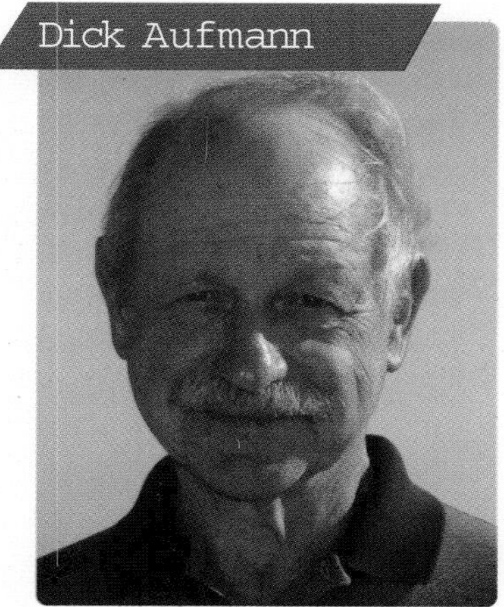

Dick Aufmann

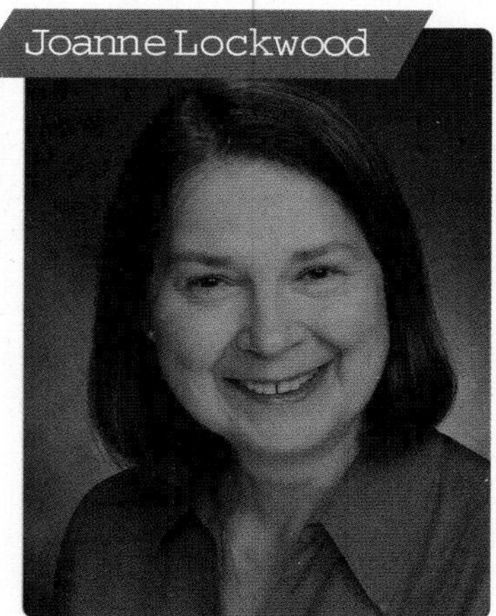

Joanne Lockwood

We have taught math for many years. During that time, we have had students ask us a number of questions about mathematics and this course. Here you find some of the questions we have been asked most often, starting with the big one.

Why do I have to take this course? You may have heard that *"Math is everywhere."* That is probably a slight exaggeration but math does find its way into many disciplines. There are obvious places like engineering, science, and medicine. There are other disciplines such as business, social science, and political science where math may be less obvious but still essential. If you are going to be an artist, writer, or musician, the direct connection to math may be even less obvious. Even so, as art historians who have studied the Mona Lisa have shown, there is a connection to math. But, suppose you find these reasons not all that compelling. **There is still a reason to learn basic math skills: You will be a better consumer and able to make better financial choices for you and your family.** For instance, is it better to buy a car or lease a car? Math can provide an answer.

I find math difficult. Why is that? It is true that some people, even very smart people, find math difficult. Some of this can be traced to previous math experiences. If your basic skills are lacking, it is more difficult to understand the math in a new math course. Some of the difficulty can be attributed to the ideas and concepts in math. They can be quite challenging to learn. Nonetheless, most of us can learn and understand the ideas in the math courses that are required for graduation. **If you want math to be less difficult, practice. When you have finished practicing, practice some more.** Ask an athlete, actor, singer, dancer, artist, doctor, skateboarder, or (name a profession) what it takes to become successful and the one common characteristic they all share is that they practiced—a lot.

Why is math important? As we mentioned earlier, math is found in many fields of study. There are, however, other reasons to take a math course. Primary among these reasons is to become a better problem solver. Math can help you learn critical thinking skills. It can help you develop a logical plan to solve a problem. Math can help you see relationships between ideas and to identify patterns. **When employers are asked what they look for in a new employee, being a problem solver is one of the highest ranked criteria.**

What do I need to do to pass this course? The most important thing you must do is to know and understand the requirements outlined by your instructor. These requirements are usually given to you in a syllabus. Once you know what is required, you can chart a course of action. Set time aside to study and do homework. If possible, choose your classes so that you have a free hour after your math class. Use this time to review your lecture notes, rework examples given by the instructor, and to begin your homework. All of us eventually need help, so know where you can get assistance with this class. This means knowing your instructor's office hours, the hours of the math help center, and how to access available online resources. And finally, do not get behind. **Try to do some math EVERY day, even if it is for only 20 minutes.**

Algebra: Introductory and Intermediate

An Applied Approach

FIFTH EDITION

Richard N. Aufmann
Palomar College

Joanne S. Lockwood
Nashua Community College

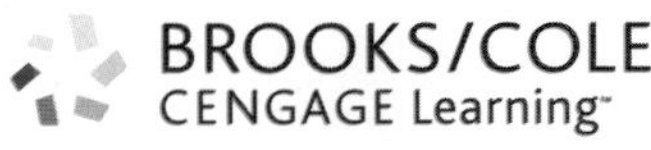

Australia • Brazil • Japan • Korea • Mexico • Singapore • Spain • United Kingdom • United States

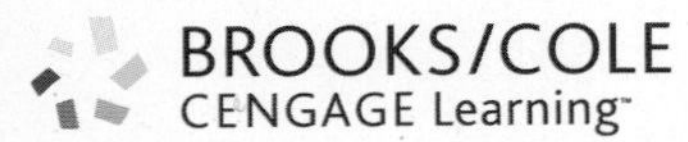

***Algebra: Introductory and Intermediate, An Applied Approach* Fifth Edition**
Richard N. Aufmann and Joanne S. Lockwood

Acquisitions Editor: Marc Bove

Developmental Editor: Erin Brown

Assistant Editor: Shaun Williams

Editorial Assistant: Kyle O'Loughlin

Media Editor: Heleny Wong

Marketing Manager: Gordon Lee

Marketing Assistant: Angela Kim

Marketing Communications Manager: Katy Malatesta

Content Project Manager: Cheryll Linthicum

Creative Director: Rob Hugel

Art Director: Vernon Boes

Print Buyer: Becky Cross

Rights Acquisitions Account Manager, Text: Roberta Broyer

Rights Acquisitions Account Manager, Image: Don Schlotman

Production Service: Graphic World Inc.

Text Designer: The Davis Group

Photo Researcher: Scott Rosen

Copy Editor: Jean Bermingham

Illustrator: Graphic World Inc.

Cover Designer: Irene Morris

Cover Image: © Richard Berenholtz/Corbis

Compositor: Graphic World Inc.

Library of Congress Control Number: 2009941953

ISBN-13: 978-1-4390-4695-1

ISBN-10: 1-4390-4695-6

Brooks/Cole
20 Davis Drive
Belmont, CA 94002-3098
USA

Cengage Learning is a leading provider of customized learning solutions with office locations around the globe, including Singapore, the United Kingdom, Australia, Mexico, Brazil, and Japan. Locate your local office at **www.cengage.com/global**

Cengage Learning products are represented in Canada by Nelson Education, Ltd.

To learn more about Brooks/Cole, visit **www.cengage.com/brookscole**
Purchase any of our products at your local college store or at our preferred online store **www.CengageBrain.com**

Printed in the United States of America
2 3 4 5 6 7 14 13 12 11 10

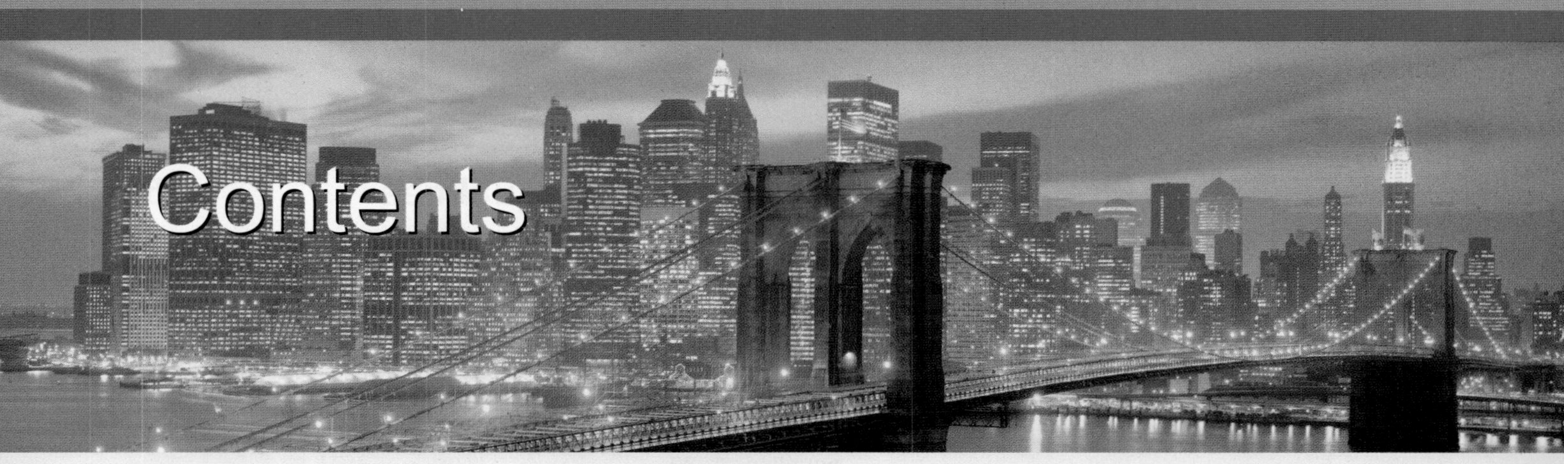
Contents

CHAPTER 2 First-Degree Equations and Inequalities 75

CHAPTER 3 Geometry 163

CHAPTER 7 Factoring 403

CHAPTER 8 Rational Expressions 457

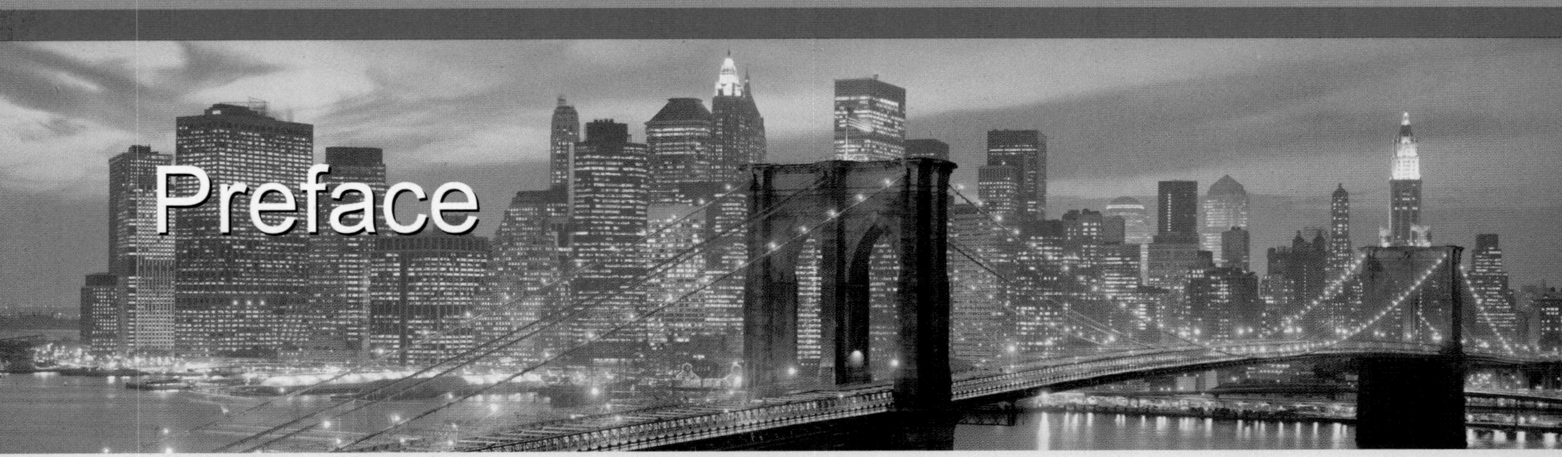

© Richard Berenholtz/Corbis

The goal in any textbook revision is to improve upon the previous edition, taking advantage of new information and new technologies, where applicable, in order to make the book more current and appealing to students and instructors. While change goes hand-in-hand with revision, a revision must be handled carefully, without compromise to valued features and pedagogy. In the fifth edition of *Algebra: Introductory and Intermediate,* we endeavored to meet these goals.

As in previous editions, the focus remains on the **Aufmann Interactive Method (AIM).** Students are encouraged to be active participants in the classroom and in their own studies as they work through the How To examples and the paired Examples and You Try It problems. The role of "active participant" is crucial to success. Providing students with worked examples, and then affording them the opportunity to immediately work similar problems, helps them build their confidence and eventually master the concepts.

To this point, simplicity plays a key factor in the organization of this edition, as in all other editions. All lessons, exercise sets, tests, and supplements are organized around a carefully constructed hierarchy of objectives. This "objective-based" approach not only serves the needs of students, in terms of helping them to clearly organize their thoughts around the content, but instructors as well, as they work to design syllabi, lesson plans, and other administrative documents.

In order to enhance the AIM and the organization of the text around objectives, we have introduced a new design. We believe students and instructors will find the page even easier to follow. Along with this change, we have introduced several new features and modifications that we believe will increase student interest and renew the appeal of presenting the content to students in the classroom, be it live or virtual.

Changes to the Fifth Edition

With the fifth edition, previous users will recognize many of the features that they have come to trust. Yet, they will notice some new additions and changes:

- Enhanced WebAssign® now accompanies the text
- Revised exercise sets with new applications
- New **In the News** applications
- New **Think About It** exercises
- Revised Chapter Review Exercises and Chapter Tests
- End-of-chapter materials now include Concept Reviews
- Revised Chapter Openers, now with Prep Tests

Take AIM and Succeed!

Algebra: Introductory and Intermediate is organized around a carefully constructed hierarchy of **OBJECTIVES**. This "objective-based" approach provides an integrated learning environment that allows students and professors to find resources such as assessment (both within the text and online), videos, tutorials, and additional exercises.

Chapter Openers are set up to help you organize your study plan for the chapter. Each opener includes: Objectives, Are You Ready? and a Prep Test.

Each Chapter Opener outlines the **OBJECTIVES** that appear in each section. The list of objectives serves as a resource to guide you in your study and review of the topics.

ARE YOU READY? outlines what you need to know to be successful in the coming chapter.

Complete each **PREP TEST** to determine which topics you may need to study more carefully, versus those you may only need to skim over to review.

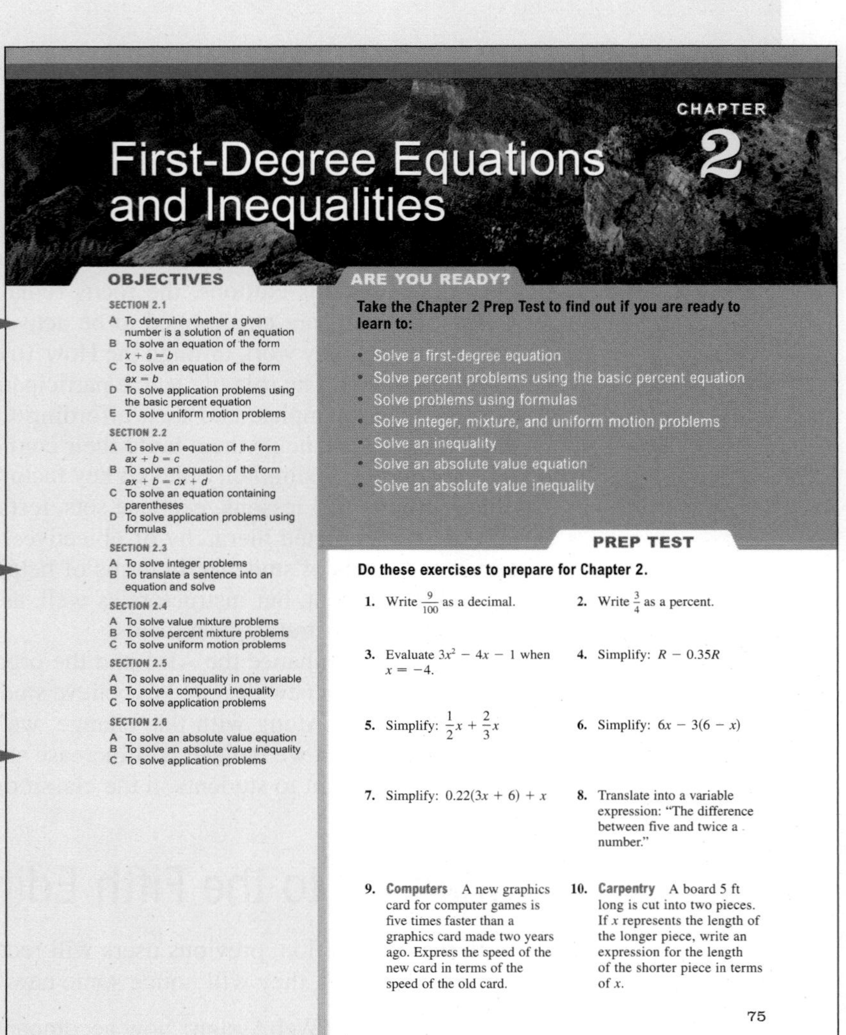

76 CHAPTER 2 First-Degree Equations and Inequalities

SECTION

2.1 Introduction to Equations

OBJECTIVE A To determine whether a given number is a solution of an equation

Point of Interest
One of the most famous equations ever stated is $E = mc^2$. This equation, stated by Albert Einstein, shows that there is a relationship between mass m and energy E. As a side note, the chemical element einsteinium was named in honor of Einstein.

An **equation** expresses the equality of two mathematical expressions. The expressions can be either numerical or variable expressions.

$9 + 3 = 12$
$3x - 2 = 10$
$y^2 + 4 = 2y - 1$
$z = 2$
Equations

The equation at the right is true if the variable is replaced by 5.

$x + 8 = 13$
$5 + 8 = 13$ A true equation

The equation is false if the variable is replaced by 7.

$7 + 8 = 13$ A false equation

A **solution of an equation** is a number that, when substituted for the variable, results in a true equation. 5 is a solution of the equation $x + 8 = 13$. 7 is not a solution of the equation $x + 8 = 13$.

HOW TO 1 Is -2 a solution of $2x + 5 = x^2 - 3$?

$2x + 5 = x^2 - 3$
$2(-2) + 5 \mid (-2)^2 - 3$ • Replace x by -2.
$-4 + 5 \mid 4 - 3$ • Evaluate the numerical expressions.
$1 = 1$ • If the results are equal, -2 is a solution of the equation. If the results are not equal, -2 is not a solution of the equation.

Yes, -2 is a solution of the equation.

Take Note
The Order of Operations Agreement applies to evaluating $2(-2) + 5$ and $(-2)^2 - 3$.

EXAMPLE 1

Is -4 a solution of $5x - 2 = 6x + 2$?

Solution
$5x - 2 = 6x + 2$
$5(-4) - 2 \mid 6(-4) + 2$
$-20 - 2 \mid -24 + 2$
$-22 = -22$

Yes, -4 is a solution.

YOU TRY IT 1

Is $\frac{1}{4}$ a solution of $5 - 4x = 8x + 2$?

Your solution

In each section, **OBJECTIVE STATEMENTS** introduce each new topic of discussion.

In each section, the **HOW TO'S** provide detailed explanations of problems related to the corresponding objectives.

The **EXAMPLE/YOU TRY IT** matched pairs are designed to actively involve you in learning the techniques presented. The You Try Its are based on the Examples. They appear side-by-side so you can easily refer to the steps in the Examples as you work through the You Try Its.

You Try It 8 $(3a - 1)5 = 15a - 5$
You Try It 9 $-8(-2a + 7b) = 16a - 56b$
You Try It 10 $3(12x^2 - x + 8) = 36x^2 - 3x + 24$
You Try It 11 $3y - 2(y - 7x) = 3y - 2y + 14x = y + 14x$

You Try It 12
$-2(x - 2y) - (-x + 3y) = -2x + 4y + x - 3y = -x + y$

You Try It 13
$3y - 2[x - 4(2 - 3y)] = 3y - 2[x - 8 + 12y] = 3y - 2x + 16 - 24y = -2x - 21y + 16$

You Try It 14 the speed of the older model: s
the speed of the new jet plane is twice the speed of the older model: $2s$

You Try It 15 the length of the longer piece: y
the length of the shorter piece: $6 - y$

SECTION 1.5
You Try It 1 $A = \{-9, -7, -5, -3, -1\}$
You Try It 2 $A = \{1, 3, 5, \ldots\}$
You Try It 3 $A \cup B = \{-2, -1, 0, 1, 2, 3, 4\}$
You Try It 4 $C \cap D = \{10, 16\}$

You Try It 11 The graph is the set of real numbers greater than -3. Use a parenthesis at -3.
-5 -4 -3 -2 -1 0 1 2 3 4 5

You Try It 12 The graph is the set of real numbers greater than -2. Use a parenthesis at -2.
-5 -4 -3 -2 -1 0 1 2 3 4 5

You Try It 13 The graph is the set of real numbers between 2 and 5, including 2 and 5. Use brackets at 2 and 5.
-5 -4 -3 -2 -1 0 1 2 3 4 5

SOLUTIONS TO CHAPTER 2 "YOU TRY IT"

SECTION 2.1
You Try It 1
$5 - 4x = 8x + 2$
$5 - 4\left(\frac{1}{4}\right) \mid 8\left(\frac{1}{4}\right) + 2$
$5 - 1 \mid 2 + 2$
$4 = 4$

Yes, $\frac{1}{4}$ is a solution.

Complete, **WORKED-OUT SOLUTIONS** to the You Try It problems are found in an appendix at the back of the text. Compare your solutions to the solutions in the appendix to obtain immediate feedback and reinforcement of the concept(s) you are studying.

Algebra: Introductory and Intermediate contains **A WIDE VARIETY OF EXERCISES** that promote skill building, skill maintenance, concept development, critical thinking, and problem solving.

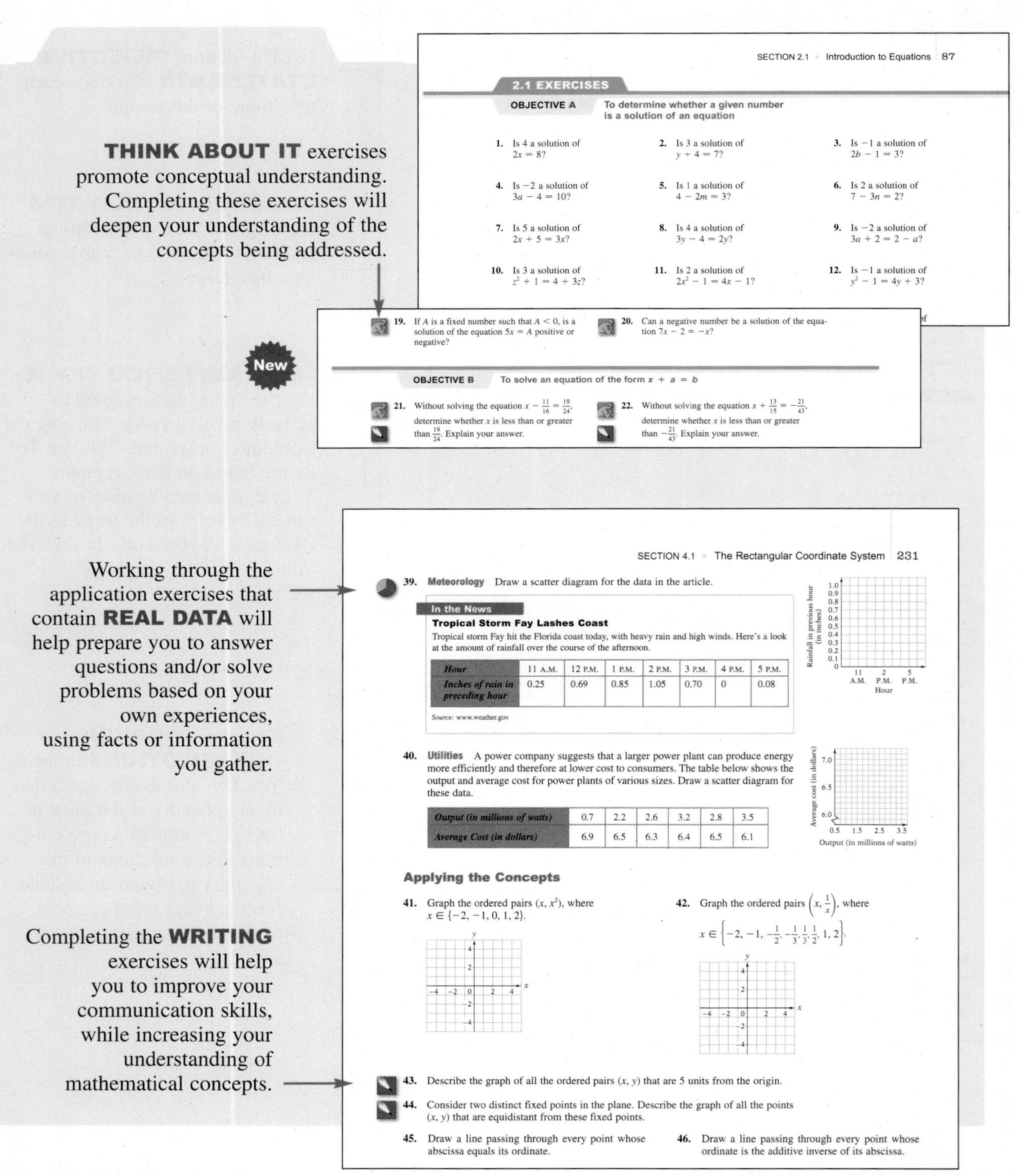

THINK ABOUT IT exercises promote conceptual understanding. Completing these exercises will deepen your understanding of the concepts being addressed.

New

SECTION 2.1 • Introduction to Equations 87

2.1 EXERCISES

OBJECTIVE A To determine whether a given number is a solution of an equation

1. Is 4 a solution of $2x = 8$?
2. Is 3 a solution of $y + 4 = 7$?
3. Is -1 a solution of $2b - 1 = 3$?
4. Is -2 a solution of $3a - 4 = 10$?
5. Is 1 a solution of $4 - 2m = 3$?
6. Is 2 a solution of $7 - 3n = 2$?
7. Is 5 a solution of $2x + 5 = 3x$?
8. Is 4 a solution of $3y - 4 = 2y$?
9. Is -2 a solution of $3a + 2 = 2 - a$?
10. Is 3 a solution of $z^2 + 1 = 4 + 3z$?
11. Is 2 a solution of $2x^2 - 1 = 4x - 1$?
12. Is -1 a solution of $y^2 - 1 = 4y + 3$?

19. If A is a fixed number such that $A < 0$, is a solution of the equation $5x = A$ positive or negative?
20. Can a negative number be a solution of the equation $7x - 2 = -x$?

OBJECTIVE B To solve an equation of the form $x + a = b$

21. Without solving the equation $x - \frac{11}{16} = \frac{19}{24}$, determine whether x is less than or greater than $\frac{19}{24}$. Explain your answer.
22. Without solving the equation $x + \frac{13}{15} = -\frac{21}{43}$, determine whether x is less than or greater than $-\frac{21}{43}$. Explain your answer.

Working through the application exercises that contain **REAL DATA** will help prepare you to answer questions and/or solve problems based on your own experiences, using facts or information you gather.

Completing the **WRITING** exercises will help you to improve your communication skills, while increasing your understanding of mathematical concepts.

SECTION 4.1 • The Rectangular Coordinate System 231

39. **Meteorology** Draw a scatter diagram for the data in the article.

In the News

Tropical Storm Fay Lashes Coast

Tropical storm Fay hit the Florida coast today, with heavy rain and high winds. Here's a look at the amount of rainfall over the course of the afternoon.

Hour	11 A.M.	12 P.M.	1 P.M.	2 P.M.	3 P.M.	4 P.M.	5 P.M.
Inches of rain in preceding hour	0.25	0.69	0.85	1.05	0.70	0	0.08

Source: www.weather.gov

40. **Utilities** A power company suggests that a larger power plant can produce energy more efficiently and therefore at lower cost to consumers. The table below shows the output and average cost for power plants of various sizes. Draw a scatter diagram for these data.

Output (in millions of watts)	0.7	2.2	2.6	3.2	2.8	3.5
Average Cost (in dollars)	6.9	6.5	6.3	6.4	6.5	6.1

Applying the Concepts

41. Graph the ordered pairs (x, x^2), where $x \in \{-2, -1, 0, 1, 2\}$.
42. Graph the ordered pairs $\left(x, \frac{1}{x}\right)$, where $x \in \left\{-2, -1, -\frac{1}{2}, -\frac{1}{3}, \frac{1}{3}, \frac{1}{2}, 1, 2\right\}$.
43. Describe the graph of all the ordered pairs (x, y) that are 5 units from the origin.
44. Consider two distinct fixed points in the plane. Describe the graph of all the points (x, y) that are equidistant from these fixed points.
45. Draw a line passing through every point whose abscissa equals its ordinate.
46. Draw a line passing through every point whose ordinate is the additive inverse of its abscissa.

264 CHAPTER 4 • Linear Functions and Inequalities in Two Variables

30. **Solar Roof** Look at the butterfly roof design shown in the article below. Which side of the roof, the left or the right, has a slope of approximately 1? Is the slope of the other side of the roof greater than 1 or less than 1? (*Note:* Consider both slopes to be positive.)

In the News

Go Green with Power Pod's Butterfly Roof

PowerHouse Enterprises has designed a modular home with a butterfly roof. The roof design combines two sections that slant toward each other at different angles, and is ideal for the use of solar panels.

Source: www.powerhouse-enterprises.com

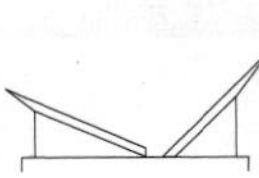

OBJECTIVE B To graph a line given a point and the slope

For Exercises 31 to 34, complete the table.

	Equation	*Value of m*	*Value of b*	*Slope*	*y-intercept*
31.	$y = -3x + 5$				
32.	$y = \frac{2}{5}x - 8$				
33.	$y = 4x$				
34.	$y = -7$				

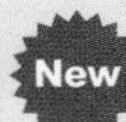

IN THE NEWS application exercises help you master the utility of mathematics in our everyday world. They are based on information found in popular media sources, including newspapers and magazines, and the Web.

APPLYING THE CONCEPTS exercises may involve further exploration of topics, or they may involve analysis. They may also integrate concepts introduced earlier in the text. **Optional** calculator exercises are included, denoted by .

Applying the Concepts

137. In your own words, explain how the signs of the last terms of the two binomial factors of a trinomial are determined.

For Exercises 138 to 143, factor.

138. $(x + 1)^2 - (x + 1) - 6$ **139.** $(x - 2)^2 + 3(x - 2) + 2$ **140.** $(y + 3)^2 - 5(y + 3) + 6$

Projects and Group Activities 391

PROJECTS AND GROUP ACTIVITIES

Astronomical Distances and Scientific Notation

Astronomers have units of measurement that are useful for measuring vast distances in space. Two of these units are the astronomical unit and the light-year. An **astronomical unit** is the average distance between Earth and the sun. A **light-year** is the distance a ray of light travels in 1 year.

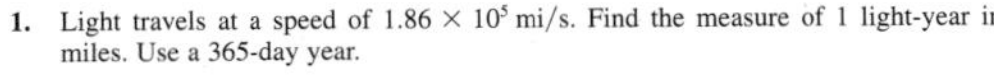

1. Light travels at a speed of 1.86×10^5 mi/s. Find the measure of 1 light-year in miles. Use a 365-day year.

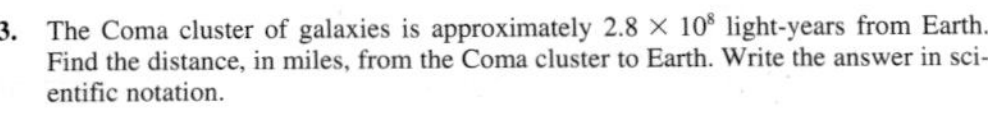

2. The distance between Earth and the star Alpha Centauri is approximately 25 trillion miles. Find the distance between Earth and Alpha Centauri in light-years. Round to the nearest hundredth.

3. The Coma cluster of galaxies is approximately 2.8×10^8 light-years from Earth. Find the distance, in miles, from the Coma cluster to Earth. Write the answer in scientific notation.

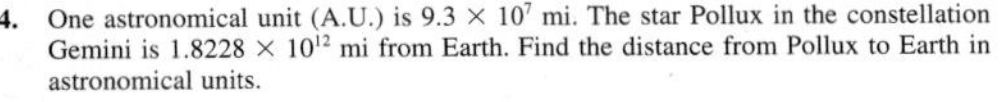

4. One astronomical unit (A.U.) is 9.3×10^7 mi. The star Pollux in the constellation Gemini is 1.8228×10^{12} mi from Earth. Find the distance from Pollux to Earth in astronomical units.

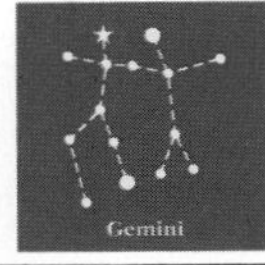

PROJECTS AND GROUP ACTIVITIES appear at the end of each chapter. Your instructor may assign these to you individually, or you may be asked to work through the activity in groups.

Algebra: Introductory and Intermediate addresses students' broad range of study styles by offering **A WIDE VARIETY OF TOOLS FOR REVIEW**.

At the end of each chapter you will find a **SUMMARY** with **KEY WORDS** and **ESSENTIAL RULES AND PROCEDURES**. Each entry includes an example of the summarized concept, an objective reference, and a page reference to show where each concept was introduced.

392 CHAPTER 6 • Polynomials

CHAPTER 6

SUMMARY

KEY WORDS	EXAMPLES
A *monomial* is a number, a variable, or a product of a number and variables. [6.1A, p. 348]	5 is a number, y is a variable. $8a^2b^2$ is a product of a number and variables. 5, y, and $8a^2b^2$ are monomials.
The *degree of a monomial* is the sum of the exponents on the variables. [6.1A, p. 348]	The degree of $8x^4y^5z$ is 10.
A *polynomial* is a variable expression in which the terms are monomials. [6.2A, p. 360]	$x^4 - 2xy - 32x + 8$ is a polynomial. The terms are x^4, $-2xy$, $-32x$, and 8.

CHAPTER 6

CONCEPT REVIEW

Test your knowledge of the concepts presented in this chapter. Answer each question. Then check your answers against the ones provided in the Answer Section.

1. Why is it important to write the terms of a polynomial in descending order before adding in a vertical format?
2. When multiplying the terms $4p^3$ and $7p^6$, what happens to the exponents?
3. How do you multiply two binomials?

CONCEPT REVIEWS actively engage you as you study and review the contents of a chapter. The **ANSWERS** to the questions are found in an appendix at the back of the text. After each answer, look for an objective reference that indicates where the concept was introduced.

By completing the chapter **REVIEW EXERCISES**, you can practice working problems that appear in an order that is different from the order they were presented in the chapter. The **ANSWERS** to these exercises include references to the section objectives upon which they are based. This will help you to quickly identify where to go to review the concepts if needed.

CHAPTER 6

REVIEW EXERCISES

1. Add: $(12y^2 + 17y - 4) + (9y^2 - 13y + 3)$
2. Divide: $\dfrac{15x^2 + 2x - 2}{3x - 2}$
3. Simplify: $(2x^{-1}y^2z^5)^4(-3x^3yz^{-3})^2$
4. Expand: $(5y - 7)^2$
5. Simplify: $\dfrac{a^{-1}b^3}{a^3b^{-3}}$
6. Use the Remainder Theorem to evaluate $P(x) = x^3 - 2x^2 + 3x - 5$ when $x = 2$.
7. Simplify: $(5x^2 - 8xy + 2y^2) - (x^2 - 3y^2)$
8. Divide: $\dfrac{12b^7 + 36b^5 - 3b^3}{3b^3}$
9. Simplify: $\dfrac{3ab^4}{-6a^2b^4}$
10. Simplify: $(-2a^2b^4)(3ab^2)$
11. Simplify: $\dfrac{8x^{12}}{12x^9}$
12. Simplify: $\dfrac{4x^3 + 27x^2 + 10x + 2}{x + 6}$

Each chapter **TEST** is designed to simulate a possible test of the concepts covered in the chapter. The **ANSWERS** include references to section objectives. References to How Tos, worked Examples, and You Try Its, that provide solutions to similar problems, are also included.

CHAPTER 6

TEST

1. Multiply: $2x(2x^2 - 3x)$
2. Use the Remainder Theorem to evaluate $P(x) = -x^3 + 4x - 8$ when $x = -2$.
3. Simplify: $\frac{12x^2}{-3x^8}$
4. Simplify: $(-2xy^2)(3x^2y^4)$
5. Divide: $(x^2 + 1) \div (x + 1)$
6. Multiply: $(x - 3)(x^2 - 4x + 5)$

CUMULATIVE REVIEW EXERCISES, which appear at the end of each chapter (beginning with Chapter 2), help you maintain skills you previously learned. The **ANSWERS** include references to the section objectives upon which the exercises are based.

CUMULATIVE REVIEW EXERCISES

1. Let $x \in \{-8, -3, 3\}$. For what values of x is the inequality $x \geq -3$ a true statement?
2. Find the additive inverse of 83.
3. Simplify: $8 - 2[-3 - (-1)]^2 \div 4$
4. Evaluate $\frac{2a - b}{b - c}$ when $a = 4$, $b = -2$, and $c = 6$.
5. Simplify: $-5\sqrt{300}$
6. Identify the property that justifies the statement $2x + (-2x) = 0$.
7. Simplify: $2x - 4[x - 2(3 - 23x) + 4]$
8. Solve: $\frac{2}{3} - y = \frac{5}{6}$
9. Solve: $8x - 3 - x = -6 + 3x - 8$
10. Solve: $3 - |2 - 3x| = -2$

A **FINAL EXAM** appears after the last chapter in the text. It is designed to simulate a possible examination of all the concepts covered in the text. The **ANSWERS** to the exam questions are provided in the answer appendix at the back of the text and include references to the section objectives upon which the questions are based.

FINAL EXAM

1. Simplify: $12 - 8[3 - (-2)]^2 \div 5 - 3$
2. Evaluate $\frac{a^2 - b^2}{a - b}$ when $a = 3$ and $b = -4$.
3. Simplify: $5 - 2[3x - 7(2 - x) - 5x]$
4. Solve: $\frac{3}{4}x - 2 = 4$
5. Solve: $8 - |5 - 3x| = 1$
6. Find the volume of a sphere with a diameter of 8 ft. Round to the nearest tenth.
7. Graph $2x - 3y = 9$ using the x- and y-intercepts.
8. Find the equation of the line containing the points $(3, -2)$ and $(1, 4)$.

Other Key Features

MARGINS Within the margins, students can find the following.

Take Note boxes alert students to concepts that require special attention.

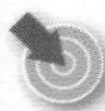

Point of Interest boxes, which may be historical in nature or be of general interest, relate to topics under discussion.

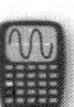

Integrating Technology boxes, which are offered as optional instruction in the proper use of the scientific calculator, appear for selected topics under discussion.

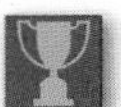

Tips for Success boxes outline good study habits.

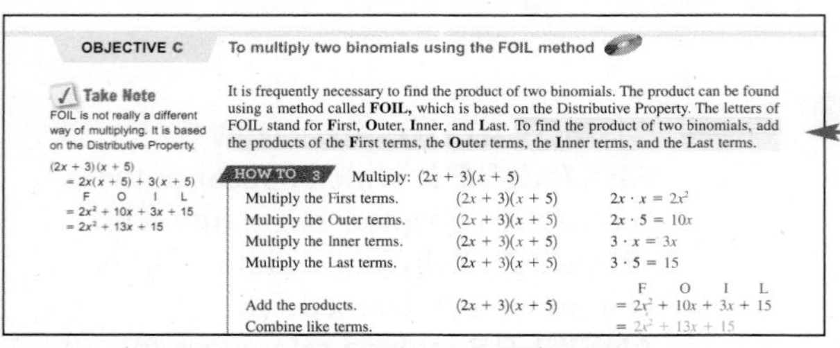

OBJECTIVE C To multiply two binomials using the FOIL method

Take Note
FOIL is not really a different way of multiplying. It is based on the Distributive Property.
$(2x + 3)(x + 5)$
$= 2x(x + 5) + 3(x + 5)$
F O I L
$= 2x^2 + 10x + 3x + 15$
$= 2x^2 + 13x + 15$

It is frequently necessary to find the product of two binomials. The product can be found using a method called **FOIL**, which is based on the Distributive Property. The letters of FOIL stand for **F**irst, **O**uter, **I**nner, and **L**ast. To find the product of two binomials, add the products of the First terms, the Outer terms, the Inner terms, and the Last terms.

HOW TO 3 Multiply: $(2x + 3)(x + 5)$

Multiply the First terms.	$(2x + 3)(x + 5)$	$2x \cdot x = 2x^2$
Multiply the Outer terms.	$(2x + 3)(x + 5)$	$2x \cdot 5 = 10x$
Multiply the Inner terms.	$(2x + 3)(x + 5)$	$3 \cdot x = 3x$
Multiply the Last terms.	$(2x + 3)(x + 5)$	$3 \cdot 5 = 15$
Add the products.	$(2x + 3)(x + 5)$	F O I L $= 2x^2 + 10x + 3x + 15$
Combine like terms.		$= 2x^2 + 13x + 15$

IMPORTANT POINTS Passages of text are now highlighted to help students recognize what is most important and to help them study more effectively.

PROBLEM-SOLVING STRATEGIES The text features a carefully developed approach to problem solving that encourages students to develop a Strategy for a problem and then to create a Solution based on the Strategy.

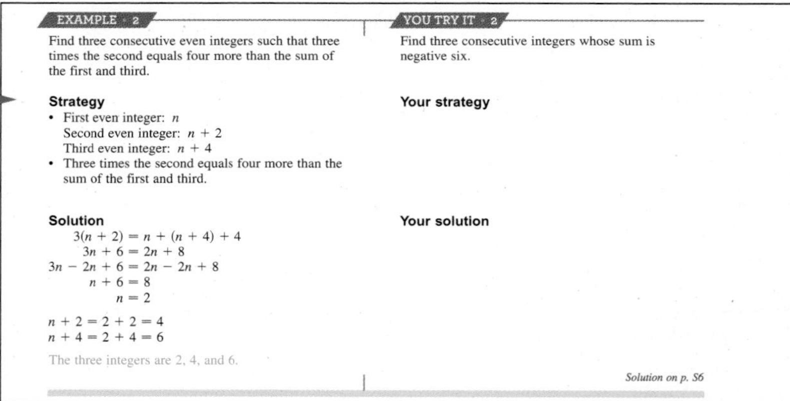

EXAMPLE 2

Find three consecutive even integers such that three times the second equals four more than the sum of the first and third.

Strategy
- First even integer: n
  Second even integer: $n + 2$
  Third even integer: $n + 4$
- Three times the second equals four more than the sum of the first and third.

Solution
$3(n + 2) = n + (n + 4) + 4$
$3n + 6 = 2n + 8$
$3n - 2n + 6 = 2n - 2n + 8$
$n + 6 = 8$
$n = 2$

$n + 2 = 2 + 2 = 4$
$n + 4 = 2 + 4 = 6$

The three integers are 2, 4, and 6.

YOU TRY IT 2

Find three consecutive integers whose sum is negative six.

Your strategy

Your solution

Solution on p. S6

FOCUS ON PROBLEM SOLVING At the end of each chapter, the Focus on Problem Solving fosters further discovery of new problem-solving strategies, such as applying solutions to other problems, working backwards, inductive reasoning, and trial and error.

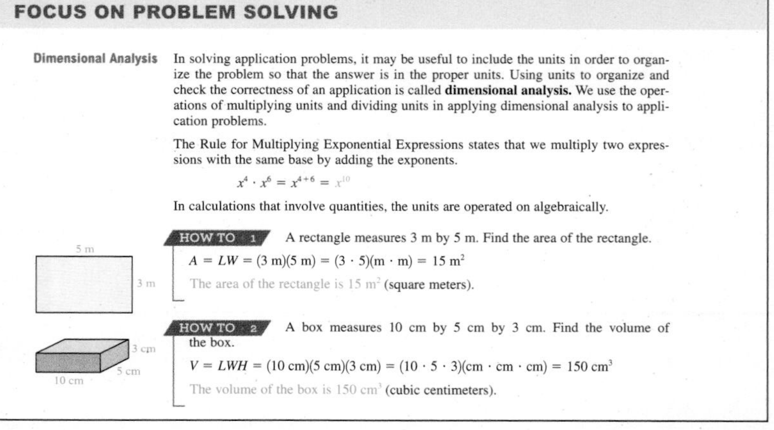

FOCUS ON PROBLEM SOLVING

Dimensional Analysis In solving application problems, it may be useful to include the units in order to organize the problem so that the answer is in the proper units. Using units to organize and check the correctness of an application is called **dimensional analysis.** We use the operations of multiplying units and dividing units in applying dimensional analysis to application problems.

The Rule for Multiplying Exponential Expressions states that we multiply two expressions with the same base by adding the exponents.

$$x^4 \cdot x^6 = x^{4+6} = x^{10}$$

In calculations that involve quantities, the units are operated on algebraically.

HOW TO 1 A rectangle measures 3 m by 5 m. Find the area of the rectangle.

$A = LW = (3 \text{ m})(5 \text{ m}) = (3 \cdot 5)(\text{m} \cdot \text{m}) = 15 \text{ m}^2$

The area of the rectangle is 15 m^2 (square meters).

HOW TO 2 A box measures 10 cm by 5 cm by 3 cm. Find the volume of the box.

$V = LWH = (10 \text{ cm})(5 \text{ cm})(3 \text{ cm}) = (10 \cdot 5 \cdot 3)(\text{cm} \cdot \text{cm} \cdot \text{cm}) = 150 \text{ cm}^3$

The volume of the box is 150 cm^3 (cubic centimeters).

General Revisions

- Chapter Openers now include Prep Tests for students to test their knowledge of prerequisite skills for the new chapter.
- Each exercise set has been thoroughly reviewed to ensure that the pace and scope of the exercises adequately cover the concepts introduced in the section.
- The variety of word problems has increased. This will appeal to instructors who teach to a range of student abilities and want to address different learning styles.
- Think About It exercises, which are conceptual in nature, have been added. They are meant to assess and strengthen a student's understanding of the material presented in an objective.
- In the News exercises have been added and are based on a media source such as a newspaper, a magazine, or the Web. The exercises demonstrate the pervasiveness and utility of mathematics in a contemporary setting.
- Concept Reviews now appear in the end-of-chapter materials to help students more actively study and review the contents of the chapter.
- The Chapter Review Exercises and Chapter Tests have been adjusted to ensure that there are questions that assess the key ideas in the chapter.
- The design has been significantly modified to make the text even easier to follow.

Acknowledgments

The authors would like to thank the people who have reviewed this manuscript and provided many valuable suggestions.

Chris Bendixen, *Lake Michigan College*
Nancy Eschen, *Florida Community College at Jacksonville*
Dorothy Fujimura, *CSU East Bay*
Oxana Grinevich, *Lourdes College*
Jean-Marie Magnier, *Springfield Technical Community College*
Rinav Mehta, *Seattle Central Community College*
Joseph Phillips, *Warren County Community College*
Melissa Rossi, *Southwestern Illinois College*
Daryl Schrader, *St. Petersburg College*
Yan Tian, *Palomar College*

The authors would also like to thank the people who reviewed the fourth edition.

Donna Foster, *Piedmont Technical College*
William Graesser, *Ivy Tech State College*
Anne Haney
Tim R. McBride, *Spartanburg Technical College*
Michael McComas, *Marshall Community and Technical College*
Linda J. Murphy, *Northern Essex Community College*

Special thanks go to Jean Bermingham for her work copyediting and proofreading, to Pat Foard and Ellena Reda for preparing the solutions manuals, and to Cindy Trimble for her work in ensuring the accuracy of the text. We would also like to thank the many people at Cengage Learning who worked to guide the manuscript from development through production.

Instructor Resources

Print Ancillaries

Complete Solutions Manual (0-538-49736-X)
Pat Foard, *South Plains College*
Ellena Reda, *Dutchess Community College*

The Complete Solutions Manual provides worked-out solutions to all of the problems in the text.

Instructor's Resource Binder (0-538-49772-6)
Maria H. Andersen, *Muskegon Community College*

New The Instructor's Resource Binder contains uniquely designed Teaching Guides, which include instruction tips, examples, activities, worksheets, overheads, and assessments, with answers to accompany them.

Appendix to accompany Instructor's Resource Binder (0-538-49772-6)
Richard N. Aufmann, *Palomar College*
Joanne S. Lockwood, *Nashua Community College*

New! The Appendix to accompany the Instructor's Resource Binder contains teacher resources that are tied directly to *Algebra: Introductory and Intermediate,* 5e. Organized by objective, the Appendix contains additional questions and short, in-class activities. The Appendix also includes answers to Writing Exercises, Focus on Problem Solving, and Projects and Group Activities found in the text.

Electronic Ancillaries

Enhanced WebAssign 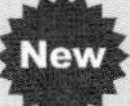

New Used by over one million students at more than 1,100 institutions, WebAssign allows you to assign, collect, grade, and record homework assignments via the Web. This proven and reliable homework system includes thousands of algorithmically generated homework problems, links to relevant textbook sections, video examples, problem-specific tutorials, and more.

Solution Builder (0-840-03616-7)

This online solutions manual allows instructors to create customizable solutions that they can print out to distribute or post as needed. This is a convenient and expedient way to deliver solutions to specific homework sets.

PowerLecture with Diploma®
(0-538-49755-6)

New This CD-ROM provides the instructor with dynamic media tools for teaching. Create, deliver, and customize tests (both print and online) in minutes with Diploma's Computerized Testing featuring algorithmic equations. Easily build solution sets for homework or exams using Solution Builder's online solutions manual. Quickly and easily update your syllabus with the new Syllabus Creator, which was created by the authors and contains the new edition's table of contents. Practice Sheets, First Day of Class PowerPoint® lecture slides, art and figures from the book, and a test bank in electronic format are also included on this CD-ROM.

Text Specific DVDs (0-538-73631-3)

Hosted by Dana Mosely and captioned for the hearing-impaired, these DVDs cover all sections in the text. Ideal for promoting individual study and review, these comprehensive DVDs also support students in online courses or those who may have missed a lecture.

Student Resources

Print Ancillaries

Student Solutions Manual (0-538-49723-8)
Pat Foard, *South Plains College*
Ellena Reda, *Dutchess Community College*

The Student Solutions Manual provides worked-out solutions to the odd-numbered problems in the textbook.

Student Workbook (0-538-49544-8)
Maria H. Andersen, *Muskegon Community College*

New Get a head-start! The Student Workbook contains assessments, activities, and worksheets from the Instructor's Resource Binder. Use them for additional practice to help you master the content.

Electronic Ancillaries

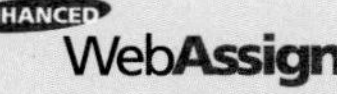

Enhanced WebAssign

New If you are looking for extra practice or additional support, Enhanced WebAssign offers practice problems, videos, and tutorials that are tied directly to the problems found in the textbook.

Text Specific DVDs (0-538-73631-3)

Hosted by Dana Mosley, an experienced mathematics instructor, the DVDs will help you to get a better handle on topics found in the textbook. A comprehensive set of DVDs for the entire course is available to order.

AIM for Success: Getting Started

Welcome to *Algebra: Introductory and Intermediate!* Students come to this course with varied backgrounds and different experiences in learning math. We are committed to your success in learning mathematics and have developed many tools and resources to support you along the way. Want to excel in this course? Read on to learn the skills you'll need and how best to use this book to get the results you want.

Motivate Yourself

You'll find many real-life problems in this book, relating to sports, money, cars, music, and more. We hope that these topics will help you understand how you will use mathematics in your real life. However, to learn all of the necessary skills and how you can apply them to your life outside this course, you need to stay motivated.

Take Note

Motivation alone won't lead to success. For example, suppose a person who cannot swim is rowed out to the middle of a lake and thrown overboard. That person has a lot of motivation to swim, but will most likely drown without some help. You'll need motivation and learning in order to succeed.

THINK ABOUT WHY YOU WANT TO SUCCEED IN THIS COURSE. LIST THE REASONS HERE (NOT IN YOUR HEAD . . . ON THE PAPER!):

____

____

We also know that this course may be a requirement for you to graduate or complete your major. That's OK. If you have a goal for the future, such as becoming a nurse or a teacher, you will need to succeed in mathematics first. Picture yourself where you want to be, and use this image to stay on track.

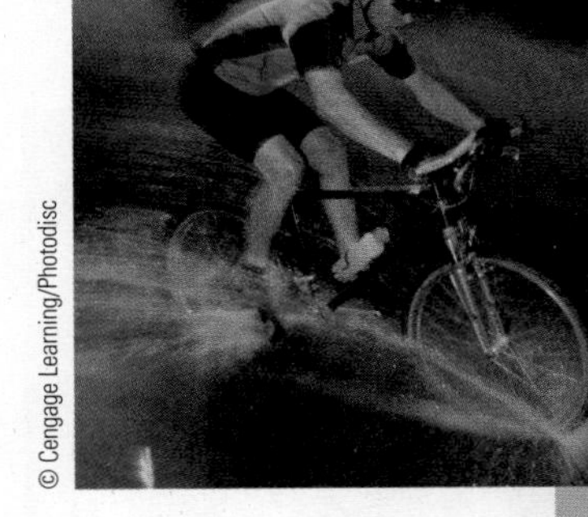

© Cengage Learning/Photodisc

Make the Commitment

Stay committed to success! With practice, you will improve your math skills. Skeptical? Think about when you first learned to ride a bike or drive a car. You probably felt self-conscious and worried that you might fail. But with time and practice, it became second nature to you.

You will also need to put in the time and practice to do well in mathematics. Think of us as your "driving" instructors. We'll lead you along the path to success, but we need you to stay focused and energized along the way.

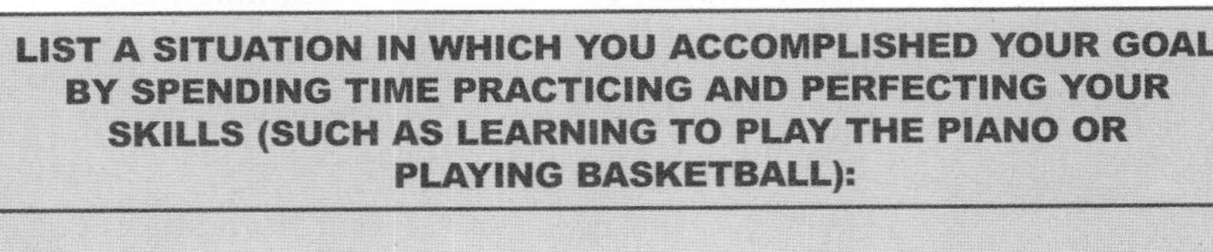

LIST A SITUATION IN WHICH YOU ACCOMPLISHED YOUR GOAL BY SPENDING TIME PRACTICING AND PERFECTING YOUR SKILLS (SUCH AS LEARNING TO PLAY THE PIANO OR PLAYING BASKETBALL):

____

____

____

____

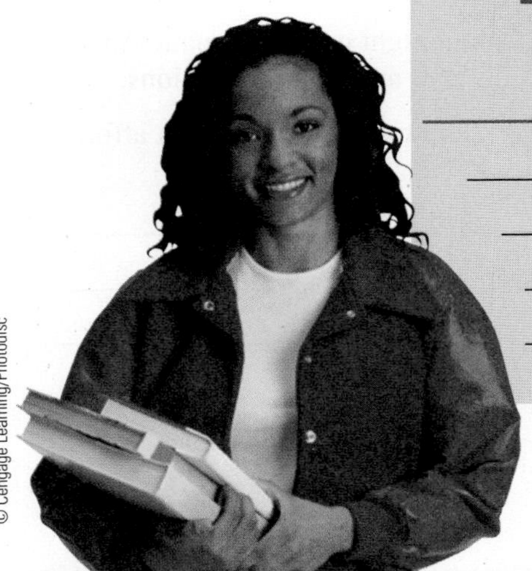

© Cengage Learning/Photodisc

If you spend time learning and practicing the skills in this book, you will also succeed in math.

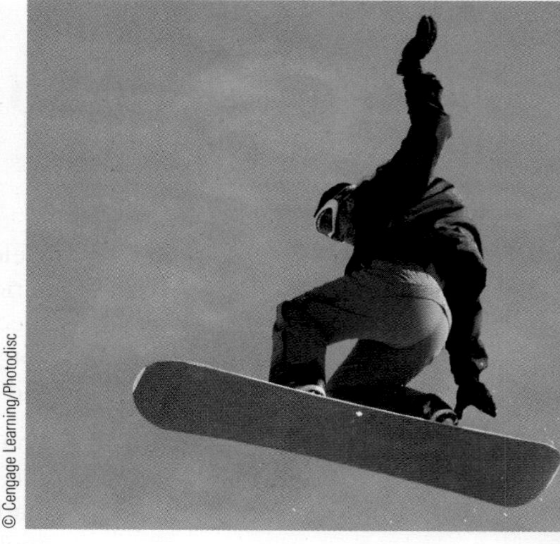
© Cengage Learning/Photodisc

Think You Can't Do Math? Think Again!

You can do math! When you first learned the skills you just listed, you may have not done them well. With practice, you got better. With practice, you will be better at math. Stay focused, motivated, and committed to success.

It is difficult for us to emphasize how important it is to overcome the "I Can't Do Math Syndrome." If you listen to interviews of very successful athletes after a particularly bad performance, you will note that they focus on the positive aspect of what they did, not the negative. Sports psychologists encourage athletes to always be positive—to have a "Can Do" attitude. Develop this attitude toward math and you will succeed.

Skills for Success

GET THE BIG PICTURE If this were an English class, we wouldn't encourage you to look ahead in the book. But this is mathematics—go right ahead! Take a few minutes to read the table of contents. Then, look through the entire book. Move quickly: scan titles, look at pictures, notice diagrams.

Getting this big picture view will help you see where this course is going. To reach your goal, it's important to get an idea of the steps you will need to take along the way.

As you look through the book, find topics that interest you. What's your preference? Horse racing? Sailing? TV? Amusement parks? Find the Index of Applications at the back of the book and pull out three subjects that interest you. Then, flip to the pages in the book where the topics are featured and read the exercises or problems where they appear.

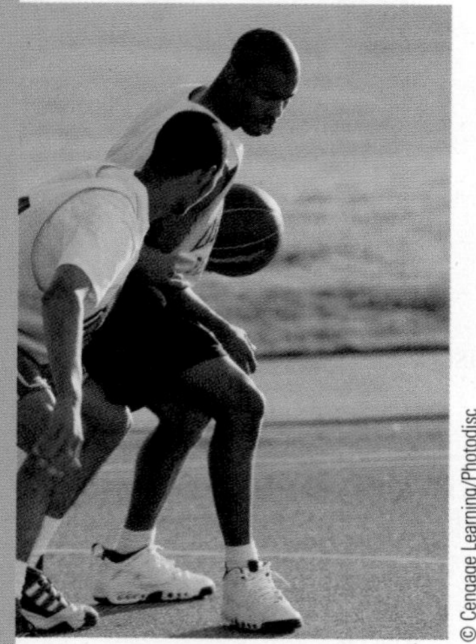
© Cengage Learning/Photodisc

WRITE THE TOPIC HERE:	WRITE THE CORRESPONDING EXERCISE/PROBLEM HERE:
__________	__________
__________	__________
__________	__________

You'll find it's easier to work at learning the material if you are interested in how it can be used in your everyday life.

Use the following activities to think about more ways you might use mathematics in your daily life. Flip open your book to the following exercises to answer the questions.

- (see p. 115, #33) I'm trying to figure out how many text messages I can afford to send in a month. I need to use algebra to . . .

____

____

____

____

- (see p. 274, #82) I just had an hour-long phone conversation. I want to know the cost. I need algebra to . . .

____

____

- (see p. 487, #21) I'm making a dessert for a party. I need enough servings for 25 people. I need algebra to . . .

____

____

____

You know that the activities you just completed are from daily life, but do you notice anything else they have in common? That's right—they are **word problems.** Try not to be intimidated by word problems. You just need a strategy. It's true that word problems can be challenging because we need to use multiple steps to solve them:

- Read the problem.
- Determine the quantity we must find.
- Think of a method to find it.
- Solve the problem.
- Check the answer.

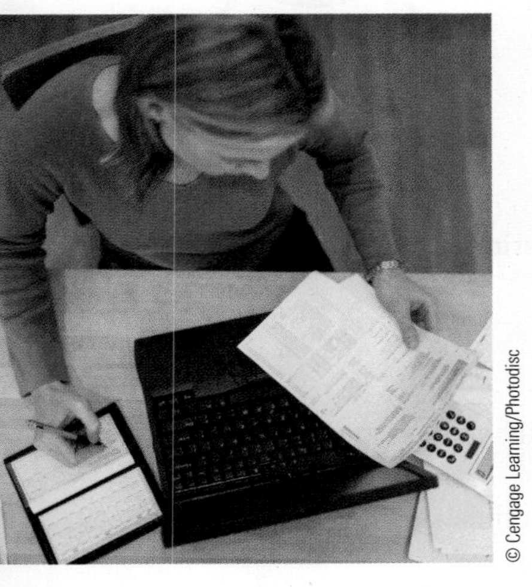

In short, we must come up with a **strategy** and then use that strategy to find the **solution.**

We'll teach you about strategies for tackling word problems that will make you feel more confident in branching out to these problems from daily life. After all, even though no one will ever come up to you on the street and ask you to solve a multiplication problem, you will need to use math every day to balance your checkbook, evaluate credit card offers, etc.

Take a look at the following example. You'll see that solving a word problem includes finding a *strategy* and using that strategy to find a *solution.* If you find yourself struggling with a word problem, try writing down the information you know about the problem. Be as specific as you can. Write out a phrase or a sentence that states what you are trying to find. Ask yourself whether there is a formula that expresses the known and unknown quantities. Then, try again!

EXAMPLE 2

The dimensions of a triangular sail are 18 ft, 11 ft, and 15 ft. What is the perimeter of the sail?

Strategy

To find the perimeter, use the formula for the perimeter of a triangle. Substitute 18 ft for a, 11 ft for b, and 15 ft for c. Solve for P.

Solution

$P = a + b + c$

$P = 18 \text{ ft} + 11 \text{ ft} + 15 \text{ ft}$

$P = 44 \text{ ft}$

The perimeter of the sail is 44 ft.

YOU TRY IT 2

What is the perimeter of a standard piece of typing paper that measures $8\frac{1}{2}$ in. by 11 in.?

Your strategy

Your solution

Page 184

 **Take Note**

Take a look at your syllabus to see if your instructor has an **attendance policy** that is part of your overall grade in the course.

The attendance policy will tell you:

- How many classes you can miss without a penalty
- What to do if you miss an exam or quiz
- If you can get the lecture notes from the professor if you miss a class

 **Take Note**

When planning your schedule, give some thought to how much time you realistically have available each week. For example, if you work 40 hours a week, take 15 units, spend the recommended study time given at the right, and sleep 8 hours a day, you will use over 80% of the available hours in a week. That leaves less than 20% of the hours in a week for family, friends, eating, recreation, and other activities.

Visit http://college.cengage.com/masterstudent/shared/content/time_chart/chart.html and use the Interactive Time Chart to see how you're spending your time—you may be surprised.

GET THE BASICS On the first day of class, your instructor will hand out a **syllabus** listing the requirements of your course. Think of this syllabus as your personal roadmap to success. It shows you the destinations (topics you need to learn) and the dates you need to arrive at those destinations (by when you need to learn the topics). Learning mathematics is a journey. But, to get the most out of this course, you'll need to know what the important stops are and what skills you'll need to learn for your arrival at those stops.

You've quickly scanned the table of contents, but now we want you to take a closer look. Flip open to the table of contents and look at it next to your syllabus. Identify when your major exams are and what material you'll need to learn by those dates. For example, if you know you have an exam in the second month of the semester, how many chapters of this text will you need to learn by then? What homework do you have to do during this time? Managing this important information will help keep you on track for success.

MANAGE YOUR TIME We know how busy you are outside of school. Do you have a full-time or a part-time job? Do you have children? Visit your family often? Play basketball or write for the school newspaper? It can be stressful to balance all of the important activities and responsibilities in your life. Making a **time management plan** will help you create a schedule that gives you enough time for everything you need to do.

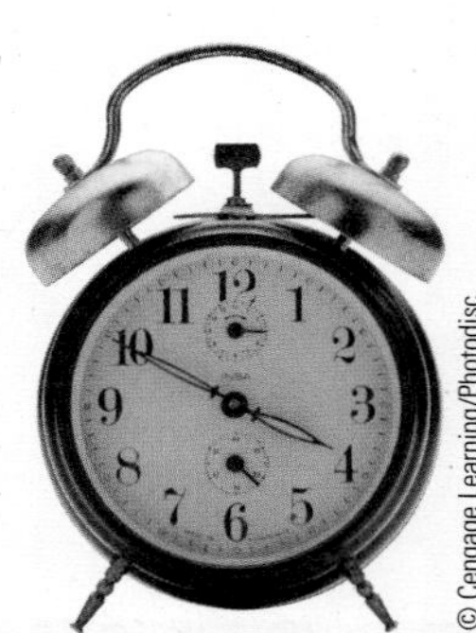

Let's get started! Create a weekly schedule.

First, list all of your responsibilities that take up certain set hours during the week. Be sure to include:

- each class you are taking
- time you spend at work
- any other commitments (child care, tutoring, volunteering, etc.)

Then, list all of your responsibilities that are more flexible. Remember to make time for:

- **STUDYING** You'll need to study to succeed, but luckily you get to choose what times work best for you. Keep in mind:
  - Most instructors ask students to spend twice as much time studying as they do in class (3 hours of class = 6 hours of study).
  - Try studying in chunks. We've found it works better to study an hour each day, rather than studying for 6 hours on one day.
  - Studying can be even more helpful if you're able to do it right after your class meets, when the material is fresh in your mind.
- **MEALS** Eating well gives you energy and stamina for attending classes and studying.
- **ENTERTAINMENT** It's impossible to stay focused on your responsibilities 100% of the time. Giving yourself a break for entertainment will reduce your stress and help keep you on track.
- **EXERCISE** Exercise contributes to overall health. You'll find you're at your most productive when you have both a healthy mind and a healthy body.

Here is a sample of what part of your schedule might look like:

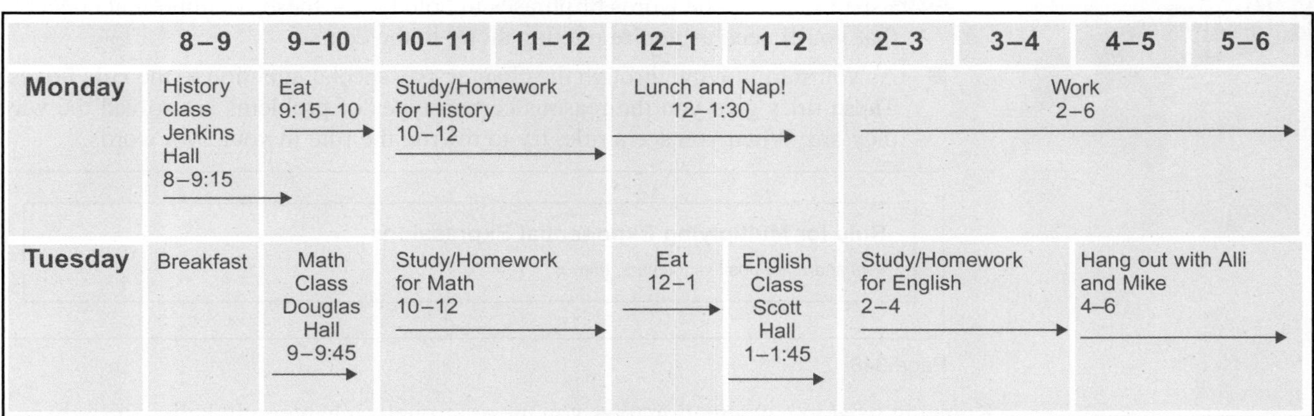

	8–9	9–10	10–11	11–12	12–1	1–2	2–3	3–4	4–5	5–6
Monday	History class Jenkins Hall 8–9:15	Eat 9:15–10	Study/Homework for History 10–12		Lunch and Nap! 12–1:30		Work 2–6			
Tuesday	Breakfast	Math Class Douglas Hall 9–9:45	Study/Homework for Math 10–12		Eat 12–1	English Class Scott Hall 1–1:45	Study/Homework for English 2–4		Hang out with Alli and Mike 4–6	

Features for Success in This Text

ORGANIZATION Let's look again at the table of contents. There are 12 chapters in this book. You'll see that every chapter is divided into **sections,** and each section contains a number of **learning objectives.** Each learning objective is labeled with a letter from A to G. Knowing how this book is organized will help you locate important topics and concepts as you're studying.

PREPARATION Ready to start a new chapter? Take a few minutes to be sure you're ready, using some of the tools in this book.

- **CUMULATIVE REVIEW EXERCISES:** You'll find these exercises after every chapter, starting with Chapter 2. The questions in the Cumulative Review Exercises are taken from the previous chapters. For example, the Cumulative Review for Chapter 3 will test all of the skills you have learned in Chapters 1, 2, and 3. Use this to refresh yourself before moving on to the next chapter, or to test what you know before a big exam.

Here's an example of how to use the Cumulative Review:

- Turn to page 219 and look at the questions for the Chapter 3 Cumulative Review, which are taken from the current chapter and the previous chapters.
- We have the answers to all of the Cumulative Review Exercises in the back of the book. Flip to page A10 to see the answers for this chapter.
- Got the answer wrong? We can tell you where to go in the book for help! For example, scroll down page A10 to find the answer for the first exercise, which is -3, 0, and 1. You'll see that after this answer, there is an **objective reference** [1.1A]. This means that the question was taken from Chapter 1, Section 1, Objective A. Go here to restudy the objective.

- **PREP TESTS:** These tests are found at the beginning of every chapter and will help you see if you've mastered all of the skills needed for the new chapter.

Here's an example of how to use the Prep Test:

- Turn to page 221 and look at the Prep Test for Chapter 4.
- All of the answers to the Prep Tests are in the back of the book. You'll find them in the first set of answers in each answer section for a chapter. Turn to page A10 to see the answers for this Prep Test.
- Restudy the objectives if you need some extra help.

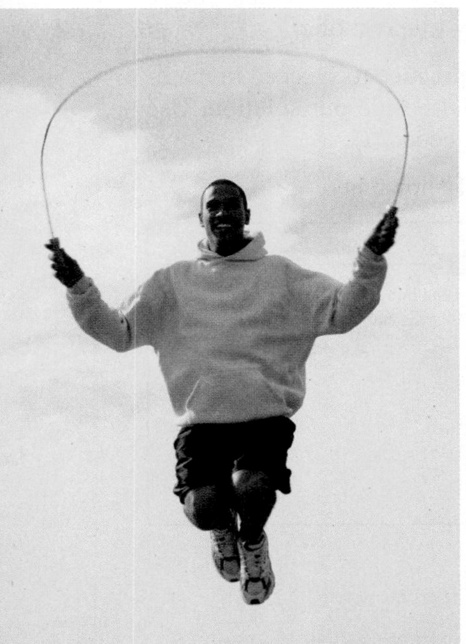

© Cengage Learning/Photodisc

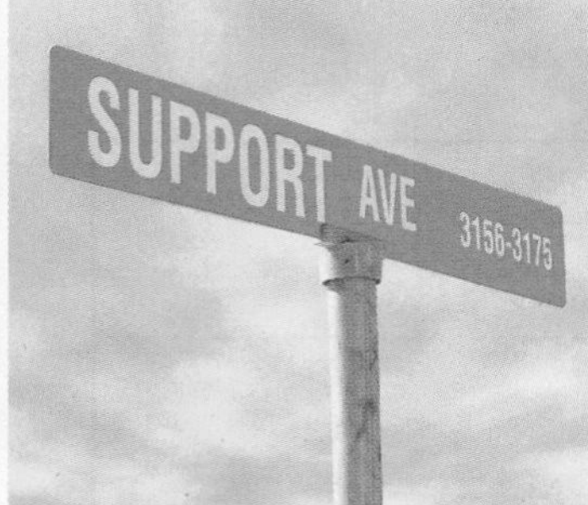

- Before you start a new section, take a few minutes to read the **Objective Statement** for that section. Then, browse through the objective material. Especially note the words or phrases in bold type—these are important concepts that you'll need as you're moving along in the course.
- As you start moving through the chapter, pay special attention to the **rule boxes.** These rules give you the reasons certain types of problems are solved the way they are. When you see a rule, try to rewrite the rule in your own words.

Rule for Multiplying Exponential Expressions

If m and n are positive integers, then $x^m \cdot x^n = x^{m+n}$.

Page 348

Knowing what to pay attention to as you move through a chapter will help you study and prepare.

INTERACTION We want you to be actively involved in learning mathematics and have given you many ways to get hands-on with this book.

- **HOW TO EXAMPLES** Take a look at page 257 shown here. See the HOW TO example? This contains an explanation by each step of the solution to a sample problem.

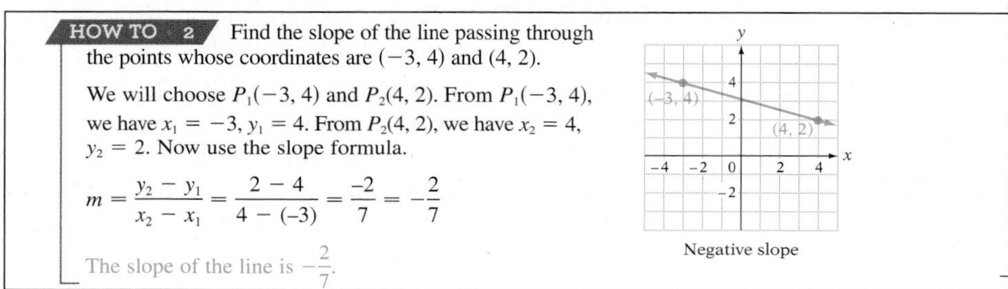

HOW TO 2 Find the slope of the line passing through the points whose coordinates are $(-3, 4)$ and $(4, 2)$.

We will choose $P_1(-3, 4)$ and $P_2(4, 2)$. From $P_1(-3, 4)$, we have $x_1 = -3$, $y_1 = 4$. From $P_2(4, 2)$, we have $x_2 = 4$, $y_2 = 2$. Now use the slope formula.

$$m = \frac{y_2 - y_1}{x_2 - x_1} = \frac{2 - 4}{4 - (-3)} = \frac{-2}{7} = -\frac{2}{7}$$

The slope of the line is $-\frac{2}{7}$.

Negative slope

Page 257

Grab a paper and pencil and work along as you're reading through each example. When you're done, get a clean sheet of paper. Write down the problem and try to complete the solution without looking at your notes or at the book. When you're done, check your answer. If you got it right, you're ready to move on.

- **EXAMPLE/YOU TRY IT PAIRS** You'll need hands-on practice to succeed in mathematics. When we show you an example, work it out beside our solution. Use the Example/You Try It pairs to get the practice you need.

  Take a look at page 259, Example 1 and You Try It 1 shown here:

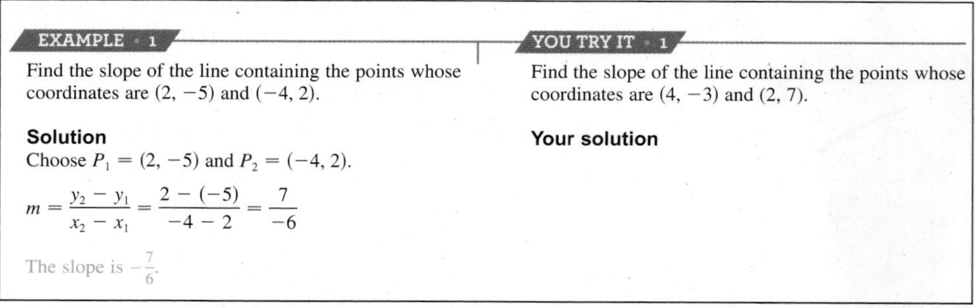

EXAMPLE 1

Find the slope of the line containing the points whose coordinates are $(2, -5)$ and $(-4, 2)$.

Solution

Choose $P_1 = (2, -5)$ and $P_2 = (-4, 2)$.

$$m = \frac{y_2 - y_1}{x_2 - x_1} = \frac{2 - (-5)}{-4 - 2} = \frac{7}{-6}$$

The slope is $-\frac{7}{6}$.

YOU TRY IT 1

Find the slope of the line containing the points whose coordinates are $(4, -3)$ and $(2, 7)$.

Your solution

Page 259

You'll see that each Example is fully worked-out. Study this Example carefully by working through each step. Then, try your hand at it by completing the You Try It. If you get stuck, the solutions to the You Try Its are provided in the back of the book. There is a page number following the You Try It, which shows you where you can find the completely worked-out solution. Use the solution to get a hint for the step on which you are stuck. Then, try again!

When you've finished the solution, check your work against the solution in the back of the book. Turn to page S12 to see the solution for You Try It 1.

Remember that sometimes there can be more than one way to solve a problem. But, your answer should always match the answers we've given in the back of the book. If you have any questions about whether your method will always work, check with your instructor.

REVIEW We have provided many opportunities for you to practice and review the skills you have learned in each chapter.

- **SECTION EXERCISES** After you're done studying a section, flip to the end of the section and complete the exercises. If you immediately practice what you've learned, you'll find it easier to master the core skills. Want to know if you answered the questions correctly? The answers to the odd-numbered exercises are given in the back of the book.
- **CHAPTER SUMMARY** Once you've completed a chapter, look at the Chapter Summary. This is divided into two sections: *Key Words* and *Essential Rules and Procedures.* Flip to page 288 to see the Chapter Summary for Chapter 4. This summary shows all of the important topics covered in the chapter. See the reference following each topic? This shows you the objective reference and the page in the text where you can find more information on the concept.
- **CONCEPT REVIEW** Following the Chapter Summary for each chapter is the Concept Review. Flip to page 291 to see the Concept Review for Chapter 4. When you read each question, jot down a reminder note on the right about whatever you feel will be most helpful to remember if you need to apply that concept during an exam. You can also use the space on the right to mark what concepts your instructor expects you to know for the next test. If you are unsure of the answer to a concept review question, flip to the answers appendix at the back of the book.
- **CHAPTER REVIEW EXERCISES** You'll find the Chapter Review Exercises after the Concept Review. Flip to page 292 to see the Chapter Review Exercises for Chapter 4. When you do the review exercises, you're giving yourself an important opportunity to test your understanding of the chapter. The answer to each review exercise is given at the back of the book, along with the objective the question relates to. When you're done with the Chapter Review Exercises, check your answers. If you had trouble with any of the questions, you can restudy the objectives and retry some of the exercises in those objectives for extra help.

- **CHAPTER TESTS** The Chapter Tests can be found after the Chapter Review Exercises and can be used to prepare for your exams. The answer to each test question is given at the back of the book, along with a reference to a How To, Example, or You Try It that the question relates to. Think of these tests as "practice runs" for your in-class tests. Take the test in a quiet place and try to work through it in the same amount of time you will be allowed for your exam.

Here are some strategies for success when you're taking your exams:

- Scan the entire test to get a feel for the questions (get the big picture).
- Read the directions carefully.
- Work the problems that are easiest for you first.
- Stay calm, and remember that you will have lots of opportunities for success in this class!

EXCEL Visit **www.cengage.com/math/aufmann** to learn about additional study tools!

- *Enhanced* ***WebAssign***® online practice exercises and homework problems match the textbook exercises.
- **DVDs** Hosted by Dana Mosley, an experienced mathematics instructor, the DVDs will help you to get a better handle on topics that may be giving you trouble. A comprehensive set of DVDs for the entire course is available to order.

Get Involved

Have a question? Ask! Your professor and your classmates are there to help. Here are some tips to help you jump in to the action:

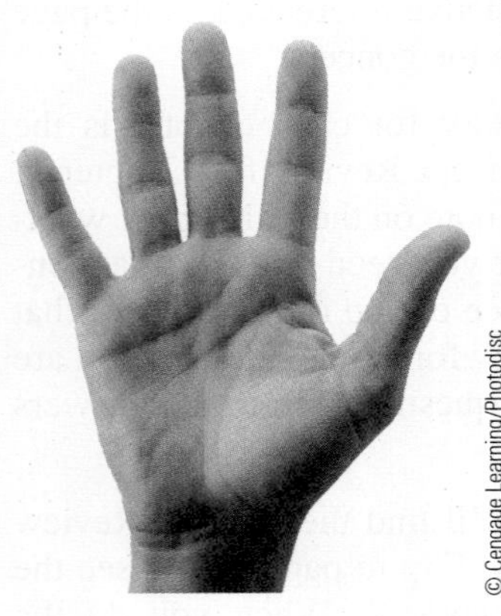

- Raise your hand in class.
- If your instructor prefers, email or call your instructor with your question. If your professor has a website where you can post your question, also look there for answers to previous questions from other students. Take advantage of these ways to get your questions answered.
- Visit a **math center.** Ask your instructor for more information about the math center services available on your campus.
- Your instructor will have **office hours** where he or she will be available to help you. Take note of where and when your instructor holds office hours. Use this time for one-on-one help, if you need it.
- Form a **study group** with students from your class. This is a great way to prepare for tests, catch up on topics you may have missed, or get extra help on problems you're struggling with. Here are a few suggestions to make the most of your study group:

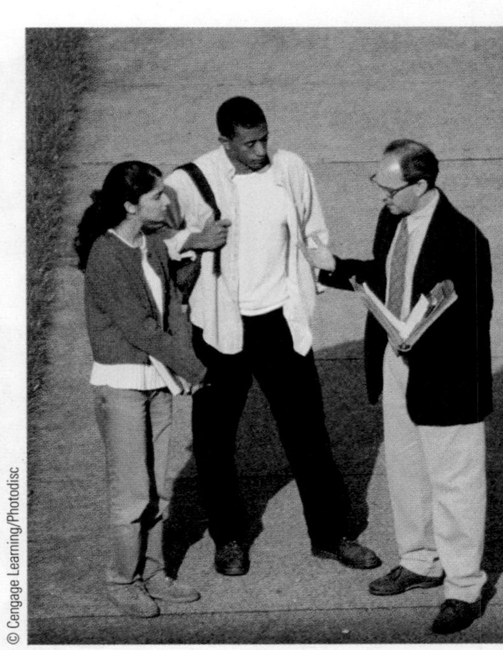

- **Test each other by asking questions.** Have each person bring a few sample questions when you get together.

- **Practice teaching each other.** We've found that you can learn a lot about what you know when you have to explain it to someone else.
- **Compare class notes.** Couldn't understand the last five minutes of class? Missed class because you were sick? Chances are someone in your group has the notes for the topics you missed.

© Cengage Learning/Photodisc

- **Brainstorm test questions.**
- **Make a plan for your meeting.** Agree on what topics you'll talk about and how long you'll be meeting. When you make a plan, you'll be sure that you make the most of your meeting.

Ready, Set, Succeed! It takes hard work and commitment to succeed, but we know you can do it! Doing well in mathematics is just one step you'll take along the path to success.

I succeeded in Introductory and Intermediate Algebra!

We are confident that if you follow our suggestions, you will succeed. Good luck!

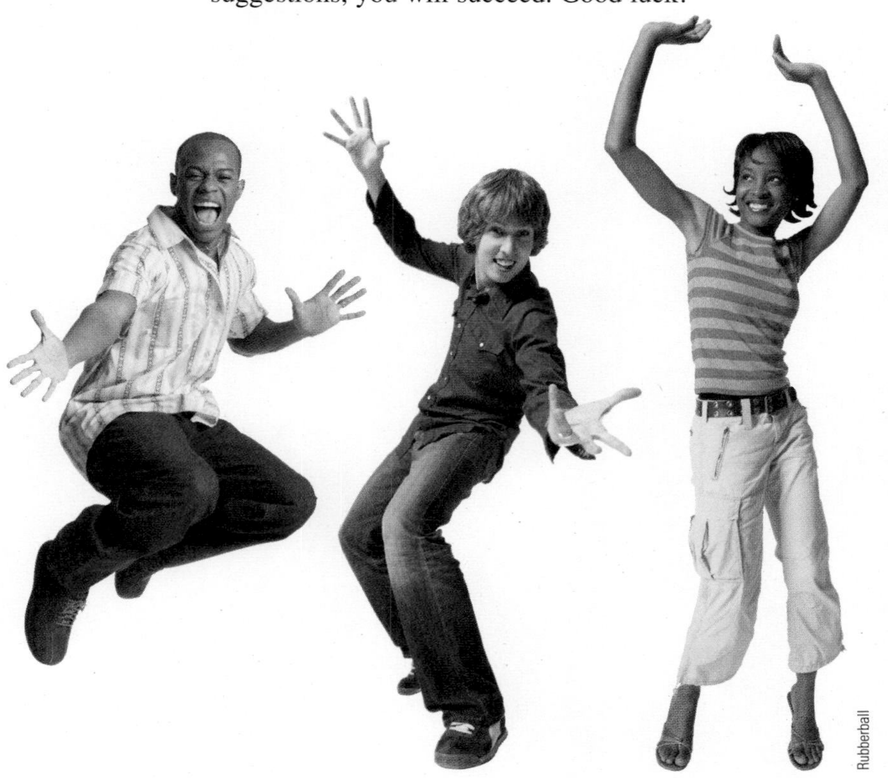

Rubberball

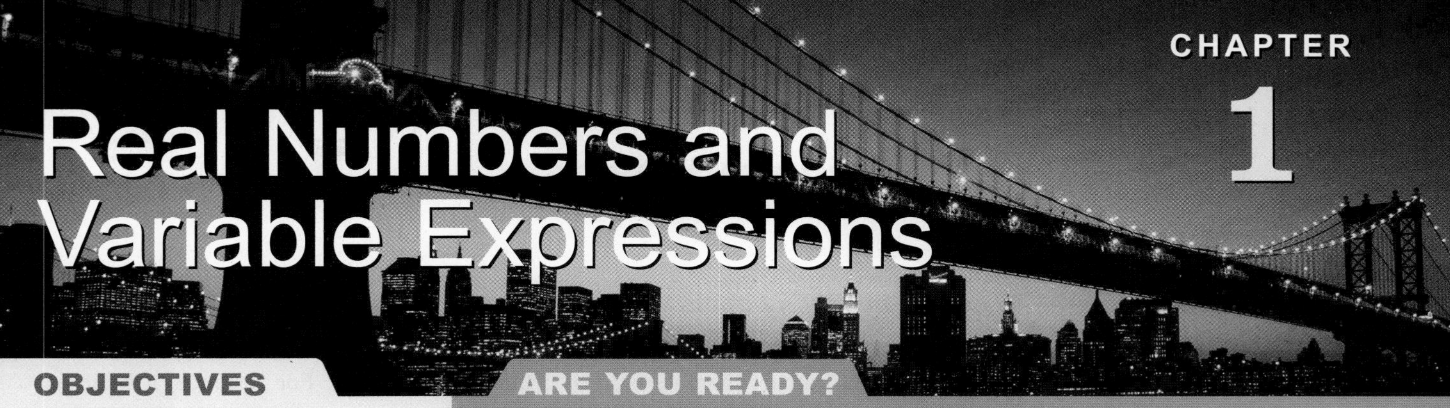

CHAPTER 1

Real Numbers and Variable Expressions

Panoramic Images/Getty Images

OBJECTIVES

SECTION 1.1

- A To use inequality symbols with integers
- B To find the additive inverse and absolute value of a number
- C To add or subtract integers
- D To multiply or divide integers
- E To solve application problems

SECTION 1.2

- A To write a rational number as a decimal
- B To convert among percents, fractions, and decimals
- C To add or subtract rational numbers
- D To multiply or divide rational numbers
- E To evaluate exponential expressions
- F To simplify numerical radical expressions
- G To solve application problems

SECTION 1.3

- A To use the Order of Operations Agreement to simplify expressions

SECTION 1.4

- A To evaluate a variable expression
- B To simplify a variable expression using the Properties of Addition
- C To simplify a variable expression using the Properties of Multiplication
- D To simplify a variable expression using the Distributive Property
- E To translate a verbal expression into a variable expression

SECTION 1.5

- A To write a set using the roster method
- B To write and graph sets of real numbers

ARE YOU READY?

Take the Chapter 1 Prep Test to find out if you are ready to learn to:

- Add, subtract, multiply, and divide integers and rational numbers
- Convert among percents, fractions, and decimals
- Evaluate numerical expressions
- Evaluate variable expressions
- Simplify variable expressions
- Write a set using the roster method, set-builder notation, and interval notation
- Graph an inequality on the number line

PREP TEST

Do these exercises to prepare for Chapter 1.

1. What is 127.1649 rounded to the nearest hundredth?

2. Add: $3416 + 42{,}561 + 537$

3. Subtract: $5004 - 487$

4. Multiply: 407×28

5. Divide: $11{,}684 \div 23$

6. What is the smallest number that both 8 and 12 divide evenly?

7. What is the greatest number that divides both 16 and 20 evenly?

8. Without using 1, write 21 as a product of two whole numbers.

9. Represent the shaded portion of the figure as a fraction in simplest form.

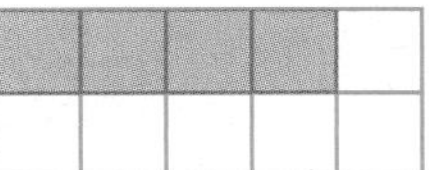

SECTION

1.1 Introduction to Integers

OBJECTIVE A To use inequality symbols with integers 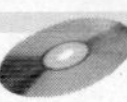

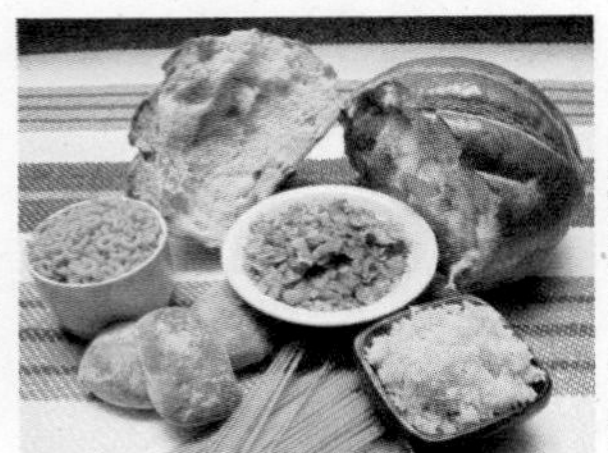
© Tony Freeman/PhotoEdit

It seems to be a human characteristic to group similar items. For instance, a biologist places similar animals in groups called *species*. Nutritionists classify foods according to *food groups;* for example, pasta, crackers, and rice are among the foods in the bread group.

Mathematicians place objects with similar properties in groups called *sets.* A **set** is a collection of objects. The objects in a set are called the **elements of the set.**

The **roster method** of writing sets encloses a list of the elements in braces. Thus the set of sections within an orchestra is written {brass, percussion, string, woodwind}. When the elements of a set are listed, each element is listed only once. For instance, if the list of numbers 1, 2, 3, 2, 3 were placed in a set, the set would be {1, 2, 3}.

The symbol $\in$ means "is an element of." $2 \in B$ is read "2 is an element of set B."

Given $C = \{3, 5, 9\}$, then $3 \in C$, $5 \in C$, and $9 \in C$. $7 \notin C$ is read "7 is not an element of set C."

The numbers that we use to count objects, such as the students in a classroom or the horses on a ranch, are the *natural numbers.*

Natural numbers = {1, 2, 3, 4, 5, 6, 7, 8, 9, 10, . . .}

The three dots mean that the list of natural numbers continues on and on and that there is no largest natural number.

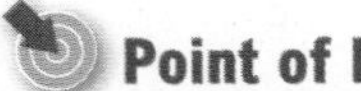
Point of Interest

The Alexandrian astronomer Ptolemy began using *omicron, O,* the first letter of the Greek word that means "nothing," as the symbol for zero in 150 A.D. It was not until the 13th century, however, that Fibonacci introduced 0 to the Western world as a placeholder so that we could distinguish, for example, 45 from 405.

Each natural number greater than 1 is a *prime* number or a *composite* number. A **prime number** is a natural number greater than 1 that is divisible (evenly) only by itself and 1. For example, 2, 3, 5, 7, 11, and 13 are the first six prime numbers. A natural number that is not a prime number is a **composite number.** The numbers 4, 6, 8, and 9 are the first four composite numbers.

The natural numbers do not have a symbol to denote the concept of none—for instance, the number of trees taller than 1000 feet. The *whole numbers* include zero and the natural numbers.

Whole numbers = {0, 1, 2, 3, 4, 5, 6, 7, 8, . . .}

The whole numbers alone do not provide all the numbers that are useful in applications. For instance, a meteorologist needs numbers below zero and above zero.

Integers = {. . . , −5, −4, −3, −2, −1, 0, 1, 2, 3, 4, 5, . . .}

Each integer can be shown on a number line. The integers to the left of zero on the number line are called **negative integers.** The integers to the right of zero are called **positive integers,** or natural numbers. Zero is neither a positive nor a negative integer.

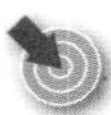

Point of Interest

There was quite a controversy over the date on which the new millennium started because of the number zero. When our current calendar was created, numbering began with the year 1 because 0 had not yet been invented. Thus at the beginning of year 2, 1 year had elapsed; at the beginning of year 3, 2 years had elapsed; and so on. This means that at the beginning of year 2000, 1999 years had elapsed. It was not until the beginning of year 2001 that 2000 years had elapsed and a new millennium began.

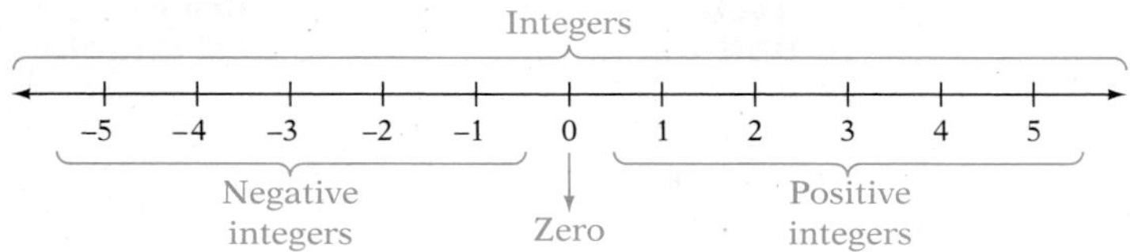

The **graph** of an integer is shown by placing a heavy dot on the number line directly above the number. The graphs of −3 and 4 are shown on the number line below.

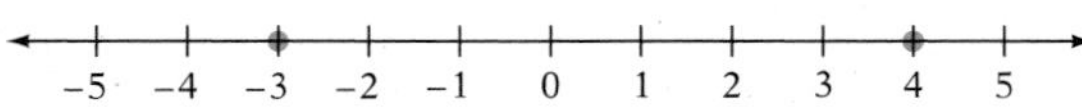

Consider the following sentences.

The quarterback threw the football and the receiver caught *it.*

A student purchased a computer and used *it* to write history papers.

In the first sentence, *it* is used to mean the football; in the second sentence, *it* means the computer. In language, the word *it* can stand for many different objects. Similarly, in mathematics, a letter of the alphabet can be used to stand for a number. Such a letter is called a **variable.** Variables are used in the following definition of inequality symbols.

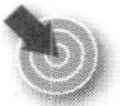

Point of Interest

The symbols for "is less than" and "is greater than" were introduced by Thomas Harriot around 1630. Before that, ⊏ and ⊐ were used for $>$ and $<$, respectively.

Inequality Symbols

If a and b are two numbers and a is to the left of b on the number line, then a **is less than** b. This is written $a < b$.

If a and b are two numbers and a is to the right of b on the number line, then a **is greater than** b. This is written $a > b$.

Here are some examples.

$$5 < 9 \qquad -4 > -10 \qquad \pi < \sqrt{17} \qquad 0 > -\frac{2}{3}$$

The inequality symbols $\leq$ (is less than or equal to) and $\geq$ (is greater than or equal to) are also important. Note the examples below.

$4 \leq 5$ is a true statement because $4 < 5$.
$5 \leq 5$ is a true statement because $5 = 5$.

EXAMPLE • 1

Let $y \in \{-7, 0, 6\}$. For which values of y is the inequality $y < 4$ a true statement?

Solution

Replace y by each of the elements of the set and determine whether the inequality is true.

$y < 4$
$-7 < 4$ True
$0 < 4$ True
$6 < 4$ False

The inequality is true for −7 and 0.

YOU TRY IT • 1

Let $z \in \{-10, -5, 6\}$. For which values of z is the inequality $z > -5$ a true statement?

Your solution

Solution on p. S1

OBJECTIVE B To find the additive inverse and absolute value of a number

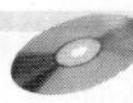

The numbers 5 and -5 are the same distance from zero on the number line but on opposite sides of zero. The numbers 5 and -5 are called **additive inverses** or **opposites** of each other.

The additive inverse (or opposite) of 5 is -5. The additive inverse of -5 is 5. The symbol for additive inverse is $-$.

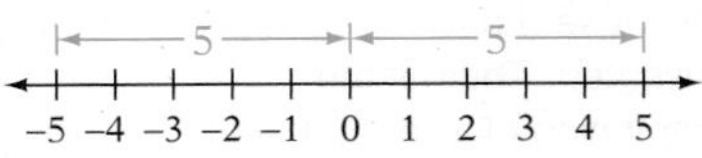

$-(2)$ means the additive inverse of *positive* 2. $\quad -(2) = -2$

$-(-5)$ means the additive inverse of *negative* 5. $\quad -(-5) = 5$

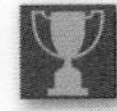

Tips for Success

One of the key instructional features of this text is the Example/You Try It pairs. Each Example is completely worked. You are to solve the You Try It problems. When you are ready, check your solution against the one given in the Solution section. The solution for You Try It 2 below is on page S1 (see the reference at the bottom right of the You Try It box). See *AIM for Success* in the Preface.

The **absolute value** of a number is its distance from zero on the number line. The symbol for absolute value is $|\ \ |$.

Note from the figure above that the distance from 0 to 5 is 5. Therefore, $|5| = 5$. That figure also shows that the distance from 0 to -5 is 5. Therefore, $|-5| = 5$.

> **Absolute Value**
>
> The absolute value of a positive number is the number itself.
> The absolute value of a negative number is the opposite of the negative number.
> The absolute value of zero is zero.

HOW TO • 1 Evaluate: $-|-12|$

$-|-12| = -12$ • **The absolute value symbol does not affect the negative sign *in front of* the absolute value symbol.**

EXAMPLE • 2

Let $y \in \{-12, 0, 4\}$.

a. Determine $-y$, the additive inverse of y, for each element of the set.

b. Evaluate $|y|$ for each element of the set.

Solution

a. Replace y in $-y$ by each element of the set and determine the value of the expression.

$$-y$$
$$-(-12) = 12$$
$$-(0) = 0$$ • **0 is neither positive nor negative.**
$$-(4) = -4$$

b. Replace y in $|y|$ by each element of the set and determine the value of the expression.

$$|y|$$
$$|-12| = 12$$
$$|0| = 0$$
$$|4| = 4$$

YOU TRY IT • 2

Let $d \in \{-11, 0, 8\}$.

a. Determine $-d$, the additive inverse of d, for each element of the set.

b. Evaluate $|d|$ for each element of the set.

Your solution

Solution on p. S1

OBJECTIVE C To add or subtract integers 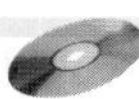

A number can be represented anywhere along the number line by an arrow. A positive number is represented by an arrow pointing to the right, and a negative number is represented by an arrow pointing to the left. The size of the number is represented by the length of the arrow.

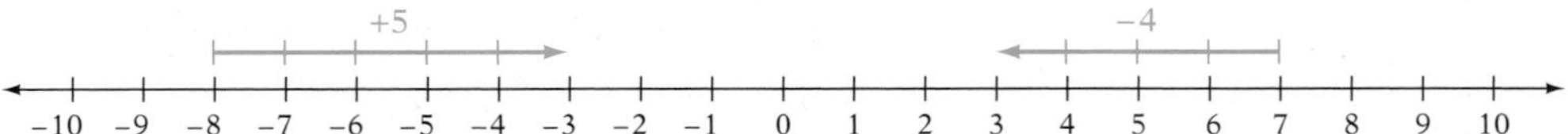

Addition is the process of finding the total of two numbers. The numbers being added are called **addends.** The total is called the **sum.** Addition of integers can be shown on the number line. To add integers, start at zero and draw, above the number line, an arrow representing the first number. At the tip of the first arrow, draw a second arrow representing the second number. The sum is below the tip of the second arrow.

$4 + 2 = 6$

$-4 + (-2) = -6$

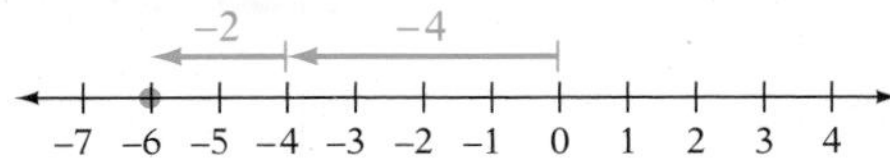

$-4 + 2 = -2$

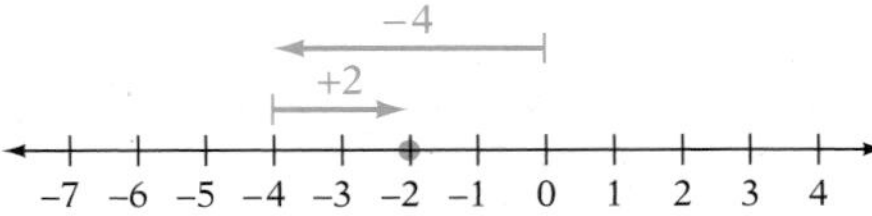

$4 + (-2) = 2$

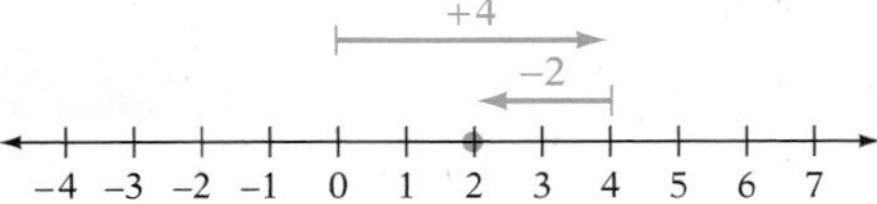

The pattern for addition shown on the number lines above is summarized in the following rules for adding integers.

Addition of Integers

To add two numbers with the same sign, add the absolute values of the numbers. Then attach the sign of the addends.

To add two numbers with different signs, find the absolute value of each number. Subtract the smaller of the two numbers from the larger. Then attach the sign of the number with the larger absolute value.

HOW TO • 2 Add: $-12 + (-26)$

$-12 + (-26) = -38$

- The signs are the same. Add the absolute values of the numbers $(12 + 26)$. Attach the sign of the addends.

Tips for Success

The HOW TO feature indicates an example with explanatory remarks. Using paper and pencil, you should work through the example. See *AIM for Success* in the Preface.

HOW TO • 3 Add: $-19 + 8$

$|-19| = 19 \quad |8| = 8$
- The signs are different. Find the absolute value of each number.

$19 - 8 = 11$
- Subtract the smaller number from the larger.

$-19 + 8 = -11$
- Attach the sign of the number with the larger absolute value.

HOW TO • 4 Find the sum of -23, 47, -18, and -10.

Recall that a sum is the answer to an addition problem.

$-23 + 47 + (-18) + (-10)$
$= 24 + (-18) + (-10)$
$= 6 + (-10) = -4$

• To add more than two numbers, add the first two numbers. Then add the sum to the third number. Continue until all the numbers are added.

Look at the expressions below. Note that each expression equals the same number.

$8 - 3 = 5$ 8 minus 3 is 5.
$8 + (-3) = 5$ 8 plus the opposite of 3 is 5.

This example suggests the following.

Subtraction of Integers

To subtract one number from another, add the opposite of the second number to the first number.

HOW TO • 5 Subtract: $-21 - (-40)$

Change this sign to plus.

$-21 - (-40) = -21 + 40 = 19$

Change -40 to the opposite of -40.

• Rewrite the subtraction as addition of the opposite. Then add.

HOW TO • 6 Subtract: $15 - 51$

Change this sign to plus.

$15 - 51 = 15 + (-51) = -36$

Change 51 to the opposite of 51.

• Rewrite the subtraction as addition of the opposite. Then add.

HOW TO • 7 Subtract: $-12 - (-21) - 15$

$-12 - (-21) - 15 = -12 + 21 + (-15)$
$= 9 + (-15) = -6$

• Rewrite each subtraction as addition of the opposite. Then add.

HOW TO • 8 Find the difference between -8 and 7.

A *difference* is the answer to a subtraction problem.

$-8 - 7 = -8 + (-7)$
$= -15$

• Rewrite the subtraction as addition of the opposite. Then add.

EXAMPLE • 3

Add: $-52 + (-39)$

Solution

$-52 + (-39) = -91$

YOU TRY IT • 3

Add: $100 + (-43)$

Your solution

Solution on p. S1

EXAMPLE • 4

Add: $37 + (-52) + (-14)$

Solution

$$37 + (-52) + (-14) = -15 + (-14) = -29$$

YOU TRY IT • 4

Add: $-51 + 42 + 17 + (-102)$

Your solution

EXAMPLE • 5

Find 11 more than -23.

Solution

$-23 + 11 = -12$

YOU TRY IT • 5

Find -8 increased by 7.

Your solution

EXAMPLE • 6

Subtract: $-14 - 18 - (-21) - 4$

Solution

$$\begin{aligned}-14 - 18 - (-21) - 4 &= -14 + (-18) + 21 + (-4)\\ &= -32 + 21 + (-4)\\ &= -11 + (-4) = -15\end{aligned}$$

YOU TRY IT • 6

Subtract: $-9 - (-12) - 17 - 4$

Your solution

EXAMPLE • 7

Find 9 less than -4.

Solution

$-4 - 9 = -4 + (-9) = -13$

YOU TRY IT • 7

Subtract -12 from -11.

Your solution

Solutions on p. S1

OBJECTIVE D To multiply or divide integers

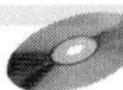

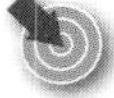

Point of Interest

The cross × was first used as a symbol for multiplication in 1631 in a book titled *The Key to Mathematics.* Also in that year, another book, *Practice of the Analytical Art,* advocated the use of a dot to indicate multiplication.

Several different symbols are used to indicate multiplication. The numbers being multiplied are called **factors;** for instance, 3 and 2 are factors in each of the examples at the right. The result is called the **product.** Note that when parentheses are used and there is no arithmetic symbol, the operation is multiplication.

$$3 \times 2 = 6$$
$$3 \cdot 2 = 6$$
$$(3)(2) = 6$$
$$3(2) = 6$$
$$(3)2 = 6$$

Multiplication is repeated addition of the same number. The product 3×5 is shown on the number line below.

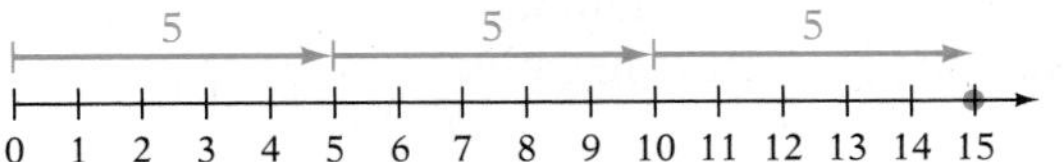

$$3 \times 5 = \underbrace{5 + 5 + 5}_{\text{5 is added 3 times.}} = 15$$

Now consider the product of a positive and a negative number.

$$3(-5) = \underbrace{(-5) + (-5) + (-5)}_{-5 \text{ is added 3 times.}} = -15$$

This suggests that the product of a positive number and a negative number is negative. Here are a few more examples.

$$4(-7) = -28 \qquad -6 \cdot 7 = -42 \qquad (-8)7 = -56$$

To find the product of two negative numbers, look at the pattern at the right. As -5 multiplies a sequence of decreasing integers, the products increase by 5.

These numbers decrease by 1. These numbers increase by 5.

$$-5 \times 3 = -15$$
$$-5 \times 2 = -10$$
$$-5 \times 1 = -5$$
$$-5 \times 0 = 0$$

The pattern can be continued by requiring that the product of two negative numbers be positive.

$$-5 \times (-1) = 5$$
$$-5 \times (-2) = 10$$
$$-5 \times (-3) =$$

Multiplication of Integers

To multiply two numbers with the same sign, multiply the absolute values of the numbers. The product is positive.

To multiply two numbers with different signs, multiply the absolute values of the numbers. The product is negative.

HOW TO • 9 Multiply: $5(-12)$

$5(-12) = -60$ • The signs are different. The product is negative.

HOW TO • 10 Find the product of -8 and -16.

A *product* is the answer to a multiplication problem.

$-8(-16) = 128$ • The signs are the same. The product is positive.

HOW TO • 11 Multiply: $-2(5)(-7)(-4)$

$$-2(5)(-7)(-4) = -10(-7)(-4)$$
$$= 70(-4) = -280$$

• To multiply more than two numbers, multiply the first two. Then multiply the product by the third number. Continue until all the numbers are multiplied.

Consider the products shown at the right. Note that when there is an even number of negative factors, the product is positive. When there is an odd number of negative factors, the product is negative.

$$(-3)(-5) = 15$$
$$(-2)(-5)(-6) = -60$$
$$(-4)(-3)(-5)(-7) = 420$$
$$(-3)(-3)(-5)(-4)(-5) = -900$$
$$(-6)(-3)(-4)(-2)(-10)(-5) = 7200$$

This idea can be summarized by the following useful rule: The product of an even number of negative factors is positive; the product of an odd number of negative factors is negative.

✓ Take Note

Think of the fraction bar as "divided by." Thus $\frac{8}{2}$ is 8 divided by 2. The number 2 is the **divisor.** The number 8 is the **dividend.** The result of the division, 4, is called the **quotient.**

For every division problem there is a related multiplication problem.

$\frac{8}{2} = 4$	because	$4 \cdot 2 = 8.$
Division		Related multiplication

Division of Integers

To divide two numbers with the same sign, divide the absolute values of the numbers. The quotient is positive.

To divide two numbers with different signs, divide the absolute values of the numbers. The quotient is negative.

HOW TO • 12 Divide: $-36 \div 9$

$-36 \div 9 = -4$ • The signs are different. The quotient is negative.

Take Note

We can denote division using, for example, $-63 \div (-7)$, $-7\overline{)-63}$, or $\frac{-63}{-7}$.

HOW TO • 13 Find the quotient of -63 and -7.

A *quotient* is the answer to a division problem.

$\frac{-63}{-7} =$ • The signs are the same. The quotient is positive.

This fact and the rules for multiplying integers can be used to illustrate the rules for dividing integers.

Note in the following examples that the quotient of two numbers with the same sign is positive.

$\frac{12}{3} =$ because $4 \cdot 3 = 12$. $\frac{-12}{-3} =$ because $4(-3) = -12$.

The next two examples illustrate that the quotient of two numbers with different signs is negative.

$\frac{12}{-3} = -4$ because $(-4)(-3) = 12$. $\frac{-12}{3} = -4$ because $(-4)3 = -12$.

HOW TO • 14 Simplify: $-\frac{-56}{7}$

$$-\frac{-56}{7} = -\left(\frac{-56}{7}\right) = -(-8) = 8$$

The properties of division are stated below. In these statements, the symbol $\neq$ is read "is not equal to."

Properties of Zero and One in Division

If $a \neq 0$, $\frac{0}{a} = 0$.	Zero divided by any number other than zero is zero.
If $a \neq 0$, $\frac{a}{a} = 1$.	Any number other than zero divided by itself is one.
$\frac{a}{1} = a$.	A number divided by one is the number.
$\frac{a}{0}$ is undefined.	Division by zero is not defined.

The fact that $\frac{-12}{3} = -4$, $\frac{12}{-3} = -4$, and $-\frac{12}{3} = -4$ suggests the following rule.

If a and b are integers, and $b \neq 0$, then $\frac{-a}{b} = \frac{a}{-b} = -\frac{a}{b}$.

EXAMPLE • 8

Multiply: $(-3)4(-5)$

Solution

$(-3)4(-5) = (-12)(-5) = 60$

YOU TRY IT • 8

Multiply: $8(-9)10$

Your solution

EXAMPLE • 9

Multiply: $12(-4)(-3)(-5)$

Solution

$$\begin{aligned} 12(-4)(-3)(-5) &= (-48)(-3)(-5) \\ &= 144(-5) \\ &= -720 \end{aligned}$$

YOU TRY IT • 9

Multiply: $(-2)3(-8)7$

Your solution

EXAMPLE • 10

Find the product of -13 and -9.

Solution

$-13(-9) = 117$

YOU TRY IT • 10

What is -9 times 34?

Your solution

EXAMPLE • 11

Divide: $(-120) \div (-8)$

Solution

$(-120) \div (-8) = 15$

YOU TRY IT • 11

Divide: $(-135) \div (-9)$

Your solution

EXAMPLE • 12

Divide: $\frac{95}{-5}$

Solution

$\frac{95}{-5} = -19$

YOU TRY IT • 12

Divide: $\frac{-72}{4}$

Your solution

EXAMPLE • 13

Simplify: $-\frac{-81}{3}$

Solution

$-\frac{-81}{3} = -(-27) = 27$

YOU TRY IT • 13

Simplify: $-\frac{36}{-12}$

Your solution

EXAMPLE • 14

Find the quotient of 98 and -14.

Solution

$98 \div (-14) = -7$

YOU TRY IT • 14

What is the ratio of -72 and -8?

Your solution

Solutions on p. S1

OBJECTIVE E To solve application problems 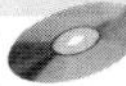

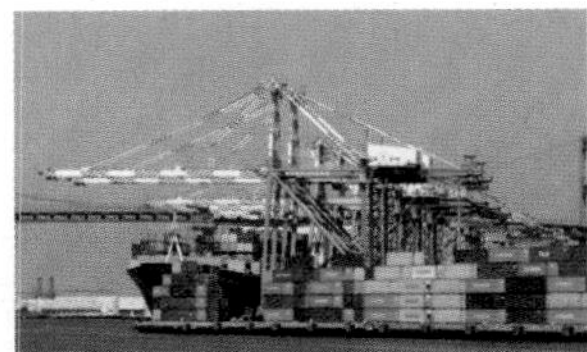
Peter Titmuss/Alamy

An **export** is a good or service produced in one's own country and sold for consumption in another country. An **import** is a good or service consumed in one's own country that was bought from another country. A nation's **balance of trade** is the difference between the value of its exports and the value of its imports over a particular period of time.

A **favorable balance of trade** exists when the value of a nation's exports is greater than the value of its imports. In this case, the balance of trade is a positive number. An **unfavorable balance of trade** exists when the value of a nation's imports is greater than the value of its exports. In this case, the balance of trade is a negative number. An unfavorable balance of trade is referred to as a **trade deficit.** A trade deficit is considered unfavorable because more money is going out of the country to pay for imported goods than is coming into the country to pay for exported goods.

The U.S. government provides data on international trade. Statistics are reported monthly, quarterly, and annually. The following table lists the U.S. balance of trade for the years 2003 through 2007. Also shown is the total balance of trade for the five years.

Year	Balance of Trade (in billions of dollars)
2003	−497
2004	−608
2005	−712
2006	−753
2007	−700
Total	−3270

Source: www.census.gov

To determine the average annual trade deficit for the years 2003 through 2007, divide the sum of the balances of trade by the number of years (5).

$$-3270 \div 5 = -654$$

The average annual trade deficit for the years 2003 through 2007 was −$654 billion.

EXAMPLE 15

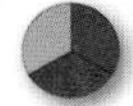

The average temperature on Mercury's sunlit side is 950°F. The average temperature on Mercury's dark side is −346°F. Find the difference between these two average temperatures.

Strategy

To find the difference, subtract the average temperature on the dark side (−346) from the average temperature on the sunlit side (950).

Solution

$$\begin{aligned} 950 - (-346) &= 950 + 346 \\ &= 1296 \end{aligned}$$

The difference between the average temperatures is 1296°F.

YOU TRY IT 15

The daily low temperatures (in degrees Celsius) during one week were recorded as −6°, −7°, 0°, −5°, −8°, −1°, and −1°. Find the average daily low temperature.

Your strategy

Your solution

Solution on p. S1

1.1 EXERCISES

OBJECTIVE A To use inequality symbols with integers

For Exercises 1 to 10, place the correct symbol, $<$ or $>$, between the two numbers.

1. 8 $\quad -6$ **2.** $-14 \quad 16$ **3.** $-12 \quad 1$ **4.** 35 $\quad$ 28 **5.** 42 $\quad$ 19

6. $-42 \quad 27$ **7.** $0 \quad -31$ **8.** $-17 \quad 0$ **9.** $53 \quad -46$ **10.** $-27 \quad -39$

For Exercises 11 to 20, answer true or false.

11. $-13 > 0$ **12.** $-20 > 3$ **13.** $12 > -31$ **14.** $9 > 7$ **15.** $-5 > -2$

16. $-44 > -21$ **17.** $-4 > -120$ **18.** $0 > -8$ **19.** $-1 > 0$ **20.** $-10 > -88$

21. Let $x \in \{-23, -18, -8, 0\}$. For which values of x is the inequality $x < -8$ a true statement?

22. Let $w \in \{-33, -24, -10, 0\}$. For which values of w is the inequality $w < -10$ a true statement?

23. Let $a \in \{-33, -15, 21, 37\}$. For which values of a is the inequality $a > -10$ a true statement?

24. Let $v \in \{-27, -14, 14, 27\}$. For which values of v is the inequality $v > -15$ a true statement?

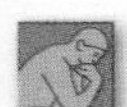
For Exercises 25 and 26, determine which of the following statements is true about n.
(i) n is positive. (ii) n is negative. (iii) n is zero. (iv) n can be positive, negative, or zero.

25. The number n is to the right of the number 5 on the number line.

26. The number n is to the left of the number 5 on the number line.

OBJECTIVE B To find the additive inverse and absolute value of a number

For Exercises 27 to 32, find the additive inverse.

27. 4 **28.** 8 **29.** -9 **30.** -12 **31.** -28 **32.** -36

For Exercises 33 to 44, evaluate.

33. $-(-14)$ **34.** $-(-40)$ **35.** $-(77)$ **36.** $-(39)$ **37.** $-(0)$ **38.** $-(-13)$

39. $|-74|$ **40.** $|-96|$ **41.** $-|-82|$ **42.** $-|-53|$ **43.** $-|81|$ **44.** $-|38|$

For Exercises 45 to 52, place the correct symbol, < or >, between the values of the two numbers.

45. $|-83| \quad |58|$ **46.** $|22| \quad |-19|$ **47.** $|43| \quad |-52|$ **48.** $|-71| \quad |-92|$

49. $|-68| \quad |-42|$ **50.** $|12| \quad |-31|$ **51.** $|-45| \quad |-61|$ **52.** $|-28| \quad |43|$

53. Let $p \in \{-19, 0, 28\}$. Evaluate $-p$ for each element of the set.

54. Let $q \in \{-34, 0, 31\}$. Evaluate $-q$ for each element of the set.

55. Let $x \in \{-45, 0, 17\}$. Evaluate $-|x|$ for each element of the set.

56. Let $y \in \{-91, 0, 48\}$. Evaluate $-|y|$ for each element of the set.

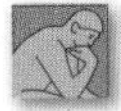

57. True or false? The absolute value of a negative number n is greater than n.

OBJECTIVE C To add or subtract integers

58. Explain how to rewrite the subtraction $8 - (-6)$ as addition of the opposite.

For Exercises 59 to 109, add or subtract.

59. $-3 + (-8)$ **60.** $-6 + (-9)$ **61.** $-8 + 3$ **62.** $-9 + 2$

63. $-3 + (-80)$ **64.** $-12 + (-1)$ **65.** $-23 + (-23)$ **66.** $-12 + (-12)$

67. $16 + (-16)$ **68.** $-17 + 17$ **69.** $48 + (-53)$ **70.** $19 + (-41)$

71. $-17 + (-3) + 29$ **72.** $13 + 62 + (-38)$ **73.** $-3 + (-8) + 12$ **74.** $-27 + (-42) + (-18)$

75. $16 - 8$ **76.** $12 - 3$ **77.** $7 - 14$ **78.** $6 - 9$

79. $-7 - 2$ **80.** $-9 - 4$ **81.** $7 - (-2)$ **82.** $3 - (-4)$

83. $-6 - (-3)$

84. $-4 - (-2)$

85. $6 - (-12)$

86. $-12 - 16$

87. $13 + (-22) + 4 + (-5)$

88. $-14 + (-3) + 7 + (-21)$

89. $-16 + (-17) + (-18) + 10$

90. $-25 + (-31) + 24 + 19$

91. $26 + (-15) + (-11) + (-12)$

92. $-32 + 40 + (-8) + (-19)$

93. $-14 + (-15) + (-11) + 40$

94. $28 + (-19) + (-8) + (-1)$

95. $-4 - 3 - 2$

96. $4 - 5 - 12$

97. $12 - (-7) - 8$

98. $-12 - (-3) - (-15)$

99. $-19 - (-19) - 18$

100. $-8 - (-8) - 14$

101. $-17 - (-8) - (-9)$

102. $7 - 8 - (-1)$

103. $-30 - (-65) - 29 - 4$

104. $42 - (-82) - 65 - 7$

105. $-16 - 47 - 63 - 12$

106. $42 - (-30) - 65 - (-11)$

107. $-47 - (-67) - 13 - 15$

108. $-18 - 49 - (-84) - 27$

109. $-19 - 17 - (-36) - 12$

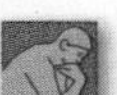

For Exercises 110 to 113, without finding the sum or difference, determine whether the sum or difference is positive or negative.

110. $812 + (-537)$

111. The sum of -57 and -31

112. $-25 - 52$

113. The difference between 8 and -5

OBJECTIVE D To multiply or divide integers

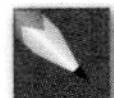

114. Name the operation in each expression. Justify your answer.
a. $8(-7)$ **b.** $8 - 7$ **c.** $8 - (-7)$ **d.** $-xy$ **e.** $x(-y)$ **f.** $-x - y$

For Exercises 115 to 172, multiply or divide.

115. $(14)3$ **116.** $(17)6$ **117.** $-7 \cdot 4$ **118.** $-8 \cdot 7$ **119.** $(-12)(-5)$ **120.** $(-13)(-9)$

121. $-11(23)$ **122.** $-8(21)$ **123.** $(-17)14$ **124.** $(-15)12$ **125.** $6(-19)$ **126.** $17(-13)$

127. $12 \div (-6)$ **128.** $18 \div (-3)$ **129.** $(-72) \div (-9)$ **130.** $(-64) \div (-8)$ **131.** $-42 \div 6$

132. $(-56) \div 8$ **133.** $(-144) \div 12$ **134.** $(-93) \div (-3)$ **135.** $48 \div (-8)$ **136.** $57 \div (-3)$

137. $\frac{-49}{7}$ **138.** $\frac{-45}{5}$ **139.** $\frac{-44}{-4}$ **140.** $\frac{-36}{-9}$ **141.** $\frac{98}{-7}$

142. $\frac{85}{-5}$ **143.** $-\frac{-120}{8}$ **144.** $-\frac{-72}{4}$ **145.** $-\frac{-80}{-5}$ **146.** $-\frac{-114}{-6}$

147. $0 \div (-9)$ **148.** $0 \div (-14)$ **149.** $\frac{-261}{9}$ **150.** $\frac{-128}{4}$ **151.** $9 \div 0$

152. $(-21) \div 0$ **153.** $\frac{132}{-12}$ **154.** $\frac{250}{-25}$ **155.** $\frac{0}{0}$ **156.** $\frac{-58}{0}$

157. $7(5)(-3)$ **158.** $(-3)(-2)8$ **159.** $9(-7)(-4)$ **160.** $(-2)(6)(-4)$

161. $16(-3)5$ **162.** $20(-4)3$ **163.** $-4(-3)8$ **164.** $-5(-9)6$

165. $-3(-8)(-9)$ **166.** $-7(-6)(-5)$ **167.** $(-9)7(5)$ **168.** $(-8)7(10)$

169. $7(-2)(5)(-6)$ **170.** $(-3)7(-2)8$ **171.** $(-14)9(-11)0$ **172.** $(-13)(15)(-19)0$

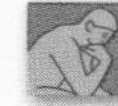

173. You multiply four positive integers and three negative integers. Is the product positive or negative?

OBJECTIVE E To solve application problems

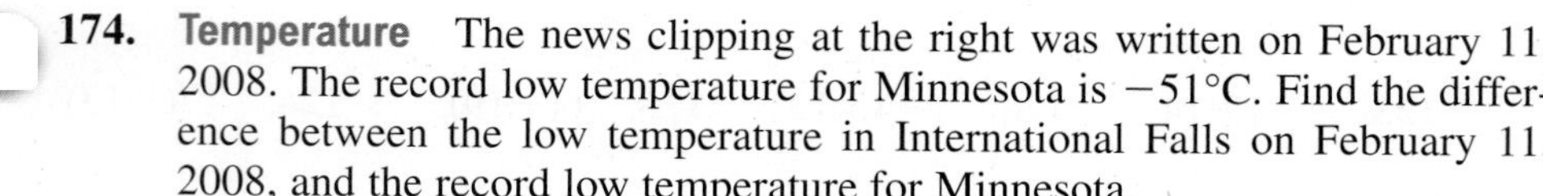

174. Temperature The news clipping at the right was written on February 11, 2008. The record low temperature for Minnesota is -51°C. Find the difference between the low temperature in International Falls on February 11, 2008, and the record low temperature for Minnesota.

In the News

Minnesota Town Named "Icebox of the Nation"

In International Falls, Minnesota, the temperature fell to -40°C just days after the citizens received word that the town had won a federal trademark naming it the "Icebox of the Nation."

Source: news.yahoo.com

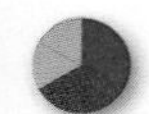

175. Temperature The record high temperature in Illinois is 117°F. The record low temperature is -36°F. Find the difference between the record high and record low temperatures in Illinois.

176. Temperature Find the temperature after a rise of 7°C from -8°C.

177. Temperature Find the temperature after a rise of 5°C from -19°C.

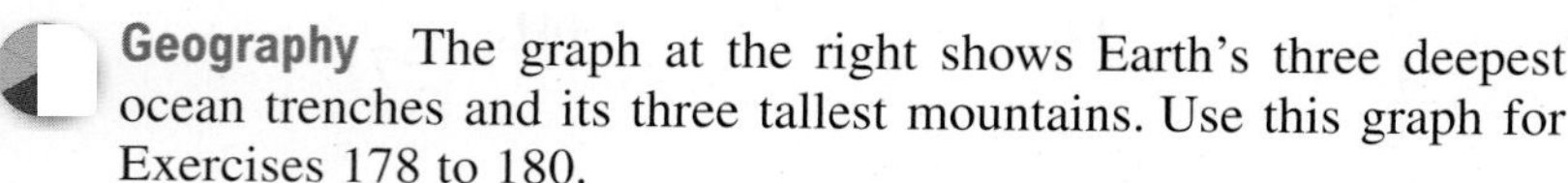

Geography The graph at the right shows Earth's three deepest ocean trenches and its three tallest mountains. Use this graph for Exercises 178 to 180.

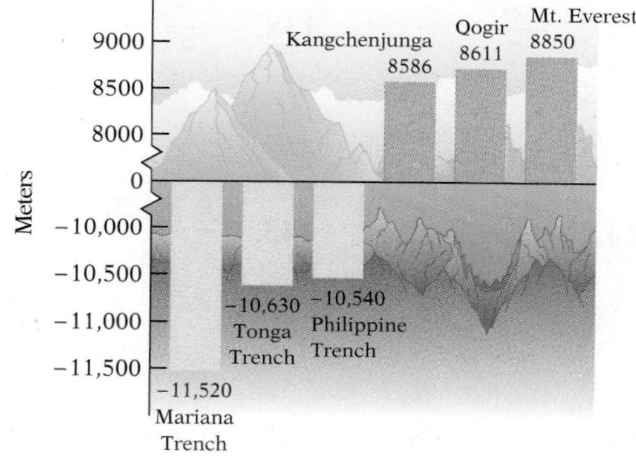

178. What is the difference between the depth of the Philippine Trench and the depth of the Mariana Trench?

179. What is the difference between the height of Mt. Everest and the depth of the Mariana Trench?

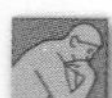

180. Could Mt. Everest fit in the Tonga Trench?

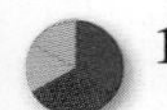

181. Golf Scores In golf, a player's score on a hole is 0 if he completes the hole in *par.* **Par** is the number of strokes in which a golfer should complete a hole. In a golf match, scores are given both as a total number of strokes taken on all holes and as a value relative to par, such as -4 ("4 under par") or $+2$ ("2 over par").

a. See the news clipping at the right. Convert each of Woods' scores for the first three days into a score relative to par.

b. In a golf tournament, players' daily scores are added. Add Woods' three daily scores to find his score, relative to par, for the first three days of the tournament.

c. Woods' score on the fourth day was 71. What was his final score, relative to par, for the four-day tournament?

In the News

Woods Leads 2008 Buick Invitational

With scores of 67, 65, and 66 on his first three days, Tiger Woods leads going into the last day of this four-day tournament. Par for the 18-hole golf course at Torrey Pines Golf Club is 72 strokes.

Source: sports.espn.go.com

182. **Meteorology** The daily low temperatures during one week were recorded as follows: 4°F, −5°F, 8°F, −1°F, −12°F, −14°F, −8°F. Find the average daily low temperature for the week.

183. **Meteorology** The daily high temperatures during one week were recorded as follows: −6°F, −11°F, 1°F, 5°F, −3°F, −9°F, −5°F. Find the average daily high temperature for the week.

184. True or false? If five temperatures are all below 0°C, then the average of the five temperatures is also below 0°C.

185. True or false? If the average of 10 temperatures is below 0°C, then all 10 temperatures are below 0°C.

186. **Chemistry** The graph at the right shows the boiling points of three chemical elements. The boiling point of neon is seven times the highest boiling point shown in the graph.
a. Without calculating the boiling point, determine whether the boiling point of neon is above 0°C or below 0°C.
b. What is the boiling point of neon?

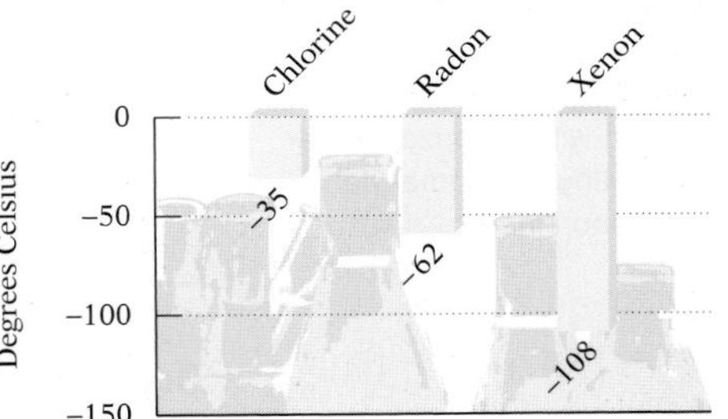

187. **Sports** The combined scores of the top 10 golfers in a tournament equaled −20 (20 under par). What was the average score of the 10 golfers?

188. **Sports** The combined scores of the top four golfers in a tournament equaled −12 (12 under par). What was the average score of the four golfers?

189. **Education** To discourage guessing on a multiple-choice exam, an instructor graded the test by giving 5 points for a correct answer, −2 points for an answer left blank, and −5 points for an incorrect answer. How many points did a student score who answered 20 questions correctly, answered 5 questions incorrectly, and left 2 questions blank?

190. **Games** During a card game of Hearts, Nick had a score of 11 points before his opponent "shot the moon," subtracting a score of 26 from Nick's total. What was Nick's score after his opponent shot the moon?

Applying the Concepts

191. **Number Problems** Consider the numbers 4, −7, −5, 13, and −9. What is the largest difference that can be obtained by subtracting one number in the list from another number in the list? Find the smallest positive difference.

192. **Number Problems** The sum of two negative integers is −8. Find the integers.

193. Determine whether the statement is *true* or *false*.
a. The product of a nonzero number and its opposite is negative.
b. The square of a negative number is a positive number.

SECTION

1.2 Rational and Irrational Numbers

OBJECTIVE A To write a rational number as a decimal

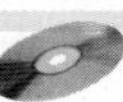

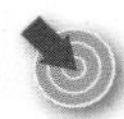

Point of Interest

As early as A.D. 630, the Hindu mathematician Brahmagupta wrote a fraction as one number over another separated by a space. The Arab mathematician al Hassar (around A.D. 1050) was the first to show a fraction with a horizontal bar separating the numerator and denominator.

A *rational number* is the quotient of two integers. A rational number written in the following way is commonly called a fraction. Here are some examples of rational numbers.

$$\frac{3}{4}, \quad \frac{-4}{9}, \quad \frac{15}{-4}, \quad \frac{8}{1}, \quad -\frac{5}{6}$$

Rational Numbers

A rational number is a number that can be written in the form $\frac{a}{b}$, where a and b are integers and $b \neq 0$.

Because an integer can be written as the quotient of the integer and 1, every integer is a rational number. For instance,

$$\frac{6}{1} = 6 \qquad \frac{-8}{1} = -8$$

Point of Interest

Simon Stevin (1548–1620) was the first to name decimal numbers. He wrote the number 2.345 as 2 0 3 1 4 2 5 3. He called the whole number part the *commencement;* the tenths digit was *prime,* the hundredths digit was *second,* the thousandths digit was *third,* and so on.

A number written in **decimal notation** is also a rational number.

three-tenths $0.3 = \frac{3}{10}$ forty-three thousandths $0.043 = \frac{43}{1000}$

A rational number written as a fraction can be rewritten in decimal notation.

HOW TO 1 Write $\frac{5}{8}$ as a decimal. Write $\frac{4}{11}$ as a decimal.

The fraction bar can be read "÷".

$$\frac{5}{8} = 5 \div 8$$

$$\begin{array}{r} 0.625 \\ 8\overline{)5.000} \\ -4\,8 \\ \hline 20 \\ -16 \\ \hline 40 \\ -40 \\ \hline 0 \end{array}$$

← This is called a terminating decimal.

← The remainder is zero.

$$\frac{5}{8} = 0.625$$

$$\begin{array}{r} 0.3636\ldots \\ 11\overline{)4.0000} \\ -3\,3 \\ \hline 70 \\ -66 \\ \hline 40 \\ -33 \\ \hline 70 \\ -66 \\ \hline 4 \end{array}$$

← This is called a repeating decimal.

← The remainder is never zero.

$$\frac{4}{11} = 0.\overline{36}$$

← The bar over the digits 3 and 6 is used to show that these digits repeat.

EXAMPLE • 1

Write $\frac{8}{11}$ as a decimal. Place a bar over the repeating digits of the decimal.

Solution

$$\frac{8}{11} = 8 \div 11 = 0.7272\ldots = 0.\overline{72}$$

YOU TRY IT • 1

Write $\frac{4}{9}$ as a decimal. Place a bar over the repeating digits of the decimal.

Your solution

Solution on p. S1

OBJECTIVE B To convert among percents, fractions, and decimals

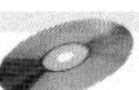

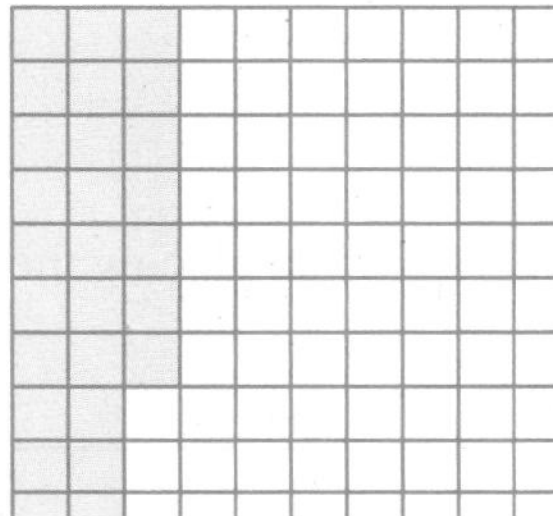

Percent means "parts of 100." Thus 27% means 27 parts of 100.

In applied problems involving percent, it may be necessary to rewrite a percent as a fraction or decimal or to rewrite a fraction or decimal as a percent.

To write a percent as a fraction, remove the percent sign and multiply by $\frac{1}{100}$.

HOW TO • 2 Write 27% as a fraction.

$$27\% = 27\left(\frac{1}{100}\right) = \frac{27}{100}$$

• Remove the percent sign and multiply by $\frac{1}{100}$.

To write a percent as a decimal, remove the percent sign and multiply by 0.01.

$$33\% = 33(0.01) = 0.33$$

Move the decimal point two places to the left and remove the percent sign.

A fraction or decimal can be written as a percent by multiplying by 100%.

$$\frac{5}{8} = \frac{5}{8}(100\%) = \frac{500}{8}\% = 62.5\%, \text{ or } 62\frac{1}{2}\%$$

$$0.82 = 0.82(100\%) = 82\%$$

Move the decimal point two places to the right and write the percent sign.

EXAMPLE • 2

Write 130% as a fraction and as a decimal.

Solution

$$130\% = 130\left(\frac{1}{100}\right) = \frac{130}{100} = \frac{13}{10}$$

$$130\% = 130(0.01) = 1.30$$

YOU TRY IT • 2

Write 125% as a fraction and as a decimal.

Your solution

Solution on p. S1

EXAMPLE • 3

Write $\frac{5}{6}$ as a percent.

Solution

$\frac{5}{6} = \frac{5}{6}(100\%) = \frac{500}{6}\% = 83\frac{1}{3}\%$

YOU TRY IT • 3

Write $\frac{1}{3}$ as a percent.

Your solution

EXAMPLE • 4

Write 0.092 as a percent.

Solution

$0.092 = 0.092(100\%) = 9.2\%$

YOU TRY IT • 4

Write 0.043 as a percent.

Your solution

Solutions on p. S2

OBJECTIVE C To add or subtract rational numbers 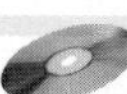

Fractions with the same denominator are added by adding the numerators and placing the sum over the common denominator.

Addition of Fractions

To add two fractions with the same denominator, add the numerators and place the sum over the common denominator.

$$\frac{a}{c} + \frac{b}{c} = \frac{a+b}{c}$$

To add fractions with different denominators, first rewrite the fractions as equivalent fractions with a common denominator. Then add the fractions.

The least common denominator is the **least common multiple** (LCM) of the denominators. This is the smallest number that is a multiple of each of the denominators.

Take Note

You can find the LCM by multiplying the denominators and then dividing by the *common factor* of the two denominators. In the case of 6 and 10, $6 \cdot 10 = 60$. Now divide by 2, the common factor of 6 and 10.

$60 \div 2 = 30$

HOW TO • 3 Add: $-\frac{5}{6} + \frac{3}{10}$

The LCM of 6 and 10 is 30. Rewrite the fractions as equivalent fractions with the denominator 30. Then add the fractions.

$$-\frac{5}{6} + \frac{3}{10} = -\frac{5}{6} \cdot \frac{5}{5} + \frac{3}{10} \cdot \frac{3}{3} = -\frac{25}{30} + \frac{9}{30} = \frac{-25+9}{30} = \frac{-16}{30} = -\frac{8}{15}$$

To subtract fractions with the same denominator, subtract the numerators and place the difference over the common denominator.

Take Note

The least common multiple of the denominators is frequently called the **least common denominator** (LCD).

HOW TO • 4 Subtract: $-\frac{4}{9} - \left(-\frac{7}{12}\right)$

Rewrite subtraction as addition of the opposite. The LCM of 9 and 12 is 36. Rewrite the fractions as equivalent fractions with the denominator 36.

$$-\frac{4}{9} - \left(-\frac{7}{12}\right) = -\frac{4}{9} + \frac{7}{12} = -\frac{16}{36} + \frac{21}{36} = \frac{-16+21}{36} = \frac{5}{36}$$

To add or subtract decimals, write the numbers so that the decimal points are in a vertical line. Then proceed as in the addition or subtraction of integers. Write the decimal point in the answer directly below the decimal points in the problem.

HOW TO • 5 Add: $-114.039 + 84.76$

$|-114.039| = 114.039$
$|84.76| = 84.76$
• The signs are different. Find the absolute value of each number.

$$\begin{array}{r} 114.039 \\ -84.76 \\ \hline 29.279 \end{array}$$
• Subtract the smaller of these numbers from the larger.

$-114.039 + 84.76 = -29.279$
• Attach the sign of the number with the larger absolute value. Because $|-114.039| > |84.76|$, use the sign of -114.039.

EXAMPLE • 5

Simplify: $-\frac{3}{4} + \frac{1}{6} - \frac{5}{8}$

Solution The LCM of 4, 6, and 8 is 24.

$$-\frac{3}{4} + \frac{1}{6} - \frac{5}{8} = -\frac{18}{24} + \frac{4}{24} - \frac{15}{24}$$
$$= \frac{-18 + 4 - 15}{24}$$
$$= \frac{-29}{24} = -\frac{29}{24}$$

YOU TRY IT • 5

Simplify: $-\frac{7}{8} - \frac{5}{6} + \frac{3}{4}$

Your solution

EXAMPLE • 6

Subtract: $42.987 - 98.61$

Solution $42.987 - 98.61$
$= 42.987 + (-98.61)$
$= -55.623$

YOU TRY IT • 6

Subtract: $16.127 - 67.91$

Your solution

Solutions on p. S2

OBJECTIVE D To multiply or divide rational numbers 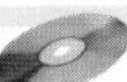

The product of two fractions is the product of the numerators divided by the product of the denominators.

$$\frac{a}{b} \cdot \frac{c}{d} = \frac{ac}{bd}$$

HOW TO • 6 Multiply: $\frac{3}{8} \cdot \frac{12}{17}$

$$\frac{3}{8} \cdot \frac{12}{17} = \frac{3 \cdot 12}{8 \cdot 17}$$
• Multiply the numerators. Multiply the denominators.

$$= \frac{3 \cdot \overset{1}{\cancel{2}} \cdot \overset{1}{\cancel{2}} \cdot 3}{2 \cdot \underset{1}{\cancel{2}} \cdot \underset{1}{\cancel{2}} \cdot 17}$$
• Write the prime factorization of each factor. Divide by the common factors.

$$= \frac{9}{34}$$
• Multiply the factors in the numerator and the factors in the denominator.

To divide fractions, invert the divisor. Then multiply the fractions.

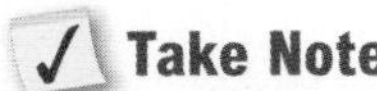

Take Note
To invert the divisor means to write its reciprocal. The reciprocal of $\frac{18}{25}$ is $\frac{25}{18}$.

HOW TO • 7 Divide: $\frac{3}{10} \div \left(-\frac{18}{25}\right)$

The signs are different. The quotient is negative.

$$\frac{3}{10} \div \left(-\frac{18}{25}\right) = -\left(\frac{3}{10} \div \frac{18}{25}\right) = -\left(\frac{3}{10} \cdot \frac{25}{18}\right) = -\left(\frac{3 \cdot 25}{10 \cdot 18}\right)$$

$$= -\left(\frac{\overset{1}{\cancel{3}} \cdot \overset{1}{\cancel{5}} \cdot 5}{2 \cdot \underset{1}{\cancel{5}} \cdot 2 \cdot \underset{1}{\cancel{3}} \cdot 3}\right) = -\frac{5}{12}$$

To multiply decimals, multiply as with integers. Write the decimal point in the product so that the number of decimal places in the product equals the sum of the numbers of decimal places in the factors.

HOW TO • 8 Multiply: $-6.89(0.00035)$

6.89	2 decimal places	
× 0.00035	5 decimal places	• Multiply the absolute values.
3445		
2067		
0.0024115	7 decimal places	

$-6.89(0.00035) = -0.0024115$ • The signs are different. The product is negative.

To divide decimals, move the decimal point in the divisor to the right to make the divisor a whole number. Move the decimal point in the dividend the same number of places to the right. Place the decimal point in the quotient directly over the decimal point in the dividend. Then divide as with whole numbers.

Take Note
The symbol ≈ is used to indicate that the quotient is an approximate value that has been rounded off.

HOW TO • 9 Divide: $1.32 \div 0.27$. Round to the nearest tenth.

```
          4.88 ≈ 4.9
0.27.)1.32.00
     -1 08
        240
       -216
        240
       -216
         24
```

• Move the decimal point 2 places to the right in the divisor and then in the dividend. Place the decimal point in the quotient above the decimal point in the dividend.

EXAMPLE • 7

Divide: $-\frac{5}{8} \div \left(-\frac{5}{40}\right)$

Solution

The quotient is positive.

$$-\frac{5}{8} \div \left(-\frac{5}{40}\right) = \frac{5}{8} \div \frac{5}{40} = \frac{5}{8} \cdot \frac{40}{5} = \frac{5 \cdot 40}{8 \cdot 5}$$

$$= \frac{\overset{1}{\cancel{5}} \cdot \overset{1}{\cancel{2}} \cdot \overset{1}{\cancel{2}} \cdot \overset{1}{\cancel{2}} \cdot 5}{\underset{1}{\cancel{2}} \cdot \underset{1}{\cancel{2}} \cdot \underset{1}{\cancel{2}} \cdot \underset{1}{\cancel{5}}} = \frac{5}{1} = 5$$

YOU TRY IT • 7

Divide: $-\frac{3}{8} \div \left(-\frac{5}{12}\right)$

Your solution

Solution on p. S2

EXAMPLE • 8

Multiply: $-4.29(8.2)$

Solution

The product is negative.

$$\begin{array}{r} 4.29 \\ \times \quad 8.2 \\ \hline 858 \\ 3432 \\ \hline 35.178 \end{array}$$

$-4.29(8.2) = -35.178$

YOU TRY IT • 8

Multiply: $-5.44(3.8)$

Your solution

Solution on p. S2

OBJECTIVE E To evaluate exponential expressions

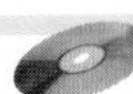

Point of Interest

René Descartes (1596–1650) was the first mathematician to use exponential notation extensively as it is used today. However, for some unknown reason, he always used xx for x^2.

Repeated multiplication of the same factor can be written using an exponent.

$2 \cdot 2 \cdot 2 \cdot 2 \cdot 2 = 2^5$ ← exponent; 2 is the base

$a \cdot a \cdot a \cdot a = a^4$ ← exponent; a is the base

The **exponent** indicates how many times the factor, called the **base,** occurs in the multiplication. The multiplication $2 \cdot 2 \cdot 2 \cdot 2 \cdot 2$ is in **factored form.** The exponential expression 2^5 is in **exponential form.**

2^1 is read "the first power of 2" or just 2. Usually the exponent 1 is not written.

2^2 is read "the second power of 2" or "2 squared."

2^3 is read "the third power of 2" or "2 cubed."

2^4 is read "the fourth power of 2."

a^4 is read "the fourth power of a."

There is a geometric interpretation of the first three natural-number powers.

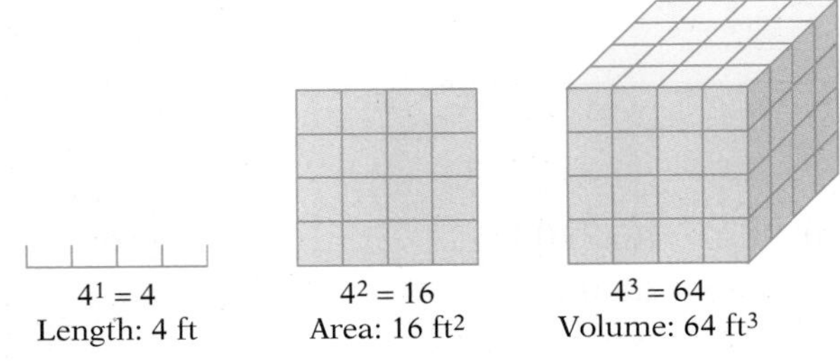

To evaluate an exponential expression, write each factor as many times as indicated by the exponent. Then multiply.

HOW TO • 10 Evaluate $(-2)^4$.

$(-2)^4 = (-2)(-2)(-2)(-2)$ • Write (-2) as a factor 4 times.

$= 16$ • Multiply.

HOW TO • 11 Evaluate -2^4.

$-2^4 = -(2 \cdot 2 \cdot 2 \cdot 2)$ • Write 2 as a factor 4 times.

$= -16$ • Multiply.

From these last two examples, note the difference between $(-2)^4$ and -2^4.

$$(-2)^4 = 16$$
$$-2^4 = -(2)^4 = -16$$

EXAMPLE • 9

Evaluate -5^3.

Solution

$-5^3 = -(5 \cdot 5 \cdot 5) = -125$

YOU TRY IT • 9

Evaluate -6^3.

Your solution

EXAMPLE • 10

Evaluate $(-4)^4$.

Solution

$(-4)^4 = (-4)(-4)(-4)(-4)$
$= 256$

YOU TRY IT • 10

Evaluate $(-3)^4$.

Your solution

EXAMPLE • 11

Evaluate $(-3)^2 \cdot 2^3$.

Solution $(-3)^2 \cdot 2^3 = (-3)(-3) \cdot (2)(2)(2)$
$= 9 \cdot 8 = 72$

YOU TRY IT • 11

Evaluate $(3^3) \cdot (-2)^3$.

Your solution

EXAMPLE • 12

Evaluate $\left(-\frac{2}{3}\right)^3$.

Solution $\left(-\frac{2}{3}\right)^3 = \left(-\frac{2}{3}\right)\left(-\frac{2}{3}\right)\left(-\frac{2}{3}\right)$

$= -\frac{2 \cdot 2 \cdot 2}{3 \cdot 3 \cdot 3} = -\frac{8}{27}$

YOU TRY IT • 12

Evaluate $\left(-\frac{2}{5}\right)^2$.

Your solution

EXAMPLE • 13

Evaluate $-4(0.7)^2$.

Solution $-4(0.7)^2 = -4(0.7)(0.7)$
$= -2.8(0.7) = -1.96$

YOU TRY IT • 13

Evaluate $-3(0.3)^3$.

Your solution

Solutions on p. S2

OBJECTIVE F To simplify numerical radical expressions 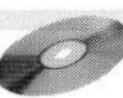

A **square root** of a positive number x is a number whose square is x.

A square root of 16 is 4 because $4^2 = 16$.
A square root of 16 is -4 because $(-4)^2 = 16$.

Every positive number has two square roots, one a positive number and one a negative number. The symbol "$\sqrt{\ }$," called a **radical sign,** is used to indicate the positive or **principal square root** of a number. For example, $\sqrt{16} = 4$ and $\sqrt{25} = 5$. The number under the radical sign is called the **radicand.**

Square Roots of Perfect Squares
$\sqrt{1} = 1$
$\sqrt{4} = 2$
$\sqrt{9} = 3$
$\sqrt{16} = 4$
$\sqrt{25} = 5$
$\sqrt{36} = 6$
$\sqrt{49} = 7$
$\sqrt{64} = 8$
$\sqrt{81} = 9$
$\sqrt{100} = 10$
$\sqrt{121} = 11$
$\sqrt{144} = 12$

When the negative square root of a number is to be found, a negative sign is placed in front of the radical. For example, $-\sqrt{16} = -4$ and $-\sqrt{25} = -5$.

The square of an integer is a **perfect square.** 49, 81, and 144 are examples of perfect squares.

$7^2 = 49$
$9^2 = 81$
$12^2 = 144$

The principal square root of an integer that is a perfect square is a positive integer.

$\sqrt{49} = 7$
$\sqrt{81} = 9$
$\sqrt{144} = 12$

If a number is not a perfect square, its square root can only be approximated. For example, 2 and 7 are not perfect squares. The square roots of these numbers are **irrational numbers.** Their decimal representations never terminate or repeat.

$\sqrt{2} \approx 1.4142135\ldots$ $\qquad$ $\sqrt{7} \approx 2.6457513\ldots$

Recall that rational numbers are fractions, such as $-\frac{6}{7}$ or $-\frac{10}{3}$, in which the numerator and denominator are integers. Rational numbers are also represented by repeating decimals, such as $0.25767676\ldots$, and by terminating decimals, such as 1.73. An irrational number is neither a repeating nor a terminating decimal. For instance, $2.45445444544445\ldots$ is an irrational number.

Real Numbers

The rational numbers and the irrational numbers taken together are called the **real numbers.**

Take Note

Recall that a factor of a number divides the number evenly. For instance, 6 is a factor of 18. The perfect square 9 is also a factor of 18. 9 is a *perfect-square factor* of 18, whereas 6 is not a perfect-square factor of 18.

Radical expressions that contain radicands that are not perfect squares are frequently written in simplest form. A radical expression is in simplest form when the radicand contains no factor greater than 1 that is a perfect square. For instance, $\sqrt{50}$ is not in simplest form because 25 is a perfect-square factor of 50. The radical expression $\sqrt{15}$ is in simplest form because there are no perfect-square factors of 15 that are greater than 1.

The Product Property of Square Roots and a knowledge of perfect squares are used to simplify radicands that are not perfect squares.

The Product Property of Square Roots

If a and b are positive real numbers, then $\sqrt{ab} = \sqrt{a} \cdot \sqrt{b}$.

Take Note

From the example at the right, $\sqrt{72} = 6\sqrt{2}$. The two expressions are different representations of the same number. Using a calculator, we find that $\sqrt{72} \approx 8.485281$ and $6\sqrt{2} \approx 8.485281$.

HOW TO • 12 Simplify: $\sqrt{72}$

$\sqrt{72} = \sqrt{36 \cdot 2}$
- Write the radicand as the product of a perfect square and a factor that does not contain a perfect square.

$= \sqrt{36}\sqrt{2}$
- Use the Product Property of Square Roots to write the expression as a product.

$= 6\sqrt{2}$
- Simplify $\sqrt{36}$.

Note that 72 must be written as the product of a perfect square and *a factor that does not contain a perfect square.* Therefore, it would not be correct to rewrite $\sqrt{72}$ as $\sqrt{9 \cdot 8}$ and simplify the expression as shown at the right. Although 9 is a perfect-square factor of 72, 8 has a perfect square factor ($8 = 4 \cdot 2$). Therefore, $\sqrt{8}$ is not in simplest form. Remember to find the *largest* perfect-square factor of the radicand.

$$\begin{aligned}\sqrt{72} &= \sqrt{9 \cdot 8}\\ &= \sqrt{9}\sqrt{8}\\ &= 3\sqrt{8}\end{aligned}$$

Not in simplest form

HOW TO • 13 Simplify: $\sqrt{-16}$

Because the square of any real number is positive, there is no real number whose square is -16. $\sqrt{-16}$ is not a real number.

EXAMPLE • 14

Simplify: $3\sqrt{90}$

Solution

$$\begin{aligned}3\sqrt{90} &= 3\sqrt{9 \cdot 10}\\ &= 3\sqrt{9}\sqrt{10}\\ &= 3 \cdot 3\sqrt{10}\\ &= 9\sqrt{10}\end{aligned}$$

YOU TRY IT • 14

Simplify: $-5\sqrt{32}$

Your solution

EXAMPLE • 15

Simplify: $\sqrt{252}$

Solution

$$\begin{aligned}\sqrt{252} &= \sqrt{36 \cdot 7}\\ &= \sqrt{36}\sqrt{7}\\ &= 6\sqrt{7}\end{aligned}$$

YOU TRY IT • 15

Simplify: $\sqrt{216}$

Your solution

Solutions on p. S2

OBJECTIVE G To solve application problems 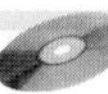

One of the applications of percent is to express a portion of a total as a percent. For instance, a recent survey of 450 mall shoppers found that 270 preferred the mall closest to their home even though it did not have as much store variety as a mall farther from home. The percent of shoppers who preferred the mall closest to home can be found by converting a fraction to a percent.

$$\frac{\text{Portion preferring mall closest to home}}{\text{Total number surveyed}} = \frac{270}{450}$$
$$= 0.60 = 60\%$$

The Congressional Budget Office projected that the total surpluses for 2001 through 2011 would be \$5.6 trillion. The number 5.6 trillion means

$$5.6 \times \underbrace{1{,}000{,}000{,}000{,}000}_{\text{1 trillion}} = 5{,}600{,}000{,}000{,}000$$

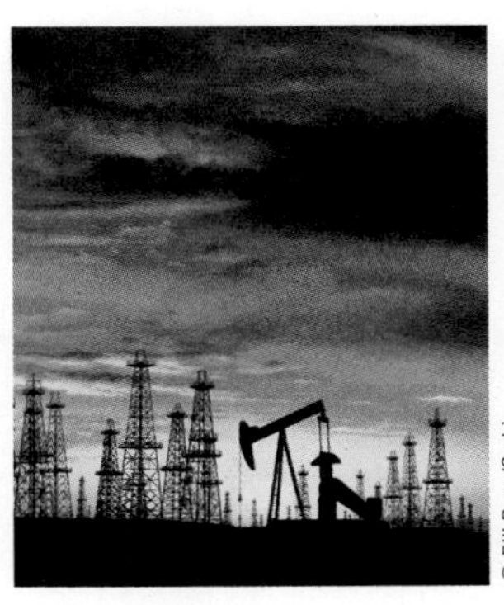

© Bill Ross/Corbis

The graph at the right shows the number of barrels of oil, in millions, produced each day in 2001 and 2007 for five regions of the world. Use this graph for Example 16 and You Try It 16.

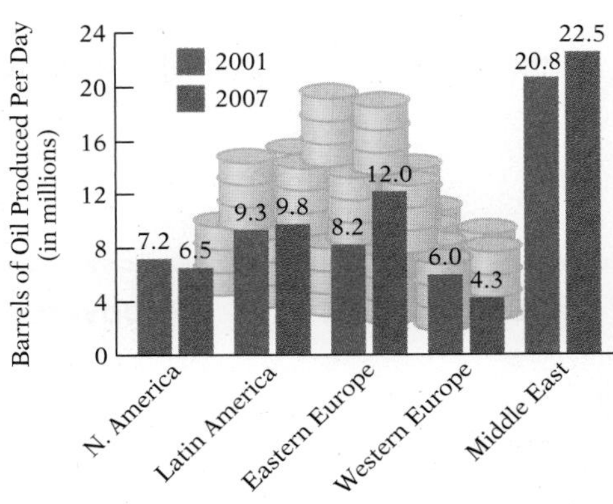

Source: www.opec.org

EXAMPLE 16

Using the graph above, find the total number of barrels of oil produced each day in 2007 for the five regions shown.

Strategy

To find the total number of barrels:

- Read the numbers from the graph that correspond to 2007 (6.5, 9.8, 12.0, 4.3, 22.5).
- Add the numbers.

Solution

$6.5 + 9.8 + 12.0 + 4.3 + 22.5 = 55.1$

In 2007, a total of 55.1 million barrels of oil were produced each day in the five regions shown.

YOU TRY IT 16

For the regions given in the graph above, find the difference between the total number of barrels of oil produced each day in 2007 and in 2001.

Your strategy

Your solution

Solution on p. S2

1.2 EXERCISES

OBJECTIVE A To write a rational number as a decimal

For Exercises 1 to 15, write as a decimal. Place a bar over the repeating digits of a repeating decimal.

1. $\frac{1}{8}$ **2.** $\frac{7}{8}$ **3.** $\frac{2}{9}$ **4.** $\frac{8}{9}$ **5.** $\frac{1}{6}$

6. $\frac{5}{6}$ **7.** $\frac{9}{16}$ **8.** $\frac{15}{16}$ **9.** $\frac{7}{12}$ **10.** $\frac{11}{12}$

11. $\frac{6}{25}$ **12.** $\frac{14}{25}$ **13.** $\frac{9}{40}$ **14.** $\frac{21}{40}$ **15.** $\frac{5}{11}$

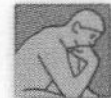

16. The denominator of a fraction that is in simplest form is a multiple of 3. You write the fraction as a decimal. Is the result a terminating or a repeating decimal?

OBJECTIVE B To convert among percents, fractions, and decimals

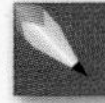

17. Explain why multiplying a number by 100% does not change the value of the number.

For Exercises 18 to 27, write as a fraction and as a decimal.

18. 75% **19.** 40% **20.** 64% **21.** 88% **22.** 125%

23. 160% **24.** 19% **25.** 87% **26.** 5% **27.** 450%

For Exercises 28 to 37, write as a fraction.

28. $11\frac{1}{9}\%$ **29.** $4\frac{2}{7}\%$ **30.** $12\frac{1}{2}\%$ **31.** $37\frac{1}{2}\%$ **32.** $66\frac{2}{3}\%$

33. $\frac{1}{4}\%$ **34.** $\frac{1}{2}\%$ **35.** $6\frac{1}{4}\%$ **36.** $83\frac{1}{3}\%$ **37.** $5\frac{3}{4}\%$

For Exercises 38 to 47, write as a decimal.

38. 7.3% **39.** 9.1% **40.** 15.8% **41.** 16.7% **42.** 0.3%

43. 0.9% **44.** 9.9% **45.** 9.15% **46.** 121.2% **47.** 18.23%

For Exercises 48 to 67, write as a percent.

48. 0.15 **49.** 0.37 **50.** 0.05 **51.** 0.02 **52.** 0.175

53. 0.125 **54.** 1.15 **55.** 1.36 **56.** 0.008 **57.** 0.004

58. $\frac{27}{50}$ **59.** $\frac{83}{100}$ **60.** $\frac{1}{3}$ **61.** $\frac{3}{8}$ **62.** $\frac{5}{11}$

63. $\frac{4}{9}$ **64.** $\frac{7}{8}$ **65.** $\frac{9}{20}$ **66.** $1\frac{2}{3}$ **67.** $2\frac{1}{2}$

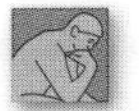

68. Does $\frac{4}{3}$ represent a percent greater than 100% or less than 100%?

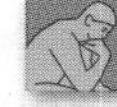

69. Does 0.055 represent a percent greater than 1% or less than 1%?

OBJECTIVE C To add or subtract rational numbers

For Exercises 70 to 94, add or subtract.

70. $-\frac{5}{6} - \frac{5}{9}$ **71.** $-\frac{6}{13} + \frac{17}{26}$ **72.** $-\frac{7}{12} + \frac{5}{8}$ **73.** $\frac{5}{8} - \left(-\frac{3}{4}\right)$

74. $\frac{3}{5} - \frac{11}{12}$ **75.** $\frac{11}{12} - \frac{5}{6}$ **76.** $-\frac{2}{3} - \left(-\frac{11}{18}\right)$ **77.** $-\frac{5}{8} - \left(-\frac{11}{12}\right)$

78. $\frac{1}{3} + \frac{5}{6} - \frac{2}{9}$

79. $\frac{1}{2} - \frac{2}{3} + \frac{1}{6}$

80. $-\frac{5}{16} + \frac{3}{4} - \frac{7}{8}$

81. $\frac{1}{2} - \frac{3}{8} - \left(-\frac{1}{4}\right)$

82. $-13.092 + 6.9$

83. $2.54 - 3.6$

84. $5.43 + 7.925$

85. $-16.92 - 6.925$

86. $-3.87 + 8.546$

87. $6.9027 - 17.692$

88. $2.09 - 6.72 - 5.4$

89. $-18.39 + 4.9 - 23.7$

90. $19 - (-3.72) - 82.75$

91. $-3.07 - (-2.97) - 17.4$

92. $16.4 - (-3.09) - 7.93$

93. $-3.09 - 4.6 - (-27.3)$

94. $2.66 - (-4.66) - 8.2$

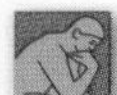
For Exercises 95 to 98, estimate each sum to the nearest integer. Do not find the exact sum.

95. $\frac{7}{8} + \frac{4}{5}$

96. $\frac{1}{3} + \left(-\frac{1}{2}\right)$

97. $-0.125 + 1.25$

98. $-1.3 + 0.2$

OBJECTIVE D To multiply or divide rational numbers

For Exercises 99 to 116, multiply or divide.

99. $\left(-\frac{3}{4}\right)\left(-\frac{8}{27}\right)$

100. $-\frac{1}{2}\left(\frac{8}{9}\right)$

101. $\frac{5}{12}\left(-\frac{8}{15}\right)$

102. $\frac{5}{8}\left(-\frac{7}{12}\right)\frac{16}{25}$

103. $\left(\frac{5}{12}\right)\left(-\frac{8}{15}\right)\left(\frac{1}{3}\right)$

104. $\frac{1}{2}\left(-\frac{3}{4}\right)\left(-\frac{5}{8}\right)$

105. $\frac{3}{8} \div \frac{1}{4}$

106. $\frac{5}{6} \div \left(-\frac{3}{4}\right)$

107. $-\frac{5}{12} \div \frac{15}{32}$

108. $\frac{1}{8} \div \left(-\frac{5}{12}\right)$

109. $-\frac{4}{9} \div \left(-\frac{2}{3}\right)$

110. $-\frac{6}{11} \div \frac{4}{9}$

111. $1.2(3.47)$

112. $(-0.8)6.2$

113. $(-1.89)(-2.3)$

114. $(6.9)(-4.2)$

115. $1.2(-0.5)(3.7)$

116. $2.3(-0.6)(0.8)$

For Exercises 117 to 122, divide. Round to the nearest hundredth.

117. $-1.27 \div (-1.7)$

118. $9.07 \div (-3.5)$

119. $0.0976 \div 0.042$

120. $-6.904 \div 1.35$

121. $-7.894 \div (-2.06)$

122. $-354.2086 \div 0.1719$

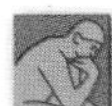

123. **a.** Without finding the product, determine whether $\frac{11}{13} \cdot \frac{50}{51}$ is greater than 1 or less than 1.

b. Without finding the quotient, determine whether $8.713 \div 7.2$ is greater than 1 or less than 1.

OBJECTIVE E To evaluate exponential expressions

For Exercises 124 to 142, evaluate.

124. 6^2

125. 7^4

126. -7^2

127. -4^3

128. $(-3)^2$

129. $(-2)^3$

130. $(-3)^4$

131. $(-5)^3$

132. $\left(\frac{1}{2}\right)^2$

133. $\left(-\frac{3}{4}\right)^3$

134. $(0.3)^2$

135. $(1.5)^3$

136. $\left(\frac{2}{3}\right)^2 \cdot 3^3$

137. $\left(-\frac{1}{2}\right)^3 \cdot 8$

138. $(0.3)^3 \cdot 2^3$

139. $(-2) \cdot (-2)^2$

140. $2^3 \cdot 3^3 \cdot (-4)$

141. $(-3)^3 \cdot 5^2 \cdot 10$

142. $(-7) \cdot 4^2 \cdot 3^2$

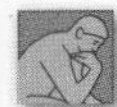
For Exercises 143 to 146, without finding the product, state whether the given expression simplifies to a positive or a negative number.

143. $(-9)^7$ **144.** -8^6 **145.** $(-9^{10})(-5^4)$ **146.** $-(3^4)(-2^5)$

OBJECTIVE F To simplify numerical radical expressions

For Exercises 147 to 170, simplify.

147. $\sqrt{16}$ **148.** $\sqrt{64}$ **149.** $\sqrt{49}$ **150.** $\sqrt{144}$ **151.** $\sqrt{32}$ **152.** $\sqrt{50}$

153. $\sqrt{8}$ **154.** $\sqrt{12}$ **155.** $6\sqrt{18}$ **156.** $-3\sqrt{48}$ **157.** $5\sqrt{40}$ **158.** $2\sqrt{28}$

159. $\sqrt{15}$ **160.** $\sqrt{21}$ **161.** $\sqrt{29}$ **162.** $\sqrt{13}$ **163.** $-9\sqrt{72}$ **164.** $11\sqrt{80}$

165. $\sqrt{45}$ **166.** $\sqrt{225}$ **167.** $\sqrt{0}$ **168.** $\sqrt{210}$ **169.** $6\sqrt{128}$ **170.** $9\sqrt{288}$

For Exercises 171 to 176, find the decimal approximation rounded to the nearest thousandth.

171. $\sqrt{240}$ **172.** $\sqrt{300}$ **173.** $\sqrt{288}$ **174.** $\sqrt{600}$ **175.** $\sqrt{256}$ **176.** $\sqrt{324}$

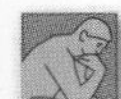
For Exercises 177 to 180, find consecutive integers *m* and *n* such that the given number is between *m* and *n*, or state that the given number is not a real number. Do not use a calculator.

177. $-\sqrt{115}$ **178.** $-\sqrt{-90}$ **179.** $\sqrt{\sqrt{64}}$ **180.** $\sqrt{200}$

OBJECTIVE G To solve application problems

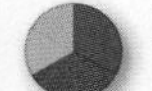
181. **Meteorology** On January 23, 1916, the temperature in Browing, Montana, was 6.67°C. On January 24, 1916, the temperature in Browing was −48.9°C. Find the difference between the temperatures in Browing on these two days.

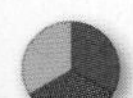
182. **Meteorology** On January 22, 1943, in Spearfish, South Dakota, the temperature fell from 12.22°C at 9 A.M. to −20°C at 9:27 A.M. How many degrees did the temperature fall during the 27-minute period?

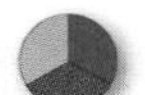

183. Temperature The date of the news clipping at the right is July 20, 2007. Find the difference between the record high and low temperatures for Slovakia.

> **In the News**
>
> **Slovakia Hits Record High**
>
> Slovakia, which became an independent country in 1993 with the peaceful division of Czechoslovakia, hit a record high temperature today of 104.5°F. The record low temperature, set on February 11, 1929, was −41.8°F.
>
> *Source:* wikipedia.org

184. If the temperature begins at 4.8°C and ends up below 0°C, is the difference between the starting and ending temperatures less than or greater than 4.8?

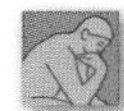

185. If the temperature rose 20.3°F during one day and ended up at a high temperature of 15.7°F, did the temperature begin above or below 0°F?

186. Chemistry The boiling point of nitrogen is −195.8°C, and the melting point is −209.86°C. Find the difference between the boiling point and the melting point of nitrogen.

187. Chemistry The boiling point of oxygen is −182.962°C. Oxygen's melting point is −218.4°C. What is the difference between the boiling point and the melting point of oxygen?

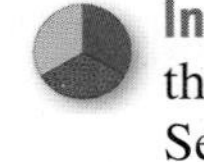

Investments The chart at the right shows the closing price of a share of stock on September 4, 2008, for each of five companies. Also shown is the change in the price from the previous day. To find the closing price on the previous day, subtract the change in price from the closing price on September 4. Use this chart for Exercises 188 and 189.

Company	*Closing Price*	*Change in Price*
Del Monte Foods Co.	8.18	−0.14
General Mills, Inc.	66.93	+0.21
Hershey Foods, Inc.	36.42	−0.41
Hormel Food Corp.	35.39	−0.07
Sara Lee Corp.	13.35	−0.07

188. **a.** Find the closing price on the previous day for General Mills.
b. Find the closing price on the previous day for Hormel Foods.

189. **a.** Find the closing price on the previous day for Sara Lee.
b. Find the closing price on the previous day for Hershey Foods.

Applying the Concepts

190. Determine whether the statement is true or false.
a. Every integer is a rational number.
b. Every whole number is an integer.
c. Every integer is a positive number.
d. Every rational number is an integer.

191. Number Problems
a. Find a rational number between 0.1 and 0.2.
b. Find a rational number between 1 and 1.1.
c. Find a rational number between 0 and 0.005.

SECTION

1.3 The Order of Operations Agreement

OBJECTIVE A To use the Order of Operations Agreement to simplify expressions 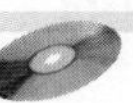

Let's evaluate $2 + 3 \cdot 5$.

There are two arithmetic operations, addition and multiplication, in this expression. The operations could be performed in different orders. We could multiply first and then add, or we could add first and then multiply. To prevent there being more than one answer for a numerical expression, an Order of Operations Agreement has been established.

The Order of Operations Agreement

Step 1 Perform operations inside grouping symbols. Grouping symbols include parentheses ( ), brackets [ ], braces { }, the absolute value symbol ||, and the fraction bar.
Step 2 Simplify exponential expressions.
Step 3 Do multiplication and division as they occur from left to right.
Step 4 Do addition and subtraction as they occur from left to right.

Integrating Technology

See the Keystroke Guide: *Basic Operations* for instruction on using a calculator to evaluate a numerical expression.

HOW TO • 1 Evaluate $12 - 24(8 - 5) \div 2^2$.

$12 - 24(8 - 5) \div 2^2 = 12 - 24(3) \div 2^2$ • Perform operations inside grouping symbols.

$= 12 - 24(3) \div 4$ • Simplify exponential expressions.

$= 12 - 72 \div 4$ • Do multiplication and division as they occur from left to right.

$= 12 - 18$

$= -6$ • Do addition and subtraction as they occur from left to right.

One or more of the steps listed above may not be needed to evaluate an expression. In that case, proceed to the next step in the Order of Operations Agreement.

HOW TO • 2 Evaluate $\frac{4 + 8}{2 + 1} - (3 - 1) + 2$.

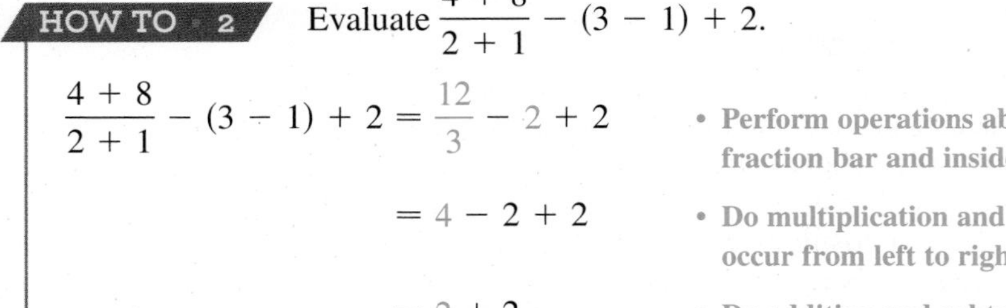

$\frac{4 + 8}{2 + 1} - (3 - 1) + 2 = \frac{12}{3} - 2 + 2$ • Perform operations above and below the fraction bar and inside parentheses.

$= 4 - 2 + 2$ • Do multiplication and division as they occur from left to right.

$= 2 + 2$ • Do addition and subtraction as they occur from left to right.

$= 4$

When an expression has grouping symbols inside grouping symbols, perform the operations inside the inner grouping symbols first.

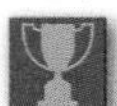

Tips for Success

The HOW TO feature gives an example with explanatory remarks. Using paper and pencil, you should work through the example. See *AIM for Success* in the Preface.

HOW TO • 3 Evaluate $6 \div [4 - (6 - 8)] + 2^2$.

$6 \div [4 - (6 - 8)] + 2^2 = 6 \div [4 - (-2)] + 2^2$ • Perform operations inside grouping symbols.

$= 6 \div 6 + 2^2$

$= 6 \div 6 + 4$ • Simplify exponential expressions.

$= 1 + 4$ • Do multiplication and division as they occur from left to right.

$= 5$ • Do addition and subtraction as they occur from left to right.

EXAMPLE • 1

Evaluate $4 - 3[4 - 2(6 - 3)] \div 2$.

Solution

$4 - 3[4 - 2(6 - 3)] \div 2$

$= 4 - 3[4 - 2 \cdot 3] \div 2$ • Perform operations inside grouping symbols.

$= 4 - 3[4 - 6] \div 2$

$= 4 - 3[-2] \div 2$

$= 4 + 6 \div 2$ • Do multiplication and division from left to right.

$= 4 + 3$

$= 7$ • Do addition and subtraction from left to right.

YOU TRY IT • 1

Evaluate $18 - 5[8 - 2(2 - 5)] \div 10$.

Your solution

EXAMPLE • 2

Evaluate $(1.75 - 1.3)^2 \div 0.025 + 6.1$.

Solution

$(1.75 - 1.3)^2 \div 0.025 + 6.1$

$= (0.45)^2 \div 0.025 + 6.1$ • Perform operations inside grouping symbols.

$= 0.2025 \div 0.025 + 6.1$ • Simplify exponential expressions.

$= 8.1 + 6.1$ • Do the division.

$= 14.2$ • Do the addition.

YOU TRY IT • 2

Evaluate $(6.97 - 4.72)^2 \cdot 4.5 \div 0.05$.

Your solution

Solutions on p. S2

1.3 EXERCISES

OBJECTIVE A **To use the Order of Operations Agreement to simplify expressions**

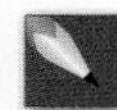

1. Why do we need an Order of Operations Agreement?

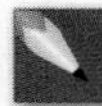

2. Describe each step in the Order of Operations Agreement.

For Exercises 3 to 35, evaluate by using the Order of Operations Agreement.

3. $4 - 8 \div 2$

4. $2^2 \cdot 3 - 3$

5. $2(3 - 4) - (-3)^2$

6. $16 - 32 \div 2^3$

7. $24 - 18 \div 3 + 2$

8. $8 - (-3)^2 - (-2)$

9. $8 - 2(3)^2$

10. $16 - 16 \cdot 2 \div 4$

11. $12 + 16 \div 4 \cdot 2$

12. $16 - 2 \cdot 4^2$

13. $27 - 18 \div (-3^2)$

14. $4 + 12 \div 3 \cdot 2$

15. $16 + 15 \div (-5) - 2$

16. $14 - 2^2 - (4 - 7)$

17. $14 - 2^2 - |4 - 7|$

18. $10 - |5 - 8| + 2^3$

19. $3 - 2[8 - (3 - 2)]$

20. $-2^2 + 4[16 \div (3 - 5)]$

21. $6 + \dfrac{16 - 4}{2^2 + 2} - 2$

22. $24 \div \dfrac{3^2}{8 - 5} - (-5)$

23. $18 \div |9 - 2^3| + (-3)$

24. $96 \div 2[12 + (6 - 2)] - 3^2$

25. $4[16 - (7 - 1)] \div 10$

26. $18 \div 2 - 4^2 - (-3)^2$

27. $20 \div (10 - 2^3) + (-5)$

28. $16 - 3(8 - 3)^2 \div 5$

29. $4(-8) \div [2(7 - 3)^2]$

30. $\dfrac{(-10) + (-2)}{6^2 - 30} \div |2 - 4|$

31. $16 - 4 \cdot \dfrac{3^3 - 7}{2^3 + 2} - (-2)^2$

32. $(0.2)^2 \cdot (-0.5) + 1.72$

33. $0.3(1.7 - 4.8) + (1.2)^2$

34. $(1.8)^2 - 2.52 \div 1.8$

35. $(1.65 - 1.05)^2 \div 0.4 + 0.8$

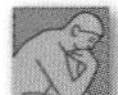

36. Which expression is equivalent to $15 + 15 \div 3 - 4^2$?
(i) $30 \div 3 - 16$ (ii) $15 + 5 - 16$ (iii) $15 + 5 + 16$ (iv) $15 + 15 \div (-1)^2$

Applying the Concepts

37. Find two fractions between $\frac{2}{3}$ and $\frac{3}{4}$. (There is more than one answer to this question.)

38. A **magic square** is one in which the numbers in every row, column, and diagonal sum to the same number. Complete the magic square at the right.

$\frac{2}{3}$		
	$\frac{1}{6}$	$\frac{5}{6}$
		$-\frac{1}{3}$

39. For each part below, find a rational number r that satisfies the condition.
a. $r^2 < r$ **b.** $r^2 = r$ **c.** $r^2 > r$

40. Find three different natural numbers a, b, and c such that $\frac{1}{a} + \frac{1}{b} + \frac{1}{c}$ is a natural number.

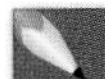

41. The following was offered as the simplification of $6 + 2(4 - 9)$.

$$\begin{aligned} 6 + 2(4 - 9) &= 6 + 2(-5) \\ &= 8(-5) \\ &= -40 \end{aligned}$$

If this is a correct simplification, write yes for the answer. If it is incorrect, write no and explain the incorrect step.

SECTION

1.4 Variable Expressions

OBJECTIVE A To evaluate a variable expression 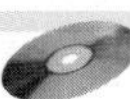

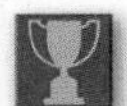

Tips for Success

Before you begin a new chapter, you should take some time to review previously learned skills. One way to do this is to complete the Prep Test. See page 75. This test focuses on the particular skills that will be required for the new chapter.

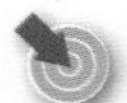

Point of Interest

Historical manuscripts indicate that mathematics is at least 4000 years old. Yet it was only 400 years ago that mathematicians started using variables to stand for numbers. The idea that a letter can stand for some number was a critical turning point in mathematics.

Today, x is used by most nations as the standard letter for a single unknown. In fact, x-rays were so named because the scientists who discovered them did not know what they were and thus labeled them the "unknown rays" or x-rays.

Often we discuss a quantity without knowing its exact value—for example, the price of gold next month, the cost of a new automobile next year, or the tuition cost for next semester. Recall that a letter of the alphabet, called a **variable,** is used to stand for a quantity that is unknown or that can change, or *vary.* An expression that contains one or more variables is called a **variable expression.**

A variable expression is shown at the right. The expression can be rewritten by writing subtraction as the addition of the opposite.

$$3x^2 - 5y + 2xy - x - 7$$
$$3x^2 + (-5y) + 2xy + (-x) + (-7)$$

Note that the expression has five addends. The **terms** of a variable expression are the addends of the expression. The expression has five terms.

Five terms

$$3x^2 \quad - \; 5y \quad + \; 2xy \quad - \; x \qquad - \; 7$$

Variable terms — Constant term

The terms $3x^2$, $-5y$, $2xy$, and $-x$ are **variable terms.**

The term -7 is a **constant term,** or simply a **constant.**

Each variable term is composed of a **numerical coefficient** and a **variable part** (the variable or variables and their exponents).

Numerical coefficient

$$3x^2 \; - \; 5y \; + \; 2xy \; - \; 1x - 7$$

Variable part

When the numerical coefficient is 1 or -1, the 1 is usually not written ($x = 1x$ and $-x = -1x$).

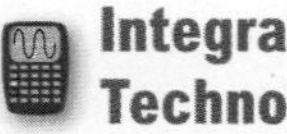

Integrating Technology

See the Keystroke Guide: *Evaluating Variable Expressions* for instructions on using a graphing calculator to evaluate variable expressions.

Replacing each variable by its value and then simplifying the resulting numerical expression is called **evaluating a variable expression.**

HOW TO • 1 Evaluate $ab - b^2$ when $a = 2$ and $b = -3$.

Replace each variable in the expression by its value. Then use the Order of Operations Agreement to simplify the resulting numerical expression.

$ab - b^2$

$2(-3) - (-3)^2 = -6 - 9$

$= -15$

When $a = 2$ and $b = -3$, the value of $ab - b^2$ is -15.

EXAMPLE • 1

Evaluate $\frac{a^2 - b^2}{a - b}$ when $a = 3$ and $b = -4$.

Solution

$$\frac{a^2 - b^2}{a - b}$$

$$\frac{3^2 - (-4)^2}{3 - (-4)} = \frac{9 - 16}{3 - (-4)}$$

$$= \frac{-7}{7} = -1$$

YOU TRY IT • 1

Evaluate $\frac{a^2 + b^2}{a + b}$ when $a = 5$ and $b = -3$.

Your solution

EXAMPLE • 2

Evaluate $x^2 - 3(x - y) - z^2$ when $x = 2$, $y = -1$, and $z = 3$.

Solution

$$x^2 - 3(x - y) - z^2$$

$$2^2 - 3[2 - (-1)] - 3^2$$

$$= 2^2 - 3(3) - 3^2$$

$$= 4 - 3(3) - 9$$

$$= 4 - 9 - 9$$

$$= -5 - 9$$

$$= -14$$

YOU TRY IT • 2

Evaluate $x^3 - 2(x + y) + z^2$ when $x = 2$, $y = -4$, and $z = -3$.

Your solution

Solutions on p. S2

OBJECTIVE B To simplify a variable expression using the Properties of Addition

Take Note

Here is an example of the Distributive Property using just numbers.

$2(5 + 9) = 2(5) + 2(9)$
$= 10 + 18 = 28$

This is the same result we would obtain using the Order of Operations Agreement.

$2(5 + 9) = 2(14) = 28$

The usefulness of the Distributive Property will become more apparent as we explore variable expressions.

Like terms of a variable expression are terms with the same variable part. (Because $x^2 = x \cdot x$, x^2 and x are not like terms.)

Constant terms are like terms. 4 and 9 are like terms.

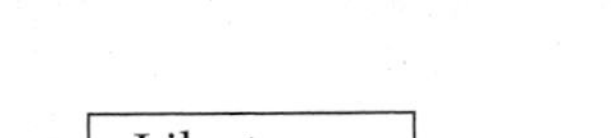

To simplify a variable expression, use the Distributive Property to combine like terms by adding the numerical coefficients. The variable part remains unchanged.

Distributive Property

If a, b, and c are real numbers, then $a(b + c) = ab + ac$.

The Distributive Property can also be written $ba + ca = (b + c)a$. This form is used to simplify a variable expression.

To simplify $2x + 3x$, use the Distributive Property to add the numerical coefficients of the like variable terms. This is called **combining like terms.**

$$2x + 3x = (2 + 3)x$$
$$= 5x$$

HOW TO • 2 Simplify: $5y - 11y$

$5y - 11y = (5 - 11)y$ • Use the **Distributive Property.**
$= -6y$

Take Note
Simplifying an expression means combining like terms. The constant term 5 and the variable term $7p$ are not like terms and therefore cannot be combined.

HOW TO • 3 Simplify: $5 + 7p$

The terms 5 and $7p$ are not like terms.

The expression $5 + 7p$ is in simplest form.

The Associative Property of Addition

If a, b, and c are real numbers, then $(a + b) + c = a + (b + c)$.

When three or more terms are added, the terms can be grouped (with parentheses, for example) in any order. The sum is the same. For example,

$$(5 + 7) + 15 = 5 + (7 + 15) \qquad (3x + 5x) + 9x = 3x + (5x + 9x)$$
$$12 + 15 = 5 + 22 \qquad 8x + 9x = 3x + 14x$$
$$27 = 27 \qquad 17x = 17x$$

The Commutative Property of Addition

If a and b are real numbers, then $a + b = b + a$.

When two like terms are added, the terms can be added in either order. The sum is the same. For example,

$$15 + (-28) = (-28) + 15 \qquad 2x + (-4x) = -4x + 2x$$
$$-13 = -13 \qquad -2x = -2x$$

The Addition Property of Zero

If a is a real number, then $a + 0 = 0 + a = a$.

The sum of a term and zero is the term. For example,

$$-9 + 0 = 0 + (-9) = -9 \qquad 5x + 0 = 0 + 5x = 5x$$

The Inverse Property of Addition

If a is a real number, then $a + (-a) = (-a) + a = 0$.

The sum of a term and its opposite is zero. Recall that the opposite of a number is called its **additive inverse.**

$$12 + (-12) = (-12) + 12 = 0 \qquad 7x + (-7x) = -7x + 7x = 0$$

HOW TO • 4 Simplify: $8x + 4y - 8x + y$

$8x + 4y - 8x + y$
$= (8x - 8x) + (4y + y)$ • Use the Commutative and Associative Properties of Addition to rearrange and group like terms.
$= 0 + 5y = 5y$ • Combine like terms.

HOW TO • 5 Simplify: $4x^2 + 5x - 6x^2 - 2x + 1$

$4x^2 + 5x - 6x^2 - 2x + 1$
$= (4x^2 - 6x^2) + (5x - 2x) + 1$ • Use the Commutative and Associative Properties of Addition to rearrange and group like terms.
$= -2x^2 + 3x + 1$ • Combine like terms.

EXAMPLE • 3

Simplify: $3x + 4y - 10x + 7y$

Solution

$3x + 4y - 10x + 7y = -7x + 11y$

YOU TRY IT • 3

Simplify: $3a - 2b - 5a + 6b$

Your solution

EXAMPLE • 4

Simplify: $x^2 - 7 + 4x^2 - 16$

Solution

$x^2 - 7 + 4x^2 - 16 = 5x^2 - 23$

YOU TRY IT • 4

Simplify: $-3y^2 + 7 + 8y^2 - 14$

Your solution

Solutions on p. S2

OBJECTIVE C To simplify a variable expression using the Properties of Multiplication

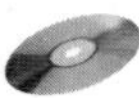

In simplifying variable expressions, the following Properties of Multiplication are used.

Take Note
The Associative Property of Multiplication allows us to multiply a coefficient by a number. Without this property, the expression $2(3x)$ could not be changed.

The Associative Property of Multiplication

If a, b, and c are real numbers, then $(ab)c = a(bc)$.

When three or more factors are multiplied, the factors can be grouped in any order. The product is the same.

$$3(5 \cdot 6) = (3 \cdot 5)6 \qquad 2(3x) = (2 \cdot 3)x$$
$$3(30) = (15)6 \qquad = 6x$$
$$90 = 90$$

Take Note
The Commutative Property of Multiplication allows us to rearrange factors. This property, along with the Associative Property of Multiplication, allows us to simplify some variable expressions.

The Commutative Property of Multiplication

If a and b are real numbers, then $ab = ba$.

Two factors can be multiplied in either order. The product is the same.

$5(-7) = -7(5)$
$-35 = -35$

$(5x) \cdot 3 = 3 \cdot (5x)$ • Commutative Property of Multiplication
$= (3 \cdot 5)x$ • Associative Property of Multiplication
$= 15x$

The Multiplication Property of One

If a is a real number, then $a \cdot 1 = 1 \cdot a = a$.

The product of a term and 1 is the term.

$$9 \cdot 1 = 1 \cdot 9 = 9 \qquad (8x) \cdot 1 = 1 \cdot (8x) = 8x$$

The Inverse Property of Multiplication

If a is a real number and a is not equal to zero, then

$$a \cdot \frac{1}{a} = \frac{1}{a} \cdot a = 1$$

$\frac{1}{a}$ is called the **reciprocal** of a. $\frac{1}{a}$ is also called the **multiplicative inverse** of a. The product of a number and its reciprocal is 1.

Take Note
We must state that $x \neq 0$ because division by zero is undefined.

$$7 \cdot \frac{1}{7} = \frac{1}{7} \cdot 7 = 1 \qquad x \cdot \frac{1}{x} = \frac{1}{x} \cdot x = 1, \quad x \neq 0$$

The multiplication properties are used to simplify variable expressions.

HOW TO 6 Simplify: $2(-x)$

$2(-x) = 2(-1 \cdot x)$
$= [2(-1)]x$ • Use the Associative Property of Multiplication to group factors.
$= -2x$

HOW TO 7 Simplify: $\frac{3}{2}\left(\frac{2x}{3}\right)$

$\frac{3}{2}\left(\frac{2x}{3}\right) = \frac{3}{2}\left(\frac{2}{3}x\right)$ • Note that $\frac{2x}{3} = \frac{2}{3}x$.

$= \left(\frac{3}{2} \cdot \frac{2}{3}\right)x$ • Use the Associative Property of Multiplication to group factors.

$= 1 \cdot x$
$= x$

HOW TO • 8 Simplify: $(16x)2$

$(16x)2 = 2(16x)$ • Use the Commutative and Associative Properties of Multiplication to rearrange and group factors.

$= (2 \cdot 16)x$

$= 32x$

EXAMPLE • 5

Simplify: $-2(3x^2)$

Solution

$-2(3x^2) = -6x^2$

YOU TRY IT • 5

Simplify: $-5(4y^2)$

Your solution

EXAMPLE • 6

Simplify: $-5(-10x)$

Solution

$-5(-10x) = 50x$

YOU TRY IT • 6

Simplify: $-7(-2a)$

Your solution

EXAMPLE • 7

Simplify: $-\frac{3}{4}\left(\frac{2}{3}x\right)$

Solution

$-\frac{3}{4}\left(\frac{2}{3}x\right) = -\frac{1}{2}x$

YOU TRY IT • 7

Simplify: $-\frac{3}{5}\left(-\frac{7}{9}a\right)$

Your solution

Solutions on p. S2

OBJECTIVE D To simplify a variable expression using the Distributive Property

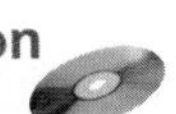

Recall that the Distributive Property states that if a, b, and c are real numbers, then

$$a(b + c) = ab + ac$$

The Distributive Property is used to remove parentheses from a variable expression.

HOW TO • 9 Simplify: $3(2x + 7)$

$3(2x + 7) = 3(2x) + 3(7)$ • Use the **Distributive Property.** Multiply each term inside the parentheses by **3.**

$= 6x + 21$

HOW TO • 10 Simplify: $-5(4x + 6)$

$-5(4x + 6) = -5(4x) + (-5)(6)$ • Use the **Distributive Property.**

$= -20x - 30$

HOW TO • 11 Simplify: $-(2x - 4)$

$$\begin{aligned} -(2x-4) &= -1(2x-4) \\ &= -1(2x) - (-1)(4) \\ &= -2x + 4 \end{aligned}$$

• Use the **Distributive Property**.

Note: When a negative sign immediately precedes the parentheses, the sign of each term inside the parentheses is changed.

HOW TO • 12 Simplify: $-\frac{1}{2}(8x - 12y)$

$$\begin{aligned} -\frac{1}{2}(8x-12y) &= -\frac{1}{2}(8x) - \left(-\frac{1}{2}\right)(12y) \\ &= -4x + 6y \end{aligned}$$

• Use the **Distributive Property**.

HOW TO • 13 Simplify: $4(x - y) - 2(-3x + 6y)$

$$\begin{aligned} &4(x-y) - 2(-3x+6y) \\ &\quad = 4x - 4y + 6x - 12y \\ &\quad = 10x - 16y \end{aligned}$$

• Use the Distributive Property.
• Combine like terms.

An extension of the Distributive Property is used when an expression contains more than two terms.

HOW TO • 14 Simplify: $3(4x - 2y - z)$

$$\begin{aligned} 3(4x - 2y - z) &= 3(4x) - 3(2y) - 3(z) \\ &= 12x - 6y - 3z \end{aligned}$$

• Use the **Distributive Property**.

EXAMPLE • 8

Simplify: $(2x - 6)2$

Solution

$(2x - 6)2 = 4x - 12$

YOU TRY IT • 8

Simplify: $(3a - 1)5$

Your solution

EXAMPLE • 9

Simplify: $-3(-5a + 7b)$

Solution

$-3(-5a + 7b) = 15a - 21b$

YOU TRY IT • 9

Simplify: $-8(-2a + 7b)$

Your solution

Solutions on p. S3

EXAMPLE 10

Simplify: $3(x^2 - x - 5)$

Solution

$3(x^2 - x - 5) = 3x^2 - 3x - 15$

YOU TRY IT 10

Simplify: $3(12x^2 - x + 8)$

Your solution

EXAMPLE 11

Simplify: $2x - 3(2x - 7y)$

Solution

$$\begin{aligned} 2x - 3(2x - 7y) &= 2x - 6x + 21y \\ &= -4x + 21y \end{aligned}$$

YOU TRY IT 11

Simplify: $3y - 2(y - 7x)$

Your solution

EXAMPLE 12

Simplify: $7(x - 2y) - (-x - 2y)$

Solution

$$\begin{aligned} &7(x - 2y) - (-x - 2y) \\ &= 7x - 14y + x + 2y \\ &= 8x - 12y \end{aligned}$$

YOU TRY IT 12

Simplify: $-2(x - 2y) - (-x + 3y)$

Your solution

EXAMPLE 13

Simplify: $2x - 3[2x - 3(x + 7)]$

Solution

$$\begin{aligned} &2x - 3[2x - 3(x + 7)] \\ &= 2x - 3[2x - 3x - 21] \\ &= 2x - 3[-x - 21] \\ &= 2x + 3x + 63 \\ &= 5x + 63 \end{aligned}$$

YOU TRY IT 13

Simplify: $3y - 2[x - 4(2 - 3y)]$

Your solution

Solutions on p. S3

OBJECTIVE E To translate a verbal expression into a variable expression

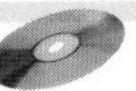

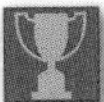

Tips for Success

Before the class meeting in which your professor begins a new section, you should read each objective statement for that section. Next, browse through the objective material. The purpose of browsing through the material is so that your brain will be prepared to accept and organize the new information when it is presented to you. See *AIM for Success* in the Preface.

One of the major skills required in applied mathematics is to translate a verbal expression into a variable expression. This requires recognizing the verbal phrases that translate into mathematical operations. A partial list of the verbal phrases used to indicate the different mathematical operations follows.

Addition	added to	6 added to y	$y + 6$
	more than	8 more than x	$x + 8$
	the sum of	the sum of x and z	$x + z$
	increased by	t increased by 9	$t + 9$
	the total of	the total of 5 and y	$5 + y$

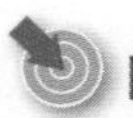

Point of Interest

The way in which expressions are symbolized has changed over time. Here are how some of the expressions shown at the right may have appeared in the early 16th century.

R p. 9 for $x + 9$. The symbol R was used for a variable to the first power. The symbol p. was used for plus.

R m. 3 for $x - 3$. The symbol R is still used for the variable. The symbol m. was used for minus.

The square of a variable was designated by Q and the cube was designated by C. The expression $x^2 + x^3$ was written Q p. C.

Subtraction	minus	x minus 2	$x - 2$
	less than	7 less than t	$t - 7$
	decreased by	m decreased by 3	$m - 3$
	the difference between	the difference between y and 4	$y - 4$
	subtract...from...	subtract 9 from z	$z - 9$
Multiplication	times	10 times t	$10t$
	twice	twice w	$2w$
	of	one-half of x	$\frac{1}{2}x$
	the product of	the product of y and z	yz
	multiplied by	y multiplied by 11	$11y$
Division	divided by	x divided by 12	$\frac{x}{12}$
	the quotient of	the quotient of y and z	$\frac{y}{z}$
	the ratio of	the ratio of t to 9	$\frac{t}{9}$
Power	the square of	the square of x	x^2
	the cube of	the cube of a	a^3

HOW TO • 15 Translate "14 less than the cube of x" into a variable expression.

14 less than the cube of x — • Identify the words that indicate the mathematical operations.

$x^3 - 14$ — • Use the identified operations to write the variable expression.

In most applications that involve translating phrases into variable expressions, the variable to be used is not given. To translate these phrases, a variable must be assigned to an unknown quantity before the variable expression can be written.

HOW TO • 16 Translate "the sum of two consecutive integers" into a variable expression. Then simplify.

the first integer: n — • Assign a variable to one of the unknown quantities.

the next consecutive integer: $n + 1$ — • Use the assigned variable to write an expression for any other unknown quantity.

$n + (n + 1)$ — • Use the assigned variable to write the variable expression.

$(n + n) + 1$ — • Simplify the variable expression.

$2n + 1$

Many applications in mathematics require that you identify the unknown quantity, assign a variable to that quantity, and then attempt to express other unknown quantities in terms of the variable.

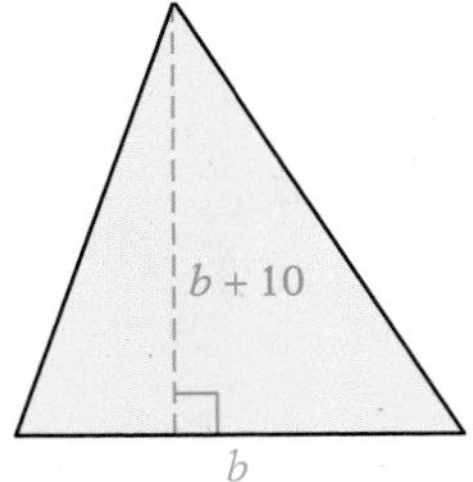

HOW TO • 17 The height of a triangle is 10 ft longer than the base of the triangle. Express the height of the triangle in terms of the base of the triangle.

the base of the triangle: b • Assign a variable to the base of the triangle.

the height is 10 more than the base: $b + 10$ • Express the height of the triangle in terms of b.

EXAMPLE • 14

The length of a swimming pool is 4 ft less than two times the width. Express the length of the pool in terms of the width.

Solution

the width of the pool: w

the length is 4 ft less than two times the width: $2w - 4$

YOU TRY IT • 14

The speed of a new jet plane is twice the speed of an older model. Express the speed of the new model in terms of the speed of the older model.

Your solution

EXAMPLE • 15

A banker divided $5000 between two accounts, one paying 10% annual interest and the second paying 8% annual interest. Express the amount invested in the 10% account in terms of the amount invested in the 8% account.

Solution

the amount invested at 8%: x

the amount invested at 10%: $5000 - x$

YOU TRY IT • 15

A guitar string 6 ft long was cut into two pieces. Express the length of the shorter piece in terms of the length of the longer piece.

Your solution

Solutions on p. S3

1.4 EXERCISES

OBJECTIVE A To evaluate a variable expression

For Exercises 1 to 12, evaluate the variable expression when $a = 2$, $b = 3$, and $c = -4$.

1. $6b \div (-a)$

2. $bc \div (2a)$

3. $b^2 - 4ac$

4. $a^2 - b^2$

5. $b^2 - c^2$

6. $(a + b)^2$

7. $a^2 + b^2$

8. $2a - (c + a)^2$

9. $(b - a)^2 + 4c$

10. $b^2 - \frac{ac}{8}$

11. $\frac{5ab}{6} - 3cb$

12. $(b - 2a)^2 + bc$

For Exercises 13 to 30, evaluate the variable expression when $a = -2$, $b = 4$, $c = -1$, and $d = 3$.

13. $\frac{b + c}{d}$

14. $\frac{d - b}{c}$

15. $\frac{2d + b}{-a}$

16. $\frac{b + 2d}{b}$

17. $\frac{b - d}{c - a}$

18. $\frac{2c - d}{-ad}$

19. $(b + d)^2 - 4a$

20. $(d - a)^2 - 3c$

21. $(d - a)^2 \div 5$

22. $3(b - a) - bc$

23. $\frac{b - 2a}{bc^2 - d}$

24. $\frac{b^2 - a}{ad + 3c}$

25. $\frac{1}{3}d^2 - \frac{3}{8}b^2$

26. $\frac{5}{8}a^4 - c^2$

27. $\frac{-4bc}{2a - b}$

28. $-\frac{3}{4}b + \frac{1}{2}(ac + bd)$

29. $-\frac{2}{3}d - \frac{1}{5}(bd - ac)$

30. $(b - a)^2 - (d - c)^2$

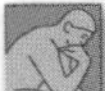

For Exercises 31 to 34, without evaluating the expression, determine whether the expression is positive or negative when $a = -25$, $b = 67$, and $c = -82$.

31. $(c - a)(-b)$

32. $(a - c) + 3b$

33. $\dfrac{b + c}{abc}$

34. $\dfrac{ac}{-b^2}$

OBJECTIVE B **To simplify a variable expression using the Properties of Addition**

For Exercises 35 to 70, simplify.

35. $6x + 8x$

36. $12x + 13x$

37. $9a - 4a$

38. $12a - 3a$

39. $4y - 10y$

40. $8y - 6y$

41. $7 - 3b$

42. $5 + 2a$

43. $-12a + 17a$

44. $-3a + 12a$

45. $5ab - 7ab$

46. $9ab - 3ab$

47. $-12xy + 17xy$

48. $-15xy + 3xy$

49. $-3ab + 3ab$

50. $-7ab + 7ab$

51. $-\dfrac{1}{2}x - \dfrac{1}{3}x$

52. $-\dfrac{2}{5}y + \dfrac{3}{10}y$

53. $2.3x + 4.2x$

54. $6.1y - 9.2y$

55. $x - 0.55x$

56. $0.65A - A$

57. $5a - 3a + 5a$

58. $10a - 17a + 3a$

59. $-5x^2 - 12x^2 + 3x^2$

60. $-y^2 - 8y^2 + 7y^2$

61. $\dfrac{3}{4}x - \dfrac{1}{3}x - \dfrac{7}{8}x$

62. $-\dfrac{2}{5}a - \left(-\dfrac{3}{10}a\right) - \dfrac{11}{15}a$

63. $7x - 3y + 10x$

64. $8y + 8x - 8y$

65. $3a + (-7b) - 5a + b$

66. $-5b + 7a - 7b + 12a$

67. $3x + (-8y) - 10x + 4x$

68. $3y + (-12x) - 7y + 2y$

69. $x^2 - 7x + (-5x^2) + 5x$

70. $3x^2 + 5x - 10x^2 - 10x$

71. Which of the following expressions are equivalent to $-10x - 10y - 10y - 10x$?
(i) 0 (ii) $-20y$ (iii) $-20x$ (iv) $-20x - 20y$ (v) $-20y - 20x$

OBJECTIVE C **To simplify a variable expression using the Properties of Multiplication**

For Exercises 72 to 111, simplify.

72. $4(3x)$ **73.** $12(5x)$ **74.** $-3(7a)$ **75.** $-2(5a)$ **76.** $-2(-3y)$

77. $-5(-6y)$ **78.** $(4x)2$ **79.** $(6x)12$ **80.** $(3a)(-2)$ **81.** $(7a)(-4)$

82. $(-3b)(-4)$ **83.** $(-12b)(-9)$ **84.** $-5(3x^2)$ **85.** $-8(7x^2)$ **86.** $\frac{1}{3}(3x^2)$

87. $\frac{1}{6}(6x^2)$ **88.** $\frac{1}{5}(5a)$ **89.** $\frac{1}{8}(8x)$ **90.** $-\frac{1}{2}(-2x)$ **91.** $-\frac{1}{4}(-4a)$

92. $-\frac{1}{7}(-7n)$ **93.** $-\frac{1}{9}(-9b)$ **94.** $(3x)\left(\frac{1}{3}\right)$ **95.** $(12x)\left(\frac{1}{12}\right)$ **96.** $(-6y)\left(-\frac{1}{6}\right)$

97. $(-10n)\left(-\frac{1}{10}\right)$ **98.** $\frac{1}{3}(9x)$ **99.** $\frac{1}{7}(14x)$ **100.** $-0.2(10x)$ **101.** $-0.25(8x)$

102. $-\frac{2}{3}(12a^2)$ **103.** $-\frac{5}{8}(24a^2)$ **104.** $-0.5(-16y)$ **105.** $-0.75(-8y)$ **106.** $(16y)\left(\frac{1}{4}\right)$

107. $(33y)\left(\frac{1}{11}\right)$ **108.** $(-6x)\left(\frac{1}{3}\right)$ **109.** $(-10x)\left(\frac{1}{5}\right)$ **110.** $(-8a)\left(-\frac{3}{4}\right)$ **111.** $(21y)\left(-\frac{3}{7}\right)$

112. After multiplying $\frac{2}{7}x^2$ by a proper fraction, is the coefficient of x^2 greater than 1 or less than 1?

OBJECTIVE D To simplify a variable expression using the Distributive Property

For Exercises 113 to 151, simplify.

113. $2(4x - 3)$

114. $5(2x - 7)$

115. $-2(a + 7)$

116. $-5(a + 16)$

117. $-3(2y - 8)$

118. $-5(3y - 7)$

119. $-(x + 2)$

120. $-(x + 7)$

121. $(5 - 3b)7$

122. $(10 - 7b)2$

123. $\frac{1}{3}(6 - 15y)$

124. $\frac{1}{2}(-8x + 4y)$

125. $3(5x^2 + 2x)$

126. $6(3x^2 + 2x)$

127. $-2(-y + 9)$

128. $-5(-2x + 7)$

129. $(-3x - 6)5$

130. $(-2x + 7)7$

131. $2(-3x^2 - 14)$

132. $5(-6x^2 - 3)$

133. $-3(2y^2 - 7)$

134. $-8(3y^2 - 12)$

135. $3(x^2 - y^2)$

136. $5(x^2 + y^2)$

137. $-\frac{2}{3}(6x - 18y)$

138. $-\frac{1}{2}(x - 4y)$

139. $-(6a^2 - 7b^2)$

140. $3(x^2 + 2x - 6)$

141. $4(x^2 - 3x + 5)$

142. $-2(y^2 - 2y + 4)$

143. $\frac{3}{4}(2x - 6y + 8)$

144. $-\frac{2}{3}(6x - 9y + 1)$

145. $4(-3a^2 - 5a + 7)$

146. $-5(-2x^2 - 3x + 7)$

147. $-3(-4x^2 + 3x - 4)$

148. $3(2x^2 + xy - 3y^2)$

149. $5(2x^2 - 4xy - y^2)$

150. $-(3a^2 + 5a - 4)$

151. $-(8b^2 - 6b + 9)$

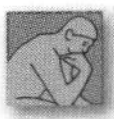

152. After the expression $17x - 31$ is multiplied by a negative integer, is the constant term positive or negative?

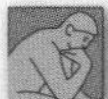

153. Which of the following expressions is equivalent to $12 - 7(y - 9)$?
(i) $5(y - 9)$ (ii) $12 - 7y - 63$ (iii) $12 - 7y + 63$ (iv) $12 - 7y - 9$

For Exercises 154 to 177, simplify.

154. $4x - 2(3x + 8)$

155. $6a - (5a + 7)$

156. $9 - 3(4y + 6)$

157. $10 - (11x - 3)$

158. $5n - (7 - 2n)$

159. $8 - (12 + 4y)$

160. $3(x + 2) - 5(x - 7)$

161. $2(x - 4) - 4(x + 2)$

162. $12(y - 2) + 3(7 - 3y)$

163. $6(2y - 7) - (3 - 2y)$

164. $3(a - b) - (a + b)$

165. $2(a + 2b) - (a - 3b)$

166. $4[x - 2(x - 3)]$

167. $2[x + 2(x + 7)]$

168. $-2[3x + 2(4 - x)]$

169. $-5[2x + 3(5 - x)]$

170. $-3[2x - (x + 7)]$

171. $-2[3x - (5x - 2)]$

172. $2x - 3[x - (4 - x)]$

173. $-7x + 3[x - (3 - 2x)]$

174. $-5x - 2[2x - 4(x + 7)] - 6$

175. $0.12(2x + 3) + x$

176. $0.05x + 0.02(4 - x)$

177. $0.03x + 0.04(1000 - x)$

OBJECTIVE E To translate a verbal expression into a variable expression

For Exercises 178 to 187, translate into a variable expression.

178. twelve minus a number

179. a number divided by eighteen

180. two-thirds of a number

181. twenty more than a number

182. the quotient of twice a number and nine

183. eight less than the product of eleven and a number

184. the quotient of fifteen and the sum of a number and twelve

185. the difference between forty and the quotient of a number and twenty

186. the quotient of five more than twice a number and the number

187. the sum of the square of a number and twice the number

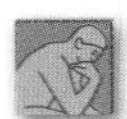

188. Which of the following phrases translate into the variable expression $32 - \frac{a}{7}$?

(i) the difference between thirty-two and the quotient of a number and seven
(ii) thirty-two decreased by the quotient of a number and seven
(iii) thirty-two minus the ratio of a number to seven

For Exercises 189 to 200, translate into a variable expression. Then simplify.

189. ten times the difference between a number and fifty

190. nine less than the total of a number and two

191. the difference between a number and three more than the number

192. four times the sum of a number and nineteen

193. a number added to the difference between twice the number and four

194. the product of five less than a number and seven

195. a number added to the product of three and the number

196. a number increased by the total of the number and nine

197. five more than the sum of a number and six

198. a number decreased by the difference between eight and the number

199. a number minus the sum of the number and ten

200. two more than the total of a number and five

201. **Museums** In a recent year, 3.8 million more people visited the Louvre in Paris than visited the Metropolitan Museum of Art in New York City. (*Sources:* The *Art Newspaper;* museums' accounts) Express the number of visitors to the Louvre in terms of the number of visitors to the Metropolitan Museum of Art.

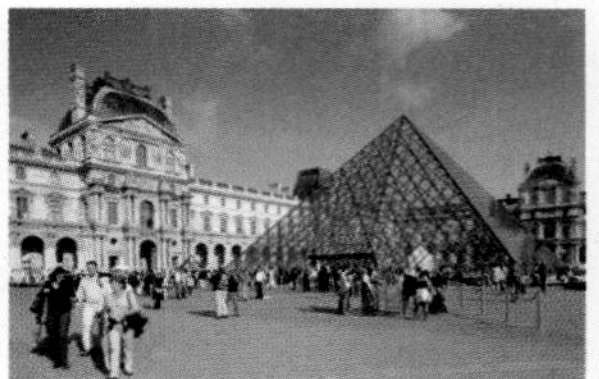
isifa Image Service s.r.o./Alamy

The Louvre

202. **Salaries** For an employee with a bachelor's degree in business, the average annual salary depends on experience. An employee with less than 5 years' experience is paid, on average, $29,100 less than an employee with 10 to 20 years' experience. (*Sources:* PayScale; *The Princeton Review*) Express the salary of an employee with less than 5 years' experience in terms of the salary of an employee with 10 to 20 years' experience.

In the News

Google Websites Most Popular

During the month of February 2008, Google websites ranked number one in the world, with the highest number of unique visitors. Microsoft websites came in second, with 63 million fewer unique visitors.

Source: comScore

203. **Websites** See the news clipping at the right. Express the number of unique visitors to Microsoft websites in terms of the number of unique visitors to Google websites.

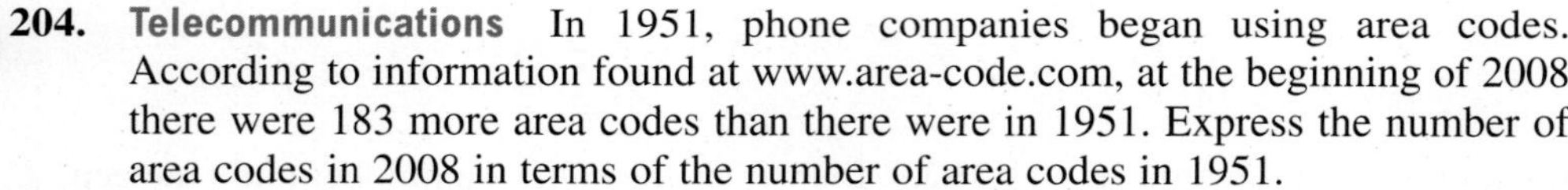

204. **Telecommunications** In 1951, phone companies began using area codes. According to information found at www.area-code.com, at the beginning of 2008 there were 183 more area codes than there were in 1951. Express the number of area codes in 2008 in terms of the number of area codes in 1951.

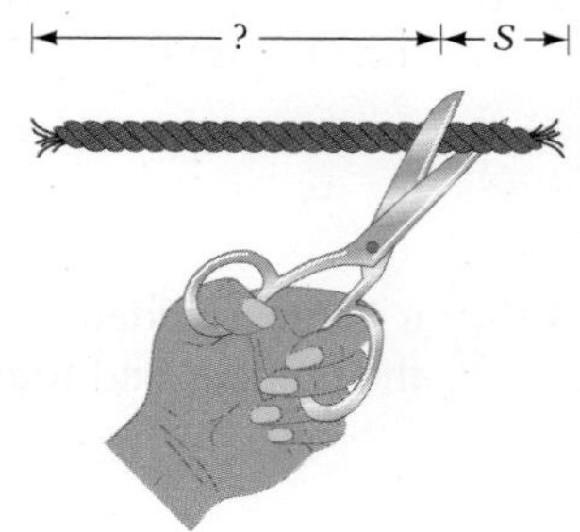

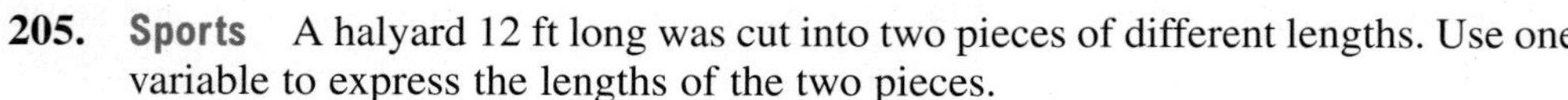

205. **Sports** A halyard 12 ft long was cut into two pieces of different lengths. Use one variable to express the lengths of the two pieces.

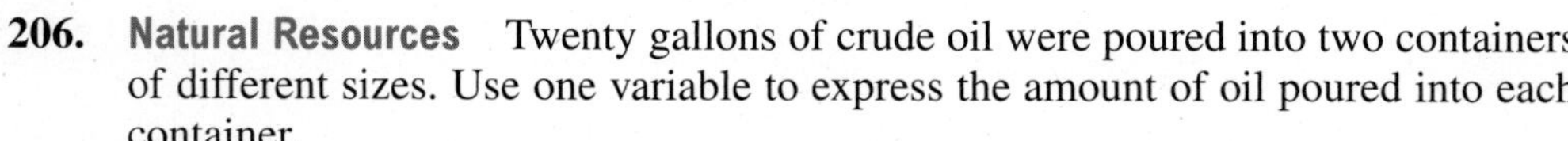

206. **Natural Resources** Twenty gallons of crude oil were poured into two containers of different sizes. Use one variable to express the amount of oil poured into each container.

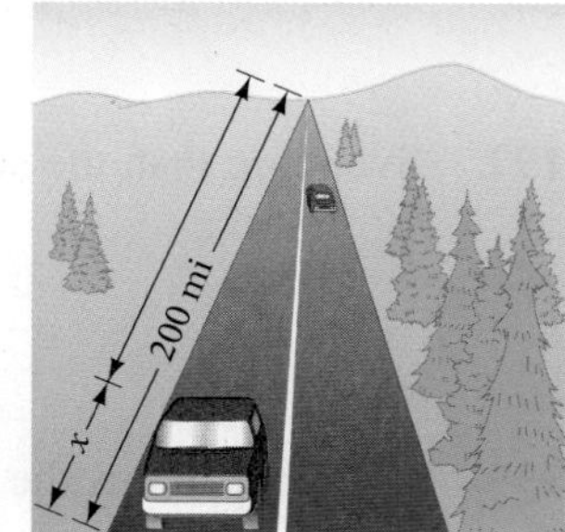

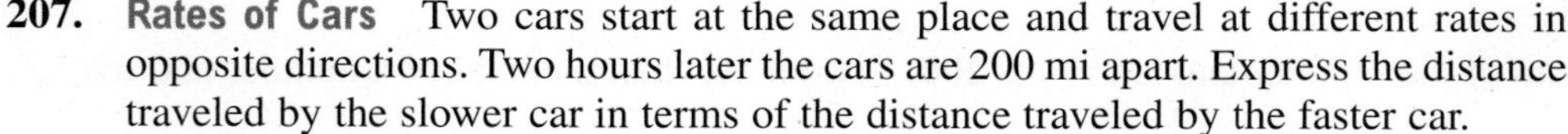

207. **Rates of Cars** Two cars start at the same place and travel at different rates in opposite directions. Two hours later the cars are 200 mi apart. Express the distance traveled by the slower car in terms of the distance traveled by the faster car.

208. **Social Networking** In June 2007, the combined number of visitors to the social networking sites Facebook and MySpace was 116,314. (*Source:* www.watblog.com) Express the number of visitors to MySpace in terms of the number of visitors to Facebook.

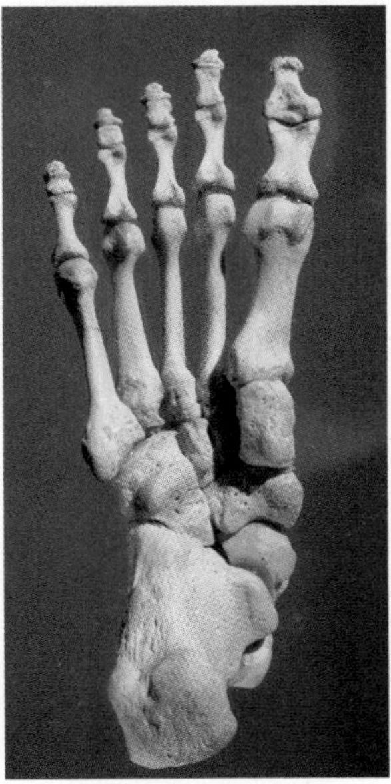
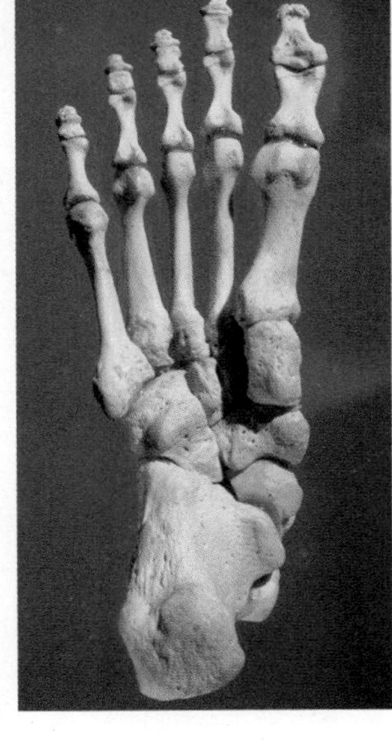

209. **Medicine** According to the American Podiatric Medical Association, the bones in your foot account for one-fourth of all the bones in your body. Express the number of bones in your foot in terms of the total number of bones in your body.

210. **Sports** The diameter of a basketball is approximately four times the diameter of a baseball. Express the diameter of a basketball in terms of the diameter of a baseball.

211. **Tax Refunds** A recent survey conducted by Turbotax.com asked, "If you receive a tax refund, what will you do?" Forty-three percent of respondents said they would pay down their debt. (*Source: USA Today,* March 27, 2008) Express the number of people who would pay down their debt in terms of the number of people surveyed.

212. **Endangered Species** Use the information in the news clipping at the right.

a. Express the number of wild tigers in India in 2007 in terms of the number of wild tigers in India in 2002.

b. Express the number of wild tigers in Tamil Nadu in 2007 in terms of the number of wild tigers in Tamil Nadu in 2002.

In the News

Endangered: Indian Wild Tigers

A recent survey of wild tiger populations in India shows that the number of wild tigers decreased by 2231 between 2002 and 2007. Only one state, in the southern part of the country, showed an increase in its number of tigers: the tiger population in Tamil Nadu rose by 16 tigers.

Source: news.oneindia.in

Applying the Concepts

213. **Metalwork** A wire whose length is given as x inches is bent into a square. Express the length of a side of the square in terms of x.

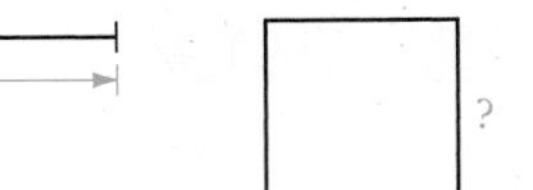

?

214. **Chemistry** The chemical formula for glucose (sugar) is $C_6H_{12}O_6$. This formula means that there are 12 hydrogen atoms for every 6 carbon atoms and 6 oxygen atoms in each molecule of glucose (see the figure at the right). If x represents the number of atoms of oxygen in a pound of sugar, express the number of hydrogen atoms in the pound of sugar in terms of x.

H C O
H − C − OH
HO − C − H
H − C − OH
H − C − OH
CH_2OH

215. Translate the expressions $5x + 8$ and $5(x + 8)$ into phrases.

SECTION

1.5 Sets

OBJECTIVE A To write a set using the roster method 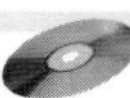

Recall that a *set* is a collection of objects, which are called the *elements* of the set. The roster method of writing a set encloses a list of the elements in braces.

The set of the positive integers less than 5 is written $\{1, 2, 3, 4\}$.

HOW TO • 1 Use the roster method to write the set of integers between 0 and 10.

$A = \{1, 2, 3, 4, 5, 6, 7, 8, 9\}$ • A set can be designated by a capital letter. Note that 0 and 10 are not elements of the set.

HOW TO • 2 Use the roster method to write the set of natural numbers.

$A = \{1, 2, 3, 4, \ldots\}$ • The three dots mean that the pattern of numbers continues without end.

The **empty set,** or **null set,** is the set that contains no elements. The symbol $\varnothing$ or $\{\ \}$ is used to represent the empty set.

The set of people who have run a 2-minute mile is an empty set.

Union of Two Sets

The **union** of two sets, written $A \cup B$, is the set of all elements that belong to either set A *or* set B.

HOW TO • 3 Find $A \cup B$, given $A = \{1, 2, 3, 4\}$ and $B = \{3, 4, 5, 6\}$.

$A \cup B = \{1, 2, 3, 4, 5, 6\}$ • The union of A and B contains all the elements of A and all the elements of B. Elements in both sets are listed only once.

Intersection of Two Sets

The **intersection** of two sets, written $A \cap B$, is the set that contains the elements that are common to both A and B.

HOW TO • 4 Find $A \cap B$, given $A = \{1, 2, 3, 4\}$ and $B = \{3, 4, 5, 6\}$.

$A \cap B = \{3, 4\}$ • The intersection of A and B contains the elements common to A and B.

EXAMPLE • 1

Use the roster method to write the set of the odd positive integers less than 12.

Solution

$A = \{1, 3, 5, 7, 9, 11\}$

YOU TRY IT • 1

Use the roster method to write the set of the odd negative integers greater than -10.

Your solution

Solution on p. S3

EXAMPLE • 2

Use the roster method to write the set of the even positive integers.

Solution

$A = \{2, 4, 6, \ldots\}$

YOU TRY IT • 2

Use the roster method to write the set of the odd positive integers.

Your solution

EXAMPLE • 3

Find $D \cup E$, given $D = \{6, 8, 10, 12\}$ and $E = \{-8, -6, 10, 12\}$.

Solution

$D \cup E = \{-8, -6, 6, 8, 10, 12\}$

YOU TRY IT • 3

Find $A \cup B$, given $A = \{-2, -1, 0, 1, 2\}$ and $B = \{0, 1, 2, 3, 4\}$.

Your solution

EXAMPLE • 4

Find $A \cap B$, given $A = \{5, 6, 9, 11\}$ and $B = \{5, 9, 13, 15\}$.

Solution

$A \cap B = \{5, 9\}$

YOU TRY IT • 4

Find $C \cap D$, given $C = \{10, 12, 14, 16\}$ and $D = \{10, 16, 20, 26\}$.

Your solution

EXAMPLE • 5

Find $A \cap B$, given $A = \{1, 2, 3, 4\}$ and $B = \{8, 9, 10, 11\}$.

Solution

$A \cap B = \varnothing$

YOU TRY IT • 5

Find $A \cap B$, given $A = \{-5, -4, -3, -2\}$ and $B = \{2, 3, 4, 5\}$.

Your solution

Solutions on p. S3

OBJECTIVE B To write and graph sets of real numbers 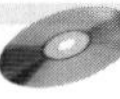

Point of Interest

The symbol $\in$ was first used in the book *Arithmeticae Principia*, published in 1889. It is the first letter of the Greek word $\epsilon\sigma\tau\iota$, which means "is." The symbols for union and intersection were also introduced around the same time.

Another method of representing sets is called **set-builder notation.** This method of writing sets uses a rule to describe the elements of the set. Using set-builder notation, we represent the set of all positive integers less than 10 as

$\{x \mid x < 10, x \in \text{positive integers}\}$, which is read "the set of all positive integers x that are less than 10."

HOW TO • 5 Use set-builder notation to write the set of integers less than or equal to 12.

$\{x \mid x \leq 12, x \in \text{integers}\}$

- **This is read "the set of all integers x that are less than or equal to 12."**

HOW TO • 6 Use set-builder notation to write the set of real numbers greater than 4.

$\{x \mid x > 4, x \in \text{real numbers}\}$

- **This is read "the set of all real numbers x that are greater than 4."**

For the remainder of this section, all variables will represent real numbers. Given this convention, $\{x|x > 4, x \in \text{real numbers}\}$ is written $\{x|x > 4\}$.

Some sets of real numbers written in set-builder notation can be written in **interval notation.** For instance, the interval notation $[-3, 2)$ represents the set of real numbers between -3 and 2. The bracket means that -3 is included in the set, and the parenthesis means that 2 is *not* included in the set. Using set-builder notation, the interval $[-3, 2)$ is written

$\{x|-3 \leq x < 2\}$ • **This is read "the set of all real numbers x between -3 and 2, including -3 but excluding 2."**

To indicate an interval that extends forever in the positive direction, we use the **infinity symbol, ∞;** to indicate an interval that extends forever in the negative direction, we use the **negative infinity symbol, −∞.**

✓ Take Note

When writing a set in interval notation, note that we always use a parenthesis to the right of ∞ and to the left of −∞. Infinity is not a real number, so it cannot be represented as belonging to the set of real numbers by using a bracket.

HOW TO • 7 Write $\{x|x > 1\}$ in interval notation.

$\{x|x > 1\}$ is the set of real numbers greater than 1. This set extends forever in the positive direction. In interval notation, this set is written $(1, \infty)$.

HOW TO • 8 Write $\{x|x \leq -2\}$ in interval notation.

$\{x|x \leq -2\}$ is the set of real numbers less than or equal to -2. This set extends forever in the negative direction. In interval notation, this set is written $(-\infty, -2]$.

HOW TO • 9 Write $[1, 3]$ in set-builder notation.

This is the set of real numbers between 1 and 3, including 1 and 3. In set-builder notation, this set is written $\{x|1 \leq x \leq 3\}$.

We can graph sets of real numbers given in set-builder notation or in interval notation.

HOW TO • 10 Graph: $(-\infty, -1)$

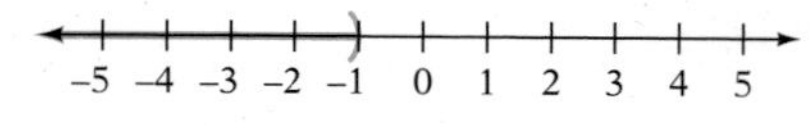

This is the set of real numbers less than -1, excluding -1. The parenthesis at -1 indicates that -1 is excluded from the set.

HOW TO • 11 Graph: $\{x|x \geq 1\}$

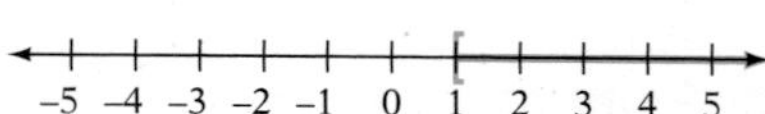

This is the set of real numbers greater than or equal to 1. The bracket at 1 indicates that 1 is included in the set.

EXAMPLE • 6

Write $\{x|x \geq 2\}$ in interval notation.

Solution

$\{x|x \geq 2\}$ is the set of real numbers greater than or equal to 2. This set extends forever in the positive direction. In interval notation, this set is written $[2, \infty)$.

YOU TRY IT • 6

Write $\{x|x \leq 3\}$ in interval notation.

Your solution

Solution on p. S3

EXAMPLE • 7

Write $\{x|0 \leq x \leq 1\}$ in interval notation.

Solution
$\{x|0 \leq x \leq 1\}$ is the set of real numbers between 0 and 1, including 0 and 1. In interval notation, this set is written $[0, 1]$.

YOU TRY IT • 7

Write $\{x|-5 \leq x \leq -3\}$ in interval notation.

Your solution

EXAMPLE • 8

Write $(-\infty, 0]$ in set-builder notation.

Solution
The interval $(-\infty, 0]$ is the set of real numbers less than or equal to 0. In set-builder notation, this set is written $\{x|x \leq 0\}$.

YOU TRY IT • 8

Write $(-3, \infty)$ in set-builder notation.

Your solution

EXAMPLE • 9

Write $(-3, 3)$ in set-builder notation.

Solution
The interval $(-3, 3)$ is the set of real numbers between -3 and 3, excluding -3 and 3. In set-builder notation, this set is written $\{x|-3 < x < 3\}$.

YOU TRY IT • 9

Write $[0, 4)$ in set-builder notation.

Your solution

EXAMPLE • 10

Graph: $\{x|-2 < x < 1\}$

Solution
The graph is the set of real numbers between -2 and 1, excluding -2 and 1. Use parentheses at -2 and 1.

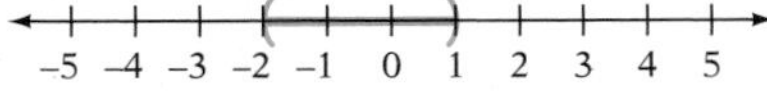

YOU TRY IT • 10

Graph: $\{x|-4 \leq x \leq 4\}$

Your solution

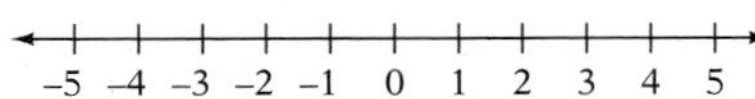

EXAMPLE • 11

Graph: $\{x|x < 4\}$

Solution
The graph is the set of real numbers less than 4. Use a parenthesis at 4.

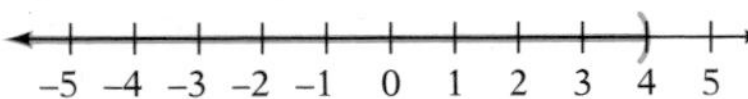

YOU TRY IT • 11

Graph: $\{x|x > -3\}$

Your solution

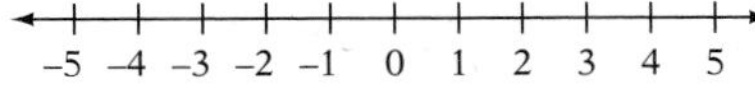

EXAMPLE • 12

Graph: $(-\infty, 5)$

Solution
The graph is the set of real numbers less than 5. Use a parenthesis at 5.

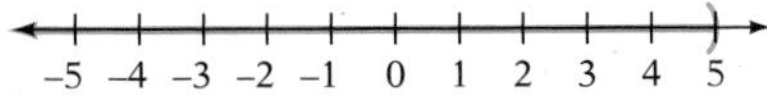

YOU TRY IT • 12

Graph: $(-2, \infty)$

Your solution

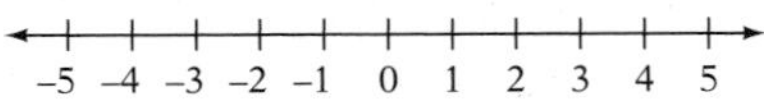

EXAMPLE • 13

Graph: $[-4, 3)$

Solution
The graph is the set of real numbers between -4 and 3, including -4 and excluding 3.

YOU TRY IT • 13

Graph: $[2, 5]$

Your solution

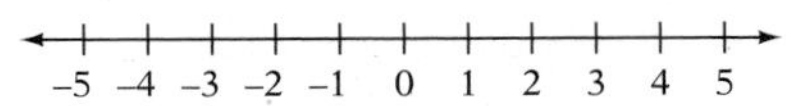

Solutions on p. S3

1.5 EXERCISES

OBJECTIVE A To write a set using the roster method

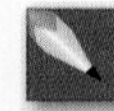
1. Explain how to find the union of two sets.

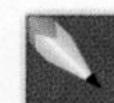
2. Explain how to find the intersection of two sets.

For Exercises 3 to 8, use the roster method to write the set.

3. The integers between 15 and 22

4. The integers between -10 and -4

5. The odd integers between 8 and 18

6. The even integers between -11 and -1

7. The letters of the alphabet between a and d

8. The letters of the alphabet between p and v

For Exercises 9 to 16, find $A \cup B$.

9. $A = \{3, 4, 5\}$ $B = \{4, 5, 6\}$

10. $A = \{-3, -2, -1\}$ $B = \{-2, -1, 0\}$

11. $A = \{-10, -9, -8\}$ $B = \{8, 9, 10\}$

12. $A = \{a, b, c\}$ $B = \{x, y, z\}$

13. $A = \{a, b, d, e\}$ $B = \{c, d, e, f\}$

14. $A = \{m, n, p, q\}$ $B = \{m, n, o\}$

15. $A = \{1, 3, 7, 9\}$ $B = \{7, 9, 11, 13\}$

16. $A = \{-3, -2, -1\}$ $B = \{-1, 1, 2\}$

For Exercises 17 to 22, find $A \cap B$.

17. $A = \{3, 4, 5\}$ $B = \{4, 5, 6\}$

18. $A = \{-4, -3, -2\}$ $B = \{-6, -5, -4\}$

19. $A = \{-4, -3, -2\}$ $B = \{2, 3, 4\}$

20. $A = \{1, 2, 3, 4\}$ $B = \{1, 2, 3, 4\}$

21. $A = \{a, b, c, d, e\}$ $B = \{c, d, e, f, g\}$

22. $A = \{m, n, o, p\}$ $B = \{k, l, m, n\}$

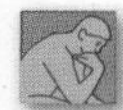
23. Make up sets A and B such that $A \cup B$ has five elements and $A \cap B$ has two elements. Write your sets using the roster method.

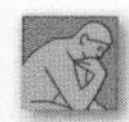
24. True or false? If $A \cup B = A$, then $A \cap B = B$.

OBJECTIVE B To write and graph sets of real numbers

For Exercises 25 to 30, use set-builder notation to write the set.

25. The negative integers greater than -5

26. The positive integers less than 5

27. The integers greater than 30

28. The integers less than -70

29. The real numbers greater than 8

30. The real numbers less than 57

For Exercises 31 to 39, write the set in interval notation.

31. $\{x|1 < x < 2\}$

32. $\{x|-2 < x \leq 4\}$

33. $\{x|x > 3\}$

34. $\{x|x \leq 0\}$

35. $\{x|-4 \leq x < 5\}$

36. $\{x|-3 \leq x \leq 0\}$

37. $\{x|x \leq 2\}$

38. $\{x|x \geq -3\}$

39. $\{x|-3 \leq x \leq 1\}$

For Exercises 40 to 48, write the interval in set-builder notation.

40. $[-4, 5]$

41. $(-5, -3)$

42. $(4, \infty)$

43. $(-\infty, -2]$

44. $(4, 9]$

45. $[-3, -2]$

46. $[0, \infty)$

47. $(-\infty, 6]$

48. $(-\infty, \infty)$

For Exercises 49 to 64, graph the set.

49. $[-5, 4]$

–5 –4 –3 –2 –1 0 1 2 3 4 5

50. $(-3, 5]$

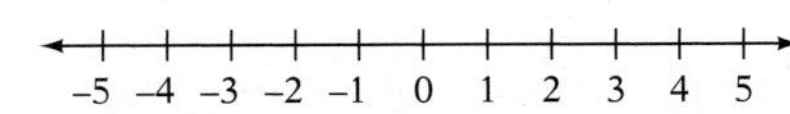

51. $\{x|x < 4\}$

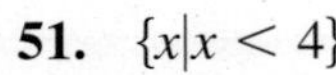

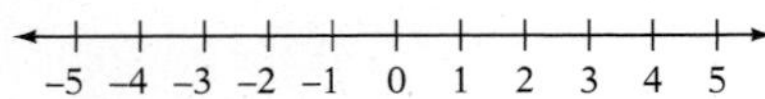

52. $\{x|x \geq -3\}$

−5 −4 −3 −2 −1 0 1 2 3 4 5

53. $\{x|x \leq -4\}$

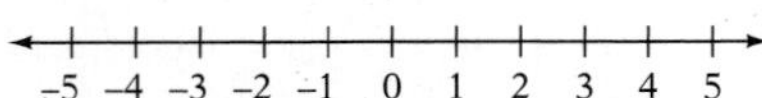

54. $\{x|x > 0\}$

−5 −4 −3 −2 −1 0 1 2 3 4 5

55. $(-\infty, 3]$

−5 −4 −3 −2 −1 0 1 2 3 4 5

56. $(4, \infty)$

−5 −4 −3 −2 −1 0 1 2 3 4 5

57. $[-1, 3)$

−5 −4 −3 −2 −1 0 1 2 3 4 5

58. $(-3, 0]$

−5 −4 −3 −2 −1 0 1 2 3 4 5

59. $\{x|-3 < x < 3\}$

−5 −4 −3 −2 −1 0 1 2 3 4 5

60. $\{x|0 \leq x < 4\}$

−5 −4 −3 −2 −1 0 1 2 3 4 5

61. $\{x|2 \leq x \leq 4\}$

−5 −4 −3 −2 −1 0 1 2 3 4 5

62. $\{x|-4 < x < 1\}$

−5 −4 −3 −2 −1 0 1 2 3 4 5

63. $\{x|-\infty < x < \infty\}$

−5 −4 −3 −2 −1 0 1 2 3 4 5

64. $(-\infty, \infty)$

−5 −4 −3 −2 −1 0 1 2 3 4 5

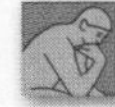

65. How many elements are in the set given in interval notation as $(4, 4)$?

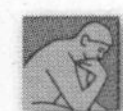

66. How many elements are in the set given by $\{x|4 \leq x \leq 4\}$?

Applying the Concepts

For Exercises 67 and 68, write an inequality that describes the situation.

67. To avoid shipping charges, one must spend a minimum m of \$250.

68. The temperature t never got above freezing (32°F).

FOCUS ON PROBLEM SOLVING

From Concrete to Abstract

In your study of algebra, you will find that the problems are less concrete than those you studied in arithmetic. Problems that are concrete provide information pertaining to a specific instance. Algebra is more abstract. Abstract problems are theoretical; they are stated without reference to a specific instance. Let's look at an example of an abstract problem.

How many minutes are in h hours?

A strategy that can be used to solve this problem is to solve the same problem after substituting a number for the variable.

How many minutes are in 5 hours?

You know that there are 60 minutes in 1 hour. To find the number of minutes in 5 hours, multiply 5 by 60.

$60 \cdot 5 = 300$ There are 300 minutes in 5 hours.

Use the same procedure to find the number of minutes in h hours: multiply h by 60.

$60 \cdot h = 60h$ There are $60h$ minutes in h hours.

This problem might be taken a step further:

If you walk 1 mile in x minutes, how far can you walk in h hours?

Consider the same problem using numbers in place of the variables.

If you walk 1 mile in 20 minutes, how far can you walk in 3 hours?

To solve this problem, you need to calculate the number of minutes in 3 hours (multiply 3 by 60) and divide the result by the number of minutes it takes to walk 1 mile (20 minutes).

$\frac{60 \cdot 3}{20} = \frac{180}{20} = 9$ If you walk 1 mile in 20 minutes, you can walk 9 miles in 3 hours.

Use the same procedure to solve the related abstract problem. Calculate the number of minutes in h hours (multiply h by 60), and divide the result by the number of minutes it takes to walk 1 mile (x minutes).

$\frac{60 \cdot h}{x} = \frac{60h}{x}$ If you walk 1 mile in x minutes, you can walk $\frac{60h}{x}$ miles in h hours.

At the heart of the study of algebra is the use of variables. It is the variables in the problems above that make them abstract. But it is variables that enable us to generalize situations and state rules about mathematics.

Try the following problems.

1. How many hours are in d days?
2. You earn d dollars an hour. What are your wages for working h hours?

3. If p is the price of one share of stock, how many shares can you purchase with d dollars?

4. A company pays a television station d dollars to air a commercial lasting s seconds. What is the cost per second?

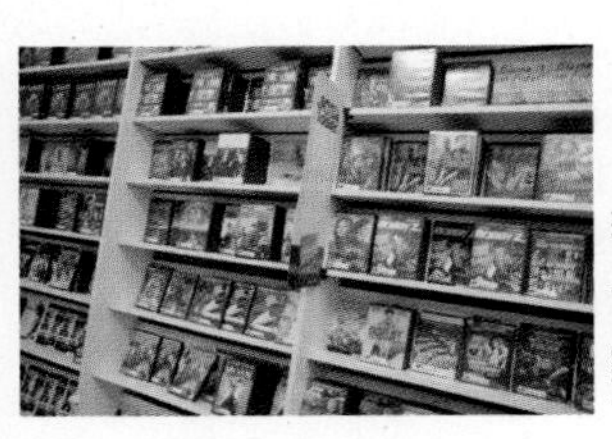
Jeff Greenberg/Alamy

5. After every D DVD rentals, you are entitled to one free rental. You have rented R DVDs, where $R < D$. How many more DVDs do you need to rent before you are entitled to a free rental?

6. Your car gets g miles per gallon. How many gallons of gasoline does your car consume traveling t miles?

7. If you drink j ounces of juice each day, how many days will q quarts of the juice last?

8. A TV station airs m minutes of commercials each hour. How many ads lasting s seconds each can be sold for each hour of programming?

PROJECTS AND GROUP ACTIVITIES

The +/- Key on a Calculator

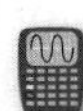

Using your calculator to simplify numerical expressions sometimes requires use of the +/- key or, on some calculators, the negative key, which is frequently shown as (-).

To enter -4:

- For those calculators with +/-, press 4 and then +/-.
- For those calculators with (-), press (-) and then 4.

Here are the keystrokes for evaluating the expression $3(-4) - (-5)$.

Calculators with +/- key: 3 x 4 +/- - 5 +/- =

Calculators with (-) key: 3 x (-) 4 - (-) 5 =

This example illustrates that calculators make a distinction between negative and minus. To perform the operation $3 - (-3)$, you cannot enter 3 - - 3. This would result in 0, which is not the correct answer. You must enter

3 - 3 +/- = or 3 - (-) 3 =

For Exercises 1 to 6, use a calculator to evaluate.

1. $-16 \div 2$
2. $3(-8)$
3. $47 - (-9)$
4. $-50 - (-14)$
5. $4 - (-3)^2$
6. $-8 + (-6)^2 - 7$

Balance of Trade

Objective 1.1E on page 11 describes the concept of balance of trade. You can find data on international trade at www.census.gov. On the next page is a portion of a table from that website.

U.S. International Trade in Goods and Services (in millions of dollars)									
Period (2008)	*Balance*			*Exports*			*Imports*		
	Total	*Goods*	*Services*	*Total*	*Goods*	*Services*	*Total*	*Goods*	*Services*
January	−58,711	−70,147	11,436	149,389	104,686	44,703	208,100	174,833	33,267
February	−61,435	−72,483	11,048	152,551	108,098	44,453	213,985	180,580	33,405
March	−56,964	−68,403	11,439	149,706	105,029	44,677	206,670	173,432	33,238

The sum of the cost of goods exported and the cost of services exported equals the total cost of exports. For January 2008:

$$104{,}686 + 44{,}703 = 149{,}389$$

The total cost of exports minus the total cost of imports equals the total balance. For January 2008:

$$149{,}389 - 208{,}100 = -58{,}711$$

1. The difference between the cost of goods exported and the cost of goods imported equals the balance of the cost of goods. Show this calculation for January 2008.

2. Use the "Balance" columns to show that, for January 2008, the balance of the cost of goods plus the balance of the cost of services equals the total balance of trade.

CHAPTER 1

SUMMARY

KEY WORDS	EXAMPLES
The set of *natural numbers* is $\{1, 2, 3, 4, 5, \ldots\}$. The set of *whole numbers* is $\{0, 1, 2, 3, 4, 5, \ldots\}$. The set of *integers* is $\{\ldots, -3, -2, -1, 0, 1, 2, 3, \ldots\}$. [1.1A, p. 2]	The number 1 is an element of the set of natural numbers, the set of whole numbers, and the set of integers.
A *prime number* is a natural number greater than 1 that is divisible (evenly) only by itself and 1. A natural number that is not a prime number is a *composite number.* [1.1A, p. 2]	2, 3, 5, 7, 11 and 13 are the first six prime numbers. 4, 6, 8, 9, 10, and 12 are the first six composite numbers.
A number *a is less than* a number *b,* written $a < b$, if *a* is to the left of *b* on a number line. A number *a is greater than* a number *b,* written $a > b$, if *a* is to the right of *b* on a number line. The symbol $\leq$ means *is less than or equal to.* The symbol $\geq$ means *is greater than or equal to.* [1.1A, p. 3]	$-5 < -3$ $9 > 0$ $3 \leq 3$ $4 \leq 7$ $5 \geq 5$ $-6 \geq -9$
Two numbers that are the same distance from zero on the number line but on opposite sides of zero are *opposite numbers* or *opposites.* The *additive inverse* of a number is the opposite of the number. [1.1B, p. 4]	7 and −7 are opposites. $-\frac{3}{4}$ and $\frac{3}{4}$ are opposites.

The *absolute value* of a number is its distance from 0 on the number line. [1.1B, p. 4]

$|5| = 5$ $|-2.3| = 2.3$ $|0| = 0$

A *rational number* (or fraction) is a number that can be written in the form $\frac{a}{b}$, where a and b are integers and $b \neq 0$. A rational number can be represented as a *terminating* or *repeating decimal.* [1.2A, p. 18]

$\frac{3}{8}$, $-\frac{9}{2}$, and 4 are rational numbers. 1.13 and $0.4\overline{73}$ are also rational numbers.

Percent means "parts of 100." [1.2B, p. 19]

72% means 72 of 100 equal parts.

An expression of the form a^n is in *exponential form.* The *base* is a and the *exponent* is n. [1.2E, p. 23]

5^4 is an exponential expression. The base is 5 and the exponent is 4.

A *square root* of a positive number x is a number whose square is x. The *principal square root* of a number is the positive square root. The symbol $\sqrt{\ }$ is called a *radical sign* and is used to indicate the principal square root of a number. The *radicand* is the number under the radical sign. [1.2F, p. 25]

$\sqrt{25} = 5$
$-\sqrt{25} = -5$

The square of an integer is a *perfect square.* If a number is not a perfect square, its square root can only be approximated. [1.2F, p. 25]

$7^2 = 49$; 49 is a perfect square.

An *irrational number* is a number that has a decimal representation that never terminates or repeats. [1.2F, p. 25]

π, $\sqrt{2}$, and 1.343343333433334 . . . are irrational numbers.

The rational numbers and the irrational numbers taken together are the *real numbers.* [1.2F, p. 25]

$\frac{3}{8}$, $-\frac{9}{2}$, 4, 1.13, $0.4\overline{73}$, π, $\sqrt{2}$, and 1.343343333433334 . . . are real numbers.

A *variable* is a letter that is used for a quantity that is unknown or that can change. A *variable expression* is an expression that contains one or more variables. [1.4A, p. 38]

$4x + 2y - 6z$ is a variable expression. It contains the variables x, y, and z.

The *terms* of a variable expression are the addends of the expression. Each term is a *variable term* or a *constant term.* [1.4A, p. 38]

The expression $2a^2 - 3b^3 + 7$ has three terms, $2a^2$, $-3b^3$, and 7. $2a^2$ and $-3b^3$ are variable terms. 7 is a constant term.

A variable term is composed of a *numerical coefficient* and a *variable part.* [1.4A, p. 38]

For the expression $-7x^3y^2$, -7 is the coefficient and x^3y^2 is the variable part.

In a variable expression, replacing each variable by its value and then simplifying the resulting numerical expression is called *evaluating the variable expression.* [1.4A, p. 38]	To evaluate $2ab - b^2$ when $a = 3$ and $b = -2$, replace a by 3 and b by -2 and then simplify the numerical expression. $2(3)(-2) - (-2)^2 = -16$
Like terms of a variable expression are terms with the same variable part. Constant terms are like terms. [1.4B, p. 39]	For the expressions $3a^2 + 2b - 3$ and $2a^2 - 3a + 4$, $3a^2$ and $2a^2$ are like terms; -3 and 4 are like terms.
To simplify the sum of like variable terms, use the Distributive Property to add the numerical coefficients. This is called *combining like terms.* [1.4B, p. 40]	$5y + 3y = (5 + 3)y$ $= 8y$
The *multiplicative inverse* of a number is the *reciprocal of the number.* [1.4C, p. 42]	$\frac{3}{4}$ is the multiplicative inverse of $\frac{4}{3}$. $-\frac{1}{4}$ is the multiplicative inverse of -4.
A *set* is a collection of objects, which are called the *elements* of the set. The *roster method* of writing a set encloses a list of the elements in braces. The *empty set* or *null set,* written $\varnothing$ or { }, is the set that contains no elements. [1.5A, p. 56]	The set of cars that can travel faster than 1000 mph is an empty set.
The *union* of two sets, written $A \cup B$, is the set that contains the elements of A and the elements of B. [1.5A, p. 56]	Let $A = \{2, 4, 6, 8\}$ and $B = \{0, 1, 2, 3, 4\}$. Then $A \cup B = \{0, 1, 2, 3, 4, 6, 8\}$.
The *intersection* of two sets, written $A \cap B$, is the set that contains the elements that are common to both A and B. [1.5A, p. 56]	Let $A = \{2, 4, 6, 8\}$ and $B = \{0, 1, 2, 3, 4\}$. Then $A \cap B = \{2, 4\}$.
Set-builder notation and *interval notation* are used to describe the elements of a set. [1.5B, pp. 57–58]	The set of real numbers greater than 2 is written in set-builder notation as $\{x \mid x > 2, x \in \text{real numbers}\}$ and in interval notation as $(2, \infty)$.

ESSENTIAL RULES AND PROCEDURES	EXAMPLES
To add two numbers with the same sign, add the absolute values of the numbers. Then attach the sign of the addends. [1.1C, p. 5]	$7 + 15 = 22$ $-7 + (-15) = -22$
To add two numbers with different signs, find the absolute value of each number. Subtract the smaller of the two numbers from the larger. Then attach the sign of the number with the larger absolute value. [1.1C, p. 5]	$7 + (-15) = -8$ $-7 + 15 = 8$
To subtract one number from another, add the opposite of the second number to the first number. [1.1C, p. 6]	$7 - 19 = 7 + (-19) = -12$ $-6 - (-13) = -6 + 13 = 7$

To multiply two numbers with the same sign, multiply the absolute values of the numbers. The product is positive. [1.1D, p. 8]	$7 \cdot 8 = 56$ $-7(-8) = 56$
To multiply two numbers with different signs, multiply the absolute values of the numbers. The product is negative. [1.1D, p. 8]	$-7 \cdot 8 = -56$ $7(-8) = -56$
To divide two numbers with the same sign, divide the absolute values of the numbers. The quotient is positive. [1.1D, p. 9]	$54 \div 9 = 6$ $(-54) \div (-9) = 6$
To divide two numbers with different signs, divide the absolute values of the numbers. The quotient is negative. [1.1D, p. 9]	$(-54) \div 9 = -6$ $54 \div (-9) = -6$
Properties of Zero and One in Division [1.1D, p. 9] If $a \neq 0$, $\frac{0}{a} = 0$. If $a \neq 0$, $\frac{a}{a} = 1$. $\frac{a}{1} = a$ $\frac{a}{0}$ is undefined.	$\frac{0}{-5} = 0$ $\frac{-12}{-12} = 1$ $\frac{7}{1} = 7$ $\frac{8}{0}$ is undefined.
To write a percent as a fraction, remove the percent sign and multiply by $\frac{1}{100}$. [1.2B, p. 19]	$60\% = 60\left(\frac{1}{100}\right) = \frac{60}{100} = \frac{3}{5}$
To write a percent as a decimal, remove the percent sign and multiply by 0.01. [1.2B, p. 19]	$73\% = 73(0.01) = 0.73$ $1.3\% = 1.3(0.01) = 0.013$
To write a decimal or a fraction as a percent, multiply by 100%. [1.2B, p. 19]	$0.3 = 0.3(100\%) = 30\%$ $\frac{5}{8} = \frac{5}{8}(100\%) = \frac{500}{8}\% = 62.5\%$
To add two fractions with the same denominator, add the numerators and place the sum over the common denominator. [1.2C, p. 20]	$\frac{7}{10} + \frac{1}{10} = \frac{7+1}{10} = \frac{8}{10} = \frac{4}{5}$
To subtract two fractions with the same denominator, subtract the numerators and place the difference over the common denominator. [1.2C, p. 20]	$\frac{7}{10} - \frac{1}{10} = \frac{7-1}{10} = \frac{6}{10} = \frac{3}{5}$
To multiply two fractions, place the product of the numerators over the product of the denominators. [1.2D, p. 21]	$-\frac{2}{3} \cdot \frac{5}{6} = -\frac{2 \cdot 5}{3 \cdot 6} = -\frac{10}{18} = -\frac{5}{9}$

To divide two fractions, multiply the dividend by the reciprocal of the divisor. [1.2D, p. 22]	$-\frac{4}{5} \div \frac{2}{3} = -\frac{4}{5} \cdot \frac{3}{2} = -\frac{2 \cdot 2 \cdot 3}{5 \cdot 2} = -\frac{6}{5}$
Product Property of Square Roots [1.2F, p. 26] $\sqrt{ab} = \sqrt{a} \cdot \sqrt{b}$	$\sqrt{50} = \sqrt{25 \cdot 2}$ $= \sqrt{25}\sqrt{2} = 5\sqrt{2}$
Order of Operations Agreement [1.3A, p. 34] **Step 1** Perform operations inside grouping symbols. Grouping symbols include parentheses ( ), brackets [ ], braces { }, and the fraction bar. **Step 2** Simplify exponential expressions. **Step 3** Do multiplication and division as they occur from left to right. **Step 4** Do addition and subtraction as they occur from left to right.	$50 \div (-5)^2 + 2(7 - 16)$ $= 50 \div (-5)^2 + 2(-9)$ $= 50 \div 25 + 2(-9)$ $= 2 + (-18)$ $= -16$
The Distributive Property [1.4B, p. 39] If a, b, and c are real numbers, then $a(b + c) = ab + ac$.	$5(4 + 7) = 5 \cdot 4 + 5 \cdot 7$ $= 20 + 35 = 55$
The Associative Property of Addition [1.4B, p. 40] If a, b, and c are real numbers, then $(a + b) + c = a + (b + c)$.	$-4 + (2 + 7) = -4 + 9 = 5$ $(-4 + 2) + 7 = -2 + 7 = 5$
The Commutative Property of Addition [1.4B, p. 40] If a and b are real numbers, then $a + b = b + a$.	$2 + 5 = 7$ and $5 + 2 = 7$
The Addition Property of Zero [1.4B, p. 40] If a is a real number, then $a + 0 = 0 + a = a$.	$-8 + 0 = -8$ and $0 + (-8) = -8$
The Inverse Property of Addition [1.4B, p. 41] If a is a real number, then $a + (-a) = (-a) + a = 0$.	$5 + (-5) = 0$ and $(-5) + 5 = 0$
The Associative Property of Multiplication [1.4C, p. 41] If a, b, and c are real numbers, then $(ab)c = a(bc)$.	$-3 \cdot (5 \cdot 4) = -3(20) = -60$ $(-3 \cdot 5) \cdot 4 = -15 \cdot 4 = -60$
The Commutative Property of Multiplication [1.4C, p. 42] If a and b are real numbers, then $ab = ba$.	$-3(7) = -21$ and $7(-3) = -21$
The Multiplication Property of One [1.4C, p. 42] If a is a real number, then $a \cdot 1 = 1 \cdot a = a$.	$-3(1) = -3$ and $1(-3) = -3$
The Inverse Property of Multiplication [1.4C, p. 42] If a is a real number and a is not equal to zero, then $a \cdot \frac{1}{a} = \frac{1}{a} \cdot a = 1$.	$-3 \cdot -\frac{1}{3} = 1$ and $-\frac{1}{3} \cdot -3 = 1$

CHAPTER 1

CONCEPT REVIEW

Test your knowledge of the concepts presented in this chapter. Answer each question. Then check your answers against the ones provided in the Answer Section.

1. How is the opposite of a number different from the absolute value of the number?

2. What is the difference between a minus sign and a negative sign?

3. Explain how to determine the sign of the product of two nonzero integers.

4. What are the steps in the Order of Operations Agreement?

5. When adding two fractions, why is it important to first find a common denominator?

6. When multiplying two fractions, do you have to first find a common denominator?

7. How do you find the reciprocal of a fraction?

8. In a term, what is the difference between the variable part and the numerical coefficient?

9. What are like terms of a variable expression?

10. How is the empty set or null set represented?

11. What do $A \cup B$ and $A \cap B$ mean?

12. What is the difference between the roster method and set-builder notation?

CHAPTER 1

REVIEW EXERCISES

1. Let $x \in \{-4, 0, 11\}$. For what values of x is the inequality $x < 1$ a true statement?

2. Find the additive inverse of -4.

3. Evaluate $-|-5|$.

4. Add: $-3 + (-12) + 6 + (-4)$

5. Subtract: $16 - (-3) - 18$

6. Multiply: $(-6)(7)$

7. Divide: $-100 \div 5$

8. Write $\frac{7}{25}$ as a decimal.

9. Write 6.2% as a decimal.

10. Write $\frac{5}{8}$ as a percent.

11. Simplify: $\frac{1}{3} - \frac{1}{6} + \frac{5}{12}$

12. Subtract: $5.17 - 6.238$

13. Divide: $-\frac{18}{35} \div \frac{17}{28}$

14. Multiply: $4.32(-1.07)$

15. Evaluate $\left(-\frac{2}{3}\right)^4$.

16. Simplify: $2\sqrt{36}$

17. Simplify: $-3\sqrt{120}$

18. Evaluate $-3^2 + 4[18 + (12 - 20)]$.

19. Evaluate $(b - a)^2 + c$ when $a = -2$, $b = 3$, and $c = 4$.

20. Simplify: $6a - 4b + 2a$

21. Simplify: $-3(-12y)$

22. Simplify: $5(2x - 7)$

23. Simplify: $-4(2x - 9) + 5(3x + 2)$

24. Simplify: $5[2 - 3(6x - 1)]$

25. Use the roster method to write the set of odd positive integers less than 8.

26. Find $A \cap B$, given $A = \{1, 5, 9, 13\}$ and $B = \{1, 3, 5, 7, 9\}$.

27. Graph $\{x | x > 3\}$.

−5 −4 −3 −2 −1 0 1 2 3 4 5

28. Graph the interval [1, 4].

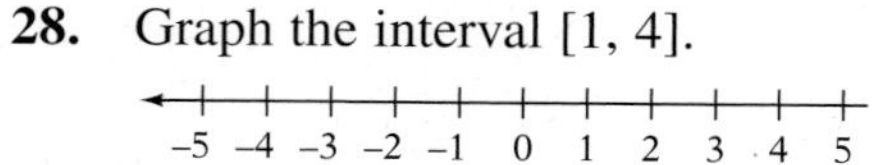

29. Write the set $\{x | x > -4\}$ in interval notation.

30. **Testing** To discourage random guessing on a multiple-choice exam, a professor assigns 6 points for a correct answer, -4 points for an incorrect answer, and -2 points for leaving a question blank. What is the score for a student who had 21 correct answers, had 5 incorrect answers, and left 4 questions blank?

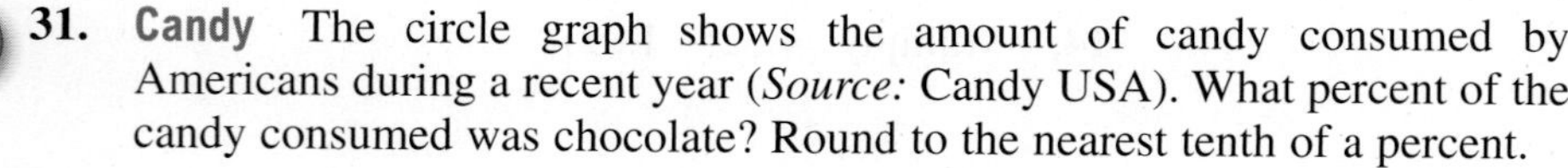

31. **Candy** The circle graph shows the amount of candy consumed by Americans during a recent year (*Source:* Candy USA). What percent of the candy consumed was chocolate? Round to the nearest tenth of a percent.

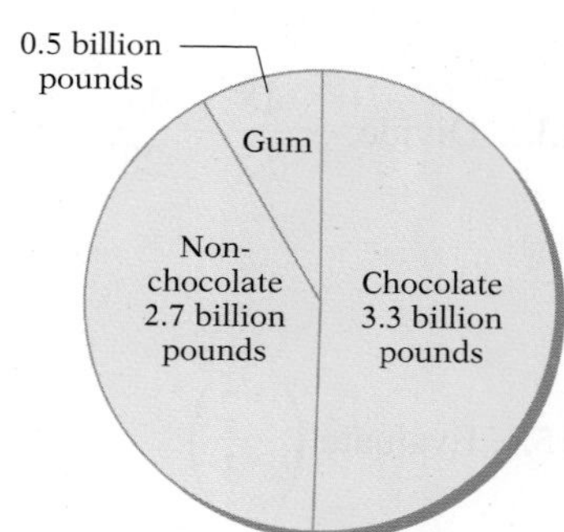

32. Translate "the difference between twice a number and one-half of the number" into a variable expression. Then simplify.

33. **Baseball Cards** A baseball card collection contains five times as many National League players' cards as American League players' cards. Express the number of National League players' cards in terms of the number of American League players' cards.

CHAPTER 1

TEST

1. Place the correct symbol, $<$ or $>$, between the two numbers.
$-2 \quad -40$

2. Find the opposite of -4.

3. Evaluate $-|-4|$.

4. Subtract: $16 - 30$

5. Add: $-22 + 14 + (-8)$

6. Subtract: $16 - (-30) - 42$

7. Divide: $-561 \div (-33)$

8. Write $\frac{7}{9}$ as a decimal. Place a bar over the repeating digit of the decimal.

9. Write 45% as a fraction and as a decimal.

10. Add: $-\frac{2}{5} + \frac{7}{15}$

11. Multiply: $6.02(-0.89)$

12. Divide: $\frac{5}{12} \div \left(-\frac{5}{6}\right)$

13. Evaluate $\frac{3}{4} \cdot (4)^2$.

14. Simplify: $-2\sqrt{45}$

15. Evaluate $16 \div 2[8 - 3(4 - 2)] + 1$.

16. Evaluate $b^2 - 3ab$ when $a = 3$ and $b = -2$.

17. Simplify: $3x - 5x + 7x$

18. Simplify: $\frac{1}{5}(10x)$

19. Simplify: $-3(2x^2 - 7y^2)$

20. Simplify: $2x - 3(x - 2)$

21. Simplify: $2x + 3[4 - (3x - 7)]$

22. Use the roster method to write the set of integers between -3 and 4.

23. Use set-builder notation to write the set of real numbers less than -3.

24. Find $A \cup B$ given $A = \{1, 3, 5, 7\}$ and $B = \{2, 4, 6, 8\}$.

25. Graph $\{x|x < 1\}$.

−5 −4 −3 −2 −1 0 1 2 3 4 5

26. Graph the interval (0, 5).

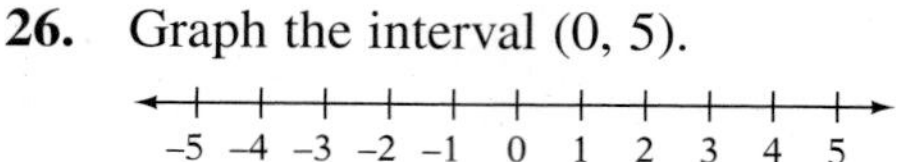

27. Translate "ten times the difference between a number and 3" into a variable expression. Then simplify.

28. **Baseball** The speed of a pitcher's fastball is twice the speed of the catcher's return throw. Express the speed of the fastball in terms of the speed of the return throw.

29. **Balance of Trade** The table at the right shows the U.S. balance of trade, in billions of dollars, for the years 1980 to 2000 (*Source:* U.S. Dept. of Commerce).

a. In which years did the trade balance increase from the previous year?

b. Calculate the difference between the trade balance in 1990 and the trade balance in 2000.

c. During which two consecutive years was the difference in the trade balance greatest?

d. How many times greater was the trade balance in 1990 than in 1980? Round to the nearest whole number.

e. Calculate the average trade balance per quarter for the year 2000.

Year	*Trade Balance*
1980	−19.4
1981	−16.2
1982	−24.2
1983	−57.8
1984	−109.2
1985	−122.1
1986	−140.6
1987	−153.3
1988	−115.9
1989	−92.2
1990	−81.1
1991	−30.7
1992	−35.7
1993	−68.9
1994	−97.0
1995	−95.9
1996	−102.1
1997	−104.7
1998	−166.9
1999	−265.0
2000	−369.7

30. **Temperature** The lowest temperature recorded in North America is −81.4°F. The highest temperature recorded is 134.0°F (*Source:* National Climatic Data Center). Find the difference beween these two extremes.

CHAPTER 2

First-Degree Equations and Inequalities

Don Smith/The Image Bank/Getty Images

OBJECTIVES

SECTION 2.1

- A To determine whether a given number is a solution of an equation
- B To solve an equation of the form $x + a = b$
- C To solve an equation of the form $ax = b$
- D To solve application problems using the basic percent equation
- E To solve uniform motion problems

SECTION 2.2

- A To solve an equation of the form $ax + b = c$
- B To solve an equation of the form $ax + b = cx + d$
- C To solve an equation containing parentheses
- D To solve application problems using formulas

SECTION 2.3

- A To solve integer problems
- B To translate a sentence into an equation and solve

SECTION 2.4

- A To solve value mixture problems
- B To solve percent mixture problems
- C To solve uniform motion problems

SECTION 2.5

- A To solve an inequality in one variable
- B To solve a compound inequality
- C To solve application problems

SECTION 2.6

- A To solve an absolute value equation
- B To solve an absolute value inequality
- C To solve application problems

ARE YOU READY?

Take the Chapter 2 Prep Test to find out if you are ready to learn to:

- Solve a first-degree equation
- Solve percent problems using the basic percent equation
- Solve problems using formulas
- Solve integer, mixture, and uniform motion problems
- Solve an inequality
- Solve an absolute value equation
- Solve an absolute value inequality

PREP TEST

Do these exercises to prepare for Chapter 2.

1. Write $\frac{9}{100}$ as a decimal.

2. Write $\frac{3}{4}$ as a percent.

3. Evaluate $3x^2 - 4x - 1$ when $x = -4$.

4. Simplify: $R - 0.35R$

5. Simplify: $\frac{1}{2}x + \frac{2}{3}x$

6. Simplify: $6x - 3(6 - x)$

7. Simplify: $0.22(3x + 6) + x$

8. Translate into a variable expression: "The difference between five and twice a number."

9. **Computers** A new graphics card for computer games is five times faster than a graphics card made two years ago. Express the speed of the new card in terms of the speed of the old card.

10. **Carpentry** A board 5 ft long is cut into two pieces. If x represents the length of the longer piece, write an expression for the length of the shorter piece in terms of x.

SECTION

2.1 Introduction to Equations

OBJECTIVE A To determine whether a given number is a solution of an equation

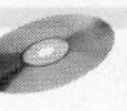

Point of Interest

One of the most famous equations ever stated is $E = mc^2$. This equation, stated by Albert Einstein, shows that there is a relationship between mass m and energy E. As a side note, the chemical element einsteinium was named in honor of Einstein.

An **equation** expresses the equality of two mathematical expressions. The expressions can be either numerical or variable expressions.

$$\left.\begin{array}{l} 9 + 3 = 12 \\ 3x - 2 = 10 \\ y^2 + 4 = 2y - 1 \\ z = 2 \end{array}\right\} \text{Equations}$$

The equation at the right is true if the variable is replaced by 5.

$x + 8 = 13$

$5 + 8 = 13$ A true equation

The equation is false if the variable is replaced by 7.

$7 + 8 = 13$ A false equation

A **solution of an equation** is a number that, when substituted for the variable, results in a true equation. 5 is a solution of the equation $x + 8 = 13$. 7 is not a solution of the equation $x + 8 = 13$.

HOW TO • 1 Is -2 a solution of $2x + 5 = x^2 - 3$?

$$\begin{array}{r|l} \multicolumn{2}{c}{2x + 5 = x^2 - 3} \\ \hline 2(-2) + 5 & (-2)^2 - 3 \\ -4 + 5 & 4 - 3 \\ \multicolumn{2}{c}{1 = 1} \end{array}$$

- Replace x by -2.
- Evaluate the numerical expressions.
- If the results are equal, -2 is a solution of the equation. If the results are not equal, -2 is not a solution of the equation.

Yes, -2 is a solution of the equation.

Take Note

The Order of Operations Agreement applies to evaluating $2(-2) + 5$ and $(-2)^2 - 3$.

EXAMPLE • 1

Is -4 a solution of $5x - 2 = 6x + 2$?

Solution

$$\begin{array}{r|l} \multicolumn{2}{c}{5x - 2 = 6x + 2} \\ \hline 5(-4) - 2 & 6(-4) + 2 \\ -20 - 2 & -24 + 2 \\ \multicolumn{2}{c}{-22 = -22} \end{array}$$

Yes, -4 is a solution.

YOU TRY IT • 1

Is $\frac{1}{4}$ a solution of $5 - 4x = 8x + 2$?

Your solution

EXAMPLE • 2

Is -4 a solution of $4 + 5x = x^2 - 2x$?

Solution

$$\begin{array}{r|l} \multicolumn{2}{c}{4 + 5x = x^2 - 2x} \\ \hline 4 + 5(-4) & (-4)^2 - 2(-4) \\ 4 + (-20) & 16 - (-8) \\ \multicolumn{2}{c}{-16 \neq 24} \end{array}$$

($\neq$ means "is not equal to")

No, -4 is not a solution.

YOU TRY IT • 2

Is 5 a solution of $10x - x^2 = 3x - 10$?

Your solution

Solutions on p. S3

OBJECTIVE B To solve an equation of the form $x + a = b$

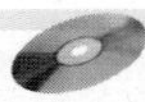

Tips for Success

To learn mathematics, you must be an active participant. Listening and watching your professor do mathematics are not enough. Take notes in class, mentally think through every question your instructor asks, and try to answer it even if you are not called on to answer it verbally. Ask questions when you have them. See *AIM for Success* in the Preface for other ways to be an active learner.

To **solve an equation** means to find a solution of the equation. The simplest equation to solve is an equation of the form *variable* = *constant,* because the constant is the solution.

The solution of the equation $x = 5$ is 5 because $5 = 5$ is a true equation.

The solution of the equation at the right is 7 because $7 + 2 = 9$ is a true equation.

$$x + 2 = 9 \qquad 7 + 2 = 9$$

Note that if 4 is added to each side of the equation $x + 2 = 9$, the solution is still 7.

$$\begin{aligned} x + 2 &= 9 \\ x + 2 + 4 &= 9 + 4 \\ x + 6 &= 13 \qquad 7 + 6 = 13 \end{aligned}$$

If -5 is added to each side of the equation $x + 2 = 9$, the solution is still 7.

$$\begin{aligned} x + 2 &= 9 \\ x + 2 + (-5) &= 9 + (-5) \\ x - 3 &= 4 \qquad 7 - 3 = 4 \end{aligned}$$

Equations that have the same solution are called **equivalent equations.** The equations $x + 2 = 9$, $x + 6 = 13$, and $x - 3 = 4$ are equivalent equations; each equation has 7 as its solution. These examples suggest that adding the same number to each side of an equation produces an equivalent equation. This is called the *Addition Property of Equations.*

Addition Property of Equations

The same number can be added to each side of an equation without changing its solution. In symbols, the equation $a = b$ has the same solution as the equation $a + c = b + c$.

In solving an equation, the goal is to rewrite the given equation in the form *variable* = *constant.* The Addition Property of Equations is used to remove a *term* from one side of the equation by adding the opposite of that term to each side of the equation.

Take Note

An equation has some properties that are similar to those of a balance scale. For instance, if a balance scale is in balance and equal weights are added to each side of the scale, then the balance scale remains in balance. If an equation is true, then adding the same number to each side of the equation produces another true equation.

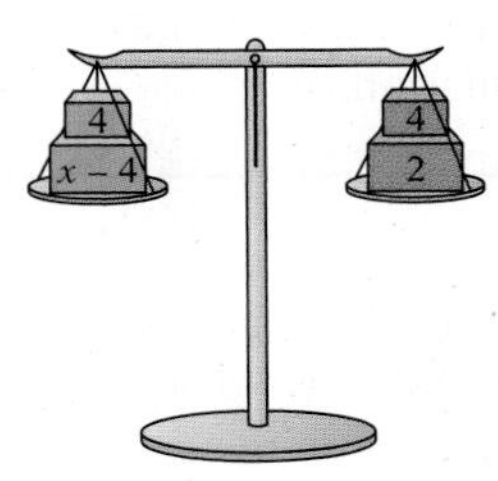

HOW TO • 2 Solve: $x - 4 = 2$

$x - 4 = 2$ • The goal is to rewrite the equation in the form *variable* = *constant.*

$x - 4 + 4 = 2 + 4$ • Add 4 to each side of the equation.

$x + 0 = 6$ • Simplify.

$x = 6$ • The equation is in the form *variable* = *constant.*

Check: $x - 4 = 2$

$6 - 4 \mid 2$

$2 = 2$ A true equation

The solution is 6.

Because subtraction is defined in terms of addition, the Addition Property of Equations also makes it possible to subtract the same number from each side of an equation without changing the solution of the equation.

HOW TO • 3 Solve: $y + \frac{3}{4} = \frac{1}{2}$

$y + \frac{3}{4} = \frac{1}{2}$ • The goal is to rewrite the equation in the form *variable* = *constant*.

$y + \frac{3}{4} - \frac{3}{4} = \frac{1}{2} - \frac{3}{4}$ • **Subtract** $\frac{3}{4}$ from each side of the equation.

$y + 0 = \frac{2}{4} - \frac{3}{4}$ • Simplify.

$y = -\frac{1}{4}$ • The equation is in the form *variable* = *constant*.

The solution is $-\frac{1}{4}$. You should check this solution.

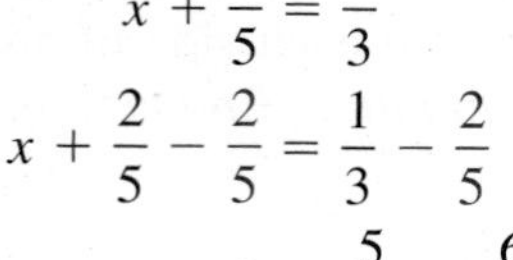

EXAMPLE • 3

Solve: $x + \frac{2}{5} = \frac{1}{3}$

Solution

$x + \frac{2}{5} = \frac{1}{3}$

$x + \frac{2}{5} - \frac{2}{5} = \frac{1}{3} - \frac{2}{5}$ • **Subtract** $\frac{2}{5}$ from each side.

$x + 0 = \frac{5}{15} - \frac{6}{15}$ • Rewrite $\frac{1}{3}$ and $\frac{2}{5}$ with a common denominator.

$x = -\frac{1}{15}$

The solution is $-\frac{1}{15}$.

YOU TRY IT • 3

Solve: $\frac{5}{6} = y - \frac{3}{8}$

Your solution

Solution on p. S3

OBJECTIVE C To solve an equation of the form $ax = b$

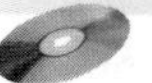

The solution of the equation at the right is 3 because $2 \cdot 3 = 6$ is a true equation.

$$2x = 6 \qquad 2 \cdot 3 = 6$$

Note that if each side of $2x = 6$ is multiplied by 5, the solution is still 3.

$$\begin{aligned} 2x &= 6 \\ 5(2x) &= 5 \cdot 6 \\ 10x &= 30 \end{aligned} \qquad 10 \cdot 3 = 30$$

If each side of $2x = 6$ is multiplied by -4, the solution is still 3.

$$\begin{aligned} 2x &= 6 \\ (-4)(2x) &= (-4)6 \\ -8x &= -24 \end{aligned} \qquad -8 \cdot 3 = -24$$

The equations $2x = 6$, $10x = 30$, and $-8x = -24$ are equivalent equations; each equation has 3 as its solution. These examples suggest that multiplying each side of an equation by the same nonzero number produces an equivalent equation.

Multiplication Property of Equations

Each side of an equation can be multiplied by the same nonzero number without changing the solution of the equation. In symbols, if $c \neq 0$, then the equation $a = b$ has the same solutions as the equation $ac = bc$.

The Multiplication Property of Equations is used to remove a coefficient by multiplying each side of the equation by the reciprocal of the coefficient.

HOW TO • 4 Solve: $\frac{3}{4}z = 9$

$\frac{3}{4}z = 9$ • The goal is to rewrite the equation in the form *variable* = *constant*.

$\frac{4}{3} \cdot \frac{3}{4}z = \frac{4}{3} \cdot 9$ • **Multiply** each side of the equation **by** $\frac{4}{3}$.

$1 \cdot z = 12$ • Simplify.

$z = 12$ • The equation is in the form *variable* = *constant*.

The solution is 12. You should check this solution.

Because division is defined in terms of multiplication, each side of an equation can be divided by the same nonzero number without changing the solution of the equation.

Take Note

Remember to check the solution.

Check: $6x = 14$

$6\left(\frac{7}{3}\right) \mid 14$

$14 = 14$

HOW TO • 5 Solve: $6x = 14$

$6x = 14$ • The goal is to rewrite the equation in the form *variable* = *constant*.

$\frac{6x}{6} = \frac{14}{6}$ • **Divide** each side of the equation **by 6.**

$x = \frac{7}{3}$ • Simplify. The equation is in the form *variable* = *constant*.

The solution is $\frac{7}{3}$.

When using the Multiplication Property of Equations, multiply each side of the equation by the reciprocal of the coefficient when the coefficient is a fraction. Divide each side of the equation by the coefficient when the coefficient is an integer or a decimal.

EXAMPLE • 4

Solve: $\frac{3x}{4} = -9$

Solution

$\frac{3x}{4} = -9$

$\frac{4}{3} \cdot \frac{3}{4}x = \frac{4}{3}(-9)$ • $\frac{3x}{4} = \frac{3}{4}$

$x = -12$

The solution is -12.

YOU TRY IT • 4

Solve: $-\frac{2x}{5} = 6$

Your solution

EXAMPLE • 5

Solve: $5x - 9x = 12$

Solution

$5x - 9x = 12$

$-4x = 12$ • Combine like terms.

$\frac{-4x}{-4} = \frac{12}{-4}$

$x = -3$

The solution is -3.

YOU TRY IT • 5

Solve: $4x - 8x = 16$

Your solution

Solutions on pp. S3–S4

OBJECTIVE D **To solve application problems using the basic percent equation** 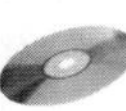

An equation that is used frequently in mathematics applications is the basic percent equation.

Basic Percent Equation

$$\text{Percent} \cdot \text{Base} = \text{Amount}$$
$$P \cdot B = A$$

In many application problems involving percent, the base follows the word *of*.

HOW TO • 6 20% of what number is 30?

$P \cdot B = A$ • Use the basic percent equation.

$0.20B = 30$ • $P = 20\% = 0.20$, $A = 30$, and B is unknown.

$\dfrac{0.20B}{0.20} = \dfrac{30}{0.20}$ • Solve for B.

$B = 150$

The number is 150.

✓ Take Note
We have written $P(80) = 70$ because that is the form of the basic percent equation. We could have written $80P = 70$. The important point is that each side of the equation is divided by 80, the coefficient of P.

HOW TO • 7 70 is what percent of 80?

$P \cdot B = A$ • Use the basic percent equation.

$P(80) = 70$ • $B = 80$, $A = 70$, and P is unknown.

$\dfrac{P(80)}{80} = \dfrac{70}{80}$ • Solve for P.

$P = 0.875$ • The question asked for a percent.

$P = 87.5\%$ • Convert the decimal to a percent.

70 is 87.5% of 80.

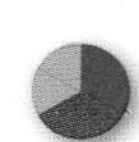

HOW TO • 8 The world's production of cocoa for a recent year was 2928 metric tons. Of this, 1969 metric tons came from Africa. (*Source:* World Cocoa Foundation) What percent of the world's cocoa production came from Africa? Round to the nearest tenth of a percent.

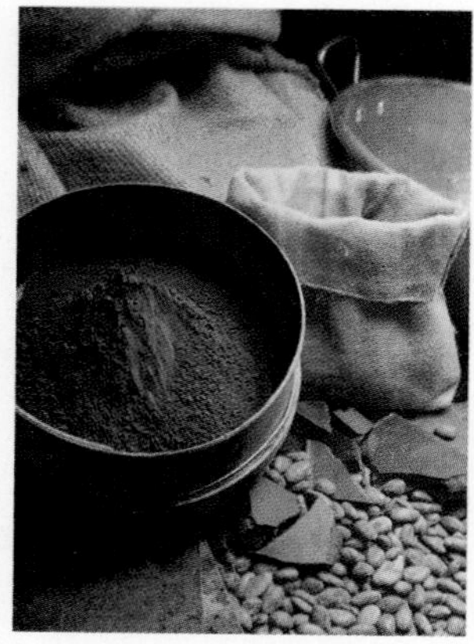
© Philippe S. Giraud/Terres du Sud/Sygma/Corbis

Strategy
To find the percent, use the basic percent equation. $B = 2928$, $A = 1969$, and P is unknown.

Solution

$P \cdot B = A$

$P(2928) = 1969$

$P = \dfrac{1969}{2928} \approx 0.672$

Approximately 67.2% of the world's cocoa production came from Africa.

The simple interest that an investment earns is given by the **simple interest equation** $I = Prt$, where I is the simple interest, P is the principal, or amount invested, r is the simple interest rate, and t is the time.

HOW TO • 9 A $1500 investment has an annual simple interest rate of 7%. Find the simple interest earned on the investment after 18 months.

The time is given in months but the interest rate is an annual rate. Therefore, we must convert 18 months to years.

$$18 \text{ months} = \frac{18}{12} \text{ years} = 1.5 \text{ years}$$

To find the interest, solve $I = Prt$ for I.

$$I = Prt$$
$$I = 1500(0.07)(1.5) \qquad \bullet\ P = 1500, r = 0.07, t = 1.5$$
$$I = 157.5$$

The investment earned $157.50.

Point of Interest

In the jewelry industry, the amount of gold in a piece of jewelry is measured by the *karat*. Pure gold is 24 karats. A necklace that is 18 karats is $\frac{18}{24} = 0.75 = 75\%$ gold.

The amount of a substance in a solution can be given as a percent of the total solution. For instance, if a certain fruit juice drink is advertised as containing 27% cranberry juice, then 27% of the contents of the bottle must be cranberry juice.

The method for solving problems involving mixtures is based on the **percent mixture equation** $Q = Ar$, where Q is the quantity of a substance in the solution, A is the amount of the solution, and r is the percent concentration of the substance.

HOW TO • 10 Part of the formula for a perfume requires that the concentration of jasmine be 1.2% of the total amount of perfume. How many ounces of jasmine are in a 2-ounce bottle of this perfume?

The amount of perfume is 2 oz. Therefore, $A = 2$. The percent concentration is 1.2%, so $r = 0.012$. To find the number of ounces of jasmine, solve $Q = Ar$ for Q.

$$Q = Ar$$
$$Q = 2(0.012) \qquad \bullet\ A = 2, r = 0.012$$
$$Q = 0.024$$

There is 0.024 oz of jasmine in the perfume.

In most cases, you should write the percent as a decimal before solving the basic percent equation. However, some percents are more easily written as a fraction. For example,

$$33\frac{1}{3}\% = \frac{1}{3} \qquad 66\frac{2}{3}\% = \frac{2}{3} \qquad 16\frac{2}{3}\% = \frac{1}{6} \qquad 83\frac{1}{3}\% = \frac{5}{6}$$

EXAMPLE • 6

12 is $33\frac{1}{3}\%$ of what number?

Solution

$P \cdot B = A$ • Use the basic percent equation.

$\frac{1}{3}B = 12$ • $33\frac{1}{3}\% = \frac{1}{3}$

$3 \cdot \frac{1}{3}B = 3 \cdot 12$

$B = 36$

12 is $33\frac{1}{3}\%$ of 36.

YOU TRY IT • 6

18 is $16\frac{2}{3}\%$ of what number?

Your solution

EXAMPLE • 7

The data in the table below show the number of households (in millions) that downloaded music files for a three-month period in a recent year. (*Source:* NPD Group)

Month	April	May	June
Downloads	14.5	12.7	10.4

For the three-month period, what percent of the files were downloaded in May? Round to the nearest percent.

Strategy

To find the percent:

- Find the total number of files downloaded for the three-month period.
- Use the basic percent equation. B is the total number of files downloaded for the three-month period; $A = 12.7$, the number of files downloaded in May; P is unknown.

Solution

$14.5 + 12.7 + 10.4 = 37.6$

$P \cdot B = A$ • Use the basic percent equation.

$P(37.6) = 12.7$ • $B = 37.6, A = 12.7$

$P = \frac{12.7}{37.6} \approx 0.34$

Approximately 34% of the files were downloaded in May.

YOU TRY IT • 7

According to AdAge.com, 97.5 million people watched Super Bowl XLII. What percent of the U.S. population watched Super Bowl XLII? Use a figure of 300 million for the U.S. population.

Your strategy

Your solution

Solutions on p. S4

EXAMPLE • 8

In April, Marshall Wardell was charged an interest fee of \$8.72 on an unpaid credit card balance of \$545. Find the annual interest rate on this credit card.

Strategy

The interest is \$8.72. Therefore, $I = 8.72$. The unpaid balance is \$545. This is the principal on which interest is calculated. Therefore, $P = 545$. The time is 1 month. Because the *annual* interest rate must be found and the time is given as 1 month, we write 1 month as $\frac{1}{12}$ year: $t = \frac{1}{12}$. To find the interest rate, solve $I = Prt$ for r.

Solution

$$I = Prt \quad \text{• Use the simple interest equation.}$$

$$8.72 = 545r\left(\frac{1}{12}\right) \quad \text{• } I = 8.72, P = 545, t = \frac{1}{12}$$

$$8.72 = \frac{545}{12}r$$

$$\frac{12}{545}(8.72) = \frac{12}{545}\left(\frac{545}{12}r\right)$$

$$0.192 = r$$

The annual interest rate is 19.2%.

YOU TRY IT • 8

Clarissa Adams purchased a municipal bond for \$1000 that earns an annual simple interest rate of 6.4%. How much must she deposit into an account that earns 8% annual simple interest so that the interest earned from each account after 1 year is the same?

Your strategy

Your solution

EXAMPLE • 9

To make a certain color of blue, 4 oz of cyan must be contained in 1 gal of paint. What is the percent concentration of cyan in the paint?

Strategy

The cyan is given in ounces and the amount of paint is given in gallons. We must convert ounces to gallons or gallons to ounces. For this problem, we will convert gallons to ounces: 1 gal = 128 oz. Solve $Q = Ar$ for r, with $Q = 4$ and $A = 128$.

Solution

$$Q = Ar \quad \text{• Use the percent mixture equation.}$$

$$4 = 128r \quad \text{• } Q = 4, A = 128$$

$$\frac{4}{128} = \frac{128r}{128}$$

$$0.03125 = r$$

The percent concentration of cyan is 3.125%.

YOU TRY IT • 9

The concentration of sugar in a certain breakfast cereal is 25%. If there are 2 oz of sugar contained in a bowl of cereal, how many ounces of cereal are in the bowl?

Your strategy

Your solution

Solutions on p. S4

OBJECTIVE E To solve uniform motion problems 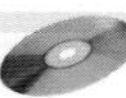

✓ Take Note
A car traveling in a *circle* at a constant speed of 45 mph is *not* in uniform motion because the direction of the car is always changing.

Any object that travels at a constant speed in a straight line is said to be in *uniform motion.* **Uniform motion** means that the speed and direction of an object do not change. For instance, a car traveling at a constant speed of 45 mph on a straight road is in uniform motion.

The solution of a uniform motion problem is based on the **uniform motion equation** $d = rt$, where d is the distance traveled, r is the rate of travel, and t is the time spent traveling. For instance, suppose a car travels at 50 mph for 3 h. Because the rate (50 mph) and time (3 h) are known, we can find the distance traveled by solving the equation $d = rt$ for d.

$$d = rt$$
$$d = 50(3) \qquad \bullet\ r = 50, t = 3$$
$$d = 150$$

The car travels a distance of 150 mi.

HOW TO • 11 A jogger runs 3 mi in 45 min. What is the rate of the jogger in miles per hour?

Strategy
- Because the answer must be in miles per *hour* and the given time is in *minutes,* convert 45 min to hours.
- To find the rate of the jogger, solve the equation $d = rt$ for r.

Solution

$$45 \text{ min} = \frac{45}{60}\text{ h} = \frac{3}{4}\text{ h}$$

$$d = rt$$
$$3 = r\left(\frac{3}{4}\right) \qquad \bullet\ d = 3, t = \frac{3}{4}$$
$$3 = \frac{3}{4}r$$
$$\left(\frac{4}{3}\right)3 = \left(\frac{4}{3}\right)\frac{3}{4}r \qquad \bullet\ \text{Multiply each side of the equation by the reciprocal of } \frac{3}{4}.$$
$$4 = r$$

The rate of the jogger is 4 mph.

If two objects are moving in opposite directions, then the rate at which the distance between them is increasing is the sum of the speeds of the two objects. For instance, in the diagram below, two cars start from the same point and travel in opposite directions. The distance between them is changing at 70 mph.

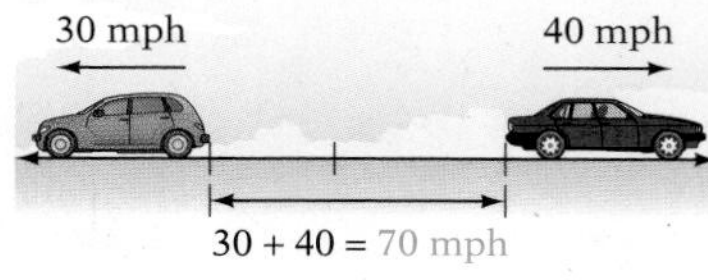

Similarly, if two objects are moving toward each other, the distance between them is decreasing at a rate that is equal to the sum of the speeds. The rate at which the two planes at the right are approaching one another is 800 mph.

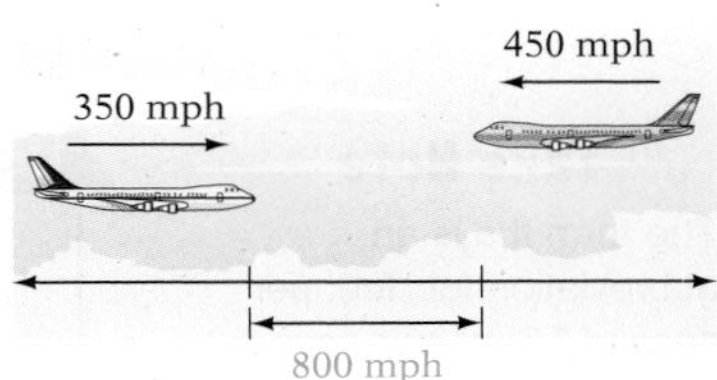

HOW TO • 12 Two cars start from the same point and move in opposite directions. The car moving west is traveling 45 mph, and the car moving east is traveling 60 mph. In how many hours will the cars be 210 mi apart?

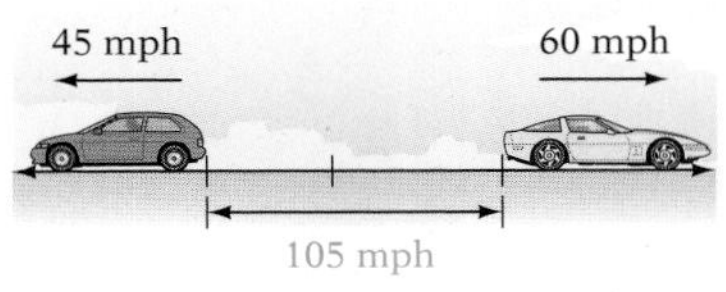

Strategy

The distance is 210 mi. Therefore, $d = 210$. The cars are moving in opposite directions, so the rate at which the distance between them is changing is the sum of the rates of each of the cars. The rate is 45 mph + 60 mph = 105 mph. Therefore, $r = 105$. To find the time, solve the equation $d = rt$ for t.

Solution

$$d = rt$$

$$210 = 105t \quad \bullet\ d = 210, r = 105$$

$$\frac{210}{105} = \frac{105t}{105} \quad \bullet\ \text{Solve for } t.$$

$$2 = t$$

In 2 h, the cars will be 210 mi apart.

If a motorboat is on a river that is flowing at a rate of 4 mph, then the boat will float down the river at a speed of 4 mph when the motor is not on. Now suppose the motor is turned on and the power adjusted so that the boat would travel 10 mph without the aid of the current. Then, if the boat is moving with the current, its effective speed is the speed of the boat using power plus the speed of the current: 10 mph + 4 mph = 14 mph. (See the figure below.)

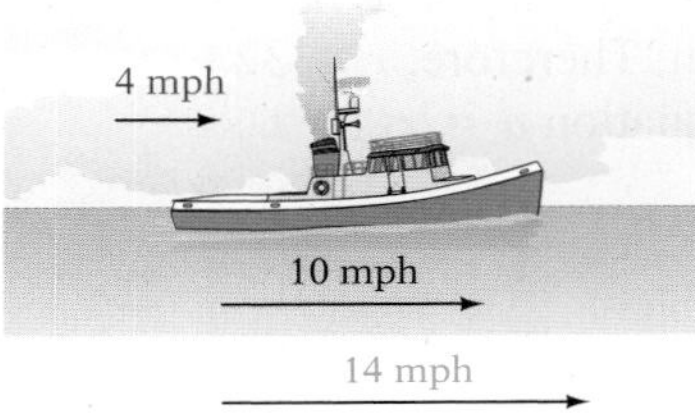

However, if the boat is moving against the current, the current slows the boat down. The effective speed of the boat is the speed of the boat using power minus the speed of the current: 10 mph − 4 mph = 6 mph. (See the figure below.)

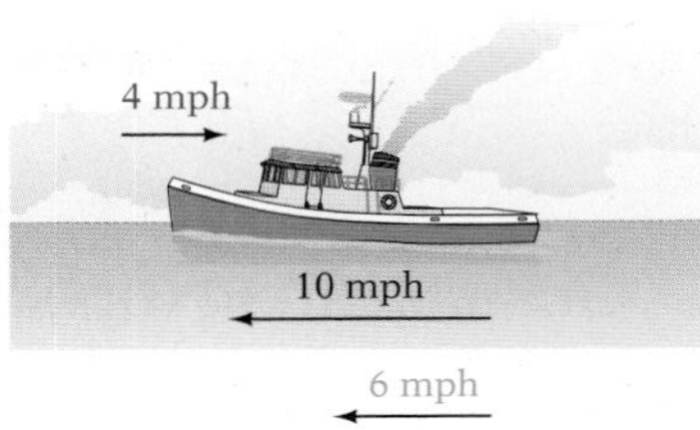

There are other situations in which the preceding concepts may be applied.

Take Note
The term ft/s is an abbreviation for "feet per second." Similarly, cm/s is "centimeters per second" and m/s is "meters per second."

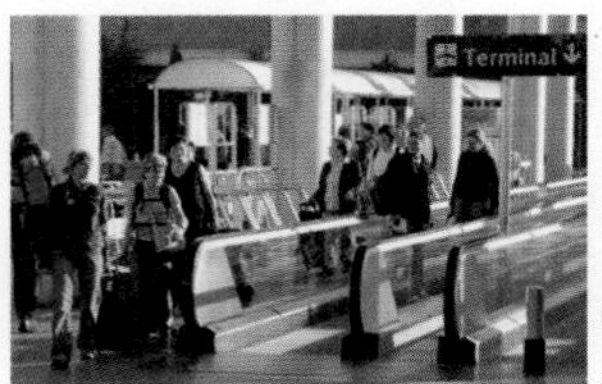

Peter Titmuss/Alamy

HOW TO • 13 An airline passenger is walking between two airline terminals and decides to get on a moving sidewalk that is 150 ft long. If the passenger walks at a rate of 7 ft/s and the moving sidewalk moves at a rate of 9 ft/s, how long, in seconds, will it take for the passenger to walk from one end of the moving sidewalk to the other? Round to the nearest thousandth.

Strategy
The distance is 150 ft. Therefore, $d = 150$. The passenger is traveling at 7 ft/s and the moving sidewalk is traveling at 9 ft/s. The rate of the passenger is the sum of the two rates, or 16 ft/s. Therefore, $r = 16$. To find the time, solve the equation $d = rt$ for t.

Solution

$d = rt$

$150 = 16t$ • $d = 150, r = 16$

$\frac{150}{16} = \frac{16t}{16}$ • Solve for t.

$9.375 = t$

It will take 9.375 s for the passenger to travel the length of the moving sidewalk.

EXAMPLE • 10

Two cyclists start at the same time at opposite ends of an 80-mile course. One cyclist is traveling 18 mph, and the second cyclist is traveling 14 mph. How long after they begin cycling will they meet?

Strategy
The distance is 80 mi. Therefore, $d = 80$. The cyclists are moving toward each other, so the rate at which the distance between them is changing is the sum of the rates of each of the cyclists. The rate is 18 mph + 14 mph = 32 mph. Therefore, $r = 32$. To find the time, solve the equation $d = rt$ for t.

Solution

$d = rt$

$80 = 32t$ • $d = 80, r = 32$

$\frac{80}{32} = \frac{32t}{32}$ • Solve for t.

$2.5 = t$

The cyclists will meet in 2.5 h.

YOU TRY IT • 10

A plane that can normally travel at 250 mph in calm air is flying into a headwind of 25 mph. How far can the plane fly in 3 h?

Your strategy

Your solution

Solution on p. S4

2.1 EXERCISES

OBJECTIVE A **To determine whether a given number is a solution of an equation**

1. Is 4 a solution of $2x = 8$?

2. Is 3 a solution of $y + 4 = 7$?

3. Is -1 a solution of $2b - 1 = 3$?

4. Is -2 a solution of $3a - 4 = 10$?

5. Is 1 a solution of $4 - 2m = 3$?

6. Is 2 a solution of $7 - 3n = 2$?

7. Is 5 a solution of $2x + 5 = 3x$?

8. Is 4 a solution of $3y - 4 = 2y$?

9. Is -2 a solution of $3a + 2 = 2 - a$?

10. Is 3 a solution of $z^2 + 1 = 4 + 3z$?

11. Is 2 a solution of $2x^2 - 1 = 4x - 1$?

12. Is -1 a solution of $y^2 - 1 = 4y + 3$?

13. Is 4 a solution of $x(x + 1) = x^2 + 5$?

14. Is 3 a solution of $2a(a - 1) = 3a + 3$?

15. Is $-\frac{1}{4}$ a solution of $8t + 1 = -1$?

16. Is $\frac{1}{2}$ a solution of $4y + 1 = 3$?

17. Is $\frac{2}{5}$ a solution of $5m + 1 = 10m - 3$?

18. Is $\frac{3}{4}$ a solution of $8x - 1 = 12x + 3$?

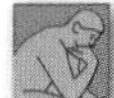

19. If A is a fixed number such that $A < 0$, is a solution of the equation $5x = A$ positive or negative?

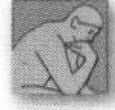

20. Can a negative number be a solution of the equation $7x - 2 = -x$?

OBJECTIVE B **To solve an equation of the form $x + a = b$**

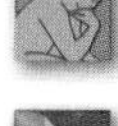

21. Without solving the equation $x - \frac{11}{16} = \frac{19}{24}$, determine whether x is less than or greater than $\frac{19}{24}$. Explain your answer.

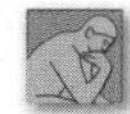

22. Without solving the equation $x + \frac{13}{15} = -\frac{21}{43}$, determine whether x is less than or greater than $-\frac{21}{43}$. Explain your answer.

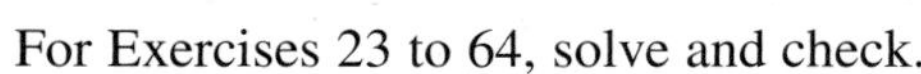

For Exercises 23 to 64, solve and check.

23. $x + 5 = 7$

24. $y + 3 = 9$

25. $b - 4 = 11$

26. $z - 6 = 10$

27. $2 + a = 8$

28. $5 + x = 12$

29. $n - 5 = -2$

30. $x - 6 = -5$

31. $b + 7 = 7$

32. $y - 5 = -5$

33. $z + 9 = 2$

34. $n + 11 = 1$

35. $10 + m = 3$

36. $8 + x = 5$

37. $9 + x = -3$

38. $10 + y = -4$

39. $2 = x + 7$

40. $-8 = n + 1$

41. $4 = m - 11$

42. $-6 = y - 5$

43. $12 = 3 + w$

44. $-9 = 5 + x$

45. $4 = -10 + b$

46. $-7 = -2 + x$

47. $m + \frac{2}{3} = -\frac{1}{3}$

48. $c + \frac{3}{4} = -\frac{1}{4}$

49. $x - \frac{1}{2} = \frac{1}{2}$

50. $x - \frac{2}{5} = \frac{3}{5}$

51. $\frac{5}{8} + y = \frac{1}{8}$

52. $\frac{4}{9} + a = -\frac{2}{9}$

53. $m + \frac{1}{2} = -\frac{1}{4}$

54. $b + \frac{1}{6} = -\frac{1}{3}$

55. $x + \frac{2}{3} = \frac{3}{4}$

56. $n + \frac{2}{5} = \frac{2}{3}$

57. $-\frac{5}{6} = x - \frac{1}{4}$

58. $-\frac{1}{4} = c - \frac{2}{3}$

59. $d + 1.3619 = 2.0148$

60. $w + 2.932 = 4.801$

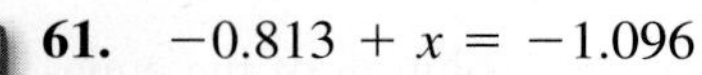

61. $-0.813 + x = -1.096$

62. $-1.926 + t = -1.042$

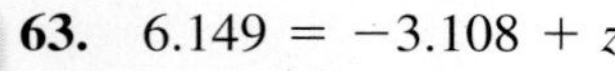

63. $6.149 = -3.108 + z$

64. $5.237 = -2.014 + x$

OBJECTIVE C To solve an equation of the form $ax = b$

For Exercises 65 to 108, solve and check.

65. $5x = -15$

66. $4y = -28$

67. $3b = 0$

68. $2a = 0$

69. $-3x = 6$

70. $-5m = 20$

71. $-3x = -27$

72. $-\frac{1}{6}n = -30$

73. $20 = \frac{1}{4}c$

74. $18 = 2t$

75. $0 = -5x$

76. $0 = -8a$

77. $49 = -7t$

78. $\frac{x}{3} = 2$

79. $\frac{x}{4} = 3$

80. $-\frac{y}{2} = 5$

81. $-\frac{b}{3} = 6$

82. $\frac{3}{4}y = 9$

83. $\frac{2}{5}x = 6$

84. $-\frac{2}{3}d = 8$

85. $-\frac{3}{5}m = 12$

86. $\frac{2n}{3} = 0$

87. $\frac{5x}{6} = 0$

88. $\frac{-3z}{8} = 9$

89. $\frac{3x}{4} = 2$

90. $\frac{3}{4}c = \frac{3}{5}$

91. $\frac{2}{9} = \frac{2}{3}y$

92. $-\frac{6}{7} = -\frac{3}{4}b$

93. $\frac{1}{5}x = -\frac{1}{10}$

94. $-\frac{2}{3}y = -\frac{8}{9}$

95. $-1 = \frac{2n}{3}$

96. $-\frac{3}{4} = \frac{a}{8}$

97. $-\frac{2}{5}m = -\frac{6}{7}$

98. $5x + 2x = 14$

99. $3n + 2n = 20$

100. $7d - 4d = 9$

101. $10y - 3y = 21$

102. $2x - 5x = 9$

103. $\frac{x}{1.46} = 3.25$

104. $\frac{z}{2.95} = -7.88$

105. $3.47a = 7.1482$

106. $2.31m = 2.4255$

107. $-3.7x = 7.881$

108. $\frac{n}{2.65} = 9.08$

For Exercises 109 to 112, suppose y is a positive integer. Determine whether x is positive or negative.

109. $15x = y$

110. $-6x = y$

111. $-\frac{1}{4}x = y$

112. $\frac{2}{9}x = -y$

OBJECTIVE D **To solve application problems using the basic percent equation**

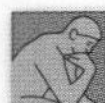

113. Without solving an equation, determine whether 40% of 80 is *less than, equal to,* or *greater than* 80% of 40.

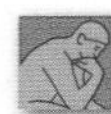

114. Without solving an equation, determine whether $\frac{1}{4}$% of 80 is *less than, equal to,* or *greater than* 25% of 80.

115. What is 35% of 80?

116. What percent of 8 is 0.5?

117. Find 1.2% of 60.

118. 8 is what percent of 5?

119. 125% of what is 80?

120. What percent of 20 is 30?

121. 12 is what percent of 50?

122. What percent of 125 is 50?

123. Find 18% of 40.

124. What is 25% of 60?

125. 12% of what is 48?

126. 45% of what is 9?

127. What is $33\frac{1}{3}$% of 27?

128. Find $16\frac{2}{3}$% of 30.

129. What percent of 12 is 3?

130. 10 is what percent of 15?

131. 12 is what percent of 6?

132. 20 is what percent of 16?

133. $5\frac{1}{4}$% of what is 21?

134. $37\frac{1}{2}$% of what is 15?

135. Find 15.4% of 50.

136. What is 18.5% of 46?

137. 1 is 0.5% of what?

138. 3 is 1.5% of what?

139. $\frac{3}{4}$% of what is 3?

140. $\frac{1}{2}$% of what is 3?

141. What is 250% of 12?

142. Government To override a presidential veto, at least $66\frac{2}{3}\%$ of the Senate must vote to override the veto. There are 100 senators in the Senate. What is the minimum number of votes needed to override a veto?

In the News

Thousands Complete Boston Marathon

This year, there were 25,283 entrants in the Boston Marathon, the world's oldest annual marathon. Of those registered, 22,377 people started the race, and 21,948 finished the 26.2-mile course.

Source: www.bostonmarathon.org

143. Boston Marathon See the news clipping at the right. What percent of the runners who started the course finished the race? Round to the nearest tenth of a percent.

144. Income According to the U.S. Census Bureau, the median income fell 1.1% between two successive years. If the median income before the decline was \$42,900, what was the median income the next year? Round to the nearest dollar.

145. School Enrollment The circle graph at the right represents the U.S. population over three years old that is enrolled in school. To answer the question "How many people are enrolled in college or graduate school?," what additional piece of information is necessary?

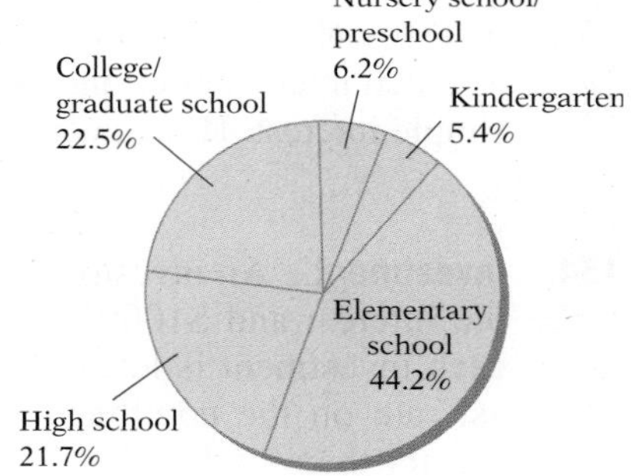

Source: U.S. Census Bureau

146. Fuel-Efficient Cars Lighter cars are more fuel-efficient than heavier cars. A report from the Energy Information Administration stated that "the average car weight in 2020 is projected to be 364 pounds lighter than the average car weight in model year 2000, a decrease of 11.8 percent." Find the average weight of a car in model year 2000.

147. Teen Smoking Use the information in the news clipping at the right.

a. Determine the approximate teen population of New York City in 2007. Round to the nearest ten thousand.

b. Suppose the smoking rate among New York City teens had not dropped from 2001 to 2007 but instead had remained unchanged. Using the 2007 teen population you found in part (a), determine how many New York City teens would have been smokers in 2007. Round to the nearest thousand.

In the News

Teen Smoking at Record Low

Data released today show that 20,000 New York City teens smoke. The smoking rate among New York City teens dropped from 17.6% in 2001 to 8.5% in 2007. The mayor links the decline to the city's efforts to reduce smoking among adults.

Source: www.nyc.gov

148. Investment If Kachina Caron invested \$1200 in a simple interest account and earned \$72 in 8 months, what is the annual interest rate?

149. Investment How much money must Andrea invest for 2 years in an account that earns an annual interest rate of 8% if she wants to earn \$300 from the investment?

150. Investment Sal Boxer decided to divide a gift of \$3000 into two different accounts. He placed \$1000 in one account that earns an annual simple interest rate of 7.5%. The remaining money was placed in an account that earns an annual simple interest rate of 8.25%. How much interest will Sal earn from the two accounts after 1 year?

151. Investment If Americo invests \$2500 at an 8% annual simple interest rate and Octavia invests \$3000 at a 7% annual simple interest rate, which of the two will earn the greater amount of interest after 1 year?

152. Investment Makana invested \$900 in a simple interest account that had an interest rate that was 1% more than that of her friend Marlys. If Marlys earned \$51 after one year from an investment of \$850, how much would Makana earn in 1 year?

153. Investment A \$2000 investment at an annual simple interest rate of 6% earned as much interest after one year as another investment in an account that earns 8% simple interest. How much was invested at 8%?

154. Investment An investor placed \$1000 in an account that earns 9% annual simple interest and \$1000 in an account that earns 6% annual simple interest. If each investment is left in the account for the same period of time, is the interest rate on the combined investment less than 6%, between 6% and 9%, or greater than 9%?

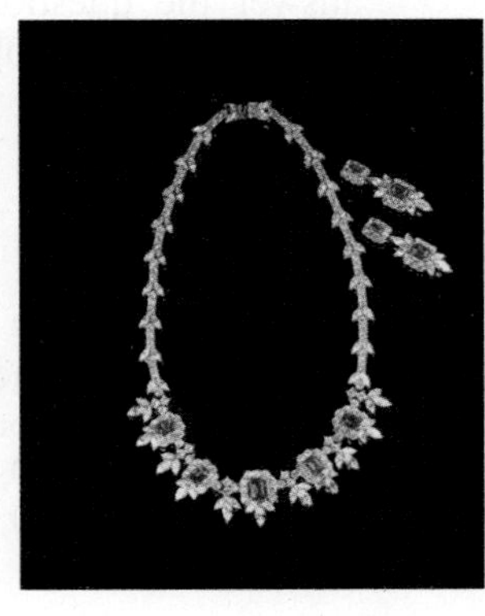

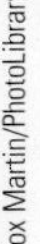
Fox Martin/PhotoLibrary

155. Metallurgy The concentration of platinum in a necklace is 15%. If the necklace weighs 12 g, find the amount of platinum in the necklace.

156. Dye Mixtures A 250-milliliter solution of a fabric dye contains 5 ml of hydrogen peroxide. What is the percent concentration of the hydrogen peroxide?

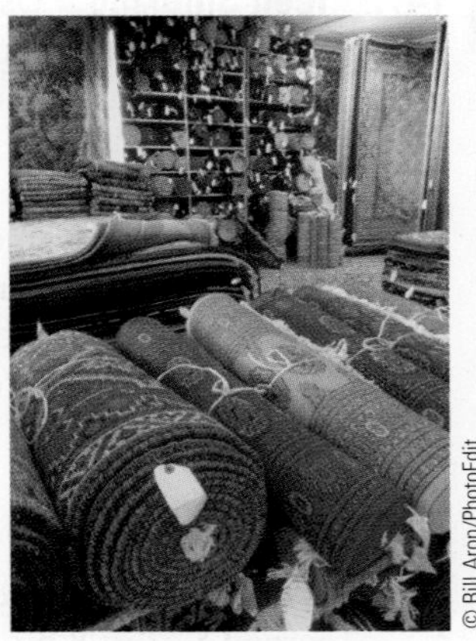
© Bill Aron/PhotoEdit

157. Fabric Mixtures A carpet is made with a blend of wool and other fibers. If the concentration of wool in the carpet is 75% and the carpet weighs 175 lb, how much wool is in the carpet?

158. Juice Mixtures Apple Dan's 32-ounce apple-flavored fruit drink contains 8 oz of apple juice. A 40-ounce generic brand of an apple-flavored fruit drink contains 9 oz of apple juice. Which of the two brands has the greater concentration of apple juice?

159. Food Mixtures Bakers use simple syrup in many of their recipes. Simple syrup is made by combining 500 g of sugar with 500 g of water and mixing it well until the sugar dissolves. What is the percent concentration of sugar in the simple syrup?

160. Pharmacology A pharmacist has 50 g of a topical cream that contains 75% glycerine. How many grams of the cream are not glycerine?

161. **Chemistry** A chemist has 100 ml of a solution that is 9% acetic acid. If the chemist adds 50 ml of pure water to this solution, what is the percent concentration of the resulting mixture?

162. **Chemistry** A 500-gram salt-and-water solution contains 50 g of salt. This mixture is left in the open air, and 100 g of water evaporates from the solution. What is the percent concentration of salt in the remaining solution?

OBJECTIVE E To solve uniform motion problems

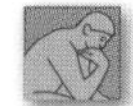

163. Joe and John live 2 mi apart. They leave their houses at the same time and walk toward each other until they meet. Joe walks faster than John does.

a. Is the distance walked by Joe less than, equal to, or greater than the distance walked by John?
b. Is the time spent walking by Joe less than, equal to, or greater than the time spent walking by John?
c. What is the total distance traveled by both Joe and John?

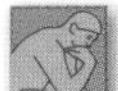

164. Morgan and Emma ride their bikes from Morgan's house to the store. Morgan begins biking 5 min before Emma begins. Emma bikes faster than Morgan and catches up with her just as they reach the store.

a. Is the distance biked by Emma less than, equal to, or greater than the distance biked by Morgan?
b. Is the time spent biking by Emma less than, equal to, or greater than the time spent biking by Morgan?

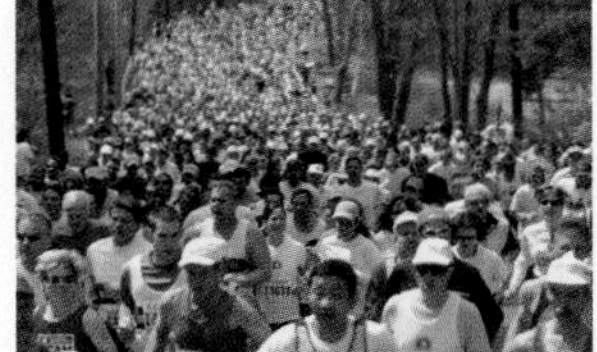

Michael Dwyer/Alamy

165. As part of a training program for the Boston Marathon, a runner wants to build endurance by running at a rate of 9 mph for 20 min. How far will the runner travel in that time period?

166. It takes a hospital dietician 40 min to drive from home to the hospital, a distance of 20 mi. What is the dietician's average rate of speed?

167. Marcella leaves home at 9:00 A.M. and drives to school, arriving at 9:45 A.M. If the distance between home and school is 27 mi, what is Marcella's average rate of speed?

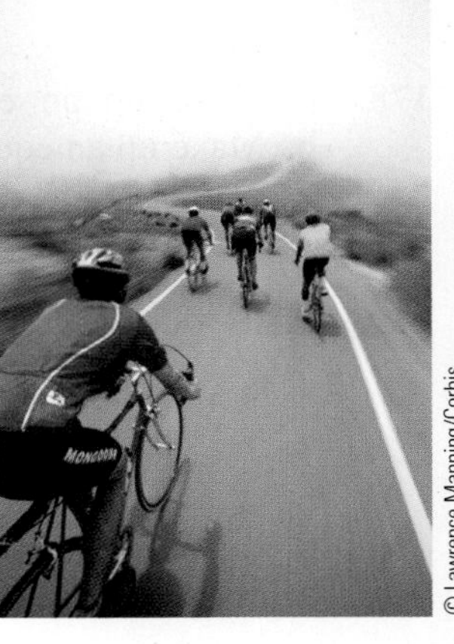

© Lawrence Manning/Corbis

168. The Ride for Health Bicycle Club has chosen a 36-mile course for this Saturday's ride. If the riders plan on averaging 12 mph while they are riding, and they have a 1-hour lunch break planned, how long will it take them to complete the trip?

169. Palmer's average running speed is 3 km/h faster than his walking speed. If Palmer can run around a 30-kilometer course in 2 h, how many hours would it take for Palmer to walk the same course?

170. A shopping mall has a moving sidewalk that takes shoppers from the shopping area to the parking garage, a distance of 250 ft. If your normal walking rate is 5 ft/s and the moving sidewalk is traveling at 3 ft/s, how many seconds would it take for you to walk from one end of the moving sidewalk to the other end?

171. Two joggers start at the same time from opposite ends of an 8-mile jogging trail and begin running toward each other. One jogger is running at a rate of 5 mph, and the other jogger is running at a rate of 7 mph. How long, in minutes, after they start will the two joggers meet?

172. Two cyclists start from the same point at the same time and move in opposite directions. One cyclist is traveling at 8 mph, and the other cyclist is traveling at 9 mph. After 30 min, how far apart are the two cyclists?

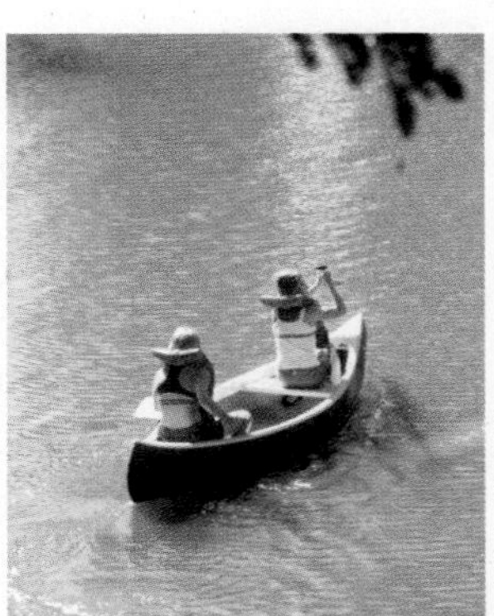
© Davis Barber/PhotoEdit

173. Petra and Celine can paddle their canoe at a rate of 10 mph in calm water. How long will it take them to travel 4 mi against the 2-mph current of the river?

174. At 8:00 A.M., a train leaves a station and travels at a rate of 45 mph. At 9:00 A.M., a second train leaves the same station on the same track and travels in the direction of the first train at a speed of 60 mph. At 10:00 A.M., how far apart are the two trains?

Applying the Concepts

175. Let r be a positive number less than 1. Is the solution of the equation $\frac{10}{9} + x = r$ positive or negative?

176. Let a be a negative number less than -5. Is the solution of the equation $a = -5b$ less than or greater than 1?

177. Why, when the Multiplication Property of Equations is used, must the quantity that multiplies each side of the equation not be zero?

178. **a.** Make up an equation of the form $x + a = b$ that has 2 as a solution.
b. Make up an equation of the form $ax = b$ that has -1 as a solution.

179. Write out the steps for solving the equation $\frac{1}{2}x = -3$. Identify each Property of Real Numbers or Property of Equations as you use it.

180. If a quantity increases by 100%, how many times its original value is the new value?

SECTION

2.2 General Equations

OBJECTIVE A To solve an equation of the form $ax + b = c$ 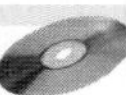

In solving an equation of the form $ax + b = c$, the goal is to rewrite the equation in the form *variable* = *constant*. This requires the application of both the Addition and the Multiplication Properties of Equations.

HOW TO 1 Solve: $\frac{3}{4}x - 2 = -11$

The goal is to write the equation in the form *variable* = *constant*.

$$\frac{3}{4}x - 2 = -11$$

$$\frac{3}{4}x - 2 + 2 = -11 + 2$$ • **Add 2** to each side of the equation.

$$\frac{3}{4}x = -9$$ • Simplify.

$$\frac{4}{3} \cdot \frac{3}{4}x = \frac{4}{3}(-9)$$ • **Multiply** each side of the equation **by** $\frac{4}{3}$.

$$x = -12$$ • The equation is in the form *variable* = *constant*.

The solution is -12.

Take Note

Check:

$\frac{3}{4}x - 2 = -11$	
$\frac{3}{4}(-12) - 2$	-11
$-9 - 2$	-11
$-11 = -11$	

A true equation

Here is an example of solving an equation that contains more than one fraction.

HOW TO 2 Solve: $\frac{2}{3}x + \frac{1}{2} = \frac{3}{4}$

$$\frac{2}{3}x + \frac{1}{2} = \frac{3}{4}$$

$$\frac{2}{3}x + \frac{1}{2} - \frac{1}{2} = \frac{3}{4} - \frac{1}{2}$$ • **Subtract** $\frac{1}{2}$ from each side of the equation.

$$\frac{2}{3}x = \frac{1}{4}$$ • Simplify.

$$\frac{3}{2}\left(\frac{2}{3}x\right) = \frac{3}{2}\left(\frac{1}{4}\right)$$ • **Multiply** each side of the equation **by** $\frac{3}{2}$, **the reciprocal of** $\frac{2}{3}$.

$$x = \frac{3}{8}$$

The solution is $\frac{3}{8}$.

It may be easier to solve an equation containing two or more fractions by multiplying each side of the equation by the least common multiple (LCM) of the denominators. For the equation above, the LCM of 3, 2, and 4 is 12. The LCM has the property that 3, 2, and 4 will divide evenly into it. Therefore, if both sides of the equation are multiplied by 12, the denominators will divide evenly into 12. The result is an equation that does not contain any fractions. Multiplying each side of an equation that contains fractions by the LCM of the denominators is called **clearing denominators.** It is an alternative method, as we show in the next example, of solving an equation that contains fractions.

Take Note

This is the same example solved on the previous page, but this time we are using the method of clearing denominators.

Observe that after we multiply both sides of the equation by the LCM of the denominators and then simplify, the equation no longer contains fractions.

HOW TO • 3 Solve: $\frac{2}{3}x + \frac{1}{2} = \frac{3}{4}$

$$\frac{2}{3}x + \frac{1}{2} = \frac{3}{4}$$

$$12\left(\frac{2}{3}x + \frac{1}{2}\right) = 12\left(\frac{3}{4}\right)$$

• **Multiply** each side of the equation **by 12, the LCM of 3, 2, and 4.**

$$12\left(\frac{2}{3}x\right) + 12\left(\frac{1}{2}\right) = 12\left(\frac{3}{4}\right)$$

• Use the Distributive Property.

$$8x + 6 = 9$$

• Simplify.

$$8x + 6 - 6 = 9 - 6$$

• **Subtract 6** from each side of the equation.

$$8x = 3$$

$$\frac{8x}{8} = \frac{3}{8}$$

• **Divide** each side of the equation **by 8.**

$$x = \frac{3}{8}$$

The solution is $\frac{3}{8}$.

Note that both methods give exactly the same solution. You may use either method to solve an equation containing fractions.

EXAMPLE • 1

Solve: $3x - 7 = -5$

Solution

$$3x - 7 = -5$$

$$3x - 7 + 7 = -5 + 7$$

• **Add 7** to each side.

$$3x = 2$$

$$\frac{3x}{3} = \frac{2}{3}$$

• **Divide** each side **by 3.**

$$x = \frac{2}{3}$$

The solution is $\frac{2}{3}$.

YOU TRY IT • 1

Solve: $5x + 7 = 10$

Your solution

EXAMPLE • 2

Solve: $5 = 9 - 2x$

Solution

$$5 = 9 - 2x$$

$$5 - 9 = 9 - 9 - 2x$$

• **Subtract 9** from each side.

$$-4 = -2x$$

$$\frac{-4}{-2} = \frac{-2x}{-2}$$

• **Divide** each side **by −2.**

$$2 = x$$

The solution is 2.

YOU TRY IT • 2

Solve: $2 = 11 + 3x$

Your solution

Solutions on p. S4

EXAMPLE • 3

Solve: $\frac{2}{3} - \frac{x}{2} = \frac{3}{4}$

Solution

$$\frac{2}{3} - \frac{x}{2} = \frac{3}{4}$$

$$\frac{2}{3} - \frac{2}{3} - \frac{x}{2} = \frac{3}{4} - \frac{2}{3}$$ • **Subtract** $\frac{2}{3}$ **from each side.**

$$-\frac{x}{2} = \frac{1}{12}$$

$$-2\left(-\frac{x}{2}\right) = -2\left(\frac{1}{12}\right)$$ • **Multiply each side by −2.**

$$x = -\frac{1}{6}$$

The solution is $-\frac{1}{6}$.

YOU TRY IT • 3

Solve: $\frac{5}{8} - \frac{2x}{3} = \frac{5}{4}$

Your solution

EXAMPLE • 4

Solve $\frac{4}{5}x - \frac{1}{2} = \frac{3}{4}$ by first clearing denominators.

Solution

The LCM of 5, 2, and 4 is 20.

$$\frac{4}{5}x - \frac{1}{2} = \frac{3}{4}$$

$$20\left(\frac{4}{5}x - \frac{1}{2}\right) = 20\left(\frac{3}{4}\right)$$ • **Multiply each side by 20.**

$$20\left(\frac{4}{5}x\right) - 20\left(\frac{1}{2}\right) = 20\left(\frac{3}{4}\right)$$ • **Use the Distributive Property.**

$$16x - 10 = 15$$

$$16x - 10 + 10 = 15 + 10$$ • **Add 10 to each side.**

$$16x = 25$$

$$\frac{16x}{16} = \frac{25}{16}$$ • **Divide each side by 16.**

$$x = \frac{25}{16}$$

The solution is $\frac{25}{16}$.

YOU TRY IT • 4

Solve $\frac{2}{3}x + 3 = \frac{7}{2}$ by first clearing denominators.

Your solution

Solutions on p. S5

EXAMPLE 5

Solve: $2x + 4 - 5x = 10$

Solution

$$2x + 4 - 5x = 10$$

$$-3x + 4 = 10$$ • Combine like terms.

$$-3x + 4 - 4 = 10 - 4$$ • Subtract 4 from each side.

$$-3x = 6$$

$$\frac{-3x}{-3} = \frac{6}{-3}$$ • Divide each side by −3.

$$x = -2$$

The solution is −2.

YOU TRY IT 5

Solve: $x - 5 + 4x = 25$

Your solution

Solution on p. S5

OBJECTIVE B To solve an equation of the form $ax + b = cx + d$ 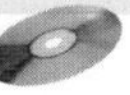

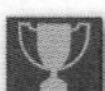

Tips for Success

Have you considered joining a study group? Getting together regularly with other students in the class to go over material and quiz each other can be very beneficial. See *AIM for Success* in the Preface.

In solving an equation of the form $ax + b = cx + d$, the goal is to rewrite the equation in the form *variable* = *constant*. Begin by rewriting the equation so that there is only one variable term in the equation. Then rewrite the equation so that there is only one constant term.

HOW TO 4 Solve: $2x + 3 = 5x - 9$

$$2x + 3 = 5x - 9$$

$$2x - 5x + 3 = 5x - 5x - 9$$ • Subtract $5x$ from each side of the equation.

$$-3x + 3 = -9$$ • Simplify. There is only one variable term.

$$-3x + 3 - 3 = -9 - 3$$ • Subtract 3 from each side of the equation.

$$-3x = -12$$ • Simplify. There is only one constant term.

$$\frac{-3x}{-3} = \frac{-12}{-3}$$ • Divide each side of the equation by −3.

$$x = 4$$ • The equation is in the form *variable* = *constant*.

The solution is 4. You should verify this by checking this solution.

EXAMPLE • 6

Solve: $4x - 5 = 8x - 7$

Solution

$$4x - 5 = 8x - 7$$

$$4x - 8x - 5 = 8x - 8x - 7$$ • **Subtract 8*x*** from each side.

$$-4x - 5 = -7$$

$$-4x - 5 + 5 = -7 + 5$$ • **Add 5** to each side.

$$-4x = -2$$

$$\frac{-4x}{-4} = \frac{-2}{-4}$$ • **Divide** each side **by −4**.

$$x = \frac{1}{2}$$

The solution is $\frac{1}{2}$.

YOU TRY IT • 6

Solve: $5x + 4 = 6 + 10x$

Your solution

EXAMPLE • 7

Solve: $3x + 4 - 5x = 2 - 4x$

Solution

$$3x + 4 - 5x = 2 - 4x$$

$$-2x + 4 = 2 - 4x$$ • Combine like terms.

$$-2x + 4x + 4 = 2 - 4x + 4x$$ • **Add 4*x*** to each side.

$$2x + 4 = 2$$

$$2x + 4 - 4 = 2 - 4$$ • **Subtract 4** from each side.

$$2x = -2$$

$$\frac{2x}{2} = \frac{-2}{2}$$ • **Divide** each side **by 2**.

$$x = -1$$

The solution is −1.

YOU TRY IT • 7

Solve: $5x - 10 - 3x = 6 - 4x$

Your solution

Solutions on p. S5

OBJECTIVE C To solve an equation containing parentheses 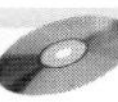

When an equation contains parentheses, one of the steps in solving the equation requires the use of the Distributive Property. The Distributive Property is used to remove parentheses from a variable expression.

HOW TO • 5 Solve: $4 + 5(2x - 3) = 3(4x - 1)$

$$
\begin{aligned}
4 + 5(2x - 3) &= 3(4x - 1) \\
4 + 10x - 15 &= 12x - 3 && \bullet \text{ Use the Distributive Property. Then simplify.} \\
10x - 11 &= 12x - 3 \\
10x - 12x - 11 &= 12x - 12x - 3 && \bullet \text{ Subtract } 12x \text{ from each side of the equation.} \\
-2x - 11 &= -3 && \bullet \text{ Simplify.} \\
-2x - 11 + 11 &= -3 + 11 && \bullet \text{ Add 11 to each side of the equation.} \\
-2x &= 8 && \bullet \text{ Simplify.} \\
\frac{-2x}{-2} &= \frac{8}{-2} && \bullet \text{ Divide each side of the equation by } -2. \\
x &= -4 && \bullet \text{ The equation is in the form } variable = constant.
\end{aligned}
$$

The solution is -4. You should verify this by checking this solution.

In the next example, we solve an equation with parentheses and decimals.

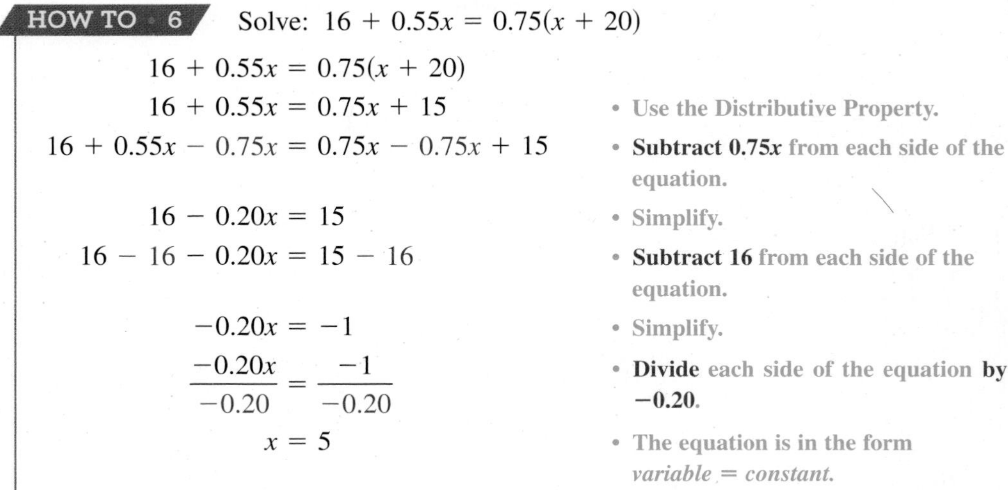

HOW TO • 6 Solve: $16 + 0.55x = 0.75(x + 20)$

$$
\begin{aligned}
16 + 0.55x &= 0.75(x + 20) \\
16 + 0.55x &= 0.75x + 15 && \bullet \text{ Use the Distributive Property.} \\
16 + 0.55x - 0.75x &= 0.75x - 0.75x + 15 && \bullet \text{ Subtract } 0.75x \text{ from each side of the equation.} \\
16 - 0.20x &= 15 && \bullet \text{ Simplify.} \\
16 - 16 - 0.20x &= 15 - 16 && \bullet \text{ Subtract 16 from each side of the equation.} \\
-0.20x &= -1 && \bullet \text{ Simplify.} \\
\frac{-0.20x}{-0.20} &= \frac{-1}{-0.20} && \bullet \text{ Divide each side of the equation by } -0.20. \\
x &= 5 && \bullet \text{ The equation is in the form } variable = constant.
\end{aligned}
$$

The solution is 5.

EXAMPLE 8

Solve: $3x - 4(2 - x) = 3(x - 2) - 4$

Solution

$$3x - 4(2 - x) = 3(x - 2) - 4$$

$$3x - 8 + 4x = 3x - 6 - 4$$ • Distributive Property

$$7x - 8 = 3x - 10$$

$$7x - 3x - 8 = 3x - 3x - 10$$ • Subtract $3x$.

$$4x - 8 = -10$$

$$4x - 8 + 8 = -10 + 8$$ • Add 8.

$$4x = -2$$

$$\frac{4x}{4} = \frac{-2}{4}$$ • Divide by 4.

$$x = -\frac{1}{2}$$

The solution is $-\frac{1}{2}$.

YOU TRY IT 8

Solve: $5x - 4(3 - 2x) = 2(3x - 2) + 6$

Your solution

EXAMPLE 9

Solve: $3[2 - 4(2x - 1)] = 4x - 10$

Solution

$$3[2 - 4(2x - 1)] = 4x - 10$$

$$3[2 - 8x + 4] = 4x - 10$$ • Distributive Property

$$3[6 - 8x] = 4x - 10$$

$$18 - 24x = 4x - 10$$ • Distributive Property

$$18 - 24x - 4x = 4x - 4x - 10$$ • Subtract $4x$.

$$18 - 28x = -10$$

$$18 - 18 - 28x = -10 - 18$$ • Subtract 18.

$$-28x = -28$$

$$\frac{-28x}{-28} = \frac{-28}{-28}$$ • Divide by -28.

$$x = 1$$

The solution is 1.

YOU TRY IT 9

Solve: $-2[3x - 5(2x - 3)] = 3x - 8$

Your solution

Solutions on p. S5

OBJECTIVE D To solve application problems using formulas 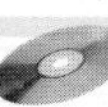

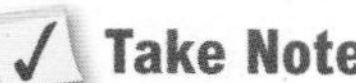

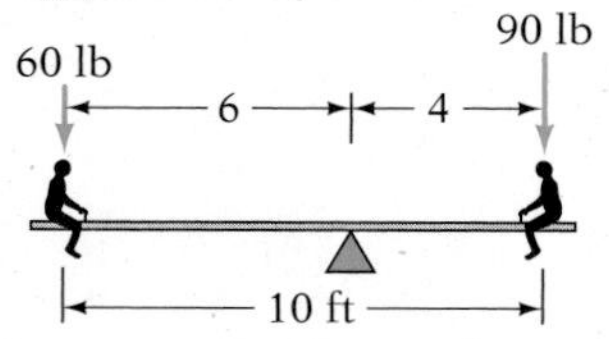

This system balances because

$$F_1x = F_2(d - x)$$
$$60(6) = 90(10 - 6)$$
$$60(6) = 90(4)$$
$$360 = 360$$

A lever system is shown at the right. It consists of a lever, or bar; a fulcrum; and two forces, F_1 and F_2. The distance d represents the length of the lever, x represents the distance from F_1 to the fulcrum, and $d - x$ represents the distance from F_2 to the fulcrum.

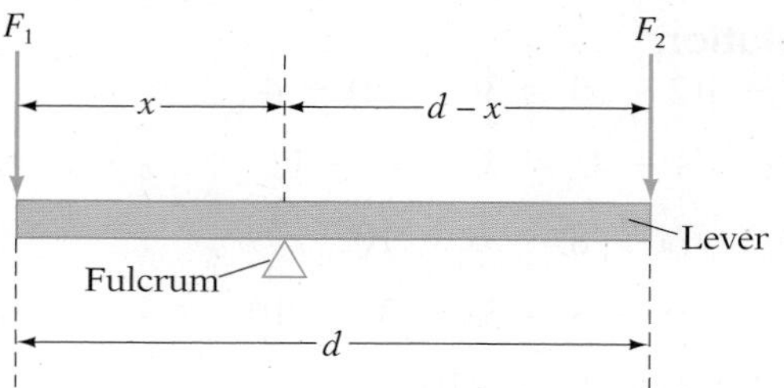

A principle of physics states that when the lever system balances, $F_1x = F_2(d - x)$.

EXAMPLE • 10

A lever is 15 ft long. A force of 50 lb is applied to one end of the lever, and a force of 100 lb is applied to the other end. Where is the fulcrum located when the system balances?

Strategy

Make a drawing.

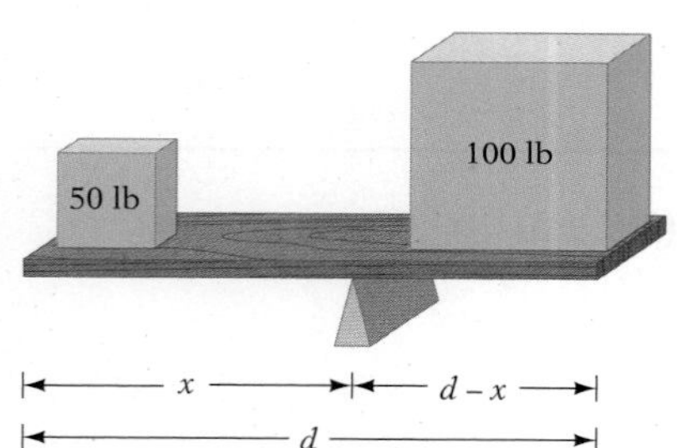

Given: $F_1 = 50$
$F_2 = 100$
$d = 15$

Unknown: x

Solution

$$F_1x = F_2(d - x)$$
$$50x = 100(15 - x)$$
$$50x = 1500 - 100x$$
$$50x + 100x = 1500 - 100x + 100x \quad \text{• Add } 100x.$$
$$150x = 1500$$
$$\frac{150x}{150} = \frac{1500}{150} \quad \text{• Divide by 150.}$$
$$x = 10$$

The fulcrum is 10 ft from the 50-pound force.

YOU TRY IT • 10

A lever is 25 ft long. A force of 45 lb is applied to one end of the lever, and a force of 80 lb is applied to the other end. Where is the location of the fulcrum when the system balances?

Your strategy

Your solution

Solution on p. S5

2.2 EXERCISES

OBJECTIVE A **To solve an equation of the form $ax + b = c$**

For Exercises 1 to 80, solve and check.

1. $3x + 1 = 10$ **2.** $4y + 3 = 11$ **3.** $2a - 5 = 7$ **4.** $5m - 6 = 9$

5. $5 = 4x + 9$ **6.** $2 = 5b + 12$ **7.** $2x - 5 = -11$ **8.** $3n - 7 = -19$

9. $4 - 3w = -2$ **10.** $5 - 6x = -13$ **11.** $8 - 3t = 2$ **12.** $12 - 5x = 7$

13. $4a - 20 = 0$ **14.** $3y - 9 = 0$ **15.** $6 + 2b = 0$ **16.** $10 + 5m = 0$

17. $-2x + 5 = -7$ **18.** $-5d + 3 = -12$ **19.** $-1.2x + 3 = -0.6$ **20.** $-1.3 = -1.1y + 0.9$

21. $2 = 7 - 5a$ **22.** $3 = 11 - 4n$ **23.** $-35 = -6b + 1$ **24.** $-8x + 3 = -29$

25. $-3m - 21 = 0$ **26.** $-5x - 30 = 0$ **27.** $-4y + 15 = 15$ **28.** $-3x + 19 = 19$

29. $9 - 4x = 6$ **30.** $3t - 2 = 0$ **31.** $9x - 4 = 0$ **32.** $7 - 8z = 0$

33. $1 - 3x = 0$ **34.** $9d + 10 = 7$ **35.** $12w + 11 = 5$ **36.** $6y - 5 = -7$

37. $8b - 3 = -9$ **38.** $5 - 6m = 2$ **39.** $7 - 9a = 4$ **40.** $9 = -12c + 5$

41. $10 = -18x + 7$

42. $2y + \frac{1}{3} = \frac{7}{3}$

43. $4a + \frac{3}{4} = \frac{19}{4}$

44. $2n - \frac{3}{4} = \frac{13}{4}$

45. $3x - \frac{5}{6} = \frac{13}{6}$

46. $5y + \frac{3}{7} = \frac{3}{7}$

47. $9x + \frac{4}{5} = \frac{4}{5}$

48. $0.8 = 7d + 0.1$

49. $0.9 = 10x - 0.6$

50. $4 = 7 - 2w$

51. $7 = 9 - 5a$

52. $8t + 13 = 3$

53. $12x + 19 = 3$

54. $-6y + 5 = 13$

55. $-4x + 3 = 9$

56. $\frac{1}{2}a - 3 = 1$

57. $\frac{1}{3}m - 1 = 5$

58. $\frac{2}{5}y + 4 = 6$

59. $\frac{3}{4}n + 7 = 13$

60. $-\frac{2}{3}x + 1 = 7$

61. $-\frac{3}{8}b + 4 = 10$

62. $\frac{x}{4} - 6 = 1$

63. $\frac{y}{5} - 2 = 3$

64. $\frac{2x}{3} - 1 = 5$

65. $\frac{2}{3}x - \frac{5}{6} = -\frac{1}{3}$

66. $\frac{5}{4}x + \frac{2}{3} = \frac{1}{4}$

67. $\frac{1}{2} - \frac{2}{3}x = \frac{1}{4}$

68. $\frac{3}{4} - \frac{3}{5}x = \frac{19}{20}$

69. $\frac{3}{2} = \frac{5}{6} + \frac{3x}{8}$

70. $-\frac{1}{4} = \frac{5}{12} + \frac{5x}{6}$

71. $\frac{11}{27} = \frac{4}{9} - \frac{2x}{3}$

72. $\frac{37}{24} = \frac{7}{8} - \frac{5x}{6}$

73. $7 = \frac{2x}{5} + 4$

74. $5 - \frac{4c}{7} = 8$

75. $7 - \frac{5}{9}y = 9$

76. $6a + 3 + 2a = 11$

77. $5y + 9 + 2y = 23$

78. $7x - 4 - 2x = 6$

79. $11z - 3 - 7z = 9$

80. $2x - 6x + 1 = 9$

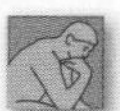

For Exercises 81 to 84, without solving the equation, determine whether the solution is positive or negative.

81. $15x + 73 = -347$

82. $17 = 25 - 40a$

83. $290 + 51n = 187$

84. $-72 = -86y + 49$

OBJECTIVE B To solve an equation of the form $ax + b = cx + d$

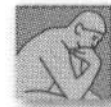

85. Describe the step that will enable you to rewrite the equation $2x - 3 = 7x + 12$ so that it has one variable term with a positive coefficient.

For Exercises 86 to 112, solve and check.

86. $8x + 5 = 4x + 13$

87. $6y + 2 = y + 17$

88. $5x - 4 = 2x + 5$

89. $13b - 1 = 4b - 19$

90. $15x - 2 = 4x - 13$

91. $7a - 5 = 2a - 20$

92. $3x + 1 = 11 - 2x$

93. $n - 2 = 6 - 3n$

94. $2x - 3 = -11 - 2x$

95. $4y - 2 = -16 - 3y$

96. $0.2b + 3 = 0.5b + 12$

97. $m + 0.4 = 3m + 0.8$

98. $4y - 8 = y - 8$

99. $5a + 7 = 2a + 7$

100. $6 - 5x = 8 - 3x$

101. $10 - 4n = 16 - n$

102. $5 + 7x = 11 + 9x$

103. $3 - 2y = 15 + 4y$

104. $2x - 4 = 6x$

105. $2b - 10 = 7b$

106. $8m = 3m + 20$

107. $9y = 5y + 16$

108. $8b + 5 = 5b + 7$

109. $6y - 1 = 2y + 2$

110. $7x - 8 = x - 3$

111. $2y - 7 = -1 - 2y$

112. $2m - 1 = -6m + 5$

113. If $5x = 3x - 8$, evaluate $4x + 2$.

114. If $7x + 3 = 5x - 7$, evaluate $3x - 2$.

115. If $2 - 6a = 5 - 3a$, evaluate $4a^2 - 2a + 1$.

116. If $1 - 5c = 4 - 4c$, evaluate $3c^2 - 4c + 2$.

OBJECTIVE C **To solve an equation containing parentheses**

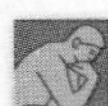

117. Without solving any of the equations, determine which of the following equations has the same solution as the equation $5 - 2(x - 1) = 8$.
(i) $3(x - 1) = 8$ (ii) $5 - 2x + 2 = 8$ (iii) $5 - 2x + 1 = 8$

For Exercises 118 to 138, solve and check.

118. $5x + 2(x + 1) = 23$

119. $6y + 2(2y + 3) = 16$

120. $9n - 3(2n - 1) = 15$

121. $12x - 2(4x - 6) = 28$

122. $7a - (3a - 4) = 12$

123. $9m - 4(2m - 3) = 11$

124. $5(3 - 2y) + 4y = 3$

125. $4(1 - 3x) + 7x = 9$

126. $5y - 3 = 7 + 4(y - 2)$

127. $0.22(x + 6) = 0.2x + 1.8$

128. $0.05(4 - x) + 0.1x = 0.32$

129. $0.3x + 0.3(x + 10) = 300$

130. $2a - 5 = 4(3a + 1) - 2$

131. $5 - (9 - 6x) = 2x - 2$

132. $7 - (5 - 8x) = 4x + 3$

133. $3[2 - 4(y - 1)] = 3(2y + 8)$

134. $5[2 - (2x - 4)] = 2(5 - 3x)$

135. $3a + 2[2 + 3(a - 1)] = 2(3a + 4)$

136. $5 + 3[1 + 2(2x - 3)] = 6(x + 5)$

137. $-2[4 - (3b + 2)] = 5 - 2(3b + 6)$

138. $-4[x - 2(2x - 3)] + 1 = 2x - 3$

139. If $4 - 3a = 7 - 2(2a + 5)$, evaluate $a^2 + 7a$.

140. If $9 - 5x = 12 - (6x + 7)$, evaluate $x^2 - 3x - 2$.

OBJECTIVE D To solve application problems using formulas

Diving Scores In a diving competition, a diver's total score for a dive is calculated using the formula $P = D(x + y + z)$, where P is the total points awarded, D is the degree of difficulty of the dive, and x, y, and z are the scores from three judges. Use this formula and the information in the article at the right for Exercises 141 to 144.

In the News

Hurricane Divers Make a Splash

University of Miami divers JJ Kinzbach and Rueben Ross took the top two spots in Men's Platform diving at the 2008 NCAA Zone B Championships. Ross also placed second in the Men's 3-meter and 1-meter events. Brittany Viola won the Women's Platform diving event and placed third in the 3-meter and 1-meter events. Statistics from some of the best dives follow.

Diver	*Event*	*Dive*	*Degree of Difficulty*	*Total Points*
Kinzbach	Platform	Inward 3½ somersault tuck	3.2	81.60
Ross	1-meter	Inward 2½ somersault tuck	3.1	77.50
Viola	Platform	Forward 3½ somersault pike	3.0	72.0
Viola	1-meter	Inward 1½ somersault pike	2.4	57.60

Source: divemeets.com

141. Two judges gave Kinzbach's platform dive scores of 8.5. Find the score given by the third judge.

142. Two judges gave Ross's 1-meter dive scores of 8 and 8.5. Find the score given by the third judge.

143. Two judges gave Viola's platform dive scores of 8. Find the score given by the third judge.

144. Two judges gave Viola's 1-meter dive scores of 8 and 8.5. Find the score given by the third judge.

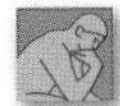

145. Physics Two people sit on a seesaw that is 8 ft long. The seesaw balances when the fulcrum is 3 ft from one of the people.
- **a.** How far is the fulcrum from the other person?
- **b.** Which person is heavier, the person who is 3 ft from the fulcrum or the other person?
- **c.** If the two people switch places, will the seesaw still balance?

Physics For Exercises 146 to 151, solve. Use the lever system equation $F_1x = F_2(d - x)$.

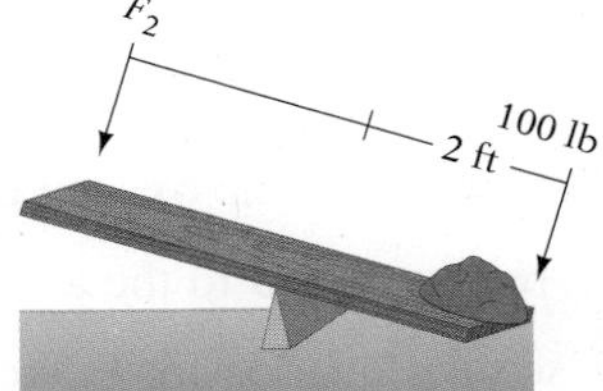

146. A lever 10 ft long is used to move a 100-pound rock. The fulcrum is placed 2 ft from the rock. What force must be applied to the other end of the lever to move the rock?

147. An adult and a child are on a seesaw 14 ft long. The adult weighs 175 lb and the child weighs 70 lb. How many feet from the child must the fulcrum be placed so that the seesaw balances?

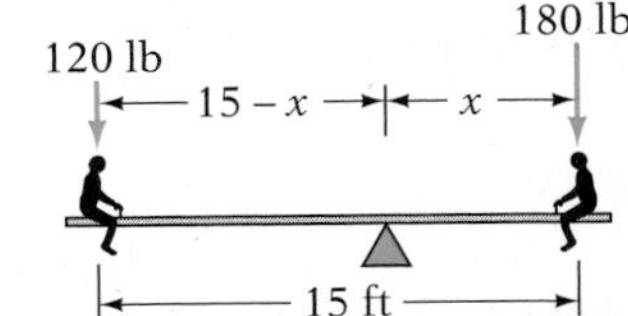

148. Two people are sitting 15 ft apart on a seesaw. One person weighs 180 lb. The second person weighs 120 lb. How far from the 180-pound person should the fulcrum be placed so that the seesaw balances?

149. Two children are sitting on a seesaw that is 12 ft long. One child weighs 60 lb. The other child weighs 90 lb. How far from the 90-pound child should the fulcrum be placed so that the seesaw balances?

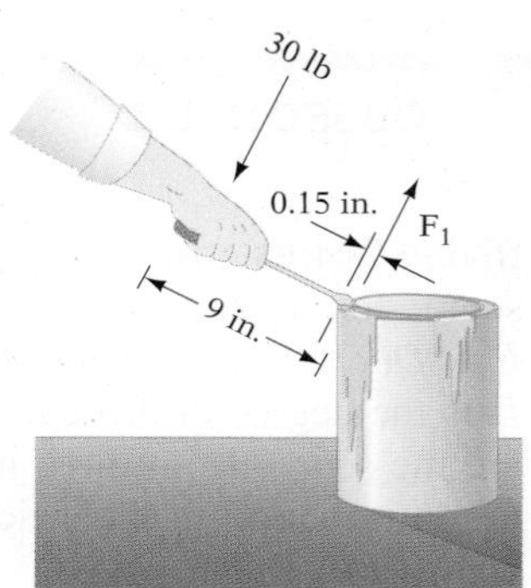

150. In preparation for a stunt, two acrobats are standing on a plank 18 ft long. One acrobat weighs 128 lb and the second acrobat weighs 160 lb. How far from the 128-pound acrobat must the fulcrum be placed so that the acrobats are balanced on the plank?

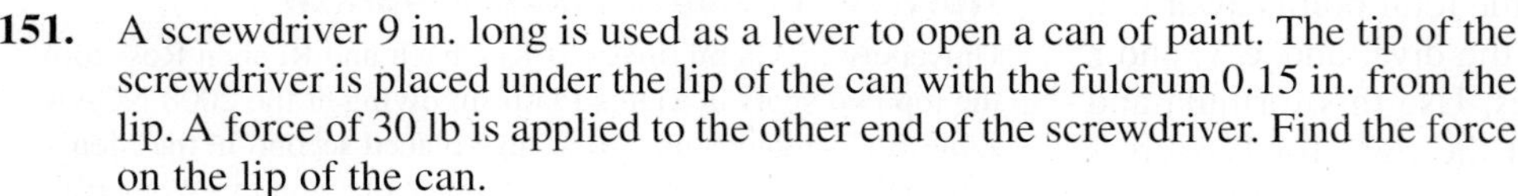

151. A screwdriver 9 in. long is used as a lever to open a can of paint. The tip of the screwdriver is placed under the lip of the can with the fulcrum 0.15 in. from the lip. A force of 30 lb is applied to the other end of the screwdriver. Find the force on the lip of the can.

Business To determine the break-even point, or the number of units that must be sold so that no profit or loss occurs, an economist uses the formula $Px = Cx + F$, where P is the selling price per unit, x is the number of units that must be sold to break even, C is the cost to make each unit, and F is the fixed cost. Use this equation for Exercises 152 to 155.

152. A business analyst has determined that the selling price per unit for a laser printer is \$1600. The cost to make one laser printer is \$950, and the fixed cost is \$211,250. Find the break-even point.

153. A business analyst has determined that the selling price per unit for a gas barbecue is \$325. The cost to make one gas barbecue is \$175, and the fixed cost is \$39,000. Find the break-even point.

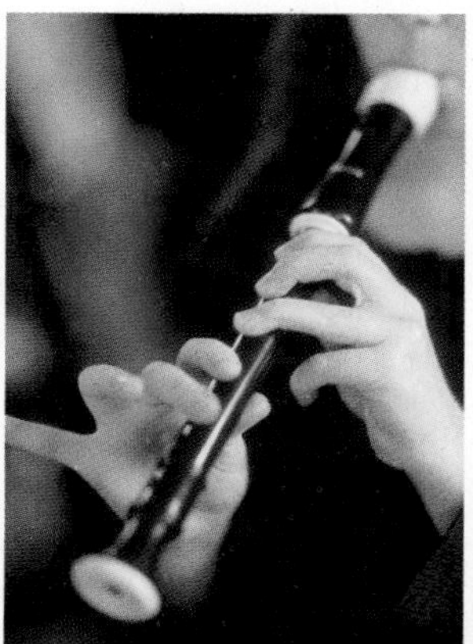
Charles Mistral/Alamy

154. A manufacturer of headphones determines that the cost per unit for a pair of headphones is \$38 and that the fixed cost is \$24,400. The selling price for the headphones is \$99. Find the break-even point.

155. A manufacturing engineer determines that the cost per unit for a soprano recorder is \$12 and that the fixed cost is \$19,240. The selling price for the recorder is \$49. Find the break-even point.

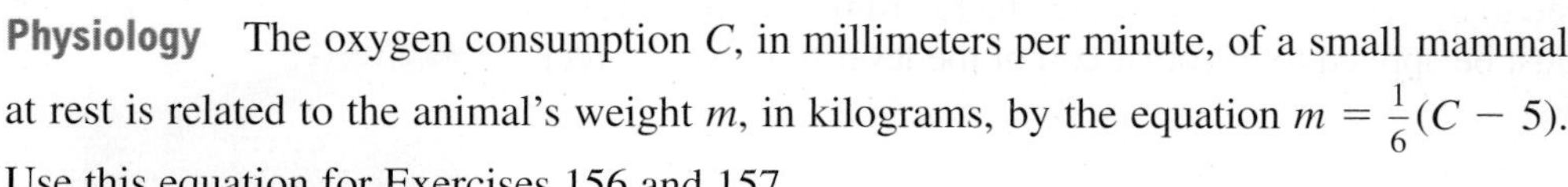

Physiology The oxygen consumption C, in millimeters per minute, of a small mammal at rest is related to the animal's weight m, in kilograms, by the equation $m = \frac{1}{6}(C - 5)$. Use this equation for Exercises 156 and 157.

156. What is the oxygen consumption of a mammal that weighs 10.4 kg?

157. What is the oxygen consumption of a mammal that weighs 8.3 kg?

Applying the Concepts

158. The equation $x = x + 1$ has no solution, whereas the solution of the equation $2x + 3 = 3$ is zero. Is there a difference between no solution and a solution of zero? Explain your answer.

SECTION 2.3 Translating Sentences into Equations

OBJECTIVE A To solve integer problems 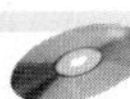

An equation states that two mathematical expressions are equal. Therefore, to **translate** a sentence into an equation requires recognition of the words or phrases that mean "equals." Some of these phrases are listed below.

equals
is
is equal to
amounts to
represents

} translate to =

Once the sentence is translated into an equation, the equation can be solved by rewriting the equation in the form *variable* = *constant*.

Take Note
You can check the solution to a translation problem.

Check:

5 less than 18 is 13

18 − 5	13
13 =	13

HOW TO 1 Translate "five less than a number is thirteen" into an equation and solve.

The unknown number: n • Assign a variable to the unknown number.

Five less than a number	is	thirteen

• Find two verbal expressions for the same value.

$n - 5 = 13$ • Write a mathematical expression for each verbal expression. Write the equals sign.

$n - 5 + 5 = 13 + 5$ • Solve the equation.

$n = 18$

The number is 18.

Recall that the integers are the numbers {..., −4, −3, −2, −1, 0, 1, 2, 3, 4, ...}. An **even integer** is an integer that is divisible by 2. Examples of even integers are −8, 0, and 22. An **odd integer** is an integer that is not divisible by 2. Examples of odd integers are −17, 1, and 39.

Consecutive integers are integers that follow one another in order. Examples of consecutive integers are shown at the right. (Assume that the variable n represents an integer.)

11, 12, 13
−8, −7, −6
$n, n + 1, n + 2$

Examples of **consecutive even integers** are shown at the right. (Assume that the variable n represents an even integer.)

24, 26, 28
−10, −8, −6
$n, n + 2, n + 4$

Take Note
Both consecutive even and consecutive odd integers are represented using n, $n + 2, n + 4, \ldots$.

Examples of **consecutive odd integers** are shown at the right. (Assume that the variable n represents an odd integer.)

19, 21, 23
−1, 1, 3
$n, n + 2, n + 4$

HOW TO • 2 The sum of three consecutive odd integers is forty-five. Find the integers.

Strategy

- First odd integer: n • Represent three consecutive odd integers.
  Second odd integer: $n + 2$
  Third odd integer: $n + 4$
- The sum of the three odd integers is 45.

Solution

$n + (n + 2) + (n + 4) = 45$ • Write an equation.

$3n + 6 = 45$ • Solve the equation.

$3n = 39$

$n = 13$ • The first odd integer is 13.

$n + 2 = 13 + 2 = 15$ • Find the second odd integer.

$n + 4 = 13 + 4 = 17$ • Find the third odd integer.

The three consecutive odd integers are 13, 15, and 17.

EXAMPLE • 1

The sum of two numbers is sixteen. The difference between four times the smaller number and two is two more than twice the larger number. Find the two numbers.

Strategy

The smaller number: n
The larger number: $16 - n$

The difference between four times the smaller and two	is	two more than twice the larger

Solution

$$\begin{aligned} 4n - 2 &= 2(16 - n) + 2 \\ 4n - 2 &= 32 - 2n + 2 \\ 4n - 2 &= 34 - 2n \\ 4n + 2n - 2 &= 34 - 2n + 2n \\ 6n - 2 &= 34 \\ 6n - 2 + 2 &= 34 + 2 \\ 6n &= 36 \\ \frac{6n}{6} &= \frac{36}{6} \\ n &= 6 \end{aligned}$$

$16 - n = 16 - 6 = 10$

The smaller number is 6.
The larger number is 10.

YOU TRY IT • 1

The sum of two numbers is twelve. The total of three times the smaller number and six amounts to seven less than the product of four and the larger number. Find the two numbers.

Your strategy

Your solution

Solution on p. S6

EXAMPLE • 2

Find three consecutive even integers such that three times the second equals four more than the sum of the first and third.

Strategy

- First even integer: n
  Second even integer: $n + 2$
  Third even integer: $n + 4$
- Three times the second equals four more than the sum of the first and third.

Solution

$$\begin{aligned} 3(n + 2) &= n + (n + 4) + 4 \\ 3n + 6 &= 2n + 8 \\ 3n - 2n + 6 &= 2n - 2n + 8 \\ n + 6 &= 8 \\ n &= 2 \end{aligned}$$

$n + 2 = 2 + 2 = 4$
$n + 4 = 2 + 4 = 6$

The three integers are 2, 4, and 6.

YOU TRY IT • 2

Find three consecutive integers whose sum is negative six.

Your strategy

Your solution

Solution on p. S6

OBJECTIVE B To translate a sentence into an equation and solve 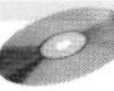

EXAMPLE • 3

A wallpaper hanger charges a fee of $25 plus $12 for each roll of wallpaper used in a room. If the total charge for hanging wallpaper is $97, how many rolls of wallpaper were used?

Strategy

To find the number of rolls of wallpaper used, write and solve an equation using n to represent the number of rolls of wallpaper used.

Solution

$25 plus $12 for each roll of wallpaper	is	$97

$$\begin{aligned} 25 + 12n &= 97 \\ 12n &= 72 \\ \frac{12n}{12} &= \frac{72}{12} \\ n &= 6 \end{aligned}$$

6 rolls of wallpaper were used.

YOU TRY IT • 3

The fee charged by a ticketing agency for a concert is $3.50 plus $17.50 for each ticket purchased. If your total charge for tickets is $161, how many tickets are you purchasing?

Your strategy

Your solution

Solution on p. S6

EXAMPLE • 4

A board 20 ft long is cut into two pieces. Five times the length of the shorter piece is 2 ft more than twice the length of the longer piece. Find the length of each piece.

Strategy

Let x represent the length of the shorter piece. Then $20 - x$ represents the length of the longer piece.

Make a drawing.

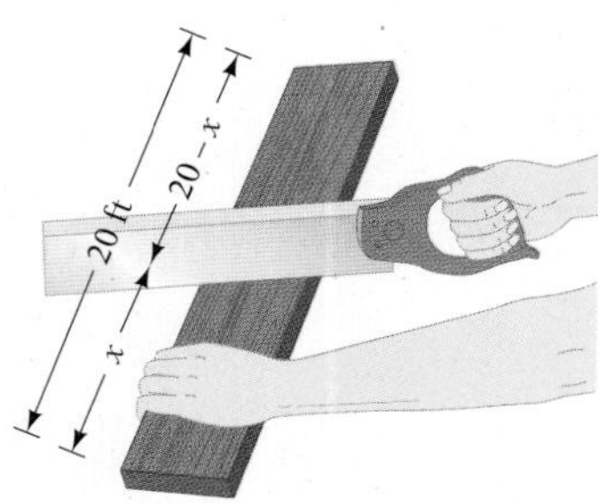

To find the lengths, write and solve an equation using x to represent the length of the shorter piece and $20 - x$ to represent the length of the longer piece.

Solution

Five times the length of the shorter piece	is	2 ft more than twice the length of the longer piece

$$
\begin{aligned}
5x &= 2(20 - x) + 2 \\
5x &= 40 - 2x + 2 \\
5x &= 42 - 2x \\
5x + 2x &= 42 - 2x + 2x \\
7x &= 42 \\
\frac{7x}{7} &= \frac{42}{7} \\
x &= 6
\end{aligned}
$$

$20 - x = 20 - 6 = 14$

The length of the shorter piece is 6 ft.
The length of the longer piece is 14 ft.

YOU TRY IT • 4

A wire 22 in. long is cut into two pieces. The length of the longer piece is 4 in. more than twice the length of the shorter piece. Find the length of each piece.

Your strategy

Your solution

Solution on p. S6

2.3 EXERCISES

OBJECTIVE A **To solve integer problems**

For Exercises 1 to 16, translate into an equation and solve.

1. The difference between a number and fifteen is seven. Find the number.

2. The sum of five and a number is three. Find the number.

3. The difference between nine and a number is seven. Find the number.

4. Three-fifths of a number is negative thirty. Find the number.

5. The difference between five and twice a number is one. Find the number.

6. Four more than three times a number is thirteen. Find the number.

7. The sum of twice a number and five is fifteen. Find the number.

8. The difference between nine times a number and six is twelve. Find the number.

9. Six less than four times a number is twenty-two. Find the number.

10. Four times the sum of twice a number and three is twelve. Find the number.

11. Three times the difference between four times a number and seven is fifteen. Find the number.

12. Twice the difference between a number and twenty-five is three times the number. Find the number.

13. The sum of two numbers is twenty. Three times the smaller is equal to two times the larger. Find the two numbers.

14. The sum of two numbers is fifteen. One less than three times the smaller is equal to the larger. Find the two numbers.

15. The sum of two numbers is fourteen. The difference between two times the smaller and the larger is one. Find the two numbers.

16. The sum of two numbers is eighteen. The total of three times the smaller and twice the larger is forty-four. Find the two numbers.

17. The sum of three consecutive odd integers is fifty-one. Find the integers.

18. Find three consecutive even integers whose sum is negative eighteen.

19. Find three consecutive odd integers such that three times the middle integer is one more than the sum of the first and third.

20. Twice the smallest of three consecutive odd integers is seven more than the largest. Find the integers.

21. Find two consecutive even integers such that three times the first equals twice the second.

22. Find two consecutive even integers such that four times the first is three times the second.

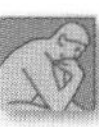

23. The sum of two numbers is seven. Twice one number is four less than the other number. Which of the following equations does *not* represent this situation?
(i) $2(7 - x) = x - 4$ (ii) $2x = (7 - x) - 4$ (iii) $2n - 4 = 7 - n$

OBJECTIVE B To translate a sentence into an equation and solve

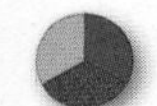

24. **Recycling** Use the information in the article at the right to find how many tons of plastic drink bottles were stocked for sale in U.S. stores.

In the News

Americans' Unquenchable Thirst

Despite efforts to increase recycling, the 2.16 million tons of plastic drink bottles that ended up in landfills this year represent four-fifths of the plastic drink bottles stocked for sale in U.S. stores.

And Americans can't seem to get enough of bottled water. Last year, stores stocked 7.5 billion gallons of bottled water, an amount that is approximately the same as the volume of water that goes over Niagara Falls every three hours.

Source: scienceline.org

25. **Robots** Kiva Systems, Inc., builds robots that companies can use to streamline order fulfillment operations in their warehouses. Salary and other benefits for one human warehouse worker can cost a company about $64,000 a year, an amount that is 103 times the company's yearly maintenance and operation costs for one robot. Find the yearly costs for a robot. Round to the nearest hundred. (*Source: The Boston Globe*)

26. **Geometry** An isosceles triangle has two sides of equal length. The length of the third side is 1 ft less than twice the length of an equal side. Find the length of each side when the perimeter is 23 ft.

27. **Geometry** An isosceles triangle has two sides of equal length. The length of one of the equal sides is 2 more than three times the length of the third side. If the perimeter is 46 m, find the length of each side.

28. **Union Dues** A union charges monthly dues of $4.00 plus $.15 for each hour worked during the month. A union member's dues for March were $29.20. How many hours did the union member work during the month of March?

29. **Technical Support** A technical information hotline charges a customer $15.00 plus $2.00 per minute to answer questions about software. How many minutes did a customer who received a bill for $37 use this service?

30. **Construction** The total cost to paint the inside of a house was $1346. This cost included $125 for materials and $33 per hour for labor. How many hours of labor were required to paint the inside of the house?

31. **Telecommunications** The cellular phone service for a business executive is $35 per month plus $.40 per minute of phone use. For a month in which the executive's cellular phone bill was $99.80, how many minutes did the executive use the phone?

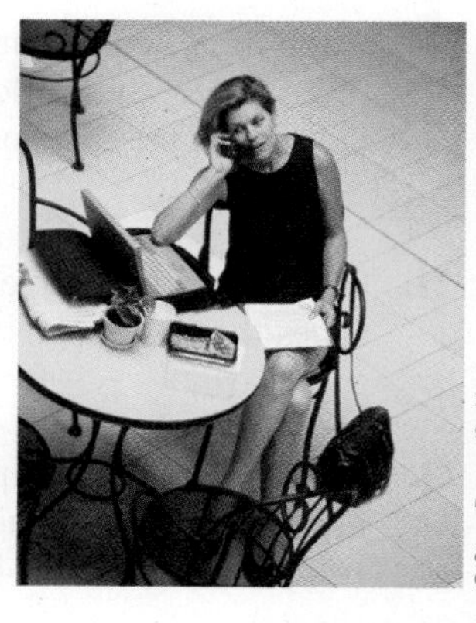

32. **Energy** The cost of electricity in a certain city is $.08 for each of the first 300 kWh (kilowatt-hours) and $.13 for each kilowatt-hour over 300 kWh. Find the number of kilowatt-hours used by a family with a $51.95 electric bill.

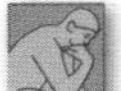

Text Messaging For Exercises 33 and 34, use the expression $2.99 + 0.15n$, which represents the total monthly text-messaging bill for n text messages over 300 in 1 month.

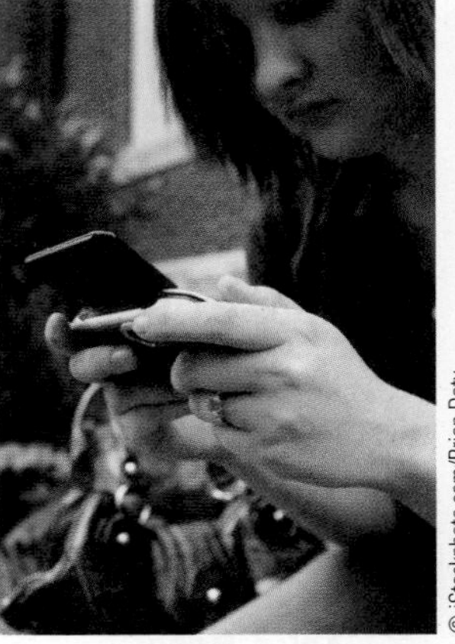

© iStockphoto.com/Brian Doty

33. How much does the customer pay per text message over 300 messages?

34. What is the fixed charge per month for the text-messaging service?

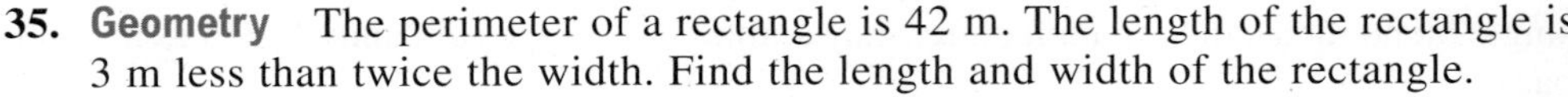

35. Geometry The perimeter of a rectangle is 42 m. The length of the rectangle is 3 m less than twice the width. Find the length and width of the rectangle.

36. Geometry A rectangular vegetable garden has a perimeter of 64 ft. The length of the garden is 20 ft. Find the width of the garden.

37. Carpentry A 12-foot board is cut into two pieces. Twice the length of the shorter piece is 3 ft less than the length of the longer piece. Find the length of each piece.

38. Sports A 14-yard fishing line is cut into two pieces. Three times the length of the longer piece is four times the length of the shorter piece. Find the length of each piece.

39. Education Seven thousand dollars is divided into two scholarships. Twice the amount of the smaller scholarship is \$1000 less than the larger scholarship. What is the amount of the larger scholarship?

40. Investing An investment of \$10,000 is divided into two accounts, one for stocks and one for mutual funds. The value of the stock account is \$2000 less than twice the value of the mutual fund account. Find the amount in each account.

Applying the Concepts

41. Make up two word problems: one that requires solving the equation $6x = 123$, and one that requires solving the equation $8x + 100 = 300$, to find the answer to the problem.

42. It is always important to check the answer to an application problem to be sure that the answer makes sense. Consider the following problem. A 4-quart juice mixture is made from apple juice and cranberry juice. There are 6 more quarts of apple juice than cranberry juice. Write and solve an equation for the number of quarts of each juice in the mixture. Does the answer to this question make sense? Explain.

SECTION

2.4 Mixture and Uniform Motion Problems

OBJECTIVE A To solve value mixture problems

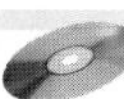

A value mixture problem involves combining two ingredients that have different prices into a single blend. For example, a coffee merchant may blend two types of coffee into a single blend, or a candy manufacturer may combine two types of candy to sell as a variety pack.

✓ Take Note

The equation $AC = V$ is used to find the value of an ingredient. For example, the value of 4 lb of cashews costing $6 per pound is

$$AC = V$$
$$4 \cdot \$6 = V$$
$$\$24 = V$$

The solution of a value mixture problem is based on the **value mixture equation** $AC = V$, where A is the amount of an ingredient, C is the cost per unit of the ingredient, and V is the value of the ingredient.

HOW TO • 1 A coffee merchant wants to make 6 lb of a blend of coffee costing $5 per pound. The blend is made using a $6-per-pound grade and a $3-per-pound grade of coffee. How many pounds of each of these grades should be used?

Strategy for Solving a Value Mixture Problem

1. For each ingredient in the mixture, write a numerical or variable expression for the amount of the ingredient used, the unit cost of the ingredient, and the value of the amount used. For the blend, write a numerical or variable expression for the amount, the unit cost of the blend, and the value of the amount. The results can be recorded in a table.

The sum of the amounts is 6 lb.

Amount of $6 coffee: x
Amount of $3 coffee: $6 - x$

✓ Take Note

Use the information given in the problem to fill in the amount and unit cost columns of the table. Fill in the value column by multiplying the two expressions you wrote in each row. Use the expressions in the last column to write the equation.

	Amount, A	·	*Unit Cost, C*	=	*Value, V*
$6 grade	x	·	6	=	$6x$
$3 grade	$6 - x$	·	3	=	$3(6 - x)$
$5 blend	6	·	5	=	$5(6)$

2. Determine how the values of the ingredients are related. Use the fact that the sum of the values of all the ingredients is equal to the value of the blend.

The sum of the values of the $6 grade and the $3 grade is equal to the value of the $5 blend.

$$6x + 3(6 - x) = 5(6)$$
$$6x + 18 - 3x = 30$$
$$3x + 18 = 30$$
$$3x = 12$$
$$x = 4$$

$6 - x = 6 - 4 = 2$ • **Find the amount of the $3 grade coffee.**

The merchant must use 4 lb of the $6 coffee and 2 lb of the $3 coffee.

EXAMPLE • 1

How many ounces of a silver alloy that costs \$4 an ounce must be mixed with 10 oz of an alloy that costs \$6 an ounce to make a mixture that costs \$4.32 an ounce?

Strategy

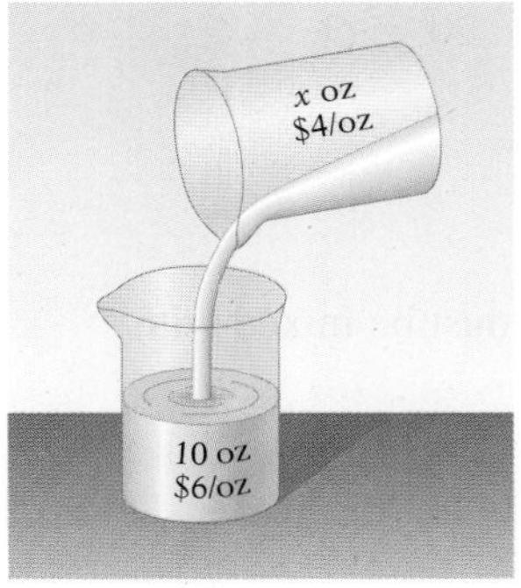

- Ounces of \$4 alloy: x

	Amount	*Cost*	*Value*
\$4 alloy	x	4	$4x$
\$6 alloy	10	6	6(10)
\$4.32 mixture	$10 + x$	4.32	$4.32(10 + x)$

- The sum of the values before mixing equals the value after mixing.

Solution

$$4x + 6(10) = 4.32(10 + x)$$
$$4x + 60 = 43.2 + 4.32x$$
$$-0.32x + 60 = 43.2$$
$$-0.32x = -16.8$$
$$x = 52.5$$

52.5 oz of the \$4 silver alloy must be used.

YOU TRY IT • 1

A gardener has 20 lb of a lawn fertilizer that costs \$.80 per pound. How many pounds of a fertilizer that costs \$.55 per pound should be mixed with this 20 lb of lawn fertilizer to produce a mixture that costs \$.75 per pound?

Your strategy

Your solution

Solution on p. S6

OBJECTIVE B To solve percent mixture problems 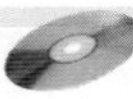

Recall from Section 2.1 that a percent mixture problem can be solved using the equation $Ar = Q$, where A is the amount of a solution, r is the percent concentration of a substance in the solution, and Q is the quantity of the substance in the solution.

For example, a 500-milliliter bottle is filled with a 4% solution of hydrogen peroxide.

$$\begin{aligned} Ar &= Q \\ 500(0.04) &= Q \\ 20 &= Q \end{aligned}$$

The bottle contains 20 ml of hydrogen peroxide.

HOW TO • 2 How many gallons of a 20% salt solution must be mixed with 6 gal of a 30% salt solution to make a 22% salt solution?

Strategy for Solving a Percent Mixture Problem

1. For each solution, write a numerical or variable expression for the amount of solution, the percent concentration, and the quantity of the substance in the solution. The results can be recorded in a table.

The unknown quantity of 20% solution: x

	Amount of Solution, A	·	*Percent Concentration, r*	=	*Quantity of Substance, Q*
20% solution	x	·	0.20	=	$0.20x$
30% solution	6	·	0.30	=	$0.30(6)$
22% solution	$x + 6$	·	0.22	=	$0.22(x + 6)$

Take Note
Use the information given in the problem to fill in the amount and percent columns of the table. Fill in the quantity column by multiplying the two expressions you wrote in each row. Use the expressions in the last column to write the equation.

2. Determine how the quantities of the substances in the solutions are related. Use the fact that the sum of the quantities of the substances being mixed is equal to the quantity of the substance after mixing.

The sum of the quantities of the substances in the 20% solution and the 30% solution is equal to the quantity of the substance in the 22% solution.

$$\begin{aligned} 0.20x + 0.30(6) &= 0.22(x + 6) \\ 0.20x + 1.80 &= 0.22x + 1.32 \\ -0.02x + 1.80 &= 1.32 \\ -0.02x &= -0.48 \\ x &= 24 \end{aligned}$$

24 gal of the 20% solution are required.

EXAMPLE • 2

A chemist wishes to make 2 L of an 8% acid solution by mixing a 10% acid solution and a 5% acid solution. How many liters of each solution should the chemist use?

Strategy

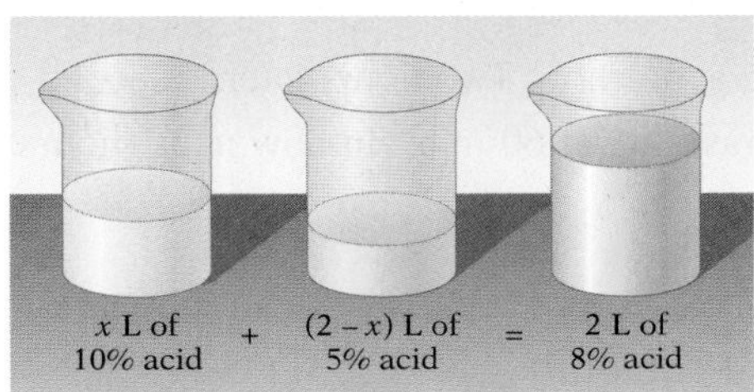

- Liters of 10% solution: x
  Liters of 5% solution: $2 - x$

	Amount	*Percent*	*Quantity*
10% solution	x	0.10	$0.10x$
5% solution	$2 - x$	0.05	$0.05(2 - x)$
8% solution	2	0.08	$0.08(2)$

- The sum of the quantities before mixing is equal to the quantity after mixing.

Solution

$$0.10x + 0.05(2 - x) = 0.08(2)$$

$$0.10x + 0.10 - 0.05x = 0.16$$

$$0.05x + 0.10 = 0.16$$

$$0.05x = 0.06$$

$$x = 1.2$$

$$2 - x = 2 - 1.2 = 0.8$$

The chemist needs 1.2 L of the 10% solution and 0.8 L of the 5% solution.

YOU TRY IT • 2

A pharmacist dilutes 5 L of a 12% solution with a 6% solution. How many liters of the 6% solution are added to make an 8% solution?

Your strategy

Your solution

Solution on p. S7

OBJECTIVE C To solve uniform motion problems 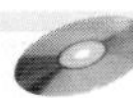

Recall from Section 2.1 that an object traveling at a constant speed in a straight line is in *uniform motion.* The solution of a uniform motion problem is based on the equation $rt = d$, where r is the rate of travel, t is the time spent traveling, and d is the distance traveled.

HOW TO • 3 A car leaves a town traveling at 40 mph. Two hours later, a second car leaves the same town, on the same road, traveling at 60 mph. In how many hours will the second car pass the first car?

Strategy for Solving a Uniform Motion Problem

1. For each object, write a numerical or variable expression for the rate, time, and distance. The results can be recorded in a table.

The first car traveled 2 h longer than the second car.

Unknown time for the second car: t
Time for the first car: $t + 2$

Take Note
Use the information given in the problem to fill in the rate and time columns of the table. Find the expression in the distance column by multiplying the two expressions you wrote in each row.

	Rate, r	·	*Time, t*	=	*Distance, d*
First car	40	·	$t + 2$	=	$40(t + 2)$
Second car	60	·	t	=	$60t$

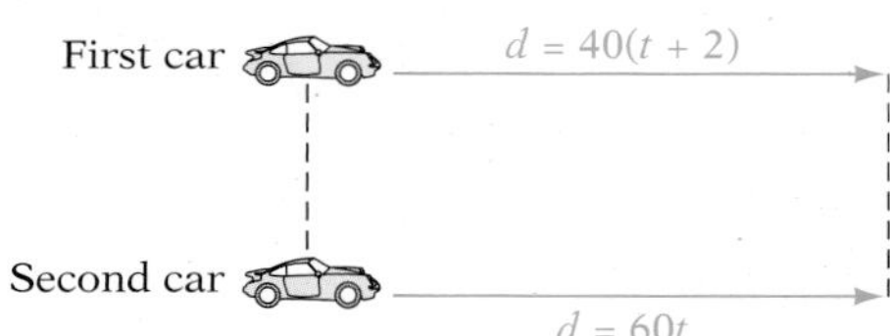

2. Determine how the distances traveled by the two objects are related. For example, the total distance traveled by both objects may be known, or it may be known that the two objects traveled the same distance.

The two cars travel the same distance.

$$40(t + 2) = 60t$$
$$40t + 80 = 60t$$
$$80 = 20t$$
$$4 = t$$

The second car will pass the first car in 4 h.

EXAMPLE • 3

Two cars, one traveling 10 mph faster than the other, start at the same time from the same point and travel in opposite directions. In 3 h they are 300 mi apart. Find the rate of each car.

Strategy

- Rate of 1st car: r
  Rate of 2nd car: $r + 10$

	Rate	*Time*	*Distance*
1st car	r	3	$3r$
2nd car	$r + 10$	3	$3(r + 10)$

- The total distance traveled by the two cars is 300 mi.

Solution

$$3r + 3(r + 10) = 300$$
$$3r + 3r + 30 = 300$$
$$6r + 30 = 300$$
$$6r = 270$$
$$r = 45$$

$r + 10 = 45 + 10 = 55$

The first car is traveling 45 mph.
The second car is traveling 55 mph.

YOU TRY IT • 3

Two trains, one traveling at twice the speed of the other, start at the same time on parallel tracks from stations that are 288 mi apart and travel toward each other. In 3 h, the trains pass each other. Find the rate of each train.

Your strategy

Your solution

EXAMPLE • 4

How far can the members of a bicycling club ride out into the country at a speed of 12 mph and return over the same road at 8 mph if they travel a total of 10 h?

Strategy

- Time spent riding out: t
  Time spent riding back: $10 - t$

	Rate	*Time*	*Distance*
Out	12	t	$12t$
Back	8	$10 - t$	$8(10 - t)$

- The distance out equals the distance back.

Solution

$$12t = 8(10 - t)$$
$$12t = 80 - 8t$$
$$20t = 80$$
$$t = 4 \quad \text{(The time is 4 h.)}$$

The distance out $= 12t = 12(4) = 48$ mi.

The club can ride 48 mi into the country.

YOU TRY IT • 4

A pilot flew out to a parcel of land and back in 5 h. The rate out was 150 mph, and the rate returning was 100 mph. How far away was the parcel of land?

Your strategy

Your solution

Solutions on p. S7

2.4 EXERCISES

OBJECTIVE A To solve value mixture problems

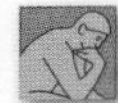

1. A grocer mixes peanuts that cost $3 per pound with almonds that cost $7 per pound. Which of the following statements could be true about the cost per pound, C, of the mixture? There may be more than one correct answer.
(i) $C = \$10$ (ii) $C > \$7$ (iii) $C < \$7$
(iv) $C < \$3$ (v) $C > \$3$ (vi) $C = \$3$

2. An herbalist has 30 oz of herbs costing $2 per ounce. How many ounces of herbs costing $1 per ounce should be mixed with the 30 oz to produce a mixture costing $1.60 per ounce?

3. The manager of a farmer's market has 500 lb of grain that costs $1.20 per pound. How many pounds of meal costing $.80 per pound should be mixed with the 500 lb of grain to produce a mixture that costs $1.05 per pound?

4. Find the cost per pound of a meatloaf mixture made from 3 lb of ground beef costing $1.99 per pound and 1 lb of ground turkey costing $1.39 per pound.

5. Find the cost per ounce of a sunscreen made from 100 oz of a lotion that costs $2.50 per ounce and 50 oz of a lotion that costs $4.00 per ounce.

6. A snack food is made by mixing 5 lb of popcorn that costs $.80 per pound with caramel that costs $2.40 per pound. How much caramel is needed to make a mixture that costs $1.40 per pound?

7. A wild birdseed mix is made by combining 100 lb of millet seed costing $.60 per pound with sunflower seeds costing $1.10 per pound. How many pounds of sunflower seeds are needed to make a mixture that costs $.70 per pound?

8. Ten cups of a restaurant's house Italian dressing are made by blending olive oil costing $1.50 per cup with vinegar that costs $.25 per cup. How many cups of each are used if the cost of the blend is $.50 per cup?

9. A high-protein diet supplement that costs $6.75 per pound is mixed with a vitamin supplement that costs $3.25 per pound. How many pounds of each should be used to make 5 lb of a mixture that costs $4.65 per pound?

10. Find the cost per ounce of a mixture of 200 oz of a cologne that costs $5.50 per ounce and 500 oz of a cologne that costs $2.00 per ounce.

11. Find the cost per pound of a trail mix made from 40 lb of raisins that cost \$4.40 per pound and 100 lb of granola that costs \$2.30 per pound.

12. The manager of a specialty food store combined almonds that cost \$4.50 per pound with walnuts that cost \$2.50 per pound. How many pounds of each were used to make a 100-pound mixture that costs \$3.24 per pound?

13. A goldsmith combined an alloy that cost \$4.30 per ounce with an alloy that cost \$1.80 per ounce. How many ounces of each were used to make a mixture of 200 oz costing \$2.50 per ounce?

14. Find the cost per pound of sugar-coated breakfast cereal made from 40 lb of sugar that costs \$1.00 per pound and 120 lb of corn flakes that cost \$.60 per pound.

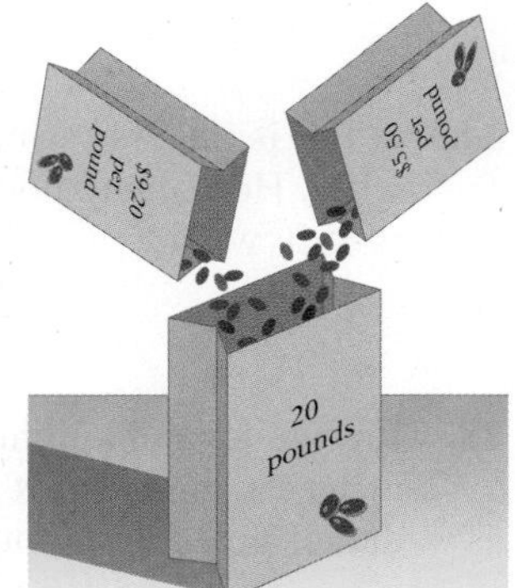

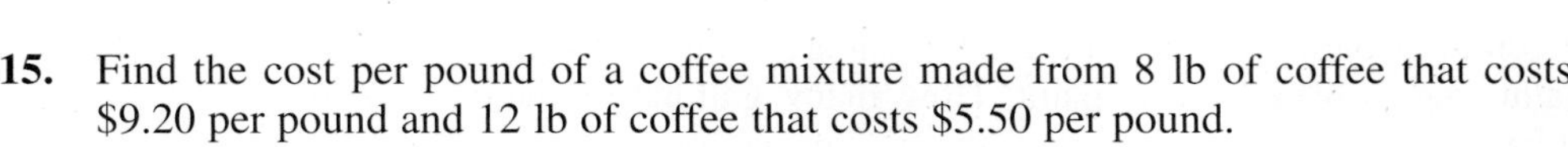

15. Find the cost per pound of a coffee mixture made from 8 lb of coffee that costs \$9.20 per pound and 12 lb of coffee that costs \$5.50 per pound.

16. Adult tickets for a play cost \$6.00, and children's tickets cost \$2.50. For one performance, 370 tickets were sold. Receipts for the performance totaled \$1723. Find the number of adult tickets sold.

17. Tickets for a piano concert sold for \$4.50 for each adult ticket. Student tickets sold for \$2.00 each. The total receipts for 1720 tickets were \$5980. Find the number of adult tickets sold.

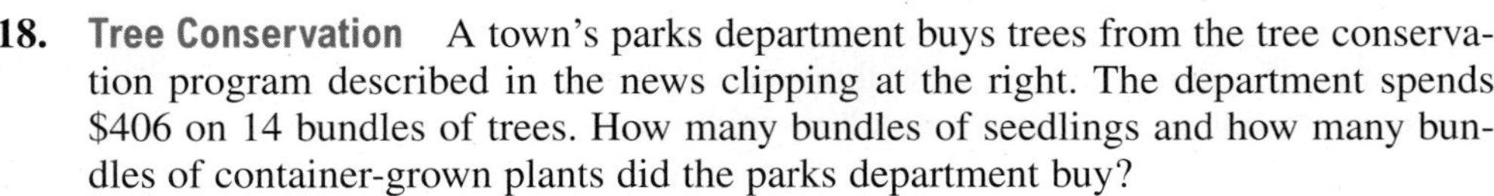

18. **Tree Conservation** A town's parks department buys trees from the tree conservation program described in the news clipping at the right. The department spends \$406 on 14 bundles of trees. How many bundles of seedlings and how many bundles of container-grown plants did the parks department buy?

In the News

Conservation Tree Planting Program Underway

The Kansas Forest Service is again offering its Conservation Tree Planting Program. Trees are sold in bundles of 25, in two sizes—seedlings cost \$17 a bundle and larger container-grown plants cost \$45 a bundle.

Source: Kansas Canopy

OBJECTIVE B To solve percent mixture problems

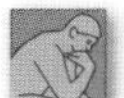

19. True or false? A 10% salt solution can be combined with some amount of a 20% salt solution to create a 30% salt solution.

20. Forty ounces of a 30% gold alloy are mixed with 60 oz of a 20% gold alloy. Find the percent concentration of the resulting gold alloy.

21. One hundred ounces of juice that is 50% tomato juice is added to 200 oz of a vegetable juice that is 25% tomato juice. What is the percent concentration of tomato juice in the resulting mixture?

22. How many gallons of a 15% acid solution must be mixed with 5 gal of a 20% acid solution to make a 16% acid solution?

23. How many pounds of a chicken feed that is 50% corn must be mixed with 400 lb of a feed that is 80% corn to make a chicken feed that is 75% corn?

24. A rug is made by weaving 20 lb of yarn that is 50% wool with a yarn that is 25% wool. How many pounds of the yarn that is 25% wool are used if the finished rug is 35% wool?

25. Five gallons of a light green latex paint that is 20% yellow paint are combined with a darker green latex paint that is 40% yellow paint. How many gallons of the darker green paint must be used to create a green paint that is 25% yellow paint?

26. How many gallons of a plant food that is 9% nitrogen must be combined with another plant food that is 25% nitrogen to make 10 gal of a solution that is 15% nitrogen?

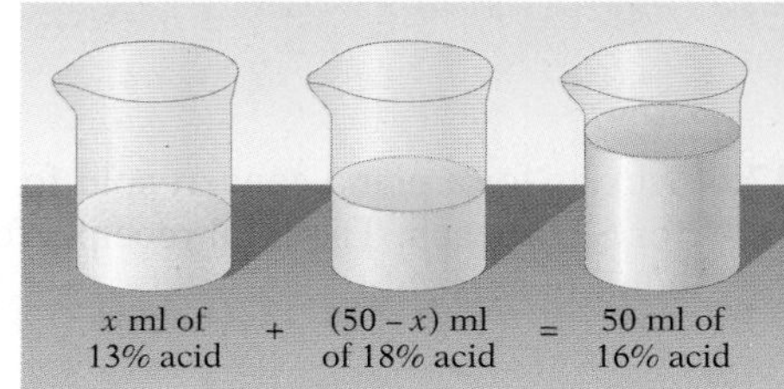

27. A chemist wants to make 50 ml of a 16% acid solution by mixing a 13% acid solution and an 18% acid solution. How many milliliters of each solution should the chemist use?

28. Five grams of sugar are added to a 45-gram serving of a breakfast cereal that is 10% sugar. What is the percent concentration of sugar in the resulting mixture?

29. A goldsmith mixes 8 oz of a 30% gold alloy with 12 oz of a 25% gold alloy. What is the percent concentration of the resulting alloy?

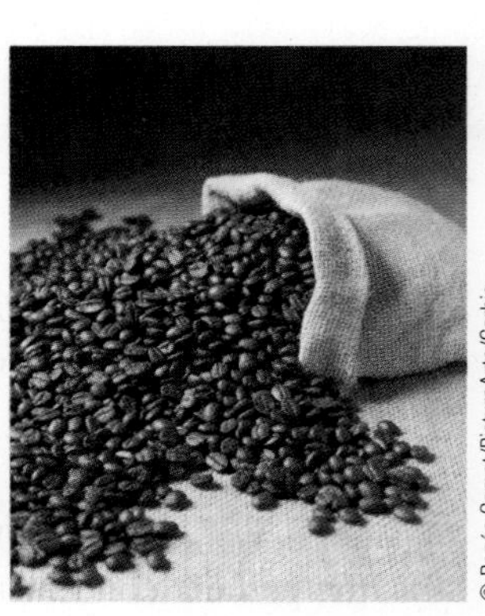

30. How many pounds of coffee that is 40% java beans must be mixed with 80 lb of coffee that is 30% java beans to make a coffee blend that is 32% java beans?

31. The manager of a garden shop mixes grass seed that is 60% rye grass with 70 lb of grass seed that is 80% rye grass to make a mixture that is 74% rye grass. How much of the 60% rye grass is used?

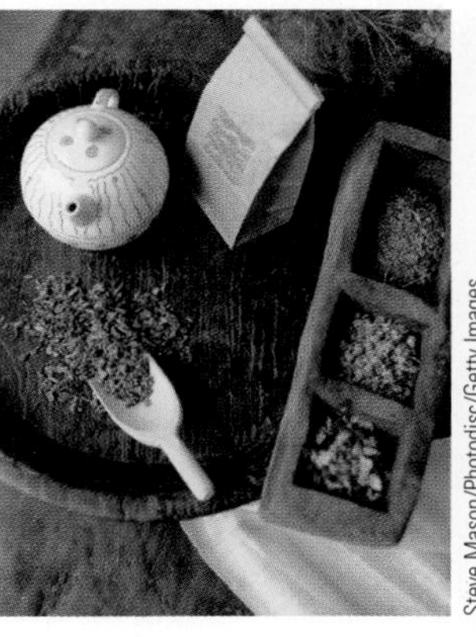
Steve Mason/Photodisc/Getty Images

32. A hair dye is made by blending a 7% hydrogen peroxide solution and a 4% hydrogen peroxide solution. How many milliliters of each are used to make a 300-milliliter solution that is 5% hydrogen peroxide?

33. A tea that is 20% jasmine is blended with a tea that is 15% jasmine. How many pounds of each tea are used to make 5 lb of tea that is 18% jasmine?

34. How many ounces of pure chocolate must be added to 150 oz of chocolate topping that is 50% chocolate to make a topping that is 75% chocolate?

35. How many ounces of pure bran flakes must be added to 50 oz of cereal that is 40% bran flakes to produce a mixture that is 50% bran flakes?

36. A clothing manufacturer has some pure silk thread and some thread that is 85% silk. How many kilograms of each must be woven together to make 75 kg of cloth that is 96% silk?

OBJECTIVE C To solve uniform motion problems

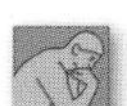

For Exercises 37 and 38, read the problem and state which of the following types of equations you would write to solve the problem.
(i) An equation showing two distances set equal to each other
(ii) An equation showing two distances added together and set equal to a total distance

37. Sam hiked up a mountain at a rate of 2.5 mph and returned along the same trail at a rate of 3 mph. His total hiking time was 11 h. How long was the hiking trail?

38. Sam hiked 16 mi. He hiked at one rate for the first 2 h of his hike, and then decreased his speed by 0.5 mph for the last 3 h of his hike. What was Sam's speed for the first 2 h?

39. Two small planes start from the same point and fly in opposite directions. The first plane is flying 25 mph slower than the second plane. In 2 h, the planes are 470 mi apart. Find the rate of each plane.

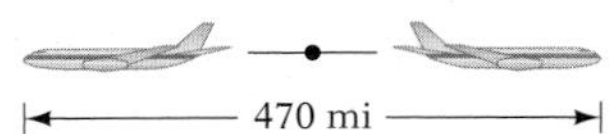

40. Two cyclists start from the same point and ride in opposite directions. One cyclist rides twice as fast as the other. In 3 h, they are 81 mi apart. Find the rate of each cyclist.

41. Two planes leave an airport at 8 A.M., one flying north at 480 km/h and the other flying south at 520 km/h. At what time will they be 3000 km apart?

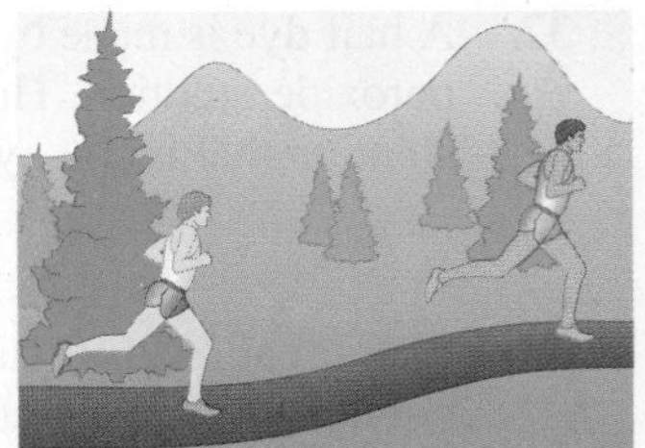

42. A long-distance runner started on a course running at an average speed of 6 mph. One-half hour later, a second runner began the same course at an average speed of 7 mph. How long after the second runner started did the second runner overtake the first runner?

43. A motorboat leaves a harbor and travels at an average speed of 9 mph toward a small island. Two hours later a cabin cruiser leaves the same harbor and travels at an average speed of 18 mph toward the same island. In how many hours after the cabin cruiser leaves the harbor will it be alongside the motorboat?

44. A 555-mile, 5-hour plane trip was flown at two speeds. For the first part of the trip, the average speed was 105 mph. For the remainder of the trip, the average speed was 115 mph. How long did the plane fly at each speed?

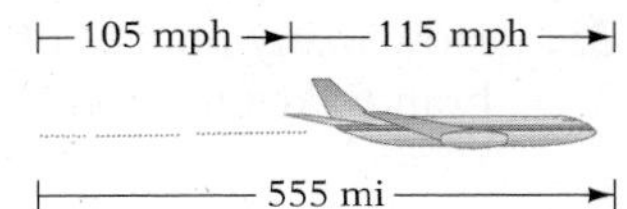

45. An executive drove from home at an average speed of 30 mph to an airport where a helicopter was waiting. The executive boarded the helicopter and flew to the corporate offices at an average speed of 60 mph. The entire distance was 150 mi. The entire trip took 3 h. Find the distance from the airport to the corporate offices.

46. After a sailboat had been on the water for 3 h, a change in the wind direction reduced the average speed of the boat by 5 mph. The entire distance sailed was 57 mi. The total time spent sailing was 6 h. How far did the sailboat travel in the first 3 h?

47. A car and a bus set out at 3 P.M. from the same point headed in the same direction. The average speed of the car is twice the average speed of the bus. In 2 h the car is 68 mi ahead of the bus. Find the rate of the car.

48. A passenger train leaves a train depot 2 h after a freight train leaves the same depot. The freight train is traveling 20 mph slower than the passenger train. Find the rate of each train if the passenger train overtakes the freight train in 3 h.

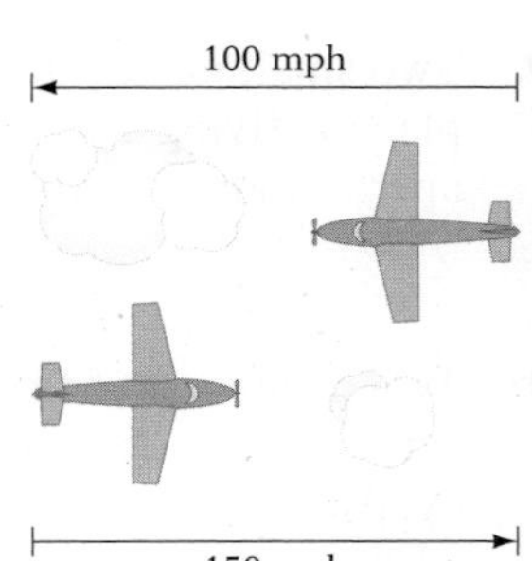

49. As part of flight training, a student pilot was required to fly to an airport and then return. The average speed on the way to the airport was 100 mph, and the average speed returning was 150 mph. Find the distance between the two airports if the total flying time was 5 h.

50. A ship traveling east at 25 mph is 10 mi from a harbor when another ship leaves the harbor traveling east at 35 mph. How long does it take the second ship to catch up to the first ship?

51. At 10 A.M. a plane leaves Boston, Massachusetts, for Seattle, Washington, a distance of 3000 mi. One hour later a plane leaves Seattle for Boston. Both planes are traveling at a speed of 500 mph. How many hours after the plane leaves Seattle will the planes pass each other?

52. At noon a train leaves Washington, D.C., headed for Charleston, South Carolina, a distance of 500 mi. The train travels at a speed of 60 mph. At 1 P.M. a second train leaves Charleston headed for Washington, D.C., traveling at 50 mph. How long after the train leaves Charleston will the two trains pass each other?

53. Two cyclists start at the same time from opposite ends of a course that is 51 mi long. One cyclist is riding at a rate of 16 mph, and the second cyclist is riding at a rate of 18 mph. How long after they begin will they meet?

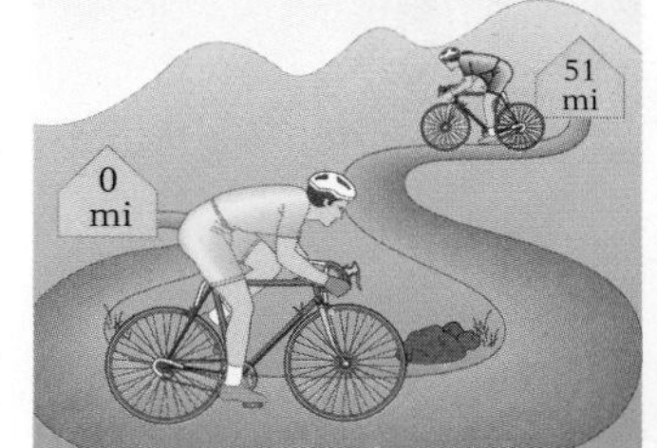

54. A bus traveling at a rate of 60 mph overtakes a car traveling at a rate of 45 mph. If the car had a 1-hour head start, how far from the starting point does the bus overtake the car?

55. A car traveling at 48 mph overtakes a cyclist who, riding at 12 mph, had a 3-hour head start. How far from the starting point does the car overtake the cyclist?

56. **sQuba** See the news clipping at the right. Two sQubas are on opposite sides of a lake 1.6 mi wide. They start toward each other at the same time, one traveling on the surface of the water and the other traveling underwater. In how many minutes after they start will the sQuba on the surface of the water be directly above the sQuba that is underwater? Assume they are traveling at top speed.

> **In the News**
>
> **Underwater Driving—Not So Fast!**
>
> Swiss company Rinspeed, Inc., presented its new car, the sQuba, at the Geneva Auto Show. The sQuba can travel on land, on water, and underwater. With a new sQuba, you can expect top speeds of 77 mph when driving on land, 3 mph when driving on the surface of the water, and 1.8 mph when driving underwater!
>
> *Source: Seattle Times*

Applying the Concepts

57. **Chemistry** How many ounces of water must be evaporated from 50 oz of a 12% salt solution to produce a 15% salt solution?

58. **Transportation** A bicyclist rides for 2 h at a speed of 10 mph and then returns at a speed of 20 mph. Find the cyclist's average speed for the trip.

59. **Travel** A car travels a 1-mile track at an average speed of 30 mph. At what average speed must the car travel the next mile so that the average speed for the 2 mi is 60 mph?

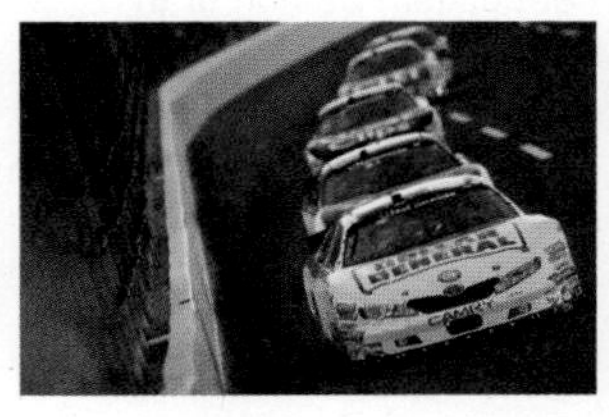

© Jared C. Tilton/Icon SMI/Corbis

SECTION

2.5 First-Degree Inequalities

OBJECTIVE A To solve an inequality in one variable

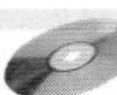

The **solution set of an inequality** is a set of numbers, each element of which, when substituted for the variable, results in a true inequality.

The inequality at the right is true if the variable is replaced by (for instance) 3, -1.98, or $\frac{2}{3}$.

$$x - 1 < 4$$
$$3 - 1 < 4$$
$$-1.98 - 1 < 4$$
$$\frac{2}{3} - 1 < 4$$

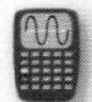

Integrating Technology

See the Keystroke Guide: *Test* for instructions on using a graphing calculator to graph the solution set of an inequality.

There are many values of the variable x that will make the inequality $x - 1 < 4$ true. The solution set of the inequality is any number less than 5. The solution set can be written in set-builder notation as $\{x \mid x < 5\}$.

The graph of the solution set of $x - 1 < 4$ is shown at the right.

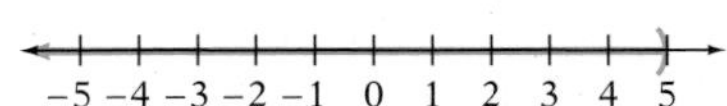

When solving an inequality, we use the **Addition and Multiplication Properties of Inequalities** to rewrite the inequality in the form *variable* $<$ *constant* or in the form *variable* $>$ *constant*.

Addition Property of Inequalities

The same term can be added to each side of an inequality without changing the solution set of the inequality. Symbolically, this is written

If $a < b$, then $a + c < b + c$.

If $a > b$, then $a + c > b + c$.

This property is also true for an inequality that contains $\leq$ or $\geq$.

The Addition Property of Inequalities is used to remove a term from one side of an inequality by adding the additive inverse of that term to each side of the inequality. Because subtraction is defined in terms of addition, the same number can be subtracted from each side of an inequality without changing the solution set of the inequality.

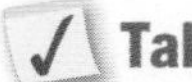

Take Note

The solution set of an inequality can be written in set-builder notation or in interval notation.

HOW TO • 1 Solve and graph the solution set: $x + 2 \geq 4$

$$x + 2 \geq 4$$
$$x + 2 - 2 \geq 4 - 2 \quad \text{• Subtract 2 from each side of the inequality.}$$
$$x \geq 2 \quad \text{• Simplify.}$$

The solution set is $\{x \mid x \geq 2\}$ or $[2, \infty)$.

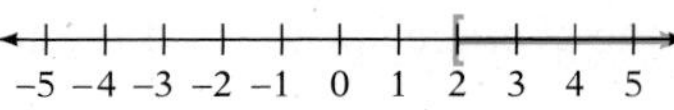

HOW TO • 2 Solve: $3x - 4 < 2x - 1$
Write the solution set in set-builder notation.

$3x - 4 < 2x - 1$

$3x - 4 - 2x < 2x - 1 - 2x$ • **Subtract $2x$** from each side of the inequality.

$x - 4 < -1$

$x - 4 + 4 < -1 + 4$ • **Add 4** to each side of the inequality.

$x < 3$

The solution set is $\{x \mid x < 3\}$.

Care must be taken when multiplying each side of an inequality by a nonzero constant. The rule is different when multiplying each side by a *positive* number than when multiplying each side by a *negative* number.

Take Note
$c > 0$ means c is a positive number. Note that the inequality symbols do not change.

$c < 0$ means c is a negative number. Note that the inequality symbols are reversed.

Multiplication Property of Inequalities

Rule 1 Each side of an inequality can be multiplied by the same *positive* constant without changing the solution set of the inequality. Symbolically, this is written

If $a < b$ and $c > 0$, then $ac < bc$.

If $a > b$ and $c > 0$, then $ac > bc$.

Rule 2 If each side of an inequality is multiplied by the same *negative* constant and the inequality symbol is reversed, then the solution set of the inequality is not changed. Symbolically, this is written

If $a < b$ and $c < 0$, then $ac > bc$.

If $a > b$ and $c < 0$, then $ac < bc$.

This property is also true for an inequality that contains $\le$ or $\ge$.

Here are examples of this property.

Rule 1: Multiply by a *positive* number.		**Rule 2:** Multiply by a *negative* number.	
$2 < 5$	$3 > 2$	$2 < 5$	$3 > 2$
$2(4) < 5(4)$	$3(4) > 2(4)$	$2(-4) > 5(-4)$	$3(-4) < 2(-4)$
$8 < 20$	$12 > 8$	$-8 > -20$	$-12 < -8$

The Multiplication Property of Inequalities is used to remove a coefficient from one side of an inequality by multiplying each side of the inequality by the reciprocal of the coefficient.

Take Note
Each side of the inequality is divided by a negative number; the inequality symbol must be reversed.

HOW TO • 3 Solve: $-3x > 9$
Write the solution set in interval notation.

$-3x > 9$

$\frac{-3x}{-3} < \frac{9}{-3}$ • **Divide** each side of the inequality by the coefficient -3. Because -3 is a negative number, the inequality symbol must be reversed.

$x < -3$

The solution set is $(-\infty, -3)$.

HOW TO • 4 Solve: $3x + 2 < -4$
Write the solution set in set-builder notation.

$$3x + 2 < -4$$
$$3x < -6$$ • Subtract 2 from each side of the inequality.
$$\frac{3x}{3} < \frac{-6}{3}$$ • Divide each side of the inequality by the coefficient **3**.
$$x < -2$$ • Because **3** is a positive number, the inequality symbol remains the same.

The solution set is $\{x \mid x < -2\}$.

HOW TO • 5 Solve: $2x - 9 > 4x + 5$
Write the solution set in set-builder notation.

$$2x - 9 > 4x + 5$$
$$-2x - 9 > 5$$ • Subtract $4x$ from each side of the inequality.
$$-2x > 14$$ • Add 9 to each side of the inequality.
$$\frac{-2x}{-2} < \frac{14}{-2}$$ • Divide each side of the inequality by the coefficient **−2**. Because **−2** is a negative number, reverse the inequality symbol.
$$x < -7$$

The solution set is $\{x \mid x < -7\}$.

HOW TO • 6 Solve: $5(x - 2) \geq 9x - 3(2x - 4)$
Write the solution set in interval notation.

$$5(x - 2) \geq 9x - 3(2x - 4)$$
$$5x - 10 \geq 9x - 6x + 12$$ • Use the Distributive Property to remove parentheses.
$$5x - 10 \geq 3x + 12$$
$$2x - 10 \geq 12$$ • Subtract $3x$ from each side of the inequality.
$$2x \geq 22$$ • Add 10 to each side of the inequality.
$$\frac{2x}{2} \geq \frac{22}{2}$$ • Divide each side of the inequality by the coefficient **2**.
$$x \geq 11$$

The solution set is $[11, \infty)$.

EXAMPLE • 1

Solve and graph the solution set: $\frac{1}{6} - \frac{3}{4}x > \frac{11}{12}$

Write the solution set in set-builder notation.

Solution

$$\frac{1}{6} - \frac{3}{4}x > \frac{11}{12}$$
$$12\left(\frac{1}{6}\right) - 12\left(\frac{3}{4}x\right) > 12\left(\frac{11}{12}\right)$$ • Clear fractions by multiplying each side of the inequality by 12.
$$2 - 9x > 11$$
$$-9x > 9$$ • Subtract 2 from each side.

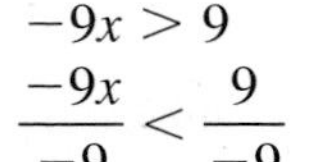

$$\frac{-9x}{-9} < \frac{9}{-9}$$ • Divide each side by −9.
$$x < -1$$

The solution set is $\{x \mid x < -1\}$.

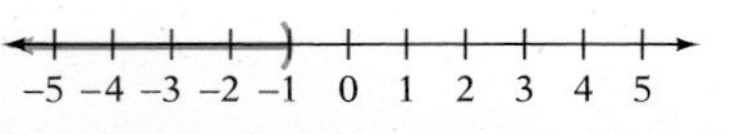

YOU TRY IT • 1

Solve and graph the solution set: $2x - 1 < 6x + 7$
Write the solution set in set-builder notation.

Your solution

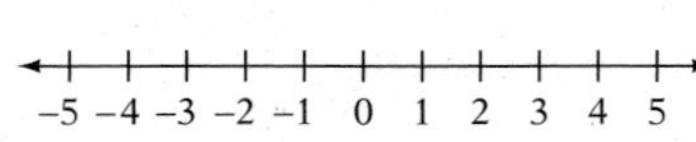

Solution on p. S7

EXAMPLE • 2

Solve: $3x - 5 \leq 3 - 2(3x + 1)$
Write the solution set in interval notation.

Solution

$3x - 5 \leq 3 - 2(3x + 1)$
$3x - 5 \leq 3 - 6x - 2$
$3x - 5 \leq 1 - 6x$
$9x - 5 \leq 1$
$9x \leq 6$
$\frac{9x}{9} \leq \frac{6}{9}$
$x \leq \frac{2}{3}$

$\left(-\infty, \frac{2}{3}\right]$

YOU TRY IT • 2

Solve: $5x - 2 \leq 4 - 3(x - 2)$
Write the solution set in interval notation.

Your solution

Solution on p. S7

OBJECTIVE B To solve a compound inequality 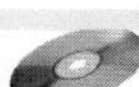

A **compound inequality** is formed by joining two inequalities with a connective word such as *and* or *or*. The inequalities at the right are compound inequalities.

$2x < 4$ and $3x - 2 > -8$

$2x + 3 > 5$ or $x + 2 < 5$

The solution set of a compound inequality with the connective word *and* is the set of all elements that are common to the solution sets of both inequalities. Therefore, it is the intersection of the solution sets of the two inequalities.

HOW TO • 7 Solve: $2x < 6$ and $3x + 2 > -4$

$2x < 6$	and	$3x + 2 > -4$	
$x < 3$		$3x > -6$	• Solve each inequality.
$\{x \mid x < 3\}$		$x > -2$	
		$\{x \mid x > -2\}$	

The solution set of a compound inequality with *and* is the intersection of the solution sets of the two inequalities.

$\{x \mid x < 3\} \cap \{x \mid x > -2\} = \{x \mid -2 < x < 3\}$ or, in interval notation, $(-2, 3)$.

HOW TO • 8 Solve: $-3 < 2x + 1 < 5$

This inequality is equivalent to the compound inequality $-3 < 2x + 1$ and $2x + 1 < 5$.

$-3 < 2x + 1$	and	$2x + 1 < 5$	
$-4 < 2x$		$2x < 4$	• Solve each inequality.
$-2 < x$		$x < 2$	
$\{x \mid x > -2\}$		$\{x \mid x < 2\}$	

The solution set of a compound inequality with *and* is the intersection of the solution sets of the two inequalities.

$\{x \mid x > -2\} \cap \{x \mid x < 2\} = \{x \mid -2 < x < 2\}$ or, in interval notation, $(-2, 2)$.

There is an alternative method for solving the inequality in the last example.

HOW TO • 9 Solve: $-3 < 2x + 1 < 5$

$$-3 < 2x + 1 < 5$$
$$-3 - 1 < 2x + 1 - 1 < 5 - 1$$
$$-4 < 2x < 4$$
$$\frac{-4}{2} < \frac{2x}{2} < \frac{4}{2}$$
$$-2 < x < 2$$

• **Subtract 1** from each of the three parts of the inequality.

• **Divide** each of the three parts of the inequality by the coefficient **2**.

The solution set is $\{x \mid -2 < x < 2\}$ or, in interval notation, $(-2, 2)$.

The solution set of a compound inequality with the connective word *or* is the union of the solution sets of the two inequalities.

HOW TO • 10 Solve: $2x + 3 > 7$ or $4x - 1 < 3$

$2x + 3 > 7$	or	$4x - 1 < 3$
$2x > 4$		$4x < 4$
$x > 2$		$x < 1$
$\{x \mid x > 2\}$		$\{x \mid x < 1\}$

• Solve each inequality.

Find the union of the solution sets.

$\{x \mid x > 2\} \cup \{x \mid x < 1\} = \{x \mid x > 2 \text{ or } x < 1\}$ or, in interval notation, $(-\infty, 1) \cup (2, \infty)$.

EXAMPLE • 3

Solve: $1 < 3x - 5 < 4$
Write the solution set in interval notation.

Solution

$$1 < 3x - 5 < 4$$
$$1 + 5 < 3x - 5 + 5 < 4 + 5$$
$$6 < 3x < 9$$
$$\frac{6}{3} < \frac{3x}{3} < \frac{9}{3}$$
$$2 < x < 3$$

• **Add 5** to each of the three parts.

• **Divide** each of the three parts by **3**.

$(2, 3)$

YOU TRY IT • 3

Solve: $-2 \le 5x + 3 \le 13$
Write the solution set in interval notation.

Your solution

EXAMPLE • 4

Solve: $11 - 2x > -3$ and $7 - 3x < 4$
Write the solution set in set-builder notation.

Solution

$11 - 2x > -3$	and	$7 - 3x < 4$
$-2x > -14$		$-3x < -3$
$x < 7$		$x > 1$
$\{x \mid x < 7\}$		$\{x \mid x > 1\}$

$\{x \mid x < 7\} \cap \{x \mid x > 1\} = \{x \mid 1 < x < 7\}$

YOU TRY IT • 4

Solve: $2 - 3x > 11$ or $5 + 2x > 7$
Write the solution set in set-builder notation.

Your solution

Solutions on p. S7

OBJECTIVE C To solve application problems 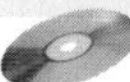

EXAMPLE • 5

A cellular phone company advertises two pricing plans. The first costs \$19.95 per month with 20 free minutes and \$.39 per minute thereafter. The second costs \$23.95 per month with 20 free minutes and \$.30 per minute thereafter. How many minutes can you talk per month for the first plan to cost less than the second?

Strategy

To find the number of minutes, write and solve an inequality using N to represent the number of minutes. Then $N - 20$ is the number of minutes for which you are charged after the first free 20 min.

Solution

$$\begin{aligned} \text{Cost of first plan} &< \text{cost of second plan} \\ 19.95 + 0.39(N - 20) &< 23.95 + 0.30(N - 20) \\ 19.95 + 0.39N - 7.8 &< 23.95 + 0.30N - 6 \\ 12.15 + 0.39N &< 17.95 + 0.30N \\ 12.15 + 0.09N &< 17.95 \\ 0.09N &< 5.8 \\ N &< 64.\overline{4} \end{aligned}$$

The first plan costs less if you talk less than 65 min.

YOU TRY IT • 5

The base of a triangle is 12 in., and the height is $(x + 2)$ in. Express as an integer the maximum height of the triangle when the area is less than 50 in^2.

Your strategy

Your solution

EXAMPLE • 6

Find three consecutive positive odd integers whose sum is between 27 and 51.

Strategy

To find the three integers, write and solve a compound inequality using n to represent the first odd integer.

Solution

$$\text{Lower limit of the sum} < \text{sum} < \text{upper limit of the sum}$$

$$27 < n + (n + 2) + (n + 4) < 51$$

$$27 < 3n + 6 < 51$$

$$27 - 6 < 3n + 6 - 6 < 51 - 6$$

$$21 < 3n < 45$$

$$\frac{21}{3} < \frac{3n}{3} < \frac{45}{3}$$

$$7 < n < 15$$

The three odd integers are 9, 11, and 13; or 11, 13, and 15; or 13, 15, and 17.

YOU TRY IT • 6

An average score of 80 to 89 in a history course receives a B. Luisa Montez has grades of 72, 94, 83, and 70 on four exams. Find the range of scores on the fifth exam that will give Luisa a B for the course.

Your strategy

Your solution

Solutions on p. S8

2.5 EXERCISES

OBJECTIVE A To solve an inequality in one variable

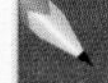

1. State the Addition Property of Inequalities, and give numerical examples of its use.

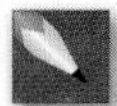

2. State the Multiplication Property of Inequalities, and give numerical examples of its use.

3. Which numbers are solutions of the inequality $x + 7 \leq -3$?
(i) -17 (ii) 8 (iii) -10 (iv) 0

4. Which numbers are solutions of the inequality $2x - 1 > 5$?
(i) 6 (ii) -4 (iii) 3 (iv) 5

For Exercises 5 to 31, solve. Write the solution in set-builder notation. For Exercises 5 to 10, graph the solution set.

5. $x - 3 < 2$

−5 −4 −3 −2 −1 0 1 2 3 4 5

6. $x + 4 \geq 2$

−5 −4 −3 −2 −1 0 1 2 3 4 5

7. $4x \leq 8$

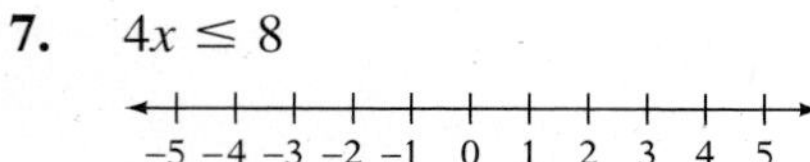

8. $6x > 12$

−5 −4 −3 −2 −1 0 1 2 3 4 5

9. $-2x > 8$

−5 −4 −3 −2 −1 0 1 2 3 4 5

10. $-3x \leq -9$

−5 −4 −3 −2 −1 0 1 2 3 4 5

11. $3x - 1 > 2x + 2$

12. $5x + 2 \geq 4x - 1$

13. $2x - 1 > 7$

14. $3x + 2 < 8$

15. $5x - 2 \leq 8$

16. $4x + 3 \leq -1$

17. $6x + 3 > 4x - 1$

18. $7x + 4 < 2x - 6$

19. $8x + 1 \geq 2x + 13$

20. $5x - 4 < 2x + 5$

21. $4 - 3x < 10$

22. $2 - 5x > 7$

23. $7 - 2x \geq 1$

24. $3 - 5x \leq 18$

25. $-3 - 4x > -11$

26. $-2 - x < 7$

27. $4x - 2 < x - 11$

28. $6x + 5 \leq x - 10$

29. $x + 7 \geq 4x - 8$

30. $3x + 1 \leq 7x - 15$

31. $3x + 2 \leq 7x + 4$

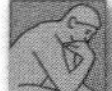

For Exercises 32 to 35, state whether the solution set of an inequality of the given form contains only negative numbers, only positive numbers, or both positive and negative numbers.

32. $x + n > a$, where both n and a are positive, and $n < a$

33. $nx > a$, where both n and a are negative

34. $nx > a$, where n is negative and a is positive

35. $x - n > -a$, where both n and a are positive, and $n < a$

For Exercises 36 to 51, solve. Write the solution in interval notation.

36. $3x - 5 \geq -2x + 5$

37. $\frac{3}{5}x - 2 < \frac{3}{10} - x$

38. $\frac{5}{6}x - \frac{1}{6} < x - 4$

39. $\frac{2}{3}x - \frac{3}{2} < \frac{7}{6} - \frac{1}{3}x$

40. $\frac{7}{12}x - \frac{3}{2} < \frac{2}{3}x + \frac{5}{6}$

41. $\frac{1}{2}x - \frac{3}{4} < \frac{7}{4}x - 2$

42. $6 - 2(x - 4) \leq 2x + 10$

43. $4(2x - 1) > 3x - 2(3x - 5)$

44. $2(1 - 3x) - 4 > 10 + 3(1 - x)$

45. $2 - 5(x + 1) \geq 3(x - 1) - 8$

46. $2 - 2(7 - 2x) < 3(3 - x)$

47. $3 + 2(x + 5) \geq x + 5(x + 1) + 1$

48. $10 - 13(2 - x) < 5(3x - 2)$

49. $3 - 4(x + 2) \leq 6 + 4(2x + 1)$

50. $3x - 2(3x - 5) \leq 2 - 5(x - 4)$

51. $12 - 2(3x - 2) \geq 5x - 2(5 - x)$

OBJECTIVE B To solve a compound inequality

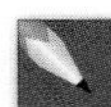

52. **a.** Which set operation is used when a compound inequality is combined with *or*?
b. Which set operation is used when a compound inequality is combined with *and*?

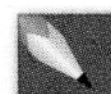

53. Explain why the inequality $-3 > x > 4$ does not make sense.

For Exercises 54 to 67, solve. Write the solution set in interval notation.

54. $3x < 6$ and $x + 2 > 1$

55. $x - 3 \leq 1$ and $2x \geq -4$

56. $x + 2 \geq 5$ or $3x \leq 3$

57. $2x < 6$ or $x - 4 > 1$

58. $-2x > -8$ and $-3x < 6$

59. $\frac{1}{2}x > -2$ and $5x < 10$

60. $\frac{1}{3}x < -1$ or $2x > 0$

61. $\frac{2}{3}x > 4$ or $2x < -8$

62. $x + 4 \geq 5$ and $2x \geq 6$

63. $3x < -9$ and $x - 2 < 2$

64. $-5x > 10$ and $x + 1 > 6$

65. $2x - 3 > 1$ and $3x - 1 < 2$

66. $7x < 14$ and $1 - x < 4$

67. $4x + 1 < 5$ and $4x + 7 > -1$

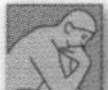

For Exercises 68 to 71, state whether the inequality describes the empty set, all real numbers, two intervals of real numbers, or one interval of real numbers.

68. $x > -3$ and $x > 2$

69. $x > -3$ or $x < 2$

70. $x < -3$ and $x > 2$

71. $x < -3$ or $x > 2$

For Exercises 72 to 91, solve. Write the solution set in set-builder notation.

72. $3x + 7 < 10$ or $2x - 1 > 5$

73. $6x - 2 < -14$ or $5x + 1 > 11$

74. $-5 < 3x + 4 < 16$

75. $5 < 4x - 3 < 21$

76. $0 < 2x - 6 < 4$

77. $-2 < 3x + 7 < 1$

78. $4x - 1 > 11$ or $4x - 1 \leq -11$

79. $3x - 5 > 10$ or $3x - 5 < -10$

80. $9x - 2 < 7$ and $3x - 5 > 10$

81. $8x + 2 \leq -14$ and $4x - 2 > 10$

82. $3x - 11 < 4$ or $4x + 9 \geq 1$

83. $5x + 12 \geq 2$ or $7x - 1 \leq 13$

84. $-6 \le 5x + 14 \le 24$

85. $3 \le 7x - 14 \le 31$

86. $3 - 2x > 7$ and $5x + 2 > -18$

87. $1 - 3x < 16$ and $1 - 3x > -16$

88. $5 - 4x > 21$ or $7x - 2 > 19$

89. $6x + 5 < -1$ or $1 - 2x < 7$

90. $3 - 7x \le 31$ and $5 - 4x > 1$

91. $9 - x \ge 7$ and $9 - 2x < 3$

OBJECTIVE C To solve application problems

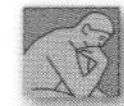

Exercises 92 to 95 make statements about temperatures t on a particular day. Match each statement with one of the following inequalities. Some inequalities may be used more than once.

$t > 21$	$t < 21$	$t \ge 21$	$t \le 21$	$21 \le t \le 42$
$t < 42$	$t > 42$	$t \le 42$	$t \ge 42$	$21 < t < 42$

92. The low temperature was 21°F.

93. The temperature did not go above 42°F.

94. The temperature ranged from 21°F to 42°F.

95. The high temperature was 42°F.

96. Integers Five times the difference between a number and two is greater than the quotient of two times the number and three. Find the smallest integer that will satisfy the inequality.

97. Integers Two times the difference between a number and eight is less than or equal to five times the sum of the number and four. Find the smallest number that will satisfy the inequality.

98. Geometry The length of a rectangle is 2 ft more than four times the width. Express as an integer the maximum width of the rectangle when the perimeter is less than 34 ft.

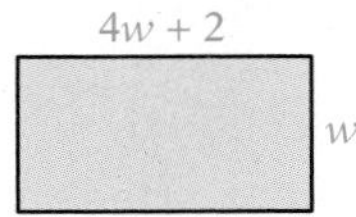

99. Geometry The length of a rectangle is 5 cm less than twice the width. Express as an integer the maximum width of the rectangle when the perimeter is less than 60 cm.

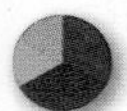

100. **Aquariums** The following is a rule-of-thumb for making sure fish kept in an aquarium are not too crowded: The surface area of the water should be at least 12 times the total length of fish kept in the aquarium. (*Source*: www.tacomapet.com) Your 10-gallon aquarium has a water surface area of 288 in^2 and houses the following fish: one 2-inch odessa barb, three 1-inch gold tetra, three 1.75-inch cobra guppies, and five 1-inch neon tetra.

    a. Find the total length of all the fish in your aquarium.

    b. Write and solve an inequality to find the greatest number n of 2-inch black hatchetfish that you can safely add to your aquarium without overcrowding the fish.

101. **Advertising** To run an advertisement on a certain website, the website owner charges a setup fee of \$250 and \$12 per day to display the advertisement. If a marketing group has a budget of \$1500 for an advertisement, what is the maximum number of days the advertisement can run on the site?

102. **Consumerism** The entry fee to a state fair is \$25 and includes five tickets for carnival rides at the fair. Additional tickets for carnival rides cost \$1.50 each. If Alisha wants to spend a maximum of \$45 for the entry fee and rides, how many additional carnival ride tickets can she purchase?

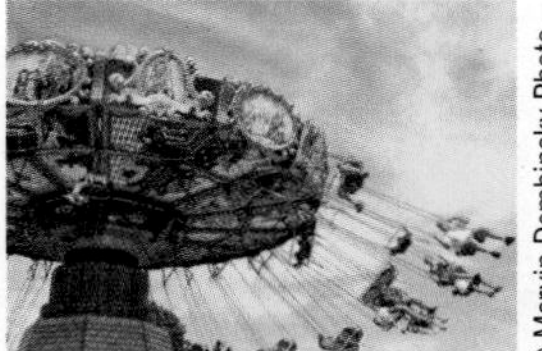

103. **Consumerism** A homeowner has a budget of \$100 to paint a room that has 320 ft^2 of wall space. Drop cloths, masking tape, and paint brushes cost \$24. If 1 gal of paint will cover 100 ft^2 of wall space, what is the maximum cost per gallon of paint that the homeowner can pay?

104. **Temperature** The temperature range for a week was between 14°F and 77°F. Find the temperature range in degrees Celsius. $F = \frac{9}{5}C + 32$

105. **Temperature** The temperature range for a week in a mountain town was between 0°C and 30°C. Find the temperature range in degrees Fahrenheit. $C = \frac{5(F - 32)}{9}$

106. **Compensation** You are a sales account executive earning \$1200 per month plus 6% commission on the amount of sales. Your goal is to earn a minimum of \$6000 per month. What amount of sales will enable you to earn \$6000 or more per month?

107. **Compensation** George Stoia earns \$1000 per month plus 5% commission on the amount of sales. George's goal is to earn a minimum of \$3200 per month. What amount of sales will enable George to earn \$3200 or more per month?

108. **Banking** Heritage National Bank offers two different checking accounts. The first charges \$3 per month and \$.50 per check after the first 10 checks. The second account charges \$8 per month with unlimited check writing. How many checks can be written per month if the first account is to be less expensive than the second account?

109. **Banking** Glendale Federal Bank offers a checking account to small businesses. The charge is \$8 per month plus \$.12 per check after the first 100 checks. A competitor is offering an account for \$5 per month plus \$.15 per check after the first 100 checks. If a business chooses the first account, how many checks does the business write monthly if it is assumed that the first account will cost less than the competitor's account?

110. **Education** An average score of 90 or above in a history class receives an A grade. You have scores of 95, 89, and 81 on three exams. Find the range of scores on the fourth exam that will give you an A grade for the course.

111. **Education** An average of 70 to 79 in a mathematics class receives a C grade. A student has scores of 56, 91, 83, and 62 on four tests. Find the range of scores on the fifth test that will give the student a C for the course.

112. **Integers** Find four consecutive integers whose sum is between 62 and 78.

113. **Integers** Find three consecutive even integers whose sum is between 30 and 51.

Applying the Concepts

114. Let $-2 \le x \le 3$ and $a \le 2x + 1 \le b$.
a. Find the largest possible value of a.
b. Find the smallest possible value of b.

115. Determine whether the following statements are always true, sometimes true, or never true.
a. If $a > b$, then $-a < -b$.
b. If $a < b$ and $a \ne 0$, $b \ne 0$, then $\frac{1}{a} < \frac{1}{b}$.
c. When dividing both sides of an inequality by an integer, we must reverse the inequality symbol.
d. If $a < 1$, then $a^2 < a$.
e. If $a < b < 0$ and $c < d < 0$, then $ac > bd$.

116. The following is offered as the solution of $2 + 3(2x - 4) < 6x + 5$.

$$2 + 3(2x - 4) < 6x + 5$$
$$2 + 6x - 12 < 6x + 5 \quad \text{• Use the Distributive Property.}$$
$$6x - 10 < 6x + 5 \quad \text{• Simplify.}$$
$$6x - 6x - 10 < 6x - 6x + 5 \quad \text{• Subtract } 6x \text{ from each side.}$$
$$-10 < 5$$

Because $-10 < 5$ is a true inequality, the solution set is all real numbers.

If this result is correct, so state. If it is not correct, explain the incorrect step and supply the correct answer.

SECTION

2.6 Absolute Value Equations and Inequalities

OBJECTIVE A To solve an absolute value equation 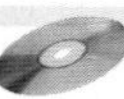

Tips for Success
Before the class meeting in which your professor begins a new section, you should read each objective statement for that section. Next, browse through the objective material. The purpose of browsing through the material is to set the stage for your brain to accept and organize new information when it is presented to you. See *AIM for Success* in the Preface.

The *absolute value* of a number is its distance from zero on the number line. Distance is always a positive number or zero. Therefore, the absolute value of a number is always a positive number or zero.

The distance from 0 to 3 or from 0 to -3 is 3 units.

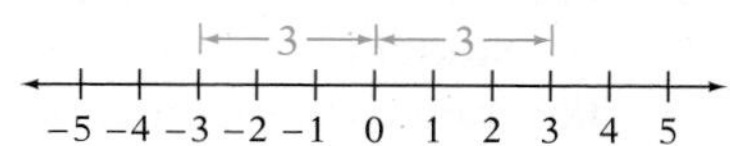

$|3| = 3$ $\qquad |-3| = 3$

Absolute value can be used to represent the distance between any two points on the number line. The **distance between two points** on the number line is the absolute value of the difference between the coordinates of the two points.

The distance between point a and point b is given by $|b - a|$.

The distance between 4 and -3 on the number line is 7 units. Note that the order in which the coordinates are subtracted does not affect the distance.

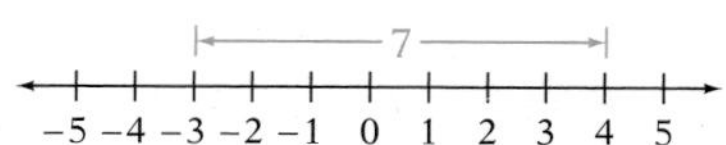

$$\text{Distance} = |-3 - 4| = |-7| = 7 \qquad \text{Distance} = |4 - (-3)| = |7| = 7$$

For any two numbers a and b, $|b - a| = |a - b|$.

An equation containing an absolute value symbol is called an **absolute value equation.** Here are three examples.

$$|x| = 3 \qquad |x + 2| = 8 \qquad |3x - 4| = 5x - 9$$

Solutions of an Absolute Value Equation
If $a \geq 0$ and $|x| = a$, then $x = a$ or $x = -a$.

For instance, given $|x| = 3$, then $x = 3$ or $x = -3$ because $|3| = 3$ and $|-3| = 3$. We can solve this equation as follows:

$|x| = 3$
$x = 3 \qquad x = -3$ • Remove the absolute value sign from $|x|$ and let x equal **3** and **the opposite of 3.**

Check:

$\lvert x\rvert = 3$		$\lvert x\rvert = 3$	
$\lvert 3\rvert$	3	$\lvert -3\rvert$	3
$3 = 3$		$3 = 3$	

The solutions are 3 and -3.

HOW TO • 1 Solve: $|x + 2| = 8$

$$|x + 2| = 8$$

$x + 2 = 8 \qquad x + 2 = -8$

$x = 6 \qquad x = -10$

• Remove the absolute value sign and rewrite as two equations.
• Solve each equation.

Check:

$\|x + 2\| = 8$	
$\|6 + 2\|$	8
$\|8\|$	8
$8 = 8$	

$\|x + 2\| = 8$	
$\|-10 + 2\|$	8
$\|-8\|$	8
$8 = 8$	

The solutions are 6 and -10.

HOW TO • 2 Solve: $|5 - 3x| - 8 = -4$

$$|5 - 3x| - 8 = -4$$

$$|5 - 3x| = 4$$

$5 - 3x = 4 \qquad 5 - 3x = -4$

$-3x = -1 \qquad -3x = -9$

$x = \frac{1}{3} \qquad x = 3$

• Solve for the absolute value.
• Remove the absolute value sign and rewrite as two equations.
• Solve each equation.

Check:

$\|5 - 3x\| - 8 = -4$	
$\left\|5 - 3\left(\frac{1}{3}\right)\right\| - 8$	-4
$\|5 - 1\| - 8$	-4
$4 - 8$	-4
$-4 = -4$	

$\|5 - 3x\| - 8 = -4$	
$\|5 - 3(3)\| - 8$	-4
$\|5 - 9\| - 8$	-4
$4 - 8$	-4
$-4 = -4$	

The solutions are $\frac{1}{3}$ and 3.

EXAMPLE • 1

Solve: $|2 - x| = 12$

Solution

$$|2 - x| = 12$$

$2 - x = 12 \qquad 2 - x = -12$

$-x = 10 \qquad -x = -14$ • Subtract 2.

$x = -10 \qquad x = 14$ • Multiply by -1.

The solutions are -10 and 14.

YOU TRY IT • 1

Solve: $|2x - 3| = 5$

Your solution

EXAMPLE • 2

Solve: $|2x| = -4$

Solution

$|2x| = -4$

There is no solution to this equation because the absolute value of a number must be nonnegative.

YOU TRY IT • 2

Solve: $|x - 3| = -2$

Your solution

Solutions on p. S8

EXAMPLE 3

Solve: $3 - |2x - 4| = -5$

Solution

$$3 - |2x - 4| = -5$$
$$-|2x - 4| = -8 \quad \text{• Subtract 3.}$$
$$|2x - 4| = 8 \quad \text{• Multiply by } -1.$$

$$2x - 4 = 8 \qquad 2x - 4 = -8$$
$$2x = 12 \qquad 2x = -4$$
$$x = 6 \qquad x = -2$$

The solutions are 6 and -2.

YOU TRY IT 3

Solve: $5 - |3x + 5| = 3$

Your solution

Solution on p. S8

OBJECTIVE B To solve an absolute value inequality

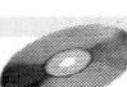

Recall that absolute value represents the distance between two points. For example, the solutions of the absolute value equation $|x - 1| = 3$ are the numbers whose distance from 1 is 3. Therefore, the solutions are -2 and 4. An **absolute value inequality** is an inequality that contains an absolute value symbol.

The solutions of the absolute value inequality $|x - 1| < 3$ are the numbers whose distance from 1 is less than 3. Therefore, the solutions are the numbers greater than -2 and less than 4. The solution set is $\{x|-2 < x < 4\}$.

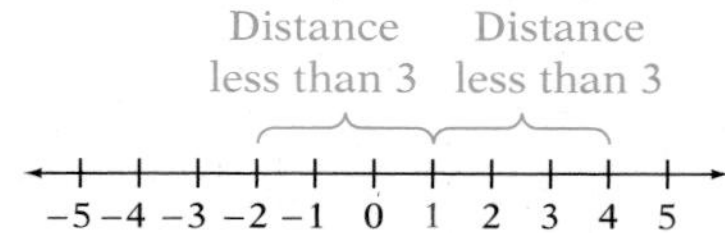

To solve an absolute value inequality of the form $|ax + b| < c$, solve the equivalent compound inequality $-c < ax + b < c$.

HOW TO 3 Solve: $|3x - 1| < 5$

$$|3x - 1| < 5$$
$$-5 < 3x - 1 < 5 \quad \text{• Solve the equivalent compound inequality.}$$
$$-5 + 1 < 3x - 1 + 1 < 5 + 1$$
$$-4 < 3x < 6$$
$$\frac{-4}{3} < \frac{3x}{3} < \frac{6}{3}$$
$$-\frac{4}{3} < x < 2$$

The solution set is $\left\{x \,\middle|\, -\frac{4}{3} < x < 2\right\}$.

Take Note
In this objective, we will write all solution sets in set-builder notation.

The solutions of the absolute value inequality $|x + 1| > 2$ are the numbers whose distance from -1 is greater than 2. Therefore, the solutions are the numbers that are less than -3 or greater than 1. The solution set of $|x + 1| > 2$ is $\{x|x < -3 \text{ or } x > 1\}$.

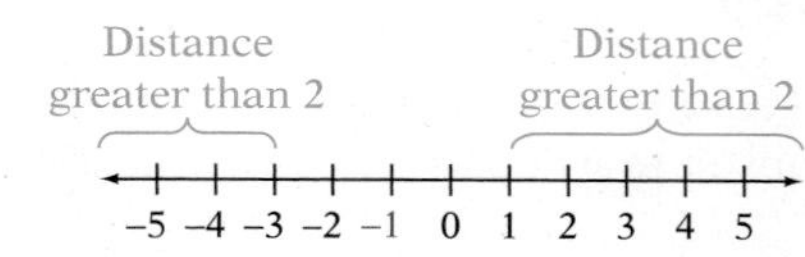

Take Note
Carefully observe the difference between the solution method of $|ax + b| > c$ shown here and that of $|ax + b| < c$ shown on the preceding page.

To solve an absolute value inequality of the form $|ax + b| > c$, solve the equivalent compound inequality $ax + b < -c$ or $ax + b > c$.

HOW TO • 4 Solve: $|3 - 2x| > 1$

$$\begin{array}{lcl} 3 - 2x < -1 & \text{or} & 3 - 2x > 1 \\ -2x < -4 & & -2x > -2 \\ x > 2 & & x < 1 \\ \{x \mid x > 2\} & & \{x \mid x < 1\} \end{array}$$

• Solve each inequality.

The solution set of a compound inequality with *or* is the union of the solution sets of the two inequalities.

$\{x \mid x > 2\} \cup \{x \mid x < 1\} = \{x \mid x > 2 \text{ or } x < 1\}$

The rules for solving absolute value inequalities are summarized below.

Solutions of Absolute Value Inequalities

To solve an absolute value inequality of the form $|ax + b| < c$, $c > 0$, solve the equivalent compound inequality $-c < ax + b < c$.

To solve an absolute value inequality of the form $|ax + b| > c$, solve the equivalent compound inequality $ax + b < -c$ or $ax + b > c$.

EXAMPLE • 4

Solve: $|4x - 3| < 5$

Solution
Solve the equivalent compound inequality.

$$-5 < 4x - 3 < 5$$
$$-5 + 3 < 4x - 3 + 3 < 5 + 3$$
$$-2 < 4x < 8$$
$$\frac{-2}{4} < \frac{4x}{4} < \frac{8}{4}$$
$$-\frac{1}{2} < x < 2$$

$\left\{x \,\middle|\, -\frac{1}{2} < x < 2\right\}$

YOU TRY IT • 4

Solve: $|3x + 2| < 8$

Your solution

EXAMPLE • 5

Solve: $|x - 3| < 0$

Solution
The absolute value of a number is greater than or equal to zero, since it measures the number's distance from zero on the number line. Therefore, the solution set of $|x - 3| < 0$ is the empty set.

YOU TRY IT • 5

Solve: $|3x - 7| < 0$

Your solution

Solutions on p. S8

EXAMPLE 6

Solve: $|x + 4| > -2$

Solution

The absolute value of a number is greater than or equal to zero. Therefore, the solution set of $|x + 4| > -2$ is the set of real numbers.

YOU TRY IT 6

Solve: $|2x + 7| \geq -1$

Your solution

EXAMPLE 7

Solve: $|2x - 1| > 7$

Solution

Solve the equivalent compound inequality.

$$\begin{aligned} 2x - 1 < -7 \quad &\text{or} \quad 2x - 1 > 7 \\ 2x < -6 \quad &\qquad\quad 2x > 8 \\ x < -3 \quad &\qquad\quad x > 4 \\ \{x \mid x < -3\} \quad &\qquad \{x \mid x > 4\} \end{aligned}$$

$$\{x \mid x < -3\} \cup \{x \mid x > 4\}$$
$$= \{x \mid x < -3 \text{ or } x > 4\}$$

YOU TRY IT 7

Solve: $|5x + 3| > 8$

Your solution

Solutions on pp. S8–S9

OBJECTIVE C To solve application problems 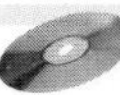

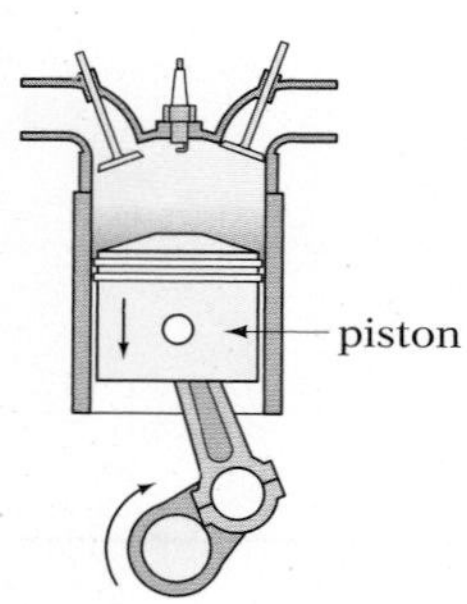

The **tolerance** of a component, or part, is the amount by which it is acceptable for the component to vary from a given measurement. For example, the diameter of a piston may vary from the given measurement of 9 cm by 0.001 cm. This is written 9 cm ± 0.001 cm and is read "9 centimeters plus or minus 0.001 centimeter." The maximum diameter, or **upper limit,** of the piston is 9 cm + 0.001 cm = 9.001 cm. The minimum diameter, or **lower limit,** is 9 cm − 0.001 cm = 8.999 cm.

The lower and upper limits of the diameter of the piston could also be found by solving the absolute value inequality $|d - 9| \leq 0.001$, where d is the diameter of the piston.

$$|d - 9| \leq 0.001$$
$$-0.001 \leq d - 9 \leq 0.001$$
$$-0.001 + 9 \leq d - 9 + 9 \leq 0.001 + 9$$
$$8.999 \leq d \leq 9.001$$

The lower and upper limits of the diameter of the piston are 8.999 cm and 9.001 cm.

EXAMPLE 8

The diameter of a piston for an automobile is $3\frac{5}{16}$ in., with a tolerance of $\frac{1}{64}$ in. Find the lower and upper limits of the diameter of the piston.

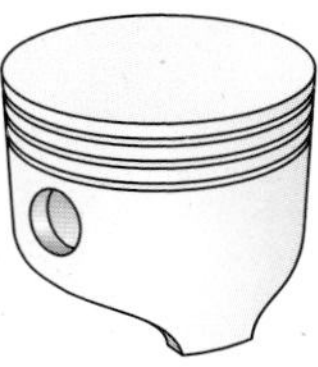

Strategy

To find the lower and upper limits of the diameter of the piston, let d represent the diameter of the piston, T the tolerance, and L the lower and upper limits of the diameter. Solve the absolute value inequality $|L - d| \le T$ for L.

Solution

$$|L - d| \le T$$

$$\left|L - 3\frac{5}{16}\right| \le \frac{1}{64}$$

$$-\frac{1}{64} \le L - 3\frac{5}{16} \le \frac{1}{64}$$

$$-\frac{1}{64} + 3\frac{5}{16} \le L - 3\frac{5}{16} + 3\frac{5}{16} \le \frac{1}{64} + 3\frac{5}{16}$$

$$3\frac{19}{64} \le L \le 3\frac{21}{64}$$

The lower and upper limits of the diameter of the piston are $3\frac{19}{64}$ in. and $3\frac{21}{64}$ in.

YOU TRY IT 8

A machinist must make a bushing that has a tolerance of 0.003 in. The diameter of the bushing is 2.55 in. Find the lower and upper limits of the diameter of the bushing.

Your strategy

Your solution

Solution on p. S9

2.6 EXERCISES

OBJECTIVE A **To solve an absolute value equation**

1. Is 2 a solution of $|x - 8| = 6$?

2. Is -2 a solution of $|2x - 5| = 9$?

3. Is -1 a solution of $|3x - 4| = 7$?

4. Is 1 a solution of $|6x - 1| = -5$?

For Exercises 5 to 64, solve.

5. $|x| = 7$

6. $|a| = 2$

7. $|b| = 4$

8. $|c| = 12$

9. $|-y| = 6$

10. $|-t| = 3$

11. $|-a| = 7$

12. $|-x| = 3$

13. $|x| = -4$

14. $|y| = -3$

15. $|-t| = -3$

16. $|-y| = -2$

17. $|x + 2| = 3$

18. $|x + 5| = 2$

19. $|y - 5| = 3$

20. $|y - 8| = 4$

21. $|a - 2| = 0$

22. $|a + 7| = 0$

23. $|x - 2| = -4$

24. $|x + 8| = -2$

25. $|3 - 4x| = 9$

26. $|2 - 5x| = 3$

27. $|2x - 3| = 0$

28. $|5x + 5| = 0$

29. $|3x - 2| = -4$

30. $|2x + 5| = -2$

31. $|x - 2| - 2 = 3$

32. $|x - 9| - 3 = 2$

33. $|3a + 2| - 4 = 4$

34. $|2a + 9| + 4 = 5$

35. $|2 - y| + 3 = 4$

36. $|8 - y| - 3 = 1$

37. $|2x - 3| + 3 = 3$

38. $|4x - 7| - 5 = -5$

39. $|2x - 3| + 4 = -4$

40. $|3x - 2| + 1 = -1$

41. $|6x - 5| - 2 = 4$

42. $|4b + 3| - 2 = 7$

43. $|3t + 2| + 3 = 4$

44. $|5x - 2| + 5 = 7$

45. $3 - |x - 4| = 5$

46. $2 - |x - 5| = 4$

47. $8 - |2x - 3| = 5$

48. $8 - |3x + 2| = 3$

49. $|2 - 3x| + 7 = 2$

50. $|1 - 5a| + 2 = 3$

51. $|8 - 3x| - 3 = 2$

52. $|6 - 5b| - 4 = 3$

53. $|2x - 8| + 12 = 2$

54. $|3x - 4| + 8 = 3$

55. $2 + |3x - 4| = 5$

56. $5 + |2x + 1| = 8$

57. $5 - |2x + 1| = 5$

58. $3 - |5x + 3| = 3$

59. $6 - |2x + 4| = 3$

60. $8 - |3x - 2| = 5$

61. $8 - |1 - 3x| = -1$

62. $3 - |3 - 5x| = -2$

63. $5 + |2 - x| = 3$

64. $6 + |3 - 2x| = 2$

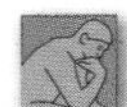

For Exercises 65 to 68, assume that a and b are positive numbers such that $a < b$. State whether the given equation has no solution, two negative solutions, two positive solutions, or one positive and one negative solution.

65. $|x - b| = a$

66. $|x - b| = -a$

67. $|x + b| = a$

68. $|x + a| = b$

OBJECTIVE B To solve an absolute value inequality

For Exercises 69 to 98, solve.

69. $|x| > 3$

70. $|x| < 5$

71. $|x + 1| > 2$

72. $|x - 2| > 1$

73. $|x - 5| \leq 1$

74. $|x - 4| \leq 3$

75. $|2 - x| \geq 3$

76. $|3 - x| \geq 2$

77. $|2x + 1| < 5$

78. $|3x - 2| < 4$

79. $|5x + 2| > 12$

80. $|7x - 1| > 13$

81. $|4x - 3| \leq -2$

82. $|5x + 1| \leq -4$

83. $|2x + 7| > -5$

84. $|3x - 1| > -4$

85. $|4 - 3x| \geq 5$

86. $|7 - 2x| > 9$

87. $|5 - 4x| \leq 13$

88. $|3 - 7x| < 17$

89. $|6 - 3x| \leq 0$

90. $|10 - 5x| \geq 0$

91. $|2 - 9x| > 20$

92. $|5x - 1| < 16$

93. $|2x - 3| + 2 < 8$

94. $|3x - 5| + 1 < 7$

95. $|2 - 5x| - 4 > -2$

96. $|4 - 2x| - 9 > -3$

97. $8 - |2x - 5| < 3$

98. $12 - |3x - 4| > 7$

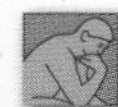
For Exercises 99 and 100, assume that a and b are positive numbers such that $a < b$. State whether the given inequality has no solution, all negative solutions, all positive solutions, or both positive and negative solutions.

99. $|x + b| < a$

100. $|x + a| < b$

OBJECTIVE C To solve application problems

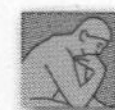
101. A dosage of medicine may safely range from 2.8 ml to 3.2 ml. What is the desired dosage of the medicine? What is the tolerance?

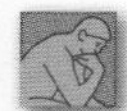
102. The tolerance, in inches, for the diameter of a piston is described by the absolute value inequality $|d - 5| \leq 0.01$. What is the desired diameter of the piston? By how much can the actual diameter of the piston vary from the desired diameter?

103. **Mechanics** The diameter of a bushing is 1.75 in. The bushing has a tolerance of 0.008 in. Find the lower and upper limits of the diameter of the bushing.

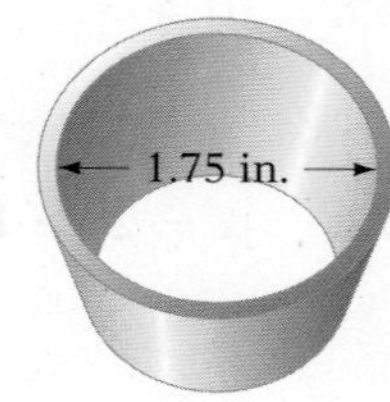

104. **Mechanics** A machinist must make a bushing that has a tolerance of 0.004 in. The diameter of the bushing is 3.48 in. Find the lower and upper limits of the diameter of the bushing.

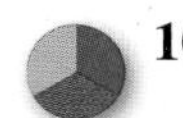

105. **Political Polling** Read the article at the right. For the poll described, the pollsters are 95% sure that the percent of American voters who felt the economy was the most important election issue lies between what lower and upper limits?

In the News

Economy Is Number-One Issue

A *Washington Post*/ABC News poll conducted between April 10 and April 13, 2008, showed that 41% of American voters felt the economy was the most important election issue. The results of the poll had a margin of error of plus or minus 3 percentage points. Note: *Margin of error* is a measure of the pollsters' confidence in their results. If the pollsters conduct this poll many times, they expect that 95% of the time they will get results that fall within the margin of error of the reported results.

Source: www.washingtonpost.com

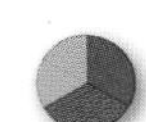

106. **Aquatic Environments** Different species of fish have different requirements for the temperature and pH of the water in which they live. The gold swordtail requires a temperature of 73°F plus or minus 9°F and a pH level of 7.65 plus or minus 0.65. Find the upper and lower limits of **a.** the temperature and **b.** the pH level for the water in which a gold swordtail lives. (*Source:* www.tacomapet.com)

107. **Automobiles** A piston rod for an automobile is $9\frac{5}{8}$ in. long, with a tolerance of $\frac{1}{32}$ in. Find the lower and upper limits of the length of the piston rod.

108. **Football Manufacturing** An NCAA football must conform to the measurements shown in the diagram, with tolerances of $\frac{1}{4}$ in. for the girth, $\frac{3}{8}$ in. for the circumference, and $\frac{5}{32}$ in. for the length. Find the upper and lower limits for **a.** the girth, **b.** the circumference, and **c.** the length of an NCAA football. (*Source:* www.ncaa.org)

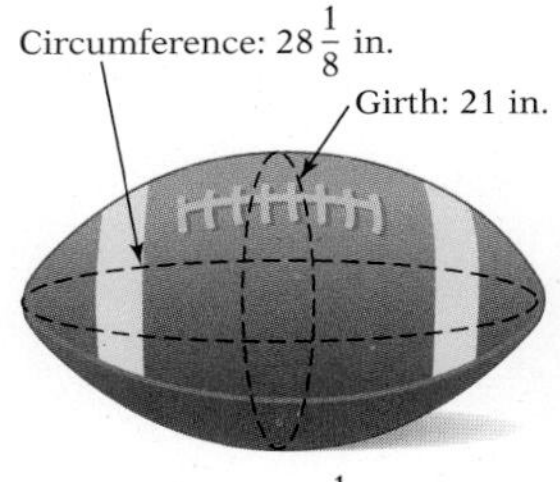

Electronics The tolerance of the resistors used in electronics is given as a percent. Use your calculator for Exercises 109 and 110.

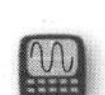

109. Find the lower and upper limits of a 29,000-ohm resistor with a 2% tolerance.

110. Find the lower and upper limits of a 15,000-ohm resistor with a 10% tolerance.

Applying the Concepts

111. For what values of the variable is the equation true? Write the solution set in set-builder notation.

a. $|x + 3| = x + 3$ **b.** $|a - 4| = 4 - a$

112. Write an absolute value inequality to represent all real numbers within 5 units of 2.

113. Replace the question mark with $\leq$, $\geq$, or $=$.

a. $|x + y| \ ? \ |x| + |y|$ **b.** $|x - y| \ ? \ |x| - |y|$

c. $||x| - |y|| \ ? \ |x| - |y|$ **d.** $|xy| \ ? \ |x||y|$

114. Let $|x| \leq 2$ and $|3x - 2| \leq a$. Find the smallest possible value of a.

FOCUS ON PROBLEM SOLVING

Trial-and-Error Approach to Problem Solving

The questions below require an answer of *always true, sometimes true,* or *never true.* These problems are best solved by the trial-and-error method. The trial-and-error method of arriving at a solution to a problem involves repeated tests or experiments.

For example, consider the following statement.

> Both sides of an equation can be divided by the same number without changing the solution of the equation.

The solution of the equation $6x = 18$ is 3. If we divide both sides of the equation by 2, the result is $3x = 9$, and the solution is still 3. So the answer "never true" has been eliminated. We still need to determine whether there is a case for which the statement is not true. Is there a number that we could divide both sides of the equation by and the result would be an equation for which the solution is not 3?

If we divide both sides of the equation by 0, the result is $\frac{6x}{0} = \frac{18}{0}$. The solution of this equation is not 3 because the expressions on either side of the equals sign are undefined. Thus the statement is true for some numbers and not true for 0. The statement is sometimes true.

For Exercises 1 to 10, determine whether the statement is *always true, sometimes true,* or *never true.*

1. Both sides of an equation can be multiplied by the same number without changing the solution of the equation.

2. For an equation of the form $ax = b$, $a \neq 0$, multiplying both sides of the equation by the reciprocal of a will result in an equation of the form $x = constant$.

3. The Multiplication Property of Equations is used to remove a term from one side of an equation.

4. Adding -3 to each side of an equation yields the same result as subtracting 3 from each side of the equation.

5. An equation contains an equals sign.

6. The same variable term can be added to both sides of an equation without changing the solution of the equation.

7. An equation of the form $ax + b = c$ cannot be solved if a is a negative number.

8. The solution of the equation $\frac{x}{0} = 0$ is 0.

9. An even integer is a multiple of 2.

10. In solving an equation of the form $ax + b = cx + d$, subtracting cx from each side of the equation results in an equation with only one variable term.

PROJECTS AND GROUP ACTIVITIES

Water Displacement When an object is placed in water, the object displaces an amount of water that is equal to the volume of the object.

HOW TO 1 A sphere with a diameter of 4 in. is placed in a rectangular tank of water that is 6 in. long and 5 in. wide. How much does the water level rise? Round to the nearest hundredth.

$V = \frac{4}{3}\pi r^3$ • Use the formula for the volume of a sphere.

$V = \frac{4}{3}\pi(2^3) = \frac{32}{3}\pi$ • $r = \frac{1}{2}d = \frac{1}{2}(4) = 2$

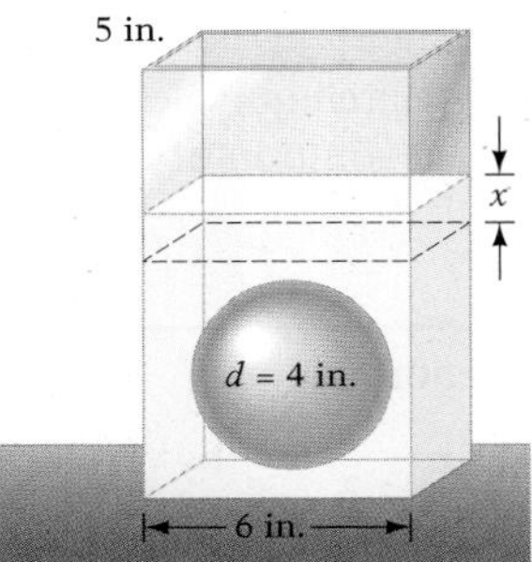

Let x represent the amount of the rise in water level. The volume of the sphere will equal the volume displaced by the water. As shown at the left, this volume is equal to the volume of a rectangular solid with width 5 in., length 6 in., and height x in.

$V = LWH$ • Use the formula for the volume of a rectangular solid.

$\frac{32}{3}\pi = (6)(5)x$ • Substitute $\frac{32}{3}\pi$ for V, 5 for W, and 6 for L.

$\frac{32}{90}\pi = x$ • The exact height that the water will fill is $\frac{32}{90}\pi$.

$1.12 \approx x$ • Use a calculator to find an approximation.

The water will rise approximately 1.12 in.

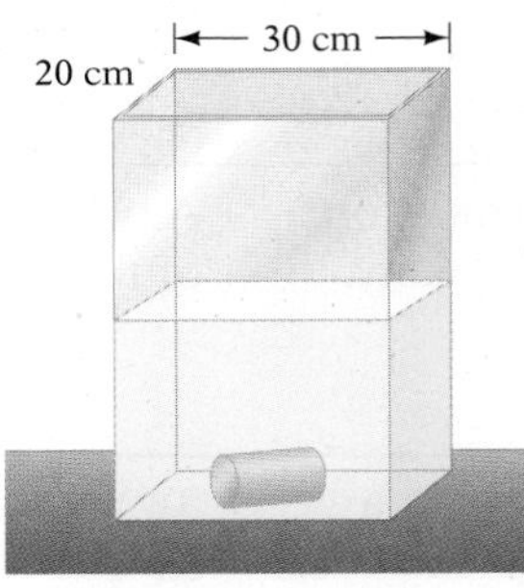

FIGURE 1

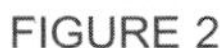

FIGURE 2

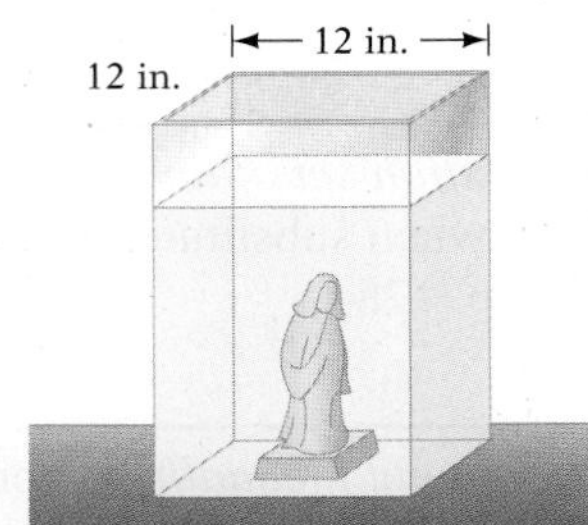

FIGURE 3

1. A cylinder with a 2-centimeter radius and a height of 10 cm is submerged in a tank of water that is 20 cm wide and 30 cm long (see Figure 1). How much does the water level rise? Round to the nearest hundredth.

2. A sphere with a radius of 6 in. is placed in a rectangular tank of water that is 16 in. wide and 20 in. long (see Figure 2). The sphere displaces water until two-thirds of the sphere is submerged. How much does the water level rise? Round to the nearest hundredth.

3. A chemist wants to know the density of a statue that weighs 15 lb. The statue is placed in a rectangular tank of water that is 12 in. long and 12 in. wide (see Figure 3). The water level rises 0.42 in. Find the density of the statue. Round to the nearest hundredth. (*Hint:* Density = weight ÷ volume)

CHAPTER 2

SUMMARY

KEY WORDS	EXAMPLES
An *equation* expresses the equality of two mathematical expressions. [2.1A, p. 76]	$3 + 2(4x - 5) = x + 4$ is an equation.
A *solution of an equation* is a number that, when substituted for the variable, results in a true equation. [2.1A, p. 76]	-2 is a solution of $2 - 3x = 8$ because $2 - 3(-2) = 8$ is a true equation.
To *solve an equation* means to find a solution of the equation. The goal is to rewrite the equation in the form variable = constant, because the constant is the solution. [2.1B, p. 77]	The equation $x = -3$ is in the form variable = constant. The constant, -3, is the solution of the equation.
Equivalent equations are equations that have the same solution. [2.1B, p. 77]	$x + 3 = 7$ and $x = 4$ are equivalent equations because the solution of each equation is 4.
Multiplying each side of an equation that contains fractions by the LCM of the denominators is called *clearing denominators*. [2.2A, p. 95]	We can clear the denominators from the equation $\frac{2}{3}x - \frac{1}{2} = \frac{3}{4}$ by multiplying each side of the equation by 12.
Consecutive integers follow one another in order. [2.3A, p. 109]	5, 6, 7 are consecutive integers. $-9, -8, -7$ are consecutive integers.
The *solution set of an inequality* is a set of numbers, each element of which, when substituted in the inequality, results in a true inequality. [2.5A, p. 128]	Any number greater than 4 is a solution of the inequality $x > 4$.
A *compound inequality* is formed by joining two inequalities with a connective word such as *and* or *or*. [2.5B, p. 131]	$3x > 6$ and $2x + 5 < 7$ $2x + 1 < 3$ or $x + 2 > 4$
An *absolute value equation* is an equation that contains an absolute value symbol. [2.6A, p. 140]	$\|x - 2\| = 3$
An *absolute value inequality* is an inequality that contains an absolute value symbol. [2.6B, p. 142]	$\|x - 4\| < 5$ $\|2x - 3\| > 6$
The *tolerance* of a component or part is the amount by which it is acceptable for the component to vary from a given measurement. The maximum measurement is the *upper limit*. The minimum measurement is the *lower limit*. [2.6C, p. 144]	The diameter of a bushing is 1.5 in. with a tolerance of 0.005 in. The lower and upper limits of the diameter of the bushing are 1.5 in. ± 0.005 in.

ESSENTIAL RULES AND PROCEDURES	EXAMPLES
Addition Property of Equations [2.1B, p. 77] The same number can be added to each side of an equation without changing the solution of the equation. If $a = b$, then $a + c = b + c$.	$x + 5 = -3$ $x + 5 - 5 = -3 - 5$ $x = -8$
Multiplication Property of Equations [2.1C, p. 78] Each side of an equation can be multiplied by the same *nonzero* number without changing the solution of the equation. If $a = b$ and $c \neq 0$, then $ac = bc$.	$\frac{2}{3}x = 4$ $\left(\frac{3}{2}\right)\left(\frac{2}{3}x\right) = \left(\frac{3}{2}\right)4$ $x = 6$
Basic Percent Equation [2.1D, p. 80] Percent · Base = Amount $P \cdot B = A$	30% of what number is 24? $PB = A$ $0.30B = 24$ $\frac{0.30B}{0.30} = \frac{24}{0.30}$ $B = 80$
Simple Interest Equation [2.1D, p. 81] Interest = Principal · Rate · Time $I = Prt$	A credit card company charges an annual interest rate of 21% on the monthly unpaid balance on a card. Find the amount of interest charged on an unpaid balance of \$232 for April. $I = Prt$ $I = 232(0.21)\left(\frac{1}{12}\right) = 4.06$ The interest charged is \$4.06.
Consecutive Integers [2.3A, p. 109] $n, n + 1, n + 2, \ldots$	The sum of three consecutive integers is 33. $n + (n + 1) + (n + 2) = 33$
Consecutive Even or Consecutive Odd Integers [2.3A, p. 109] $n, n + 2, n + 4, \ldots$	The sum of three consecutive odd integers is 33. $n + (n + 2) + (n + 4) = 33$
Value Mixture Equation [2.4A, p. 116] Amount · Unit Cost = Value $AC = V$	A merchant combines coffee that costs \$6 per pound with coffee that costs \$3.20 per pound. How many pounds of each should be used to make 60 lb of a blend that costs \$4.50 per pound? $6x + 3.20(60 - x) = 4.50(60)$

Percent Mixture Problems [2.4B, p. 118]

$$\text{Amount of solution} \cdot \text{Percent of concentration} = \text{Quantity of substance}$$

$$Ar = Q$$

A silversmith mixed 120 oz of an 80% silver alloy with 240 oz of a 30% silver alloy. Find the percent concentration of the resulting silver alloy.

$$0.80(120) + 0.30(240) = x(360)$$

Uniform Motion Equation [2.4C, p. 120]

$$\text{Rate} \cdot \text{Time} = \text{Distance}$$

$$rt = d$$

Two planes are 1640 mi apart and are traveling toward each other. One plane is traveling 60 mph faster than the other plane. The planes meet in 2 h. Find the speed of each plane.

$$2r + 2(r + 60) = 1640$$

Addition Property of Inequalities [2.5A, p. 128]

If $a > b$, then $a + c > b + c$.
If $a < b$, then $a + c < b + c$.

$$x + 3 > -2$$
$$x + 3 - 3 > -2 - 3$$
$$x > -5$$

Multiplication Property of Inequalities [2.5A, p. 129]

Rule 1 If $a > b$ and $c > 0$, then $ac > bc$.
If $a < b$ and $c > 0$, then $ac < bc$.

$$\frac{3}{4}x > 12$$
$$\frac{4}{3}\left(\frac{3}{4}x\right) > \frac{4}{3}(12)$$
$$x > 16$$

Rule 2 If $a > b$ and $c < 0$, then $ac < bc$.
If $a < b$ and $c < 0$, then $ac > bc$.

$$-2x < 8$$
$$\frac{-2x}{-2} > \frac{8}{-2}$$
$$x > -4$$

Solutions of an Absolute Value Equation [2.6A, p. 140]

If $a \geq 0$ and $|x| = a$, then $x = a$ or $x = -a$.

$$|x - 3| = 7$$
$$x - 3 = 7 \qquad x - 3 = -7$$
$$x = 10 \qquad x = -4$$

Solutions of Absolute Value Inequalities [2.6B, p. 143]

To solve an absolute value inequality of the form $|ax + b| < c$, $c > 0$, solve the equivalent compound inequality $-c < ax + b < c$.

$$|x - 5| < 9$$
$$-9 < x - 5 < 9$$
$$-9 + 5 < x - 5 + 5 < 9 + 5$$
$$-4 < x < 14$$

To solve an absolute value inequality of the form $|ax + b| > c$, solve the equivalent compound inequality $ax + b < -c$ or $ax + b > c$.

$$|x - 5| > 9$$
$$x - 5 < -9 \quad \text{or} \quad x - 5 > 9$$
$$x < -4 \quad \text{or} \quad x > 14$$

CHAPTER 2

CONCEPT REVIEW

Test your knowledge of the concepts presented in this chapter. Answer each question. Then check your answers against the ones provided in the Answer Section.

1. How is the Addition Property of Equations used to solve an equation?

2. How do you check the solution of an equation?

3. How do you solve an equation containing parentheses?

4. What is the difference between consecutive integers and consecutive even integers?

5. In a percent mixture problem, when mixing a 15% solution with a 20% solution, what percent concentration should the resulting solution be: (1) greater than 20%, (2) between 15% and 20%, or (3) less than 15%?

6. What formula is used in solving a percent mixture problem?

7. What formula is used to solve a uniform motion problem?

8. How is the Multiplication Property of Inequalities different from the Multiplication Property of Equations?

9. What compound inequality is equivalent to an absolute value inequality of the form $|ax + b| < c, c > 0$?

10. What compound inequality is equivalent to an absolute value inequality of the form $|ax + b| > c$?

11. How do you check the solution to an absolute value equation?

CHAPTER 2

REVIEW EXERCISES

1. Solve: $3t - 3 + 2t = 7t - 15$

2. Solve: $3x - 7 > -2$
Write the solution set in set-builder notation.

3. Is 3 a solution of $5x - 2 = 4x + 5$?

4. Solve: $x + 4 = -5$

5. Solve: $3x < 4$ and $x + 2 > -1$
Write the solution set in set-builder notation.

6. Solve: $\frac{3}{5}x - 3 = 2x + 5$

7. Solve: $-\frac{2}{3}x = \frac{4}{9}$

8. Solve: $|x - 4| - 8 = -3$

9. Solve: $|2x - 5| < 3$

10. Solve: $\frac{2x - 3}{3} + 2 = \frac{2 - 3x}{5}$

11. Solve: $2(a - 3) = 5(4 - 3a)$

12. Solve: $5x - 2 > 8$ or $3x + 2 < -4$
Write the solution set in set-builder notation.

13. Solve: $|4x - 5| \geq 3$

14. 30 is what percent of 12?

15. Solve: $\frac{1}{2}x - \frac{5}{8} = \frac{3}{4}x + \frac{3}{2}$

16. Solve: $6 + |3x - 3| = 2$

17. Solve: $3x - 2 > x - 4$ or $7x - 5 < 3x + 3$

18. Solve: $2x - (3 - 2x) = 4 - 3(4 - 2x)$

19. Solve: $x + 9 = -6$

20. Solve: $\frac{2}{3} = x + \frac{3}{4}$

21. Solve: $-3x = -21$

22. Solve: $\frac{2}{3}a = \frac{4}{9}$

23. Solve: $3y - 5 = 3 - 2y$

24. Solve: $4x - 5 + x = 6x - 8$

25. Solve: $3(x - 4) = -5(6 - x)$

26. Solve: $\frac{3x - 2}{4} + 1 = \frac{2x - 3}{2}$

27. Solve: $5x - 8 < -3$
Write the solution set in interval notation.

28. Solve: $2x - 9 \leq 8x + 15$
Write the solution set in interval notation.

29. Solve: $\frac{2}{3}x - \frac{5}{8} \geq \frac{3}{4}x + 1$
Write the solution set in set-builder notation.

30. Solve: $2 - 3(2x - 4) \leq 4x - 2(1 - 3x)$
Write the solution set in interval notation.

31. Solve: $-5 < 4x - 1 < 7$
Write the solution set in set-builder notation.

32. Solve: $|2x - 3| = 8$

33. Solve: $|5x + 8| = 0$

34. Solve: $|5x - 4| < -2$

35. **Uniform Motion** A ferry leaves a dock and travels to an island at an average speed of 16 mph. On the return trip, the ferry travels at an average speed of 12 mph. The total time for the trip is $2\frac{1}{3}$ h. How far is the island from the dock?

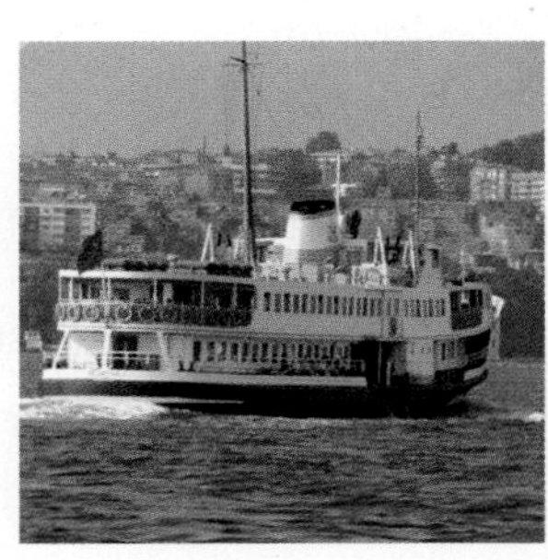

Alan Oddie/PhotoEdit

36. **Mixtures** A grocer mixed apple juice that costs \$4.20 per gallon with 40 gal of cranberry juice that costs \$6.50 per gallon. How much apple juice was used to make cranapple juice costing \$5.20 per gallon?

37. **Compensation** A sales executive earns \$800 per month plus 4% commission on the amount of sales. The executive's goal is to earn \$3000 per month. What amount of sales will enable the executive to earn \$3000 or more per month?

38. **Integers** Translate "four less than the product of five and a number is sixteen" into an equation and solve.

39. **Mechanics** The diameter of a bushing is 2.75 in. The bushing has a tolerance of 0.003 in. Find the lower and upper limits of the diameter of the bushing.

40. **Integers** The sum of two integers is twenty. Five times the smaller integer is two more than twice the larger integer. Find the two integers.

41. **Education** An average score of 80 to 90 in a psychology class receives a B grade. A student has scores of 92, 66, 72, and 88 on four tests. Find the range of scores on the fifth test that will give the student a B for the course.

42. **Uniform Motion** Two planes are 1680 mi apart and are traveling toward each other. One plane is traveling 80 mph faster than the other plane. The planes meet in 1.75 h. Find the speed of each plane.

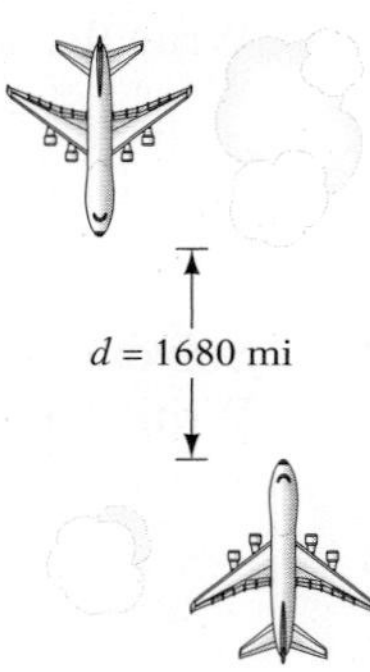

43. **Mixtures** An alloy containing 30% tin is mixed with an alloy containing 70% tin. How many pounds of each were used to make 500 lb of an alloy containing 40% tin?

44. **Automobiles** A piston rod for an automobile is $10\frac{3}{8}$ in. long with a tolerance of $\frac{1}{32}$ in. Find the lower and upper limits of the length of the piston rod.

CHAPTER 2

TEST

1. Solve: $x - 2 = -4$

2. Solve: $b + \frac{3}{4} = \frac{5}{8}$

3. Solve: $-\frac{3}{4}y = -\frac{5}{8}$

4. Solve: $3x - 5 = 7$

5. Solve: $\frac{3}{4}y - 2 = 6$

6. Solve: $2x - 3 - 5x = 8 + 2x - 10$

7. Solve: $2[a - (2 - 3a) - 4] = a - 5$

8. Is -2 a solution of $x^2 - 3x = 2x - 6$?

9. Solve: $3x - 6 = 10x - 18$

10. Solve: $3x - 2 \geq 6x + 7$
    Write the solution set in set-builder notation.

11. What is 0.5% of 8?

12. Solve: $4x - 1 > 5$ or $2 - 3x < 8$
    Write the solution set in set-builder notation.

13. Solve: $4 - 3x \geq 7$ and $2x + 3 \geq 7$

14. Solve: $|3 - 5x| = 12$

15. Solve: $2 - |2x - 5| = -7$

16. Solve: $|3x - 5| \leq 4$

17. Solve: $|4x - 3| > 5$

18. **Consumerism** Gambelli Agency rents cars for \$12 per day plus 10¢ for every mile driven. McDougal Rental rents cars for \$24 per day with unlimited mileage. How many miles a day can you drive a Gambelli Agency car if it is to cost you less than a McDougal Rental car?

19. **Mechanics** A machinist must make a bushing that has a tolerance of 0.002 in. The diameter of the bushing is 2.65 in. Find the lower and upper limits of the diameter of the bushing.

20. **Integers** The sum of two integers is fifteen. Eight times the smaller integer is one less than three times the larger integer. Find the integers.

21. **Mixtures** How many gallons of water must be mixed with 5 gal of a 20% salt solution to make a 16% salt solution?

22. **Mixtures** A butcher combines 100 lb of hamburger that costs \$2.10 per pound with 60 lb of hamburger that costs \$3.70 per pound. Find the cost of the hamburger mixture.

23. **Uniform Motion** A jogger runs a distance at a speed of 8 mph and returns the same distance running at a speed of 6 mph. Find the total distance that the jogger ran if the total time running was 1 hour and 45 minutes.

24. **Uniform Motion** Two trains are 250 mi apart and are traveling toward each other. One train is traveling 5 mph faster than the other train. The trains pass each other in 2 h. Find the speed of each train.

25. **Mixtures** How many ounces of pure water must be added to 60 oz of an 8% salt solution to make a 3% salt solution?

CUMULATIVE REVIEW EXERCISES

1. Subtract: $-6 - (-20) - 8$

2. Multiply: $(-2)(-6)(-4)$

3. Subtract: $-\frac{5}{6} - \left(-\frac{7}{16}\right)$

4. Simplify: $-4^2 \cdot \left(-\frac{3}{2}\right)^3$

5. Simplify: $\frac{5}{8} - \left(\frac{1}{2}\right)^2 \div \left(\frac{1}{3} - \frac{3}{4}\right)$

6. Evaluate $3(a - c) - 2ab$ when $a = 2$, $b = 3$, and $c = -4$.

7. Simplify: $3x - 8x + (-12x)$

8. Simplify: $2a - (-b) - 7a - 5b$

9. Simplify: $(16x)\left(\frac{1}{8}\right)$

10. Simplify: $-4(-9y)$

11. Simplify: $-2(-x^2 - 3x + 2)$

12. Simplify: $-2(x - 3) + 2(4 - x)$

13. Simplify: $-3[2x - 4(x - 3)] + 2$

14. Find $A \cap B$ given $A = \{-4, -2, 0, 2\}$ and $B = \{-4, 0, 4, 8\}$.

15. Graph: $\{x \mid x < 3\} \cap \{x \mid x > -2\}$

−5 −4 −3 −2 −1 0 1 2 3 4 5

16. Is -3 a solution of $x^2 + 6x + 9 = x + 3$?

17. Solve: $\frac{3}{5}x = -15$

18. Solve: $7x - 8 = -29$

19. Solve: $13 - 9x = -14$

20. Solve: $5x - 8 = 12x + 13$

21. Solve: $11 - 4x = 2x + 8$

22. Solve: $8x - 3(4x - 5) = -2x - 11$

23. Solve: $3 - 2(2x - 1) \geq 3(2x - 2) + 1$

24. Solve: $3x + 2 \leq 5$ and $x + 5 \geq 1$

25. Solve: $|3 - 2x| = 5$

26. Solve: $|3x - 1| > 5$

27. Write 55% as a fraction.

28. Write 1.03 as a percent.

29. 25% of what number is 30?

30. **Integers** Translate "the sum of six times a number and thirteen is five less than the product of three and the number" into an equation and solve.

31. **Mixtures** How many pounds of an oat flour that costs \$.80 per pound must be mixed with 40 lb of a wheat flour that costs \$.50 per pound to make a blend that costs \$.60 per pound?

32. **Mixtures** How many grams of pure gold must be added to 100 g of a 20% gold alloy to make an alloy that is 36% gold?

33. **Uniform Motion** A sprinter ran to the end of a track at an average rate of 8 m/s and then jogged back to the starting point at an average rate of 3 m/s. The sprinter took 55 s to run to the end of the track and jog back. Find the length of the track.

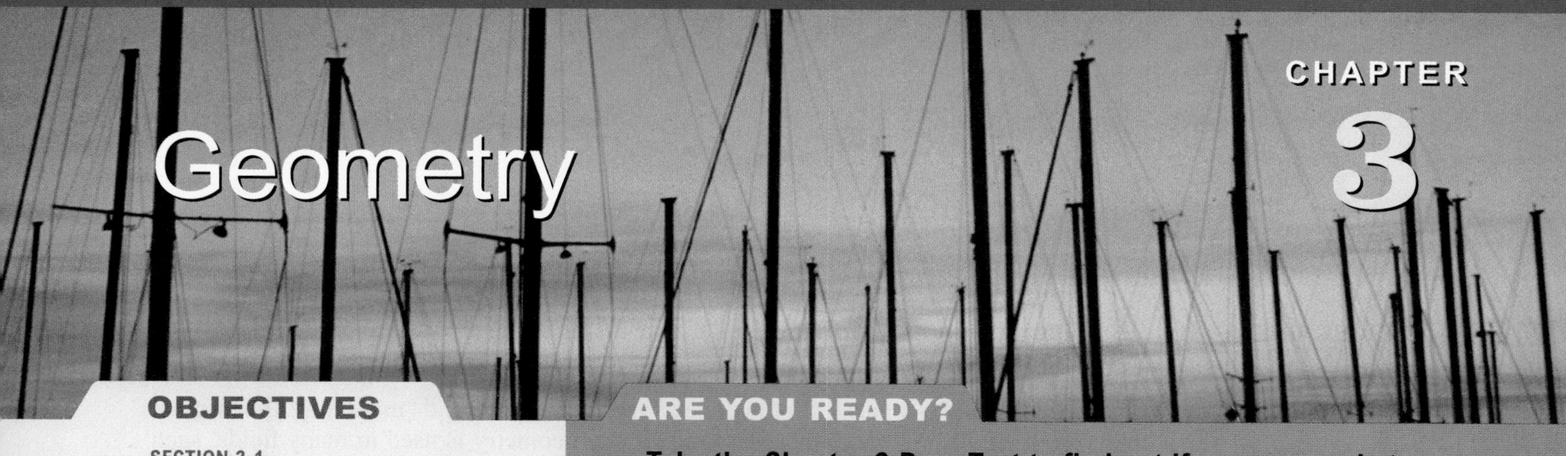

Tony Ise, Photodisc

OBJECTIVES

SECTION 3.1

A To solve problems involving lines and angles
B To solve problems involving angles formed by intersecting lines
C To solve problems involving the angles of a triangle

SECTION 3.2

A To solve problems involving the perimeter of a geometric figure
B To solve problems involving the area of a geometric figure

SECTION 3.3

A To solve problems involving the volume of a solid
B To solve problems involving the surface area of a solid

ARE YOU READY?

Take the Chapter 3 Prep Test to find out if you are ready to learn to:

- Solve problems involving lines and angles
- Find the perimeter and area of geometric figures
- Find the volume and surface area of geometric solids

PREP TEST

Do these exercises to prepare for Chapter 3.

1. Solve: $x + 47 = 90$

2. Solve: $32 + 97 + x = 180$

3. Simplify: $2(18) + 2(10)$

4. Evaluate abc when $a = 2$, $b = 3.14$, and $c = 9$.

5. Evaluate xyz^3 when $x = \frac{4}{3}$, $y = 3.14$, and $z = 3$.

6. Evaluate $\frac{1}{2}a(b + c)$ when $a = 6$, $b = 25$, and $c = 15$.

SECTION

3.1 Introduction to Geometry

OBJECTIVE A To solve problems involving lines and angles 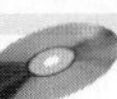

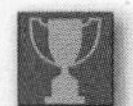

Tips for Success
Before you begin a new chapter, you should take some time to review previously learned skills. One way to do this is to complete the Prep Test. See page 163. This test focuses on the particular skills that will be required for the new chapter.

The word *geometry* comes from the Greek words for "earth" and "measure." The original purpose of geometry was to measure land. Today geometry is used in many fields, such as physics, medicine, and geology, and is applied in such areas as mechanical drawing and astronomy. Geometric forms are also used in art and design.

Three basic concepts of geometry are the point, line, and plane. A **point** is symbolized by drawing a dot. A **line** is determined by two distinct points and extends indefinitely in both directions, as the arrows on the line shown at the right indicate. This line contains points A and B and is represented by $\overleftrightarrow{AB}$. A line can also be represented by a single letter, such as ℓ.

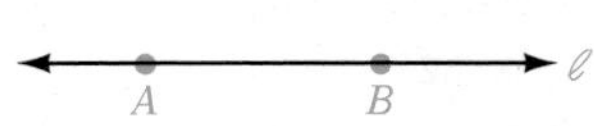

A **ray** starts at a point and extends indefinitely in *one* direction. The point at which a ray starts is called the **endpoint** of the ray. The ray shown at the right is denoted by $\overrightarrow{AB}$. Point A is the endpoint of the ray.

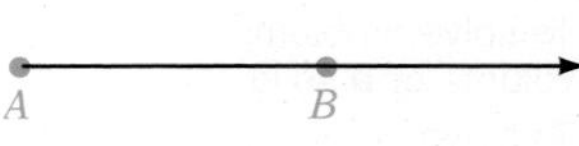

A **line segment** is part of a line and has two endpoints. The line segment shown at the right is denoted by $\overline{AB}$.

The distance between the endpoints of $\overline{AC}$ is denoted by AC. If B is a point on $\overline{AC}$, then AC (the distance from A to C) is the sum of AB (the distance from A to B) and BC (the distance from B to C).

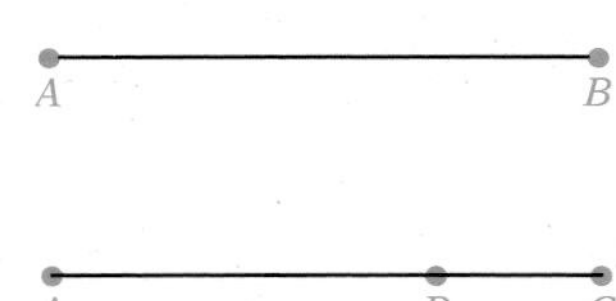

HOW TO • 1 Given $AB = 22$ cm and $AC = 31$ cm, find BC.

$AC = AB + BC$ • Write an equation for the distances between points on the line segment.

$31 = 22 + BC$ • Substitute the given distances for AB and AC into the equation.

$9 = BC$ • Solve for BC.

$BC = 9$ cm

In this section we will be discussing figures that lie in a plane. A **plane** is a flat surface and can be pictured as a table top or whiteboard that extends in all directions. Figures that lie in a plane are called **plane figures.**

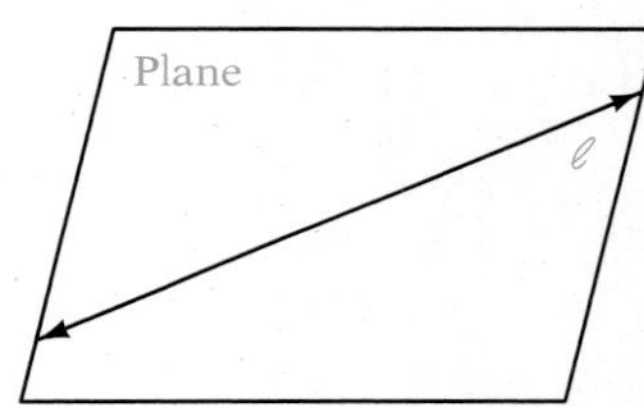

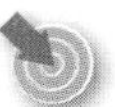

Point of Interest

Geometry is one of the oldest branches of mathematics. Around 350 B.C., the Greek mathematician Euclid wrote the *Elements*, which contained all of the known concepts of geometry. Euclid's contribution was to unify various concepts into a single deductive system that was based on a set of axioms.

Lines in a plane can be intersecting or parallel. **Intersecting lines** cross at a point in the plane. **Parallel lines** never meet. The distance between them is always the same.

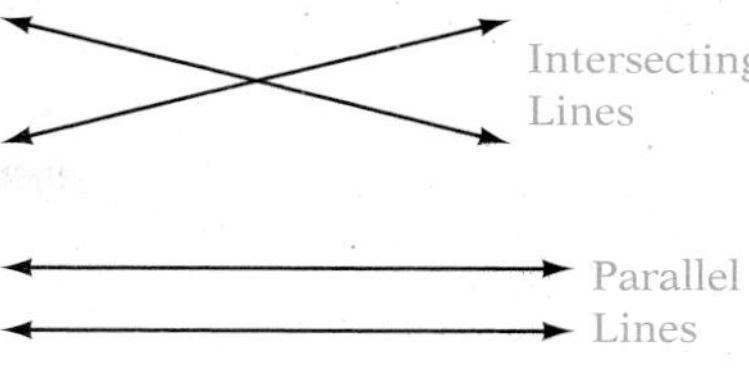

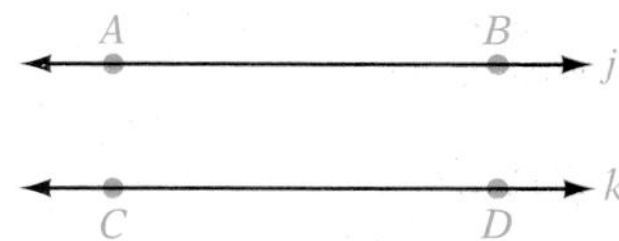

The symbol $\parallel$ means "is parallel to." In the figure at the right, $j \parallel k$ and $\overline{AB} \parallel \overline{CD}$. Note that j contains $\overline{AB}$ and k contains $\overline{CD}$. Parallel lines contain parallel line segments.

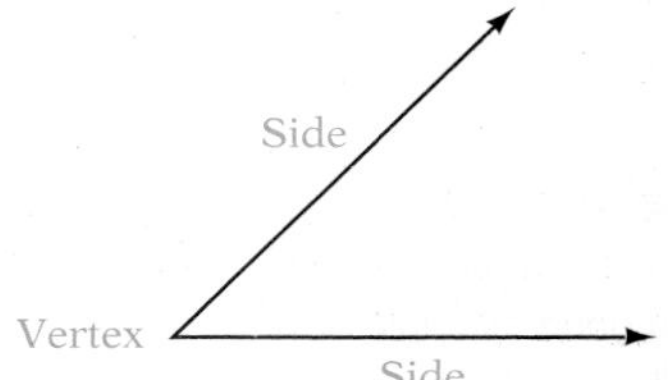

An **angle** is formed by two rays with the same endpoint. The **vertex** of the angle is the point at which the two rays meet. The rays are called the **sides** of the angle.

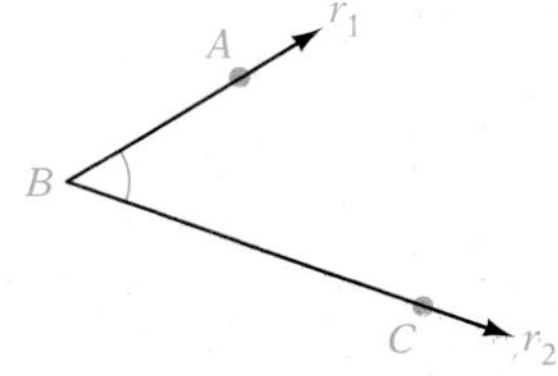

If A and C are points on rays r_1 and r_2, and B is the vertex, then the angle is called $\angle B$ or $\angle ABC$, where $\angle$ is the symbol for angle. Note that either the angle is named by the vertex, or the vertex is the second point listed when the angle is named by giving three points. $\angle ABC$ could also be called $\angle CBA$.

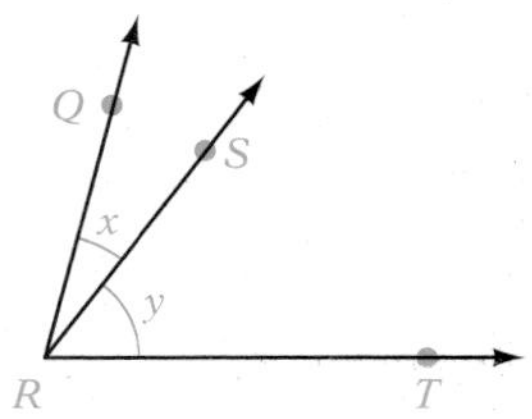

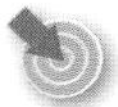

Point of Interest

The first woman mathematician for whom documented evidence exists is Hypatia (370–415). She lived in Alexandria, Egypt, and lectured at the Museum, the forerunner of our modern university. She made important contributions in mathematics, astronomy, and philosophy.

An angle can also be named by a variable written between the rays close to the vertex. In the figure at the right, $\angle x = \angle QRS$ and $\angle y = \angle SRT$. Note that in this figure, more than two rays meet at R. In this case, the vertex cannot be used to name an angle.

An angle is measured in **degrees.** The symbol for degrees is a small raised circle, °. Probably because early Babylonians believed that Earth revolves around the sun in approximately 360 days, the angle formed by a circle has a measure of 360° (360 degrees).

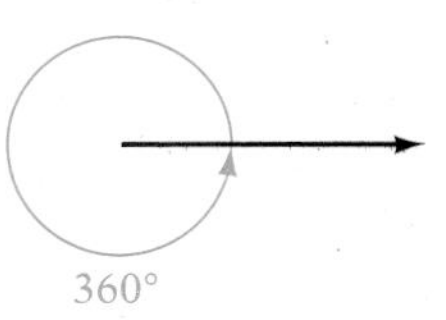

A **protractor** is used to measure an angle. Place the center of the protractor at the vertex of the angle with the edge of the protractor along a side of the angle. The angle shown in the figure below measures 58°.

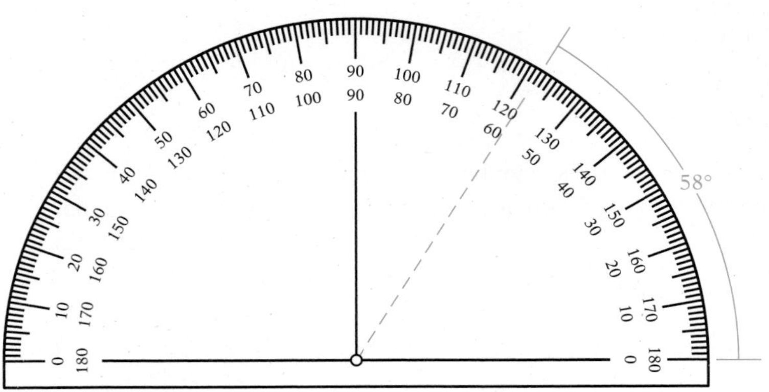

Take Note

The corner of a page of this book is a good example of a 90° angle.

A 90° angle is called a **right angle.** The symbol ∟ represents a right angle.

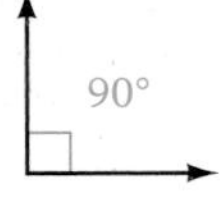

Perpendicular lines are intersecting lines that form right angles.

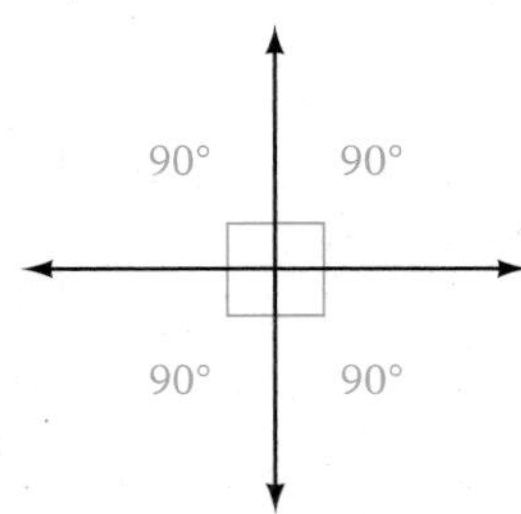

The symbol ⊥ means "is perpendicular to." In the figure at the right, $p \perp q$ and $\overline{AB} \perp \overline{CD}$. Note that line p contains $\overline{AB}$ and line q contains $\overline{CD}$. Perpendicular lines contain perpendicular line segments.

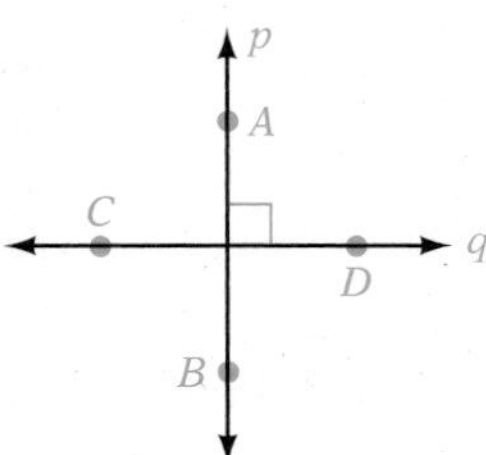

Complementary angles are two angles whose measures have the sum 90°.

$$\angle A + \angle B = 70° + 20° = 90°$$

$\angle A$ and $\angle B$ are complementary angles.

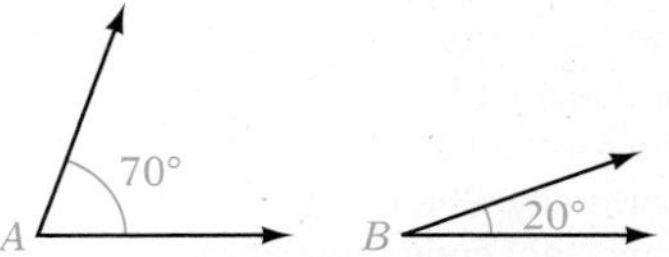

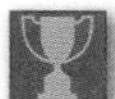

Tips for Success
A great many new vocabulary words are introduced in this chapter. All of these terms are in **bold type.** The bold type indicates that these are concepts you must know to learn the material. Be sure to study each new term as it is presented.

A 180° angle is called a **straight angle.**

$\angle AOB$ is a straight angle.

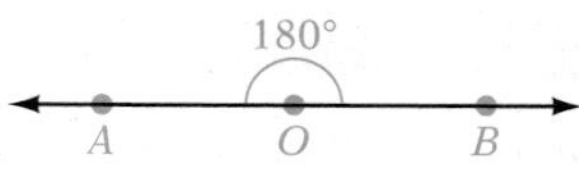

Supplementary angles are two angles whose measures have the sum 180°.

$$\angle A + \angle B = 130° + 50° = 180°$$

$\angle A$ and $\angle B$ are supplementary angles.

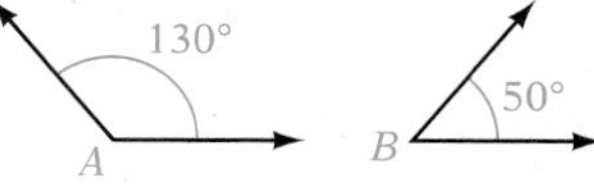

An **acute angle** is an angle whose measure is between 0° and 90°. $\angle B$ above is an acute angle. An **obtuse angle** is an angle whose measure is between 90° and 180°. $\angle A$ above is an obtuse angle.

Two angles that share a common side are **adjacent angles.** In the figure at the right, $\angle DAC$ and $\angle CAB$ are adjacent angles. $\angle DAC = 45°$ and $\angle CAB = 55°$.

$$\begin{aligned}\angle DAB &= \angle DAC + \angle CAB \\ &= 45° + 55° = 100°\end{aligned}$$

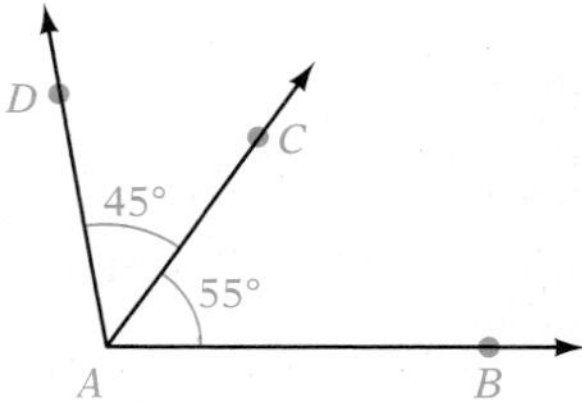

HOW TO • 2 In the figure at the right, $\angle EDG = 80°$. $\angle FDG$ is three times the measure of $\angle EDF$. Find the measure of $\angle EDF$.

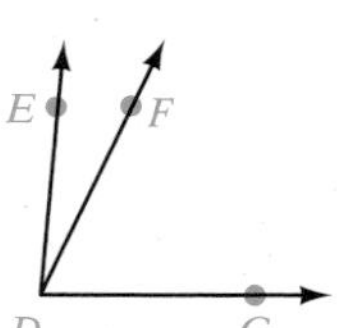

Let x = the measure of $\angle EDF$. Then $3x$ = the measure of $\angle FDG$. Write an equation and solve for x, the measure of $\angle EDF$.

$$\begin{aligned}\angle EDF + \angle FDG &= \angle EDG \\ x + 3x &= 80 \\ 4x &= 80 \\ x &= 20\end{aligned}$$

$\angle EDF = 20°$

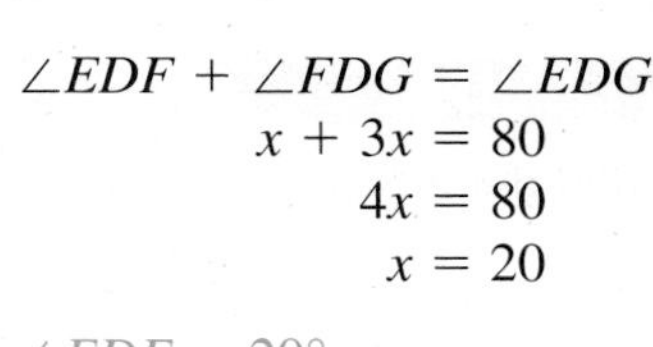

Take Note
Answers to application problems must have units, such as degrees, feet, dollars, or hours.

EXAMPLE • 1

Given $MN = 15$ mm, $NO = 18$ mm, and $MP = 48$ mm, find OP.

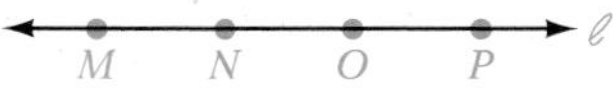

Solution

$$\begin{aligned}MN + NO + OP &= MP \\ 15 + 18 + OP &= 48 \quad \bullet\ MN = 15, NO = 18, MP = 48 \\ 33 + OP &= 48 \\ OP &= 15\end{aligned}$$

$OP = 15$ mm

YOU TRY IT • 1

Given $QR = 24$ cm, $ST = 17$ cm, and $QT = 62$ cm, find RS.

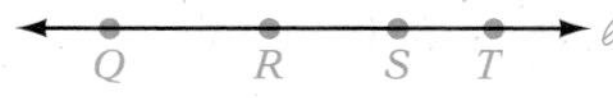

Your solution

Solution on p. S9

EXAMPLE • 2

Given $XY = 9$ m and YZ is twice XY, find XZ.

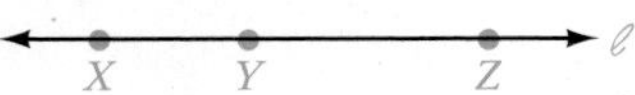

Solution

$XZ = XY + YZ$
$XZ = XY + 2(XY)$ • YZ is twice XY.
$XZ = 9 + 2(9)$ • $XY = 9$.
$XZ = 9 + 18$
$XZ = 27$

$XZ = 27$ m

YOU TRY IT • 2

Given $BC = 16$ ft and $AB = \frac{1}{4}(BC)$, find AC.

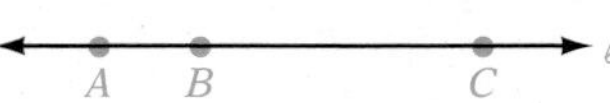

Your solution

EXAMPLE • 3

Find the complement of a 38° angle.

Strategy

Complementary angles are two angles whose sum is 90°. To find the complement, let x represent the complement of a 38° angle. Write an equation and solve for x.

Solution

$x + 38° = 90°$
$x = 52°$

The complement of a 38° angle is a 52° angle.

YOU TRY IT • 3

Find the supplement of a 129° angle.

Your strategy

Your solution

EXAMPLE • 4

Find the measure of $\angle x$.

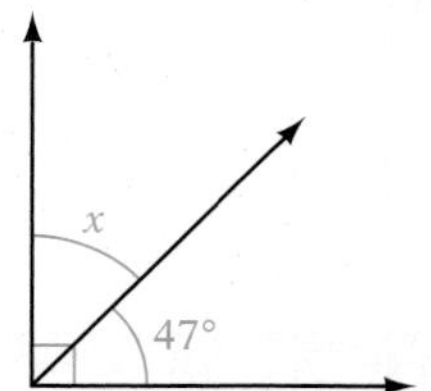

Strategy

To find the measure of $\angle x$, write an equation using the fact that the sum of the measure of $\angle x$ and 47° is 90°. Solve for $\angle x$.

Solution

$\angle x + 47° = 90°$
$\angle x = 43°$

The measure of $\angle x$ is 43°.

YOU TRY IT • 4

Find the measure of $\angle a$.

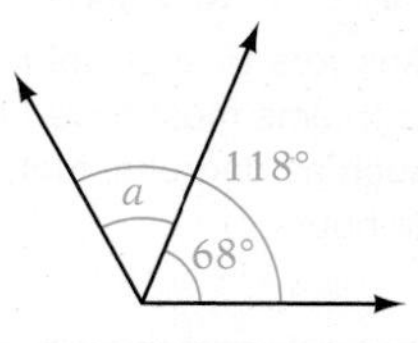

Your strategy

Your solution

Solutions on p. S9

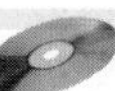

OBJECTIVE B To solve problems involving angles formed by intersecting lines

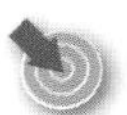

Point of Interest

Many cities in the New World, unlike those in Europe, were designed using rectangular street grids. Washington, D.C. was planned that way except that diagonal avenues were added, primarily for the purpose of enabling quick troop movement in the event the city required defense. As an added precaution, monuments of statuary were constructed at major intersections so that attackers would not have a straight shot down a boulevard.

Four angles are formed by the intersection of two lines. If the two lines are perpendicular, then each of the four angles is a right angle.

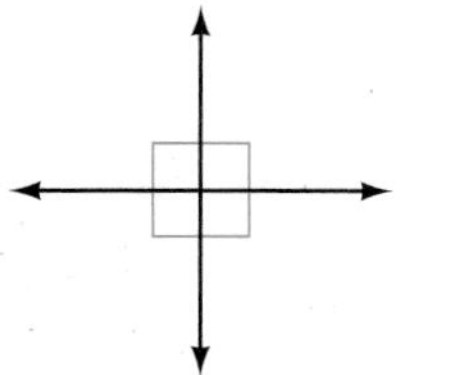

If the two lines are not perpendicular, then two of the angles formed are acute angles and two of the angles are obtuse angles. The two acute angles are always opposite each other, and the two obtuse angles are always opposite each other. In the figure at the right, $\angle w$ and $\angle y$ are acute angles. $\angle x$ and $\angle z$ are obtuse angles.

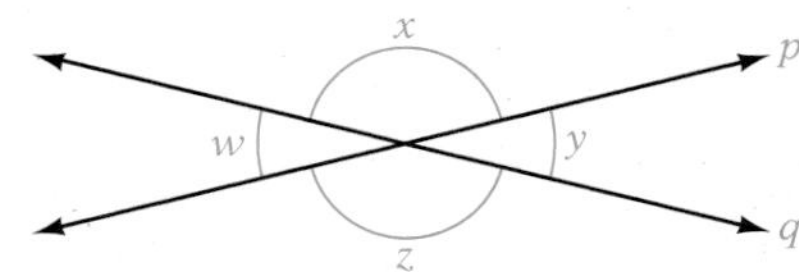

Two angles that are on opposite sides of the intersection of two lines are called **vertical angles.** Vertical angles have the same measure. $\angle w$ and $\angle y$ are vertical angles. $\angle x$ and $\angle z$ are vertical angles.

Vertical angles have the same measure.

$$\angle w = \angle y$$
$$\angle x = \angle z$$

Two angles that share a common side are called **adjacent angles.** For the figure shown above, $\angle x$ and $\angle y$ are adjacent angles, as are $\angle y$ and $\angle z$, $\angle z$ and $\angle w$, and $\angle w$ and $\angle x$. Adjacent angles of intersecting lines are supplementary angles.

Adjacent angles of intersecting lines are supplementary angles.

$$\angle x + \angle y = 180^\circ$$
$$\angle y + \angle z = 180^\circ$$
$$\angle z + \angle w = 180^\circ$$
$$\angle w + \angle x = 180^\circ$$

HOW TO • 3 Given that $\angle c = 65^\circ$, find the measures of angles a, b, and d.

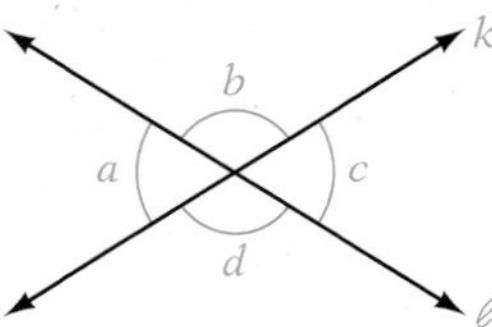

$\angle a = 65^\circ$

• $\angle a = \angle c$ because $\angle a$ and $\angle c$ are vertical angles.

$\angle b + \angle c = 180^\circ$
$\angle b + 65^\circ = 180^\circ$
$\angle b = 115^\circ$

• $\angle b$ is supplementary to $\angle c$ because $\angle b$ and $\angle c$ are adjacent angles of intersecting lines.

$\angle d = 115^\circ$

• $\angle d = \angle b$ because $\angle d$ and $\angle b$ are vertical angles.

A line that intersects two other lines at different points is called a **transversal.**

If the lines cut by a transversal t are parallel lines and the transversal is perpendicular to the parallel lines, all eight angles formed are right angles.

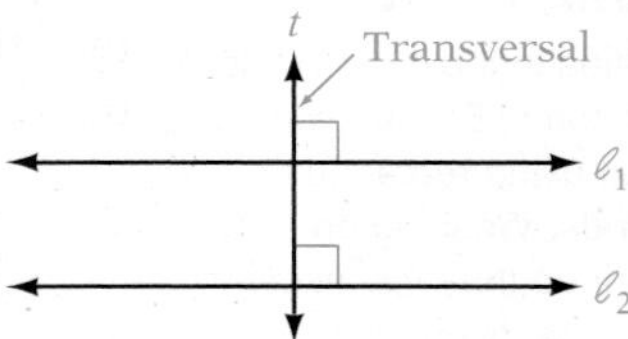

If the lines cut by a transversal t are parallel lines and the transversal is not perpendicular to the parallel lines, all four acute angles have the same measure and all four obtuse angles have the same measure. For the figure at the right,

$$\angle b = \angle d = \angle x = \angle z$$
$$\angle a = \angle c = \angle w = \angle y$$

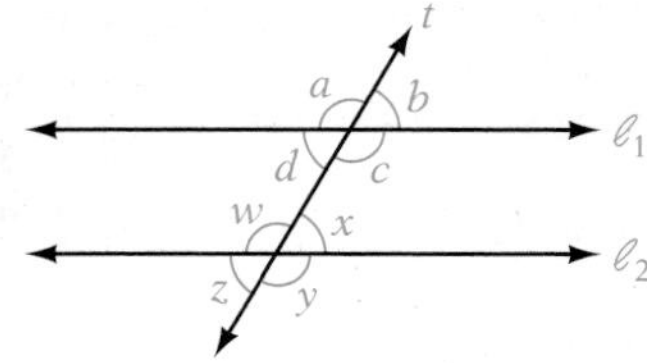

Alternate interior angles are two nonadjacent angles that are on opposite sides of the transversal and lie between the parallel lines. In the figure above, $\angle c$ and $\angle w$ are alternate interior angles; $\angle d$ and $\angle x$ are alternate interior angles. Alternate interior angles have the same measure.

Alternate interior angles have the same measure.

$$\angle c = \angle w$$
$$\angle d = \angle x$$

Alternate exterior angles are two nonadjacent angles that are on opposite sides of the transversal and lie outside the parallel lines. In the figure above, $\angle a$ and $\angle y$ are alternate exterior angles; $\angle b$ and $\angle z$ are alternate exterior angles. Alternate exterior angles have the same measure.

Alternate exterior angles have the same measure.

$$\angle a = \angle y$$
$$\angle b = \angle z$$

Corresponding angles are two angles that are on the same side of the transversal and are both acute angles or are both obtuse angles. For the figure above, the following pairs of angles are corresponding angles: $\angle a$ and $\angle w$, $\angle d$ and $\angle z$, $\angle b$ and $\angle x$, and $\angle c$ and $\angle y$. Corresponding angles have the same measure.

Corresponding angles have the same measure.

$$\angle a = \angle w$$
$$\angle d = \angle z$$
$$\angle b = \angle x$$
$$\angle c = \angle y$$

HOW TO • 4 Given that $\ell_1 \parallel \ell_2$ and $\angle c = 58°$, find the measures of $\angle f$, $\angle h$, and $\angle g$.

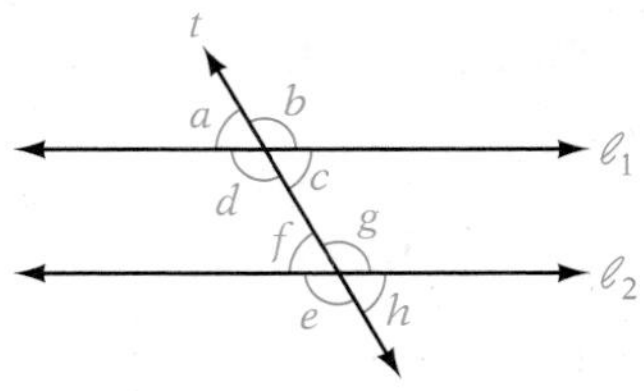

$\angle f = \angle c = 58°$ • $\angle c$ and $\angle f$ are alternate interior angles.

$\angle h = \angle c = 58°$ • $\angle c$ and $\angle h$ are corresponding angles.

$\angle g + \angle h = 180°$ • $\angle g$ is supplementary to $\angle h$.
$\angle g + 58° = 180°$
$\angle g = 122°$

EXAMPLE • 5

Find x.

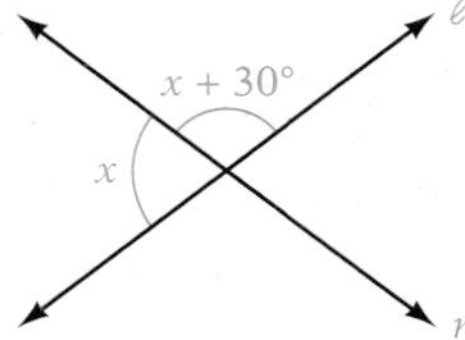

Strategy
The angles labeled are adjacent angles of intersecting lines and are, therefore, supplementary angles. To find x, write an equation and solve for x.

Solution
$x + (x + 30°) = 180°$
$2x + 30° = 180°$
$2x = 150°$
$x = 75°$

YOU TRY IT • 5

Find x.

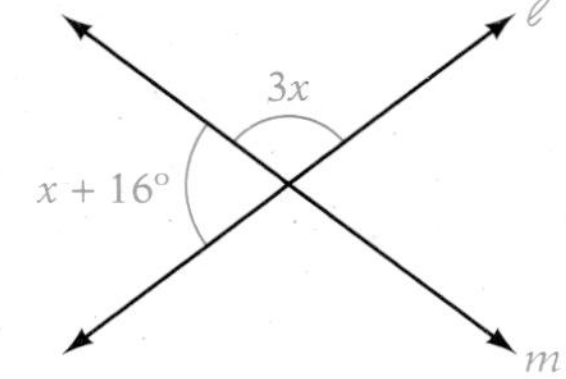

Your strategy

Your solution

EXAMPLE • 6

Given $\ell_1 \parallel \ell_2$, find x.

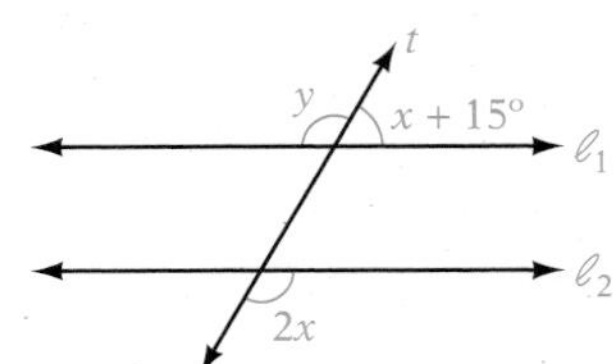

Strategy
$2x = y$ because alternate exterior angles have the same measure. $(x + 15°) + y = 180°$ because adjacent angles of intersecting lines are supplementary angles. Substitute $2x$ for y and solve for x.

Solution
$(x + 15°) + 2x = 180°$
$3x + 15° = 180°$
$3x = 165°$
$x = 55°$

YOU TRY IT • 6

Given $\ell_1 \parallel \ell_2$, find x.

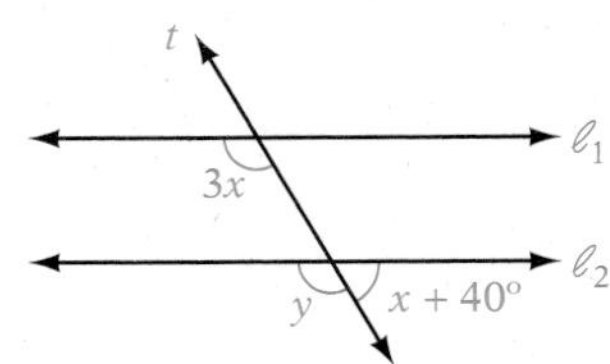

Your strategy

Your solution

Solutions on p. S9

OBJECTIVE C To solve problems involving the angles of a triangle

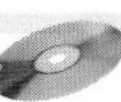

If the lines cut by a transversal are not parallel lines, the three lines will intersect at three points. In the figure at the right, the transversal t intersects lines p and q. The three lines intersect at points A, B, and C. These three points define three line segments: $\overline{AB}$, $\overline{BC}$, and $\overline{AC}$. The plane figure formed by these three line segments is called a **triangle.**

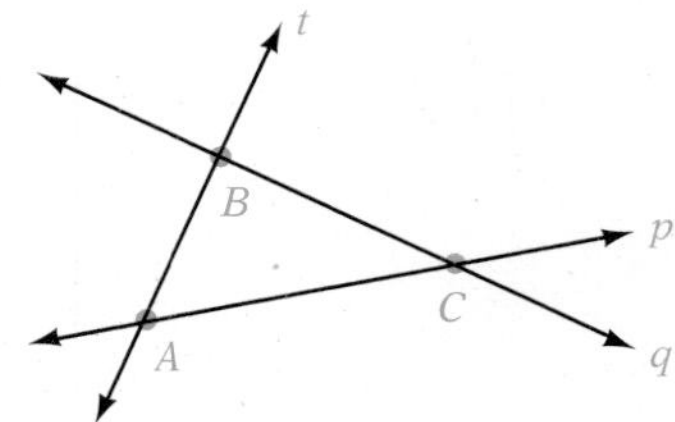

Each of the three points of intersection is the vertex of four angles. The angles within the region enclosed by the triangle are called **interior angles.** In the figure at the right, angles a, b, and c are interior angles. The sum of the measures of the interior angles of a triangle is 180°.

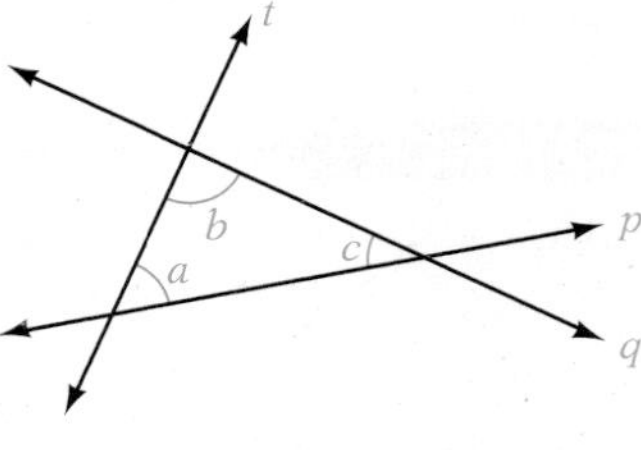

$\angle a + \angle b + \angle c = 180°$

> **The Sum of the Measures of the Interior Angles of a Triangle**
>
> The sum of the measures of the interior angles of a triangle is 180°.

An angle adjacent to an interior angle is an **exterior angle.** In the figure at the right, angles m and n are exterior angles for angle a. The sum of the measures of an interior and an exterior angle is 180°.

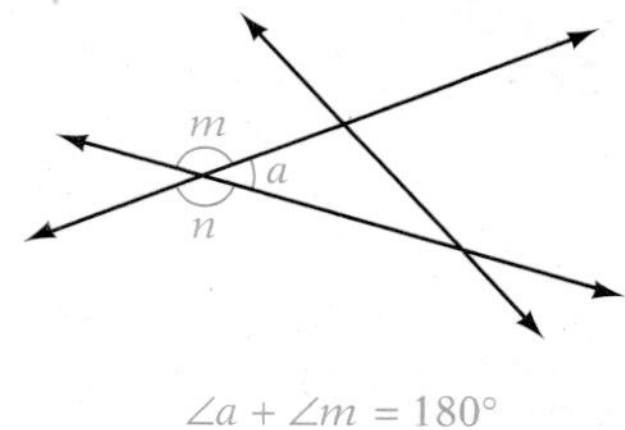

$\angle a + \angle m = 180°$
$\angle a + \angle n = 180°$

HOW TO • 5 Given that $\angle c = 40°$ and $\angle d = 100°$, find the measure of $\angle e$.

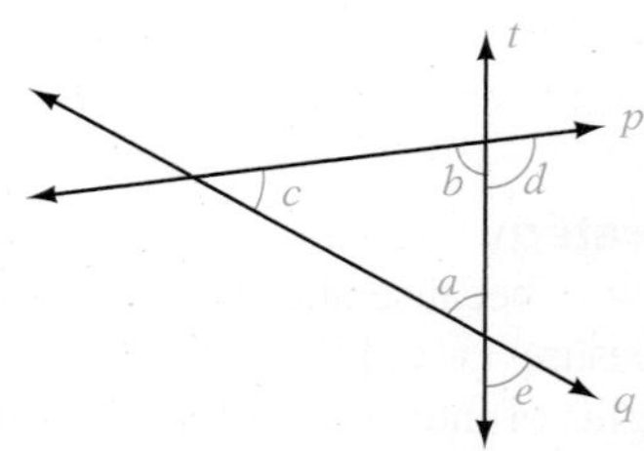

$\angle d$ and $\angle b$ are supplementary angles.

$$\angle d + \angle b = 180°$$
$$100° + \angle b = 180°$$
$$\angle b = 80°$$

The sum of the interior angles is 180°.

$$\angle c + \angle b + \angle a = 180°$$
$$40° + 80° + \angle a = 180°$$
$$120° + \angle a = 180°$$
$$\angle a = 60°$$

$\angle a$ and $\angle e$ are vertical angles.

$$\angle e = \angle a = 60°$$

EXAMPLE • 7

Given that $\angle y = 55°$, find the measures of angles a, b, and d.

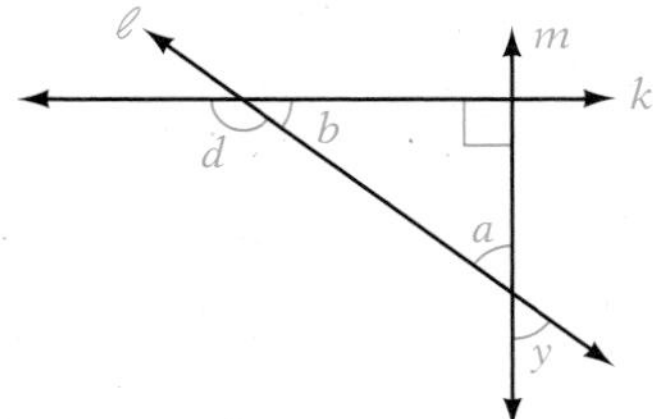

Strategy

- To find the measure of angle a, use the fact that $\angle a$ and $\angle y$ are vertical angles.
- To find the measure of angle b, use the fact that the sum of the measures of the interior angles of a triangle is 180°.
- To find the measure of angle d, use the fact that the sum of an interior and an exterior angle is 180°.

Solution

$\angle a = \angle y = 55°$

$$\begin{aligned} \angle a + \angle b + 90° &= 180° \\ 55° + \angle b + 90° &= 180° \\ \angle b + 145° &= 180° \\ \angle b &= 35° \end{aligned}$$

$$\begin{aligned} \angle d + \angle b &= 180° \\ \angle d + 35° &= 180° \\ \angle d &= 145° \end{aligned}$$

YOU TRY IT • 7

Given that $\angle a = 45°$ and $\angle x = 100°$, find the measures of angles b, c, and y.

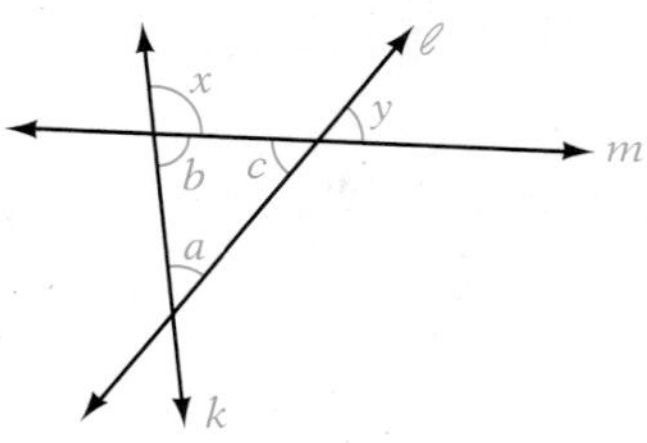

Your strategy

Your solution

EXAMPLE • 8

Two angles of a triangle measure 53° and 78°. Find the measure of the third angle.

Strategy

To find the measure of the third angle, use the fact that the sum of the measures of the interior angles of a triangle is 180°. Write an equation using x to represent the measure of the third angle. Solve the equation for x.

Solution

$$\begin{aligned} x + 53° + 78° &= 180° \\ x + 131° &= 180° \\ x &= 49° \end{aligned}$$

The measure of the third angle is 49°.

YOU TRY IT • 8

One angle in a triangle is a right angle, and one angle measures 34°. Find the measure of the third angle.

Your strategy

Your solution

Solutions on pp. S9–S10

3.1 EXERCISES

OBJECTIVE A To solve problems involving lines and angles

For Exercises 1 to 6, use a protractor to measure the angle. State whether the angle is acute, obtuse, or right.

1. 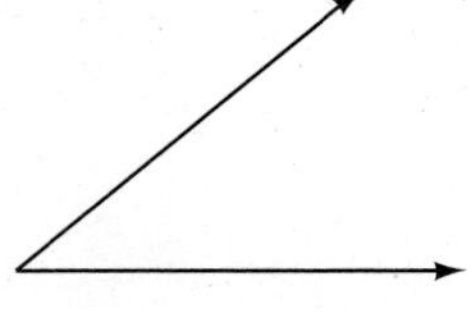

2. 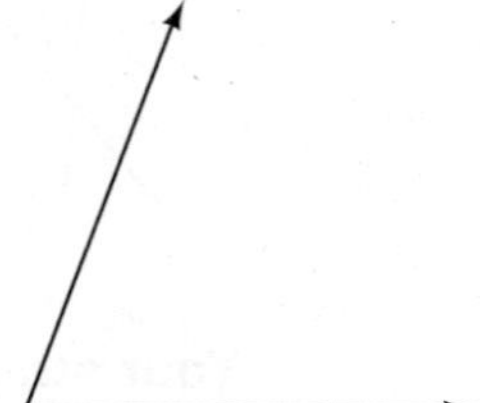

3. 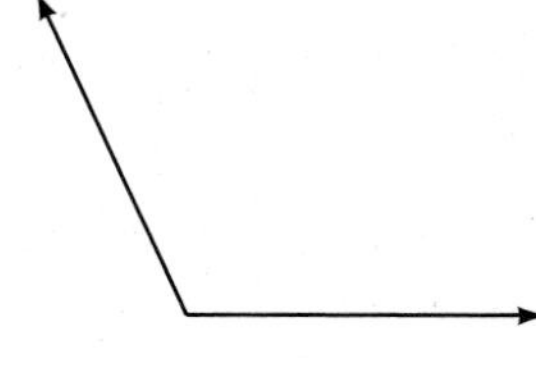

4. 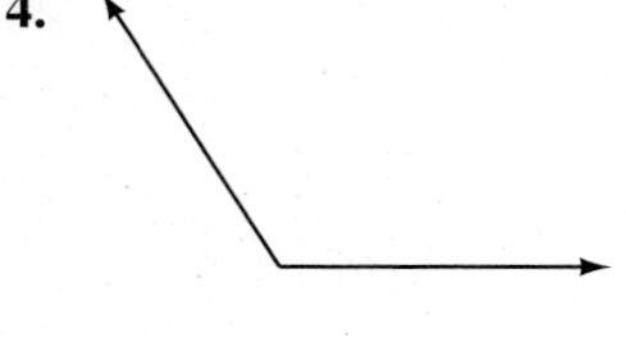

5. 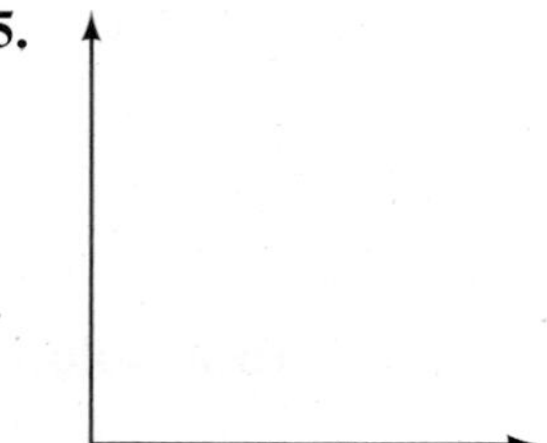

6. 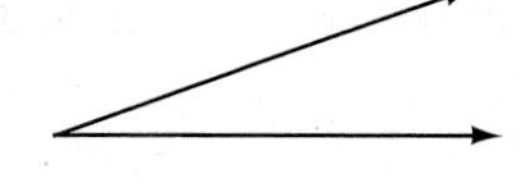

7. Find the complement of a 62° angle.

8. Find the complement of a 31° angle.

9. Find the supplement of a 162° angle.

10. Find the supplement of a 72° angle.

11. Given $AB = 12$ cm, $CD = 9$ cm, and $AD = 35$ cm, find BC.

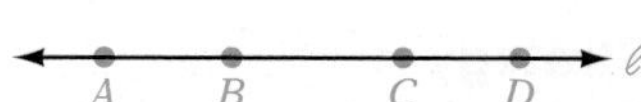

12. Given $AB = 21$ mm, $BC = 14$ mm, and $AD = 54$ mm, find CD.

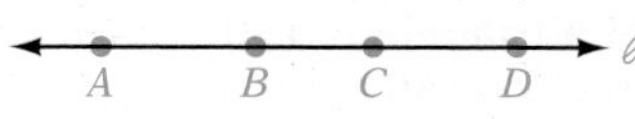

13. Given $QR = 7$ ft and RS is three times the length of QR, find QS.

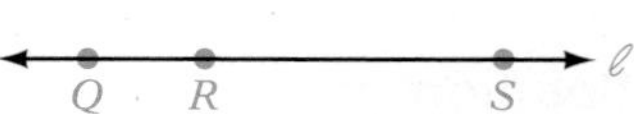

14. Given $QR = 15$ in. and RS is twice the length of QR, find QS.

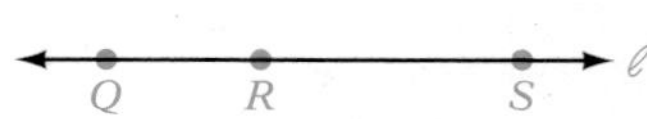

15. Given $EF = 20$ m and FG is one-half the length of EF, find EG.

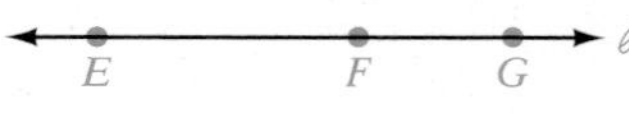

16. Given $EF = 18$ cm and FG is one-third the length of EF, find EG.

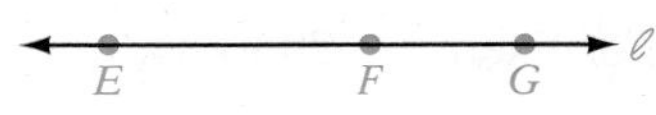

17. Given $\angle LOM = 53°$ and $\angle LON = 139°$, find the measure of $\angle MON$.

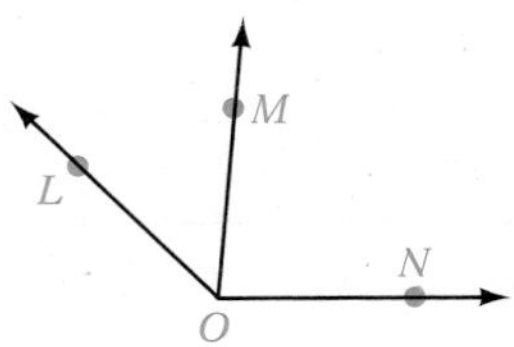

18. Given $\angle MON = 38°$ and $\angle LON = 85°$, find the measure of $\angle LOM$.

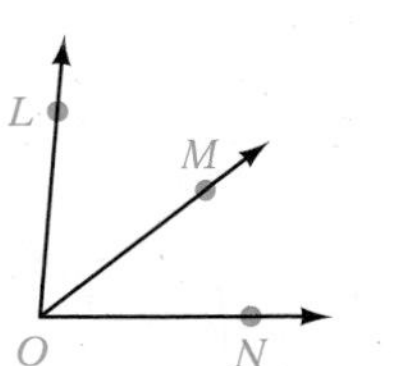

For Exercises 19 and 20, find the measure of $\angle x$.

19.

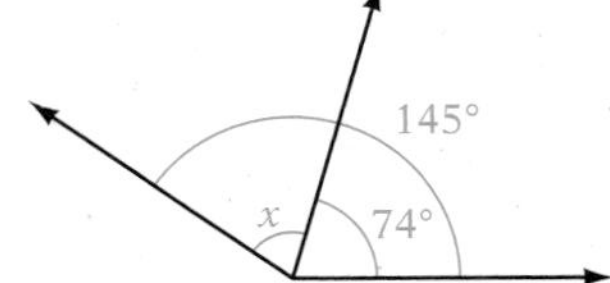

20.

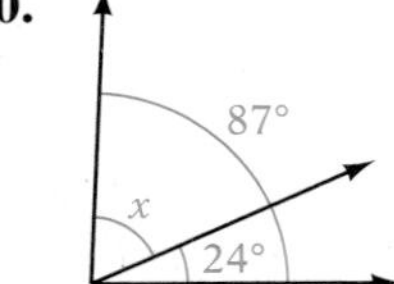

For Exercises 21 to 24, given that $\angle LON$ is a right angle, find the measure of $\angle x$.

21.

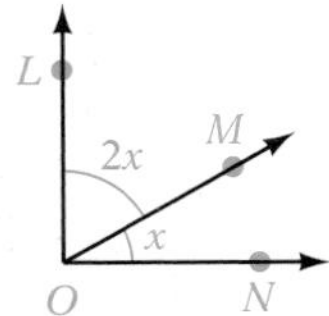

22.

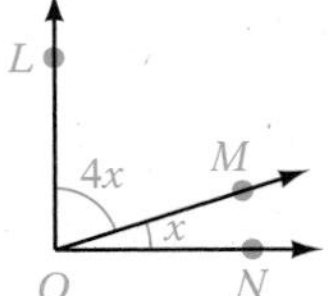

23.

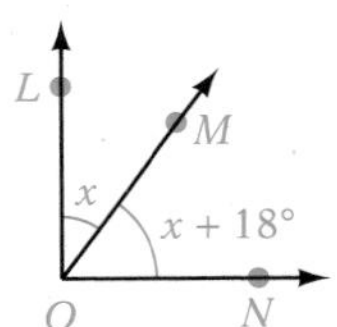

24.

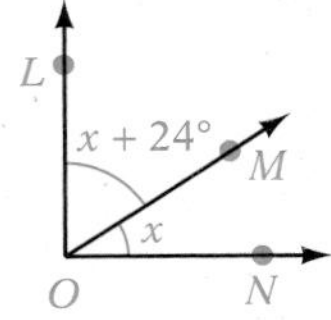

For Exercises 25 to 28, find the measure of $\angle a$.

25.

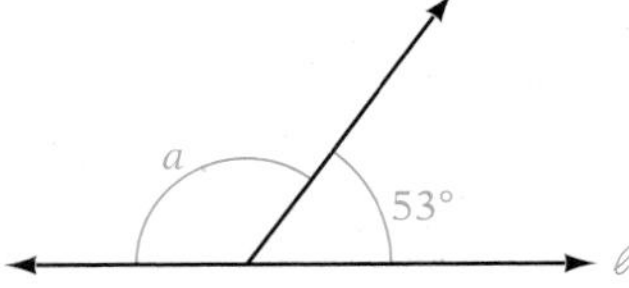

26.

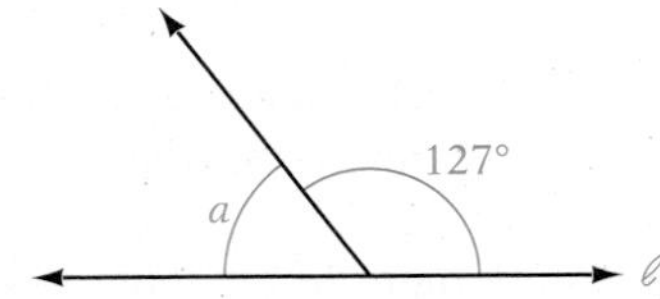

27.

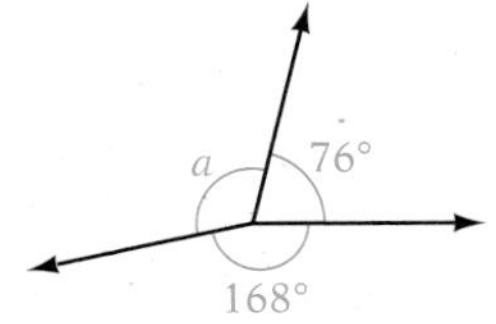

28.

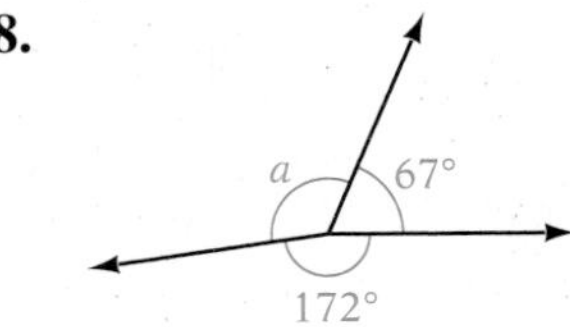

For Exercises 29 to 34, find x.

29.

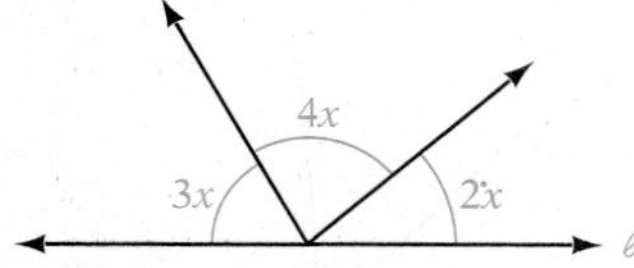

30.

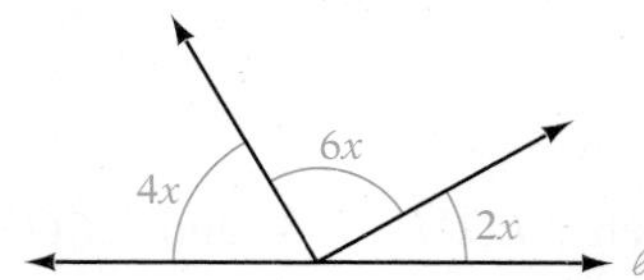

31.

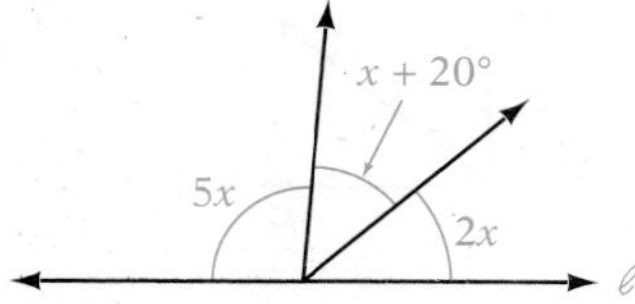

32.

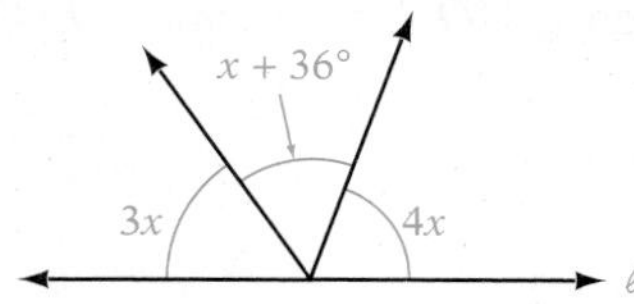

33.

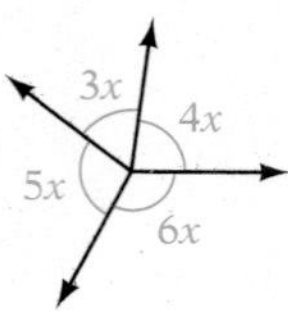

34.

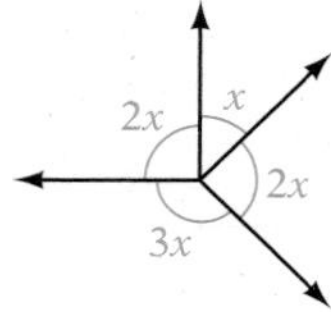

35. Given $\angle a = 51°$, find the measure of $\angle b$.

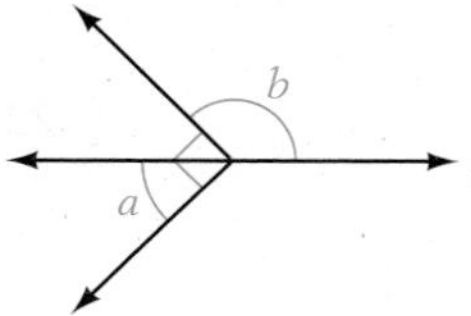

36. Given $\angle a = 38°$, find the measure of $\angle b$.

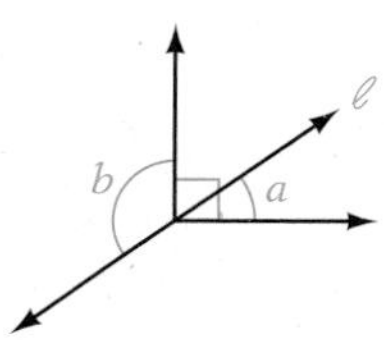

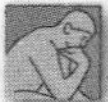

For Exercises 37 to 40, determine whether the described angle is an acute angle, is a right angle, is an obtuse angle, or does not exist.

37. The complement of an acute angle

38. The supplement of a right angle

39. The supplement of an acute angle

40. The complement of an obtuse angle

OBJECTIVE B To solve problems involving angles formed by intersecting lines

For Exercises 41 and 42, find the measure of $\angle x$.

41. 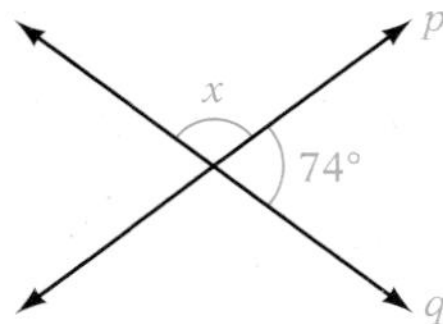

42. 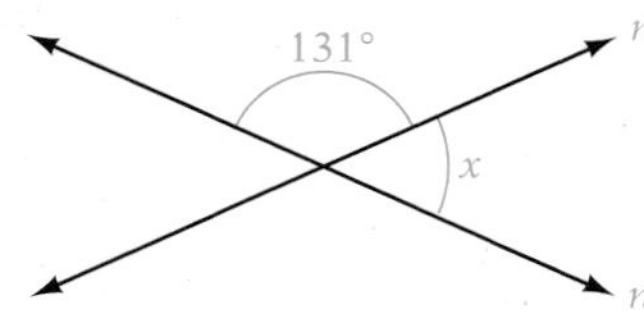

For Exercises 43 and 44, find x.

43. 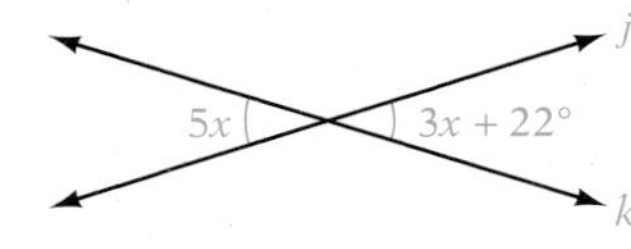

44. 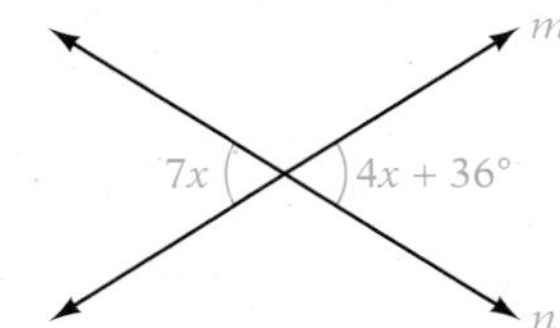

For Exercises 45 to 48, given that $\ell_1 \parallel \ell_2$, find the measures of angles a and b.

45. 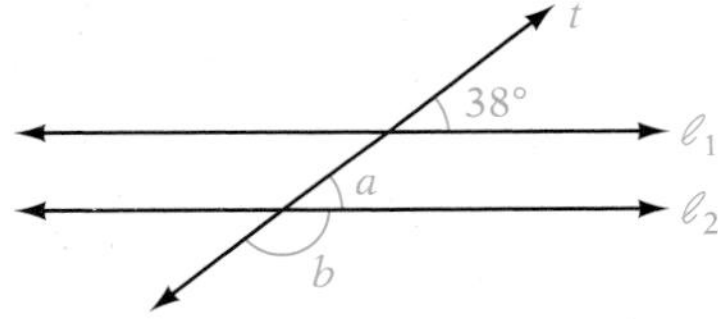

46. 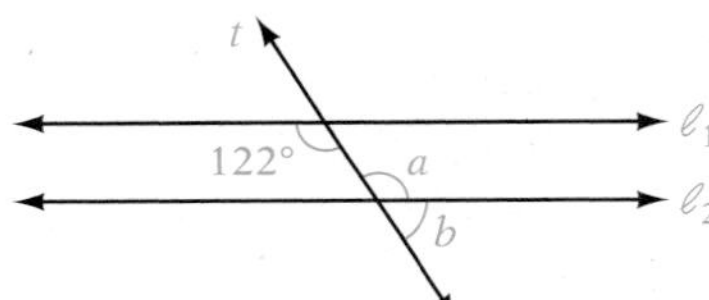

47. 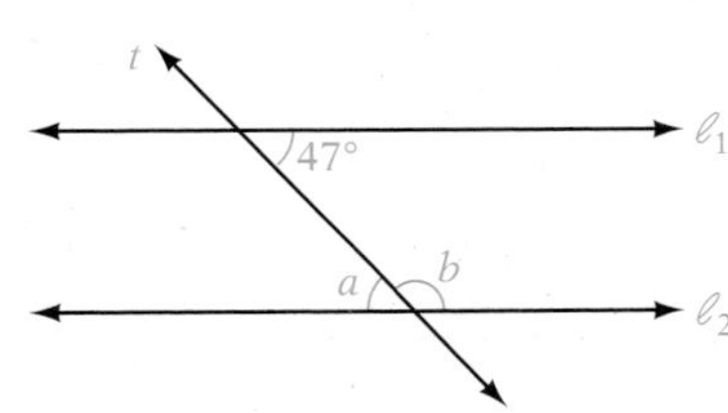

48. 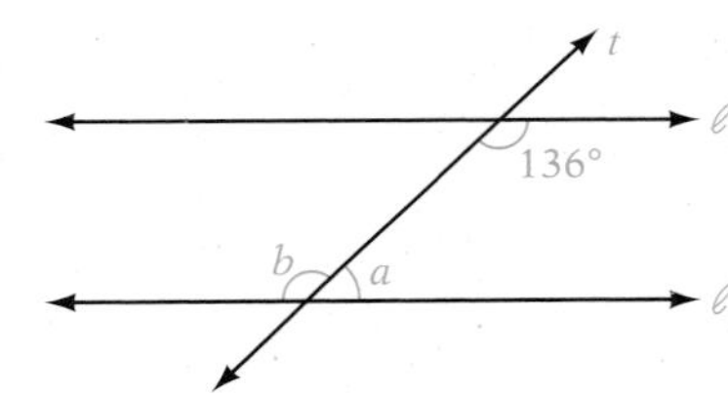

For Exercises 49 to 52, given that $\ell_1 \parallel \ell_2$, find x.

49. 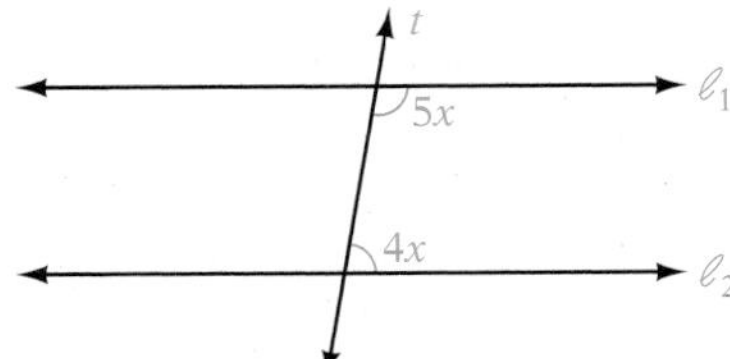

50. 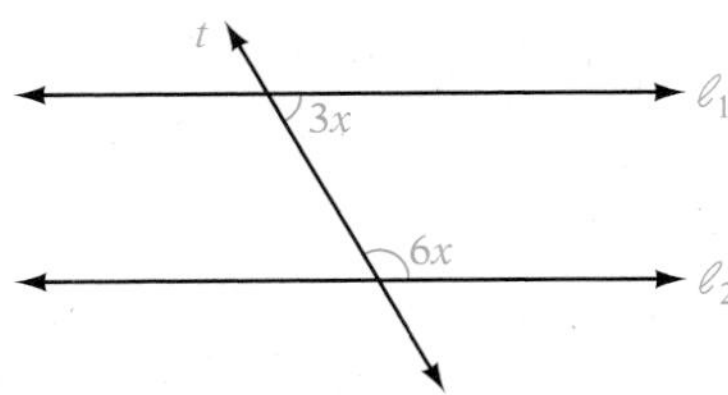

51. 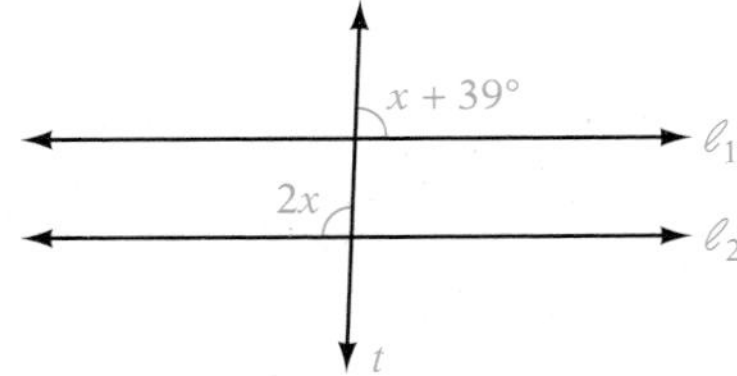

52. 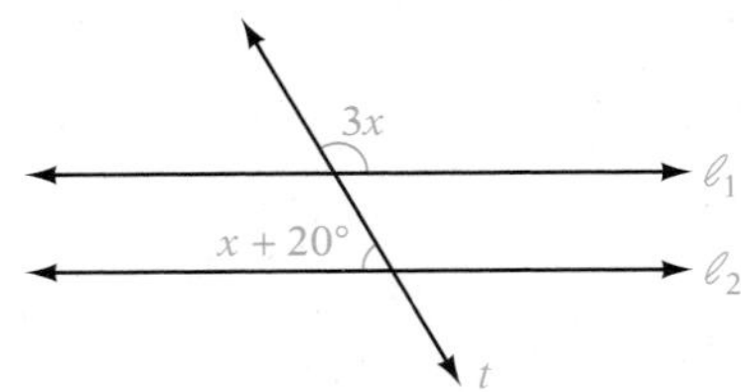

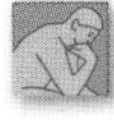

For Exercises 53 and 54, use the diagram at the right. Determine whether the given statement is *true* or *false*.

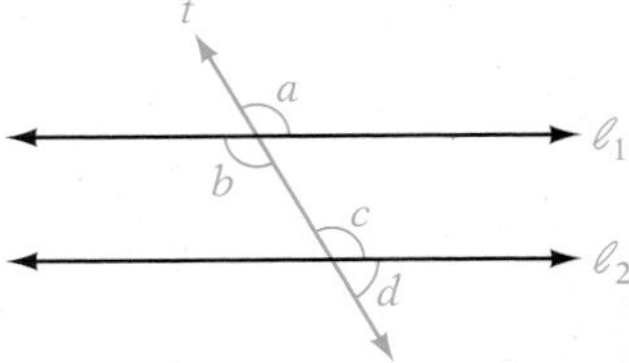

53. $\angle a$ and $\angle b$ have the same measure even if ℓ_1 and ℓ_2 are not parallel.

54. If $\angle a$ is greater than $\angle c$, then ℓ_1 and ℓ_2 are not parallel.

OBJECTIVE C **To solve problems involving the angles of a triangle**

55. Given that $\angle a = 95°$ and $\angle b = 70°$, find the measures of angles x and y.

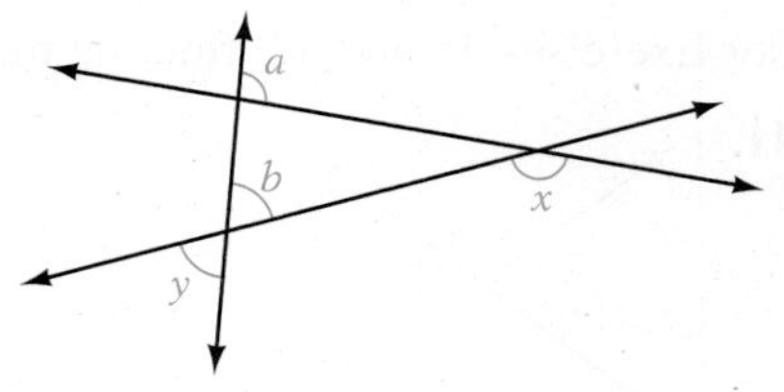

56. Given that $\angle a = 35°$ and $\angle b = 55°$, find the measures of angles x and y.

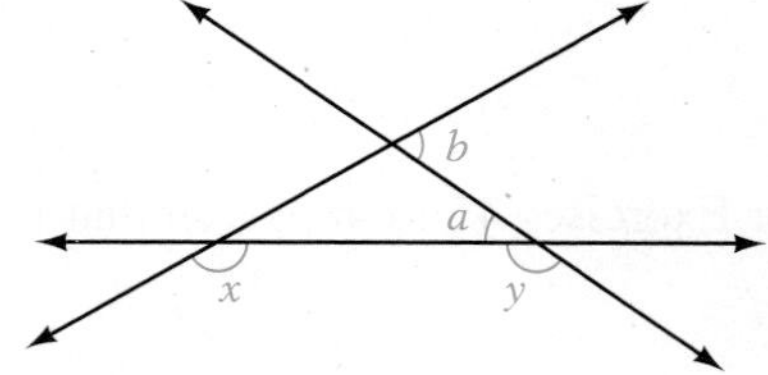

57. Given that $\angle y = 45°$, find the measures of angles a and b.

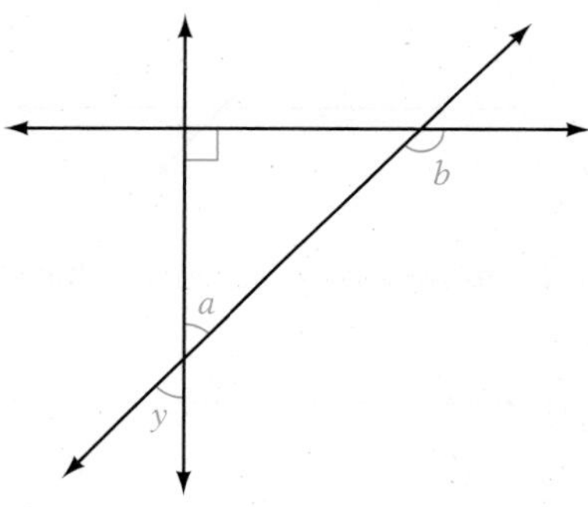

58. Given that $\angle y = 130°$, find the measures of angles a and b.

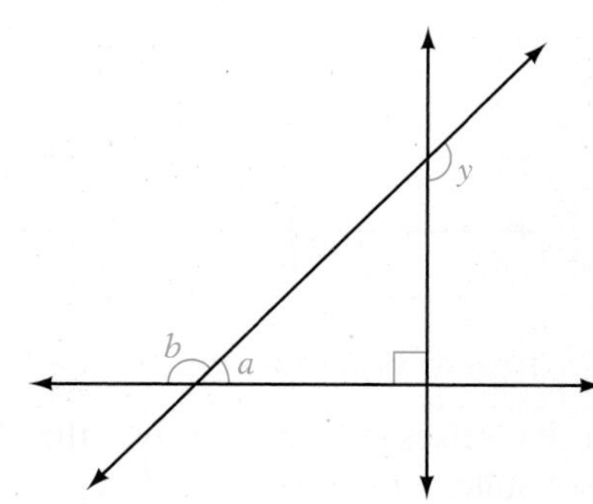

59. Given that $\overline{AO} \perp \overline{OB}$, express in terms of x the number of degrees in $\angle BOC$.

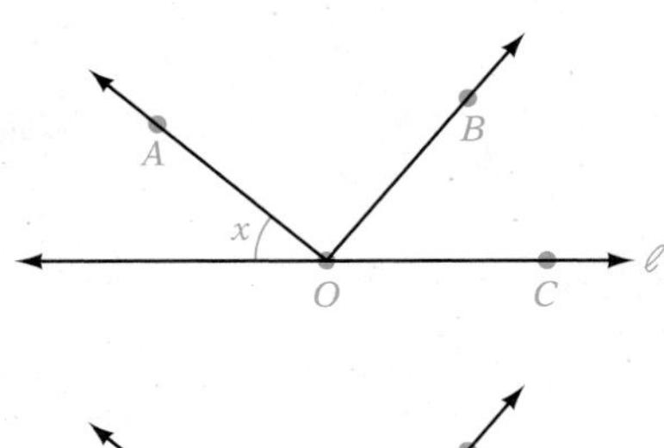

60. Given that $\overline{AO} \perp \overline{OB}$, express in terms of x the number of degrees in $\angle AOC$.

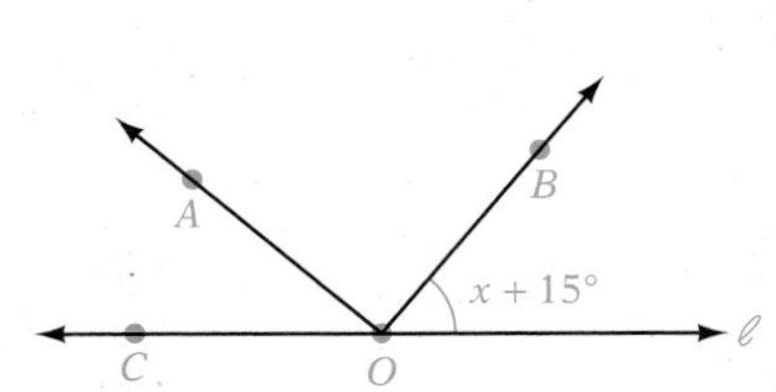

61. One angle in a triangle is a right angle, and one angle is equal to 30°. What is the measure of the third angle?

62. A triangle has a 45° angle and a right angle. Find the measure of the third angle.

63. Two angles of a triangle measure 42° and 103°. Find the measure of the third angle.

64. Two angles of a triangle measure 62° and 45°. Find the measure of the third angle.

65. A triangle has a 13° angle and a 65° angle. What is the measure of the third angle?

66. A triangle has a 105° angle and a 32° angle. What is the measure of the third angle?

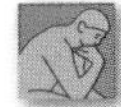

For Exercises 67 and 68, determine whether the statement is *true* or *false*.

67. A triangle can have two obtuse angles.

68. If the sum of two angles of a triangle is less than 90°, then the third angle is an obtuse angle.

Applying the Concepts

69. Cut out a triangle and then tear off two of the angles, as shown at the right. Position the pieces you tore off so that angle a is adjacent to angle b and angle c is adjacent to angle b (on the other side). Describe what you observe. What does this demonstrate?

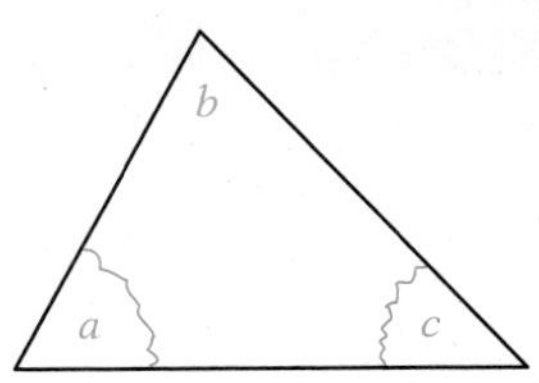

70. Determine whether the statement is always true, sometimes true, or never true.
a. Two lines that are parallel to a third line are parallel to each other.
b. A triangle contains two acute angles.
c. Vertical angles are complementary angles.

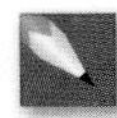

71. If $\overline{AB}$ and $\overline{CD}$ intersect at point O, and $\angle AOC = \angle BOC$, explain why $\overline{AB} \perp \overline{CD}$.

SECTION

3.2 Plane Geometric Figures

OBJECTIVE A To solve problems involving the perimeter of a geometric figure

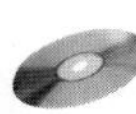

A **polygon** is a closed figure determined by three or more line segments that lie in a plane. The **sides of a polygon** are the line segments that form the polygon. The figures below are examples of polygons.

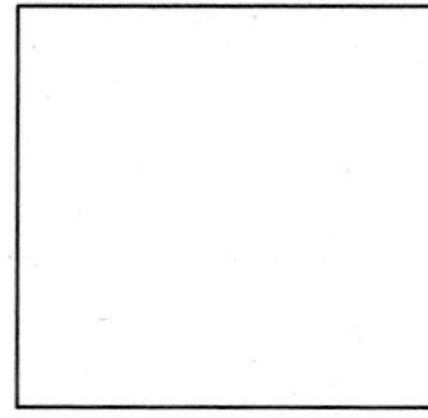
A

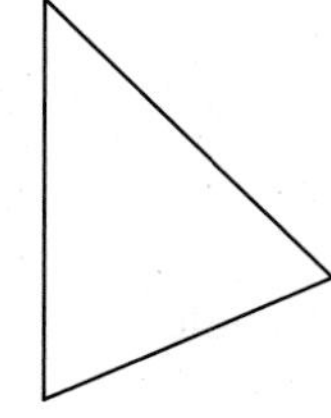
B

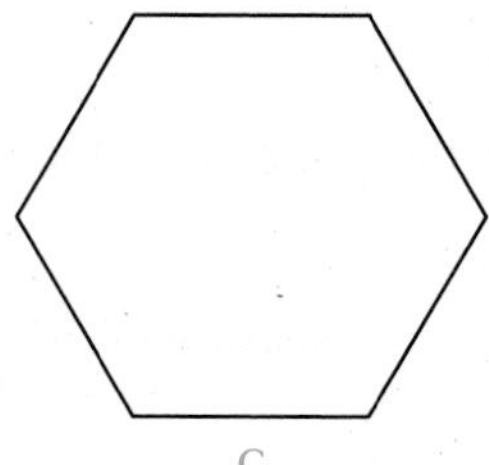
C

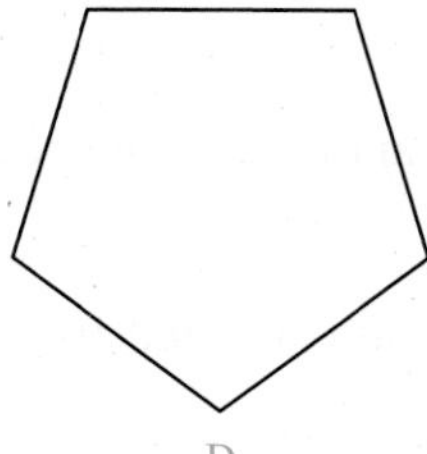
D

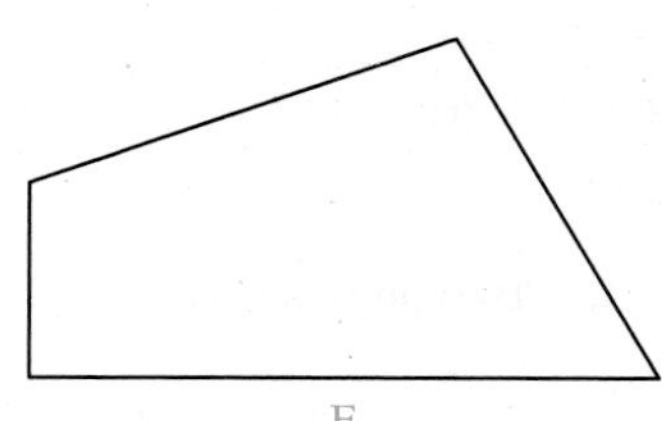
E

Point of Interest

Although a polygon is defined in terms of its *sides* (see the definition above), the word actually comes from the Latin word *polygonum,* which means "having many *angles.*" This is certainly the case for some polygons.

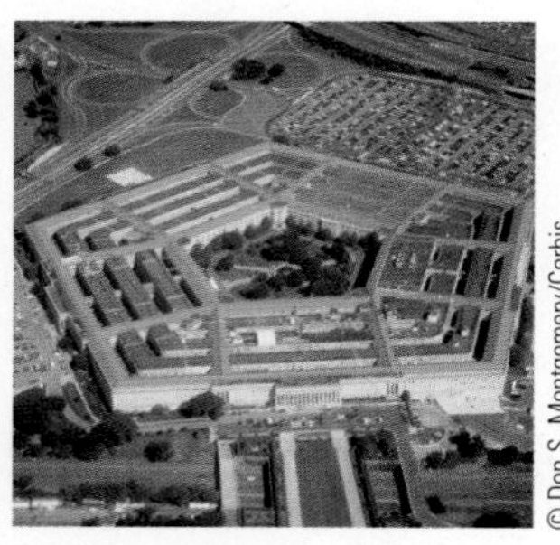
© Don S. Montgomery/Corbis

The Pentagon in Arlington, Virginia

A **regular polygon** is one in which each side has the same length and each angle has the same measure. The polygons in Figures A, C, and D above are regular polygons.

The name of a polygon is based on the number of its sides. The table below lists the names of polygons that have from 3 to 10 sides.

Number of Sides	*Name of the Polygon*
3	Triangle
4	Quadrilateral
5	Pentagon
6	Hexagon
7	Heptagon
8	Octagon
9	Nonagon
10	Decagon

Triangles and quadrilaterals are two of the most common types of polygons. Triangles are distinguished by the number of equal sides and also by the measures of their angles.

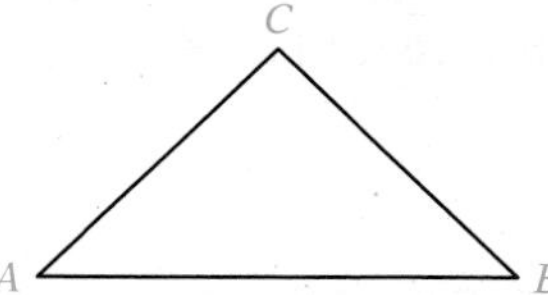

An **isosceles triangle** has two sides of equal length. The angles opposite the equal sides are of equal measure.
$AC = BC$
$\angle A = \angle B$

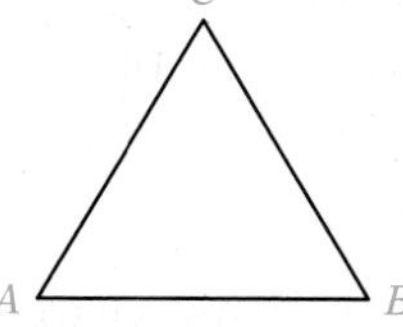

The three sides of an **equilateral triangle** are of equal length. The three angles are of equal measure.
$AB = BC = AC$
$\angle A = \angle B = \angle C$

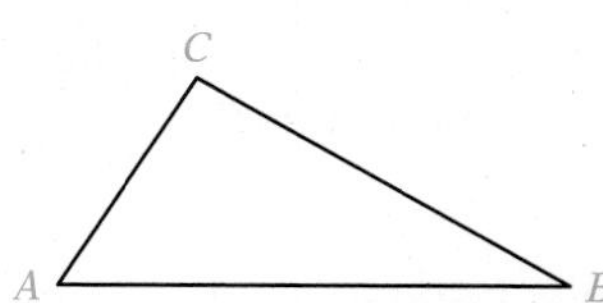

A **scalene triangle** has no two sides of equal length. No two angles are of equal measure.

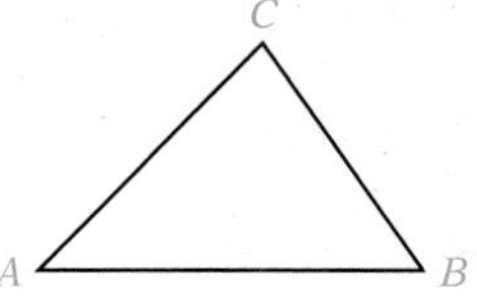

An **acute triangle** has three acute angles.

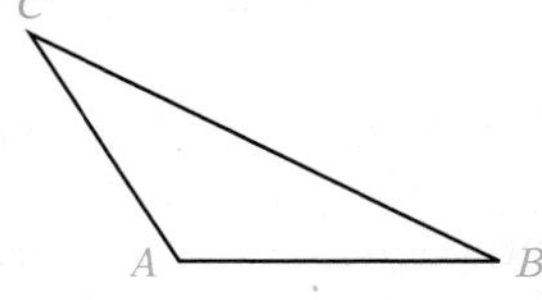

An **obtuse triangle** has one obtuse angle.

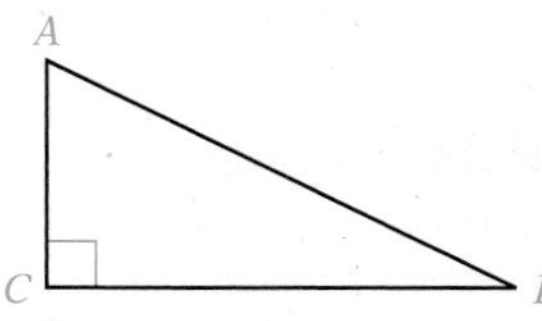

A **right triangle** has a right angle.

Quadrilaterals are also distinguished by their sides and angles, as shown below. Note that a rectangle, a square, and a rhombus are different forms of a parallelogram.

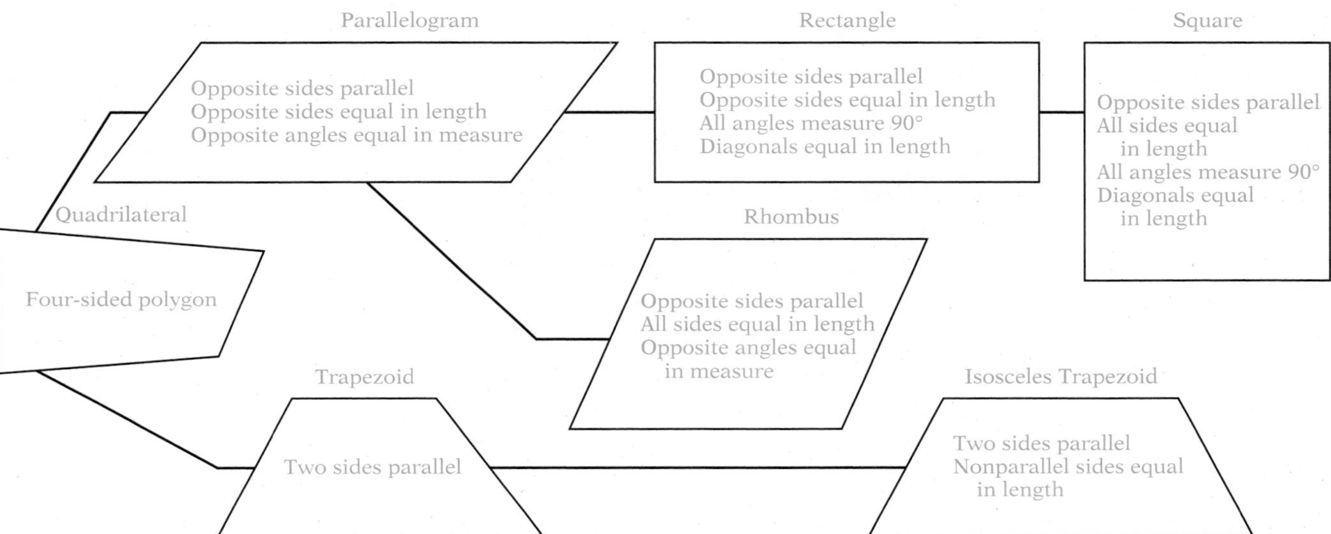

The **perimeter** of a plane geometric figure is a measure of the distance around the figure. Perimeter is used in buying fencing for a lawn or determining how much baseboard is needed for a room.

The perimeter of a triangle is the sum of the lengths of the three sides.

Perimeter of a Triangle

Let a, b, and c be the lengths of the sides of a triangle. The perimeter, P, of the triangle is given by $P = a + b + c$.

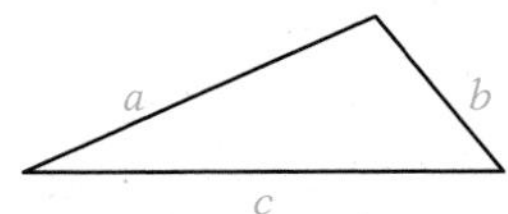

HOW TO • 1 Find the perimeter of the triangle shown at the right.

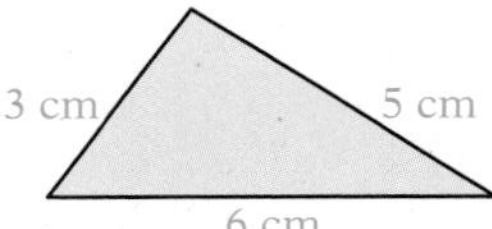

$$\begin{aligned} P &= a + b + c \\ &= 3 \text{ cm} + 5 \text{ cm} + 6 \text{ cm} \\ &= 14 \text{ cm} \end{aligned}$$

The perimeter of the triangle is 14 cm.

The perimeter of a quadrilateral is the sum of the lengths of its four sides.

Point of Interest

Leonardo DaVinci painted the Mona Lisa on a rectangular canvas whose height was approximately 1.6 times its width. Rectangles with these proportions, called golden rectangles, were used extensively in Renaissance art.

A rectangle is a quadrilateral with opposite sides of equal length. Usually the length, L, of a rectangle refers to the length of one of the longer sides of the rectangle, and the width, W, refers to the length of one of the shorter sides. The perimeter can then be represented by $P = L + W + L + W$.

$P = L + W + L + W$

The formula for the perimeter of a rectangle is derived by combining like terms.

$P = 2L + 2W$

Perimeter of a Rectangle

Let L represent the length and W the width of a rectangle. The perimeter, P, of the rectangle is given by $P = 2L + 2W$.

HOW TO • 2 Find the perimeter of the rectangle shown at the right.

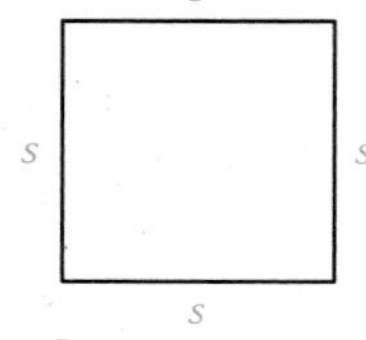

$$\begin{aligned} P &= 2L + 2W \\ &= 2(6\text{ m}) + 2(3\text{ m}) \quad \bullet\ L = 6\text{ m};\ W = 3\text{ m} \\ &= 12\text{ m} + 6\text{ m} \\ &= 18\text{ m} \end{aligned}$$

The perimeter of the rectangle is 18 m.

A square is a rectangle in which each side has the same length. Let s represent the length of each side of a square. Then the perimeter of the square can be represented by $P = s + s + s + s$.

$P = s + s + s + s$

The formula for the perimeter of a square is derived by combining like terms.

$P = 4s$

Perimeter of a Square

Let s represent the length of a side of a square. The perimeter, P, of the square is given by $P = 4s$.

HOW TO • 3 Find the perimeter of the square shown at the right.

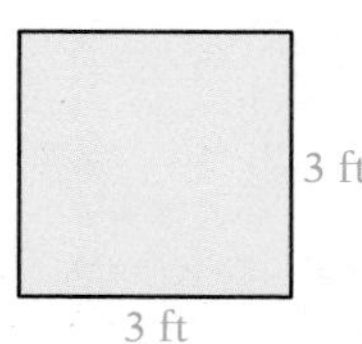

$$\begin{aligned} P &= 4s \\ &= 4(3\text{ ft}) \quad \bullet\ s = 3\text{ ft} \\ &= 12\text{ ft} \end{aligned}$$

The perimeter of the square is 12 ft.

A **circle** is a plane figure in which all points are the same distance from point *O*, which is called the **center** of the circle.

The **diameter** of a circle is a line segment across the circle through point *O*. *AB* is a diameter of the circle at the right. The variable *d* is used to designate the diameter of a circle.

The **radius** of a circle is a line segment from the center of the circle to a point on the circle. *OC* is a radius of the circle at the right. The variable *r* is used to designate a radius of a circle.

The length of the diameter of a circle is twice the length of the radius.

$d = 2r$ or $r = \frac{1}{2}d$

Point of Interest

Archimedes (c. 287–212 B.C.) was the mathematician who gave us the approximate value of π as $\frac{22}{7} = 3\frac{1}{7}$. He actually showed that π was between $3\frac{10}{71}$ and $3\frac{1}{7}$. The approximation $3\frac{10}{71}$ is closer to the exact value of π, but it is more difficult to use.

The distance around a circle is called the **circumference.** The circumference, *C*, of a circle is equal to the product of π (pi) and the diameter.

$C = \pi d$

Because $d = 2r$, the formula for the circumference can be written in terms of *r*.

$C = 2\pi r$

The Circumference of a Circle

The circumference, *C*, of a circle with diameter *d* and radius *r* is given by $C = \pi d$ or $C = 2\pi r$.

The formula for circumference uses the number π, which is an irrational number. The value of π can be approximated by a fraction or by a decimal.

$\pi \approx \frac{22}{7}$ or $\pi \approx 3.14$

The π key on a scientific calculator gives a closer approximation of π than 3.14. Use a scientific calculator to find approximate values in calculations involving π.

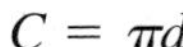

Find the circumference of a circle with a diameter of 6 in.

$C = \pi d$

$C = \pi(6 \text{ in.})$

• The diameter of the circle is given. Use the circumference formula that involves the diameter. $d = 6$ **in.**

$C = 6\pi$ in.

• The exact circumference of the circle is 6π in.

$C \approx 18.85$ in.

• An approximate measure is found by using the π key on a calculator.

The circumference is approximately 18.85 in.

Integrating Technology

The π key on your calculator can be used to find decimal approximations for expressions that contain π. To perform the calculation at the right, enter $6 \times \pi =$.

EXAMPLE • 1

A carpenter is designing a square patio with a perimeter of 44 ft. What is the length of each side?

Strategy
To find the length of each side, use the formula for the perimeter of a square. Substitute 44 for P and solve for s.

Solution
$P = 4s$
$44 = 4s$
$11 = s$

The length of each side of the patio is 11 ft.

YOU TRY IT • 1

The infield of a softball field is a square with each side of length 60 ft. Find the perimeter of the infield.

Your strategy

Your solution

EXAMPLE • 2

The dimensions of a triangular sail are 18 ft, 11 ft, and 15 ft. What is the perimeter of the sail?

Strategy
To find the perimeter, use the formula for the perimeter of a triangle. Substitute 18 ft for a, 11 ft for b, and 15 ft for c. Solve for P.

Solution
$P = a + b + c$
$P = 18 \text{ ft} + 11 \text{ ft} + 15 \text{ ft}$
$P = 44 \text{ ft}$

The perimeter of the sail is 44 ft.

YOU TRY IT • 2

What is the perimeter of a standard piece of typing paper that measures $8\frac{1}{2}$ in. by 11 in.?

Your strategy

Your solution

EXAMPLE • 3

Find the circumference of a circle with a radius of 15 cm. Round to the nearest hundredth.

Strategy
To find the circumference, use the circumference formula that involves the radius. An approximation is asked for; use the π key on a calculator. $r = 15$ cm.

Solution
$C = 2\pi r = 2\pi(15 \text{ cm}) = 30\pi \text{ cm} \approx 94.25 \text{ cm}$

The circumference is approximately 94.25 cm.

YOU TRY IT • 3

Find the circumference of a circle with a diameter of 9 in. Give the exact measure.

Your strategy

Your solution

Solutions on p. S10

OBJECTIVE B To solve problems involving the area of a geometric figure 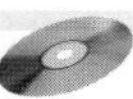

Area is a measure of the amount of surface in a region. Area can be used to describe the size of a rug, a parking lot, a farm, or a national park. Area is measured in square units.

A square that measures 1 in. on each side has an area of 1 square inch, which is written 1 in^2.

A square that measures 1 cm on each side has an area of 1 square centimeter, which is written 1 cm^2.

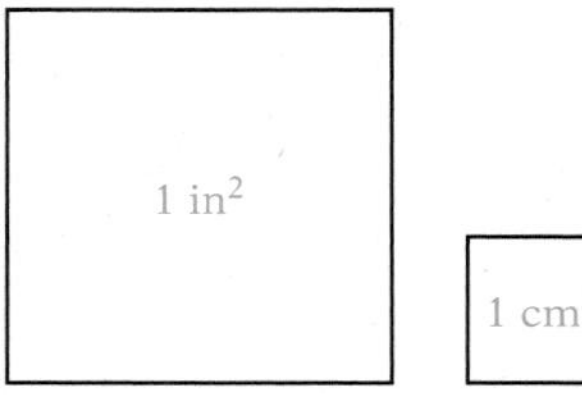

Larger areas can be measured in square feet (ft^2), square meters (m^2), square miles (mi^2), acres (43,560 ft^2), or any other square unit.

The area of a geometric figure is the number of squares that are necessary to cover the figure. In the figures below, two rectangles have been drawn and covered with squares. In the figure on the left, 12 squares, each of area 1 cm^2, were used to cover the rectangle. The area of the rectangle is 12 cm^2. In the figure on the right, 6 squares, each of area 1 in^2, were used to cover the rectangle. The area of the rectangle is 6 in^2.

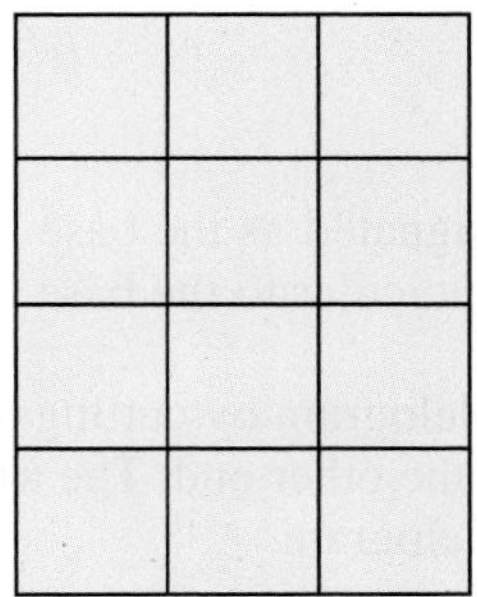

The area of the rectangle is 12 cm^2.

The area of the rectangle is 6 in^2.

Note from the above figures that the area of a rectangle can be found by multiplying the length of the rectangle by its width.

Area of a Rectangle

Let L represent the length and W the width of a rectangle. The area, A, of the rectangle is given by $A = LW$.

HOW TO 5 Find the area of the rectangle shown at the right.

$A = LW$
$= (8\text{ ft})(5\text{ ft})$ • $L = 8$ ft, $W = 5$ ft
$= 40\text{ ft}^2$

The area of the rectangle is 40 ft^2.

A square is a rectangle in which all sides are the same length. Therefore, both the length and the width of a square can be represented by s, and $A = LW = s \cdot s = s^2$.

Area of a Square

Let s represent the length of a side of a square. The area, A, of the square is given by $A = s^2$.

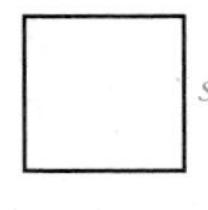

$A = s \cdot s = s^2$

HOW TO • 6 Find the area of the square shown at the right.

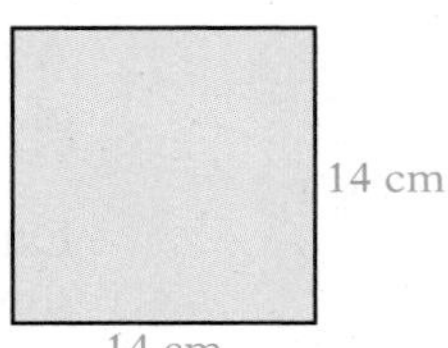

$$A = s^2$$
$$= (14 \text{ cm})^2 \quad \bullet\ s = 14 \text{ cm}$$
$$= 196 \text{ cm}^2$$

The area of the square is 196 cm^2.

Figure $ABCD$ is a parallelogram. BC is the **base,** b, of the parallelogram. AE, perpendicular to the base, is the **height,** h, of the parallelogram.

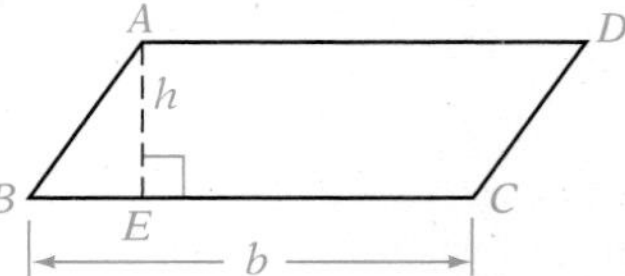

Any side of a parallelogram can be designated as the base. The corresponding height is found by drawing a line segment perpendicular to the base from the opposite side.

A rectangle can be formed from a parallelogram by cutting a right triangle from one end of the parallelogram and attaching it to the other end. The area of the resulting rectangle will equal the area of the original parallelogram.

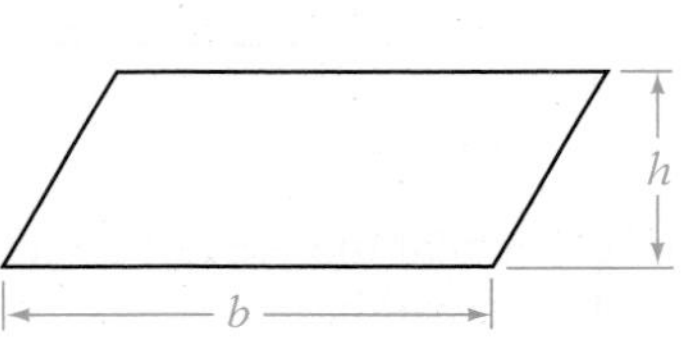

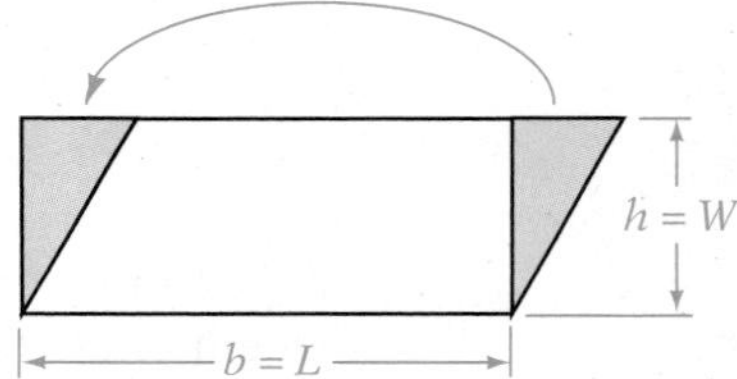

Area of a Parallelogram

Let b represent the length of the base and h the height of a parallelogram. The area, A, of the parallelogram is given by $A = bh$.

HOW TO • 7 Find the area of the parallelogram shown at the right.

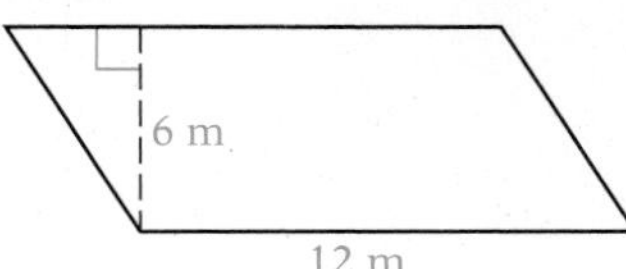

$$A = bh = (12 \text{ m})(6 \text{ m}) = 72 \text{ m}^2$$

The area is 72 m^2.

Figure *ABC* is a triangle. *AB* is the **base,** *b*, of the triangle. *CD*, perpendicular to the base, is the **height,** *h*, of the triangle.

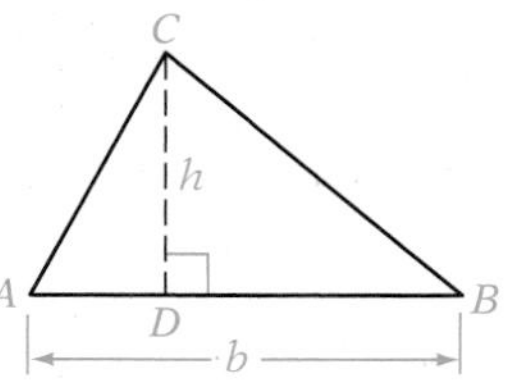

Any side of a triangle can be designated as the base. The corresponding height is found by drawing a line segment perpendicular to the base from the vertex opposite the base.

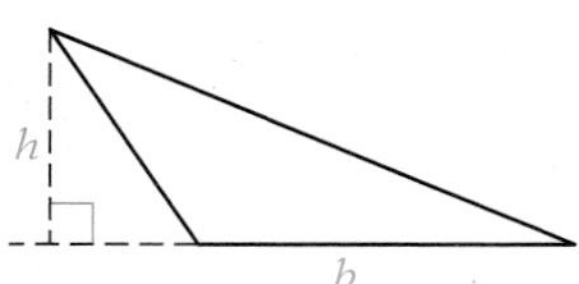

Consider the triangle with base *b* and height *h* shown at the right. By extending a line from *C* parallel to the base *AB* and equal in length to the base, and extending a line from *B* parallel to *AC* and equal in length to *AC*, a parallelogram is formed. The area of the parallelogram is *bh* and is twice the area of the triangle. Therefore, the area of the triangle is one-half the area of the parallelogram, or $\frac{1}{2}bh$.

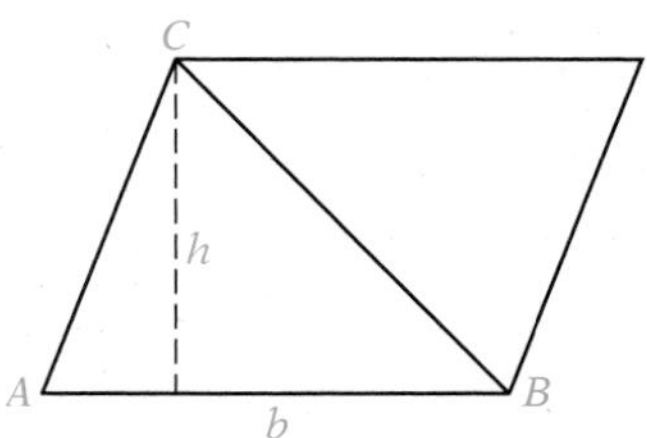

Area of a Triangle

Let b represent the length of the base and h the height of a triangle. The area, A, of the triangle is given by $A = \frac{1}{2}bh$.

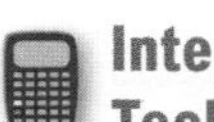

Integrating Technology

To calculate the area of the triangle shown at the right, you can enter

20 × 5 ÷ 2 =

or

.5 × 20 × 5 =

HOW TO • 8 Find the area of the triangle shown at the right.

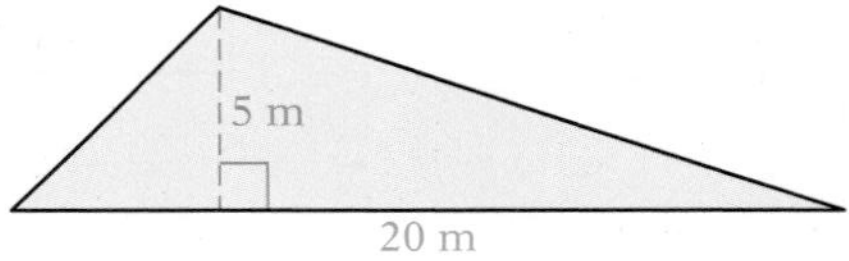

$$A = \frac{1}{2}bh$$

$$= \frac{1}{2}(20 \text{ m})(5 \text{ m}) \quad \bullet\ b = 20 \text{ m}, h = 5 \text{ m}$$

$$= 50 \text{ m}^2$$

The area of the triangle is 50 m^2.

Figure *ABCD* is a trapezoid. *AB* is one **base,** b_1, of the trapezoid, and *CD* is the other base, b_2. *AE,* perpendicular to the two bases, is the **height,** *h*.

In the trapezoid at the right, the line segment *BD* divides the trapezoid into two triangles, *ABD* and *BCD*. In triangle *ABD*, b_1 is the base and *h* is the height. In triangle *BCD*, b_2 is the base and *h* is the height. The area of the trapezoid is the sum of the areas of the two triangles.

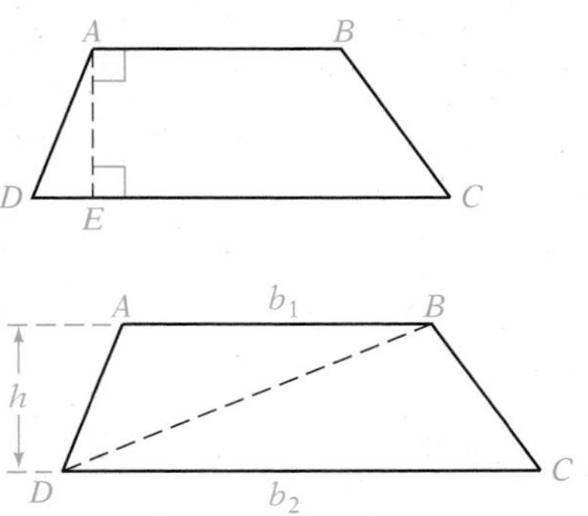

$$\text{Area of trapezoid } ABCD = \text{area of triangle } ABD + \text{area of triangle } BCD$$

$$= \frac{1}{2}b_1h + \frac{1}{2}b_2h = \frac{1}{2}h(b_1 + b_2)$$

Area of a Trapezoid

Let b_1 and b_2 represent the lengths of the bases and h the height of a trapezoid. The area, A, of the trapezoid is given by $A = \frac{1}{2}h(b_1 + b_2)$.

HOW TO • 9 Find the area of a trapezoid that has bases measuring 15 in. and 5 in. and a height of 8 in.

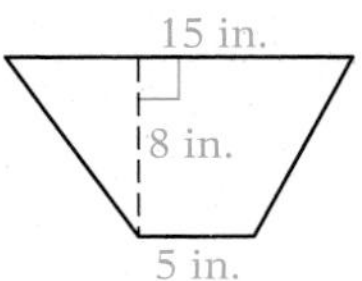

$$A = \frac{1}{2}h(b_1 + b_2)$$
$$= \frac{1}{2}(8 \text{ in.})(15 \text{ in.} + 5 \text{ in.}) = (4 \text{ in.})(20 \text{ in.}) = 80 \text{ in}^2$$

The area is 80 in^2.

The area of a circle is equal to the product of π and the square of the radius.

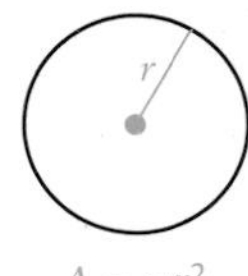

$A = \pi r^2$

Area of a Circle

The area, A, of a circle with radius r is given by $A = \pi r^2$.

HOW TO • 10 Find the area of a circle that has a radius of 6 cm.

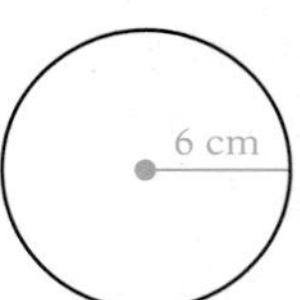

$A = \pi r^2$
$A = \pi(6 \text{ cm})^2$
$A = \pi(36 \text{ cm}^2)$

• Use the formula for the area of a circle. $r = 6$ cm

$A = 36\pi \text{ cm}^2$
$A \approx 113.10 \text{ cm}^2$

• The exact area of the circle is $36\pi \text{ cm}^2$.
• An approximate measure is found by using the π key on a calculator.

The approximate area of the circle is 113.10 cm^2.

Integrating Technology

To approximate 36π on your calculator, enter $36 \times \pi =$.

For your reference, all of the formulas for the perimeters and areas of the geometric figures presented in this section are listed in the Chapter Summary located at the end of this chapter.

EXAMPLE • 4

The Parks and Recreation Department of a city plans to plant grass seed in a playground that has the shape of a trapezoid, as shown below. Each bag of grass seed will seed 1500 ft^2. How many bags of grass seed should the department purchase?

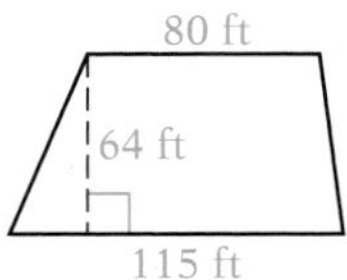

Strategy

To find the number of bags to be purchased:

- Use the formula for the area of a trapezoid to find the area of the playground.
- Divide the area of the playground by the area one bag will seed (1500).

Solution

$$A = \frac{1}{2}h(b_1 + b_2)$$

$$A = \frac{1}{2} \cdot (64 \text{ ft})(80 \text{ ft} + 115 \text{ ft})$$

$$A = 6240 \text{ ft}^2$$ • The area of the playground is 6240 ft^2.

$$6240 \div 1500 = 4.16$$

Because a portion of a fifth bag is needed, 5 bags of grass seed should be purchased.

YOU TRY IT • 4

An interior designer decides to wallpaper two walls of a room. Each roll of wallpaper will cover 30 ft^2. Each wall measures 8 ft by 12 ft. How many rolls of wallpaper should be purchased?

Your strategy

Your solution

EXAMPLE • 5

Find the area of a circle with a diameter of 5 ft. Give the exact measure.

Strategy

To find the area:

- Find the radius of the circle.
- Use the formula for the area of a circle. Leave the answer in terms of π.

Solution

$$r = \frac{1}{2}d = \frac{1}{2}(5 \text{ ft}) = 2.5 \text{ ft}$$

$$A = \pi r^2 = \pi(2.5 \text{ ft})^2 = \pi(6.25 \text{ ft}^2) = 6.25\pi \text{ ft}^2$$

The area of the circle is $6.25\pi \text{ ft}^2$.

YOU TRY IT • 5

Find the area of a circle with a radius of 11 cm. Round to the nearest hundredth.

Your strategy

Your solution

Solutions on p. S10

3.2 EXERCISES

OBJECTIVE A **To solve problems involving the perimeter of a geometric figure**

For Exercises 1 to 4, name each polygon.

1. 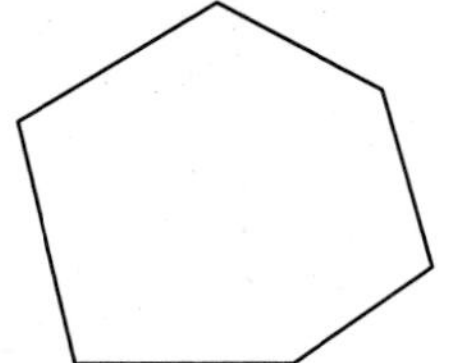

2. 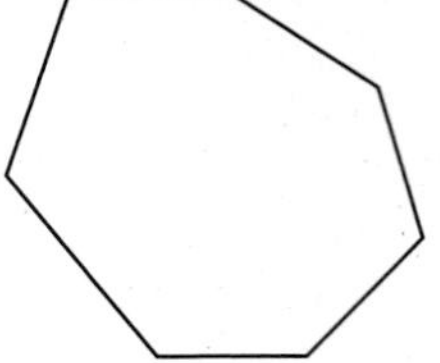

3. 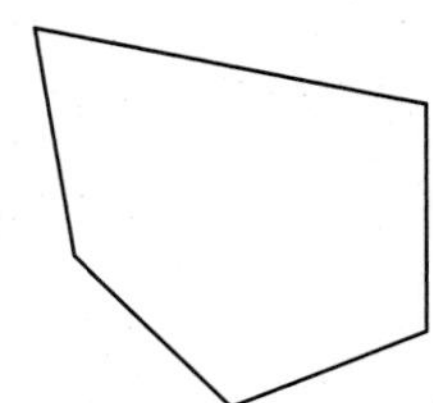

4. 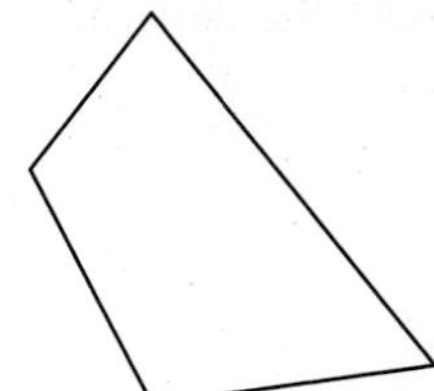

For Exercises 5 to 8, classify the triangle as isosceles, equilateral, or scalene.

5. 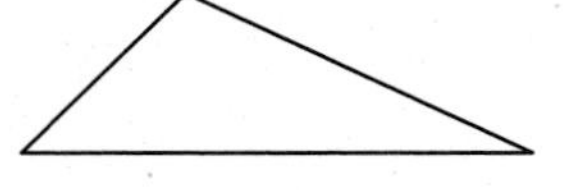

6. 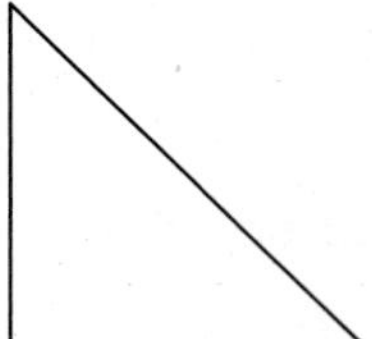

7. 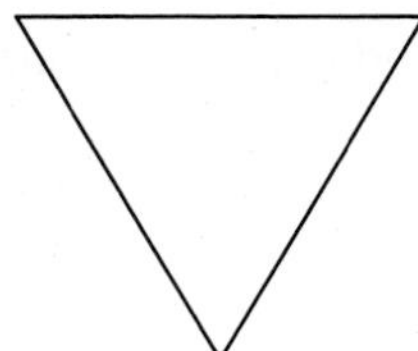

8. 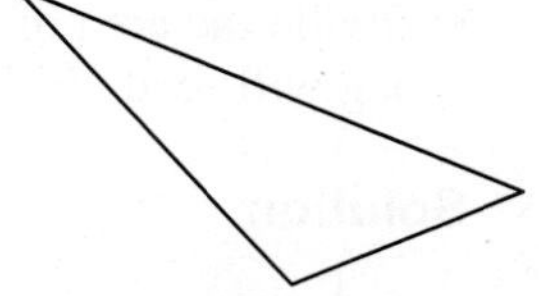

For Exercises 9 to 12, classify the triangle as acute, obtuse, or right.

9. 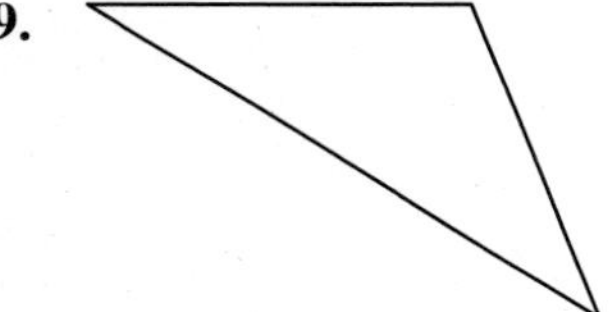

10. 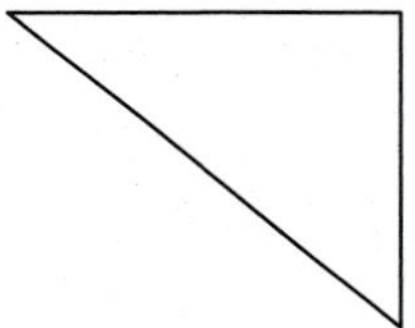

11. 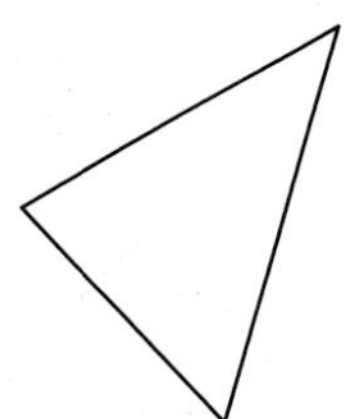

12. 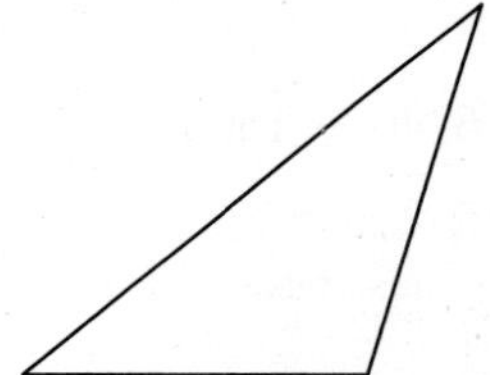

For Exercises 13 to 18, find the perimeter of the figure.

13. 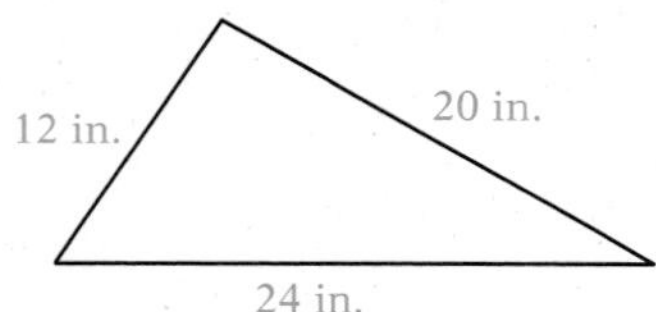

14. 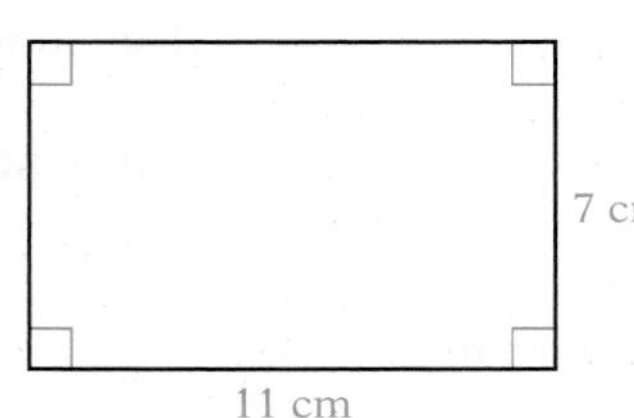

15. 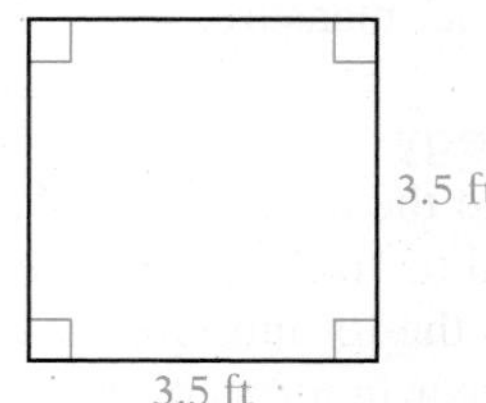

16. 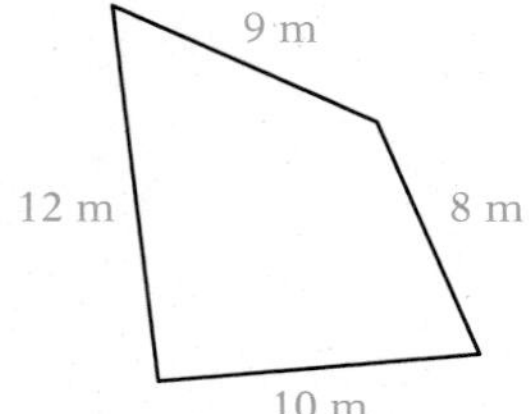

17. 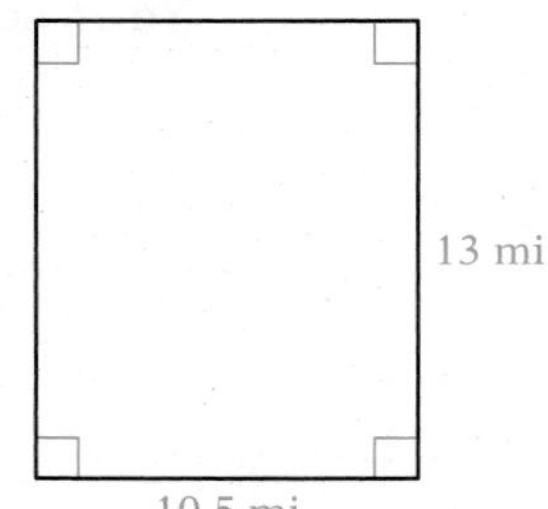

18. 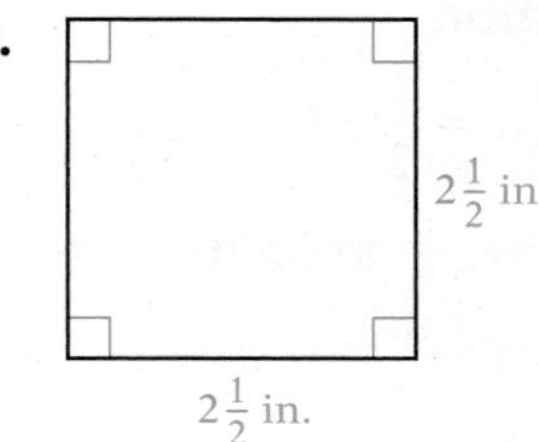

For Exercises 19 to 24, find the circumference of the figure. Give both the exact value and an approximation to the nearest hundredth.

19.

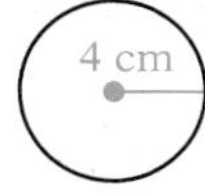

20.

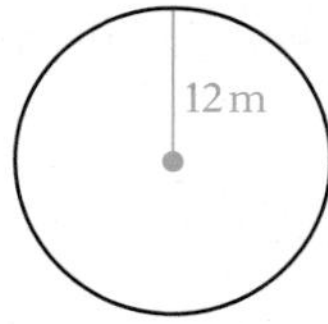

21.

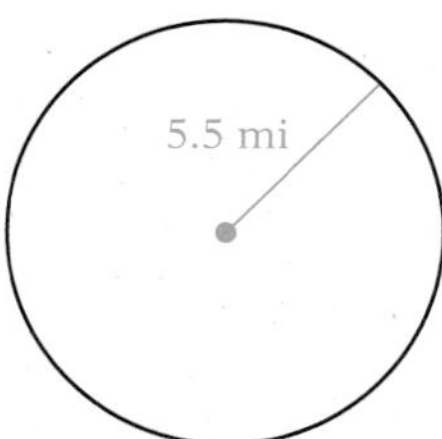

22.

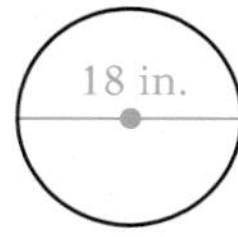

23.

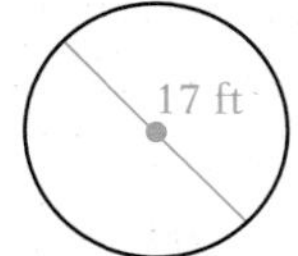

24.

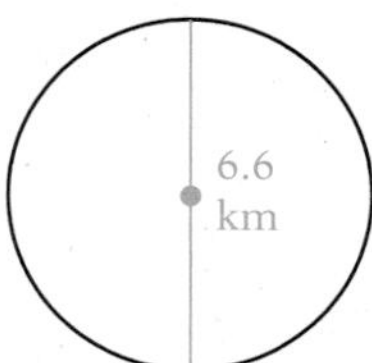

25. The lengths of the three sides of a triangle are 3.8 cm, 5.2 cm, and 8.4 cm. Find the perimeter of the triangle.

26. The lengths of the three sides of a triangle are 7.5 m, 6.1 m, and 4.9 m. Find the perimeter of the triangle.

27. The length of each of two sides of an isosceles triangle is $2\frac{1}{2}$ cm. The third side measures 3 cm. Find the perimeter of the triangle.

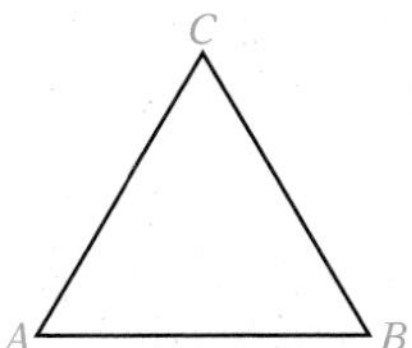

28. The length of each side of an equilateral triangle is $4\frac{1}{2}$ in. Find the perimeter of the triangle.

29. A rectangle has a length of 8.5 m and a width of 3.5 m. Find the perimeter of the rectangle.

30. Find the perimeter of a rectangle that has a length of $5\frac{1}{2}$ ft and a width of 4 ft.

31. The length of each side of a square is 12.2 cm. Find the perimeter of the square.

32. Find the perimeter of a square that measures 0.5 m on each side.

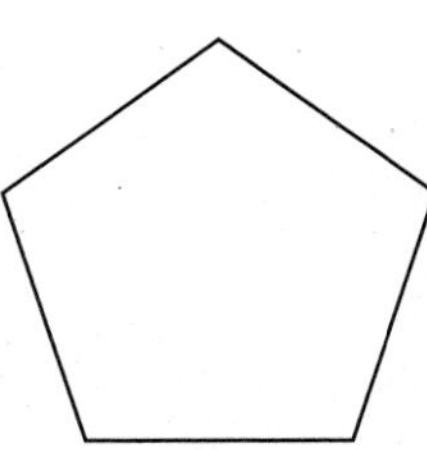

33. Find the perimeter of a regular pentagon that measures 3.5 in. on each side.

34. What is the perimeter of a regular hexagon that measures 8.5 cm on each side?

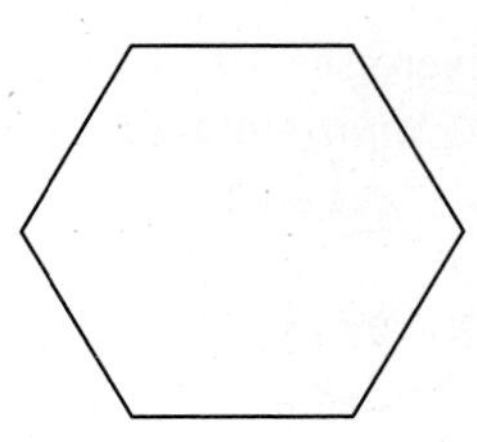

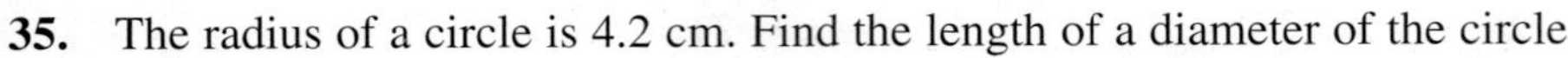

35. The radius of a circle is 4.2 cm. Find the length of a diameter of the circle.

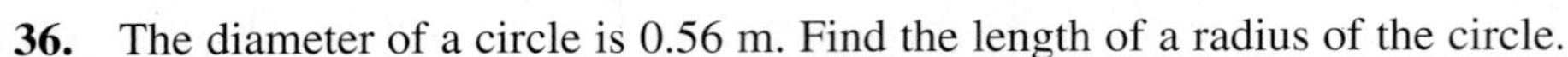

36. The diameter of a circle is 0.56 m. Find the length of a radius of the circle.

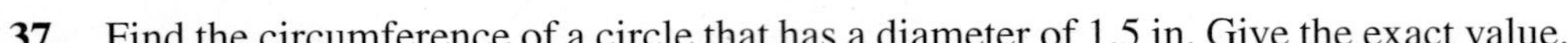

37. Find the circumference of a circle that has a diameter of 1.5 in. Give the exact value.

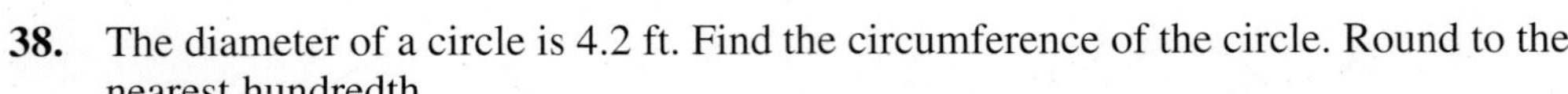

38. The diameter of a circle is 4.2 ft. Find the circumference of the circle. Round to the nearest hundredth.

39. The radius of a circle is 36 cm. Find the circumference of the circle. Round to the nearest hundredth.

40. Find the circumference of a circle that has a radius of 2.5 m. Give the exact value.

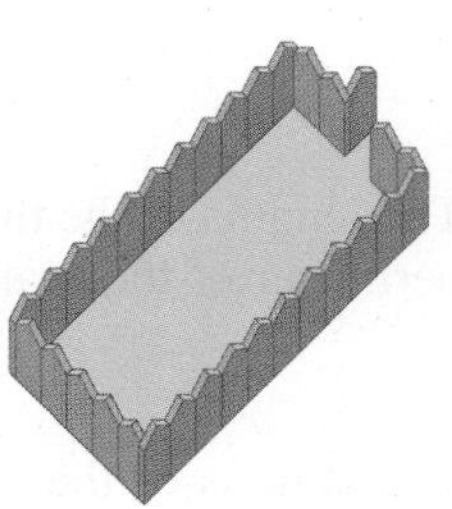

41. **Fencing** How many feet of fencing should be purchased for a rectangular garden that is 18 ft long and 12 ft wide?

42. **Quilting** How many meters of binding are required to bind the edge of a rectangular quilt that measures 3.5 m by 8.5 m?

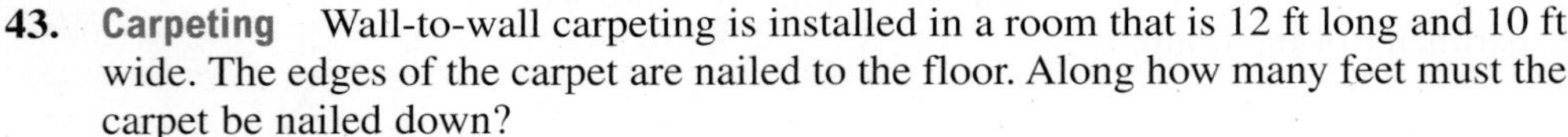

43. **Carpeting** Wall-to-wall carpeting is installed in a room that is 12 ft long and 10 ft wide. The edges of the carpet are nailed to the floor. Along how many feet must the carpet be nailed down?

44. **Fencing** The length of a rectangular park is 55 yd. The width is 47 yd. How many yards of fencing are needed to surround the park?

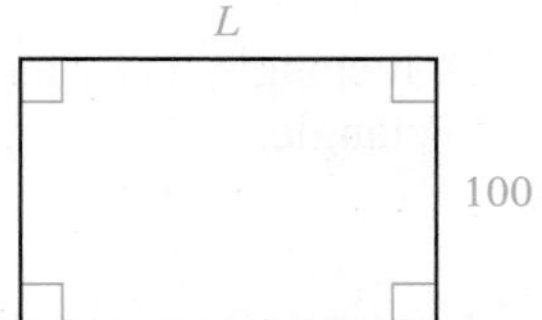

45. **Playgrounds** The perimeter of a rectangular playground is 440 ft. If the width is 100 ft, what is the length of the playground?

46. **Gardens** A rectangular vegetable garden has a perimeter of 64 ft. The length of the garden is 20 ft. What is the width of the garden?

47. **Banners** Each of two sides of a triangular banner measures 18 in. If the perimeter of the banner is 46 in., what is the length of the third side of the banner?

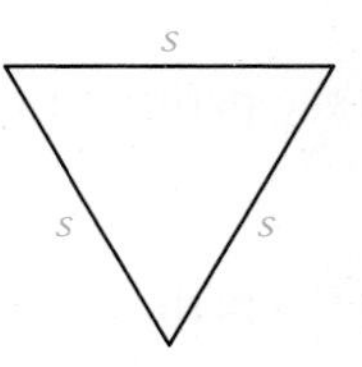

48. The perimeter of an equilateral triangle is 13.2 cm. What is the length of each side of the triangle?

49. **Framing** The perimeter of a square picture frame is 48 in. Find the length of each side of the frame.

50. **Carpeting** A square rug has a perimeter of 32 ft. Find the length of each edge of the rug.

51. The circumference of a circle is 8 cm. Find the length of a diameter of the circle. Round to the nearest hundredth.

52. The circumference of a circle is 15 in. Find the length of a radius of the circle. Round to the nearest hundredth.

53. **Carpentry** Find the length of molding needed to put around a circular table that is 4.2 ft in diameter. Round to the nearest hundredth.

54. **Carpeting** How much binding is needed to bind the edge of a circular rug that is 3 m in diameter? Round to the nearest hundredth.

55. **Cycling** A bicycle tire has a diameter of 24 in. How many feet does the bicycle travel when the wheel makes eight revolutions? Round to the nearest hundredth.

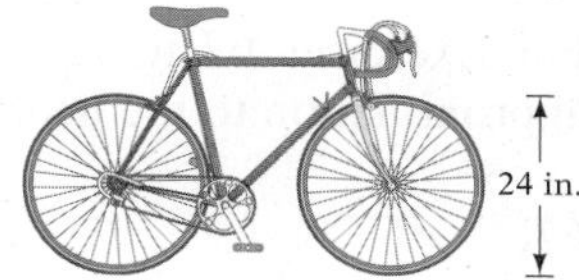

56. **Cycling** A tricycle tire has a diameter of 12 in. How many feet does the tricycle travel when the wheel makes twelve revolutions? Round to the nearest hundredth.

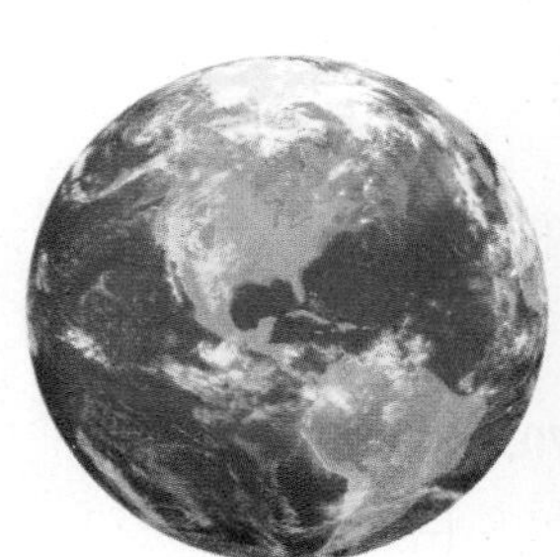

57. **Earth Science** The distance from the surface of Earth to its center is 6356 km. What is the circumference of Earth? Round to the nearest hundredth.

58. **Sewing** Bias binding is to be sewed around the edge of a rectangular tablecloth measuring 72 in. by 45 in. If the bias binding comes in packages containing 15 ft of binding, how many packages of bias binding are needed for the tablecloth?

59. The length of a side of a square is equal to the diameter of a circle. Which is greater, the perimeter of the square or the circumference of the circle?

60. The length of a rectangle is equal to the diameter of a circle, and the width of the rectangle is equal to the radius of the same circle. Which is greater, the perimeter of the rectangle or the circumference of the circle?

OBJECTIVE B To solve problems involving the area of a geometric figure

For Exercises 61 to 66, find the area of the figure.

61. 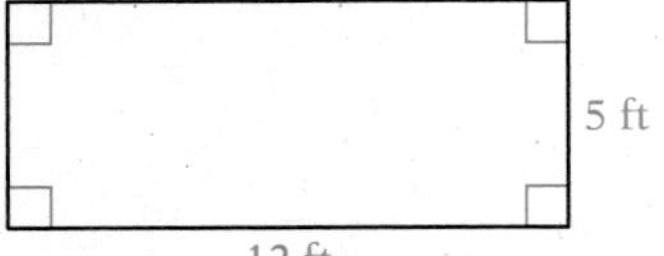

62. 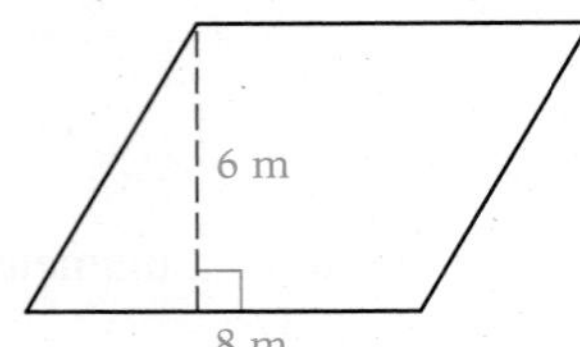

63. 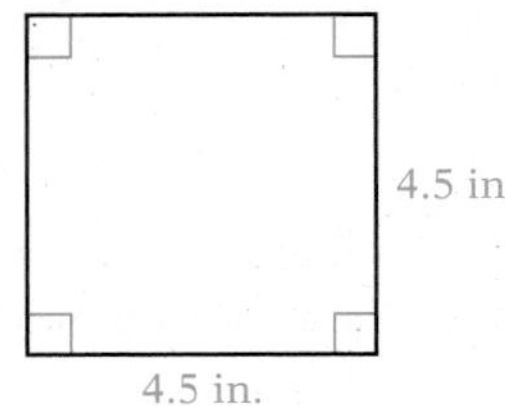

64. 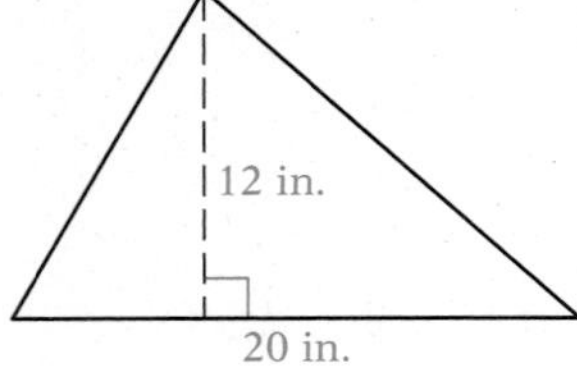

65. 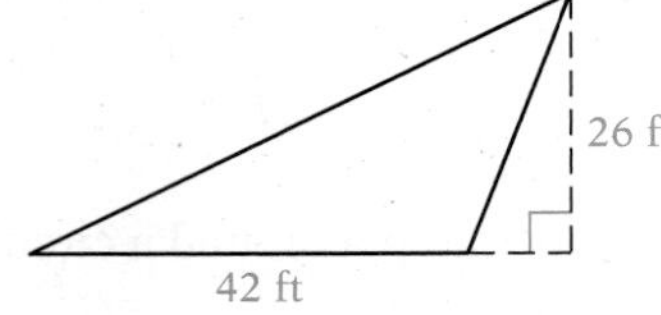

66. 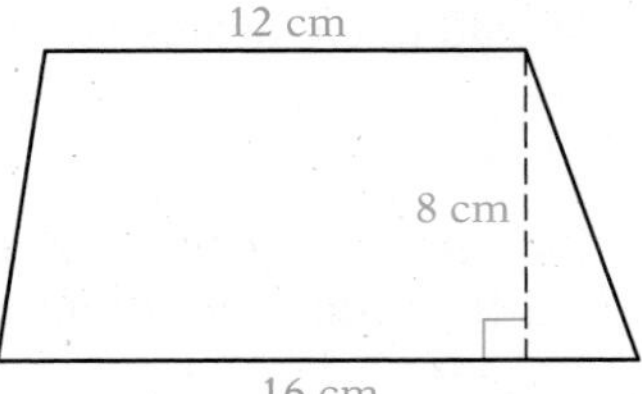

For Exercises 67 to 72, find the area of the figure. Give both the exact value and an approximation to the nearest hundredth.

67. 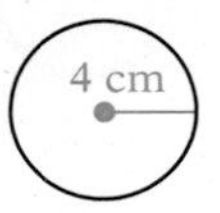

68. 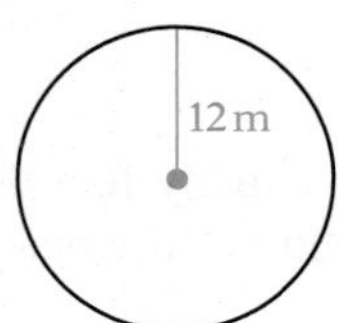

69. 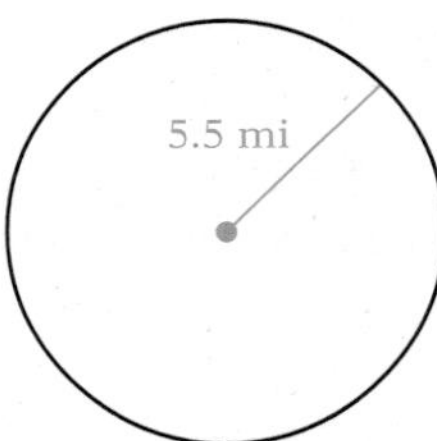

70. 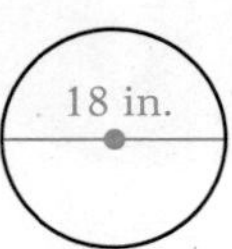

71. 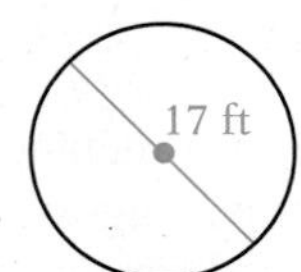

72. 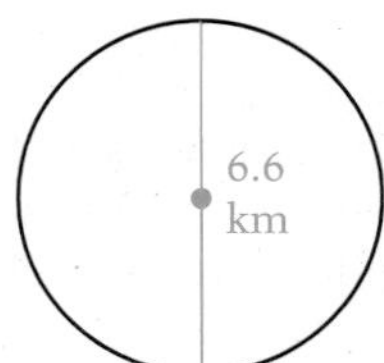

73. The length of a side of a square is 12.5 cm. Find the area of the square.

74. Each side of a square measures $3\frac{1}{2}$ in. Find the area of the square.

75. The length of a rectangle is 38 in., and the width is 15 in. Find the area of the rectangle.

76. Find the area of a rectangle that has a length of 6.5 m and a width of 3.8 m.

77. The length of the base of a parallelogram is 16 in., and the height is 12 in. Find the area of the parallelogram.

78. The height of a parallelogram is 3.4 m, and the length of the base is 5.2 m. Find the area of the parallelogram.

79. The length of the base of a triangle is 6 ft. The height is 4.5 ft. Find the area of the triangle.

80. The height of a triangle is 4.2 cm. The length of the base is 5 cm. Find the area of the triangle.

81. The length of one base of a trapezoid is 35 cm, and the length of the other base is 20 cm. If the height is 12 cm, what is the area of the trapezoid?

82. The height of a trapezoid is 5 in. The bases measure 16 in. and 18 in. Find the area of the trapezoid.

83. The radius of a circle is 5 in. Find the area of the circle. Give the exact value.

84. Find the area of a circle with a radius of 14 m. Round to the nearest hundredth.

85. Find the area of a circle that has a diameter of 3.4 ft. Round to the nearest hundredth.

86. The diameter of a circle is 6.5 m. Find the area of the circle. Give the exact value.

87. **Telescopes** The dome of the Hale telescope at Mount Palomar, California, has a diameter of 200 in. Find the area across the dome. Give the exact value.

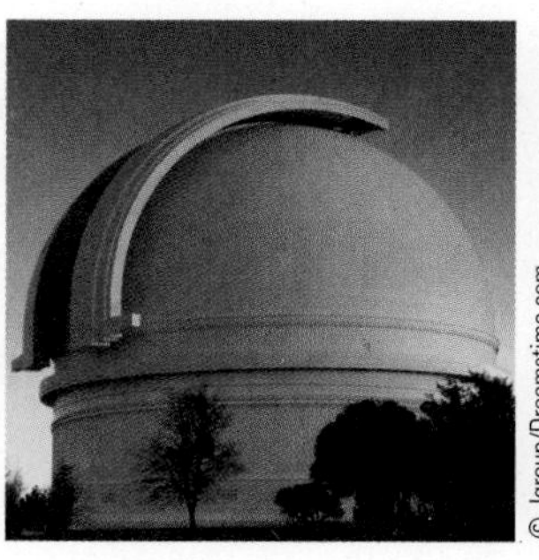

© Jgroup/Dreamstime.com

88. **Patios** What is the area of a square patio that measures 8.5 m on each side?

89. **Gardens** Find the area of a rectangular flower garden that measures 14 ft by 9 ft.

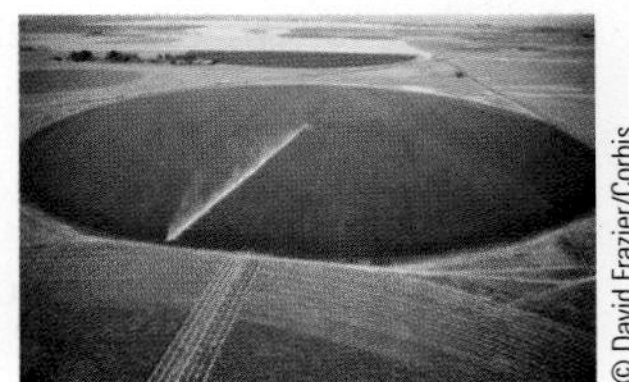

90. **Irrigation** An irrigation system waters a circular field that has a 50-foot radius. Find the area watered by the irrigation system. Give the exact value.

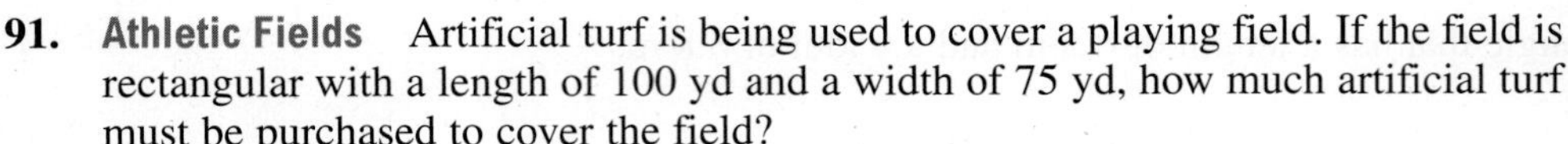

91. **Athletic Fields** Artificial turf is being used to cover a playing field. If the field is rectangular with a length of 100 yd and a width of 75 yd, how much artificial turf must be purchased to cover the field?

92. **Interior Decorating** A fabric wall hanging is to fill a space that measures 5 m by 3.5 m. Allowing for 0.1 m of the fabric to be folded back along each edge, how much fabric must be purchased for the wall hanging?

93. The area of a rectangle is 300 in^2. If the length of the rectangle is 30 in., what is the width?

30 in.

W

94. The width of a rectangle is 12 ft. If the area is 312 ft^2, what is the length of the rectangle?

95. The height of a triangle is 5 m. The area of the triangle is 50 m^2. Find the length of the base of the triangle.

96. The area of a parallelogram is 42 m^2. If the height of the parallelogram is 7 m, what is the length of the base?

97. **Home Maintenance** You plan to stain the wooden deck attached to your house. The deck measures 10 ft by 8 ft. If a quart of stain will cover 50 ft^2, how many quarts of stain should you buy?

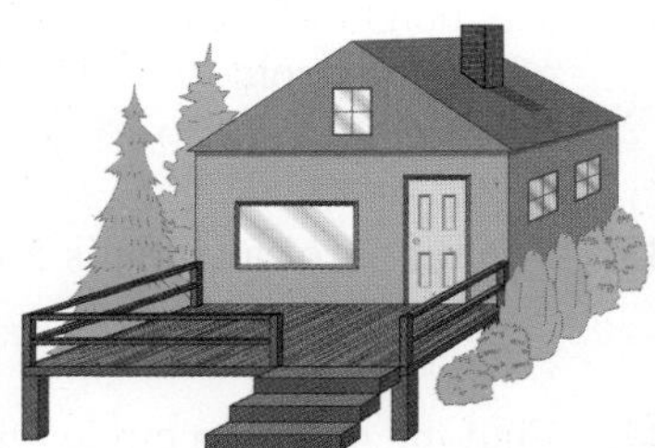

98. **Flooring** You want to tile your kitchen floor. The floor measures 12 ft by 9 ft. How many tiles, each a square with side $1\frac{1}{2}$ ft, should you purchase for the job?

99. **Interior Decorating** You are wallpapering two walls of a child's room, one measuring 9 ft by 8 ft and the other measuring 11 ft by 8 ft. The wallpaper costs \$24.50 per roll, and each roll of the wallpaper will cover 40 ft^2. What will it cost to wallpaper the two walls?

100. **Parks** An urban renewal project involves reseeding a park that is in the shape of a square 60 ft on each side. Each bag of grass seed costs \$9.75 and will seed 1200 ft^2. How much money should be budgeted for buying grass seed for the park?

101. A circle has a radius of 8 in. Find the increase in area when the radius is increased by 2 in. Round to the nearest hundredth.

102. A circle has a radius of 6 cm. Find the increase in area when the radius is doubled. Round to the nearest hundredth.

103. **Landscaping** A walkway 2 m wide surrounds a rectangular plot of grass. The plot is 30 m long and 20 m wide. What is the area of the walkway?

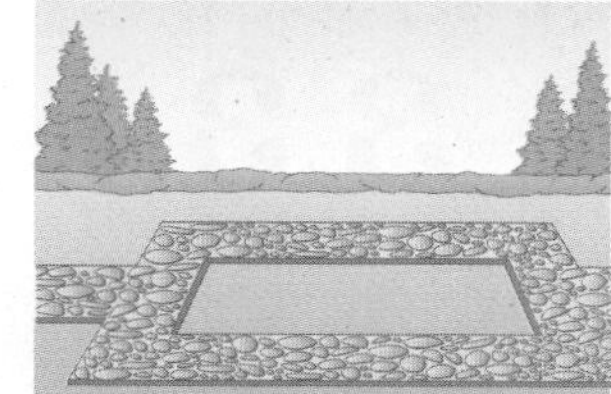

104. **Interior Decorating** You want to paint the walls of your bedroom. Two walls measure 15 ft by 9 ft, and the other two walls measure 12 ft by 9 ft. The paint you wish to purchase costs $28.98 per gallon, and each gallon will cover 400 ft^2 of wall. Find the total amount you will spend on paint.

105. **Carpeting** You want to install wall-to-wall carpeting in your living room, which measures 15 ft by 24 ft. If the cost of the carpet you would like to purchase is $21.95 per square yard, what will be the cost of the carpeting for your living room? (*Hint:* $9 \text{ ft}^2 = 1 \text{ yd}^2$)

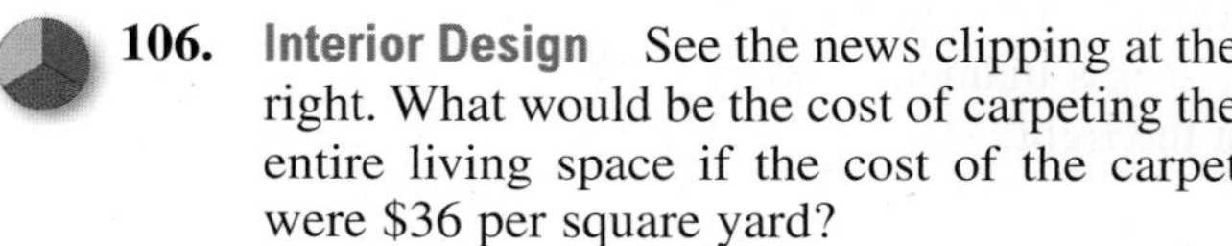

106. **Interior Design** See the news clipping at the right. What would be the cost of carpeting the entire living space if the cost of the carpet were $36 per square yard?

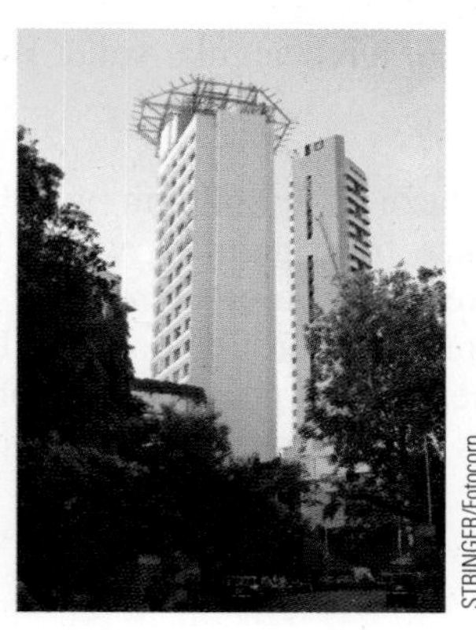

STRINGER/Fotocorp

In the News

Billion-Dollar Home Built in Mumbai

The world's first billion-dollar home is a 27-story skyscraper in downtown Mumbai, India (formerly known as Bombay). It is 550 ft high with 400,000 square feet of living space.

Source: Forbes.com

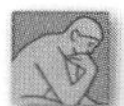

For Exercises 107 and 108, determine whether the area of the first figure is *less than, equal to,* or *greater than* the area of the second figure.

107.

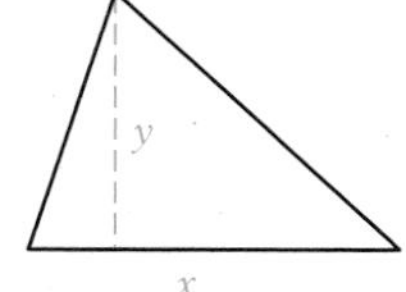

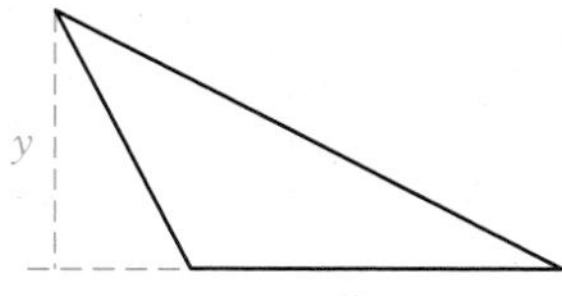

108.

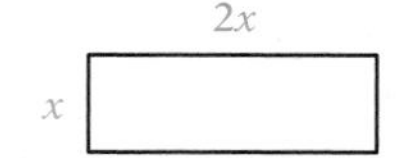

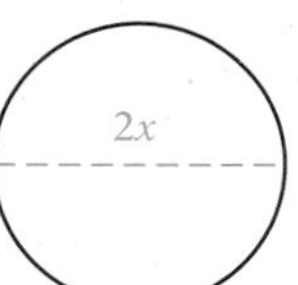

Applying the Concepts

109. If both the length and the width of a rectangle are doubled, how many times larger is the area of the resulting rectangle?

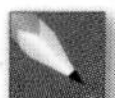

110. Suppose a circle is cut into 16 equal pieces, which are then arranged as shown at the right. The figure formed resembles a parallelogram. What variable expression could describe the base of the parallelogram? What variable could describe its height? Explain how the formula for the area of a circle is derived from this approach.

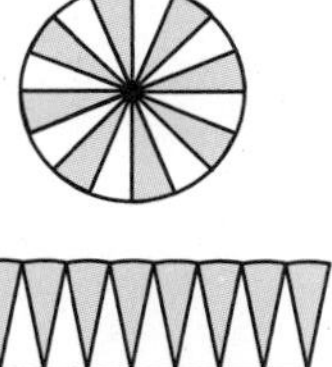

SECTION

3.3 Solids

OBJECTIVE A To solve problems involving the volume of a solid 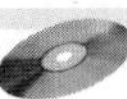

Geometric solids are figures in space. Five common geometric solids are the rectangular solid, the sphere, the cylinder, the cone, and the pyramid.

A **rectangular solid** is one in which all six sides, called faces, are rectangles. The variable L is used to represent the length of a rectangular solid, W its width, and H its height.

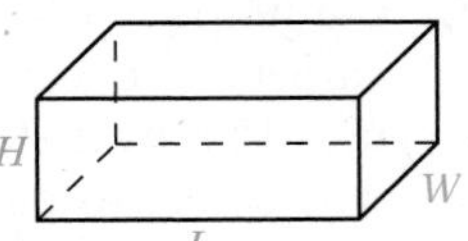

A **sphere** is a solid in which all points are the same distance from point O, which is called the **center** of the sphere. The **diameter**, d, of a sphere is a line across the sphere going through point O. The **radius,** r, is a line from the center to a point on the sphere. AB is a diameter and OC is a radius of the sphere shown at the right.

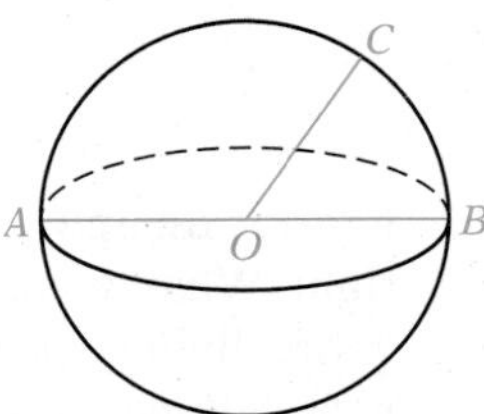

$$d = 2r \text{ or } r = \frac{1}{2}d$$

The most common cylinder, called a **right circular cylinder,** is one in which the bases are circles and are perpendicular to the height of the cylinder. The variable r is used to represent the radius of a base of a cylinder, and h represents the height. In this text, only right circular cylinders are discussed.

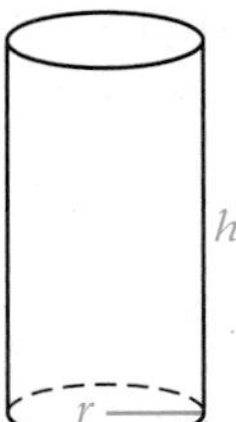

A **right circular cone** is obtained when one base of a right circular cylinder is shrunk to a point, called a **vertex,** V. The variable r is used to represent the radius of the base of the cone, and h represents the height. The variable ℓ is used to represent the **slant height,** which is the distance from a point on the circumference of the base to the vertex. In this text, only right circular cones are discussed.

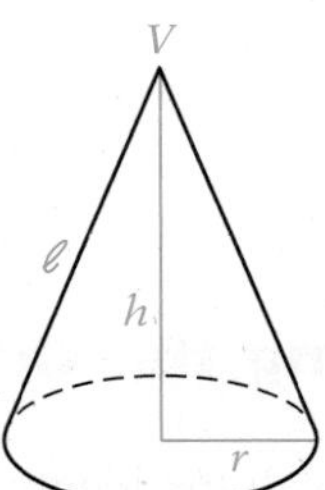

The base of a **regular pyramid** is a regular polygon, and the sides are isosceles triangles. The height, h, is the distance from the vertex, V, to the base and is perpendicular to the base. The variable ℓ is used to represent the **slant height,** which is the height of one of the isosceles triangles on the face of the pyramid. The regular square pyramid at the right has a square base. This is the only type of pyramid discussed in this text.

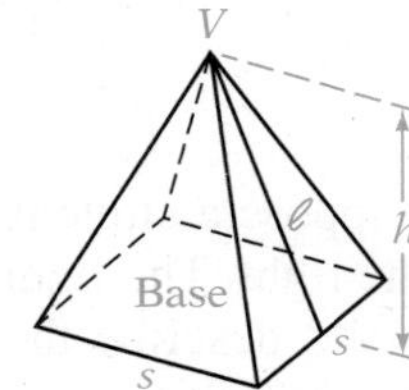

A **cube** is a special type of rectangular solid. Each of the six faces of a cube is a square. The variable s is used to represent the length of one side of a cube.

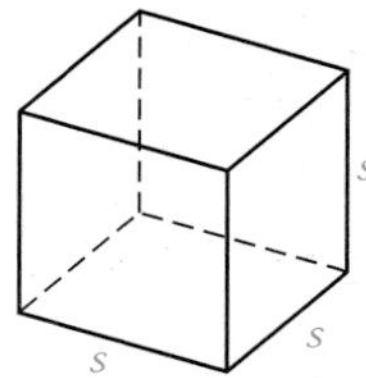

Volume is a measure of the amount of space inside a figure in space. Volume can be used to describe the amount of heating gas used for cooking, the amount of concrete delivered for the foundation of a house, or the amount of water in storage for a city's water supply.

A cube that is 1 ft on each side has a volume of 1 cubic foot, which is written 1 ft^3. A cube that measures 1 cm on each side has a volume of 1 cubic centimeter, which is written 1 cm^3.

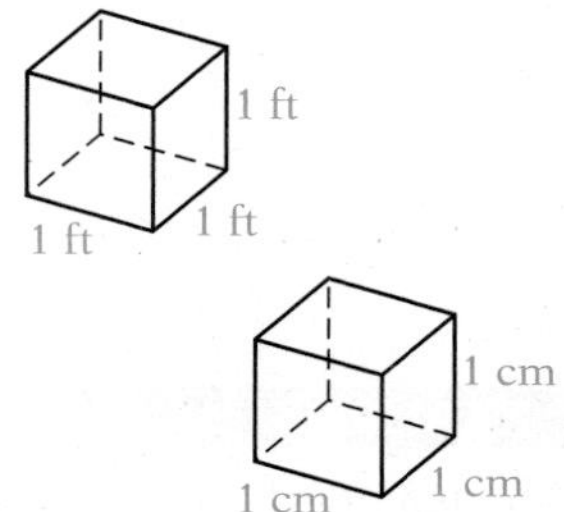

The volume of a solid is the number of cubes that are necessary to exactly fill the solid. The volume of the rectangular solid at the right is 24 cm^3 because it will hold exactly 24 cubes, each 1 cm on a side. Note that the volume can be found by multiplying the length times the width times the height.

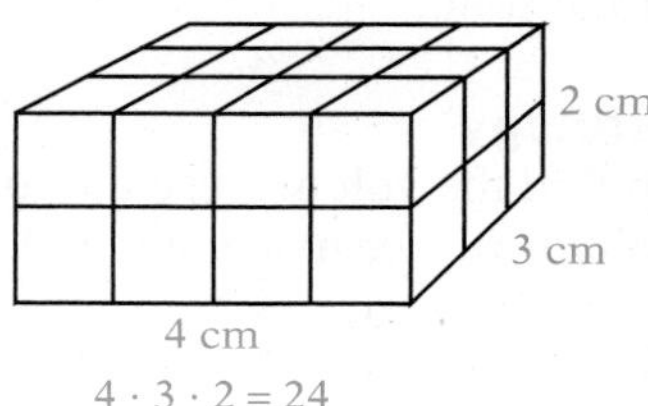

$4 \cdot 3 \cdot 2 = 24$

The formulas for the volumes of the geometric solids described above are given below.

Volumes of Geometric Solids

The volume, V, of a **rectangular solid** with length L, width W, and height H is given by $V = LWH$.

The volume, V, of a **cube** with side s is given by $V = s^3$.

The volume, V, of a **sphere** with radius r is given by $V = \frac{4}{3}\pi r^3$.

The volume, V, of a **right circular cylinder** is given by $V = \pi r^2 h$, where r is the radius of the base and h is the height.

The volume, V, of a **right circular cone** is given by $V = \frac{1}{3}\pi r^2 h$, where r is the radius of the circular base and h is the height.

The volume, V, of a **regular square pyramid** is given by $V = \frac{1}{3}s^2 h$, where s is the length of a side of the base and h is the height.

HOW TO • 1 Find the volume of a sphere with a diameter of 6 in.

$r = \frac{1}{2}d = \frac{1}{2}(6 \text{ in.}) = 3 \text{ in.}$ • First find the radius of the sphere.

$V = \frac{4}{3}\pi r^3$ • Use the formula for the volume of a sphere.

$V = \frac{4}{3}\pi(3 \text{ in.})^3$

$V = \frac{4}{3}\pi(27 \text{ in}^3)$

$V = 36\pi \text{ in}^3$ • The exact volume of the sphere is $36\pi \text{ in}^3$.

$V \approx 113.10 \text{ in}^3$ • An approximate measure can be found by using the π key on a calculator.

The approximate volume is 113.10 in^3.

EXAMPLE • 1

The length of a rectangular solid is 5 m, the width is 3.2 m, and the height is 4 m. Find the volume of the solid.

Strategy

To find the volume, use the formula for the volume of a rectangular solid. $L = 5$ m, $W = 3.2$ m, $H = 4$ m.

Solution

$V = LWH = (5 \text{ m})(3.2 \text{ m})(4 \text{ m}) = 64 \text{ m}^3$

The volume of the rectangular solid is 64 m^3.

YOU TRY IT • 1

Find the volume of a cube that measures 2.5 m on a side.

Your strategy

Your solution

EXAMPLE • 2

The radius of the base of a cone is 8 cm. The height is 12 cm. Find the volume of the cone. Round to the nearest hundredth.

Strategy

To find the volume, use the formula for the volume of a cone. An approximation is asked for; use the π key on a calculator. $r = 8$ cm, $h = 12$ cm.

Solution

$V = \frac{1}{3}\pi r^2 h$

$V = \frac{1}{3}\pi(8 \text{ cm})^2(12 \text{ cm}) = \frac{1}{3}\pi(64 \text{ cm}^2)(12 \text{ cm})$

$= 256\pi \text{ cm}^3 \approx 804.25 \text{ cm}^3$

The volume is approximately 804.25 cm^3.

YOU TRY IT • 2

The diameter of the base of a cylinder is 8 ft. The height of the cylinder is 22 ft. Find the exact volume of the cylinder.

Your strategy

Your solution

Solutions on p. S10

OBJECTIVE B To solve problems involving the surface area of a solid 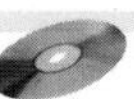

The **surface area** of a solid is the total area on the surface of the solid.

When a rectangular solid is cut open and flattened out, each face is a rectangle. The surface area, SA, of the rectangular solid is the sum of the areas of the six rectangles:

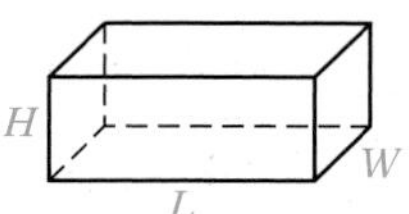

$$SA = LW + LH + WH + LW + WH + LH$$

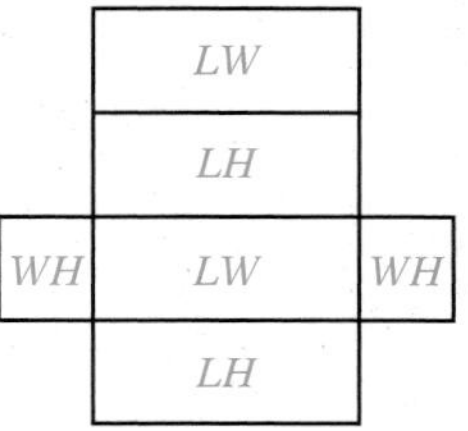

which simplifies to

$$\mathbf{SA = 2LW + 2LH + 2WH}$$

The surface area of a cube is the sum of the areas of the six faces of the cube. The area of each face is s^2. Therefore, the surface area, SA, of a cube is given by the formula $\mathbf{SA = 6s^2}$.

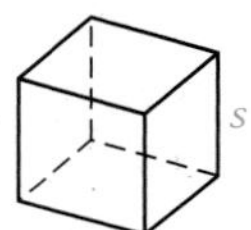

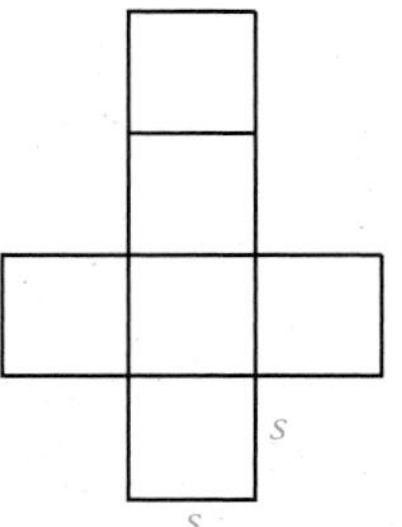

When a cylinder is cut open and flattened out, the top and bottom of the cylinder are circles. The side of the cylinder flattens out to a rectangle. The length of the rectangle is the circumference of the base, which is $2\pi r$; the width is h, the height of the cylinder. Therefore, the area of the rectangle is $2\pi rh$. The surface area, SA, of the cylinder is

$$SA = \pi r^2 + 2\pi rh + \pi r^2$$

which simplifies to

$$\mathbf{SA = 2\pi r^2 + 2\pi rh}$$

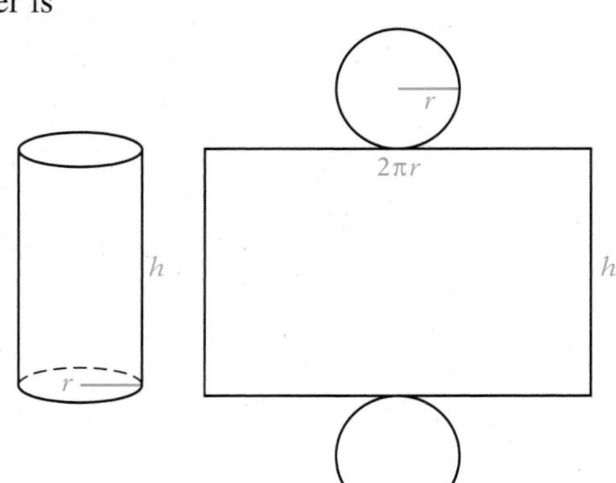

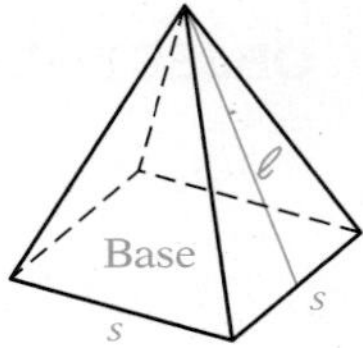

The surface area of a regular square pyramid is the area of the base plus the area of the four isosceles triangles. A side of the square base is s; therefore, the area of the base is s^2. The slant height, ℓ, is the height of each triangle, and s is the base of each triangle. The surface area, SA, of a pyramid is

$$SA = s^2 + 4\left(\frac{1}{2}s\ell\right)$$

which simplifies to

$$\mathbf{SA = s^2 + 2s\ell}$$

Formulas for the surface areas of geometric solids are given below.

> **Surface Areas of Geometric Solids**
>
> The surface area, SA, of a **rectangular solid** with length L, width W, and height H is given by $SA = 2LW + 2LH + 2WH$.
>
> The surface area, SA, of a **cube** with side s is given by $SA = 6s^2$.
>
> The surface area, SA, of a **sphere** with radius r is given by $SA = 4\pi r^2$.
>
> The surface area, SA, of a **right circular cylinder** is given by $SA = 2\pi r^2 + 2\pi rh$, where r is the radius of the base and h is the height.
>
> The surface area, SA, of a **right circular cone** is given by $SA = \pi r^2 + \pi r\ell$, where r is the radius of the circular base and ℓ is the slant height.
>
> The surface area, SA, of a **regular square pyramid** is given by $SA = s^2 + 2s\ell$, where s is the length of a side of the base and ℓ is the slant height.

HOW TO • 2 Find the surface area of a sphere with a diameter of 18 cm.

$r = \frac{1}{2}d = \frac{1}{2}(18 \text{ cm}) = 9 \text{ cm}$ • First find the radius of the sphere.

$SA = 4\pi r^2$ • Use the formula for the surface area of a sphere.
$SA = 4\pi(9 \text{ cm})^2$
$SA = 4\pi(81 \text{ cm}^2)$
$SA = 324\pi \text{ cm}^2$ • The exact surface area of the sphere is 324π cm².

$SA \approx 1017.88 \text{ cm}^2$ • An approximate measure can be found by using the π key on a calculator.

The approximate surface area is 1017.88 cm².

EXAMPLE • 3

The diameter of the base of a cone is 5 m, and the slant height is 4 m. Find the surface area of the cone. Give the exact measure.

Strategy

To find the surface area of the cone:

- Find the radius of the base of the cone.
- Use the formula for the surface area of a cone. Leave the answer in terms of π.

Solution

$r = \frac{1}{2}d = \frac{1}{2}(5\text{ m}) = 2.5\text{ m}$

$SA = \pi r^2 + \pi r\ell$

$SA = \pi(2.5\text{ m})^2 + \pi(2.5\text{ m})(4\text{ m})$

$SA = 6.25\pi\text{ m}^2 + 10\pi\text{ m}^2$

$SA = 16.25\pi\text{ m}^2$

The surface area of the cone is $16.25\pi\text{ m}^2$.

YOU TRY IT • 3

The diameter of the base of a cylinder is 6 ft, and the height is 8 ft. Find the surface area of the cylinder. Round to the nearest hundredth.

Your strategy

Your solution

EXAMPLE • 4

Find the area of a label used to cover a soup can that has a radius of 4 cm and a height of 12 cm. Round to the nearest hundredth.

Strategy

To find the area of the label, use the fact that the surface area of the sides of a cylinder is given by $2\pi rh$. An approximation is asked for; use the π key on a calculator. $r = 4$ cm, $h = 12$ cm.

Solution

Area of the label $= 2\pi rh$

Area of the label $= 2\pi(4\text{ cm})(12\text{ cm})$

$= 96\pi\text{ cm}^2 \approx 301.59\text{ cm}^2$

The area is approximately 301.59 cm^2.

YOU TRY IT • 4

Which has a larger surface area, a cube with a side measuring 10 cm or a sphere with a diameter measuring 8 cm?

Your strategy

Your solution

Solutions on p. S11

3.3 EXERCISES

OBJECTIVE A To solve problems involving the volume of a solid

For Exercises 1 to 6, find the volume of the figure. For calculations involving π, give both the exact value and an approximation to the nearest hundredth.

1.

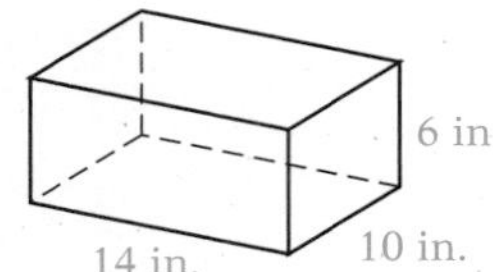

2.

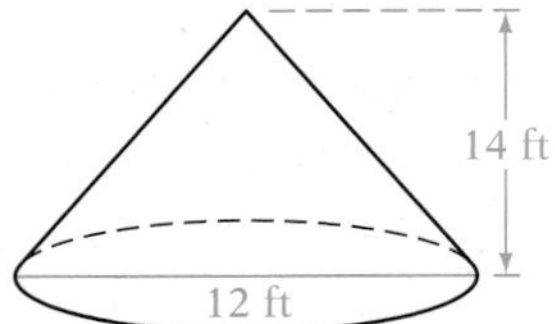

3.

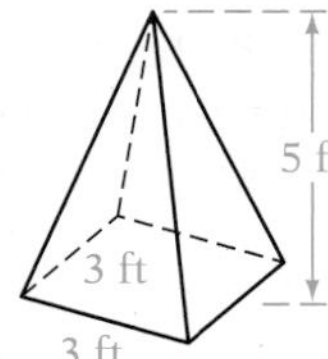

4.

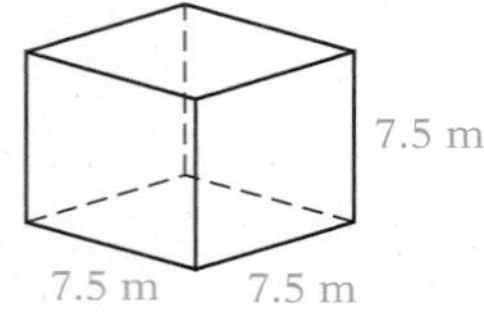

5.

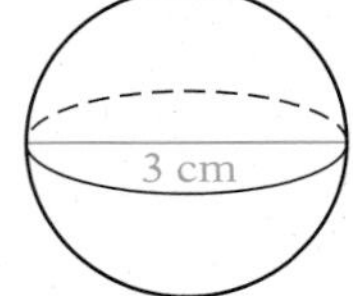

6.

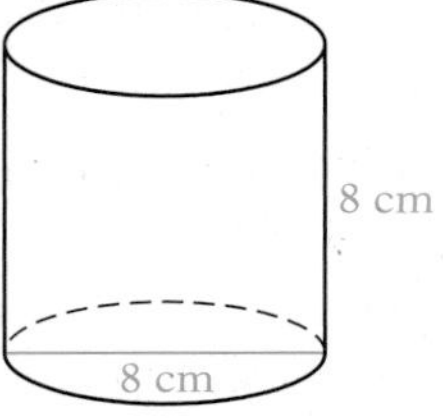

7. **Storage Units** A rectangular storage unit has a length of 6.8 m, a width of 2.5 m, and a height of 2 m. Find the volume of the storage unit.

8. **Fish Hatchery** A rectangular tank at a fish hatchery is 9 m long, 3 m wide, and 1.5 m deep. Find the volume of the water in the tank when the tank is full.

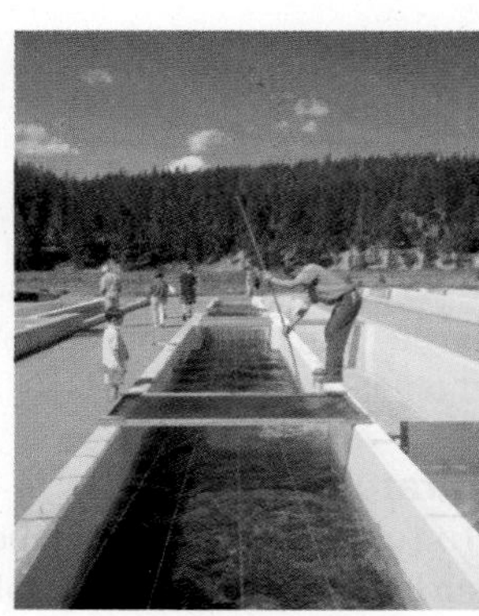

9. Find the volume of a cube whose side measures 3.5 in.

10. The length of a side of a cube is 7 cm. Find the volume of the cube.

11. The diameter of a sphere is 6 ft. Find the volume of the sphere. Give the exact measure.

12. Find the volume of a sphere that has a radius of 1.2 m. Round to the nearest tenth.

13. The diameter of the base of a cylinder is 24 cm. The height of the cylinder is 18 cm. Find the volume of the cylinder. Round to the nearest hundredth.

14. The height of a cylinder is 7.2 m. The radius of the base is 4 m. Find the volume of the cylinder. Give the exact measure.

15. The radius of the base of a cone is 5 in. The height of the cone is 9 in. Find the volume of the cone. Give the exact measure.

16. The height of a cone is 15 cm. The diameter of the cone is 10 cm. Find the volume of the cone. Round to the nearest hundredth.

17. The length of a side of the base of a pyramid is 6 in., and the height is 10 in. Find the volume of the pyramid.

18. The height of a pyramid is 8 m, and the length of a side of the base is 9 m. What is the volume of the pyramid?

19. **Appliances** The volume of a freezer with a length of 7 ft and a height of 3 ft is 52.5 ft^3. Find the width of the freezer.

20. **Aquariums** The length of an aquarium is 18 in., and the width is 12 in. If the volume of the aquarium is 1836 in^3, what is the height of the aquarium?

21. The volume of a cylinder is 502.4 in^3. The diameter of the base is 10 in. Find the height of the cylinder. Round to the nearest hundredth.

22. The diameter of the base of a cylinder is 14 cm. If the volume of the cylinder is 2310 cm^3, find the height of the cylinder. Round to the nearest hundredth.

23. **Guacamole Consumption** See the news clipping at the right. What is the volume of the guacamole in cubic feet?

24. **Guacamole Consumption** See the news clipping at the right. Assuming that each person eats 1 c of guacamole, how many people could be fed from the covered football field? (1 ft^3 = 59.84 pt)

25. The length of a side of a cube is equal to the radius of a sphere. Which solid has the greater volume?

26. A sphere and a cylinder have the same radius. The height of the cylinder is equal to the radius of its base. Which solid has the greater volume?

In the News

Super Bowl Win for Guacamole

Guacamole has become the dish of choice at Super Bowl parties. If all the guacamole eaten during the Super Bowl were piled onto a football field—that's a football field which, including endzones, is 360 ft long and 160 ft wide—it would cover the field to a depth of 3 ft!

Source: www.azcentral.com

OBJECTIVE B To solve problems involving the surface area of a solid

For Exercises 27 to 29, find the surface area of the figure.

27.

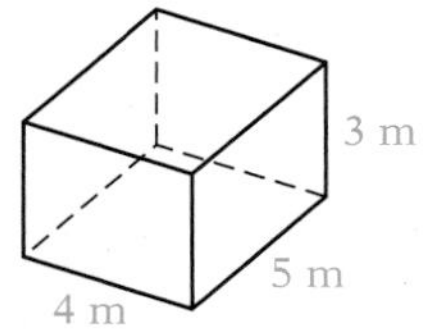

28.

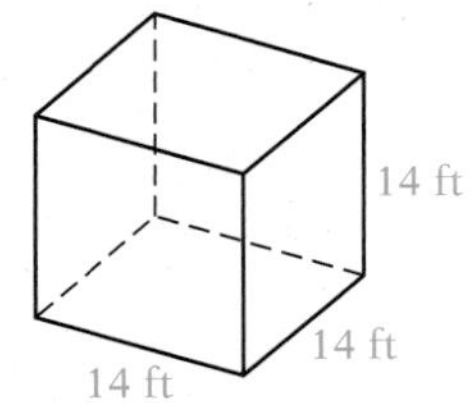

29.

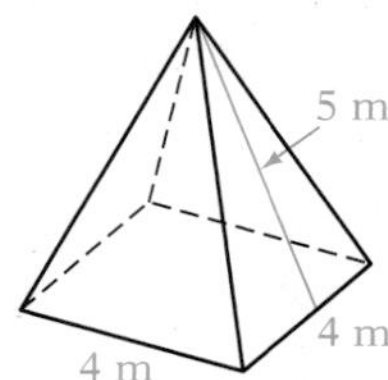

For Exercises 30 to 32, find the surface area of the figure. Give both the exact value and an approximation to the nearest hundredth.

30.

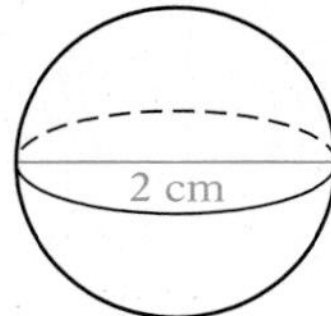

31.

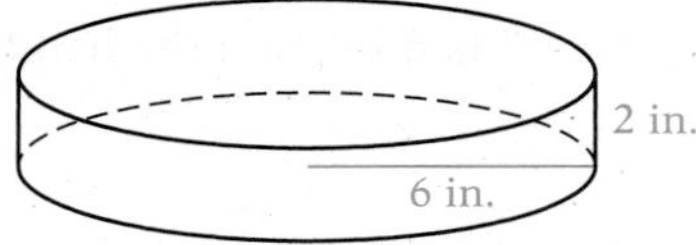

32.

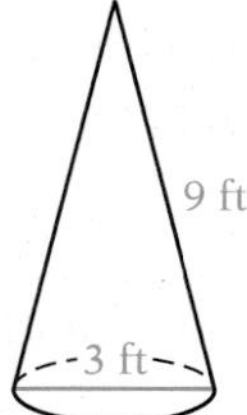

33. The height of a rectangular solid is 5 ft. The length is 8 ft, and the width is 4 ft. Find the surface area of the solid.

34. The width of a rectangular solid is 32 cm. The length is 60 cm, and the height is 14 cm. What is the surface area of the solid?

35. The side of a cube measures 3.4 m. Find the surface area of the cube.

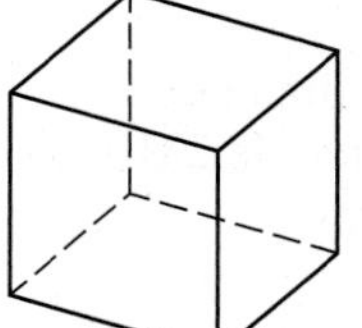

36. Find the surface area of a cube that has a side measuring 1.5 in.

37. Find the surface area of a sphere with a diameter of 15 cm. Give the exact value.

38. The radius of a sphere is 2 in. Find the surface area of the sphere. Round to the nearest hundredth.

39. The radius of the base of a cylinder is 4 in. The height of the cylinder is 12 in. Find the surface area of the cylinder. Round to the nearest hundredth.

40. The diameter of the base of a cylinder is 1.8 m. The height of the cylinder is 0.7 m. Find the surface area of the cylinder. Give the exact value.

41. The slant height of a cone is 2.5 ft. The radius of the base is 1.5 ft. Find the surface area of the cone. Give the exact value.

42. The diameter of the base of a cone is 21 in. The slant height is 16 in. What is the surface area of the cone? Round to the nearest hundredth.

43. The length of a side of the base of a pyramid is 9 in., and the slant height is 12 in. Find the surface area of the pyramid.

44. The slant height of a pyramid is 18 m, and the length of a side of the base is 16 m. What is the surface area of the pyramid?

45. The surface area of a rectangular solid is 108 cm^2. The height of the solid is 4 cm, and the length is 6 cm. Find the width of the rectangular solid.

46. The length of a rectangular solid is 12 ft. The width is 3 ft. If the surface area is 162 ft^2, find the height of the rectangular solid.

47. **Paint** A can of paint will cover 300 ft^2. How many cans of paint should be purchased in order to paint a cylinder that has a height of 30 ft and a radius of 12 ft?

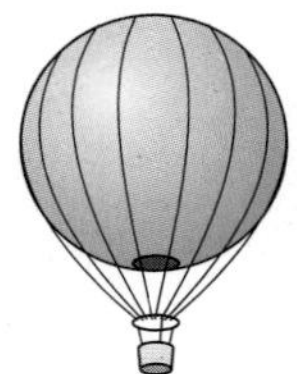

48. **Ballooning** A hot air balloon is in the shape of a sphere. Approximately how much fabric was used to construct the balloon if its diameter is 32 ft? Round to the nearest whole number.

49. **Aquariums** How much glass is needed to make a fish tank that is 12 in. long, 8 in. wide, and 9 in. high? The fish tank is open at the top.

50. **Packaging** Find the area of a label used to cover a can of juice that has a diameter of 16.5 cm and a height of 17 cm. Round to the nearest hundredth.

51. The length of a side of the base of a pyramid is 5 cm, and the slant height is 8 cm. How much larger is the surface area of this pyramid than the surface area of a cone with a diameter of 5 cm and a slant height of 8 cm? Round to the nearest hundredth.

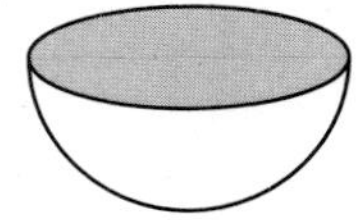

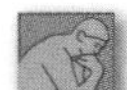

52. Half of a sphere is called a **hemisphere.** Derive formulas for the volume and surface area of a hemisphere.

53. Determine whether the statement is always true, sometimes true, or never true.
a. The slant height of a regular pyramid is longer than the height.
b. The slant height of a cone is shorter than the height.
c. The four triangular faces of a regular pyramid are equilateral triangles.

Applying the Concepts

54. **a.** What is the effect on the surface area of a rectangular solid when the width and height are doubled?
b. What is the effect on the volume of a rectangular solid when both the length and the width are doubled?
c. What is the effect on the volume of a cube when the length of each side of the cube is doubled?
d. What is the effect on the surface area of a cylinder when the radius and height are doubled?

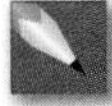

55. Explain how you could cut through a cube so that the face of the resulting solid is
a. a square
b. an equilateral triangle
c. a trapezoid
d. a hexagon

FOCUS ON PROBLEM SOLVING

More on the Trial-and-Error Approach to Problem Solving

Some problems in mathematics are solved by using **trial and error.** The trial-and-error method of arriving at a solution to a problem involves performing repeated tests or experiments until a satisfactory conclusion is reached.

Many of the Applying the Concepts exercises in this text require a trial-and-error method of solution. For example, an exercise in Section 3 of this chapter reads:

Explain how you could cut through a cube so that the face of the resulting solid is **(a)** a square, **(b)** an equilateral triangle, **(c)** a trapezoid, **(d)** a hexagon.

There is no formula to apply to these problems; there is no computation to perform. These problems require that you picture a cube and the results after it is cut through at different places on its surface and at different angles. For part (a), cutting perpendicular to the top and bottom of the cube and parallel to two of its sides will result in a square. The other shapes may prove more difficult.

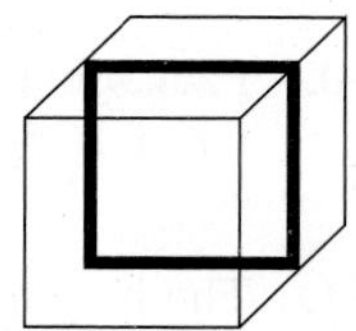

When solving problems of this type, keep an open mind. Sometimes when using the trial-and-error method, we are hampered by our narrowness of vision; we cannot expand our thinking to include other possibilities. Then, when we see someone else's solution, it appears so obvious to us! For example, for the Applying the Concepts question above, it is necessary to conceive of cutting through the cube at places other than the top surface; we need to be open to the idea of beginning the cut at one of the corner points of the cube.

One topic of the Projects and Group Activities in this chapter is symmetry. Here again, the trial-and-error method is used to determine the lines of symmetry inherent in an object. For example, in determining lines of symmetry for a square, begin by drawing a square. The horizontal line of symmetry and the vertical line of symmetry may be immediately obvious to you.

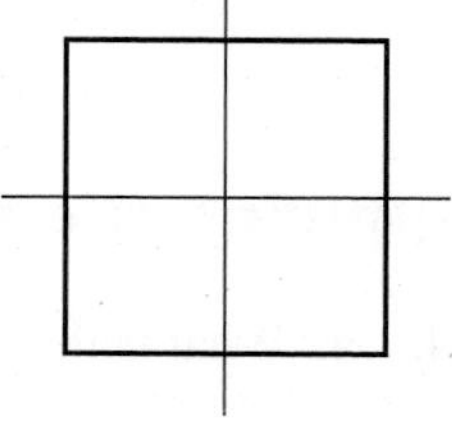

But there are two others. Do you see that a line drawn through opposite corners of the square is also a line of symmetry?

Many of the questions in this text that require an answer of "always true, sometimes true, or never true" are best solved by the trial-and-error method. For example, consider the following statement.

If two rectangles have the same area, then they have the same perimeter.

Try some numbers. Each of two rectangles, one measuring 6 units by 2 units and another measuring 4 units by 3 units, has an area of 12 square units, but the perimeter of the first is 16 units and the perimeter of the second is 14 units, so the answer "always true" has been eliminated. We still need to determine whether there is a case when the statement is true. After experimenting with a lot of numbers, you may come to realize that we are trying to determine whether it is possible for two different pairs of factors of a number to have the same sum. Is it?

Don't be afraid to make many experiments, and remember that *errors,* or tests that "don't work," are a part of the trial-and-*error* process.

PROJECTS AND GROUP ACTIVITIES

Investigating Perimeter

The perimeter of the square at the right is 4 units.

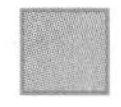

If two squares are joined along one of the sides, the perimeter is 6 units. Note that it does not matter which sides are joined; the perimeter is still 6 units.

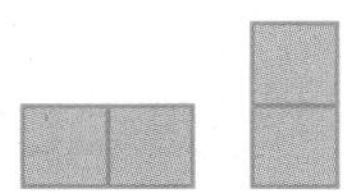

If three squares are joined, the perimeter of the resulting figure is 8 units for each possible placement of the squares.

Four squares can be joined in five different ways as shown. There are two possible perimeters: 10 units for A, B, C, and D, and 8 for E.

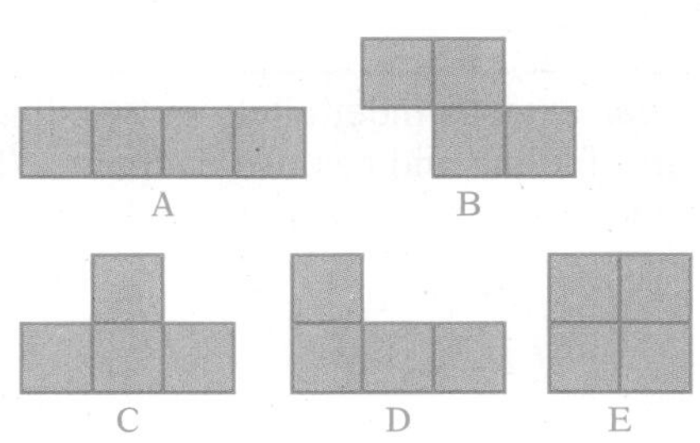

1. If five squares are joined, what is the maximum perimeter possible?
2. If five squares are joined, what is the minimum perimeter possible?
3. If six squares are joined, what is the maximum perimeter possible?
4. If six squares are joined, what is the minimum perimeter possible?

Symmetry Look at the letter A printed at the left. If the letter were folded along line ℓ, the two sides of the letter would match exactly. This letter has **symmetry** with respect to line ℓ. Line ℓ is called the **axis of symmetry.**

Now consider the letter H printed below at the left. Both lines ℓ_1 and ℓ_2 are axes of symmetry for this letter; the letter could be folded along either line and the two sides would match exactly.

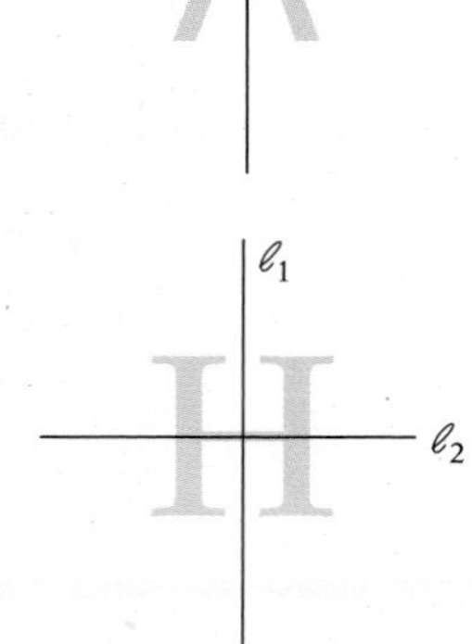

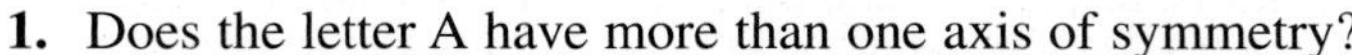

1. Does the letter A have more than one axis of symmetry?
2. Find axes of symmetry for other capital letters of the alphabet.
3. Which lowercase letters have one axis of symmetry?
4. Do any of the lowercase letters have more than one axis of symmetry?
5. Find the number of axes of symmetry for each of the plane geometric figures presented in this chapter.
6. There are other types of symmetry. Look up the meaning of *point symmetry* and *rotational symmetry.* Which plane geometric figures provide examples of these types of symmetry?
7. Find examples of symmetry in nature, art, and architecture.

CHAPTER 3 SUMMARY

KEY WORDS	EXAMPLES
A *line* extends indefinitely in two directions. A *line segment* is part of a line and has two endpoints. The length of a line segment is the distance between the endpoints of the line segment. [3.1A, p. 164]	Line A B Line segment
Parallel lines never meet; the distance between them is always the same. The symbol $\parallel$ means "is parallel to." *Intersecting lines* cross at a point in the plane. *Perpendicular lines* are intersecting lines that form right angles. The symbol $\perp$ means "is perpendicular to." [3.1A, pp. 165–166]	Parallel lines Perpendicular lines

A *ray* starts at a point and extends indefinitely in one direction. An *angle* is formed when two rays start from the same point. The common point is called the *vertex* of the angle. An angle is measured in *degrees.* A 90° angle is a *right angle.* A 180° angle is a *straight angle.* [3.1A, pp. 164–167]

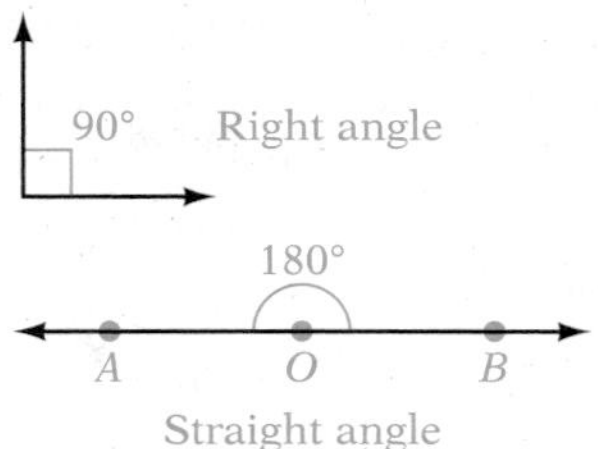

Complementary angles are two angles whose measures have the sum 90°. *Supplementary angles* are two angles whose measures have the sum 180°. [3.1A, pp. 166–167]

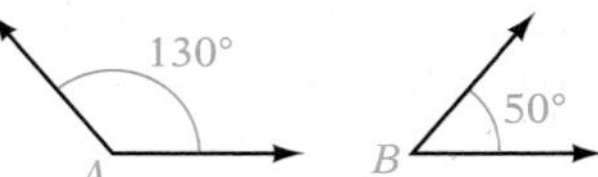

An *acute angle* is an angle whose measure is between 0° and 90°. An *obtuse angle* is an angle whose measure is between 90° and 180°. [3.1A, p. 167]

$\angle A$ above is an obtuse angle.
$\angle B$ above is an acute angle.

Two angles that are on opposite sides of the intersection of two lines are *vertical angles;* vertical angles have the same measure. Two angles that share a common side are *adjacent angles;* adjacent angles of intersecting lines are *supplementary angles.* [3.1B, p. 169]

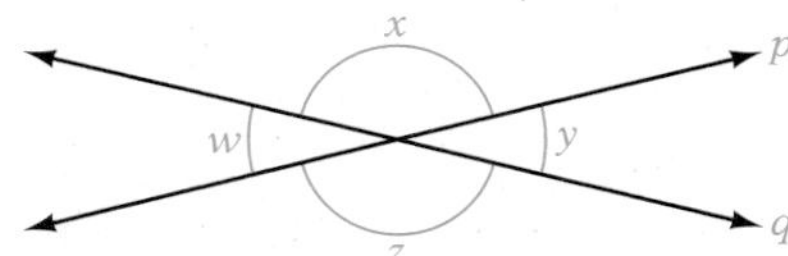

Angles w and y are vertical angles.
Angles x and y are adjacent angles.

A line that intersects two other lines at two different points is a *transversal.* If the lines cut by a transversal are parallel lines, equal angles are formed: *alternate interior angles, alternate exterior angles,* and *corresponding angles.* [3.1B, p. 170]

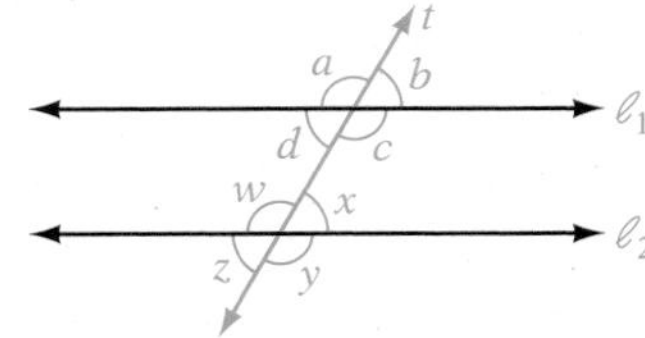

Parallel lines ℓ_1 and ℓ_2 are cut by transversal t. All four acute angles have the same measure. All four obtuse angles have the same measure.

A *quadrilateral* is a four-sided polygon. A *parallelogram,* a *rectangle,* and a *square* are quadrilaterals. [3.2A, pp. 180–181]

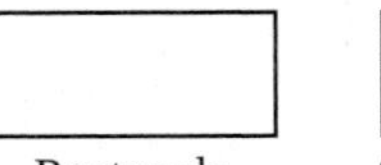

A *polygon* is a closed figure determined by three or more line segments. The line segments that form the polygon are its *sides.* A *regular polygon* is one in which each side has the same length and each angle has the same measure. Polygons are classified by the number of sides. [3.2A, p. 180]

Number of Sides	Name of the Polygon
3	Triangle
4	Quadrilateral
5	Pentagon
6	Hexagon
7	Heptagon
8	Octagon
9	Nonagon
10	Decagon

A *triangle* is a closed, three-sided plane figure. [3.1C, p. 172]

An *isosceles triangle* has two sides of equal length. The three sides of an *equilateral triangle* are of equal length. A *scalene triangle* has no two sides of equal length. An *acute triangle* has three acute angles. An *obtuse triangle* has one obtuse angle. A *right triangle* contains a right angle. [3.2A, pp. 180–181]

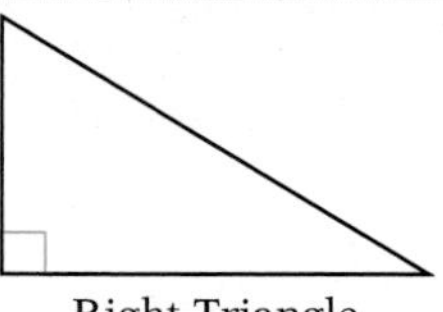

Right Triangle

A *circle* is a plane figure in which all points are the same distance from the center of the circle. A *diameter* of a circle is a line segment across the circle through the center. A *radius* of a circle is a line segment from the center of the circle to a point on the circle. [3.2A, p. 183]

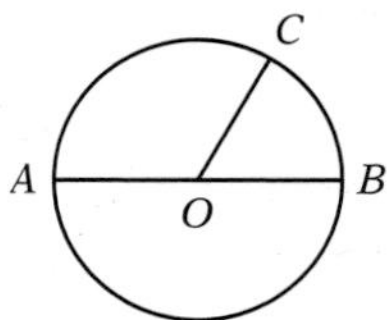

AB is a diameter of the circle.
OC is a radius.

Geometric solids are figures in space. Six common space figures are the rectangular solid, cube, sphere, cylinder, cone, and pyramid. A *rectangular solid* is a solid in which all six faces are rectangles. A *cube* is a rectangular solid in which all six faces are squares. A *sphere* is a solid in which all points on the sphere are the same distance from the center of the sphere. The most common *cylinder* is one in which the bases are circles and are perpendicular to the height. In this text the only types of cones and pyramids discussed are the *right circular cone* and the *regular square pyramid.* [3.3A, pp. 198–199]

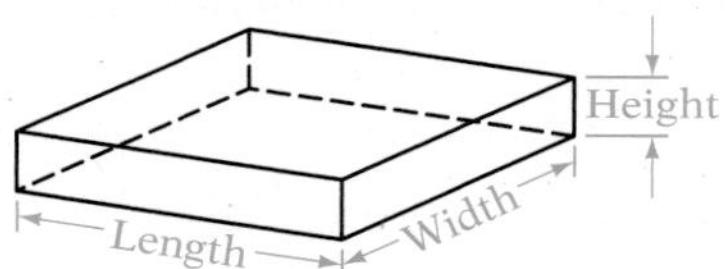

Rectangular Solid

ESSENTIAL RULES AND PROCEDURES

EXAMPLES

Triangles [3.1C, p. 172]

Sum of the measures of the interior angles = 180°

Two angles of a triangle measure 32° and 48°. Find the measure of the third angle.

$$\angle A + \angle B + \angle C = 180^\circ$$
$$\angle A + 32^\circ + 48^\circ = 180^\circ$$
$$\angle A + 80^\circ = 180^\circ$$
$$\angle A + 80^\circ - 80^\circ = 180^\circ - 80^\circ$$
$$\angle A = 100^\circ$$

The measure of the third angle is 100°.

Formulas for Perimeter (the distance around a figure)
[3.2A, pp. 181–183]
Triangle: $P = a + b + c$
Rectangle: $P = 2L + 2W$
Square: $P = 4s$
Circumference of a circle: $C = \pi d$ or $C = 2\pi r$

The length of a rectangle is 8 m. The width is 5.5 m. Find the perimeter of the rectangle.
$P = 2L + 2W$
$P = 2(8 \text{ m}) + 2(5.5 \text{ m})$
$P = 16 \text{ m} + 11 \text{ m}$
$P = 27 \text{ m}$
The perimeter is 27 m.

Formulas for Area (the amount of surface in a region)
[3.2B, pp. 185–188]
Rectangle: $A = LW$
Square: $A = s^2$
Parallelogram: $A = bh$
Triangle: $A = \frac{1}{2}bh$
Trapezoid: $A = \frac{1}{2}h(b_1 + b_2)$
Circle: $A = \pi r^2$

Find the area of a circle with a radius of 4 cm. Round to the nearest hundredth.
$A = \pi r^2$
$A = \pi(4 \text{ cm})^2$
$A = 16\pi \text{ cm}^2$
$A \approx 50.27 \text{ cm}^2$
The area is approximately 50.27 cm^2.

Formulas for Volume (the amount of space inside a figure in space)
[3.3A, p. 199]
Rectangular solid: $V = LWH$
Cube: $V = s^3$
Sphere: $V = \frac{4}{3}\pi r^3$
Cylinder: $V = \pi r^2 h$
Right circular cone: $V = \frac{1}{3}\pi r^2 h$
Regular square pyramid: $V = \frac{1}{3}s^2 h$

Find the volume of a cube that measures 3 in. on a side.
$V = s^3$
$V = (3 \text{ in.})^3$
$V = 27 \text{ in}^3$
The volume is 27 in^3.

Formulas for Surface Area (the total area on the surface of a solid)
[3.3B, p. 202]
Rectangular solid: $SA = 2LW + 2LH + 2WH$
Cube: $SA = 6s^2$
Sphere: $SA = 4\pi r^2$
Right circular cylinder: $SA = 2\pi r^2 + 2\pi rh$
Right circular cone: $SA = \pi r^2 + \pi r\ell$
Regular square pyramid: $SA = s^2 + 2s\ell$

Find the surface area of a sphere that has a radius of 3 in. Round to the nearest hundredth.
$SA = 4\pi r^2$
$SA = 4\pi(3 \text{ in.})^2$
$SA = 4\pi(9 \text{ in}^2)$
$SA = 36\pi \text{ in}^2$
$SA \approx 113.10 \text{ in}^2$
The surface area is approximately 113.10 in^2.

CHAPTER 3

CONCEPT REVIEW

Test your knowledge of the concepts presented in this chapter. Answer each question. Then check your answers against the ones provided in the Answer Section.

1. What are perpendicular lines?

2. When are two angles complementary?

3. For the figure at the right, name the obtuse angles that have the same measure.

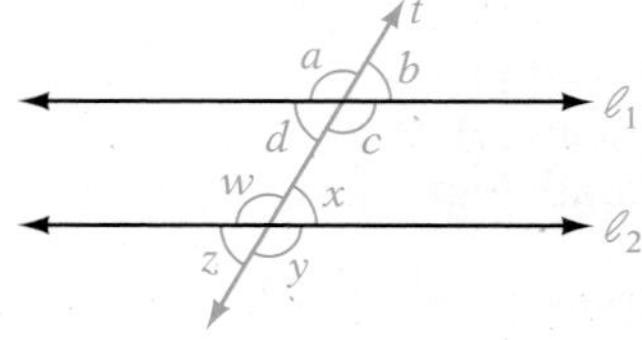

4. How are the angles in a triangle related?

5. If you know the diameter of a circle, how can you find the radius?

6. What is the formula for the perimeter of a rectangle?

7. How do you find the circumference of a circle?

8. What is the formula for the area of a triangle?

9. What three dimensions are needed to find the volume of a rectangular solid?

10. Explain the difference between perimeter and area.

11. The surface area of a rectangular solid is the sum of the areas of how many rectangles?

CHAPTER 3

REVIEW EXERCISES

1. Given that $\angle a = 74°$ and $\angle b = 52°$, find the measures of angles x and y.

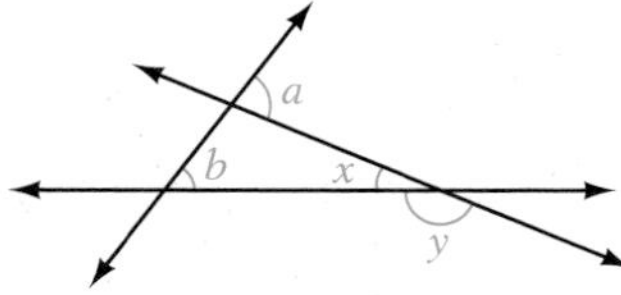

2. Find the measure of $\angle x$.

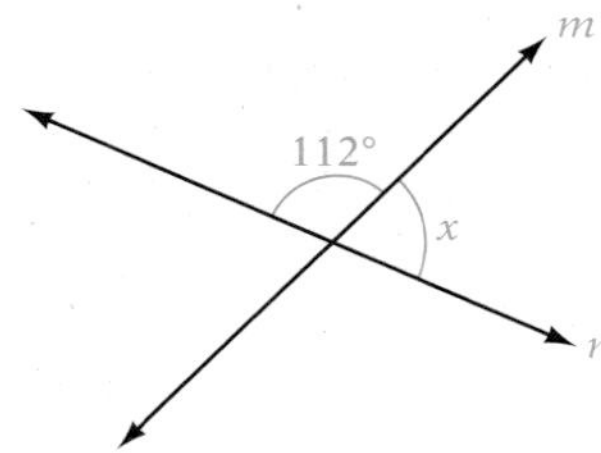

3. Given that $BC = 11$ cm and AB is three times BC, find AC.

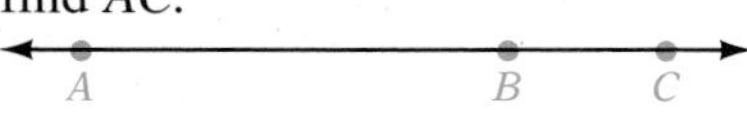

4. Find x.

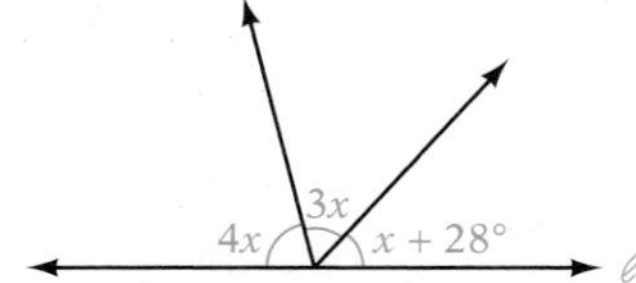

5. Find the volume of the figure.

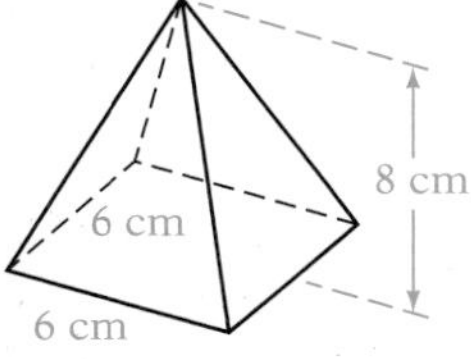

6. Given that $\ell_1 \parallel \ell_2$, find the measures of angles a and b.

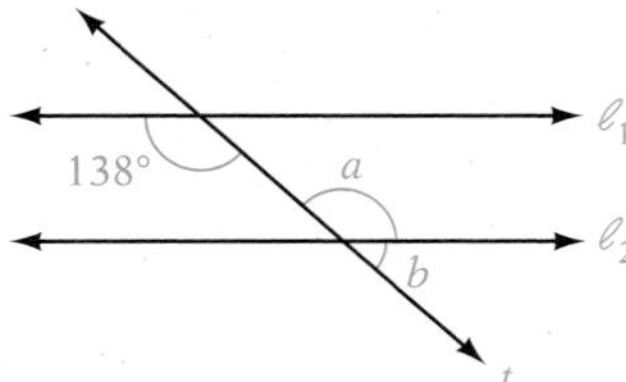

7. Find the surface area of the figure.

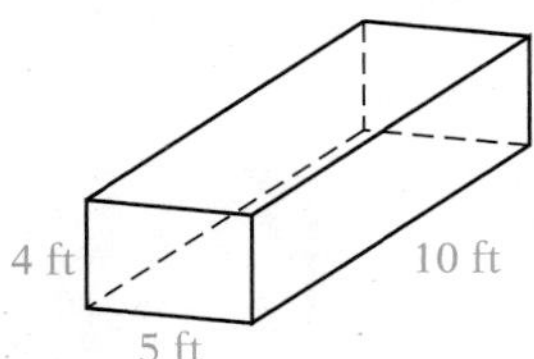

8. Find the supplement of a 32° angle.

9. Determine the area of a rectangle with a length of 12 cm and a width of 6.5 cm.

10. Determine the area of a triangle whose base is 9 m and whose height is 14 m.

11. Find the volume of a rectangular solid with a length of 6.5 ft, a width of 2 ft, and a height of 3 ft.

12. Two angles of a triangle measure 37° and 48°. Find the measure of the third angle.

13. The height of a triangle is 7 cm. The area of the triangle is 28 cm^2. Find the length of the base of the triangle.

14. Find the volume of a sphere that has a diameter of 12 mm. Find the exact value.

15. Determine the exact volume of a right circular cone whose radius is 7 cm and whose height is 16 cm.

16. **Framing** The perimeter of a square picture frame is 86 cm. Find the length of each side of the frame.

17. **Paint** A can of paint will cover 200 ft^2. How many cans of paint should be purchased in order to paint a cylinder that has a height of 15 ft and a radius of 6 ft?

18. **Parks** The length of a rectangular park is 56 yd. The width is 48 yd. How many yards of fencing are needed to surround the park?

19. **Alcatraz** Inmate cells at Alcatraz were 9 ft long and 5 ft wide. The height of a cell was 7 ft.
a. Find the area of the floor of a cell at Alcatraz.
b. Find the volume of a cell at Alcatraz.

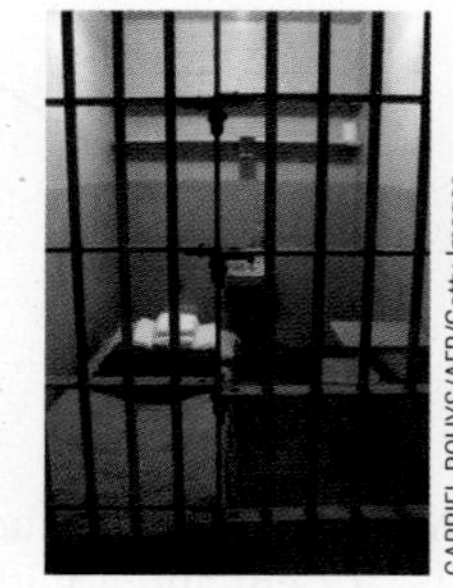
GABRIEL BOUYS/AFP/Getty Images

20. **Landscaping** A walkway 2 m wide surrounds a rectangular plot of grass. The plot is 40 m long and 25 m wide. What is the area of the walkway?

CHAPTER 3

TEST

1. The diameter of a sphere is 1.5 m. Find the radius of the sphere.

2. Find the circumference of a circle with a radius of 5 cm. Round to the nearest hundredth.

3. Find the perimeter of the rectangle in the figure below.

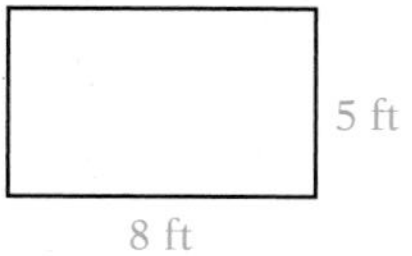

4. Given $AB = 15$, $CD = 6$, and $AD = 24$, find BC.

A B C D ℓ

5. Find the volume of a sphere with a diameter of 8 ft. Round to the nearest hundredth.

6. Find the area of the circle shown below. Round to the nearest hundredth.

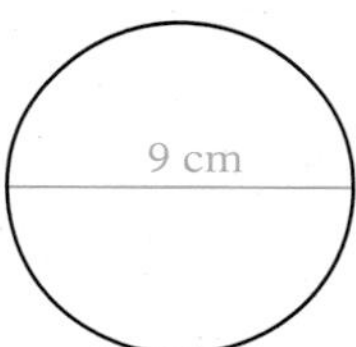

7. Given that $\ell_1 \parallel \ell_2$, find the measures of angles a and b.

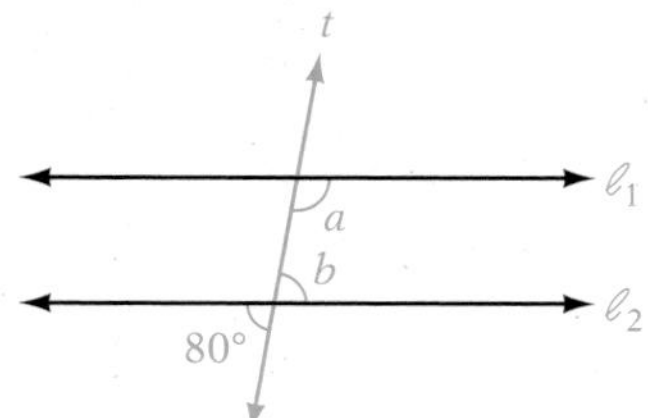

8. Find the supplement of a 105° angle.

9. Given that $\ell_1 \parallel \ell_2$, find the measures of angles a and b.

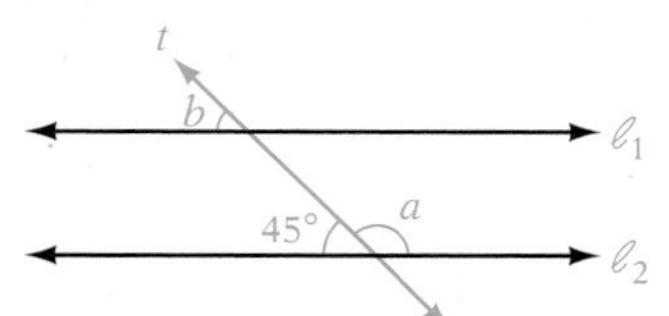

10. Find the area of the rectangle shown below.

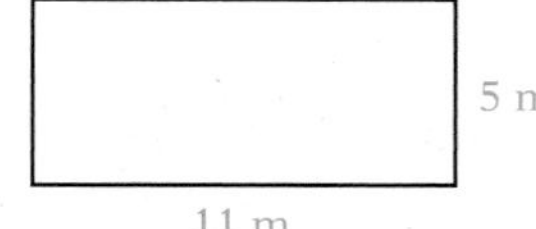

11. Find the volume of a cylinder with a height of 6 m and a radius of 3 m. Round to the nearest hundredth.

12. Find the perimeter of a rectangle that has a length of 2 m and a width of 1.4 m.

13. Find the complement of a 32° angle.

14. Find the surface area of the figure. Round to the nearest hundredth.

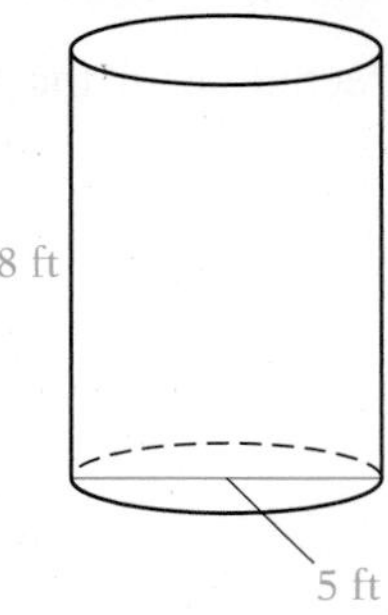

15. **Pizza** How much more pizza is contained in a pizza with radius 10 in. than in one with radius 8 in.? Round to the nearest hundredth.

16. **Triangles** A right triangle has a 32° angle. Find the measures of the other two angles.

17. **Cycling** A bicycle tire has a diameter of 28 in. How many feet does the bicycle travel if the wheel makes 10 revolutions? Round to the nearest tenth.

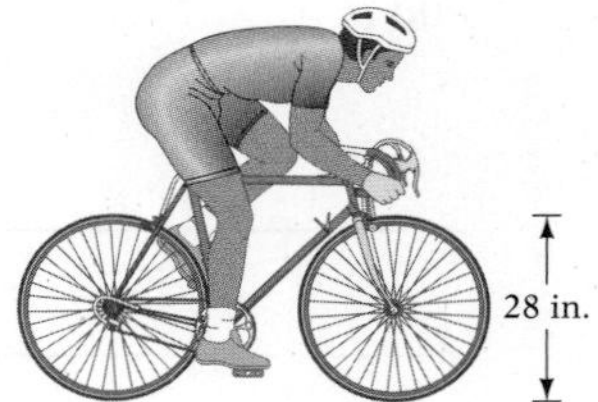

18. **Carpeting** New carpet is installed in a room measuring 18 ft by 14 ft. Find the area of the room in square yards. ($9 \text{ ft}^2 = 1 \text{ yd}^2$)

19. **Agriculture** A silo, which is in the shape of a cylinder, is 9 ft in diameter and has a height of 18 ft. Find the volume of the silo. Round to the nearest hundredth.

20. **Triangles** Find the area of a right triangle with a base of 8 m and a height of 2.75 m.

CUMULATIVE REVIEW EXERCISES

1. Let $x \in \{-3, 0, 1\}$. For what values of x is the inequality $x \leq 1$ a true statement?

2. Write 8.9% as a decimal.

3. Write $\frac{7}{20}$ as a percent.

4. Divide: $-\frac{4}{9} \div \frac{2}{3}$

5. Multiply: $5.7(-4.3)$

6. Simplify: $-\sqrt{125}$

7. Evaluate $5 - 3[10 + (5 - 6)^2]$.

8. Evaluate $a(b - c)^3$ when $a = -1$, $b = -2$, and $c = -4$.

9. Simplify: $5m + 3n - 8m$

10. Simplify: $-7(-3y)$

11. Simplify: $4(3x + 2) - (5x - 1)$

12. Use the roster method to write the set of negative integers greater than or equal to -2.

13. Find $C \cup D$, given $C = \{0, 10, 20, 30\}$ and $D = \{-10, 0, 10\}$.

14. Graph: $x \leq 1$

    −5 −4 −3 −2 −1 0 1 2 3 4 5

15. Solve: $4x + 2 = 6x - 8$

16. Solve: $3(2x + 5) = 18$

17. Solve: $4y - 3 \geq 6y + 5$
    Write the solution set in interval notation.

18. Solve: $8 - 4(3x + 5) \leq 6(x - 8)$
    Write the solution set in set-builder notation.

19. Solve: $2x - 3 > 5$ or $x + 4 < 1$

20. Solve: $-3 \le 2x - 7 \le 5$

21. Solve: $|3x - 1| = 2$

22. Solve: $|x - 8| \le 2$

23. Find the measure of $\angle x$.

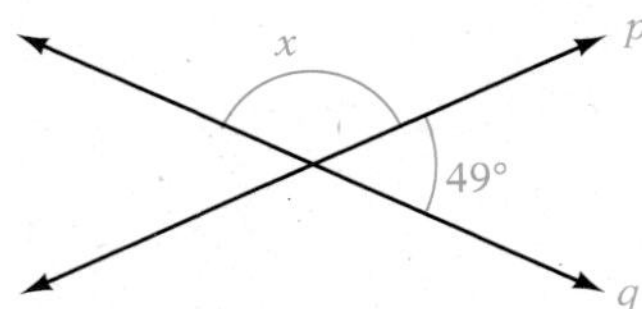

24. Translate "the difference between four times a number and ten is two" into an equation and solve.

25. **Triangles** Two angles of a triangle measure 37° and 21°. Find the measure of the third angle of the triangle.

26. **Investments** Michael deposits \$5000 in an account that earns an annual simple interest rate of 4.5% and \$2500 in an account that earns an annual simple interest rate of 3.5%. How much interest will Michael earn from the two accounts in one year?

27. **Triangles** Two sides of an isosceles triangle measure 7.5 m. The perimeter of the triangle is 19.5 m. Find the measure of the third side of the triangle.

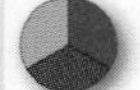

28. **Annual Earnings** According to the Census Bureau, the median annual earnings of a man with a bachelor's degree is \$49,982, and the median annual earnings of a woman with a bachelor's degree is \$35,408. What percent of the men's median annual earnings is the women's median annual earnings? Round to the nearest tenth of a percent. (Median is a type of average.)

29. Find the exact area of a circle that has a diameter of 9 cm.

30. The volume of a box is 144 ft^3. The length of the box is 12 ft, and the width is 4 ft. Find the height of the box.

CHAPTER 4

Linear Functions and Inequalities in Two Variables

Murat Taner/zefa Value/Corbis

OBJECTIVES

SECTION 4.1

- A To graph points in a rectangular coordinate system
- B To determine ordered-pair solutions of an equation in two variables
- C To graph a scatter diagram

SECTION 4.2

- A To evaluate a function

SECTION 4.3

- A To graph a linear function
- B To graph an equation of the form $Ax + By = C$
- C To find the x- and y-intercepts of a straight line
- D To solve application problems

SECTION 4.4

- A To find the slope of a line given two points
- B To graph a line given a point and the slope

SECTION 4.5

- A To find the equation of a line given a point and the slope
- B To find the equation of a line given two points
- C To solve application problems

SECTION 4.6

- A To find parallel and perpendicular lines

SECTION 4.7

- A To graph the solution set of an inequality in two variables

ARE YOU READY?

Take the Chapter 4 Prep Test to find out if you are ready to learn to:

- Evaluate a function
- Graph equations of the form $y = mx + b$ and of the form $Ax + By = C$
- Find the slope of a line
- Find the equation of a line given a point and the slope or given two points
- Find parallel and perpendicular lines
- Graph the solution set of an inequality in two variables

PREP TEST

Do these exercises to prepare for Chapter 4.

For Exercises 1 to 3, simplify.

1. $-4(x - 3)$

2. $\sqrt{(-6)^2 + (-8)^2}$

3. $\dfrac{3 - (-5)}{2 - 6}$

4. Evaluate $-2x + 5$ for $x = -3$.

5. Evaluate $\dfrac{2r}{r - 1}$ for $r = 5$.

6. Evaluate $2p^3 - 3p + 4$ for $p = -1$.

7. Evaluate $\dfrac{x_1 + x_2}{2}$ for $x_1 = 7$ and $x_2 = -5$.

8. Given $3x - 4y = 12$, find the value of x when $y = 0$.

SECTION

4.1 The Rectangular Coordinate System

OBJECTIVE A To graph points in a rectangular coordinate system

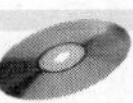

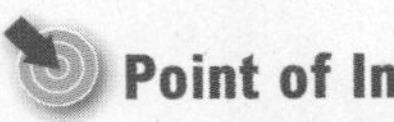

Point of Interest

A rectangular coordinate system is also called a **Cartesian coordinate system,** in honor of Descartes.

Before the 15th century, geometry and algebra were considered separate branches of mathematics. That all changed when René Descartes, a French mathematician who lived from 1596 to 1650, founded analytic geometry. In this geometry, a *coordinate system* is used to study relationships between variables.

A **rectangular coordinate system** is formed by two number lines, one horizontal and one vertical, that intersect at the zero point of each line. The point of intersection is called the **origin.** The two lines are called **coordinate axes,** or simply **axes.**

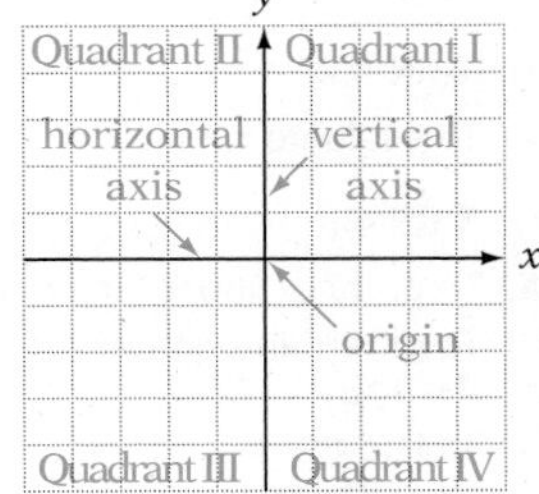

The axes determine a **plane,** which can be thought of as a large, flat sheet of paper. The two axes divide the plane into four regions called **quadrants.** The quadrants are numbered counterclockwise from I to IV.

Point of Interest

Gottfried Leibnitz introduced the words *abscissa* and *ordinate. Abscissa* is from Latin, meaning "to cut off." Originally, Leibnitz used the phrase *abscissa linea,* "cut off a line" (axis). The root of *ordinate* is also a Latin word used to suggest a sense of order.

Each point in the plane can be identified by a pair of numbers called an **ordered pair.** The first number of the pair measures a horizontal distance and is called the **abscissa.** The second number of the pair measures a vertical distance and is called the **ordinate.** The **coordinates of a point** are the numbers in the ordered pair associated with the point. The abscissa is also called the **first coordinate** of the ordered pair, and the ordinate is also called the **second coordinate** of the ordered pair.

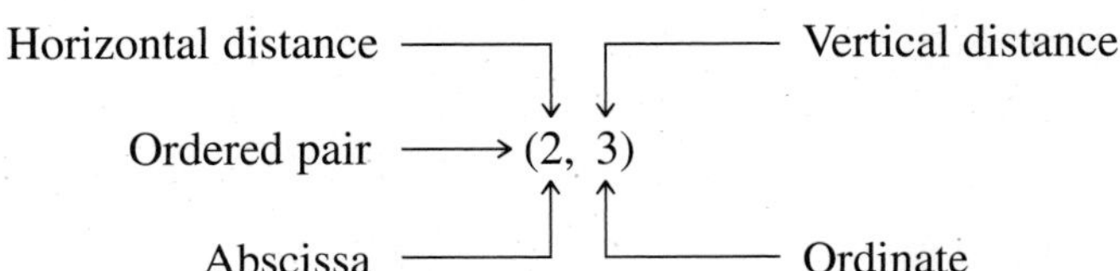

Graphing, or plotting, an ordered pair in the plane means placing a dot at the location given by the ordered pair. The **graph of an ordered pair** is the dot drawn at the coordinates of the point in the plane. The points whose coordinates are (3, 4) and (−2.5, −3) are graphed in the figure at the right.

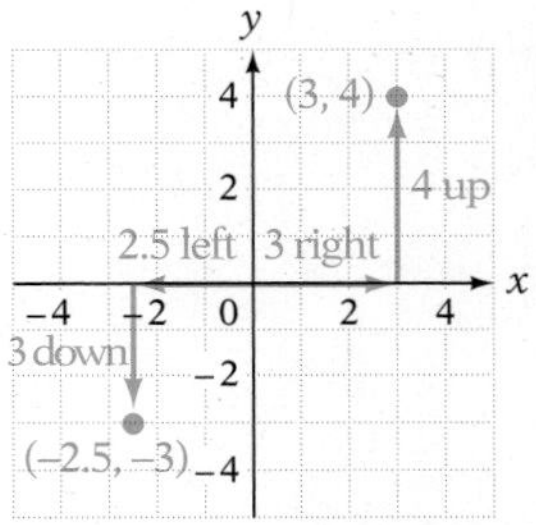

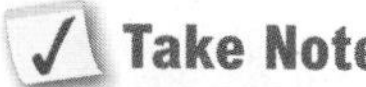

Take Note

This concept is very important. An **ordered pair** is a *pair* of coordinates, and the *order* in which the coordinates appear is crucial.

The points whose coordinates are $(3, -1)$ and $(-1, 3)$ are graphed at the right. Note that the graphed points are in different locations. *The order of the coordinates of an ordered pair is important.*

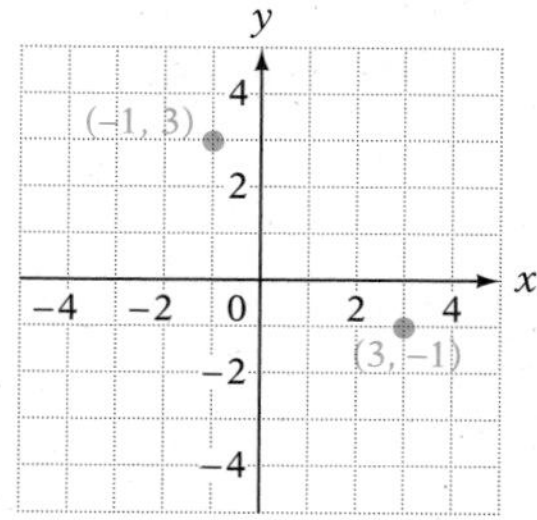

Each point in the plane is associated with an ordered pair, and each ordered pair is associated with a point in the plane. Although only the labels for integers are given on a coordinate grid, the graph of any ordered pair can be approximated. For example, the points whose coordinates are $(-2.3, 4.1)$ and $(\pi, 1)$ are shown on the graph at the right.

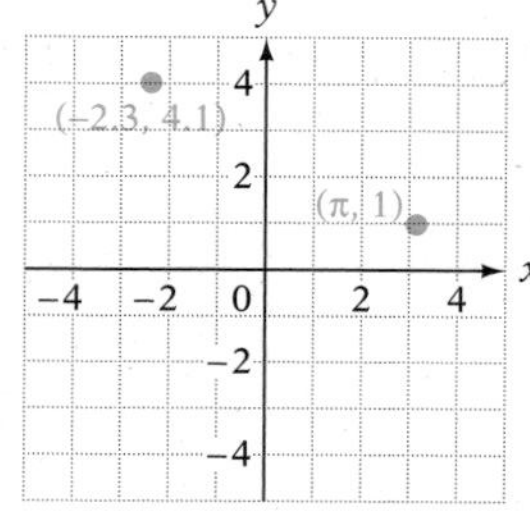

EXAMPLE • 1

Graph the ordered pairs $(-2, -3)$, $(3, -2)$, $(0, -2)$, and $(3, 0)$.

Solution

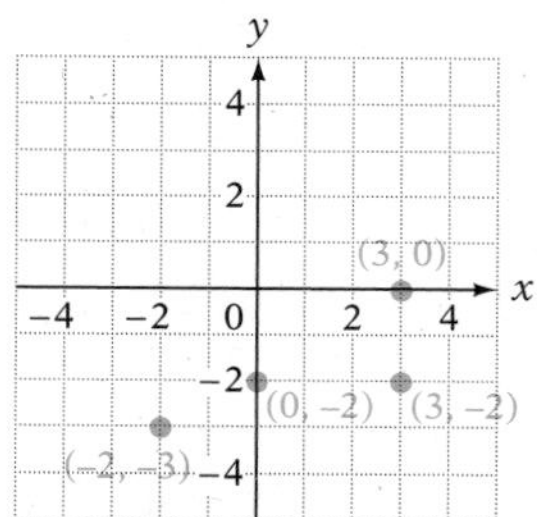

YOU TRY IT • 1

Graph the ordered pairs $(-4, 1)$, $(3, -3)$, $(0, 4)$, and $(-3, 0)$.

Your solution

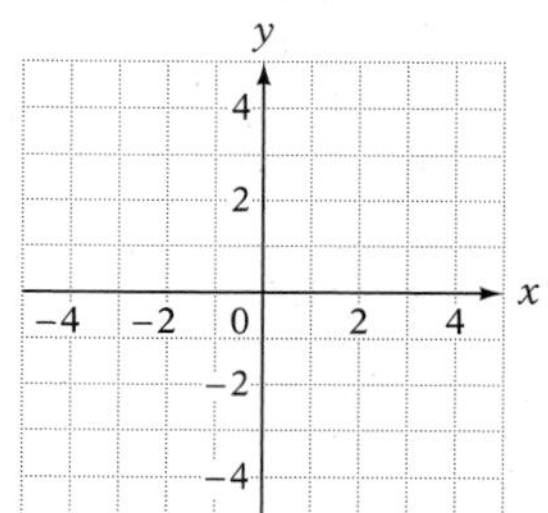

EXAMPLE • 2

Give the coordinates of the points labeled A and B. Give the abscissa of point C and the ordinate of point D.

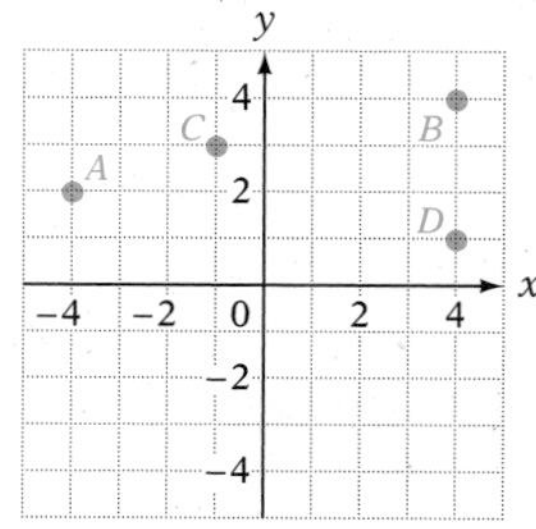

Solution

The coordinates of A are $(-4, 2)$.
The coordinates of B are $(4, 4)$.
The abscissa of C is -1.
The ordinate of D is 1.

YOU TRY IT • 2

Give the coordinates of the points labeled A and B. Give the abscissa of point D and the ordinate of point C.

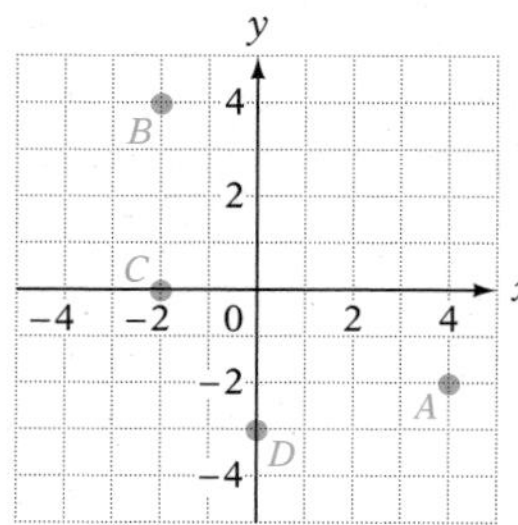

Your solution

Solutions on p. S11

OBJECTIVE B To determine ordered-pair solutions of an equation in two variables 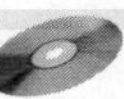

When drawing a rectangular coordinate system, we often label the horizontal axis x and the vertical axis y. In this case, the coordinate system is called an ***xy*-coordinate system.** The coordinates of the points are given by ordered pairs (x, y), where the abscissa is called the ***x*-coordinate** and the ordinate is called the ***y*-coordinate.**

An xy-coordinate system is used to study the relationship between two variables. Frequently this relationship is given by an equation. Examples of equations in two variables include

$$y = 2x - 3 \qquad 3x + 2y = 6 \qquad x^2 - y = 0$$

A **solution of an equation in two variables** is an ordered pair (x, y) whose coordinates make the equation a true statement.

HOW TO 1 Is the ordered pair $(-3, 7)$ a solution of the equation $y = -2x + 1$?

y	$= -2x + 1$	
7	$-2(-3) + 1$	• Replace x by -3 and y by 7.
7	$6 + 1$	• Simplify.
$7 = 7$		• Compare the results. If the resulting equation is a true statement, the ordered pair is a solution of the equation. If it is not a true statement, the ordered pair is not a solution of the equation.

Yes, the ordered pair $(-3, 7)$ is a solution of the equation.

Besides $(-3, 7)$, there are many other ordered-pair solutions of $y = -2x + 1$. For example, $(0, 1)$, $\left(-\frac{3}{2}, 4\right)$, and $(4, -7)$ are also solutions.

In general, an equation in two variables has an infinite number of solutions. By choosing any value of x and substituting that value into the equation, we can calculate a corresponding value of y.

HOW TO 2 Find the ordered-pair solution of $y = \frac{2}{3}x - 3$ that corresponds to $x = 6$.

$$y = \frac{2}{3}x - 3$$

$$= \frac{2}{3}(6) - 3 \quad \text{• Replace } x \text{ by } 6.$$

$$= 4 - 3 = 1 \quad \text{• Simplify.}$$

The ordered-pair solution is $(6, 1)$.

The solutions of an equation in two variables can be graphed in an xy-coordinate system.

HOW TO • 3 Graph the ordered-pair solutions of $y = -2x + 1$ when $x = -2, -1, 0, 1$, and 2.

Use the values of x to determine ordered-pair solutions of the equation. It is convenient to record these in a table.

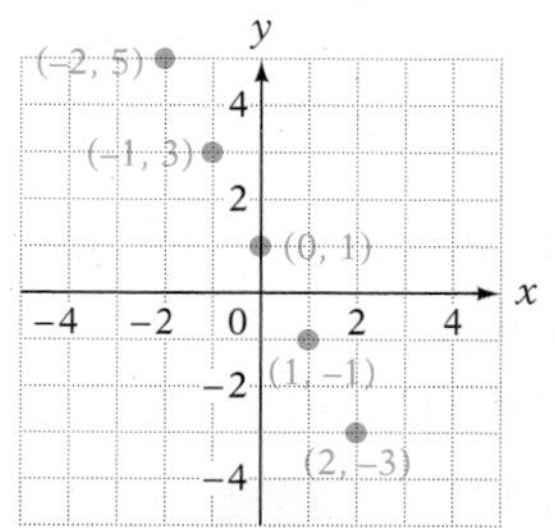

x	$y = -2x + 1$	y	(x, y)
−2	$-2(-2) + 1$	5	(−2, 5)
−1	$-2(-1) + 1$	3	(−1, 3)
0	$-2(0) + 1$	1	(0, 1)
1	$-2(1) + 1$	−1	(1, −1)
2	$-2(2) + 1$	−3	(2, −3)

EXAMPLE • 3

Is $(3, -2)$ a solution of $3x - 4y = 15$?

Solution

$$\begin{array}{r|l} 3x - 4y & = 15 \\ \hline 3(3) - 4(-2) & 15 \\ 9 + 8 & \\ 17 & \neq 15 \end{array}$$

• Replace x by 3 and y by −2.

No. $(3, -2)$ is not a solution of $3x - 4y = 15$.

YOU TRY IT • 3

Is $(-2, 4)$ a solution of $x - 3y = -14$?

Your solution

$-2 - 3(4) = -14$

EXAMPLE • 4

Find the ordered-pair solution of $y = \frac{x}{x - 2}$ corresponding to $x = 4$.

Solution

Replace x by 4 and solve for y.

$$y = \frac{x}{x - 2} = \frac{4}{4 - 2} = \frac{4}{2} = 2$$

The ordered-pair solution is (4, 2).

YOU TRY IT • 4

Find the ordered-pair solution of $y = \frac{3x}{x + 1}$ corresponding to $x = -2$.

Your solution

$y = \frac{3(-2)}{-2+1} = \frac{-6}{-1}$

$(-2, 6)$

Solutions on p. S11

EXAMPLE • 5

Graph the ordered-pair solutions of $y = \frac{2}{3}x - 2$ when $x = -3, 0, 3, 6$.

Solution

Replace x in $y = \frac{2}{3}x - 2$ by $-3, 0, 3$, and 6. For each value of x, determine the value of y.

x	$y = \frac{2}{3}x - 2$	y	(x, y)
-3	$\frac{2}{3}(-3) - 2$	-4	$(-3, -4)$
0	$\frac{2}{3}(0) - 2$	-2	$(0, -2)$
3	$\frac{2}{3}(3) - 2$	0	$(3, 0)$
6	$\frac{2}{3}(6) - 2$	2	$(6, 2)$

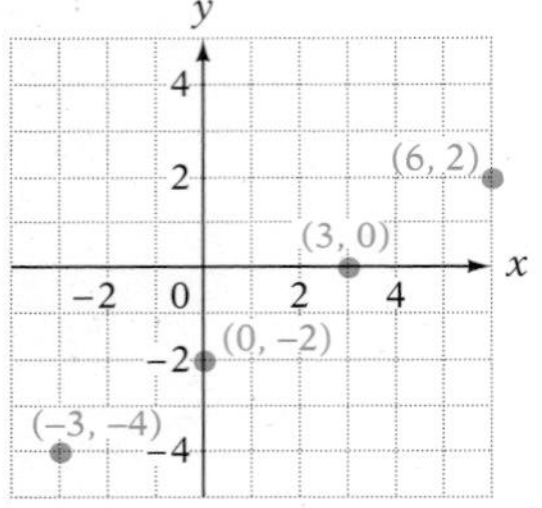

YOU TRY IT • 5

Graph the ordered-pair solutions of $y = -\frac{1}{2}x + 2$ when $x = -4, -2, 0, 2$.

Your solution

y
4
2
−4 −2 0 2 4 x
−2
−4

Solution on p. S11

OBJECTIVE C **To graph a scatter diagram** 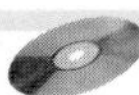

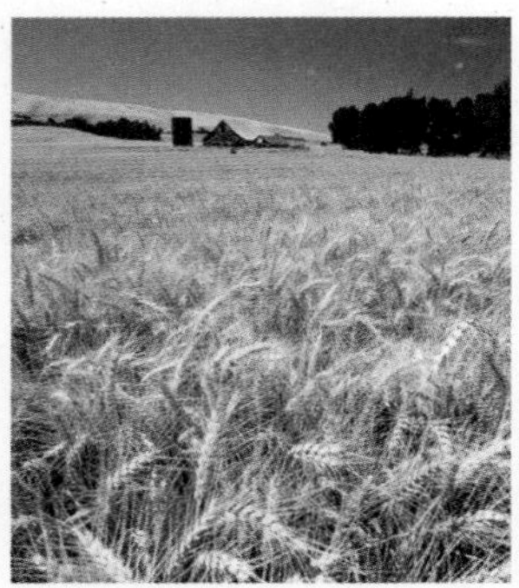

Finding a relationship between two variables is an important task in the study of mathematics. These relationships occur in many forms and in a wide variety of applications. Here are some examples.

- A botanist wants to know the relationship between the number of bushels of wheat yielded per acre and the amount of watering per acre.
- An environmental scientist wants to know the relationship between the incidence of skin cancer and the amount of ozone in the atmosphere.
- A business analyst wants to know the relationship between the price of a product and the number of products that are sold at that price.

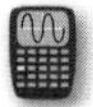

Integrating Technology

See the Keystroke Guide: *Scatter Diagrams* for instructions on using a graphing calculator to create a scatter diagram.

A researcher may investigate the relationship between two variables by means of *regression analysis,* which is a branch of statistics. The study of the relationship between the two variables may begin with a **scatter diagram,** which is a graph of the ordered pairs of the known data.

The following table shows randomly selected data for a recent Boston Marathon. It shows times (in minutes) for various participants 40 years old and older.

Age (x)	55	46	53	40	40	44	54	44	41	50
Time (y)	254	204	243	194	281	197	238	300	232	216

Take Note

The jagged portion of the horizontal axis in the figure at the right indicates that the numbers between 0 and 40 are missing.

The scatter diagram for these data is shown at the right. Each ordered pair represents the age and time for a participant. For instance, the ordered pair (53, 243) indicates that a 53-year-old participant ran the marathon in 243 min.

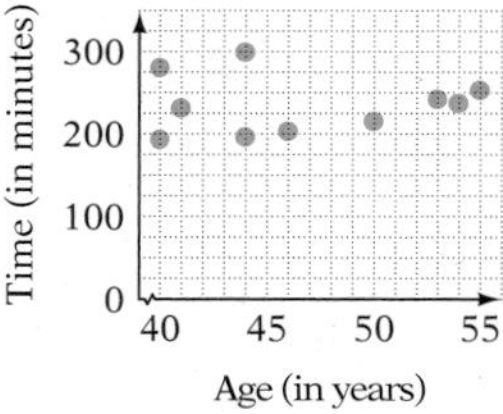

EXAMPLE 6

Life expectancy depends on a number of factors, including age. The table below, based on data from the Census Bureau, shows the life expectancy (in years) for people up to age 70 in the United States. Draw a scatter diagram for these data.

Age	0	10	20	30	40	50	60	70
Life Expectancy	77.8	78.6	78.8	79.3	79.9	80.9	82.5	85.2

Strategy

To draw a scatter diagram:

- Draw a coordinate grid with the horizontal axis representing the age of the person and the vertical axis representing life expectancy. Because the life expectancies start at 77.8, it is more convenient to start labeling the vertical axis at 77 (any number less than 77.8 could be used). A jagged line on the vertical axis indicates that the graph does not start at zero.
- Graph the ordered pairs (0, 77.8), (10, 78.6), (20, 78.8), (30, 79.3), (40, 79.9), (50, 80.9), (60, 82.5), and (70, 85.2).

Solution

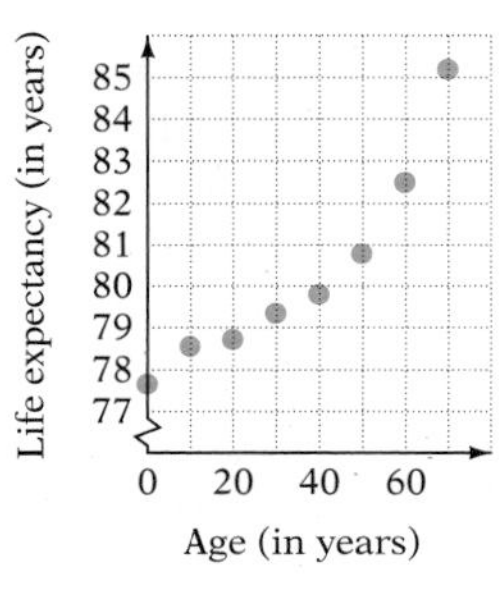

YOU TRY IT 6

A cup of tea is placed in a microwave oven and its temperature (in °F) recorded at 10-second intervals. The results are shown in the table below. Draw a scatter diagram for these data.

Time (in seconds)	0	10	20	30	40	50	60
Temperature (in °F)	72	76	88	106	126	140	160

Your strategy

Your solution

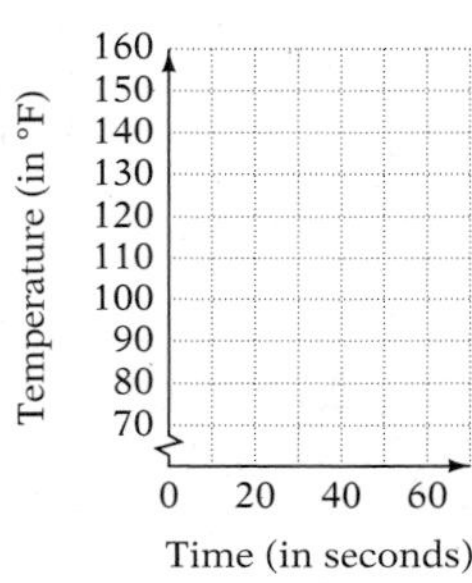

Solution on p. S11

4.1 EXERCISES

OBJECTIVE A **To graph points in a rectangular coordinate system**

1. Graph $(-2, 1)$, $(3, -5)$, $(-2, 4)$, and $(0, 3)$.

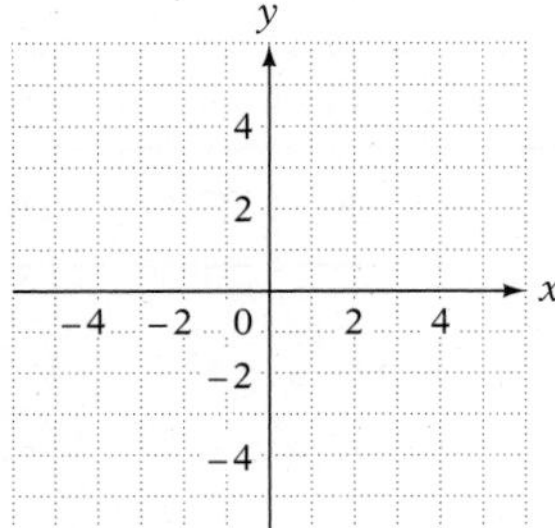

2. Graph $(5, -1)$, $(-3, -3)$, $(-1, 0)$, and $(1, -1)$.

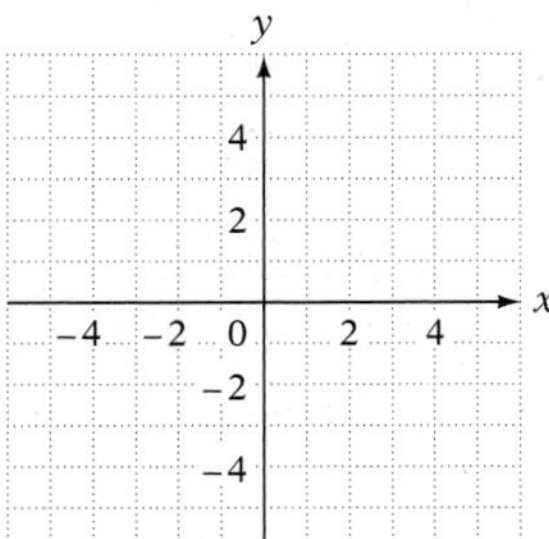

3. Graph $(0, 0)$, $(0, -5)$, $(-3, 0)$, and $(0, 2)$.

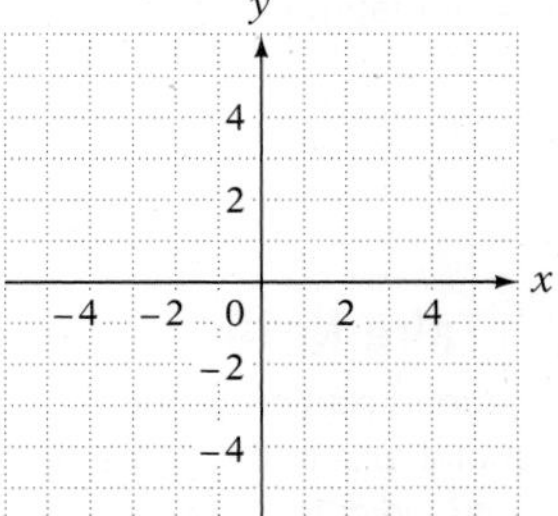

4. Graph $(-4, 5)$, $(-3, 1)$, $(3, -4)$, and $(5, 0)$.

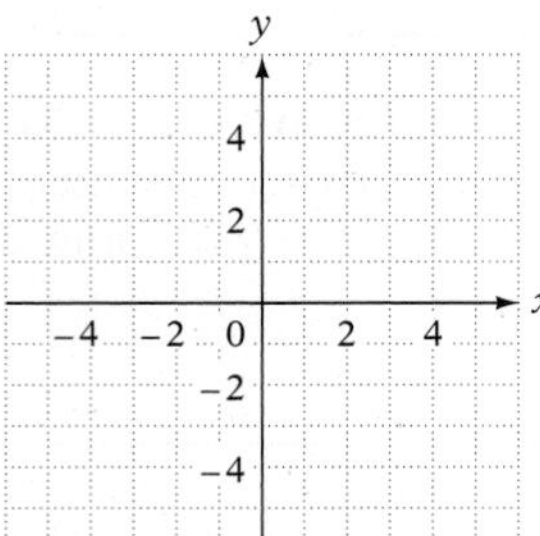

5. Graph $(-1, 4)$, $(-2, -3)$, $(0, 2)$, and $(4, 0)$.

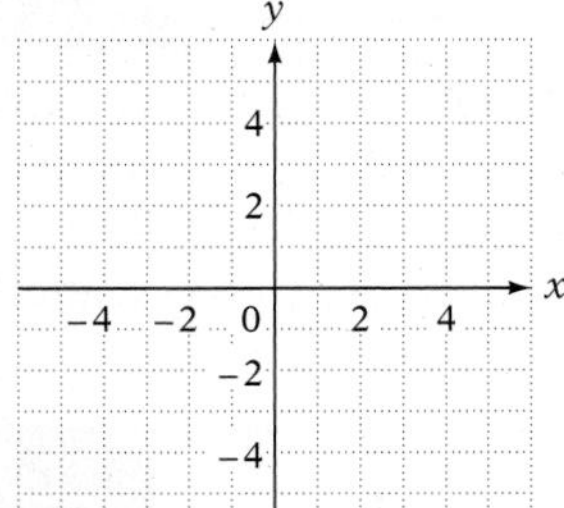

6. Graph $(5, 2)$, $(-4, -1)$, $(0, 0)$, and $(0, 3)$.

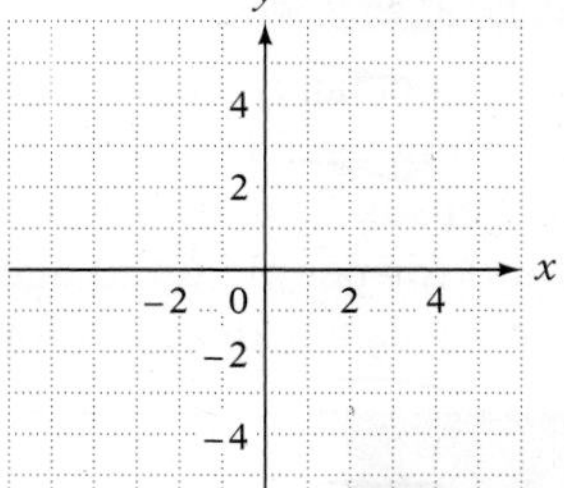

7. Find the coordinates of each of the points.

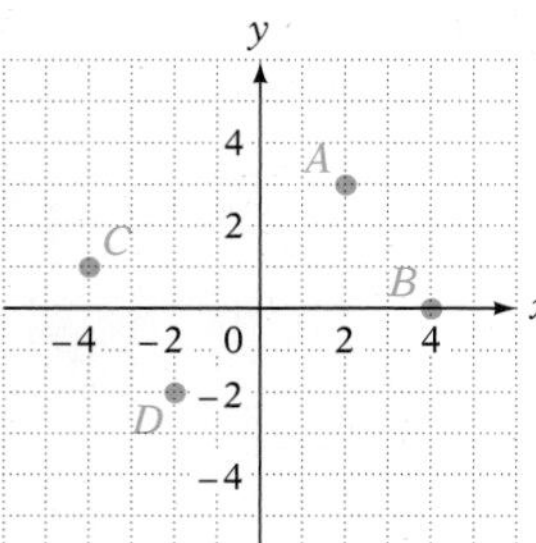

8. Find the coordinates of each of the points.

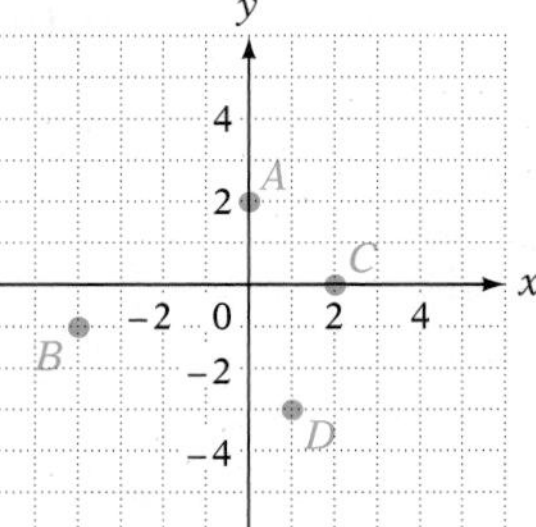

9. Find the coordinates of each of the points.

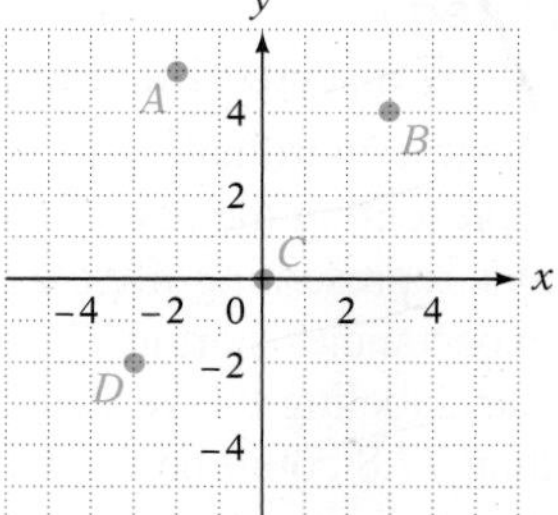

10. Find the coordinates of each of the points.

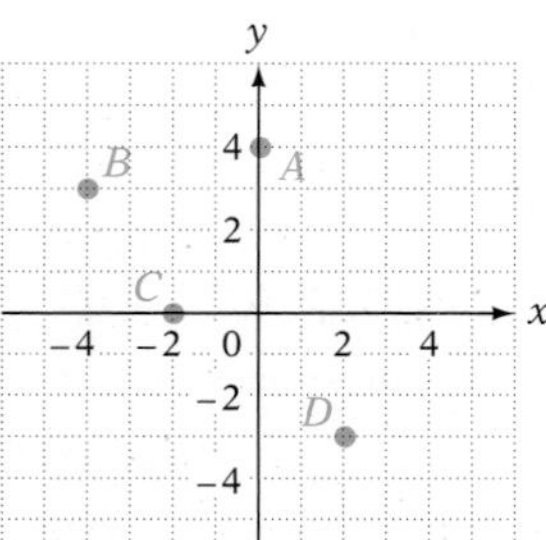

11. **a.** Name the abscissas of points *A* and *C*.
b. Name the ordinates of points *B* and *D*.

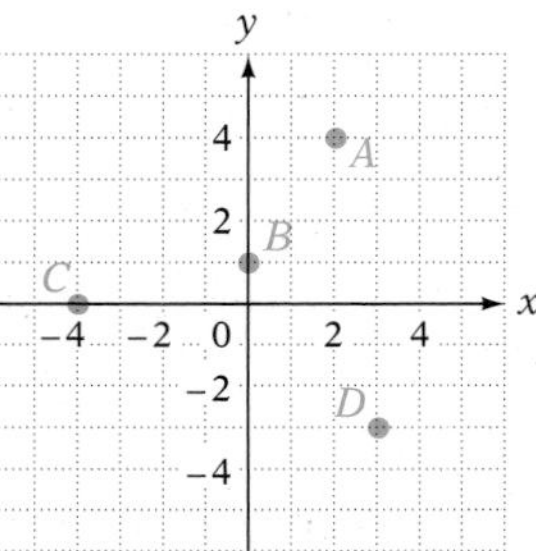

12. **a.** Name the abscissas of points *A* and *C*.
b. Name the ordinates of points *B* and *D*.

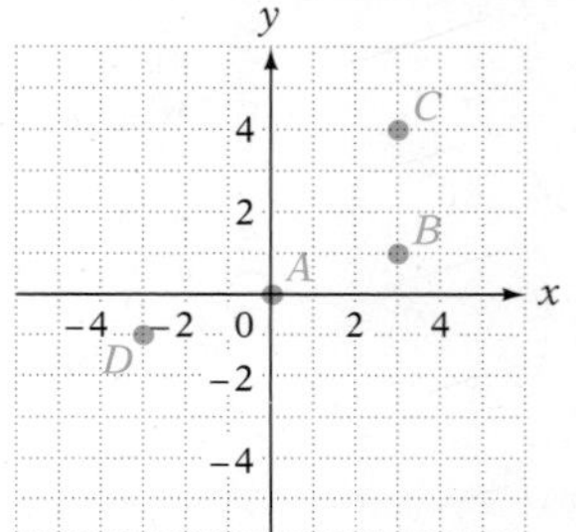

13. **a.** On an xy-coordinate system, what is the name of the axis for which all the x-coordinates are zero?
b. On an xy-coordinate system, what is the name of the axis for which all the y-coordinates are zero?

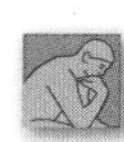

14. Let a and b be positive numbers such that $a < b$. In which quadrant is each point located?
a. (a, b) **b.** $(-a, b)$ **c.** $(-a, -b)$ **d.** $(b - a, -b)$

OBJECTIVE B To determine ordered-pair solutions of an equation in two variables

15. Is (3, 4) a solution of $y = -x + 7$?

16. Is (2, −3) a solution of $y = x + 5$?

17. Is (−1, 2) a solution of $y = \frac{1}{2}x - 1$?

18. Is (1, −3) a solution of $y = -2x - 1$?

19. Is (4, 1) a solution of $2x - 5y = 4$?

20. Is (−5, 3) a solution of $3x - 2y = 9$?

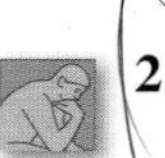

21. Suppose (x, y) is a solution of the equation $y = -3x + 6$, where $x > 2$. Is y positive or negative?

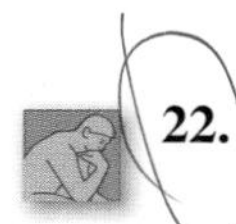

22. Suppose (x, y) is a solution of the equation $y = 4x - 8$, where $y > 0$. Is x less than or greater than 2?

23. Find the ordered-pair solution of $y = 3x - 2$ corresponding to $x = 3$.

24. Find the ordered-pair solution of $y = 4x + 1$ corresponding to $x = -1$.

25. Find the ordered-pair solution of $y = \frac{2}{3}x - 1$ corresponding to $x = 6$.

26. Find the ordered-pair solution of $y = \frac{3}{4}x - 2$ corresponding to $x = 4$.

27. Find the ordered-pair solution of $y = -3x + 1$ corresponding to $x = 0$.

28. Find the ordered-pair solution of $y = \frac{2}{5}x - 5$ corresponding to $x = 0$.

29. Find the ordered-pair solution of $y = \frac{2}{5}x + 2$ corresponding to $x = -5$.

30. Find the ordered-pair solution of $y = -\frac{1}{6}x - 2$ corresponding to $x = 12$.

For Exercises 31 to 36, graph the ordered-pair solutions of the equation for the given values of x.

31. $y = 2x; x = -2, -1, 0, 2$

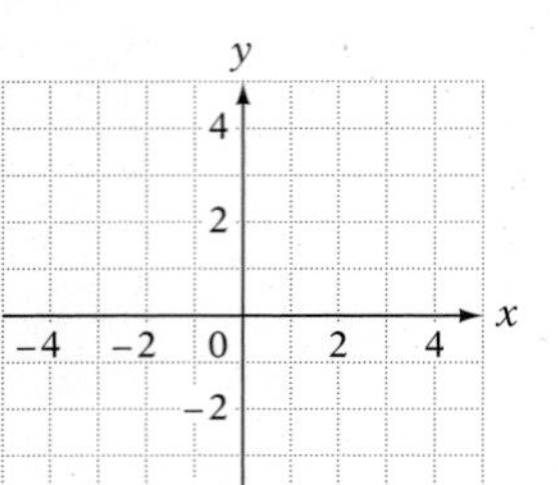

32. $y = -2x; x = -2, -1, 0, 2$

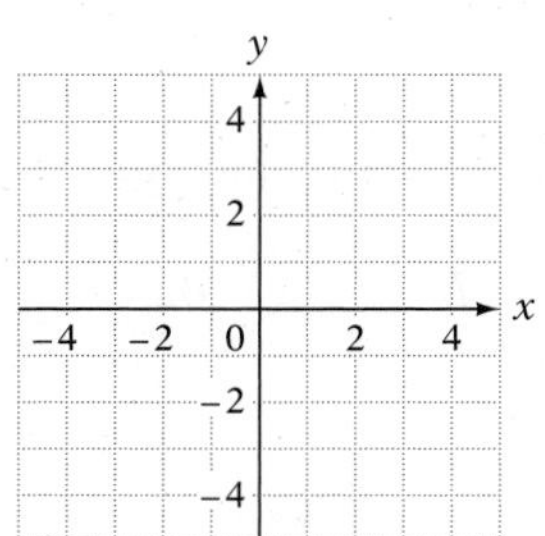

33. $y = \frac{2}{3}x + 1; x = -3, 0, 3$

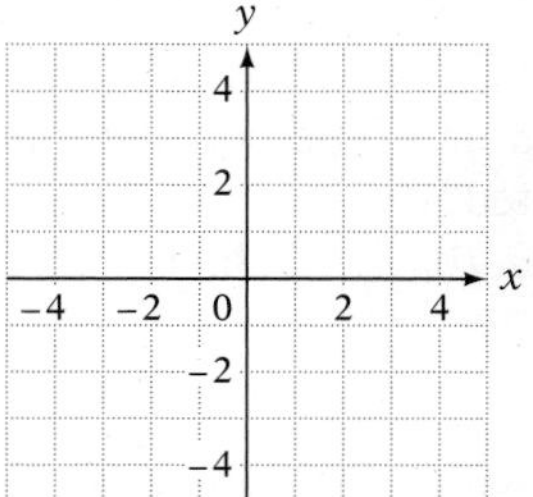

34. $y = -\frac{1}{3}x - 2; x = -3, 0, 3$

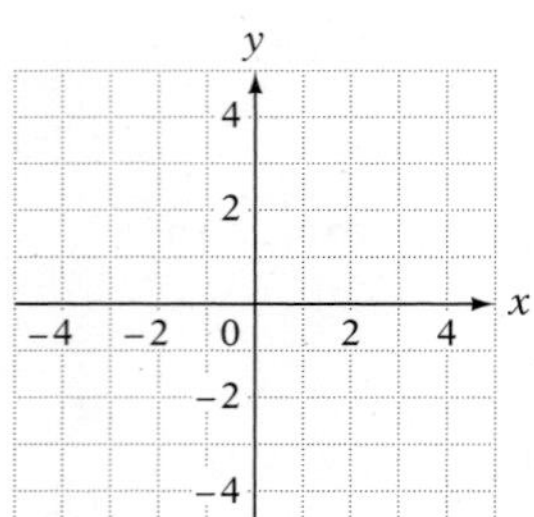

35. $2x + 3y = 6; x = -3, 0, 3$

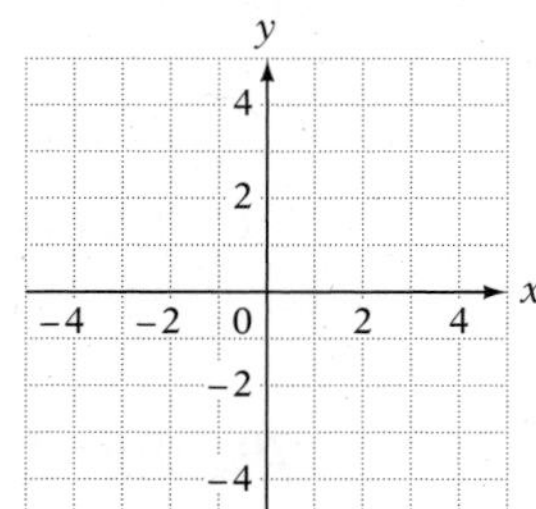

36. $x - 2y = 4; x = -2, 0, 2$

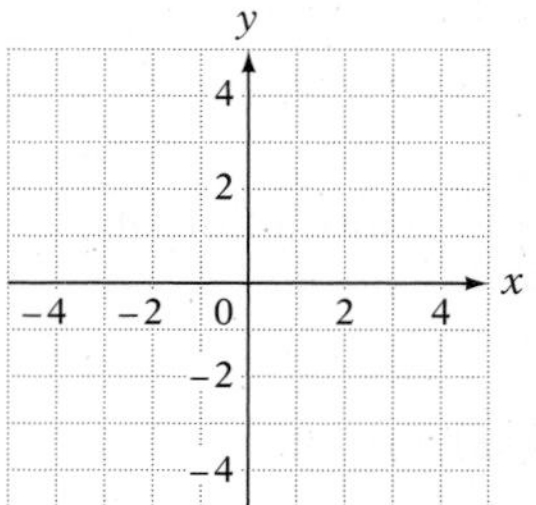

OBJECTIVE C To graph a scatter diagram

37. **Chemistry** The amount of a substance that can be dissolved in a fixed amount of water usually increases as the temperature of the water increases. Cerium selenate, however, does not behave in this manner. The graph at the right shows the number of grams of cerium selenate that will dissolve in 100 mg of water for various temperatures, in degrees Celsius.

a. Determine the temperature at which 25 g of cerium selenate will dissolve.

b. Determine the number of grams of cerium selenate that will dissolve when the temperature is 80°C.

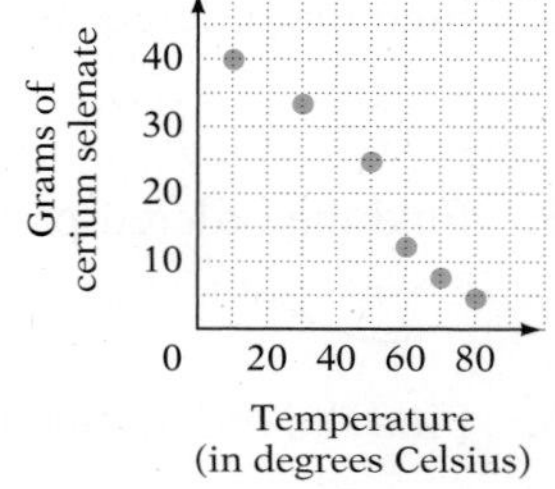

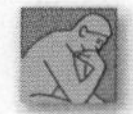

38. The scatter diagram at the right shows the quarterly profits for two fast food restaurants, Quick Eats and Pizza Your Way.

a. Which company had the greater profit in the first quarter of 2010?

b. If the trend shown in the scatter diagram were to continue, which restaurant would have the larger profit in 2011?

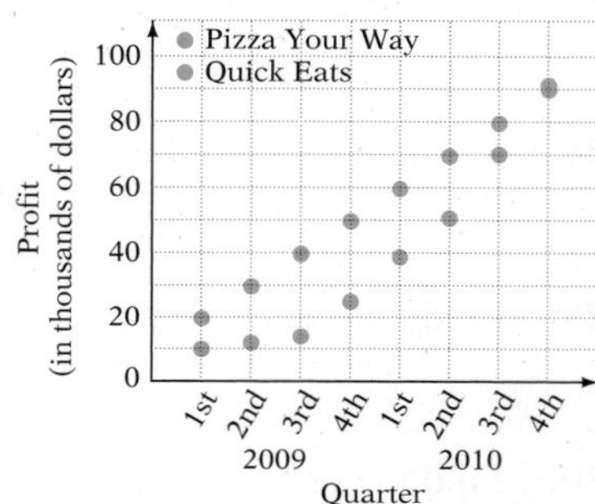

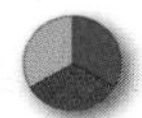

39. Meteorology Draw a scatter diagram for the data in the article.

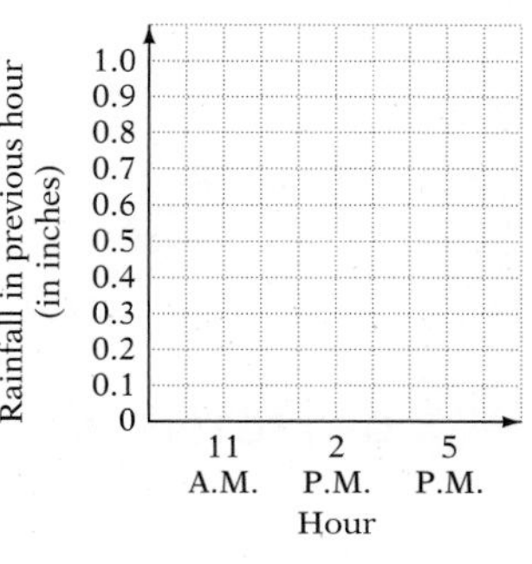

In the News

Tropical Storm Fay Lashes Coast

Tropical storm Fay hit the Florida coast today, with heavy rain and high winds. Here's a look at the amount of rainfall over the course of the afternoon.

Hour	11 A.M.	12 P.M.	1 P.M.	2 P.M.	3 P.M.	4 P.M.	5 P.M.
Inches of rain in preceding hour	0.25	0.69	0.85	1.05	0.70	0	0.08

Source: www.weather.gov

40. Utilities A power company suggests that a larger power plant can produce energy more efficiently and therefore at lower cost to consumers. The table below shows the output and average cost for power plants of various sizes. Draw a scatter diagram for these data.

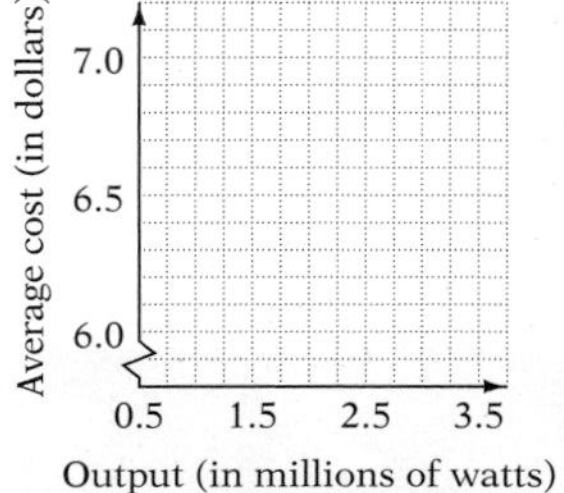

Output (in millions of watts)	0.7	2.2	2.6	3.2	2.8	3.5
Average Cost (in dollars)	6.9	6.5	6.3	6.4	6.5	6.1

Applying the Concepts

41. Graph the ordered pairs (x, x^2), where $x \in \{-2, -1, 0, 1, 2\}$.

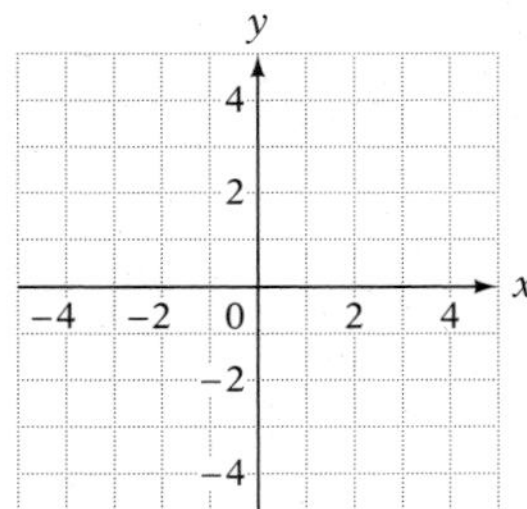

42. Graph the ordered pairs $\left(x, \frac{1}{x}\right)$, where $x \in \left\{-2, -1, -\frac{1}{2}, -\frac{1}{3}, \frac{1}{3}, \frac{1}{2}, 1, 2\right\}$.

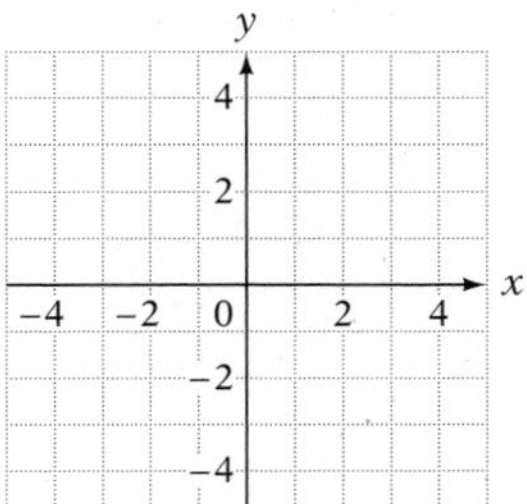

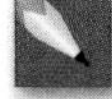

43. Describe the graph of all the ordered pairs (x, y) that are 5 units from the origin.

44. Consider two distinct fixed points in the plane. Describe the graph of all the points (x, y) that are equidistant from these fixed points.

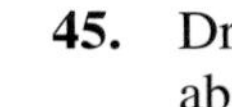

45. Draw a line passing through every point whose abscissa equals its ordinate.

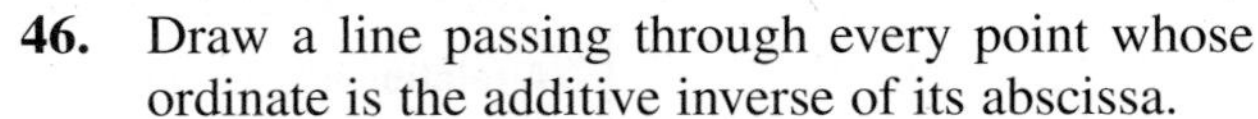

46. Draw a line passing through every point whose ordinate is the additive inverse of its abscissa.

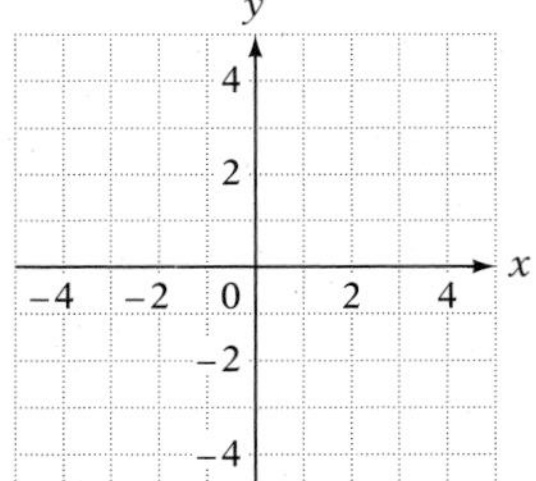

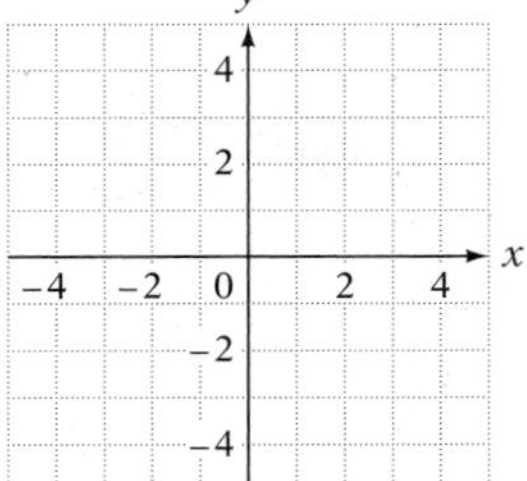

SECTION

4.2 Introduction to Functions

OBJECTIVE A To evaluate a function 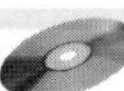

In mathematics and its applications, there are many times when it is necessary to investigate a relationship between two quantities. Here is a financial application: Consider a person who is planning to finance the purchase of a car. If the current interest rate for a 5-year loan is 5%, the equation that describes the relationship between the amount that is borrowed B and the monthly payment P is $P = 0.018871B$.

For each amount the purchaser may borrow (B), there is a certain monthly payment (P). The relationship between the amount borrowed and the payment can be recorded as ordered pairs, where the first coordinate is the amount borrowed and the second coordinate is the monthly payment. Some of these ordered pairs are shown at the right.

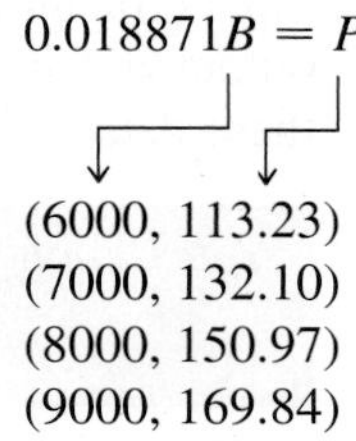

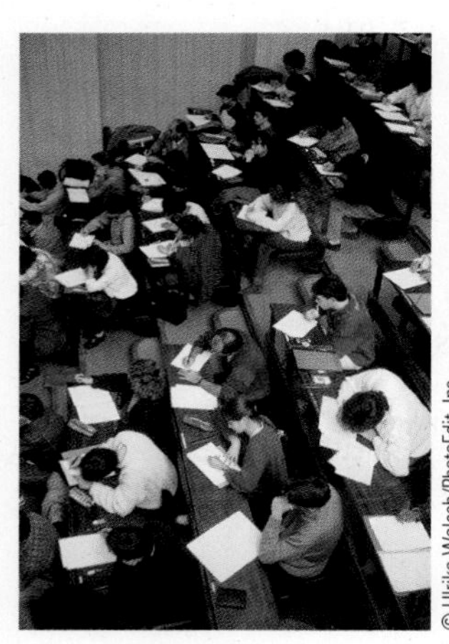
© Ulrike Welsch/PhotoEdit, Inc.

A relationship between two quantities is not always given by an equation. The table at the right describes a grading scale that defines a relationship between a score on a test and a letter grade. For each score, the table assigns only one letter grade. The ordered pair (84, B) indicates that a score of 84 receives a letter grade of B.

Score	*Grade*
90–100	A
80–89	B
70–79	C
60–69	D
0–59	F

The graph at the right also shows a relationship between two quantities. It is a graph of the viscosity V of SAE 40 motor oil at various temperatures T. Ordered pairs can be approximated from the graph. The ordered pair (120, 250) indicates that the viscosity of the oil at 120°F is 250 units.

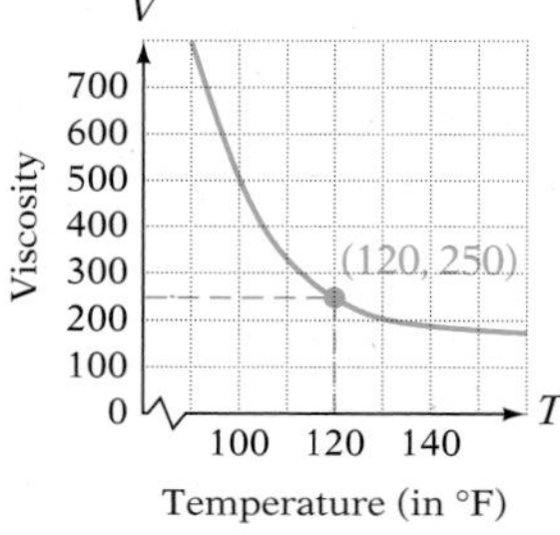

In each of these examples, there is a rule (an equation, a table, or a graph) that determines a certain set of ordered pairs.

Definition of Relation

A **relation** is a set of ordered pairs.

Here are some of the ordered pairs for the relations given above.

Relation	*Some of the Ordered Pairs of the Relation*
Car Payment	(7500, 141.53), (8750, 165.12), (9390, 177.20)
Grading Scale	(78, C), (98, A), (70, C), (81, B), (94, A)
Oil Viscosity	(100, 500), (120, 250), (130, 200), (150, 180)

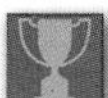

Tips for Success

Have you considered joining a study group? Getting together regularly with other students in the class to go over material and quiz each other can be very beneficial. See *AIM for Success* in the Preface.

Each of these three relations is actually a special type of relation called a *function*. Functions play an important role in mathematics and its applications.

Definition of Function

A **function** is a relation in which no two ordered pairs have the same first coordinate and different second coordinates.

The **domain** of a function is the set of the first coordinates of all the ordered pairs of the function. The **range** is the set of the second coordinates of all the ordered pairs of the function.

For the function defined by the ordered pairs

$$\{(2, 3), (4, 5), (6, 7), (8, 9)\}$$

the domain is $\{2, 4, 6, 8\}$ and the range is $\{3, 5, 7, 9\}$.

HOW TO • 1 Find the domain and range of the function $\{(2, 3), (4, 6), (6, 8), (10, 6)\}$.

The domain is $\{2, 4, 6, 10\}$.
- **The domain of the function is the set of the first coordinates of the ordered pairs.**

The range is $\{3, 6, 8\}$.
- **The range of the function is the set of the second coordinates of the ordered pairs.**

For each element of the domain of a function there is a corresponding element in the range of the function. A possible diagram for the function above is

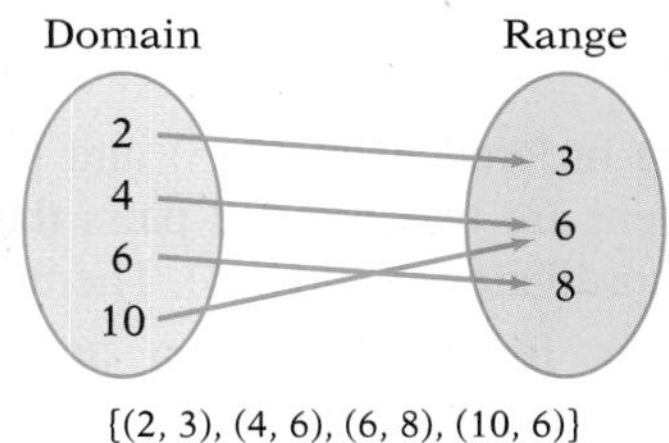

$\{(2, 3), (4, 6), (6, 8), (10, 6)\}$

Functions defined by tables or graphs, such as those described at the beginning of this section, have important applications. However, a major focus of this text is functions defined by equations in two variables.

The **square function,** which pairs each real number with its square, can be defined by the equation

$$y = x^2$$

This equation states that for a given value of x in the domain, the corresponding value of y in the range is the square of x. For instance, if $x = 6$, then $y = 36$ and if $x = -7$, then $y = 49$. Because the value of y *depends* on the value of x, y is called the **dependent variable** and x is called the **independent variable.**

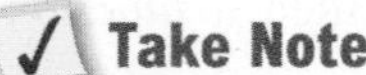

A pictorial representation of the square function is shown at the right. The function acts as a machine that changes a number from the domain into the square of the number.

A function can be thought of as a rule that pairs one number with another number. For instance, the *square function* pairs a number with its square. The ordered pairs for the values shown at the right are $(-5, 25)$, $\left(\frac{3}{5}, \frac{9}{25}\right)$, $(0, 0)$, and $(3, 9)$. For this function, the second coordinate is the square of the first coordinate. If we let x represent the first coordinate, then the second coordinate is x^2 and we have the ordered pair (x, x^2).

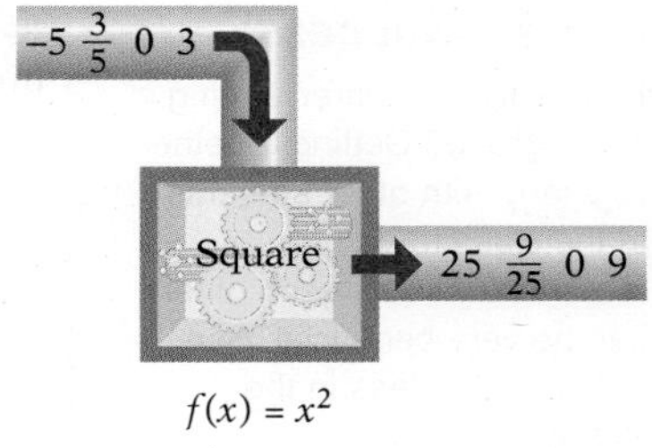

$f(x) = x^2$

A function cannot have two ordered pairs with *different* second coordinates and the same first coordinate. However, a function may contain ordered pairs with the *same* second coordinate. For instance, the square function has the ordered pairs $(-3, 9)$ and $(3, 9)$; the second coordinates are the same but the first coordinates are different.

The **double function** pairs a number with twice that number. The ordered pairs for the values shown at the right are $(-5, -10)$, $\left(\frac{3}{5}, \frac{6}{5}\right)$, $(0, 0)$, and $(3, 6)$. For this function, the second coordinate is twice the first coordinate. If we let x represent the first coordinate, then the second coordinate is $2x$ and we have the ordered pair $(x, 2x)$.

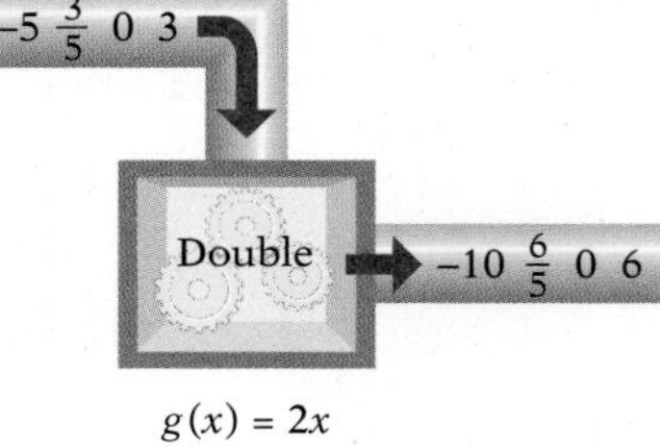

$g(x) = 2x$

Not every equation in two variables defines a function. For instance, consider the equation

$$y^2 = x^2 + 9$$

Because

$$5^2 = 4^2 + 9 \quad \text{and} \quad (-5)^2 = 4^2 + 9$$

the ordered pairs $(4, 5)$ and $(4, -5)$ are both solutions of the equation. Consequently, there are two ordered pairs that have the same first coordinate (4) but *different* second coordinates (5 and -5). Therefore, the equation does not define a function. Other ordered pairs for this equation are $(0, -3)$, $(0, 3)$, $\left(\sqrt{7}, -4\right)$, and $\left(\sqrt{7}, 4\right)$. A graphical representation of these ordered pairs is shown below.

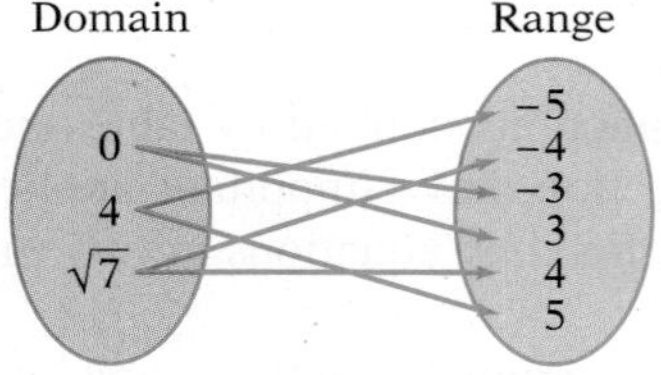

Note from this graphical representation that each element from the domain has two arrows pointing to two different elements in the range. Any time this occurs, the situation does not represent a function. However, this diagram *does* represent a relation. The relation for the values shown is $\{(0, -3), (0, 3), (4, -5), (4, 5), \left(\sqrt{7}, -4\right), \left(\sqrt{7}, 4\right)\}$.

The phrase "y is a function of x," or the same phrase with different variables, is used to describe an equation in two variables that defines a function. To emphasize that the equation represents a function, **function notation** is used.

Just as the variable x is commonly used to represent a number, the letter f is commonly used to name a function. The square function is written in function notation as follows:

Take Note

The dependent variable y and the notation $f(x)$ can be used interchangeably.

This is the value of the function. It is the number that is paired with x.

$$f(x) = x^2$$

The name of the function is f.

This is an algebraic expression that defines the relationship between the dependent and independent variables.

The symbol $f(x)$ is read "the value of f at x" or "f of x."

It is important to note that $f(x)$ does *not* mean f times x. The symbol $f(x)$ is the **value of the function** and represents the value of the dependent variable for a given value of the independent variable. We often write $y = f(x)$ to emphasize the relationship between the independent variable x and the dependent variable y. Remember that y and $f(x)$ are different symbols for the same number.

The letters used to represent a function are somewhat arbitrary. All of the following equations represent the same function.

$$\left.\begin{aligned} f(x) &= x^2 \\ s(t) &= t^2 \\ P(v) &= v^2 \end{aligned}\right\} \text{Each equation represents the square function.}$$

The process of determining $f(x)$ for a given value of x is called **evaluating a function.** For instance, to evaluate $f(x) = x^2$ when $x = 4$, replace x by 4 and simplify.

$$\begin{aligned} f(x) &= x^2 \\ f(4) &= 4^2 = 16 \end{aligned}$$

The *value* of the function is 16 when $x = 4$. An ordered pair of the function is (4, 16).

Integrating Technology

See the Projects and Group Activities at the end of this chapter for instructions on using a graphing calculator to evaluate a function. Instructions are also provided in the Keystroke Guide: *Evaluating Functions.*

HOW TO 2 Evaluate $g(t) = 3t^2 - 5t + 1$ when $t = -2$.

$$\begin{aligned} g(t) &= 3t^2 - 5t + 1 \\ g(-2) &= 3(-2)^2 - 5(-2) + 1 \\ &= 3(4) - 5(-2) + 1 \\ &= 12 + 10 + 1 = 23 \end{aligned}$$

• Replace t by -2 and then simplify.

When t is -2, the value of the function is 23.

Therefore, an ordered pair of the function is $(-2, 23)$.

It is possible to evaluate a function for a variable expression.

HOW TO 3 Evaluate $P(z) = 3z - 7$ when $z = 3 + h$.

$$\begin{aligned} P(z) &= 3z - 7 \\ P(3 + h) &= 3(3 + h) - 7 \\ &= 9 + 3h - 7 \\ &= 3h + 2 \end{aligned}$$

• Replace z by $3 + h$ and then simplify.

When z is $3 + h$, the value of the function is $3h + 2$.

Therefore, an ordered pair of the function is $(3 + h, 3h + 2)$.

Recall that the range of a function is found by applying the function to each element of the domain. If the domain contains an infinite number of elements, it may be difficult to find the range. However, if the domain has a finite number of elements, then the range can be found by evaluating the function for each element in the domain.

HOW TO • 4 Find the range of $f(x) = x^3 + x$ if the domain is $\{-2, -1, 0, 1, 2\}$.

$$\begin{aligned} f(x) &= x^3 + x \\ f(-2) &= (-2)^3 + (-2) = -10 \\ f(-1) &= (-1)^3 + (-1) = -2 \\ f(0) &= 0^3 + 0 = 0 \\ f(1) &= 1^3 + 1 = 2 \\ f(2) &= 2^3 + 2 = 10 \end{aligned}$$

• Replace x by each member of the domain. The range includes the values of $f(-2)$, $f(-1)$, $f(0)$, $f(1)$, and $f(2)$.

The range is $\{-10, -2, 0, 2, 10\}$.

When a function is represented by an equation, the domain of the function is all real numbers for which the value of the function is a real number. For instance:

- The domain of $f(x) = x^2$ is all real numbers, because the square of every real number is a real number.
- The domain of $g(x) = \frac{1}{x - 2}$ is all real numbers except 2, because when $x = 2$, $g(2) = \frac{1}{2 - 2} = \frac{1}{0}$, which is not a real number.

The domain of the grading-scale function is the set of whole numbers from 0 to 100. In set-builder notation, this is written $\{x \mid 0 \le x \le 100, x \in \text{whole numbers}\}$. The range is $\{A, B, C, D, F\}$.

Score	Grade
90–100	A
80–89	B
70–79	C
60–69	D
0–59	F

HOW TO • 5 What values, if any, are excluded from the domain of $f(x) = 2x^2 - 7x + 1$?

Because the value of $2x^2 - 7x + 1$ is a real number for any value of x, the domain of the function is all real numbers. No values are excluded from the domain of $f(x) = 2x^2 - 7x + 1$.

EXAMPLE • 1

Find the domain and range of the function $\{(5, 3), (9, 7), (13, 7), (17, 3)\}$.

Solution

Domain: $\{5, 9, 13, 17\}$;
Range: $\{3, 7\}$

• The domain is the set of first coordinates.

YOU TRY IT • 1

Find the domain and range of the function $\{(-1, 5), (3, 5), (4, 5), (6, 5)\}$.

Your solution

Solution on p. S11

EXAMPLE • 2

Given $p(r) = 5r^3 - 6r - 2$, find $p(-3)$.

Solution

$$\begin{aligned} p(r) &= 5r^3 - 6r - 2 \\ p(-3) &= 5(-3)^3 - 6(-3) - 2 \\ &= 5(-27) + 18 - 2 \\ &= -135 + 18 - 2 = -119 \end{aligned}$$

YOU TRY IT • 2

Evaluate $G(x) = \frac{3x}{x+2}$ when $x = -4$.

Your solution

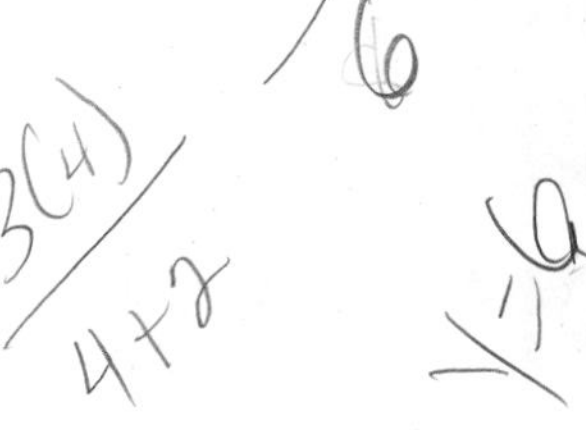

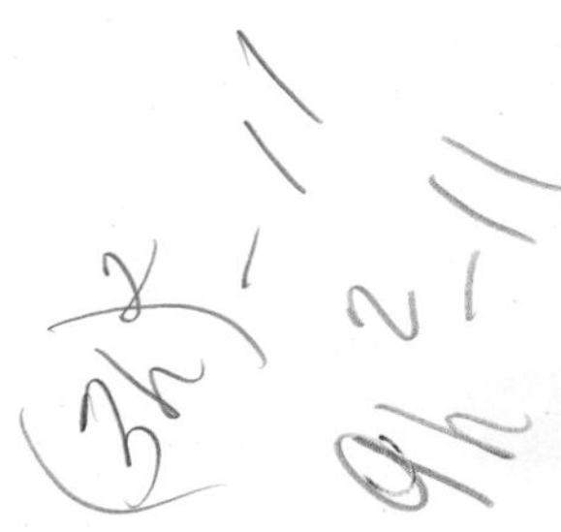

EXAMPLE • 3

Evaluate $Q(r) = 2r + 5$ when $r = h + 3$.

Solution

$$\begin{aligned} Q(r) &= 2r + 5 \\ Q(h+3) &= 2(h+3) + 5 \\ &= 2h + 6 + 5 \\ &= 2h + 11 \end{aligned}$$

YOU TRY IT • 3

Evaluate $f(x) = x^2 - 11$ when $x = 3h$.

Your solution

EXAMPLE • 4

Find the range of $f(x) = x^2 - 1$ if the domain is $\{-2, -1, 0, 1, 2\}$.

Solution

To find the range, evaluate the function at each element of the domain.

$$\begin{aligned} f(x) &= x^2 - 1 \\ f(-2) &= (-2)^2 - 1 = 4 - 1 = 3 \\ f(-1) &= (-1)^2 - 1 = 1 - 1 = 0 \\ f(0) &= 0^2 - 1 = 0 - 1 = -1 \\ f(1) &= 1^2 - 1 = 1 - 1 = 0 \\ f(2) &= 2^2 - 1 = 4 - 1 = 3 \end{aligned}$$

The range is $\{-1, 0, 3\}$. Note that 0 and 3 are listed only once.

YOU TRY IT • 4

Find the range of $h(z) = 3z + 1$ if the domain is $\left\{0, \frac{1}{3}, \frac{2}{3}, 1\right\}$.

Your solution

EXAMPLE • 5

What is the domain of $f(x) = 3x^2 - 5x + 2$?

Solution

Because $3x^2 - 5x + 2$ evaluates to a real number for any value of x, the domain of the function is all real numbers.

YOU TRY IT • 5

What value is excluded from the domain of $f(x) = \frac{2}{x-5}$?

Your solution

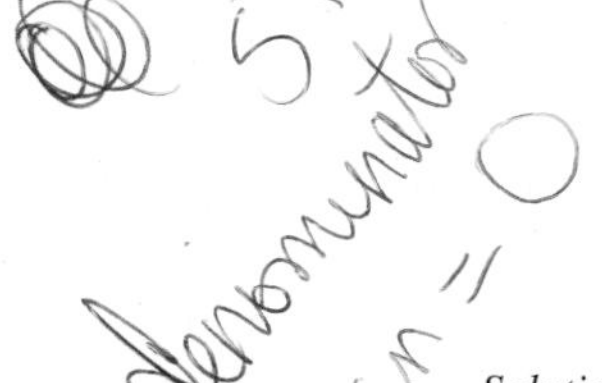

Solutions on p. S12

4.2 EXERCISES

OBJECTIVE A To evaluate a function

1. In your own words, explain what a function is.

2. What is the domain of a function? What is the range of a function?

3. Does the diagram below represent a function? Explain your answer.

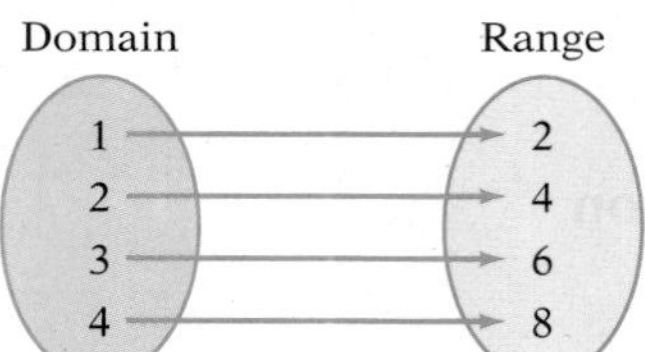

4. Does the diagram below represent a function? Explain your answer.

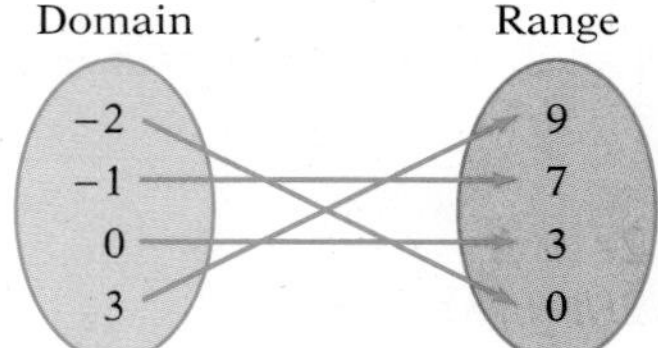

5. Does the diagram below represent a function? Explain your answer.

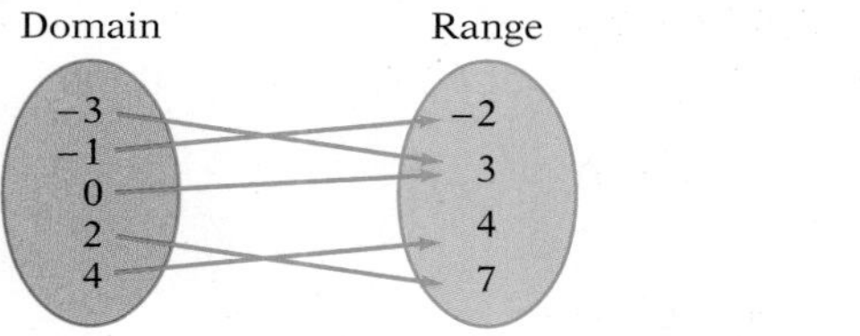

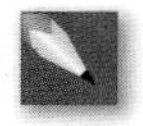

6. Does the diagram below represent a function? Explain your answer.

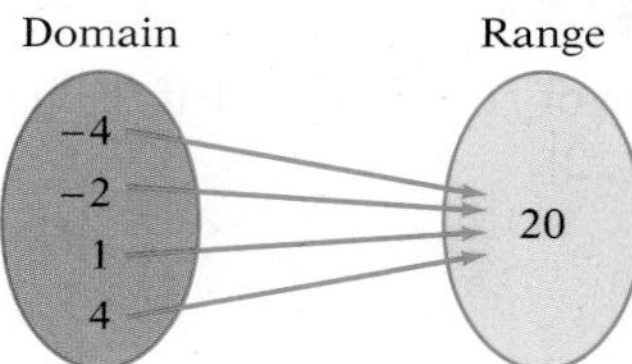

7. Does the diagram below represent a function? Explain your answer.

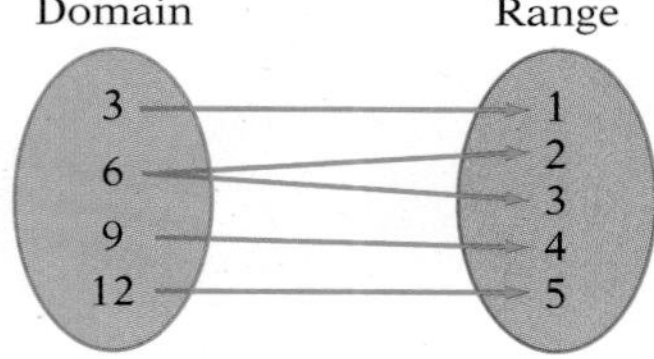

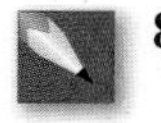

8. Does the diagram below represent a function? Explain your answer.

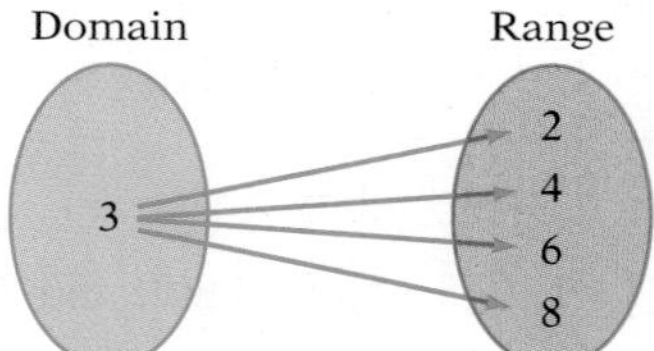

For Exercises 9 to 16, state whether the relation is a function.

9. $\{(0, 0), (2, 4), (3, 6), (4, 8), (5, 10)\}$

10. $\{(1, 3), (3, 5), (5, 7), (7, 9)\}$

11. $\{(-2, -1), (-4, -5), (0, -1), (3, 5)\}$

12. $\{(-3, -1), (-1, -1), (0, 1), (2, 6)\}$

13. $\{(-2, 3), (-1, 3), (0, -3), (1, 3), (2, 3)\}$

14. $\{(0, 0), (1, 0), (2, 0), (3, 0), (4, 0)\}$

15. $\{(1, 1), (4, 2), (9, 3), (1, -1), (4, -2)\}$

16. $\{(3, 1), (3, 2), (3, 3), (3, 4)\}$

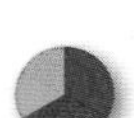

17. Traffic Safety The table in the article at the right shows the speeding fines that went into effect in the state of Nebraska in 2008.
a. Does this table define a function?
b. Given $x = 18$ mph, find y.

In the News

State Raises Speeding Fines

Beginning next week, you will pay more to speed in Nebraska.

Miles per hour over limit, x	*Fine (in dollars), y*
$1 \le x \le 5$	10
$5 < x \le 10$	25
$10 < x \le 15$	75
$15 < x \le 20$	125
$20 < x \le 35$	200
$x > 35$	300

Source: The Grand Island Independent

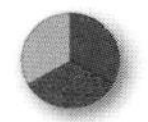

18. Shipping The table at the right shows the cost to send a Priority Mail package using the U.S. Postal Service.
a. Does this table define a function?
b. Given $x = 2$ lb, find y.

Weight in pounds (x)	*Cost (y)*
$0 < x \le 1$	\$4.95
$1 < x \le 2$	\$5.20
$2 < x \le 3$	\$6.25
$3 < x \le 4$	\$7.10
$4 < x \le 5$	\$8.15

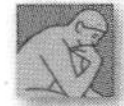

19. True or false? If f is a function, then it is possible that $f(0) = -2$ and $f(3) = -2$.

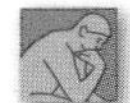

20. True or false? If f is a function, then it is possible that $f(4) = 3$ and $f(4) = 2$.

For Exercises 21 to 24, given $f(x) = 5x - 4$, evaluate.

21. $f(3)$ **22.** $f(-2)$ **23.** $f(0)$ **24.** $f(-1)$

For Exercises 25 to 28, given $G(t) = 4 - 3t$, evaluate.

25. $G(0)$ **26.** $G(-3)$ **27.** $G(-2)$ **28.** $G(4)$

For Exercises 29 to 32, given $q(r) = r^2 - 4$, evaluate.

29. $q(3)$ **30.** $q(4)$ **31.** $q(-2)$ **32.** $q(-5)$

For Exercises 33 to 36, given $F(x) = x^2 + 3x - 4$, evaluate.

33. $F(4)$ **34.** $F(-4)$ **35.** $F(-3)$ **36.** $F(-6)$

For Exercises 37 to 40, given $H(p) = \dfrac{3p}{p + 2}$, evaluate.

37. $H(1)$ **38.** $H(-3)$ **39.** $H(t)$ **40.** $H(v)$

For Exercises 41 to 44, given $s(t) = t^3 - 3t + 4$, evaluate.

41. $s(-1)$ **42.** $s(2)$ **43.** $s(a)$ **44.** $s(w)$

45. Given $P(x) = 4x + 7$, write $P(-2 + h) - P(-2)$ in simplest form.

46. Given $G(t) = 9 - 2t$, write $G(-3 + h) - G(-3)$ in simplest form.

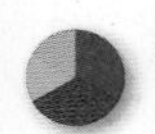

47. Automotive Technology Read the article below. If you change the tires on your car to low-rolling-resistance tires that increase your car's fuel efficiency by 5%, what is your annual cost savings? Round to the nearest cent.

In the News

Lower Rolling Resistance Means Lower Cost

Carmakers, seeking ways to meet consumer demands for better gas mileage, are using low-rolling-resistance tires that increase fuel efficiency by as much as 10%. If you drive 12,000 miles a year and the price of gas is $4.00 per gallon, increasing your fuel efficiency by p percent can give you an annual cost savings of $S = \dfrac{2400p}{1 + p}$ dollars.

Source: The Detroit News

48. Airports Airport administrators have a tendency to price airport parking at a rate that discourages people from using the parking lot for long periods of time. The rate structure for an airport is given in the table at the right.

a. Evaluate this function when $t = 2.5$ h.
b. Evaluate this function when $t = 7$ h.

Hours Parked	*Cost*
$0 < t \leq 1$	\$1.00
$1 < t \leq 2$	\$3.00
$2 < t \leq 4$	\$6.50
$4 < t \leq 7$	\$10.00
$7 < t \leq 12$	\$14.00

49. Business Game Engineering has just completed the programming and testing for a new computer game. The cost to manufacture and package the game depends on the number of units Game Engineering plans to sell. The table at the right shows the cost per game for packaging various quantities.

a. Evaluate this function when $x = 7000$.
b. Evaluate this function when $x = 20{,}000$.

Number of Games Manufactured	*Cost to Manufacture One Game*
$0 < x \leq 2500$	\$6.00
$2500 < x \leq 5000$	\$5.50
$5000 < x \leq 10{,}000$	\$4.75
$10{,}000 < x \leq 20{,}000$	\$4.00
$20{,}000 < x \leq 40{,}000$	\$3.00

50. Real Estate A real estate appraiser charges a fee that depends on the estimated value V of the property. A table giving the fees charged for various estimated values of real estate appears at the right.

a. Evaluate this function when $V = \$5{,}000{,}000$.
b. Evaluate this function when $V = \$767{,}000$.

Value of Property	*Appraisal Fee*
$V < 100{,}000$	\$350
$100{,}000 \leq V < 500{,}000$	\$525
$500{,}000 \leq V < 1{,}000{,}000$	\$950
$1{,}000{,}000 \leq V < 5{,}000{,}000$	\$2500
$5{,}000{,}000 \leq V < 10{,}000{,}000$	\$3000

For Exercises 51 to 60, find the domain and range of the function.

51. $\{(1, 1), (2, 4), (3, 7), (4, 10), (5, 13)\}$

52. $\{(2, 6), (4, 18), (6, 38), (8, 66), (10, 102)\}$

53. $\{(0, 1), (2, 2), (4, 3), (6, 4)\}$

54. $\{(0, 1), (1, 2), (4, 3), (9, 4)\}$

55. $\{(1, 0), (3, 0), (5, 0), (7, 0), (9, 0)\}$

56. $\{(-2, -4), (2, 4), (-1, 1), (1, 1), (-3, 9), (3, 9)\}$

57. $\{(0, 0), (1, 1), (-1, 1), (2, 2), (-2, 2)\}$

58. $\{(0, -5), (5, 0), (10, 5), (15, 10)\}$

59. $\{(-2, -3), (-1, 6), (0, 7), (2, 3), (1, 9)\}$

60. $\{(-8, 0), (-4, 2), (-2, 4), (0, -4), (4, 4)\}$

For Exercises 61 to 78, what values are excluded from the domain of the function?

61. $f(x) = \dfrac{1}{x - 1}$

62. $g(x) = \dfrac{1}{x + 4}$

63. $h(x) = \dfrac{x + 3}{x + 8}$

64. $F(x) = \dfrac{2x - 5}{x - 4}$

65. $f(x) = 3x + 2$

66. $g(x) = 4 - 2x$

67. $G(x) = x^2 + 1$

68. $H(x) = \dfrac{1}{2}x^2$

69. $f(x) = \dfrac{x - 1}{x}$

70. $g(x) = \dfrac{2x + 5}{7}$

71. $H(x) = x^2 - x + 1$

72. $f(x) = 3x^2 + x + 4$

73. $f(x) = \dfrac{2x - 5}{3}$

74. $g(x) = \dfrac{3 - 5x}{5}$

75. $H(x) = \dfrac{x - 2}{x + 2}$

76. $h(x) = \dfrac{3 - x}{6 - x}$

77. $f(x) = \dfrac{x - 2}{2}$

78. $G(x) = \dfrac{2}{x - 2}$

For Exercises 79 to 92, find the range of the function defined by the equation and the given domain.

79. $f(x) = 4x - 3$; domain $= \{0, 1, 2, 3\}$

80. $G(x) = 3 - 5x$; domain $= \{-2, -1, 0, 1, 2\}$

81. $g(x) = 5x - 8$; domain $= \{-3, -1, 0, 1, 3\}$

82. $h(x) = 3x - 7$; domain $= \{-4, -2, 0, 2, 4\}$

83. $h(x) = x^2$; domain $= \{-2, -1, 0, 1, 2\}$

84. $H(x) = 1 - x^2$; domain $= \{-2, -1, 0, 1, 2\}$

85. $f(x) = 2x^2 - 2x + 2$; domain $= \{-4, -2, 0, 4\}$

86. $G(x) = -2x^2 + 5x - 2$; domain $= \{-3, -1, 0, 1, 3\}$

87. $H(x) = \dfrac{5}{1 - x}$; domain $= \{-2, 0, 2\}$

88. $g(x) = \dfrac{4}{4 - x}$; domain $= \{-5, 0, 3\}$

89. $f(x) = \dfrac{2}{x - 4}$; domain $= \{-2, 0, 2, 6\}$

90. $g(x) = \dfrac{x}{3 - x}$; domain $= \{-2, -1, 0, 1, 2\}$

91. $H(x) = 2 - 3x - x^2$; domain $= \{-5, 0, 5\}$

92. $G(x) = 4 - 3x - x^3$; domain $= \{-3, 0, 3\}$

Applying the Concepts

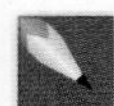

93. Explain the meanings of the words *relation* and *function*. Include in your explanation how the meanings of the two words differ.

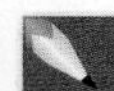

94. Give a real-world example of a relation that is not a function. Is it possible to give an example of a function that is not a relation? If so, give one. If not, explain why it is not possible.

95. **a.** Find the set of ordered pairs (x, y) determined by the equation $y = x^3$, where $x \in \{-2, -1, 0, 1, 2\}$.

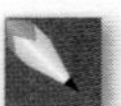

b. Does the set of ordered pairs define a function? Why or why not?

96. **a.** Find the set of ordered pairs (x, y) determined by the equation $|y| = x$, where $x \in \{0, 1, 2, 3\}$.

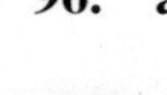

b. Does the set of ordered pairs define a function? Why or why not?

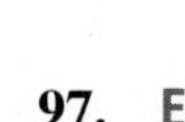

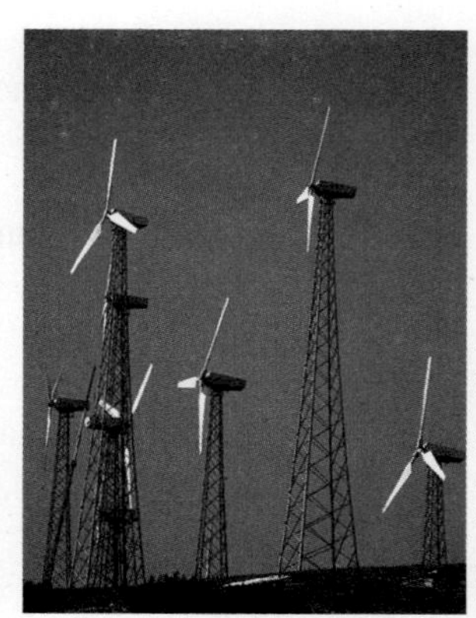

97. **Energy** The power a windmill can generate is a function of the velocity of the wind. The function can be approximated by $P = f(v) = 0.015v^3$, where P is the power in watts and v is the velocity of the wind in meters per second. How much power will be produced by a windmill when the velocity of the wind is 15 m/s?

98. **Automotive Technology** The distance s (in feet) a car will skid on a certain road surface after the brakes are applied is a function of the car's velocity v (in miles per hour). The function can be approximated by $s = f(v) = 0.017v^2$. How far will a car skid after its brakes are applied if it is traveling 60 mph?

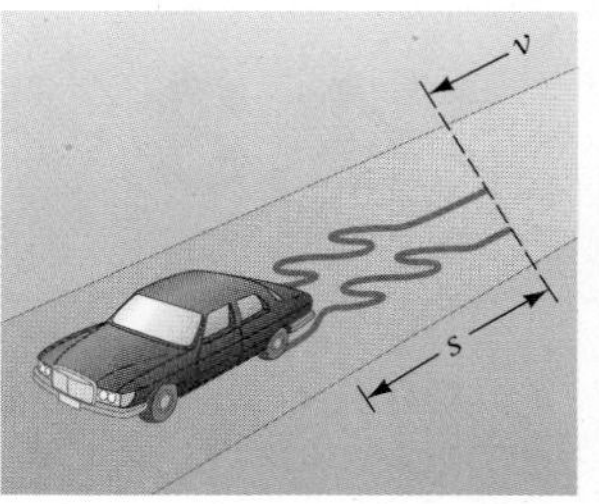

For Exercises 99 to 102, each graph defines a function.

99. **Parachuting** The graph at the right shows the descent speed of a paratrooper after making a low-level training jump.

a. The ordered pair (11.5, 36.3) gives the coordinates of a point on the graph. Write a sentence that explains the meaning of this ordered pair.

b. Estimate the speed at which the paratrooper is falling 1 s after jumping from the plane.

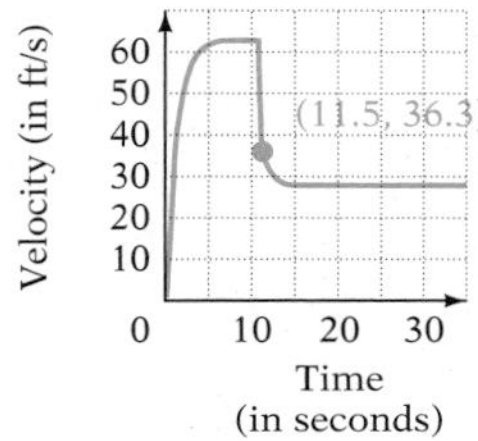

100. **Parachuting** The graph at the right shows the distance above ground of a paratrooper after making a low-level training jump.

a. The ordered pair (11.5, 590.2) gives the coordinates of a point on the graph. Write a sentence that explains the meaning of this ordered pair.

b. Estimate the time from the beginning of the jump to the end of the jump.

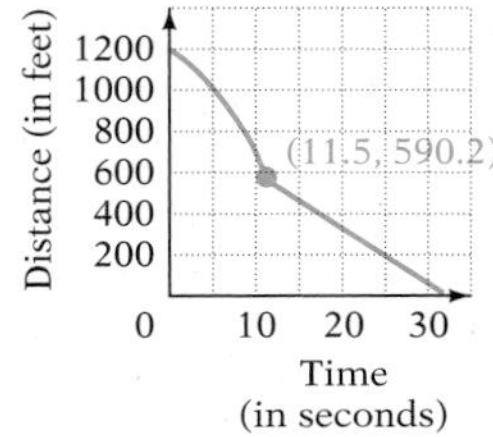

101. **Computer Science** The graph at the right is based on data from *PC Magazine* and shows the trend in the number of malware (malicious software) attacks, in thousands, for some recent years. (*Source: PC Magazine,* June 2008)

a. Estimate the number of malware attacks in 2005.

b. Estimate the number of malware attacks in 2008.

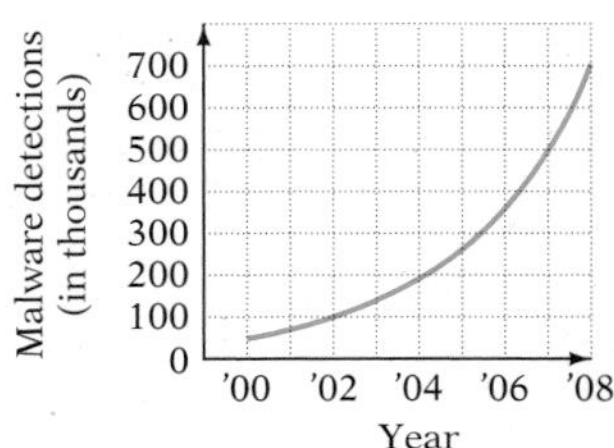

102. **Athletics** The graph at the right shows the decrease in the heart rate r of a runner (in beats per minute) t minutes after the completion of a race.

a. Estimate the heart rate of a runner when $t = 5$ min.

b. Estimate the heart rate of a runner when $t = 20$ min.

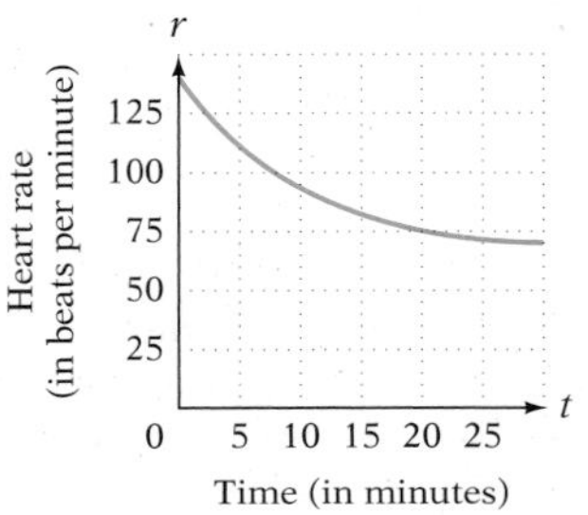

SECTION

4.3 Linear Functions

OBJECTIVE A To graph a linear function 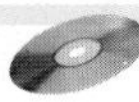

Recall that the ordered pairs of a function can be written as $(x, f(x))$ or (x, y). The **graph of a function** is a graph of the ordered pairs (x, y) that belong to the function. Certain functions have characteristic graphs. A function that can be written in the form $f(x) = mx + b$ (or $y = mx + b$) is called a **linear function** because its graph is a straight line.

Examples of linear functions are shown at the right. Note that the exponent on each variable is 1.

$f(x) = 2x + 5 \quad (m = 2, b = 5)$

$P(t) = 3t - 2 \quad (m = 3, b = -2)$

$y = -2x \quad (m = -2, b = 0)$

$y = -\frac{2}{3}x + 1 \quad \left(m = -\frac{2}{3}, b = 1\right)$

$g(z) = z - 2 \quad (m = 1, b = -2)$

The equation $y = x^2 + 4x + 3$ is not a linear function because it includes a term with a variable squared. The equation $f(x) = \frac{3}{x-2}$ is not a linear function because a variable occurs in the denominator. Another example of an equation that is not a linear function is $y = \sqrt{x} + 4$; this equation contains a variable within a radical and so is not a linear function.

Consider $f(x) = 2x + 1$. Evaluating the linear function when $x = -2$, $-1, 0, 1$, and 2 produces some of the ordered pairs of the function. It is convenient to record the results in a table similar to the one at the right. The graph of the ordered pairs is shown in the leftmost figure below.

x	$f(x) = 2x + 1$	y	(x, y)
−2	2(−2) + 1	−3	(−2, −3)
−1	2(−1) + 1	−1	(−1, −1)
0	2(0) + 1	1	(0, 1)
1	2(1) + 1	3	(1, 3)
2	2(2) + 1	5	(2, 5)

Evaluating the function when x is not an integer produces more ordered pairs to graph, such as $\left(-\frac{5}{2}, -4\right)$ and $\left(\frac{3}{2}, 4\right)$, as shown in the middle figure below. Evaluating the function for still other values of x would result in more and more ordered pairs being graphed. The result would be so many dots that the graph would look like the straight line shown in the rightmost figure, which is the graph of $f(x) = 2x + 1$.

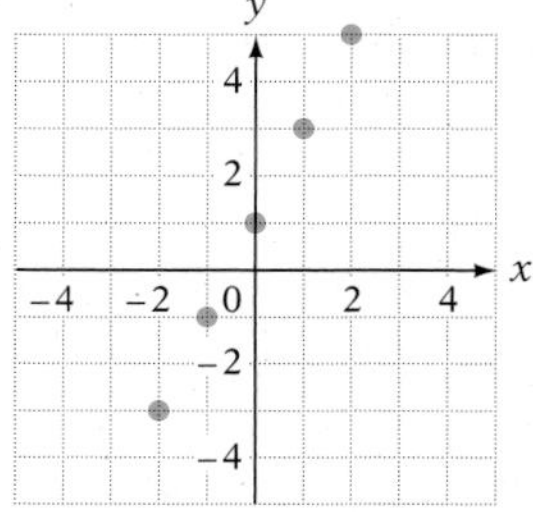

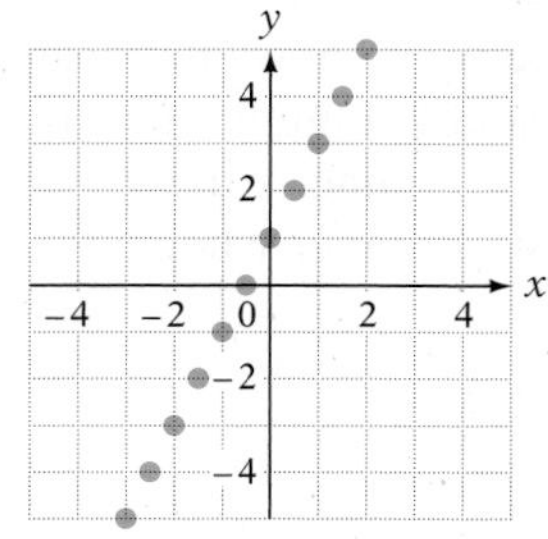

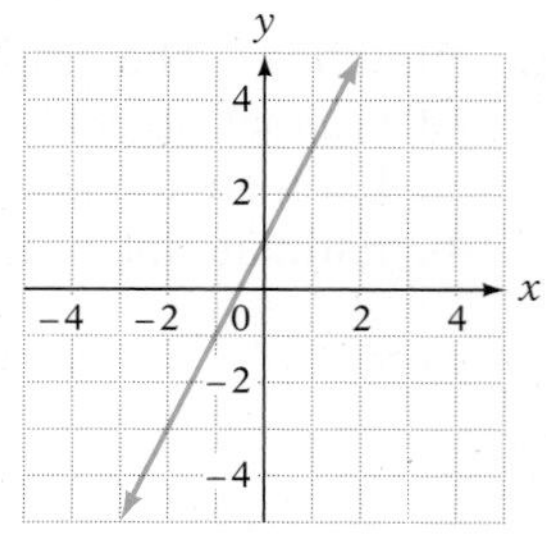

No matter what value of x is chosen, $2x + 1$ is a real number. This means the domain of $f(x) = 2x + 1$ is all real numbers. Therefore, we can use any real number when evaluating the function. Normally, however, values such as π or $\sqrt{5}$ are not used because it is difficult to graph the resulting ordered pairs.

Note from the graph of $f(x) = 2x + 1$ shown at the right that $(-1.5, -2)$ and $(3, 7)$ are the coordinates of points on the graph and that $f(-1.5) = -2$ and $f(3) = 7$. Note also that the point whose coordinates are $(2, 1)$ is not a point on the graph and that $f(2) \neq 1$. Every point on the graph is an ordered pair that belongs to the function, and every ordered pair that belongs to the function corresponds to a point on the graph.

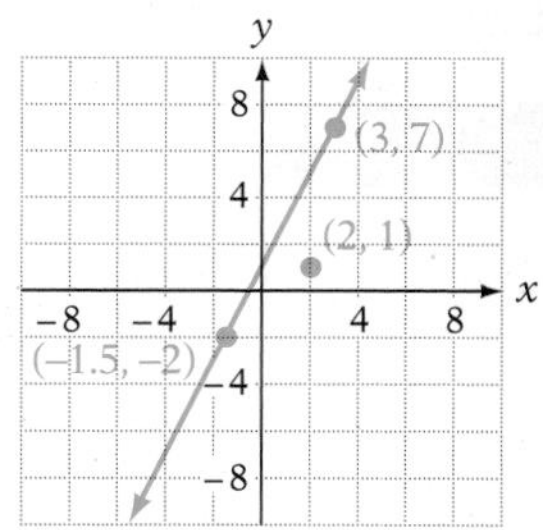

Integrating Technology

See the Projects and Group Activities at the end of this chapter for instructions on using a graphing calculator to graph a linear function. Instructions are also provided in the Keystroke Guide: *Graph*.

Whether an equation is written as $f(x) = mx + b$ or as $y = mx + b$, the equation represents a linear function, and the graph of the equation is a straight line.

Because the graph of a linear function is a straight line, and a straight line is determined by two points, the graph of a linear function can be drawn by finding only two of the ordered pairs of the function. However, it is recommended that you find at least *three* ordered pairs to ensure accuracy.

Take Note

When the coefficient of x is a fraction, choose values of x that are multiples of the denominator of the fraction. This will result in coordinates that are integers.

HOW TO • 1 Graph: $f(x) = -\frac{1}{2}x + 3$

x	$y = f(x)$
−4	5
0	3
2	2

• Find at least three ordered pairs. When the coefficient of x is a fraction, choose values of x that will simplify the calculations. The ordered pairs can be displayed in a table.

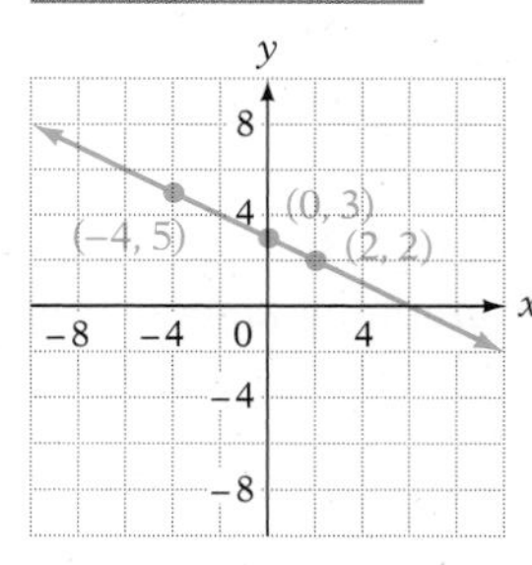

• Graph the ordered pairs and draw a line through the points.

EXAMPLE • 1

Graph: $f(x) = -\frac{3}{2}x - 3$

Solution

x	$y = f(x)$
0	−3
−2	0
−4	3

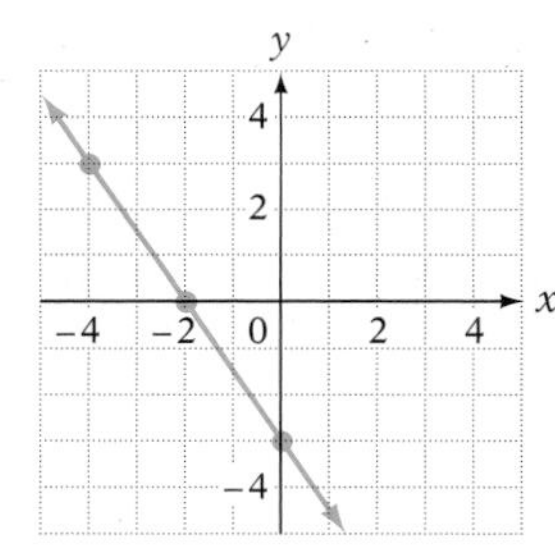

YOU TRY IT • 1

Graph: $f(x) = \frac{3}{5}x - 4$

Your solution

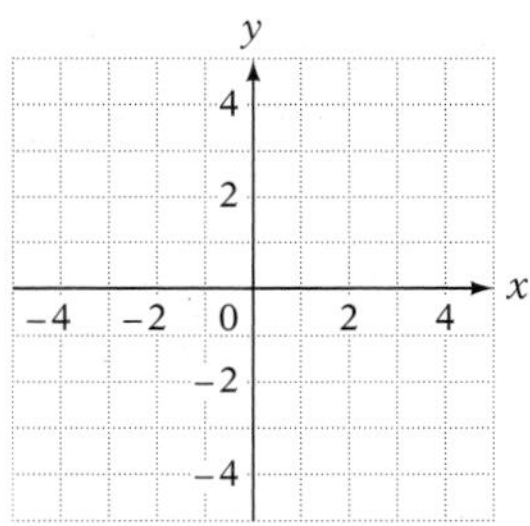

Solution on p. S12

EXAMPLE 2

Graph: $y = \frac{2}{3}x$

Solution

x	y
0	0
3	2
−3	−2

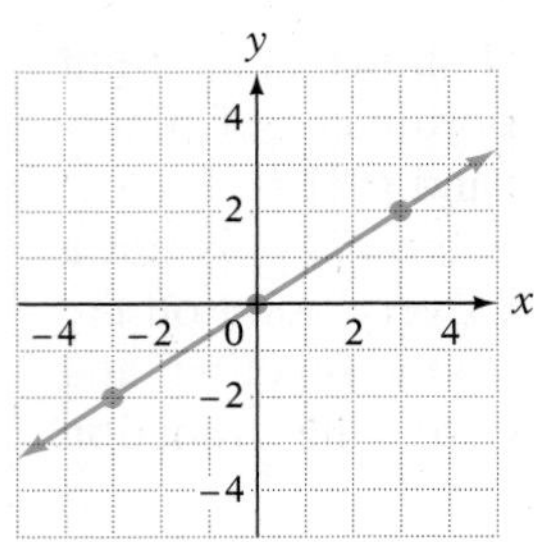

YOU TRY IT 2

Graph: $y = -\frac{3}{4}x$

Your solution

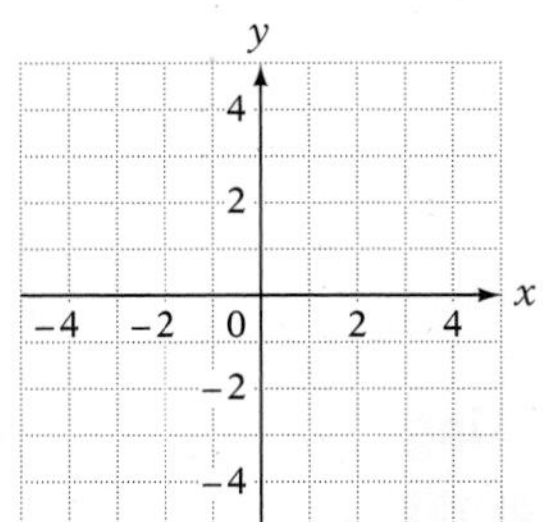

Solution on p. S12

OBJECTIVE B To graph an equation of the form $Ax + By = C$

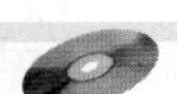

An equation of the form $Ax + By = C$, where A and B are coefficients and C is a constant, is also a *linear equation in two variables.* This equation can be written in the form $y = mx + b$.

HOW TO 2 Write $4x - 3y = 6$ in the form $y = mx + b$.

$$4x - 3y = 6$$
$$-3y = -4x + 6$$

• Subtract $4x$ from each side of the equation.

$$y = \frac{4}{3}x - 2$$

• Divide each side of the equation by −3. This is the form $y = mx + b$. $m = \frac{4}{3}$, $b = -2$

To graph an equation of the form $Ax + By = C$, first solve the equation for y. Then follow the same procedure used for graphing an equation of the form $y = mx + b$.

HOW TO 3 Graph: $3x + 2y = 6$

$$3x + 2y = 6$$
$$2y = -3x + 6$$
$$y = -\frac{3}{2}x + 3$$

• Solve the equation for y.

x	y
0	3
2	0
4	−3

• Find at least three solutions.

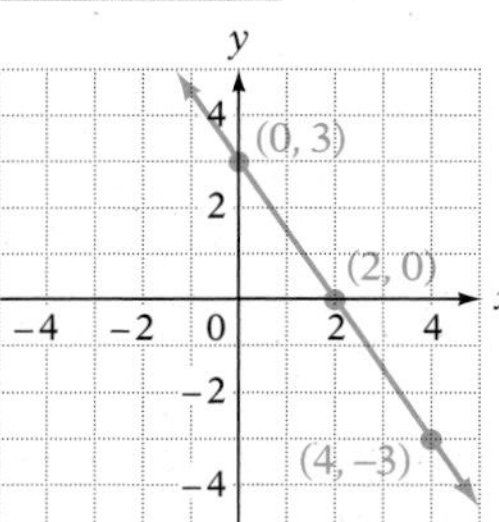

• Graph the ordered pairs in a rectangular coordinate system. Draw a straight line through the points.

If one of the coefficients A or B is zero, the graph of $Ax + By = C$ is a horizontal or vertical line. Consider the equation $y = -2$, where A, the coefficient of x, is 0. We can write this equation in *two* variables as $0 \cdot x + y = -2$. No matter what value of x is selected, $0 \cdot x = 0$. Therefore, y equals -2 for all values of x. The table below shows some of the ordered-pair solutions of $y = -2$. The graph is shown to the right of the table.

x	$0 \cdot x + y = -2$	y	(x, y)
-1	$0 \cdot (-1) + y = -2$	-2	$(-1, -2)$
0	$0 \cdot 0 + y = -2$	-2	$(0, -2)$
3	$0 \cdot 3 + y = -2$	-2	$(3, -2)$

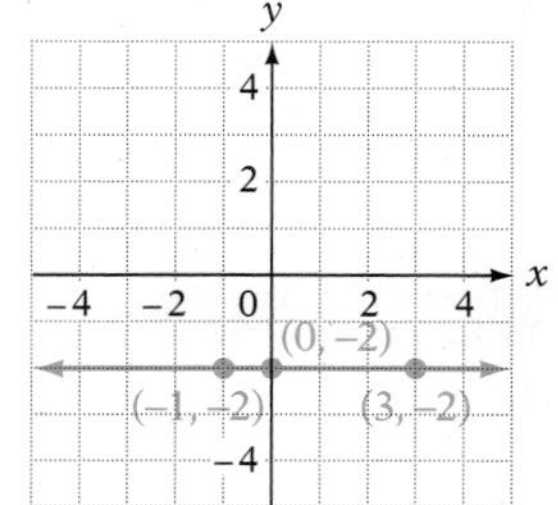

> **Graph of $y = b$**
>
> The **graph of $y = b$** is a horizontal line passing through the point $(0, b)$.

HOW TO 4 Graph: $y + 4 = 0$

Solve for y.

$$y + 4 = 0$$
$$y = -4$$

The graph of $y = -4$ is a horizontal line passing through $(0, -4)$.

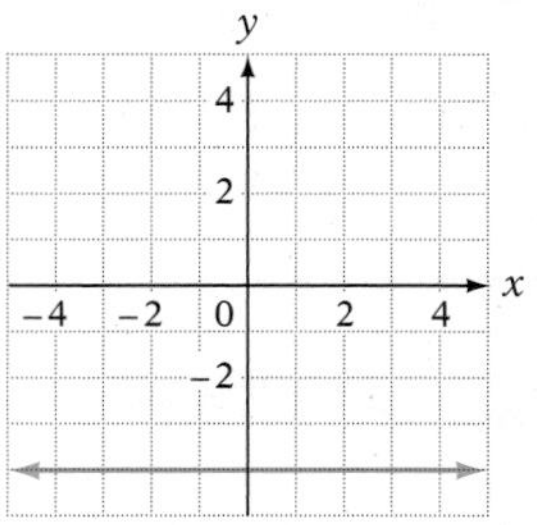

The equation $y = -2$ represents a function. Some of the ordered pairs of this function are $(-1, -2)$, $(0, -2)$, and $(3, -2)$. In function notation, we write $f(x) = -2$. This function is an example of a *constant function.* For every value of x, the value of the function is the constant -2. For instance, we have $f(17) = -2$, $f(\sqrt{2}) = -2$, and $f(\pi) = -2$.

> **Constant Function**
>
> A function given by $f(x) = b$, where b is a constant, is a **constant function.** The graph of a constant function is a horizontal line passing through $(0, b)$.

Now consider the equation $x = 2$, where B, the coefficient of y, is zero. We write this equation in two variables by writing $x + 0 \cdot y = 2$, where the coefficient of y is 0. No matter what value of y is selected, $0 \cdot y = 0$. Therefore, x equals 2 for all values of y. The following table shows some of the ordered-pair solutions of $x = 2$. The graph is shown to the right of the table.

y	$x + 0 \cdot y = 2$	x	(x, y)
6	$x + 0 \cdot 6 = 2$	2	$(2, 6)$
1	$x + 0 \cdot 1 = 2$	2	$(2, 1)$
-4	$x + 0 \cdot (-4) = 2$	2	$(2, -4)$

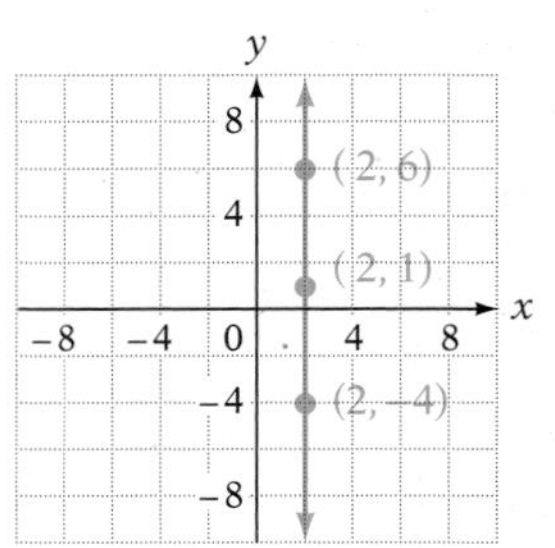

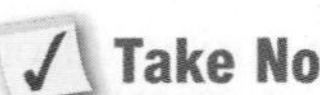

The equation $y = b$ represents a function. The equation $x = a$ does *not* represent a function. Remember, not all equations represent functions.

Graph of $x = a$

The **graph of $x = a$** is a vertical line passing through the point $(a, 0)$.

Recall that a function is a set of ordered pairs in which no two ordered pairs have the same first coordinate and different second coordinates. Because (2, 6), (2, 1), and (2, −4) are ordered pairs belonging to the equation $x = 2$, this equation does not represent a function, and the graph is not the graph of a function.

EXAMPLE • 3

Graph: $2x + 3y = 9$

Solution Solve the equation for y.

$$2x + 3y = 9$$
$$3y = -2x + 9$$
$$y = -\frac{2}{3}x + 3$$

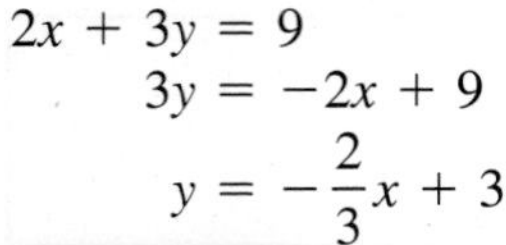

x	y
−3	5
0	3
3	1

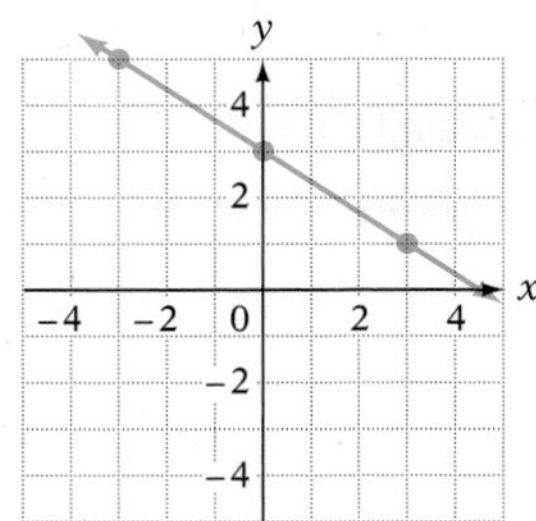

YOU TRY IT • 3

Graph: $-3x + 2y = 4$

Your solution

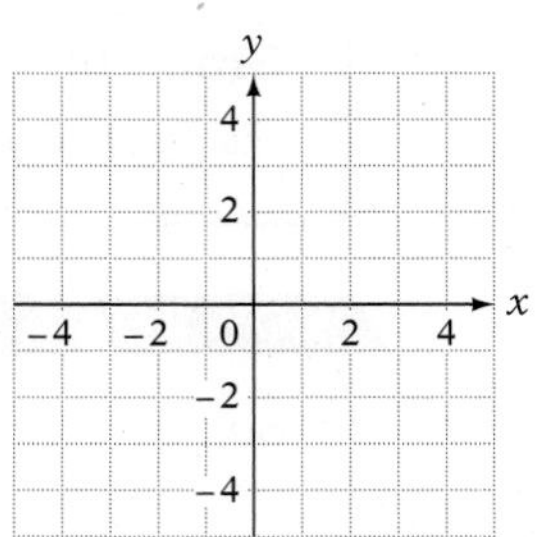

EXAMPLE • 4

Graph: $x = -4$

Solution

The graph of an equation of the form $x = a$ is a vertical line passing through the point whose coordinates are $(a, 0)$.

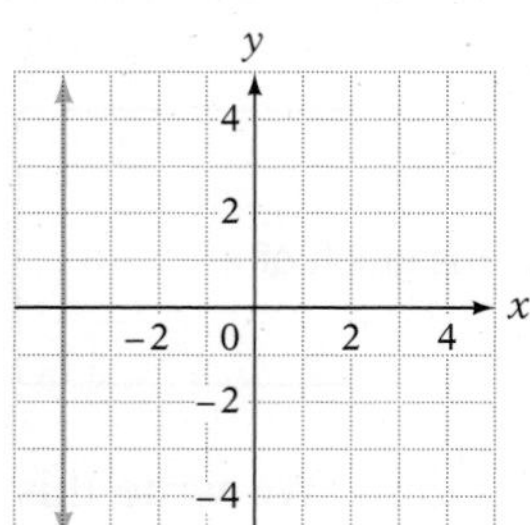

• The graph of $x = -4$ goes through the point $(-4, 0)$.

YOU TRY IT • 4

Graph: $y - 3 = 0$

Your solution

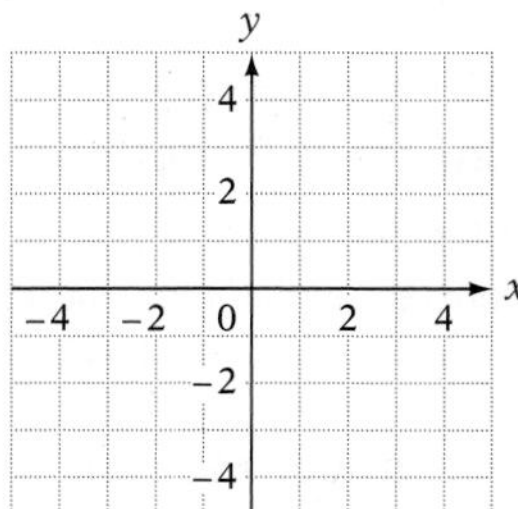

Solutions on p. S12

OBJECTIVE C To find the x- and y-intercepts of a straight line

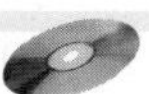

The graph of the equation $x - 2y = 4$ is shown at the right. The graph crosses the x-axis at the point (4, 0). This point is called the ***x*-intercept.** The graph also crosses the y-axis at the point (0, −2). This point is called the ***y*-intercept.**

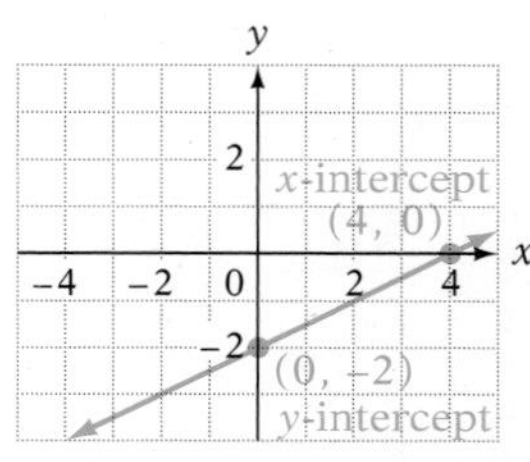

✓ Take Note

The x-intercept occurs when $y = 0$.

The y-intercept occurs when $x = 0$.

HOW TO • 5 Find the x- and y-intercepts of the graph of the equation $3x + 4y = -12$.

To find the x-intercept, let $y = 0$. (Any point on the x-axis has y-coordinate 0.)

$$\begin{aligned} 3x + 4y &= -12 \\ 3x + 4(0) &= -12 \\ 3x &= -12 \\ x &= -4 \end{aligned}$$

The x-intercept is $(-4, 0)$.

To find the y-intercept, let $x = 0$. (Any point on the y-axis has x-coordinate 0.)

$$\begin{aligned} 3x + 4y &= -12 \\ 3(0) + 4y &= -12 \\ 4y &= -12 \\ y &= -3 \end{aligned}$$

The y-intercept is $(0, -3)$.

A linear equation can be graphed by finding the x- and y-intercepts and then drawing a line through the two points.

HOW TO • 6 Graph $3x - 2y = 6$ by using the x- and y-intercepts.

$$\begin{aligned} 3x - 2y &= 6 \\ 3x - 2(0) &= 6 \\ 3x &= 6 \\ x &= 2 \end{aligned}$$

• To find the x-intercept, let $y = 0$. Then solve for x.

$$\begin{aligned} 3x - 2y &= 6 \\ 3(0) - 2y &= 6 \\ -2y &= 6 \\ y &= -3 \end{aligned}$$

• To find the y-intercept, let $x = 0$. Then solve for y.

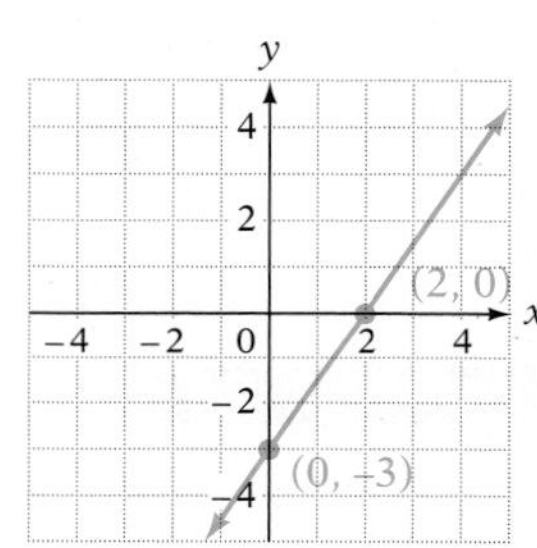

The x-intercept is $(2, 0)$. The y-intercept is $(0, -3)$.

EXAMPLE • 5

Graph $4x - y = 4$ by using the x- and y-intercepts.

Solution

x-intercept:

$$\begin{aligned} 4x - y &= 4 \\ 4x - 0 &= 4 \\ 4x &= 4 \\ x &= 1 \end{aligned}$$

• Let $y = 0$.

$(1, 0)$

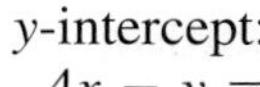

y-intercept:

$$\begin{aligned} 4x - y &= 4 \\ 4(0) - y &= 4 \\ -y &= 4 \\ y &= -4 \end{aligned}$$

• Let $x = 0$.

$(0, -4)$

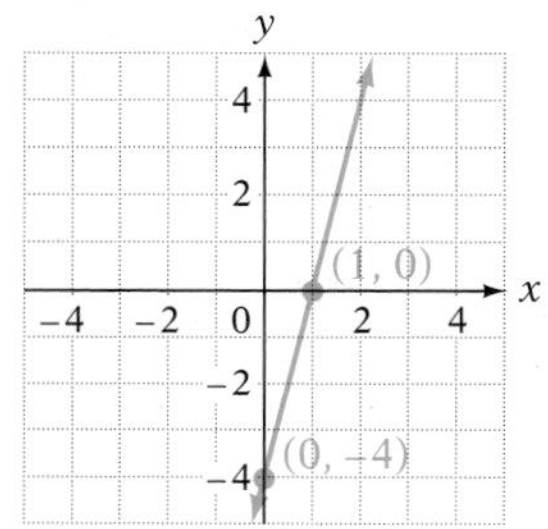

YOU TRY IT • 5

Graph $3x - y = 2$ by using the x- and y-intercepts.

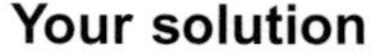

Your solution

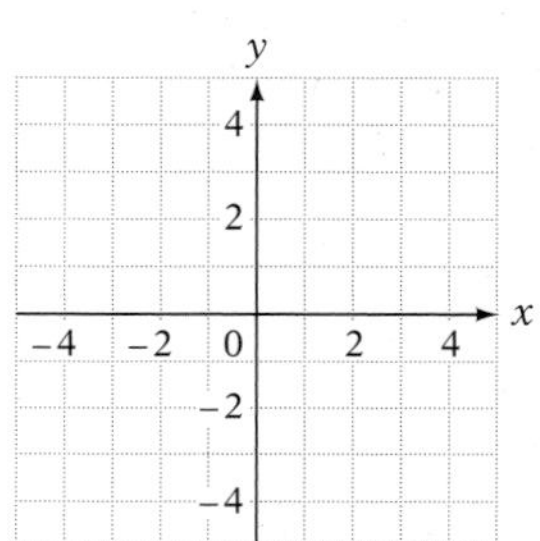

Solution on p. S12

OBJECTIVE D To solve application problems 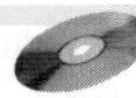

There are a variety of applications of linear functions.

Take Note
In many applications, the domain of the variable is given such that the equation makes sense. For this application, it would not be sensible to have values of t that are less than 0. This would indicate negative time! The number 10 is somewhat arbitrary, but after 10 min most people's heart rates would level off, and a linear function would no longer apply.

HOW TO • 7 The heart rate R after t minutes for a person taking a brisk walk can be approximated by the equation $R = 2t + 72$.

a. Graph this equation for $0 \le t \le 10$.
b. The point whose coordinates are (5, 82) is on the graph. Write a sentence that describes the meaning of this ordered pair.

a.

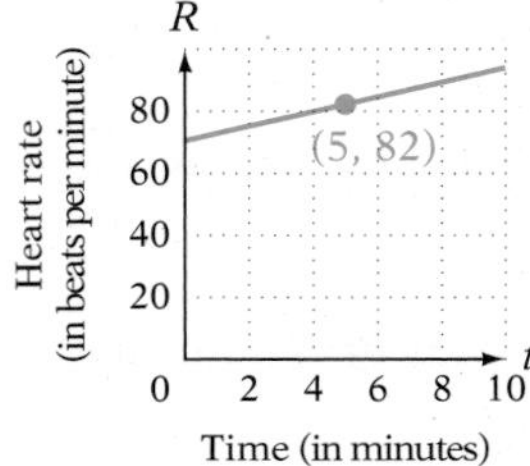

b. The ordered pair (5, 82) means that after 5 min the person's heart rate is 82 beats per minute.

EXAMPLE • 6

An electronics technician charges $45 plus $1 per minute to repair defective wiring in a home or apartment. The equation that describes the total cost C to have defective wiring repaired is given by $C = t + 45$, where t is the number of minutes the technician works. Graph this equation for $0 \le t \le 60$. The point whose coordinates are (15, 60) is on this graph. Write a sentence that describes the meaning of this ordered pair.

Solution

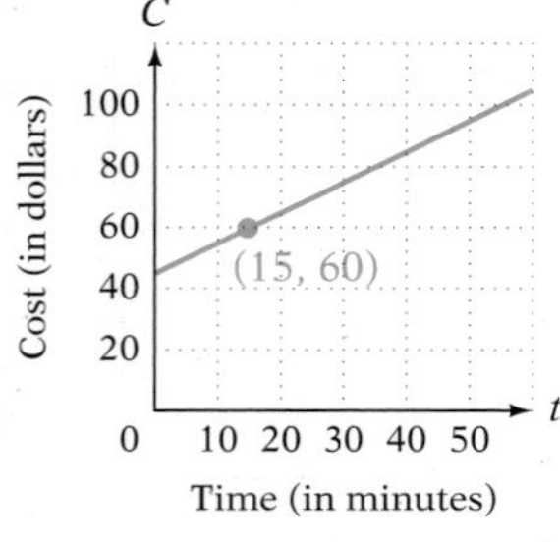

- Graph $C = t + 45$.
  When $t = 0$, $C = 45$.
  When $t = 50$, $C = 95$.

The ordered pair (15, 60) means that it costs $60 for the technician to work 15 min.

YOU TRY IT • 6

The height h (in inches) of a person and the length L (in inches) of that person's stride while walking are related. The equation $h = \frac{3}{4}L + 50$ approximates this relationship. Graph this equation for $15 \le L \le 40$. The point whose coordinates are (32, 74) is on this graph. Write a sentence that describes the meaning of this ordered pair.

Your solution

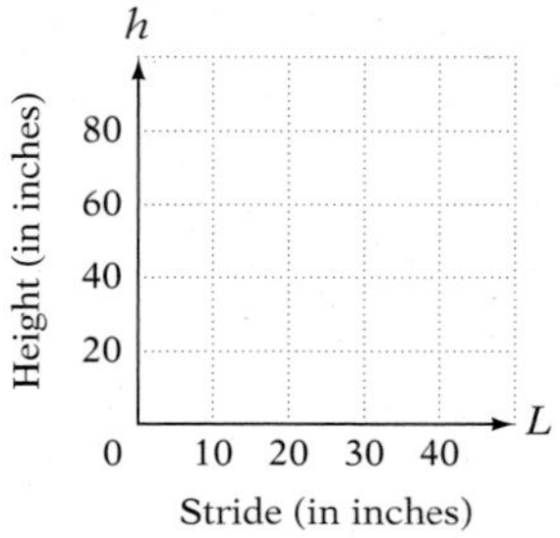

Solution on p. S12

4.3 EXERCISES

OBJECTIVE A To graph a linear function

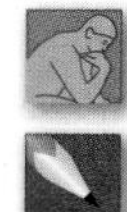
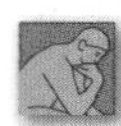

1. When finding ordered pairs to graph a line, why do we recommend that you find three ordered pairs?

2. To graph points on the graph of $y = \frac{5}{3}x + 4$, it is helpful to choose values of x that are divisible by what number?

For Exercises 3 to 14, graph.

3. $y = x - 3$

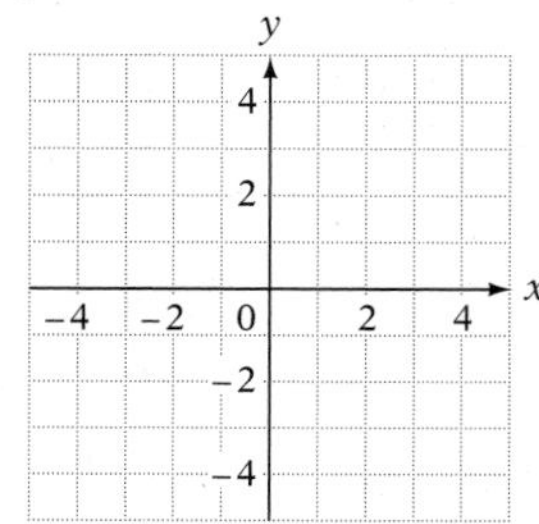

4. $y = -x + 2$

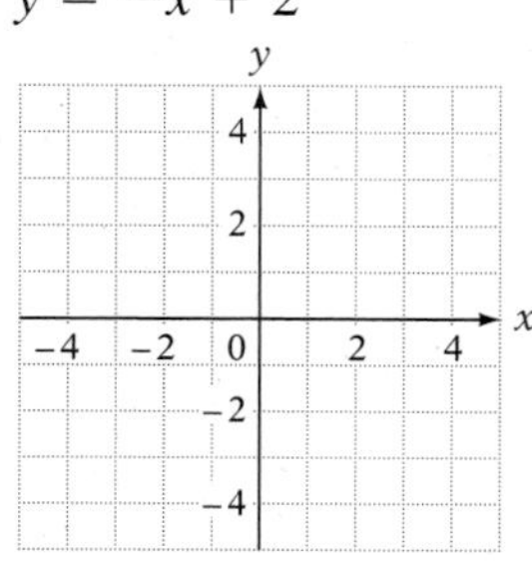

5. $y = -3x + 2$

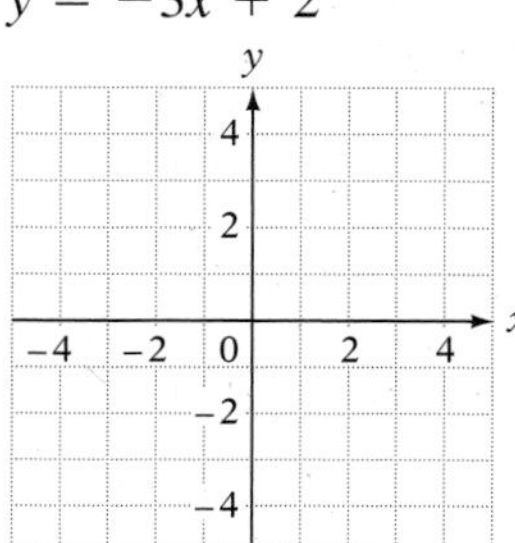

6. $y = -2x + 3$

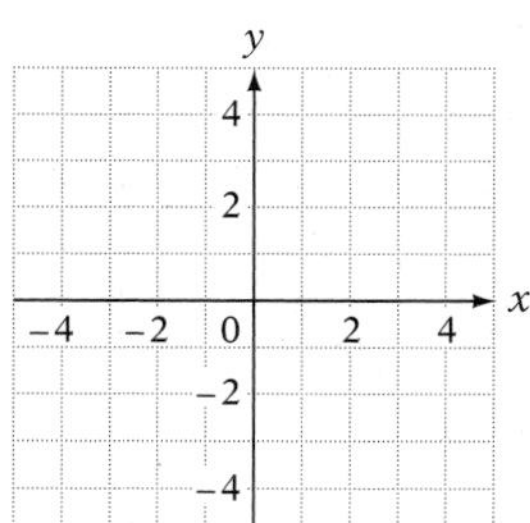

7. $f(x) = 3x - 4$

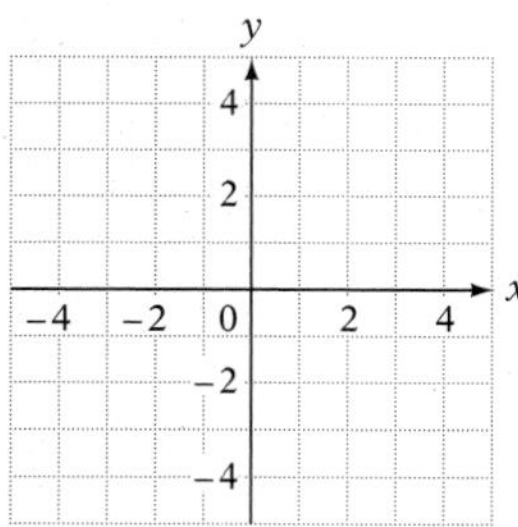

8. $f(x) = \frac{3}{2}x$

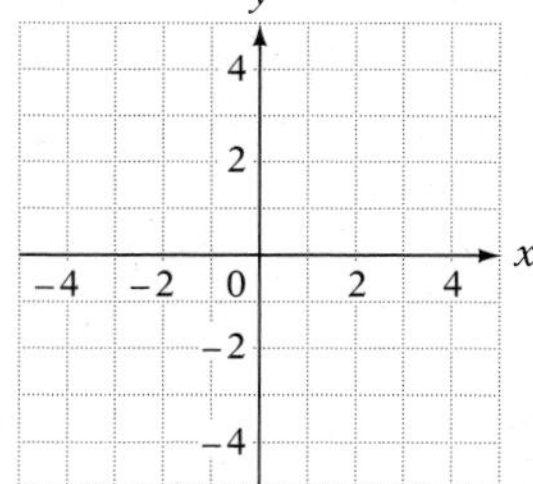

9. $f(x) = -\frac{2}{3}x$

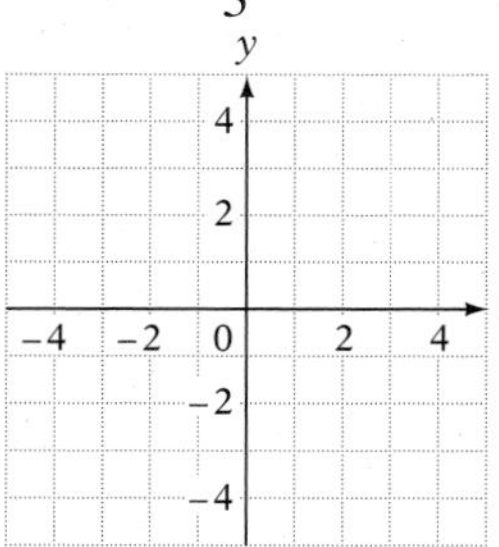

10. $f(x) = \frac{3}{4}x + 2$

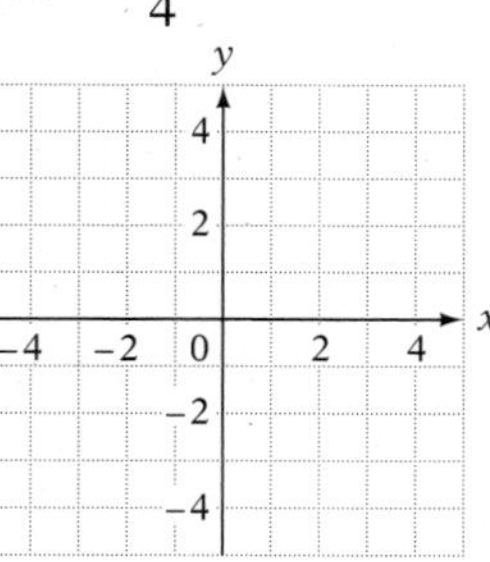

11. $y = \frac{2}{3}x - 4$

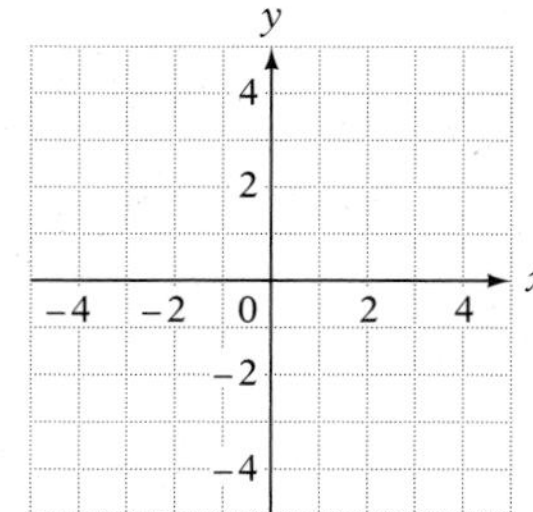

12. $y = -\frac{3}{2}x - 3$

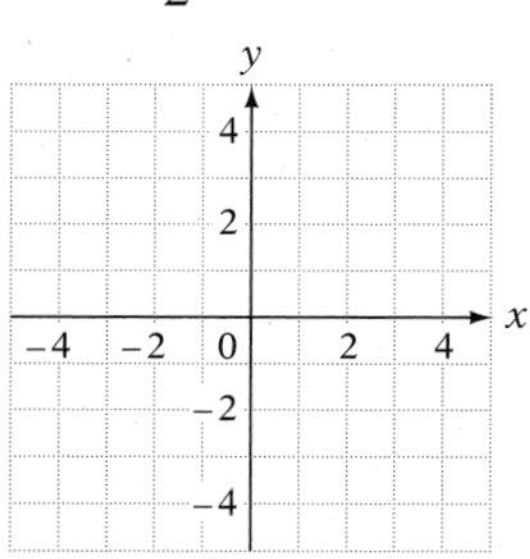

13. $f(x) = -\frac{1}{3}x + 2$

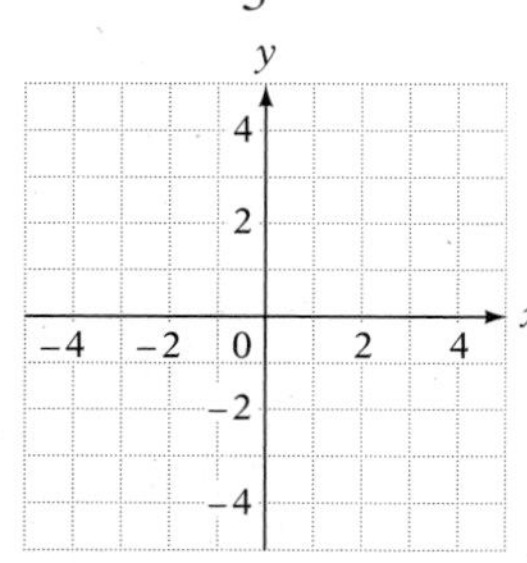

14. $f(x) = \frac{3}{5}x - 1$

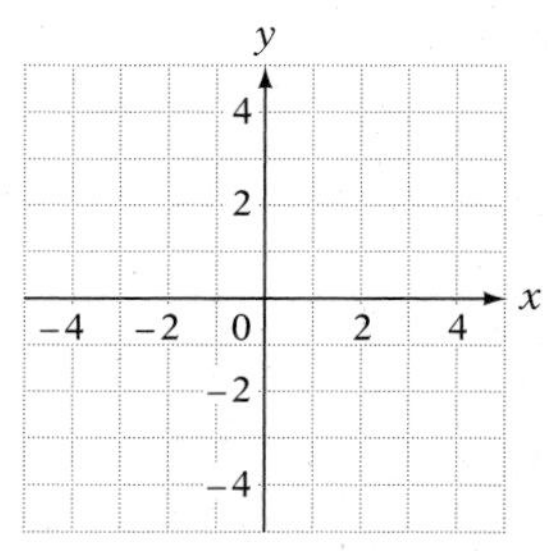

OBJECTIVE B To graph an equation of the form $Ax + By = C$

For Exercises 15 to 29, graph.

15. $2x + y = -3$

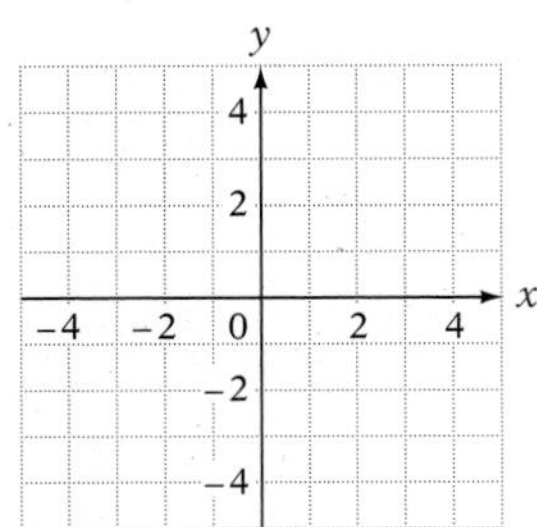

16. $2x - y = 3$

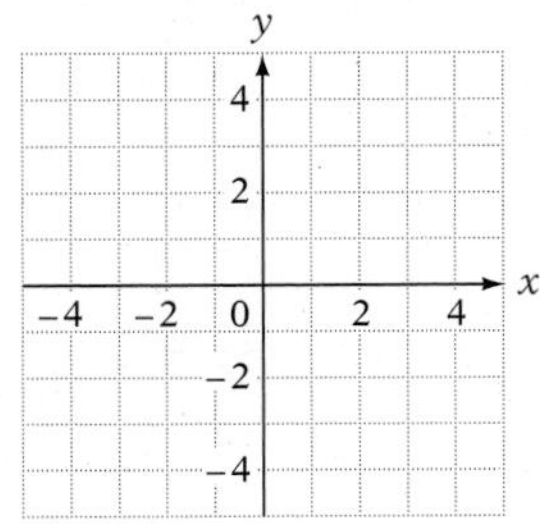

17. $x - 4y = 8$

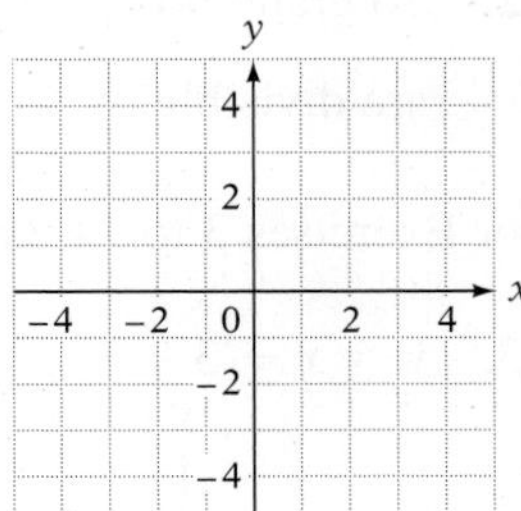

18. $2x + 5y = 10$

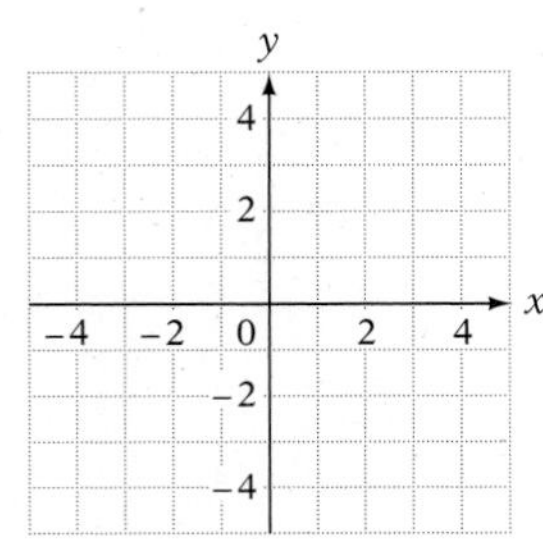

19. $4x + 3y = 12$

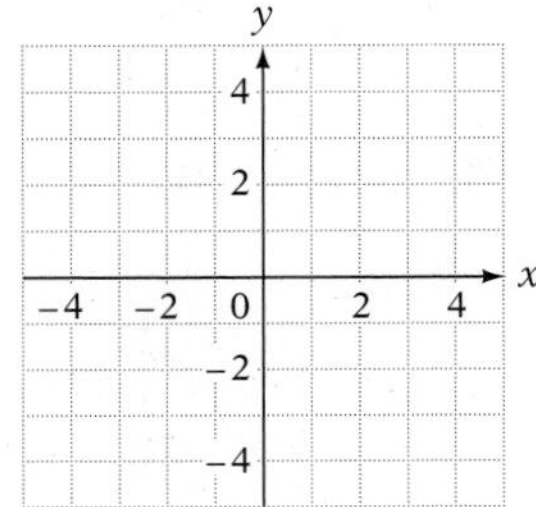

20. $2x - 5y = 10$

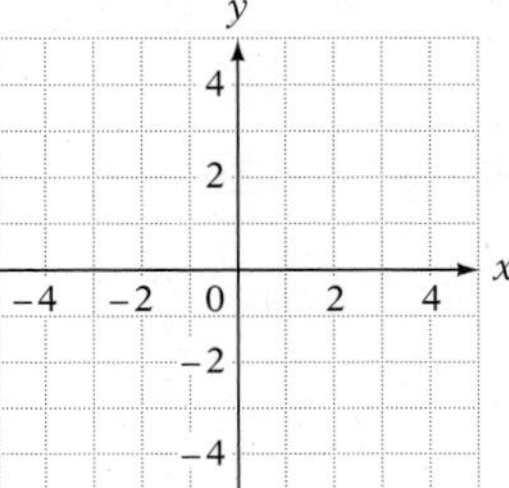

21. $x - 3y = 0$

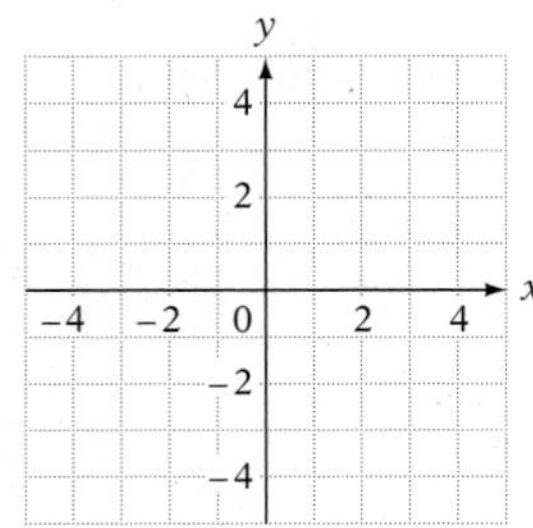

22. $x + 3y = 9$

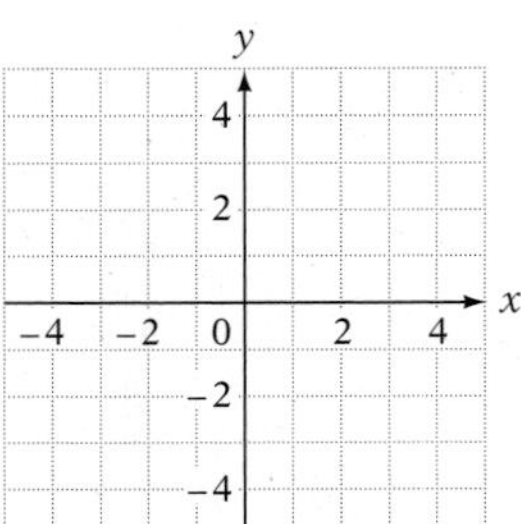

23. $y = -2$

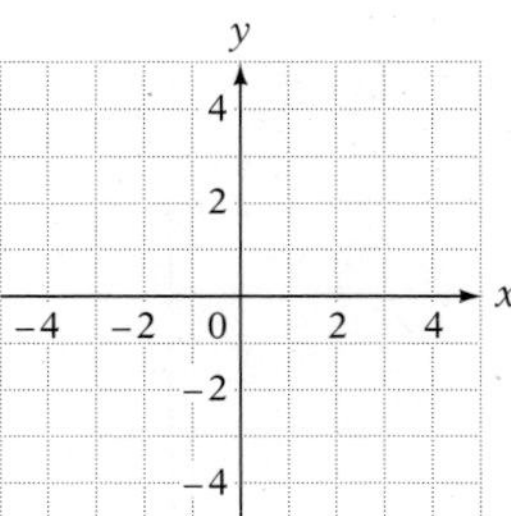

24. $f(x) = 3$

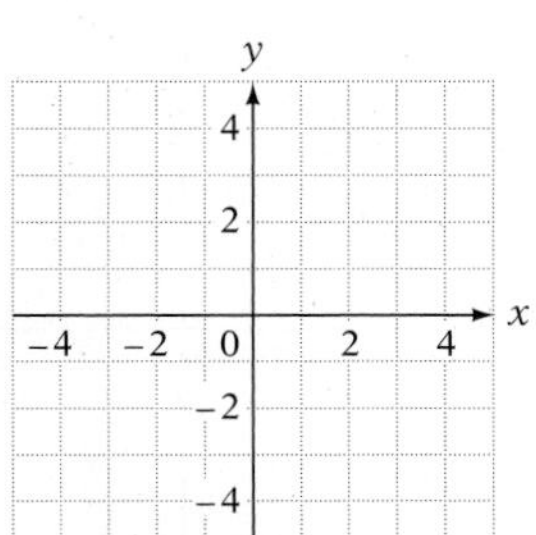

25. $x = -3$

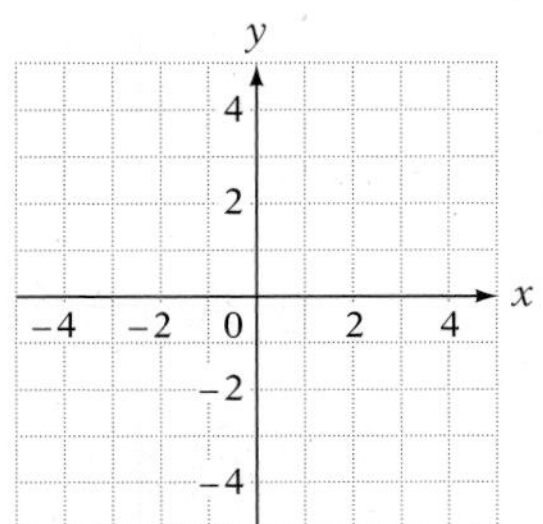

26. $x = 1$

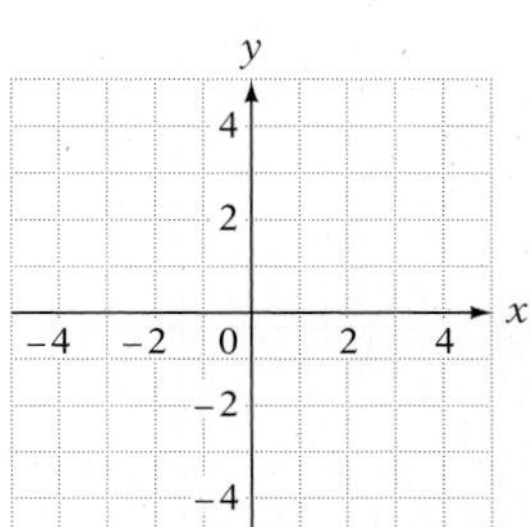

27. $3x - y = -2$

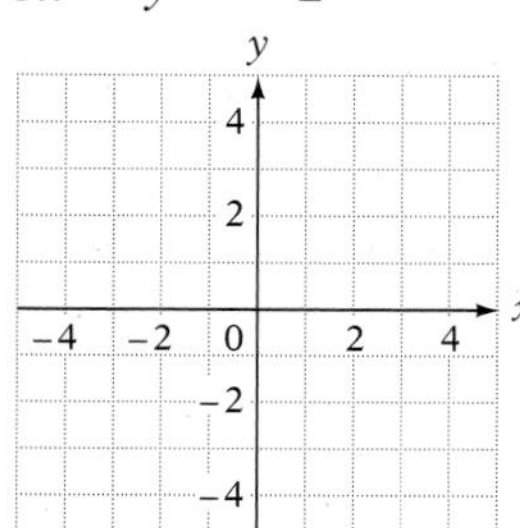

28. $2x - 3y = 12$

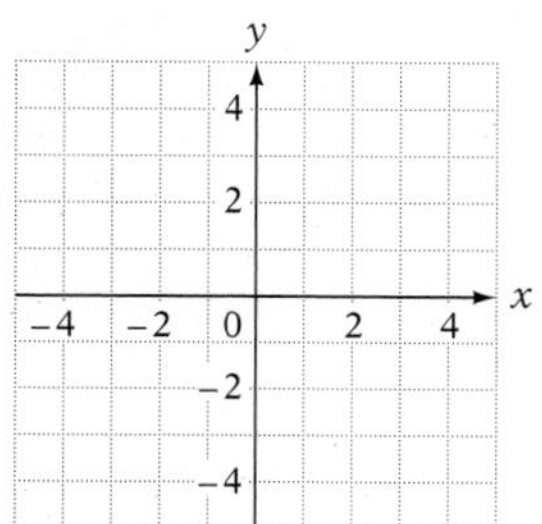

29. $3x - 2y = 8$

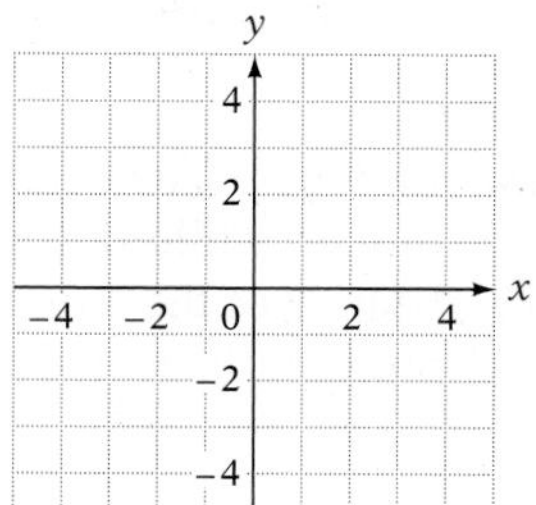

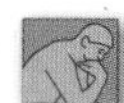

30. Suppose the graph of $Ax + By = C$ is a horizontal line. Which of the numbers A, B, or C must be zero?

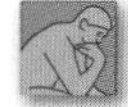

31. Is it always possible to solve $Ax + By = C$ for y and write the equation in the form $y = mx + b$? If not, explain.

OBJECTIVE C To find the x- and y-intercepts of a straight line

For Exercises 32 to 40, find the x- and y-intercepts and graph.

32. $3x + y = 3$

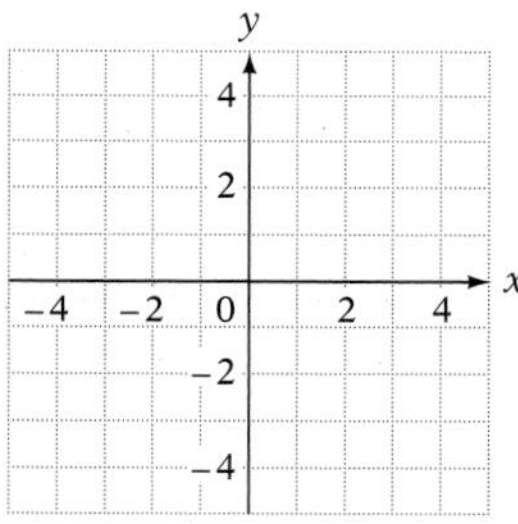

33. $x - 2y = -4$

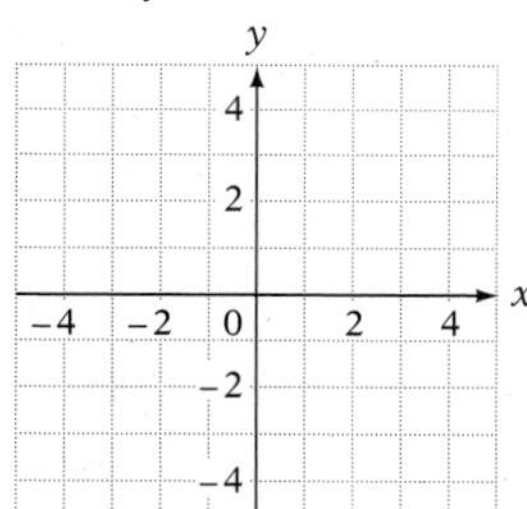

34. $2x - y = 4$

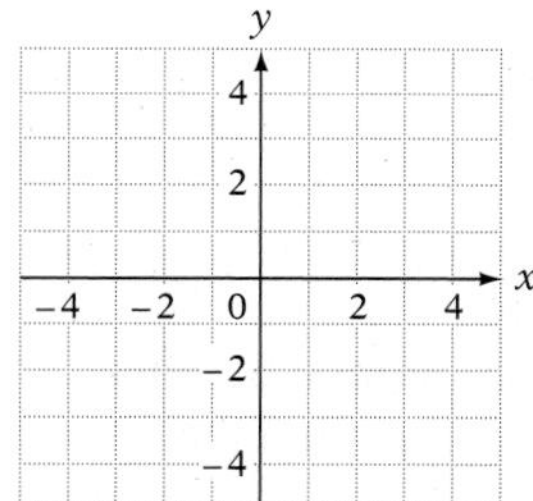

35. $2x - 3y = 9$

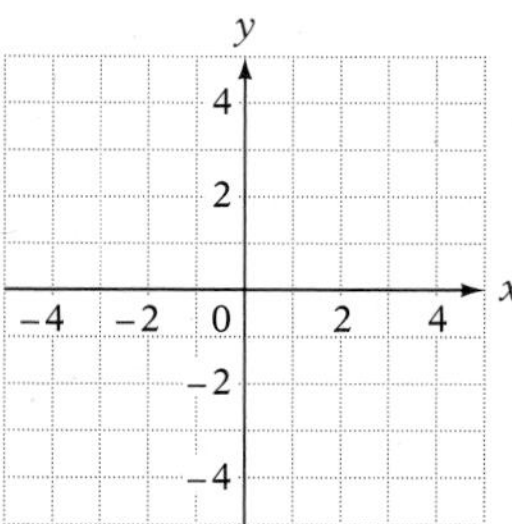

36. $4x - 3y = 8$

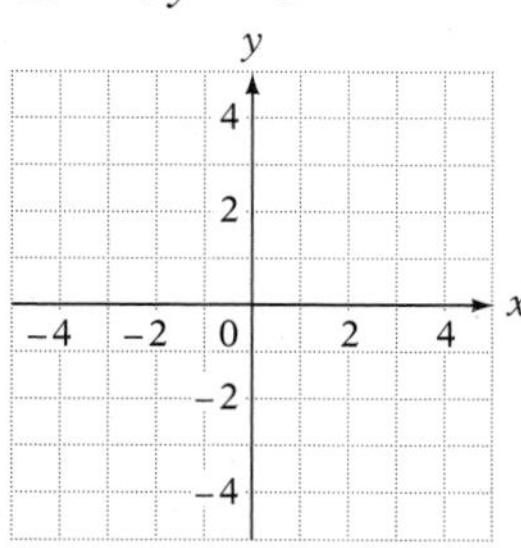

37. $2x + y = 3$

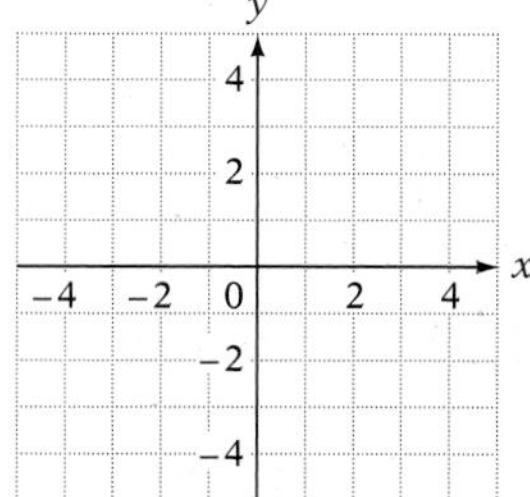

38. $2x - 3y = 4$

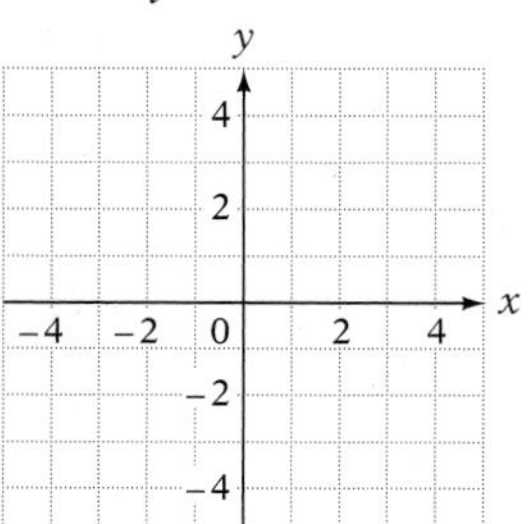

39. $3x + 2y = 4$

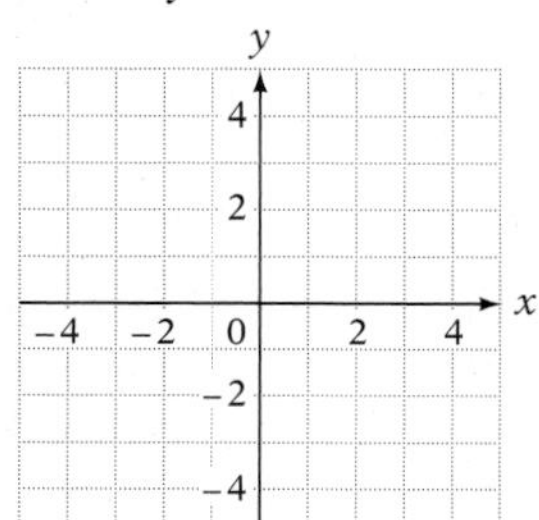

40. $4x - 3y = 6$

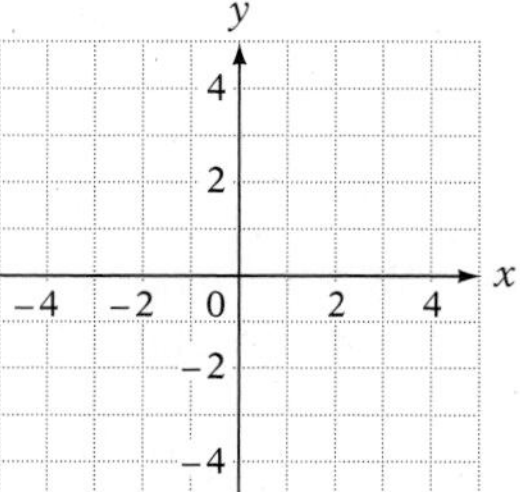

41. Does the graph of every straight line have a y-intercept? Explain.

42. Why is it not possible to graph an equation of the form $Ax + By = 0$ by using only the x- and y-intercepts?

OBJECTIVE D To solve application problems

43. Biology The heart of a ruby-throated hummingbird beats about 1200 times per minute while in flight. The equation $B = 1200t$ gives the total number of heartbeats B of the hummingbird in t minutes of flight. How many times will this hummingbird's heart beat in 7 min of flight?

44. Telecommunications The monthly cost of a wireless telephone plan is $39.99 for up to 450 min of calling time plus $.45 per minute for each minute over 450 min. The equation that describes the cost of this plan is $C = 0.45x + 39.99$, where x is the number of minutes over 450. What is the cost of this plan if a person uses **a.** 398 min and **b.** 475 min of calling time?

45. Compensation Marlys receives $11 per hour as a mathematics department tutor. The equation that describes Marlys's wages is $W = 11t$, where t is the number of hours she spends tutoring. Graph this equation for $0 \leq t \leq 20$. The ordered pair (15, 165) is on the graph. Write a sentence that describes the meaning of this ordered pair.

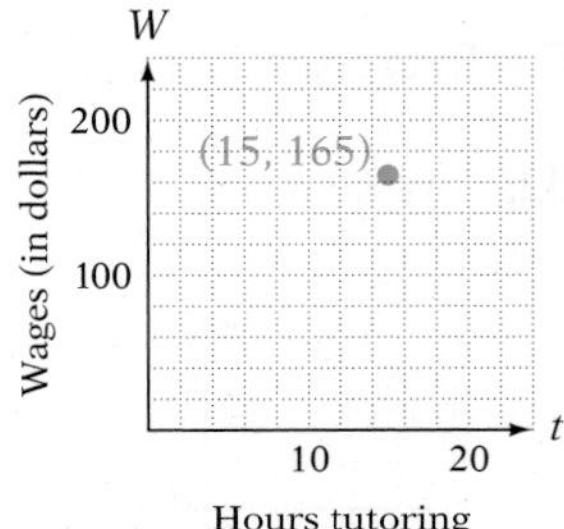

46. Animal Science A bee beats its wings approximately 100 times per second. The equation that describes the total number of times a bee beats its wings is given by $B = 100t$. Graph this equation for $0 \leq t \leq 60$. The point (35, 3500) is on this graph. Write a sentence that describes the meaning of this ordered pair.

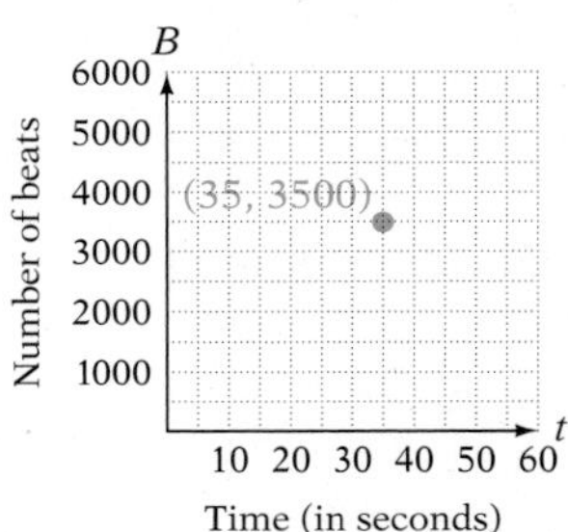

47. Manufacturing The cost of manufacturing skis is $5000 for startup costs and $80 per pair of skis manufactured. The equation that describes the cost of manufacturing n pairs of skis is $C = 80n + 5000$. Graph this equation for $0 \leq n \leq 2000$. The point (50, 9000) is on the graph. Write a sentence that describes the meaning of this ordered pair.

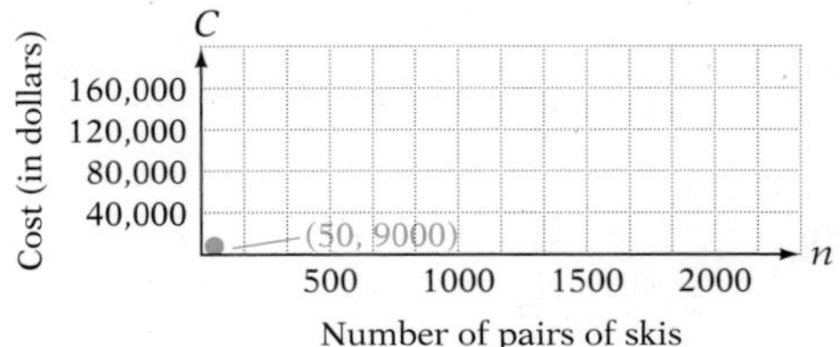

48. Atomic Physics The Large Hadron Collider, or LHC, is a machine that is capable of accelerating a proton to a velocity that is 99.99% the speed of light. At this speed, a proton will travel approximately 0.98 ft in a billionth of a second (one nanosecond). (*Source:* news.yahoo.com) The equation $d = 0.98t$ gives the distance d, in feet, traveled by a proton in t nanoseconds. Graph this equation for $0 \leq t \leq 10$. The ordered pair (4, 3.92) is on the graph. Write a sentence that explains the meaning of this ordered pair.

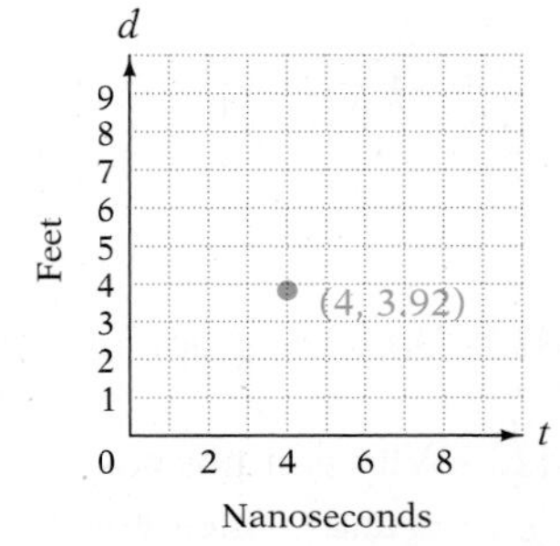

Applying the Concepts

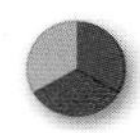

Oceanography For Exercises 49–51, read the article below about the small submarine *Alvin*, used by scientists to explore the ocean for more than 40 years.

In the News

Alvin, First Viewer of the Titanic, to Retire

Woods Hole Oceanographic Institute announced plans for a replacement for *Alvin*, the original Human Occupied Vehicle (HOV), built in 1964 for deep-sea exploration. *Alvin* can descend at a rate of 30 m/min to a maximum depth of 4500 m. The replacement HOV will be able to descend at a rate of 48 m/min to a maximum depth of 6500 m.

Source: Woods Hole Oceanographic Institute

49. **a.** The equation that describes *Alvin's* depth D, in meters, is $D = -30t$, where t is the number of minutes *Alvin* has been descending. Graph this equation for $0 \le t \le 150$. The ordered pair $(65, -1950)$ is on the graph. Write a sentence that describes the meaning of this ordered pair.

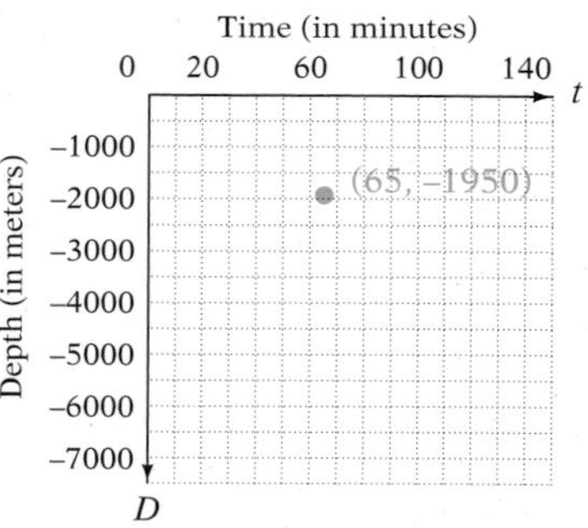

b. The equation that describes the replacement HOV's depth D, in meters, is $D = -48t$, where t is the number of minutes the HOV has been descending. On the same set of axes you used in part (a), graph this equation for $0 \le t \le 150$. Is the ordered pair with x-coordinate 65 above or below the ordered pair $(65, -1950)$ from part (a)?

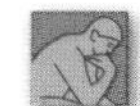

50. Does your graph from Exercise 49(a) include any points that do not represent depths that *Alvin* can descend to? If so, describe them.

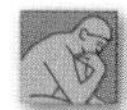

51. Does your graph from Exercise 49(b) include any points that do not represent depths that the replacement HOV can descend to? If so, describe them.

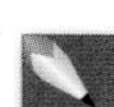

52. Explain what the graph of an equation represents.

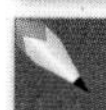

53. Explain how to graph the equation of a straight line by plotting points.

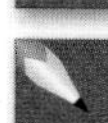

54. Explain how to graph the equation of a straight line by using its x- and y-intercepts.

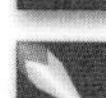

55. Explain why you cannot graph the equation $4x + 3y = 0$ by using just its intercepts.

An equation of the form $\frac{x}{a} + \frac{y}{b} = 1$, where $a \neq 0$ and $b \neq 0$, is called the *intercept form of a straight line* because $(a, 0)$ and $(0, b)$ are the x- and y-intercepts of the graph of the equation. Graph the equations in Exercises 56 to 58.

56. $\frac{x}{3} + \frac{y}{5} = 1$

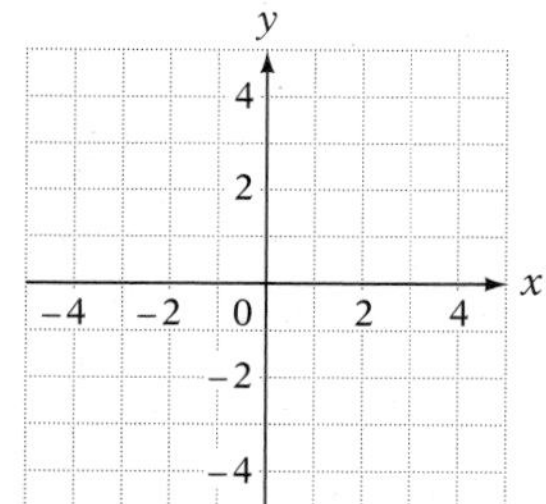

57. $\frac{x}{2} + \frac{y}{3} = 1$

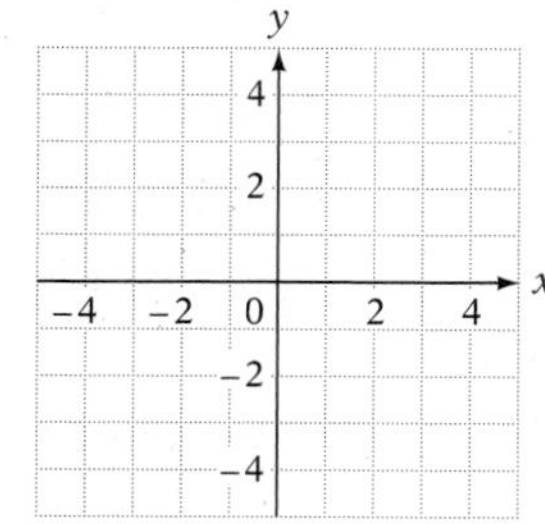

58. $\frac{x}{3} - \frac{y}{2} = 1$

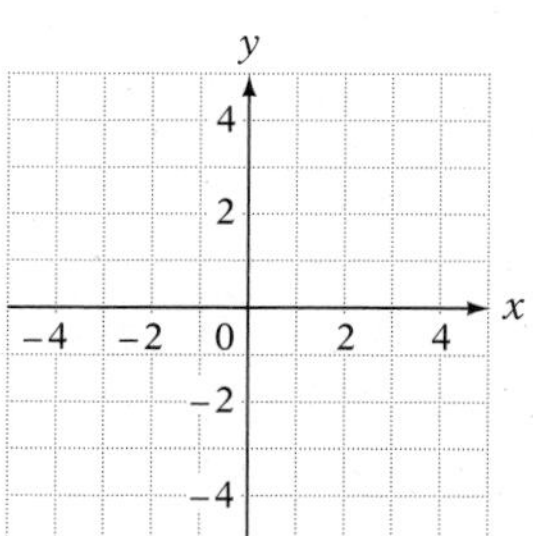

SECTION

4.4 Slope of a Straight Line

OBJECTIVE A To find the slope of a line given two points 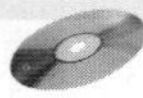

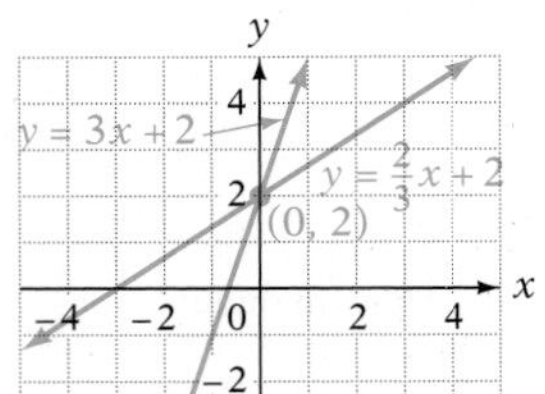

The graphs of $y = 3x + 2$ and $y = \frac{2}{3}x + 2$ are shown at the left. Each graph crosses the y-axis at the point $P(0, 2)$, but the graphs have different slants. The **slope** of a line is a measure of the slant of the line. The symbol for slope is m.

The slope of a line containing two points is the ratio of the change in the y values of the two points to the change in the x values. The line containing the points whose coordinates are $(-1, -3)$ and $(5, 2)$ is shown below.

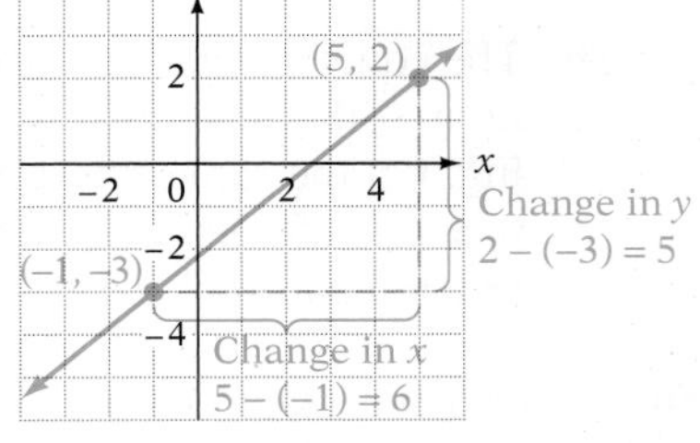

The change in the y values is the difference between the y-coordinates of the two points.

$$\text{Change in } y = 2 - (-3) = 5$$

The change in the x values is the difference between the x-coordinates of the two points.

$$\text{Change in } x = 5 - (-1) = 6$$

The slope of the line between the two points is the ratio of the change in y to the change in x.

$$\text{Slope} = m = \frac{\text{change in } y}{\text{change in } x} = \frac{5}{6} \qquad m = \frac{2 - (-3)}{5 - (-1)} = \frac{5}{6}$$

In general, if $P_1(x_1, y_1)$ and $P_2(x_2, y_2)$ are two points on a line, then

$$\text{Change in } y = y_2 - y_1 \qquad \text{Change in } x = x_2 - x_1$$

Using these ideas, we can state a formula for slope.

Slope Formula

The slope of the line containing the two points $P_1(x_1, y_1)$ and $P_2(x_2, y_2)$ is given by

$$m = \frac{y_2 - y_1}{x_2 - x_1}, \; x_1 \neq x_2$$

Frequently, the Greek letter Δ is used to designate the change in a variable. Using this notation, we can write equations for the change in y and the change in x as follows:

$$\text{Change in } y = \Delta y = y_2 - y_1 \qquad \text{Change in } x = \Delta x = x_2 - x_1$$

Using this notation, the slope formula is written $m = \frac{\Delta y}{\Delta x}$.

Take Note

When asked to find the slope of a line between two points on a graph, it does not matter which is selected as P_1 and which is selected as P_2. For the graph at the far right, we could have labeled the points $P_1(4, 5)$ and $P_2(-2, 0)$, where we have reversed the naming of P_1 and P_2. Then

$$m = \frac{y_2 - y_1}{x_2 - x_1}$$

$$m = \frac{0 - 5}{-2 - 4} = \frac{-5}{-6} = \frac{5}{6}$$

The value of the slope is the same.

HOW TO • 1 Find the slope of the line passing through the points whose coordinates are $(-2, 0)$ and $(4, 5)$.

Choose P_1 (point 1) and P_2 (point 2). We will choose $P_1(-2, 0)$ and $P_2(4, 5)$. From $P_1(-2, 0)$, we have $x_1 = -2$, $y_1 = 0$. From $P_2(4, 5)$, we have $x_2 = 4$, $y_2 = 5$. Now use the slope formula.

$$m = \frac{y_2 - y_1}{x_2 - x_1} = \frac{5 - 0}{4 - (-2)} = \frac{5}{6}$$

The slope of the line is $\frac{5}{6}$.

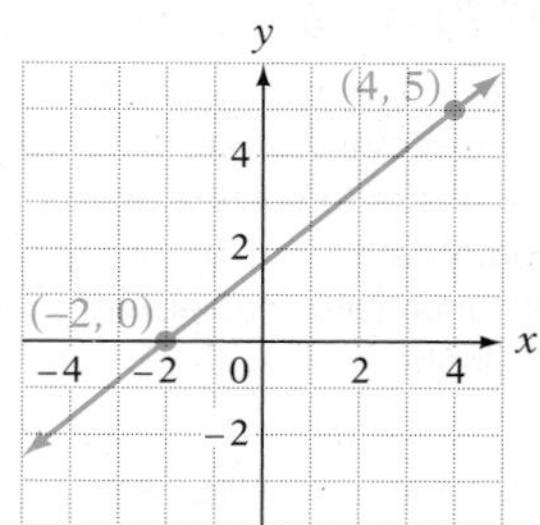

Positive slope

A line that slants upward to the right has a **positive slope.**

HOW TO • 2 Find the slope of the line passing through the points whose coordinates are $(-3, 4)$ and $(4, 2)$.

We will choose $P_1(-3, 4)$ and $P_2(4, 2)$. From $P_1(-3, 4)$, we have $x_1 = -3$, $y_1 = 4$. From $P_2(4, 2)$, we have $x_2 = 4$, $y_2 = 2$. Now use the slope formula.

$$m = \frac{y_2 - y_1}{x_2 - x_1} = \frac{2 - 4}{4 - (-3)} = \frac{-2}{7} = -\frac{2}{7}$$

The slope of the line is $-\frac{2}{7}$.

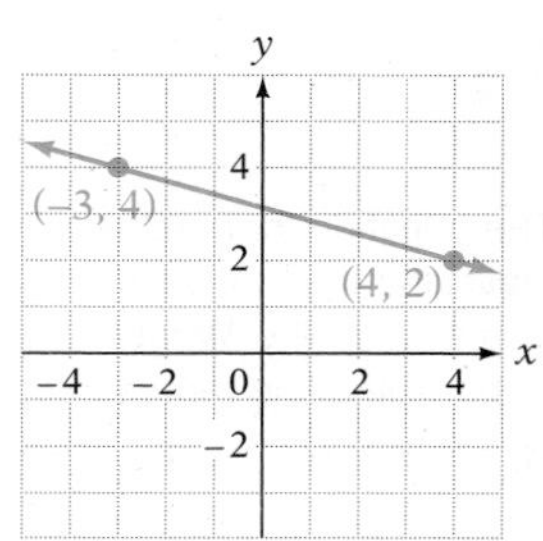

Negative slope

A line that slants downward to the right has a **negative slope.**

HOW TO • 3 Find the slope of the line passing through the points whose coordinates are $(-2, 2)$ and $(4, 2)$.

We will choose $P_1(-2, 2)$ and $P_2(4, 2)$. From $P_1(-2, 2)$, we have $x_1 = -2$, $y_1 = 2$. From $P_2(4, 2)$, we have $x_2 = 4$, $y_2 = 2$. Now use the slope formula.

$$m = \frac{y_2 - y_1}{x_2 - x_1} = \frac{2 - 2}{4 - (-2)} = \frac{0}{6} = 0$$

The slope of the line is 0.

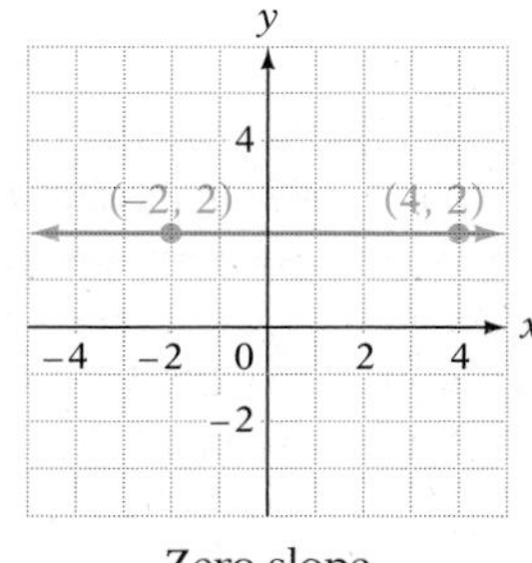

Zero slope

A horizontal line has **zero slope.**

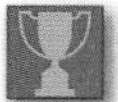

Tips for Success

To learn mathematics, you must be an active participant. Listening to and watching your professor do mathematics is not enough. Take notes in class, mentally think through every question your instructor asks, and try to answer it even if you are not called on to answer it verbally. Ask questions when you have them. See *AIM for Success* in the Preface for other ways to be an active learner.

HOW TO • 4 Find the slope of the line passing through the points whose coordinates are $(1, -2)$ and $(1, 3)$.

We will choose $P_1(1, -2)$ and $P_2(1, 3)$. From $P_1(1, -2)$, we have $x_1 = 1$, $y_1 = -2$. From $P_2(1, 3)$, we have $x_2 = 1$, $y_2 = 3$. Now use the slope formula.

$$m = \frac{y_2 - y_1}{x_2 - x_1} = \frac{3 - (-2)}{1 - 1} = \frac{5}{0} \quad \text{Not a real number}$$

The slope of the line is undefined.

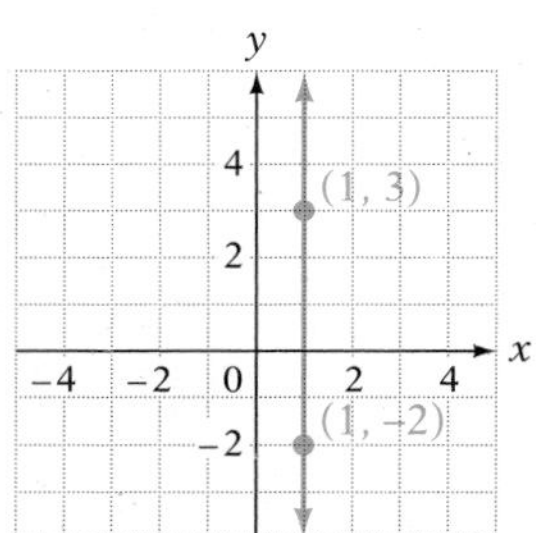

Undefined

A vertical line has **undefined slope.** Sometimes a vertical line is said to have **no slope.**

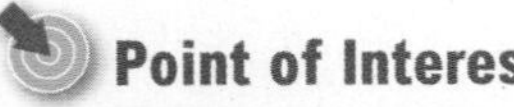

Point of Interest

One of the motivations for the discovery of calculus was the desire to solve a more complicated version of the distance-rate problem at the right.

You may be familiar with twirling a ball on the end of a string. If you release the string, the ball flies off in a path as shown below.

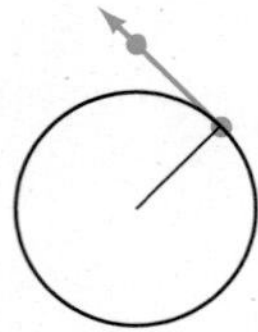

The question that mathematicians tried to answer was essentially, "What is the slope of the line represented by the arrow?"

Answering questions similar to this led to the development of one aspect of calculus.

There are many applications of slope. Here are two examples.

The first record for the 1-mile run was recorded in 1865 in England. Richard Webster ran the mile in 4 min 36.5 s. His average speed was approximately 19 ft/s.

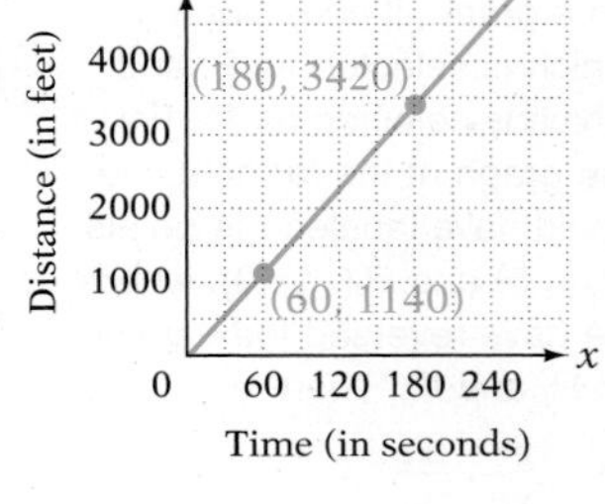

The graph at the right shows the distance Webster ran during that run. From the graph, note that after 60 s (1 min) he had traveled 1140 ft and that after 180 s (3 min) he had traveled 3420 ft.

We will choose $P_1(60, 1140)$ and $P_2(180, 3420)$. The slope of the line between these two points is

$$m = \frac{y_2 - y_1}{x_2 - x_1}$$
$$= \frac{3420 - 1140}{180 - 60} = \frac{2280}{120} = 19$$

Note that the slope of the line is the same as Webster's average speed, 19 ft/s.

Average speed is related to slope.

The following example is related to the automotive industry.

The resale value of a 2006 Chevrolet Corvette declines as the number of miles the car is driven increases. (*Source*: Edmunds.com, July 2008)

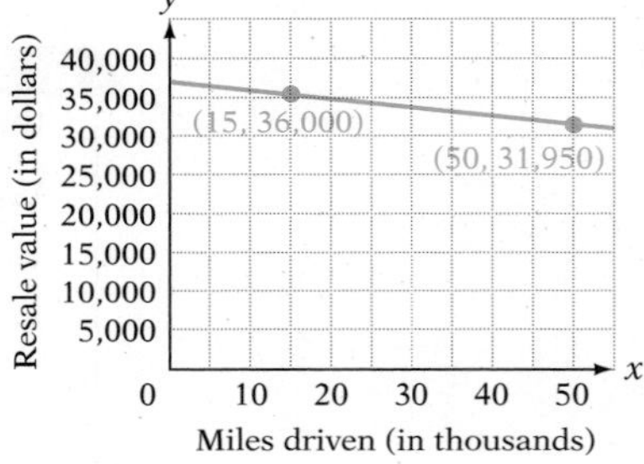

We will choose $P_1(15, 36{,}000)$ and $P_2(50, 31{,}950)$. The slope of the line between the two points is

$$m = \frac{y_2 - y_1}{x_2 - x_1}$$
$$= \frac{31{,}950 - 36{,}000}{50 - 15} = \frac{-4050}{35} \approx -115.71$$

If we interpret negative slope as decreasing value, then the slope of the line represents the dollar decline of the value of the car for each 1000 mi driven. Thus the value of the car declines approximately \$115.71 for each 1000 mi driven.

In general, any quantity that is expressed by using the word *per* is represented mathematically as slope. In the first example, the slope represented the average speed, 19 ft/s. In the second example, the slope represented the rate at which the value of the car was decreasing, \$115.71 for each 1000 mi driven.

EXAMPLE • 1

Find the slope of the line containing the points whose coordinates are (2, −5) and (−4, 2).

Solution

Choose $P_1 = (2, -5)$ and $P_2 = (-4, 2)$.

$$m = \frac{y_2 - y_1}{x_2 - x_1} = \frac{2 - (-5)}{-4 - 2} = \frac{7}{-6}$$

The slope is $-\frac{7}{6}$.

YOU TRY IT • 1

Find the slope of the line containing the points whose coordinates are (4, −3) and (2, 7).

Your solution

EXAMPLE • 2

Find the slope of the line containing the points whose coordinates are (−3, 4) and (5, 4).

Solution

Choose $P_1 = (-3, 4)$ and $P_2 = (5, 4)$.

$$m = \frac{y_2 - y_1}{x_2 - x_1} = \frac{4 - 4}{5 - (-3)} = \frac{0}{8} = 0$$

The slope of the line is zero.

YOU TRY IT • 2

Find the slope of the line containing the points whose coordinates are (6, −1) and (6, 7).

Your solution

EXAMPLE • 3

The graph below shows the relationship between the cost of an item and the sales tax. Find the slope of the line between the two points shown on the graph. Write a sentence that states the meaning of the slope.

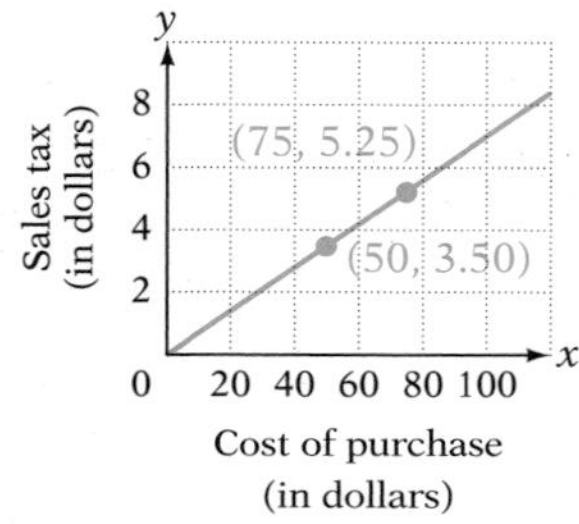

Solution

$$m = \frac{5.25 - 3.50}{75 - 50} = \frac{1.75}{25} = 0.07$$

• Choose $P_1(50, 3.50)$ and $P_2(75, 5.25)$.

A slope of 0.07 means that the sales tax is \$.07 per dollar.

YOU TRY IT • 3

The graph below shows the decrease in the value of a recycling truck for a period of 6 years. Find the slope of the line between the two points shown on the graph. Write a sentence that states the meaning of the slope.

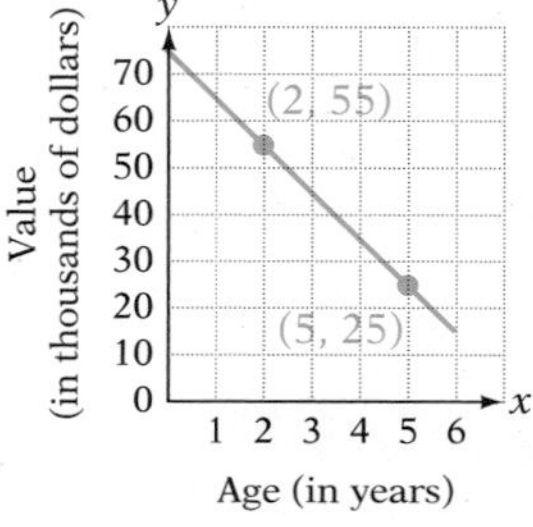

Your solution

Solutions on pp. S12–S13

OBJECTIVE B To graph a line given a point and the slope 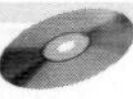

The graph of the equation $y = -\frac{3}{4}x + 4$ is shown at the right. The points with coordinates $(-4, 7)$ and $(4, 1)$ are on the graph. The slope of the line is

$$m = \frac{7 - 1}{-4 - 4} = \frac{6}{-8} = -\frac{3}{4}$$

The slope of the line has the same value as the coefficient of x.

Recall that the y-intercept is found by replacing x by zero and solving for y.

$$y = -\frac{3}{4}x + 4 \qquad y = -\frac{3}{4}(0) + 4 = 4$$

The y-intercept is $(0, 4)$. The y-coordinate of the y-intercept is the constant term of the equation.

> **Slope-Intercept Form of a Straight Line**
>
> The equation $y = mx + b$ is called the **slope-intercept form of a straight line.** The slope of the line is m, the coefficient of x. The y-intercept is $(0, b)$.

When the equation of a straight line is in the form $y = mx + b$, its graph can be drawn by using the slope and y-intercept. First locate the y-intercept. Use the slope to find a second point on the line. Then draw a line through the two points.

✓ Take Note
When graphing a line by using its slope and y-intercept, *always* start at the y-intercept.

HOW TO 5 Graph $y = \frac{5}{3}x - 4$ by using the slope and y-intercept.

The slope is the coefficient of x:

$$m = \frac{5}{3} = \frac{\text{change in } y}{\text{change in } x}. \qquad \text{The } y\text{-intercept is } (0, -4).$$

Beginning at the y-intercept $(0, -4)$, move right 3 units (change in x) and then up 5 units (change in y).

The point whose coordinates are $(3, 1)$ is a second point on the graph. Draw a line through the points whose coordinates are $(0, -4)$ and $(3, 1)$.

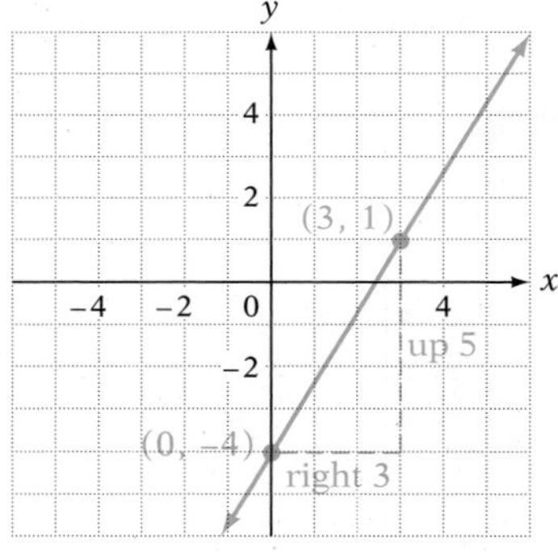

HOW TO 6 Graph $x + 2y = 4$ by using the slope and y-intercept.

Solve the equation for y.

$$x + 2y = 4$$
$$2y = -x + 4 \qquad \bullet\ m = -\frac{1}{2} = \frac{-1}{2},$$
$$y = -\frac{1}{2}x + 2 \qquad y\text{-intercept} = (0, 2)$$

Beginning at the y-intercept $(0, 2)$, move right 2 units (change in x) and then down 1 unit (change in y). The point whose coordinates are $(2, 1)$ is a second point on the graph. Draw a line through the points whose coordinates are $(0, 2)$ and $(2, 1)$.

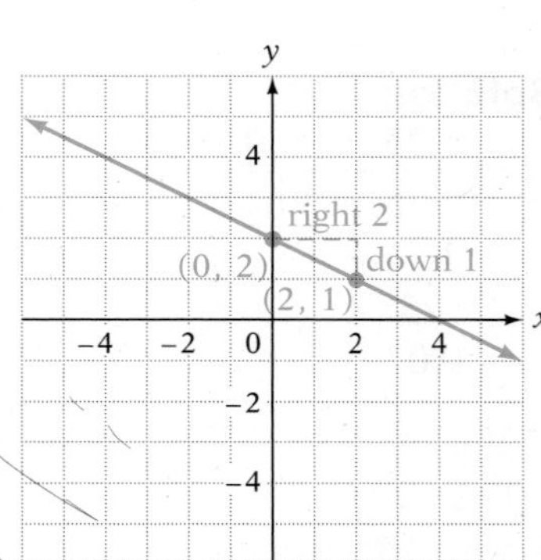

The graph of a line can be drawn when any point on the line and the slope of the line are given.

Take Note

This example differs from the preceding two in that a point other than the y-intercept is used. In this case, start at the given point.

HOW TO • 7 Graph the line that passes through the point whose coordinates are $(-4, -4)$ and has slope 2.

When the slope is an integer, write it as a fraction with denominator 1.

$$m = 2 = \frac{2}{1} = \frac{\text{change in } y}{\text{change in } x}$$

Locate $(-4, -4)$ in the coordinate plane. Beginning at that point, move right 1 unit (change in x) and then up 2 units (change in y).

The point whose coordinates are $(-3, -2)$ is a second point on the graph. Draw a line through the points $(-4, -4)$ and $(-3, -2)$.

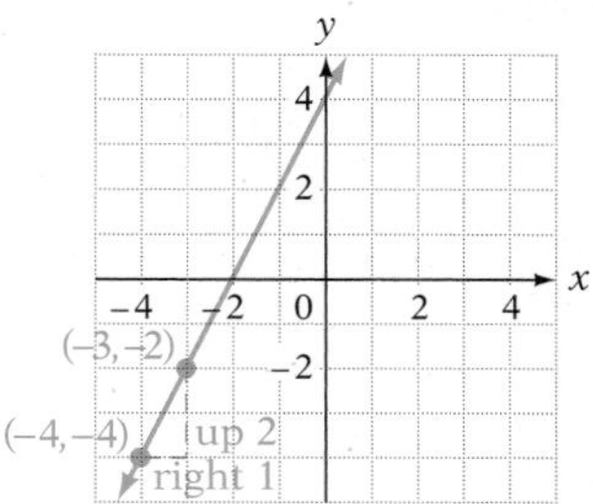

EXAMPLE • 4

Graph $y = -\frac{3}{2}x + 4$ by using the slope and y-intercept.

Solution $m = -\frac{3}{2} = \frac{-3}{2}$

y-intercept $= (0, 4)$

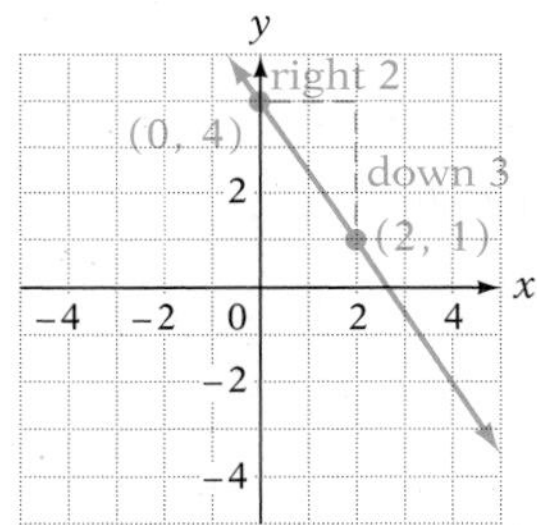

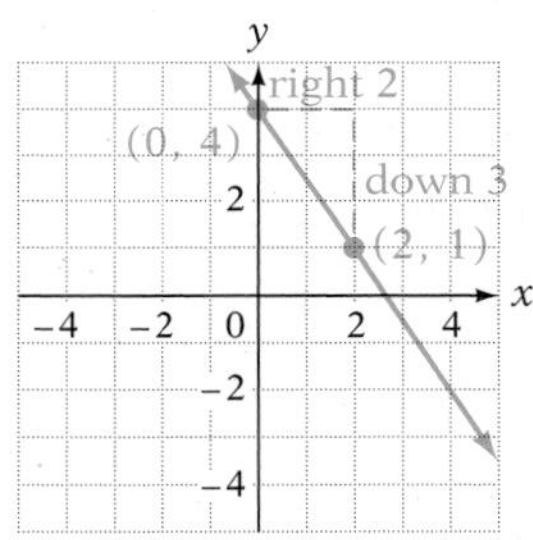

YOU TRY IT • 4

Graph $2x + 3y = 6$ by using the slope and y-intercept.

Your solution

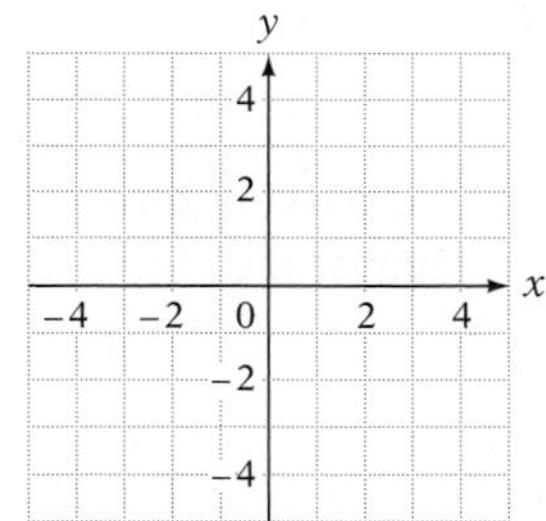

EXAMPLE • 5

Graph the line that passes through the point whose coordinates are $(-2, 3)$ and has slope $-\frac{4}{3}$.

Solution Locate $(-2, 3)$.

$m = -\frac{4}{3} = \frac{-4}{3}$

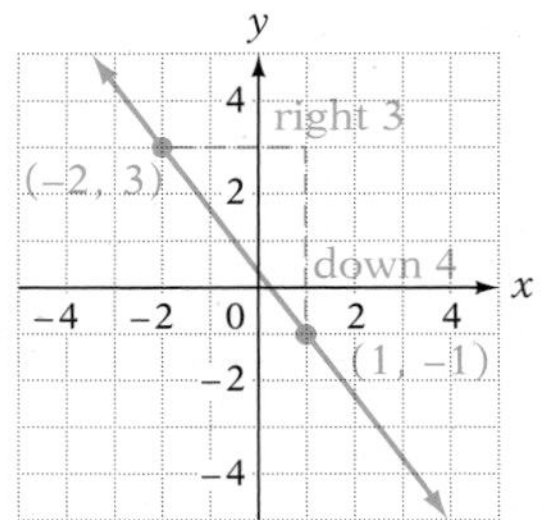

YOU TRY IT • 5

Graph the line that passes through the point whose coordinates are $(-3, -2)$ and has slope 3.

Your solution

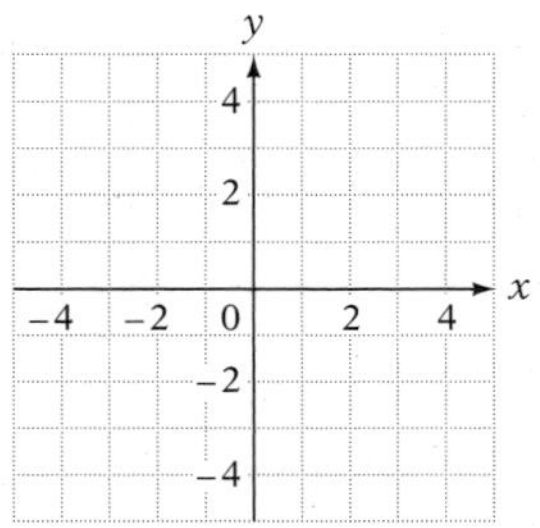

Solutions on p. S13

4.4 EXERCISES

OBJECTIVE A To find the slope of a line given two points

For Exercises 1 to 18, find the slope of the line containing the points with the given coordinates.

1. (1, 3), (3, 1)

2. (2, 3), (5, 1)

3. $(-1, 4), (2, 5)$

4. $(3, -2), (1, 4)$

5. $(-1, 3), (-4, 5)$

6. $(-1, -2), (-3, 2)$

7. (0, 3), (4, 0)

8. $(-2, 0), (0, 3)$

9. $(2, 4), (2, -2)$

10. $(4, 1), (4, -3)$

11. $(2, 5), (-3, -2)$

12. $(4, 1), (-1, -2)$

13. $(2, 3), (-1, 3)$

14. (3, 4), (0, 4)

15. $(0, 4), (-2, 5)$

16. $(-2, 3), (-2, 5)$

17. $(-3, -1), (-3, 4)$

18. $(-2, -5), (-4, -1)$

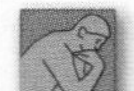

19. Let l be a line passing through the points (a, b) and (c, d). Which two of a, b, c, and d are equal if the slope of l is undefined?

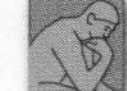

20. Let l be a line passing through the points (a, b) and (c, d). Which two of a, b, c, and d are equal if the slope of l is zero?

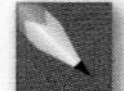

21. **Travel** The graph below shows the relationship between the distance traveled by a motorist and the time of travel. Find the slope of the line between the two points shown on the graph. Write a sentence that states the meaning of the slope.

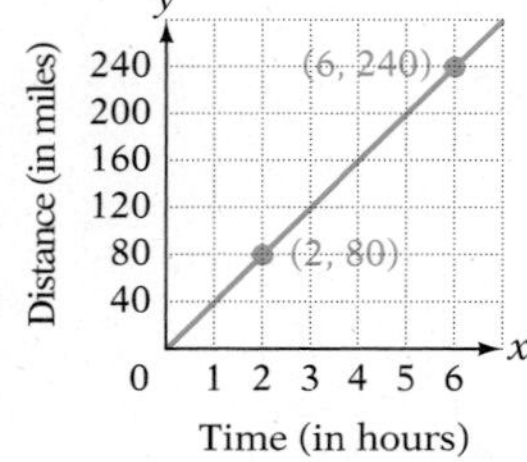

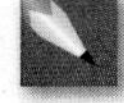

22. **Media** The graph below shows the number of people subscribing to a sports magazine of increasing popularity. Find the slope of the line between the two points shown on the graph. Write a sentence that states the meaning of the slope.

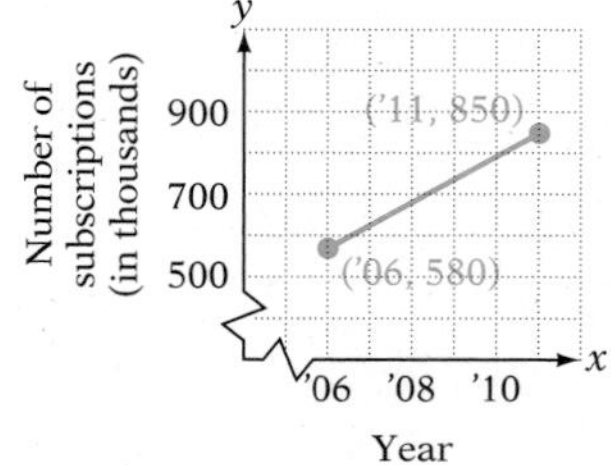

23. **Temperature** The graph below shows the relationship between the temperature inside an oven and the time since the oven was turned off. Find the slope of the line. Write a sentence that states the meaning of the slope.

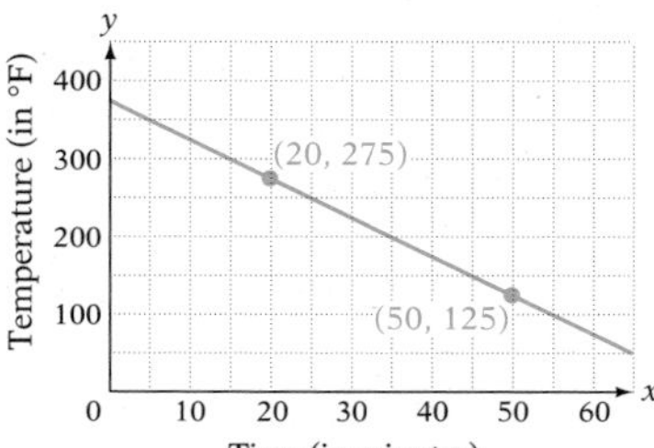

24. **Home Maintenance** The graph below shows the number of gallons of water remaining in a pool x minutes after a valve is opened to drain the pool. Find the slope of the line. Write a sentence that states the meaning of the slope.

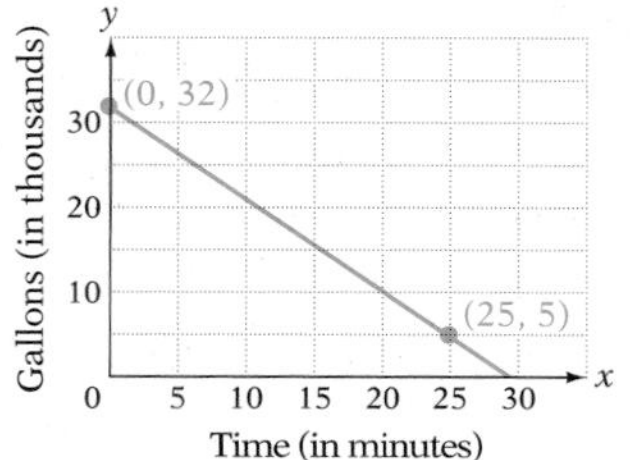

25. **Fuel Consumption** The graph below shows how the amount of gas in the tank of a car decreases as the car is driven. Find the slope of the line. Write a sentence that states the meaning of the slope.

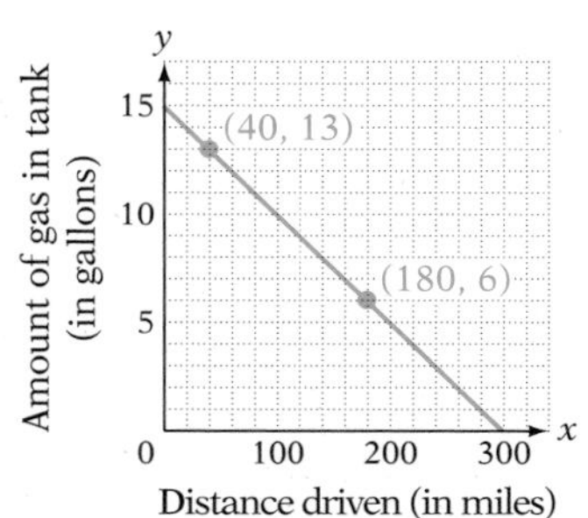

26. **Meteorology** The troposphere extends from Earth's surface to an elevation of about 11 km. The graph below shows the decrease in the temperature of the troposphere as altitude increases. Find the slope of the line. Write a sentence that states the meaning of the slope.

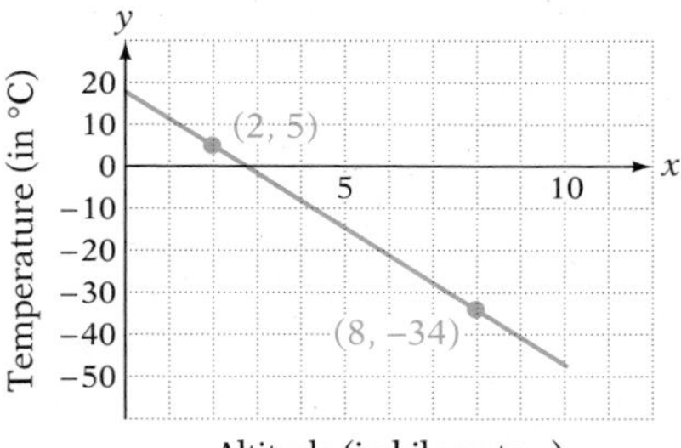

27. **Sports** The graph below shows the relationship between distance and time for the 5000-meter run for the world record by Tirunesh Dibaba in 2008. Find the slope of the line between the two points shown on the graph. Round to the nearest tenth. Write a sentence that states the meaning of the slope.

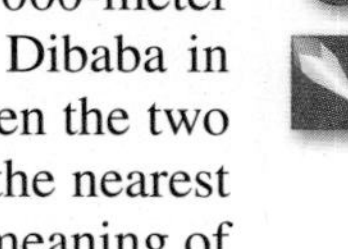

28. **Sports** The graph below shows the relationship between distance and time for the 10,000-meter run for the Olympic record by Kenenisa Bekele in 2008. Find the slope of the line between the two points shown on the graph. Round to the nearest tenth. Write a sentence that states the meaning of the slope.

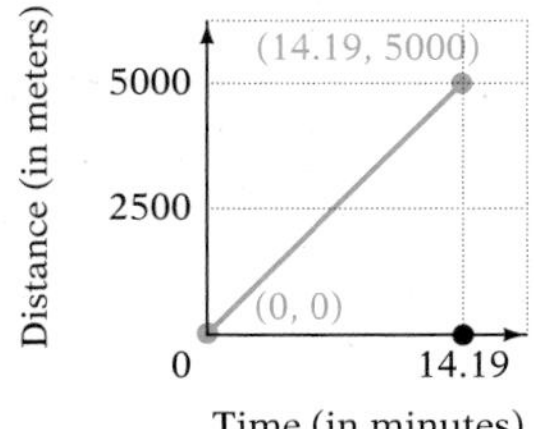

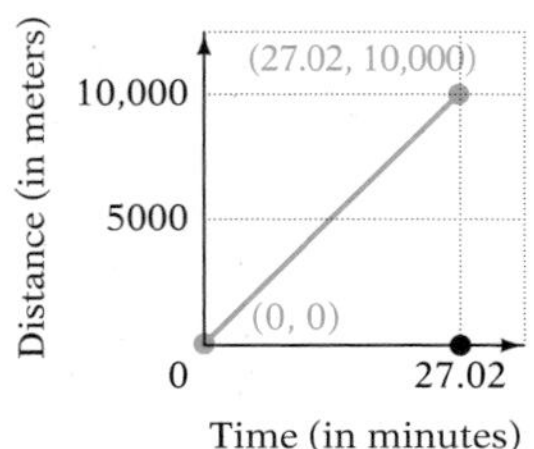

29. **Construction** The American National Standards Institute (ANSI) states that the slope for a wheelchair ramp must not exceed $\frac{1}{12}$.

a. Does a ramp that is 6 in. high and 5 ft long meet the requirements of ANSI?
b. Does a ramp that is 12 in. high and 170 in. long meet the requirements of ANSI?

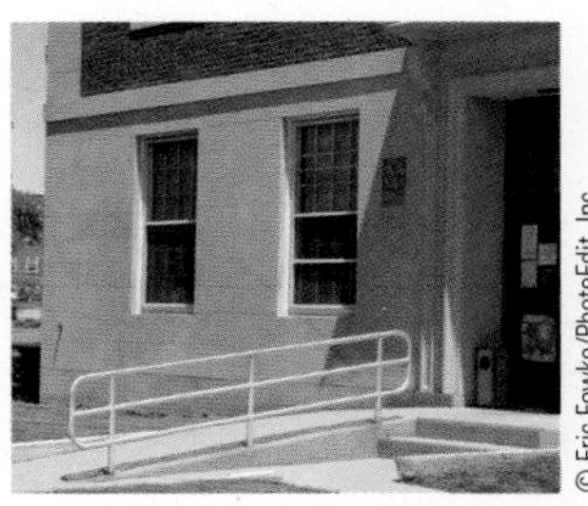

30. **Solar Roof** Look at the butterfly roof design shown in the article below. Which side of the roof, the left or the right, has a slope of approximately 1? Is the slope of the other side of the roof greater than 1 or less than 1? (*Note:* Consider both slopes to be positive.)

In the News

Go Green with Power Pod's Butterfly Roof

PowerHouse Enterprises has designed a modular home with a butterfly roof. The roof design combines two sections that slant toward each other at different angles, and is ideal for the use of solar panels.

Source: www.powerhouse-enterprises.com

OBJECTIVE B To graph a line given a point and the slope

For Exercises 31 to 34, complete the table.

	Equation	*Value of m*	*Value of b*	*Slope*	*y-intercept*
31.	$y = -3x + 5$				
32.	$y = \frac{2}{5}x - 8$				
33.	$y = 4x$				
34.	$y = -7$				

For Exercises 35 to 46, graph by using the slope and the *y*-intercept.

35. $y = \frac{1}{2}x + 2$

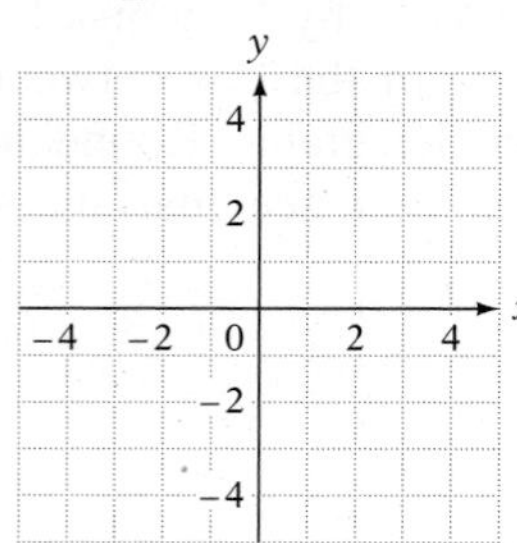

36. $y = \frac{2}{3}x - 3$

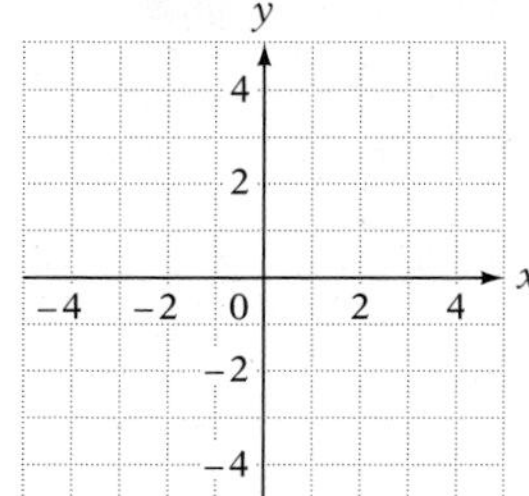

37. $y = -\frac{3}{2}x$

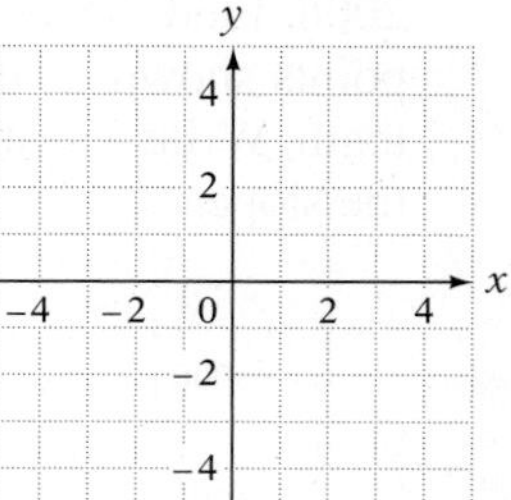

38. $y = \frac{3}{4}x$

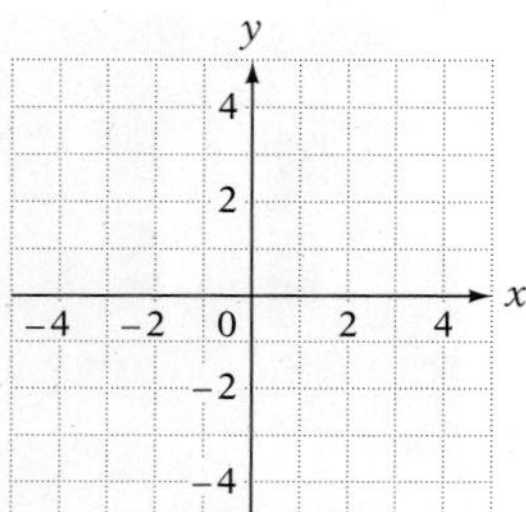

39. $y = -\frac{1}{2}x + 2$

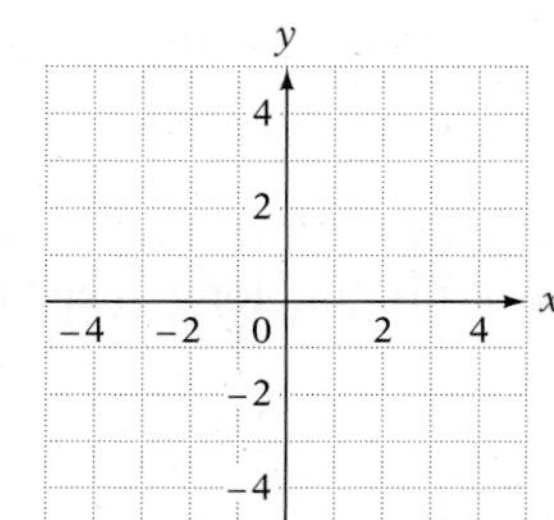

40. $y = \frac{2}{3}x - 1$

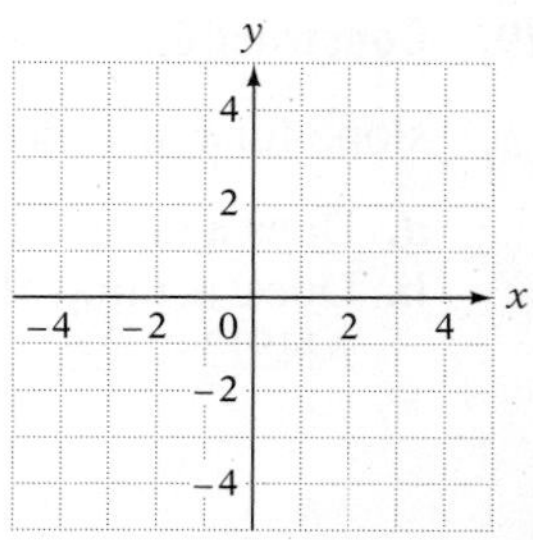

41. $y = 2x - 4$

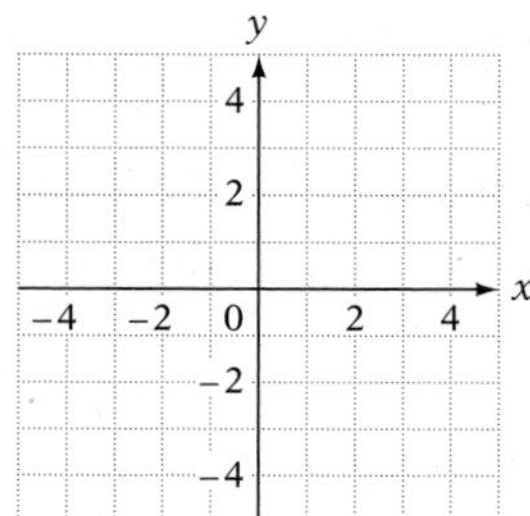

42. $y = -3x + 1$

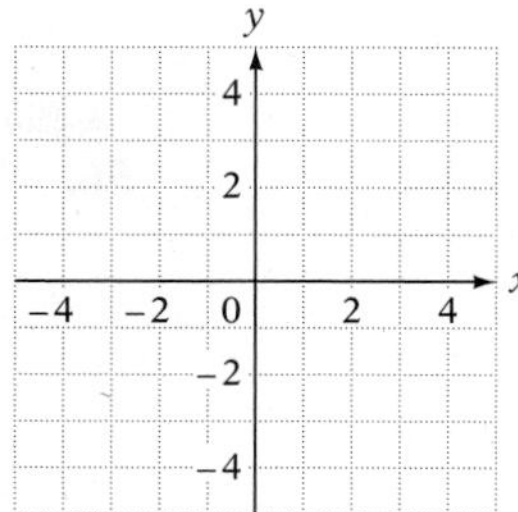

43. $4x - y = 1$

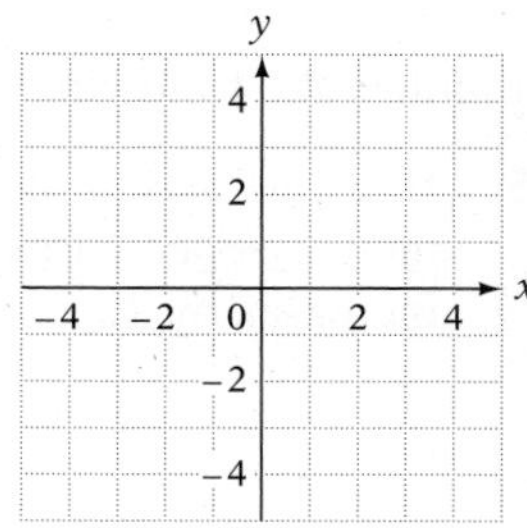

44. $4x + y = 2$

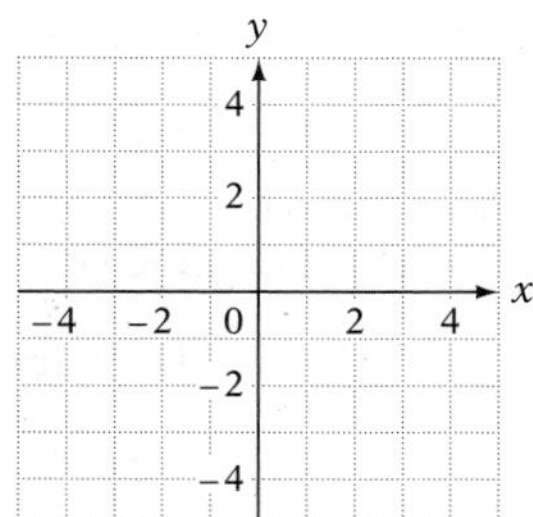

45. $x - 3y = 3$

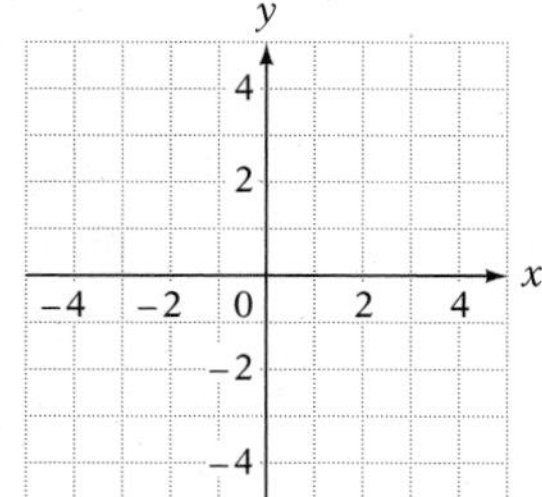

46. $3x + 2y = 8$

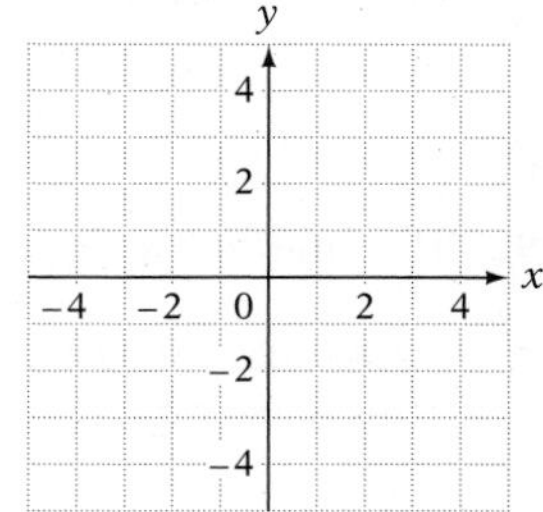

47. Graph the line that passes through the point $(-1, -3)$ and has slope $\frac{4}{3}$.

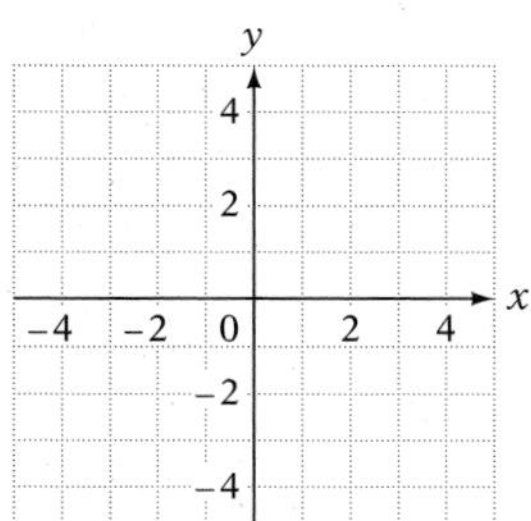

48. Graph the line that passes through the point $(-2, -3)$ and has slope $\frac{5}{4}$.

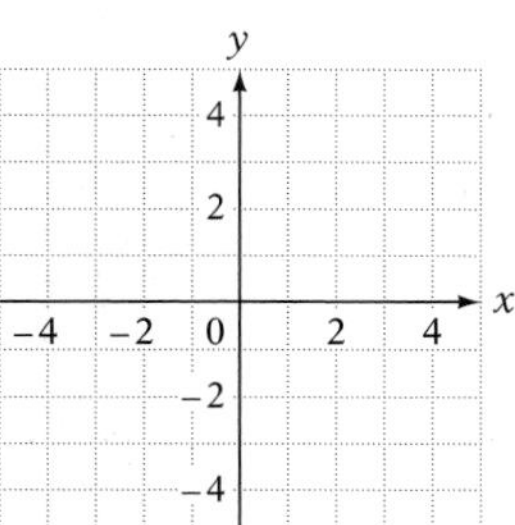

49. Graph the line that passes through the point $(-3, 0)$ and has slope -3.

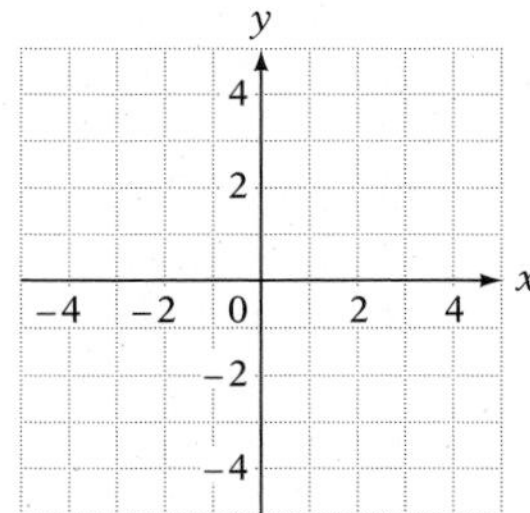

50. Graph the line that passes through the point $(2, -4)$ and has slope $-\frac{1}{2}$.

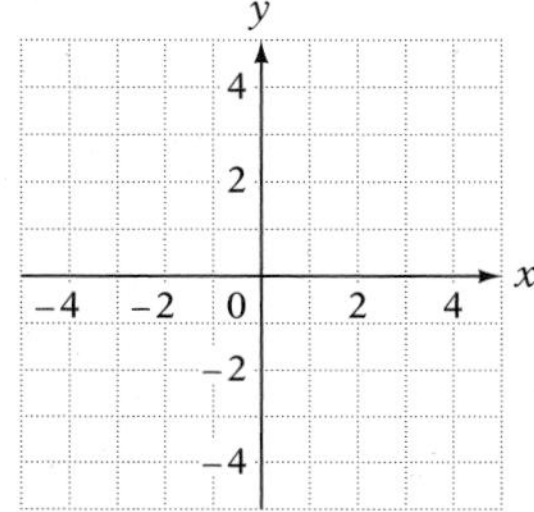

51. Graph the line that passes through the point $(-4, 1)$ and has slope $\frac{2}{3}$.

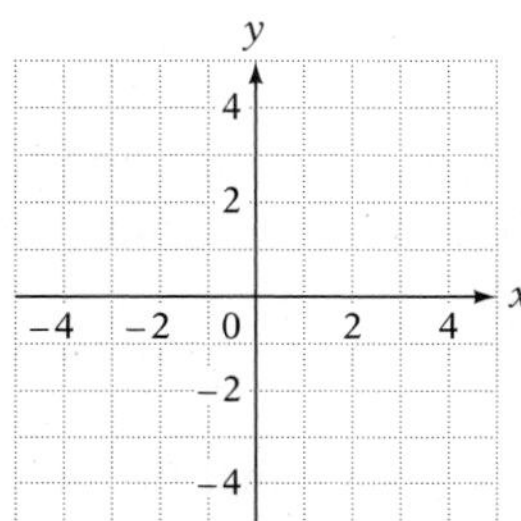

52. Graph the line that passes through the point $(1, 5)$ and has slope -4.

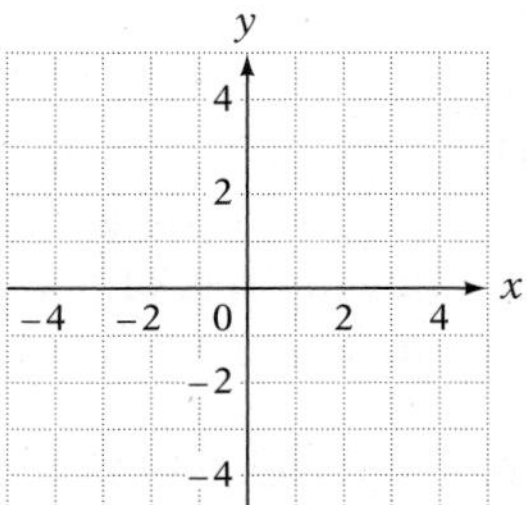

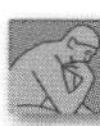

For Exercises 53 and 54, for the given conditions, state **a.** whether the y-intercept of the graph of $Ax + By = C$ lies above or below the x-axis, and **b.** whether the graph of $Ax + By = C$ has positive or negative slope.

53. A and B are positive numbers, and C is a negative number.

54. A and C are positive numbers, and B is a negative number.

Applying the Concepts

Complete the following sentences.

55. If a line has a slope of 2, then the value of y <u>increases/decreases</u> by _______ as the value of x increases by 1.

56. If a line has a slope of -3, then the value of y <u>increases/decreases</u> by _______ as the value of x increases by 1.

57. If a line has a slope of -2, then the value of y <u>increases/decreases</u> by _______ as the value of x decreases by 1.

58. If a line has a slope of 3, then the value of y <u>increases/decreases</u> by _______ as the value of x decreases by 1.

59. If a line has a slope of $-\frac{2}{3}$, then the value of y <u>increases/decreases</u> by _______ as the value of x increases by 1.

60. If a line has a slope of $\frac{1}{2}$, then the value of y <u>increases/decreases</u> by _______ as the value of x increases by 1.

61. Match each equation with its graph.

i. $y = -2x + 4$

ii. $y = 2x - 4$

iii. $y = 2$

iv. $2x + 4y = 0$

v. $y = \frac{1}{2}x + 4$

vi. $y = -\frac{1}{4}x - 2$

A. **B.**

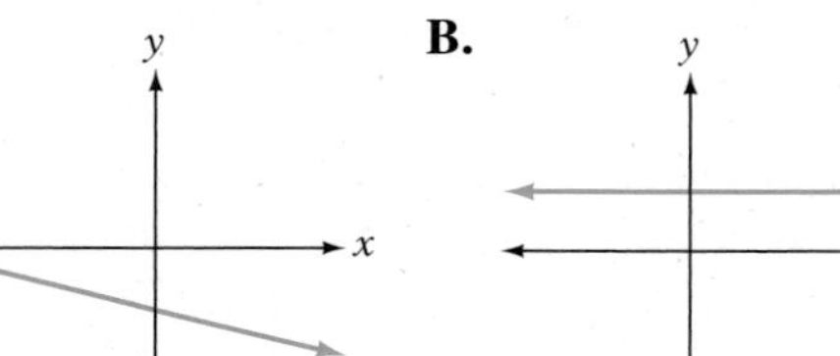

C.

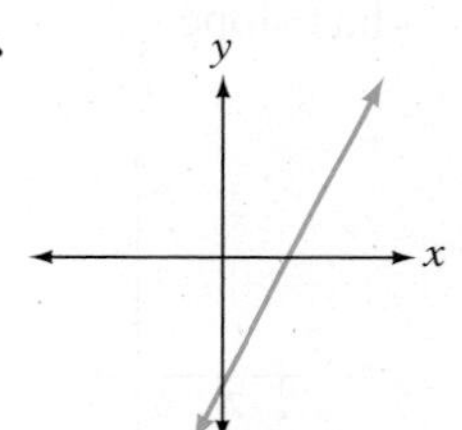

D.

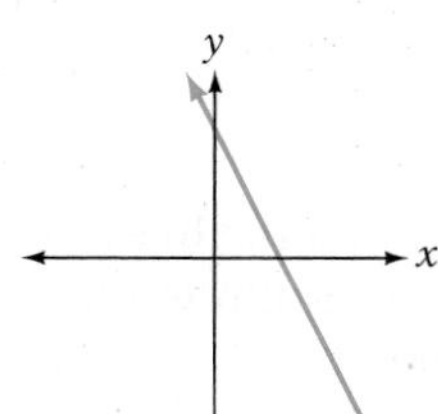

E. **F.**

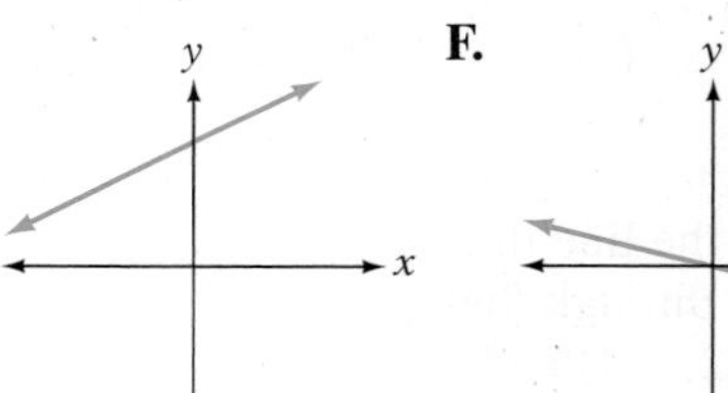

62. Explain how you can use the slope of a line to determine whether three given points lie on the same line. Then use your procedure to determine whether each of the following sets of points lie on the same line.
a. $(2, 5), (-1, -1), (3, 7)$ **b.** $(-1, 5), (0, 3), (-3, 4)$

For Exercises 63 to 66, determine the value of k such that the points whose coordinates are given lie on the same line.

63. $(3, 2), (4, 6), (5, k)$

64. $(-2, 3), (1, 0), (k, 2)$

65. $(k, 1), (0, -1), (2, -2)$

66. $(4, -1), (3, -4), (k, k)$

SECTION

4.5 Finding Equations of Lines

OBJECTIVE A To find the equation of a line given a point and the slope 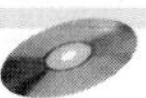

When the slope of a line and a point on the line are known, the equation of the line can be determined. If the particular point is the y-intercept, use the slope-intercept form, $y = mx + b$, to find the equation.

HOW TO 1 Find the equation of the line that contains the point $P(0, 3)$ and has slope $\frac{1}{2}$.

The known point is the y-intercept, $P(0, 3)$.

$y = mx + b$ • Use the slope-intercept form.

$y = \frac{1}{2}x + 3$ • Replace m with $\frac{1}{2}$, the given slope. Replace b with 3, the y-coordinate of the y-intercept.

The equation of the line is $y = \frac{1}{2}x + 3$.

One method of finding the equation of a line when the slope and *any* point on the line are known involves using the *point-slope formula.* This formula is derived from the formula for the slope of a line as follows.

Let $P_1(x_1, y_1)$ be the given point on the line, and let $P(x, y)$ be any other point on the line. See the graph at the left.

$\frac{y - y_1}{x - x_1} = m$ • Use the formula for the slope of a line.

$\frac{y - y_1}{x - x_1}(x - x_1) = m(x - x_1)$ • Multiply each side by $(x - x_1)$.

$y - y_1 = m(x - x_1)$ • Simplify.

Point-Slope Formula

Let m be the slope of a line, and let $P_1(x_1, y_1)$ be the coordinates of a point on the line. The equation of the line can be found from the **point-slope formula:**

$$y - y_1 = m(x - x_1)$$

HOW TO 2 Find the equation of the line that contains the point whose coordinates are $P(4, -1)$ and has slope $-\frac{3}{4}$.

$y - y_1 = m(x - x_1)$ • Use the point-slope formula.

$y - (-1) = \left(-\frac{3}{4}\right)(x - 4)$ • $m = -\frac{3}{4}, (x_1, y_1) = (4, -1)$

$y + 1 = -\frac{3}{4}x + 3$ • Simplify.

$y = -\frac{3}{4}x + 2$ • Write the equation in the form $y = mx + b$.

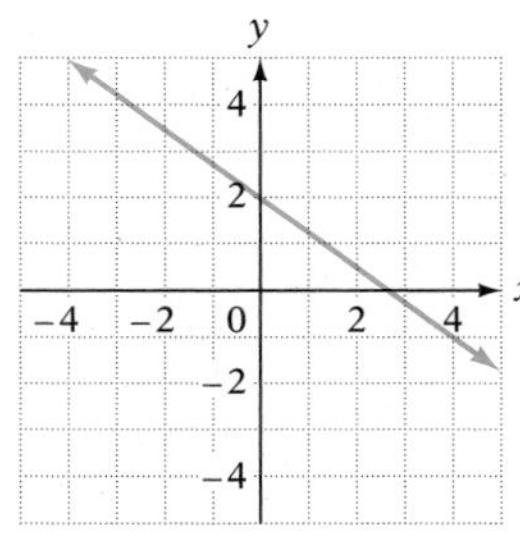

$y = -\frac{3}{4}x + 2$

HOW TO • 3 Find the equation of the line that passes through the point whose coordinates are (4, 3) and whose slope is undefined.

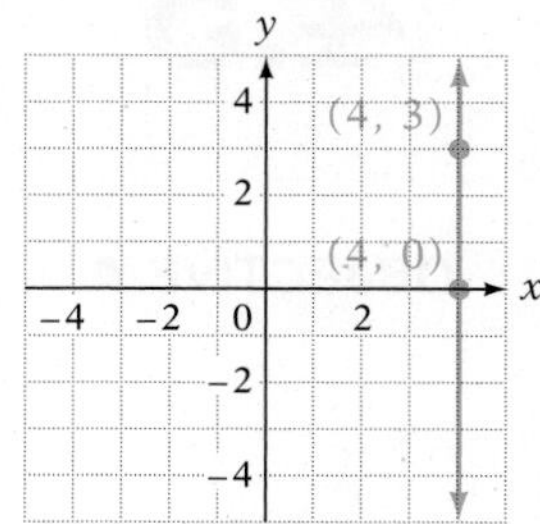

Because the slope is undefined, the point-slope formula cannot be used to find the equation. Instead, recall that when the slope of a line is undefined, the line is vertical. The equation of a vertical line is $x = a$, where a is the x-coordinate of the x-intercept. Because the line is vertical and passes through (4, 3), the x-intercept is (4, 0).
The equation of the line is $x = 4$.

EXAMPLE • 1

Find the equation of the line that contains the point $P(3, 0)$ and has slope -4.

Solution

$m = -4 \qquad (x_1, y_1) = (3, 0)$

$$y - y_1 = m(x - x_1)$$
$$y - 0 = -4(x - 3)$$
$$y = -4x + 12$$

The equation of the line is $y = -4x + 12$.

YOU TRY IT • 1

Find the equation of the line that contains the point $P(-3, -2)$ and has slope $-\frac{1}{3}$.

Your solution

EXAMPLE • 2

Find the equation of the line that contains the point $P(-2, 4)$ and has slope 2.

Solution

$m = 2 \qquad (x_1, y_1) = (-2, 4)$

$$y - y_1 = m(x - x_1)$$
$$y - 4 = 2[x - (-2)]$$
$$y - 4 = 2(x + 2)$$
$$y - 4 = 2x + 4$$
$$y = 2x + 8$$

The equation of the line is $y = 2x + 8$.

YOU TRY IT • 2

Find the equation of the line that contains the point $P(4, -3)$ and has slope -3.

Your solution

Solutions on p. S13

OBJECTIVE B To find the equation of a line given two points

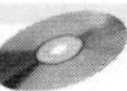

The point-slope formula and the formula for slope are used to find the equation of a line when two points are known.

HOW TO • 4 Find the equation of the line containing the points whose coordinates are (3, 2) and (−5, 6).

To use the point-slope formula, we must know the slope. Use the formula for slope to determine the slope of the line between the two given points. We will choose $P_1(3, 2)$ and $P_2(-5, 6)$.

$$m = \frac{y_2 - y_1}{x_2 - x_1} = \frac{6 - 2}{-5 - 3} = \frac{4}{-8} = -\frac{1}{2}$$

Now use the point-slope formula with $m = -\frac{1}{2}$ and $(x_1, y_1) = (3, 2)$.

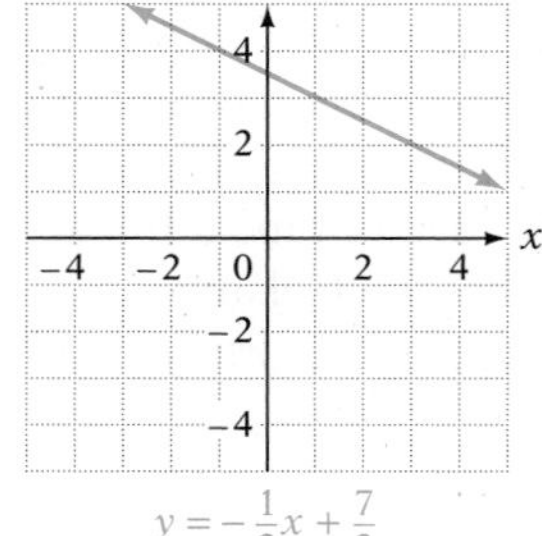

$y = -\frac{1}{2}x + \frac{7}{2}$

$y - y_1 = m(x - x_1)$ • Use the point-slope formula.

$y - 2 = \left(-\frac{1}{2}\right)(x - 3)$ • $m = -\frac{1}{2}$, $(x_1, y_1) = (3, 2)$

$y - 2 = -\frac{1}{2}x + \frac{3}{2}$ • Simplify.

$y = -\frac{1}{2}x + \frac{7}{2}$

The equation of the line is $y = -\frac{1}{2}x + \frac{7}{2}$.

EXAMPLE • 3

Find the equation of the line passing through the points whose coordinates are (2, 3) and (4, 1).

Solution

Choose $P_1(2, 3)$ and $P_2(4, 1)$.

$$m = \frac{y_2 - y_1}{x_2 - x_1} = \frac{1 - 3}{4 - 2} = \frac{-2}{2} = -1$$

$y - y_1 = m(x - x_1)$

$y - 3 = -1(x - 2)$

$y - 3 = -x + 2$

$y = -x + 5$

The equation of the line is $y = -x + 5$.

YOU TRY IT • 3

Find the equation of the line passing through the points whose coordinates are (2, 0) and (5, 3).

Your solution

EXAMPLE • 4

Find the equation of the line containing the points whose coordinates are (2, −3) and (2, 5).

Solution

Choose $P_1(2, -3)$ and $P_2(2, 5)$.

$$m = \frac{y_2 - y_1}{x_2 - x_1} = \frac{5 - (-3)}{2 - 2} = \frac{8}{0}$$

The slope is undefined, so the graph of the line is vertical.

The equation of the line is $x = 2$.

YOU TRY IT • 4

Find the equation of the line containing the points whose coordinates are (2, 3) and (−5, 3).

Your solution

Solutions on p. S13

OBJECTIVE C To solve application problems 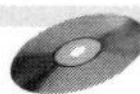

Linear functions can be used to model a variety of applications in science and business. For each application, data are collected and the independent and dependent variables are selected. Then a linear function is determined that models the data.

EXAMPLE • 5

Suppose a manufacturer has determined that at a price of $150, consumers will purchase 1 million 8 GB digital media players and that at a price of $125, consumers will purchase 1.25 million 8 GB digital media players. Describe this situation with a linear function. Use this function to predict how many 8 GB digital media players consumers will purchase if the price is $80.

Strategy

- Select the independent and dependent variables. Because you are trying to determine the number of 8 GB digital media players, that quantity is the *dependent* variable, y. The price of the 8 GB digital media players is the *independent* variable, x.
- From the given data, two ordered pairs are (150, 1) and (125, 1.25). (The ordinates are in millions of units.) Use these ordered pairs to determine the linear function.
- Evaluate this function when $x = 80$ to predict how many 8 GB digital media players consumers will purchase if the price is $80.

Solution

Choose $P_1(150, 1)$ and $P_2(125, 1.25)$.

$$m = \frac{y_2 - y_1}{x_2 - x_1} = \frac{1.25 - 1}{125 - 150} = \frac{0.25}{-25} = -0.01$$

$$y - y_1 = m(x - x_1)$$

$$y - 1 = -0.01(x - 150)$$

$$y = -0.01x + 2.50$$

The linear function is $f(x) = -0.01x + 2.50$.

$f(80) = -0.01(80) + 2.50 = 1.7$

Consumers will purchase 1.7 million 8 GB digital media players at a price of $80.

YOU TRY IT • 5

Gabriel Daniel Fahrenheit invented the mercury thermometer in 1717. In terms of readings on this thermometer, water freezes at 32ºF and boils at 212ºF. In 1742, Anders Celsius invented the Celsius temperature scale. On this scale, water freezes at 0ºC and boils at 100ºC. Determine a linear function that can be used to predict the Celsius temperature when the Fahrenheit temperature is known.

Your strategy

Your solution

Solution on p. S13

4.5 EXERCISES

OBJECTIVE A To find the equation of a line given a point and the slope

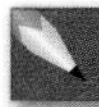
1. Explain how to find the equation of a line given its slope and its y-intercept.

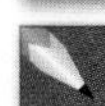
2. What is the point-slope formula and how is it used?

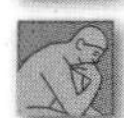
3. After you find an equation of a line given the coordinates of a point on the line and its slope, how can you determine whether you have the correct equation?

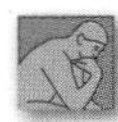
4. What point must the graph of the equation $y = mx$ pass through?

For Exercises 5 to 40, find the equation of the line that contains the given point and has the given slope.

5. $P(0, 5)$, $m = 2$
6. $P(0, 3)$, $m = 1$
7. $P(2, 3)$, $m = \frac{1}{2}$
8. $P(5, 1)$, $m = \frac{2}{3}$
9. $P(-1, 4)$, $m = \frac{5}{4}$
10. $P(-2, 1)$, $m = \frac{3}{2}$
11. $P(3, 0)$, $m = -\frac{5}{3}$
12. $P(-2, 0)$, $m = \frac{3}{2}$
13. $P(2, 3)$, $m = -3$
14. $P(1, 5)$, $m = -\frac{4}{5}$
15. $P(-1, 7)$, $m = -3$
16. $P(-2, 4)$, $m = -4$
17. $P(-1, -3)$, $m = \frac{2}{3}$
18. $P(-2, -4)$, $m = \frac{1}{4}$
19. $P(0, 0)$, $m = \frac{1}{2}$
20. $P(0, 0)$, $m = \frac{3}{4}$
21. $P(2, -3)$, $m = 3$
22. $P(4, -5)$, $m = 2$
23. $P(3, 5)$, $m = -\frac{2}{3}$
24. $P(5, 1)$, $m = -\frac{4}{5}$
25. $P(0, -3)$, $m = -1$
26. $P(2, 0)$, $m = \frac{5}{6}$
27. $P(1, -4)$, $m = \frac{7}{5}$
28. $P(3, 5)$, $m = -\frac{3}{7}$

29. $P(4, -1)$, $m = -\frac{2}{5}$

30. $P(-3, 5)$, $m = -\frac{1}{4}$

31. $P(3, -4)$, slope is undefined

32. $P(-2, 5)$, slope is undefined

33. $P(-2, -5)$, $m = -\frac{5}{4}$

34. $P(-3, -2)$, $m = -\frac{2}{3}$

35. $P(-2, -3)$, $m = 0$

36. $P(-3, -2)$, $m = 0$

37. $P(4, -5)$, $m = -2$

38. $P(-3, 5)$, $m = 3$

39. $P(-5, -1)$, slope is undefined

40. $P(0, 4)$, slope is undefined

OBJECTIVE B **To find the equation of a line given two points**

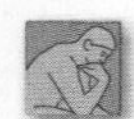

41. After you find an equation of a line given the coordinates of two points on the line, how can you determine whether you have the correct equation?

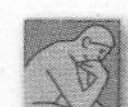

42. If you are asked to find the equation of a line through two given points, does it matter which point is selected as (x_1, y_1) and which point is selected as (x_2, y_2)?

For Exercises 43 to 78, find the equation of the line that contains the points with the given coordinates.

43. $(0, 2), (3, 5)$

44. $(0, 4), (1, 5)$

45. $(0, -3), (-4, 5)$

46. $(0, -2), (-3, 4)$

47. $(2, 3), (5, 5)$

48. $(4, 1), (6, 3)$

49. $(-1, 3), (2, 4)$

50. $(-1, 1), (4, 4)$

51. $(-1, -2), (3, 4)$

52. $(-3, -1), (2, 4)$

53. $(0, 3), (2, 0)$

54. $(0, 4), (2, 0)$

55. $(-3, -1), (2, -1)$

56. $(-3, -5), (4, -5)$

57. $(-2, -3), (-1, -2)$

58. (4, 1), (3, −2)

59. (−2, 3), (2, −1)

60. (3, 1), (−3, −2)

61. (2, 3), (5, −5)

62. (7, 2), (4, 4)

63. (2, 0), (0, −1)

64. (0, 4), (−2, 0)

65. (3, −4), (−2, −4)

66. (−3, 3), (−2, 3)

67. (0, 0), (4, 3)

68. (2, −5), (0, 0)

69. (2, −1), (−1, 3)

70. (3, −5), (−2, 1)

71. (−2, 5), (−2, −5)

72. (3, 2), (3, −4)

73. (2, 1), (−2, −3)

74. (−3, −2), (1, −4)

75. (−4, −3), (2, 5)

76. (4, 5), (−4, 3)

77. (0, 3), (3, 0)

78. (1, −3), (−2, 4)

OBJECTIVE C To solve application problems

79. **Aviation** The pilot of a Boeing 747 jet takes off from Boston's Logan Airport, which is at sea level, and climbs to a cruising altitude of 32,000 ft at a constant rate of 1200 ft/min.
a. Write a linear function for the height of the plane in terms of the time after takeoff.
b. Use your function to find the height of the plane 11 min after takeoff.

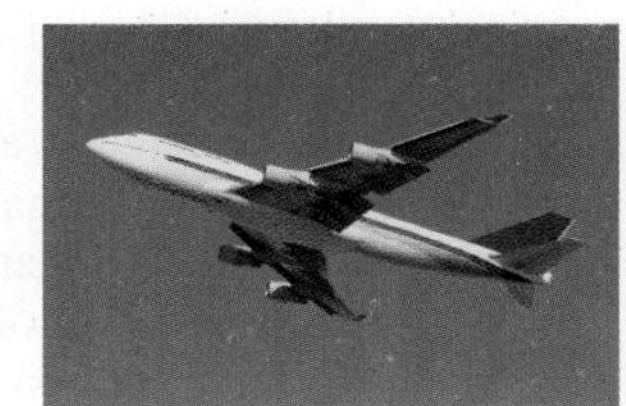
© David Frazier/Corbis

80. **Calories** A jogger running at 9 mph burns approximately 14 Calories per minute.
a. Write a linear function for the number of Calories burned by the jogger in terms of the number of minutes run.
b. Use your function to find the number of Calories that the jogger has burned after jogging for 32 min.

81. **Ecology** Use the information in the article at the right.
   a. Determine a linear function for the percent of trees at 2600 ft that are hardwoods in terms of the year.
   b. Use your function from part (a) to predict the percent of trees at 2600 ft that will be hardwoods in 2012.

82. **Telecommunications** A cellular phone company offers several different service options. One option, for people who plan on using the phone only in emergencies, costs the user $4.95 per month plus $.59 per minute for each minute the phone is used.
   a. Write a linear function for the monthly cost of the phone in terms of the number of minutes the phone is used.
   b. Use your function to find the cost of using the cellular phone for 13 min in 1 month.

83. **Fuel Consumption** The gas tank of a certain car contains 16 gal when the driver of the car begins a trip. Each mile driven by the driver decreases the amount of gas in the tank by 0.032 gal.
   a. Write a linear function for the number of gallons of gas in the tank in terms of the number of miles driven.
   b. Use your function to find the number of gallons in the tank after 150 mi are driven.

84. **Boiling Points** At sea level, the boiling point of water is 100°C. At an altitude of 2 km, the boiling point of water is 93°C.
   a. Write a linear function for the boiling point of water in terms of the altitude above sea level.
   b. Use your function to predict the boiling point of water on top of Mount Everest, which is approximately 8.85 km above sea level. Round to the nearest degree.

85. **Business** A manufacturer of motorcycles has determined that 50,000 motorcycles per month can be sold at a price of $9000. At a price of $8750, the number of motorcycles sold per month would increase to 55,000.
   a. Determine a linear function that will predict the number of motorcycles that would be sold each month at a given price.
   b. Use this model to predict the number of motorcycles that would be sold at a price of $8500.

86. **Business** A manufacturer of graphing calculators has determined that 10,000 calculators per week will be sold at a price of $95. At a price of $90, it is estimated that 12,000 calculators would be sold.
   a. Determine a linear function that will predict the number of calculators that would be sold each week at a given price.
   b. Use this model to predict the number of calculators that would be sold each week at a price of $75.

87. **Calories** There are approximately 126 Calories in a 2-ounce serving of lean hamburger and approximately 189 Calories in a 3-ounce serving.
   a. Determine a linear function for the number of Calories in lean hamburger in terms of the size of the serving.
   b. Use your function to estimate the number of Calories in a 5-ounce serving of lean hamburger.

88. **Compensation** An account executive receives a base salary plus a commission. On $20,000 in monthly sales, the account executive receives $1800. On $50,000 in monthly sales, the account executive receives $3000.
   a. Determine a linear function that will yield the compensation of the sales executive for a given amount of monthly sales.
   b. Use this model to determine the account executive's compensation for $85,000 in monthly sales.

In the News

Is Global Warming Moving Mountains?

In the mountains of Vermont, maples, beeches, and other hardwood trees that thrive in warm climates are gradually taking over areas that once supported more cold-loving trees, such as balsam and fir. Ecologists report that in 2004, 82% of the trees at an elevation of 2600 ft were hardwoods, as compared to only 57% in 1964.

Source: The Boston Globe

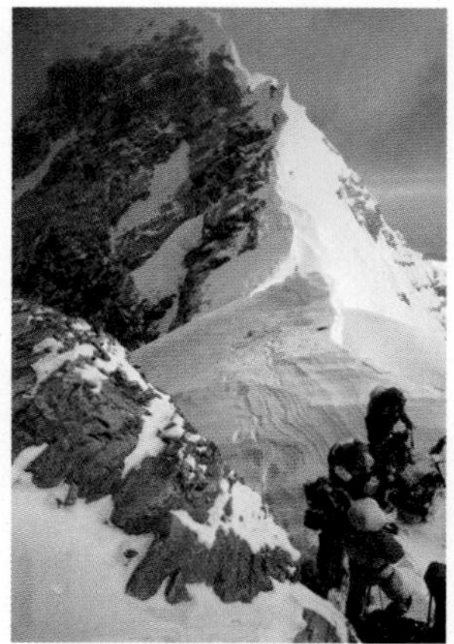

Mt. Everest

89. Refer to Exercise 86. Describe how you could use the linear function found in part (a) to find the price at which the manufacturer should sell the calculators in order to sell 15,000 calculators a week.

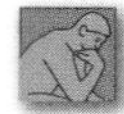

90. Refer to Exercise 88. Describe how you could use the linear function found in part (a) to find the monthly sales the executive would need to make in order to earn a commission of \$6000 a month.

91. Let f be a linear function. If $f(2) = 5$ and $f(0) = 3$, find $f(x)$.

92. Let f be a linear function. If $f(-3) = 4$ and $f(1) = -8$, find $f(x)$.

93. Given that f is a linear function for which $f(1) = 3$ and $f(-1) = 5$, determine $f(4)$.

94. Given that f is a linear function for which $f(-3) = 2$ and $f(2) = 7$, determine $f(0)$.

95. A line with slope $\frac{4}{3}$ passes through the point $(3, 2)$.

a. What is y when $x = -6$?
b. What is x when $y = 6$?

96. A line with slope $-\frac{3}{4}$ passes through the point $(8, -2)$.

a. What is y when $x = -4$?
b. What is x when $y = 1$?

Applying the Concepts

97. Explain why the point-slope formula cannot be used to find the equation of a line that is parallel to the y-axis.

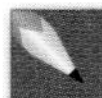

98. Refer to Example 5 in this section for each of the following.

a. Explain the meaning of the slope of the graph of the linear function given in the example.
b. Explain the meaning of the y-intercept.
c. Explain the meaning of the x-intercept.

99. A line contains the points $(-3, 6)$ and $(6, 0)$. Find the coordinates of three other points that are on this line.

100. A line contains the points $(4, -1)$ and $(2, 1)$. Find the coordinates of three other points that are on this line.

101. Find the equation of the line that passes through the midpoint of the line segment between $P_1(2, 5)$ and $P_2(-4, 1)$ and has slope -2.

SECTION

4.6 Parallel and Perpendicular Lines

OBJECTIVE A To find parallel and perpendicular lines 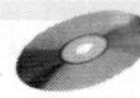

Two lines that have the same slope do not intersect and are called **parallel lines.**

The slope of each of the lines at the right is $\frac{2}{3}$.

The lines are parallel.

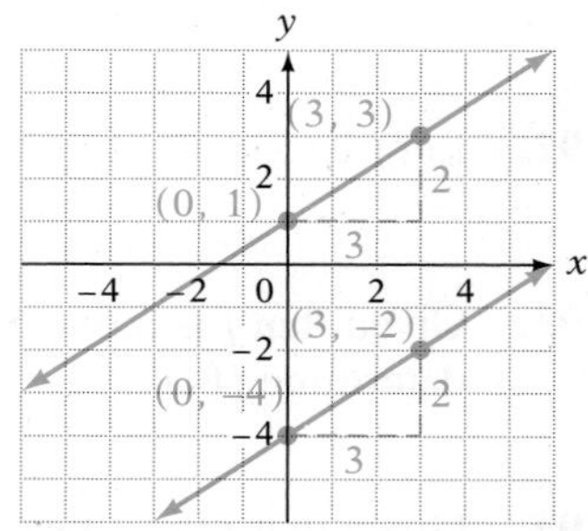

Slopes of Parallel Lines

Two nonvertical lines with slopes of m_1 and m_2 are parallel if and only if $m_1 = m_2$. Any two vertical lines are parallel.

HOW TO • 1 Is the line containing the points whose coordinates are $(-2, 1)$ and $(-5, -1)$ parallel to the line that contains the points whose coordinates are $(1, 0)$ and $(4, 2)$?

$$m_1 = \frac{-1 - 1}{-5 - (-2)} = \frac{-2}{-3} = \frac{2}{3}$$
• Find the slope of the line through $(-2, 1)$ and $(-5, -1)$.

$$m_2 = \frac{2 - 0}{4 - 1} = \frac{2}{3}$$
• Find the slope of the line through $(1, 0)$ and $(4, 2)$.

Because $m_1 = m_2$, the lines are parallel.

HOW TO • 2 Find the equation of the line that contains the point whose coordinates are $(2, 3)$ and is parallel to the graph of $y = \frac{1}{2}x - 4$.

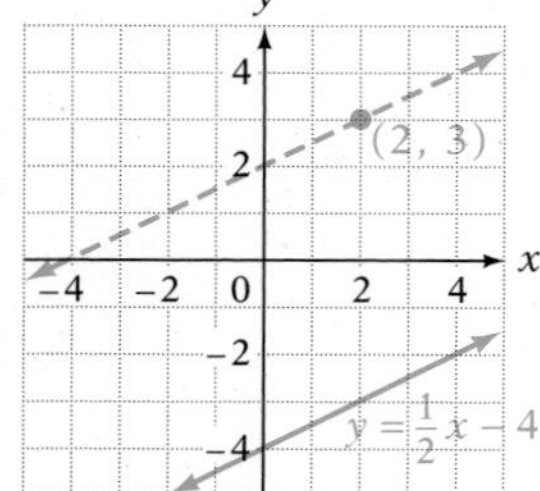

The slope of the given line is $\frac{1}{2}$. Because parallel lines have the same slope, the slope of the unknown line is also $\frac{1}{2}$.

$$y - y_1 = m(x - x_1)$$
• Use the point-slope formula.

$$y - 3 = \frac{1}{2}(x - 2)$$
• $m = \frac{1}{2}$, $(x_1, y_1) = (2, 3)$

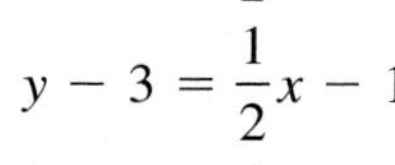

$$y - 3 = \frac{1}{2}x - 1$$
• Simplify.

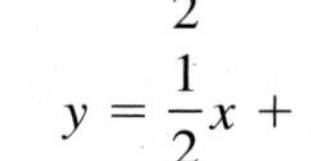

$$y = \frac{1}{2}x + 2$$
• Write the equation in the form $y = mx + b$.

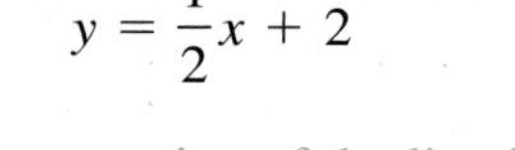

The equation of the line is $y = \frac{1}{2}x + 2$.

HOW TO • 3 Find the equation of the line that contains the point whose coordinates are $(-1, 4)$ and is parallel to the graph of $2x - 3y = 5$.

Because the lines are parallel, the slope of the unknown line is the same as the slope of the given line. Solve $2x - 3y = 5$ for y and determine its slope.

$$2x - 3y = 5$$
$$-3y = -2x + 5$$
$$y = \frac{2}{3}x - \frac{5}{3}$$

The slope of the given line is $\frac{2}{3}$. Because the lines are parallel, this is the slope of the unknown line. Use the point-slope formula to determine the equation.

$$y - y_1 = m(x - x_1)$$ • Use the point-slope formula.

$$y - 4 = \frac{2}{3}[x - (-1)]$$ • $m = \frac{2}{3}, (x_1, y_1) = (-1, 4)$

$$y - 4 = \frac{2}{3}x + \frac{2}{3}$$ • Simplify.

$$y = \frac{2}{3}x + \frac{14}{3}$$ • Write the equation in the form $y = mx + b$.

The equation of the line is $y = \frac{2}{3}x + \frac{14}{3}$.

Two lines that intersect at right angles are **perpendicular lines.**

Any horizontal line is perpendicular to any vertical line. For example, $x = 3$ is perpendicular to $y = -2$.

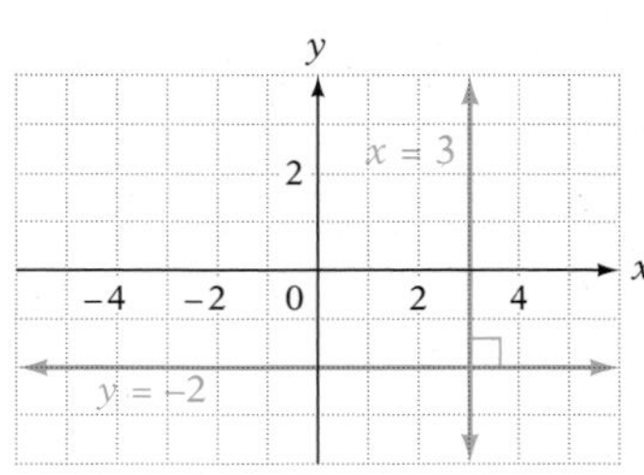

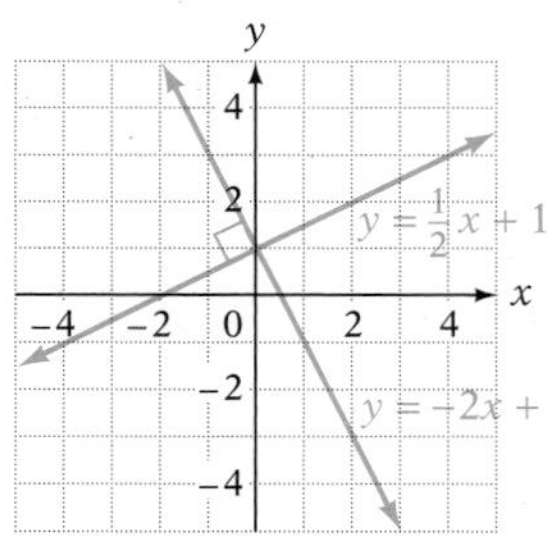

Slopes of Perpendicular Lines

If m_1 and m_2 are the slopes of two lines, neither of which is vertical, then the lines are perpendicular if and only if $m_1 \cdot m_2 = -1$.

A vertical line is perpendicular to a horizontal line.

Solving $m_1 \cdot m_2 = -1$ for m_1 gives $m_1 = -\frac{1}{m_2}$. This last equation states that the slopes of perpendicular lines are **negative reciprocals** of each other.

HOW TO • 4 Is the line that contains the points whose coordinates are $(4, 2)$ and $(-2, 5)$ perpendicular to the line that contains the points whose coordinates are $(-4, 3)$ and $(-3, 5)$?

$$m_1 = \frac{5 - 2}{-2 - 4} = \frac{3}{-6} = -\frac{1}{2}$$ • Find the slope of the line through $(4, 2)$ and $(-2, 5)$.

$$m_2 = \frac{5 - 3}{-3 - (-4)} = \frac{2}{1} = 2$$ • Find the slope of the line through $(-4, 3)$ and $(-3, 5)$.

$$m_1 \cdot m_2 = -\frac{1}{2}(2) = -1$$ • Find the product of the two slopes.

Because $m_1 \cdot m_2 = -1$, the lines are perpendicular.

HOW TO • 5 Are the graphs of the lines whose equations are $3x + 4y = 8$ and $8x + 6y = 5$ perpendicular?

To determine whether the lines are perpendicular, solve each equation for y and find the slope of each line. Then use the equation $m_1 \cdot m_2 = -1$.

$$\begin{aligned} 3x + 4y &= 8 \\ 4y &= -3x + 8 \\ y &= -\frac{3}{4}x + 2 \end{aligned} \qquad \begin{aligned} 8x + 6y &= 5 \\ 6y &= -8x + 5 \\ y &= -\frac{4}{3}x + \frac{5}{6} \end{aligned}$$

$$m_1 = -\frac{3}{4} \qquad m_2 = -\frac{4}{3}$$

$$m_1 \cdot m_2 = \left(-\frac{3}{4}\right)\left(-\frac{4}{3}\right) = 1$$

Because $m_1 \cdot m_2 = 1 \neq -1$, the lines are not perpendicular.

HOW TO • 6 Find the equation of the line that contains the point whose coordinates are $(-2, 1)$ and is perpendicular to the graph of $y = -\frac{2}{3}x + 2$.

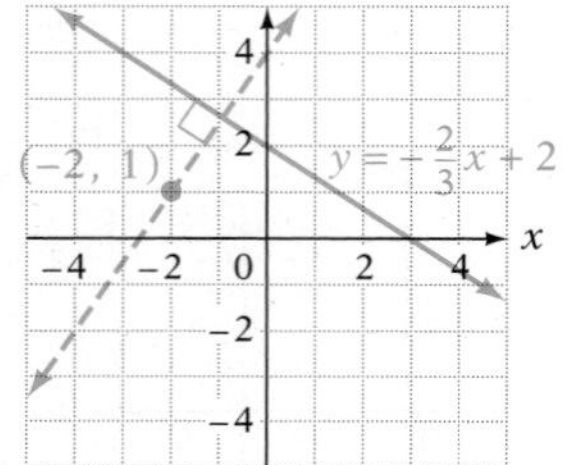

The slope of the given line is $-\frac{2}{3}$. The slope of the line perpendicular to the given line is the negative reciprocal of $-\frac{2}{3}$, which is $\frac{3}{2}$. Substitute this slope and the coordinates of the given point, $(-2, 1)$, into the point-slope formula.

$y - y_1 = m(x - x_1)$ • The point-slope formula

$y - 1 = \frac{3}{2}[x - (-2)]$ • $m = \frac{3}{2}, (x_1, y_1) = (-2, 1)$

$y - 1 = \frac{3}{2}x + 3$ • Simplify.

$y = \frac{3}{2}x + 4$ • Write the equation in the form $y = mx + b$.

The equation of the perpendicular line is $y = \frac{3}{2}x + 4$.

HOW TO • 7 Find the equation of the line that contains the point whose coordinates are $(3, -4)$ and is perpendicular to the graph of $2x - y = -3$.

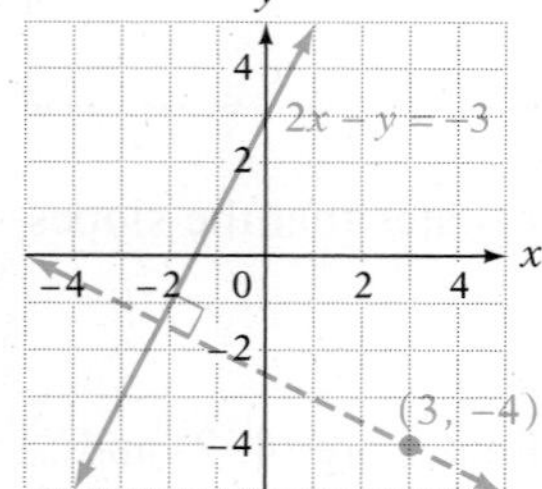

$2x - y = -3$ • Determine the slope of the given line by solving the equation for y.

$-y = -2x - 3$

$y = 2x + 3$ • The slope is 2.

The slope of the line perpendicular to the given line is $-\frac{1}{2}$, the negative reciprocal of 2. Now use the point-slope formula to find the equation of the line.

$y - y_1 = m(x - x_1)$ • The point-slope formula

$y - (-4) = -\frac{1}{2}(x - 3)$ • $m = -\frac{1}{2}, (x_1, y_1) = (3, -4)$

$y + 4 = -\frac{1}{2}x + \frac{3}{2}$ • Simplify.

$y = -\frac{1}{2}x - \frac{5}{2}$ • Write the equation in the form $y = mx + b$.

The equation of the perpendicular line is $y = -\frac{1}{2}x - \frac{5}{2}$.

EXAMPLE • 1

Is the line that contains the points whose coordinates are (−4, 2) and (1, 6) parallel to the line that contains the points whose coordinates are (2, −4) and (7, 0)?

Solution

$$m_1 = \frac{6-2}{1-(-4)} = \frac{4}{5} \qquad \bullet\ (x_1, y_1) = (-4, 2),\ (x_2, y_2) = (1, 6)$$

$$m_2 = \frac{0-(-4)}{7-2} = \frac{4}{5} \qquad \bullet\ (x_1, y_1) = (2, -4),\ (x_2, y_2) = (7, 0)$$

$$m_1 = m_2 = \frac{4}{5}$$

The lines are parallel.

YOU TRY IT • 1

Is the line that contains the points whose coordinates are (−2, −3) and (7, 1) perpendicular to the line that contains the points whose coordinates are (4, 1) and (6, −5)?

Your solution

EXAMPLE • 2

Are the lines $4x - y = -2$ and $x + 4y = -12$ perpendicular?

Solution

$$4x - y = -2$$
$$-y = -4x - 2$$
$$y = 4x + 2$$
$$m_1 = 4$$

$$x + 4y = -12$$
$$4y = -x - 12$$
$$y = -\frac{1}{4}x - 3$$
$$m_2 = -\frac{1}{4}$$

$$m_1 \cdot m_2 = 4\left(-\frac{1}{4}\right) = -1$$

The lines are perpendicular.

YOU TRY IT • 2

Are the lines $5x + 2y = 2$ and $5x + 2y = -6$ parallel?

Your solution

EXAMPLE • 3

Find the equation of the line that contains the point whose coordinates are (3, −1) and is parallel to the graph of $3x - 2y = 4$.

Solution

$$3x - 2y = 4$$
$$-2y = -3x + 4$$
$$y = \frac{3}{2}x - 2 \qquad \bullet\ m = \frac{3}{2}$$

$$y - y_1 = m(x - x_1)$$
$$y - (-1) = \frac{3}{2}(x - 3) \qquad \bullet\ (x_1, y_1) = (3, -1)$$
$$y + 1 = \frac{3}{2}x - \frac{9}{2}$$
$$y = \frac{3}{2}x - \frac{11}{2}$$

The equation of the line is $y = \frac{3}{2}x - \frac{11}{2}$.

YOU TRY IT • 3

Find the equation of the line that contains the point whose coordinates are (−2, 2) and is perpendicular to the graph of $x - 4y = 3$.

Your solution

Solutions on p. S14

4.6 EXERCISES

OBJECTIVE A To find parallel and perpendicular lines

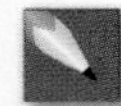

1. Explain how to determine whether the graphs of two lines are parallel.

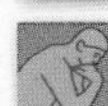

2. Explain how to determine whether the graphs of two lines are perpendicular.

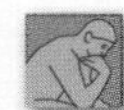

3. Is it possible for two lines to be perpendicular and for the slope of each to be a positive number? Explain.

4. Is it possible for two lines to be parallel and for the slope of one line to be positive and the slope of the other line to be negative? Explain.

5. The slope of a line is -5. What is the slope of any line parallel to this line?

6. The slope of a line is $\frac{3}{2}$. What is the slope of any line parallel to this line?

7. The slope of a line is 4. What is the slope of any line perpendicular to this line?

8. The slope of a line is $-\frac{4}{5}$. What is the slope of any line perpendicular to this line?

9. Is the graph of $x = -2$ perpendicular to the graph of $y = 3$?

10. Is the graph of $y = \frac{1}{2}$ perpendicular to the graph of $y = -2$?

11. Is the graph of $x = -3$ parallel to the graph of $y = \frac{1}{3}$?

12. Is the graph of $x = 4$ parallel to the graph of $x = -4$?

13. Is the graph of $y = \frac{2}{3}x - 4$ parallel to the graph of $y = -\frac{3}{2}x - 4$?

14. Is the graph of $y = -2x + \frac{2}{3}$ parallel to the graph of $y = -2x + 3$?

15. Is the graph of $y = \frac{4}{3}x - 2$ perpendicular to the graph of $y = -\frac{3}{4}x + 2$?

16. Is the graph of $y = \frac{1}{2}x + \frac{3}{2}$ perpendicular to the graph of $y = -\frac{1}{2}x + \frac{3}{2}$?

17. Are the graphs of $2x + 3y = 2$ and $2x + 3y = -4$ parallel?

18. Are the graphs of $2x - 4y = 3$ and $2x + 4y = -3$ parallel?

19. Are the graphs of $x - 4y = 2$ and $4x + y = 8$ perpendicular?

20. Are the graphs of $4x - 3y = 2$ and $4x + 3y = -7$ perpendicular?

21. Is the line that contains the points whose coordinates are (3, 2) and (1, 6) parallel to the line that contains the points whose coordinates are $(-1, 3)$ and $(-1, -1)$?

22. Is the line that contains the points whose coordinates are $(4, -3)$ and (2, 5) parallel to the line that contains the points whose coordinates are $(-2, -3)$ and $(-4, 1)$?

23. Is the line that contains the points whose coordinates are $(-3, 2)$ and $(4, -1)$ perpendicular to the line that contains the points whose coordinates are (1, 3) and $(-2, -4)$?

24. Is the line that contains the points whose coordinates are $(-1, 2)$ and (3, 4) perpendicular to the line that contains the points whose coordinates are $(-1, 3)$ and $(-4, 1)$?

25. Find the equation of the line that contains the point with coordinates $(3, -2)$ and is parallel to the graph of $y = 2x + 1$.

26. Find the equation of the line that contains the point with coordinates $(-1, 3)$ and is parallel to the graph of $y = -x + 3$.

27. Find the equation of the line that contains the point with coordinates $(-2, -1)$ and is perpendicular to the graph of $y = -\frac{2}{3}x - 2$.

28. Find the equation of the line that contains the point with coordinates $(-4, 1)$ and is perpendicular to the graph of $y = 2x - 5$.

29. Find the equation of the line containing the point whose coordinates are $(-2, -4)$ and parallel to the graph of $2x - 3y = 2$.

30. Find the equation of the line containing the point whose coordinates are $(3, 2)$ and parallel to the graph of $3x + y = -3$.

31. Find the equation of the line containing the point whose coordinates are $(4, 1)$ and perpendicular to the graph of $y = -3x + 4$.

32. Find the equation of the line containing the point whose coordinates are $(2, -5)$ and perpendicular to the graph of $y = \frac{5}{2}x - 4$.

33. Find the equation of the line containing the point whose coordinates are $(-1, -3)$ and perpendicular to the graph of $3x - 5y = 2$.

34. Find the equation of the line containing the point whose coordinates are $(-1, 3)$ and perpendicular to the graph of $2x + 4y = -1$.

Applying the Concepts

Physics For Exercises 35 and 36, suppose a ball is being twirled at the end of a string and the center of rotation is the origin of a coordinate system. If the string breaks, the initial path of the ball is on a line that is perpendicular to the radius of the circle.

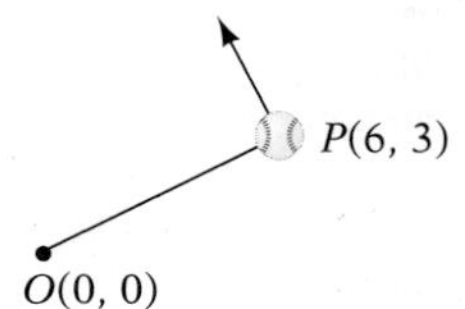

35. Suppose the string breaks when the ball is at the point $P(6, 3)$. Find the equation of the line on which the initial path lies.

36. Suppose the string breaks when the ball is at the point $P(2, 8)$. Find the equation of the line on which the initial path lies.

37. If the graphs of $A_1x + B_1y = C_1$ and $A_2x + B_2y = C_2$ are parallel, express $\frac{A_1}{B_1}$ in terms of A_2 and B_2.

38. If the graphs of $A_1x + B_1y = C_1$ and $A_2x + B_2y = C_2$ are perpendicular, express $\frac{A_1}{B_1}$ in terms of A_2 and B_2.

SECTION

4.7 Inequalities in Two Variables

OBJECTIVE A To graph the solution set of an inequality in two variables

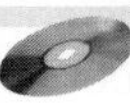

The graph of the linear equation $y = x - 1$ separates the plane into three sets: the set of points on the line, the set of points above the line, and the set of points below the line.

The point whose coordinates are (2, 1) is a solution of $y = x - 1$ and is a point on the line.

The point whose coordinates are (2, 4) is a solution of $y > x - 1$ and is a point above the line.

The point whose coordinates are (2, −2) is a solution of $y < x - 1$ and is a point below the line.

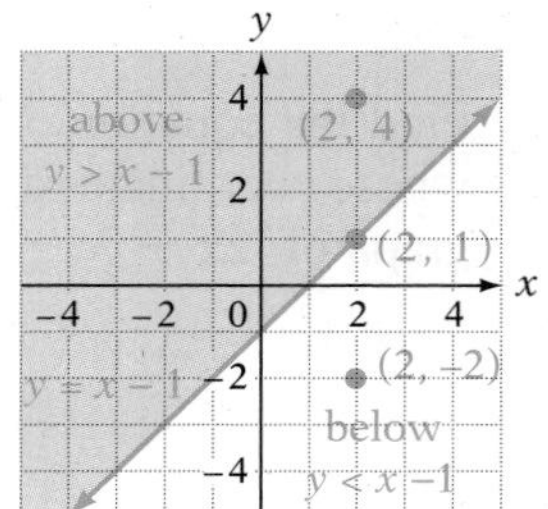

The set of points on the line is the solution of the equation $y = x - 1$. The set of points above the line is the solution of the inequality $y > x - 1$. These points form a **half-plane.** The set of points below the line is the solution of the inequality $y < x - 1$. These points also form a half-plane.

An inequality of the form $y > mx + b$ or $Ax + By > C$ is a **linear inequality in two variables.** (The inequality symbol $>$ could be replaced by $\geq$, $<$, or +.) The solution set of a linear inequality in two variables is a half-plane.

The following illustrates the procedure for graphing the solution set of a linear inequality in two variables.

Take Note

When solving the inequality at the right for y, both sides of the inequality are divided by −4, so the inequality symbol must be reversed.

HOW TO 1 Graph the solution set of $3x - 4y < 12$.

$$3x - 4y < 12$$
$$-4y < -3x + 12$$
$$y > \frac{3}{4}x - 3$$

• Solve the inequality for y.

Change the inequality $y > \frac{3}{4}x - 3$ to the equality $y = \frac{3}{4}x - 3$, and graph the line.

If the inequality contains + or $\geq$, the line belongs to the solution set and is shown by a *solid line.* If the inequality contains $<$ or $>$, the line is not part of the solution set and is shown by a *dashed line.*

If the inequality contains $>$ or $\geq$, shade the upper half-plane. If the inequality contains $<$ or +, shade the lower half-plane.

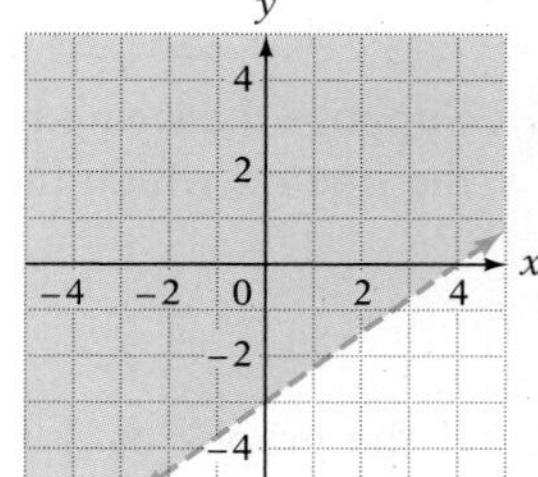

Take Note

As shown below, (0, 0) is a solution of the inequality in the example at the right.

$$y > \frac{3}{4}x - 3$$
$$0 > \frac{3}{4}(0) - 3$$
$$0 > 0 - 3$$
$$0 > -3$$

Because (0, 0) is a solution of the inequality, (0, 0) should be in the shaded region. The solution set as graphed is correct.

As a check, use the ordered pair (0, 0) to determine whether the correct region of the plane has been shaded. If (0, 0) is a solution of the inequality, then (0, 0) should be in the shaded region. If (0, 0) is not a solution of the inequality, then (0, 0) should not be in the shaded region.

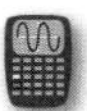

Integrating Technology

See the Keystroke Guide: *Graphing Inequalities* for instructions on using a graphing calculator to graph the solution set of an inequality in two variables.

If the line passes through the point (0, 0), then another point, such as (0, 1), must be used as a check.

From the graph of $y > \frac{3}{4}x - 3$, note that for a given value of x, more than one value of y can be paired with that value of x. For instance, (4, 1), (4, 3), (5, 1), and $\left(5, \frac{9}{4}\right)$ are all ordered pairs that belong to the graph. Because there are ordered pairs with the same first coordinate and different second coordinates, the inequality does not represent a function. The inequality is a relation but not a function.

EXAMPLE • 1

Graph the solution set of $x + 2y \leq 4$.

Solution

$$x + 2y \leq 4$$
$$2y \leq -x + 4$$
$$y \leq -\frac{1}{2}x + 2$$

Graph $y = -\frac{1}{2}x + 2$ as a solid line.

Shade the lower half-plane.

Check:

$y \leq -\frac{1}{2}x + 2$

$0 \leq -\frac{1}{2}(0) + 2$

$0 \leq 0 + 2$

$0 \leq 2$

The point (0, 0) should be in the shaded region.

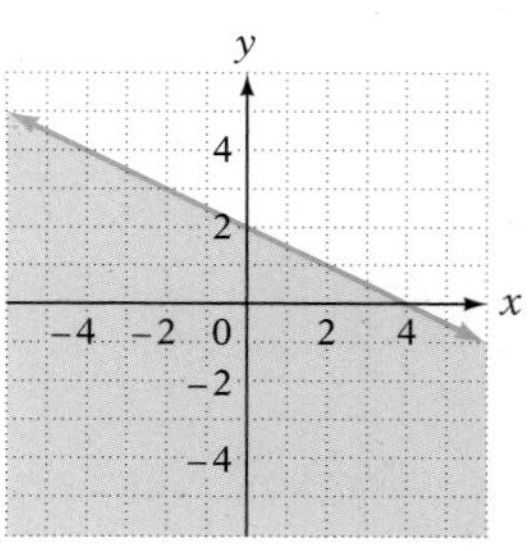

YOU TRY IT • 1

Graph the solution set of $x + 3y > 6$.

Your solution

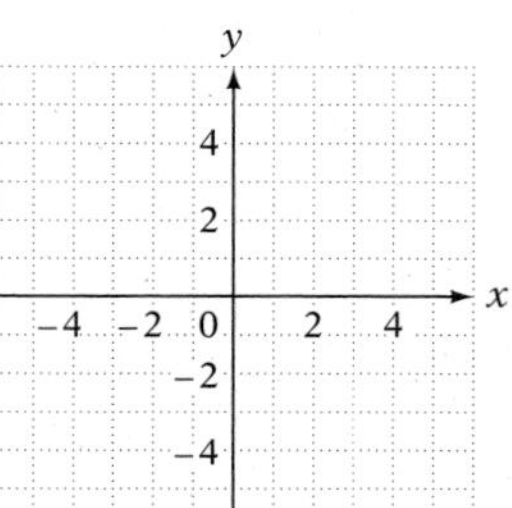

EXAMPLE • 2

Graph the solution set of $x \geq -1$.

Solution Graph $x = -1$ as a solid line.

Shade the half-plane to the right of the line.

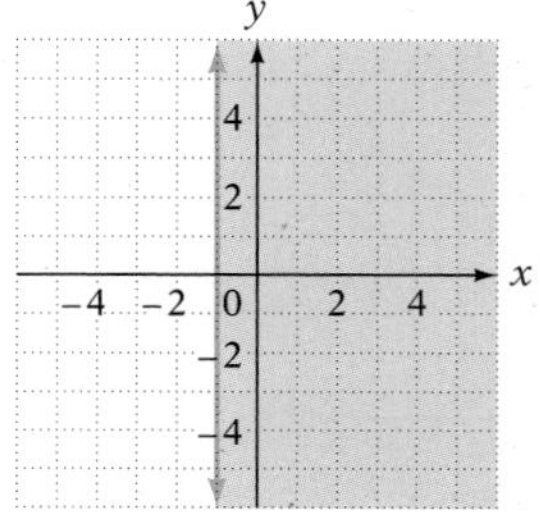

YOU TRY IT • 2

Graph the solution set of $y < 2$.

Your solution

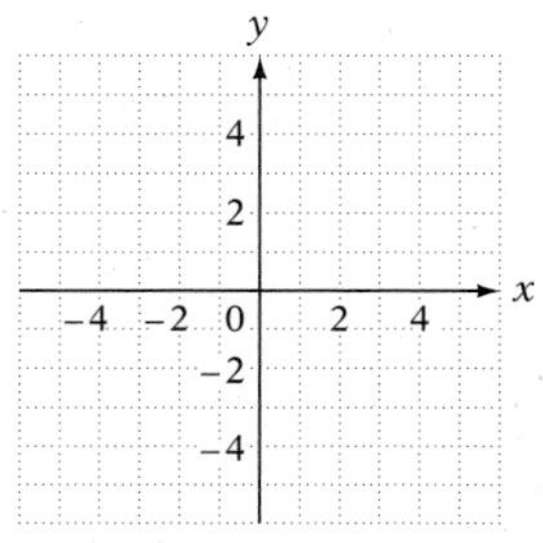

Solutions on p. S14

4.7 EXERCISES

OBJECTIVE A **To graph the solution set of an inequality in two variables**

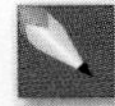

1. What is a half-plane?

2. Explain a method you can use to check that the graph of a linear inequality in two variables has been shaded correctly.

3. Is (0, 0) a solution of $y > 2x - 7$?

4. Is (0, 0) a solution of $y < 5x + 3$?

5. Is (0, 0) a solution of $y \leq -\frac{2}{3}x - 8$?

6. Is (0, 0) a solution of $y \geq -\frac{3}{4}x + 9$?

For Exercises 7 to 24, graph the solution set.

7. $y \leq \frac{3}{2}x - 3$

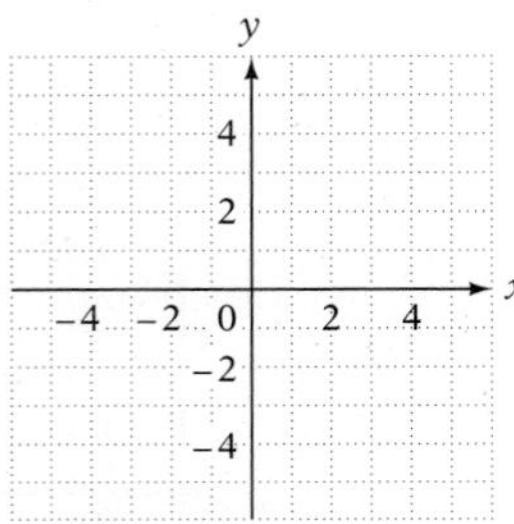

8. $y \geq \frac{4}{3}x - 4$

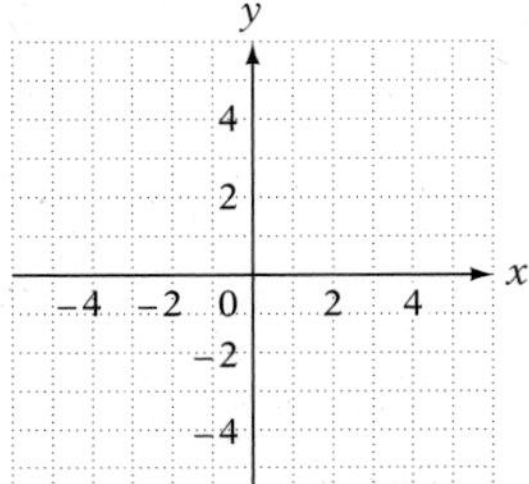

9. $y < -\frac{1}{3}x + 1$

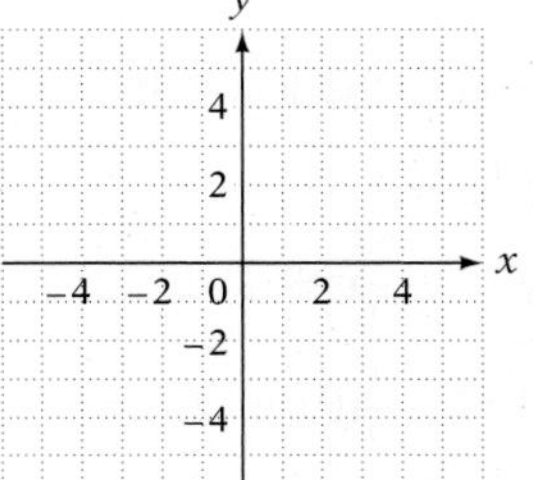

10. $y < \frac{3}{5}x - 3$

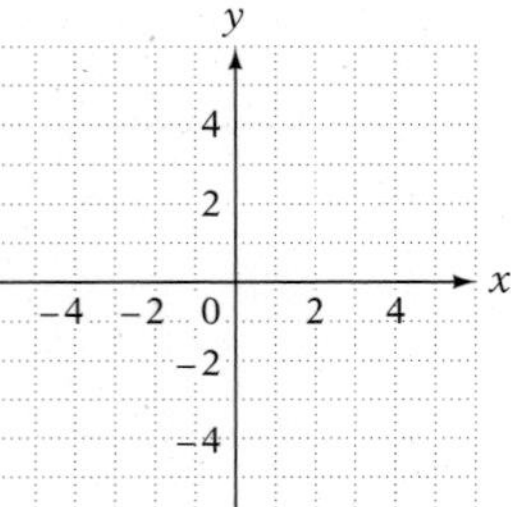

11. $4x - 5y > 10$

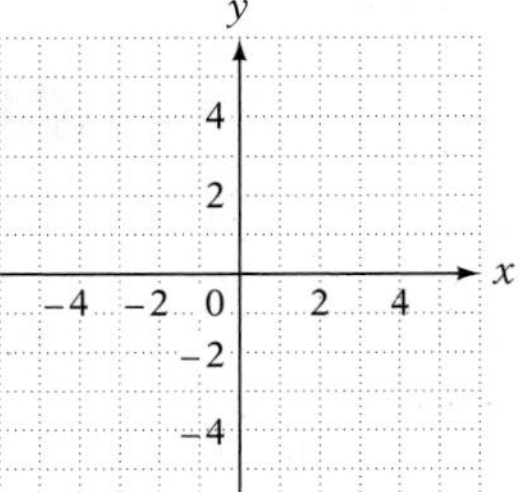

12. $4x + 3y < 9$

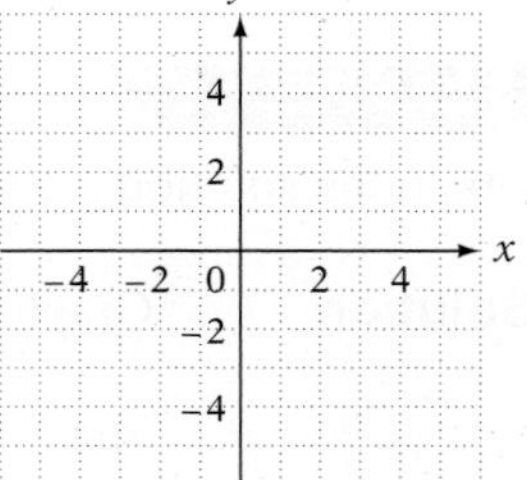

13. $x + 3y < 6$

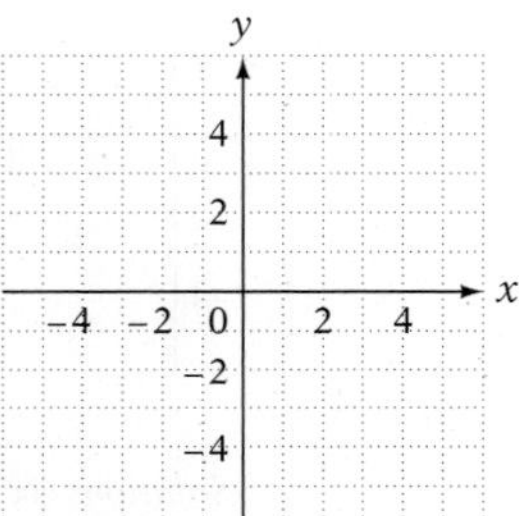

14. $2x - 5y \leq 10$

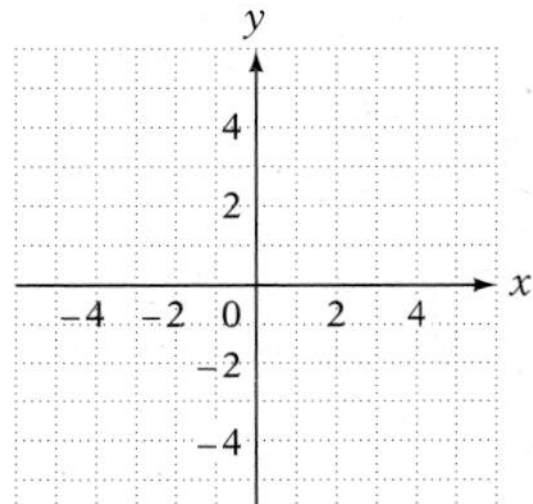

15. $2x + 3y \geq 6$

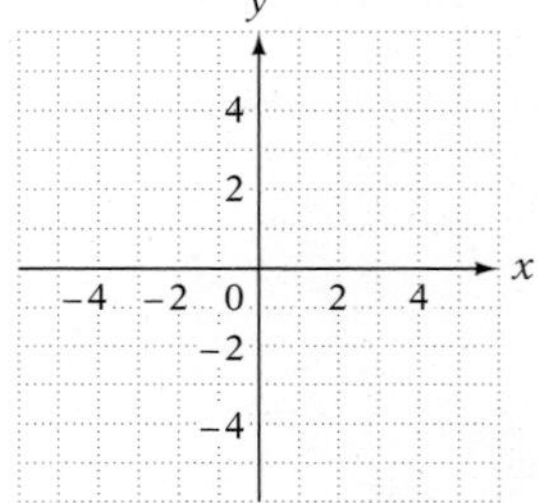

16. $3x + 2y < 4$

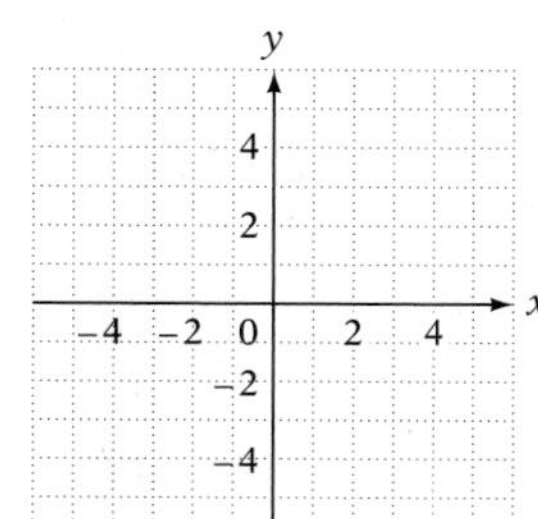

17. $-x + 2y > -8$

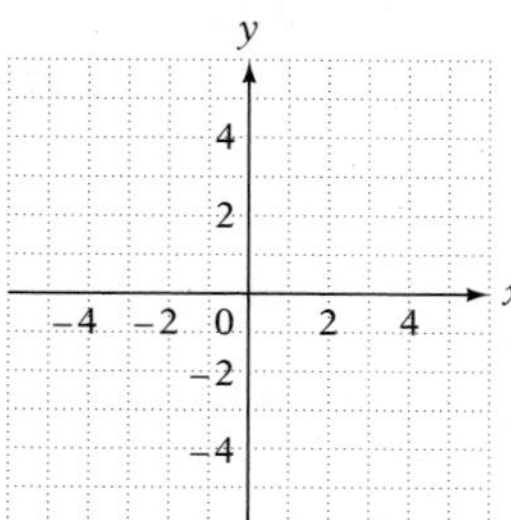

18. $-3x + 2y > 2$

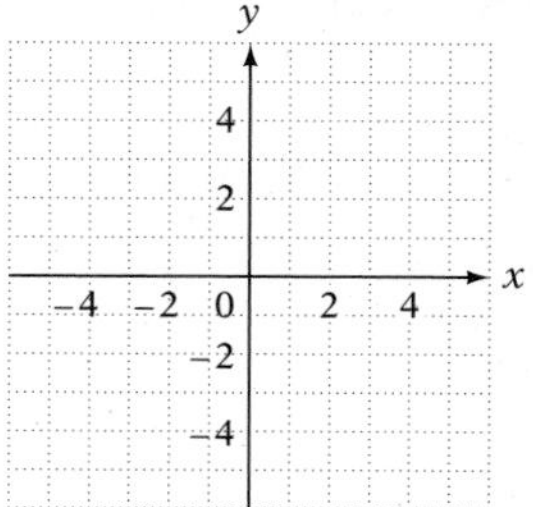

19. $y - 4 < 0$

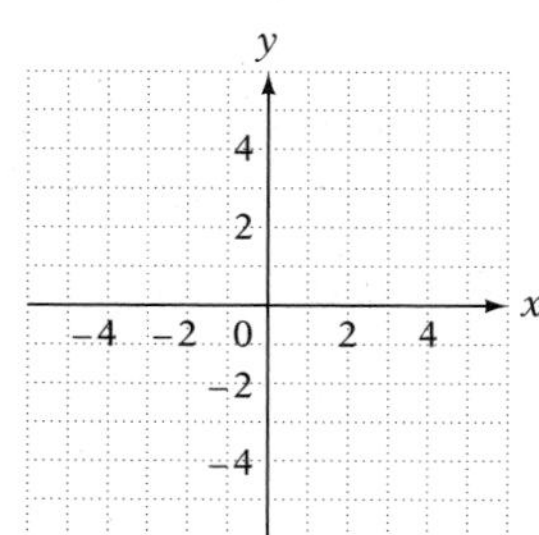

20. $x + 2 \geq 0$

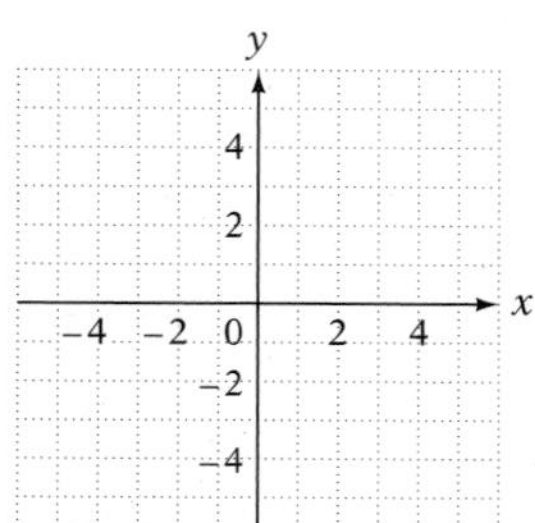

21. $6x + 5y < 15$

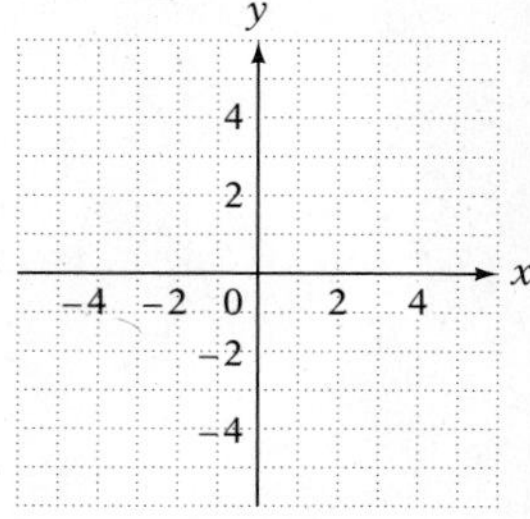

22. $3x - 5y < 10$

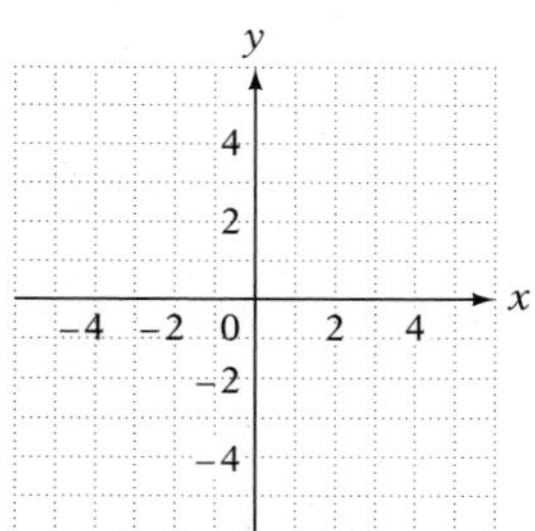

23. $-5x + 3y \geq -12$

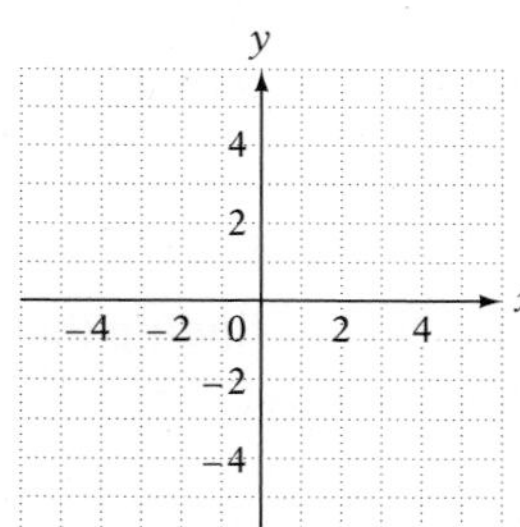

24. $3x + 4y \geq 12$

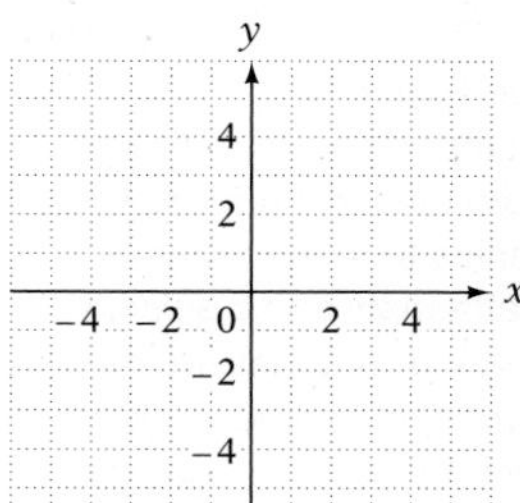

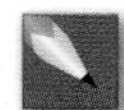

25. What quadrant is represented by the two linear inequalities $x > 0$ and $y > 0$?

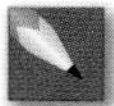

26. What quadrant is represented by the two linear inequalities $x < 0$ and $y < 0$?

Applying the Concepts

27. Does the inequality $y < 3x - 1$ represent a function? Explain your answer.

28. Are there any points whose coordinates satisfy both $y \leq x + 3$ and $y \geq -\frac{1}{2}x + 1$? If so, give the coordinates of three such points. If not, explain why not.

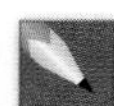

29. Are there any points whose coordinates satisfy both $y \leq x - 1$ and $y \geq x + 2$? If so, give the coordinates of three such points. If not, explain why not.

FOCUS ON PROBLEM SOLVING

Find a Pattern

Polya's four recommended problem-solving steps are stated below.

1. Understand the problem.
2. Devise a plan.
3. Carry out the plan.
4. Review the solution.

One of the several ways of devising a plan is first to try to find a pattern. Karl Friedrich Gauss supposedly used this method to solve a problem that was given to his math class when he was in elementary school. As the story goes, his teacher wanted to grade some papers while the class worked on a math problem. The problem given to the class was to find the sum

$$1 + 2 + 3 + 4 + \cdots + 100$$

Gauss quickly solved the problem by seeing a pattern. Here is what he saw.

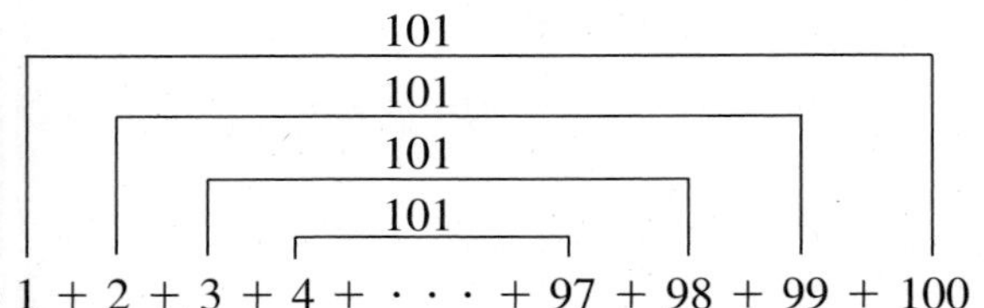

Note that

$1 + 100 = 101$
$2 + 99 = 101$
$3 + 98 = 101$
$4 + 97 = 101$

Gauss noted that there were 50 sums of 101. Therefore, the sum of the first 100 natural numbers is

$$1 + 2 + 3 + 4 + \cdots + 97 + 98 + 99 + 100 = 50(101) = 5050$$

Try to solve Exercises 1 to 6 by finding a pattern.

1. Find the sum $2 + 4 + 6 + \cdots + 96 + 98 + 100$.
2. Find the sum $1 + 3 + 5 + \cdots + 97 + 99 + 101$.
3. Find another method of finding the sum $1 + 3 + 5 + \cdots + 97 + 99 + 101$ given in the preceding exercise.
4. Find the sum $\frac{1}{1 \cdot 2} + \frac{1}{2 \cdot 3} + \frac{1}{3 \cdot 4} + \cdots + \frac{1}{49 \cdot 50}$.

   Hint: $\frac{1}{1 \cdot 2} = \frac{1}{2}, \frac{1}{1 \cdot 2} + \frac{1}{2 \cdot 3} = \frac{2}{3}, \frac{1}{1 \cdot 2} + \frac{1}{2 \cdot 3} + \frac{1}{3 \cdot 4} = \frac{3}{4}$
5. A *polynomial number* is a number that can be represented by arranging that number of dots in rows to form a geometric figure such as a triangle, square, pentagon, or hexagon. For instance, the first four *triangular numbers,* 3, 6, 10, and 15, are shown below. What are the next two triangular numbers?

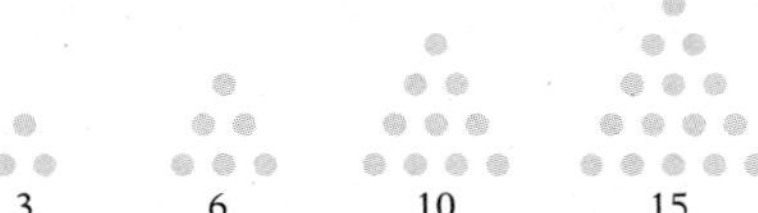

6. The following problem shows that checking a few cases does not always result in a conjecture that is true for *all* cases. Select any two points on a circle (see the drawing in the left margin) and draw a *chord,* a line connecting the points. The chord divides the circle into two regions. Now select three different points and draw chords connecting each of the three points with every other point. The chords divide the circle into four regions. Now select four points and connect each of the points with every other point. Make a conjecture as to the relationship between the number of regions and the number of points on the circle. Does your conjecture work for five points? Six points?

Point of Interest

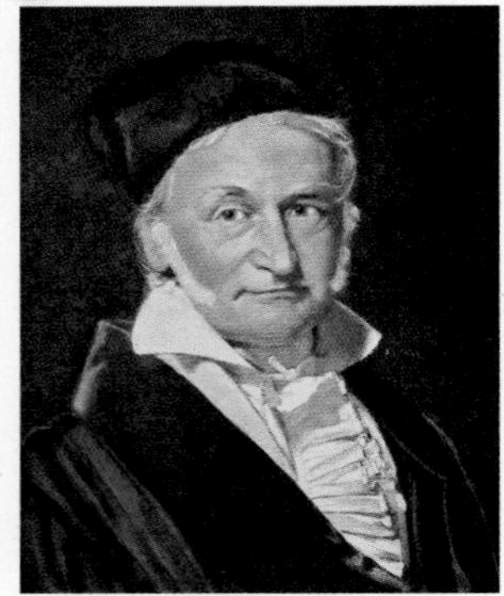

The Granger Collection, New York

Karl Friedrich Gauss

Karl Friedrich Gauss (1777–1855) has been called the "Prince of Mathematicians" by some historians. He applied his genius to many areas of mathematics and science. A unit of magnetism, the *gauss*, is named in his honor. Some electronic equipment (televisions, for instance) contains a *degausser* that controls magnetic fields.

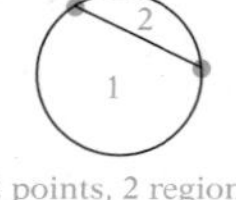

2 points, 2 regions

3 points, 4 regions

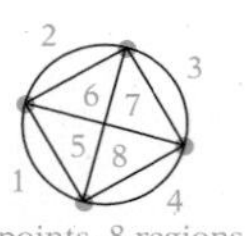

4 points, 8 regions

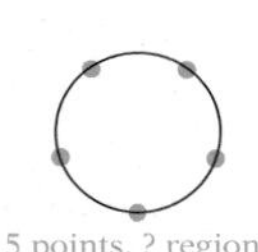

5 points, ? regions

PROJECTS AND GROUP ACTIVITIES

Evaluating a Function with a Graphing Calculator

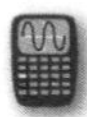

You can use a graphing calculator to evaluate some functions. Shown below are the keystrokes needed to evaluate $f(x) = 3x^2 + 2x - 1$ for $x = -2$ on a TI-84. (There are other methods of evaluating functions. This is just one of them.) Try these keystrokes. The calculator should display 7 as the value of the function.

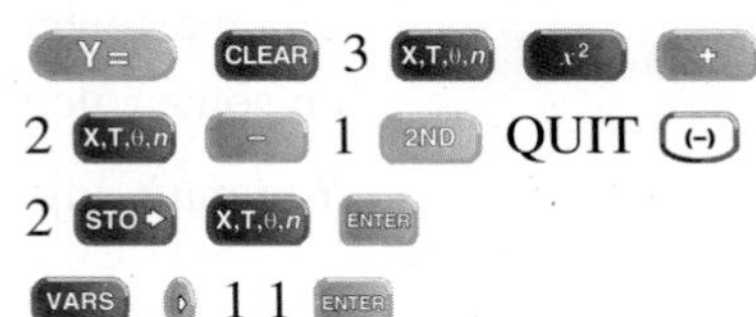

Introduction to Graphing Calculators

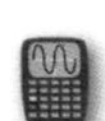

There are a variety of computer programs and calculators that can graph an equation. A computer or graphing calculator screen is divided into pixels. Depending on the computer or calculator, there are approximately 6000 to 790,000 pixels available on the screen. The greater the number of pixels, the smoother the graph will appear. A portion of a screen is shown at the left. Each little rectangle represents one pixel.

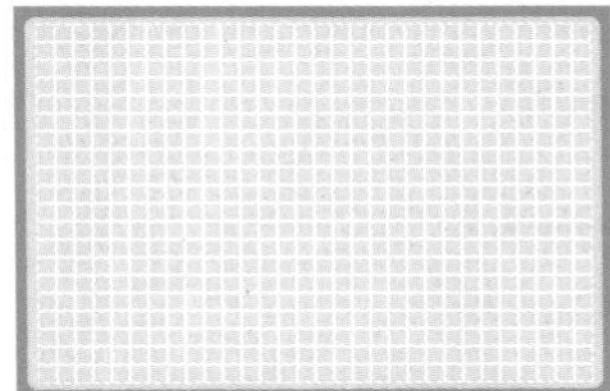

Calculator Screen

A graphing calculator draws a graph in a manner similar to the method we have used in this chapter. Values of x are chosen and ordered pairs calculated. Then a graph is drawn through those points by illuminating pixels (an abbreviation for "picture element") on the screen.

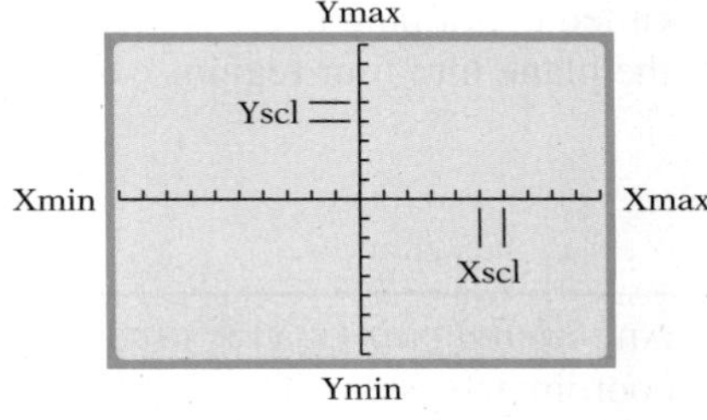

Graphing utilities can display only a portion of the xy-plane, called a window. The window [Xmin, Xmax] by [Ymin, Ymax] consists of those points (x, y) that satisfy both of the following inequalities:

$$\text{Xmin} \le x \le \text{Xmax} \quad \text{and} \quad \text{Ymin} \le y \le \text{Ymax}$$

The user sets these values before a graph is drawn.

The numbers Xscl and Yscl are the distances between the tick marks that are drawn on the x- and y-axes. If you do not want tick marks on the axes, set Xscl = 0 and Yscl = 0.

The graph at the right is a portion of the graph of $y = \frac{1}{2}x + 1$ as it was drawn with a graphing calculator. The window is

Xmin = −4.7, Xmax = 4.7
Ymin = −3.1, Ymax = 3.1
Xscl = 1 and Yscl = 1

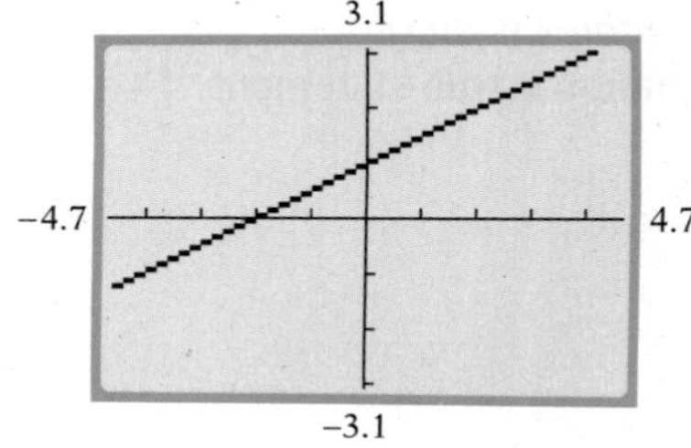

Graph the equations in Exercises 1 to 6 by using a graphing calculator.

1. $y = 2x + 1$ — For $2x$, you may enter $2 \times x$ or just $2x$. Use of the times sign $\times$ is not necessary on many graphing calculators.

2. $y = -x + 2$ — Many calculators use the (–) key to enter a negative sign.

3. $3x + 2y = 6$ — Solve for y. Then enter the equation.

4. $y = 50x$ — You must adjust the viewing window. Try the window $[-4.7, 4.7]$ by $[-250, 250]$, with Yscl = 50.

5. $y = \frac{2}{3}x - 3$ — You may enter $\frac{2}{3}x$ as $2x/3$ or $(2/3)x$. Although entering $2/3x$ works on some calculators, it is not recommended.

6. $4x + 3y = 75$ — You must adjust the viewing window.

CHAPTER 4
SUMMARY

KEY WORDS	EXAMPLES
A *rectangular coordinate system* is formed by two number lines, one horizontal and one vertical, that intersect at the zero point of each line. The point of intersection is called the *origin.* The number lines that make up a rectangular coordinate system are called *coordinate axes.* A rectangular coordinate system divides the plane into four regions called *quadrants.* [4.1A, p. 222]	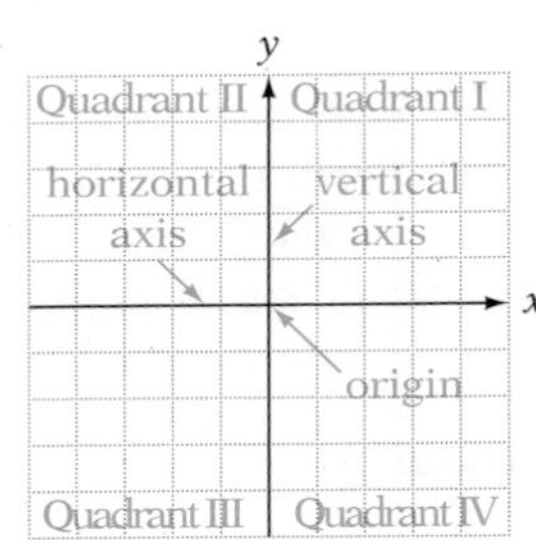
An *ordered pair* (x, y) is used to locate a point in a rectangular coordinate system. The first number of the pair measures a horizontal distance and is called the *abscissa* or *x-coordinate.* The second number of the pair measures a vertical distance and is called the *ordinate* or *y-coordinate.* The *coordinates* of the point are the numbers in the ordered pair associated with the point. To *graph,* or *plot,* a point in the plane, place a dot at the location given by the ordered pair. The *graph of an ordered pair* is the dot drawn at the coordinates of the point in the plane. [4.1A, 4.1B, pp. 222–224]	(3, 4) is an ordered pair. 3 is the abscissa. 4 is the ordinate. The graph of (3, 4) is shown below. 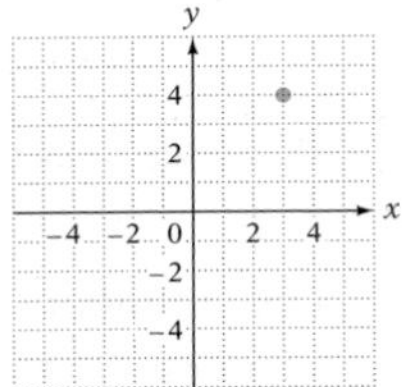
An equation of the form $y = mx + b$, where m and b are constants, is an equation in two variables. A *solution of an equation in two variables* is an ordered pair (x, y) whose coordinates make the equation a true statement. [4.1B, p. 224]	$y = 3x + 2$ is an equation in two variables. Ordered-pair solutions of $y = 3x + 2$ are shown below, along with the graph of the equation.

x	y
1	5
0	2
−1	−1

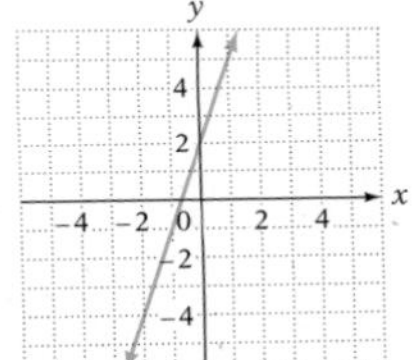

A *scatter diagram* is a graph of ordered-pair data. [4.1C, p. 227]	
A *relation* is any set of ordered pairs. [4.2A, p. 232]	$\{(2, 3), (2, 4), (3, 4), (5, 7)\}$
A *function* is a relation in which no two ordered pairs have the same first coordinate and different second coordinates. The *domain* of a function is the set of the first coordinates of all the ordered pairs of the function. The *range* is the set of the second coordinates of all the ordered pairs of the function. [4.2A, p. 233]	$\{(2, 3), (3, 5), (5, 7), (6, 9)\}$ The domain is $\{2, 3, 5, 6\}$. The range is $\{3, 5, 7, 9\}$.
Function notation is used for those equations that represent functions. For the equation at the right, x is the *independent variable* and y is the *dependent variable.* The symbol $f(x)$ is the *value of the function* and represents the value of the dependent variable for a given value of the independent variable. [4.2A, pp. 233–235]	In function notation, $y = 3x + 7$ is written as $f(x) = 3x + 7$.
The process of determining $f(x)$ for a given value of x is called *evaluating the function.* [4.2A, p. 235]	Evaluate $f(x) = 2x - 3$ when $x = 4$. $f(x) = 2x - 3$ $f(4) = 2(4) - 3$ $f(4) = 5$
The *graph of a function* is a graph of the ordered pairs (x, y) that belong to the function. A function that can be written in the form $f(x) = mx + b$ (or $y = mx + b$) is a *linear function* because its graph is a straight line. [4.3A, p. 244]	$f(x) = -\frac{2}{3}x + 3$ is an example of a linear function. In this case, $m = -\frac{2}{3}$ and $b = 3$.
The point at which a graph crosses the x-axis is called the *x-intercept,* and the point at which a graph crosses the y-axis is called the *y-intercept.* [4.3C, p. 248]	The x-intercept of $x + y = 4$ is $(4, 0)$. The y-intercept of $x + y = 4$ is $(0, 4)$.
The *slope* of a line is a measure of the slant, or tilt, of the line. The symbol for slope is m. A line that slants upward to the right has a *positive slope,* and a line that slants downward to the right has a *negative slope.* A horizontal line has *zero slope.* The slope of a vertical line is *undefined.* [4.4A, pp. 256–257]	The line $y = 2x - 3$ has a slope of 2 and slants upward to the right. The line $y = -5x + 2$ has a slope of -5 and slants downward to the right. The line $y = 4$ has a slope of 0.
An inequality of the form $y > mx + b$ or of the form $Ax + By > C$ is a *linear inequality in two variables.* (The symbol $>$ can be replaced by $\geq$, $<$, or $\leq$.) The solution set of an inequality in two variables is a *half-plane.* [4.7A, p. 282]	$4x - 3y < 12$ and $y \geq 2x + 6$ are linear inequalities in two variables.

ESSENTIAL RULES AND PROCEDURES	EXAMPLES
Graph of $y = b$ [4.3B, p. 247] The graph of $y = b$ is a horizontal line passing through the point $(0, b)$.	The graph of $y = -5$ is a horizontal line passing through the point $(0, -5)$.

Graph of a Constant Function [4.3B, p. 247]
A function given by $f(x) = b$, where b is a constant, is a *constant function.* The graph of the constant function is a horizontal line passing through $(0, b)$.

The graph of $f(x) = -5$ is a horizontal line passing through the point $(0, -5)$. Note that this is the same as the graph of $y = -5$.

Graph of $x = a$ [4.3B, p. 248]
The graph of $x = a$ is a vertical line passing through the point $(a, 0)$.

The graph of $x = 4$ is a vertical line passing through the point $(4, 0)$.

Finding Intercepts of Graphs of Linear Equations [4.3C, p. 249]
To find the x-intercept, let $y = 0$.
To find the y-intercept, let $x = 0$.
For any equation of the form $y = mx + b$, the y-intercept is $(0, b)$.

$3x + 4y = 12$

Let $y = 0$:
$$3x + 4(0) = 12$$
$$3x = 12$$
$$x = 4$$
The x-intercept is $(4, 0)$.

Let $x = 0$:
$$3(0) + 4y = 12$$
$$4y = 12$$
$$y = 3$$
The y-intercept is $(0, 3)$.

Slope Formula [4.4A, p. 256]
The slope of the line containing the two points $P_1(x_1, y_1)$ and $P_2(x_2, y_2)$ is given by $m = \dfrac{y_2 - y_1}{x_2 - x_1}$, $x_1 \neq x_2$.

$(x_1, y_1) = (-3, 2), (x_2, y_2) = (1, 4)$
$$m = \frac{y_2 - y_1}{x_2 - x_1} = \frac{4 - 2}{1 - (-3)} = \frac{1}{2}$$
The slope of the line through the points $(-3, 2)$ and $(1, 4)$ is $\frac{1}{2}$.

Slope-Intercept Form of a Straight Line [4.4B, p. 260]
The equation $y = mx + b$ is called the *slope-intercept form* of a straight line. The slope of the line is m, the coefficient of x. The y-intercept is $(0, b)$.

For the equation $y = -3x + 2$, the slope is -3 and the y-intercept is $(0, 2)$.

Point-Slope Formula [4.5A, p. 267]
Let m be the slope of a line, and let (x_1, y_1) be the coordinates of a point on the line. The equation of the line can be found from the point-slope formula: $y - y_1 = m(x - x_1)$.

The equation of the line that passes through the point $(4, 2)$ and has slope -3 is
$$y - y_1 = m(x - x_1)$$
$$y - 2 = -3(x - 4)$$
$$y - 2 = -3x + 12$$
$$y = -3x + 14$$

Slopes of Parallel Lines [4.6A, p. 276]
Two nonvertical lines with slopes of m_1 and m_2 are parallel if and only if $m_1 = m_2$. Any two vertical lines are parallel.

$y = 3x - 4, \quad m_1 = 3$
$y = 3x + 2, \quad m_2 = 3$
Because $m_1 = m_2$, the lines are parallel.

Slopes of Perpendicular Lines [4.6A, p. 277]
If m_1 and m_2 are the slopes of two lines, neither of which is vertical, then the lines are perpendicular if and only if $m_1 \cdot m_2 = -1$. This states that the slopes of perpendicular lines are *negative reciprocals* of each other. A vertical line is perpendicular to a horizontal line.

$y = \frac{1}{2}x - 1, \quad m_1 = \frac{1}{2}$
$y = -2x + 2, \quad m_2 = -2$
Because $m_1 \cdot m_2 = -1$, the lines are perpendicular.

CHAPTER 4

CONCEPT REVIEW

Test your knowledge of the concepts presented in this chapter. Answer each question. Then check your answers against the ones provided in the Answer Section.

1. How is the ordinate different from the abscissa?

2. How many ordered-pair solutions are there for a linear equation in two variables?

3. When is a relation a function?

4. What is the difference between an independent variable and a dependent variable?

5. In the general equation $y = mx + b$, what do m and b represent?

6. How many ordered-pair solutions of a linear function should be found to ensure the accuracy of a graph?

7. How is the equation of a vertical line different from the equation of a horizontal line?

8. How are the ordered pairs different for an x-intercept and a y-intercept?

9. What does it mean for a line to have an undefined slope?

10. Given two ordered pairs on a line, how do you find the slope of the line?

11. What is the difference between parallel and perpendicular lines?

12. What is the point-slope formula?

CHAPTER 4

REVIEW EXERCISES

1. Determine the ordered-pair solution of $y = \frac{x}{x-2}$ that corresponds to $x = 4$.

2. Given $P(x) = 3x + 4$, evaluate $P(-2)$ and $P(a)$.

3. Graph the ordered-pair solutions of $y = 2x^2 - 5$ when $x = -2, -1, 0, 1,$ and 2.

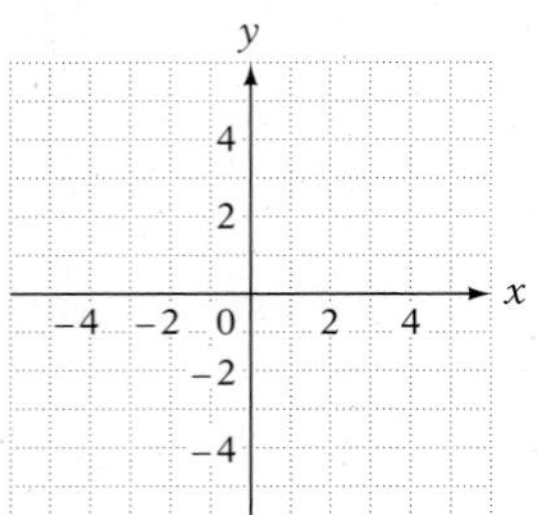

4. Draw a line through all the points with an abscissa of -3.

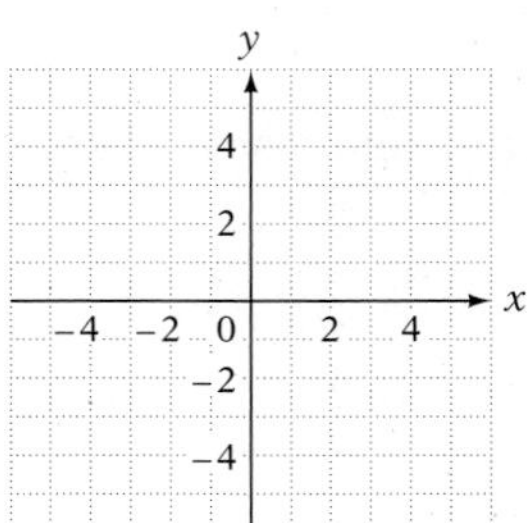

5. Find the range of $f(x) = x^2 + x - 1$ if the domain is $\{-2, -1, 0, 1, 2\}$.

6. Find the domain and range of the function $\{(-1, 0), (0, 2), (1, 2), (5, 4)\}$.

7. Is the line that passes through the points $(-2, 3)$ and $(3, 7)$ parallel to the line that passes through the points $(1, -4)$ and $(6, 0)$?

8. What value of x is excluded from the domain of $f(x) = \frac{x}{x+4}$?

9. Find the x- and y-intercepts and graph

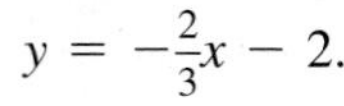

$y = -\frac{2}{3}x - 2$.

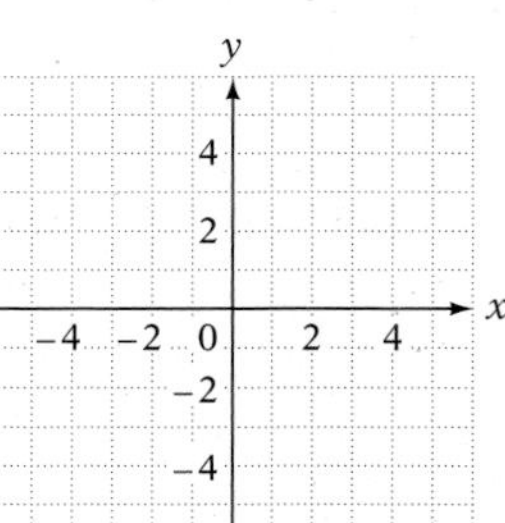

10. Graph $3x + 2y = -6$ by using the x- and y-intercepts.

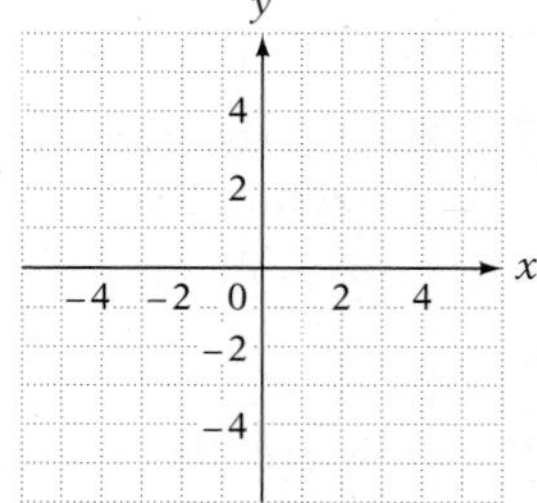

11. Graph: $y = -2x + 2$

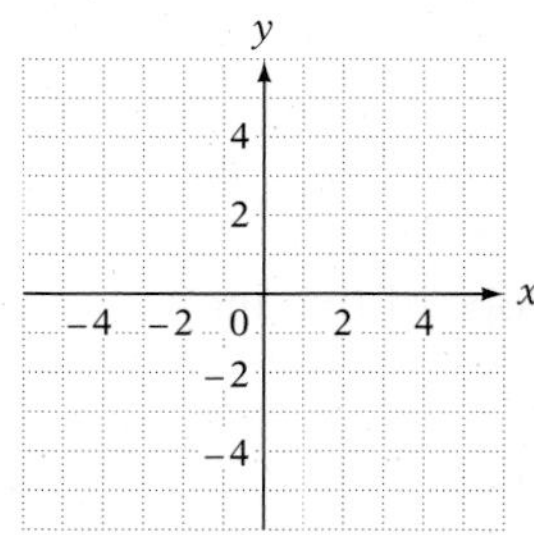

12. Graph: $4x - 3y = 12$

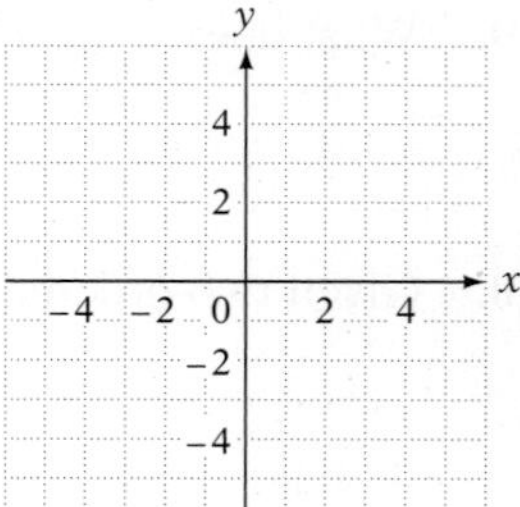

13. Find the slope of the line that contains the ordered pairs $(3, -2)$ and $(-1, 2)$.

14. Find the equation of the line that contains the ordered pair $(-3, 4)$ and has slope $\frac{5}{2}$.

15. Draw a line through all points with an ordinate of -2.

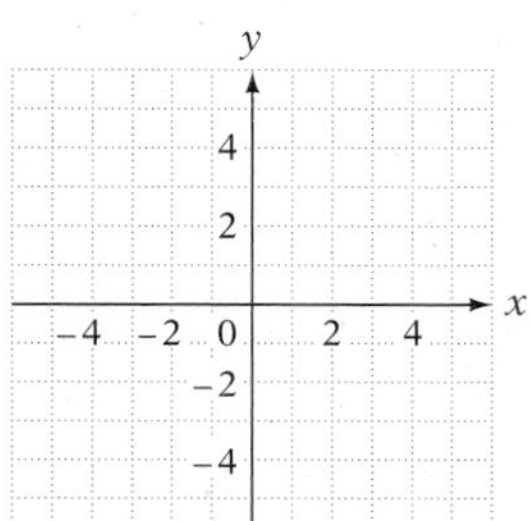

16. Graph the line that passes through the point $(-2, 3)$ and has slope $-\frac{1}{4}$.

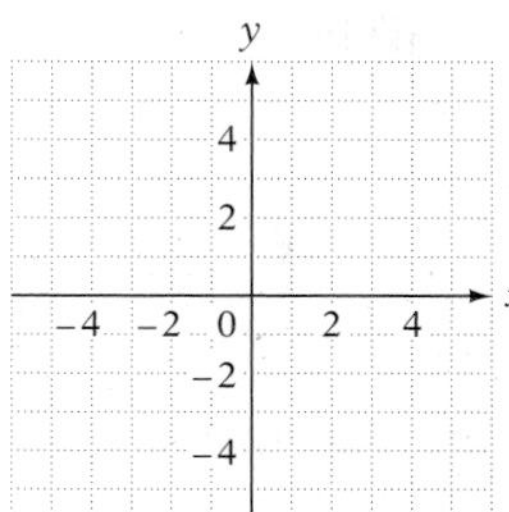

17. Find the range of $f(x) = x^2 - 2$ if the domain is $\{-2, -1, 0, 1, 2\}$.

18. **The Hospitality Industry** The manager of a hotel determines that 200 rooms will be occupied if the rate is \$95 per night. For each \$10 increase in the rate, 10 fewer rooms will be occupied.
   a. Determine a linear function that will predict the number of rooms that will be occupied at a given rate.
   b. Use the model to predict occupancy when the rate is \$120.

19. Find the equation of the line that contains the point $(-2, 3)$ and is parallel to the line $y = -4x + 3$.

20. Find the equation of the line that contains the point $(-2, 3)$ and is perpendicular to the line $y = -\frac{2}{5}x - 3$.

21. Graph $y = 1$.

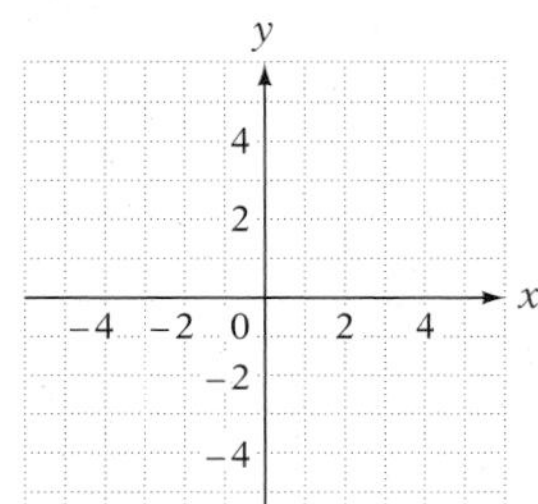

22. Graph $x = -1$.

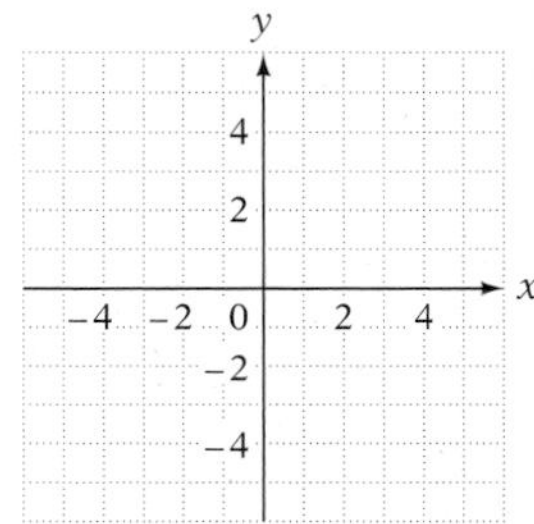

23. Find the equation of the line that contains the ordered pair $(-3, 3)$ and has slope $-\frac{2}{3}$.

24. Find the equation of the line that contains the ordered pairs $(-8, 2)$ and $(4, 5)$.

25. Find the x- and y-intercepts of $3x - 2y = -12$.

26. Find the ordered-pair solution of $y = 3x - 4$ corresponding to $x = -1$.

27. **Life Expectancy** The table below shows the average number of years that women and men were estimated to live beyond age 65 in 1950, 1960, 1970, 1980, 1990, and 2000. (*Source:* 2000 Social Security Trustees Report) Figures are rounded to the nearest whole number. Graph the scatter diagram for these data.

Women, x	15	16	17	18	19	19
Men, y	13	13	13	14	15	16

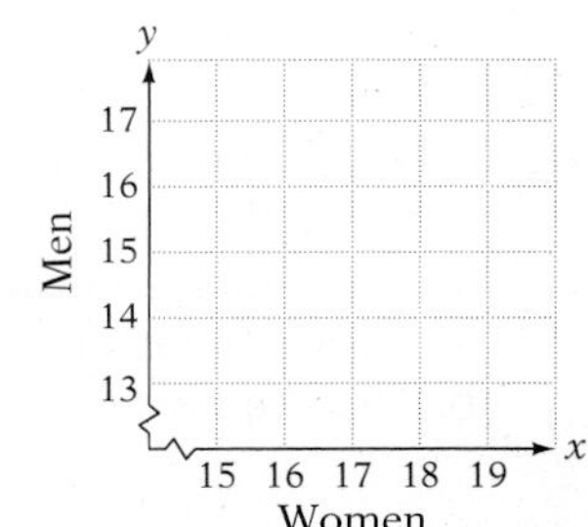

28. Graph the solution set of $y \geq 2x - 3$.

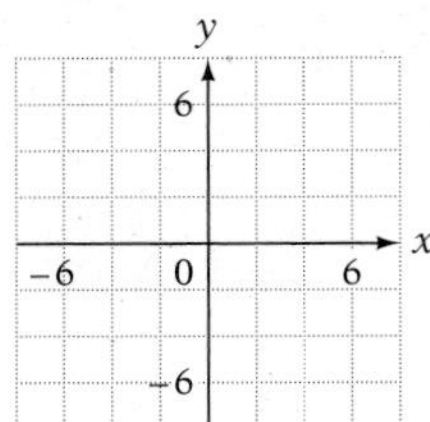

29. Graph the solution set of $3x - 2y < 6$.

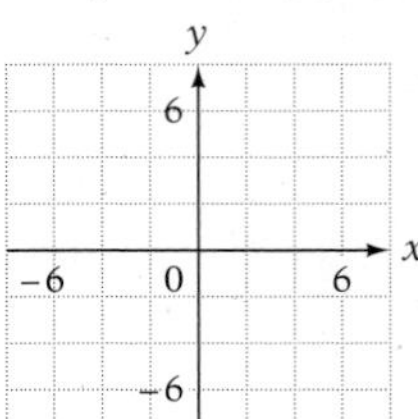

30. Find the equation of the line that contains the ordered pairs $(-2, 4)$ and $(4, -3)$.

31. Find the equation of the line that contains the ordered pair $(-2, -4)$ and is parallel to the graph of $4x - 2y = 7$.

32. Find the equation of the line that contains the ordered pair $(3, -2)$ and is parallel to the graph of $y = -3x + 4$.

33. Find the equation of the line that contains the ordered pair $(2, 5)$ and is perpendicular to the graph of the line $y = -\frac{2}{3}x + 6$.

34. Graph the line that passes through the point whose coordinates are $(-1, 4)$ and has slope $-\frac{1}{3}$.

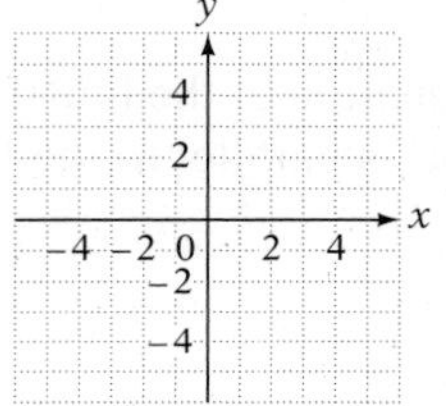

35. Travel A car is traveling at 55 mph. The equation that describes the distance traveled is $d = 55t$. Graph this equation for $0 \leq t \leq 6$. The point whose coordinates are $(4, 220)$ is on the graph. Write a sentence that explains the meaning of this ordered pair.

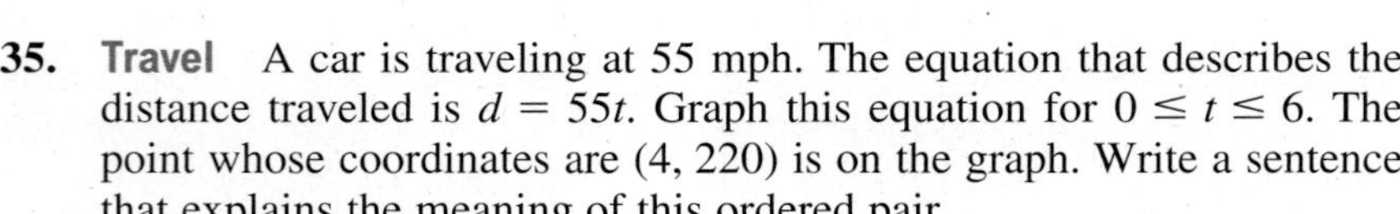

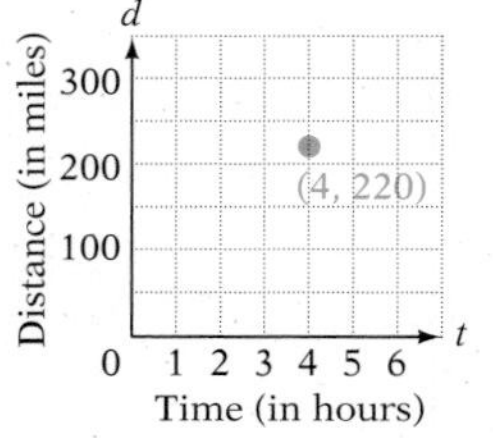

36. Manufacturing The graph at the right shows the relationship between the cost of manufacturing calculators and the number of calculators manufactured. Find the slope of the line between the two points shown on the graph. Write a sentence that states the meaning of the slope.

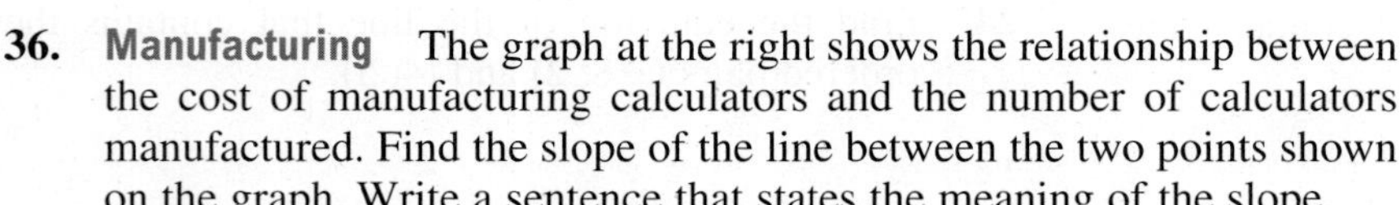

12,000
10,000
8000
6000
4000
2000
Cost (in dollars)
(500, 12,000)
(200, 6000)
0
100 200 300 400 500
Calculators manufactured

37. Construction A building contractor estimates that the cost to build a new home is \$25,000 plus \$80 for each square foot of floor space.

a. Determine a linear function that will give the cost to build a house that contains a given number of square feet.

b. Use the model to determine the cost to build a house that contains 2000 ft^2.

CHAPTER 4

TEST

1. Graph the ordered-pair solutions of $P(x) = 2 - x^2$ when $x = -2, -1, 0, 1,$ and 2.

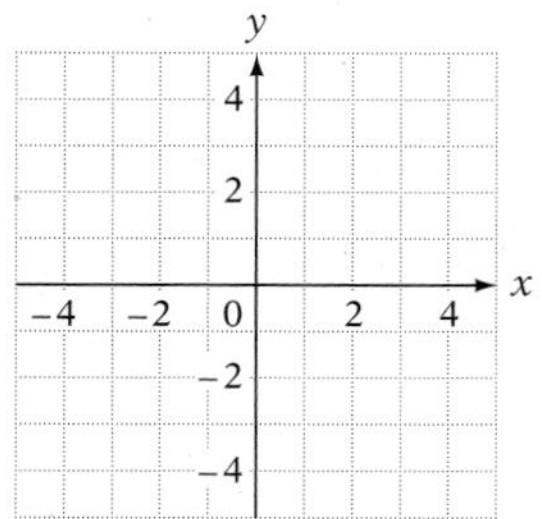

2. Find the ordered-pair solution of $y = 2x + 6$ that corresponds to $x = -3$.

3. Graph: $y = \frac{2}{3}x - 4$

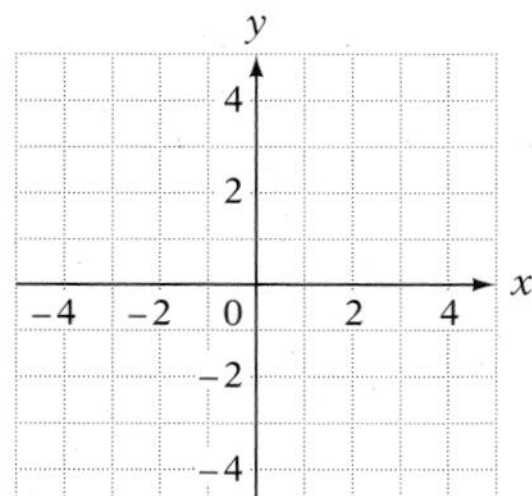

4. Graph: $2x + 3y = -3$

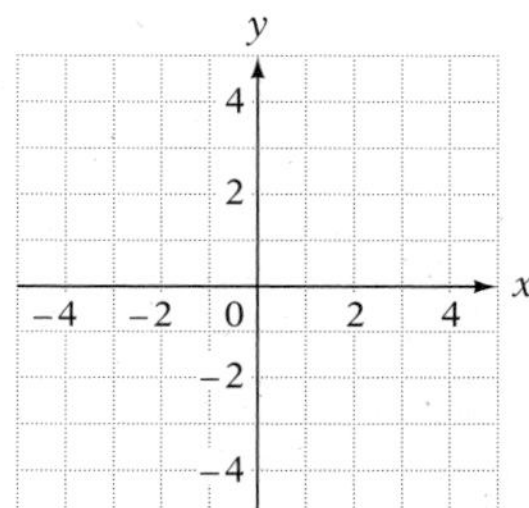

5. Find the equation of the vertical line that contains the point $(-2, 3)$.

6. Is the line that passes through the points $(-3, 1)$ and $(2, 5)$ perpendicular to the line that passes through the points $(-4, 1)$ and $(0, 6)$?

7. Find the slope of the line that contains the points $(-2, 3)$ and $(4, 2)$.

8. Given $P(x) = 3x^2 - 2x + 1$, evaluate $P(2)$.

9. Graph $2x - 3y = 6$ by using the x- and y-intercepts.

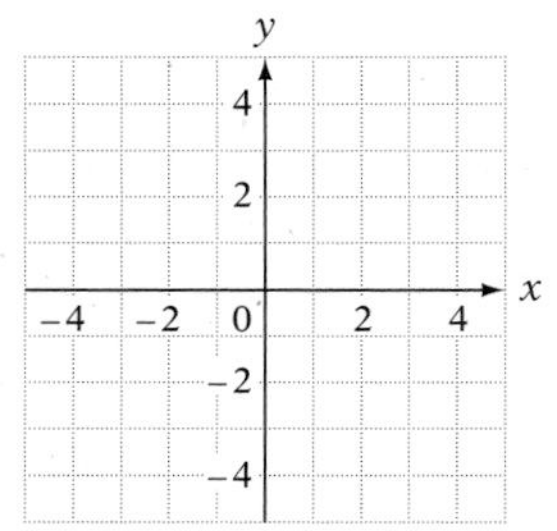

10. Graph the line that passes through the point $(-2, 3)$ and has slope $-\frac{3}{2}$.

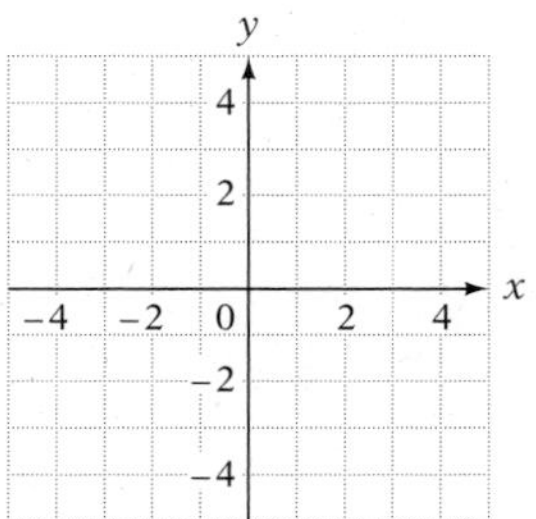

11. Find the equation of the line that contains the point $(-5, 2)$ and has slope $\frac{2}{5}$.

12. What value of x is excluded from the domain of $f(x) = \frac{2x + 1}{x}$?

13. Find the equation of the line that contains the points $(3, -4)$ and $(-2, 3)$.

14. Find the equation of the horizontal line that contains the point $(4, -3)$.

15. Find the domain and range of the function $\{(-4, 2), (-2, 2), (0, 0), (3, 5)\}$.

16. Find the equation of the line that contains the point $(1, 2)$ and is parallel to the line $y = -\frac{3}{2}x - 6$.

17. Find the equation of the line that contains the point $(-2, -3)$ and is perpendicular to the line $y = -\frac{1}{2}x - 3$.

18. Graph the solution set of $3x - 4y > 8$.

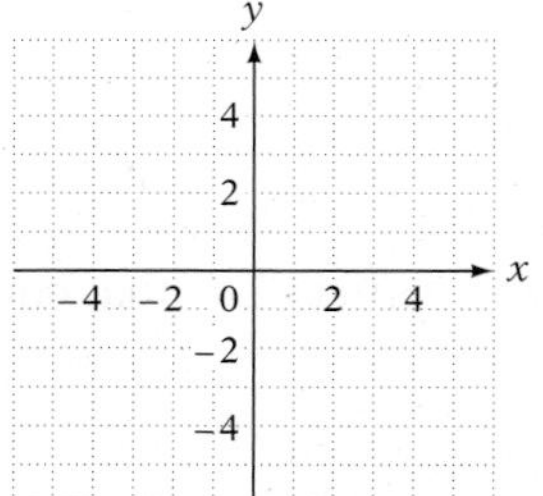

19. Depreciation The graph below shows the relationship between the cost of a rental house and the depreciation allowed for income tax purposes. Find the slope between the two points shown on the graph. Write a sentence that states the meaning of the slope.

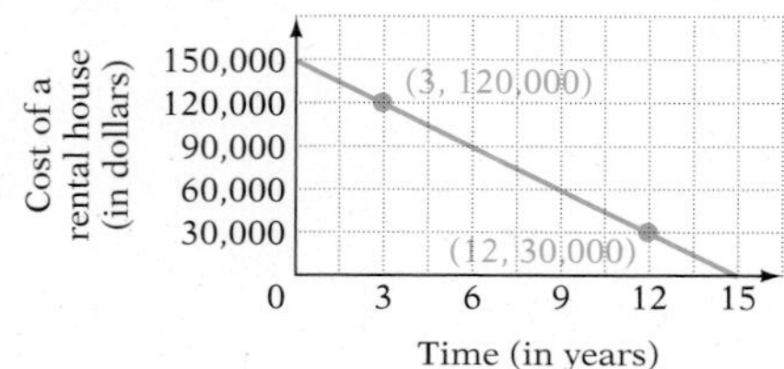

20. Summer Camp The director of a baseball camp estimates that 100 students will enroll if the tuition is \$250. For each \$20 increase in tuition, six fewer students will enroll.

a. Determine a linear function that will predict the number of students who will enroll at a given tuition.

b. Use this model to predict enrollment when the tuition is \$300.

CUMULATIVE REVIEW EXERCISES

1. Let $x \in \{-5, -3, -1\}$. For what values of x is the inequality $x \leq -3$ a true statement?

2. Write $\frac{17}{20}$ as a decimal.

3. Simplify: $3\sqrt{45}$

4. Simplify: $12 - 18 \div 3(-2)^2$

5. Evaluate $\frac{a - b}{a^2 - c}$ when $a = -2$, $b = 3$, and $c = -4$.

6. Simplify: $3d - 9 - 7d$

7. Simplify: $4(-8z)$

8. Simplify: $2(x + y) - 5(3x - y)$

9. Graph: $\{x|x < -2\} \cup \{x|x > 0\}$

   −5 −4 −3 −2 −1 0 1 2 3 4 5

10. Solve: $2x - \frac{2}{3} = \frac{7}{3}$

11. Solve: $3x - 2(10x - 6) = x - 6$

12. Solve: $4x - 3 < 9x + 2$

13. Solve: $3x - 1 < 4$ and $x - 2 > 2$

14. Solve: $|3x - 5| < 5$

15. Given $f(t) = t^2 + t$, find $f(2)$.

16. Find the slope of the line containing the points $(2, -3)$ and $(4, 1)$.

17. Graph $y = 3x + 1$.

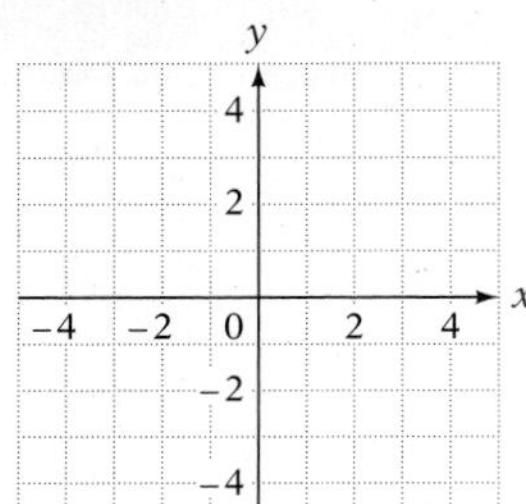

18. Graph $x = -4$.

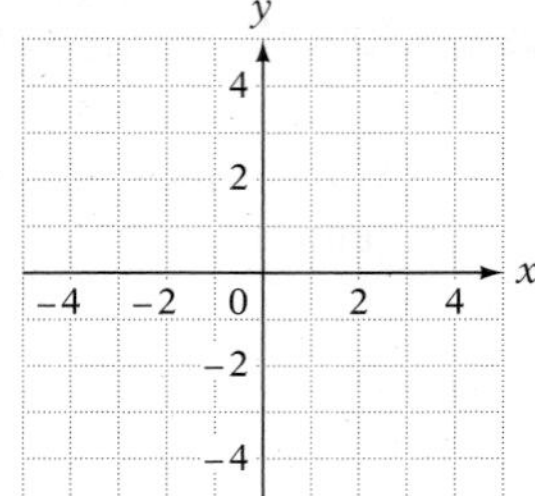

19. Graph the line that has slope $\frac{1}{2}$ and y-intercept $(0, -1)$.

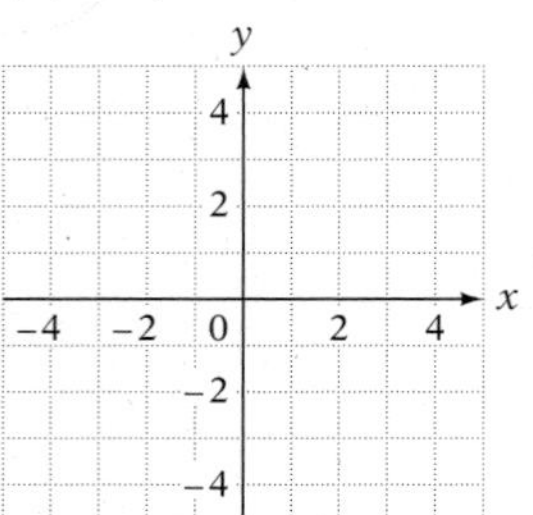

20. Graph the solution set of $3x - 2y \geq 6$.

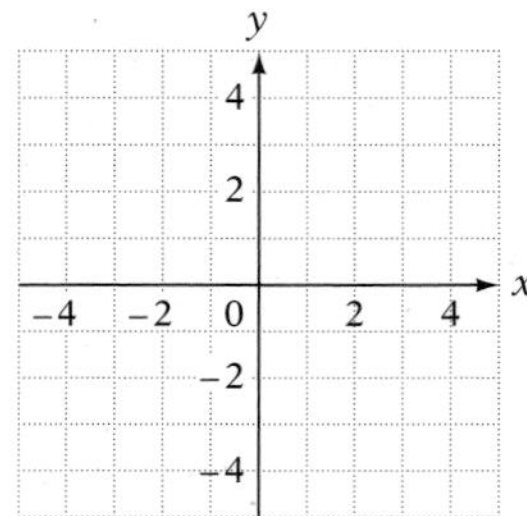

21. Find the equation of the line that passes through the points $(6, -4)$ and $(-3, -1)$.

22. Find the equation of the line that contains the point $(2, 4)$ and is parallel to the line $y = -\frac{3}{2}x + 2$.

23. Travel Two planes are 1800 mi apart and traveling toward each other. The first plane is traveling at twice the speed of the second plane. The planes meet in 3 h. Find the speed of each plane.

24. Mixtures A grocer combines coffee that costs \$9 per pound with coffee that costs \$6 per pound. How many pounds of each should be used to make 60 lb of a blend that costs \$8 per pound?

25. Depreciation The graph at the right shows the relationship between the cost of a backhoe loader and the depreciation allowed for income tax purposes. Find the slope of the line between the two points on the graph. Write a sentence that states the meaning of the slope.

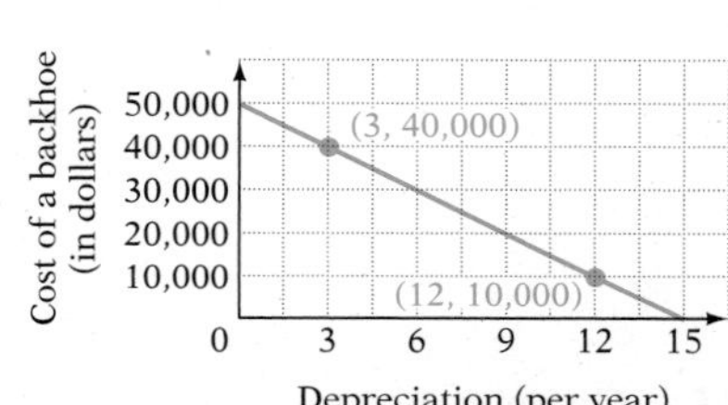

CHAPTER 5

Systems of Linear Equations and Inequalities

OBJECTIVES

SECTION 5.1

A To solve a system of linear equations by graphing

B To solve a system of linear equations by the substitution method

C To solve investment problems

SECTION 5.2

A To solve a system of two linear equations in two variables by the addition method

B To solve a system of three linear equations in three variables by the addition method

SECTION 5.3

A To solve rate-of-wind or rate-of-current problems

B To solve application problems

SECTION 5.4

A To graph the solution set of a system of linear inequalities

ARE YOU READY?

Take the Chapter 5 Prep Test to find out if you are ready to learn to:

- Solve a system of two linear equations in two variables by graphing, by the substitution method, and by the addition method
- Solve a system of three linear equations in three variables by the addition method
- Solve investment problems and rate-of-wind or rate-of-current problems
- Graph the solution set of a system of linear inequalities

PREP TEST

Do these exercises to prepare for Chapter 5.

1. Simplify: $10\left(\frac{3}{5}x + \frac{1}{2}y\right)$

2. Evaluate $3x + 2y - z$ for $x = -1$, $y = 4$, and $z = -2$.

3. Given $3x - 2z = 4$, find the value of x when $z = -2$.

4. Solve: $3x + 4(-2x - 5) = -5$

5. Solve: $0.45x + 0.06(-x + 4000) = 630$

6. Graph: $3x - 2y = 6$

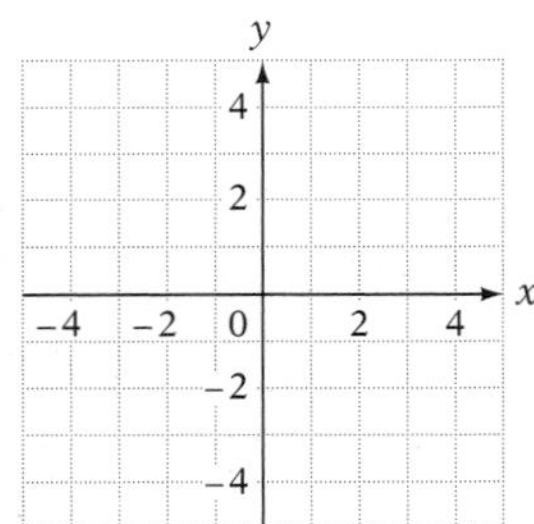

7. Graph: $y > -\frac{3}{5}x + 1$

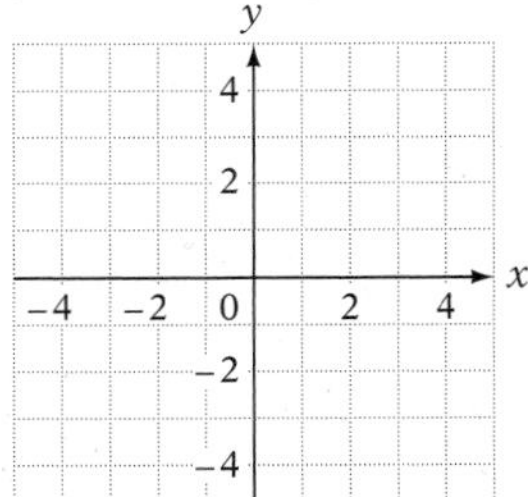

SECTION

5.1 Solving Systems of Linear Equations by Graphing and by the Substitution Method

OBJECTIVE A To solve a system of linear equations by graphing

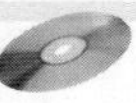

A **system of equations** is two or more equations considered together. The system at the right is a system of two linear equations in two variables. The graphs of the equations are straight lines.

$$3x + 4y = 7$$
$$2x - 3y = 6$$

A **solution of a system of equations in two variables** is an ordered pair that is a solution of each equation of the system.

HOW TO 1 Is $(3, -2)$ a solution of the system

$$2x - 3y = 12$$
$$5x + 2y = 11?$$

$2x - 3y = 12$		$5x + 2y = 11$	
$2(3) - 3(-2)$	12	$5(3) + 2(-2)$	11
$6 - (-6)$	12	$15 + (-4)$	11
	$12 = 12$		$11 = 11$

• Replace x by 3 and y by -2.

Yes, because $(3, -2)$ is a solution of each equation, it is a solution of the system of equations.

A solution of a system of linear equations can be found by graphing the lines of the system on the same set of coordinate axes. Three examples of linear equations in two variables are shown below, along with the graphs of the equations of the systems.

System I
$x + 2y = 4$
$2x + y = -1$

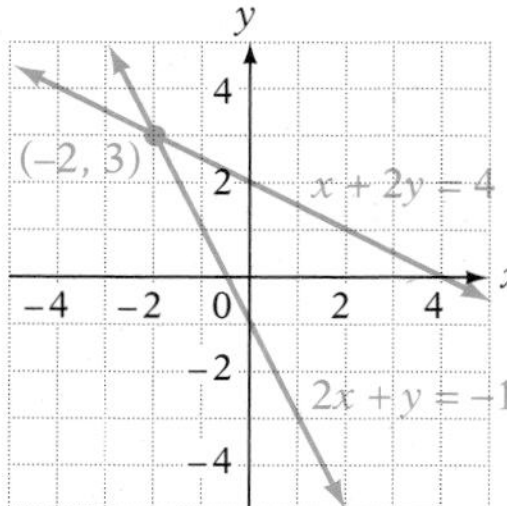

System II
$2x + 3y = 6$
$4x + 6y = -12$

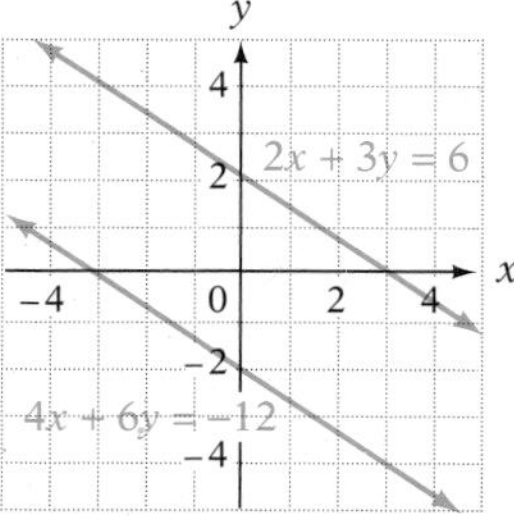

System III
$x - 2y = 4$
$2x - 4y = 8$

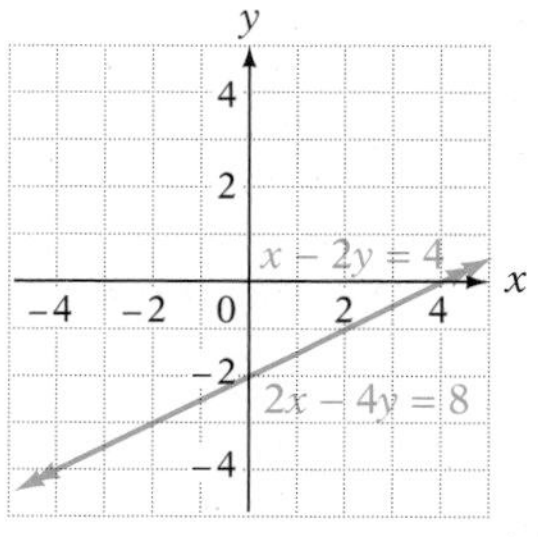

Check:

$x + 2y = 4$	
$-2 + 2(3)$	4
$-2 + 6$	4
	$4 = 4$

$2x + y = -1$	
$2(-2) + 3$	-1
$-4 + 3$	-1
	$-1 = -1$

In System I, the two lines intersect at a single point whose coordinates are $(-2, 3)$. Because this point lies on both lines, it is a solution of each equation of the system of equations. We can check this by replacing x by -2 and y by 3. The check is shown at the left. The ordered pair $(-2, 3)$ is a solution of System I.

When the graphs of a system of equations intersect at only one point, the system is called an **independent system of equations.** System I is an independent system of equations.

System II from the preceding page and the graph of the equations of that system are shown again at the right. Note in this case that the graphs of the lines are parallel and do not intersect. Since the graphs do not intersect, there is no point that is on both lines. Therefore, the system of equations has no solution.

$$2x + 3y = 6$$
$$4x + 6y = -12$$

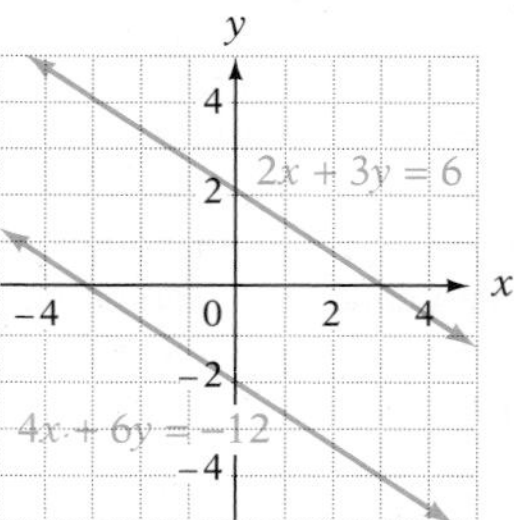

When a system of equations has no solution, it is called an **inconsistent system of equations.** System II is an inconsistent system of equations.

System III from the preceding page and the graph of the equations of that system are shown again at the right. Note that the graph of $x - 2y = 4$ lies directly on top of the graph of $2x - 4y = 8$. Thus the two lines intersect at an infinite number of points. Because the graphs intersect at an infinite number of points, there is an infinite number of solutions of this system of equations. Since each equation represents the same set of points, the solutions of the system of equations can be stated by using the ordered pairs of either one of the equations. Therefore, we can say, "The solutions are the ordered pairs that satisfy $x - 2y = 4$," or we can solve the equation for y and say, "The solutions are the ordered pairs that satisfy $y = \frac{1}{2}x - 2$." We normally state this solution using ordered pairs—"The solutions are the ordered pairs $\left(x, \frac{1}{2}x - 2\right)$."

$$x - 2y = 4$$
$$2x - 4y = 8$$

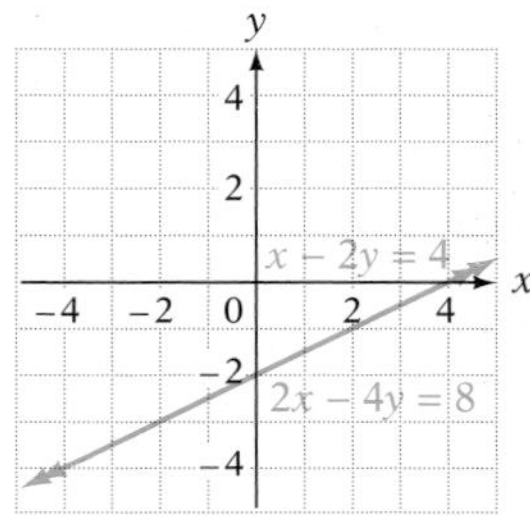

When the two equations in a system of equations represent the same line, the system is called a **dependent system of equations.** System III is a dependent system of equations.

Take Note

Keep in mind the differences among independent, dependent, and inconsistent systems of equations. You should be able to express your understanding of these terms by using graphs.

Summary of the Three Possibilities for a System of Linear Equations in Two Variables

1. The graphs intersect at one point.
   The solution of the system of equations is the ordered pair (x, y) whose coordinates name the point of intersection.
   The system of equations is independent.

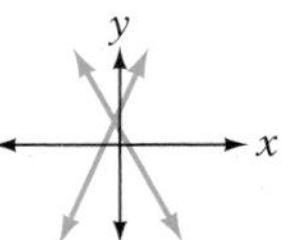

2. The lines are parallel and never intersect.
   There is no solution of the system of equations.
   The system of equations is inconsistent.

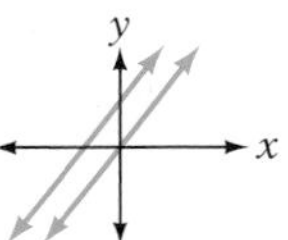

3. The graphs are the same line, and they intersect at infinitely many points.
   There are infinitely many solutions of the system of equations.
   The system of equations is dependent.

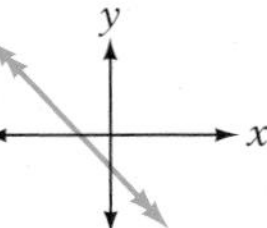

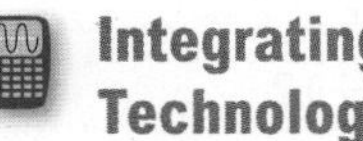

Integrating Technology

See the Projects and Group Activities at the end of this chapter for instructions on using a graphing calculator to solve a system of equations. Instructions are also provided in the Keystroke Guide: *Intersect.*

HOW TO • 2 Solve by graphing: $2x - y = 3$
$4x - 2y = 6$

Graph each line.
The system of equations is dependent.
Solve one of the equations for y.

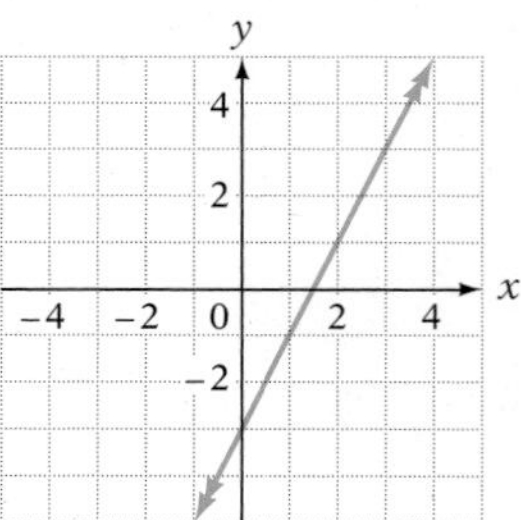

$$2x - y = 3$$
$$-y = -2x + 3$$
$$y = 2x - 3$$

The solutions are the ordered pairs $(x, 2x - 3)$.

EXAMPLE • 1

Solve by graphing:
$2x - y = 3$
$3x + y = 2$

Solution

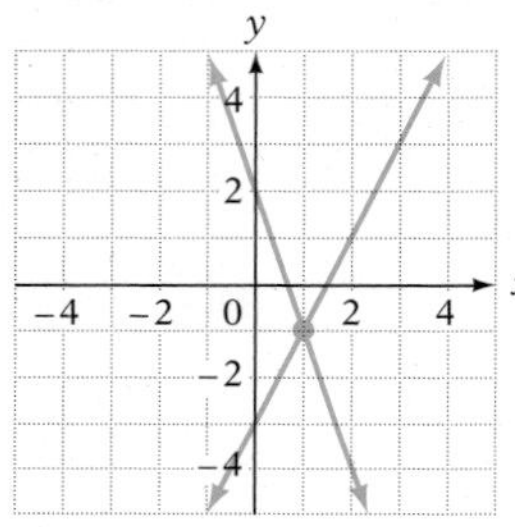

• Find the point of intersection of the graphs of the equations.

The solution is $(1, -1)$.

YOU TRY IT • 1

Solve by graphing:
$x + y = 1$
$2x + y = 0$

Your solution

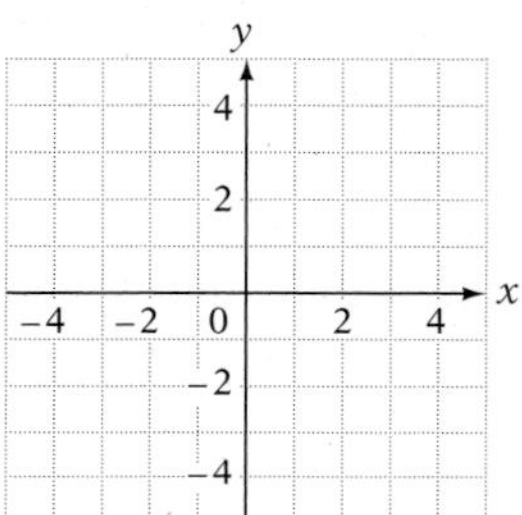

EXAMPLE • 2

Solve by graphing:
$2x + 3y = 6$
$y = -\frac{2}{3}x + 1$

Solution

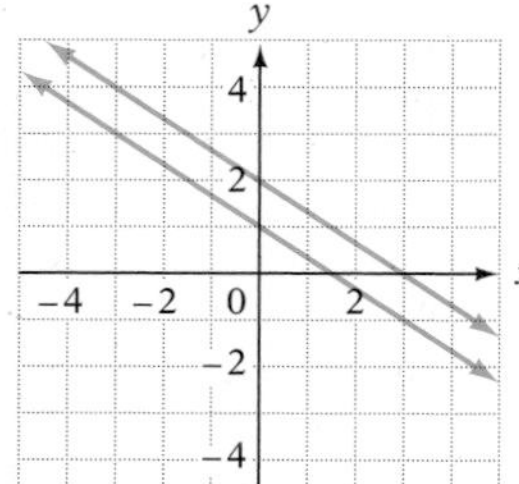

• Graph the two equations.

The lines are parallel and therefore do not intersect. The system of equations has no solution. The system of equations is inconsistent.

YOU TRY IT • 2

Solve by graphing:
$2x + 5y = 10$
$y = -\frac{2}{5}x - 2$

Your solution

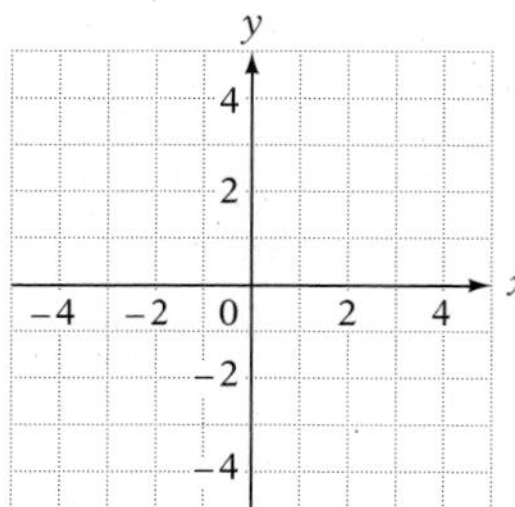

Solutions on p. S14

EXAMPLE • 3

Solve by graphing:
$x - 2y = 6$
$y = \frac{1}{2}x - 3$

Solution

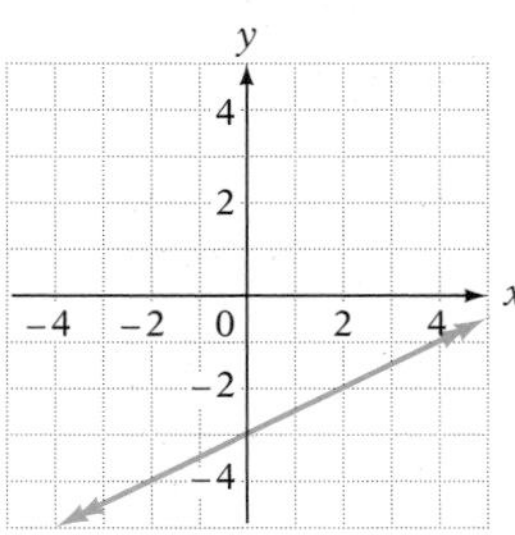

• Graph the two equations.

The system of equations is dependent. The solutions are the ordered pairs $\left(x, \frac{1}{2}x - 3\right)$.

YOU TRY IT • 3

Solve by graphing:
$3x - 4y = 12$
$y = \frac{3}{4}x - 3$

Your solution

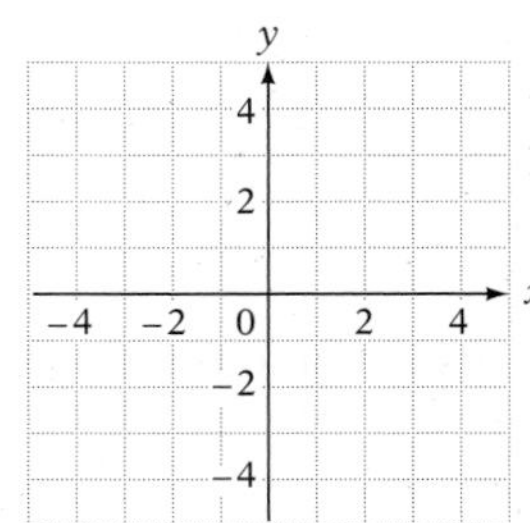

Solution on p. S14

OBJECTIVE B **To solve a system of linear equations by the substitution method** 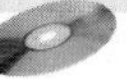

The graphical solution of a system of equations is based on approximating the coordinates of a point of intersection. An algebraic method called the **substitution method** can be used to find an exact solution of a system of equations. To use the substitution method, we must write one of the equations of the system in terms of x or in terms of y.

HOW TO • 3 Solve by the substitution method: (1) $3x + y = 5$
(2) $4x + 5y = 3$

$3x + y = 5$
(3) $y = -3x + 5$
• Solve Equation (1) for y. This is Equation (3).

(2) $4x + 5y = 3$
• This is Equation (2).

$4x + 5(-3x + 5) = 3$
• Equation (3) states that $y = -3x + 5$. Substitute $-3x + 5$ for y in Equation (2).

$4x - 15x + 25 = 3$
$-11x + 25 = 3$
$-11x = -22$
$x = 2$
• Solve for x.

(3) $y = -3x + 5$
$y = -3(2) + 5$
$y = -6 + 5$
$y = -1$
• Substitute the value of x into Equation (3) and solve for y.

The solution is the ordered pair $(2, -1)$.

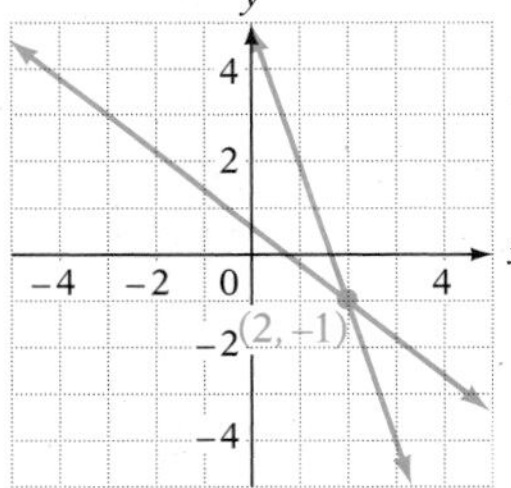

The graph of the system of equations is shown at the left. Note that the graphs intersect at the point whose coordinates are $(2, -1)$, the solution of the system of equations.

HOW TO • 4 Solve by the substitution method: (1) $6x + 2y = 8$
(2) $3x + y = 2$

$3x + y = 2$ • We will solve Equation (2) for y.

(3) $y = -3x + 2$ • This is Equation (3).

(1) $6x + 2y = 8$ • This is Equation (1).

$6x + 2(-3x + 2) = 8$ • Equation (3) states that $y = -3x + 2$. Substitute $-3x + 2$ for y in Equation (1).

$6x - 6x + 4 = 8$ • Solve for x.

$0x + 4 = 8$

$4 = 8$

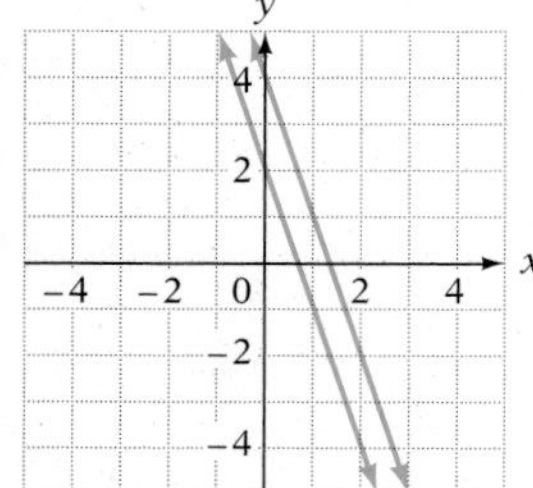

This is not a true equation.
The system of equations has no solution.
The system of equations is inconsistent.

The graph of the system of equations is shown at the left. Note that the lines are parallel.

EXAMPLE • 4

Solve by substitution:
(1) $3x - 2y = 4$
(2) $-x + 4y = -3$

Solution Solve Equation (2) for x.

$$-x + 4y = -3$$
$$-x = -4y - 3$$
$$x = 4y + 3 \quad \text{• Equation (3)}$$

Substitute $4y + 3$ for x in Equation (1).

$$3x - 2y = 4 \quad \text{• Equation (1)}$$
$$3(4y + 3) - 2y = 4 \quad \text{• } x = 4y + 3$$
$$12y + 9 - 2y = 4$$
$$10y + 9 = 4$$
$$10y = -5$$
$$y = -\frac{5}{10} = -\frac{1}{2}$$

Substitute the value of y into Equation (3).

$$x = 4y + 3 \quad \text{• Equation (3)}$$
$$= 4\left(-\frac{1}{2}\right) + 3 \quad \text{• } y = -\frac{1}{2}$$
$$= -2 + 3 = 1$$

The solution is $\left(1, -\frac{1}{2}\right)$.

YOU TRY IT • 4

Solve by substitution:
$3x - y = 3$
$6x + 3y = -4$

Your solution

Solution on p. S15

EXAMPLE 5

Solve by substitution and graph:

$3x - 3y = 2$

$y = x + 2$

Solution

$$\begin{aligned} 3x - 3y &= 2 \\ 3x - 3(x + 2) &= 2 \\ 3x - 3x - 6 &= 2 \\ -6 &= 2 \end{aligned}$$

• $y = x + 2$

This is not a true equation. The lines are parallel, so the system is inconsistent. The system does not have a solution.

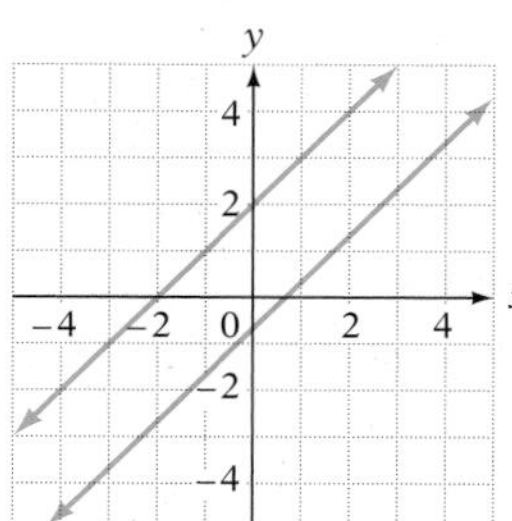

• **Graph the two equations.**

YOU TRY IT 5

Solve by substitution and graph:

$y = 2x - 3$

$3x - 2y = 6$

Your solution

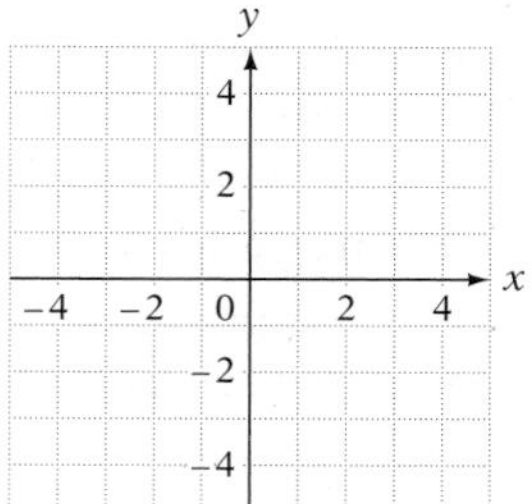

EXAMPLE 6

Solve by substitution and graph:

$9x + 3y = 12$

$y = -3x + 4$

Solution

$$\begin{aligned} 9x + 3y &= 12 \\ 9x + 3(-3x + 4) &= 12 \\ 9x - 9x + 12 &= 12 \\ 12 &= 12 \end{aligned}$$

• $y = -3x + 4$

This is a true equation. The system is dependent. The solutions are the ordered pairs $(x, -3x + 4)$.

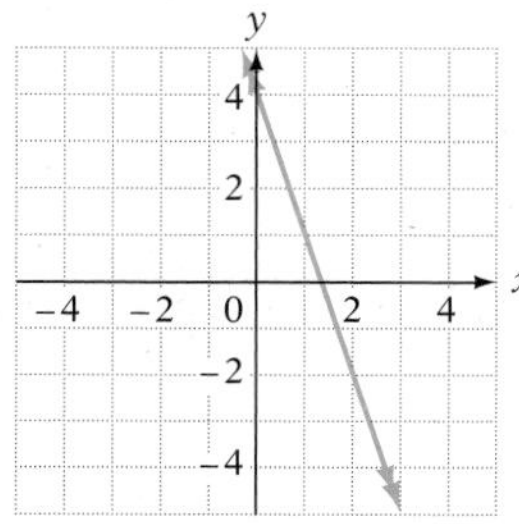

• **Graph the two equations.**

YOU TRY IT 6

Solve by substitution and graph:

$6x - 3y = 6$

$2x - y = 2$

Your solution

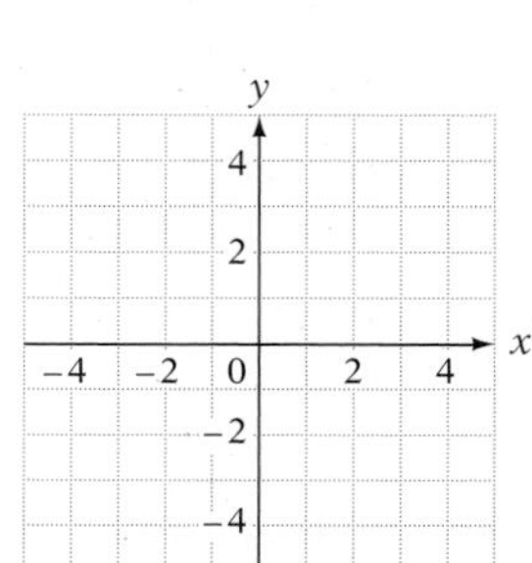

Solutions on p. S15

OBJECTIVE C To solve investment problems 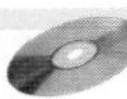

The annual simple interest that an investment earns is given by the equation $Pr = I$, where P is the principal, or the amount invested, r is the simple interest rate, and I is the simple interest.

For instance, if you invest \$500 at a simple interest rate of 5%, then the interest earned after one year is calculated as follows:

$$Pr = I$$
$$500(0.05) = I \quad \text{• Replace } P \text{ by 500 and } r \text{ by 0.05 (5\%).}$$
$$25 = I \quad \text{• Simplify.}$$

The amount of interest earned is \$25.

Tips for Success

Note that solving a word problem includes stating a strategy and using the strategy to find a solution. If you have difficulty with a word problem, write down the known information. Be very specific. Write out a phrase or sentence that states what you are trying to find. See *AIM for Success* in the Preface.

HOW TO • 5 You have a total of \$5000 to invest in two simple interest accounts. On one account, the money market fund, the annual simple interest rate is 3.5%. On the second account, the bond fund, the annual simple interest rate is 7.5%. If you earn \$245 per year from these two investments, how much do you have invested in each account?

Strategy for Solving Simple-Interest Investment Problems

1. For each amount invested, use the equation $Pr = I$. Write a numerical or variable expression for the principal, the interest rate, and the interest earned.

Amount invested at 3.5%: x
Amount invested at 7.5%: y

	Principal, P	·	*Interest rate, r*	=	*Interest earned, I*
Amount at 3.5%	x	·	0.035	=	$0.035x$
Amount at 7.5%	y	·	0.075	=	$0.075y$

2. Write a system of equations. One equation will express the relationship between the amounts invested. The second equation will express the relationship between the amounts of interest earned by the investments.

The total amount invested is \$5000: $x + y = 5000$
The total annual interest earned is \$245: $0.035x + 0.075y = 245$

Solve the system of equations.
$$(1) \quad x + y = 5000$$
$$(2) \quad 0.035x + 0.075y = 245$$

Solve Equation (1) for y:
$$(3) \quad y = -x + 5000$$

Substitute into Equation (2):
$$(2) \quad 0.035x + 0.075(-x + 5000) = 245$$
$$0.035x - 0.075x + 375 = 245$$
$$-0.04x = -130$$
$$x = 3250$$

Substitute the value of x into Equation (3) and solve for y.

$$y = -x + 5000$$
$$y = -3250 + 5000 = 1750$$

The amount invested at 3.5% is \$3250.
The amount invested at 7.5% is \$1750.

EXAMPLE • 7

An investment of \$4000 is made at an annual simple interest rate of 4.9%. How much additional money must be invested at an annual simple interest rate of 7.4% so that the total interest earned is 6.4% of the total investment?

Strategy

- Amount invested at 4.9%: 4000
  Amount invested at 7.4%: x
  Amount invested at 6.4%: y

	Principal	*Rate*	*Interest*
Amount at 4.9%	4000	0.049	0.049(4000)
Amount at 7.4%	x	0.074	$0.074x$
Amount at 6.4%	y	0.064	$0.064y$

- The amount invested at 6.4% (y) is \$4000 more than the amount invested at 7.4% (x):
  $y = x + 4000$
- The sum of the interest earned at 4.9% and the interest earned at 7.4% equals the interest earned at 6.4%:
  $0.049(4000) + 0.074x = 0.064y$

Solution

$$y = x + 4000 \qquad (1)$$
$$0.049(4000) + 0.074x = 0.064y \qquad (2)$$

Replace y in Equation (2) by $x + 4000$ from Equation (1). Then solve for x.

$$0.049(4000) + 0.074x = 0.064(x + 4000)$$
$$196 + 0.074x = 0.064x + 256$$
$$0.01x = 60$$
$$x = 6000$$

\$6000 must be invested at an annual simple interest rate of 7.4%.

YOU TRY IT • 7

An investment club invested \$13,600 in two simple interest accounts. On one account, the annual simple interest rate is 4.2%. On the other, the annual simple interest rate is 6%. How much should be invested in each account so that both accounts earn the same annual interest?

Your strategy

Your solution

Solution on p. S15

5.1 EXERCISES

OBJECTIVE A To solve a system of linear equations by graphing

For Exercises 1 to 4, determine whether the ordered pair is a solution of the system of equations.

1. $(0, -1)$
$3x - 2y = 2$
$x + 2y = 6$

2. $(2, 1)$
$x + y = 3$
$2x - 3y = 1$

3. $(-3, -5)$
$x + y = -8$
$2x + 5y = -31$

4. $(1, -1)$
$3x - y = 4$
$7x + 2y = -5$

For Exercises 5 to 8, state whether the system of equations is independent, inconsistent, or dependent.

5.

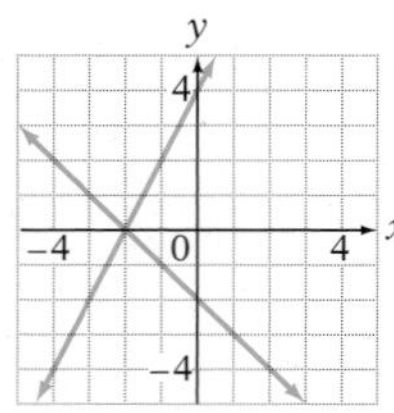

6.

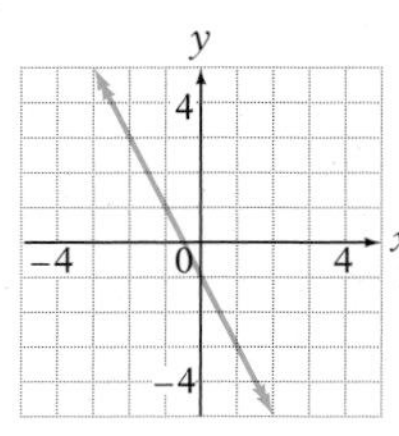

7.

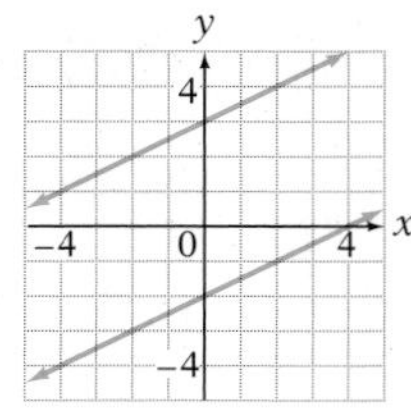

8.

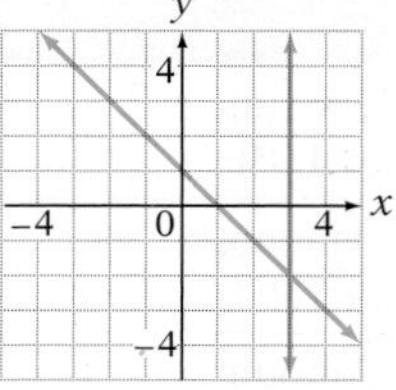

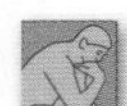

9. In a system of linear equations in two variables, the slope of one line is twice the slope of the other line. Is the system of equations independent, dependent, or inconsistent? If there is not enough information to answer the question, so state.

10. In an inconsistent system of linear equations in two variables, what is the relationship between the slopes of the lines? If the graphs of the lines are not vertical, what is the relationship between the y-intercepts of the two lines?

For Exercises 11 to 28, solve by graphing.

11. $x + y = 2$
$x - y = 4$

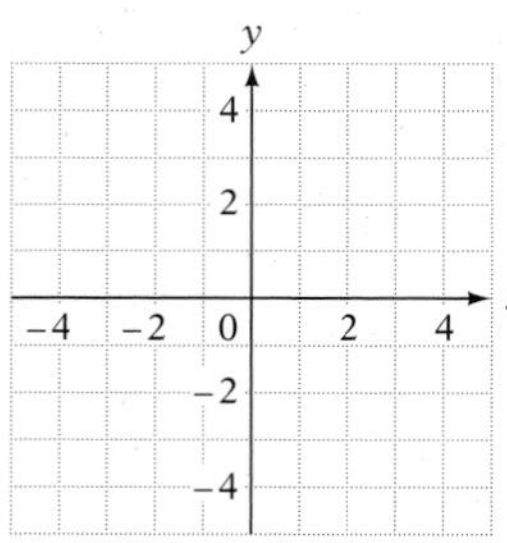

12. $x + y = 1$
$3x - y = -5$

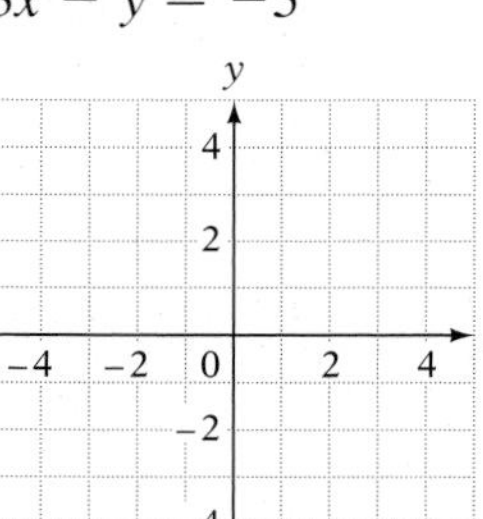

13. $x - y = -2$
$x + 2y = 10$

14. $2x - y = 5$
$3x + y = 5$

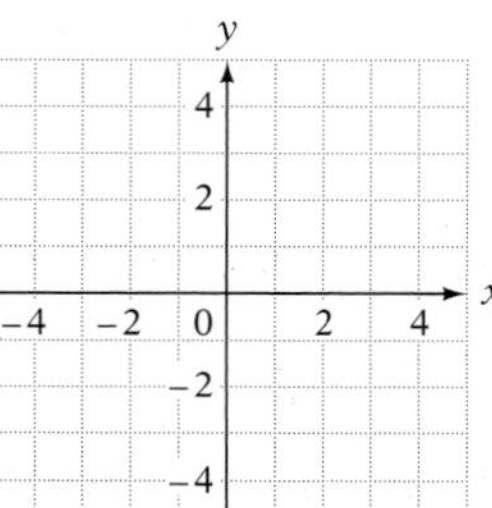

15. $3x - 2y = 6$
$y = 3$

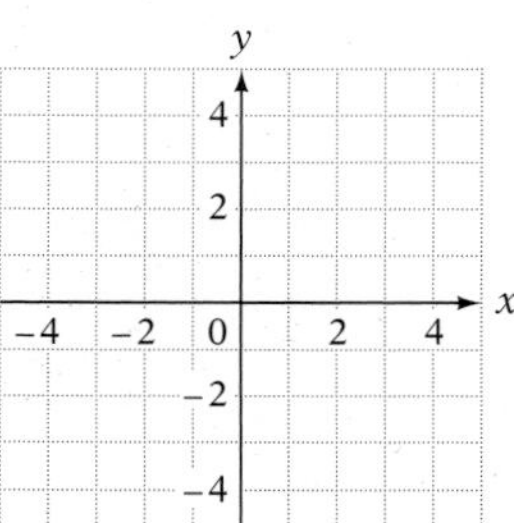

16. $x = 4$
$3x - 2y = 4$

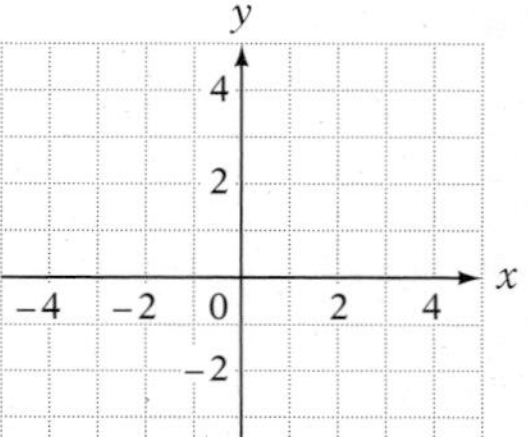

17. 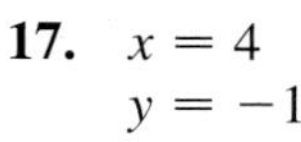
$x = 4$
$y = -1$

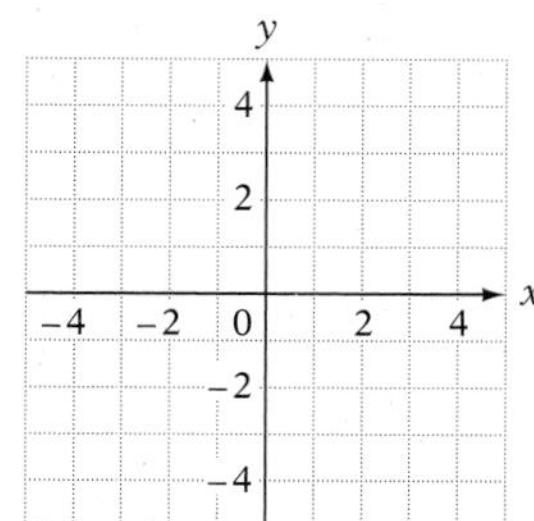

18. $x + 2 = 0$
$y - 1 = 0$

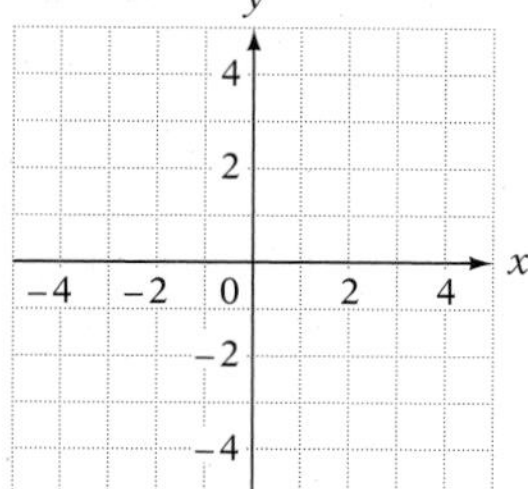

19. $2x + y = 3$
$x - 2 = 0$

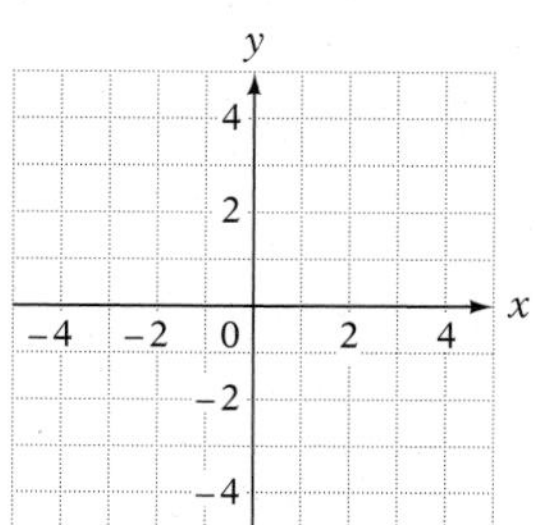

20. $x - 3y = 6$
$y + 3 = 0$

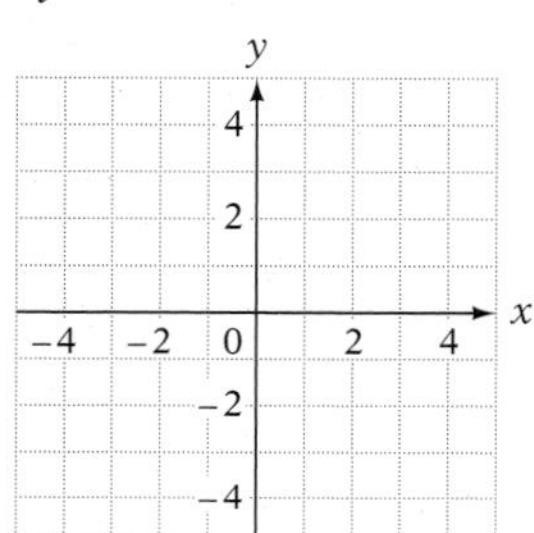

21. $x - y = 6$
$x + y = 2$

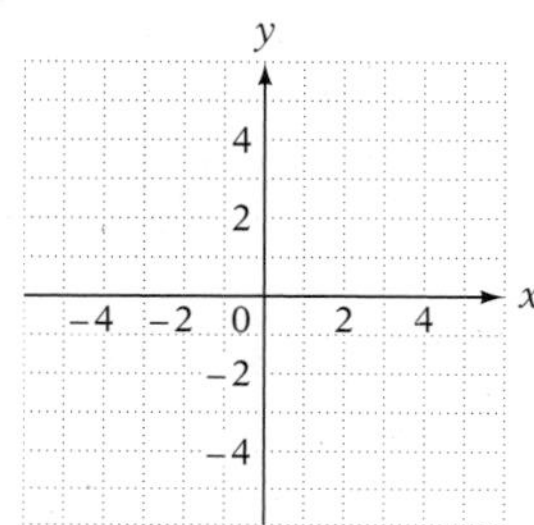

22. $2x + y = 2$
$-x + y = 5$

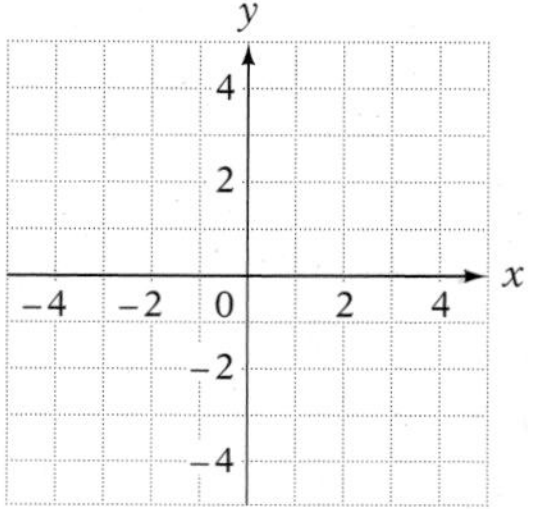

23. $y = x - 5$
$2x + y = 4$

24. $2x - 5y = 4$
$y = x + 1$

25. $y = \frac{1}{2}x - 2$
$x - 2y = 8$

26. $2x + 3y = 6$
$y = -\frac{2}{3}x + 1$

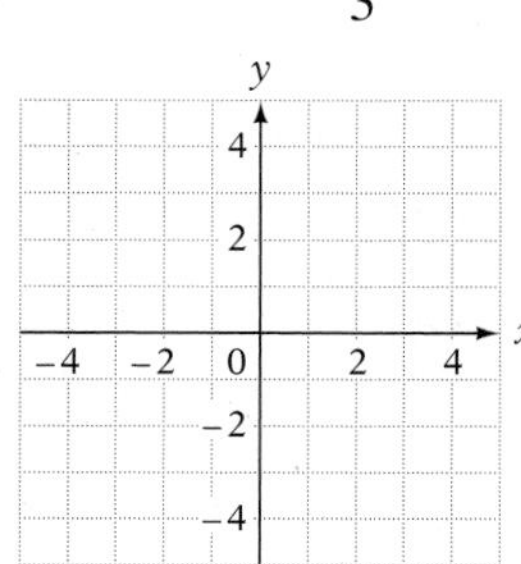

27. $2x - 5y = 10$
$y = \frac{2}{5}x - 2$

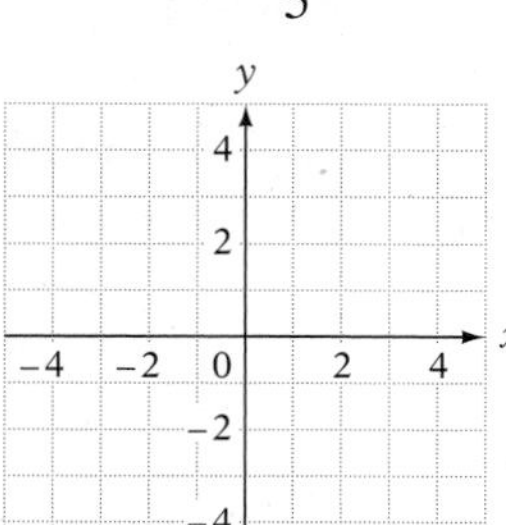

28. $3x - 2y = 6$
$y = \frac{3}{2}x - 3$

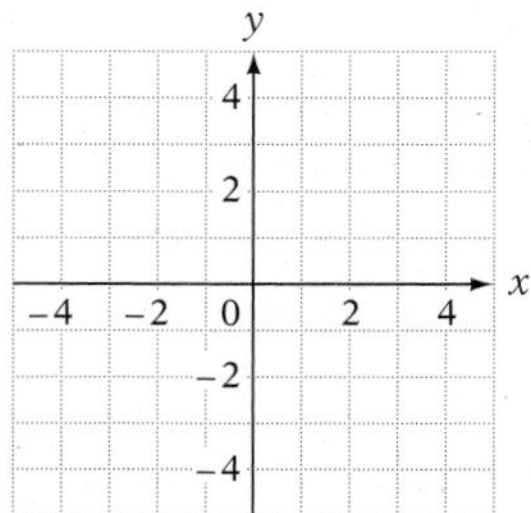

OBJECTIVE B **To solve a system of linear equations by the substitution method**

For Exercises 29 to 58, solve by the substitution method.

29. $y = -x + 1$
$2x - y = 5$

30. $x = 3y + 1$
$x - 2y = 6$

31. $x = 2y - 3$
$3x + y = 5$

32. $4x - 3y = 5$
$y = 2x - 3$

33. $3x + 5y = -1$
$y = 2x - 8$

34. $5x - 2y = 9$
$y = 3x - 4$

35. $4x - 3y = 2$
$y = 2x + 1$

36. $x = 2y + 4$
$4x + 3y = -17$

37. $3x - 2y = -11$
$x = 2y - 9$

38. $5x + 4y = -1$
$y = 2 - 2x$

39. $3x + 2y = 4$
$y = 1 - 2x$

40. $2x - 5y = -9$
$y = 9 - 2x$

41. $5x + 2y = 15$
$x = 6 - y$

42. $7x - 3y = 3$
$x = 2y + 2$

43. $3x - 4y = 6$
$x = 3y + 2$

44. $2x + 2y = 7$
$y = 4x + 1$

45. $3x + 7y = -5$
$y = 6x - 5$

46. $3x + y = 5$
$2x + 3y = 8$

47. $3x - y = 10$
$6x - 2y = 5$

48. $6x - 4y = 3$
$3x - 2y = 9$

49. $3x + 4y = 14$
$2x + y = 1$

50. $5x + 3y = 8$
$3x + y = 8$

51. $3x + 5y = 0$
$x - 4y = 0$

52. $2x - 7y = 0$
$3x + y = 0$

53. $2x - 4y = 16$
$-x + 2y = -8$

54. $3x - 12y = -24$
$-x + 4y = 8$

55. $y = 3x + 2$
$y = 2x + 3$

56. $y = 3x - 7$
$y = 2x - 5$

57. $y = 3x + 1$
$y = 6x - 1$

58. $y = 2x - 3$
$y = 4x - 4$

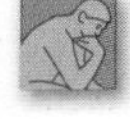

59. The system of equations at the right is inconsistent. What is the value of $\frac{a}{b}$?

$4x - 6y = 7$
$ax - by = 9$

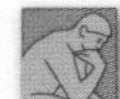

60. Give an example of a system of linear equations in two variables that has (0, 0) as its only solution.

OBJECTIVE C **To solve investment problems**

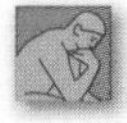

For Exercises 61 and 62, use the system of equations shown at the right. The system models the investment of x dollars in one simple interest account and y dollars in a second simple interest account.

$$x + y = 6000$$
$$0.055x + 0.072y = 391.20$$

61. What are the interest rates on the two accounts?

62. What is the total amount of money invested?

63. The Community Relief Charity Group is earning 3.5% simple interest on the $2800 it invested in a savings account. It also earns 4.2% simple interest on an insured bond fund. The annual interest earned from both accounts is $329. How much is invested in the insured bond fund?

64. Two investments earn an annual income of $575. One investment earns an annual simple interest rate of 8.5%, and the other investment earns an annual simple interest rate of 6.4%. The total amount invested is $8000. How much is invested in each account?

65. An investment club invested $6000 at an annual simple interest rate of 4.0%. How much additional money must be invested at an annual simple interest rate of 6.5% so that the total annual interest earned will be 5% of the total investment?

66. A company invested $30,000, putting part of it into a savings account that earned 3.2% annual simple interest and the remainder in a stock fund that earned 12.6% annual simple interest. If the investments earned $1665 annually, how much was in each account?

67. An account executive divided $42,000 between two simple interest accounts. On the tax-free account the annual simple interest rate is 3.5%, and on the money market fund the annual simple interest rate is 4.5%. How much should be invested in each account so that both accounts earn the same annual interest?

68. An investment club placed $33,000 into two simple interest accounts. On one account, the annual simple interest rate is 6.5%. On the other, the annual simple interest rate is 4.5%. How much should be invested in each account so that both accounts earn the same annual interest?

69. The Cross Creek Investment Club decided to invest $16,000 in two bond funds. The first, a mutual bond fund, earns 4.5% annual simple interest. The second, a corporate bond fund, earns 8% annual simple interest. If the club earned $1070 from these two accounts, how much was invested in the mutual bond fund?

70. Cabin Financial Service Group recommends that a client purchase for $10,000 a corporate bond that earns 5% annual simple interest. How much additional money must be placed in U.S. government securities that earn a simple interest rate of 3.5% so that the total annual interest earned from the two investments is 4% of the total investment?

Applying the Concepts

For Exercises 71 to 73, use a graphing calculator to estimate the solution to the system of equations. Round coordinates to the nearest hundredth. See the Projects and Group Activities at the end of this chapter for assistance.

71. $y = -\frac{1}{2}x + 2$
$y = 2x - 1$

72. $y = \sqrt{2}x - 1$
$y = -\sqrt{3}x + 1$

73. $y = \pi x - \frac{2}{3}$
$y = -x + \frac{\pi}{2}$

SECTION

5.2 Solving Systems of Linear Equations by the Addition Method

OBJECTIVE A To solve a system of two linear equations in two variables by the addition method 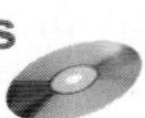

The **addition method** is an alternative method for solving a system of equations. This method is based on the Addition Property of Equations. Use the addition method when it is not convenient to solve one equation for one variable in terms of another variable.

Note for the system of equations at the right the effect of adding Equation (2) to Equation (1). Because $-3y$ and $3y$ are additive inverses, adding the equations results in an equation with only one variable.

$$\begin{aligned} (1)\ 5x - 3y &= 14 \\ (2)\ 2x + 3y &= -7 \\ 7x + 0y &= 7 \\ 7x &= 7 \end{aligned}$$

The solution of the resulting equation is the first coordinate of the ordered-pair solution of the system.

$$\begin{aligned} 7x &= 7 \\ x &= 1 \end{aligned}$$

The second coordinate is found by substituting the value of x into Equation (1) or (2) and then solving for y. Equation (1) is used here.

$$\begin{aligned} (1)\quad 5x - 3y &= 14 \\ 5(1) - 3y &= 14 \\ 5 - 3y &= 14 \\ -3y &= 9 \\ y &= -3 \end{aligned}$$

The solution is $(1, -3)$.

Point of Interest

There are records of Babylonian mathematicians solving systems of equations 3600 years ago. Here is a system of equations from that time (in our modern notation):

$$\frac{2}{3}x = \frac{1}{2}y + 500$$

$$x + y = 1800$$

We say *modern notation* for many reasons. Foremost is the fact that the use of variables did not become widespread until the 17th century. There are many other reasons, however. The equals sign had not been invented, 2 and 3 did not look like they do today, and zero had not even been considered as a possible number.

Sometimes each equation of the system of equations must be multiplied by a constant so that the coefficients of one of the variables are opposites.

HOW TO • 1 Solve by the addition method:

$$\begin{aligned} (1)\ 3x + 4y &= 2 \\ (2)\ 2x + 5y &= -1 \end{aligned}$$

To eliminate x, multiply Equation (1) by 2 and Equation (2) by -3. Note at the right how the constants are chosen.

$$\begin{aligned} 2(3x + 4y) &= 2 \cdot 2 \\ -3(2x + 5y) &= -3(-1) \end{aligned}$$

• The negative is used so that the coefficients will be opposites.

$$\begin{aligned} 6x + 8y &= 4 \\ -6x - 15y &= 3 \\ -7y &= 7 \\ y &= -1 \end{aligned}$$

• 2 times Equation (1).
• -3 times Equation (2).
• Add the equations.
• Solve for y.

Substitute the value of y into Equation (1) or Equation (2) and solve for x. Equation (1) will be used here.

$$\begin{aligned} (1)\quad 3x + 4y &= 2 \\ 3x + 4(-1) &= 2 \\ 3x - 4 &= 2 \\ 3x &= 6 \\ x &= 2 \end{aligned}$$

• Substitute -1 for y.
• Solve for x.

The solution is $(2, -1)$.

Tips for Success

Always check the proposed solution of a system of equations. For the system at the right:

$$\begin{array}{r|l} 3x + 4y & 2 \\ \hline 3(2) + 4(-1) & 2 \\ 6 - 4 & 2 \\ 2 = 2 & \end{array}$$

$$\begin{array}{r|l} 2x + 5y & -1 \\ \hline 2(2) + 5(-1) & -1 \\ 4 - 5 & -1 \\ -1 = -1 & \end{array}$$

The solution checks.

HOW TO • 2 Solve by the addition method:

$$(1)\quad 2x - y = 3$$
$$(2)\quad 4x - 2y = 6$$

Eliminate y. Multiply Equation (1) by -2.

$(1)\quad -2(2x - y) = -2(3)$ • -2 times Equation (1).

$(3)\quad -4x + 2y = -6$ • This is Equation (3).

Take Note
The result of adding Equations (3) and (2) is $0 = 0$. It is not $x = 0$ and it is not $y = 0$. There is no variable in the equation $0 = 0$. This result does not indicate that the solution is (0, 0); rather, it indicates a dependent system of equations.

Add Equation (3) to Equation (2).

$$\begin{array}{lr} (2) & 4x - 2y = 6 \\ (3) & -4x + 2y = -6 \\ \hline & 0 = 0 \end{array}$$

The equation $0 = 0$ indicates that the system of equations is dependent. This means that the graphs of the two lines are the same. Therefore, the solutions of the system of equations are the ordered-pair solutions of the equation of the line. Solve Equation (1) for y.

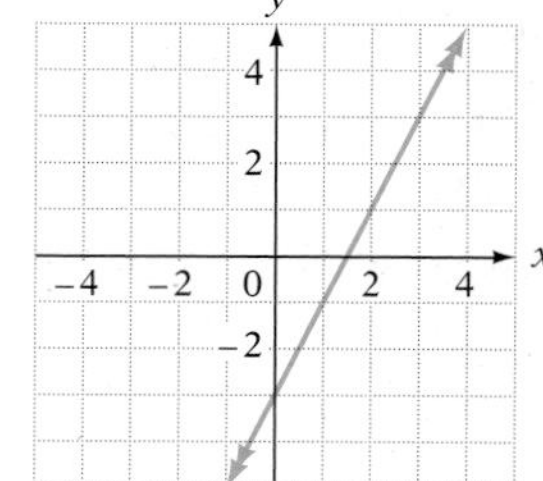

$$2x - y = 3$$
$$-y = -2x + 3$$
$$y = 2x - 3$$

The ordered-pair solutions are $(x, 2x - 3)$, where $2x - 3$ is substituted for y.

HOW TO • 3 Solve by the addition method:

$$(1)\quad \frac{2}{3}x + \frac{1}{2}y = 4$$
$$(2)\quad \frac{1}{4}x - \frac{3}{8}y = -\frac{3}{4}$$

Clear fractions. Multiply each equation by the LCM of the denominators.

$6\left(\frac{2}{3}x + \frac{1}{2}y\right) = 6(4)$ • The LCM of 3 and 2 is 6.

$8\left(\frac{1}{4}x - \frac{3}{8}y\right) = 8\left(-\frac{3}{4}\right)$ • The LCM of 4 and 8 is 8.

$$\begin{array}{r} 4x + 3y = 24 \\ 2x - 3y = -6 \\ \hline 6x = 18 \\ x = 3 \end{array}$$

• Eliminate y. Add the equations. Then solve for x.

$\frac{2}{3}x + \frac{1}{2}y = 4$ • This is Equation (1).

$\frac{2}{3}(3) + \frac{1}{2}y = 4$ • Substitute $x = 3$ into Equation (1) and solve for y.

$$2 + \frac{1}{2}y = 4$$
$$\frac{1}{2}y = 2$$
$$y = 4$$

The solution is (3, 4).

EXAMPLE • 1

Solve by the addition method:

(1) $3x - 2y = 2x + 5$
(2) $2x + 3y = -4$

Solution

Write Equation (1) in the form $Ax + By = C$.

$$3x - 2y = 2x + 5$$
$$x - 2y = 5$$

Solve the system:

$$x - 2y = 5$$
$$2x + 3y = -4$$

Eliminate x.

$$-2(x - 2y) = -2(5)$$
$$2x + 3y = -4$$

$$-2x + 4y = -10$$
$$2x + 3y = -4$$

$$7y = -14$$ • Add the equations.
$$y = -2$$ • Solve for y.

Replace y in Equation (2).

$$2x + 3y = -4$$
$$2x + 3(-2) = -4$$
$$2x - 6 = -4$$
$$2x = 2$$
$$x = 1$$

The solution is $(1, -2)$.

YOU TRY IT • 1

Solve by the addition method:

$2x + 5y = 6$
$3x - 2y = 6x + 2$

Your solution

EXAMPLE • 2

Solve by the addition method:

(1) $4x - 8y = 36$
(2) $3x - 6y = 27$

Solution

Eliminate x.

$$3(4x - 8y) = 3(36)$$
$$-4(3x - 6y) = -4(27)$$

$$12x - 24y = 108$$
$$-12x + 24y = -108$$

$$0 = 0$$ • Add the equations.

The system of equations is dependent. Solve Equation (1) for y.

$$4x - 8y = 36$$
$$-8y = -4x + 36$$
$$y = \frac{1}{2}x - \frac{9}{2}$$

The solutions are the ordered pairs $\left(x, \frac{1}{2}x - \frac{9}{2}\right)$.

YOU TRY IT • 2

Solve by the addition method:

$2x + y = 5$
$4x + 2y = 6$

Your solution

Solutions on p. S16

OBJECTIVE B To solve a system of three linear equations in three variables by the addition method 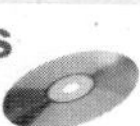

An equation of the form $Ax + By + Cz = D$, where A, B, and C are the coefficients of the variables and D is a constant, is a **linear equation in three variables.** Examples of this type of equation are shown at the right.

$$2x + 4y - 3z = 7$$
$$x - 6y + z = -3$$

Graphing an equation in three variables requires a third coordinate axis perpendicular to the xy-plane. The third axis is commonly called the z-axis. The result is a three-dimensional coordinate system called the ***xyz*-coordinate system.** To help visualize a three-dimensional coordinate system, think of a corner of a room: The floor is the xy-plane, one wall is the yz-plane, and the other wall is the xz-plane. A three-dimensional coordinate system is shown at the right.

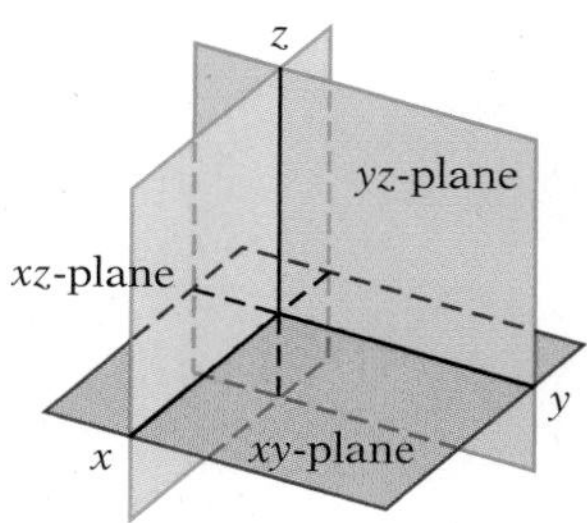

The graph of a point in an xyz-coordinate system is an **ordered triple** (x, y, z). Graphing an ordered triple requires three moves, the first in the direction of the x-axis, the second in the direction of the y-axis, and the third in the direction of the z-axis. The graph of the points $(-4, 2, 3)$ and $(3, 4, -2)$ is shown at the right.

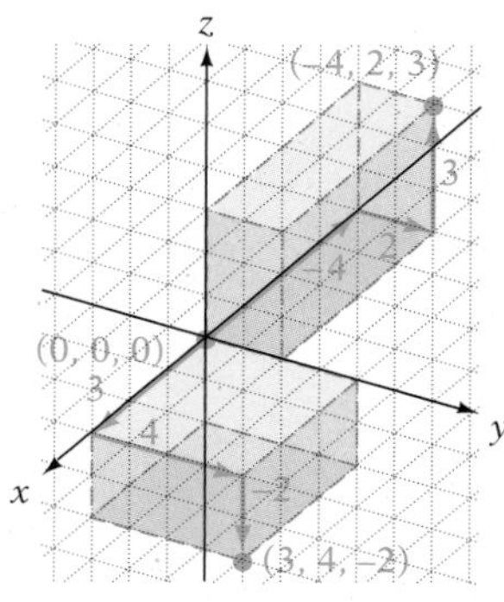

The graph of a linear equation in three variables is a plane. That is, if all the solutions of a linear equation in three variables were plotted in an xyz-coordinate system, the graph would look like a large piece of paper extending infinitely. The graph of $x + y + z = 3$ is shown at the right.

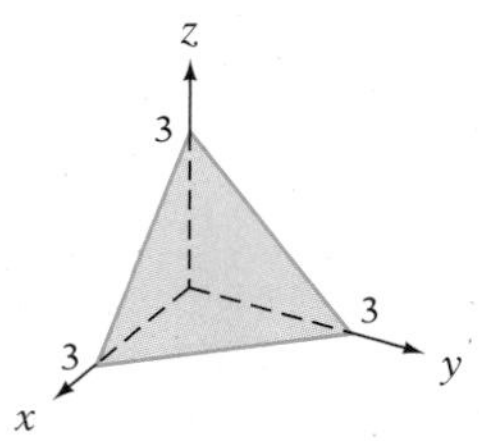

There are different ways in which three planes can be oriented in an *xyz*-coordinate system. The systems of equations represented by the planes below are inconsistent.

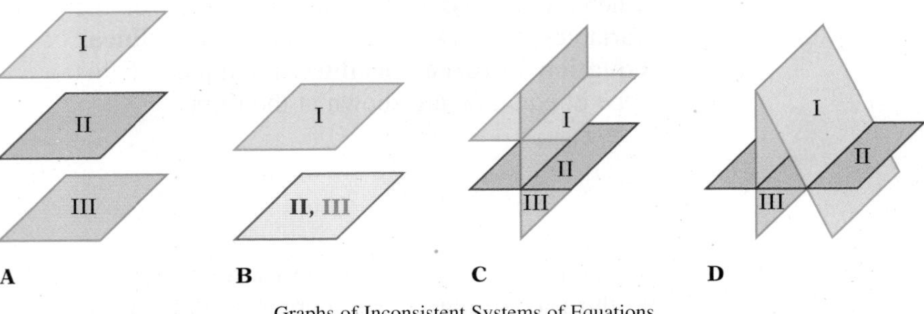

Graphs of Inconsistent Systems of Equations

For a system of three equations in three variables to have a solution, the graphs of the planes must intersect at a single point, they must intersect along a common line, or all equations must have a graph that is the same plane. These situations are shown in the figures below.

The three planes shown in Figure E intersect at a point. A system of equations represented by planes that intersect at a point is independent.

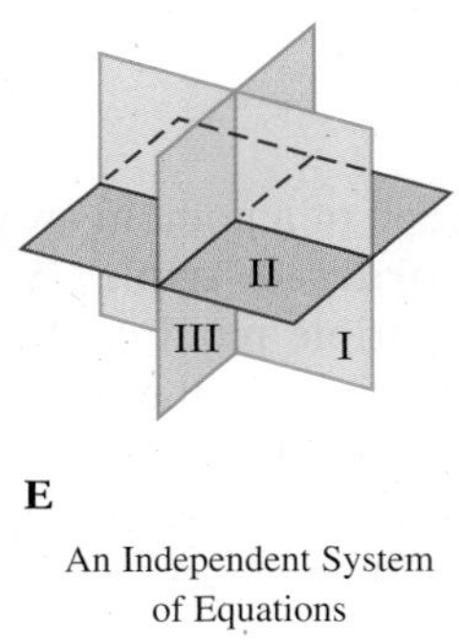

An Independent System of Equations

The planes shown in Figures F and G intersect along a common line. The system of equations represented by the planes in Figure H has a graph that is the same plane. The systems of equations represented by the graphs below are dependent.

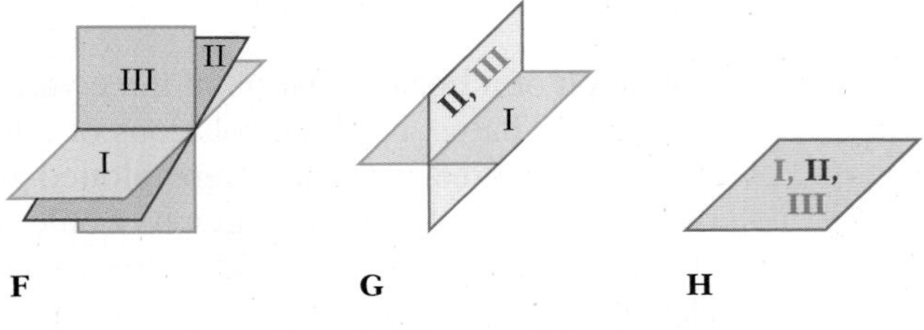

Dependent Systems of Equations

Point of Interest

In the early 1980s, Stephen Hoppe became interested in winning Monopoly strategies. Finding these strategies required solving a system that contained 123 equations in 123 variables!

Just as a solution of an equation in two variables is an ordered pair (x, y), a **solution of an equation in three variables** is an ordered triple (x, y, z). For example, $(2, 1, -3)$ is a solution of the equation $2x - y - 2z = 9$. The ordered triple $(1, 3, 2)$ is not a solution.

A **system of linear equations in three variables** is shown at the right. A **solution of a system of equations in three variables** is an ordered triple that is a solution of each equation of the system.

$$\begin{aligned} x - 2y + z &= 6 \\ 3x + y - 2z &= 2 \\ 2x - 3y + 5z &= 1 \end{aligned}$$

A system of linear equations in three variables can be solved by using the addition method. First, eliminate one variable from any two of the given equations. Then eliminate the same variable from any other two equations. The result will be a system of two equations in two variables. Solve this system by the addition method.

HOW TO • 4 Solve:

$$\begin{aligned} (1) \quad x + 4y - z &= 10 \\ (2) \quad 3x + 2y + z &= 4 \\ (3) \quad 2x - 3y + 2z &= -7 \end{aligned}$$

Eliminate z from Equations (1) and (2) by adding the two equations.

$$\begin{aligned} x + 4y - z &= 10 \\ 3x + 2y + z &= 4 \\ \hline (4) \quad 4x + 6y &= 14 \end{aligned}$$

• Add the equations.

Eliminate z from Equations (1) and (3). Multiply Equation (1) by 2 and add to Equation (3).

$$\begin{aligned} 2x + 8y - 2z &= 20 \\ 2x - 3y + 2z &= -7 \\ \hline (5) \quad 4x + 5y &= 13 \end{aligned}$$

• 2 times Equation (1).
• This is Equation (3).
• Add the equations.

Using Equations (4) and (5), solve the system of two equations in two variables.

$$\begin{aligned} (4) \quad 4x + 6y &= 14 \\ (5) \quad 4x + 5y &= 13 \end{aligned}$$

Eliminate x. Multiply Equation (5) by -1 and add to Equation (4).

$$\begin{aligned} 4x + 6y &= 14 \\ -4x - 5y &= -13 \\ \hline y &= 1 \end{aligned}$$

• This is Equation (4).
• -1 times Equation (5).
• Add the equations.

Substitute the value of y into Equation (4) or Equation (5) and solve for x. Equation (4) is used here.

$$\begin{aligned} 4x + 6y &= 14 \\ 4x + 6(1) &= 14 \\ 4x + 6 &= 14 \\ 4x &= 8 \\ x &= 2 \end{aligned}$$

• This is Equation (4).
• $y = 1$
• Solve for x.

Substitute the value of y and the value of x into one of the equations in the original system. Equation (2) is used here.

$$\begin{aligned} 3x + 2y + z &= 4 \\ 3(2) + 2(1) + z &= 4 \\ 6 + 2 + z &= 4 \\ 8 + z &= 4 \\ z &= -4 \end{aligned}$$

• $x = 2, y = 1$

The solution is $(2, 1, -4)$.

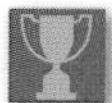

Tips for Success

Always check the proposed solution of a system of equations. For the system at the right:

$$\begin{array}{r|l} x + 4y - z = 10 & \\ \hline 2 + 4(1) - (-4) & 10 \\ 10 = 10 & \end{array}$$

$$\begin{array}{r|l} 3x + 2y + z = 4 & \\ \hline 3(2) + 2(1) + (-4) & 4 \\ 4 = 4 & \end{array}$$

$$\begin{array}{r|l} 2x - 3y + 2z = -7 & \\ \hline 2(2) - 3(1) + 2(-4) & -7 \\ -7 = -7 & \end{array}$$

The solution checks.

HOW TO • 5 Solve:

$$\begin{aligned} (1)\quad 2x - 3y - z &= 1 \\ (2)\quad x + 4y + 3z &= 2 \\ (3)\quad 4x - 6y - 2z &= 5 \end{aligned}$$

Eliminate x from Equations (1) and (2).

$$\begin{aligned} 2x - 3y - z &= 1 \\ -2x - 8y - 6z &= -4 \\ \hline -11y - 7z &= -3 \end{aligned}$$

- This is Equation (1).
- −2 times Equation (2).
- Add the equations.

Eliminate x from Equations (1) and (3).

$$\begin{aligned} -4x + 6y + 2z &= -2 \\ 4x - 6y - 2z &= 5 \\ \hline 0 &= 3 \end{aligned}$$

- −2 times Equation (1).
- This is Equation (3).
- Add the equations.

The equation $0 = 3$ is not a true equation. The system of equations is inconsistent and therefore has no solution.

HOW TO • 6 Solve:

$$\begin{aligned} (1)\quad 3x - z &= -1 \\ (2)\quad 2y - 3z &= 10 \\ (3)\quad x + 3y - z &= 7 \end{aligned}$$

Eliminate x from Equations (1) and (3). Multiply Equation (3) by -3 and add to Equation (1).

$$\begin{aligned} 3x \qquad - z &= -1 \\ -3x - 9y + 3z &= -21 \\ \hline (4)\qquad -9y + 2z &= -22 \end{aligned}$$

- This is Equation (1).
- −3 times Equation (3).
- Add the equations.

Use Equations (2) and (4) to form a system of equations in two variables.

$$\begin{aligned} (2)\quad 2y - 3z &= 10 \\ (4)\quad -9y + 2z &= -22 \end{aligned}$$

Eliminate z. Multiply Equation (2) by 2 and Equation (4) by 3.

$$\begin{aligned} 4y - 6z &= 20 \\ -27y + 6z &= -66 \\ \hline -23y &= -46 \\ y &= 2 \end{aligned}$$

- 2 times Equation (2).
- 3 times Equation (4).
- Add the equations.
- Solve for y.

Substitute the value of y into Equation (2) or Equation (4) and solve for z. Equation (2) is used here.

$$\begin{aligned} (2)\quad 2y - 3z &= 10 \\ 2(2) - 3z &= 10 \\ 4 - 3z &= 10 \\ -3z &= 6 \\ z &= -2 \end{aligned}$$

- This is Equation (2).
- $y = 2$
- Solve for z.

Substitute the value of z into Equation (1) and solve for x.

$$\begin{aligned} (1)\quad 3x - z &= -1 \\ 3x - (-2) &= -1 \\ 3x + 2 &= -1 \\ 3x &= -3 \\ x &= -1 \end{aligned}$$

- This is Equation (1).
- $z = -2$
- Solve for x.

The solution is $(-1, 2, -2)$.

EXAMPLE 3

Solve:
(1) $3x - y + 2z = 1$
(2) $2x + 3y + 3z = 4$
(3) $x + y - 4z = -9$

Solution

Eliminate y. Add Equations (1) and (3).

$$\begin{aligned} 3x - y + 2z &= 1 \\ x + y - 4z &= -9 \\ \hline 4x - 2z &= -8 \end{aligned}$$

Multiply each side of the equation by $\frac{1}{2}$.

$2x - z = -4$ • Equation (4)

Multiply Equation (1) by 3 and add to Equation (2).

$$\begin{aligned} 9x - 3y + 6z &= 3 \\ 2x + 3y + 3z &= 4 \\ \hline 11x + 9z &= 7 \end{aligned}$$ • Equation (5)

Solve the system of two equations.

(4) $2x - z = -4$
(5) $11x + 9z = 7$

Multiply Equation (4) by 9 and add to Equation (5).

$$\begin{aligned} 18x - 9z &= -36 \\ 11x + 9z &= 7 \\ \hline 29x &= -29 \\ x &= -1 \end{aligned}$$

Replace x by -1 in Equation (4).

$$\begin{aligned} 2x - z &= -4 \\ 2(-1) - z &= -4 \\ -2 - z &= -4 \\ -z &= -2 \\ z &= 2 \end{aligned}$$

Replace x by -1 and z by 2 in Equation (3).

$$\begin{aligned} x + y - 4z &= -9 \\ -1 + y - 4(2) &= -9 \\ -9 + y &= -9 \\ y &= 0 \end{aligned}$$

The solution is $(-1, 0, 2)$.

YOU TRY IT 3

Solve:
$x - y + z = 6$
$2x + 3y - z = 1$
$x + 2y + 2z = 5$

Your solution

Solution on p. S16

5.2 EXERCISES

OBJECTIVE A **To solve a system of two linear equations in two variables by the addition method**

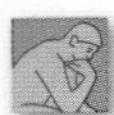

For Exercises 1 and 2, use the system of equations at the right.

$$(1)\ 5x - 7y = 9$$
$$(2)\ 6x + 3y = 12$$

1. If you were using the addition method to eliminate x, you could multiply Equation (1) by ______ and Equation (2) by ______ and then add the resulting equations.

2. If you were using the addition method to eliminate y, you could multiply Equation (1) by ______ and Equation (2) by ______ and then add the resulting equations.

For Exercises 3 to 44, solve by the addition method.

3. $x - y = 5$
$x + y = 7$

4. $x + y = 1$
$2x - y = 5$

5. $3x + y = 4$
$x + y = 2$

6. $x - 3y = 4$
$x + 5y = -4$

7. $3x + y = 7$
$x + 2y = 4$

8. $x - 2y = 7$
$3x - 2y = 9$

9. $2x + 3y = -1$
$x + 5y = 3$

10. $x + 5y = 7$
$2x + 7y = 8$

11. $3x - y = 4$
$6x - 2y = 8$

12. $x - 2y = -3$
$-2x + 4y = 6$

13. $2x + 5y = 9$
$4x - 7y = -16$

14. $8x - 3y = 21$
$4x + 5y = -9$

15. $4x - 6y = 5$
$2x - 3y = 7$

16. $3x + 6y = 7$
$2x + 4y = 5$

17. $3x - 5y = 7$
$x - 2y = 3$

18. $3x + 4y = 25$
$2x + y = 10$

19. $x + 3y = 7$
$-2x + 3y = 22$

20. $2x - 3y = 14$
$5x - 6y = 32$

21. $3x + 2y = 16$
$2x - 3y = -11$

22. $2x - 5y = 13$
$5x + 3y = 17$

23. $4x + 4y = 5$
$2x - 8y = -5$

24. $3x + 7y = 16$
$4x - 3y = 9$

25. $5x + 4y = 0$
$3x + 7y = 0$

26. $3x - 4y = 0$
$4x - 7y = 0$

27. $5x + 2y = 1$
$2x + 3y = 7$

28. $3x + 5y = 16$
$5x - 7y = -4$

29. $3x - 6y = 6$
$9x - 3y = 8$

30. $\frac{2}{3}x - \frac{1}{2}y = 3$
$\frac{1}{3}x - \frac{1}{4}y = \frac{3}{2}$

31. $\frac{3}{4}x + \frac{1}{3}y = -\frac{1}{2}$
$\frac{1}{2}x - \frac{5}{6}y = -\frac{7}{2}$

32. $\frac{2}{5}x - \frac{1}{3}y = 1$
$\frac{3}{5}x + \frac{2}{3}y = 5$

33. $\frac{5x}{6} + \frac{y}{3} = \frac{4}{3}$
$\frac{2x}{3} - \frac{y}{2} = \frac{11}{6}$

34. $\frac{3x}{4} + \frac{2y}{5} = -\frac{3}{20}$
$\frac{3x}{2} - \frac{y}{4} = \frac{3}{4}$

35. $\frac{2x}{5} - \frac{y}{2} = \frac{13}{2}$
$\frac{3x}{4} - \frac{y}{5} = \frac{17}{2}$

36. $\frac{x}{2} + \frac{y}{3} = \frac{5}{12}$
$\frac{x}{2} - \frac{y}{3} = \frac{1}{12}$

37. $\frac{3x}{2} - \frac{y}{4} = -\frac{11}{12}$
$\frac{x}{3} - y = -\frac{5}{6}$

38. $\frac{3x}{4} - \frac{2y}{3} = 0$
$\frac{5x}{4} - \frac{y}{3} = \frac{7}{12}$

39. $4x - 5y = 3y + 4$
$2x + 3y = 2x + 1$

40. $5x - 2y = 8x - 1$
$2x + 7y = 4y + 9$

41. $2x + 5y = 5x + 1$
$3x - 2y = 3y + 3$

42. $4x - 8y = 5$
$8x + 2y = 1$

43. $5x + 2y = 2x + 1$
$2x - 3y = 3x + 2$

44. $3x + 3y = y + 1$
$x + 3y = 9 - x$

OBJECTIVE B To solve a system of three linear equations in three variables by the addition method

For Exercises 45 to 68, solve by the addition method.

45. $x + 2y - z = 1$
$2x - y + z = 6$
$x + 3y - z = 2$

46. $x + 3y + z = 6$
$3x + y - z = -2$
$2x + 2y - z = 1$

47. $2x - y + 2z = 7$
$x + y + z = 2$
$3x - y + z = 6$

48. $x - 2y + z = 6$
$x + 3y + z = 16$
$3x - y - z = 12$

49. $3x + y = 5$
$3y - z = 2$
$x + z = 5$

50. $2y + z = 7$
$2x - z = 3$
$x - y = 3$

51. $x - y + z = 1$
$2x + 3y - z = 3$
$-x + 2y - 4z = 4$

52. $2x + y - 3z = 7$
$x - 2y + 3z = 1$
$3x + 4y - 3z = 13$

53. $2x + 3z = 5$
$3y + 2z = 3$
$3x + 4y = -10$

54. $3x + 4z = 5$
$2y + 3z = 2$
$2x - 5y = 8$

55. $2x + 4y - 2z = 3$
$x + 3y + 4z = 1$
$x + 2y - z = 4$

56. $x - 3y + 2z = 1$
$x - 2y + 3z = 5$
$2x - 6y + 4z = 3$

57. $2x + y - z = 5$
$x + 3y + z = 14$
$3x - y + 2z = 1$

58. $3x - y - 2z = 11$
$2x + y - 2z = 11$
$x + 3y - z = 8$

59. $3x + y - 2z = 2$
$x + 2y + 3z = 13$
$2x - 2y + 5z = 6$

60. $\begin{aligned} 4x + 5y + z &= 6 \\ 2x - y + 2z &= 11 \\ x + 2y + 2z &= 6 \end{aligned}$

61. $\begin{aligned} 2x - y + z &= 6 \\ 3x + 2y + z &= 4 \\ x - 2y + 3z &= 12 \end{aligned}$

62. $\begin{aligned} 3x + 2y - 3z &= 8 \\ 2x + 3y + 2z &= 10 \\ x + y - z &= 2 \end{aligned}$

63. $\begin{aligned} 3x - 2y + 3z &= -4 \\ 2x + y - 3z &= 2 \\ 3x + 4y + 5z &= 8 \end{aligned}$

64. $\begin{aligned} 3x - 3y + 4z &= 6 \\ 4x - 5y + 2z &= 10 \\ x - 2y + 3z &= 4 \end{aligned}$

65. $\begin{aligned} 3x - y + 2z &= 2 \\ 4x + 2y - 7z &= 0 \\ 2x + 3y - 5z &= 7 \end{aligned}$

66. $\begin{aligned} 2x + 2y + 3z &= 13 \\ -3x + 4y - z &= 5 \\ 5x - 3y + z &= 2 \end{aligned}$

67. $\begin{aligned} 2x - 3y + 7z &= 0 \\ x + 4y - 4z &= -2 \\ 3x + 2y + 5z &= 1 \end{aligned}$

68. $\begin{aligned} 5x + 3y - z &= 5 \\ 3x - 2y + 4z &= 13 \\ 4x + 3y + 5z &= 22 \end{aligned}$

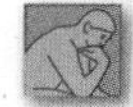

69. In the following sentences, fill in the blanks with one of the following phrases: (i) exactly one point, (ii) more than one point, or (iii) no points.

a. For an inconsistent system of linear equations in three variables, the planes representing the equations intersect at _______.

b. For a dependent system of linear equations in three variables, the planes representing the equations intersect at _______.

c. For an independent system of linear equations in three variables, the planes representing the equations intersect at _______.

Applying the Concepts

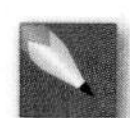

70. Describe the graph of each of the following equations in an *xyz*-coordinate system.

a. $x = 3$ **b.** $y = 4$ **c.** $z = 2$ **d.** $y = x$

In Exercises 71 to 74, the systems are not systems of linear equations. However, they can be solved by using a modification of the addition method. Solve each system of equations.

71. $\begin{aligned} \frac{1}{x} - \frac{2}{y} &= 3 \\ \frac{2}{x} + \frac{3}{y} &= -1 \end{aligned}$

72. $\begin{aligned} \frac{1}{x} + \frac{2}{y} &= 3 \\ \frac{1}{x} - \frac{3}{y} &= -2 \end{aligned}$

73. $\begin{aligned} \frac{3}{x} + \frac{2}{y} &= 1 \\ \frac{2}{x} + \frac{4}{y} &= -2 \end{aligned}$

74. $\begin{aligned} \frac{3}{x} - \frac{5}{y} &= -\frac{3}{2} \\ \frac{1}{x} - \frac{2}{y} &= -\frac{2}{3} \end{aligned}$

SECTION

5.3 Application Problems

OBJECTIVE A To solve rate-of-wind or rate-of-current problems 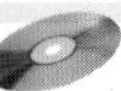

Solving motion problems that involve an object moving with or against a wind or current normally requires two variables.

HOW TO • 1 A motorboat traveling with the current can go 24 mi in 2 h. Against the current, it takes 3 h to go the same distance. Find the rate of the motorboat in calm water and the rate of the current.

Strategy for Solving Rate-of-Wind or Rate-of-Current Problems

1. Choose one variable to represent the rate of the object in calm conditions and a second variable to represent the rate of the wind or current. Using these variables, express the rate of the object with and against the wind or current. Use the equation $rt = d$ to write expressions for the distance traveled by the object. The results can be recorded in a table.

Rate of the boat in calm water: x
Rate of the current: y

	Rate	·	Time	=	Distance
With the current	$x + y$	·	2	=	$2(x + y)$
Against the current	$x - y$	·	3	=	$3(x - y)$

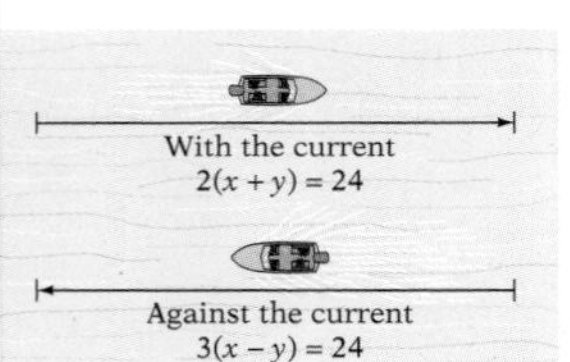

2. Determine how the expressions for distance are related.

The distance traveled with the current is 24 mi: $2(x + y) = 24$
The distance traveled against the current is 24 mi: $3(x - y) = 24$

Solve the system of equations.

$$2(x + y) = 24 \xrightarrow{\text{Multiply by } \frac{1}{2}.} \frac{1}{2} \cdot 2(x + y) = \frac{1}{2} \cdot 24 \longrightarrow x + y = 12$$

$$3(x - y) = 24 \xrightarrow{\text{Multiply by } \frac{1}{3}.} \frac{1}{3} \cdot 3(x - y) = \frac{1}{3} \cdot 24 \longrightarrow x - y = 8$$

$$2x = 20$$
$$x = 10$$

• Add the equations.

Replace x by 10 in the equation $x + y = 12$.
Solve for y.

$$x + y = 12$$
$$10 + y = 12$$
$$y = 2$$

The rate of the boat in calm water is 10 mph.
The rate of the current is 2 mph.

EXAMPLE • 1

Flying with the wind, a plane flew 1000 mi in 5 h. Flying against the wind, the plane could fly only 500 mi in the same amount of time. Find the rate of the plane in calm air and the rate of the wind.

Strategy

- Rate of the plane in still air: p
  Rate of the wind: w

	Rate	*Time*	*Distance*
With wind	$p + w$	5	$5(p + w)$
Against wind	$p - w$	5	$5(p - w)$

- The distance traveled with the wind is 1000 mi.
  The distance traveled against the wind is 500 mi.

$$5(p + w) = 1000$$
$$5(p - w) = 500$$

Solution

$$5(p + w) = 1000 \qquad \frac{1}{5} \cdot 5(p + w) = \frac{1}{5} \cdot 1000$$

$$5(p - w) = 500 \qquad \frac{1}{5} \cdot 5(p - w) = \frac{1}{5} \cdot 500$$

$$\begin{aligned} p + w &= 200 \\ p - w &= 100 \\ \hline 2p &= 300 \\ p &= 150 \end{aligned}$$

$$\begin{aligned} p + w &= 200 \\ 150 + w &= 200 \\ w &= 50 \end{aligned}$$

- Substitute **150** for p.

The rate of the plane in calm air is 150 mph. The rate of the wind is 50 mph.

YOU TRY IT • 1

A rowing team rowing with the current traveled 18 mi in 2 h. Against the current, the team rowed 10 mi in 2 h. Find the rate of the rowing team in calm water and the rate of the current.

Your strategy

Your solution

Solution on p. S16

OBJECTIVE B To solve application problems 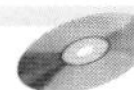

The application problems in this section are varieties of problems solved earlier in the text. Each of the strategies for the problems in this section will result in a system of equations.

HOW TO • 2 A store owner purchased twenty 60-watt light bulbs and 30 fluorescent bulbs for a total cost of \$80. A second purchase, at the same prices, included thirty 60-watt light bulbs and 10 fluorescent bulbs for a total cost of \$50. Find the cost of a 60-watt bulb and that of a fluorescent bulb.

Strategy for Solving Application Problems

1. Choose a variable to represent each of the unknown quantities. Write numerical or variable expressions for all the remaining quantities. These results may be recorded in tables, one for each condition.

Cost of 60-watt bulb: b
Cost of fluorescent bulb: f

First Purchase

	Amount	·	*Unit Cost*	=	*Value*
60-watt	20	·	b	=	$20b$
Fluorescent	30	·	f	=	$30f$

Second Purchase

	Amount	·	*Unit Cost*	=	*Value*
60-watt	30	·	b	=	$30b$
Fluorescent	10	·	f	=	$10f$

2. Determine a system of equations. The strategies presented in the chapter on First-Degree Equations and Inequalities can be used to determine the relationships among the expressions in the tables. Each table will give one equation of the system of equations.

The total of the first purchase was \$80: $20b + 30f = 80$
The total of the second purchase was \$50: $30b + 10f = 50$

Solve the system of equations: (1) $20b + 30f = 80$
(2) $30b + 10f = 50$

$$
\begin{aligned}
60b + 90f &= 240 \\
-60b - 20f &= -100 \\
\hline
70f &= 140 \\
f &= 2
\end{aligned}
$$

• **3 times Equation (1).**
• **−2 times Equation (2).**

Replace f by 2 in Equation (1) and solve for b.

$$
\begin{aligned}
20b + 30f &= 80 \\
20b + 30(2) &= 80 \\
20b + 60 &= 80 \\
20b &= 20 \\
b &= 1
\end{aligned}
$$

The cost of a 60-watt bulb was \$1.00.
The cost of a fluorescent bulb was \$2.00.

Some application problems may require more than two variables, as shown in Example 2 and You Try It 2 on the next page.

EXAMPLE 2

An investor has a total of \$20,000 deposited in three different accounts, which earn annual interest rates of 9%, 7%, and 5%. The amount deposited in the 9% account is twice the amount in the 7% account. If the total annual interest earned for the three accounts is \$1300, how much is invested in each account?

Strategy

- Amount invested at 9%: x
  Amount invested at 7%: y
  Amount invested at 5%: z

	Principal	*Rate*	*Interest*
Amount at 9%	x	0.09	$0.09x$
Amount at 7%	y	0.07	$0.07y$
Amount at 5%	z	0.05	$0.05z$

- The amount invested at 9% (x) is twice the amount invested at 7% (y): $x = 2y$
  The sum of the interest earned for all three accounts is \$1300:
  $0.09x + 0.07y + 0.05z = 1300$
  The total amount invested is \$20,000:
  $x + y + z = 20{,}000$

Solution

(1) $x = 2y$
(2) $0.09x + 0.07y + 0.05z = 1300$
(3) $x + y + z = 20{,}000$

Solve the system of equations. Substitute $2y$ for x in Equation (2) and Equation (3).

$0.09(2y) + 0.07y + 0.05z = 1300$
$2y + y + z = 20{,}000$

(4) $0.25y + 0.05z = 1300$ • $0.09(2y) + 0.07y = 0.25y$
(5) $3y + z = 20{,}000$ • $2y + y = 3y$

Solve the system of equations in two variables by multiplying Equation (5) by -0.05 and adding to Equation (4).

$$\begin{aligned} 0.25y + 0.05z &= 1300 \\ -0.15y - 0.05z &= -1000 \\ \hline 0.10y &= 300 \\ y &= 3000 \end{aligned}$$

Substituting the value of y into Equation (1), $x = 6000$.
Substituting the values of x and y into Equation (3), $z = 11{,}000$.

The investor placed \$6000 in the 9% account, \$3000 in the 7% account, and \$11,000 in the 5% account.

YOU TRY IT 2

A coin bank contains only nickels, dimes, and quarters. The value of the 19 coins in the bank is \$2. If there are twice as many nickels as dimes, find the number of each type of coin in the bank.

Your strategy

Your solution

Solution on p. S17

5.3 EXERCISES

OBJECTIVE A To solve rate-of-wind or rate-of-current problems

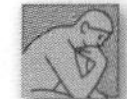

1. Traveling with the wind, a plane flies m miles in h hours. Traveling against the wind, the plane flies n miles in h hours. Is n less than, equal to, or greater than m?

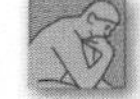

2. Traveling against the current, it takes a boat h hours to go m miles. Traveling with the current, the boat takes k hours to go m miles. Is h less than, equal to, or greater than k?

3. A motorboat traveling with the current went 36 mi in 2 h. Against the current, it took 3 h to travel the same distance. Find the rate of the boat in calm water and the rate of the current.

4. A cabin cruiser traveling with the current went 45 mi in 3 h. Against the current, it took 5 h to travel the same distance. Find the rate of the cabin cruiser in calm water and the rate of the current.

5. A jet plane flying with the wind went 2200 mi in 4 h. Against the wind, the plane could fly only 1820 mi in the same amount of time. Find the rate of the plane in calm air and the rate of the wind.

6. Flying with the wind, a small plane flew 300 mi in 2 h. Against the wind, the plane could fly only 270 mi in the same amount of time. Find the rate of the plane in calm air and the rate of the wind.

7. A rowing team rowing with the current traveled 20 km in 2 h. Rowing against the current, the team rowed 12 km in the same amount of time. Find the rate of the team in calm water and the rate of the current.

8. A motorboat traveling with the current went 72 km in 3 h. Against the current, the boat could go only 48 km in the same amount of time. Find the rate of the boat in calm water and the rate of the current.

9. A turboprop plane flying with the wind flew 800 mi in 4 h. Flying against the wind, the plane required 5 h to travel the same distance. Find the rate of the wind and the rate of the plane in calm air.

10. Flying with the wind, a pilot flew 600 mi between two cities in 4 h. The return trip against the wind took 5 h. Find the rate of the plane in calm air and the rate of the wind.

11. A plane flying with a tailwind flew 600 mi in 5 h. Against the wind, the plane required 6 h to fly the same distance. Find the rate of the plane in calm air and the rate of the wind.

12. Flying with the wind, a plane flew 720 mi in 3 h. Against the wind, the plane required 4 h to fly the same distance. Find the rate of the plane in calm air and the rate of the wind.

OBJECTIVE B To solve application problems

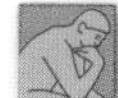

13. A coffee merchant's house blend contains 3 lb of dark roast coffee and 1 lb of light roast coffee. The merchant's breakfast blend contains 1 lb of the dark roast coffee and 3 lb of the light roast coffee. If the cost per pound of the house blend is greater than the cost per pound of the breakfast blend, is the cost per pound of the dark roast coffee less than, equal to, or greater than the cost per pound of the light roast coffee?

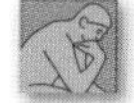

14. The total value of dimes and quarters in a bank is V dollars. If the dimes were quarters and the quarters were dimes, the total value would be more than V dollars. Is the number of quarters in the bank less than, equal to, or greater than the number of dimes in the bank?

15. Coins A coin bank contains only nickels and dimes. The total value of the coins in the bank is \$2.50. If the nickels were dimes and the dimes were nickels, the total value of the coins would be \$3.50. Find the number of nickels in the bank.

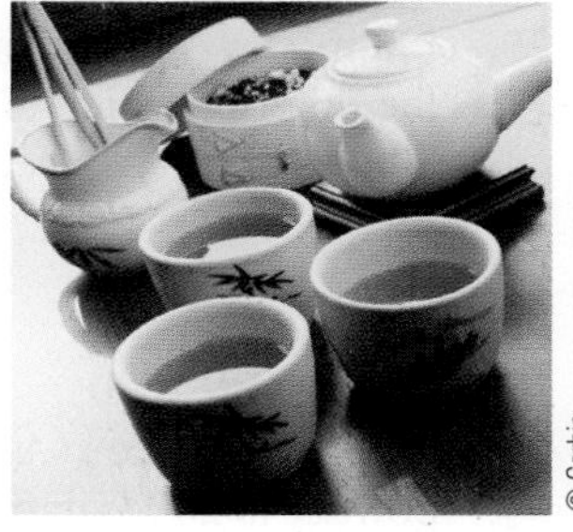

© Corbis

16. Business A merchant mixed 10 lb of cinnamon tea with 5 lb of spice tea. The 15-pound mixture cost \$40. A second mixture included 12 lb of the cinnamon tea and 8 lb of the spice tea. The 20-pound mixture cost \$54. Find the cost per pound of the cinnamon tea and of the spice tea.

17. Purchasing A carpenter purchased 60 ft of redwood and 80 ft of pine for a total cost of \$286. A second purchase, at the same prices, included 100 ft of redwood and 60 ft of pine for a total cost of \$396. Find the cost per foot of redwood and of pine.

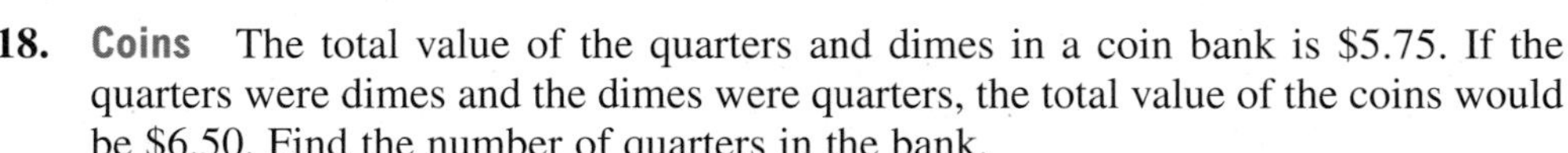

18. Coins The total value of the quarters and dimes in a coin bank is \$5.75. If the quarters were dimes and the dimes were quarters, the total value of the coins would be \$6.50. Find the number of quarters in the bank.

19. Purchasing A contractor buys 16 yd of nylon carpet and 20 yd of wool carpet for \$1840. A second purchase, at the same prices, includes 18 yd of nylon carpet and 25 yd of wool carpet for \$2200. Find the cost per yard of the wool carpet.

20. Finances During one month, a homeowner used 500 units of electricity and 100 units of gas for a total cost of \$352. The next month, 400 units of electricity and 150 units of gas were used for a total cost of \$304. Find the cost per unit of gas.

21. Manufacturing A company manufactures both mountain bikes and trail bikes. The cost of materials for a mountain bike is \$70, and the cost of materials for a trail bike is \$50. The cost of labor to manufacture a mountain bike is \$80, and the cost of labor to manufacture a trail bike is \$40. During a week in which the company has budgeted \$2500 for materials and \$2600 for labor, how many mountain bikes does the company plan to manufacture?

22. Manufacturing A company manufactures both LCD and plasma televisions. The cost for materials for an LCD television is $125, and the cost of materials for a plasma TV is $150. The cost of labor to manufacture one LCD television is $80, and the cost of labor for one plasma television is $85. How many of each television can a manufacturer produce during a week in which $18,000 has been budgeted for materials and $10,750 has been budgeted for labor?

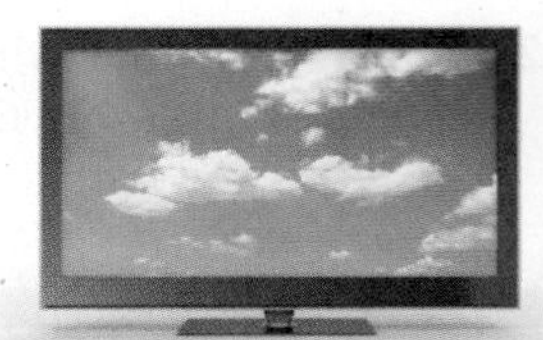

Fuel Economy Use the information in the article at the right for Exercises 23 and 24.

23. One week, the owner of a hybrid car drove 394 mi and spent $34.74 on gasoline. How many miles did the owner drive in the city? On the highway?

24. Gasoline for one week of driving cost the owner of a hybrid car $26.50. The owner would have spent $51.50 for gasoline to drive the same number of miles in a traditional car. How many miles did the owner drive in the city? On the highway?

In the News

Hybrids Easier on the Pocketbook?

A hybrid car can make up for its high sticker price with savings at the pump. At current gas prices, here's a look at the cost per mile for one company's hybrid and traditional cars.

Gasoline Cost per Mile

Car Type	*City ($/mi)*	*Highway ($/mi)*
Hybrid	0.09	0.08
Traditional	0.18	0.13

Source: www.fueleconomy.gov

25. Chemistry A chemist has two alloys, one of which is 10% gold and 15% lead, and the other of which is 30% gold and 40% lead. How many grams of each of the two alloys should be used to make an alloy that contains 60 g of gold and 88 g of lead?

26. Health Science A pharmacist has two vitamin-supplement powders. The first powder is 20% vitamin B1 and 10% vitamin B2. The second is 15% vitamin B1 and 20% vitamin B2. How many milligrams of each powder should the pharmacist use to make a mixture that contains 130 mg of vitamin B1 and 80 mg of vitamin B2?

27. Business On Monday, a computer manufacturing company sent out three shipments. The first order, which contained a bill for $114,000, was for 4 Model II, 6 Model VI, and 10 Model IX computers. The second shipment, which contained a bill for $72,000, was for 8 Model II, 3 Model VI, and 5 Model IX computers. The third shipment, which contained a bill for $81,000, was for 2 Model II, 9 Model VI, and 5 Model IX computers. What does the manufacturer charge for each Model VI computer?

28. Purchasing A relief organization supplies blankets, cots, and lanterns to victims of fires, floods, and other natural disasters. One week the organization purchased 15 blankets, 5 cots, and 10 lanterns for a total cost of $1250. The next week, at the same prices, the organization purchased 20 blankets, 10 cots, and 15 lanterns for a total cost of $2000. The next week, at the same prices, the organization purchased 10 blankets, 15 cots, and 5 lanterns for a total cost of $1625. Find the cost of one blanket, the cost of one cot, and the cost of one lantern.

29. **Investments** An investor has a total of \$25,000 deposited in three different accounts, which earn annual interest rates of 8%, 6%, and 4%. The amount deposited in the 8% account is twice the amount in the 6% account. If the three accounts earn total annual interest of \$1520, how much money is deposited in each account?

Applying the Concepts

30. **Geometry** Two angles are complementary. The measure of the larger angle is 9° more than eight times the measure of the smaller angle. Find the measures of the two angles. (Complementary angles are two angles whose sum is 90°.)

31. **Geometry** Two angles are supplementary. The measure of the larger angle is 40° more than three times the measure of the smaller angle. Find the measures of the two angles. (Supplementary angles are two angles whose sum is 180°.)

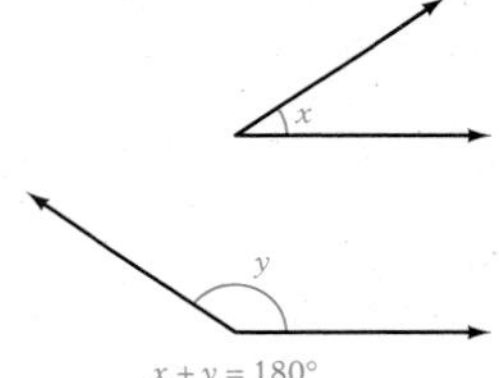

32. **Coins** The sum of the ages of a gold coin and a silver coin is 75 years. The age of the gold coin 10 years from now will be 5 years less than the age of the silver coin 10 years ago. Find the present ages of the two coins.

33. **Art** The difference between the ages of an oil painting and a watercolor is 35 years. The age of the oil painting 5 years from now will be twice the age of the watercolor 5 years ago. Find the present ages of each.

34. **Coins** A coin bank contains only dimes and quarters. The total value of all the coins is \$1. How many of each type of coin are in the bank? (*Hint:* There is more than one answer to this problem.)

SECTION

5.4 Solving Systems of Linear Inequalities

OBJECTIVE A To graph the solution set of a system of linear inequalities

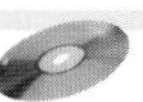

Point of Interest

Large systems of linear inequalities containing over 100 inequalities have been used to solve application problems in such diverse areas as providing health care and hardening a nuclear missile silo.

Two or more inequalities considered together are called a **system of inequalities.** The **solution set of a system of inequalities** is the intersection of the solution sets of the individual inequalities. To graph the solution set of a system of inequalities, first graph the solution set of each inequality. The solution set of the system of inequalities is the region of the plane represented by the intersection of the shaded areas.

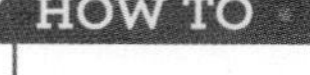

HOW TO 1 Graph the solution set: $2x - y \leq 3$, $3x + 2y > 8$

Solve each inequality for y.

$$2x - y \leq 3$$
$$-y \leq -2x + 3$$
$$y \geq 2x - 3$$

$$3x + 2y > 8$$
$$2y > -3x + 8$$
$$y > -\frac{3}{2}x + 4$$

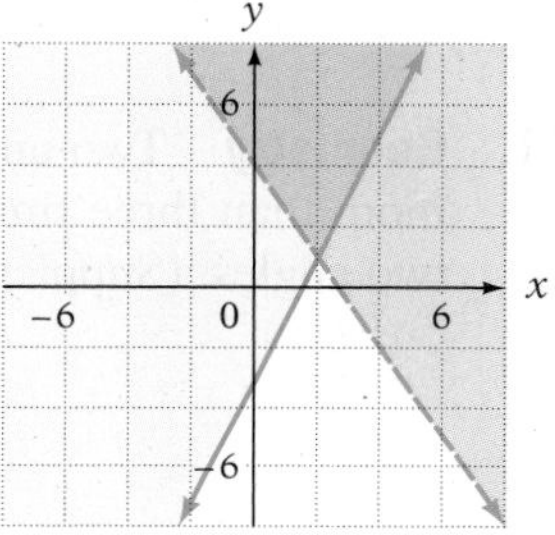

Graph $y = 2x - 3$ as a solid line. Because the inequality is $\geq$, shade above the line.

Graph $y = -\frac{3}{2}x + 4$ as a dashed line. Because the inequality is $>$, shade above the line.

The solution set of the system is the region of the plane represented by the intersection of the solution sets of the individual inequalities.

Take Note

You can use a test point to check that the correct region has been denoted as the solution set. We can see from the graph that the point (2, 4) is in the solution set and, as shown below, it is a solution of each inequality in the system. This indicates that the solution set as graphed is correct.

$$2x - y \leq 3$$
$$2(2) - (4) \leq 3$$
$$0 \leq 3 \quad \text{True}$$

$$3x + 2y > 8$$
$$3(2) + 2(4) > 8$$
$$14 > 8 \quad \text{True}$$

HOW TO 2 Graph the solution set: $-x + 2y \geq 4$, $x - 2y \geq 6$

Solve each inequality for y.

$$-x + 2y \geq 4$$
$$2y \geq x + 4$$
$$y \geq \frac{1}{2}x + 2$$

$$x - 2y \geq 6$$
$$-2y \geq -x + 6$$
$$y \leq \frac{1}{2}x - 3$$

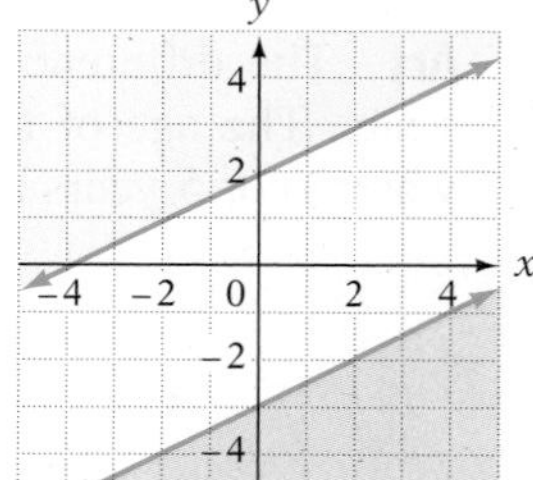

Shade above the solid line $y = \frac{1}{2}x + 2$.

Shade below the solid line $y = \frac{1}{2}x - 3$.

Because the solution sets of the two inequalities do not intersect, the solution of the system is the empty set.

Integrating Technology

See the Keystroke Guide: *Graphing Inequalities* for instructions on using a graphing calculator to graph the solution set of a system of inequalities.

EXAMPLE • 1

Graph the solution set: $y \geq x - 1$

$y < -2x$

Solution

Shade above the solid line $y = x - 1$.
Shade below the dashed line $y = -2x$.

The solution of the system is the intersection of the solution sets of the individual inequalities.

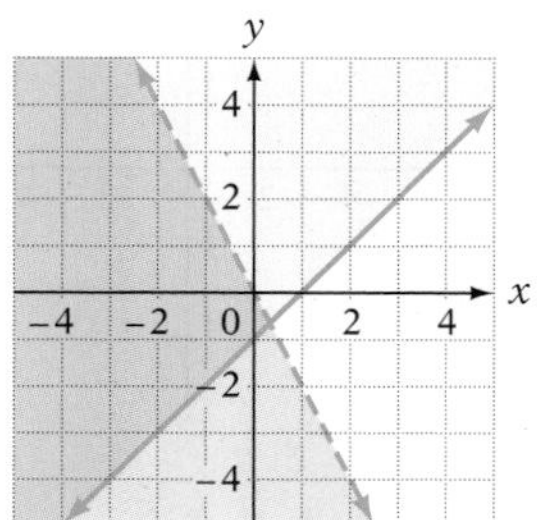

YOU TRY IT • 1

Graph the solution set: $y \geq 2x - 3$

$y > -3x$

Your solution

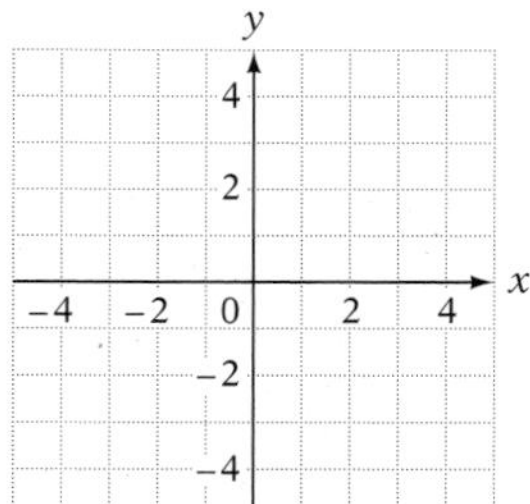

EXAMPLE • 2

Graph the solution set: $2x + 3y > 9$

$$y < -\frac{2}{3}x + 1$$

Solution

$2x + 3y > 9$

$3y > -2x + 9$

$y > -\frac{2}{3}x + 3$

Graph above the dashed line $y = -\frac{2}{3}x + 3$.

Graph below the dashed line $y = -\frac{2}{3}x + 1$.

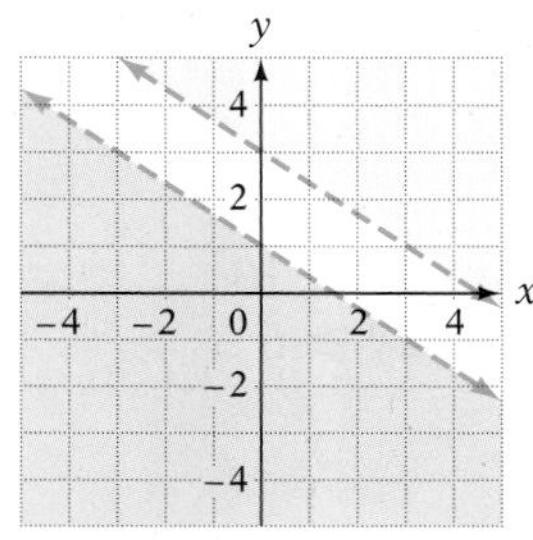

The intersection of the system is the empty set because the solution sets of the two inequalities do not intersect.

YOU TRY IT • 2

Graph the solution set: $3x + 4y > 12$

$$y < \frac{3}{4}x - 1$$

Your solution

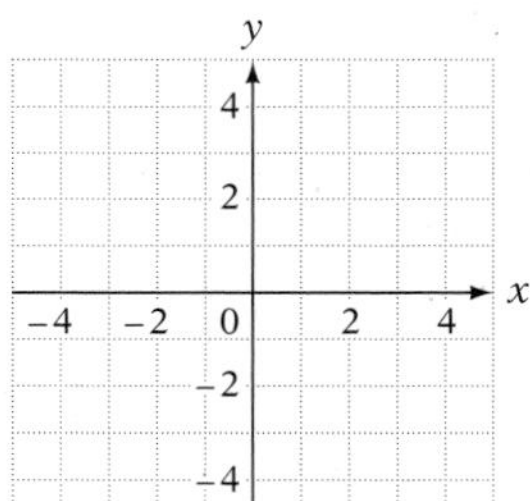

Solutions on p. S17

5.4 EXERCISES

OBJECTIVE A **To graph the solution set of a system of linear inequalities**

For Exercises 1 to 18, graph the solution set.

1. $x - y \geq 3$
$x + y \leq 5$

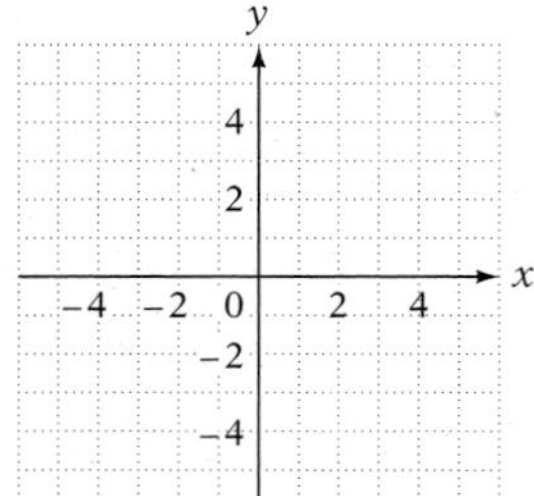

2. $2x - y < 4$
$x + y < 5$

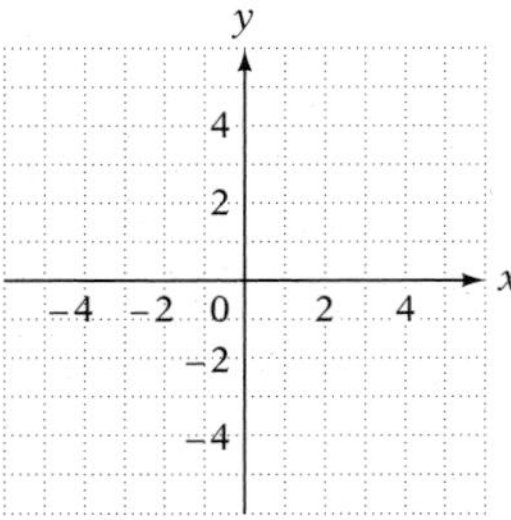

3. $3x - y < 3$
$2x + y \geq 2$

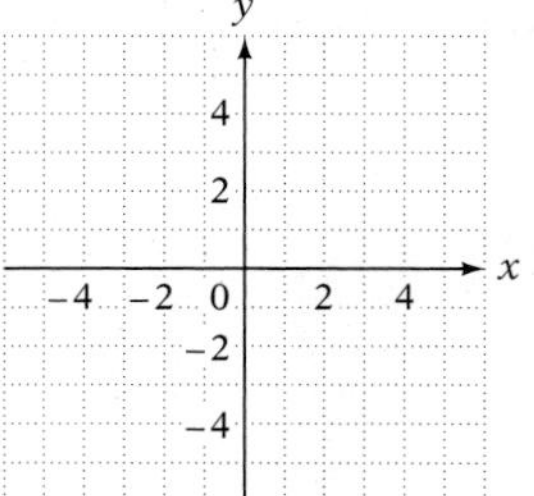

4. $x + 2y \leq 6$
$x - y \leq 3$

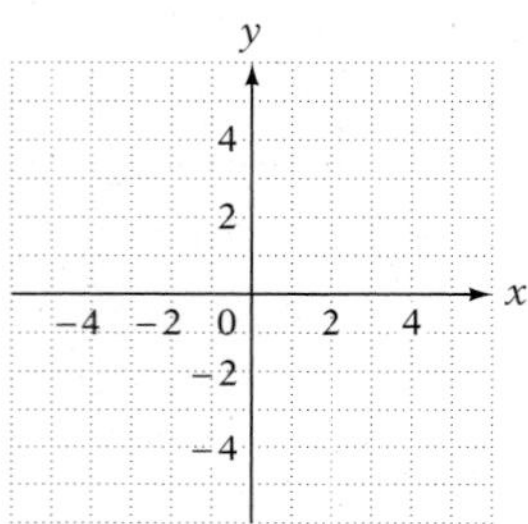

5. $2x + y \geq -2$
$6x + 3y \leq 6$

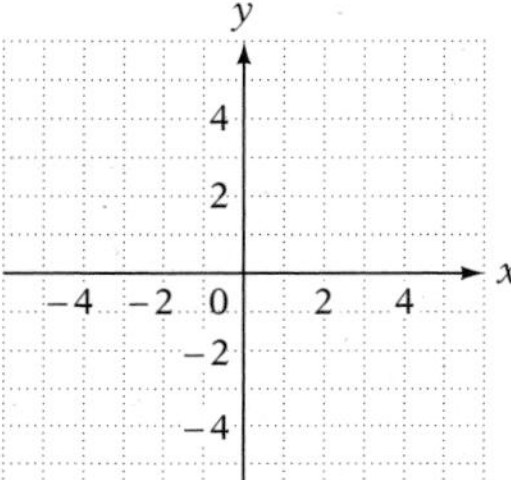

6. $x + y \geq 5$
$3x + 3y \leq 6$

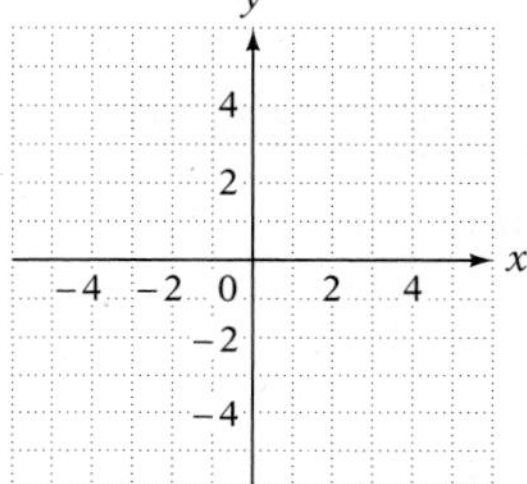

7. $3x - 2y < 6$
$y \leq 3$

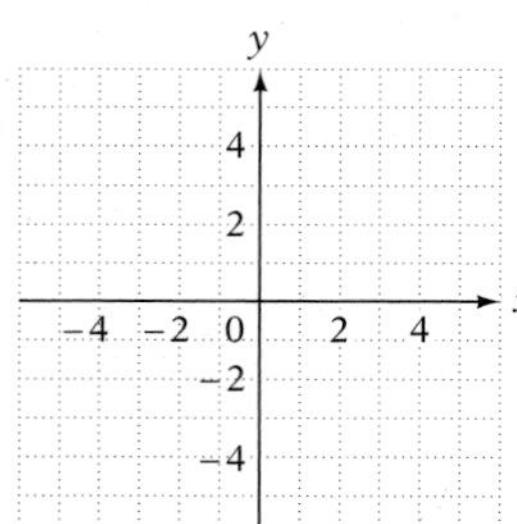

8. $x \leq 2$
$3x + 2y > 4$

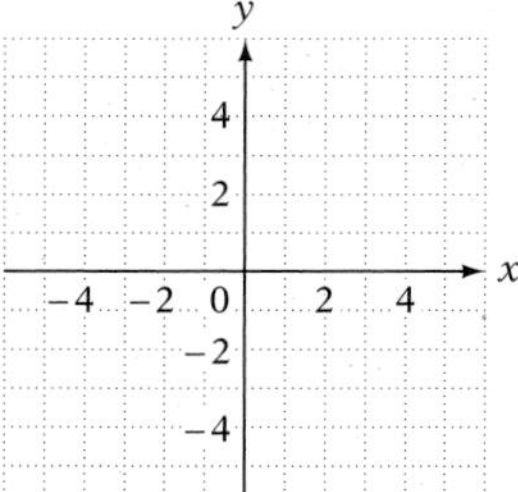

9. $y > 2x - 6$
$x + y < 0$

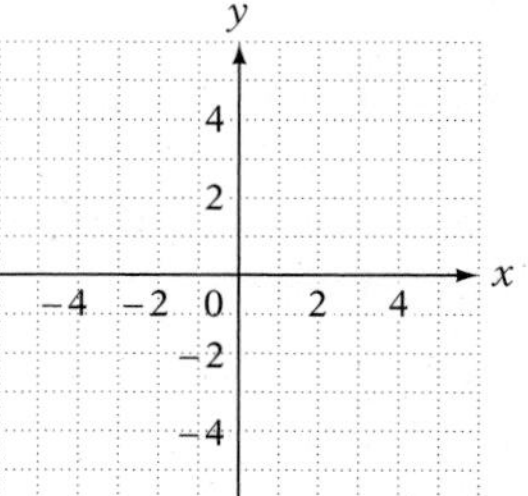

10. $x < 3$
$y < -2$

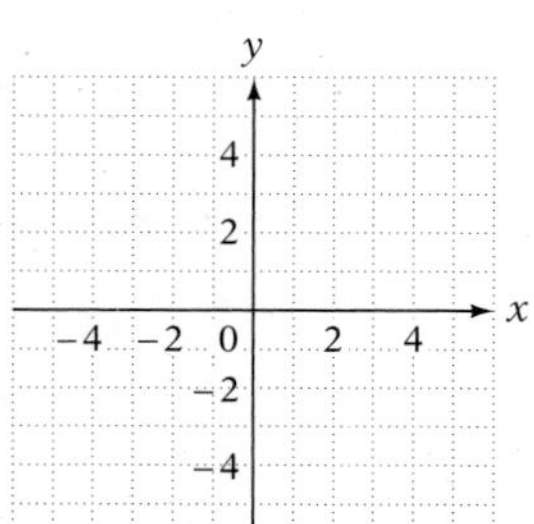

11. $x + 1 \geq 0$
$y - 3 \leq 0$

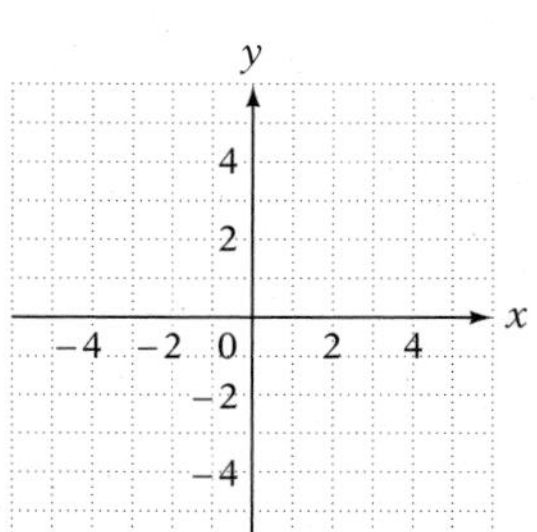

12. $5x - 2y \geq 10$
$3x + 2y \geq 6$

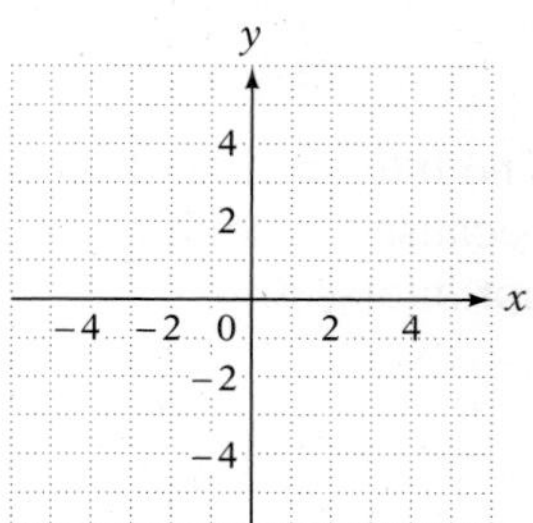

13. $2x + y \geq 4$
$3x - 2y < 6$

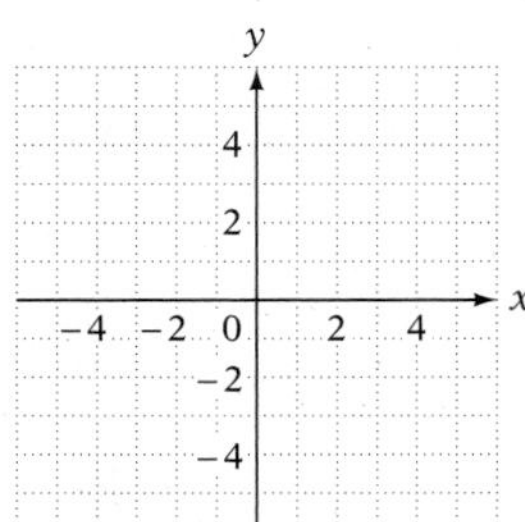

14. $3x - 4y < 12$
$x + 2y < 6$

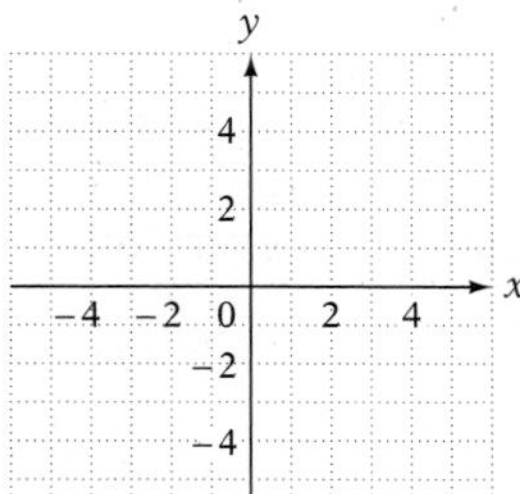

15. $x - 2y \leq 6$
$2x + 3y \leq 6$

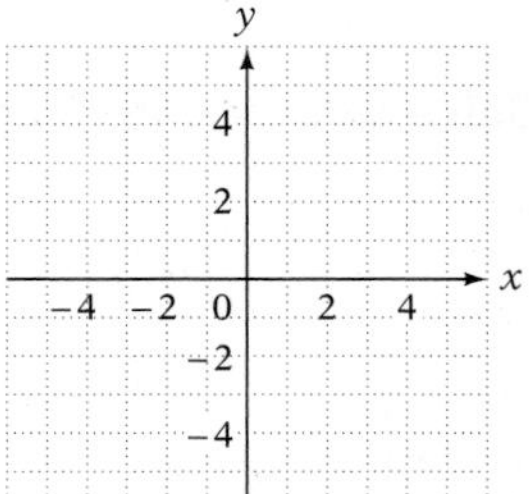

16. $x - 3y > 6$
$2x + y > 5$

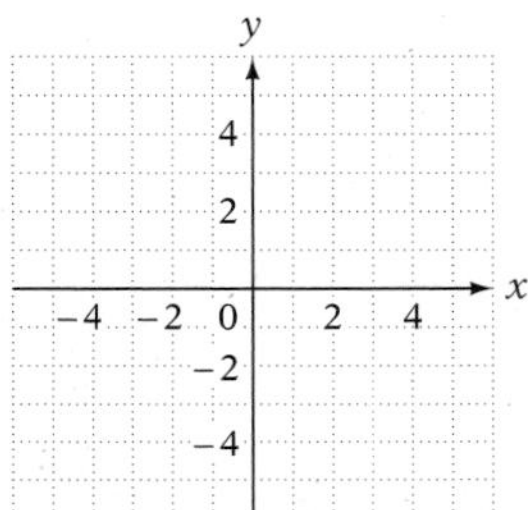

17. $x - 2y \leq 4$
$3x + 2y \leq 8$

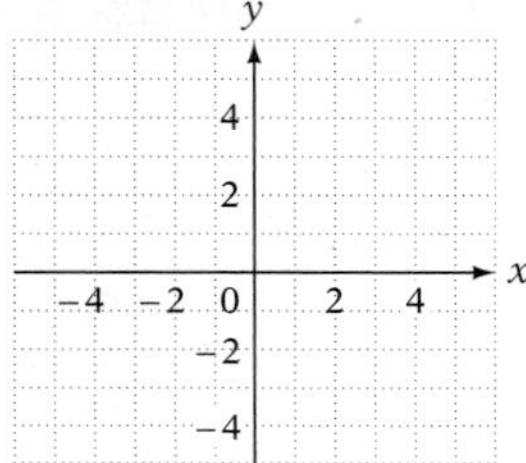

18. $3x - 2y < 0$
$5x + 3y > 9$

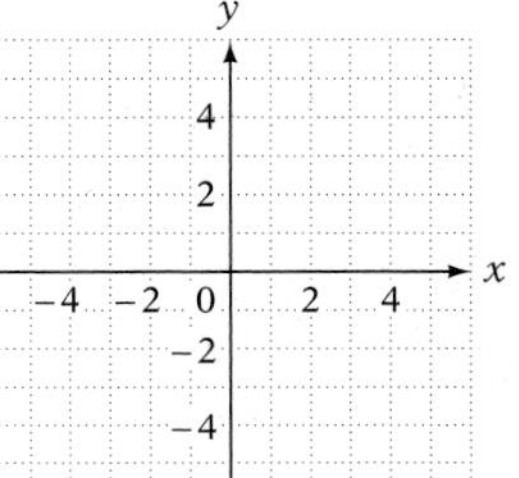

For Exercises 19 to 22, assume that a and b are positive numbers such that $a > b$. Describe the solution set of each system of inequalities.

19. $x + y < a$
$x + y < b$

20. $x + y > a$
$x + y < b$

21. $x + y < a$
$x + y > b$

22. $x + y > a$
$x + y > b$

Applying the Concepts

For Exercises 23 to 25, graph the solution set.

23. $2x + 3y \leq 15$
$3x - y \leq 6$
$y \geq 0$

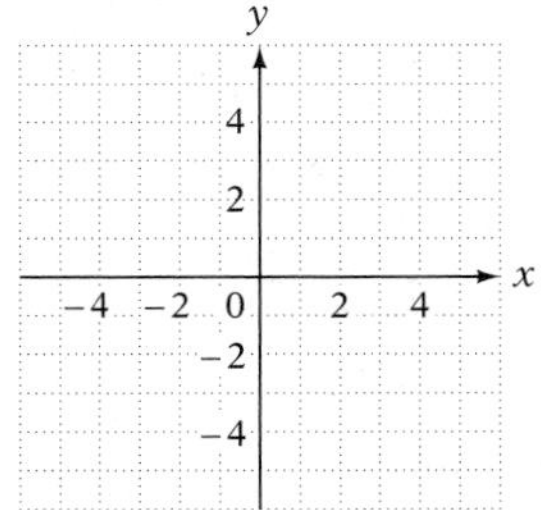

24. $x + y \leq 6$
$x - y \leq 2$
$x \geq 0$

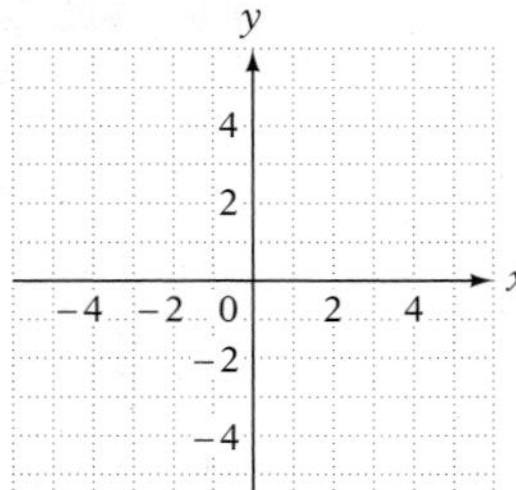

25. $2x - y \leq 4$
$3x + y < 1$
$y \leq 0$

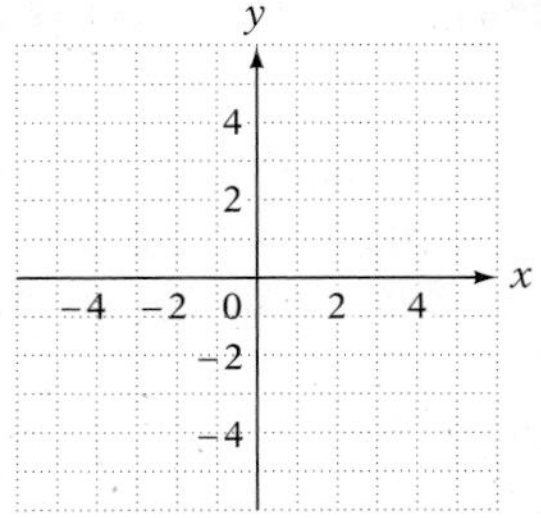

FOCUS ON PROBLEM SOLVING

Solve an Easier Problem

One approach to problem solving is to try to solve an easier problem. Suppose you are in charge of your softball league, which consists of 15 teams. You must devise a schedule in which each team plays every other team once. How many games must be scheduled?

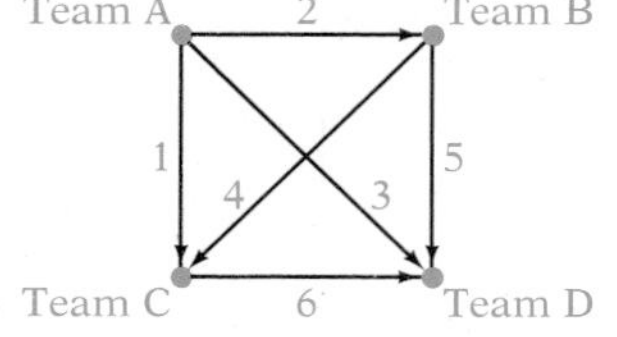

To solve this problem, we will attempt an easier problem first. Suppose that your league contains only a small number of teams. For instance, if there were only1 team, you would schedule 0 games. If there were 2 teams, you would schedule 1 game. If there were 3 teams, you would schedule 3 games. The diagram at the left shows that 6 games must be scheduled when there are 4 teams in the league.

Here is a table of our results so far. (Remember that making a table is another strategy to be used in problem solving.)

Number of Teams	*Number of Games*	*Possible Pattern*
1	0	0
2	1	1
3	3	1 + 2
4	6	1 + 2 + 3

1. Draw a diagram with five dots to represent the teams. Draw lines from each dot to a second dot, and determine the number of games required.

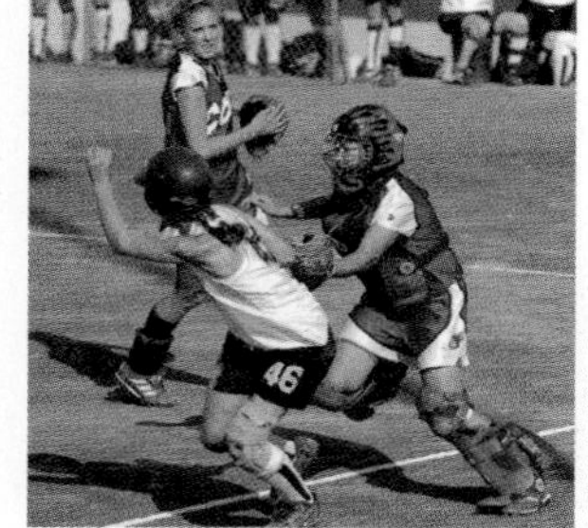
AP/World Wide Photos

2. What is the apparent pattern for the number of games required?
3. Assuming that the pattern continues, how many games must be scheduled for the 15 teams of the original problem?

After solving a problem, good problem solvers ask whether it is possible to solve the problem in a different manner. Here is a possible alternative method of solving the scheduling problem.

Begin with one of the 15 teams (say team A) and ask, "How many games must this team play?" Because there are 14 teams left to play, you must schedule 14 games. Now move to team B. It is already scheduled to play team A, and it does not play itself, so there are 13 teams left for it to play. Consequently, you must schedule 14 + 13 games.

4. Continue this reasoning for the remaining teams and determine the number of games that must be scheduled. Does this answer correspond to the answer you obtained using the first method?

PROJECTS AND GROUP ACTIVITIES

Using a Graphing Calculator to Solve a System of Equations

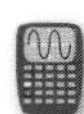

A graphing calculator can be used to solve a system of equations. For this procedure to work on most calculators, it is necessary that the point of intersection be on the screen. This means that you may have to experiment with Xmin, Xmax, Ymin, and Ymax values until the graphs intersect on the screen.

To solve a system of equations graphically, solve each equation for *y*. Then graph the equations of the system. Their point of intersection is the solution.

For instance, to solve the system of equations

$$4x - 3y = 7$$
$$5x + 4y = 2$$

first solve each equation for y.

$$4x - 3y = 7 \Rightarrow y = \frac{4}{3}x - \frac{7}{3}$$
$$5x + 4y = 2 \Rightarrow y = -\frac{5}{4}x + \frac{1}{2}$$

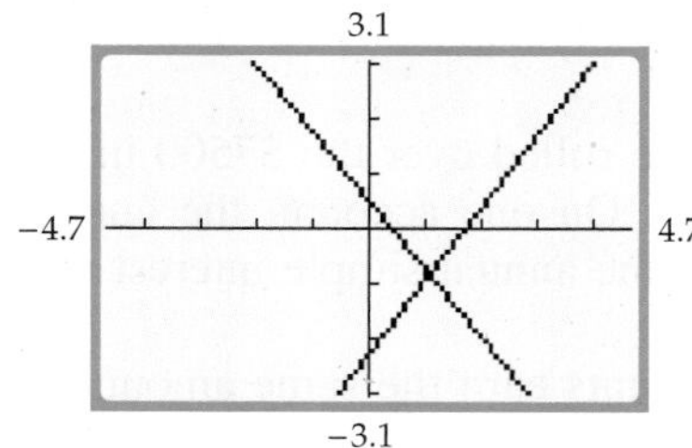

The keystrokes needed to solve this system using a TI-84 are given below. We are using a viewing window of $[-4.7, 4.7]$ by $[-3.1, 3.1]$. The approximate solution is $(1.096774, -0.870968)$.

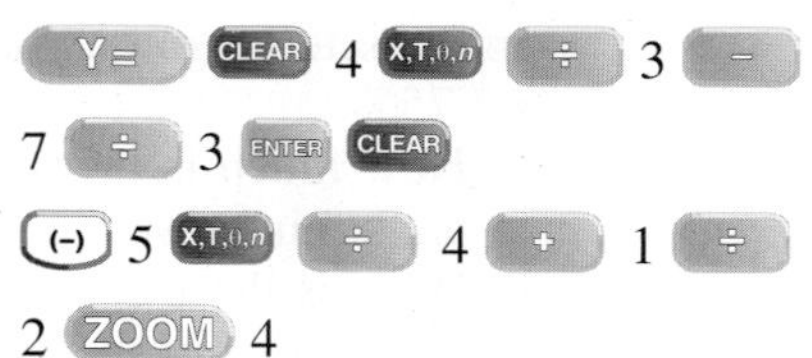

Once the calculator has drawn the graphs, use the following keystrokes to determine the point of intersection.

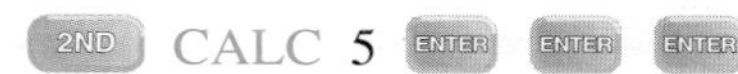

Here is an example of using a graphing calculator to solve an investment problem.

Take Note

As shown on the calculator screen in the example at the right, at the point of intersection, $y = 157.5$. This is the interest earned, \$157.50, on \$4500 invested at 3.5%. It is also the interest earned on \$3500 invested at 4.5%. In other words, it is the interest earned on each account when both accounts earn the same interest.

HOW TO 1 A marketing manager deposited \$8000 in two simple interest accounts, one with an interest rate of 3.5% and the other with an interest rate of 4.5%. How much is deposited in each account if both accounts earn the same interest?

Amounted invested at 3.5%: x
Amount invested at 4.5%: $8000 - x$

Interest earned on the 3.5% account: $0.035x$
Interest earned on the 4.5% account: $0.045(8000 - x)$

Enter $0.035x$ into Y_1. Enter $0.045(8000 - x)$ into Y_2. Graph the equations. (We used a window of Xmin = 0, Xmax = 8000, Ymin = 0, Ymax = 400.) Use the intersect feature to find the point of intersection.

400
Intersection
X=4500 Y=157.5
0 8000
0

At the point of intersection, $x = 4500$. This is the amount in the 3.5% account.
The amount in the 4.5% account is $8000 - x = 8000 - 4500 = 3500$.

\$4500 is invested at 3.5%. \$3500 is invested at 4.5%.

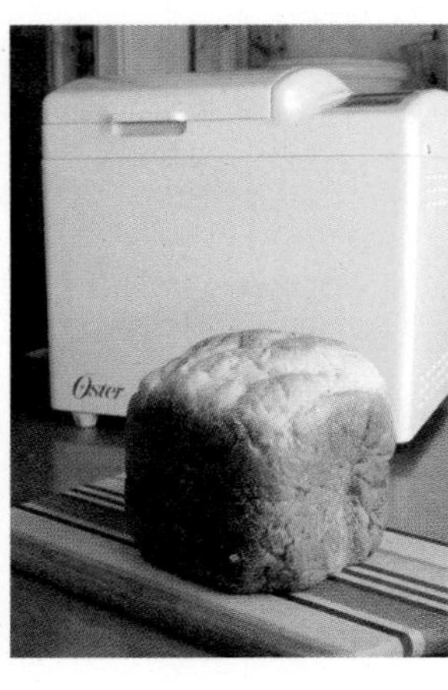

Michael Newman/PhotoEdit

For Exercises 1 to 3, solve using a graphing calculator.

1. **Finances** Suppose that a breadmaker costs \$190, and that the ingredients and electricity needed to make one loaf of bread cost \$.95. If a comparable loaf of bread at a grocery store costs \$1.98, how many loaves of bread must you make before the breadmaker pays for itself?
2. **Finances** Suppose a natural gas clothes dryer costs \$520 and uses \$.80 of gas to dry a load of clothes for 1 h. The laundromat charges \$3.50 to use a dryer for 1 h.
   a. How many loads of clothes must you dry before the gas dryer purchase becomes the more economical choice?
   b. What is the y-coordinate of the point of intersection? What does this represent in the context of the problem?
3. **Investments** When Mitch Deerfield changed jobs, he rolled over the \$7500 in his retirement account into two simple interest accounts. On one account, the annual simple interest rate is 6.25%; on the second account, the annual simple interest rate is 5.75%.
   a. How much is invested in each account if the accounts earn the same amount of annual interest?
   b. What is the y-coordinate of the point of intersection? What does this represent in the context of the problem?

CHAPTER 5 SUMMARY

KEY WORDS	EXAMPLES
A *system of equations* is two or more equations considered together. A *solution of a system of equations in two variables* is an ordered pair that is a solution of each equation in the system. [5.1A, p. 300]	The solution of the system $x + y = 2$, $x - y = 4$ is the ordered pair $(3, -1)$. $(3, -1)$ is the only ordered pair that is a solution of both equations.
When the graphs of a system of equations intersect at only one point, the system is called an *independent system of equations.* [5.1A, p. 300]	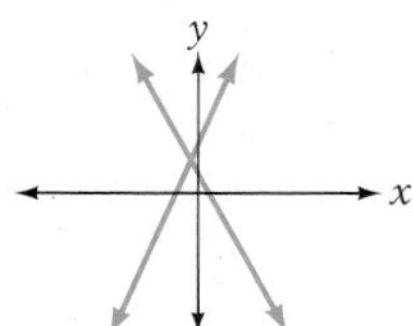
When the graphs of a system of equations do not intersect, the system has no solution and is called an *inconsistent system of equations.* [5.1A, p. 301]	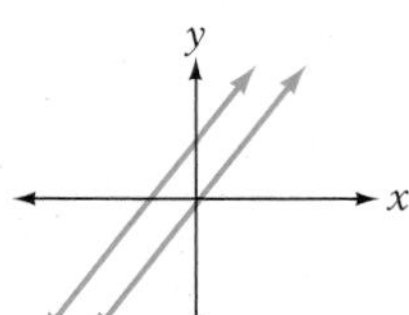
When the graphs of a system of equations coincide, the system is called a *dependent system of equations.* [5.1A, p. 301]	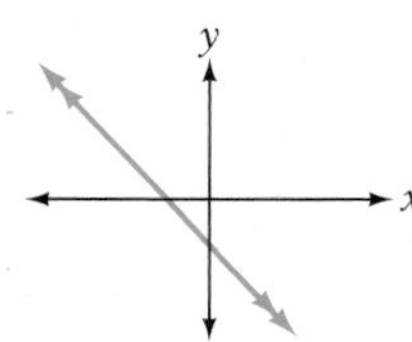

An equation of the form $Ax + By + Cz = D$, where A, B, and C are coefficients of the variables and D is a constant, is a *linear equation in three variables.* A *solution of an equation in three variables* is an *ordered triple* (x, y, z). [5.2B, p. 315]	$3x + 2y - 5z = 12$ is a linear equation in three variables. One solution of this equation is the ordered triple $(0, 1, -2)$.
A *solution of a system of equations in three variables* is an ordered triple that is a solution of each equation of the system. [5.2B, p. 317]	The solution of the system $\begin{aligned} 3x + y - 3z &= 2 \\ -x + 2y + 3z &= 6 \\ 2x + 2y - 2z &= 4 \end{aligned}$ is the ordered triple $(1, 2, 1)$. $(1, 2, 1)$ is the only ordered triple that is a solution of all three equations.
Two or more inequalities considered together are called a *system of inequalities.* The *solution set of a system of inequalities* is the intersection of the solution sets of the individual inequalities. [5.4A, p. 332]	$x + y > 3$ $x - y > -2$ 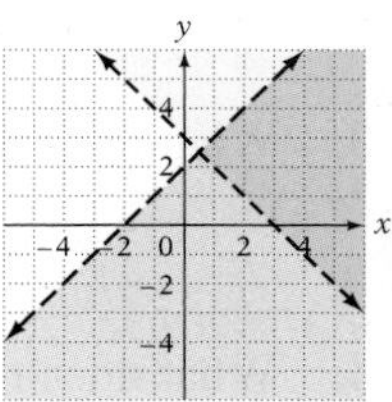

ESSENTIAL RULES AND PROCEDURES	EXAMPLES
Solving Systems of Equations A system of equations can be solved by: **a.** *Graphing* [5.1A, p. 300]	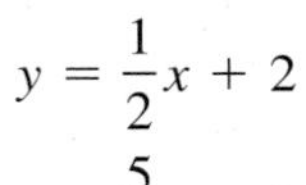 $y = \frac{1}{2}x + 2$ $y = \frac{5}{2}x - 2$ 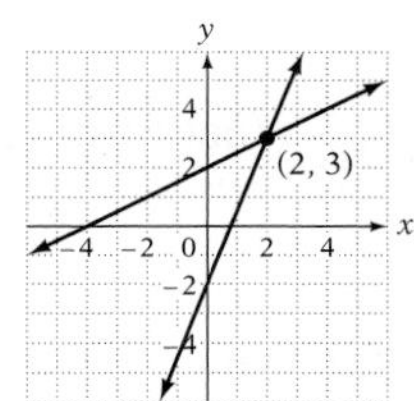
b. *The substitution method* [5.1B, p. 303]	(1) $2x - 3y = 4$ (2) $y = -x + 2$ Substitute the value of y into equation (1). $2x - 3(-x + 2) = 4$
c. *The addition method* [5.2A, p. 312]	$-2x + 3y = 7$ $2x - 5y = 2$ $-2y = 9$ • Add the equations.
Annual Simple Interest Equation [5.1C, p. 306] Principal · simple interest rate = simple interest $Pr = I$	You have a total of \$10,000 to invest in two simple interest accounts, one earning 4.5% annual simple interest and the other earning 5% annual simple interest. If you earn \$485 per year in interest from these two investments, how much do you have invested in each account? $x + y = 10{,}000$ $0.045x + 0.05y = 485$

CHAPTER 5

CONCEPT REVIEW

Test your knowledge of the concepts presented in this chapter. Answer each question. Then check your answers against the ones provided in the Answer Section.

1. After graphing a system of linear equations, how is the solution determined?

2. What is the difference between an independent system and a dependent system of equations?

3. What does the graph of an inconsistent system of equations look like?

4. What does the graph of a dependent system of equations look like?

5. What steps are used to solve a system of two linear equations by the substitution method?

6. What formula is used to solve a simple interest problem?

7. What steps are used to solve a system of two linear equations by the addition method?

8. When using the addition method, after adding the two equations in a system of two equations, what type of resulting equation tells you that the system of equations is dependent?

9. In a rate-of-wind problem, what do the expressions $p + w$ and $p - w$ represent?

10. In application problems in two variables, why are two equations written?

11. How do you check the solution to a system of three equations in three variables?

12. How is intersection used to solve a system of linear inequalities?

CHAPTER 5

REVIEW EXERCISES

1. Solve by substitution: $2x - 6y = 15$
$x = 4y + 8$

2. Solve by the addition method: $3x + 2y = 2$
$x + y = 3$

3. Solve by graphing: $x + y = 3$
$3x - 2y = -6$

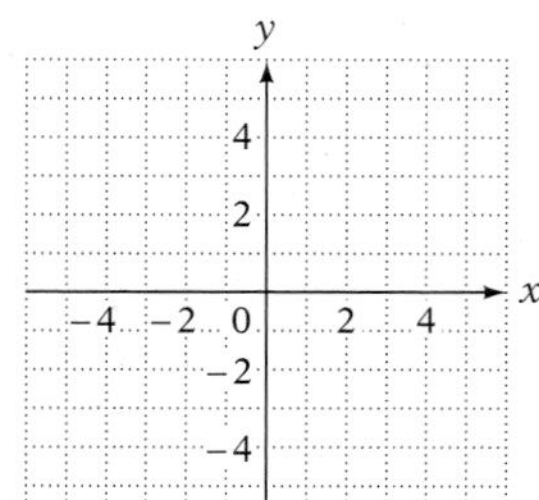

4. Solve by graphing: $2x - y = 4$
$y = 2x - 4$

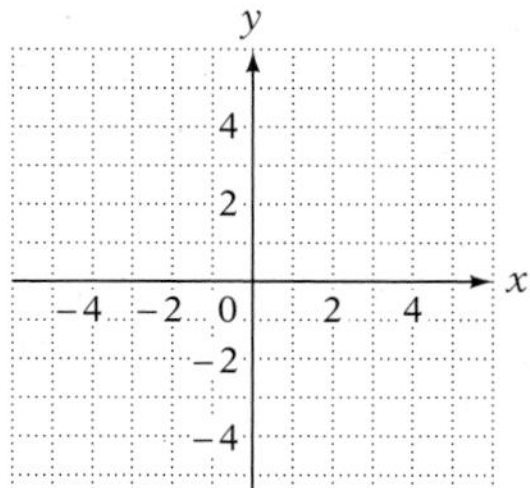

5. Solve by substitution: $3x + 12y = 18$
$x + 4y = 6$

6. Solve by the addition method: $5x - 15y = 30$
$x - 3y = 6$

7. Solve:
$3x - 4y - 2z = 17$
$4x - 3y + 5z = 5$
$5x - 5y + 3z = 14$

8. Solve:
$3x + y = 13$
$2y + 3z = 5$
$x + 2z = 11$

9. Is $(1, -2)$ a solution of the system of equations?
$6x + y = 4$
$2x - 5y = 12$

10. Solve by substitution:
$2x - 4y = 11$
$y = 3x - 4$

11. Solve by substitution:
$$2x - y = 7$$
$$3x + 2y = 7$$

12. Solve by the addition method:
$$3x - 4y = 1$$
$$2x + 5y = 16$$

13. Solve:
$$x + y + z = 0$$
$$x + 2y + 3z = 5$$
$$2x + y + 2z = 3$$

14. Solve:
$$x + 3y + z = 6$$
$$2x + y - z = 12$$
$$x + 2y - z = 13$$

15. Graph the solution set:
$$x + 3y \leq 6$$
$$2x - y \geq 4$$

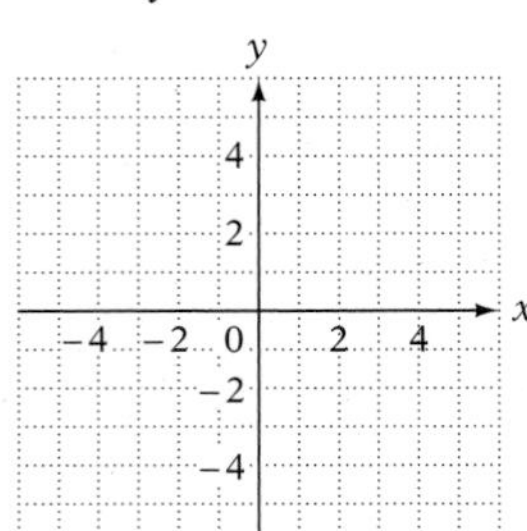

16. Graph the solution set:
$$2x + 4y \geq 8$$
$$x + y \leq 3$$

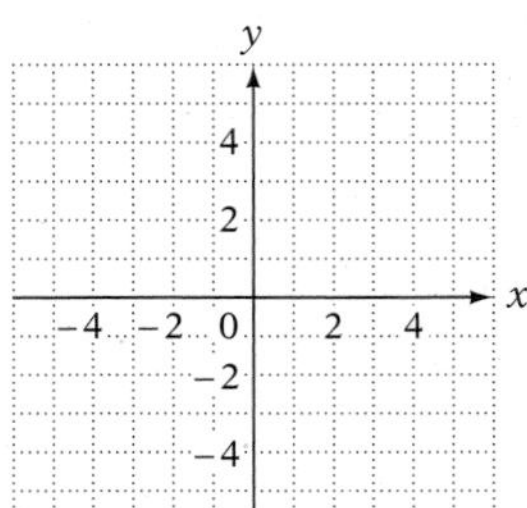

17. **Boating** A cabin cruiser traveling with the current went 60 mi in 3 h. Against the current, it took 5 h to travel the same distance. Find the rate of the cabin cruiser in calm water and the rate of the current.

18. **Aeronautics** A pilot flying with the wind flew 600 mi in 3 h. Flying against the wind, the pilot required 4 h to travel the same distance. Find the rate of the plane in calm air and the rate of the wind.

19. **Ticket Sales** At a movie theater, admission tickets are \$5 for children and \$8 for adults. The receipts for one Friday evening were \$2500. The next day, there were three times as many children as the preceding evening and only half the number of adults as the night before, yet the receipts were still \$2500. Find the number of children who attended on Friday evening.

20. **Investments** A trust administrator divides \$20,000 between two accounts. One account earns an annual simple interest rate of 3%, and a second account earns an annual simple interest rate of 7%. The total annual income from the two accounts is \$1200. How much is invested in each account?

CHAPTER 5

TEST

1. Solve by graphing: $2x - 3y = -6$
   $2x - y = 2$

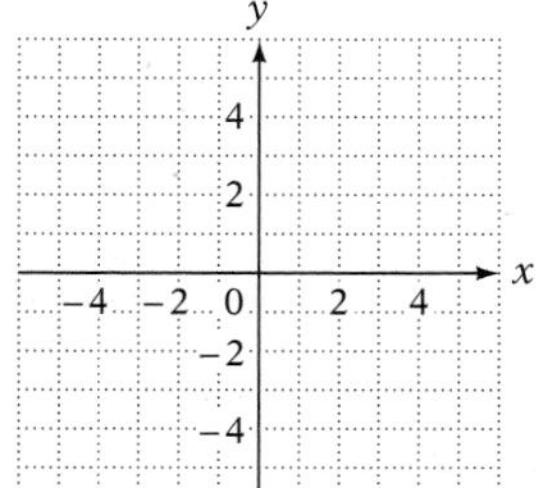

2. Solve by graphing: $x - 2y = -5$
   $3x + 4y = -15$

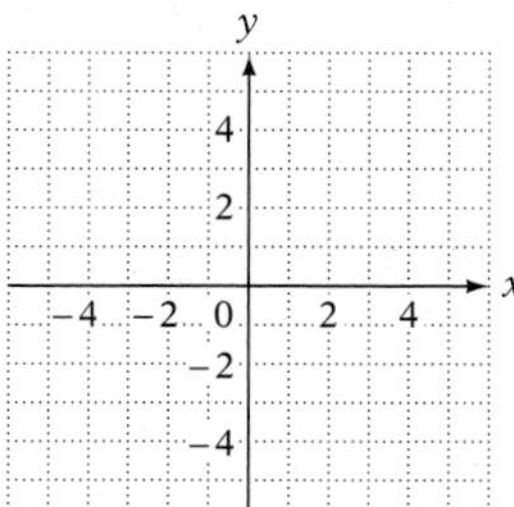

3. Graph the solution set: $2x - y < 3$
   $4x + 3y < 11$

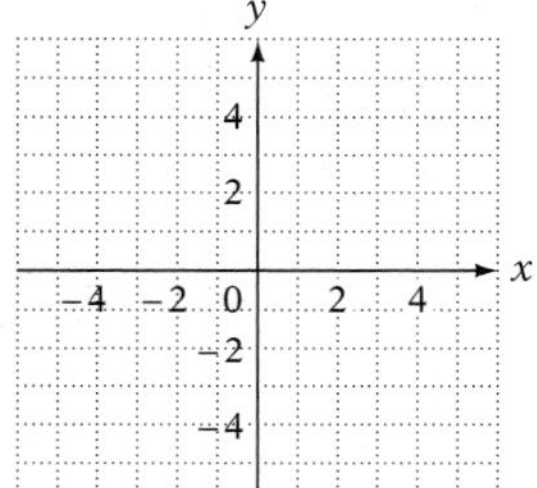

4. Graph the solution set: $x + y > 2$
   $2x - y < -1$

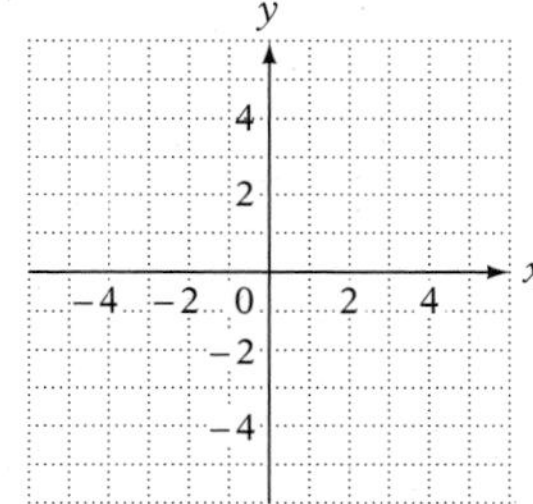

5. Solve by substitution: $3x + 2y = 4$
   $x = 2y - 1$

6. Solve by substitution: $5x + 2y = -23$
   $2x + y = -10$

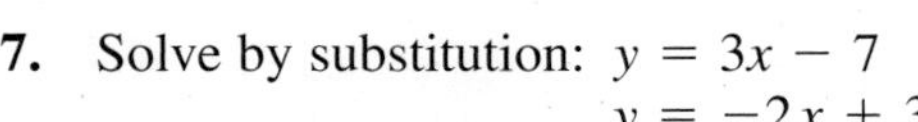

7. Solve by substitution: $y = 3x - 7$
   $y = -2x + 3$

8. Solve by the addition method:
   $3x + 4y = -2$
   $2x + 5y = 1$

9. Solve by the addition method:
   $4x - 6y = 5$
   $6x - 9y = 4$

10. Solve by the addition method:
    $3x - y = 2x + y - 1$
    $5x + 2y = y + 6$

11. Solve:
$$2x + 4y - z = 3$$
$$x + 2y + z = 5$$
$$4x + 8y - 2z = 7$$

12. Solve:
$$x - y - z = 5$$
$$2x + z = 2$$
$$3y - 2z = 1$$

13. Solve by substitution:
$$x - y = 3$$
$$2x + y = -4$$

14. Is $(2, -2)$ a solution of the system of equations?
$$5x + 2y = 6$$
$$3x + 5y = -4$$

15. Solve:
$$x - y + z = 2$$
$$2x - y - z = 1$$
$$x + 2y - 3z = -4$$

16. **Aeronautics** A plane flying with the wind went 350 mi in 2 h. The return trip, flying against the wind, took 2.8 h. Find the rate of the plane in calm air and the rate of the wind.

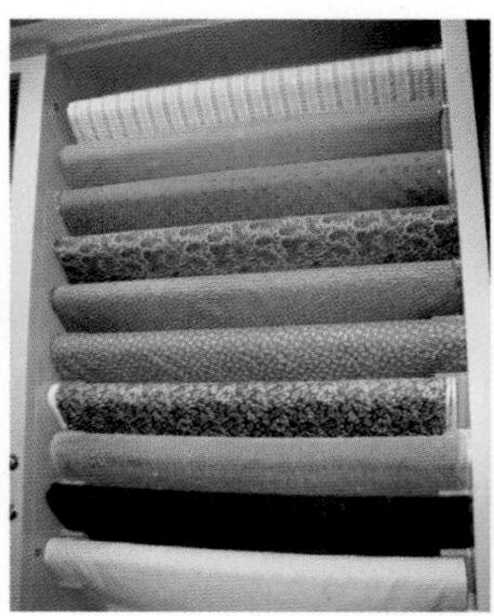
Susan Van Etten/PhotoEdit

17. **Purchasing** A clothing manufacturer purchased 60 yd of cotton and 90 yd of wool for a total cost of $1800. Another purchase, at the same prices, included 80 yd of cotton and 20 yd of wool for a total cost of $1000. Find the cost per yard of the cotton and the wool.

18. **Investments** The annual interest earned on two investments is $549. One investment is in a 2.7% tax-free annual simple interest account, and the other investment is in a 5.1% annual simple interest CD. The total amount invested is $15,000. How much is invested in each account?

CUMULATIVE REVIEW EXERCISES

1. Simplify: $-2\sqrt{90}$

2. Solve: $3(x - 5) = 2x + 7$

3. Simplify: $3[x - 2(5 - 2x) - 4x] + 6$

4. Evaluate $a + bc \div 2$ when $a = 4$, $b = 8$, and $c = -2$.

5. Solve: $2x - 3 < 9$ or $5x - 1 < 4$

6. Solve: $|x - 2| - 4 < 2$

7. Solve: $|2x - 3| > 5$

8. Given $F(x) = x^2 - 3$, find $F(2)$.

9. Graph the solution set of $\{x|x \leq 2\} \cap \{x|x > -3\}$.

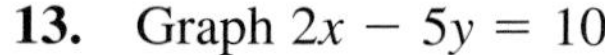

10. Find the equation of the line that contains the point $(-2, 3)$ and has slope $-\frac{2}{3}$.

11. Find the equation of the line that contains the points $(2, -1)$ and $(3, 4)$.

12. Find the equation of the line that contains the point $(-2, 2)$ and is perpendicular to the line $2x - 3y = 6$.

13. Graph $2x - 5y = 10$.

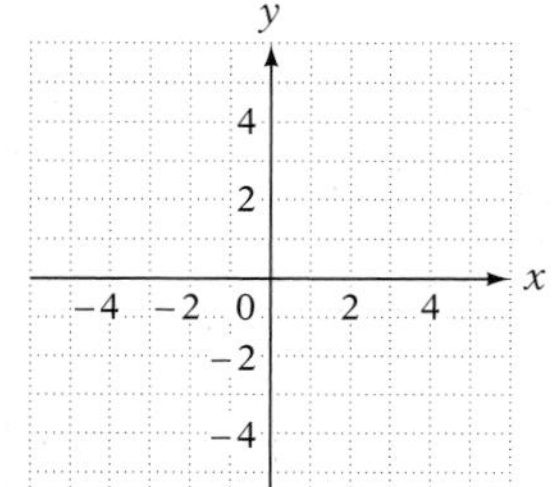

14. Graph the solution set of $3x - 4y \geq 8$.

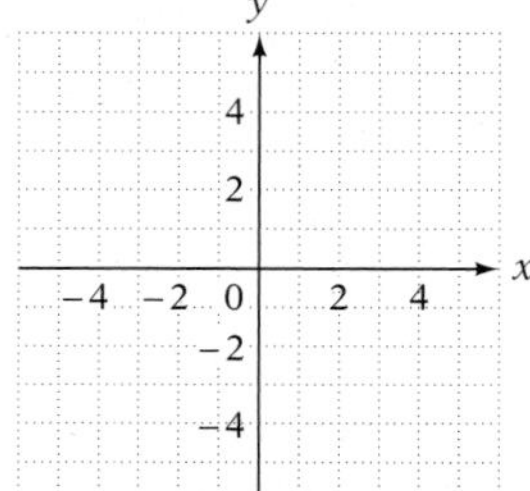

15. Solve by graphing.
$5x - 2y = 10$
$3x + 2y = 6$

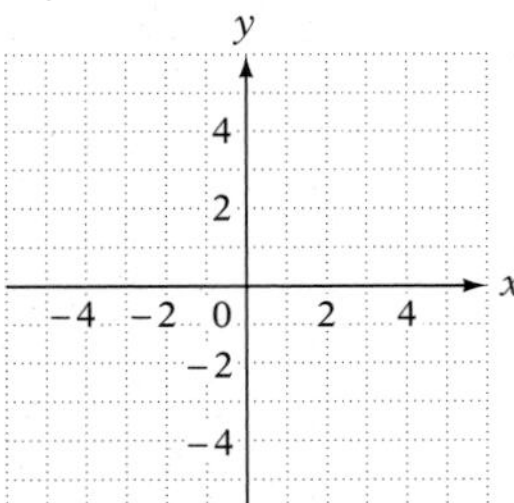

16. Graph the solution set.
$3x - 2y \geq 4$
$x + y < 3$

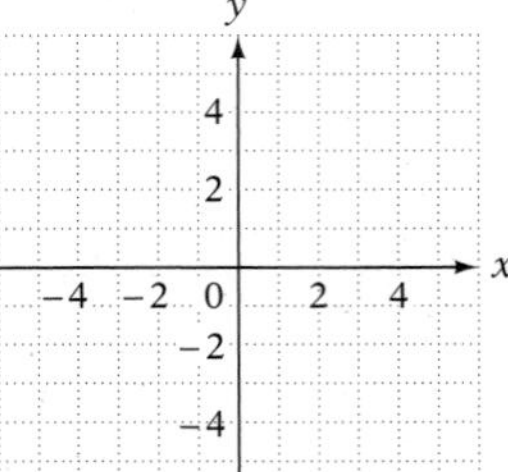

17. Solve:
$3x + 2z = 1$
$2y - z = 1$
$x + 2y = 1$

18. Solve:
$2x - y + z = 2$
$3x + y + 2z = 5$
$3x - y + 4z = 1$

19. Solve by the addition method:
$4x - 3y = 17$
$3x - 2y = 12$

20. Solve by substitution:
$3x - 2y = 7$
$y = 2x - 1$

21. Mixtures How many milliliters of pure water must be added to 100 ml of a 4% salt solution to make a 2.5% salt solution?

22. Travel Flying with the wind, a small plane required 2 h to fly 150 mi. Against the wind, it took 3 h to fly the same distance. Find the rate of the wind.

23. Purchasing A restaurant manager buys 100 lb of hamburger and 50 lb of steak for a total cost of \$980. A second purchase, at the same prices, includes 150 lb of hamburger and 100 lb of steak. The total cost is \$1720. Find the price of 1 lb of steak.

24. Electronics Find the lower and upper limits of a 12,000-ohm resistor with a 15% tolerance.

25. Compensation The graph shows the relationship between the monthly income and the sales of an account executive. Find the slope of the line between the two points shown on the graph. Write a sentence that states the meaning of the slope.

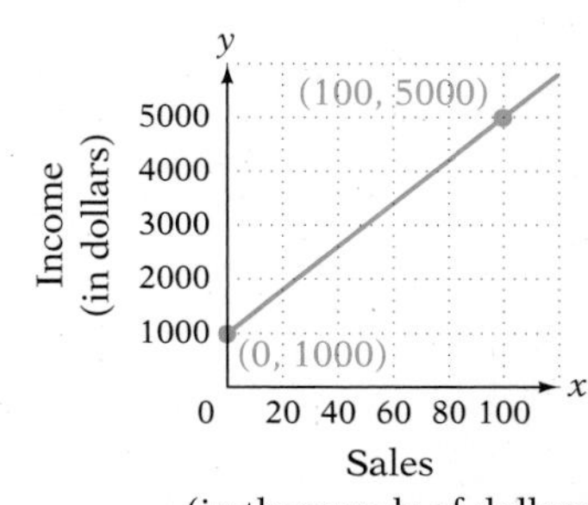

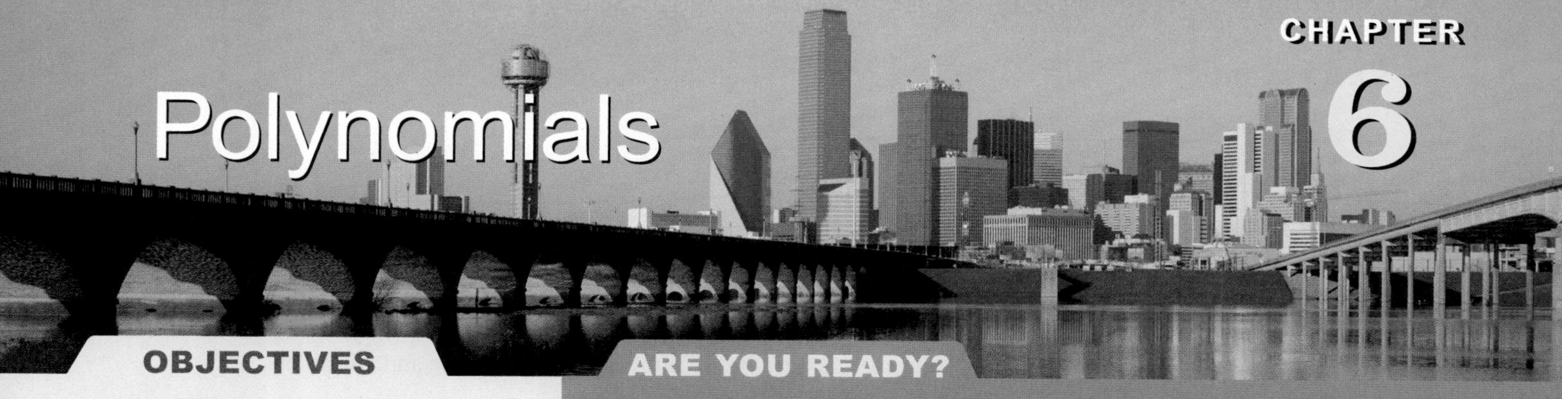

CHAPTER 6
Polynomials

photodisc/First Light

OBJECTIVES

SECTION 6.1

A To multiply monomials
B To divide monomials and simplify expressions with negative exponents
C To write a number using scientific notation
D To solve application problems

SECTION 6.2

A To evaluate polynomial functions
B To add or subtract polynomials

SECTION 6.3

A To multiply a polynomial by a monomial
B To multiply polynomials
C To multiply polynomials that have special products
D To solve application problems

SECTION 6.4

A To divide a polynomial by a monomial
B To divide polynomials
C To divide polynomials by using synthetic division
D To evaluate a polynomial function using synthetic division

ARE YOU READY?

Take the Chapter 6 Prep Test to find out if you are ready to learn to:

- Add, subtract, multiply, and divide polynomials
- Simplify expressions with negative exponents
- Write a number in scientific notation

PREP TEST

Do these exercises to prepare for Chapter 6.

1. Subtract: $-2 - (-3)$

2. Multiply: $-3(6)$

3. Simplify: $-\frac{24}{-36}$

4. Evaluate $3n^4$ when $n = -2$.

5. If $\frac{a}{b}$ is a fraction in simplest form, what number is not a possible value of b?

6. Are $2x^2$ and $2x$ like terms?

7. Simplify: $3x^2 - 4x + 1 + 2x^2 - 5x - 7$

8. Simplify: $-4y + 4y$

9. Simplify: $-3(2x - 8)$

10. Simplify: $3xy - 4y - 2(5xy - 7y)$

SECTION

6.1 Exponential Expressions

OBJECTIVE A To multiply monomials

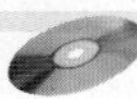

Point of Interest

Around A.D. 250, the monomial $3x^2$ shown at the right would have been written $\Delta^Y 3$, or at least approximately like that. In A.D. 250, the symbol for 3 was not the one we use today.

A **monomial** is a number, a variable, or a product of a number and variables.

The examples at the right are monomials. The **degree of a monomial** is the sum of the exponents on the variables.

x degree 1 ($x = x^1$)
$3x^2$ degree 2
$4x^2y$ degree 3
$6x^3y^4z^2$ degree 9

In this chapter, the variable n is considered a positive integer when used as an exponent.

x^n degree n

The degree of a nonzero constant term is zero.

6 degree 0

The expression $5\sqrt{x}$ is not a monomial because $\sqrt{x}$ cannot be written as a product of variables. The expression $\frac{x}{y}$ is not a monomial because it is a quotient of variables.

The expression x^4 is an exponential expression. The exponent, 4, indicates the number of times the base, x, occurs as a factor.

The product of exponential expressions with the *same* base can be simplified by writing each expression in factored form and writing the result with an exponent.

$$x^3 \cdot x^4 = \overbrace{(x \cdot x \cdot x)}^{3 \text{ factors}} \cdot \overbrace{(x \cdot x \cdot x \cdot x)}^{4 \text{ factors}}$$
(7 factors)
$$= x^7$$

Note that adding the exponents results in the same product.

$$x^3 \cdot x^4 = x^{3+4} = x^7$$

Rule for Multiplying Exponential Expressions

If m and n are positive integers, then $x^m \cdot x^n = x^{m+n}$.

HOW TO • 1 Simplify: $(-4x^5y^3)(3xy^2)$

$(-4x^5y^3)(3xy^2) = (-4 \cdot 3)(x^5 \cdot x)(y^3 \cdot y^2)$
• Use the Commutative and Associative Properties of Multiplication to rearrange and group factors.

$= -12(x^{5+1})(y^{3+2})$
• Multiply variables with the same base by adding their exponents.

$= -12x^6y^5$
• Simplify.

As shown below, the power of a monomial can be simplified by writing the power in factored form and then using the Rule for Multiplying Exponential Expressions. It can also be simplified by multiplying each exponent inside the parentheses by the exponent outside the parentheses.

$(a^2)^3 = a^2 \cdot a^2 \cdot a^2$
$= a^{2+2+2}$
$= a^6$

$(x^3y^4)^2 = (x^3y^4)(x^3y^4)$
$= x^{3+3}y^{4+4}$
$= x^6y^8$

• Write in factored form. Then use the Rule for Multiplying Exponential Expressions.

$(a^2)^3 = a^{2\cdot 3} = a^6$

$(x^3y^4)^2 = x^{3\cdot 2}y^{4\cdot 2} = x^6y^8$

• Multiply each exponent inside the parentheses by the exponent outside the parentheses.

Rule for Simplifying the Power of an Exponential Expression

If m and n are positive integers, then $(x^m)^n = x^{mn}$.

Rule for Simplifying Powers of Products

If m, n, and p are positive integers, then $(x^my^n)^p = x^{mp}y^{np}$.

HOW TO • 2 Simplify: $(x^4)^5$

$(x^4)^5 = x^{4\cdot 5}$
$= x^{20}$

• Use the Rule for Simplifying Powers of Exponential Expressions to multiply the exponents.

HOW TO • 3 Simplify: $(2a^3b^4)^3$

$(2a^3b^4)^3 = 2^{1\cdot 3}a^{3\cdot 3}b^{4\cdot 3}$
$= 2^3a^9b^{12}$
$= 8a^9b^{12}$

• Use the Rule for Simplifying Powers of Products to multiply each exponent inside the parentheses by the exponent outside the parentheses.

EXAMPLE • 1

Simplify: $(2xy^2)(-3xy^4)^3$

Solution $(2xy^2)(-3xy^4)^3 = (2xy^2)[(-3)^3x^3y^{12}]$
$= (2xy^2)(-27x^3y^{12})$
$= -54x^4y^{14}$

YOU TRY IT • 1

Simplify: $(-3a^2b^4)(-2ab^3)^4$

Your solution

EXAMPLE • 2

Simplify: $(-3x^3y)^2(-2x^3y^5)^3$

Solution

$(-3x^3y)^2(-2x^3y^5)^3 = [(-3)^2x^6y^2][(-2)^3x^9y^{15}]$
$= [9x^6y^2][-8x^9y^{15}]$
$= -72x^{15}y^{17}$

YOU TRY IT • 2

Simplify: $(-4ab^4)^2(2a^4b^2)^4$

Your solution

EXAMPLE • 3

Simplify: $(x^{n+2})^5$

Solution $(x^{n+2})^5 = x^{5n+10}$ • Multiply the exponents.

YOU TRY IT • 3

Simplify: $(y^{n-3})^2$

Your solution

Solutions on p. S17

OBJECTIVE B

To divide monomials and simplify expressions with negative exponents

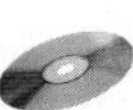

The quotient of two exponential expressions with the same base can be simplified by writing each expression in factored form, dividing by the common factors, and then writing the result with an exponent.

$$\frac{x^5}{x^2} = \frac{\overset{1}{\cancel{x}} \cdot \overset{1}{\cancel{x}} \cdot x \cdot x \cdot x}{\underset{1}{\cancel{x}} \cdot \underset{1}{\cancel{x}}} = x^3$$

Note that subtracting the exponents gives the same result.

$$\frac{x^5}{x^2} = x^{5-2} = x^3$$

To divide two monomials with the same base, subtract the exponents of the like bases.

HOW TO 4 Simplify: $\frac{z^8}{z^2}$

$$\frac{z^8}{z^2} = z^{8-2}$$
$$= z^6$$

• The bases are the same. Subtract the exponents.

HOW TO 5 Simplify: $\frac{a^5b^9}{a^4b}$

$$\frac{a^5b^9}{a^4b} = a^{5-4}b^{9-1}$$
$$= ab^8$$

• Subtract the exponents of the like bases.

Consider the expression $\frac{x^4}{x^4}$, $x \neq 0$. This expression can be simplified, as shown below, by subtracting exponents or dividing by common factors.

$$\frac{x^4}{x^4} = x^{4-4} = x^0 \qquad \frac{x^4}{x^4} = \frac{\overset{1}{\cancel{x}} \cdot \overset{1}{\cancel{x}} \cdot \overset{1}{\cancel{x}} \cdot \overset{1}{\cancel{x}}}{\underset{1}{\cancel{x}} \cdot \underset{1}{\cancel{x}} \cdot \underset{1}{\cancel{x}} \cdot \underset{1}{\cancel{x}}} = 1$$

The equations $\frac{x^4}{x^4} = x^0$ and $\frac{x^4}{x^4} = 1$ suggest the following definition of x^0.

Definition of Zero as an Exponent

If $x \neq 0$, then $x^0 = 1$. The expression 0^0 is not defined.

Take Note

In the example at the right, we indicated that $z \neq 0$. If we try to evaluate $(16z^5)^0$ when $z = 0$, we have $[16(0)^5]^0 = [16(0)]^0 = 0^0$. However, 0^0 is not defined, so we must assume that $z \neq 0$. To avoid stating this for every example or exercise, we will assume that variables cannot take on values that result in the expression 0^0.

HOW TO 6 Simplify: $(16z^5)^0$, $z \neq 0$

$$(16z^5)^0 = 1$$

• Any nonzero expression to the zero power is 1.

HOW TO 7 Simplify: $-(7x^4y^3)^0$

$$-(7x^4y^3)^0 = -(1) = -1$$

• The negative sign outside the parentheses is not affected by the exponent.

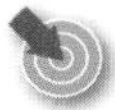

Point of Interest

In the 15th century, the expression $12^{2\bar{m}}$ was used to mean $12x^{-2}$. The use of $\bar{m}$ reflects an Italian influence, where m was used for minus and p was used for plus. It was understood that $2\bar{m}$ referred to an unnamed variable. Isaac Newton, in the 17th century, advocated the use of a negative exponent, the symbol we use today.

Consider the expression $\frac{x^4}{x^6}$, $x \neq 0$. This expression can be simplified, as shown below, by subtracting exponents or by dividing by common factors.

$$\frac{x^4}{x^6} = x^{4-6} = x^{-2} \qquad \frac{x^4}{x^6} = \frac{\overset{1}{\cancel{x}} \cdot \overset{1}{\cancel{x}} \cdot \overset{1}{\cancel{x}} \cdot \overset{1}{\cancel{x}}}{\underset{1}{\cancel{x}} \cdot \underset{1}{\cancel{x}} \cdot \underset{1}{\cancel{x}} \cdot \underset{1}{\cancel{x}} \cdot x \cdot x} = \frac{1}{x^2}$$

The equations $\frac{x^4}{x^6} = x^{-2}$ and $\frac{x^4}{x^6} = \frac{1}{x^2}$ suggest that $x^{-2} = \frac{1}{x^2}$.

Definition of a Negative Exponent

If $x \neq 0$ and n is a positive integer, then

$$x^{-n} = \frac{1}{x^n} \quad \text{and} \quad \frac{1}{x^{-n}} = x^n$$

Take Note

Note from the example at the right that 2^{-4} is a *positive* number. A negative exponent does not change the sign of a number.

HOW TO • 8 Evaluate: 2^{-4}

$2^{-4} = \frac{1}{2^4}$ • Use the Definition of a Negative Exponent.

$= \frac{1}{16}$ • Evaluate the expression.

The expression $\left(\frac{x^3}{y^4}\right)^2$, $y \neq 0$, can be simplified by squaring $\frac{x^3}{y^4}$ or by multiplying each exponent in the quotient by the exponent outside the parentheses.

$$\left(\frac{x^3}{y^4}\right)^2 = \left(\frac{x^3}{y^4}\right)\left(\frac{x^3}{y^4}\right) = \frac{x^3 \cdot x^3}{y^4 \cdot y^4} = \frac{x^{3+3}}{y^{4+4}} = \frac{x^6}{y^8} \qquad \left(\frac{x^3}{y^4}\right)^2 = \frac{x^{3\cdot 2}}{y^{4\cdot 2}} = \frac{x^6}{y^8}$$

Rule for Simplifying Powers of Quotients

If m, n, and p are integers and $y \neq 0$, then $\left(\frac{x^m}{y^n}\right)^p = \frac{x^{mp}}{y^{np}}$.

HOW TO • 9 Simplify: $\left(\frac{a^2}{b^3}\right)^{-2}$

$\left(\frac{a^2}{b^3}\right)^{-2} = \frac{a^{2(-2)}}{b^{3(-2)}}$ • Use the Rule for Simplifying Powers of Quotients.

$= \frac{a^{-4}}{b^{-6}} = \frac{b^6}{a^4}$ • Use the Definition of a Negative Exponent to write the expression with positive exponents.

The preceding example suggests the following rule.

Rule for Negative Exponents on Fractional Expressions

If $a \neq 0$, $b \neq 0$, and n is a positive integer, then $\left(\frac{a}{b}\right)^{-n} = \left(\frac{b}{a}\right)^{n}$.

Take Note

The exponent on d is -5 (negative 5). The d^{-5} is written in the denominator as d^5. The exponent on 7 is 1 (positive 1). The 7 remains in the numerator.

Also, note that we indicated $d \neq 0$. This is necessary because division by zero is not defined. In this textbook, we will assume that values of the variables are chosen such that division by zero does not occur.

An exponential expression is in simplest form when it is written with only positive exponents.

HOW TO • 10 Simplify: $7d^{-5}$, $d \neq 0$

$7d^{-5} = 7 \cdot \frac{1}{d^5} = \frac{7}{d^5}$

- Use the Definition of a Negative Exponent to rewrite the expression with a positive exponent.

HOW TO • 11 Simplify: $\frac{2}{5a^{-4}}$

$\frac{2}{5a^{-4}} = \frac{2}{5} \cdot \frac{1}{a^{-4}} = \frac{2}{5} \cdot a^4 = \frac{2a^4}{5}$

- Use the Definition of a Negative Exponent to rewrite the expression with a positive exponent.

Now that zero as an exponent and negative exponents have been defined, a rule for dividing exponential expressions can be stated.

Rule for Dividing Exponential Expressions

If m and n are integers and $x \neq 0$, then $\frac{x^m}{x^n} = x^{m-n}$.

HOW TO • 12 Simplify: $\frac{x^4}{x^9}$

$\frac{x^4}{x^9} = x^{4-9}$

- Use the Rule for Dividing Exponential Expressions.

$= x^{-5}$

- Subtract the exponents.

$= \frac{1}{x^5}$

- Use the Definition of a Negative Exponent to rewrite the expression with a positive exponent.

The rules for simplifying exponential expressions and powers of exponential expressions are true for all integers. These rules are restated here for convenience.

Rules of Exponents

If m, n, and p are integers, then

$x^m \cdot x^n = x^{m+n}$ $\qquad (x^m)^n = x^{mn}$ $\qquad (x^m y^n)^p = x^{mp} y^{np}$

$\frac{x^m}{x^n} = x^{m-n}, x \neq 0$ $\qquad \left(\frac{x^m}{y^n}\right)^p = \frac{x^{mp}}{y^{np}}, y \neq 0$ $\qquad x^{-n} = \frac{1}{x^n}, x \neq 0$

$x^0 = 1, x \neq 0$

HOW TO • 13 Simplify: $(3ab^{-4})(-2a^{-3}b^7)$

$$\begin{aligned}(3ab^{-4})(-2a^{-3}b^7) &= [3 \cdot (-2)](a^{1+(-3)}b^{-4+7}) \\ &= -6a^{-2}b^3 \\ &= -\frac{6b^3}{a^2}\end{aligned}$$

• When multiplying expressions, add the exponents on like bases.

HOW TO • 14 Simplify: $\frac{4a^{-2}b^5}{6a^5b^2}$

$$\frac{4a^{-2}b^5}{6a^5b^2} = \frac{\cancel{2} \cdot 2a^{-2}b^5}{\cancel{2} \cdot 3a^5b^2} = \frac{2a^{-2}b^5}{3a^5b^2}$$

• Divide the coefficients by their common factor.

$$= \frac{2a^{-2-5}b^{5-2}}{3}$$

• Use the Rule for Dividing Exponential Expressions.

$$= \frac{2a^{-7}b^3}{3} = \frac{2b^3}{3a^7}$$

• Use the Definition of a Negative Exponent to rewrite the expression with a positive exponent.

HOW TO • 15 Simplify: $\left[\frac{6m^2n^3}{8m^7n^2}\right]^{-3}$

$$\left[\frac{6m^2n^3}{8m^7n^2}\right]^{-3} = \left[\frac{3m^{2-7}n^{3-2}}{4}\right]^{-3}$$

• Simplify inside the brackets.

$$= \left[\frac{3m^{-5}n}{4}\right]^{-3}$$

• Subtract the exponents.

$$= \frac{3^{-3}m^{15}n^{-3}}{4^{-3}}$$

• Use the Rule for Simplifying Powers of Quotients.

$$= \frac{4^3m^{15}}{3^3n^3} = \frac{64m^{15}}{27n^3}$$

• Use the Definition of a Negative Exponent to rewrite the expression with positive exponents. Then simplify.

EXAMPLE • 4

Simplify: $\frac{-28x^6z^{-3}}{42x^{-1}z^4}$

Solution

$$\begin{aligned}\frac{-28x^6z^{-3}}{42x^{-1}z^4} &= -\frac{14 \cdot 2x^{6-(-1)}z^{-3-4}}{14 \cdot 3} \\ &= -\frac{2x^7z^{-7}}{3} = -\frac{2x^7}{3z^7}\end{aligned}$$

YOU TRY IT • 4

Simplify: $\frac{20r^{-2}t^{-5}}{-16r^{-3}s^{-2}}$

Your solution

EXAMPLE • 5

Simplify: $\frac{(3a^{-1}b^4)^{-3}}{(6^{-1}a^{-3}b^{-4})^3}$

Solution

$$\begin{aligned}\frac{(3a^{-1}b^4)^{-3}}{(6^{-1}a^{-3}b^{-4})^3} &= \frac{3^{-3}a^3b^{-12}}{6^{-3}a^{-9}b^{-12}} = 3^{-3} \cdot 6^3a^{12}b^0 \\ &= \frac{6^3a^{12}}{3^3} = \frac{216a^{12}}{27} = 8a^{12}\end{aligned}$$

YOU TRY IT • 5

Simplify: $\frac{(9u^{-6}v^4)^{-1}}{(6u^{-3}v^{-2})^{-2}}$

Your solution

Solutions on p. S17

EXAMPLE 6

Simplify: $\dfrac{x^{4n-2}}{x^{2n-5}}$

Solution $\dfrac{x^{4n-2}}{x^{2n-5}} = x^{4n-2-(2n-5)}$ • Subtract the exponents.

$= x^{4n-2-2n+5}$

$= x^{2n+3}$

YOU TRY IT 6

Simplify: $\dfrac{a^{2n+1}}{a^{n+3}}$

Your solution

Solution on p. S17

OBJECTIVE C To write a number using scientific notation

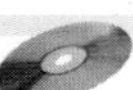

Point of Interest

Astronomers measure the distance of some stars by using a unit called the parsec. One parsec is approximately 1.91×10^{13} mi.

Integer exponents are used to represent the very large and very small numbers encountered in the fields of science and engineering. For example, the mass of an electron is 0.0000000000000000000000000000009 g. Numbers such as this are difficult to read and write, so a more convenient system for writing such numbers has been developed. It is called **scientific notation.**

To express a number in scientific notation, write the number as the product of a number between 1 and 10 and a power of 10. The form for scientific notation is $a \times 10^n$, where $1 \leq a < 10$.

For numbers greater than 10, move the decimal point to the right of the first digit. The exponent n is positive and equal to the number of places the decimal point has been moved.

$965{,}000 = 9.65 \times 10^5$

$3{,}600{,}000 = 3.6 \times 10^6$

$92{,}000{,}000{,}000 = 9.2 \times 10^{10}$

For numbers less than 1, move the decimal point to the right of the first nonzero digit. The exponent n is negative. The absolute value of the exponent is equal to the number of places the decimal point has been moved.

$0.0002 = 2 \times 10^{-4}$

$0.0000000974 = 9.74 \times 10^{-8}$

$0.000000000086 = 8.6 \times 10^{-11}$

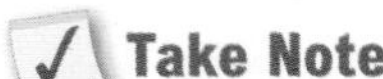

Take Note

There are two steps involved in writing a number in scientific notation:
(1) determine the number between 1 and 10, and
(2) determine the exponent on 10.

Converting a number written in scientific notation to decimal notation requires moving the decimal point.

When the exponent is positive, move the decimal point to the right the same number of places as the exponent.

$1.32 \times 10^4 = 13{,}200$

$1.4 \times 10^8 = 140{,}000{,}000$

When the exponent is negative, move the decimal point to the left the same number of places as the absolute value of the exponent.

$1.32 \times 10^{-2} = 0.0132$

$1.4 \times 10^{-4} = 0.00014$

Numerical calculations involving numbers that have more digits than a hand-held calculator is able to handle can be performed using scientific notation.

Integrating Technology

See the Keystroke Guide: *Scientific Notation* for instructions on entering a number that is in scientific notation into a graphing calculator.

HOW TO 16 Simplify: $\dfrac{220{,}000 \times 0.000000092}{0.0000011}$

$$\frac{220{,}000 \times 0.000000092}{0.0000011} = \frac{2.2 \times 10^5 \times 9.2 \times 10^{-8}}{1.1 \times 10^{-6}}$$ • Write the numbers in scientific notation.

$$= \frac{(2.2)(9.2) \times 10^{5+(-8)-(-6)}}{1.1}$$ • Simplify.

$$= 18.4 \times 10^3 = 18{,}400$$

EXAMPLE • 7

Write 0.000041 in scientific notation.

Solution $0.000041 = 4.1 \times 10^{-5}$

YOU TRY IT • 7

Write 942,000,000 in scientific notation.

Your solution

EXAMPLE • 8

Write 3.3×10^7 in decimal notation.

Solution $3.3 \times 10^7 = 33{,}000{,}000$

YOU TRY IT • 8

Write 2.7×10^{-5} in decimal notation.

Your solution

EXAMPLE • 9

Simplify:
$$\frac{2{,}400{,}000{,}000 \times 0.0000063}{0.00009 \times 480}$$

Solution
$$\frac{2{,}400{,}000{,}000 \times 0.0000063}{0.00009 \times 480}$$
$$= \frac{2.4 \times 10^9 \times 6.3 \times 10^{-6}}{9 \times 10^{-5} \times 4.8 \times 10^2}$$
$$= \frac{(2.4)(6.3) \times 10^{9+(-6)-(-5)-2}}{(9)(4.8)}$$
$$= 0.35 \times 10^6 = 350{,}000$$

YOU TRY IT • 9

Simplify:
$$\frac{5{,}600{,}000 \times 0.000000081}{900 \times 0.000000028}$$

Your solution

Solutions on p. S18

OBJECTIVE D To solve application problems

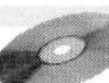

EXAMPLE • 10

How many miles does light travel in 1 day? The speed of light is 186,000 mi/s. Write the answer in scientific notation.

Strategy

To find the distance traveled:

- Write the speed of light in scientific notation.
- Write the number of seconds in 1 day in scientific notation.
- Use the equation $d = rt$, where r is the speed of light and t is the number of seconds in 1 day.

Solution

$r = 186{,}000 = 1.86 \times 10^5$

$t = 24 \cdot 60 \cdot 60 = 86{,}400 = 8.64 \times 10^4$

$d = rt$

$d = (1.86 \times 10^5)(8.64 \times 10^4)$

$= 1.86 \times 8.64 \times 10^9$

$= 16.0704 \times 10^9$

$= 1.60704 \times 10^{10}$

Light travels 1.60704×10^{10} mi in 1 day.

YOU TRY IT • 10

The Roadrunner supercomputer from IBM can perform one arithmetic operation, called a FLOP (**FL**oating-point **OP**eration), in 9.74×10^{-16} s. In scientific notation, how many arithmetic operations can be performed in 1 min?

Your strategy

Your solution

Solution on p. S18

6.1 EXERCISES

OBJECTIVE A To multiply monomials

For Exercises 1 to 37, simplify.

1. $(ab^3)(a^3b)$

2. $(-2ab^4)(-3a^2b^4)$

3. $(9xy^2)(-2x^2y^2)$

4. $(x^2y)^2$

5. $(x^2y^4)^4$

6. $(-2ab^2)^3$

7. $(-3x^2y^3)^4$

8. $(4a^2b^3)^3$

9. $(27a^5b^3)^2$

10. $[(2ab)^3]^2$

11. $[(2a^4b^3)^3]^2$

12. $(xy)(x^2y)^4$

13. $(x^2y^2)(xy^3)^3$

14. $(-4x^3y)^2(2x^2y)$

15. $(-5ab)(3a^3b^2)^2$

16. $(-4r^2s^3)^3(2s^2)$

17. $(3x^5y)(-4x^3)^3$

18. $(4x^3z)(-3y^4z^5)$

19. $(-6a^4b^2)(-7a^2c^5)$

20. $(4ab)^2(-2ab^2c^3)^3$

21. $(-2ab^2)(-3a^4b^5)^3$

22. $(2a^2b)^3(-3ab^4)^2$

23. $(-3ab^3)^3(-2^2a^2b)^2$

24. $(-c^3)(-2a^2bc)(3a^2b)$

25. $(-2x^2y^3z)(3x^2yz^4)$

26. $(x^2z^4)(2xyz^4)(-3x^3y^2)$

27. $(2xy)(-3x^2yz)(x^2y^3z^3)$

28. $(2a^2b)(-3ab^2c)(-b^3c^5)$

29. $(3b^5)(2ab^2)(-2ab^2c^2)$

30. $y^n \cdot y^{2n}$

31. $x^n \cdot x^{n+1}$

32. $y^{2n} \cdot y^{4n+1}$

33. $y^{3n} \cdot y^{3n-2}$

34. $(a^n)^{2n}$

35. $(a^{n-3})^{2n}$

36. $(y^{2n-1})^3$

37. $(x^{3n+2})^5$

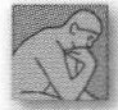

38. If $x^m \cdot x^n = x^9$ and $n = 3$, what is the value of m?

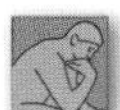

39. What is the value of n in the equation $2^{32} + 2^{32} = 2^n$?

OBJECTIVE B To divide monomials and simplify expressions with negative exponents

For Exercises 40 to 90, simplify.

40. 2^{-3}

41. $\frac{1}{3^{-5}}$

42. $\frac{1}{x^{-4}}$

43. $\frac{1}{y^{-3}}$

44. $\dfrac{2x^{-2}}{y^4}$

45. $\dfrac{a^3}{4b^{-2}}$

46. $x^{-3}y$

47. xy^{-4}

48. $-5x^0$

49. $\dfrac{1}{2x^0}$

50. $\dfrac{(2x)^0}{-2^3}$

51. $\dfrac{-3^{-2}}{(2y)^0}$

52. $(x^2y^{-4})^2$

53. $(x^3y^5)^{-2}$

54. $(2x^3y^{-2})(-3x^{-4}y^3)$

55. $(-3a^{-4}b^{-5})(-5a^{-2}b^4)$

56. $(5m^{-3}n^2)^{-2}(10m^2n)$

57. $(4y^{-3}z^{-4})(-3y^3z^{-3})^{-2}$

58. $(-6mn^{-2})^3(-4m^{-3}n^{-1})^{-2}$

59. $(4x^{-3}y^2)^{-3}(2xy^{-3})^4$

60. $\dfrac{6a^4}{4a^6}$

61. $\dfrac{9x^5}{12x^8}$

62. $\dfrac{x^{17}y^5}{-x^7y^{10}}$

63. $\dfrac{-6x^2y}{12x^4y}$

64. $\dfrac{y^{-7}}{y^{-8}}$

65. $\dfrac{y^{-2}}{y^6}$

66. $\dfrac{x^{-2}y^{-11}}{xy^{-2}}$

67. $\dfrac{x^4y^3}{x^{-1}y^{-2}}$

68. $\dfrac{a^{-1}b^{-3}}{a^4b^{-5}}$

69. $\dfrac{a^6b^{-4}}{a^{-2}b^5}$

70. $\dfrac{2x^2y^4}{(3xy^2)^3}$

71. $\dfrac{-3ab^2}{(9a^2b^4)^3}$

72. $\dfrac{(4x^2y)^2}{(2xy^3)^3}$

73. $\dfrac{(3a^2b)^3}{(-6ab^3)^2}$

74. $\dfrac{(-4xy^3)^3}{(-2x^7y)^4}$

75. $\dfrac{(-8x^2y^2)^4}{(16x^3y^7)^2}$

76. $\dfrac{(-3x^3y^{-2})^3}{(2xy^{-3})^{-2}}$

77. $\dfrac{(3a^4b^{-2})^{-2}}{(2a^{-3}b)^3}$

78. $\dfrac{(2m^{-2}n^3)^{-2}}{(6m^{-1}n^{-2})^{-3}}$

79. $\dfrac{(-2x^{-5}y^2)^{-3}}{(4xy^{-2})^{-4}}$

80. $\dfrac{a^{5n}}{a^{3n}}$

81. $\dfrac{b^{6n}}{b^{10n}}$

82. $\dfrac{-x^{5n}}{x^{2n}}$

83. $\dfrac{y^{2n}}{-y^{8n}}$

84. $\dfrac{x^{2n-1}}{x^{n-3}}$

85. $\dfrac{y^{3n+2}}{y^{2n+4}}$

86. $\dfrac{a^{3n-2}b^{n+1}}{a^{2n+1}b^{2n+2}}$

87. $\dfrac{x^{2n-1}y^{n-3}}{x^{n+4}y^{n+3}}$

88. $\left(\dfrac{4^{-2}xy^{-3}}{x^{-3}y}\right)^3\left(\dfrac{8^{-1}x^{-2}y}{x^4y^{-1}}\right)^{-2}$

89. $\left(\dfrac{9ab^{-2}}{8a^{-2}b}\right)^{-2}\left(\dfrac{3a^{-2}b}{2a^2b^{-2}}\right)^3$

90. $\left(\dfrac{2ab^{-1}}{ab}\right)^{-1}\left(\dfrac{3a^{-2}b}{a^2b^2}\right)^{-2}$

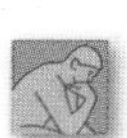

91. If $\dfrac{x^p}{x^q} = 1$, what is the value of $p - q$?

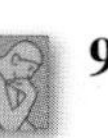

92. If $m < n$, is the value of $\dfrac{2^m}{2^n}$ less than 1 or greater than 1?

OBJECTIVE C To write a number using scientific notation

For Exercises 93 to 98, write in scientific notation.

93. 0.00000467 **94.** 0.00000005 **95.** 0.00000000017

96. 4,300,000 **97.** 200,000,000,000 **98.** 9,800,000,000

For Exercises 99 to 104, write in decimal notation.

99. 1.23×10^{-7} **100.** 6.2×10^{-12} **101.** 8.2×10^{15}

102. 6.34×10^{5} **103.** 3.9×10^{-2} **104.** 4.35×10^{9}

For Exercises 105 to 116, simplify. Write the answer in decimal notation.

105. $(3 \times 10^{-12})(5 \times 10^{16})$ **106.** $(8.9 \times 10^{-5})(3.2 \times 10^{-6})$

107. $(0.0000065)(3{,}200{,}000{,}000{,}000)$ **108.** $(480{,}000)(0.0000000096)$

109. $\dfrac{9 \times 10^{-3}}{6 \times 10^{5}}$ **110.** $\dfrac{2.7 \times 10^{4}}{3 \times 10^{-6}}$ **111.** $\dfrac{0.0089}{500{,}000{,}000}$ **112.** $\dfrac{0.000000346}{0.0000005}$

113. $\dfrac{(3.3 \times 10^{-11})(2.7 \times 10^{15})}{8.1 \times 10^{-3}}$ **114.** $\dfrac{(6.9 \times 10^{27})(8.2 \times 10^{-13})}{4.1 \times 10^{15}}$

115. $\dfrac{(0.00000004)(84{,}000)}{(0.0003)(1{,}400{,}000)}$ **116.** $\dfrac{(720)(0.0000000039)}{(26{,}000{,}000{,}000)(0.018)}$

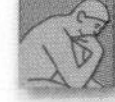

117. Is 5.27×10^{-6} less than zero or greater than zero?

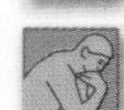

118. Place the correct symbol, <, =, or >, between the two numbers: 4.61×10^{5} ____ 46.1×10^{4}.

OBJECTIVE D To solve application problems

For Exercises 119 to 127, solve. Write the answer in scientific notation.

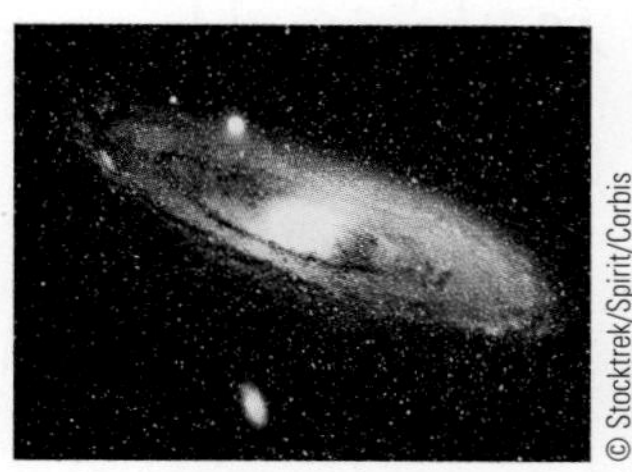

The Milky Way

© Stocktrek/Spirit/Corbis

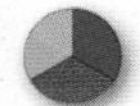

119. Astronomy Our galaxy is estimated to be 5.6×10^{19} mi across. How many years would it take a spaceship to cross the galaxy traveling at 25,000 mph?

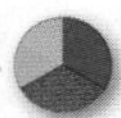

120. Astronomy How long does it take light to travel to Earth from the sun? The sun is 9.3×10^{7} mi from Earth, and light travels 1.86×10^{5} mi/s.

121. **Physics** The mass of an electron is 9.109×10^{-31} kg. The mass of a proton is 1.673×10^{-27} kg. How many times larger is the mass of a proton than the mass of an electron?

122. **The Federal Government** In 2010, the gross national debt was approximately 1.1×10^{13} dollars. How much would each American have to pay in order to pay off the debt? Use 3×10^8 as the number of citizens.

123. **Geology** The mass of Earth is 5.9×10^{24} kg. The mass of the sun is 2×10^{30} kg. How many times larger is the mass of the sun than the mass of Earth?

124. **Astronomy** Use the information in the article below to determine the average number of miles traveled per day by *Phoenix* on its trip to Mars.

In the News

A Mars Landing for *Phoenix*

At 7:53 P.M., a safe landing on the surface of Mars brought an end to the *Phoenix* spacecraft's 296-day, 422-million-mile journey to the Red Planet.

Source: The Los Angeles Times

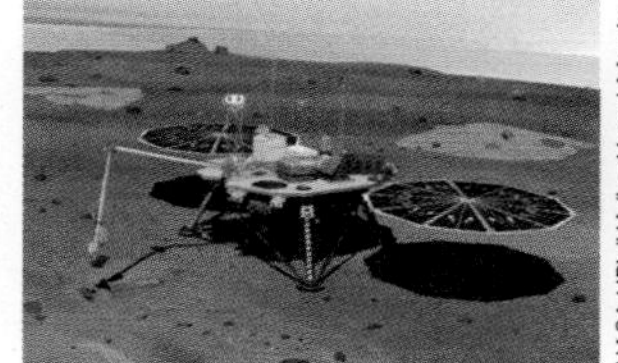
NASA/JPL/UA/Lockheed Martin

125. **Astronomy** It took 11 min for the commands from a computer on Earth to travel to the *Phoenix* Mars Lander, a distance of 119 million miles. How fast did the signals from Earth to Mars travel?

126. **Forestry** Use the information in the article at the right. If every burned acre of Yellowstone Park had 12,000 lodgepole pine seedlings growing on it 1 year after the fire, how many new seedlings would be growing?

127. **Forestry** Use the information in the article at the right. Find the number of seeds released by the lodgepole pine trees for each surviving seedling.

In the News

Forest Fires Spread Seeds

Forest fires may be feared by humans, but not by the lodgepole pine, a tree that uses the intense heat of a fire to release its seeds from their cones. After a blaze that burned 12,000,000 acres of Yellowstone National Park, scientists counted 2 million lodgepole pine seeds on a single acre of the park. One year later, they returned to find 12,000 lodgepole pine seedlings growing.

Source: National Public Radio

128. One light-year is approximately 5.9×10^{12} mi and is defined as the distance light can travel in a vacuum in 1 year. *Voyager 1* is approximately 15 light-hours away from Earth and took about 30 years to travel that distance. One light-hour is approximately 6.7×10^8 mi. True or false: $\dfrac{5.9 \times 10^{12}}{6.7 \times 10^8} \approx$ number of hours in 1 year.

Applying the Concepts

129. Correct the error in each of the following expressions. Explain which rule or property was used incorrectly.
a. $x^0 = 0$ **b.** $(x^4)^5 = x^9$ **c.** $x^2 \cdot x^3 = x^6$

130. Simplify. **a.** $1 + [1 + (1 + 2^{-1})^{-1}]^{-1}$ **b.** $2 - [2 - (2 - 2^{-1})^{-1}]^{-1}$

SECTION

6.2 Introduction to Polynomial Functions

OBJECTIVE A To evaluate polynomial functions

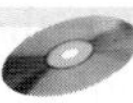

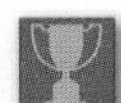

Tips for Success
A great many new vocabulary words are introduced in this chapter. All of these terms are in **boldface type.** The bold type indicates that these are concepts you must know to learn the material. Be sure to study each new term as it is presented.

A **polynomial** is a variable expression in which the terms are monomials.

A polynomial of one term is a **monomial.** $5x$

A polynomial of two terms is a **binomial.** $5x^2y + 6x$

A polynomial of three terms is a **trinomial.** $3x^2 + 9xy - 5y$

Polynomials with more than three terms do not have special names.

The **degree of a polynomial** is the greatest of the degrees of any of its terms.

$3x + 2$ degree 1
$3x^2 + 2x - 4$ degree 2
$4x^3y^2 + 6x^4$ degree 5
$3x^{2n} - 5x^n - 2$ degree $2n$

The terms of a polynomial in one variable are usually arranged so that the exponents on the variable decrease from left to right. This is called **descending order.**

$2x^2 - x + 8$

$3y^3 - 3y^2 + y - 12$

For a polynomial in more than one variable, descending order may refer to any one of the variables.

The polynomial at the right is shown first in descending order of the x variable and then in descending order of the y variable.

$2x^2 + 3xy + 5y^2$

$5y^2 + 3xy + 2x^2$

Polynomial functions have many applications in mathematics. In general, a **polynomial function** is an expression whose terms are monomials. The **linear function** given by $f(x) = mx + b$ is an example of a polynomial function. It is a polynomial function of degree 1. A second-degree polynomial function, called a **quadratic function,** is given by the equation $f(x) = ax^2 + bx + c$, $a \neq 0$. A third-degree polynomial function is called a **cubic function.**

To **evaluate a polynomial function,** replace the variable by its value and simplify.

HOW TO • 1 Given $P(x) = x^3 - 3x^2 + 4$, evaluate $P(-3)$.

$$P(x) = x^3 - 3x^2 + 4$$
$$P(-3) = (-3)^3 - 3(-3)^2 + 4$$ • Substitute −3 for x and simplify.
$$= -27 - 27 + 4$$
$$= -50$$

Integrating Technology
See the Keystroke Guide: *Evaluating Functions* for instructions on using a graphing calculator to evaluate a function.

The **leading coefficient** of a polynomial function is the coefficient of the variable with the largest exponent. The **constant term** is the term without a variable.

HOW TO • 2 Find the leading coefficient, the constant term, and the degree of the polynomial function $P(x) = 7x^4 - 3x^2 + 2x - 4$.

The leading coefficient is 7, the constant term is -4, and the degree is 4.

The three equations below do not represent polynomial functions.

$f(x) = 3x^2 + 2x^{-1}$ A polynomial function does not have a variable raised to a negative power.

$g(x) = 2\sqrt{x} - 3$ A polynomial function does not have a variable expression within a radical.

$h(x) = \dfrac{x}{x - 1}$ A polynomial function does not have a variable in the denominator of a fraction.

The graph of a linear function is a straight line and can be found by plotting just two points. The graph of a polynomial function of degree greater than 1 is a curve. Consequently, many points may have to be found before an accurate graph can be drawn.

Evaluating the quadratic function given by the equation $f(x) = x^2 - x - 6$ when $x = -3, -2, -1, 0, 1, 2, 3,$ and 4 gives the points shown in Figure 1 below. For instance, $f(-3) = 6$, so $(-3, 6)$ is graphed; $f(2) = -4$, so $(2, -4)$ is graphed; and $f(4) = 6$, so $(4, 6)$ is graphed. Evaluating the function when x is not an integer, such as when $x = -\frac{3}{2}$ and $x = \frac{5}{2}$, produces more points to graph, as shown in Figure 2. Connecting the points with a smooth curve results in Figure 3, which is the graph of f.

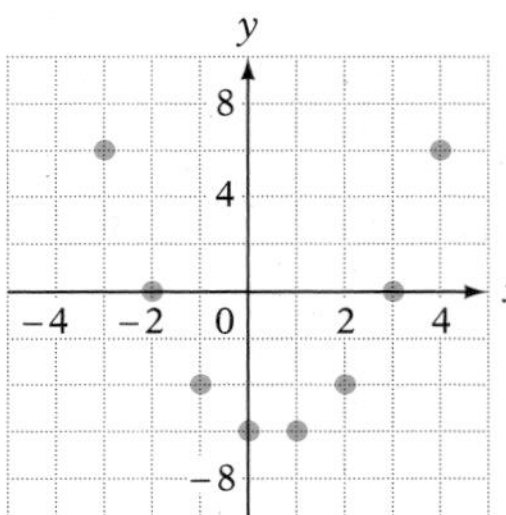

FIGURE 1

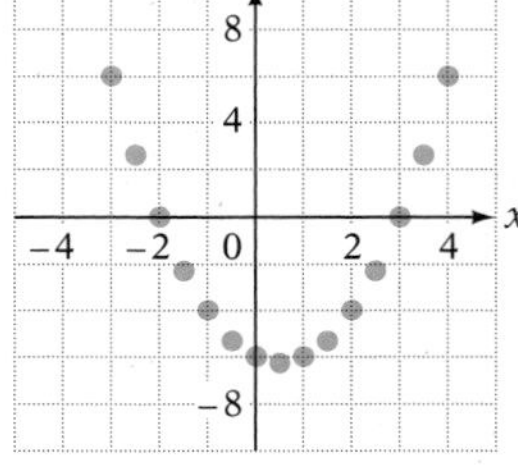

FIGURE 2

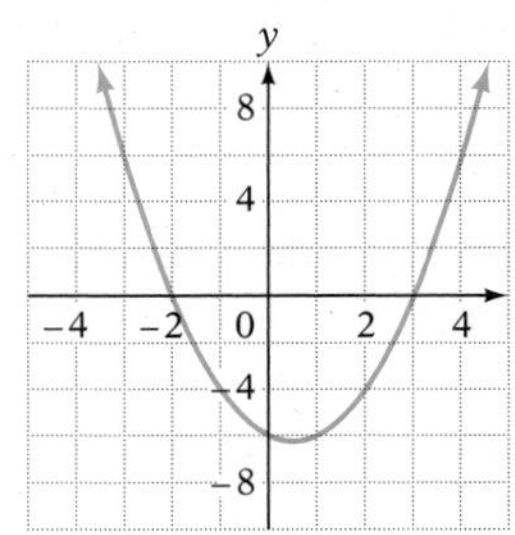

FIGURE 3

Here is an example of graphing a cubic function, $P(x) = x^3 - 2x^2 - 5x + 6$. Evaluating the function when $x = -2, -1, 0, 1, 2, 3,$ and 4 gives the graph in Figure 4 below. Evaluating for some noninteger values gives the graph in Figure 5. Finally, connecting the dots with a smooth curve gives the graph in Figure 6.

Integrating Technology

You can verify the graphs of these polynomial functions by using a graphing calculator. See the Keystroke Guide: *Graph* for instructions on using a graphing calculator to graph a function.

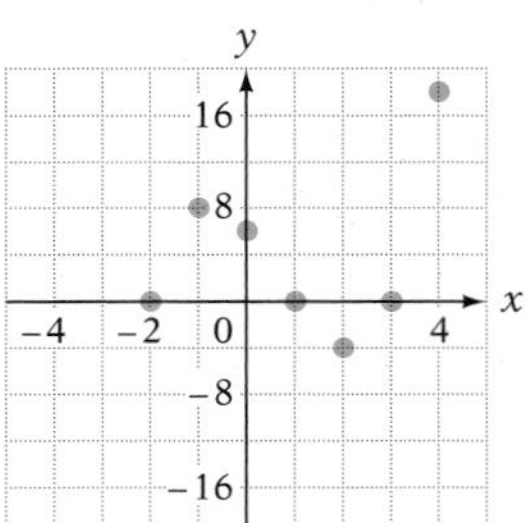

FIGURE 4

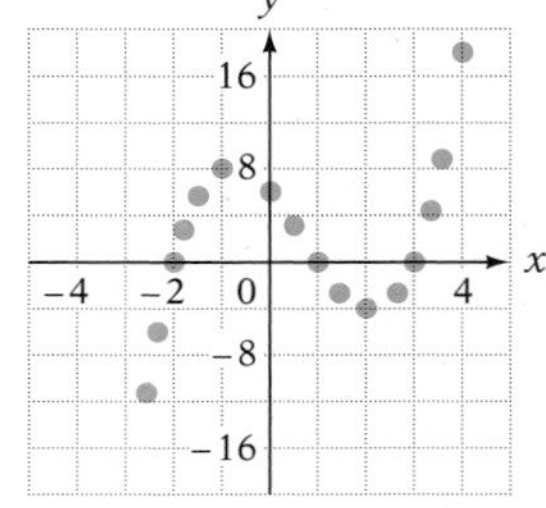

FIGURE 5

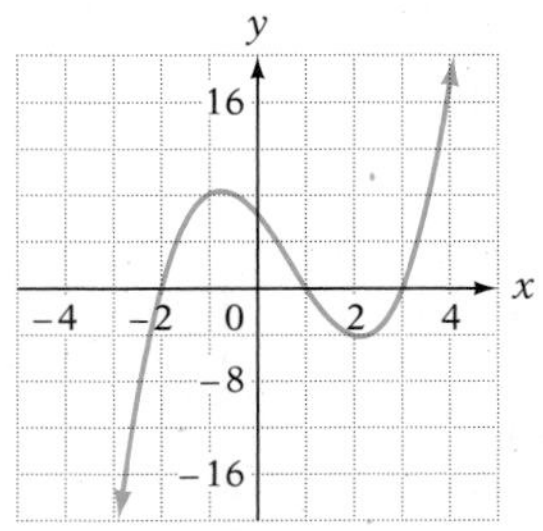

FIGURE 6

EXAMPLE • 1

Given $P(x) = x^3 + 3x^2 - 2x + 8$, evaluate $P(-2)$.

Solution

$$\begin{aligned} P(x) &= x^3 + 3x^2 - 2x + 8 \\ P(-2) &= (-2)^3 + 3(-2)^2 - 2(-2) + 8 \\ &= (-8) + 3(4) + 4 + 8 \\ &= -8 + 12 + 4 + 8 \\ &= 16 \end{aligned}$$

• Replace x by -2. Simplify.

YOU TRY IT • 1

Given $R(x) = -2x^4 - 5x^3 + 2x - 8$, evaluate $R(2)$.

Your solution

EXAMPLE • 2

Find the leading coefficient, the constant term, and the degree of the polynomial.
$P(x) = 5x^6 - 4x^5 - 3x^2 + 7$

Solution

The leading coefficient is 5, the constant term is 7, and the degree is 6.

YOU TRY IT • 2

Find the leading coefficient, the constant term, and the degree of the polynomial.
$r(x) = -3x^4 + 3x^3 + 3x^2 - 2x - 12$

Your solution

EXAMPLE • 3

Which of the following is a polynomial function?

a. $P(x) = 3x^{\frac{1}{2}} + 2x^2 - 3$
b. $T(x) = 3\sqrt{x} - 2x^2 - 3x + 2$
c. $R(x) = 14x^3 - \pi x^2 + 3x + 2$

Solution

a. This is not a polynomial function. A polynomial function does not have a variable raised to a fractional power.
b. This is not a polynomial function. A polynomial function does not have a variable expression within a radical.
c. This is a polynomial function.

YOU TRY IT • 3

Which of the following is a polynomial function?

a. $R(x) = 5x^{14} - 5$
b. $V(x) = -x^{-1} + 2x - 7$
c. $P(x) = 2x^4 - 3\sqrt{x} - 3$

Your solution

EXAMPLE • 4

Graph $f(x) = x^2 - 2$.

Solution

x	$y = f(x)$
−3	7
−2	2
−1	−1
0	−2
1	−1
2	2
3	7

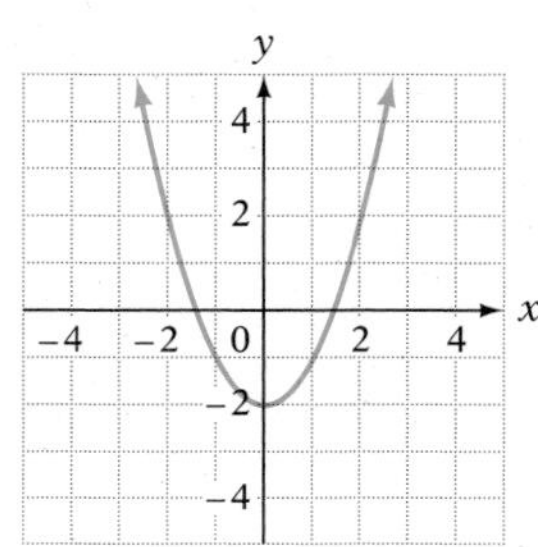

YOU TRY IT • 4

Graph $f(x) = x^2 + 2x - 3$.

Your solution

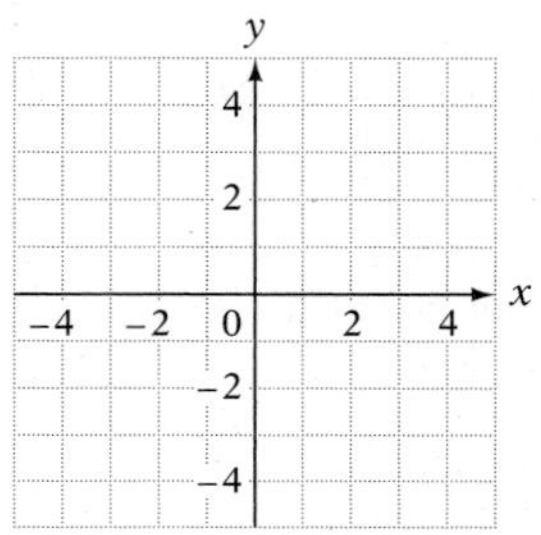

Solutions on p. S18

EXAMPLE 5

Graph $f(x) = x^3 - 1$.

Solution

x	$y = f(x)$
−2	−9
−1	−2
0	−1
1	0
2	7

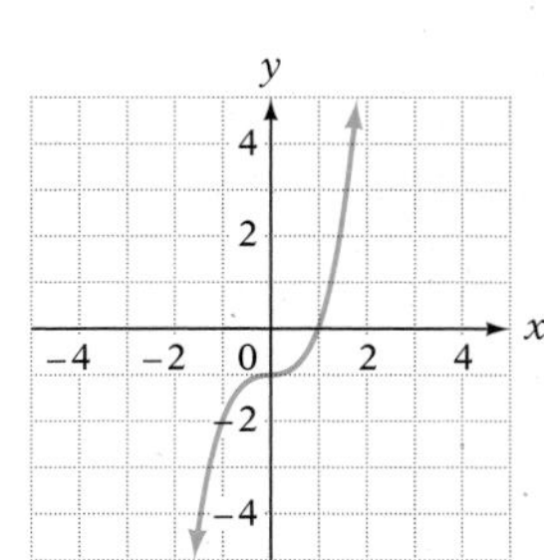

YOU TRY IT 5

Graph $f(x) = -x^3 + 1$.

Your solution

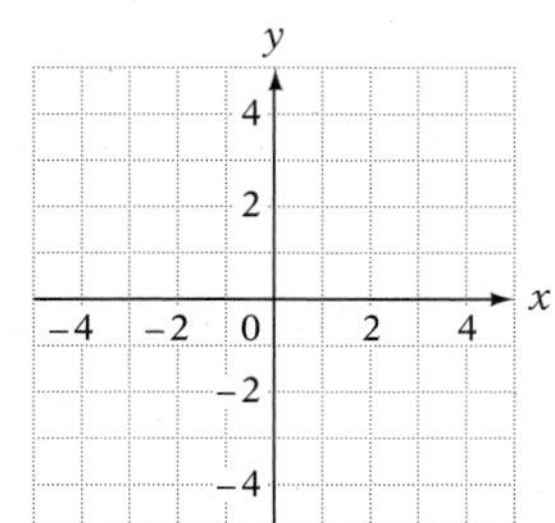

Solution on p. S18

OBJECTIVE B To add or subtract polynomials 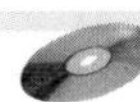

Polynomials can be added by combining like terms. Either a vertical or a horizontal format can be used.

HOW TO 3 Add $(3x^2 + 2x - 7) + (7x^3 - 3 + 4x^2)$. Use a horizontal format.

Use the Commutative and Associative Properties of Addition to rearrange and group like terms.

$$\begin{aligned}&(3x^2 + 2x - 7) + (7x^3 - 3 + 4x^2)\\&= 7x^3 + (3x^2 + 4x^2) + 2x + (-7 - 3)\\&= 7x^3 + 7x^2 + 2x - 10\end{aligned}$$

• Combine like terms.

HOW TO 4 Add $(4x^2 + 5x - 3) + (7x^3 - 7x + 1) + (2x - 3x^2 + 4x^3 + 1)$. Use a vertical format.

Arrange the terms of each polynomial in descending order, with like terms in the same column.

$$\begin{array}{rrrr} & 4x^2 & +\ 5x & -\ 3\\ 7x^3 & & -\ 7x & +\ 1\\ 4x^3 & -\ 3x^2 & +\ 2x & +\ 1\\ \hline 11x^3 & +\ x^2 & & -\ 1\end{array}$$

• Add the terms in each column.

Take Note

The additive inverse of a polynomial is that polynomial with the sign of every term changed.

The **additive inverse of the polynomial** $x^2 + 5x - 4$ is $-(x^2 + 5x - 4)$.

To simplify the additive inverse of a polynomial, change the sign of every term inside the parentheses.

$$-(x^2 + 5x - 4) = -x^2 - 5x + 4$$

Take Note
This is the same definition used for subtraction of integers: subtraction is addition of the opposite.

To subtract two polynomials, add the additive inverse of the second polynomial to the first.

HOW TO • 5 Subtract $(3x^2 - 7xy + y^2) - (-4x^2 + 7xy - 3y^2)$. Use a horizontal format.

Rewrite the subtraction as addition of the additive inverse.

$$\begin{aligned}&(3x^2 - 7xy + y^2) - (-4x^2 + 7xy - 3y^2)\\ &\quad = (3x^2 - 7xy + y^2) + (4x^2 - 7xy + 3y^2)\\ &\quad = 7x^2 - 14xy + 4y^2\end{aligned}$$

• Combine like terms.

HOW TO • 6 Subtract $(6x^3 - 3x + 7) - (3x^2 - 5x + 12)$. Use a vertical format.

Rewrite the subtraction as addition of the additive inverse.

$$(6x^3 - 3x + 7) - (3x^2 - 5x + 12) = (6x^3 - 3x + 7) + (-3x^2 + 5x - 12)$$

Arrange the terms of each polynomial in descending order, with like terms in the same column.

$$\begin{array}{rrrr} 6x^3 & & -\ 3x & +\ \ 7 \\ & -\ 3x^2 & +\ 5x & -\ 12 \\ \hline 6x^3 & -\ 3x^2 & +\ 2x & -\ \ 5 \end{array}$$

• Combine the terms in each column.

Function notation can be used when adding or subtracting polynomials.

HOW TO • 7 Given $P(x) = 3x^2 - 2x + 4$ and $R(x) = -5x^3 + 4x + 7$, find $P(x) + R(x)$.

$$\begin{aligned}P(x) + R(x) &= (3x^2 - 2x + 4) + (-5x^3 + 4x + 7)\\ &= -5x^3 + 3x^2 + 2x + 11\end{aligned}$$

HOW TO • 8 Given $P(x) = -5x^2 + 8x - 4$ and $R(x) = -3x^2 - 5x + 9$, find $P(x) - R(x)$.

$$\begin{aligned}P(x) - R(x) &= (-5x^2 + 8x - 4) - (-3x^2 - 5x + 9)\\ &= (-5x^2 + 8x - 4) + (3x^2 + 5x - 9)\\ &= -2x^2 + 13x - 13\end{aligned}$$

HOW TO • 9 Given $P(x) = 3x^2 - 5x + 6$ and $R(x) = 2x^2 - 5x - 7$, find $S(x)$, the sum of the two polynomials.

$$\begin{aligned}S(x) = P(x) + R(x) &= (3x^2 - 5x + 6) + (2x^2 - 5x - 7)\\ &= 5x^2 - 10x - 1\end{aligned}$$

Note from the preceding example that evaluating $P(x) = 3x^2 - 5x + 6$ and $R(x) = 2x^2 - 5x - 7$ at, for example, $x = 3$ and then adding the values is the same as evaluating $S(x) = 5x^2 - 10x - 1$ at 3.

$$\begin{aligned}P(3) &= 3(3)^2 - 5(3) + 6 = 27 - 15 + 6 = 18\\ R(3) &= 2(3)^2 - 5(3) - 7 = 18 - 15 - 7 = -4\end{aligned}$$

$$P(3) + R(3) = 18 + (-4) = 14$$

$$S(3) = 5(3)^2 - 10(3) - 1 = 45 - 30 - 1 = 14$$

EXAMPLE • 6

Add:
$(4x^2 - 3xy + 7y^2) + (-3x^2 + 7xy + y^2)$
Use a vertical format.

Solution

$$\begin{array}{r} 4x^2 - 3xy + 7y^2 \\ -3x^2 + 7xy + y^2 \\ \hline x^2 + 4xy + 8y^2 \end{array}$$

YOU TRY IT • 6

Add:
$(-3x^2 - 4x + 9) + (-5x^2 - 7x + 1)$
Use a vertical format.

Your solution

EXAMPLE • 7

Subtract:
$(3x^2 - 2x + 4) - (7x^2 + 3x - 12)$
Use a vertical format.

Solution

Add the additive inverse of $7x^2 + 3x - 12$ to $3x^2 - 2x + 4$.

$$\begin{array}{r} 3x^2 - 2x + 4 \\ -7x^2 - 3x + 12 \\ \hline -4x^2 - 5x + 16 \end{array}$$

YOU TRY IT • 7

Subtract:
$(-5x^2 + 2x - 3) - (6x^2 + 3x - 7)$
Use a vertical format.

Your solution

EXAMPLE • 8

Given $P(x) = -3x^2 + 2x - 6$ and $R(x) = 4x^3 - 3x + 4$, find $S(x) = P(x) + R(x)$. Evaluate $S(-2)$.

Solution

$$\begin{aligned} S(x) &= P(x) + R(x) \\ S(x) &= (-3x^2 + 2x - 6) + (4x^3 - 3x + 4) \\ &= 4x^3 - 3x^2 - x - 2 \end{aligned}$$

$$\begin{aligned} S(-2) &= 4(-2)^3 - 3(-2)^2 - (-2) - 2 \\ &= 4(-8) - 3(4) + 2 - 2 \\ &= -44 \end{aligned}$$

YOU TRY IT • 8

Given $P(x) = 4x^3 - 3x^2 + 2$ and $R(x) = -2x^2 + 2x - 3$, find $S(x) = P(x) + R(x)$. Evaluate $S(-1)$.

Your solution

EXAMPLE • 9

Given $P(x) = (2x^{2n} - 3x^n + 7)$ and $R(x) = (3x^{2n} + 3x^n + 5)$, find $D(x) = P(x) - R(x)$.

Solution

$$\begin{aligned} D(x) &= P(x) - R(x) \\ D(x) &= (2x^{2n} - 3x^n + 7) - (3x^{2n} + 3x^n + 5) \\ &= (2x^{2n} - 3x^n + 7) + (-3x^{2n} - 3x^n - 5) \\ &= -x^{2n} - 6x^n + 2 \end{aligned}$$

YOU TRY IT • 9

Given $P(x) = (5x^{2n} - 3x^n - 7)$ and $R(x) = (-2x^{2n} - 5x^n + 8)$, find $D(x) = P(x) - R(x)$.

Your solution

Solutions on p. S18

6.2 EXERCISES

OBJECTIVE A To evaluate polynomial functions

1. Given $P(x) = 3x^2 - 2x - 8$, evaluate $P(3)$.

2. Given $P(x) = -3x^2 - 5x + 8$, evaluate $P(-5)$.

3. Given $R(x) = 2x^3 - 3x^2 + 4x - 2$, evaluate $R(2)$.

4. Given $R(x) = -x^3 + 2x^2 - 3x + 4$, evaluate $R(-1)$.

5. Given $f(x) = x^4 - 2x^2 - 10$, evaluate $f(-1)$.

6. Given $f(x) = x^5 - 2x^3 + 4x$, evaluate $f(2)$.

In Exercises 7 to 18, indicate whether the expression defines a polynomial function. For those that are polynomial functions: **a.** Identify the leading coefficient. **b.** Identify the constant term. **c.** State the degree.

7. $P(x) = -x^2 + 3x + 8$

8. $P(x) = 3x^4 - 3x - 7$

9. $f(x) = \sqrt{x} - x^2 + 2$

10. $f(x) = x^2 - \sqrt{x + 2} - 8$

11. $R(x) = \dfrac{x}{x + 1}$

12. $R(x) = \dfrac{3x^2 - 2x + 1}{x}$

13. $g(x) = 3x^5 - 2x^2 + \pi$

14. $g(x) = -4x^5 - 3x^2 + x - \sqrt{7}$

15. $P(x) = 3x^2 - 5x^3 + 2$

16. $P(x) = x^2 - 5x^4 - x^6$

17. $R(x) = 14$

18. $R(x) = \dfrac{1}{x} + 2$

For Exercises 19 to 24, graph.

19. $P(x) = x^2 - 1$

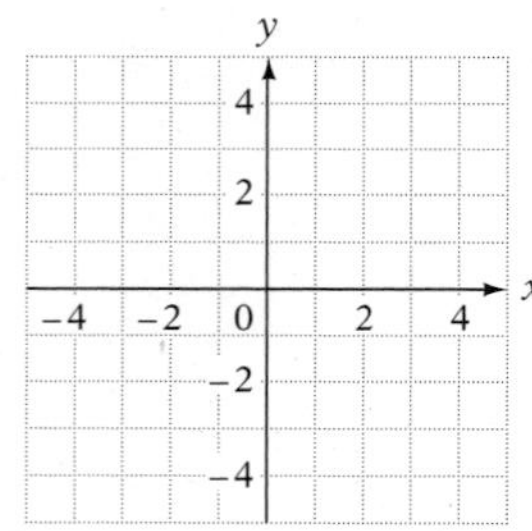

20. $P(x) = 2x^2 + 3$

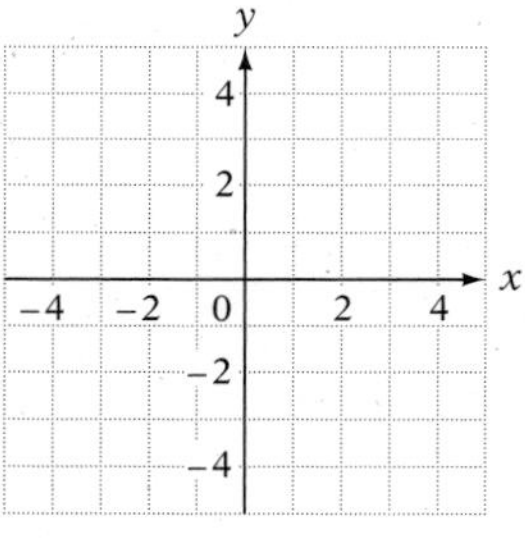

21. $R(x) = x^3 + 2$

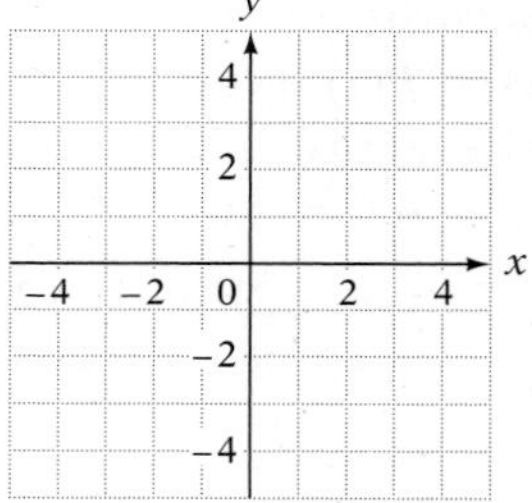

22. $R(x) = x^4 + 1$

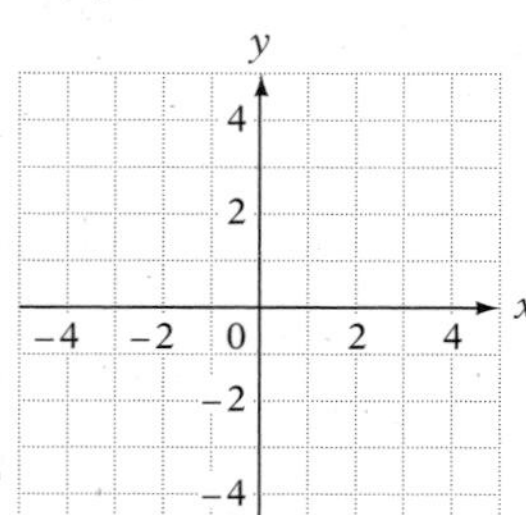

23. $f(x) = x^3 - 2x$

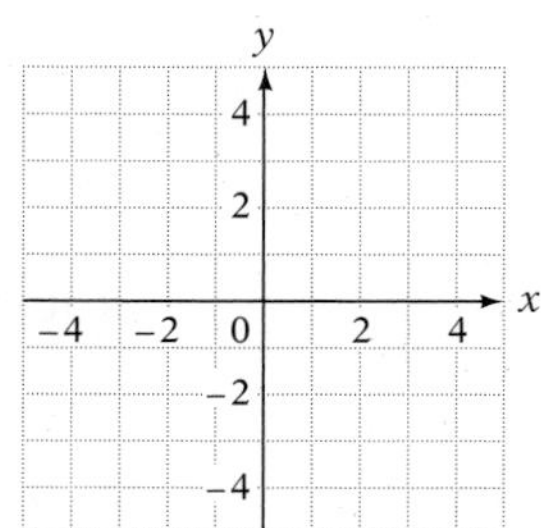

24. $f(x) = x^2 - x - 2$

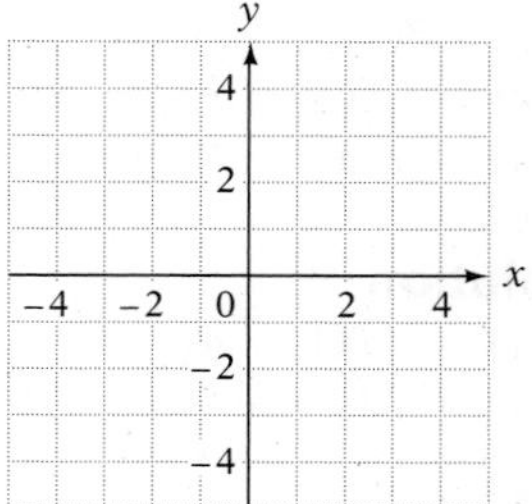

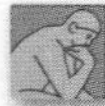

25. Suppose $f(x) = x^2$ and $g(x) = x^3$. For any number c between 0 and 1, is $f(c) - g(c) > 0$ or is $f(c) - g(c) < 0$? Suppose c is between -1 and 0. Is $f(c) - g(c) > 0$ or is $f(c) - g(c) < 0$?

OBJECTIVE B To add or subtract polynomials

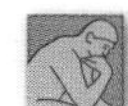

26. If $P(x)$ and $Q(x)$ are polynomials of degree 3, is it possible for the sum of the two polynomials to have degree 2? If so, give an example. If not, explain why.

For Exercises 27 to 30, simplify. Use a vertical format.

27. $(5x^2 + 2x - 7) + (x^2 - 8x + 12)$

28. $(3x^2 - 2x + 7) + (-3x^2 + 2x - 12)$

29. $(x^2 - 3x + 8) - (2x^2 - 3x + 7)$

30. $(2x^2 + 3x - 7) - (5x^2 - 8x - 1)$

For Exercises 31 to 34, simplify. Use a horizontal format.

31. $(3y^2 - 7y) + (2y^2 - 8y + 2)$

32. $(-2y^2 - 4y - 12) + (5y^2 - 5y)$

33. $(2a^2 - 3a - 7) - (-5a^2 - 2a - 9)$

34. $(3a^2 - 9a) - (-5a^2 + 7a - 6)$

35. Given $P(x) = x^2 - 3xy + y^2$ and $R(x) = 2x^2 - 3y^2$, find $P(x) + R(x)$.

36. Given $P(x) = x^{2n} + 7x^n - 3$ and $R(x) = -x^{2n} + 2x^n + 8$, find $P(x) + R(x)$.

37. Given $P(x) = 3x^2 + 2y^2$ and $R(x) = -5x^2 + 2xy - 3y^2$, find $P(x) - R(x)$.

38. Given $P(x) = 2x^{2n} - x^n - 1$ and $R(x) = 5x^{2n} + 7x^n + 1$, find $P(x) - R(x)$.

39. Given $P(x) = 3x^4 - 3x^3 - x^2$ and $R(x) = 3x^3 - 7x^2 + 2x$, find $S(x) = P(x) + R(x)$. Evaluate $S(2)$.

40. Given $P(x) = 3x^4 - 2x + 1$ and $R(x) = 3x^5 - 5x - 8$, find $S(x) = P(x) + R(x)$. Evaluate $S(-1)$.

Applying the Concepts

41. For what value of k is the given equation an identity?
    a. $(2x^3 + 3x^2 + kx + 5) - (x^3 + 2x^2 + 3x + 7) = x^3 + x^2 + 5x - 2$
    b. $(6x^3 + kx^2 - 2x - 1) - (4x^3 - 3x^2 + 1) = 2x^3 - x^2 - 2x - 2$

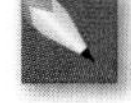

42. If $P(x)$ is a third-degree polynomial and $Q(x)$ is a fourth-degree polynomial, what can be said about the degree of $P(x) + Q(x)$? Give some examples of polynomials that support your answer.

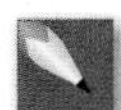

43. If $P(x)$ is a fifth-degree polynomial and $Q(x)$ is a fourth-degree polynomial, what can be said about the degree of $P(x) - Q(x)$? Give some examples of polynomials that support your answer.

44. **Sports** The deflection D (in inches) of a beam that is uniformly loaded is given by the polynomial function $D(x) = 0.005x^4 - 0.1x^3 + 0.5x^2$, where x is the distance from one end of the beam. See the figure at the right. The maximum deflection occurs when x is the midpoint of the beam. Determine the maximum deflection for the beam in the diagram.

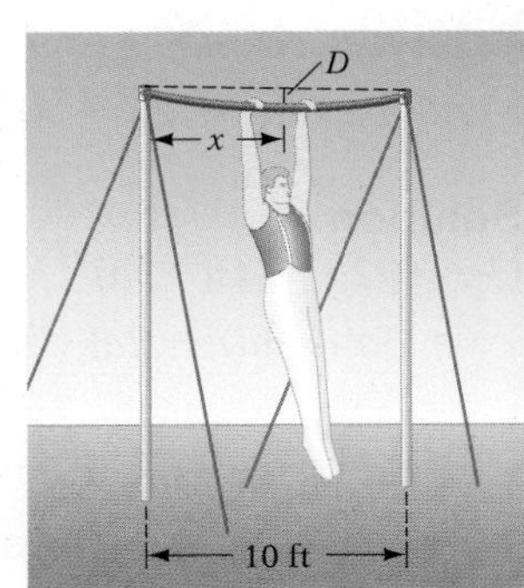

SECTION 6.3 Multiplication of Polynomials

OBJECTIVE A To multiply a polynomial by a monomial 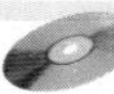

To multiply a polynomial by a monomial, use the Distributive Property and the Rule for Multiplying Exponential Expressions.

HOW TO • 1 Multiply: $-3x^2(2x^2 - 5x + 3)$

$-3x^2(2x^2 - 5x + 3)$

$= -3x^2(2x^2) - (-3x^2)(5x) + (-3x^2)(3)$ • Use the Distributive Property.

$= -6x^4 + 15x^3 - 9x^2$ • Use the Rule for Multiplying Exponential Expressions.

HOW TO • 2 Simplify: $5x(3x - 6) + 3(4x - 2)$

$5x(3x - 6) + 3(4x - 2)$

$= 5x(3x) - 5x(6) + 3(4x) - 3(2)$ • Use the Distributive Property.

$= 15x^2 - 30x + 12x - 6$

$= 15x^2 - 18x - 6$ • Simplify.

HOW TO • 3 Simplify: $2x^2 - 3x[2 - x(4x + 1) + 2]$

$2x^2 - 3x[2 - x(4x + 1) + 2]$

$= 2x^2 - 3x[2 - 4x^2 - x + 2]$ • Use the Distributive Property to remove the parentheses.

$= 2x^2 - 3x[-4x^2 - x + 4]$ • Simplify.

$= 2x^2 + 12x^3 + 3x^2 - 12x$ • Use the Distributive Property to remove the brackets.

$= 12x^3 + 5x^2 - 12x$ • Simplify.

EXAMPLE • 1

Multiply: $(3a^2 - 2a + 4)(-3a)$

Solution

$(3a^2 - 2a + 4)(-3a)$

$= 3a^2(-3a) - 2a(-3a) + 4(-3a)$ • Use the Distributive Property.

$= -9a^3 + 6a^2 - 12a$

YOU TRY IT • 1

Multiply: $(2b^2 - 7b - 8)(-5b)$

Your solution

Solution on p. S18

EXAMPLE 2

Simplify: $y - 3y[y - 2(3y - 6) + 2]$

Solution

$$\begin{aligned} y - 3y[y - 2(3y - 6) + 2] &= y - 3y[y - 6y + 12 + 2] \\ &= y - 3y[-5y + 14] \\ &= y + 15y^2 - 42y \\ &= 15y^2 - 41y \end{aligned}$$

YOU TRY IT 2

Simplify: $x^2 - 2x[x - x(4x - 5) + x^2]$

Your solution

Solution on p. S18

OBJECTIVE B To multiply polynomials

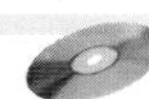

Multiplication of polynomials requires repeated application of the Distributive Property.

HOW TO 4 Multiply: $(2x^2 - 5x + 1)(3x + 2)$

Use the Distributive Property to multiply the trinomial by each term of the binomial.

$$\begin{aligned} (2x^2 - 5x + 1)(3x + 2) &= (2x^2 - 5x + 1)3x + (2x^2 - 5x + 1)2 \\ &= (6x^3 - 15x^2 + 3x) + (4x^2 - 10x + 2) \\ &= 6x^3 - 11x^2 - 7x + 2 \end{aligned}$$

A convenient method of multiplying two polynomials is to use a vertical format similar to that used for multiplication of whole numbers.

HOW TO 5 Multiply: $(2x^2 - 5x + 1)(3x + 2)$

$$\begin{array}{rrrrl} & 2x^2 & - 5x & + 1 & \\ & & 3x & + 2 & \\ \hline & 4x^2 & - 10x & + 2 & = (2x^2 - 5x + 1)2 \\ 6x^3 & - 15x^2 & + 3x & & = (2x^2 - 5x + 1)3x \\ \hline 6x^3 & - 11x^2 & - 7x & + 2 & \end{array}$$

Take Note

FOIL is not really a different way of multiplying. It is based on the Distributive Property.

$$\begin{aligned} &(3x - 2)(2x + 5) \\ &= 3x(2x + 5) - 2(2x + 5) \\ &= 6x^2 + 15x - 4x - 10 \\ &= 6x^2 + 11x - 10 \end{aligned}$$

FOIL is an efficient way of remembering how to do binomial multiplication.

It is frequently necessary to find the product of two binomials. The product can be found by using a method called **FOIL,** which is based on the Distributive Property. The letters of FOIL stand for **F**irst, **O**uter, **I**nner, and **L**ast.

Multiply: $(3x - 2)(2x + 5)$

Multiply the First terms.	$(3x - 2)(2x + 5)$	$3x \cdot 2x = 6x^2$
Multiply the Outer terms.	$(3x - 2)(2x + 5)$	$3x \cdot 5 = 15x$
Multiply the Inner terms.	$(3x - 2)(2x + 5)$	$-2 \cdot 2x = -4x$
Multiply the Last terms.	$(3x - 2)(2x + 5)$	$-2 \cdot 5 = -10$
		F O I L
Add the products.	$(3x - 2)(2x + 5)$	$= 6x^2 + 15x - 4x - 10$
Combine like terms.		$= 6x^2 + 11x - 10$

HOW TO • 6 Multiply: $(6x - 5)(3x - 4)$

$$\begin{aligned}(6x - 5)(3x - 4) &= 6x(3x) + 6x(-4) + (-5)(3x) + (-5)(-4)\\ &= 18x^2 - 24x - 15x + 20\\ &= 18x^2 - 39x + 20\end{aligned}$$

Take Note

The product of $x^2 - x - 12$ and $x + 2$ could have been found using a vertical format.

$$\begin{array}{rrrr} & x^2 & -\ x & -\ 12\\ & & x & +\ 2\\ \hline & 2x^2 & -\ 2x & -\ 24\\ x^3 & -\ x^2 & -\ 12x & \\ \hline x^3 & +\ x^2 & -\ 14x & -\ 24\end{array}$$

HOW TO • 7 Multiply: $(x + 3)(x - 4)(x + 2)$

$(x + 3)(x - 4)(x + 2)$

$= (x^2 - x - 12)(x + 2)$ • Multiply $(x + 3)(x - 4)$.

$= x^2(x + 2) - x(x + 2) - 12(x + 2)$ • Use the Distributive Property.

$= x^3 + 2x^2 - x^2 - 2x - 12x - 24$

$= x^3 + x^2 - 14x - 24$ • Simplify.

EXAMPLE • 3

Multiply: $(4a^3 - 3a + 7)(a - 5)$

Solution Keep like terms in the same columns.

$$\begin{array}{rrrrr} & 4a^3 & & -\ 3a & +\ 7\\ & & & a & -\ 5\\ \hline & -\ 20a^3 & & +\ 15a & -\ 35\\ 4a^4 & & -\ 3a^2 & +\ 7a & \\ \hline 4a^4 & -\ 20a^3 & -\ 3a^2 & +\ 22a & -\ 35\end{array}$$

• $-5(4a^3 - 3a + 7)$

• $a(4a^3 - 3a + 7)$

YOU TRY IT • 3

Multiply: $(-2b^2 + 5b - 4)(-3b + 2)$

Your solution

EXAMPLE • 4

Multiply: $(5a - 3b)(2a + 7b)$

Solution

$(5a - 3b)(2a + 7b)$

$= 10a^2 + 35ab - 6ab - 21b^2$ • FOIL

$= 10a^2 + 29ab - 21b^2$

YOU TRY IT • 4

Multiply: $(3x - 4)(2x - 3)$

Your solution

EXAMPLE • 5

Multiply: $(2x^2 - 3)(4x^2 + 1)$

Solution

$(2x^2 - 3)(4x^2 + 1)$

$= 8x^4 + 2x^2 - 12x^2 - 3$

$= 8x^4 - 10x^2 - 3$

YOU TRY IT • 5

Multiply: $(3ab + 4)(5ab - 3)$

Your solution

Solutions on p. S18

OBJECTIVE C To multiply polynomials that have special products 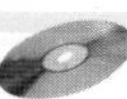

Using FOIL, a pattern can be found for the **product of the sum and difference of two terms** [that is, a polynomial that can be expressed in the form $(a + b)(a - b)$] and for the **square of a binomial** [that is, a polynomial that can be expressed in the form $(a + b)^2$].

The Sum and Difference of Two Terms

$$(a + b)(a - b) = a^2 - ab + ab - b^2$$
$$= a^2 - b^2$$

Square of the first term (a^2)
Square of the second term (b^2)

The Square of a Binomial

$$(a + b)^2 = (a + b)(a + b) = a^2 + ab + ab + b^2$$
$$= a^2 + 2ab + b^2$$

Square of the first term (a^2)
Twice the product of the two terms ($2ab$)
Square of the second term (b^2)

HOW TO • 8 Multiply: $(4x + 3)(4x - 3)$

$(4x + 3)(4x - 3)$ is the sum and difference of the same two terms. The product is the difference of the squares of the terms.

$$(4x + 3)(4x - 3) = (4x)^2 - 3^2$$
$$= 16x^2 - 9$$

Take Note
The word *expand* is sometimes used to mean "multiply," especially when referring to a power of a binomial.

HOW TO • 9 Expand: $(2x - 3y)^2$

$(2x - 3y)^2$ is the square of a binomial.

$$(2x - 3y)^2 = (2x)^2 + 2(2x)(-3y) + (-3y)^2$$
$$= 4x^2 - 12xy + 9y^2$$

EXAMPLE • 6

Multiply: $(2a - 3)(2a + 3)$

Solution

$(2a - 3)(2a + 3) = 4a^2 - 9$ • The sum and difference of two terms

YOU TRY IT • 6

Multiply: $(3x - 7)(3x + 7)$

Your solution

EXAMPLE • 7

Multiply: $(5x + y)(5x - y)$

Solution

$(5x + y)(5x - y) = 25x^2 - y^2$ • The sum and difference of two terms

YOU TRY IT • 7

Multiply: $(2ab + 7)(2ab - 7)$

Your solution

Solutions on p. S19

EXAMPLE • 8

Expand: $(2x + 7y)^2$

Solution

$(2x + 7y)^2 = 4x^2 + 28xy + 49y^2$ • The square of a binomial

YOU TRY IT • 8

Expand: $(3x - 4y)^2$

Your solution

EXAMPLE • 9

Expand: $(3a^2 - b)^2$

Solution

$(3a^2 - b)^2 = 9a^4 - 6a^2b + b^2$ • The square of a binomial

YOU TRY IT • 9

Expand: $(5xy + 4)^2$

Your solution

Solutions on p. S19

OBJECTIVE D To solve application problems

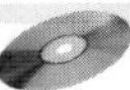

EXAMPLE • 10

The length of a rectangle is $(2x + 3)$ ft. The width is $(x - 5)$ ft. Find the area of the rectangle in terms of the variable x.

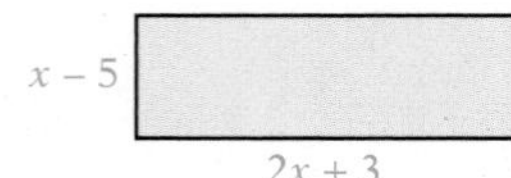

Strategy

To find the area, replace the variables L and W in the equation $A = L \cdot W$ by the given values, and solve for A.

Solution

$$A = L \cdot W$$
$$A = (2x + 3)(x - 5)$$
$$= 2x^2 - 10x + 3x - 15 \quad \text{• FOIL}$$
$$= 2x^2 - 7x - 15$$

The area is $(2x^2 - 7x - 15)$ ft^2.

YOU TRY IT • 10

The base of a triangle is $(2x + 6)$ ft. The height is $(x - 4)$ ft. Find the area of the triangle in terms of the variable x.

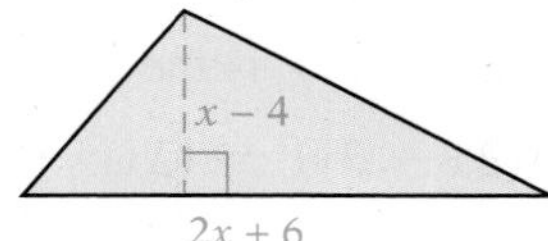

Your strategy

Your solution

Solution on p. S19

EXAMPLE • 11

The corners are cut from a rectangular piece of cardboard measuring 8 in. by 12 in. The sides are folded up to make a box. Find the volume of the box in terms of the variable x, where x is the length of a side of the square cut from each corner of the rectangle.

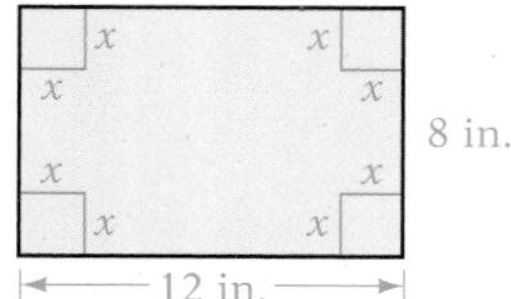

Strategy

Length of the box: $12 - 2x$
Width of the box: $8 - 2x$
Height of the box: x
To find the volume, replace the variables L, W, and H in the equation $V = L \cdot W \cdot H$, and solve for V.

Solution

$V = L \cdot W \cdot H$
$V = (12 - 2x)(8 - 2x)x$
$= (96 - 24x - 16x + 4x^2)x$ • FOIL
$= (96 - 40x + 4x^2)x$
$= 96x - 40x^2 + 4x^3$
$= 4x^3 - 40x^2 + 96x$

The volume is $(4x^3 - 40x^2 + 96x)$ in^3.

YOU TRY IT • 11

Find the volume of the solid shown in the diagram below. All dimensions are in feet.

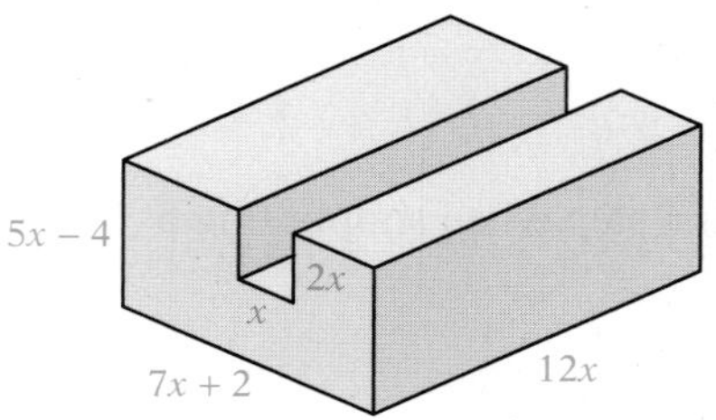

Your strategy

Your solution

EXAMPLE • 12

The radius of a circle is $(3x - 2)$ cm. Find the area of the circle in terms of the variable x. Use 3.14 for π.

Strategy

To find the area, replace the variable r in the equation $A = \pi r^2$ by the given value, and solve for A.

Solution

$A = \pi r^2$
$A \approx 3.14(3x - 2)^2$
$= 3.14(9x^2 - 12x + 4)$
$= 28.26x^2 - 37.68x + 12.56$

The area is $(28.26x^2 - 37.68x + 12.56)$ cm^2.

YOU TRY IT • 12

The radius of a circle is $(2x + 3)$ cm. Find the area of the circle in terms of the variable x. Use 3.14 for π.

Your strategy

Your solution

Solutions on p. S19

6.3 EXERCISES

OBJECTIVE A To multiply a polynomial by a monomial

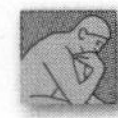

1. What is the name of the property that is used to remove parentheses from the expression $3x(4x^2 - 5x - 7)$?

2. What is the first step when simplifying $5 + 2(2x + 1)$? Why?

For Exercises 3 to 28, simplify.

3. $2x(x - 3)$

4. $2a(2a + 4)$

5. $3x^2(2x^2 - x)$

6. $-4y^2(4y - 6y^2)$

7. $3xy(2x - 3y)$

8. $-4ab(5a - 3b)$

9. $x^n(x + 1)$

10. $y^n(y^{2n} - 3)$

11. $x^n(x^n + y^n)$

12. $x - 2x(x - 2)$

13. $2b + 4b(2 - b)$

14. $-2y(3 - y) + 2y^2$

15. $-2a^2(3a^2 - 2a + 3)$

16. $4b(3b^3 - 12b^2 - 6)$

17. $(-3y^2 - 4y + 2)(y^2)$

18. $(6b^4 - 5b^2 - 3)(-2b^3)$

19. $-5x^2(4 - 3x + 3x^2 + 4x^3)$

20. $-2y^2(3 - 2y - 3y^2 + 2y^3)$

21. $-2x^2y(x^2 - 3xy + 2y^2)$

22. $3ab^2(3a^2 - 2ab + 4b^2)$

23. $5x^3 - 4x(2x^2 + 3x - 7)$

24. $7a^3 - 2a(6a^2 - 5a - 3)$

25. $2y^2 - y[3 - 2(y - 4) - y]$

26. $3x^2 - x[x - 2(3x - 4)]$

27. $2y - 3[y - 2y(y - 3) + 4y]$

28. $4a^2 - 2a[3 - a(2 - a + a^2)]$

29. Given $P(b) = 3b$ and $Q(b) = 3b^4 - 3b^2 + 8$, find $P(b) \cdot Q(b)$.

30. Given $P(x) = -2x^2$ and $Q(x) = 2x^2 - 3x - 7$, find $P(x) \cdot Q(x)$.

OBJECTIVE B **To multiply polynomials**

For Exercises 31 to 60, multiply.

31. $(x - 2)(x + 7)$

32. $(y + 8)(y + 3)$

33. $(2y - 3)(4y + 7)$

34. $(5x - 7)(3x - 8)$

35. $(a + 3c)(4a - 5c)$

36. $(2m - 3n)(5m + 4n)$

37. $(5x - 7)(5x - 7)$

38. $(5r + 2t)(5r - 2t)$

39. $2(2x - 3y)(2x + 5y)$

40. $-3(7x - 3y)(2x - 9y)$

41. $(xy + 4)(xy - 3)$

42. $(xy - 5)(2xy + 7)$

43. $(2x^2 - 5)(x^2 - 5)$

44. $(x^2 - 4)(x^2 - 6)$

45. $(5x^2 - 5y)(2x^2 - y)$

46. $(x^2 - 2y^2)(x^2 + 4y^2)$

47. $(x + 5)(x^2 - 3x + 4)$

48. $(a + 2)(a^2 - 3a + 7)$

49. $(2a - 3b)(5a^2 - 6ab + 4b^2)$

50. $(3a + b)(2a^2 - 5ab - 3b^2)$

51. $(2x^3 + 3x^2 - 2x + 5)(2x - 3)$

52. $(3a^3 + 4a - 7)(4a - 2)$

53. $(2x - 5)(2x^4 - 3x^3 - 2x + 9)$

54. $(2a - 5)(3a^4 - 3a^2 + 2a - 5)$

55. $(x^2 + 2x - 3)(x^2 - 5x + 7)$

56. $(x^2 - 3x + 1)(x^2 - 2x + 7)$

57. $(a - 2)(2a - 3)(a + 7)$

58. $(b - 3)(3b - 2)(b - 1)$

59. $(2x + 3)(x - 4)(3x + 5)$

60. $(3a - 5)(2a + 1)(a - 3)$

61. Given $P(y) = 2y^2 - 1$ and $Q(y) = y^3 - 5y^2 - 3$, find $P(y) \cdot Q(y)$.

62. Given $P(b) = 2b^2 - 3$ and $Q(b) = 3b^2 - 3b + 6$, find $P(b) \cdot Q(b)$.

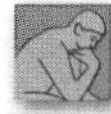

63. If $P(x)$ is a polynomial of degree 3 and $Q(x)$ is a polynomial of degree 2, what is the degree of the product of the two polynomials?

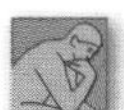

64. Do all polynomials of degree 2 factor over the integers? If not, give an example of a polynomial of degree 2 that does not factor over the integers.

OBJECTIVE C To multiply polynomials that have special products

For Exercises 65 to 88, simplify or expand.

65. $(3x - 2)(3x + 2)$

66. $(4y + 1)(4y - 1)$

67. $(6 - x)(6 + x)$

68. $(10 + b)(10 - b)$

69. $(2a - 3b)(2a + 3b)$

70. $(5x - 7y)(5x + 7y)$

71. $(3ab + 4)(3ab - 4)$

72. $(5xy - 8)(5xy + 8)$

73. $(x^2 + 1)(x^2 - 1)$

74. $(x^2 + y^2)(x^2 - y^2)$

75. $(x - 5)^2$

76. $(y + 2)^2$

77. $(3a + 5b)^2$

78. $(5x - 4y)^2$

79. $(x^2 - 3)^2$

80. $(x^2 + y^2)^2$

81. $(2x^2 - 3y^2)^2$

82. $(2xy + 3)^2$

83. $(3mn - 5)^2$

84. $(2 - 7xy)^2$

85. $y^2 - (x - y)^2$

86. $a^2 + (a + b)^2$

87. $(x - y)^2 - (x + y)^2$

88. $(a + b)^2 + (a - b)^2$

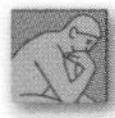

89. True or false? $a^2 + b^2 = (a + b)(a + b)$

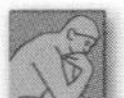

90. If $P(x)$ is a polynomial of degree 2 that factors as the difference of squares, is the coefficient of x in $P(x)$ **(i)** less than zero, **(ii)** equal to zero, **(iii)** greater than zero, or **(iv)** either less than or greater than zero?

OBJECTIVE D To solve application problems

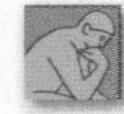

91. If the measures of the width and length of the floor of a room are given in feet, what is the unit of measure of the area?

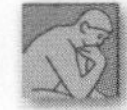

92. If the measure of the width, length, and height of a box are given in meters, what is the unit of measure of the volume of the box?

93. **Geometry** The length of a rectangle is $(3x - 2)$ ft. The width is $(x + 4)$ ft. Find the area of the rectangle in terms of the variable x.

94. **Geometry** The base of a triangle is $(x - 4)$ ft. The height is $(3x + 2)$ ft. Find the area of the triangle in terms of the variable x.

95. **Geometry** Find the area of the figure shown below. All dimensions given are in meters.

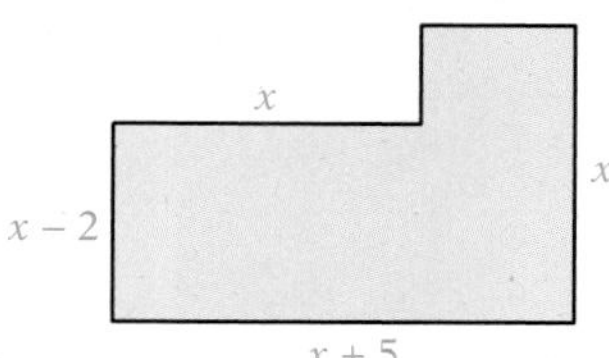

96. **Geometry** Find the area of the figure shown below. All dimensions given are in feet.

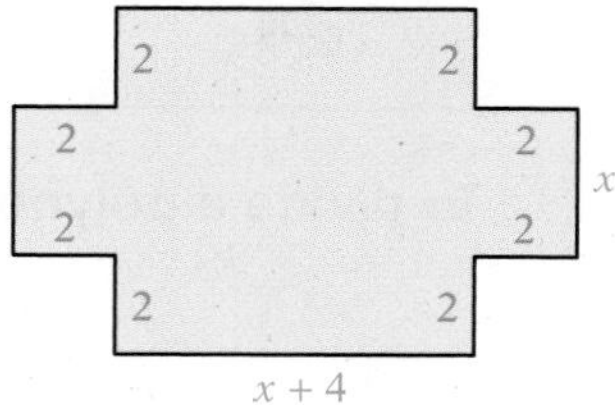

97. **Geometry** The length of the side of a cube is $(x + 3)$ cm. Find the volume of the cube in terms of the variable x.

98. **Geometry** The length of a box is $(3x + 2)$ cm, the width is $(x - 4)$ cm, and the height is x cm. Find the volume of the box in terms of the variable x.

99. **Geometry** Find the volume of the figure shown below. All dimensions given are in inches.

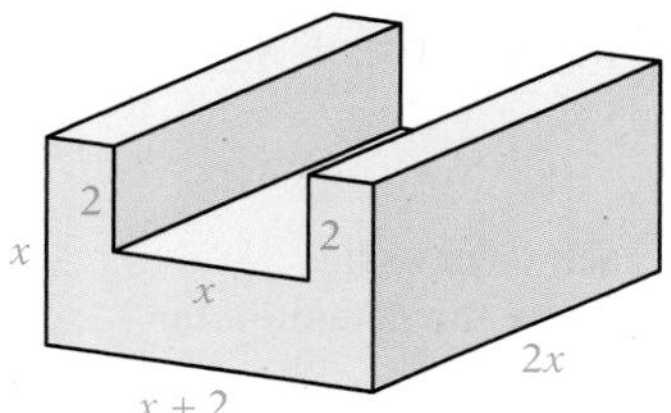

100. **Geometry** Find the volume of the figure shown below. All dimensions given are in centimeters.

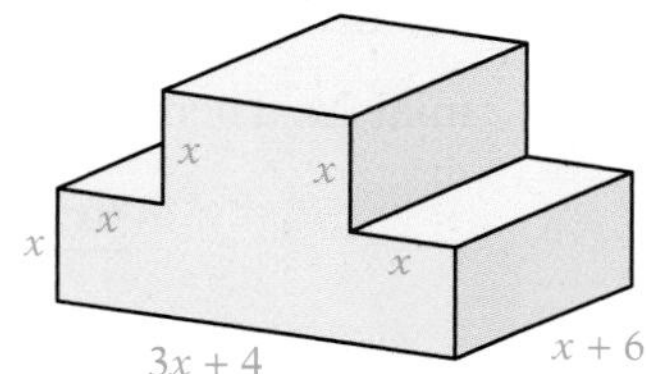

101. **Geometry** The radius of a circle is $(5x + 4)$ in. Find the area of the circle in terms of the variable x. Use 3.14 for π.

102. **Geometry** The radius of a circle is $(x - 2)$ in. Find the area of the circle in terms of the variable x. Use 3.14 for π.

Applying the Concepts

103. Find the product.

a. $(a - b)(a^2 + ab + b^2)$

b. $(x + y)(x^2 - xy + y^2)$

104. Correct the error in each of the following.

a. $(x + 3)^2 = x^2 + 9$

b. $(a - b)^2 = a^2 - b^2$

105. For what value of k is the given equation an identity?

a. $(3x - k)(2x + k) = 6x^2 + 5x - k^2$

b. $(4x + k)^2 = 16x^2 + 8x + k^2$

106. Complete.

a. If $m = n + 1$, then $\dfrac{a^m}{a^n} =$ ____.

b. If $m = n + 2$, then $\dfrac{a^m}{a^n} =$ ____.

107. Subtract the product of $4a + b$ and $2a - b$ from $9a^2 - 2ab$.

108. Subtract the product of $5x - y$ and $x + 3y$ from $6x^2 + 12xy - 2y^2$.

SECTION

6.4 Division of Polynomials

OBJECTIVE A To divide a polynomial by a monomial 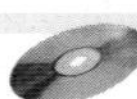

As shown below, $\frac{6+4}{2}$ can be simplified by first adding the terms in the numerator and then dividing the result by the denominator. It can also be simplified by first dividing each term in the numerator by the denominator and then adding the results.

$$\frac{6+4}{2} = \frac{10}{2} = 5 \qquad \frac{6+4}{2} = \frac{6}{2} + \frac{4}{2} = 3 + 2 = 5$$

It is this second method that is used to divide a polynomial by a monomial: Divide each term in the numerator by the denominator, and then write the sum of the quotients.

To divide $\frac{6x^2 + 4x}{2x}$, divide each term of the polynomial $6x^2 + 4x$ by the monomial $2x$. Then simplify each quotient.

Take Note
Recall that the fraction bar can be read "divided by."

$$\frac{6x^2 + 4x}{2x} = \frac{6x^2}{2x} + \frac{4x}{2x}$$ • Divide each term in the numerator by the denominator.

$$= 3x + 2$$ • Simplify each quotient.

We can check this quotient by multiplying it by the divisor.

$2x(3x + 2) = 6x^2 + 4x$ • The product is the dividend. The quotient checks.

HOW TO • 1 Divide and check: $\frac{16x^5 - 8x^3 + 4x}{2x}$

$$\frac{16x^5 - 8x^3 + 4x}{2x} = \frac{16x^5}{2x} - \frac{8x^3}{2x} + \frac{4x}{2x}$$ • Divide each term in the numerator by the denominator.

$$= 8x^4 - 4x^2 + 2$$ • Simplify each quotient.

Check:
$2x(8x^4 - 4x^2 + 2) = 16x^5 - 8x^3 + 4x$ • The quotient checks.

EXAMPLE • 1

Divide and check: $\frac{6x^3 - 3x^2 + 9x}{3x}$

Solution

$\frac{6x^3 - 3x^2 + 9x}{3x}$

$= \frac{6x^3}{3x} - \frac{3x^2}{3x} + \frac{9x}{3x}$ • Divide each term in the numerator by the denominator.

$= 2x^2 - x + 3$ • Simplify each quotient.

Check: $3x(2x^2 - x + 3) = 6x^3 - 3x^2 + 9x$

YOU TRY IT • 1

Divide and check: $\frac{4x^3y + 8x^2y^2 - 4xy^3}{2xy}$

Your solution

Solution on p. S19

OBJECTIVE B To divide polynomials 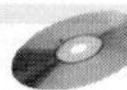

The division method illustrated in Objective A is appropriate only when the divisor is a monomial. To divide two polynomials in which the divisor is not a monomial, use a method similar to that used for division of whole numbers.

To check division of polynomials, use

Dividend = (quotient × divisor) + remainder

HOW TO • 2 Divide: $(x^2 + 5x - 7) \div (x + 3)$

Step 1

$$\begin{array}{r} x\phantom{{}+5x-7} \\ x+3\overline{)x^2 + 5x - 7} \\ \underline{x^2 + 3x}\phantom{{}-7}\downarrow \\ 2x - 7 \end{array}$$

Think: $x\overline{)x^2} = \dfrac{x^2}{x} = x$

Multiply: $x(x + 3) = x^2 + 3x$

Subtract: $(x^2 + 5x) - (x^2 + 3x) = 2x$

Step 2

$$\begin{array}{r} x + 2 \\ x+3\overline{)x^2 + 5x - 7} \\ \underline{x^2 + 3x}\phantom{{}-7} \\ 2x - 7 \\ \underline{2x + 6} \\ -13 \end{array}$$

Think: $x\overline{)2x} = \dfrac{2x}{x} = 2$

Multiply: $2(x + 3) = 2x + 6$

Subtract: $(2x - 7) - (2x + 6) = -13$

The remainder is -13.

Check: $(x + 2)(x + 3) + (-13) = x^2 + 3x + 2x + 6 - 13 = x^2 + 5x - 7$

$$(x^2 + 5x - 7) \div (x + 3) = x + 2 - \frac{13}{x + 3}$$

HOW TO • 3 Divide: $\dfrac{6 - 6x^2 + 4x^3}{2x + 3}$

Arrange the terms in descending order. Note that there is no term containing x in $4x^3 - 6x^2 + 6$. Insert a zero for the missing term so that like terms will be in the same columns.

$$\begin{array}{r} 2x^2 - 6x + 9 \\ 2x+3\overline{)4x^3 - 6x^2 + 0x + 6} \\ \underline{4x^3 + 6x^2}\phantom{{}+0x+6} \\ -12x^2 + 0x\phantom{{}+6} \\ \underline{-12x^2 - 18x}\phantom{{}+6} \\ 18x + 6 \\ \underline{18x + 27} \\ -21 \end{array}$$

$$\frac{4x^3 - 6x^2 + 6}{2x + 3} = 2x^2 - 6x + 9 - \frac{21}{2x + 3}$$

EXAMPLE • 2

Divide: $\dfrac{12x^2 - 11x + 10}{4x - 5}$

Solution

$$\begin{array}{r} 3x + 1 \\ 4x - 5\overline{)12x^2 - 11x + 10} \\ \underline{12x^2 - 15x} \quad\quad \\ 4x + 10 \\ \underline{4x - 5} \\ 15 \end{array}$$

$$\frac{12x^2 - 11x + 10}{4x - 5} = 3x + 1 + \frac{15}{4x - 5}$$

YOU TRY IT • 2

Divide: $\dfrac{15x^2 + 17x - 20}{3x + 4}$

Your solution

EXAMPLE • 3

Divide: $\dfrac{x^3 + 1}{x + 1}$

Solution

$$\begin{array}{r} x^2 - x + 1 \\ x + 1\overline{)x^3 + 0x^2 + 0x + 1} \\ \underline{x^3 + x^2} \quad\quad\quad\quad \\ -x^2 + 0x \quad\quad \\ \underline{-x^2 - x} \quad\quad \\ x + 1 \\ \underline{x + 1} \\ 0 \end{array}$$

• Insert zeros for the missing terms.

$$\frac{x^3 + 1}{x + 1} = x^2 - x + 1$$

YOU TRY IT • 3

Divide: $\dfrac{3x^3 + 8x^2 - 6x + 2}{3x - 1}$

Your solution

EXAMPLE • 4

Divide:
$(2x^4 - 7x^3 + 3x^2 + 4x - 5) \div (x^2 - 2x - 2)$

Solution

$$\begin{array}{r} 2x^2 - 3x + 1 \\ x^2 - 2x - 2\overline{)2x^4 - 7x^3 + 3x^2 + 4x - 5} \\ \underline{2x^4 - 4x^3 - 4x^2} \quad\quad\quad\quad \\ -3x^3 + 7x^2 + 4x \quad\quad \\ \underline{-3x^3 + 6x^2 + 6x} \quad\quad \\ x^2 - 2x - 5 \\ \underline{x^2 - 2x - 2} \\ -3 \end{array}$$

$(2x^4 - 7x^3 + 3x^2 + 4x - 5) \div (x^2 - 2x - 2)$
$= 2x^2 - 3x + 1 - \dfrac{3}{x^2 - 2x - 2}$

YOU TRY IT • 4

Divide:
$(3x^4 - 11x^3 + 16x^2 - 16x + 8) \div (x^2 - 3x + 2)$

Your solution

Solutions on p. S19

OBJECTIVE C To divide polynomials by using synthetic division 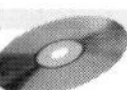

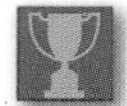

Tips for Success
An important element of success is practice. We cannot do anything well if we do not practice it repeatedly. Practice is crucial to success in mathematics. In this objective you are learning a new procedure, synthetic division. You will need to practice this procedure in order to be successful at it.

Synthetic division is a shorter method of dividing a polynomial by a binomial of the form $x - a$.

Divide $(3x^2 - 4x + 6) \div (x - 2)$ by using long division.

$$\begin{array}{r}
3x + 2 \\
x - 2\overline{)3x^2 - 4x + 6} \\
\underline{3x^2 - 6x} \\
2x + 6 \\
\underline{2x - 4} \\
10
\end{array}$$

$$(3x^2 - 4x + 6) \div (x - 2) = 3x + 2 + \frac{10}{x - 2}$$

The variables can be omitted because the position of a term indicates the power of the term.

$$\begin{array}{r}
3 \quad\; 2 \\
-2\overline{)3 \quad -4 \quad\; 6} \\
\underline{3 \quad -6} \\
2 \quad\; 6 \\
\underline{2 \quad -4} \\
10
\end{array}$$

Each number shown in color is exactly the same as the number above it. Removing the colored numbers condenses the vertical spacing.

$$\begin{array}{r}
3 \quad\; 2 \\
-2\overline{)3 \quad -4 \quad\; 6} \\
\underline{-6 \quad -4} \\
2 \quad 10
\end{array}$$

The number in color in the top row is the same as the one in the bottom row. Writing the 3 from the top row in the bottom row allows the spacing to be condensed even further.

$$\begin{array}{r|rrr}
-2 & 3 & -4 & 6 \\
 & & -6 & -4 \\
\hline
 & 3 & 2 & 10
\end{array}$$

Terms of the quotient (3, 2)   Remainder (10)

Because the degree of the dividend $(3x^2 - 4x + 6)$ is 2 and the degree of the divisor $(x - 2)$ is 1, the degree of the quotient is $2 - 1 = 1$. This means that, using the terms of the quotient given above, that quotient is $3x + 2$. The remainder is 10.

In general, the degree of the quotient of two polynomials is the difference between the degree of the dividend and the degree of the divisor.

By replacing the constant term in the divisor by its additive inverse, we may add rather than subtract terms. This is illustrated in the following example.

HOW TO • 4 Divide: $(3x^3 + 6x^2 - x - 2) \div (x + 3)$

The additive inverse of the binomial constant ↓ ; Coefficients of the polynomial

$$\begin{array}{r|rrrr} -3 & 3 & 6 & -1 & -2 \\ & \downarrow & & & \\ \hline & 3 & & & \end{array}$$

• Bring down the 3.

$$\begin{array}{r|rrrr} -3 & 3 & 6 & -1 & -2 \\ & & -9 & & \\ \hline & 3 & -3 & & \end{array}$$

• Multiply $-3(3)$ and add the product to 6.

$$\begin{array}{r|rrrr} -3 & 3 & 6 & -1 & -2 \\ & & -9 & 9 & \\ \hline & 3 & -3 & 8 & \end{array}$$

• Multiply $-3(-3)$ and add the product to -1.

$$\begin{array}{r|rrrr} -3 & 3 & 6 & -1 & -2 \\ & & -9 & 9 & -24 \\ \hline & 3 & -3 & 8 & -26 \end{array}$$

• Multiply $-3(8)$ and add the product to -2.

Terms of the quotient; Remainder

The degree of the dividend is 3 and the degree of the divisor is 1. Therefore, the degree of the quotient is $3 - 1 = 2$.

$$(3x^3 + 6x^2 - x - 2) \div (x + 3) = 3x^2 - 3x + 8 - \frac{26}{x + 3}$$

HOW TO • 5 Divide: $(2x^3 - x + 2) \div (x - 2)$

The additive inverse of the binomial constant ↓ ; Coefficients of the polynomial

$$\begin{array}{r|rrrr} 2 & 2 & 0 & -1 & 2 \\ & \downarrow & & & \\ \hline & 2 & & & \end{array}$$

• Insert a **0** for the missing term and bring down the 2.

$$\begin{array}{r|rrrr} 2 & 2 & 0 & -1 & 2 \\ & & 4 & & \\ \hline & 2 & 4 & & \end{array}$$

• Multiply 2(2) and add the product to 0.

$$\begin{array}{r|rrrr} 2 & 2 & 0 & -1 & 2 \\ & & 4 & 8 & \\ \hline & 2 & 4 & 7 & \end{array}$$

• Multiply 2(4) and add the product to -1.

$$\begin{array}{r|rrrr} 2 & 2 & 0 & -1 & 2 \\ & & 4 & 8 & 14 \\ \hline & 2 & 4 & 7 & 16 \end{array}$$

• Multiply 2(7) and add the product to 2.

Terms of the quotient; Remainder

$$(2x^3 - x + 2) \div (x - 2) = 2x^2 + 4x + 7 + \frac{16}{x - 2}$$

EXAMPLE • 5

Divide: $(7 - 3x + 5x^2) \div (x - 1)$

Solution
Arrange the coefficients in decreasing powers of x.

1	5	−3	7
		5	2
	5	2	9

$$(5x^2 - 3x + 7) \div (x - 1) = 5x + 2 + \frac{9}{x - 1}$$

YOU TRY IT • 5

Divide: $(6x^2 + 8x - 5) \div (x + 2)$

Your solution

EXAMPLE • 6

Divide: $(2x^3 + 4x^2 - 3x + 12) \div (x + 4)$

Solution

−4	2	4	−3	12
		−8	16	−52
	2	−4	13	−40

$$(2x^3 + 4x^2 - 3x + 12) \div (x + 4)$$
$$= 2x^2 - 4x + 13 - \frac{40}{x + 4}$$

YOU TRY IT • 6

Divide: $(5x^3 - 12x^2 - 8x + 16) \div (x - 2)$

Your solution

EXAMPLE • 7

Divide: $(3x^4 - 8x^2 + 2x + 1) \div (x + 2)$

Solution
Insert a zero for the missing term.

−2	3	0	−8	2	1
		−6	12	−8	12
	3	−6	4	−6	13

$$(3x^4 - 8x^2 + 2x + 1) \div (x + 2)$$
$$= 3x^3 - 6x^2 + 4x - 6 + \frac{13}{x + 2}$$

YOU TRY IT • 7

Divide: $(2x^4 - 3x^3 - 8x^2 - 2) \div (x - 3)$

Your solution

Solutions on pp. S19–S20

OBJECTIVE D To evaluate a polynomial function using synthetic division 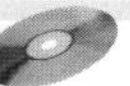

A polynomial can be evaluated by using synthetic division. Consider the polynomial $P(x) = 2x^4 - 3x^3 + 4x^2 - 5x + 1$. One way to evaluate the polynomial when $x = 2$ is to replace x by 2 and then simplify the numerical expression.

$$P(x) = 2x^4 - 3x^3 + 4x^2 - 5x + 1$$

$$\begin{aligned} P(2) &= 2(2)^4 - 3(2)^3 + 4(2)^2 - 5(2) + 1 \\ &= 2(16) - 3(8) + 4(4) - 5(2) + 1 \\ &= 32 - 24 + 16 - 10 + 1 \\ &= 15 \end{aligned}$$

Now use synthetic division to divide $(2x^4 - 3x^3 + 4x^2 - 5x + 1) \div (x - 2)$.

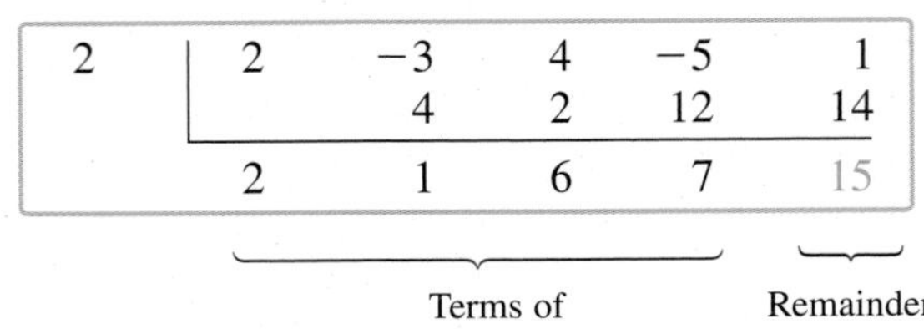

Note that the remainder is 15, which is the same value as $P(2)$. This is not a coincidence. The following theorem states that this situation is always true.

Remainder Theorem

If the polynomial $P(x)$ is divided by $x - a$, the remainder is $P(a)$.

HOW TO • 6 Evaluate $P(x) = x^4 - 3x^2 + 4x - 5$ when $x = -2$ by using the Remainder Theorem.

The value at which the polynomial is evaluated

−2	1	0	−3	4	−5
		−2	4	−2	−4
	1	−2	1	2	−9

• A **0** is inserted for the x^3 term.

← The remainder

$P(-2) = -9$

EXAMPLE • 8

Evaluate $P(x) = x^2 - 6x + 4$ when $x = 3$ by using the Remainder Theorem.

Solution

3	1	−6	4
		3	−9
	1	−3	−5

$P(3) = -5$

YOU TRY IT • 8

Evaluate $P(x) = 2x^2 - 3x - 5$ when $x = 2$ by using the Remainder Theorem.

Your solution

EXAMPLE • 9

Evaluate $P(x) = -x^4 + 3x^3 + 2x^2 - x - 5$ when $x = -2$ by using the Remainder Theorem.

Solution

−2	−1	3	2	−1	−5
		2	−10	16	−30
	−1	5	−8	15	−35

$P(-2) = -35$

YOU TRY IT • 9

Evaluate $P(x) = 2x^3 - 5x^2 + 7$ when $x = -3$ by using the Remainder Theorem.

Your solution

Solutions on p. S20

6.4 EXERCISES

OBJECTIVE A To divide a polynomial by a monomial

For Exercises 1 to 12, divide and check.

1. $\dfrac{3x^2 - 6x}{3x}$

2. $\dfrac{10y^2 - 6y}{2y}$

3. $\dfrac{5x^2 - 10x}{-5x}$

4. $\dfrac{3y^2 - 27y}{-3y}$

5. $\dfrac{5x^2y^2 + 10xy}{5xy}$

6. $\dfrac{8x^2y^2 - 24xy}{8xy}$

7. $\dfrac{x^3 + 3x^2 - 5x}{x}$

8. $\dfrac{a^3 - 5a^2 + 7a}{a}$

9. $\dfrac{9b^5 + 12b^4 + 6b^3}{3b^2}$

10. $\dfrac{a^8 - 5a^5 - 3a^3}{a^2}$

11. $\dfrac{a^5b - 6a^3b + ab}{ab}$

12. $\dfrac{5c^3d + 10c^2d^2 - 15cd^3}{5cd}$

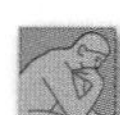

13. If $\dfrac{P(x)}{3x} = 2x^2 + 7x - 5$, what is $P(x)$?

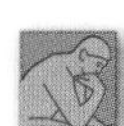

14. If $\dfrac{6x^3 + 15x^2 - 24x}{ax} = 2x^2 + 5x - 8$, what is the value of a?

OBJECTIVE B To divide polynomials

For Exercises 15 to 36, divide by using long division.

15. $(x^2 + 3x - 40) \div (x - 5)$

16. $(x^2 - 14x + 24) \div (x - 2)$

17. $(x^3 - 3x + 2) \div (x - 3)$

18. $(x^3 + 4x^2 - 8) \div (x + 4)$

19. $(6x^2 + 13x + 8) \div (2x + 1)$

20. $(12x^2 + 13x - 14) \div (3x - 2)$

21. $(10x^2 + 9x - 5) \div (2x - 1)$

22. $(18x^2 - 3x + 2) \div (3x + 2)$

23. $(8x^3 - 9) \div (2x - 3)$

24. $(64x^3 + 4) \div (4x + 2)$

25. $(6x^4 - 13x^2 - 4) \div (2x^2 - 5)$

26. $(12x^4 - 11x^2 + 10) \div (3x^2 + 1)$

27. $\dfrac{-10 - 33x + 3x^3 - 8x^2}{3x + 1}$

28. $\dfrac{10 - 49x + 38x^2 - 8x^3}{1 - 4x}$

29. $\dfrac{x^3 - 5x^2 + 7x - 4}{x - 3}$

30. $\dfrac{2x^3 - 3x^2 + 6x + 4}{2x + 1}$

31. $\dfrac{16x^2 - 13x^3 + 2x^4 + 20 - 9x}{x - 5}$

32. $\dfrac{x - x^2 + 5x^3 + 3x^4 - 2}{x + 2}$

33. $\dfrac{2x^3 + 4x^2 - x + 2}{x^2 + 2x - 1}$

34. $\dfrac{3x^3 - 2x^2 + 5x - 4}{x^2 - x + 3}$

35. $\dfrac{x^4 + 2x^3 - 3x^2 - 6x + 2}{x^2 - 2x - 1}$

36. $\dfrac{x^4 - 3x^3 + 4x^2 - x + 1}{x^2 + x - 3}$

37. Given $Q(x) = 2x + 1$ and $P(x) = 2x^3 + x^2 + 8x + 7$, find $\dfrac{P(x)}{Q(x)}$.

38. Given $Q(x) = 3x - 2$ and $P(x) = 3x^3 - 2x^2 + 3x - 5$, find $\dfrac{P(x)}{Q(x)}$.

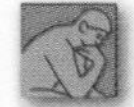

39. True or false? When a tenth-degree polynomial is divided by a second-degree polynomial, the quotient is a fifth-degree polynomial.

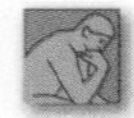

40. Let $p(x)$ be a polynomial of degree 5, and let $q(x)$ be a polynomial of degree 2. If $r(x)$ is the remainder of $\dfrac{p(x)}{q(x)}$, is the degree of $r(x)$ (i) greater than 5, (ii) between 2 and 5, or (iii) less than 2?

OBJECTIVE C **To divide polynomials by using synthetic division**

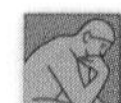

41. The display below shows the beginning of a synthetic division. What is the degree of the dividend?

4	2	−5	4	1

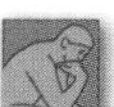

42. For the synthetic division shown below, what is the quotient and what is the remainder?

−2	3	7	4	−3
		−6	−2	−4
	3	1	2	−7

For Exercises 43 to 58, divide by using synthetic division.

43. $(2x^2 - 6x - 8) \div (x + 1)$

44. $(3x^2 + 19x + 20) \div (x + 5)$

45. $(3x^2 - 14x + 16) \div (x - 2)$

46. $(4x^2 - 23x + 28) \div (x - 4)$

47. $(3x^2 - 4) \div (x - 1)$

48. $(4x^2 - 8) \div (x - 2)$

49. $(2x^3 - x^2 + 6x + 9) \div (x + 1)$

50. $(3x^3 + 10x^2 + 6x - 4) \div (x + 2)$

51. $(18 + x - 4x^3) \div (2 - x)$

52. $(12 - 3x^2 + x^3) \div (x + 3)$

53. $(2x^3 + 5x^2 - 5x + 20) \div (x + 4)$

54. $(5x^3 + 3x^2 - 17x + 6) \div (x + 2)$

55. $\dfrac{5 + 5x - 8x^2 + 4x^3 - 3x^4}{2 - x}$

56. $\dfrac{3 - 13x - 5x^2 + 9x^3 - 2x^4}{3 - x}$

57. $\dfrac{3x^4 + 3x^3 - x^2 + 3x + 2}{x + 1}$

58. $\dfrac{4x^4 + 12x^3 - x^2 - x + 2}{x + 3}$

59. Given $Q(x) = x - 2$ and $P(x) = 3x^2 - 5x + 6$, find $\dfrac{P(x)}{Q(x)}$.

60. Given $Q(x) = x + 5$ and $P(x) = 2x^2 + 7x - 12$, find $\dfrac{P(x)}{Q(x)}$.

OBJECTIVE D **To evaluate a polynomial function using synthetic division**

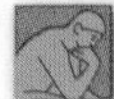

61. The result of a synthetic division is shown below. What is a first-degree polynomial factor of the dividend $p(x)$?

$$\begin{array}{r|rrrr} 3 & 1 & 1 & -7 & -15 \\ & & 3 & 12 & 15 \\ \hline & 1 & 4 & 5 & 0 \end{array}$$

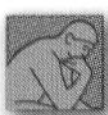

62. The result of a synthetic division of $p(x)$ is shown below. What is $p(-3)$?

$$\begin{array}{r|rrrr} -3 & 1 & 2 & -7 & -4 \\ & & -3 & 3 & 12 \\ \hline & 1 & -1 & -4 & 8 \end{array}$$

For Exercises 63 to 80, use the Remainder Theorem to evaluate the polynomial function.

63. $P(x) = 2x^2 - 3x - 1;\ P(3)$

64. $Q(x) = 3x^2 - 5x - 1;\ Q(2)$

65. $R(x) = x^3 - 2x^2 + 3x - 1;\ R(4)$

66. $F(x) = x^3 + 4x^2 - 3x + 2;\ F(3)$

67. $P(z) = 2z^3 - 4z^2 + 3z - 1;\ P(-2)$

68. $R(t) = 3t^3 + t^2 - 4t + 2;\ R(-3)$

69. $Z(p) = 2p^3 - p^2 + 3;\ Z(-3)$

70. $P(y) = 3y^3 + 2y^2 - 5;\ P(-2)$

71. $Q(x) = x^4 + 3x^3 - 2x^2 + 4x - 9;\ Q(2)$

72. $Y(z) = z^4 - 2z^3 - 3z^2 - z + 7;\ Y(3)$

73. $F(x) = 2x^4 - x^3 + 2x - 5;\ F(-3)$

74. $Q(x) = x^4 - 2x^3 + 4x - 2;\ Q(-2)$

75. $P(x) = x^3 - 3;\ P(5)$

76. $S(t) = 4t^3 + 5;\ S(-4)$

77. $R(t) = 4t^4 - 3t^2 + 5;\ R(-3)$

78. $P(z) = 2z^4 + z^2 - 3;\ P(-4)$

79. $Q(x) = x^5 - 4x^3 - 2x^2 + 5x - 2;\ Q(2)$

80. $R(x) = 2x^5 - x^3 + 4x - 1;\ R(-2)$

Applying the Concepts

81. Divide by using long division.

a. $\dfrac{a^3 + b^3}{a + b}$ **b.** $\dfrac{x^5 + y^5}{x + y}$ **c.** $\dfrac{x^6 - y^6}{x + y}$

82. For what value of k will the remainder be zero?

a. $(x^3 - x^2 - 3x + k) \div (x + 3)$ **b.** $(2x^3 - x + k) \div (x - 1)$

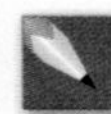

83. Show how synthetic division can be modified so that the divisor can be of the form $ax + b$.

FOCUS ON PROBLEM SOLVING

Dimensional Analysis

In solving application problems, it may be useful to include the units in order to organize the problem so that the answer is in the proper units. Using units to organize and check the correctness of an application is called **dimensional analysis.** We use the operations of multiplying units and dividing units in applying dimensional analysis to application problems.

The Rule for Multiplying Exponential Expressions states that we multiply two expressions with the same base by adding the exponents.

$$x^4 \cdot x^6 = x^{4+6} = x^{10}$$

In calculations that involve quantities, the units are operated on algebraically.

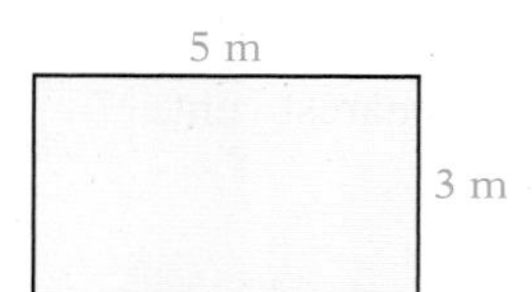

HOW TO 1 A rectangle measures 3 m by 5 m. Find the area of the rectangle.

$A = LW = (3 \text{ m})(5 \text{ m}) = (3 \cdot 5)(\text{m} \cdot \text{m}) = 15 \text{ m}^2$

The area of the rectangle is 15 m^2 (square meters).

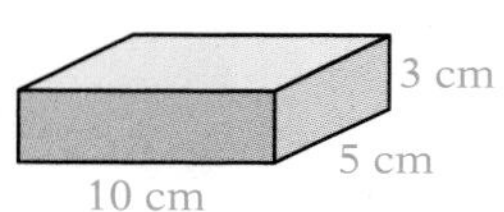

HOW TO 2 A box measures 10 cm by 5 cm by 3 cm. Find the volume of the box.

$V = LWH = (10 \text{ cm})(5 \text{ cm})(3 \text{ cm}) = (10 \cdot 5 \cdot 3)(\text{cm} \cdot \text{cm} \cdot \text{cm}) = 150 \text{ cm}^3$

The volume of the box is 150 cm^3 (cubic centimeters).

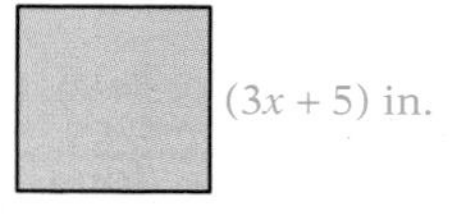

HOW TO 3 Find the area of a square whose side measures $(3x + 5)$ in.

$A = s^2 = [(3x + 5) \text{ in.}]^2 = (3x + 5)^2 \text{ in}^2 = (9x^2 + 30x + 25) \text{ in}^2$

The area of the square is $(9x^2 + 30x + 25) \text{ in}^2$ (square inches).

Dimensional analysis is used in the conversion of units.

1 mi
5280 ft

The following example converts the unit miles to feet. The equivalent measures 1 mi = 5280 ft are used to form the following rates, which are called conversion factors: $\frac{1 \text{ mi}}{5280 \text{ ft}}$ and $\frac{5280 \text{ ft}}{1 \text{ mi}}$. Because 1 mi = 5280 ft, both of the conversion factors $\frac{1 \text{ mi}}{5280 \text{ ft}}$ and $\frac{5280 \text{ ft}}{1 \text{ mi}}$ are equal to 1.

To convert 3 mi to feet, multiply 3 mi by the conversion factor $\frac{5280 \text{ ft}}{1 \text{ mi}}$.

3 mi
15,840 ft

$$3 \text{ mi} = 3 \text{ mi} \cdot 1 = \frac{3 \text{ mi}}{1} \cdot \frac{5280 \text{ ft}}{1 \text{ mi}} = \frac{3 \cancel{\text{mi}} \cdot 5280 \text{ ft}}{1 \cancel{\text{mi}}} = 3 \cdot 5280 \text{ ft} = 15{,}840 \text{ ft}$$

There are two important points in the above illustration. First, you can think of dividing the numerator and denominator by the common unit "mile" just as you would divide the numerator and denominator of a fraction by a common factor.

Second, the conversion factor $\frac{5280 \text{ ft}}{1 \text{ mi}}$ is equal to 1, and multiplying an expression by 1 does not change the value of the expression.

In the application problem that follows, the units are kept in the problem while the problem is worked.

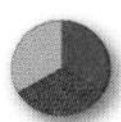

In 2008, a horse named Big Brown ran a 1.25-mile race in 2.02 min. Find Big Brown's average speed for that race in miles per hour. Round to the nearest tenth.

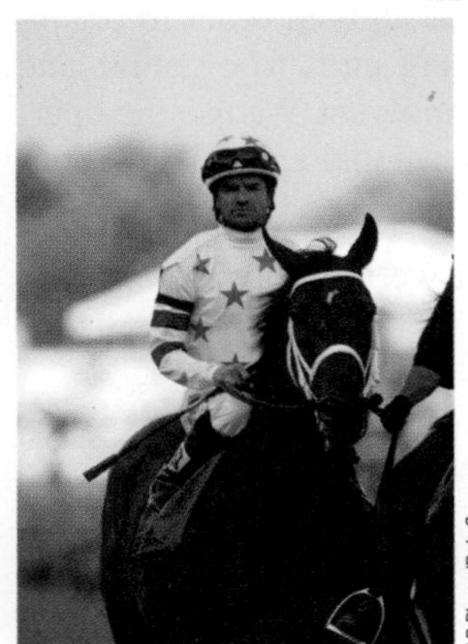

AP Photo/Rob Carr

"Big Brown"

Strategy To find the average speed, use the formula $r = \frac{d}{t}$, where r is the speed, d is the distance, and t is the time. Use the conversion factor $\frac{60\text{ min}}{1\text{ h}}$.

Solution

$$r = \frac{d}{t} = \frac{1.25\text{ mi}}{2.02\text{ min}} = \frac{1.25\text{ mi}}{2.02\text{ min}} \cdot \frac{60\text{ min}}{1\text{ h}}$$

$$= \frac{75\text{ mi}}{2.02\text{ h}} \approx 37.1\text{ mph}$$

Big Brown's average speed was 37.1 mph.

Try each of the following problems. Round to the nearest tenth or nearest cent.

1. Convert 88 ft/s to miles per hour.

2. Convert 8 m/s to kilometers per hour (1 km = 1000 m).

3. A carpet is to be placed in a meeting hall that is 36 ft wide and 80 ft long. At $21.50 per square yard, how much will it cost to carpet the meeting hall?

4. A carpet is to be placed in a room that is 20 ft wide and 30 ft long. At $22.25 per square yard, how much will it cost to carpet the area?

5. Find the number of gallons of water in a fish tank that is 36 in. long and 24 in. wide and is filled to a depth of 16 in. (1 gal = 231 in^3).

6. Find the number of gallons of water in a fish tank that is 24 in. long and 18 in. wide and is filled to a depth of 12 in. (1 gal = 231 in^3).

7. A $\frac{1}{4}$-acre commercial lot is on sale for $2.15 per square foot. Find the sale price of the commercial lot (1 acre = 43,560 ft^2).

8. A 0.75-acre industrial parcel was sold for $98,010. Find the parcel's price per square foot (1 acre = 43,560 ft^2).

9. A new driveway will require 800 ft^3 of concrete. Concrete is ordered by the cubic yard. How much concrete should be ordered?

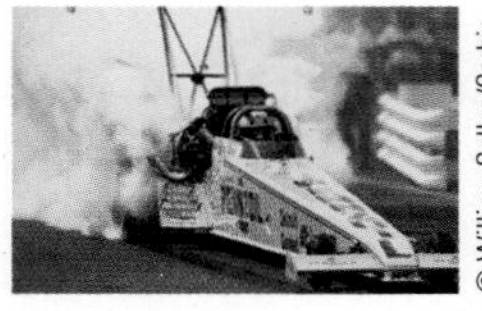

© William Sallaz/Corbis

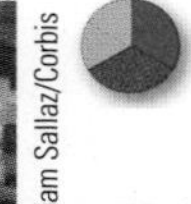

10. A piston-engined dragster traveled 440 yd in 4.936 s at Ennis, Texas, on October 9, 1988. Find the average speed of the dragster in miles per hour.

11. The Marianas Trench in the Pacific Ocean is the deepest part of the ocean. Its depth is 6.85 mi. The speed of sound under water is 4700 ft/s. Find the time it takes sound to travel from the surface of the ocean to the bottom of the Marianas Trench and back.

PROJECTS AND GROUP ACTIVITIES

Astronomical Distances and Scientific Notation

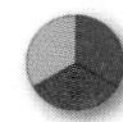

Astronomers have units of measurement that are useful for measuring vast distances in space. Two of these units are the astronomical unit and the light-year. An **astronomical unit** is the average distance between Earth and the sun. A **light-year** is the distance a ray of light travels in 1 year.

1. Light travels at a speed of 1.86×10^5 mi/s. Find the measure of 1 light-year in miles. Use a 365-day year.

2. The distance between Earth and the star Alpha Centauri is approximately 25 trillion miles. Find the distance between Earth and Alpha Centauri in light-years. Round to the nearest hundredth.

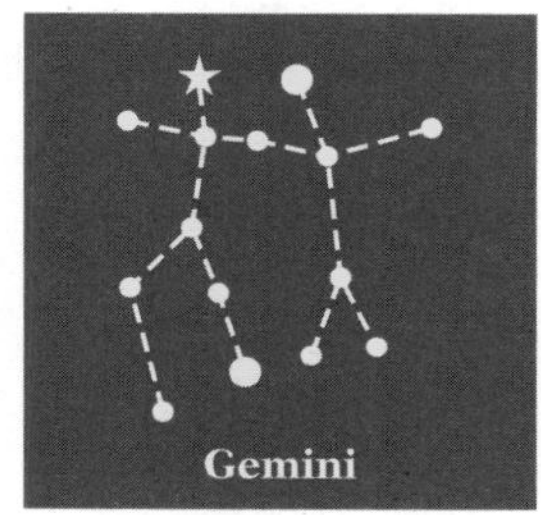

3. The Coma cluster of galaxies is approximately 2.8×10^8 light-years from Earth. Find the distance, in miles, from the Coma cluster to Earth. Write the answer in scientific notation.

4. One astronomical unit (A.U.) is 9.3×10^7 mi. The star Pollux in the constellation Gemini is 1.8228×10^{12} mi from Earth. Find the distance from Pollux to Earth in astronomical units.

5. One light-year is equal to approximately how many astronomical units? Round to the nearest thousand.

Shown below are data on the planets in our solar system. The planets are listed in alphabetical order.

Point of Interest

In November 2001, the Hubble Space Telescope took photos of the atmosphere of a planet orbiting a star 150 light-years from Earth in the constellation Pegasus. The planet is about the size of Jupiter and orbits close to the star HD209458. It was the first discovery of an atmosphere around a planet outside our solar system.

NASA/Roger Ressmeyer/Corbis

Jupiter

Planet	*Distance from the Sun (in kilometers)*	*Mass (in kilograms)*
Earth	1.50×10^8	5.97×10^{24}
Jupiter	7.79×10^8	1.90×10^{27}
Mars	2.28×10^8	6.42×10^{23}
Mercury	5.79×10^7	3.30×10^{23}
Neptune	4.50×10^9	1.02×10^{26}
Saturn	1.43×10^9	5.68×10^{26}
Uranus	2.87×10^9	8.68×10^{25}
Venus	1.08×10^8	4.87×10^{24}

6. Arrange the planets in order from closest to the sun to farthest from the sun.

7. Arrange the planets in order from the one with the greatest mass to the one with the least mass.

8. Write a rule for ordering numbers written in scientific notation.

CHAPTER 6

SUMMARY

KEY WORDS	EXAMPLES
A *monomial* is a number, a variable, or a product of a number and variables. [6.1A, p. 348]	5 is a number, y is a variable. $8a^2b^2$ is a product of a number and variables. 5, y, and $8a^2b^2$ are monomials.
The *degree of a monomial* is the sum of the exponents on the variables. [6.1A, p. 348]	The degree of $8x^4y^5z$ is 10.
A *polynomial* is a variable expression in which the terms are monomials. [6.2A, p. 360]	$x^4 - 2xy - 32x + 8$ is a polynomial. The terms are x^4, $-2xy$, $-32x$, and 8.
A polynomial of one term is a *monomial*, a polynomial of two terms is a *binomial*, and a polynomial of three terms is a *trinomial*. [6.2A, p. 360]	$5x^4$ is a monomial. $6y^3 - 2y$ is a binomial. $2x^2 - 5x + 3$ is a trinomial.
The *degree of a polynomial* is the greatest of the degrees of any of its terms. [6.2A, p. 360]	The degree of the polynomial $x^3 + 3x^2y^2 - 4xy - 3$ is 4.
The terms of a polynomial in one variable are usually arranged so that the exponents of the variable decrease from left to right. This is called *descending order*. [6.2A, p. 360]	The polynomial $4x^3 + 5x^2 - x + 7$ is written in descending order.
A *polynomial function* is an expression whose terms are monomials. Polynomial functions include the *linear function* given by $f(x) = mx + b$; the *quadratic function* given by $f(x) = ax^2 + bx + c$, $a \neq 0$; and the *cubic function*, which is a third-degree polynomial function. The *leading coefficient* of a polynomial function is the coefficient of the variable with the largest exponent. The *constant term* is the term without a variable. [6.2A, p. 360]	$f(x) = 5x - 4$ is a linear function. $f(x) = 3x^2 - 2x + 1$ is a quadratic function. 3 is the leading coefficient, and 1 is the constant term. $f(x) = x^3 - 1$ is a cubic function.

ESSENTIAL RULES AND PROCEDURES	EXAMPLES
Rule for Multiplying Exponential Expressions [6.1A, p. 348] $x^m \cdot x^n = x^{m+n}$	$b^5 \cdot b^4 = b^{5+4} = b^9$

Rule for Simplifying the Power of an Exponential Expression [6.1A, p. 349]

$(x^m)^n = x^{mn}$

$(y^3)^7 = y^{3(7)} = y^{21}$

Rule for Simplifying Powers of Products [6.1A, p. 349]

$(x^m y^n)^p = x^{mp} y^{np}$

$(x^6 y^4 z^5)^2 = x^{6(2)} y^{4(2)} z^{5(2)} = x^{12} y^8 z^{10}$

Definition of Zero as an Exponent [6.1B, p. 350]

For $x \neq 0$, $x^0 = 1$. The expression 0^0 is not defined.

$17^0 = 1$
$(5y)^0 = 1, y \neq 0$

Definition of a Negative Exponent [6.1B, p. 351]

For $x \neq 0$, $x^{-n} = \dfrac{1}{x^n}$ and $\dfrac{1}{x^{-n}} = x^n$.

$x^{-6} = \dfrac{1}{x^6}$ and $\dfrac{1}{x^{-6}} = x^6$

Rule for Simplifying Powers of Quotients [6.1B, p. 351]

For $y \neq 0$, $\left(\dfrac{x^m}{y^n}\right)^p = \dfrac{x^{mp}}{y^{np}}$.

$\left(\dfrac{x^2}{y^4}\right)^5 = \dfrac{x^{2\cdot 5}}{y^{4\cdot 5}} = \dfrac{x^{10}}{y^{20}}$

Rule for Negative Exponents on Fractional Expressions [6.1B, p. 352]

For $a \neq 0$, $b \neq 0$, $\left(\dfrac{a}{b}\right)^{-n} = \left(\dfrac{b}{a}\right)^n$.

$\left(\dfrac{3}{8}\right)^{-4} = \left(\dfrac{8}{3}\right)^4$

Rule for Dividing Exponential Expressions [6.1B, p. 352]

For $x \neq 0$, $\dfrac{x^m}{x^n} = x^{m-n}$.

$\dfrac{y^8}{y^3} = y^{8-3} = y^5$

Scientific Notation [6.1C, p. 354]

To express a number in scientific notation, write it in the form $a \times 10^n$, where a is a number between 1 and 10 and n is an integer.

If the number is greater than 10, the exponent on 10 will be positive.

$367{,}000{,}000 = 3.67 \times 10^8$

If the number is less than 1, the exponent on 10 will be negative.

$0.0000059 = 5.9 \times 10^{-6}$

To change a number in scientific notation to decimal notation, move the decimal point to the right if the exponent on 10 is positive and to the left if the exponent on 10 is negative. Move the decimal point the same number of places as the absolute value of the exponent on 10.

$2.418 \times 10^7 = 24{,}180{,}000$
$9.06 \times 10^{-5} = 0.0000906$

To write the additive inverse of a polynomial, change the sign of every term of the polynomial. [6.2B, p. 363]

The additive inverse of $-y^2 + 4y - 5$ is $y^2 - 4y + 5$.

To add polynomials, combine like terms, which means to add the coefficients of the like terms. [6.2B, p. 363]

$$\begin{aligned}&(8x^2 + 2x - 9) + (-3x^2 + 5x - 7)\\&\quad= (8x^2 - 3x^2) + (2x + 5x) + (-9 - 7)\\&\quad= 5x^2 + 7x - 16\end{aligned}$$

To subtract two polynomials, add the additive inverse of the second polynomial to the first polynomial. [6.2B, p. 364]

$$\begin{aligned}&(3y^2 - 8y + 6) - (-y^2 + 4y - 5)\\&\quad= (3y^2 - 8y + 6) + (y^2 - 4y + 5)\\&\quad= 4y^2 - 12y + 11\end{aligned}$$

To multiply a polynomial by a monomial, use the Distributive Property and the Rule for Multiplying Exponential Expressions. [6.3A, p. 368]

$$\begin{aligned}&-2x^3(4x^2 + 5x - 1)\\&\quad= -8x^5 - 10x^4 + 2x^3\end{aligned}$$

The FOIL Method [6.3B, p. 369]
The product of two binomials can be found by adding the products of the **F**irst terms, the **O**uter terms, the **I**nner terms, and the **L**ast terms.

$$\begin{aligned}&(4x + 3)(2x - 5)\\&\quad= (4x)(2x) + (4x)(-5)\\&\qquad+ (3)(2x) + (3)(-5)\\&\quad= 8x^2 - 20x + 6x - 15\\&\quad= 8x^2 - 14x - 15\end{aligned}$$

To divide a polynomial by a monomial, divide each term of the polynomial by the monomial. [6.4A, p. 378]

$$\frac{12x^5 + 8x^3 - 6x}{4x^2}$$

$$= \frac{12x^5}{4x^2} + \frac{8x^3}{4x^2} - \frac{6x}{4x^2} = 3x^3 + 2x - \frac{3}{2x}$$

Synthetic Division [6.4C, pp. 381–382]
Synthetic division is a shorter method of dividing a polynomial by a binomial of the form $x - a$. This method uses only the coefficients of the variable terms.

$(3x^3 - 9x - 5) \div (x - 2)$

2	3	0	−9	−5
		6	12	6
	3	6	3	1

$(3x^3 - 9x - 5) \div (x - 2)$

$$= 3x^2 + 6x + 3 + \frac{1}{x - 2}$$

Remainder Theorem [6.4D, p. 384]
If the polynomial $P(x)$ is divided by $x - a$, the remainder is $P(a)$.

$P(x) = x^3 - x^2 + x - 1$

−2	1	−1	1	−1
		−2	6	−14
	1	−3	7	−15

$P(-2) = -15$

CHAPTER 6

CONCEPT REVIEW

Test your knowledge of the concepts presented in this chapter. Answer each question. Then check your answers against the ones provided in the Answer Section.

1. Why is it important to write the terms of a polynomial in descending order before adding in a vertical format?

2. When multiplying the terms $4p^3$ and $7p^6$, what happens to the exponents?

3. How do you multiply two binomials?

4. How do you square a binomial?

5. How do you write a very large number in scientific notation? How do you write a very small number in scientific notation?

6. After you divide a polynomial by a binomial, how do you check your answer?

7. How do you know if a binomial is a factor of a polynomial?

8. What type of divisor is necessary to use synthetic division for division of a polynomial?

9. How do you write a polynomial with a missing term when using synthetic division?

10. How do you use synthetic division to evaluate a polynomial function?

CHAPTER 6

REVIEW EXERCISES

1. Add: $(12y^2 + 17y - 4) + (9y^2 - 13y + 3)$

2. Divide: $\dfrac{15x^2 + 2x - 2}{3x - 2}$

3. Simplify: $(2x^{-1}y^2z^5)^4(-3x^3yz^{-3})^2$

4. Expand: $(5y - 7)^2$

5. Simplify: $\dfrac{a^{-1}b^3}{a^3b^{-3}}$

6. Use the Remainder Theorem to evaluate $P(x) = x^3 - 2x^2 + 3x - 5$ when $x = 2$.

7. Simplify: $(5x^2 - 8xy + 2y^2) - (x^2 - 3y^2)$

8. Divide: $\dfrac{12b^7 + 36b^5 - 3b^3}{3b^3}$

9. Simplify: $\dfrac{3ab^4}{-6a^2b^4}$

10. Simplify: $(-2a^2b^4)(3ab^2)$

11. Simplify: $\dfrac{8x^{12}}{12x^9}$

12. Simplify: $\dfrac{4x^3 + 27x^2 + 10x + 2}{x + 6}$

13. Given $P(x) = 2x^3 - x + 7$, evaluate $P(-2)$.

14. Subtract: $(13y^3 - 7y - 2) - (12y^2 - 2y - 1)$

15. Divide: $(b^3 - 2b^2 - 33b - 7) \div (b - 7)$

16. Multiply: $4x^2y(3x^3y^2 + 2xy - 7y^3)$

17. Multiply: $(2a - b)(x - 2y)$

18. Multiply: $(2b - 3)(4b + 5)$

19. Simplify: $5x^2 - 4x[x - 3(3x + 2) + x]$

20. Simplify: $(xy^5z^3)(x^3y^3z)$

21. Expand: $(4x - 3y)^2$

22. Simplify: $\dfrac{x^4 - 4}{x - 4}$

23. Add: $(3x^2 - 2x - 6) + (-x^2 - 3x + 4)$

24. Multiply: $(5x^2yz^4)(2xy^3z^{-1})(7x^{-2}y^{-2}z^3)$

25. Simplify: $\dfrac{3x^4yz^{-1}}{-12xy^3z^2}$

26. Write 948,000,000 in scientific notation.

27. Simplify: $\dfrac{3 \times 10^{-3}}{15 \times 10^2}$

28. Use the Remainder Theorem to evaluate $P(x) = -2x^3 + 2x^2 - 4$ when $x = -3$.

29. Divide: $\dfrac{16x^5 - 8x^3 + 20x}{4x}$

30. Divide: $\dfrac{12x^2 - 16x - 7}{6x + 1}$

31. Multiply: $a^{2n+3}(a^n - 5a + 2)$

32. Multiply: $(x + 6)(x^3 - 3x^2 - 5x + 1)$

33. Multiply: $-2x(4x^2 + 7x - 9)$

34. Multiply: $(3y^2 + 4y - 7)(2y + 3)$

35. Simplify: $(-2u^3v^4)^4$

36. Add: $(2x^3 + 7x^2 + x) + (2x^2 - 4x - 12)$

37. Subtract: $(5x^2 - 2x - 1) - (3x^2 - 5x + 7)$

38. Multiply: $(a + 7)(a - 7)$

39. Simplify: $(5a^7b^6)^2(4ab)$

40. Write 1.46×10^7 in decimal notation.

41. Simplify: $(-2x^3)^2(-3x^4)^3$

42. Divide: $(6y^2 - 35y + 36) \div (3y - 4)$

43. Evaluate: -4^{-2}

44. Multiply: $(5a - 7)(2a + 9)$

45. Divide: $\dfrac{7 - x - x^2}{x + 3}$

46. Write 0.000000127 in scientific notation.

47. Divide: $\dfrac{16y^2 - 32y}{-4y}$

48. Simplify: $\dfrac{(2a^4b^{-3}c^2)^3}{(2a^3b^2c^{-1})^4}$

49. Multiply: $(x - 4)(3x + 2)(2x - 3)$

50. Simplify: $(-3x^{-2}y^{-3})^{-2}$

51. Simplify: $(2a^{12}b^3)(-9b^2c^6)(3ac)$

52. Multiply: $(5a + 2b)(5a - 2b)$

53. Write 2.54×10^{-3} in decimal notation.

54. Multiply: $2ab^3(4a^2 - 2ab + 3b^2)$

55. Graph $y = x^2 + 1$.

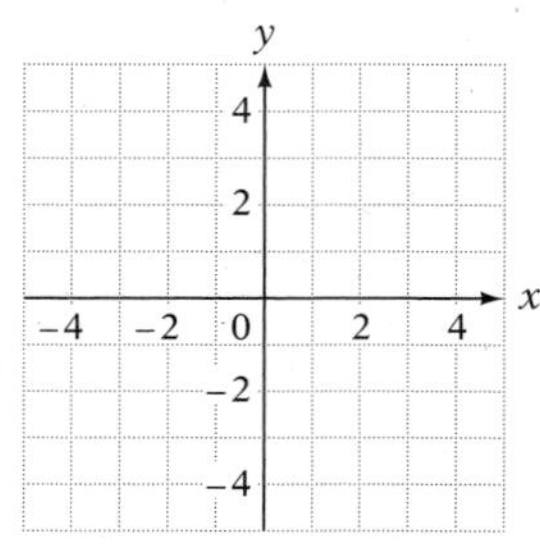

56. For the polynomial $P(x) = 3x^5 - 6x^2 + 7x + 8$:
a. Identify the leading coefficient.
b. Identify the constant term.
c. State the degree.

57. **Physics** The mass of the moon is 3.7×10^{-8} times the mass of the sun. The mass of the sun is 2.19×10^{27} tons. Find the mass of the moon. Write the answer in scientific notation.

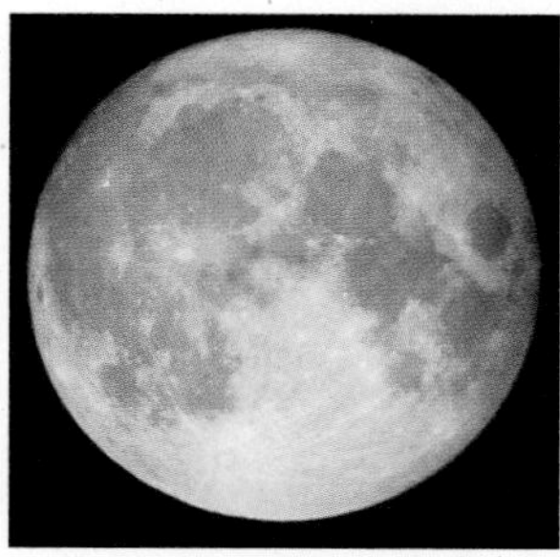
Susan Van Etten/PhotoEdit

The Moon

58. **Geometry** The side of a checkerboard is $(3x - 2)$ inches. Express the area of the checkerboard in terms of the variable x.

59. **Astronomy** The most distant object visible from Earth without the aid of a telescope is the Great Galaxy of Andromeda. It takes light from the Great Galaxy of Andromeda 2.2×10^6 years to travel to Earth. Light travels about 6.7×10^8 mph. How far from Earth is the Great Galaxy of Andromeda? Use a 365-day year.

60. **Geometry** The length of a rectangle is $(5x + 3)$ centimeters. The width is $(2x - 7)$ centimeters. Find the area of the rectangle in terms of the variable x.

CHAPTER 6

TEST

1. Multiply: $2x(2x^2 - 3x)$

2. Use the Remainder Theorem to evaluate $P(x) = -x^3 + 4x - 8$ when $x = -2$.

3. Simplify: $\dfrac{12x^2}{-3x^8}$

4. Simplify: $(-2xy^2)(3x^2y^4)$

5. Divide: $(x^2 + 1) \div (x + 1)$

6. Multiply: $(x - 3)(x^2 - 4x + 5)$

7. Simplify: $(-2a^2b)^3$

8. Simplify: $\dfrac{(3x^{-2}y^3)^3}{3x^4y^{-1}}$

9. Multiply: $(a - 2b)(a + 5b)$

10. Given $P(x) = 3x^2 - 8x + 1$, evaluate $P(2)$.

11. Divide: $(x^2 + 6x - 7) \div (x - 1)$

12. Multiply: $-3y^2(-2y^2 + 3y - 6)$

13. Multiply: $(-2x^3 + x^2 - 7)(2x - 3)$

14. Simplify: $(4y - 3)(4y + 3)$

15. Divide: $\dfrac{18x^5 + 9x^4 - 6x^3}{3x^2}$

16. Simplify: $\dfrac{2a^{-1}b}{2^{-2}a^{-2}b^{-3}}$

17. Simplify: $\dfrac{(2a^{-4}b^2)^3}{4a^{-2}b^{-1}}$

18. Subtract: $(3a^2 - 2a - 7) - (5a^3 + 2a - 10)$

19. Simplify: $(2x - 5)^2$

20. Divide: $\dfrac{x^3 - 2x^2 - 5x + 7}{x + 3}$

21. Multiply: $(2x - 7y)(5x - 4y)$

22. Add: $(3x^3 - 2x^2 - 4) + (8x^2 - 8x + 7)$

23. Write 0.00000000302 in scientific notation.

24. Write the number of seconds in 10 weeks in scientific notation.

25. **Geometry** The radius of a circle is $(x - 5)$ meters. Use the equation $A = \pi r^2$, where r is the radius, to find the area of the circle in terms of the variable x. Leave the answer in terms of π.

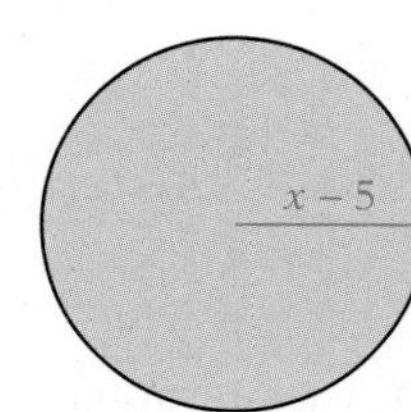

CUMULATIVE REVIEW EXERCISES

1. Let $x \in \{-8, -3, 3\}$. For what values of x is the inequality $x \geq -3$ a true statement?

2. Find the additive inverse of 83.

3. Simplify: $8 - 2[-3 - (-1)]^2 \div 4$

4. Evaluate $\frac{2a - b}{b - c}$ when $a = 4$, $b = -2$, and $c = 6$.

5. Simplify: $-5\sqrt{300}$

6. Identify the property that justifies the statement $2x + (-2x) = 0$.

7. Simplify: $2x - 4[x - 2(3 - 23x) + 4]$

8. Solve: $\frac{2}{3} - y = \frac{5}{6}$

9. Solve: $8x - 3 - x = -6 + 3x - 8$

10. Solve: $3 - |2 - 3x| = -2$

11. Given $P(x) = 3x^2 - 2x + 2$, find $P(-2)$.

12. Is the relation $\{(-1, 0), (0, 0), (1, 0)\}$ a function?

13. Find the slope of the line containing the points $(-2, 3)$ and $(4, 2)$.

14. Find the equation of the line that contains the point $(-1, 2)$ and has slope $-\frac{3}{2}$.

15. Find the equation of the line that contains the point $(-2, 4)$ and is perpendicular to the line $3x + 2y = 4$.

16. Solve by substitution:
$$2x - 3y = 4$$
$$x + y = -3$$

17. Solve by the addition method:
$$x - y + z = 0$$
$$2x + y - 3z = -7$$
$$-x + 2y + 2z = 5$$

18. Simplify: $-2x - (-xy) + 7x - 4xy$

19. Multiply: $(2x + 3)(2x^2 - 3x + 1)$

20. Write the number 0.00000501 in scientific notation.

21. Graph: $3x - 4y = 12$

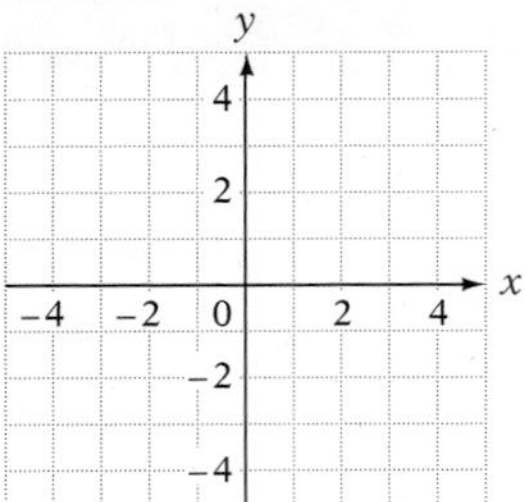

22. Graph: $-3x + 2y < 6$

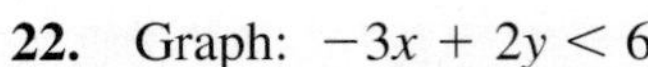

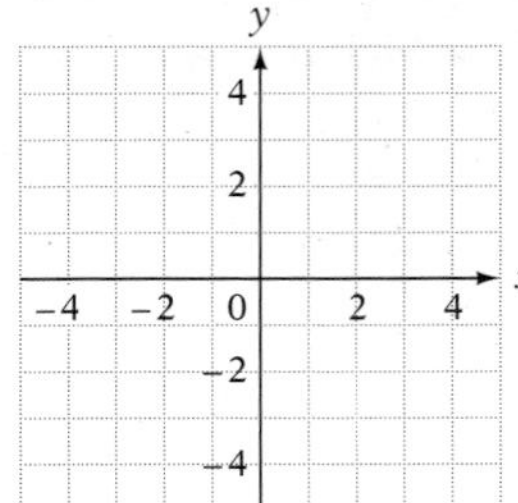

23. Solve by graphing:
$$x - 2y = 3$$
$$-2x + y = -3$$

24. Graph the solution set:
$$2x + y < 2$$
$$-6x + 3y \geq 6$$

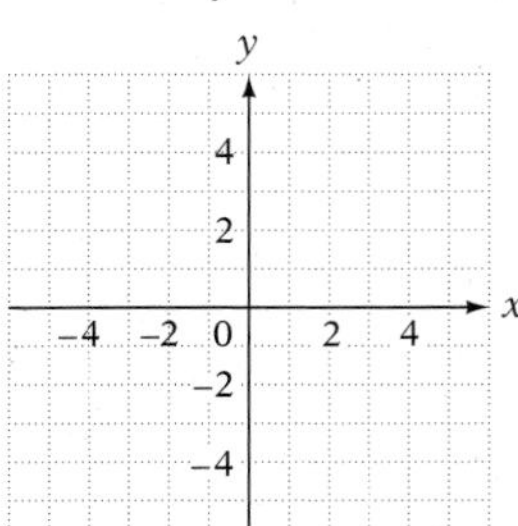

25. Simplify: $(4a^{-2}b^3)(2a^3b^{-1})^{-2}$

26. Simplify: $\dfrac{(5x^3y^{-3}z)^{-2}}{y^4z^{-2}}$

27. **Integers** The sum of two integers is twenty-four. The difference between four times the smaller integer and nine is three less than twice the larger integer. Find the integers.

28. **Mixtures** How many ounces of pure gold that costs \$360 per ounce must be mixed with 80 oz of an alloy that costs \$120 per ounce to make a mixture that costs \$200 per ounce?

29. **Travel** Two bicyclists are 25 mi apart and traveling toward each other. One cyclist is traveling at 1.5 times the rate of the other cyclist. They meet in 2 h. Find the rate of each cyclist.

30. **Investments** An investor has a total of \$12,000 invested in two simple interest accounts. On one account, the annual simple interest rate is 4%. On the second account, the annual simple interest rate is 4.5%. How much is invested in the 4% account if the total annual interest earned is \$530?

31. **Travel** The graph shows the relationship between the distance traveled and the time of travel. Find the slope of the line between the two points on the graph. Write a sentence that states the meaning of the slope.

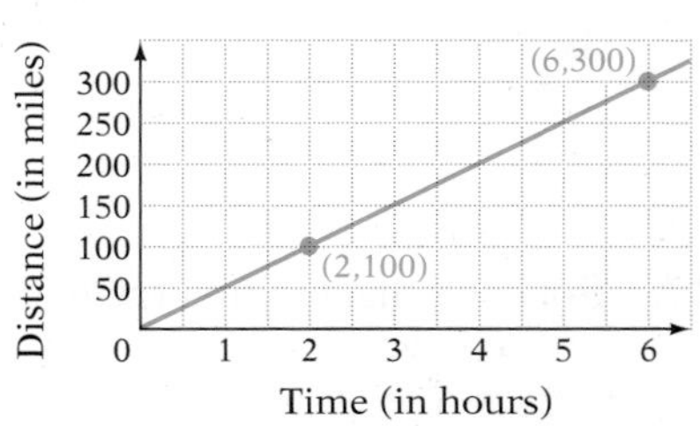

32. **Geometry** The width of a rectangle is 40% of the length. The perimeter of the rectangle is 42 m. Find the length and width of the rectangle.

33. **Geometry** The length of a side of a square is $(2x + 3)$ meters. Find the area of the square in terms of the variable x.

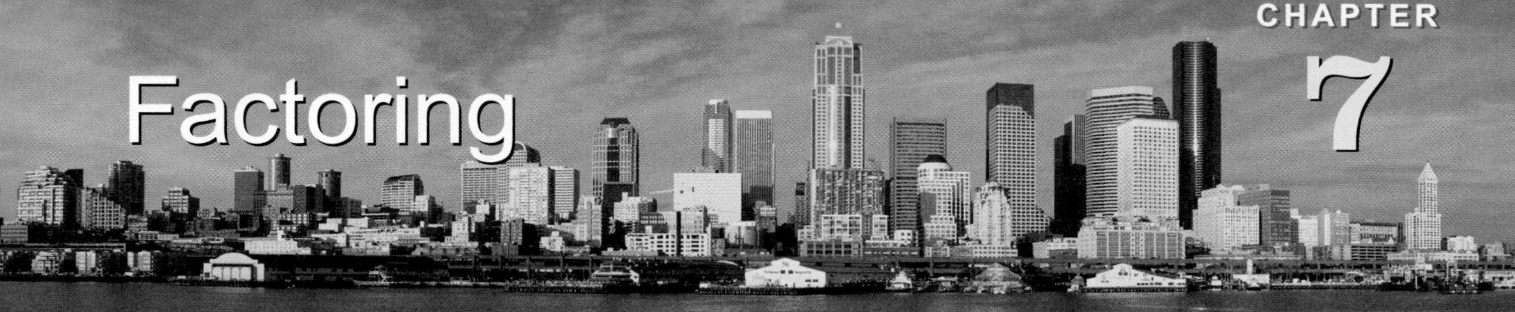

Chapter 7: Factoring

OBJECTIVES

SECTION 7.1

A To factor a monomial from a polynomial

B To factor by grouping

SECTION 7.2

A To factor a trinomial of the form $x^2 + bx + c$

B To factor completely

SECTION 7.3

A To factor a trinomial of the form $ax^2 + bx + c$ by using trial factors

B To factor a trinomial of the form $ax^2 + bx + c$ by grouping

SECTION 7.4

A To factor the difference of two perfect squares or a perfect-square trinomial

B To factor the sum or difference of two perfect cubes

C To factor a trinomial that is quadratic in form

D To factor completely

SECTION 7.5

A To solve equations by factoring

B To solve application problems

ARE YOU READY?

Take the Chapter 7 Prep Test to find out if you are ready to learn to:

- Factor a monomial from a polynomial
- Factor by grouping
- Factor trinomials
- Factor the difference of two squares, perfect-square trinomials, and the sum or difference of two cubes
- Solve equations by factoring

PREP TEST

Do these exercises to prepare for Chapter 7.

1. Write 30 as a product of prime numbers.

2. Simplify: $-3(4y - 5)$

3. Simplify: $-(a - b)$

4. Simplify: $2(a - b) - 5(a - b)$

5. Solve: $4x = 0$

6. Solve: $2x + 1 = 0$

7. Multiply: $(x + 4)(x - 6)$

8. Multiply: $(2x - 5)(3x + 2)$

9. Simplify: $\dfrac{x^5}{x^2}$

10. Simplify: $\dfrac{6x^4y^3}{2xy^2}$

SECTION 7.1 Common Factors

OBJECTIVE A To factor a monomial from a polynomial 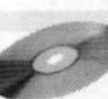

The **greatest common factor (GCF) of two or more numbers** is the largest common factor of the numbers. For example, the GCF of 9 and 12 is 3. The **GCF of two or more monomials** is the product of the GCF of the coefficients and the common variable factors.

$6x^3y = 2 \cdot 3 \cdot x \cdot x \cdot x \cdot y$
$8x^2y^2 = 2 \cdot 2 \cdot 2 \cdot x \cdot x \cdot y \cdot y$
$\text{GCF} = 2 \cdot x \cdot x \cdot y = 2x^2y$

Note that the exponent on each variable in the GCF is the same as the *smallest* exponent on that variable in either of the monomials.

The GCF of $6x^3y$ and $8x^2y^2$ is $2x^2y$.

HOW TO • 1 Find the GCF of $12a^4b$ and $18a^2b^2c$.

The common variable factors are a^2 and b; c is not a common variable factor.

$12a^4b = 2 \cdot 2 \cdot 3 \cdot a^4 \cdot b$
$18a^2b^2c = 2 \cdot 3 \cdot 3 \cdot a^2 \cdot b^2 \cdot c$
$\text{GCF} = 2 \cdot 3 \cdot a^2 \cdot b = 6a^2b$

To **factor a polynomial** means to write the polynomial as a product of other polynomials. In the example at the right, $2x$ is the GCF of the terms $2x^2$ and $10x$.

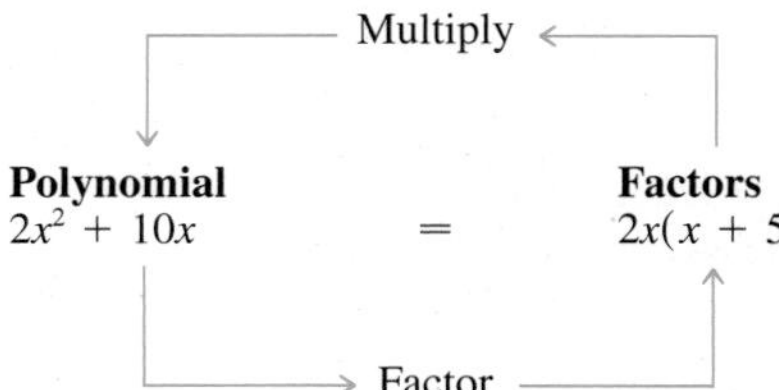

HOW TO • 2 Factor: $5x^3 - 35x^2 + 10x$

Find the GCF of the terms of the polynomial.

$5x^3 = 5 \cdot x^3$
$35x^2 = 5 \cdot 7 \cdot x^2$
$10x = 2 \cdot 5 \cdot x$

The GCF is $5x$.

Rewrite the polynomial, expressing each term as a product with the GCF as one of the factors.

$$5x^3 - 35x^2 + 10x = 5x(x^2) + 5x(-7x) + 5x(2)$$
$$= 5x(x^2 - 7x + 2)$$

• Use the Distributive Property to write the polynomial as a product of factors.

Take Note

At the right, the factors in parentheses are determined by dividing each term of the trinomial by the GCF, $5x$.

$\frac{5x^3}{5x} = x^2$, $\frac{-35x^2}{5x} = -7x$, and $\frac{10x}{5x} = 2$

HOW TO • 3 Factor: $21x^2y^3 - 6xy^5 + 15x^4y^2$

Find the GCF of the terms of the polynomial.

$$21x^2y^3 = 3 \cdot 7 \cdot x^2 \cdot y^3$$
$$6xy^5 = 2 \cdot 3 \cdot x \cdot y^5$$
$$15x^4y^2 = 3 \cdot 5 \cdot x^4 \cdot y^2$$

The GCF is $3xy^2$.

Rewrite the polynomial, expressing each term as a product with the GCF as one of the factors.

$$21x^2y^3 - 6xy^5 + 15x^4y^2$$
$$= 3xy^2(7xy) + 3xy^2(-2y^3) + 3xy^2(5x^3)$$
$$= 3xy^2(7xy - 2y^3 + 5x^3)$$

• Use the Distributive Property to write the polynomial as a product of factors.

EXAMPLE • 1

Factor: $8x^2 + 2xy$

Solution

The GCF is $2x$.

$$8x^2 + 2xy = 2x(4x) + 2x(y)$$
$$= 2x(4x + y)$$

YOU TRY IT • 1

Factor: $14a^2 - 21a^4b$

Your solution

EXAMPLE • 2

Factor: $n^3 - 5n^2 + 2n$

Solution

The GCF is n.

$$n^3 - 5n^2 + 2n = n(n^2) + n(-5n) + n(2)$$
$$= n(n^2 - 5n + 2)$$

YOU TRY IT • 2

Factor: $27b^2 + 18b + 9$

Your solution

EXAMPLE • 3

Factor: $16x^2y + 8x^4y^2 - 12x^4y^5$

Solution

The GCF is $4x^2y$.

$$16x^2y + 8x^4y^2 - 12x^4y^5$$
$$= 4x^2y(4) + 4x^2y(2x^2y) + 4x^2y(-3x^2y^4)$$
$$= 4x^2y(4 + 2x^2y - 3x^2y^4)$$

YOU TRY IT • 3

Factor: $6x^4y^2 - 9x^3y^2 + 12x^2y^4$

Your solution

Solutions on p. S20

OBJECTIVE B To factor by grouping 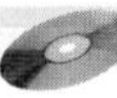

A factor that has two terms is called a **binomial factor**. In the examples at the right, the binomials $a + b$ and $x - y$ are binomial factors.

$2a(a + b)^2$
$3xy(x - y)$

The Distributive Property is used to factor a common binomial factor from an expression.

The common binomial factor of the expression $6(x - 3) + y(x - 3)$ is $(x - 3)$. To factor the expression, use the Distributive Property to write the expression as a product of factors.

$$6(x - 3) + y(x - 3) = (x - 3)(6 + y)$$

Consider the following simplification of $-(a - b)$.

$$-(a - b) = -1(a - b) = -a + b = b - a$$

Thus $$b - a = -(a - b)$$

This equation is sometimes used to factor a common binomial from an expression.

HOW TO • 4 Factor: $2x(x - y) + 5(y - x)$

$$2x(x - y) + 5(y - x) = 2x(x - y) - 5(x - y)$$
$$= (x - y)(2x - 5)$$

• $5(y - x) = 5[(-1)(x - y)]$

A polynomial can be **factored by grouping** if its terms can be grouped and factored in such a way that a common binomial factor is found.

HOW TO • 5 Factor: $ax + bx - ay - by$

$$ax + bx - ay - by = (ax + bx) - (ay + by)$$

• Group the first two terms and the last two terms. Note that $-ay - by = -(ay + by)$.

$$= x(a + b) - y(a + b)$$

• Factor each group.

$$= (a + b)(x - y)$$

• Factor the GCF, $(a + b)$, from each group.

Check: $(a + b)(x - y) = ax - ay + bx - by$
$= ax + bx - ay - by$

HOW TO • 6 Factor: $6x^2 - 9x - 4xy + 6y$

$$6x^2 - 9x - 4xy + 6y = (6x^2 - 9x) - (4xy - 6y)$$

• Group the first two terms and the last two terms. Note that $-4xy + 6y = -(4xy - 6y)$.

$$= 3x(2x - 3) - 2y(2x - 3)$$

• Factor each group.

$$= (2x - 3)(3x - 2y)$$

• Factor the GCF, $(2x - 3)$, from each group.

EXAMPLE • 4

Factor: $4x(3x - 2) - 7(3x - 2)$

Solution

$4x(3x - 2) - 7(3x - 2)$ • **$3x - 2$ is the common binomial factor.**

$= (3x - 2)(4x - 7)$

YOU TRY IT • 4

Factor: $2y(5x - 2) - 3(2 - 5x)$

Your solution

EXAMPLE • 5

Factor: $9x^2 - 15x - 6xy + 10y$

Solution

$9x^2 - 15x - 6xy + 10y$

$= (9x^2 - 15x) - (6xy - 10y)$ • **$-6xy + 10y = -(6xy - 10y)$**

$= 3x(3x - 5) - 2y(3x - 5)$ • **$3x - 5$ is the common factor.**

$= (3x - 5)(3x - 2y)$

YOU TRY IT • 5

Factor: $a^2 - 3a + 2ab - 6b$

Your solution

EXAMPLE • 6

Factor: $3x^2y - 4x - 15xy + 20$

Solution

$3x^2y - 4x - 15xy + 20$

$= (3x^2y - 4x) - (15xy - 20)$ • **$-15xy + 20 = -(15xy - 20)$**

$= x(3xy - 4) - 5(3xy - 4)$ • **$3xy - 4$ is the common factor.**

$= (3xy - 4)(x - 5)$

YOU TRY IT • 6

Factor: $2mn^2 - n + 8mn - 4$

Your solution

EXAMPLE • 7

Factor: $4ab - 6 + 3b - 2ab^2$

Solution

$4ab - 6 + 3b - 2ab^2$

$= (4ab - 6) + (3b - 2ab^2)$

$= 2(2ab - 3) + b(3 - 2ab)$

$= 2(2ab - 3) - b(2ab - 3)$ • **$3 - 2ab = -(2ab - 3)$**

$= (2ab - 3)(2 - b)$ • **$2ab - 3$ is the common factor.**

YOU TRY IT • 7

Factor: $3xy - 9y - 12 + 4x$

Your solution

Solutions on p. S20

7.1 EXERCISES

OBJECTIVE A To factor a monomial from a polynomial

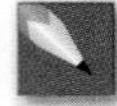

1. Explain the meaning of "a common monomial factor of a polynomial."

2. Explain the meaning of "a factor" and the meaning of "to factor."

For Exercises 3 to 41, factor.

3. $5a + 5$ **4.** $7b - 7$ **5.** $16 - 8a^2$ **6.** $12 + 12y^2$ **7.** $8x + 12$

8. $16a - 24$ **9.** $30a - 6$ **10.** $20b + 5$ **11.** $7x^2 - 3x$ **12.** $12y^2 - 5y$

13. $3a^2 + 5a^5$ **14.** $9x - 5x^2$ **15.** $14y^2 + 11y$ **16.** $6b^3 - 5b^2$ **17.** $2x^4 - 4x$

18. $3y^4 - 9y$ **19.** $10x^4 - 12x^2$ **20.** $12a^5 - 32a^2$ **21.** $8a^8 - 4a^5$ **22.** $16y^4 - 8y^7$

23. $x^2y^2 - xy$ **24.** $a^2b^2 + ab$ **25.** $3x^2y^4 - 6xy$ **26.** $12a^2b^5 - 9ab$ **27.** $x^2y - xy^3$

28. $3x^3 + 6x^2 + 9x$ **29.** $5y^3 - 20y^2 + 5y$ **30.** $2x^4 - 4x^3 + 6x^2$ **31.** $3y^4 - 9y^3 - 6y^2$

32. $2x^3 + 6x^2 - 14x$ **33.** $3y^3 - 9y^2 + 24y$ **34.** $2y^5 - 3y^4 + 7y^3$ **35.** $6a^5 - 3a^3 - 2a^2$

36. $x^3y - 3x^2y^2 + 7xy^3$ **37.** $2a^2b - 5a^2b^2 + 7ab^2$ **38.** $5y^3 + 10y^2 - 25y$

39. $4b^5 + 6b^3 - 12b$ **40.** $3a^2b^2 - 9ab^2 + 15b^2$ **41.** $8x^2y^2 - 4x^2y + x^2$

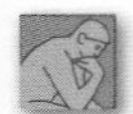

42. What is the GCF of the terms of the polynomial $x^a + x^b + x^c$ given that a, b, and c are all positive integers, and $a > b > c$?

OBJECTIVE B To factor by grouping

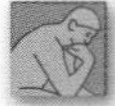

43. Use the three expressions at the right.
a. Which expressions are equivalent to $x^2 - 5x + 6$?
b. Which expression can be factored by grouping?

(i) $x^2 - 15x + 10x + 6$
(ii) $x^2 - x - 4x + 6$
(iii) $x^2 - 2x - 3x + 6$

For Exercises 44 to 70, factor.

44. $x(b + 4) + 3(b + 4)$

45. $y(a + z) + 7(a + z)$

46. $a(y - x) - b(y - x)$

47. $3r(a - b) + s(a - b)$

48. $x(x - 2) + y(2 - x)$

49. $t(m - 7) + 7(7 - m)$

50. $8c(2m - 3n) + (3n - 2m)$

51. $2y(4a + b) - (b + 4a)$

52. $2x(7 + b) - y(b + 7)$

53. $x^2 + 2x + 2xy + 4y$

54. $x^2 - 3x + 4ax - 12a$

55. $p^2 - 2p - 3rp + 6r$

56. $t^2 + 4t - st - 4s$

57. $ab + 6b - 4a - 24$

58. $xy - 5y - 2x + 10$

59. $2z^2 - z + 2yz - y$

60. $2y^2 - 10y + 7xy - 35x$

61. $8v^2 - 12vy + 14v - 21y$

62. $21x^2 + 6xy - 49x - 14y$

63. $2x^2 - 5x - 6xy + 15y$

64. $4a^2 + 5ab - 10b - 8a$

65. $3y^2 - 6y - ay + 2a$

66. $2ra + a^2 - 2r - a$

67. $3xy - y^2 - y + 3x$

68. $2ab - 3b^2 - 3b + 2a$

69. $3st + t^2 - 2t - 6s$

70. $4x^2 + 3xy - 12y - 16x$

Applying the Concepts

71. **Geometry** In the equation $P = 2L + 2W$, what is the effect on P when the quantity $L + W$ doubles?

72. **Geometry** Write an expression in factored form for the shaded portion in each of the following diagrams. Use the equation for the area of a rectangle ($A = LW$) and the equation for the area of a circle ($A = \pi r^2$).

a.

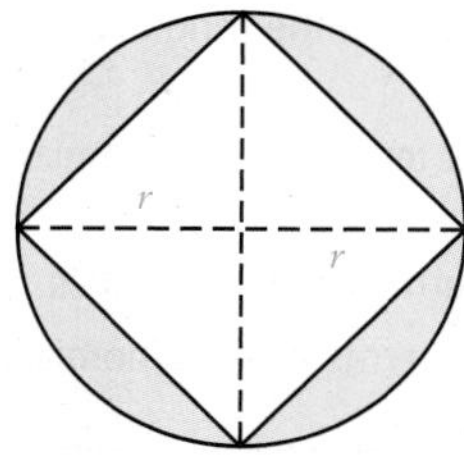

b.

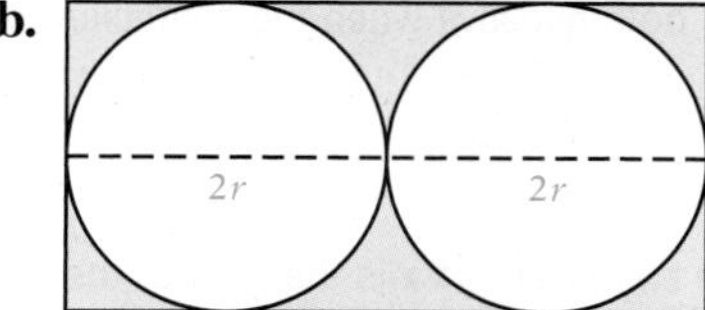

c.

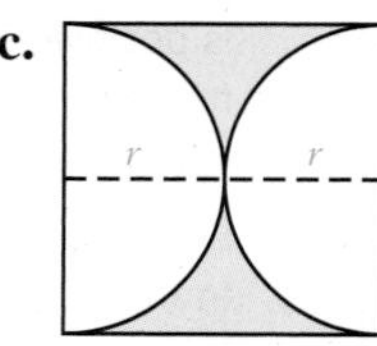

SECTION

7.2 Factoring Polynomials of the Form $x^2 + bx + c$

OBJECTIVE A To factor a trinomial of the form $x^2 + bx + c$ 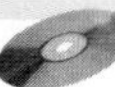

Trinomials of the form $x^2 + bx + c$, where b and c are integers, are shown at the right.

$x^2 + 8x + 12;\ b = 8,\ c = 12$
$x^2 - 7x + 12;\ b = -7,\ c = 12$
$x^2 - 2x - 15;\ b = -2,\ c = -15$

To factor a trinomial of this form means to express the trinomial as the product of two binomials.

Trinomials expressed as the product of binomials are shown at the right.

$x^2 + 8x + 12 = (x + 6)(x + 2)$
$x^2 - 7x + 12 = (x - 3)(x - 4)$
$x^2 - 2x - 15 = (x + 3)(x - 5)$

The method by which factors of a trinomial are found is based on FOIL. Consider the following binomial products, noting the relationship between the constant terms of the binomials and the terms of the trinomials.

The signs in the binomial factors are the same.

$$(x + 6)(x + 2) = x^2 + 2x + 6x + (6)(2) = x^2 + 8x + 12$$

Sum of 6 and 2 (→ 8)
Product of 6 and 2 (→ 12)

$$(x - 3)(x - 4) = x^2 - 4x - 3x + (-3)(-4) = x^2 - 7x + 12$$

Sum of −3 and −4 (→ −7)
Product of −3 and −4 (→ 12)

The signs in the binomial factors are opposites.

$$(x + 3)(x - 5) = x^2 - 5x + 3x + (3)(-5) = x^2 - 2x - 15$$

Sum of 3 and −5 (→ −2)
Product of 3 and −5 (→ −15)

$$(x - 4)(x + 6) = x^2 + 6x - 4x + (-4)(6) = x^2 + 2x - 24$$

Sum of −4 and 6 (→ 2)
Product of −4 and 6 (→ −24)

Factoring $x^2 + bx + c$: IMPORTANT RELATIONSHIPS

1. When the constant term of the trinomial is positive, the constant terms of the binomials have the same sign. They are both positive when the coefficient of the x term in the trinomial is positive. They are both negative when the coefficient of the x term in the trinomial is negative.
2. When the constant term of the trinomial is negative, the constant terms of the binomials have opposite signs.
3. In the trinomial, the coefficient of x is the sum of the constant terms of the binomials.
4. In the trinomial, the constant term is the product of the constant terms of the binomials.

HOW TO 1 Factor: $x^2 - 7x + 10$

Because the constant term is positive and the coefficient of x is negative, the binomial constants will be negative. Find two negative factors of 10 whose sum is -7. The results can be recorded in a table.

Negative Factors of 10	*Sum*
$-1, -10$	-11
$-2, -5$	-7

• These are the correct factors.

$x^2 - 7x + 10 = (x - 2)(x - 5)$ • Write the trinomial as a product of its factors.

You can check the proposed factorization by multiplying the two binomials.

Check: $(x - 2)(x - 5) = x^2 - 5x - 2x + 10 = x^2 - 7x + 10$

Take Note
Always check your proposed factorization to ensure accuracy.

HOW TO 2 Factor: $x^2 - 9x - 36$

The constant term is negative. The binomial constants will have opposite signs. Find two factors of -36 whose sum is -9.

Factors of −36	*Sum*
$+1, -36$	-35
$-1, +36$	35
$+2, -18$	-16
$-2, +18$	16
$+3, -12$	-9

• Once the correct factors are found, it is not necessary to try the remaining factors.

$x^2 - 9x - 36 = (x + 3)(x - 12)$ • Write the trinomial as a product of its factors.

For some trinomials it is not possible to find integer factors of the constant term whose sum is the coefficient of the middle term. A polynomial that does not factor using only integers is **nonfactorable over the integers.**

HOW TO 3 Factor: $x^2 + 7x + 8$

The constant term is positive and the coefficient of x is positive. The binomial constants will be positive. Find two positive factors of 8 whose sum is 7.

Positive Factors of 8	*Sum*
1, 8	9
2, 4	6

• There are no positive integer factors of 8 whose sum is 7.

$x^2 + 7x + 8$ is nonfactorable over the integers.

Take Note
Just as 17 is a prime number, $x^2 + 7x + 8$ is a **prime polynomial.** Binomials of the form $x - a$ and $x + a$ are also prime polynomials.

EXAMPLE 1

Factor: $x^2 - 8x + 15$

Solution

Find two negative factors of 15 whose sum is -8.

Factors	*Sum*
$-1, -15$	-16
$-3, -5$	-8

$x^2 - 8x + 15 = (x - 3)(x - 5)$

YOU TRY IT 1

Factor: $x^2 + 9x + 20$

Your solution

Solution on p. S20

EXAMPLE • 2

Factor: $x^2 + 6x - 27$

Solution
Find two factors of -27 whose sum is 6.

Factors	*Sum*
$+1, -27$	-26
$-1, +27$	26
$+3, -9$	-6
$-3, +9$	6

$x^2 + 6x - 27 = (x - 3)(x + 9)$

YOU TRY IT • 2

Factor: $x^2 + 7x - 18$

Your solution

Solution on p. S20

OBJECTIVE B To factor completely

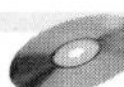

A polynomial is **factored completely** when it is written as a product of factors that are nonfactorable over the integers.

✓ Take Note
The first step in *any* factoring problem is to determine whether the terms of the polynomial have a *common factor.* If they do, factor it out first.

HOW TO • 4 Factor: $4y^3 - 4y^2 - 24y$

$4y^3 - 4y^2 - 24y = 4y(y^2) - 4y(y) - 4y(6)$ • The GCF is **4y**.

$= 4y(y^2 - y - 6)$ • Use the Distributive Property to factor out the GCF.

$= 4y(y + 2)(y - 3)$ • Factor $y^2 - y - 6$. The two factors of -6 whose sum is -1 are 2 and -3.

It is always possible to check a proposed factorization by multiplying the polynomials. Here is the check for the last example.

Check: $4y(y + 2)(y - 3) = 4y(y^2 - 3y + 2y - 6)$
$= 4y(y^2 - y - 6)$
$= 4y^3 - 4y^2 - 24y$ • This is the original polynomial.

HOW TO • 5 Factor: $x^2 + 12xy + 20y^2$

There is no common factor.
Note that the variable part of the middle term is xy, and the variable part of the last term is y^2.

$x^2 + 12xy + 20y^2 = (x + 2y)(x + 10y)$ • The two factors of 20 whose sum is 12 are 2 and 10.

✓ Take Note
The terms $2y$ and $10y$ are placed in the binomials. This is necessary so that the middle term of the trinomial contains xy and the last term contains y^2.

Note that the terms $2y$ and $10y$ are placed in the binomials. The following check shows why this is necessary.

Check: $(x + 2y)(x + 10y) = x^2 + 10xy + 2xy + 20y^2$
$= x^2 + 12xy + 20y^2$ • This is the original polynomial.

Take Note

When the coefficient of the highest power in a polynomial is negative, consider factoring out a negative GCF. Example 3 is another example of this technique.

HOW TO • 6 Factor: $15 - 2x - x^2$

Because the coefficient of x^2 is -1, factor -1 from the trinomial and then write the resulting trinomial in descending order.

$15 - 2x - x^2 = -(x^2 + 2x - 15)$ • $15 - 2x - x^2 = -1(-15 + 2x + x^2) = -(x^2 + 2x - 15)$

$= -(x + 5)(x - 3)$ • Factor $x^2 + 2x - 15$. The two factors of -15 whose sum is 2 are 5 and -3.

Check: $-(x + 5)(x - 3) = -(x^2 + 2x - 15)$
$= -x^2 - 2x + 15$
$= 15 - 2x - x^2$ • This is the original polynomial.

EXAMPLE • 3

Factor: $-3x^3 + 9x^2 + 12x$

Solution

The GCF is $-3x$.
$-3x^3 + 9x^2 + 12x = -3x(x^2 - 3x - 4)$
Factor the trinomial $x^2 - 3x - 4$. Find two factors of -4 whose sum is -3.

Factors	*Sum*
$+1, -4$	-3

$-3x^3 + 9x^2 + 12x = -3x(x + 1)(x - 4)$

YOU TRY IT • 3

Factor: $-2x^3 + 14x^2 - 12x$

Your solution

EXAMPLE • 4

Factor: $4x^2 - 40xy + 84y^2$

Solution

The GCF is 4.
$4x^2 - 40xy + 84y^2 = 4(x^2 - 10xy + 21y^2)$
Factor the trinomial $x^2 - 10xy + 21y^2$. Find two negative factors of 21 whose sum is -10.

Factors	*Sum*
$-1, -21$	-22
$-3, -7$	-10

$4x^2 - 40xy + 84y^2 = 4(x - 3y)(x - 7y)$

YOU TRY IT • 4

Factor: $3x^2 - 9xy - 12y^2$

Your solution

Solutions on pp. S20–S21

7.2 EXERCISES

OBJECTIVE A **To factor a trinomial of the form $x^2 + bx + c$**

For Exercises 1 to 73, factor.

1. $x^2 + 3x + 2$

2. $x^2 + 5x + 6$

3. $x^2 - x - 2$

4. $x^2 + x - 6$

5. $a^2 + a - 12$

6. $a^2 - 2a - 35$

7. $a^2 - 3a + 2$

8. $a^2 - 5a + 4$

9. $a^2 + a - 2$

10. $a^2 - 2a - 3$

11. $b^2 - 6b + 9$

12. $b^2 + 8b + 16$

13. $b^2 + 7b - 8$

14. $y^2 - y - 6$

15. $y^2 + 6y - 55$

16. $z^2 - 4z - 45$

17. $y^2 - 5y + 6$

18. $y^2 - 8y + 15$

19. $z^2 - 14z + 45$

20. $z^2 - 14z + 49$

21. $z^2 - 12z - 160$

22. $p^2 + 2p - 35$

23. $p^2 + 12p + 27$

24. $p^2 - 6p + 8$

25. $x^2 + 20x + 100$

26. $x^2 + 18x + 81$

27. $b^2 + 9b + 20$

28. $b^2 + 13b + 40$

29. $x^2 - 11x - 42$

30. $x^2 + 9x - 70$

31. $b^2 - b - 20$

32. $b^2 + 3b - 40$

33. $y^2 - 14y - 51$

34. $y^2 - y - 72$

35. $p^2 - 4p - 21$

36. $p^2 + 16p + 39$

37. $y^2 - 8y + 32$

38. $y^2 - 9y + 81$

39. $x^2 - 20x + 75$

40. $x^2 - 12x + 11$

41. $p^2 + 24p + 63$

42. $x^2 - 15x + 56$

43. $x^2 + 21x + 38$

44. $x^2 + x - 56$

45. $x^2 + 5x - 36$

46. $a^2 - 21a - 72$

47. $a^2 - 7a - 44$

48. $a^2 - 15a + 36$

49. $a^2 - 21a + 54$

50. $z^2 - 9z - 136$

51. $z^2 + 14z - 147$

52. $c^2 - c - 90$

53. $c^2 - 3c - 180$

54. $z^2 + 15z + 44$

55. $p^2 + 24p + 135$

56. $c^2 + 19c + 34$

57. $c^2 + 11c + 18$

58. $x^2 - 4x - 96$

59. $x^2 + 10x - 75$

60. $x^2 - 22x + 112$

61. $x^2 + 21x - 100$

62. $b^2 + 8b - 105$

63. $b^2 - 22b + 72$

64. $a^2 - 9a - 36$

65. $a^2 + 42a - 135$

66. $b^2 - 23b + 102$

67. $b^2 - 25b + 126$

68. $a^2 + 27a + 72$

69. $z^2 + 24z + 144$

70. $x^2 + 25x + 156$

71. $x^2 - 29x + 100$

72. $x^2 - 10x - 96$

73. $x^2 + 9x - 112$

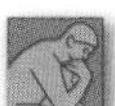

For Exercises 74 and 75, $x^2 + bx + c = (x + n)(x + m)$, where b and c are nonzero and n and m are positive integers.

74. Is c positive or negative?

75. Is b positive or negative?

OBJECTIVE B To factor completely

For Exercises 76 to 129, factor.

76. $2x^2 + 6x + 4$

77. $3x^2 + 15x + 18$

78. $18 + 7x - x^2$

79. $12 - 4x - x^2$

80. $ab^2 + 2ab - 15a$

81. $ab^2 + 7ab - 8a$

82. $xy^2 - 5xy + 6x$

83. $xy^2 + 8xy + 15x$

84. $z^3 - 7z^2 + 12z$

85. $-2a^3 - 6a^2 - 4a$

86. $-3y^3 + 15y^2 - 18y$

87. $4y^3 + 12y^2 - 72y$

88. $3x^2 + 3x - 36$

89. $2x^3 - 2x^2 + 4x$

90. $5z^2 - 15z - 140$

91. $6z^2 + 12z - 90$

92. $2a^3 + 8a^2 - 64a$

93. $3a^3 - 9a^2 - 54a$

94. $x^2 - 5xy + 6y^2$

95. $x^2 + 4xy - 21y^2$

96. $a^2 - 9ab + 20b^2$

97. $a^2 - 15ab + 50b^2$

98. $x^2 - 3xy - 28y^2$

99. $s^2 + 2st - 48t^2$

100. $y^2 - 15yz - 41z^2$

101. $x^2 + 85xy + 36y^2$

102. $z^4 - 12z^3 + 35z^2$

103. $z^4 + 2z^3 - 80z^2$

104. $b^4 - 22b^3 + 120b^2$

105. $b^4 - 3b^3 - 10b^2$

106. $2y^4 - 26y^3 - 96y^2$

107. $3y^4 + 54y^3 + 135y^2$

108. $-x^4 - 7x^3 + 8x^2$

109. $-x^4 + 11x^3 + 12x^2$

110. $4x^2y + 20xy - 56y$

111. $3x^2y - 6xy - 45y$

112. $c^3 + 18c^2 - 40c$

113. $-3x^3 + 36x^2 - 81x$

114. $-4x^3 - 4x^2 + 24x$

115. $x^2 - 8xy + 15y^2$

116. $y^2 - 7xy - 8x^2$

117. $a^2 - 13ab + 42b^2$

118. $y^2 + 4yz - 21z^2$

119. $y^2 + 8yz + 7z^2$

120. $y^2 - 16yz + 15z^2$

121. $3x^2y + 60xy - 63y$

122. $4x^2y - 68xy - 72y$

123. $3x^3 + 3x^2 - 36x$

124. $4x^3 + 12x^2 - 160x$

125. $2t^2 - 24ts + 70s^2$

126. $4a^2 - 40ab + 100b^2$

127. $3a^2 - 24ab - 99b^2$

128. $4x^3 + 8x^2y - 12xy^2$

129. $5x^3 + 30x^2y + 40xy^2$

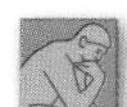

130. State whether the trinomial has a factor of $x + 3$.
a. $3x^2 - 3x - 36$ **b.** $x^2y - xy - 12y$

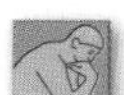

131. State whether the trinomial has a factor of $x + y$.
a. $2x^2 - 2xy - 4y^2$ **b.** $2x^2y - 4xy - 4y$

Applying the Concepts

For Exercises 132 to 134, find all integers k such that the trinomial can be factored over the integers.

132. $x^2 + kx + 35$

133. $x^2 + kx + 18$

134. $x^2 + kx + 21$

For Exercises 135 to 140, determine the positive integer values of k for which the polynomial is factorable over the integers.

135. $y^2 + 4y + k$

136. $z^2 + 7z + k$

137. $a^2 - 6a + k$

138. $c^2 - 7c + k$

139. $x^2 - 3x + k$

140. $y^2 + 5y + k$

141. In Exercises 135 to 140, there was the stated requirement that $k > 0$. If k is allowed to be any integer, how many different values of k are possible for each polynomial?

SECTION

7.3 Factoring Polynomials of the Form $ax^2 + bx + c$

OBJECTIVE A To factor a trinomial of the form $ax^2 + bx + c$ by using trial factors

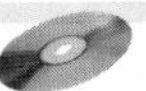

Trinomials of the form $ax^2 + bx + c$, where a, b, and c are integers, are shown at the right.

$3x^2 - x + 4$; $a = 3$, $b = -1$, $c = 4$
$6x^2 + 2x - 3$; $a = 6$, $b = 2$, $c = -3$

These trinomials differ from those in the preceding section in that the coefficient of x^2 is not 1. There are various methods of factoring these trinomials. The method described in this objective is factoring polynomials using trial factors.

To reduce the number of trial factors that must be considered, remember the following:

1. Use the signs of the constant term and the coefficient of x in the trinomial to determine the signs of the binomial factors. If the constant term is positive, the signs of the binomial factors will be the same as the sign of the coefficient of x in the trinomial. If the sign of the constant term is negative, the constant terms in the binomials have opposite signs.
2. If the terms of the trinomial do not have a common factor, then the terms of each binomial factor will not have a common factor.

HOW TO 1 Factor: $2x^2 - 7x + 3$

The terms have no common factor. The constant term is positive. The coefficient of x is negative. The binomial constants will be negative.

Positive Factors of 2 (coefficient of x^2)	Negative Factors of 3 (constant term)
1, 2	−1, −3

Write trial factors. Use the **O**uter and **I**nner products of FOIL to determine the middle term, $-7x$, of the trinomial.

Trial Factors	Middle Term
$(x - 1)(2x - 3)$	$-3x - 2x = -5x$
$(x - 3)(2x - 1)$	$-x - 6x = -7x$

Write the factors of the trinomial.

$2x^2 - 7x + 3 = (x - 3)(2x - 1)$

HOW TO 2 Factor: $3x^2 + 14x + 15$

The terms have no common factor. The constant term is positive. The coefficient of x is positive. The binomial constants will be positive.

Positive Factors of 3 (coefficient of x^2)	Positive Factors of 15 (constant term)
1, 3	1, 15
	3, 5

Write trial factors. Use the **O**uter and **I**nner products of FOIL to determine the middle term, $14x$, of the trinomial.

Trial Factors	Middle Term
$(x + 1)(3x + 15)$	Common factor
$(x + 15)(3x + 1)$	$x + 45x = 46x$
$(x + 3)(3x + 5)$	$5x + 9x = 14x$
$(x + 5)(3x + 3)$	Common factor

Write the factors of the trinomial.

$3x^2 + 14x + 15 = (x + 3)(3x + 5)$

HOW TO 3 Factor: $6x^3 + 14x^2 - 12x$

Factor the GCF, $2x$, from the terms.

$$6x^3 + 14x^2 - 12x = 2x(3x^2 + 7x - 6)$$

Factor the trinomial. The constant term is negative. The binomial constants will have opposite signs.

Positive Factors of 3	Factors of −6
1, 3	1, −6
	−1, 6
	2, −3
	−2, 3

Take Note
For this example, all the trial factors were listed. Once the correct factors have been found, however, the remaining trial factors can be omitted. For the examples and solutions in this text, all trial factors except those that have a common factor will be listed.

Write trial factors. Use the **O**uter and **I**nner products of FOIL to determine the middle term, $7x$, of the trinomial.

It is not necessary to test trial factors that have a common factor.

Trial Factors	Middle Term
$(x + 1)(3x - 6)$	Common factor
$(x - 6)(3x + 1)$	$x - 18x = -17x$
$(x - 1)(3x + 6)$	Common factor
$(x + 6)(3x - 1)$	$-x + 18x = 17x$
$(x + 2)(3x - 3)$	Common factor
$(x - 3)(3x + 2)$	$2x - 9x = -7x$
$(x - 2)(3x + 3)$	Common factor
$(x + 3)(3x - 2)$	$-2x + 9x = 7x$

Write the factors of the trinomial.

$$6x^3 + 14x^2 - 12x = 2x(x + 3)(3x - 2)$$

EXAMPLE 1

Factor: $3x^2 + x - 2$

Solution

Positive factors of 3: 1, 3

Factors of −2: 1, −2
−1, 2

Trial Factors	Middle Term
$(x + 1)(3x - 2)$	$-2x + 3x = x$
$(x - 2)(3x + 1)$	$x - 6x = -5x$
$(x - 1)(3x + 2)$	$2x - 3x = -x$
$(x + 2)(3x - 1)$	$-x + 6x = 5x$

$3x^2 + x - 2 = (x + 1)(3x - 2)$

YOU TRY IT 1

Factor: $2x^2 - x - 3$

Your solution

EXAMPLE 2

Factor: $-12x^3 - 32x^2 + 12x$

Solution

$-12x^3 - 32x^2 + 12x = -4x(3x^2 + 8x - 3)$

Factor the trinomial.

Positive factors of 3: 1, 3

Factors of −3: 1, −3
−1, 3

Trial Factors	Middle Term
$(x - 3)(3x + 1)$	$x - 9x = -8x$
$(x + 3)(3x - 1)$	$-x + 9x = 8x$

$-12x^3 - 32x^2 + 12x = -4x(x + 3)(3x - 1)$

YOU TRY IT 2

Factor: $-45y^3 + 12y^2 + 12y$

Your solution

Solutions on p. S21

OBJECTIVE B To factor a trinomial of the form $ax^2 + bx + c$ by grouping

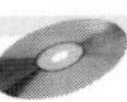

In the preceding objective, trinomials of the form $ax^2 + bx + c$ were factored by using trial factors. In this objective, these trinomials will be factored by grouping.

To factor $ax^2 + bx + c$, first find two factors of $a \cdot c$ whose sum is b. Then use factoring by grouping to write the factorization of the trinomial.

HOW TO 4 Factor: $2x^2 + 13x + 15$

Find two positive factors of 30 $(a \cdot c = 2 \cdot 15 = 30)$ whose sum is 13.

Positive Factors of 30	*Sum*
1, 30	31
2, 15	17
3, 10	13

• Once the required sum has been found, the remaining factors need not be checked.

$$2x^2 + 13x + 15 = 2x^2 + 3x + 10x + 15$$

• Use the factors of 30 whose sum is 13 to write $13x$ as $3x + 10x$.

$$= (2x^2 + 3x) + (10x + 15)$$

• Factor by grouping.

$$= x(2x + 3) + 5(2x + 3)$$
$$= (2x + 3)(x + 5)$$

Check: $(2x + 3)(x + 5) = 2x^2 + 10x + 3x + 15$
$= 2x^2 + 13x + 15$

HOW TO 5 Factor: $6x^2 - 11x - 10$

Find two factors of -60 $[a \cdot c = 6(-10) = -60]$ whose sum is -11.

Factors of −60	*Sum*
1, −60	−59
−1, 60	59
2, −30	−28
−2, 30	28
3, −20	−17
−3, 20	17
4, −15	−11

$$6x^2 - 11x - 10 = 6x^2 + 4x - 15x - 10$$

• Use the factors of -60 whose sum is -11 to write $-11x$ as $4x - 15x$.

$$= (6x^2 + 4x) - (15x + 10)$$

• Factor by grouping. Recall that $-15x - 10 = -(15x + 10)$.

$$= 2x(3x + 2) - 5(3x + 2)$$
$$= (3x + 2)(2x - 5)$$

Check: $(3x + 2)(2x - 5) = 6x^2 - 15x + 4x - 10$
$= 6x^2 - 11x - 10$

HOW TO • 6 Factor: $3x^2 - 2x - 4$

Find two factors of -12 $[a \cdot c = 3(-4) = -12]$ whose sum is -2.

Factors of −12	*Sum*
1, −12	−11
−1, 12	11
2, −6	−4
−2, 6	4
3, −4	−1
−3, 4	1

Take Note
$3x^2 - 2x - 4$ is a prime polynomial.

Because no integer factors of -12 have a sum of -2, $3x^2 - 2x - 4$ is nonfactorable over the integers.

EXAMPLE • 3

Factor: $2x^2 + 19x - 10$

Solution

Factors of −20 [2(−10)]	*Sum*
−1, 20	19

$$\begin{aligned}2x^2 + 19x - 10 &= 2x^2 - x + 20x - 10\\ &= (2x^2 - x) + (20x - 10)\\ &= x(2x - 1) + 10(2x - 1)\\ &= (2x - 1)(x + 10)\end{aligned}$$

YOU TRY IT • 3

Factor: $2a^2 + 13a - 7$

Your solution

EXAMPLE • 4

Factor: $24x^2y - 76xy + 40y$

Solution
The GCF is $4y$.
$24x^2y - 76xy + 40y = 4y(6x^2 - 19x + 10)$

Negative Factors of 60 [6(10)]	*Sum*
−1, −60	−61
−2, −30	−32
−3, −20	−23
−4, −15	−19

$$\begin{aligned}6x^2 - 19x + 10 &= 6x^2 - 4x - 15x + 10\\ &= (6x^2 - 4x) - (15x - 10)\\ &= 2x(3x - 2) - 5(3x - 2)\\ &= (3x - 2)(2x - 5)\end{aligned}$$

$$\begin{aligned}24x^2y - 76xy + 40y &= 4y(6x^2 - 19x + 10)\\ &= 4y(3x - 2)(2x - 5)\end{aligned}$$

YOU TRY IT • 4

Factor: $15x^3 + 40x^2 - 80x$

Your solution

Solutions on p. S21

7.3 EXERCISES

OBJECTIVE A **To factor a trinomial of the form $ax^2 + bx + c$ by using trial factors**

For Exercises 1 to 70, factor by using trial factors.

1. $2x^2 + 3x + 1$

2. $5x^2 + 6x + 1$

3. $2y^2 + 7y + 3$

4. $3y^2 + 7y + 2$

5. $2a^2 - 3a + 1$

6. $3a^2 - 4a + 1$

7. $2b^2 - 11b + 5$

8. $3b^2 - 13b + 4$

9. $2x^2 + x - 1$

10. $4x^2 - 3x - 1$

11. $2x^2 - 5x - 3$

12. $3x^2 + 5x - 2$

13. $2t^2 - t - 10$

14. $2t^2 + 5t - 12$

15. $3p^2 - 16p + 5$

16. $6p^2 + 5p + 1$

17. $12y^2 - 7y + 1$

18. $6y^2 - 5y + 1$

19. $6z^2 - 7z + 3$

20. $9z^2 + 3z + 2$

21. $6t^2 - 11t + 4$

22. $10t^2 + 11t + 3$

23. $8x^2 + 33x + 4$

24. $7x^2 + 50x + 7$

25. $5x^2 - 62x - 7$

26. $9x^2 - 13x - 4$

27. $12y^2 + 19y + 5$

28. $5y^2 - 22y + 8$

29. $7a^2 + 47a - 14$

30. $11a^2 - 54a - 5$

31. $3b^2 - 16b + 16$

32. $6b^2 - 19b + 15$

33. $2z^2 - 27z - 14$

34. $4z^2 + 5z - 6$

35. $3p^2 + 22p - 16$

36. $7p^2 + 19p + 10$

37. $4x^2 + 6x + 2$

38. $12x^2 + 33x - 9$

39. $15y^2 - 50y + 35$

40. $30y^2 + 10y - 20$

41. $2x^3 - 11x^2 + 5x$

42. $2x^3 - 3x^2 - 5x$

43. $3a^2b - 16ab + 16b$

44. $2a^2b - ab - 21b$

45. $3z^2 + 95z + 10$

46. $8z^2 - 36z + 1$

47. $36x - 3x^2 - 3x^3$

48. $-2x^3 + 2x^2 + 4x$

49. $80y^2 - 36y + 4$

50. $24y^2 - 24y - 18$

51. $8z^3 + 14z^2 + 3z$

52. $6z^3 - 23z^2 + 20z$

53. $6x^2y - 11xy - 10y$

54. $8x^2y - 27xy + 9y$

55. $10t^2 - 5t - 50$

56. $16t^2 + 40t - 96$

57. $3p^3 - 16p^2 + 5p$

58. $6p^3 + 5p^2 + p$

59. $26z^2 + 98z - 24$

60. $30z^2 - 87z + 30$

61. $10y^3 - 44y^2 + 16y$

62. $14y^3 + 94y^2 - 28y$

63. $4yz^3 + 5yz^2 - 6yz$

64. $12a^3 + 14a^2 - 48a$

65. $42a^3 + 45a^2 - 27a$

66. $36p^2 - 9p^3 - p^4$

67. $9x^2y - 30xy^2 + 25y^3$

68. $8x^2y - 38xy^2 + 35y^3$

69. $9x^3y - 24x^2y^2 + 16xy^3$

70. $9x^3y + 12x^2y + 4xy$

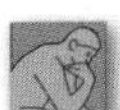

For Exercises 71 and 72, let $(nx + p)$ and $(mx + q)$ be prime factors of the trinomial $ax^2 + bx + c$.

71. If n is even, must p be even or odd?

72. If p is even, must n be even or odd?

OBJECTIVE B To factor a trinomial of the form $ax^2 + bx + c$ by grouping

For Exercises 73 to 132, factor by grouping.

73. $6x^2 - 17x + 12$

74. $15x^2 - 19x + 6$

75. $5b^2 + 33b - 14$

76. $8x^2 - 30x + 25$

77. $6a^2 + 7a - 24$

78. $14a^2 + 15a - 9$

79. $4z^2 + 11z + 6$

80. $6z^2 - 25z + 14$

81. $22p^2 + 51p - 10$

82. $14p^2 - 41p + 15$

83. $8y^2 + 17y + 9$

84. $12y^2 - 145y + 12$

85. $18t^2 - 9t - 5$

86. $12t^2 + 28t - 5$

87. $6b^2 + 71b - 12$

88. $8b^2 + 65b + 8$

89. $9x^2 + 12x + 4$

90. $25x^2 - 30x + 9$

91. $6b^2 - 13b + 6$

92. $20b^2 + 37b + 15$

93. $33b^2 + 34b - 35$

94. $15b^2 - 43b + 22$

95. $18y^2 - 39y + 20$

96. $24y^2 + 41y + 12$

97. $15a^2 + 26a - 21$

98. $6a^2 + 23a + 21$

99. $8y^2 - 26y + 15$

100. $18y^2 - 27y + 4$

101. $8z^2 + 2z - 15$

102. $10z^2 + 3z - 4$

103. $15x^2 - 82x + 24$

104. $13z^2 + 49z - 8$

105. $10z^2 - 29z + 10$

106. $15z^2 - 44z + 32$

107. $36z^2 + 72z + 35$

108. $16z^2 + 8z - 35$

109. $3x^2 + xy - 2y^2$

110. $6x^2 + 10xy + 4y^2$

111. $3a^2 + 5ab - 2b^2$

112. $2a^2 - 9ab + 9b^2$

113. $4y^2 - 11yz + 6z^2$

114. $2y^2 + 7yz + 5z^2$

115. $28 + 3z - z^2$

116. $15 - 2z - z^2$

117. $8 - 7x - x^2$

118. $12 + 11x - x^2$

119. $9x^2 + 33x - 60$

120. $16x^2 - 16x - 12$

121. $24x^2 - 52x + 24$

122. $60x^2 + 95x + 20$

123. $35a^4 + 9a^3 - 2a^2$

124. $15a^4 + 26a^3 + 7a^2$

125. $15b^2 - 115b + 70$

126. $25b^2 + 35b - 30$

127. $3x^2 - 26xy + 35y^2$

128. $4x^2 + 16xy + 15y^2$

129. $216y^2 - 3y - 3$

130. $360y^2 + 4y - 4$

131. $21 - 20x - x^2$

132. $18 + 17x - x^2$

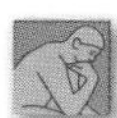

For Exercises 133 to 136, information is given about the signs of b and c in the trinomial $ax^2 + bx + c$, where $a > 0$. If you want to factor $ax^2 + bx + c$ by grouping, you look for factors of ac whose sum is b. In each case, state whether the factors of ac should be two positive numbers, two negative numbers, or one positive and one negative number.

133. $b > 0$ and $c > 0$

134. $b < 0$ and $c < 0$

135. $b < 0$ and $c > 0$

136. $b > 0$ and $c < 0$

Applying the Concepts

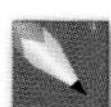

137. In your own words, explain how the signs of the last terms of the two binomial factors of a trinomial are determined.

For Exercises 138 to 143, factor.

138. $(x + 1)^2 - (x + 1) - 6$

139. $(x - 2)^2 + 3(x - 2) + 2$

140. $(y + 3)^2 - 5(y + 3) + 6$

141. $2(y + 2)^2 - (y + 2) - 3$

142. $3(a + 2)^2 - (a + 2) - 4$

143. $4(y - 1)^2 - 7(y - 1) - 2$

SECTION 7.4 Special Factoring

OBJECTIVE A To factor the difference of two perfect squares or a perfect-square trinomial

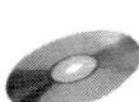

The product of a term and itself is called a **perfect square.** The exponents on variables of perfect squares are always even numbers.

Term		*Perfect Square*
5	$5 \cdot 5 =$	25
x	$x \cdot x =$	x^2
$3y^4$	$3y^4 \cdot 3y^4 =$	$9y^8$
x^n	$x^n \cdot x^n =$	x^{2n}

The **square root of a perfect square** is one of the two equal factors of the perfect square. $\sqrt{}$ is the symbol for square root. To find the exponent on the square root of a variable term, multiply the exponent by $\frac{1}{2}$.

$\sqrt{25} = 5$
$\sqrt{x^2} = x$
$\sqrt{9y^8} = 3y^4$
$\sqrt{x^{2n}} = x^n$

The **difference of two perfect squares** is the **product of the sum and difference of two terms.** The factors of the difference of two perfect squares are the sum and difference of the square roots of the perfect squares.

Factors of the Difference of Two Perfect Squares

$a^2 - b^2 = (a + b)(a - b)$

The **sum of two perfect squares,** $a^2 + b^2$, is nonfactorable over the integers.

HOW TO • 1 Factor: $4x^2 - 81y^2$

Write the binomial as the difference of two perfect squares. $4x^2 - 81y^2 = (2x)^2 - (9y)^2$

The factors are the sum and difference of the square roots of the perfect squares. $= (2x + 9y)(2x - 9y)$

A **perfect-square trinomial** is the square of a binomial.

Factors of a Perfect-Square Trinomial

$a^2 + 2ab + b^2 = (a + b)^2$

$a^2 - 2ab + b^2 = (a - b)^2$

In factoring a perfect-square trinomial, remember that the terms of the binomial are the square roots of the perfect squares of the trinomial. The sign of the binomial is the sign of the middle term of the trinomial.

HOW TO • 2 Factor: $4x^2 + 12x + 9$

Because $4x^2$ is a perfect square $[4x^2 = (2x)^2]$ and 9 is a perfect square $(9 = 3^2)$, try factoring $4x^2 + 12x + 9$ as the square of a binomial.

$4x^2 + 12x + 9 \stackrel{?}{=} (2x + 3)^2$

Check: $(2x + 3)^2 = (2x + 3)(2x + 3) = 4x^2 + 6x + 6x + 9 = 4x^2 + 12x + 9$

The check verifies that $4x^2 + 12x + 9 = (2x + 3)^2$.

It is important to check a proposed factorization as we did above. The next example illustrates the importance of this check.

HOW TO • 3 Factor: $x^2 + 13x + 36$

Because x^2 is a perfect square and 36 is a perfect square, try factoring $x^2 + 13x + 36$ as the square of a binomial.

$x^2 + 13x + 36 \stackrel{?}{=} (x + 6)^2$

Check: $(x + 6)^2 = (x + 6)(x + 6) = x^2 + 6x + 6x + 36 = x^2 + 12x + 36$

In this case, the proposed factorization of $x^2 + 13x + 36$ does *not* check.
Try another factorization. The numbers 4 and 9 are factors of 36 whose sum is 13.

$x^2 + 13x + 36 = (x + 4)(x + 9)$

EXAMPLE • 1

Factor: $25x^2 - 1$

Solution

$25x^2 - 1 = (5x)^2 - (1)^2$
$= (5x + 1)(5x - 1)$

• Difference of two squares

YOU TRY IT • 1

Factor: $x^2 - 36y^4$

Your solution

EXAMPLE • 2

Factor: $4x^2 - 20x + 25$

Solution

$4x^2 - 20x + 25 = (2x - 5)^2$

• Perfect-square trinomial

YOU TRY IT • 2

Factor: $9x^2 + 12x + 4$

Your solution

EXAMPLE • 3

Factor: $(x + y)^2 - 4$

Solution

$(x + y)^2 - 4 = (x + y)^2 - (2)^2$
$= (x + y)^2 - (2)^2$
$= (x + y + 2)(x + y - 2)$

• Difference of two squares

YOU TRY IT • 3

Factor: $(a + b)^2 - (a - b)^2$

Your solution

Solutions on p. S21

OBJECTIVE B To factor the sum or difference of two perfect cubes 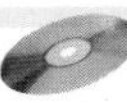

The product of the same three factors is called a **perfect cube.** The first seven perfect cube integers are:

$$1 = 1^3,\ 8 = 2^3,\ 27 = 3^3,\ 64 = 4^3,\ 125 = 5^3,\ 216 = 6^3,\ 343 = 7^3$$

A variable term is a perfect cube if the coefficient is a perfect cube and the exponent on each variable is divisible by 3. The table at the right shows some perfect-cube variable terms. Note that each exponent of the perfect cube is divisible by 3.

Term		*Perfect Cube*
x	$x \cdot x \cdot x = x^3 =$	x^3
$2y$	$2y \cdot 2y \cdot 2y = (2y)^3 =$	$8y^3$
$4x^2$	$4x^2 \cdot 4x^2 \cdot 4x^2 = (4x^2)^3 =$	$64x^6$
$3x^4y^3$	$3x^4y^3 \cdot 3x^4y^3 \cdot 3x^4y^3 = (3x^4y^3)^3 =$	$27x^{12}y^9$

The **cube root** of a perfect cube is one of the three equal factors of the perfect cube. $\sqrt[3]{\ }$ is the symbol for cube root. To find the exponents on the cube root of a perfect-cube variable expression, multiply the exponents on the variables by $\frac{1}{3}$.

$\sqrt[3]{x^3} = x$
$\sqrt[3]{8y^3} = 2y$
$\sqrt[3]{64x^6} = 4x^2$
$\sqrt[3]{27x^{12}y^9} = 3x^4y^3$

✓ Take Note

The factoring formulas at the right are the result of finding, for instance, the quotient

$$\frac{a^3 + b^3}{a + b} = a^2 - ab + b^2$$

Similarly,

$$\frac{a^3 - b^3}{a - b} = a^2 + ab + b^2.$$

The following rules are used to factor the sum or difference of two perfect cubes.

Factoring the Sum or Difference of Two Perfect Cubes

$a^3 + b^3 = (a + b)(a^2 - ab + b^2)$

$a^3 - b^3 = (a - b)(a^2 + ab + b^2)$

To factor $27x^3 - 1$:

Write the binomial as the difference of two perfect cubes.

The terms of the binomial factor are the cube roots of the perfect cubes. The sign of the binomial factor is the same as the sign of the given binomial. The trinomial factor is obtained from the binomial factor.

$$27x^3 - 1 = (3x)^3 - 1^3$$
$$= (3x - 1)(9x^2 + 3x + 1)$$

Square of the first term ($9x^2$)

Opposite of the product of the two terms ($3x$)

Square of the last term (1)

✓ Take Note

Note the placement of the signs. The sign of the binomial factor is the same as the sign of the sum or difference of the perfect cubes. The first sign of the trinomial factor is the opposite of the sign of the binomial factor.

HOW TO • 4 Factor: $m^3 + 64n^3$

$m^3 + 64n^3 = (m)^3 + (4n)^3$ • Write as the sum of two perfect cubes.

$(m)^2$ $\quad m(4n)$ $\quad (4n)^2$

$= (m + 4n)(m^2 - 4mn + 16n^2)$ • Use $a^3 + b^3 = (a + b)(a^2 - ab + b^2)$.

HOW TO • 5 Factor: $8x^3 - 27$

$8x^3 - 27 = (2x)^3 - 3^3$ • Write as the difference of two perfect cubes.

$(2x)^2$ $\quad 2x(3) \quad$ 3^2

$= (2x - 3)(4x^2 + 6x + 9)$ • Use $a^3 - b^3 = (a - b)(a^2 + ab + b^2)$.

HOW TO • 6 Factor: $64y^4 - 125y$

$64y^4 - 125y = y(64y^3 - 125)$ • Factor out y, the GCF.

$= y[(4y)^3 - 5^3]$ • Write the binomial as the difference of two perfect cubes.

$= y(4y - 5)(16y^2 + 20y + 25)$ • Factor.

EXAMPLE • 4

Factor: $x^3y^3 - 1$

Solution

$x^3y^3 - 1 = (xy)^3 - 1^3$ • Difference of two cubes

$= (xy - 1)(x^2y^2 + xy + 1)$

YOU TRY IT • 4

Factor: $a^3b^3 - 27$

Your solution

EXAMPLE • 5

Factor: $64c^3 + 8d^3$

Solution

$64c^3 + 8d^3$

$= 8(8c^3 + d^3)$ • GCF

$= 8[(2c)^3 + d^3]$ • Sum of two cubes

$= 8(2c + d)(4c^2 - 2cd + d^2)$

YOU TRY IT • 5

Factor: $8x^3 + y^3z^3$

Your solution

EXAMPLE • 6

Factor: $(x + y)^3 - x^3$

Solution

$(x + y)^3 - x^3$ • Difference of two cubes

$= [(x + y) - x][(x + y)^2 + x(x + y) + x^2]$

$= y(x^2 + 2xy + y^2 + x^2 + xy + x^2)$

$= y(3x^2 + 3xy + y^2)$

YOU TRY IT • 6

Factor: $(x - y)^3 + (x + y)^3$

Your solution

Solutions on pp. S21–S22

OBJECTIVE C To factor a trinomial that is quadratic in form 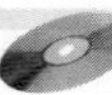

Certain trinomials that are not quadratic can be expressed as quadratic trinomials by making suitable variable substitutions. A trinomial is **quadratic in form** if it can be written as $au^2 + bu + c$.

As shown below, the trinomials $x^4 + 5x^2 + 6$ and $2x^2y^2 + 3xy - 9$ are quadratic in form.

	$x^4 + 5x^2 + 6$		$2x^2y^2 + 3xy - 9$
	$(x^2)^2 + 5(x^2) + 6$		$2(xy)^2 + 3(xy) - 9$
Let $u = x^2$.	$u^2 + 5u + 6$	Let $u = xy$.	$2u^2 + 3u - 9$

When we use this method to factor a trinomial that is quadratic in form, the variable part of the first term in each binomial will be u.

Take Note
The trinomial $x^4 + 5x^2 + 6$ was shown above to be quadratic in form.

HOW TO • 7 Factor: $x^4 + 5x^2 + 6$

$x^4 + 5x^2 + 6 = u^2 + 5u + 6$ • Let $u = x^2$.

$= (u + 3)(u + 2)$ • Factor.

$= (x^2 + 3)(x^2 + 2)$ • Replace u by x^2.

Here is an example in which $u = \sqrt{x}$.

HOW TO • 8 Factor: $x - 2\sqrt{x} - 15$

$x - 2\sqrt{x} - 15 = u^2 - 2u - 15$ • Let $u = \sqrt{x}$. Then $u^2 = x$.

$= (u - 5)(u + 3)$ • Factor.

$= (\sqrt{x} - 5)(\sqrt{x} + 3)$ • Replace u by $\sqrt{x}$.

EXAMPLE • 7

Factor: $6x^2y^2 - xy - 12$

Solution

Let $u = xy$.

$6x^2y^2 - xy - 12$
$= 6u^2 - u - 12$
$= (3u + 4)(2u - 3)$
$= (3xy + 4)(2xy - 3)$ • Replace u by xy.

YOU TRY IT • 7

Factor: $3x^4 + 4x^2 - 4$

Your solution

Solution on p. S22

OBJECTIVE D To factor completely 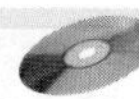

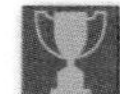

Tips for Success

You now have completed all the lessons on factoring polynomials. You will need to be able to recognize all of the factoring patterns. To test yourself, try the Chapter 7 Review Exercises.

Take Note

Remember that you may have to factor more than once in order to write the polynomial as a product of *prime* factors.

General Factoring Strategy

1. Is there a common factor? If so, factor out the GCF.
2. If the polynomial is a binomial, is it the difference of two perfect squares, the sum of two perfect cubes, or the difference of two perfect cubes? If so, factor.
3. If the polynomial is a trinomial, is it a perfect-square trinomial or the product of two binomials? If so, factor.
4. Can the polynomial be factored by grouping? If so, factor.
5. Is each factor nonfactorable over the integers? If not, factor.

EXAMPLE 8

Factor: $6a^3 + 15a^2 - 36a$

Solution

$6a^3 + 15a^2 - 36a = 3a(2a^2 + 5a - 12)$ • GCF

$= 3a(2a - 3)(a + 4)$

YOU TRY IT 8

Factor: $18x^3 - 6x^2 - 60x$

Your solution

EXAMPLE 9

Factor: $x^2y + 2x^2 - y - 2$

Solution

$x^2y + 2x^2 - y - 2$

$= (x^2y + 2x^2) - (y + 2)$ • Factor by grouping.

$= x^2(y + 2) - (y + 2)$

$= (y + 2)(x^2 - 1)$

$= (y + 2)(x + 1)(x - 1)$

YOU TRY IT 9

Factor: $4x - 4y - x^3 + x^2y$

Your solution

EXAMPLE 10

Factor: $x^{4n} - y^{4n}$

Solution

$x^{4n} - y^{4n} = (x^{2n})^2 - (y^{2n})^2$ • Difference of two squares

$= (x^{2n} + y^{2n})(x^{2n} - y^{2n})$

$= (x^{2n} + y^{2n})[(x^n)^2 - (y^n)^2]$

$= (x^{2n} + y^{2n})(x^n + y^n)(x^n - y^n)$

YOU TRY IT 10

Factor: $x^{4n} - x^{2n}y^{2n}$

Your solution

Solutions on p. S22

7.4 EXERCISES

OBJECTIVE A To factor the difference of two perfect squares or a perfect-square trinomial

For Exercises 1 and 2, determine which expressions are perfect squares.

1. $4; 8; 25x^6; 12y^{10}; 100x^4y^4$

2. $9; 18; 15a^8; 49b^{12}; 64a^{16}b^2$

For Exercises 3 to 6, find the square root of the expression.

3. $16z^8$

4. $36d^{10}$

5. $81a^4b^6$

6. $25m^2n^{12}$

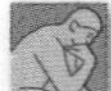

7. Which of the following polynomials, if any, is a difference of squares?
(i) $x^2 - 12$ (ii) $a^2 + 25$ (iii) $x^3 - 4x$ (iv) $m^4 - n^2$

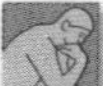

8. Determine if each statement is true or false. Explain.
a. $a^2 + b^2 = (a + b)^2$
b. $a^2 - b^2 = (a - b)(a - b)$
c. $a^2 + 10a + 9$ is a perfect-square trinomial.
d. $a^2 + 3a - 4$ is a perfect-square trinomial.

For Exercises 9 to 44, factor.

9. $x^2 - 16$

10. $y^2 - 49$

11. $4x^2 - 1$

12. $81x^2 - 4$

13. $16x^2 - 121$

14. $49y^2 - 36$

15. $1 - 9a^2$

16. $16 - 81y^2$

17. $x^2y^2 - 100$

18. $a^2b^2 - 25$

19. $x^2 + 4$

20. $a^2 + 16$

21. $25 - a^2b^2$

22. $64 - x^2y^2$

23. $a^{2n} - 1$

24. $b^{2n} - 16$

25. $x^2 - 12x + 36$

26. $y^2 - 6y + 9$

27. $b^2 - 2b + 1$

28. $a^2 + 14a + 49$

29. $16x^2 - 40x + 25$

30. $49x^2 + 28x + 4$

31. $4a^2 + 4a - 1$

32. $9x^2 + 12x - 4$

33. $b^2 + 7b + 14$

34. $y^2 - 5y + 25$

35. $x^2 + 6xy + 9y^2$

36. $4x^2y^2 + 12xy + 9$

37. $25a^2 - 40ab + 16b^2$

38. $4a^2 - 36ab + 81b^2$

39. $x^{2n} + 6x^n + 9$

40. $y^{2n} - 16y^n + 64$

41. $(x - 4)^2 - 9$

42. $16 - (a - 3)^2$

43. $(x - y)^2 - (a + b)^2$

44. $(x - 2y)^2 - (x + y)^2$

OBJECTIVE B To factor the sum or difference of two perfect cubes

For Exercises 45 and 46, determine which expressions are perfect cubes.

45. 4; 8; x^9; a^8b^8; $27c^{15}d^{18}$

46. 9; 27; y^{12}; m^3n^6; $64mn^9$

For Exercises 47 to 50, find the cube root of the expression.

47. $8x^9$

48. $27y^{15}$

49. $64a^6b^{18}$

50. $125c^{12}d^3$

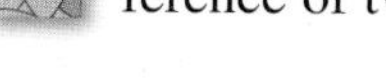

For Exercises 51 to 54, state whether the polynomial can be written as either the sum or difference of two cubes.

51. $x^6 + 64$

52. $x^9 - 32$

53. $a^6x^3 - b^3y^6$

54. $a^3x^3 - b^3y^3$

For Exercises 55 to 78, factor.

55. $x^3 - 27$

56. $y^3 + 125$

57. $8x^3 - 1$

58. $64a^3 + 27$

59. $x^3 - y^3$

60. $x^3 - 8y^3$

61. $m^3 + n^3$

62. $27a^3 + b^3$

63. $64x^3 + 1$

64. $1 - 125b^3$

65. $27x^3 - 8y^3$

66. $64x^3 + 27y^3$

67. $x^3y^3 + 64$

68. $8x^3y^3 + 27$

69. $16x^3 - y^3$

70. $27x^3 - 8y^2$

71. $8x^3 - 9y^3$

72. $27a^3 - 16$

73. $(a - b)^3 - b^3$

74. $a^3 + (a + b)^3$

75. $x^{6n} + y^{3n}$

76. $x^{3n} + y^{3n}$

77. $x^{3n} + 8$

78. $a^{3n} + 64$

OBJECTIVE C To factor a trinomial that is quadratic in form

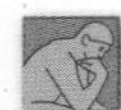

79. The expression $x - \sqrt{x} - 6$ is quadratic in form. Is the expression a polynomial? Explain.

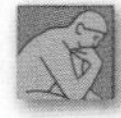

80. The polynomial $x^4 - 2x^2 - 3$ is quadratic in form. Is the polynomial a quadratic polynomial? Explain.

For Exercises 81 to 101, factor.

81. $x^2y^2 - 8xy + 15$

82. $x^2y^2 - 8xy - 33$

83. $x^2y^2 - 17xy + 60$

84. $a^2b^2 + 10ab + 24$

85. $x^4 - 9x^2 + 18$

86. $y^4 - 6y^2 - 16$

87. $b^4 - 13b^2 - 90$

88. $a^4 + 14a^2 + 45$

89. $x^4y^4 - 8x^2y^2 + 12$

90. $a^4b^4 + 11a^2b^2 - 26$

91. $x^{2n} + 3x^n + 2$

92. $a^{2n} - a^n - 12$

93. $3x^2y^2 - 14xy + 15$

94. $5x^2y^2 - 59xy + 44$

95. $6a^2b^2 - 23ab + 21$

96. $10a^2b^2 + 3ab - 7$

97. $2x^4 - 13x^2 - 15$

98. $3x^4 + 20x^2 + 32$

99. $2x^{2n} - 7x^n + 3$

100. $4x^{2n} + 8x^n - 5$

101. $6a^{2n} + 19a^n + 10$

OBJECTIVE D To factor completely

For Exercises 102 to 135, factor.

102. $5x^2 + 10x + 5$

103. $12x^2 - 36x + 27$

104. $3x^4 - 81x$

105. $27a^4 - a$

106. $7x^2 - 28$

107. $20x^2 - 5$

108. $y^4 - 10y^3 + 21y^2$

109. $y^5 + 6y^4 - 55y^3$

110. $x^4 - 16$

111. $16x^4 - 81$

112. $8x^5 - 98x^3$

113. $16a - 2a^4$

114. $x^3y^3 - x^3$

115. $a^3b^6 - b^3$

116. $x^6y^6 - x^3y^3$

117. $8x^4 - 40x^3 + 50x^2$

118. $6x^5 + 74x^4 + 24x^3$

119. $x^4 - y^4$

120. $16a^4 - b^4$

121. $x^6 + y^6$

122. $x^4 - 5x^2 - 4$

123. $a^4 - 25a^2 - 144$

124. $3b^5 - 24b^2$

125. $16a^4 - 2a$

126. $x^4y^2 - 5x^3y^3 + 6x^2y^4$

127. $a^4b^2 - 8a^3b^3 - 48a^2b^4$

128. $x^3 - 2x^2 - x + 2$

129. $x^3 - 2x^2 - 4x + 8$

130. $4x^3 + 8x^2 - 9x - 18$

131. $2x^3 + x^2 - 32x - 16$

132. $4x^2y^2 - 4x^2 - 9y^2 + 9$

133. $4x^4 - x^2 - 4x^2y^2 + y^2$

134. $2x^{n+2} - 7x^{n+1} + 3x^n$

135. $3b^{n+2} + 4b^{n+1} - 4b^n$

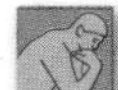

136. What is the degree of the polynomial whose factored form is $3x^3(x^2 + 4)(x - 2)(x + 3)$?

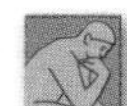

137. What is the coefficient of x^6 when $4x^2(x^2 + 3)(x + 4)(2x - 5)$ is expanded and written as a polynomial?

Applying the Concepts

138. Factor: $x^2(x - 3) - 3x(x - 3) + 2(x - 3)$

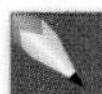

139. Given that $(x - 3)$ and $(x + 4)$ are factors of $x^3 + 6x^2 - 7x - 60$, explain how you can find a third *first-degree* factor of $x^3 + 6x^2 - 7x - 60$. Then find the factor.

SECTION

7.5 Solving Equations

OBJECTIVE A To solve equations by factoring 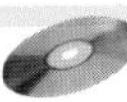

The Multiplication Property of Zero states that the product of a number and zero is zero. This property is stated below.

If a is a real number, then $a \cdot 0 = 0 \cdot a = 0$.

Now consider $a \cdot b = 0$. For this to be a true equation, then either $a = 0$ or $b = 0$.

> **Principle of Zero Products**
>
> If the product of two factors is zero, then at least one of the factors must be zero.
>
> If $a \cdot b = 0$, then $a = 0$ or $b = 0$.

The Principle of Zero Products is used to solve some equations.

HOW TO • 1 Solve: $(x - 2)(x - 3) = 0$

$(x - 2)(x - 3) = 0$

$x - 2 = 0 \qquad x - 3 = 0$ • Let each factor equal zero (the Principle of Zero Products).

$x = 2 \qquad x = 3$ • Solve each equation for x.

Check:

$(x - 2)(x - 3) = 0$

$(2 - 2)(2 - 3) \mid 0$

$0(-1) \mid 0$

$0 = 0$ • A true equation

$(x - 2)(x - 3) = 0$

$(3 - 2)(3 - 3) \mid 0$

$(1)(0) \mid 0$

$0 = 0$ • A true equation

The solutions are 2 and 3.

An equation that can be written in the form $ax^2 + bx + c = 0$, $a \neq 0$, is a **quadratic equation.** A quadratic equation is in **standard form** when the polynomial is in descending order and equal to zero. The quadratic equations at the right are in standard form.

$3x^2 + 2x + 1 = 0$
$a = 3, b = 2, c = 1$

$4x^2 - 3x + 2 = 0$
$a = 4, b = -3, c = 2$

HOW TO • 2 Solve: $2x^2 + x = 6$

$$2x^2 + x = 6$$

$$2x^2 + x - 6 = 0$$ • Write the equation in standard form.

$$(2x - 3)(x + 2) = 0$$ • Factor.

$$2x - 3 = 0 \qquad x + 2 = 0$$ • Use the Principle of Zero Products.

$$2x = 3 \qquad x = -2$$ • Solve each equation for x.

$$x = \frac{3}{2}$$

Check: $\frac{3}{2}$ and -2 check as solutions.

The solutions are $\frac{3}{2}$ and -2.

EXAMPLE • 1

Solve: $x(x - 3) = 0$

Solution

$$x(x - 3) = 0$$

$$x = 0 \qquad x - 3 = 0$$
$$x = 3$$

• Use the Principle of Zero Products.

The solutions are 0 and 3.

YOU TRY IT • 1

Solve: $2x(x + 7) = 0$

Your solution

EXAMPLE • 2

Solve: $2x^2 - 50 = 0$

Solution

$$2x^2 - 50 = 0$$

$$2(x^2 - 25) = 0$$ • Factor out the GCF, 2.

$$2(x + 5)(x - 5) = 0$$ • Factor the difference of two squares.

$$x + 5 = 0 \qquad x - 5 = 0$$
$$x = -5 \qquad x = 5$$

• Use the Principle of Zero Products.

The solutions are -5 and 5.

YOU TRY IT • 2

Solve: $4x^2 - 9 = 0$

Your solution

EXAMPLE • 3

Solve: $(x - 3)(x - 10) = -10$

Solution

$$(x - 3)(x - 10) = -10$$

$$x^2 - 13x + 30 = -10$$ • Multiply $(x - 3)(x - 10)$.

$$x^2 - 13x + 40 = 0$$ • Add 10 to each side of the equation. The equation is now in standard form.

$$(x - 8)(x - 5) = 0$$

$$x - 8 = 0 \qquad x - 5 = 0$$
$$x = 8 \qquad x = 5$$

The solutions are 8 and 5.

YOU TRY IT • 3

Solve: $(x + 2)(x - 7) = 52$

Your solution

Solutions on p. S22

OBJECTIVE B To solve application problems 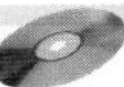

EXAMPLE • 4

The sum of the squares of two consecutive positive even integers is equal to 100. Find the two integers.

Strategy

First positive even integer: n
Second positive even integer: $n + 2$

The sum of the square of the first positive even integer and the square of the second positive even integer is 100.

Solution

$$\begin{aligned} n^2 + (n+2)^2 &= 100 \\ n^2 + n^2 + 4n + 4 &= 100 \\ 2n^2 + 4n + 4 &= 100 \\ 2n^2 + 4n - 96 &= 0 \\ 2(n^2 + 2n - 48) &= 0 \\ 2(n-6)(n+8) &= 0 \end{aligned}$$

• Quadratic equation in standard form

$n - 6 = 0 \qquad n + 8 = 0$
$n = 6 \qquad n = -8$

• Principle of Zero Products

Because -8 is not a positive even integer, it is not a solution.

$n = 6$
$n + 2 = 6 + 2 = 8$

The two integers are 6 and 8.

YOU TRY IT • 4

The sum of the squares of two consecutive positive integers is 61. Find the two integers.

Your strategy

Your solution

Solution on p. S22

EXAMPLE • 5

A stone is thrown into a well with an initial speed of 4 ft/s. The well is 420 ft deep. How many seconds later will the stone hit the bottom of the well? Use the equation $d = vt + 16t^2$, where d is the distance in feet that the stone travels in t seconds when its initial speed is v feet per second.

Strategy

To find the time for the stone to drop to the bottom of the well, replace the variables d and v by their given values and solve for t.

Solution

$$d = vt + 16t^2$$
$$420 = 4t + 16t^2$$
$$0 = -420 + 4t + 16t^2$$
$$0 = 16t^2 + 4t - 420 \quad \text{• Quadratic equation in standard form}$$
$$0 = 4(4t^2 + t - 105)$$
$$0 = 4(4t + 21)(t - 5)$$

$$4t + 21 = 0 \qquad t - 5 = 0 \quad \text{• Principle of Zero Products}$$
$$4t = -21 \qquad t = 5$$
$$t = -\frac{21}{4}$$

Because the time cannot be a negative number, $-\frac{21}{4}$ is not a solution.

The stone will hit the bottom of the well 5 s later.

YOU TRY IT • 5

The length of a rectangle is 4 in. longer than twice the width. The area of the rectangle is 96 in^2. Find the length and width of the rectangle.

Your strategy

Your solution

Solution on p. S22

7.5 EXERCISES

OBJECTIVE A **To solve equations by factoring**

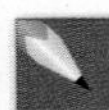

1. In your own words, explain the Principle of Zero Products.

2. Fill in the blanks. If $(x + 5)(2x - 7) = 0$, then ________ $= 0$ or ________ $= 0$.

For Exercises 3 to 60, solve.

3. $(y + 3)(y + 2) = 0$

4. $(y - 3)(y - 5) = 0$

5. $(z - 7)(z - 3) = 0$

6. $(z + 8)(z - 9) = 0$

7. $x(x - 5) = 0$

8. $x(x + 2) = 0$

9. $a(a - 9) = 0$

10. $a(a + 12) = 0$

11. $y(2y + 3) = 0$

12. $t(4t - 7) = 0$

13. $2a(3a - 2) = 0$

14. $4b(2b + 5) = 0$

15. $(b + 2)(b - 5) = 0$

16. $(b - 8)(b + 3) = 0$

17. $x^2 - 81 = 0$

18. $x^2 - 121 = 0$

19. $4x^2 - 49 = 0$

20. $16x^2 - 1 = 0$

21. $9x^2 - 1 = 0$

22. $16x^2 - 49 = 0$

23. $x^2 + 6x + 8 = 0$

24. $x^2 - 8x + 15 = 0$

25. $z^2 + 5z - 14 = 0$

26. $z^2 + z - 72 = 0$

27. $2a^2 - 9a - 5 = 0$

28. $3a^2 + 14a + 8 = 0$

29. $6z^2 + 5z + 1 = 0$

30. $6y^2 - 19y + 15 = 0$

31. $x^2 - 3x = 0$

32. $a^2 - 5a = 0$

33. $x^2 - 7x = 0$

34. $2a^2 - 8a = 0$

35. $a^2 + 5a = -4$

36. $a^2 - 5a = 24$

37. $y^2 - 5y = -6$

38. $y^2 - 7y = 8$

39. $2t^2 + 7t = 4$

40. $3t^2 + t = 10$

41. $3t^2 - 13t = -4$

42. $5t^2 - 16t = -12$

43. $x(x - 12) = -27$

44. $x(x - 11) = 12$

45. $y(y - 7) = 18$

46. $y(y + 8) = -15$

47. $p(p + 3) = -2$

48. $p(p - 1) = 20$

49. $y(y + 4) = 45$

50. $y(y - 8) = -15$

51. $x(x + 3) = 28$

52. $p(p - 14) = 15$

53. $(x + 8)(x - 3) = -30$

54. $(x + 4)(x - 1) = 14$

55. $(z - 5)(z + 4) = 52$

56. $(z - 8)(z + 4) = -35$

57. $(z - 6)(z + 1) = -10$

58. $(a + 3)(a + 4) = 72$

59. $(a - 4)(a + 7) = -18$

60. $(2x + 5)(x + 1) = -1$

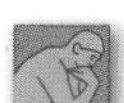

For Exercises 61 and 62, the equation $ax^2 + bx + c = 0$, $a > 0$, is a quadratic equation that can be solved by factoring and then using the Principle of Zero Products.

61. If $ax^2 + bx + c = 0$ has one positive solution and one negative solution, is c greater than, less than, or equal to zero?

62. If zero is one solution of $ax^2 + bx + c = 0$, is c greater than, less than, or equal to zero?

OBJECTIVE B To solve application problems

63. **Number Sense** The square of a positive number is six more than five times the positive number. Find the number.

64. **Number Sense** The square of a negative number is fifteen more than twice the negative number. Find the number.

65. **Number Sense** The sum of two numbers is six. The sum of the squares of the two numbers is twenty. Find the two numbers.

66. **Number Sense** The sum of two numbers is eight. The sum of the squares of the two numbers is thirty-four. Find the two numbers.

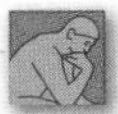

For Exercises 67 and 68, use the following problem situation: The sum of the squares of two consecutive positive integers is 113. Find the two integers.

67. Which equation could be used to solve this problem?
(i) $x^2 + x^2 + 1 = 113$ (ii) $x^2 + (x + 1)^2 = 113$ (iii) $(x + x + 1)^2 = 113$

68. Suppose the solutions of the correct equation in Exercise 67 are -8 and 7. Which solution should be eliminated, and why?

69. **Number Sense** The sum of the squares of two consecutive positive integers is forty-one. Find the two integers.

70. **Number Sense** The sum of the squares of two consecutive positive even integers is one hundred. Find the two integers.

71. **Number Sense** The sum of two numbers is ten. The product of the two numbers is twenty-one. Find the two numbers.

72. **Number Sense** The sum of two numbers is thirteen. The product of the two numbers is forty. Find the two numbers.

Sum of Natural Numbers The formula $S = \frac{n^2 + n}{2}$ gives the sum S of the first n natural numbers. Use this formula for Exercises 73 and 74.

73. How many consecutive natural numbers beginning with 1 will give a sum of 78?

74. How many consecutive natural numbers beginning with 1 will give a sum of 171?

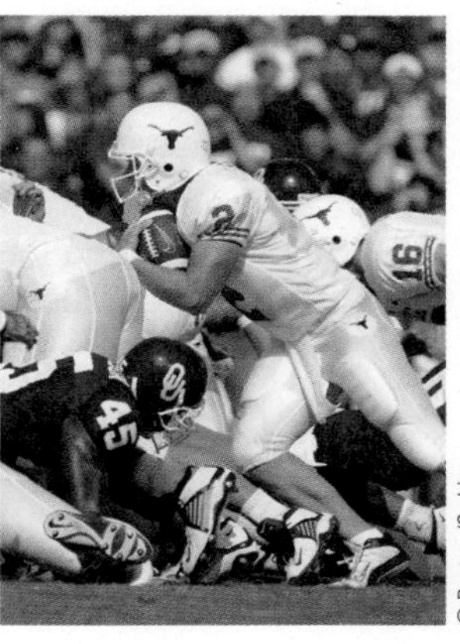
© Reuters/Corbis

Sports The formula $N = \frac{t^2 - t}{2}$ gives the number N of football games that must be scheduled in a league with t teams if each team is to play every other team once. Use this formula for Exercises 75 and 76.

75. How many teams are in a league that schedules 15 games in such a way that each team plays every other team once?

76. How many teams are in a league that schedules 45 games in such a way that each team plays every other team once?

Physics The distance s, in feet, that an object will fall (neglecting air resistance) in t seconds is given by $s = vt + 16t^2$, where v is the initial velocity of the object in feet per second. Use this formula for Exercises 77 and 78.

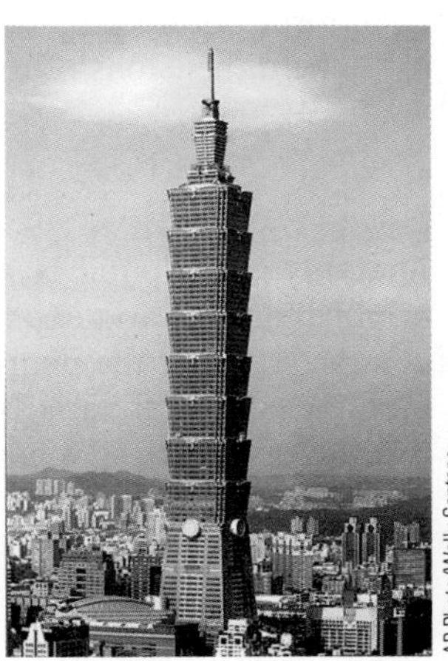
AP Photo/Wally Santana

77. An object is released from the top of a building 192 ft high. The initial velocity is 16 ft/s, and air resistance is neglected. How many seconds later will the object hit the ground?

78. Taipei 101 in Taipei, Taiwan, is the world's tallest inhabited building. The top of the spire is 1667 ft above ground. If an object is released from this building at a point 640 ft above the ground at an initial velocity of 48 ft/s, assuming no air resistance, how many seconds later will the object reach the ground?

Sports The height h, in feet, an object will attain (neglecting air resistance) in t seconds is given by $h = vt - 16t^2$, where v is the initial velocity of the object in feet per second. Use this formula for Exercises 79 and 80.

79. A golf ball is thrown onto a cement surface and rebounds straight up. The initial velocity of the rebound is 60 ft/s. How many seconds later will the golf ball return to the ground?

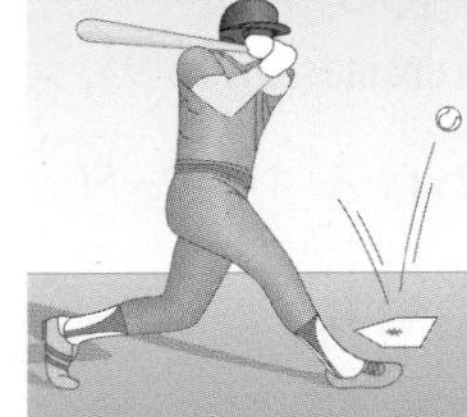

80. A foul ball leaves a bat, hits home plate, and travels straight up with an initial velocity of 64 ft/s. How many seconds later will the ball be 64 ft above the ground?

81. Geometry The length of a rectangle is 5 in. more than twice its width. Its area is 75 in^2. Find the length and width of the rectangle.

82. Geometry The width of a rectangle is 5 ft less than the length. The area of the rectangle is 176 ft^2. Find the length and width of the rectangle.

83. Geometry The height of a triangle is 4 m more than twice the length of the base. The area of the triangle is 35 m². Find the height of the triangle.

In the News

New Lane for Basketball Court

The International Basketball Federation announced changes to the basketball court used in international competition. The 3-second lane, currently a trapezoid, will be a rectangle 3 ft longer than it is wide, similar to the one used in NBA games.

Source: The New York Times

84. Geometry The length of each side of a square is extended 5 in. The area of the resulting square is 64 in². Find the length of a side of the original square.

85. Basketball See the news clipping at the right. If the area of the rectangular 3-second lane is 304 ft², find the width of the lane.

86. Gardening A small garden measures 8 ft by 10 ft. A uniform border around the garden increases the total area to 143 ft². What is the width of the border?

87. Publishing The page of a book measures 6 in. by 9 in. A uniform border around the page leaves 28 in² for type. What are the dimensions of the type area?

Anatomy The pupil is the opening in the iris that lets light into the eye. In bright light, the iris expands so that the pupil is smaller; in low light, the iris contracts so that the pupil is larger. If x is the width, in millimeters, of the iris, then the area of the iris is given by $A = (12\pi x - \pi x^2)$ mm². Use this formula for Exercises 88 and 89.

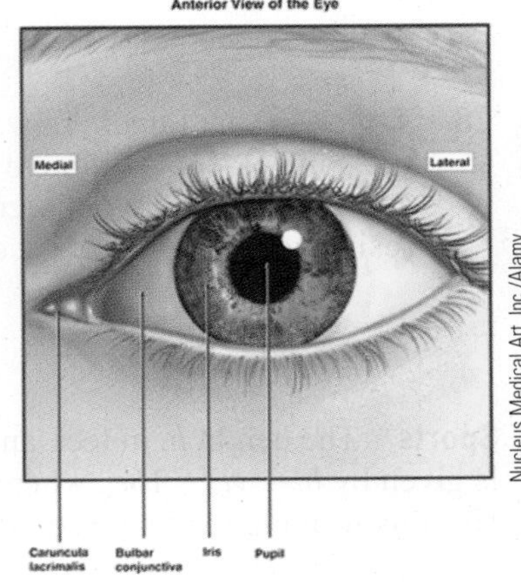

Nucleus Medical Art, Inc./Alamy

88. Find the width of the iris if the area of the iris is 20π mm².

89. Find the width of the iris if the area of the iris is 27π mm².

Applying the Concepts

90. Find $3n^2$ if $n(n + 5) = -4$.

91. Find $2n^2$ if $n(n + 3) = 4$.

For Exercises 92 to 95, solve.

92. $2y(y + 4) = -5(y + 3)$

93. $(b + 5)^2 = 16$

94. $p^3 = 9p^2$

95. $(x + 3)(2x - 1) = (3 - x)(5 - 3x)$

96. Explain the error made in solving the equation at the right. Solve the equation correctly.

$$(x + 2)(x - 3) = 6$$
$$x + 2 = 6 \qquad x - 3 = 6$$
$$x = 4 \qquad x = 9$$

97. Explain the error made in solving the equation at the right. Solve the equation correctly.

$$x^2 = x$$
$$\frac{x^2}{x} = \frac{x}{x}$$
$$x = 1$$

FOCUS ON PROBLEM SOLVING

Making a Table

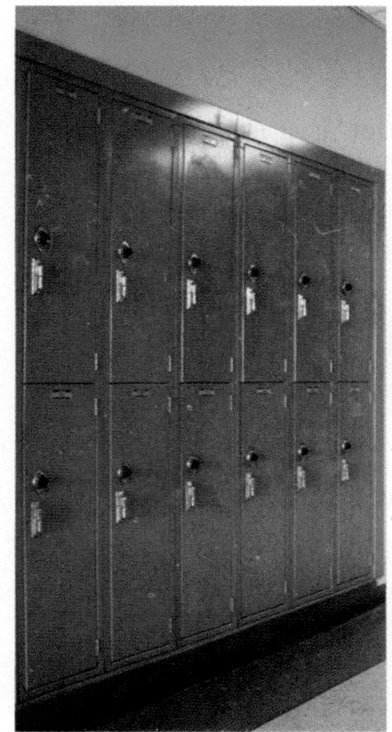

There are six students using a gym. The wall on the gym has six lockers that are numbered 1, 2, 3, 4, 5, and 6. After a practice, the first student goes by and opens all the lockers. The second student shuts every second locker, the third student changes every third locker (opens a locker if it is shut and shuts a locker if it is open), the fourth student changes every fourth locker, the fifth student changes every fifth locker, and the sixth student changes every sixth locker. After the sixth student makes changes, which lockers are open?

One method of solving this problem would be to create a table, as shown below.

Student / *Locker*	1	2	3	4	5	6
1	O	O	O	O	O	O
2	O	C	C	C	C	C
3	O	O	C	C	C	C
4	O	C	C	O	O	O
5	O	O	O	O	C	C
6	O	C	O	O	O	C

From this table, lockers 1 and 4 are open after the sixth student passes through.

Now extend this scenario to more lockers and students. In each case, the *n*th student changes multiples of the *n*th locker. For instance, the 8th student would change the 8th, 16th, 24th, . . .

1. Suppose there are 10 lockers and 10 students. Which lockers will remain open?
2. Suppose there are 16 lockers and 16 students. Which lockers will remain open?
3. Suppose there are 25 lockers and 25 students. Which lockers will remain open?
4. Suppose there are 40 lockers and 40 students. Which lockers will remain open?
5. Suppose there are 50 lockers and 50 students. Which lockers will remain open?
6. Make a conjecture as to which lockers would be open if there were 100 lockers and 100 students.
7. Give a reason why your conjecture should be true. [*Hint:* Consider how many factors there are for the door numbers that remain open and for those that remain closed. For instance, with 40 lockers and 40 students, locker 36 (which remains open) has factors 1, 2, 3, 4, 6, 9, 12, 18, and 36—an odd number of factors. Locker 28, a closed locker, has factors 1, 2, 4, 7, 14, and 28—an even number of factors.]

Find a Counterexample

When you are faced with an assertion, it may be that the assertion is false. For instance, consider the statement "Every prime number is an odd number." This assertion is false because the prime number 2 is an even number.

Finding an example that illustrates that an assertion is false is called finding a *counterexample.* The number 2 is a counterexample to the assertion that every prime number is an odd number.

If you are given an unfamiliar problem, one strategy to consider as a means of solving the problem is to try to find a counterexample. For each of the following problems, answer *true* if the assertion is always true. If the assertion is not true, answer *false* and give a counterexample. If there are terms used that you do not understand, consult a reference to find the meaning of the term.

1. If x is a real number, then x^2 is always positive.

2. The product of an odd integer and an even integer is an even integer.

3. If m is a positive integer, then $2m + 1$ is always a positive odd integer.

4. If $x < y$, then $x^2 < y^2$.

5. Given any three positive numbers a, b, and c, it is possible to construct a triangle whose sides have lengths a, b, and c.

6. The product of two irrational numbers is an irrational number.

7. If n is a positive integer greater than 2, then $1 \cdot 2 \cdot 3 \cdot 4 \cdot \cdots \cdot n + 1$ is a prime number.

8. Draw a polygon with more than three sides. Select two different points inside the polygon and join the points with a line segment. The line segment always lies completely inside the polygon.

9. Let A, B, and C be three points in the plane that are not collinear. Let d_1 be the distance from A to B, and let d_2 be the distance from A to C. Then the distance between B and C is less than $d_1 + d_2$.

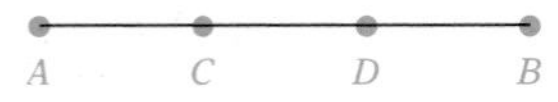

10. Consider the line segment AB shown at the left. Two points, C and D, are randomly selected on the line segment and three new segments are formed: AC, CD, and DB. The three new line segments can always be connected to form a triangle.

It may not be easy to establish that an assertion is true or to find a counterexample to the assertion. For instance, consider the assertion that every positive integer greater than 3 can be written as the sum of two primes. For example, $6 = 3 + 3$, $8 = 3 + 5$, $9 = 2 + 7$. Is this assertion always true? (*Note:* This assertion, called Goldbach's conjecture, has never been proved, nor has a counterexample been found!)

PROJECTS AND GROUP ACTIVITIES

Exploring Integers

Number theory is a branch of mathematics that focuses on integers and the relationships that exist among the integers. Some of the results from this field of study have important, practical applications for sending sensitive information such as credit card numbers over the Internet. In this project, you will be asked to discover some of those relationships.

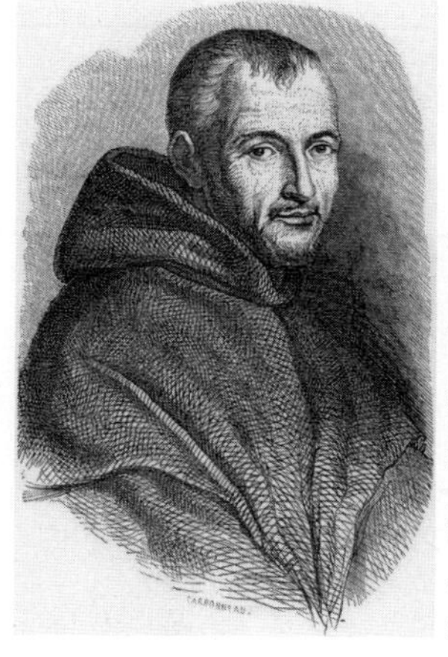

Mary Evans Picture Library/Alamy

Marin Mersenne (1588–1648) was a French mathematician, scientist, and philosopher known for his development of the Mersenne primes.

1. If n is an integer, explain why the product $n(n + 1)$ is always an even integer.
2. If n is an integer, explain why $2n$ is always an even integer.
3. If n is an integer, explain why $2n + 1$ is always an odd integer.
4. Select any odd integer greater than 1, square it, and then subtract 1. Try this for various odd integers greater than 1. Is the result always evenly divisible by 8?
5. Prove the assertion in Exercise 4. [*Suggestion:* From Exercise 3, an odd integer can be represented as $2n + 1$. Therefore, the assertion in Exercise 4 can be stated "$(2n + 1)^2 - 1$ is evenly divisible by 8." Expand this expression and explain why the result must be divisible by 8. You will need to use the result from Exercise 1.]
6. The integers 2 and 3 are consecutive prime numbers. Are there any other consecutive prime numbers? Why?
7. If n is a positive integer, for what values of n is $n^2 - 1$ a prime number?
8. A *Mersenne prime number* is a prime number that can be written in the form $2^n - 1$, where n is also a prime number. For instance, $2^5 - 1 = 32 - 1 = 31$. Because 5 and 31 are prime numbers, 31 is a Mersenne prime number. On the other hand, $2^{11} - 1 = 2048 - 1 = 2047$. In this case, although 11 is a prime number, $2047 = 23 \cdot 89$ and so is not a prime number. Find two Mersenne prime numbers other than 31.

CHAPTER 7 SUMMARY

KEY WORDS	EXAMPLES
The *greatest common factor (GCF) of two or more monomials* is the product of the GCF of the coefficients and the common variable factors. [7.1A, p. 404]	The GCF of $8x^2y$ and $12xyz$ is $4xy$.
To *factor a polynomial* means to write the polynomial as a product of other polynomials. [7.1A, p. 404]	To factor $x^2 + 3x + 2$ means to write it as the product $(x + 1)(x + 2)$.

A factor that has two terms is called a *binomial factor.* [7.1B, p. 406]

$(x + 1)$ is a binomial factor of $3x(x + 1)$.

A polynomial that does not factor using only integers is *nonfactorable over the integers.* [7.2A, p. 411]

The trinomial $x^2 + x + 4$ is nonfactorable over the integers. There are no integers whose product is 4 and whose sum is 1.

A polynomial is *factored completely* if it is written as a product of factors that are nonfactorable over the integers. [7.2B, p. 412]

The polynomial $3y^3 + 9y^2 - 12y$ is factored completely as $3y(y + 4)(y - 1)$.

The product of a term and itself is called a *perfect square*. The *square root* of a perfect square is one of the two equal factors of the perfect square. [7.4A, p. 426]

$(5x)(5x) = 25x^2$; $25x^2$ is a perfect square.
$\sqrt{25x^2} = 5x$

The product of the same three factors is called a *perfect cube.* The *cube root* of a perfect cube is one of the three equal factors of the perfect cube. [7.4B, p. 428]

$(2x)(2x)(2x) = 8x^3$; $8x^3$ is a perfect cube.
$\sqrt[3]{8x^3} = 2x$

A trinomial is *quadratic in form* if it can be written as $au^2 + bu + c$. [7.4C, p. 430]

$6x^4 - 5x^2 - 4$
$6(x^2)^2 - 5(x^2) - 4$
$6u^2 - 5u - 4$

An equation that can be written in the form $ax^2 + bx + c = 0$, $a \neq 0$, is a *quadratic equation.* A quadratic equation is in *standard form* when the polynomial is written in descending order and equal to zero. [7.5A, p. 436]

The equation $2x^2 - 3x + 7 = 0$ is a quadratic equation in standard form.

ESSENTIAL RULES AND PROCEDURES — EXAMPLES

Factoring $x^2 + bx + c$: IMPORTANT RELATIONSHIPS
[7.2A, p. 410]

1. When the constant term of the trinomial is positive, the constant terms of the binomials have the same sign. They are both positive when the coefficient of the x term in the trinomial is positive. They are both negative when the coefficient of the x term in the trinomial is negative.

$x^2 + 6x + 8 = (x + 4)(x + 2)$

$x^2 - 6x + 5 = (x - 5)(x - 1)$

2. When the constant term of the trinomial is negative, the constant terms of the binomials have opposite signs.

$x^2 - 4x - 21 = (x + 3)(x - 7)$

3. In the trinomial, the coefficient of x is the sum of the constant terms of the binomials.

In the three examples above, note that $6 = 4 + 2$, $-6 = -5 + (-1)$, and $-4 = 3 + (-7)$.

4. In the trinomial, the constant term is the product of the constant terms of the binomials.

In the three examples above, note that $8 = 4 \cdot 2$, $5 = -5(-1)$, and $-21 = 3(-7)$.

Factoring by Grouping [7.1B, p. 406]
A polynomial can be factored by grouping if its terms can be grouped and factored in such a way that a common binomial factor is found.

$$3a^2 - a - 15ab + 5b$$
$$= (3a^2 - a) - (15ab - 5b)$$
$$= a(3a - 1) - 5b(3a - 1)$$
$$= (3a - 1)(a - 5b)$$

To factor $ax^2 + bx + c$ by grouping [7.3B, p. 420]
First find two factors of $a \cdot c$ whose sum is b. Then use factoring by grouping to write the factorization of the trinomial.

$3x^2 - 11x - 20$
$a \cdot c = 3(-20) = -60$
The product of 4 and -15 is -60.
The sum of 4 and -15 is -11.
$$3x^2 + 4x - 15x - 20$$
$$= (3x^2 + 4x) - (15x + 20)$$
$$= x(3x + 4) - 5(3x + 4)$$
$$= (3x + 4)(x - 5)$$

Factoring the Difference of Two Perfect Squares [7.4A, p. 426]
The difference of two perfect squares factors as the sum and difference of the same terms.

$a^2 - b^2 = (a + b)(a - b)$

$$x^2 - 64 = (x + 8)(x - 8)$$
$$4x^2 - 81 = (2x)^2 - 9^2$$
$$= (2x + 9)(2x - 9)$$

Factoring a Perfect-Square Trinomial [7.4A, p. 426]
A perfect-square trinomial is the square of a binomial.

$a^2 + 2ab + b^2 = (a + b)^2$
$a^2 - 2ab + b^2 = (a - b)^2$

$$x^2 + 14x + 49 = (x + 7)^2$$
$$x^2 - 10x + 25 = (x - 5)^2$$

Factoring the Sum or Difference of Two Cubes [7.4B, p. 428]
$a^3 + b^3 = (a + b)(a^2 - ab + b^2)$
$a^3 - b^3 = (a - b)(a^2 + ab + b^2)$

$$x^3 + 64 = (x + 4)(x^2 - 4x + 16)$$
$$8b^3 - 1 = (2b - 1)(4b^2 + 2b + 1)$$

To Factor Completely [7.4D, p. 431]
When factoring a polynomial completely, ask the following questions about the polynomial.

1. Is there a common factor? If so, factor out the GCF.
2. If the polynomial is a binomial, is it the difference of two perfect squares, the sum of two perfect cubes, or the difference of two perfect cubes? If so, factor.
3. If the polynomial is a trinomial, is it a perfect-square trinomial or the product of two binomials? If so, factor.
4. Can the polynomial be factored by grouping? If so, factor.
5. Is each factor nonfactorable over the integers? If not, factor.

$$54x^3 - 6x = 6x(9x^2 - 1)$$
$$= 6x(3x + 1)(3x - 1)$$

Principle of Zero Products [7.5A, p. 436]
If the product of two factors is zero, then at least one of the factors must be zero.

If $ab = 0$, then $a = 0$ or $b = 0$.

$$(x - 4)(x + 2) = 0$$
$$x - 4 = 0 \qquad x + 2 = 0$$

CHAPTER 7

CONCEPT REVIEW

Test your knowledge of the concepts presented in this chapter. Answer each question. Then check your answers against the ones provided in the Answer Section.

1. What does the GCF have to do with factoring?

2. When factoring a polynomial, do the terms of the polynomial have to be like terms?

3. After factoring a polynomial, how do you check your answer?

4. How is the GCF used in factoring by grouping?

5. When is a polynomial nonfactorable over the integers?

6. What does it mean to factor a polynomial completely?

7. When factoring a trinomial of the form $x^2 + bx + c$, why do we begin by finding the possible factors of c?

8. What are trial factors?

9. What is the middle term of a trinomial?

10. What are the binomial factors of the difference of two perfect squares?

11. What is an example of a perfect-square trinomial?

12. To solve an equation using factoring, why must the equation be set equal to zero?

CHAPTER 7

REVIEW EXERCISES

1. Factor: $b^2 - 13b + 30$

2. Factor: $4x(x - 3) - 5(3 - x)$

3. Factor $2x^2 - 5x + 6$ by using trial factors.

4. Factor: $21x^4y^4 + 23x^2y^2 + 6$

5. Factor: $14y^9 - 49y^6 + 7y^3$

6. Factor: $y^2 + 5y - 36$

7. Factor $6x^2 - 29x + 28$ by using trial factors.

8. Factor: $12a^2b + 3ab^2$

9. Factor: $a^6 - 100$

10. Factor: $n^4 - 2n^3 - 3n^2$

11. Factor $12y^2 + 16y - 3$ by using trial factors.

12. Factor: $12b^3 - 58b^2 + 56b$

13. Factor: $9y^4 - 25z^2$

14. Factor: $c^2 + 8c + 12$

15. Factor $18a^2 - 3a - 10$ by grouping.

16. Solve: $4x^2 + 27x = 7$

17. Factor: $4x^3 - 20x^2 - 24x$

18. Factor: $64a^3 - 27b^3$

19. Factor $2a^2 - 19a - 60$ by grouping.

20. Solve: $(x + 1)(x - 5) = 16$

21. Factor: $21ax - 35bx - 10by + 6ay$

22. Factor: $36x^8 - 36x^4 + 5$

23. Factor: $10x^2 + 25x + 4xy + 10y$

24. Factor: $5x^2 - 5x - 30$

25. Factor: $3x^2 + 36x + 108$

26. Factor $3x^2 - 17x + 10$ by grouping.

27. **Sports** The length of the field in field hockey is 20 yd less than twice the width of the field. The area of the field in field hockey is 6000 yd^2. Find the length and width of the field.

28. **Image Projection** The size S of an image from a slide projector depends on the distance d of the screen from the projector and is given by $S = d^2$. Find the distance between the projector and the screen when the size of the picture is 400 ft^2.

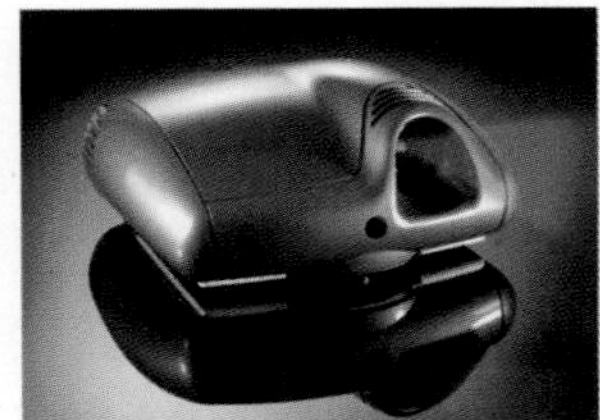
Tony Cordoza/Alamy

29. **Photography** A rectangular photograph has dimensions 15 in. by 12 in. A picture frame around the photograph increases the total area to 270 in^2. What is the width of the frame?

30. **Gardening** The length of each side of a square garden plot is extended 4 ft. The area of the resulting square is 576 ft^2. Find the length of a side of the original garden plot.

CHAPTER 7

TEST

1. Factor: $ab + 6a - 3b - 18$

2. Factor: $2y^4 - 14y^3 - 16y^2$

3. Factor $8x^2 + 20x - 48$ by grouping.

4. Factor $6x^2 + 19x + 8$ by using trial factors.

5. Factor: $a^2 - 19a + 48$

6. Factor: $6x^3 - 8x^2 + 10x$

7. Factor: $x^2 + 2x - 15$

8. Solve: $4x^2 - 1 = 0$

9. Factor: $5x^2 - 45x - 15$

10. Factor: $p^2 + 12p + 36$

11. Solve: $x(x - 8) = -15$

12. Factor: $3x^2 + 12xy + 12y^2$

13. Factor: $b^2 - 16$

14. Factor $6x^2y^2 + 9xy^2 + 3y^2$ by grouping.

15. Factor: $27x^3 - 8$

16. Factor: $6a^4 - 13a^2 - 5$

17. Factor: $x(p + 1) - (p + 1)$

18. Factor: $3a^2 - 75$

19. Factor $2x^2 + 4x - 5$ by using trial factors.

20. Factor: $x^2 - 9x - 36$

21. Factor: $4a^2 - 12ab + 9b^2$

22. Factor: $4x^2 - 49y^2$

23. Solve: $(2a - 3)(a + 7) = 0$

24. **Number Sense** The sum of two numbers is ten. The sum of the squares of the two numbers is fifty-eight. Find the two numbers.

25. **Geometry** The length of a rectangle is 3 cm longer than twice its width. The area of the rectangle is 90 cm². Find the length and width of the rectangle.

$2W + 3$

W

CUMULATIVE REVIEW EXERCISES

1. Subtract: $-2 - (-3) - 5 - (-11)$

2. Simplify: $(3 - 7)^2 \div (-2) - 3 \cdot (-4)$

3. Evaluate $-2a^2 \div (2b) - c$ when $a = -4$, $b = 2$, and $c = -1$.

4. Simplify: $-\frac{3}{4}(-20x^2)$

5. Simplify: $-2[4x - 2(3 - 2x) - 8x]$

6. Solve: $-\frac{5}{7}x = -\frac{10}{21}$

7. Solve: $3x - 2 = 12 - 5x$

8. Solve: $-2 + 4[3x - 2(4 - x) - 3] = 4x + 2$

9. 120% of what number is 54?

10. 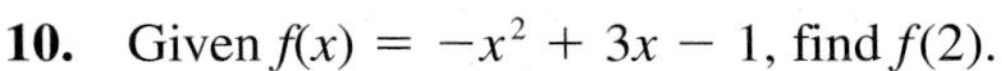
Given $f(x) = -x^2 + 3x - 1$, find $f(2)$.

11. 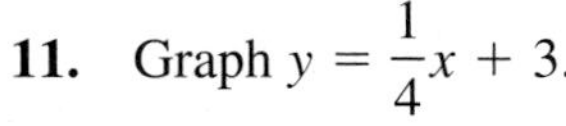
Graph $y = \frac{1}{4}x + 3$.

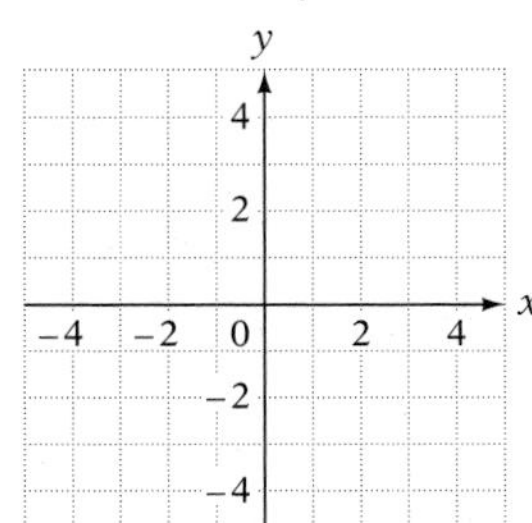

12. Graph $5x + 3y = 15$.

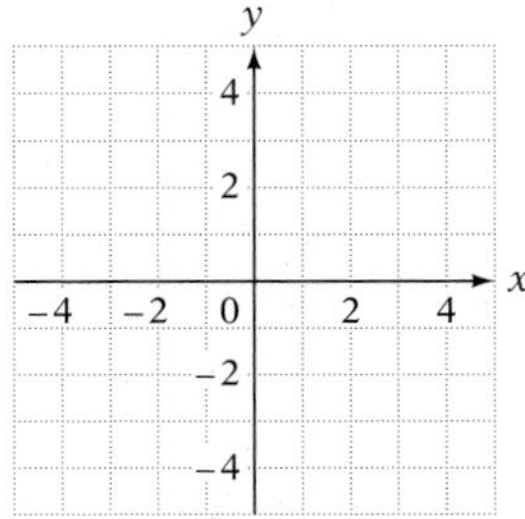

13. Find the equation of the line that contains the point $(-3, 4)$ and has slope $\frac{2}{3}$.

14. Solve by substitution: $8x - y = 2$
$y = 5x + 1$

15. Solve by the addition method:
$5x + 2y = -9$
$12x - 7y = 2$

16. Simplify: $(-3a^3b^2)^2$

17. Multiply: $(x + 2)(x^2 - 5x + 4)$

18. Divide: $(8x^2 + 4x - 3) \div (2x - 3)$

19. Simplify: $(x^{-4}y^3)^2$

20. Factor: $3a - 3b - ax + bx$

21. Factor: $15xy^2 - 20xy^4$

22. Factor: $x^2 - 5xy - 14y^2$

23. Solve: $3x^2 + 19x - 14 = 0$

24. Solve: $6x^2 + 60 = 39x$

25. **Geometry** A triangle has a 31° angle and a right angle. Find the measure of the third angle.

26. **Gardening** A rectangular flower garden has a perimeter of 86 ft. The length of the garden is 28 ft. What is the width of the garden?

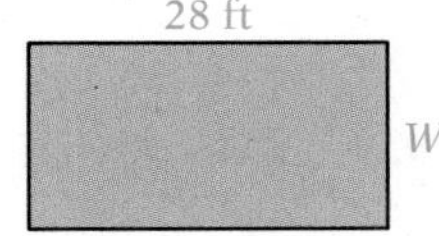

27. **Carpentry** A board 10 ft long is cut into two pieces. Four times the length of the shorter piece is 2 ft less than three times the length of the longer piece. Find the length of each piece.

28. **Investments** An investor has a total of $15,000 invested in two simple interest accounts. On one account, the annual simple interest rate is 4.5%. On the second account, the annual simple interest rate is 4%. How much is invested in the 4% account if the total annual interest earned is $635?

29. **Travel** A family drove to a resort at an average speed of 42 mph and later returned over the same road at an average speed of 56 mph. Find the distance to the resort if the total driving time was 7 h.

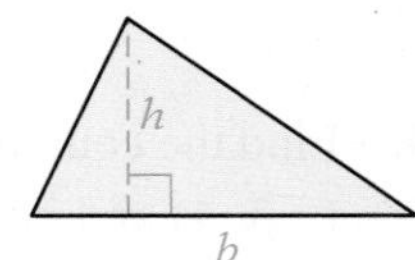

30. **Geometry** The length of the base of a triangle is three times the height. The area of the triangle is 24 in^2. Find the length of the base of the triangle.

CHAPTER 8

Rational Expressions

A.G.E. Foto Stock/First Light

OBJECTIVES

SECTION 8.1

A To simplify a rational expression
B To multiply rational expressions
C To divide rational expressions

SECTION 8.2

A To rewrite rational expressions in terms of a common denominator
B To add or subtract rational expressions

SECTION 8.3

A To simplify a complex fraction

SECTION 8.4

A To solve an equation containing fractions

SECTION 8.5

A To solve a proportion
B To solve application problems
C To solve problems involving similar triangles

SECTION 8.6

A To solve a literal equation for one of the variables

SECTION 8.7

A To solve work problems
B To use rational expressions to solve uniform motion problems

SECTION 8.8

A To solve variation problems

ARE YOU READY?

Take the Chapter 8 Prep Test to find out if you are ready to learn to:

- Simplify a rational expression
- Add, subtract, multiply, and divide rational expressions
- Solve an equation containing fractions
- Solve a proportion and use proportions to solve problems involving similar triangles
- Solve a literal equation for one of the variables
- Use rational expressions to solve work problems and uniform motion problems
- Solve variation problems

PREP TEST

Do these exercises to prepare for Chapter 8.

1. Find the LCM of 10 and 25.

For Exercises 2 to 5, add, subtract, multiply, or divide.

2. $-\frac{3}{8} \cdot \frac{4}{9}$

3. $-\frac{4}{5} \div \frac{8}{15}$

4. $-\frac{5}{6} + \frac{7}{8}$

5. $-\frac{3}{8} - \left(-\frac{7}{12}\right)$

6. Evaluate $\frac{2x - 3}{x^2 - x + 1}$ for $x = 2$.

7. Solve: $4(2x + 1) = 3(x - 2)$.

8. Solve: $10\left(\frac{t}{2} + \frac{t}{5}\right) = 10(1)$

9. **Travel** Two planes start from the same point and fly in opposite directions. The first plane is flying 20 mph slower than the second plane. In 2 h, the planes are 480 mi apart. Find the rate of each plane.

SECTION

8.1 Multiplication and Division of Rational Expressions

OBJECTIVE A To simplify a rational expression 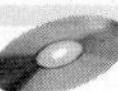

A fraction in which the numerator and denominator are polynomials is called a **rational expression.** Examples of rational expressions are shown at the right.

$$\frac{5}{z}, \quad \frac{x^2+1}{2x-1}, \quad \frac{y^2+y-1}{4y^2+1}$$

Care must be exercised with a rational expression to ensure that when the variables are replaced with numbers, the resulting denominator is not zero. Consider the rational expression at the right. The value of x cannot be 3 because the denominator would then be zero.

$$\frac{4x^2-9}{2x-6}$$

$$\frac{4(3)^2-9}{2(3)-6} = \frac{27}{0} \quad \text{Not a real number}$$

In the **simplest form of a rational expression,** the numerator and denominator have no common factors. The Multiplication Property of One is used to write a rational expression in simplest form.

HOW TO 1 Simplify: $\frac{x^2-4}{x^2-2x-8}$

$$\frac{x^2-4}{x^2-2x-8} = \frac{(x-2)(x+2)}{(x-4)(x+2)}$$

• Factor the numerator and denominator.

$$= \frac{x-2}{x-4} \cdot \frac{x+2}{x+2} = \frac{x-2}{x-4} \cdot 1$$

$$= \frac{x-2}{x-4}, x \neq -2, 4$$

• The restrictions, $x \neq -2$ or 4, are necessary to prevent division by zero.

This simplification is usually shown with slashes through the common factors:

$$\frac{x^2-4}{x^2-2x-8} = \frac{(x-2)\overset{1}{\cancel{(x+2)}}}{(x-4)\underset{1}{\cancel{(x+2)}}}$$

• Factor the numerator and denominator.

$$= \frac{x-2}{x-4}, x \neq -2, 4$$

• Divide by the common factors. The restrictions, $x \neq -2$ or 4, are necessary to prevent division by zero.

In summary, to simplify a rational expression, factor the numerator and denominator. Then divide the numerator and denominator by the common factors.

HOW TO 2 Simplify: $\frac{10+3x-x^2}{x^2-4x-5}$

$$\frac{10+3x-x^2}{x^2-4x-5} = \frac{-(x^2-3x-10)}{x^2-4x-5}$$

• Because the coefficient of x^2 in the numerator is -1, factor -1 from the numerator.

$$= \frac{-\overset{1}{\cancel{(x-5)}}(x+2)}{\underset{1}{\cancel{(x-5)}}(x+1)}$$

• Factor the numerator and denominator. Divide by the common factors.

$$= -\frac{x+2}{x+1}, x \neq -1, 5$$

For the remaining examples, we will omit the restrictions on the variables that prevent division by zero and assume the values of the variables are such that division by zero is not possible.

EXAMPLE 1

Simplify: $\dfrac{4x^3y^4}{6x^4y}$

Solution

$\dfrac{4x^3y^4}{6x^4y} = \dfrac{2y^3}{3x}$ • Use the rules of exponents.

YOU TRY IT 1

Simplify: $\dfrac{6x^5y}{12x^2y^3}$

Your solution

EXAMPLE 2

Simplify: $\dfrac{x^2 + 2x - 15}{x^2 - 7x + 12}$

Solution

$\dfrac{x^2 + 2x - 15}{x^2 - 7x + 12} = \dfrac{(x + 5)\overset{1}{\cancel{(x - 3)}}}{\underset{1}{\cancel{(x - 3)}}(x - 4)} = \dfrac{x + 5}{x - 4}$

YOU TRY IT 2

Simplify: $\dfrac{x^2 + 4x - 12}{x^2 - 3x + 2}$

Your solution

EXAMPLE 3

Simplify: $\dfrac{9 - x^2}{x^2 + x - 12}$

Solution

$\dfrac{9 - x^2}{x^2 + x - 12} = \dfrac{\overset{-1}{\cancel{(3 - x)}}(3 + x)}{\underset{1}{\cancel{(x - 3)}}(x + 4)}$ • $\dfrac{3 - x}{x - 3} = \dfrac{-1(x - 3)}{x - 3} = -1$

$= -\dfrac{x + 3}{x + 4}$

YOU TRY IT 3

Simplify: $\dfrac{x^2 + 2x - 24}{16 - x^2}$

Your solution

Solutions on pp. S22–S23

OBJECTIVE B To multiply rational expressions 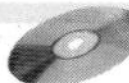

The product of two fractions is a fraction whose numerator is the product of the numerators of the two fractions and whose denominator is the product of the denominators of the two fractions.

Multiplying Rational Expressions

Multiply the numerators.
Multiply the denominators.

$$\frac{a}{b} \cdot \frac{c}{d} = \frac{ac}{bd}$$

$$\frac{2}{3} \cdot \frac{4}{5} = \frac{8}{15} \qquad \frac{3x}{y} \cdot \frac{2}{z} = \frac{6x}{yz} \qquad \frac{x + 2}{x} \cdot \frac{3}{x - 2} = \frac{3x + 6}{x^2 - 2x}$$

HOW TO • 3 Multiply: $\dfrac{x^2 + 3x}{x^2 - 3x - 4} \cdot \dfrac{x^2 - 5x + 4}{x^2 + 2x - 3}$

$$\frac{x^2 + 3x}{x^2 - 3x - 4} \cdot \frac{x^2 - 5x + 4}{x^2 + 2x - 3}$$

$$= \frac{x(x + 3)}{(x - 4)(x + 1)} \cdot \frac{(x - 4)(x - 1)}{(x + 3)(x - 1)}$$

• Factor the numerator and denominator of each fraction.

$$= \frac{x\overset{1}{\cancel{(x + 3)}}\overset{1}{\cancel{(x - 4)}}\overset{1}{\cancel{(x - 1)}}}{\underset{1}{\cancel{(x - 4)}}(x + 1)\underset{1}{\cancel{(x + 3)}}\underset{1}{\cancel{(x - 1)}}}$$

• Multiply. Then divide by the common factors.

$$= \frac{x}{x + 1}$$

• Write the answer in simplest form.

EXAMPLE • 4

Multiply: $\dfrac{10x^2 - 15x}{12x - 8} \cdot \dfrac{3x - 2}{20x - 25}$

Solution

$$\frac{10x^2 - 15x}{12x - 8} \cdot \frac{3x - 2}{20x - 25}$$

$$= \frac{5x(2x - 3)}{4(3x - 2)} \cdot \frac{(3x - 2)}{5(4x - 5)}$$

• Factor.

$$= \frac{\overset{1}{\cancel{5}}x(2x - 3)\overset{1}{\cancel{(3x - 2)}}}{4\underset{1}{\cancel{(3x - 2)}}\underset{1}{\cancel{5}}(4x - 5)}$$

• Divide by the common factors.

$$= \frac{x(2x - 3)}{4(4x - 5)}$$

YOU TRY IT • 4

Multiply: $\dfrac{12x^2 + 3x}{10x - 15} \cdot \dfrac{8x - 12}{9x + 18}$

Your solution

EXAMPLE • 5

Multiply: $\dfrac{x^2 + x - 6}{x^2 + 7x + 12} \cdot \dfrac{x^2 + 3x - 4}{4 - x^2}$

Solution

$$\frac{x^2 + x - 6}{x^2 + 7x + 12} \cdot \frac{x^2 + 3x - 4}{4 - x^2}$$

$$= \frac{(x + 3)(x - 2)}{(x + 3)(x + 4)} \cdot \frac{(x + 4)(x - 1)}{(2 - x)(2 + x)}$$

• Factor.

$$= \frac{\overset{1}{\cancel{(x + 3)}}\overset{-1}{\cancel{(x - 2)}}\overset{1}{\cancel{(x + 4)}}(x - 1)}{\underset{1}{\cancel{(x + 3)}}\underset{1}{\cancel{(x + 4)}}\underset{1}{\cancel{(2 - x)}}(2 + x)}$$

• Divide by the common factors.

$$= -\frac{x - 1}{x + 2}$$

YOU TRY IT • 5

Multiply: $\dfrac{x^2 + 2x - 15}{9 - x^2} \cdot \dfrac{x^2 - 3x - 18}{x^2 - 7x + 6}$

Your solution

Solutions on p. S23

OBJECTIVE C To divide rational expressions 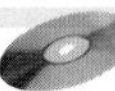

The **reciprocal of a rational expression** is the rational expression with the numerator and denominator interchanged.

Fraction	Reciprocal
$\frac{a}{b}$	$\frac{b}{a}$
$x^2 = \frac{x^2}{1}$	$\frac{1}{x^2}$
$\frac{x+2}{x}$	$\frac{x}{x+2}$

Dividing Rational Expressions

Multiply the dividend by the reciprocal of the divisor. $\frac{a}{b} \div \frac{c}{d} = \frac{a}{b} \cdot \frac{d}{c} = \frac{ad}{bc}$

$$\frac{4}{x} \div \frac{y}{5} = \frac{4}{x} \cdot \frac{5}{y} = \frac{20}{xy} \qquad \frac{x+4}{x} \div \frac{x-2}{4} = \frac{x+4}{x} \cdot \frac{4}{x-2} = \frac{4(x+4)}{x(x-2)}$$

The basis for the division rule is shown at the right.

$$\frac{a}{b} \div \frac{c}{d} = \frac{\frac{a}{b}}{\frac{c}{d}} = \frac{\frac{a}{b}}{\frac{c}{d}} \cdot \frac{\frac{d}{c}}{\frac{d}{c}} = \frac{\frac{a}{b} \cdot \frac{d}{c}}{1} = \frac{a}{b} \cdot \frac{d}{c}$$

EXAMPLE 6

Divide: $\frac{xy^2 - 3x^2y}{z^2} \div \frac{6x^2 - 2xy}{z^3}$

Solution

$$\frac{xy^2 - 3x^2y}{z^2} \div \frac{6x^2 - 2xy}{z^3}$$

$$= \frac{xy^2 - 3x^2y}{z^2} \cdot \frac{z^3}{6x^2 - 2xy}$$ • Multiply by the reciprocal.

$$= \frac{xy\overset{-1}{(y - 3x)} \cdot z^3}{z^2 \cdot 2x\underset{1}{(3x - y)}} = -\frac{yz}{2}$$

YOU TRY IT 6

Divide: $\frac{a^2}{4bc^2 - 2b^2c} \div \frac{a}{6bc - 3b^2}$

Your solution

EXAMPLE 7

Divide: $\frac{2x^2 + 5x + 2}{2x^2 + 3x - 2} \div \frac{3x^2 + 13x + 4}{2x^2 + 7x - 4}$

Solution

$$\frac{2x^2 + 5x + 2}{2x^2 + 3x - 2} \div \frac{3x^2 + 13x + 4}{2x^2 + 7x - 4}$$

$$= \frac{2x^2 + 5x + 2}{2x^2 + 3x - 2} \cdot \frac{2x^2 + 7x - 4}{3x^2 + 13x + 4}$$ • Multiply by the reciprocal.

$$= \frac{(2x+1)\overset{1}{(x+2)} \cdot \overset{1}{(2x-1)}\overset{1}{(x+4)}}{\underset{1}{(2x-1)}\underset{1}{(x+2)} \cdot (3x+1)\underset{1}{(x+4)}} = \frac{2x+1}{3x+1}$$

YOU TRY IT 7

Divide: $\frac{3x^2 + 26x + 16}{3x^2 - 7x - 6} \div \frac{2x^2 + 9x - 5}{x^2 + 2x - 15}$

Your solution

Solutions on p. S23

8.1 EXERCISES

OBJECTIVE A **To simplify a rational expression**

1. Explain the procedure for writing a rational expression in simplest form.

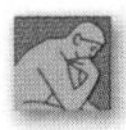

2. Why is the simplification at the right incorrect? $\frac{x+3}{x} = \frac{\overset{1}{\cancel{x}}+3}{\underset{1}{\cancel{x}}} = 4$

For Exercises 3 to 30, simplify.

3. $\frac{9x^3}{12x^4}$

4. $\frac{16x^2y}{24xy^3}$

5. $\frac{(x+3)^2}{(x+3)^3}$

6. $\frac{(2x-1)^5}{(2x-1)^4}$

7. $\frac{3n-4}{4-3n}$

8. $\frac{5-2x}{2x-5}$

9. $\frac{6y(y+2)}{9y^2(y+2)}$

10. $\frac{12x^2(3-x)}{18x(3-x)}$

11. $\frac{6x(x-5)}{8x^2(5-x)}$

12. $\frac{14x^3(7-3x)}{21x(3x-7)}$

13. $\frac{a^2+4a}{ab+4b}$

14. $\frac{x^2-3x}{2x-6}$

15. $\frac{4-6x}{3x^2-2x}$

16. $\frac{5xy-3y}{9-15x}$

17. $\frac{y^2-3y+2}{y^2-4y+3}$

18. $\frac{x^2+5x+6}{x^2+8x+15}$

19. $\frac{x^2+3x-10}{x^2+2x-8}$

20. $\frac{a^2+7a-8}{a^2+6a-7}$

21. $\frac{x^2+x-12}{x^2-6x+9}$

22. $\frac{x^2+8x+16}{x^2-2x-24}$

23. $\frac{x^2-3x-10}{25-x^2}$

24. $\frac{4-y^2}{y^2-3y-10}$

25. $\frac{2x^3+2x^2-4x}{x^3+2x^2-3x}$

26. $\frac{3x^3-12x}{6x^3-24x^2+24x}$

27. $\frac{6x^2-7x+2}{6x^2+5x-6}$

28. $\frac{2n^2-9n+4}{2n^2-5n-12}$

29. $\frac{x^2+3x-28}{24-2x-x^2}$

30. $\frac{x^2+7x-8}{1+x-2x^2}$

OBJECTIVE B To multiply rational expressions

For Exercises 31 to 54, multiply.

31. $\dfrac{8x^2}{9y^3} \cdot \dfrac{3y^2}{4x^3}$

32. $\dfrac{14a^2b^3}{15x^5y^2} \cdot \dfrac{25x^3y}{16ab}$

33. $\dfrac{12x^3y^4}{7a^2b^3} \cdot \dfrac{14a^3b^4}{9x^2y^2}$

34. $\dfrac{18a^4b^2}{25x^2y^3} \cdot \dfrac{50x^5y^6}{27a^6b^2}$

35. $\dfrac{3x - 6}{5x - 20} \cdot \dfrac{10x - 40}{27x - 54}$

36. $\dfrac{8x - 12}{14x + 7} \cdot \dfrac{42x + 21}{32x - 48}$

37. $\dfrac{3x^2 + 2x}{2xy - 3y} \cdot \dfrac{2xy^3 - 3y^3}{3x^3 + 2x^2}$

38. $\dfrac{4a^2x - 3a^2}{2by + 5b} \cdot \dfrac{2b^3y + 5b^3}{4ax - 3a}$

39. $\dfrac{x^2 + 5x + 4}{x^3y^2} \cdot \dfrac{x^2y^3}{x^2 + 2x + 1}$

40. $\dfrac{x^2 + x - 2}{xy^2} \cdot \dfrac{x^3y}{x^2 + 5x + 6}$

41. $\dfrac{x^4y^2}{x^2 + 3x - 28} \cdot \dfrac{x^2 - 49}{xy^4}$

42. $\dfrac{x^5y^3}{x^2 + 13x + 30} \cdot \dfrac{x^2 + 2x - 3}{x^7y^2}$

43. $\dfrac{2x^2 - 5x}{2xy + y} \cdot \dfrac{2xy^2 + y^2}{5x^2 - 2x^3}$

44. $\dfrac{3a^3 + 4a^2}{5ab - 3b} \cdot \dfrac{3b^3 - 5ab^3}{3a^2 + 4a}$

45. $\dfrac{x^2 - 2x - 24}{x^2 - 5x - 6} \cdot \dfrac{x^2 + 5x + 6}{x^2 + 6x + 8}$

46. $\dfrac{x^2 - 8x + 7}{x^2 + 3x - 4} \cdot \dfrac{x^2 + 3x - 10}{x^2 - 9x + 14}$

47. $\dfrac{x^2 + 2x - 35}{x^2 + 4x - 21} \cdot \dfrac{x^2 + 3x - 18}{x^2 + 9x + 18}$

48. $\dfrac{y^2 + y - 20}{y^2 + 2y - 15} \cdot \dfrac{y^2 + 4y - 21}{y^2 + 3y - 28}$

49. $\dfrac{x^2 - 3x - 4}{x^2 + 6x + 5} \cdot \dfrac{x^2 + 5x + 6}{8 + 2x - x^2}$

50. $\dfrac{25 - n^2}{n^2 - 2n - 35} \cdot \dfrac{n^2 - 8n - 20}{n^2 - 3n - 10}$

51. $\dfrac{16 + 6x - x^2}{x^2 - 10x - 24} \cdot \dfrac{x^2 - 6x - 27}{x^2 - 17x + 72}$

52. $\dfrac{x^2 - 11x + 28}{x^2 - 13x + 42} \cdot \dfrac{x^2 + 7x + 10}{20 - x - x^2}$

53. $\dfrac{2x^2 + 5x + 2}{2x^2 + 7x + 3} \cdot \dfrac{x^2 - 7x - 30}{x^2 - 6x - 40}$

54. $\dfrac{x^2 - 4x - 32}{x^2 - 8x - 48} \cdot \dfrac{3x^2 + 17x + 10}{3x^2 - 22x - 16}$

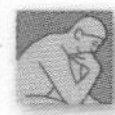

For Exercises 55 to 57, use the product $\dfrac{x^a}{y^b} \cdot \dfrac{y^c}{x^d}$, where a, b, c, and d are all positive integers.

55. If $a > d$ and $c > b$, what is the denominator of the simplified product?

56. If $a > d$ and $b > c$, which variable appears in the denominator of the simplified product?

57. If $a < d$ and $b = c$, what is the numerator of the simplified product?

OBJECTIVE C **To divide rational expressions**

58. What is the reciprocal of a rational expression?

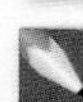

59. Explain how to divide rational expressions.

For Exercises 60 to 79, divide.

60. $\dfrac{4x^2y^3}{15a^2b^3} \div \dfrac{6xy}{5a^3b^5}$

61. $\dfrac{9x^3y^4}{16a^4b^2} \div \dfrac{45x^4y^2}{14a^7b}$

62. $\dfrac{6x - 12}{8x + 32} \div \dfrac{18x - 36}{10x + 40}$

63. $\dfrac{28x + 14}{45x - 30} \div \dfrac{14x + 7}{30x - 20}$

64. $\dfrac{6x^3 + 7x^2}{12x - 3} \div \dfrac{6x^2 + 7x}{36x - 9}$

65. $\dfrac{5a^2y + 3a^2}{2x^3 + 5x^2} \div \dfrac{10ay + 6a}{6x^3 + 15x^2}$

66. $\dfrac{x^2 + 4x + 3}{x^2y} \div \dfrac{x^2 + 2x + 1}{xy^2}$

67. $\dfrac{x^3y^2}{x^2 - 3x - 10} \div \dfrac{xy^4}{x^2 - x - 20}$

68. $\dfrac{x^2 - 49}{x^4y^3} \div \dfrac{x^2 - 14x + 49}{x^4y^3}$

69. $\dfrac{x^2y^5}{x^2 - 11x + 30} \div \dfrac{xy^6}{x^2 - 7x + 10}$

70. $\dfrac{4ax - 8a}{c^2} \div \dfrac{2y - xy}{c^3}$

71. $\dfrac{3x^2y - 9xy}{a^2b} \div \dfrac{3x^2 - x^3}{ab^2}$

72. $\dfrac{x^2 - 5x + 6}{x^2 - 9x + 18} \div \dfrac{x^2 - 6x + 8}{x^2 - 9x + 20}$

73. $\dfrac{x^2 + 3x - 40}{x^2 + 2x - 35} \div \dfrac{x^2 + 2x - 48}{x^2 + 3x - 18}$

74. $\dfrac{x^2 + 2x - 15}{x^2 - 4x - 45} \div \dfrac{x^2 + x - 12}{x^2 - 5x - 36}$

75. $\dfrac{y^2 - y - 56}{y^2 + 8y + 7} \div \dfrac{y^2 - 13y + 40}{y^2 - 4y - 5}$

76. $\dfrac{8 + 2x - x^2}{x^2 + 7x + 10} \div \dfrac{x^2 - 11x + 28}{x^2 - x - 42}$

77. $\dfrac{x^2 - x - 2}{x^2 - 7x + 10} \div \dfrac{x^2 - 3x - 4}{40 - 3x - x^2}$

78. $\dfrac{2x^2 - 3x - 20}{2x^2 - 7x - 30} \div \dfrac{2x^2 - 5x - 12}{4x^2 + 12x + 9}$

79. $\dfrac{6n^2 + 13n + 6}{4n^2 - 9} \div \dfrac{6n^2 + n - 2}{4n^2 - 1}$

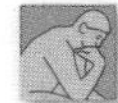

For Exercises 80 to 83, state whether the given division is equivalent to $\frac{x^2 - 3x - 4}{x^2 + 5x - 6}$.

80. $\dfrac{x - 4}{x + 6} \div \dfrac{x - 1}{x + 1}$

81. $\dfrac{x + 1}{x + 6} \div \dfrac{x - 1}{x - 4}$

82. $\dfrac{x + 1}{x - 1} \div \dfrac{x + 6}{x - 4}$

83. $\dfrac{x - 1}{x + 1} \div \dfrac{x - 4}{x + 6}$

Applying the Concepts

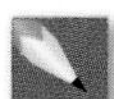

84. Given the expression $\frac{9}{x^2 + 1}$, choose some values of x and evaluate the expression for those values. Is it possible to choose a value of x for which the value of the expression is greater than 10? If so, what is that value of x? If not, explain why it is not possible.

Geometry For Exercises 85 and 86, write in simplest form the ratio of the shaded area of the figure to the total area of the figure.

85.

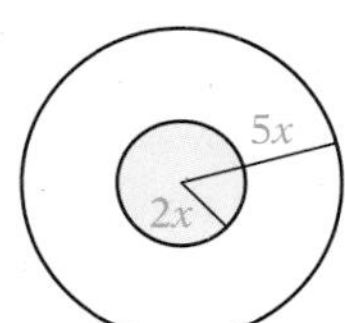

86.

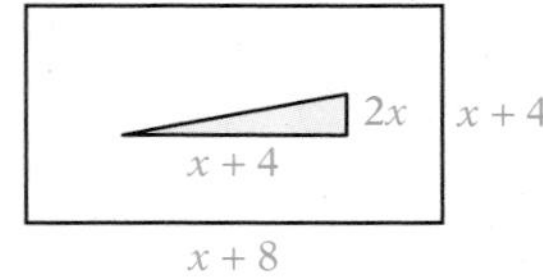

SECTION

8.2 Addition and Subtraction of Rational Expressions

OBJECTIVE A To rewrite rational expressions in terms of a common denominator

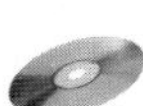

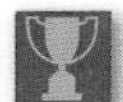

Tips for Success

As you know, often in mathematics you learn one skill in order to perform another. This is true of this objective. You are learning to rewrite rational expressions in terms of a common denominator in order to add and subtract rational expressions in the next objective. To ensure success, be certain you understand this lesson before studying the next.

In adding or subtracting rational expressions, it is frequently necessary to express the rational expressions in terms of a common denominator. This common denominator is the least common multiple (LCM) of the denominators.

The **least common multiple (LCM) of two or more polynomials** is the simplest polynomial that contains the factors of each polynomial. To find the LCM, first factor each polynomial completely. The LCM is the product of each factor the greatest number of times it occurs in any one factorization.

HOW TO 1 Find the LCM of $3x^2 + 15x$ and $6x^4 + 24x^3 - 30x^2$.

Factor each polynomial.

$$3x^2 + 15x = 3x(x + 5)$$
$$6x^4 + 24x^3 - 30x^2 = 6x^2(x^2 + 4x - 5) = 6x^2(x - 1)(x + 5)$$

The LCM is the product of the LCM of the numerical coefficients and each variable factor the greatest number of times it occurs in any one factorization.

LCM $= 6x^2(x - 1)(x + 5)$

HOW TO 2 Write the fractions $\frac{5}{6x^2y}$ and $\frac{a}{9xy^3}$ in terms of the LCM of the denominators.

The LCM is $18x^2y^3$.

• Find the LCM of the denominators.

$$\frac{5}{6x^2y} = \frac{5}{6x^2y} \cdot \frac{3y^2}{3y^2} = \frac{15y^2}{18x^2y^3}$$

$$\frac{a}{9xy^3} = \frac{a}{9xy^3} \cdot \frac{2x}{2x} = \frac{2ax}{18x^2y^3}$$

• For each fraction, multiply the numerator and denominator by the factor whose product with the denominator is the LCM.

HOW TO 3 Write the fractions $\frac{x+2}{x^2-2x}$ and $\frac{5x}{3x-6}$ in terms of the LCM of the denominators.

The LCM is $3x(x - 2)$.

• Find the LCM of the denominators.

$$\frac{x+2}{x^2-2x} = \frac{x+2}{x(x-2)} \cdot \frac{3}{3} = \frac{3x+6}{3x(x-2)}$$

$$\frac{5x}{3x-6} = \frac{5x}{3(x-2)} \cdot \frac{x}{x} = \frac{5x^2}{3x(x-2)}$$

• For each fraction, multiply the numerator and denominator by the factor whose product with the denominator is the LCM.

Take Note

$x^2 - 2x = x(x - 2)$
$3x - 6 = 3(x - 2)$
The LCM of $x(x - 2)$ and $3(x - 2)$ is $3x(x - 2)$.

EXAMPLE • 1

Write the fractions $\frac{3x}{x-1}$ and $\frac{4}{2x+5}$ in terms of the LCM of the denominators.

Solution

The LCM of $x - 1$ and $2x + 5$ is $(x - 1)(2x + 5)$.

$$\frac{3x}{x-1} = \frac{3x}{x-1} \cdot \frac{2x+5}{2x+5} = \frac{6x^2 + 15x}{(x-1)(2x+5)}$$

$$\frac{4}{2x+5} = \frac{4}{2x+5} \cdot \frac{x-1}{x-1} = \frac{4x-4}{(x-1)(2x+5)}$$

YOU TRY IT • 1

Write the fractions $\frac{2x}{2x-5}$ and $\frac{3}{x+4}$ in terms of the LCM of the denominators.

Your solution

EXAMPLE • 2

Write the fractions $\frac{2a-3}{a^2-2a}$ and $\frac{a+1}{2a^2-a-6}$ in terms of the LCM of the denominators.

Solution

$a^2 - 2a = a(a - 2)$;
$2a^2 - a - 6 = (2a + 3)(a - 2)$

The LCM is $a(a - 2)(2a + 3)$.

$$\frac{2a-3}{a^2-2a} = \frac{2a-3}{a(a-2)} \cdot \frac{2a+3}{2a+3} = \frac{4a^2-9}{a(a-2)(2a+3)}$$

$$\frac{a+1}{2a^2-a-6} = \frac{a+1}{(a-2)(2a+3)} \cdot \frac{a}{a} = \frac{a^2+a}{a(a-2)(2a+3)}$$

YOU TRY IT • 2

Write the fractions $\frac{3x}{2x^2-11x+15}$ and $\frac{x-2}{x^2-3x}$ in terms of the LCM of the denominators.

Your solution

EXAMPLE • 3

Write the fractions $\frac{2x-3}{3x-x^2}$ and $\frac{3x}{x^2-4x+3}$ in terms of the LCM of the denominators.

Solution

$3x - x^2 = x(3 - x) = -x(x - 3)$;
$x^2 - 4x + 3 = (x - 3)(x - 1)$

The LCM is $x(x - 3)(x - 1)$.

$$\frac{2x-3}{3x-x^2} = -\frac{2x-3}{x(x-3)} \cdot \frac{x-1}{x-1} = -\frac{2x^2-5x+3}{x(x-3)(x-1)}$$

$$\frac{3x}{x^2-4x+3} = \frac{3x}{(x-3)(x-1)} \cdot \frac{x}{x} = \frac{3x^2}{x(x-3)(x-1)}$$

YOU TRY IT • 3

Write the fractions $\frac{2x-7}{2x-x^2}$ and $\frac{3x-2}{3x^2-5x-2}$ in terms of the LCM of the denominators.

Your solution

Solutions on p. S23

OBJECTIVE B To add or subtract rational expressions 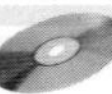

When adding rational expressions in which the denominators are the same, add the numerators. The denominator of the sum is the common denominator.

$$\frac{4x}{15} + \frac{8x}{15} = \frac{4x + 8x}{15} = \frac{12x}{15} = \frac{4x}{5}$$ • Note that the sum is written in simplest form.

$$\frac{a}{a^2 - b^2} + \frac{b}{a^2 - b^2} = \frac{a + b}{a^2 - b^2} = \frac{a + b}{(a - b)(a + b)} = \frac{\overset{1}{\cancel{(a + b)}}}{(a - b)\underset{1}{\cancel{(a + b)}}} = \frac{1}{a - b}$$

When subtracting rational expressions in which the denominators are the same, subtract the numerators. The denominator of the difference is the common denominator. Write the answer in simplest form.

$$\frac{7x - 12}{2x^2 + 5x - 12} - \frac{3x - 6}{2x^2 + 5x - 12} = \frac{(7x - 12) - (3x - 6)}{2x^2 + 5x - 12} = \frac{4x - 6}{2x^2 + 5x - 12}$$

$$= \frac{2\overset{1}{\cancel{(2x - 3)}}}{\underset{1}{\cancel{(2x - 3)}}(x + 4)} = \frac{2}{x + 4}$$

Before two rational expressions with *different* denominators can be added or subtracted, each rational expression must be expressed in terms of a common denominator. This common denominator is the LCM of the denominators of the rational expressions.

✓ Take Note

Note the steps involved in adding or subtracting rational expressions with different denominators:

1. Find the LCM of the denominators.
2. Rewrite each fraction in terms of the common denominator.
3. Add or subtract the rational expressions.
4. Simplify the resulting sum or difference.

HOW TO • 4 Simplify: $\frac{x}{x - 3} - \frac{x + 1}{x - 2}$

The LCM is $(x - 3)(x - 2)$. • Find the LCM of the denominators.

$$\frac{x}{x - 3} - \frac{x + 1}{x - 2} = \frac{x}{x - 3} \cdot \frac{x - 2}{x - 2} - \frac{x + 1}{x - 2} \cdot \frac{x - 3}{x - 3}$$ • Express each fraction in terms of the LCM.

$$= \frac{x(x - 2) - (x + 1)(x - 3)}{(x - 3)(x - 2)}$$ • Subtract the fractions.

$$= \frac{(x^2 - 2x) - (x^2 - 2x - 3)}{(x - 3)(x - 2)}$$

$$= \frac{3}{(x - 3)(x - 2)}$$ • Simplify.

HOW TO • 5 Simplify: $\frac{3x}{2x - 3} + \frac{3x + 6}{2x^2 + x - 6}$

The LCM of $2x - 3$ and $2x^2 + x - 6$ is $(2x - 3)(x + 2)$. • Find the LCM of the denominators.

$$\frac{3x}{2x - 3} + \frac{3x + 6}{2x^2 + x - 6} = \frac{3x}{2x - 3} \cdot \frac{x + 2}{x + 2} + \frac{3x + 6}{(2x - 3)(x + 2)}$$ • Express each fraction in terms of the LCM.

$$= \frac{3x(x + 2) + (3x + 6)}{(2x - 3)(x + 2)}$$ • Add the fractions.

$$= \frac{(3x^2 + 6x) + (3x + 6)}{(2x - 3)(x + 2)}$$

$$= \frac{3x^2 + 9x + 6}{(2x - 3)(x + 2)} = \frac{3\overset{1}{\cancel{(x + 2)}}(x + 1)}{(2x - 3)\underset{1}{\cancel{(x + 2)}}}$$ • Simplify.

$$= \frac{3(x + 1)}{2x - 3}$$

EXAMPLE 4

Simplify: $\frac{2}{x} - \frac{3}{x^2} + \frac{1}{xy}$

Solution

The LCM is x^2y.

$$\frac{2}{x} - \frac{3}{x^2} + \frac{1}{xy} = \frac{2}{x} \cdot \frac{xy}{xy} - \frac{3}{x^2} \cdot \frac{y}{y} + \frac{1}{xy} \cdot \frac{x}{x}$$
$$= \frac{2xy}{x^2y} - \frac{3y}{x^2y} + \frac{x}{x^2y}$$
$$= \frac{2xy - 3y + x}{x^2y}$$

YOU TRY IT 4

Simplify: $\frac{2}{b} - \frac{1}{a} + \frac{4}{ab}$

Your solution

EXAMPLE 5

Simplify: $\frac{x}{2x - 4} - \frac{4 - x}{x^2 - 2x}$

Solution

$2x - 4 = 2(x - 2)$; $x^2 - 2x = x(x - 2)$

The LCM is $2x(x - 2)$.

$$\frac{x}{2x - 4} - \frac{4 - x}{x^2 - 2x} = \frac{x}{2(x - 2)} \cdot \frac{x}{x} - \frac{4 - x}{x(x - 2)} \cdot \frac{2}{2}$$
$$= \frac{x^2 - (4 - x)2}{2x(x - 2)}$$
$$= \frac{x^2 - (8 - 2x)}{2x(x - 2)} = \frac{x^2 + 2x - 8}{2x(x - 2)}$$
$$= \frac{(x + 4)(x - 2)}{2x(x - 2)} = \frac{(x + 4)\overset{1}{\cancel{(x - 2)}}}{2x\underset{1}{\cancel{(x - 2)}}}$$
$$= \frac{x + 4}{2x}$$

YOU TRY IT 5

Simplify: $\frac{a - 3}{a^2 - 5a} + \frac{a - 9}{a^2 - 25}$

Your solution

EXAMPLE 6

Simplify: $\frac{x}{x + 1} - \frac{2}{x - 2} - \frac{3}{x^2 - x - 2}$

Solution

The LCM is $(x + 1)(x - 2)$.

$$\frac{x}{x + 1} - \frac{2}{x - 2} - \frac{3}{x^2 - x - 2}$$
$$= \frac{x}{x + 1} \cdot \frac{x - 2}{x - 2} - \frac{2}{x - 2} \cdot \frac{x + 1}{x + 1} - \frac{3}{(x + 1)(x - 2)}$$
$$= \frac{x(x - 2) - 2(x + 1) - 3}{(x + 1)(x - 2)}$$
$$= \frac{x^2 - 2x - 2x - 2 - 3}{(x + 1)(x - 2)}$$
$$= \frac{x^2 - 4x - 5}{(x + 1)(x - 2)} = \frac{\overset{1}{\cancel{(x + 1)}}(x - 5)}{\underset{1}{\cancel{(x + 1)}}(x - 2)} = \frac{x - 5}{x - 2}$$

YOU TRY IT 6

Simplify: $\frac{2x}{x - 4} - \frac{x - 1}{x + 1} + \frac{2}{x^2 - 3x - 4}$

Your solution

Solutions on p. S23

8.2 EXERCISES

OBJECTIVE A **To rewrite rational expressions in terms of a common denominator**

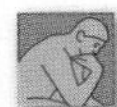

1. **a.** How many factors of a are in the LCM of $(a^2b)^3$ and a^4b^4?
b. How many factors of b are in the LCM of $(a^2b)^3$ and a^4b^4?

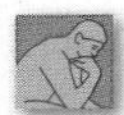

2. **a.** How many factors of $x - 4$ are in the LCM of $x^2 - x - 12$ and $x^2 - 8x + 16$?
b. How many factors of $x - 4$ are in the LCM of $x^2 - x - 12$ and $x^2 - 16$?

For Exercises 3 to 25, write each fraction in terms of the LCM of the denominators.

3. $\frac{3}{4x^2y}, \frac{17}{12xy^4}$

4. $\frac{5}{16a^3b^3}, \frac{7}{30a^5b}$

5. $\frac{x-2}{3x(x-2)}, \frac{3}{6x^2}$

6. $\frac{5x-1}{4x(2x+1)}, \frac{2}{5x^3}$

7. $\frac{3x-1}{2x(x-5)}, -3x$

8. $\frac{4x-3}{3x(x-2)}, 2x$

9. $\frac{3x}{2x-3}, \frac{5x}{2x+3}$

10. $\frac{2}{7y-3}, \frac{-3}{7y+3}$

11. $\frac{2x}{x^2-9}, \frac{x+1}{x-3}$

12. $\frac{3x}{16-x^2}, \frac{2x}{16-4x}$

13. $\frac{3}{3x^2-12y^2}, \frac{5}{6x-12y}$

14. $\frac{2x}{x^2-36}, \frac{x-1}{6x-36}$

15. $\frac{3x}{x^2-1}, \frac{5x}{x^2-2x+1}$

16. $\frac{x^2+2}{x^3-1}, \frac{3}{x^2+x+1}$

17. $\frac{x-3}{8-x^3}, \frac{2}{4+2x+x^2}$

18. $\frac{2x}{x^2+x-6}, \frac{-4x}{x^2+5x+6}$

19. $\frac{2x}{x^2+2x-3}, \frac{-x}{x^2+6x+9}$

20. $\dfrac{3x}{2x^2 - x - 3}, \dfrac{-2x}{2x^2 - 11x + 12}$

21. $\dfrac{-4x}{4x^2 - 16x + 15}, \dfrac{3x}{6x^2 - 19x + 10}$

22. $\dfrac{3}{2x^2 + 5x - 12}, \dfrac{2x}{3 - 2x}, \dfrac{3x - 1}{x + 4}$

23. $\dfrac{5}{6x^2 - 17x + 12}, \dfrac{2x}{4 - 3x}, \dfrac{x + 1}{2x - 3}$

24. $\dfrac{3x}{x - 4}, \dfrac{4}{x + 5}, \dfrac{x + 2}{20 - x - x^2}$

25. $\dfrac{2x}{x - 3}, \dfrac{-2}{x + 5}, \dfrac{x - 1}{15 - 2x - x^2}$

OBJECTIVE B To add or subtract rational expressions

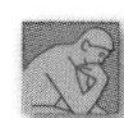

26. True or false? $\dfrac{1}{2x} + \dfrac{1}{3x} = \dfrac{1}{5x}$

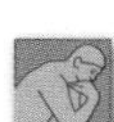

27. True or false? $\dfrac{1}{x - 3} + \dfrac{1}{3 - x} = 0$

For Exercises 28 to 83, simplify.

28. $\dfrac{3}{2xy} - \dfrac{7}{2xy} - \dfrac{9}{2xy}$

29. $-\dfrac{3}{4x^2} + \dfrac{8}{4x^2} - \dfrac{3}{4x^2}$

30. $\dfrac{x}{x^2 - 3x + 2} - \dfrac{2}{x^2 - 3x + 2}$

31. $\dfrac{3x}{3x^2 + x - 10} - \dfrac{5}{3x^2 + x - 10}$

32. $\dfrac{3}{2x^2y} - \dfrac{8}{5x} - \dfrac{9}{10xy}$

33. $\dfrac{2}{5ab} - \dfrac{3}{10a^2b} + \dfrac{4}{15ab^2}$

34. $\dfrac{2}{3x} - \dfrac{3}{2xy} + \dfrac{4}{5xy} - \dfrac{5}{6x}$

35. $\dfrac{3}{4ab} - \dfrac{2}{5a} + \dfrac{3}{10b} - \dfrac{5}{8ab}$

36. $\dfrac{2x - 1}{12x} - \dfrac{3x + 4}{9x}$

37. $\dfrac{3x - 4}{6x} - \dfrac{2x - 5}{4x}$

38. $\dfrac{3x + 2}{4x^2y} - \dfrac{y - 5}{6xy^2}$

39. $\dfrac{2y - 4}{5xy^2} + \dfrac{3 - 2x}{10x^2y}$

40. $\dfrac{2x}{x - 3} - \dfrac{3x}{x - 5}$

41. $\dfrac{3a}{a - 2} - \dfrac{5a}{a + 1}$

42. $\dfrac{3}{2a - 3} + \dfrac{2a}{3 - 2a}$

43. $\dfrac{x}{2x-5} - \dfrac{2}{5x-2}$

44. $\dfrac{1}{x+h} - \dfrac{1}{h}$

45. $\dfrac{1}{a-b} + \dfrac{1}{b}$

46. $\dfrac{2}{x} - 3 - \dfrac{10}{x-4}$

47. $\dfrac{6a}{a-3} - 5 + \dfrac{3}{a}$

48. $\dfrac{1}{2x-3} - \dfrac{5}{2x} + 1$

49. $\dfrac{5}{x} - \dfrac{5x}{5-6x} + 2$

50. $\dfrac{3}{x^2-1} + \dfrac{2x}{x^2+2x+1}$

51. $\dfrac{1}{x^2-6x+9} - \dfrac{1}{x^2-9}$

52. $\dfrac{x}{x+3} - \dfrac{3-x}{x^2-9}$

53. $\dfrac{1}{x+2} - \dfrac{3x}{x^2+4x+4}$

54. $\dfrac{2x-3}{x+5} - \dfrac{x^2-4x-19}{x^2+8x+15}$

55. $\dfrac{-3x^2+8x+2}{x^2+2x-8} - \dfrac{2x-5}{x+4}$

56. $\dfrac{x^n}{x^{2n}-1} - \dfrac{2}{x^n+1}$

57. $\dfrac{2}{x^n-1} + \dfrac{x^n}{x^{2n}-1}$

58. $\dfrac{2}{x^n-1} - \dfrac{6}{x^{2n}+x^n-2}$

59. $\dfrac{2x^n-6}{x^{2n}-x^n-6} + \dfrac{x^n}{x^n+2}$

60. $\dfrac{2x-2}{4x^2-9} - \dfrac{5}{3-2x}$

61. $\dfrac{x^2+4}{4x^2-36} - \dfrac{13}{x+3}$

62. $\dfrac{x-2}{x+1} - \dfrac{3-12x}{2x^2-x-3}$

63. $\dfrac{3x-4}{4x+1} + \dfrac{3x+6}{4x^2+9x+2}$

64. $\dfrac{x+1}{x^2+x-6} - \dfrac{x+2}{x^2+4x+3}$

65. $\dfrac{x+1}{x^2+x-12} - \dfrac{x-3}{x^2+7x+12}$

66. $\dfrac{x^2+6x}{x^2+3x-18} - \dfrac{2x-1}{x+6} + \dfrac{x-2}{3-x}$

67. $\dfrac{2x^2-2x}{x^2-2x-15} - \dfrac{2}{x+3} + \dfrac{x}{5-x}$

68. $\dfrac{7-4x}{2x^2-9x+10}+\dfrac{x-3}{x-2}-\dfrac{x+1}{2x-5}$

69. $\dfrac{x}{3x+4}+\dfrac{3x+2}{x-5}-\dfrac{7x^2+24x+28}{3x^2-11x-20}$

70. $\dfrac{32x-9}{2x^2+7x-15}+\dfrac{x-2}{3-2x}+\dfrac{3x+2}{x+5}$

71. $\dfrac{x+1}{1-2x}-\dfrac{x+3}{4x-3}+\dfrac{10x^2+7x-9}{8x^2-10x+3}$

72. $\dfrac{x^2}{x^3-8}-\dfrac{x+2}{x^2+2x+4}$

73. $\dfrac{2x}{4x^2+2x+1}+\dfrac{4x+1}{8x^3-1}$

74. $\dfrac{2x^2}{x^4-1}-\dfrac{1}{x^2-1}+\dfrac{1}{x^2+1}$

75. $\dfrac{x^2-12}{x^4-16}+\dfrac{1}{x^2-4}-\dfrac{1}{x^2+4}$

76. $\left(\dfrac{x+8}{4}+\dfrac{4}{x}\right)\div\dfrac{x+4}{16x^2}$

77. $\left(\dfrac{a-3}{a^2}-\dfrac{a-3}{9}\right)\div\dfrac{a^2-9}{3a}$

78. $\dfrac{3}{x-2}-\dfrac{x^2+x}{2x^3+3x^2}\cdot\dfrac{2x^2+x-3}{x^2+3x+2}$

79. $\dfrac{x^2-4x+4}{2x+1}\cdot\dfrac{2x^2+x}{x^3-4x}-\dfrac{3x-2}{x+1}$

80. $\left(\dfrac{x-y}{x^2}-\dfrac{x-y}{y^2}\right)\div\dfrac{x^2-y^2}{xy}$

81. $\left(\dfrac{a-2b}{b}+\dfrac{b}{a}\right)\div\left(\dfrac{b+a}{a}-\dfrac{2a}{b}\right)$

82. $\dfrac{2}{x-3}-\dfrac{x}{x^2-x-6}\cdot\dfrac{x^2-2x-3}{x^2-x}$

83. $\dfrac{2x}{x^2-x-6}-\dfrac{6x-6}{2x^2-9x+9}\div\dfrac{x^2+x-2}{2x-3}$

Applying the Concepts

84. Correct the right hand side of the following equations.

a. $\dfrac{3}{4}+\dfrac{x}{5}=\dfrac{3+x}{4+5}$ **b.** $\dfrac{4x+5}{4}=x+5$ **c.** $\dfrac{1}{x}+\dfrac{1}{y}=\dfrac{1}{x+y}$

85. Let $f(x)=\dfrac{x}{x+2}$, $g(x)=\dfrac{4}{x-3}$, and $S(x)=\dfrac{x^2+x+8}{x^2-x-6}$.

a. Evaluate $f(4)$, $g(4)$, and $S(4)$. **b.** Does $f(4)+g(4)=S(4)$?

SECTION

8.3 Complex Fractions

OBJECTIVE A To simplify a complex fraction 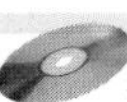

A **complex fraction** is a fraction in which the numerator or denominator contains one or more fractions. Examples of complex fractions are shown at the right.

$$\frac{5}{2+\frac{1}{2}} \qquad \frac{5+\frac{1}{y}}{5-\frac{1}{y}} \qquad \frac{x+4+\frac{1}{x+2}}{x-2+\frac{1}{x+2}}$$

If the LCMs of the denominators in the numerator and denominator of a complex fraction are the same, the complex fraction can be simplified by multiplying the numerator and denominator by that expression.

Take Note
Begin with the numerator of the complex fraction. The LCM of the denominators is x^2. Now consider the denominator of the complex fraction. The LCM of these denominators is also x^2.

HOW TO 1 Simplify: $\dfrac{1+\frac{5}{x}-\frac{6}{x^2}}{1+\frac{8}{x}+\frac{12}{x^2}}$

$$\frac{1+\frac{5}{x}-\frac{6}{x^2}}{1+\frac{8}{x}+\frac{12}{x^2}} = \frac{1+\frac{5}{x}-\frac{6}{x^2}}{1+\frac{8}{x}+\frac{12}{x^2}} \cdot \frac{x^2}{x^2}$$

• Multiply the numerator and denominator of the complex fraction by x^2.

$$= \frac{1 \cdot x^2 + \frac{5}{x} \cdot x^2 - \frac{6}{x^2} \cdot x^2}{1 \cdot x^2 + \frac{8}{x} \cdot x^2 + \frac{12}{x^2} \cdot x^2}$$

• Simplify.

$$= \frac{x^2+5x-6}{x^2+8x+12} = \frac{(x+6)(x-1)}{(x+6)(x+2)} = \frac{x-1}{x+2}$$

If the LCMs of the denominators in the numerator and denominator of a complex fraction are different, it may be easier to simplify the complex fraction by using a different approach.

Take Note
Begin with the numerator of the complex fraction. The LCM of the denominators is $x - 3$. Now consider the denominator of the complex fraction. The LCM of the denominators is $x + 1$. These expressions are different.

Take Note
Either method of simplifying a complex fraction will always work. With experience, you will be able to decide which method works best for a particular complex fraction.

HOW TO 2 Simplify: $\dfrac{3+\frac{10}{x-3}}{3-\frac{2}{x+1}}$

$$\frac{3+\frac{10}{x-3}}{3-\frac{2}{x+1}} = \frac{\frac{3(x-3)}{x-3}+\frac{10}{x-3}}{\frac{3(x+1)}{x+1}-\frac{2}{x+1}}$$

• Simplify the numerator and denominator of the complex fraction by rewriting each as a single fraction.

$$= \frac{\frac{3x-9}{x-3}+\frac{10}{x-3}}{\frac{3x+3}{x+1}-\frac{2}{x+1}} = \frac{\frac{3x+1}{x-3}}{\frac{3x+1}{x+1}}$$

• Both the numerator and denominator of the complex fraction are now written as single fractions.

$$= \frac{3x+1}{x-3} \div \frac{3x+1}{x+1} = \frac{3x+1}{x-3} \cdot \frac{x+1}{3x+1} = \frac{x+1}{x-3}$$

• Divide the numerator by the denominator.

EXAMPLE • 1

Simplify: $\dfrac{2x - 1 + \dfrac{7}{x + 4}}{3x - 8 + \dfrac{17}{x + 4}}$

Solution

The LCM of $x + 4$ and $x + 4$ is $x + 4$.

$$\frac{2x - 1 + \dfrac{7}{x + 4}}{3x - 8 + \dfrac{17}{x + 4}} = \frac{2x - 1 + \dfrac{7}{x + 4}}{3x - 8 + \dfrac{17}{x + 4}} \cdot \frac{x + 4}{x + 4}$$

$$= \frac{(2x - 1)(x + 4) + \dfrac{7}{x + 4}(x + 4)}{(3x - 8)(x + 4) + \dfrac{17}{x + 4}(x + 4)}$$

$$= \frac{2x^2 + 7x - 4 + 7}{3x^2 + 4x - 32 + 17}$$

$$= \frac{2x^2 + 7x + 3}{3x^2 + 4x - 15} = \frac{(2x + 1)(x + 3)}{(3x - 5)(x + 3)}$$

$$= \frac{(2x + 1)\cancel{(x + 3)}}{(3x - 5)\cancel{(x + 3)}} = \frac{2x + 1}{3x - 5}$$

YOU TRY IT • 1

Simplify: $\dfrac{2x + 5 + \dfrac{14}{x - 3}}{4x + 16 + \dfrac{49}{x - 3}}$

Your solution

EXAMPLE • 2

Simplify: $1 + \dfrac{a}{2 + \dfrac{1}{a}}$

Solution

The LCM of the denominators of 2 and $\frac{1}{a}$ is a.

$$1 + \frac{a}{2 + \dfrac{1}{a}} = 1 + \frac{a}{2 + \dfrac{1}{a}} \cdot \frac{a}{a}$$

$$= 1 + \frac{a \cdot a}{2 \cdot a + \dfrac{1}{a} \cdot a} = 1 + \frac{a^2}{2a + 1}$$

The LCM of the denominators of 1 and $\frac{a^2}{2a + 1}$ is $2a + 1$.

$$1 + \frac{a^2}{2a + 1} = 1 \cdot \frac{2a + 1}{2a + 1} + \frac{a^2}{2a + 1}$$

$$= \frac{2a + 1}{2a + 1} + \frac{a^2}{2a + 1} = \frac{2a + 1 + a^2}{2a + 1}$$

$$= \frac{a^2 + 2a + 1}{2a + 1} = \frac{(a + 1)^2}{2a + 1}$$

YOU TRY IT • 2

Simplify: $2 - \dfrac{1}{2 - \dfrac{1}{x}}$

Your solution

Solutions on p. S24

8.3 EXERCISES

OBJECTIVE A To simplify a complex fraction

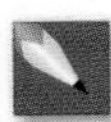 1. What is a complex fraction?

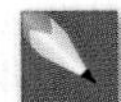 2. What is the general goal of simplifying a complex fraction?

For Exercises 3 to 46, simplify.

3. $\dfrac{2 - \dfrac{1}{3}}{4 + \dfrac{11}{3}}$

4. $\dfrac{3 + \dfrac{5}{2}}{8 - \dfrac{3}{2}}$

5. $\dfrac{3 - \dfrac{2}{3}}{5 + \dfrac{5}{6}}$

6. $\dfrac{5 - \dfrac{3}{4}}{2 + \dfrac{1}{2}}$

7. $\dfrac{1 + \dfrac{1}{x}}{1 - \dfrac{1}{x^2}}$

8. $\dfrac{\dfrac{1}{y^2} - 1}{1 + \dfrac{1}{y}}$

9. $\dfrac{a - 2}{\dfrac{4}{a} - a}$

10. $\dfrac{\dfrac{25}{a} - a}{5 + a}$

11. $\dfrac{\dfrac{1}{x + h} - \dfrac{1}{x}}{h}$

12. $\dfrac{\dfrac{1}{(x + h)^2} - \dfrac{1}{x^2}}{h}$

13. $\dfrac{x - \dfrac{1}{x}}{x + \dfrac{1}{x}}$

14. $\dfrac{a - \dfrac{1}{a}}{\dfrac{1}{a} + a}$

15. $\dfrac{\dfrac{1}{a^2} - \dfrac{1}{a}}{\dfrac{1}{a^2} + \dfrac{1}{a}}$

16. $\dfrac{\dfrac{1}{b} + \dfrac{1}{2}}{\dfrac{4}{b^2} - 1}$

17. $\dfrac{2 - \dfrac{4}{x + 2}}{5 - \dfrac{10}{x + 2}}$

18. $\dfrac{4 + \dfrac{12}{2x - 3}}{5 + \dfrac{15}{2x - 3}}$

19. $\dfrac{\dfrac{3}{2a - 3} + 2}{\dfrac{-6}{2a - 3} - 4}$

20. $\dfrac{\dfrac{-5}{b - 5} - 3}{\dfrac{10}{b - 5} + 6}$

21. $\dfrac{1 - \dfrac{1}{x - 4}}{1 - \dfrac{6}{x + 1}}$

22. $\dfrac{1 + \dfrac{3}{x + 2}}{1 + \dfrac{6}{x - 1}}$

23. $\dfrac{1 - \dfrac{2}{x - 3}}{1 + \dfrac{3}{2 - x}}$

24. $\dfrac{1 + \dfrac{x}{x + 1}}{1 + \dfrac{x - 1}{x + 2}}$

25. $\dfrac{x - 4 + \dfrac{9}{2x + 3}}{x + 3 - \dfrac{5}{2x + 3}}$

26. $\dfrac{2x - 3 - \dfrac{10}{4x - 5}}{3x + 2 + \dfrac{11}{4x - 5}}$

27. $\dfrac{x - 3 + \dfrac{10}{x + 4}}{x + 7 + \dfrac{16}{x - 3}}$

28. $\dfrac{x + 9 + \dfrac{30}{x - 2}}{x - 1 + \dfrac{8}{x + 5}}$

29. $\dfrac{1 - \dfrac{1}{x} - \dfrac{6}{x^2}}{1 - \dfrac{4}{x} + \dfrac{3}{x^2}}$

30. $\dfrac{1 - \dfrac{3}{x} - \dfrac{10}{x^2}}{1 + \dfrac{11}{x} + \dfrac{18}{x^2}}$

31. $\dfrac{1 + \dfrac{1}{x} - \dfrac{12}{x^2}}{\dfrac{9}{x^2} + \dfrac{3}{x} - 2}$

32. $\dfrac{\dfrac{15}{x^2} - \dfrac{2}{x} - 1}{\dfrac{4}{x^2} - \dfrac{5}{x} + 4}$

33. $\dfrac{\dfrac{1}{y^2} - \dfrac{1}{xy} - \dfrac{2}{x^2}}{\dfrac{1}{y^2} - \dfrac{3}{xy} + \dfrac{2}{x^2}}$

34. $\dfrac{\dfrac{2}{b^2} - \dfrac{5}{ab} - \dfrac{3}{a^2}}{\dfrac{2}{b^2} + \dfrac{7}{ab} + \dfrac{3}{a^2}}$

35. $\dfrac{\dfrac{x}{x + 1} - \dfrac{1}{x}}{\dfrac{x}{x + 1} + \dfrac{1}{x}}$

36. $\dfrac{\dfrac{2a}{a - 1} - \dfrac{3}{a}}{\dfrac{1}{a - 1} + \dfrac{2}{a}}$

37. $\dfrac{\dfrac{1}{a} - \dfrac{3}{a - 2}}{\dfrac{2}{a} + \dfrac{5}{a - 2}}$

38. $\dfrac{\dfrac{2}{b} - \dfrac{5}{b + 3}}{\dfrac{3}{b} + \dfrac{3}{b + 3}}$

39. $\dfrac{\dfrac{x - 1}{x + 1} - \dfrac{x + 1}{x - 1}}{\dfrac{x - 1}{x + 1} + \dfrac{x + 1}{x - 1}}$

40. $\dfrac{\dfrac{y}{y + 2} - \dfrac{y}{y - 2}}{\dfrac{y}{y + 2} + \dfrac{y}{y - 2}}$

41. $a + \dfrac{a}{a + \dfrac{1}{a}}$

42. $4 - \dfrac{2}{2 - \dfrac{3}{x}}$

43. $a - \dfrac{a}{1 - \dfrac{a}{1 - a}}$

44. $3 - \dfrac{3}{3 - \dfrac{3}{3 - x}}$

45. $3 - \dfrac{2}{1 - \dfrac{2}{3 - \dfrac{2}{x}}}$

46. $a + \dfrac{a}{2 + \dfrac{1}{1 - \dfrac{2}{a}}}$

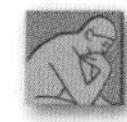

47. In simplest form, what is the reciprocal of the complex fraction $\dfrac{1}{1 - \dfrac{1}{a}}$?

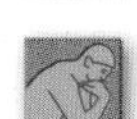

48. The denominator of a complex fraction is the reciprocal of its numerator. Which of the following is the simplified form of the complex fraction?
(i) 1 (ii) the square of the numerator of the complex fraction
(iii) the square of the denominator of the complex fraction

Applying the Concepts

For Exercises 49 to 52, simplify.

49. $(3 + 3^{-1})^{-1}$

50. $[x + (x + 1)^{-1}]^{-1}$

51. $\dfrac{x^{-1}}{x^{-1} + 2^{-1}}$

52. $\dfrac{1 + x^{-1}}{1 - x^{-1}}$

SECTION

8.4 Solving Equations Containing Fractions

OBJECTIVE A To solve an equation containing fractions 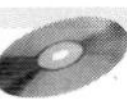

Recall that to solve an equation containing fractions, clear denominators by multiplying each side of the equation by the LCM of the denominators. Then solve for the variable.

HOW TO • 1 Solve: $\frac{3x-1}{4} + \frac{2}{3} = \frac{7}{6}$

$$\frac{3x-1}{4} + \frac{2}{3} = \frac{7}{6}$$

$$12\left(\frac{3x-1}{4} + \frac{2}{3}\right) = 12 \cdot \frac{7}{6}$$

• The LCM is 12. To clear denominators, multiply each side of the equation by the LCM.

$$12\left(\frac{3x-1}{4}\right) + 12 \cdot \frac{2}{3} = 12 \cdot \frac{7}{6}$$

• Simplify using the Distributive Property and the Properties of Fractions.

$$\frac{\overset{3}{\cancel{12}}}{1}\left(\frac{3x-1}{\underset{1}{\cancel{4}}}\right) + \frac{\overset{4}{\cancel{12}}}{1} \cdot \frac{2}{\underset{1}{\cancel{3}}} = \frac{\overset{2}{\cancel{12}}}{1} \cdot \frac{7}{\underset{1}{\cancel{6}}}$$

$$9x - 3 + 8 = 14$$

• Solve for x.

$$9x + 5 = 14$$

$$9x = 9$$

$$x = 1$$

1 checks as a solution. The solution is 1.

Occasionally, a value of the variable that appears to be a solution of an equation will make one of the denominators zero. In such a case, that value is not a solution of the equation.

HOW TO • 2 Solve: $\frac{2x}{x-2} = 1 + \frac{4}{x-2}$

$$\frac{2x}{x-2} = 1 + \frac{4}{x-2}$$

$$(x-2)\frac{2x}{x-2} = (x-2)\left(1 + \frac{4}{x-2}\right)$$

• The LCM is $x - 2$. Multiply each side of the equation by the LCM.

$$(x-2)\frac{2x}{x-2} = (x-2) \cdot 1 + (x-2)\frac{4}{x-2}$$

• Simplify using the Distributive Property and the Properties of Fractions.

$$\frac{\overset{1}{\cancel{(x-2)}}}{1} \cdot \frac{2x}{\underset{1}{\cancel{x-2}}} = (x-2) \cdot 1 + \frac{\overset{1}{\cancel{(x-2)}}}{1} \cdot \frac{4}{\underset{1}{\cancel{x-2}}}$$

$$2x = x - 2 + 4$$

• Solve for x.

$$2x = x + 2$$

$$x = 2$$

When x is replaced by 2, the denominators of $\frac{2x}{x-2}$ and $\frac{4}{x-2}$ are zero. Therefore, the equation has no solution.

✓ Take Note

The example at the right illustrates the importance of checking a solution of a rational equation when each side is multiplied by a variable expression. As shown in this example, a proposed solution may not check when it is substituted into the original equation.

EXAMPLE • 1

Solve: $\dfrac{x}{x+4} = \dfrac{2}{x}$

Solution

The LCM is $x(x+4)$.

$$\frac{x}{x+4} = \frac{2}{x}$$

$$x(x+4)\left(\frac{x}{x+4}\right) = x(x+4)\left(\frac{2}{x}\right)$$ • Multiply by the LCM.

$$\frac{\overset{1}{\cancel{x(x+4)}}}{1} \cdot \frac{x}{\underset{1}{\cancel{x+4}}} = \frac{\overset{1}{\cancel{x}}(x+4)}{1} \cdot \frac{2}{\underset{1}{\cancel{x}}}$$ • Divide by the common factors.

$$x^2 = (x+4)2$$ • Simplify.

$$x^2 = 2x + 8$$

Solve the quadratic equation by factoring.

$$x^2 - 2x - 8 = 0$$ • Write in standard form.

$$(x-4)(x+2) = 0$$ • Factor.

$$x - 4 = 0 \qquad x + 2 = 0$$ • Principle of Zero Products

$$x = 4 \qquad x = -2$$

Both 4 and −2 check as solutions.

The solutions are 4 and −2.

YOU TRY IT • 1

Solve: $\dfrac{x}{x+6} = \dfrac{3}{x}$

Your solution

EXAMPLE • 2

Solve: $\dfrac{3x}{x-4} = 5 + \dfrac{12}{x-4}$

Solution

The LCM is $x - 4$.

$$\frac{3x}{x-4} = 5 + \frac{12}{x-4}$$

$$(x-4)\left(\frac{3x}{x-4}\right) = (x-4)\left(5 + \frac{12}{x-4}\right)$$ • Clear denominators.

$$\frac{\overset{1}{\cancel{(x-4)}}}{1} \cdot \frac{3x}{\underset{1}{\cancel{x-4}}} = (x-4)5 + \frac{\overset{1}{\cancel{(x-4)}}}{1} \cdot \frac{12}{\underset{1}{\cancel{x-4}}}$$

$$3x = (x-4)5 + 12$$ • Solve for x.

$$3x = 5x - 20 + 12$$

$$3x = 5x - 8$$

$$-2x = -8$$

$$x = 4$$

4 does not check as a solution.

The equation has no solution.

YOU TRY IT • 2

Solve: $\dfrac{5x}{x+2} = 3 - \dfrac{10}{x+2}$

Your solution

Solutions on p. S24

8.4 EXERCISES

OBJECTIVE A To solve an equation containing fractions

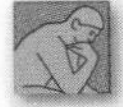

When a proposed solution of a rational equation does not check in the original equation, it is because the proposed solution results in an expression that involves division by zero. For Exercises 1 to 3, state the values of x that would result in division by zero when substituted into the original equation.

1. $\frac{6x}{x+1} - \frac{x}{x-2} = 4$

2. $\frac{1}{x+5} = \frac{x}{x-3} + \frac{2}{x^2+2x-15}$

3. $\frac{3}{x-9} = \frac{1}{x^2-9x} + 2$

For Exercises 4 to 36, solve.

4. $\frac{2x}{3} - \frac{5}{2} = -\frac{1}{2}$

5. $\frac{x}{3} - \frac{1}{4} = \frac{1}{12}$

6. $\frac{x}{3} - \frac{1}{4} = \frac{x}{4} - \frac{1}{6}$

7. $\frac{2y}{9} - \frac{1}{6} = \frac{y}{9} + \frac{1}{6}$

8. $\frac{2x-5}{8} + \frac{1}{4} = \frac{x}{8} + \frac{3}{4}$

9. $\frac{3x+4}{12} - \frac{1}{3} = \frac{5x+2}{12} - \frac{1}{2}$

10. $\frac{6}{2a+1} = 2$

11. $\frac{12}{3x-2} = 3$

12. $\frac{9}{2x-5} = -2$

13. $\frac{6}{4-3x} = 3$

14. $2 + \frac{5}{x} = 7$

15. $3 + \frac{8}{n} = 5$

16. $1 - \frac{9}{x} = 4$

17. $3 - \frac{12}{x} = 7$

18. $\frac{2}{y} + 5 = 9$

19. $\frac{6}{x} + 3 = 11$

20. $\frac{3}{x-2} = \frac{4}{x}$

21. $\frac{5}{x+3} = \frac{3}{x-1}$

22. $\frac{2}{3x-1} = \frac{3}{4x+1}$

23. $\frac{5}{3x-4} = \frac{-3}{1-2x}$

24. $\frac{-3}{2x+5} = \frac{2}{x-1}$

25. $\frac{4}{5y - 1} = \frac{2}{2y - 1}$

26. $\frac{4x}{x - 4} + 5 = \frac{5x}{x - 4}$

27. $\frac{2x}{x + 2} - 5 = \frac{7x}{x + 2}$

28. $2 + \frac{3}{a - 3} = \frac{a}{a - 3}$

29. $\frac{x}{x + 4} = 3 - \frac{4}{x + 4}$

30. $\frac{x}{x - 1} = \frac{8}{x + 2}$

31. $\frac{x}{x + 12} = \frac{1}{x + 5}$

32. $\frac{2x}{x + 4} = \frac{3}{x - 1}$

33. $\frac{5}{3n - 8} = \frac{n}{n + 2}$

34. $\frac{x}{x + 4} = \frac{11}{x^2 - 16} + 2$

35. $x - \frac{6}{x - 3} = \frac{2x}{x - 3}$

36. $\frac{8}{r} + \frac{3}{r - 1} = 3$

Applying the Concepts

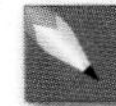

37. Explain the procedure for solving an equation containing fractions. Include in your discussion an explanation of how the LCM of the denominators is used to eliminate fractions in the equation.

For Exercises 38 to 43, solve.

38. $\frac{3}{5}y - \frac{1}{3}(1 - y) = \frac{2y - 5}{15}$

39. $\frac{3}{4}a = \frac{1}{2}(3 - a) + \frac{a - 2}{4}$

40. $\frac{b + 2}{5} = \frac{1}{4}b - \frac{3}{10}(b - 1)$

41. $\frac{x}{2x^2 - x - 1} = \frac{3}{x^2 - 1} + \frac{3}{2x + 1}$

42. $\frac{x + 1}{x^2 + x - 2} = \frac{x + 2}{x^2 - 1} + \frac{3}{x + 2}$

43. $\frac{y + 2}{y^2 - y - 2} + \frac{y + 1}{y^2 - 4} = \frac{1}{y + 1}$

SECTION

8.5 Ratio and Proportion

OBJECTIVE A To solve a proportion 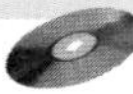

Point of Interest

The Women's Restroom Equity Bill was signed by New York City Mayor Michael Bloomberg and approved unanimously by the NYC Council. This bill requires that women's and men's bathroom stalls in bars, sports arenas, theaters, and highway service areas be in a ratio of 2 to 1. Nicknamed "potty parity," this legislation attempts to shorten the long lines at ladies rooms throughout the city.

Quantities such as 4 meters, 15 seconds, and 8 gallons are number quantities written with units. In these examples the units are meters, seconds, and gallons.

A **ratio** is the quotient of two quantities that have the same unit.

The length of a living room is 16 ft and the width is 12 ft. The ratio of the length to the width is written

$\frac{16 \text{ ft}}{12 \text{ ft}} = \frac{16}{12} = \frac{4}{3}$ A ratio is in simplest form when the two numbers do not have a common factor. Note that the units are not written.

A **rate** is the quotient of two quantities that have different units.

There are 2 lb of salt in 8 gal of water. The salt-to-water rate is

$\frac{2 \text{ lb}}{8 \text{ gal}} = \frac{1 \text{ lb}}{4 \text{ gal}}$ A rate is in simplest form when the two numbers do not have a common factor. The units are written as part of the rate.

A **proportion** is an equation that states the equality of two ratios or rates. Examples of proportions are shown at the right.

$\frac{30 \text{ mi}}{4 \text{ h}} = \frac{15 \text{ mi}}{2 \text{ h}}$ $\frac{4}{6} = \frac{8}{12}$ $\frac{3}{4} = \frac{x}{8}$

HOW TO • 1 Solve the proportion $\frac{4}{x} = \frac{2}{3}$.

$$\frac{4}{x} = \frac{2}{3}$$

$$3x\left(\frac{4}{x}\right) = 3x\left(\frac{2}{3}\right)$$ • The LCM of the denominators is $3x$. To clear denominators, multiply each side of the proportion by the LCM.

$$12 = 2x$$ • Solve the equation.

$$6 = x$$

The solution is 6.

EXAMPLE • 1

Solve: $\frac{8}{x+3} = \frac{4}{x}$

Solution

$$\frac{8}{x+3} = \frac{4}{x}$$

$$x(x+3)\frac{8}{x+3} = x(x+3)\frac{4}{x}$$ • Clear denominators.

$$8x = 4(x+3)$$ • Solve for x.

$$8x = 4x + 12$$

$$4x = 12$$

$$x = 3$$

The solution is 3.

YOU TRY IT • 1

Solve: $\frac{2}{x+3} = \frac{6}{5x+5}$

Your solution

Solution on p. S24

OBJECTIVE B To solve application problems 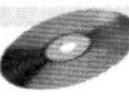

EXAMPLE 2

The monthly loan payment for a car is $28.35 for each $1000 borrowed. At this rate, find the monthly payment for a $6000 car loan.

Strategy

To find the monthly payment, write and solve a proportion, using P to represent the monthly car payment.

Solution

$$\frac{28.35}{1000} = \frac{P}{6000}$$ • Write a proportion.

$$6000\left(\frac{28.35}{1000}\right) = 6000\left(\frac{P}{6000}\right)$$ • Clear denominators.

$$170.10 = P$$

The monthly payment is $170.10.

YOU TRY IT 2

Sixteen ceramic tiles are needed to tile a 9-square-foot area. At this rate, how many square feet can be tiled using 256 ceramic tiles?

Your strategy

Your solution

Solution on p. S24

OBJECTIVE C To solve problems involving similar triangles 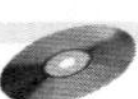

Similar objects have the same shape but not necessarily the same size. A tennis ball is similar to a basketball. A model ship is similar to an actual ship.

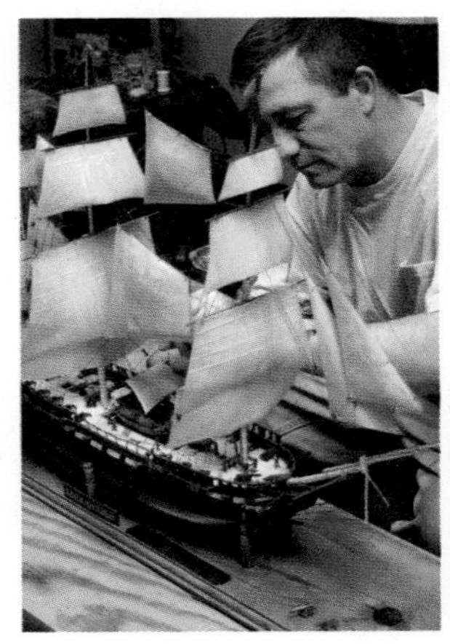
© Clayton Sharrard/PhotoEdit

Similar objects have corresponding parts; for example, the rudder on the model ship corresponds to the rudder on the actual ship. The relationship between the sizes of each of the corresponding parts can be written as a ratio, and each ratio will be the same. If the rudder on the model ship is $\frac{1}{100}$ the size of the rudder on the actual ship, then the model wheelhouse is $\frac{1}{100}$ the size of the actual wheelhouse, the width of the model is $\frac{1}{100}$ the width of the actual ship, and so on.

The two triangles ABC and DEF shown at the right are similar. Side AB corresponds to DE, side BC corresponds to EF, and side AC corresponds to DF. The height CH corresponds to the height FK. The ratios of corresponding parts of similar triangles are equal.

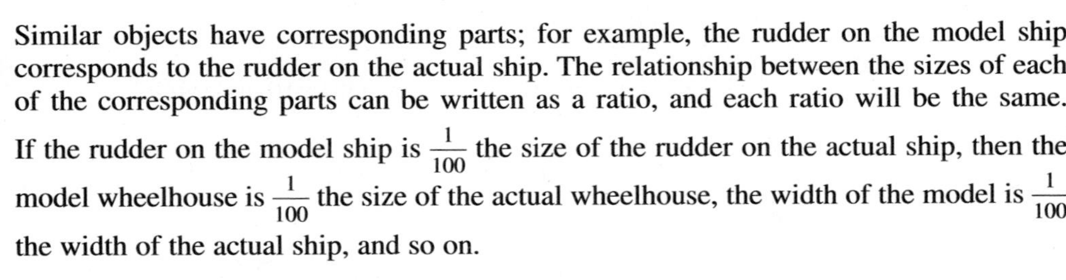

$$\frac{AB}{DE} = \frac{4}{8} = \frac{1}{2}, \quad \frac{AC}{DF} = \frac{3}{6} = \frac{1}{2}, \quad \frac{BC}{EF} = \frac{2}{4} = \frac{1}{2}, \quad \text{and} \quad \frac{CH}{FK} = \frac{1.5}{3} = \frac{1}{2}$$

Because the ratios of corresponding parts are equal, three proportions can be formed using the sides of the triangles.

$$\frac{AB}{DE} = \frac{AC}{DF}, \quad \frac{AB}{DE} = \frac{BC}{EF}, \quad \text{and} \quad \frac{AC}{DF} = \frac{BC}{EF}$$

Three proportions can also be formed by using the sides and heights of the triangles.

$$\frac{AB}{DE} = \frac{CH}{FK}, \quad \frac{AC}{DF} = \frac{CH}{FK}, \quad \text{and} \quad \frac{BC}{EF} = \frac{CH}{FK}$$

The measures of the corresponding angles in similar triangles are equal. Therefore,

$$m\angle A = m\angle D, \quad m\angle B = m\angle E, \quad \text{and} \quad m\angle C = m\angle F$$

It is also true that if the measures of the three angles of one triangle are equal, respectively, to the measures of the three angles of another triangle, then the two triangles are similar.

Take Note
Vertical angles of intersecting lines, corresponding angles of parallel lines, and angles of a triangle are discussed in Section 3.1.

A line DE is drawn parallel to the base AB in the triangle at the right. $m\angle x = m\angle m$ and $m\angle y = m\angle n$ because corresponding angles are equal. $m\angle C = m\angle C$; thus the measures of the three angles of triangle DEC are equal, respectively, to the measures of the three angles of triangle ABC. Triangle DEC is similar to triangle ABC.

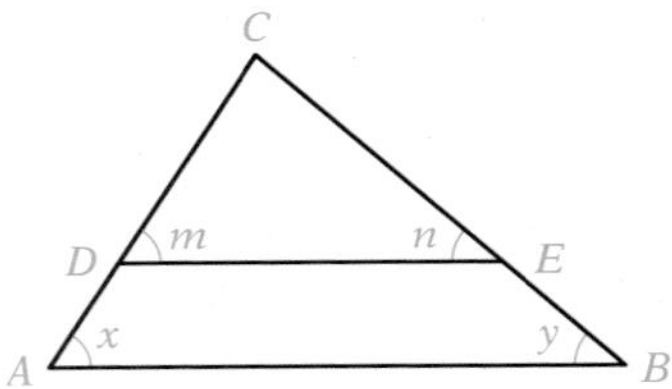

The sum of the measures of the three angles of a triangle is 180°. If two angles of one triangle are equal in measure to two angles of another triangle, then the two third angles must be equal in measure. Thus we can say that if two angles of one triangle are equal in measure to two angles of another triangle, then the two triangles are similar.

HOW TO • 2 The line segments AB and CD intersect at point O in the figure at the right. Angles C and D are right angles. Find the length of DO.

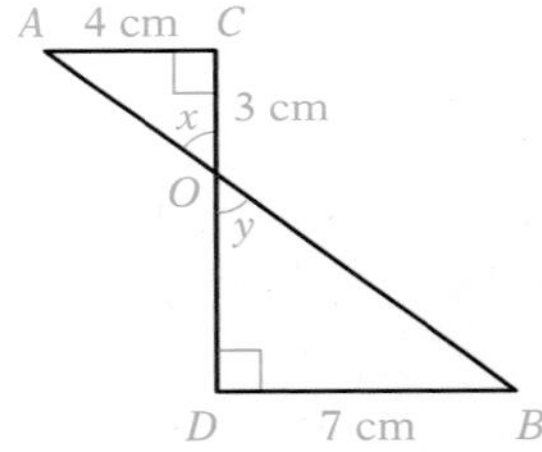

First we must determine whether triangle AOC is similar to triangle BOD.

$m\angle C = m\angle D$ because they are right angles.

$m\angle x = m\angle y$ because they are vertical angles.

Triangle AOC is similar to triangle BOD because two angles of one triangle are equal in measure to two angles of the other triangle.

$$\frac{AC}{DB} = \frac{CO}{DO}$$
• Use a proportion to find the length of the unknown side.

$$\frac{4}{7} = \frac{3}{DO}$$
• $AC = 4$, $CO = 3$, and $DB = 7$.

$$7(DO)\frac{4}{7} = 7(DO)\frac{3}{DO}$$
• To clear denominators, multiply each side of the proportion by $7(DO)$.

$$4(DO) = 7(3)$$
• Solve for DO.

$$4(DO) = 21$$
$$DO = 5.25$$

The length of DO is 5.25 cm.

HOW TO • 3 Triangles ABC and DEF at the right are similar. Find the area of triangle ABC.

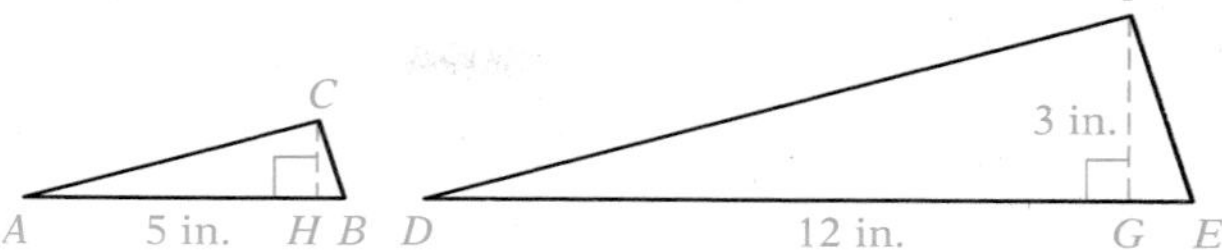

$$\frac{AB}{DE} = \frac{CH}{FG}$$ • Solve a proportion to find the height of triangle ABC.

$$\frac{5}{12} = \frac{CH}{3}$$ • $AB = 5$, $DE = 12$, and $FG = 3$.

$$12 \cdot \frac{5}{12} = 12 \cdot \frac{CH}{3}$$ • To clear denominators, multiply each side of the proportion by 12.

$$5 = 4(CH)$$ • Solve for CH.

$$1.25 = CH$$ • The height is 1.25 in. The base is 5 in.

$$A = \frac{1}{2}bh = \frac{1}{2}(5)(1.25) = 3.125$$ • Use the formula for the area of a triangle.

The area of triangle ABC is 3.125 in^2.

EXAMPLE • 3

In the figure below, AB is parallel to DC, and angles B and D are right angles. $AB = 12$ m, $DC = 4$ m, and $AC = 18$ m. Find the length of CO.

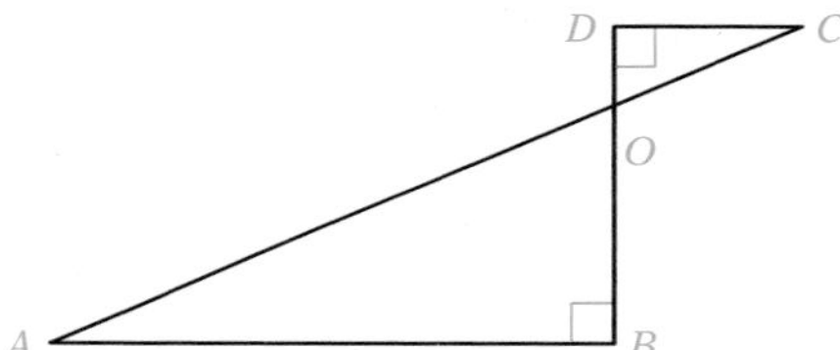

Strategy

Triangle AOB is similar to triangle COD. Solve a proportion to find the length of CO. Let x represent the length of CO and $18 - x$ represent the length of AO.

Solution

$$\frac{DC}{AB} = \frac{CO}{AO}$$ • Write a proportion.

$$\frac{4}{12} = \frac{x}{18 - x}$$ • Substitute.

$$12(18 - x) \cdot \frac{4}{12} = 12(18 - x) \cdot \frac{x}{18 - x}$$ • Clear denominators.

$$4(18 - x) = 12x$$ • Solve for x.

$$72 - 4x = 12x$$

$$72 = 16x$$

$$4.5 = x$$

The length of CO is 4.5 m.

YOU TRY IT • 3

In the figure below, AB is parallel to DC, and angles A and D are right angles. $AB = 10$ cm, $CD = 4$ cm, and $DO = 3$ cm. Find the area of triangle AOB.

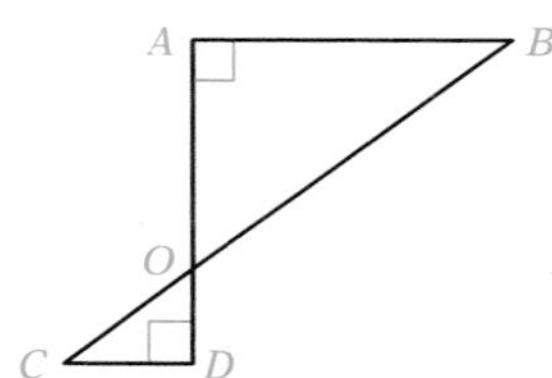

Your strategy

Your solution

Solution on p. S25

8.5 EXERCISES

OBJECTIVE A To solve a proportion

For Exercises 1 to 15, solve.

1. $\dfrac{x}{12} = \dfrac{3}{4}$

2. $\dfrac{6}{x} = \dfrac{2}{3}$

3. $\dfrac{4}{9} = \dfrac{x}{27}$

4. $\dfrac{16}{9} = \dfrac{64}{x}$

5. $\dfrac{x+3}{12} = \dfrac{5}{6}$

6. $\dfrac{3}{5} = \dfrac{x-4}{10}$

7. $\dfrac{18}{x+4} = \dfrac{9}{5}$

8. $\dfrac{2}{11} = \dfrac{20}{x-3}$

9. $\dfrac{2}{x} = \dfrac{4}{x+1}$

10. $\dfrac{16}{x-2} = \dfrac{8}{x}$

11. $\dfrac{x+3}{4} = \dfrac{x}{8}$

12. $\dfrac{x-6}{3} = \dfrac{x}{5}$

13. $\dfrac{2}{x-1} = \dfrac{6}{2x+1}$

14. $\dfrac{9}{x+2} = \dfrac{3}{x-2}$

15. $\dfrac{2x}{7} = \dfrac{x-2}{14}$

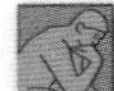

16. True or false? (Assume that a, b, c, and d do not equal zero.)

a. If $\dfrac{a}{b} = \dfrac{c}{d}$, then $\dfrac{d}{b} = \dfrac{c}{a}$.

b. If $\dfrac{a}{b} = \dfrac{c}{d}$, then $\dfrac{b}{a} = \dfrac{d}{c}$.

OBJECTIVE B To solve application problems

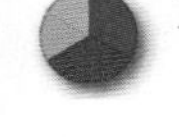

17. Health Insurance See the news clipping at the right. How many Americans do not have health insurance? Use a figure of 300 million for the population of the United States.

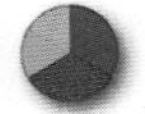
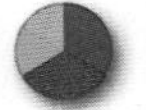

18. Poverty See the news clipping at the right. How many American children live in poverty? Use a figure of 75 million for the number of children living in the United States.

19. Surveys An exit poll survey showed that 4 out of every 7 voters cast a ballot in favor of an amendment to a city charter. At this rate, how many voters voted in favor of the amendment if 35,000 people voted?

20. Business A company decides to accept a large shipment of 10,000 computer chips if there are 2 or fewer defects in a sample of 100 randomly chosen chips. Assuming that there are 300 defective chips in the shipment and that the rate of defective chips in the sample is the same as the rate in the shipment, will the shipment be accepted?

In the News

Room for Improvement

According to U.N. publications, the United States ranks 12th in the world in the area of human development. With regard to health, 1 in 6 Americans does not have health insurance. With respect to standard of living, 1 in 5 American children lives in poverty.

Source: Time, July 28, 2008

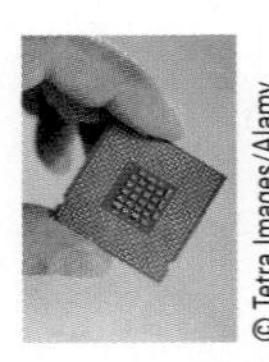
© Tetra Images/Alamy

21. Cooking Simple syrup used in making some desserts requires 2 c of sugar for every $\frac{2}{3}$ c of boiling water. At this rate, how many cups of sugar are required for 2 c boiling water?

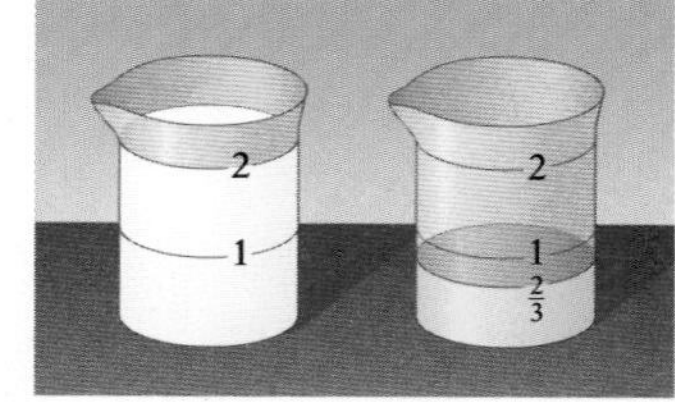

22. Cartography On a map, two cities are $2\frac{5}{8}$ in. apart. If $\frac{3}{8}$ in. on the map represents 25 mi, find the number of miles between the two cities.

23. Conservation As part of a conservation effort for a lake, 40 fish are caught, tagged, and then released. Later 80 fish are caught. Four of the 80 fish are found to have tags. Estimate the number of fish in the lake.

24. Conservation In a wildlife preserve, 10 elk are captured, tagged, and then released. Later 15 elk are captured and 2 are found to have tags. Estimate the number of elk in the preserve.

25. Art Leonardo da Vinci measured various distances on the human body in order to make accurate drawings. He determined that generally the ratio of the kneeling height of a person to the standing height of that person was $\frac{3}{4}$. Using this ratio, determine how tall a person is who has a kneeling height of 48 in.

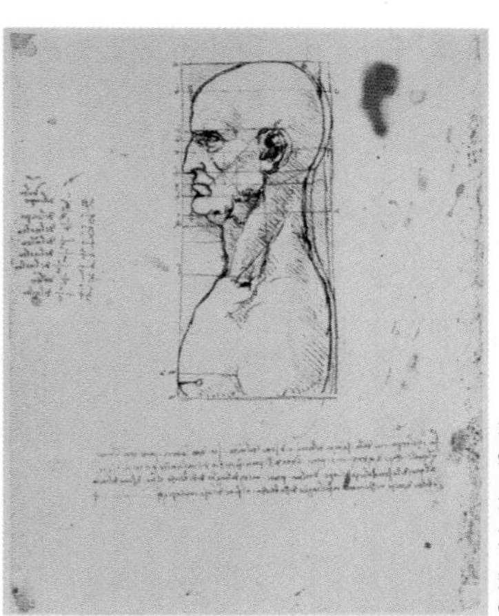

26. Art In one of Leonardo da Vinci's notebooks, he wrote that "... from the top to the bottom of the chin is the sixth part of a face, and it is the fifty-fourth part of the man." Suppose the distance from the top to the bottom of the chin of a person is 1.25 in. Using da Vinci's measurements, find the height of this person.

27. Taxes In February 2008, *USA Today* reported that if a person who uses his or her vehicle for business drives 1000 mi this year, that person will be allowed to deduct \$505 on his or her 2008 tax return. At this rate, how much will a person who drives her car 2200 mi for business in 2008 be able to deduct on her tax return?

In the News

390-Million-Year-Old Scorpion Fossil Found

Scientists have announced the unearthing of the largest fossil sea scorpion claw ever discovered. Based on the 18-inch claw length, scientists estimate that the scorpion would have measured 8.2 ft in length. The longest previously known prehistoric sea scorpion was estimated to be 6.7 ft long.

Source: news.nationalgeographic.com

Fossils For Exercises 28 and 29, use the information in the article at the right. Assume that all scorpions have approximately the same ratio of claw length to body length.

28. Estimate the length, in feet, of the longest previously known prehistoric sea scorpion's claw. Round to the nearest hundredth.

29. Today, scorpions range in length from about 0.5 in. to about 8 in. Estimate the length of a claw of a 7-inch scorpion. Round to the nearest hundredth. (*Hint:* Convert 8.2 ft to inches.)

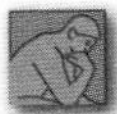

30. The scale on a map shows that a distance of 3 cm on the map represents an actual distance of 10 mi. Would a distance of 8 cm on the map represent an actual distance that is greater than 30 mi or less than 30 mi?

OBJECTIVE C To solve problems involving similar triangles

Triangles *ABC* and *DEF* in Exercises 31 to 38 are similar. Round answers to the nearest tenth.

31. Find side *AC*.

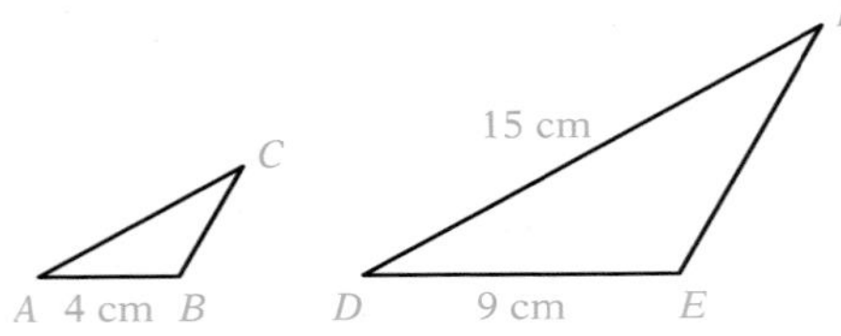

32. Find side *DE*.

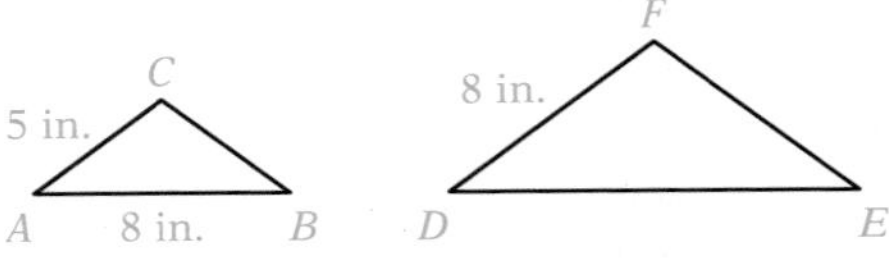

33. Find the height of triangle *ABC*.

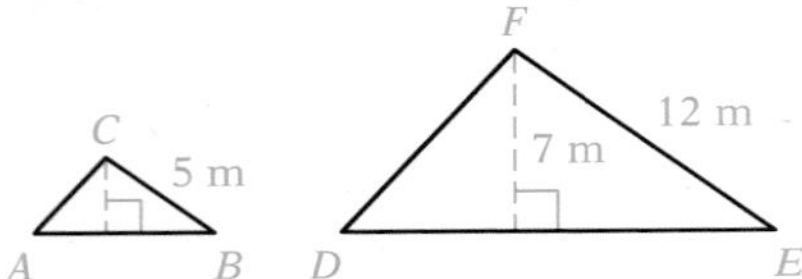

34. Find the height of triangle *DEF*.

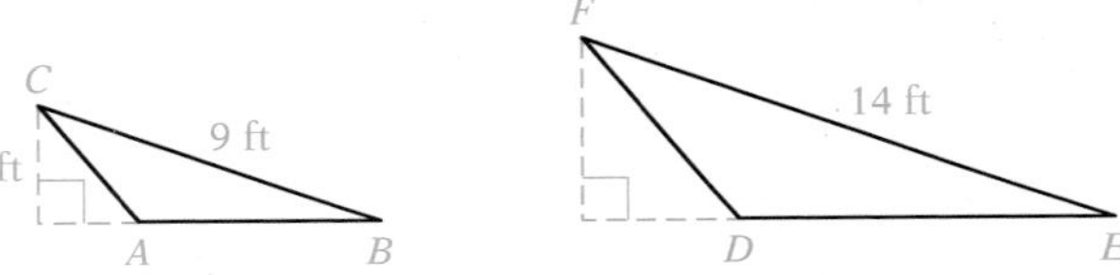

35. Find the perimeter of triangle *DEF*.

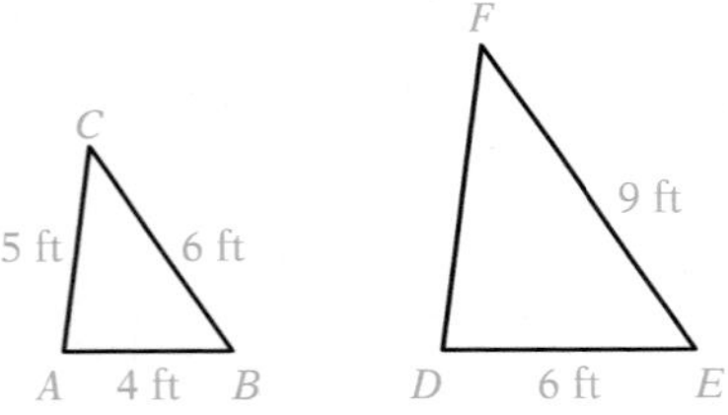

36. Find the perimeter of triangle *ABC*.

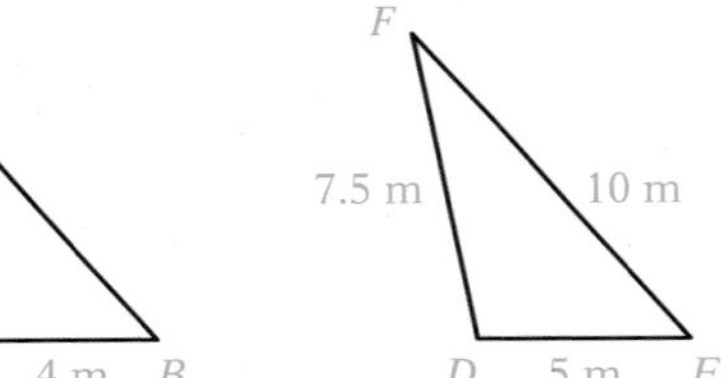

37. Find the area of triangle *ABC*.

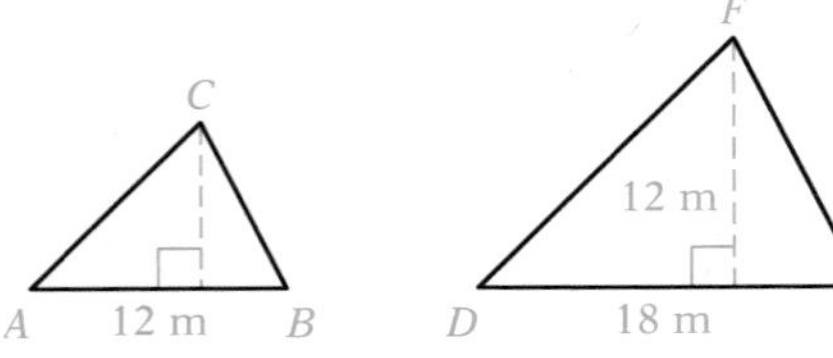

38. Find the area of triangle *ABC*.

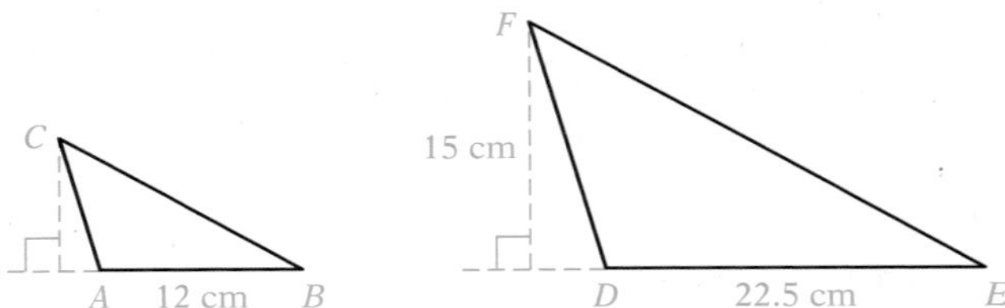

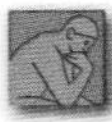

39. True or false? The ratio of the perimeters of two similar triangles is the same as the ratio of their corresponding sides.

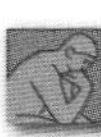

40. True or false? The ratio of the areas of two similar triangles is the same as the ratio of their corresponding sides.

41. Given $BD \parallel AE$, BD measures 5 cm, AE measures 8 cm, and AC measures 10 cm, find the length of BC.

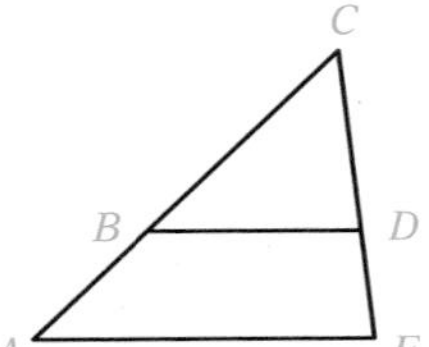

42. Given $AC \parallel DE$, BD measures 8 m, AD measures 12 m, and BE measures 6 m, find the length of BC.

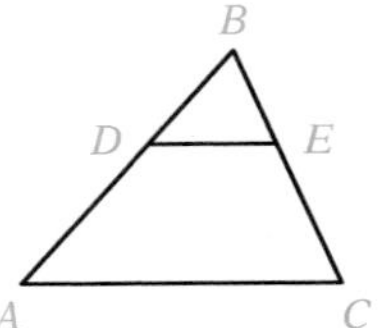

43. Given MP and NQ intersect at O, NO measures 25 ft, MO measures 20 ft, and PO measures 8 ft, find the length of QO.

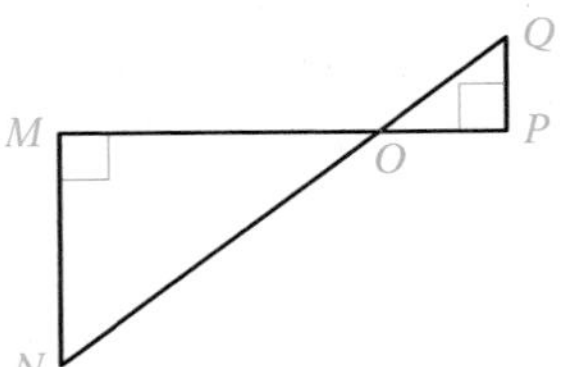

44. Given MP and NQ intersect at O, NO measures 24 cm, MP measures 39 cm, and QO measures 12 cm, find the length of OP.

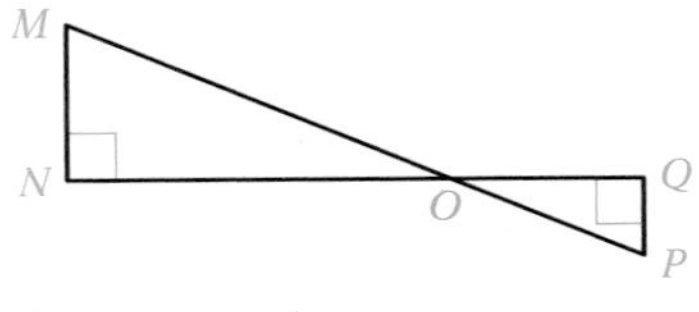

45. **Indirect Measurement** Similar triangles can be used as an indirect way of measuring inaccessible distances. The diagram at the right represents a river of width DC. The triangles AOB and DOC are similar. The distances AB, BO, and OC can be measured. Find the width of the river.

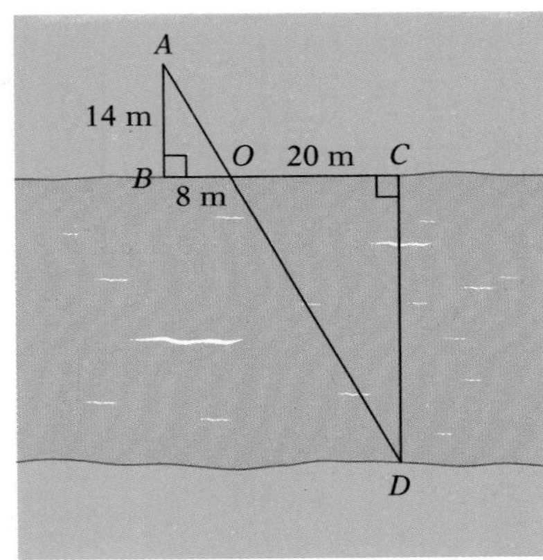

46. **Indirect Measurement** The sun's rays cast a shadow as shown in the diagram at the right. Find the height of the flagpole. Write the answer in terms of feet.

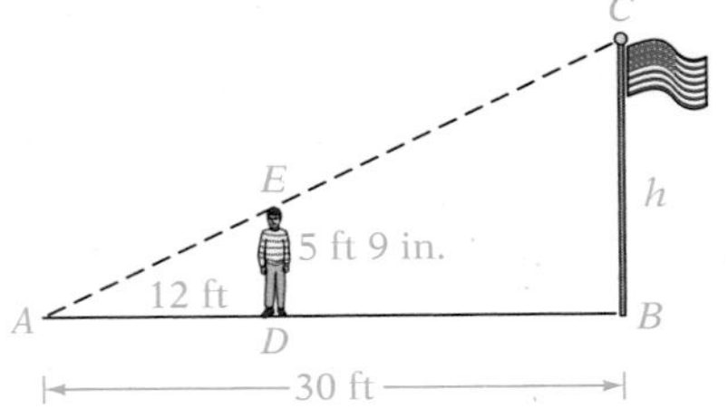

Applying the Concepts

47. **Lottery Tickets** Three people put their money together to buy lottery tickets. The first person put in \$25, the second person put in \$30, and the third person put in \$35. One of their tickets was a winning ticket. If they won \$4.5 million, what was the first person's share of the winnings?

SECTION

8.6 Literal Equations

OBJECTIVE A To solve a literal equation for one of the variables 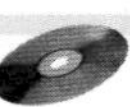

A **literal equation** is an equation that contains more than one variable. Examples of literal equations are shown at the right.

$2x + 3y = 6$
$4w - 2x + z = 0$

Formulas are used to express a relationship among physical quantities. A **formula** is a literal equation that states a rule about measurements. Examples of formulas are shown at the right.

$\frac{1}{R_1} + \frac{1}{R_2} = \frac{1}{R}$ (Physics)

$s = a + (n - 1)d$ (Mathematics)

$A = P + Prt$ (Business)

The Addition and Multiplication Properties can be used to solve a literal equation for one of the variables. The goal is to rewrite the equation so that the variable being solved for is alone on one side of the equation and all the other numbers and variables are on the other side.

HOW TO • 1 Solve $A = P(1 + i)$ for i.

The goal is to rewrite the equation so that i is on one side of the equation and all other variables are on the other side.

$A = P(1 + i)$

$A = P + Pi$ • Use the **Distributive Property** to remove parentheses.

$A - P = P - P + Pi$ • **Subtract** P from each side of the equation.

$A - P = Pi$

$\frac{A - P}{P} = \frac{Pi}{P}$ • Divide each side of the equation by P.

$\frac{A - P}{P} = i$

EXAMPLE • 1

Solve $3x - 4y = 12$ for y.

Solution

$3x - 4y = 12$

$3x - 3x - 4y = -3x + 12$ • Subtract $3x$.

$-4y = -3x + 12$

$\frac{-4y}{-4} = \frac{-3x + 12}{-4}$ • Divide by -4.

$y = \frac{3}{4}x - 3$

YOU TRY IT • 1

Solve $5x - 2y = 10$ for y.

Your solution

Solution on p. S25

EXAMPLE 2

Solve $I = \dfrac{E}{R + r}$ for R.

Solution

$$I = \frac{E}{R + r}$$

$$(R + r)I = (R + r)\frac{E}{R + r} \qquad \text{• Multiply by } (R + r).$$

$$RI + rI = E$$

$$RI + rI - rI = E - rI \qquad \text{• Subtract } rI.$$

$$RI = E - rI$$

$$\frac{RI}{I} = \frac{E - rI}{I} \qquad \text{• Divide by } I.$$

$$R = \frac{E - rI}{I}$$

YOU TRY IT 2

Solve $s = \dfrac{A + L}{2}$ for L.

Your solution

EXAMPLE 3

Solve $L = a(1 + ct)$ for c.

Solution

$$L = a(1 + ct)$$

$$L = a + act \qquad \text{• Distributive Property}$$

$$L - a = a - a + act \qquad \text{• Subtract } a.$$

$$L - a = act$$

$$\frac{L - a}{at} = \frac{act}{at} \qquad \text{• Divide by } at.$$

$$\frac{L - a}{at} = c$$

YOU TRY IT 3

Solve $S = a + (n - 1)d$ for n.

Your solution

EXAMPLE 4

Solve $S = C - rC$ for C.

Solution

$$S = C - rC$$

$$S = (1 - r)C \qquad \text{• Factor.}$$

$$\frac{S}{1 - r} = \frac{(1 - r)C}{1 - r} \qquad \text{• Divide by } (1 - r).$$

$$\frac{S}{1 - r} = C$$

YOU TRY IT 4

Solve $S = rS + C$ for S.

Your solution

Solutions on p. S25

8.6 EXERCISES

OBJECTIVE A **To solve a literal equation for one of the variables**

For Exercises 1 to 15, solve for y.

1. $3x + y = 10$

2. $2x + y = 5$

3. $4x - y = 3$

4. $5x - y = 7$

5. $3x + 2y = 6$

6. $2x + 3y = 9$

7. $2x - 5y = 10$

8. $5x - 2y = 4$

9. $2x + 7y = 14$

10. $6x - 5y = 10$

11. $x + 3y = 6$

12. $x + 2y = 8$

13. $y - 2 = 3(x + 2)$

14. $y + 4 = -2(x - 3)$

15. $y - 1 = -\frac{2}{3}(x + 6)$

For Exercises 16 to 23, solve for x.

16. $x + 3y = 6$

17. $x + 6y = 10$

18. $3x - y = 3$

19. $2x - y = 6$

20. $2x + 5y = 10$

21. $4x + 3y = 12$

22. $x - 2y + 1 = 0$

23. $x - 4y - 3 = 0$

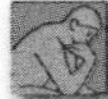

24. Two students are working with the equation $A = P(1 + i)$. State whether the two students' answers are equivalent.

a. When asked to solve the equation for i, one student answered $i = \frac{A}{P} - 1$ and the other student answered $i = \frac{A - P}{P}$.

b. When asked to solve the equation for i, one student answered $i = -\frac{P - A}{P}$ and the other student answered $i = \frac{A - P}{P}$.

For Exercises 25 to 40, solve the formula for the given variable.

25. $d = rt$; t (Physics)

26. $E = IR$; R (Physics)

27. $PV = nRT$; T (Chemistry)

28. $A = bh$; h (Geometry)

29. $P = 2l + 2w; l$ (Geometry)

30. $F = \frac{9}{5}C + 32; C$ (Temperature conversion)

31. $A = \frac{1}{2}h(b_1 + b_2); b_1$ (Geometry)

32. $s = a(x - vt); t$ (Physics)

33. $V = \frac{1}{3}Ah; h$ (Geometry)

34. $P = R - C; C$ (Business)

35. $R = \frac{C - S}{t}; S$ (Business)

36. $P = \frac{R - C}{n}; R$ (Business)

37. $A = P + Prt; P$ (Business)

38. $T = fm - gm; m$ (Engineering)

39. $A = Sw + w; w$ (Physics)

40. $a = S - Sr; S$ (Mathematics)

Applying the Concepts

Business Break-even analysis is a method used to determine the sales volume required for a company to break even, or experience neither a profit nor a loss on the sale of a product. The break-even point represents the number of units that must be made and sold for income from sales to equal the cost of the product. The break-even point can be calculated using the formula $B = \frac{F}{S - V}$, where F is the fixed costs, S is the selling price per unit, and V is the variable costs per unit. Use this information for Exercise 41.

41. **a.** Solve the formula $B = \frac{F}{S - V}$ for S.

    b. Use your answer to part (a) to find the selling price per unit required for a company to break even. The fixed costs are \$20,000, the variable costs per unit are \$80, and the company plans to make and sell 200 underwater cameras.

    c. Use your answer to part (a) to find the selling price per unit required for a company to break even. The fixed costs are \$15,000, the variable costs per unit are \$50, and the company plans to make and sell 600 pen scanners.

SECTION

8.7 Application Problems

OBJECTIVE A To solve work problems 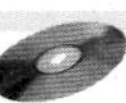

If a painter can paint a room in 4 h, then in 1 h the painter can paint $\frac{1}{4}$ of the room. The painter's rate of work is $\frac{1}{4}$ of the room each hour. The **rate of work** is the part of a task that is completed in 1 unit of time.

A pipe can fill a tank in 30 min. This pipe can fill $\frac{1}{30}$ of the tank in 1 min. The rate of work is $\frac{1}{30}$ of the tank each minute. If a second pipe can fill the tank in x min, the rate of work for the second pipe is $\frac{1}{x}$ of the tank each minute.

In solving a work problem, the goal is to determine the time it takes to complete a task. The basic equation that is used to solve work problems is

Rate of work × time worked = part of task completed

For example, if a faucet can fill a sink in 6 min, then in 5 min the faucet will fill $\frac{1}{6} \times 5 = \frac{5}{6}$ of the sink. In 5 min the faucet completes $\frac{5}{6}$ of the task.

Tips for Success

Note in the examples in this section that solving a word problem includes stating a strategy and using the strategy to find a solution. If you have difficulty with a word problem, write down the known information. Be very specific. Write out a phrase or sentence that states what you are trying to find. See *AIM for Success* in the Preface.

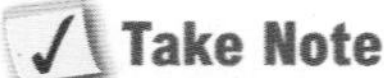

Use the information given in the problem to fill in the "Rate" and "Time" columns of the table. Fill in the "Part Completed" column by multiplying the two expressions you wrote in each row.

HOW TO 1 A painter can paint a wall in 20 min. The painter's apprentice can paint the same wall in 30 min. How long will it take them to paint the wall when they work together?

Strategy for Solving a Work Problem

1. For each person or machine, write a numerical or variable expression for the rate of work, the time worked, and the part of the task completed. The results can be recorded in a table.

Unknown time to paint the wall working together: t

	Rate of Work	·	*Time Worked*	=	*Part of Task Completed*
Painter	$\frac{1}{20}$	·	t	=	$\frac{t}{20}$
Apprentice	$\frac{1}{30}$	·	t	=	$\frac{t}{30}$

2. Determine how the parts of the task completed are related. Use the fact that the sum of the parts of the task completed must equal 1, the complete task.

$$\frac{t}{20} + \frac{t}{30} = 1$$
• The sum of the part of the task completed by the painter and the part of the task completed by the apprentice is 1.

$$60\left(\frac{t}{20} + \frac{t}{30}\right) = 60 \cdot 1$$
• Multiply by the LCM of 20 and 30.

$$3t + 2t = 60$$
• Distributive Property

$$5t = 60$$
$$t = 12$$

Working together, they will paint the wall in 12 min.

EXAMPLE • 1

A small water pipe takes three times longer to fill a tank than does a large water pipe. With both pipes open it takes 4 h to fill the tank. Find the time it would take the small pipe, working alone, to fill the tank.

Strategy

- Time for large pipe to fill the tank: t
  Time for small pipe to fill the tank: $3t$

Fills tank in $3t$ hours

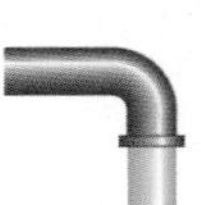

Fills $\frac{4}{3t}$ of the tank in 4 hours

Fills tank in t hours

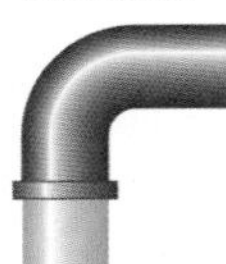

Fills $\frac{4}{t}$ of the tank in 4 hours

	Rate	Time	Part
Small pipe	$\frac{1}{3t}$	4	$\frac{4}{3t}$
Large pipe	$\frac{1}{t}$	4	$\frac{4}{t}$

- The sum of the parts of the task completed by each pipe must equal 1.

Solution

$$\frac{4}{3t} + \frac{4}{t} = 1$$

$$3t\left(\frac{4}{3t} + \frac{4}{t}\right) = 3t \cdot 1$$ • Multiply by the LCM of $3t$ and t.

$$4 + 12 = 3t$$ • Distributive Property

$$16 = 3t$$

$$\frac{16}{3} = t$$

$$3t = 3\left(\frac{16}{3}\right) = 16$$

The small pipe, working alone, takes 16 h to fill the tank.

YOU TRY IT • 1

Two computer printers that work at the same rate are working together to print the payroll checks for a large corporation. After they work together for 2 h, one of the printers quits. The second printer requires 3 h more to complete the payroll checks. Find the time it would take one printer, working alone, to print the payroll.

Your strategy

Your solution

Solution on p. S25

OBJECTIVE B To use rational expressions to solve uniform motion problems

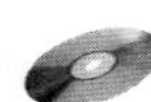

A car that travels constantly in a straight line at 30 mph is in uniform motion. **Uniform motion** means that the speed or direction of an object does not change.

The basic equation used to solve uniform motion problems is

Distance = rate × time

An alternative form of this equation can be written by solving the equation for time.

$$\frac{\textbf{Distance}}{\textbf{Rate}} = \textbf{time}$$

This form of the equation is useful when the total time of travel for two objects or the time of travel between two points is known.

HOW TO • 2 The speed of a boat in still water is 20 mph. The boat traveled 75 mi down a river in the same amount of time it took to travel 45 mi up the river. Find the rate of the river's current.

Strategy for Solving a Uniform Motion Problem

1. For each object, write a numerical or variable expression for the distance, rate, and time. The results can be recorded in a table.

The unknown rate of the river's current: r

Take Note
Use the information given in the problem to fill in the "Distance" and "Rate" columns of the table. Fill in the "Time" column by dividing the two expressions you wrote in each row.

	Distance	÷	*Rate*	=	*Time*
Down river	75	÷	$20 + r$	=	$\frac{75}{20 + r}$
Up river	45	÷	$20 - r$	=	$\frac{45}{20 - r}$

2. Determine how the times traveled by each object are related. For example, it may be known that the times are equal, or the total time may be known.

$$\frac{75}{20 + r} = \frac{45}{20 - r}$$

• The time down the river is equal to the time up the river.

$$(20 + r)(20 - r)\frac{75}{20 + r} = (20 + r)(20 - r)\frac{45}{20 - r}$$

• Multiply by the **LCM**.

$$(20 - r)75 = (20 + r)45$$

$$1500 - 75r = 900 + 45r$$

• Distributive Property

$$-120r = -600$$

$$r = 5$$

The rate of the river's current is 5 mph.

EXAMPLE • 2

A cyclist rode the first 20 mi of a trip at a constant rate. For the next 16 mi, the cyclist reduced the speed by 2 mph. The total time for the 36 mi was 4 h. Find the rate of the cyclist for each leg of the trip.

Strategy

- Rate for the first 20 mi: r
  Rate for the next 16 mi: $r - 2$

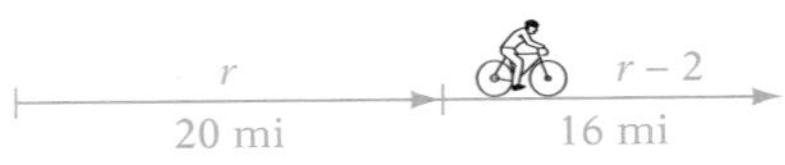

	Distance	*Rate*	*Time*
First 20 mi	20	r	$\frac{20}{r}$
Next 16 mi	16	$r - 2$	$\frac{16}{r-2}$

- The total time for the trip was 4 h.

Solution

$$\frac{20}{r} + \frac{16}{r-2} = 4$$ • The total time was 4 h.

$$r(r-2)\left[\frac{20}{r} + \frac{16}{r-2}\right] = r(r-2) \cdot 4$$ • Multiply by the LCM.

$$(r-2)20 + 16r = 4r^2 - 8r$$ • Distributive Property

$$20r - 40 + 16r = 4r^2 - 8r$$

$$36r - 40 = 4r^2 - 8r$$

Solve the quadratic equation by factoring.

$$0 = 4r^2 - 44r + 40$$ • Standard form

$$0 = 4(r^2 - 11r + 10)$$

$$0 = 4(r - 10)(r - 1)$$ • Factor.

$$r - 10 = 0 \qquad r - 1 = 0$$ • Principle of Zero Products

$$r = 10 \qquad r = 1$$

The solution $r = 1$ mph is not possible, because the rate on the last 16 mi would then be -1 mph.

10 mph was the rate for the first 20 mi.
8 mph was the rate for the next 16 mi.

YOU TRY IT • 2

The total time it took for a sailboat to sail back and forth across a lake 6 km wide was 2 h. The rate sailing back was three times the rate sailing across. Find the rate sailing out across the lake.

Your strategy

Your solution

Solution on p. S26

8.7 EXERCISES

OBJECTIVE A To solve work problems

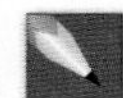

1. Explain the meaning of the phrase "rate of work."

2. If $\frac{2}{5}$ of a room can be painted in 1 h, what is the rate of work? At the same rate, how long will it take to paint the entire room?

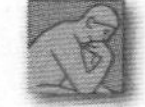

3. It takes Sam h hours to rake the yard, and it takes Emma k hours to rake the yard, where $h > k$. Let t be the amount of time it takes Sam and Emma to rake the yard working together. Is t less than k, between k and h, or greater than k?

4. One grocery clerk can stock a shelf in 20 min, whereas a second clerk requires 30 min to stock the same shelf. How long would it take to stock the shelf if the two clerks worked together?

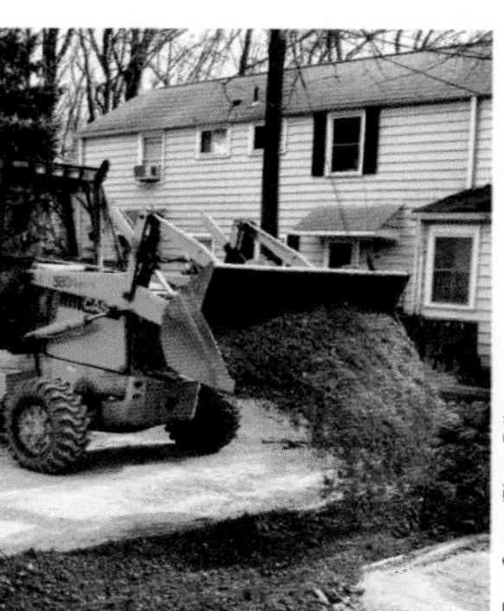
© Tom Carter/PhotoEdit

5. One person with a skiploader requires 12 h to remove a large quantity of earth. A second, larger skiploader can remove the same amount of earth in 4 h. How long would it take to remove the earth with both skiploaders working together?

6. An experienced painter can paint a fence twice as fast as an inexperienced painter. Working together, the painters require 4 h to paint the fence. How long would it take the experienced painter, working alone, to paint the fence?

7. A new machine can make 10,000 aluminum cans three times faster than an older machine. With both machines working, 10,000 cans can be made in 9 h. How long would it take the new machine, working alone, to make the 10,000 cans?

8. A small air conditioner can cool a room 5° in 75 min. A larger air conditioner can cool the room 5° in 50 min. How long would it take to cool the room 5° with both air conditioners working?

9. One printing press can print the first edition of a book in 55 min, whereas a second printing press requires 66 min to print the same number of copies. How long would it take to print the first edition with both presses operating?

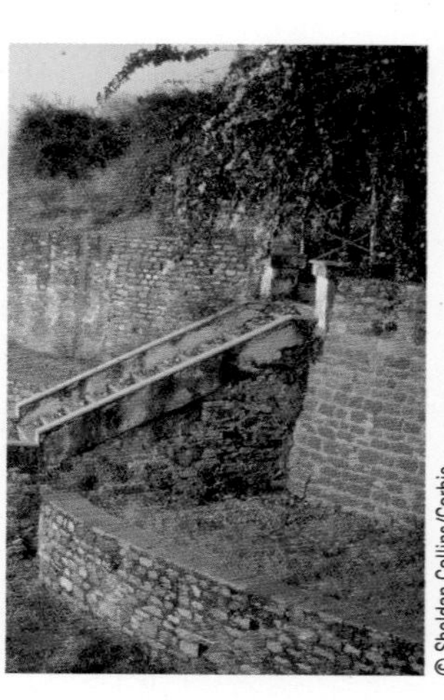
© Sheldan Collins/Corbis

10. Two oil pipelines can fill a small tank in 30 min. One of the pipelines would require 45 min to fill the tank. How long would it take the second pipeline, working alone, to fill the tank?

11. A mason can construct a retaining wall in 10 h. With the mason's apprentice assisting, the task takes 6 h. How long would it take the apprentice, working alone, to construct the wall?

12. A mechanic requires 2 h to repair a transmission, whereas an apprentice requires 6 h to make the same repairs. The mechanic worked alone for 1 h and then stopped. How long will it take the apprentice, working alone, to complete the repairs?

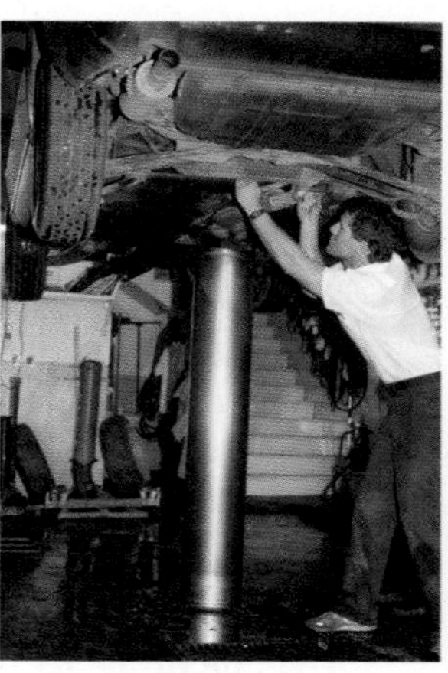
© David Frazier/Photo Edit

13. One technician can wire a security alarm in 4 h, whereas it takes 6 h for a second technician to do the same job. After working alone for 2 h, the first technician quit. How long will it take the second technician to complete the wiring?

14. A wallpaper hanger requires 2 h to hang the wallpaper on one wall of a room. A second wallpaper hanger requires 4 h to hang the same amount of paper. The first wallpaper hanger worked alone for 1 h and then quit. How long will it take the second wallpaper hanger, working alone, to complete the wall?

15. Two welders who work at the same rate are welding the girders of a building. After they work together for 10 h, one of the welders quits. The second welder requires 20 more hours to complete the welds. Find the time it would have taken one of the welders, working alone, to complete the welds.

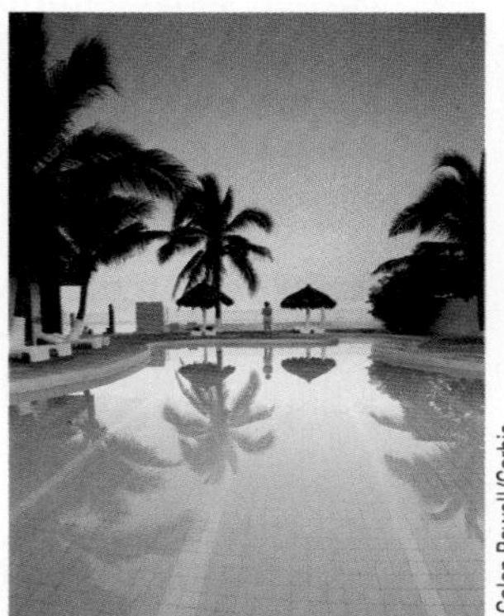
© Galen Rowell/Corbis

16. A large and a small heating unit are being used to heat the water of a pool. The large unit, working alone, requires 8 h to heat the pool. After both units have been operating for 2 h, the large unit is turned off. The small unit requires 9 h more to heat the pool. How long would it take the small unit, working alone, to heat the pool?

17. Two machines that fill cereal boxes work at the same rate. After they work together for 7 h, one machine breaks down. The second machine requires 14 h more to finish filling the boxes. How long would it have taken one of the machines, working alone, to fill the boxes?

18. A large and a small drain are opened to drain a pool. The large drain can empty the pool in 6 h. After both drains have been open for 1 h, the large drain becomes clogged and is closed. The smaller drain remains open and requires 9 h more to empty the pool. How long would it have taken the small drain, working alone, to empty the pool?

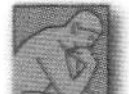

19. Zachary and Eli picked a row of peas together in m minutes. It would have taken Zachary n minutes to pick the row of peas by himself. What fraction of the row of peas did Zachary pick? What fraction of the row of peas did Eli pick?

OBJECTIVE B **To use rational expressions to solve uniform motion problems**

20. Running at a constant speed, a jogger ran 24 mi in 3 h. How far did the jogger run in 2 h?

21. For uniform motion, distance = rate · time. How is time related to distance and rate? How is rate related to distance and time?

22. Commuting from work to home, a lab technician traveled 10 mi at a constant rate through congested traffic. On reaching the expressway, the technician increased the speed by 20 mph. An additional 20 mi was traveled at the increased speed. The total time for the trip was 1 h. Find the rate of travel through the congested traffic.

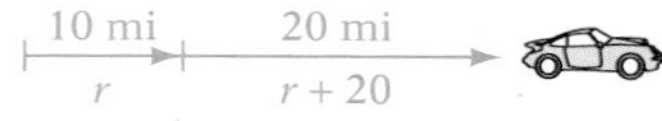

23. The president of a company traveled 1800 mi by jet and 300 mi on a prop plane. The rate of the jet was four times the rate of the prop plane. The entire trip took a total of 5 h. Find the rate of the jet plane.

24. As part of a conditioning program, a jogger ran 8 mi in the same amount of time a cyclist rode 20 mi. The rate of the cyclist was 12 mph faster than the rate of the jogger. Find the rate of the jogger and that of the cyclist.

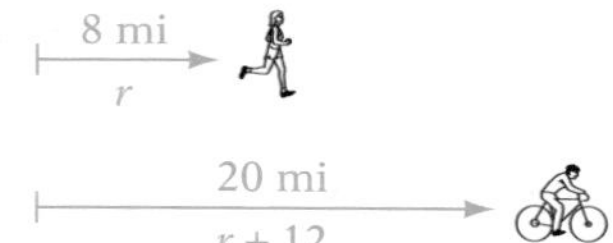

25. An express train travels 600 mi in the same amount of time it takes a freight train to travel 360 mi. The rate of the express train is 20 mph faster than that of the freight train. Find the rate of each train.

26. To assess the damage done by a fire, a forest ranger traveled 1080 mi by jet and then an additional 180 mi by helicopter. The rate of the jet was four times the rate of the helicopter. The entire trip took a total of 5 h. Find the rate of the jet.

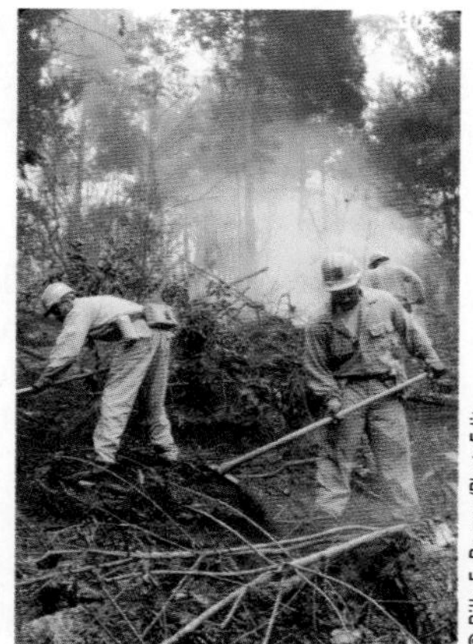

27. As part of an exercise plan, Camille Ellison walked for 40 min and then ran for 20 min. If Camille runs 3 mph faster than she walks and covered 5 mi during the 1-hour exercise period, what is her walking speed?

28. A car and a bus leave a town at 1 P.M. and head for a town 300 mi away. The rate of the car is twice the rate of the bus. The car arrives 5 h ahead of the bus. Find the rate of the car.

29. A car is traveling at a rate that is 36 mph faster than the rate of a cyclist. The car travels 384 mi in the same amount of time it takes the cyclist to travel 96 mi. Find the rate of the car.

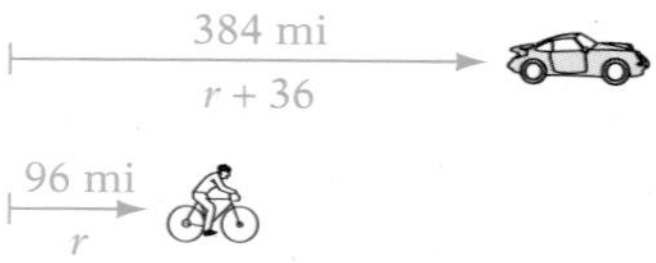

30. On a recent trip, a trucker traveled 330 mi at a constant rate. Because of road construction, the trucker then had to reduce the speed by 25 mph. An additional 30 mi was traveled at the reduced rate. The total time for the entire trip was 7 h. Find the rate of the trucker for the first 330 mi.

31. A backpacker hiking into a wilderness area walked 9 mi at a constant rate and then reduced this rate by 1 mph. Another 4 mi was hiked at the reduced rate. The time required to hike the 4 mi was 1 h less than the time required to walk the 9 mi. Find the rate at which the hiker walked the first 9 mi.

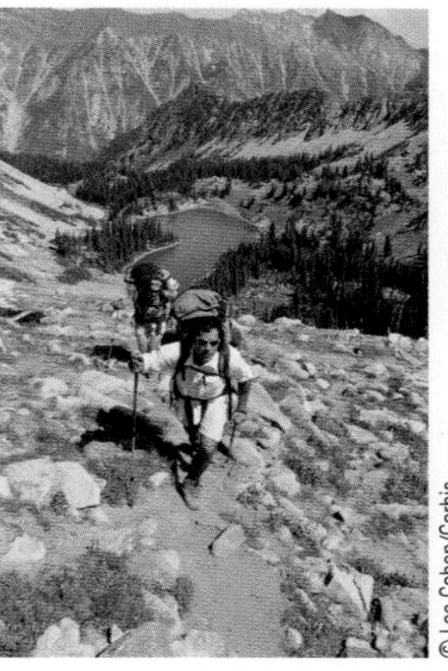

32. A plane can fly 180 mph in calm air. Flying with the wind, the plane can fly 600 mi in the same amount of time it takes to fly 480 mi against the wind. Find the rate of the wind.

33. A commercial jet can fly 550 mph in calm air. Traveling with the jet stream, the plane flew 2400 mi in the same amount of time it takes to fly 2000 mi against the jet stream. Find the rate of the jet stream.

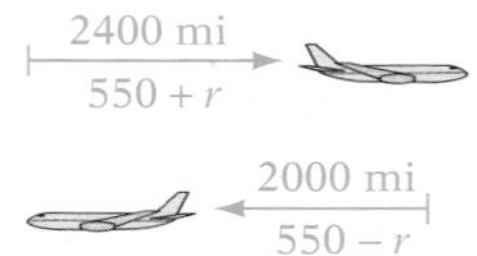

34. A cruise ship can sail at 28 mph in calm water. Sailing with the gulf current, the ship can sail 170 mi in the same amount of time that it can sail 110 mi against the gulf current. Find the rate of the gulf current.

35. Rowing with the current of a river, a rowing team can row 25 mi in the same amount of time it takes to row 15 mi against the current. The rate of the rowing team in calm water is 20 mph. Find the rate of the current.

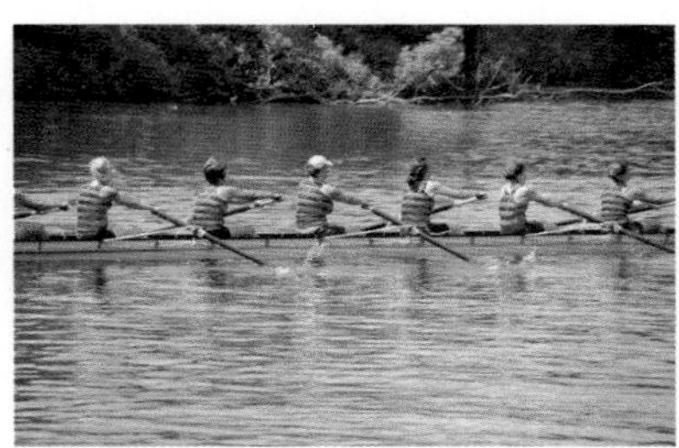

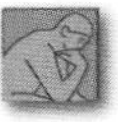
For Exercises 36 and 37, use the following problem situation: A plane can fly 380 mph in calm air. In the time it takes the plane to fly 1440 mi against a headwind, it could fly 1600 mi with the wind. Use the equation $\frac{1440}{380 - r} = \frac{1600}{380 + r}$ to find the rate r of the wind.

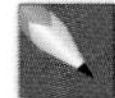
36. Explain the meaning of $380 - r$ and $380 + r$ in terms of the problem situation.

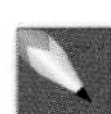
37. Explain the meaning of $\frac{1440}{380 - r}$ and $\frac{1600}{380 + r}$ in terms of the problem situation.

Applying the Concepts

38. **Work** One pipe can fill a tank in 2 h, a second pipe can fill the tank in 4 h, and a third pipe can fill the tank in 5 h. How long will it take to fill the tank with all three pipes working?

39. **Transportation** Because of bad weather, a bus driver reduced the usual speed along a 150-mile bus route by 10 mph. The bus arrived only 30 min later than its usual arrival time. How fast does the bus usually travel?

SECTION

8.8 Variation

OBJECTIVE A To solve variation problems 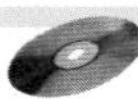

Direct variation is a special function that can be expressed as the equation $y = kx^n$, where k is a constant and n is a positive number. The equation $y = kx^n$ is read "y varies directly as x to the nth" or "y is directly proportional to x to the nth." The constant k is called the **constant of variation** or the **constant of proportionality.**

A surveyor earns \$43 per hour. The total wage w of the surveyor varies directly as the number of hours worked h. The direct variation equation is $w = 43h$. The constant of variation is 43, and the value of n is 1.

The distance s, in feet, an object falls varies directly as the square of the time t, in seconds, that it falls. The direct variation equation is $s = kt^2$. The constant of variation is k, and the value of n is 2. If the object is dropped on Earth, $k = 16$; if the object is dropped on the moon, $k = 2.7$.

Many geometry formulas are expressed as direct variation. The circumference C of a circle is directly proportional to its diameter d. The direct variation equation is $C = \pi d$. The constant of proportionality is π, and the value of n is 1.

The area A of a circle varies directly as the square of its radius r. The direct variation equation is $A = \pi r^2$. The constant of proportionality is π, and the value of n is 2.

HOW TO 1 Given that V varies directly as r and that $V = 20$ when $r = 4$, find the constant of variation and the variation equation.

$V = kr$ • Write the basic direct variation equation.

$20 = k \cdot 4$ • Replace V and r by the given values. Then solve for k.

$5 = k$ • This is the constant of variation.

$V = 5r$ • Write the direct variation equation by substituting the value of k into the basic direct variation equation.

HOW TO 2 The tension T in a spring varies directly as the distance x it is stretched. If $T = 8$ lb when $x = 2$ in., find T when $x = 4$ in.

$T = kx$ • Write the basic direct variation equation.

$8 = k \cdot 2$ • Replace T and x by the given values.

$4 = k$ • Solve for the constant of variation.

$T = 4x$ • Write the direct variation equation.

To find T when $x = 4$ in., substitute 4 for x in the equation and solve for T.

$T = 4x$

$T = 4 \cdot 4 = 16$

The tension is 16 lb.

Inverse variation is a function that can be written as the equation $y = \frac{k}{x^n}$, where k is a constant and n is a positive number. The equation $y = \frac{k}{x^n}$ is read "y varies inversely as x to the nth" or "y is inversely proportional to x to the nth."

The time t it takes a car to travel 100 mi varies inversely as the speed r of the car. The inverse variation equation is $t = \frac{100}{r}$. The variation constant is 100, and the value of n is 1.

The intensity I of a light source varies inversely as the square of the distance d from the source. The inverse variation equation is $I = \frac{k}{d^2}$. The constant of variation depends on the medium through which the light travels (air, water, glass), and the value of n is 2.

Point of Interest

This equation is important to concert sound engineers. Without additional speakers and reverberation, the sound intensity for someone about 20 rows back at this concert would be about 25 dB, the sound intensity of normal conversation.

HOW TO • 3 The intensity I, in decibels (dB), of sound varies inversely as the square of the distance d from the source. If the intensity of an open-air concert is 110 dB in the front row, 10 ft from the band, find the variation equation.

$I = \frac{k}{d^2}$ • Write the basic inverse variation equation.

$110 = \frac{k}{10^2}$ • Replace I and d by the given values.

$11{,}000 = k$ • This is the constant of variation.

$I = \frac{11{,}000}{d^2}$ • This is the variation equation.

HOW TO • 4 The length L of a rectangle with fixed area is inversely proportional to the width w. If $L = 6$ ft when $w = 2$ ft, find L when $w = 3$ ft.

$L = \frac{k}{w}$ • Write the basic inverse variation equation.

$6 = \frac{k}{2}$ • Replace L and w by the given values.

$12 = k$ • Solve for the constant of variation.

$L = \frac{12}{w}$ • Write the inverse variation equation.

To find L when $w = 3$ ft, substitute 3 for w in the equation and solve for L.

$$L = \frac{12}{w} = \frac{12}{3} = 4$$

The length is 4 ft.

Joint variation is variation in which a variable varies directly as the product of two or more other variables. A joint variation can be expressed as the equation $z = kxy$, where k is a constant. The equation $z = kxy$ is read "z varies jointly as x and y."

The area A of a triangle varies jointly as the base b and the height h. The joint variation equation is written $A = \frac{1}{2}bh$. The constant of variation is $\frac{1}{2}$.

A **combined variation** is a variation in which two or more types of variation occur at the same time. For example, in physics, the volume V of a gas varies directly as the temperature T and inversely as the pressure P. This combined variation is written $V = \frac{kT}{P}$.

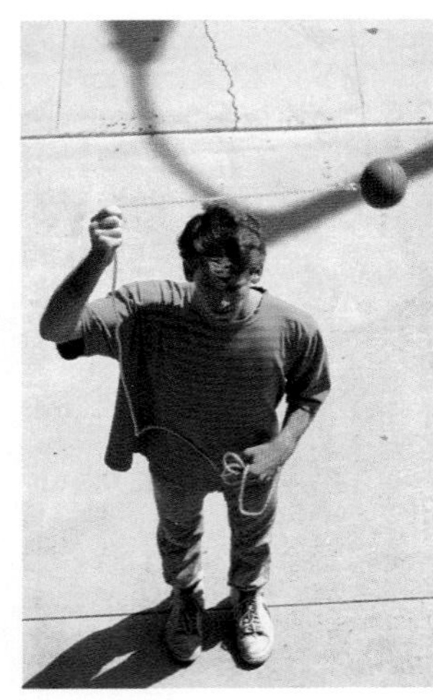
© Tony Freeman/PhotoEdit, Inc.

HOW TO • 5 A ball is being twirled on the end of a string. The tension T in the string is directly proportional to the square of the speed v of the ball and inversely proportional to the length r of the string. If the tension is 96 lb when the length of the string is 0.5 ft and the speed is 4 ft/s, find the tension when the length of the string is 1 ft and the speed is 5 ft/s.

$T = \frac{kv^2}{r}$ • Write the basic combined variation equation.

$96 = \frac{k \cdot 4^2}{0.5}$ • Replace T, v, and r by the given values.

$96 = \frac{k \cdot 16}{0.5}$

$96 = k \cdot 32$ • Solve for the constant of variation.

$3 = k$

$T = \frac{3v^2}{r}$ • Write the combined variation equation.

To find T when $r = 1$ ft and $v = 5$ ft/s, substitute 1 for r and 5 for v, and solve for T.

$$T = \frac{3v^2}{r} = \frac{3 \cdot 5^2}{1} = 3 \cdot 25 = 75$$

The tension is 75 lb.

EXAMPLE • 1

The amount A of medication prescribed for a person is directly related to the person's weight W. For a 50-kilogram person, 2 ml of medication are prescribed. How many milliliters of medication are required for a person who weighs 75 kg?

Strategy

To find the required amount of medication:

- Write the basic direct variation equation, replace the variables by the given values, and solve for k.
- Write the direct variation equation, replacing k by its value. Substitute 75 for W and solve for A.

Solution

$A = kW$ • Direct variation equation

$2 = k \cdot 50$ • Replace A by 2 and W by 50.

$\frac{1}{25} = k$

$A = \frac{1}{25}W = \frac{1}{25} \cdot 75 = 3$ • $k = \frac{1}{25}$, $W = 75$

The required amount of medication is 3 ml.

YOU TRY IT • 1

The distance s a body falls from rest varies directly as the square of the time t of the fall. An object falls 64 ft in 2 s. How far will it fall in 5 s?

Your strategy

Your solution

Solution on p. S26

EXAMPLE • 2

A company that produces personal computers has determined that the number of computers it can sell s is inversely proportional to the price P of the computer. Two thousand computers can be sold when the price is \$2500. How many computers can be sold when the price of a computer is \$2000?

Strategy

To find the number of computers:

- Write the basic inverse variation equation, replace the variables by the given values, and solve for k.
- Write the inverse variation equation, replacing k by its value. Substitute 2000 for P and solve for s.

Solution

$$s = \frac{k}{P}$$ • Inverse variation equation

$$2000 = \frac{k}{2500}$$ • Replace s by 2000 and P by 2500.

$$5{,}000{,}000 = k$$

$$s = \frac{5{,}000{,}000}{P} = \frac{5{,}000{,}000}{2000} = 2500$$ • $k = 5{,}000{,}000$, $P = 2000$

At \$2000 each, 2500 computers can be sold.

YOU TRY IT • 2

The resistance R to the flow of electric current in a wire of fixed length is inversely proportional to the square of the diameter d of the wire. If a wire of diameter 0.01 cm has a resistance of 0.5 ohm, what is the resistance in a wire that is 0.02 cm in diameter?

Your strategy

Your solution

EXAMPLE • 3

The pressure P of a gas varies directly as the temperature T and inversely as the volume V. When $T = 50°$ and $V = 275$ in^3, $P = 20$ lb/in^2. Find the pressure of a gas when $T = 60°$ and $V = 250$ in^3.

Strategy

To find the pressure:

- Write the basic combined variation equation, replace the variables by the given values, and solve for k.
- Write the combined variation equation, replacing k by its value. Substitute 60 for T and 250 for V, and solve for P.

Solution

$$P = \frac{kT}{V}$$ • Combined variation equation

$$20 = \frac{k \cdot 50}{275}$$ • Replace P by 20, T by 50, and V by 275.

$$110 = k$$

$$P = \frac{110T}{V} = \frac{110 \cdot 60}{250} = 26.4$$ • $k = 110$, $T = 60$, $V = 250$

The pressure is 26.4 lb/in^2.

YOU TRY IT • 3

The strength s of a rectangular beam varies jointly as its width W and as the square of its depth d and inversely as its length L. If the strength of a beam 2 in. wide, 12 in. deep, and 12 ft long is 1200 lb, find the strength of a beam 4 in. wide, 8 in. deep, and 16 ft long.

Your strategy

Your solution

Solutions on p. S26

8.8 EXERCISES

OBJECTIVE A To solve variation problems

1. **Business** The profit P realized by a company varies directly as the number of products it sells s. If a company makes a profit of \$4000 on the sale of 250 products, what is the profit when the company sells 5000 products?

2. **Compensation** The income I of a computer analyst varies directly as the number of hours h worked. If the analyst earns \$336 for working 8 h, how much will the analyst earn for working 36 h?

3. **Recreation** The pressure p on a diver in the water varies directly as the depth d. If the pressure is 4.5 lb/in^2 when the depth is 10 ft, what is the pressure when the depth is 15 ft?

4. **Physics** The distance d that a spring will stretch varies directly as the force f applied to the spring. If a force of 6 lb is required to stretch a spring 3 in., what force is required to stretch the spring 4 in.?

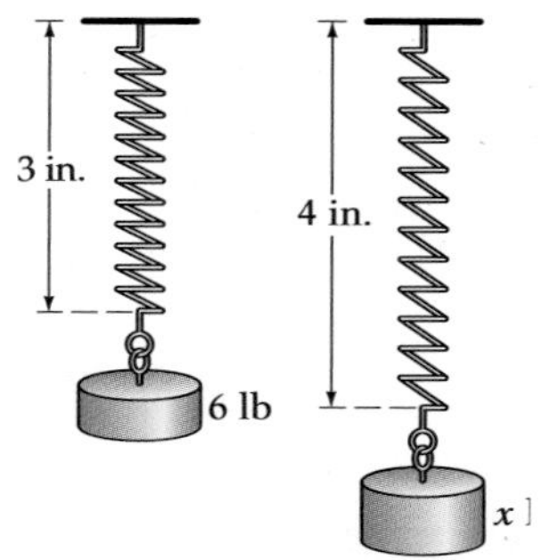

5. **Physics** The distance d an object will fall is directly proportional to the square of the time t of the fall. If an object falls 144 ft in 3 s, how far will the object fall in 10 s?

6. **Physics** The period p of a pendulum, or the time it takes the pendulum to make one complete swing, varies directly as the square root of the length L of the pendulum. If the period of a pendulum is 1.5 s when the length is 2 ft, find the period when the length is 5 ft. Round to the nearest hundredth.

7. **Computer Science** Parallel processing is the use of more than one computer to solve a particular problem. The time T it takes to solve a certain problem is inversely proportional to the number of computers n that are used. If it takes one computer 500 s to solve a problem, how long would it take five computers to solve the same problem?

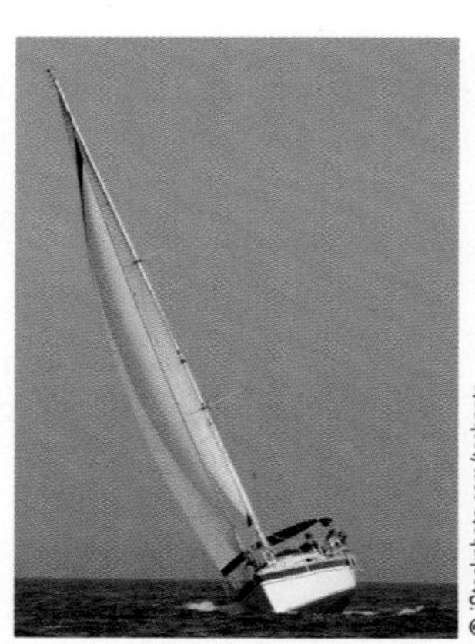
© iStockphoto.com/technotr

8. **Safety** The stopping distance s of a car varies directly as the square of its speed v. If a car traveling 30 mph requires 63 ft to stop, find the stopping distance for a car traveling 55 mph.

9. **Sailing** The load L, in pounds, on a certain sail varies directly as the square of the wind speed v, in miles per hour. If the load on a sail is 640 lb when the wind speed is 20 mph, what is the load on the sail when the wind speed is 15 mph?

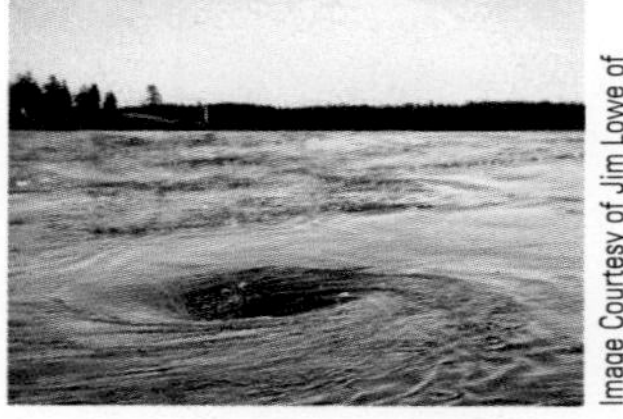
Image Courtesy of Jim Lowe of Eastport, Maine

10. **Whirlpools** The speed v of the current in a whirlpool varies inversely as the distance d from the whirlpool's center. The Old Sow whirlpool, located off the coast of eastern Canada, is one of the most powerful whirlpools on Earth. At a distance of 10 ft from the center of the whirlpool, the speed of the current is about 2.5 ft/s. What is the speed of the current 2 ft from the center?

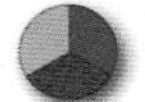

11. Real Estate Real estate agents receive income in the form of a commission, which is a portion of the selling price of the house. The commission c on the sale of a house varies directly as the selling price p of the house. Read the article at the right. Find the commission the agency referred to in the article would receive for selling an average-priced home in January 2008.

In the News

Realtors' Incomes Hit Hard

With lower house prices come lower commissions for real estate agents. One agent said, "In January 2007, my agency received a \$15,600 commission on the sale of an average-priced home (\$260,000). Now, in January 2008, that average-priced home is selling at \$230,000, and we will receive less than \$14,000 in commission."

Source: The Arizona Republic

12. Architecture The heat loss H, in watts, through a single-pane glass window varies jointly as the area A, in square meters, and the difference between the inside and outside temperatures T, in degrees Celsius. If the heat loss is 6 watts for a window with an area of 1.5 m^2 and a temperature difference of 2°C, what would be the heat loss for a window with an area of 2 m^2 and a temperature difference of 5°C?

13. Electronics The current I in a wire varies directly as the voltage v and inversely as the resistance r. If the current is 10 amps when the voltage is 110 volts and the resistance is 11 ohms, find the current when the voltage is 180 volts and the resistance is 24 ohms.

14. Magnetism The repulsive force f between the north poles of two magnets is inversely proportional to the square of the distance d between them. If the repulsive force is 20 lb when the distance is 4 in., find the repulsive force when the distance is 2 in.

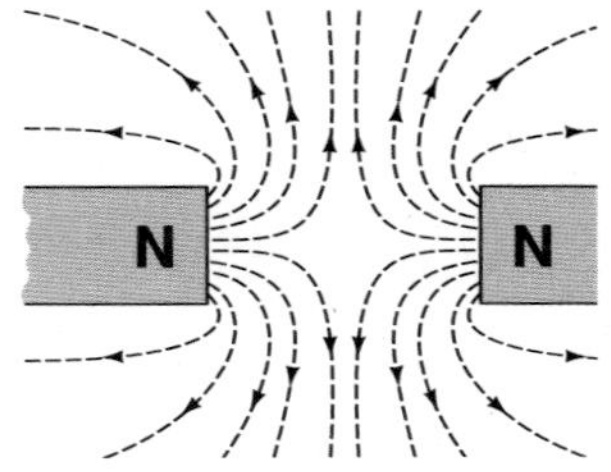

15. Light The intensity I of a light source is inversely proportional to the square of the distance d from the source. If the intensity is 12 foot-candles at a distance of 10 ft, what is the intensity when the distance is 5 ft?

16. Mechanics The speed v of a gear varies inversely as the number of teeth t. If a gear that has 45 teeth makes 24 revolutions per minute (rpm), how many revolutions per minute will a gear that has 36 teeth make?

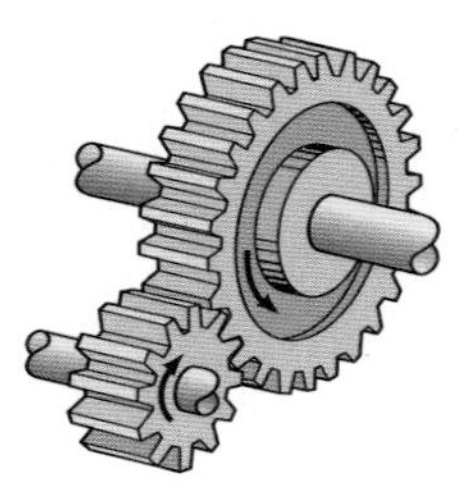

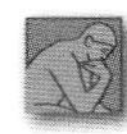

17. In the direct variation equation $y = kx$, what is the effect on y when x is doubled?

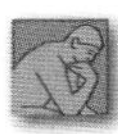

18. In the inverse variation equation $y = \frac{k}{x}$, what is the effect on x when y is doubled?

Applying the Concepts

For Exercises 19 to 22, complete using the word *directly* or *inversely*.

19. If a varies ____________ as b and c, then abc is constant.

20. If a varies directly as b and inversely as c, then c varies ____________ as b and ____________ as a.

21. If the area of a rectangle is held constant, the length of the rectangle varies ____________ as the width.

22. If the length of a rectangle is held constant, the area of the rectangle varies ____________ as the width.

FOCUS ON PROBLEM SOLVING

Implication

Sentences that are constructed using "If. . . , then. . ." occur frequently in problem-solving situations. These sentences are called **implications.** The sentence "If it rains, then I will stay home" is an implication. The phrase *it rains* is called the **antecedent** of the implication. The phrase *I will stay home* is the **consequent** of the implication.

The sentence "If $x = 4$, then $x^2 = 16$" is a true sentence. The **contrapositive** of an implication is formed by switching the antecedent and the consequent and then negating each one. The contrapositive of "If $x = 4$, then $x^2 = 16$" is "If $x^2 \neq 16$, then $x \neq 4$." This sentence is also true. It is a principle of logic that an implication and its contrapositive are either both true or both false.

The **converse** of an implication is formed by switching the antecedent and the consequent. The converse of "If $x = 4$, then $x^2 = 16$" is "If $x^2 = 16$, then $x = 4$." Note that the converse is not a true statement because if $x^2 = 16$, then x could equal -4. The converse of an implication may or may not be true.

Those statements for which the implication and the converse are both true are very important. They can be stated in the form "x if and only if y." For instance, a number is divisible by 5 if and only if the last digit of the number is 0 or 5. The implication "If a number is divisible by 5, then it ends in 0 or 5" and the converse "If a number ends in 0 or 5, then it is divisible by 5" are both true.

For Exercises 1 to 14, state the contrapositive and the converse of the implication. If the converse and the implication are both true statements, write a sentence using the phrase *if and only if.*

1. If I live in Chicago, then I live in Illinois.
2. If today is June 1, then yesterday was May 31.
3. If today is not Thursday, then tomorrow is not Friday.
4. If a number is divisible by 8, then it is divisible by 4.
5. If a number is an even number, then it is divisible by 2.
6. If a number is a multiple of 6, then it is a multiple of 3.
7. If $4z = 20$, then $z = 5$.
8. If an angle measures 90°, then it is a right angle.
9. If p is a prime number greater than 2, then p is an odd number.
10. If the equation of a graph is $y = mx + b$, then the graph of the equation is a straight line.
11. If $a = 0$ or $b = 0$, then $ab = 0$.
12. If the coordinates of a point are $(5, 0)$, then the point is on the x-axis.
13. If a quadrilateral is a square, then the quadrilateral has four sides of equal length.
14. If $x = y$, then $x^2 = y^2$.

PROJECTS AND GROUP ACTIVITIES

Intensity of Illumination

You are already aware that the standard unit of length in the metric system is the meter (m) and that the standard unit of mass in the metric system is the gram (g). You may not know that the standard unit of light intensity is the **candela (cd).**

The rate at which light falls on a 1-square-unit area of surface is called the **intensity of illumination.** Intensity of illumination is measured in **lumens (lm).** A lumen is defined in the following paragraph.

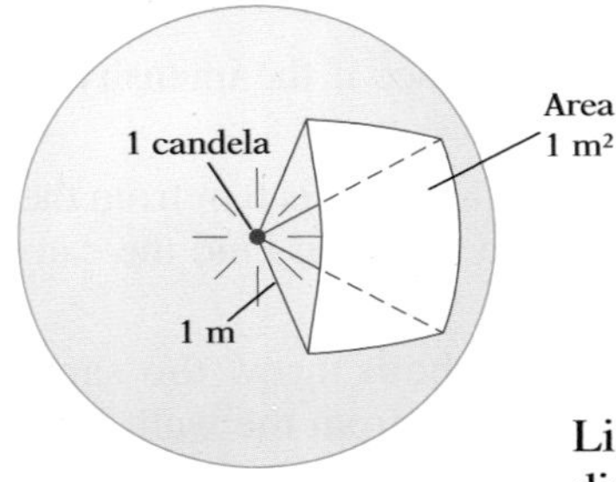

Picture a source of light equal to 1 cd positioned at the center of a hollow sphere that has a radius of 1 m. The rate at which light falls on 1 m^2 of the inner surface of the sphere is equal to 1 lm. If a light source equal to 4 cd is positioned at the center of the sphere, each square meter of the inner surface receives four times as much illumination, or 4 lm.

Light rays diverge as they leave a light source. The light that falls on an area of 1 m^2 at a distance of 1 m from the source of light spreads out over an area of 4 m^2 when it is 2 m from the source. The same light spreads out over an area of 9 m^2 when it is 3 m from the light source and over an area of 16 m^2 when it is 4 m from the light source. Therefore, as a surface moves farther away from the source of light, the intensity of illumination on the surface decreases from its value at 1 m to $\left(\frac{1}{2}\right)^2$, or $\frac{1}{4}$, that value at 2 m; to $\left(\frac{1}{3}\right)^2$, or $\frac{1}{9}$, that value at 3 m; and to $\left(\frac{1}{4}\right)^2$, or $\frac{1}{16}$, that value at 4 m.

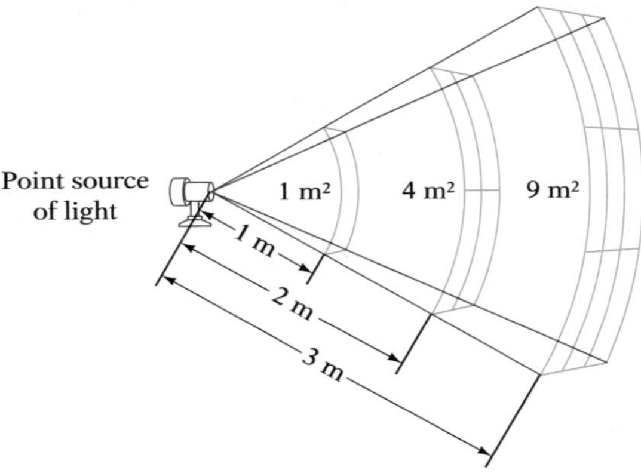

The formula for the intensity of illumination is

$$I = \frac{s}{r^2}$$

where I is the intensity of illumination in lumens, s is the strength of the light source in candelas, and r is the distance in meters between the light source and the illuminated surface.

A 30-candela lamp is 0.5 m above a desk. Find the illumination on the desk.

$$I = \frac{s}{r^2}$$

$$I = \frac{30}{(0.5)^2} = 120$$

The illumination on the desk is 120 lm.

Solve.

1. A 100-candela light is hanging 5 m above a floor. What is the intensity of the illumination on the floor beneath it?
2. A 25-candela source of light is 2 m above a desk. Find the intensity of illumination on the desk.
3. How strong a light source is needed to cast 20 lm of light on a surface 4 m from the source?
4. How strong a light source is needed to cast 80 lm of light on a surface 5 m from the source?
5. How far from the desk surface must a 40-candela light source be positioned if the desired intensity of illumination is 10 lm?
6. Find the distance between a 36-candela light source and a surface if the intensity of illumination on the surface is 0.01 lm.
7. Two lights cast the same intensity of illumination on a wall. One light is 6 m from the wall and has a rating of 36 cd. The second light is 8 m from the wall. Find the candela rating of the second light.
8. A 40-candela light source and a 10-candela light source both throw the same intensity of illumination on a wall. The 10-candela light is 6 m from the wall. Find the distance from the 40-candela light to the wall.

CHAPTER 8

SUMMARY

KEY WORDS	EXAMPLES
A *rational expression* is a fraction in which the numerator and denominator are polynomials. A rational expression is in *simplest form* when the numerator and denominator have no common factors. [8.1A, p. 458]	$\frac{2x+1}{x^2+4}$ is a rational expression in simplest form.
The *reciprocal of a rational expression* is the rational expression with the numerator and denominator interchanged. [8.1C, p. 461]	The reciprocal of $\frac{3x-y}{x+4}$ is $\frac{x+4}{3x-y}$.
The *least common multiple (LCM) of two or more polynomials* is the polynomial of least degree that contains all the factors of each polynomial. [8.2A, p. 466]	The LCM of $3x^2 - 6x$ and $x^2 - 4$ is $3x(x-2)(x+2)$, because it contains the factors of $3x^2 - 6x = 3x(x-2)$ and the factors of $x^2 - 4 = (x-2)(x+2)$.
A *complex fraction* is a fraction whose numerator or denominator contains one or more fractions. [8.3A, p. 474]	$\dfrac{x - \frac{2}{x+1}}{1 - \frac{4}{x}}$ is a complex fraction.
A *ratio* is the quotient of two quantities that have the same unit. A *rate* is the quotient of two quantities that have different units. [8.5A, p. 482]	$\frac{9}{4}$ is a ratio. $\frac{60\text{ m}}{12\text{ s}}$ is a rate.

A *proportion* is an equation that states the equality of two ratios or rates. [8.5A, p. 482]

$\frac{3}{8} = \frac{12}{32}$ and $\frac{x \text{ ft}}{12 \text{ s}} = \frac{15 \text{ ft}}{160 \text{ s}}$ are proportions.

A *literal equation* is an equation that contains more than one variable. A *formula* is a literal equation that states a rule about measurements. [8.6A, p. 490]

$3x - 4y = 12$ is a literal equation.
$A = LW$ is a literal equation that is also the formula for the area of a rectangle.

Direct variation is a special function that can be expressed as the equation $y = kx^n$, where k is a constant called the *constant of variation* or the *constant of proportionality*. [8.8A, p. 502]

$E = mc^2$ is Einstein's formula relating energy and mass. c^2 is the constant of proportionality.

Inverse variation is a function that can be expressed as the equation $y = \frac{k}{x^n}$, where k is a constant. [8.8A, p. 503]

$I = \frac{k}{d^2}$ gives the intensity of a light source at a distance d from the source.

A *joint variation* is a variation in which a variable varies directly as the product of two or more variables. A joint variation can be expressed as the equation $z = kxy$, where k is a constant. [8.8A, p. 503]

$C = kAT$ is a formula for the cost of insulation, where A is the area to be insulated and T is the thickness of the insulation.

A *combined variation* is a variation in which two or more types of variation occur at the same time. [8.8A, p. 503]

$V = \frac{kT}{P}$ is a formula that states that the volume of a gas is directly proportional to the temperature and inversely proportional to the pressure.

ESSENTIAL RULES AND PROCEDURES

EXAMPLES

Simplifying Rational Expressions [8.1A, p. 458]
Factor the numerator and denominator. Divide the numerator and denominator by the common factors.

$$\frac{x^2 - 3x - 10}{x^2 - 25} = \frac{(x + 2)(x - 5)}{(x + 5)(x - 5)} = \frac{x + 2}{x + 5}$$

Multiplying Rational Expressions [8.1B, p. 459]
Multiply the numerators. Multiply the denominators. Write the answer in simplest form.

$$\frac{a}{b} \cdot \frac{c}{d} = \frac{ac}{bd}$$

$$\frac{x^2 - 3x}{x^2 + x} \cdot \frac{x^2 + 5x + 4}{x^2 - 4x + 3}$$
$$= \frac{x(x - 3)}{x(x + 1)} \cdot \frac{(x + 1)(x + 4)}{(x - 3)(x - 1)}$$
$$= \frac{x(x - 3)(x + 1)(x + 4)}{x(x + 1)(x - 3)(x - 1)}$$
$$= \frac{x + 4}{x - 1}$$

Dividing Rational Expressions [8.1C, p. 461]

Multiply the dividend by the reciprocal of the divisor. Write the answer in simplest form.

$$\frac{a}{b} \div \frac{c}{d} = \frac{a}{b} \cdot \frac{d}{c} = \frac{ad}{bc}$$

$$\frac{4x+16}{3x-6} \div \frac{x^2+6x+8}{x^2-4}$$
$$= \frac{4x+16}{3x-6} \cdot \frac{x^2-4}{x^2+6x+8}$$
$$= \frac{4(x+4)}{3(x-2)} \cdot \frac{(x-2)(x+2)}{(x+4)(x+2)}$$
$$= \frac{4}{3}$$

Adding and Subtracting Rational Expressions [8.2B, p. 468]

1. Find the LCM of the denominators.
2. Write each fraction as an equivalent fraction using the LCM as the denominator.
3. Add or subtract the numerators and place the result over the common denominator.
4. Write the answer in simplest form.

$$\frac{a}{b} + \frac{c}{b} = \frac{a+c}{b} \qquad \frac{a}{b} - \frac{c}{b} = \frac{a-c}{b}$$

$$\frac{x}{x+1} - \frac{x+3}{x-2}$$
$$= \frac{x}{x+1} \cdot \frac{x-2}{x-2} - \frac{x+3}{x-2} \cdot \frac{x+1}{x+1}$$
$$= \frac{x(x-2)}{(x+1)(x-2)} - \frac{(x+3)(x+1)}{(x+1)(x-2)}$$
$$= \frac{x(x-2)-(x+3)(x+1)}{(x+1)(x-2)}$$
$$= \frac{(x^2-2x)-(x^2+4x+3)}{(x+1)(x-2)}$$
$$= \frac{-6x-3}{(x+1)(x-2)} = \frac{-3(2x+1)}{(x+1)(x-2)}$$

Simplifying Complex Fractions [8.3A, p. 474]

Method 1: Multiply by 1 in the form $\frac{\text{LCM}}{\text{LCM}}$.

1. Determine the LCM of the denominators of the fractions in the numerator and denominator of the complex fraction.
2. Multiply the numerator and denominator of the complex fraction by the LCM.
3. Simplify.

Method 1: $$\frac{\frac{1}{x}+\frac{1}{y}}{\frac{1}{x}-\frac{1}{y}} = \frac{\frac{1}{x}+\frac{1}{y}}{\frac{1}{x}-\frac{1}{y}} \cdot \frac{xy}{xy}$$
$$= \frac{\frac{1}{x}\cdot xy + \frac{1}{y}\cdot xy}{\frac{1}{x}\cdot xy - \frac{1}{y}\cdot xy}$$
$$= \frac{y+x}{y-x}$$

Method 2: Multiply the numerator by the reciprocal of the denominator.

1. Simplify the numerator to a single fraction and simplify the denominator to a single fraction.
2. Using the definition for dividing fractions, multiply the numerator by the reciprocal of the denominator.
3. Simplify.

Method 2: $$\frac{\frac{1}{x}+\frac{1}{y}}{\frac{1}{x}-\frac{1}{y}} = \frac{\frac{y+x}{xy}}{\frac{y-x}{xy}}$$
$$= \frac{y+x}{xy} \cdot \frac{xy}{y-x}$$
$$= \frac{y+x}{y-x}$$

Solving Equations Containing Fractions [8.4A, p. 478]
Clear denominators by multiplying each side of the equation by the LCM of the denominators. Then solve for the variable.

$$\frac{1}{2a} = \frac{2}{a} - \frac{3}{8}$$

$$8a\left(\frac{1}{2a}\right) = 8a\left(\frac{2}{a}\right) - 8a\left(\frac{3}{8}\right)$$

$$4 = 16 - 3a$$

$$-12 = -3a$$

$$4 = a$$

Similar Triangles [8.5C, pp. 483–484]
Similar triangles have the same shape but not necessarily the same size. The ratios of corresponding parts of similar triangles are equal. The measures of the corresponding angles of similar triangles are equal.

Triangles ABC and DFE are similar triangles. The ratios of corresponding parts are equal to $\frac{2}{3}$.

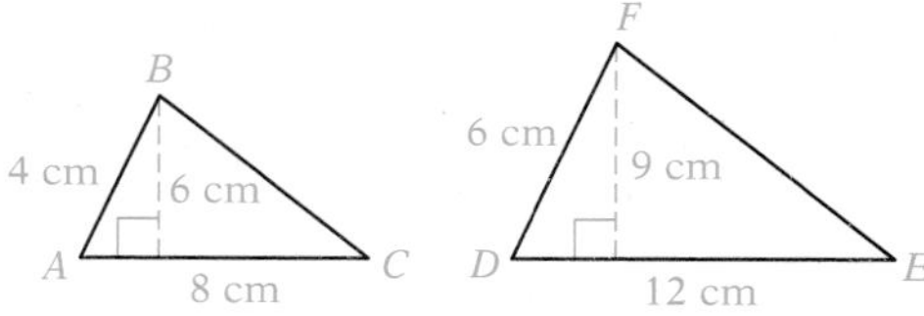

If two angles of one triangle are equal in measure to two angles of another triangle, then the two triangles are similar.

Triangles AOB and COD are similar because $m\angle AOB = m\angle COD$ and $m\angle B = m\angle D$.

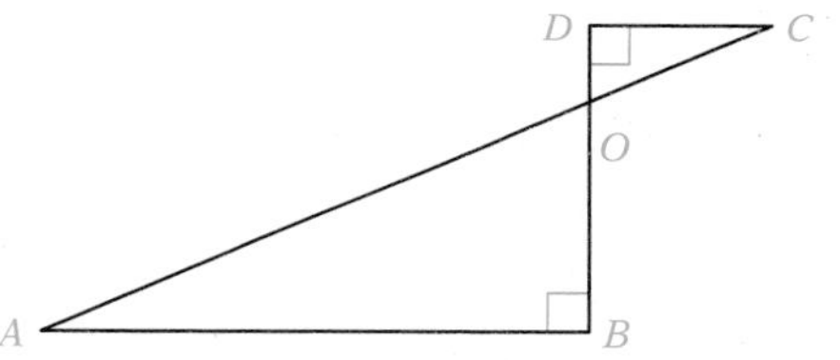

Solving Literal Equations [8.6A, p. 490]
Rewrite the equation so that the letter being solved for is alone on one side of the equation and all numbers and other variables are on the other side.

Solve $2x + ax = 5$ for x.

$$2x + ax = 5$$

$$x(2 + a) = 5$$

$$\frac{x(2 + a)}{2 + a} = \frac{5}{2 + a}$$

$$x = \frac{5}{2 + a}$$

Work Problems [8.7A, p. 494]
Rate of work × time worked = part of task completed

Pat can do a certain job in 3 h. Chris can do the same job in 5 h. How long would it take them, working together, to get the job done?

$$\frac{t}{3} + \frac{t}{5} = 1$$

Uniform Motion Problems with Rational Expressions [8.7B, p. 496]

$$\frac{\text{Distance}}{\text{Rate}} = \text{time}$$

Train A's speed is 15 mph faster than train B's speed. Train A travels 150 mi in the same amount of time it takes train B to travel 120 mi. Find the rate of train B.

$$\frac{120}{r} = \frac{150}{r + 15}$$

CHAPTER 8

CONCEPT REVIEW

Test your knowledge of the concepts presented in this chapter. Answer each question. Then check your answers against the ones provided in the Answer Section.

1. When is a rational expression in simplest form?

2. How is the reciprocal useful when dividing rational expressions?

3. How do you find the LCM of two polynomials?

4. When subtracting two rational expressions, what must be the same about both expressions before subtraction can take place?

5. What are the steps for adding rational expressions?

6. What are the steps used to simplify a complex fraction? Use either method.

7. When solving an equation that contains fractions, why do we first clear the denominators?

8. If the units in a comparison are different, is it a ratio or a rate?

9. How can you use a proportion to solve similar triangles?

10. What is the goal when you solve a literal equation for a particular variable?

11. What is the rate of work if the job is completed in x hours?

CHAPTER 8

REVIEW EXERCISES

1. Divide: $\dfrac{6a^2b^7}{25x^3y} \div \dfrac{12a^3b^4}{5x^2y^2}$

2. Add: $\dfrac{x+7}{15x} + \dfrac{x-2}{20x}$

3. Simplify: $\dfrac{x - \dfrac{16}{5x-2}}{3x - 4 - \dfrac{88}{5x-2}}$

4. Simplify: $\dfrac{x^2 + x - 30}{15 + 2x - x^2}$

5. Simplify: $\dfrac{16x^5y^3}{24xy^{10}}$

6. Solve: $\dfrac{20}{x+2} = \dfrac{5}{16}$

7. Divide: $\dfrac{10 - 23y + 12y^2}{6y^2 - y - 5} \div \dfrac{4y^2 - 13y + 10}{18y^2 + 3y - 10}$

8. Multiply: $\dfrac{8ab^2}{15x^3y} \cdot \dfrac{5xy^4}{16a^2b}$

9. Simplify: $\dfrac{1 - \dfrac{1}{x}}{1 - \dfrac{8x-7}{x^2}}$

10. Write each fraction in terms of the LCM of the denominators.

$\dfrac{x}{12x^2 + 16x - 3}, \dfrac{4x^2}{6x^2 + 7x - 3}$

11. Solve $T = 2(ab + bc + ca)$ for a.

12. Solve: $\dfrac{5}{7} + \dfrac{x}{2} = 2 - \dfrac{x}{7}$

13. Solve $i = \dfrac{100m}{c}$ for c.

14. Solve: $\dfrac{x+8}{x+4} = 1 + \dfrac{5}{x+4}$

15. Divide: $\dfrac{20x^2 - 45x}{6x^3 + 4x^2} \div \dfrac{40x^3 - 90x^2}{12x^2 + 8x}$

16. Add: $\dfrac{2y}{5y-7} + \dfrac{3}{7-5y}$

17. Subtract: $\dfrac{5x + 3}{2x^2 + 5x - 3} - \dfrac{3x + 4}{2x^2 + 5x - 3}$

18. Find the LCM of $10x^2 - 11x + 3$ and $20x^2 - 17x + 3$.

19. Solve $4x + 9y = 18$ for y.

20. Multiply: $\dfrac{24x^2 - 94x + 15}{12x^2 - 49x + 15} \cdot \dfrac{24x^2 + 7x - 5}{4 - 27x + 18x^2}$

21. Solve: $\dfrac{20}{2x + 3} = \dfrac{17x}{2x + 3} - 5$

22. Add: $\dfrac{x - 1}{x + 2} + \dfrac{3x - 2}{5 - x} + \dfrac{5x^2 + 15x - 11}{x^2 - 3x - 10}$

23. Solve: $\dfrac{6}{x - 7} = \dfrac{8}{x - 6}$

24. Solve: $\dfrac{3}{20} = \dfrac{x}{80}$

25. Triangles *ABC* and *DEF* are similar. Find the perimeter of triangle *ABC*.

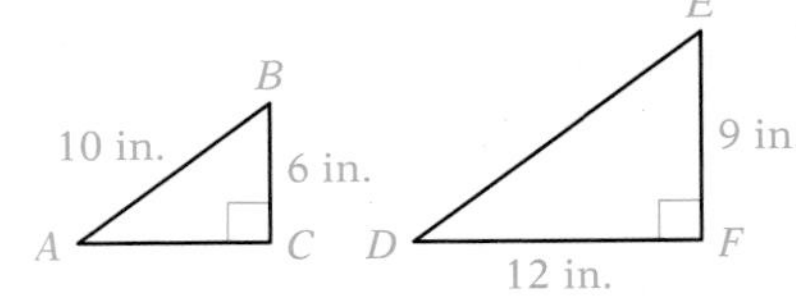

26. **Work** One hose can fill a pool in 15 h. The second hose can fill the pool in 10 h. How long would it take to fill the pool using both hoses?

27. **Travel** A car travels 315 mi in the same amount of time that a bus travels 245 mi. The rate of the car is 10 mph greater than that of the bus. Find the rate of the car.

28. **Travel** The rate of a jet is 400 mph in calm air. Traveling with the wind, the jet can fly 2100 mi in the same amount of time it takes to fly 1900 mi against the wind. Find the rate of the wind.

29. **Baseball** A pitcher's earned run average (ERA) is the average number of runs allowed in nine innings of pitching. If a pitcher allows 15 runs in 100 innings, find the pitcher's ERA.

30. **Electronics** The current (I) in an electric circuit varies inversely as the resistance (R). If the current in the circuit is 4 amps when the resistance is 50 ohms, find the current in the circuit when the resistance is 100 ohms.

CHAPTER 8

TEST

1. Simplify: $\dfrac{16x^5y}{24x^2y^4}$

2. Simplify: $\dfrac{x^2 + 4x - 5}{1 - x^2}$

3. Multiply: $\dfrac{x^3y^4}{x^2 - 4x + 4} \cdot \dfrac{x^2 - x - 2}{x^6y^4}$

4. Multiply: $\dfrac{x^2 + 2x - 3}{x^2 + 6x + 9} \cdot \dfrac{2x^2 - 11x + 5}{2x^2 + 3x - 5}$

5. Divide: $\dfrac{x^2 + 3x + 2}{x^2 + 5x + 4} \div \dfrac{x^2 - x - 6}{x^2 + 2x - 15}$

6. Find the LCM of $6x - 3$ and $2x^2 + x - 1$.

7. Write each fraction in terms of the LCM of the denominators.
$\dfrac{3}{x^2 - 2x}, \dfrac{x}{x^2 - 4}$

8. Subtract: $\dfrac{2x}{x^2 + 3x - 10} - \dfrac{4}{x^2 + 3x - 10}$

9. Subtract: $\dfrac{2}{2x - 1} - \dfrac{3}{3x + 1}$

10. Subtract: $\dfrac{x}{x + 3} - \dfrac{2x - 5}{x^2 + x - 6}$

11. Simplify: $\dfrac{1 + \dfrac{1}{x} - \dfrac{12}{x^2}}{1 + \dfrac{2}{x} - \dfrac{8}{x^2}}$

12. Solve: $\dfrac{6}{x} - 2 = 1$

13. Solve: $\dfrac{2x}{x+1} - 3 = \dfrac{-2}{x+1}$

14. Solve: $\dfrac{3}{x+4} = \dfrac{5}{x+6}$

15. Triangles ABC and DEF are similar. Find the area of triangle DEF.

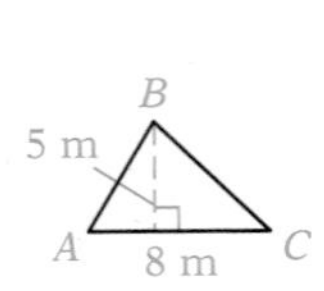

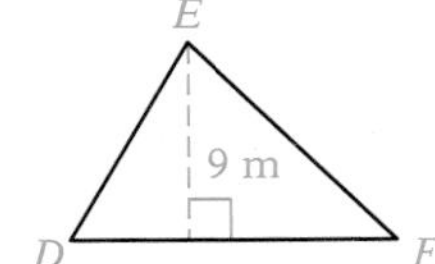

16. Solve $d = s + rt$ for t.

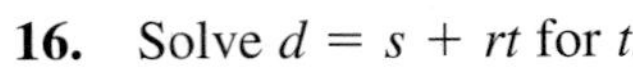

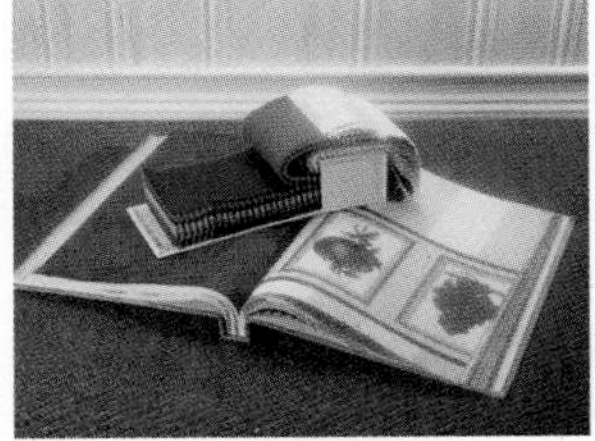

17. **Interior Design** An interior designer uses two rolls of wallpaper for every 45 ft^2 of wall space in an office. At this rate, how many rolls of wallpaper are needed for an office that has 315 ft^2 of wall space?

18. **Landscaping** One landscaper can till the soil for a lawn in 30 min, whereas it takes a second landscaper 15 min to do the same job. How long would it take to till the soil for the lawn with both landscapers working together?

19. **Travel** A cyclist travels 20 mi in the same amount of time that it takes a hiker to walk 6 mi. The rate of the cyclist is 7 mph faster than the rate of the hiker. Find the rate of the cyclist.

20. **Electronics** The electrical resistance (r) of a cable varies directly as its length (l) and inversely as the square of its diameter (d). If a cable 16,000 ft long and $\frac{1}{4}$ in. in diameter has a resistance of 3.2 ohms, what is the resistance of a cable that is 8000 ft long and $\frac{1}{2}$ in. in diameter?

CUMULATIVE REVIEW EXERCISES

1. Simplify: $\left(\frac{2}{3}\right)^2 \div \left(\frac{3}{2} - \frac{2}{3}\right) + \frac{1}{2}$

2. Evaluate $-a^2 + (a - b)^2$ when $a = -2$ and $b = 3$.

3. Simplify: $-2x - (-3y) + 7x - 5y$

4. Simplify: $2[3x - 7(x - 3) - 8]$

5. Solve: $4 - \frac{2}{3}x = 7$

6. Solve: $3[x - 2(x - 3)] = 2(3 - 2x)$

7. Find $16\frac{2}{3}\%$ of 60.

8. Solve: $x - 3(1 - 2x) \geq 1 - 4(3 - 2x)$

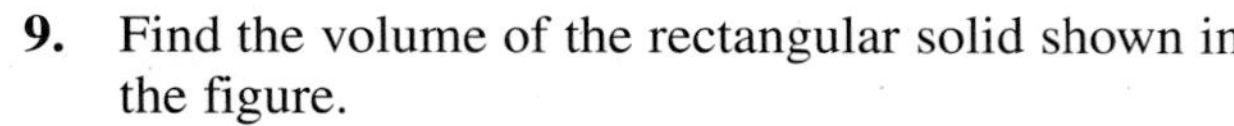

9. Find the volume of the rectangular solid shown in the figure.

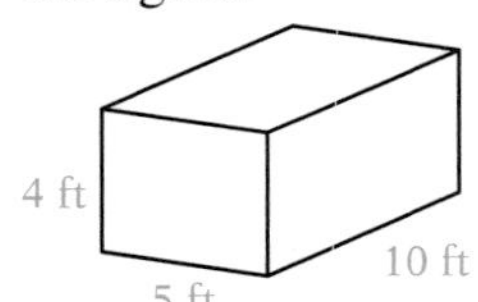

10. Graph: $x - 2y = 2$

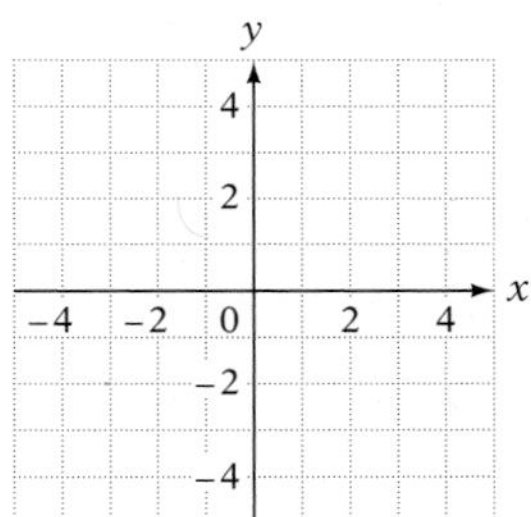

11. Given $P(x) = \frac{x - 1}{2x - 3}$, find $P(-2)$.

12. Find the equation of the line that contains the point $(-2, -1)$ and is parallel to the line $3x - 2y = 6$.

13. Solve:
$$\begin{aligned} 2x - y + z &= 2 \\ 3x + y - 2z &= 9 \\ x - y + z &= 0 \end{aligned}$$

14. Multiply: $(a^2b^5)(ab^2)$

15. Simplify: $\frac{(2a^{-2}b^3)^{-2}}{(4a)^{-1}}$

16. Write 0.000000035 in scientific notation.

17. Multiply: $(2a^2 - 3a + 1)(-2a^2)$

18. Multiply: $(a - 3b)(a + 4b)$

19. Divide: $(x^3 - 8) \div (x - 2)$

20. Factor: $y^2 - 7y + 6$

21. Factor: $12x^2 - x - 1$

22. Factor: $2a^3 + 7a^2 - 15a$

23. Factor: $4b^2 - 100$

24. Solve: $(x + 3)(2x - 5) = 0$

25. Simplify: $\dfrac{12x^4y^2}{18xy^7}$

26. Simplify: $\dfrac{x^2 - 7x + 10}{25 - x^2}$

27. Divide: $\dfrac{x^2 - x - 56}{x^2 + 8x + 7} \div \dfrac{x^2 - 13x + 40}{x^2 - 4x - 5}$

28. Subtract: $\dfrac{2}{2x - 1} - \dfrac{1}{x + 1}$

29. Simplify: $\dfrac{1 - \dfrac{2}{x} - \dfrac{15}{x^2}}{1 - \dfrac{25}{x^2}}$

30. Solve: $\dfrac{3x}{x - 3} - 2 = \dfrac{10}{x - 3}$

31. **Mixtures** A silversmith mixes 60 g of an alloy that is 40% silver with 120 g of another silver alloy. The resulting alloy is 60% silver. Find the percent of silver in the 120 g of alloy.

32. **Insurance** A life insurance policy costs $32 for every $1000 of coverage. At this rate, how much money would a policy of $5000 cost?

33. **Work** One water pipe can fill a tank in 9 min, whereas a second pipe requires 18 min to fill the tank. How long would it take both pipes, working together, to fill the tank?

CHAPTER 9

Exponents and Radicals

Panstock/First Light

OBJECTIVES

SECTION 9.1

- A To simplify expressions with rational exponents
- B To write exponential expressions as radical expressions and to write radical expressions as exponential expressions
- C To simplify radical expressions that are roots of perfect powers

SECTION 9.2

- A To simplify radical expressions
- B To add or subtract radical expressions
- C To multiply radical expressions
- D To divide radical expressions

SECTION 9.3

- A To solve a radical equation
- B To solve application problems

SECTION 9.4

- A To simplify a complex number
- B To add or subtract complex numbers
- C To multiply complex numbers
- D To divide complex numbers

ARE YOU READY?

Take the Chapter 9 Prep Test to find out if you are ready to learn to:

- Simplify expressions with rational exponents
- Simplify radical expressions
- Add, subtract, multiply, and divide radical expressions
- Solve a radical equation
- Add, subtract, multiply, and divide complex numbers

PREP TEST

Do these exercises to prepare for Chapter 9.

1. Complete: $48 = ? \cdot 3$

For Exercises 2 to 7, simplify.

2. 2^5

3. $6\left(\frac{3}{2}\right)$

4. $\frac{1}{2} - \frac{2}{3} + \frac{1}{4}$

5. $(3 - 7x) - (4 - 2x)$

6. $\frac{3x^5y^6}{12x^4y}$

7. $(3x - 2)^2$

For Exercises 8 and 9, multiply.

8. $(2 + 4x)(5 - 3x)$

9. $(6x - 1)(6x + 1)$

10. Solve: $x^2 - 14x - 5 = 10$

SECTION

9.1 Rational Exponents and Radical Expressions

OBJECTIVE A To simplify expressions with rational exponents

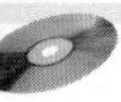

Point of Interest

Nicolas Chuquet (c. 1475), a French physician, wrote an algebra text in which he used a notation for expressions with fractional exponents. He wrote R^26 to mean $6^{1/2}$ and R^315 to mean $15^{1/3}$. This was an improvement over earlier notations that used words for these expressions.

In this section, the definition of an exponent is extended beyond integers so that any rational number can be used as an exponent. The definition is expressed in such a way that the Rules of Exponents hold true for rational exponents.

Consider the expression $(a^{1/n})^n$ for $a > 0$ and n a positive integer. Now simplify, assuming that the Rule for Simplifying the Power of an Exponential Expression is true.

$$(a^{1/n})^n = a^{\frac{1}{n}\cdot n} = a^1 = a$$

Because $(a^{1/n})^n = a$, the number $a^{1/n}$ is the number whose nth power is a.

If $a > 0$ and n is a positive number, then $a^{1/n}$ is called the nth root of a.

$25^{1/2} = 5$ because $(5)^2 = 25$.

$8^{1/3} = 2$ because $(2)^3 = 8$.

Integrating Technology

A calculator can be used to evaluate expressions with rational exponents. For example, to evaluate the expression at the right, press 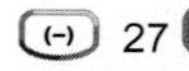 27 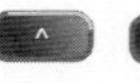 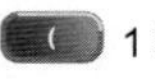 1 ÷ 3 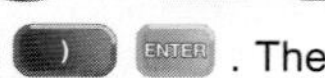. The display reads −3.

In the expression $a^{1/n}$, if a is a negative number and n is a positive even integer, then $a^{1/n}$ is not a real number.

$(-4)^{1/2}$ is not a real number, because there is no real number whose second power is -4.

When n is a positive odd integer, a can be a positive or a negative number.

$(-27)^{1/3} = -3$ because $(-3)^3 = -27$.

Using the definition of $a^{1/n}$ and the Rules of Exponents, it is possible to define any exponential expression that contains a rational exponent.

Rule for Rational Exponents

If m and n are positive integers and $a^{1/n}$ is a real number, then

$$a^{m/n} = (a^{1/n})^m$$

The expression $a^{m/n}$ can also be written $a^{m/n} = a^{m\cdot\frac{1}{n}} = (a^m)^{1/n}$. However, rewriting $a^{m/n}$ as $(a^m)^{1/n}$ is not as useful as rewriting it as $(a^{1/n})^m$. See the Take Note at the top of the next page.

As shown above, expressions that contain rational exponents do not always represent real numbers when the base of the exponential expression is a negative number. For this reason, **all variables in this chapter represent positive numbers unless otherwise stated.**

Take Note

Although we can simplify an expression by rewriting $a^{m/n}$ in the form $(a^m)^{1/n}$, it is usually easier to simplify the form $(a^{1/n})^m$. For instance, simplifying $(27^{1/3})^2$ is easier than simplifying $(27^2)^{1/3}$.

HOW TO • 1 Simplify: $27^{2/3}$

$27^{2/3} = (3^3)^{2/3}$ • Rewrite 27 as 3^3.

$= 3^{3(2/3)} = 3^2$ • Multiply the exponents.

$= 9$

Take Note

Note that $32^{-2/5} = \frac{1}{4}$, a positive number. The negative exponent does not affect the sign of a number.

HOW TO • 2 Simplify: $32^{-2/5}$

$32^{-2/5} = (2^5)^{-2/5}$ • Rewrite 32 as 2^5.

$= 2^{-2} = \frac{1}{2^2}$ • Multiply the exponents. Then use the Rule of Negative Exponents.

$= \frac{1}{4}$ • Simplify.

HOW TO • 3 Simplify: $a^{1/2} \cdot a^{2/3} \cdot a^{-1/4}$

$a^{1/2} \cdot a^{2/3} \cdot a^{-1/4} = a^{1/2 + 2/3 - 1/4}$ • Use the Rule for Multiplying Exponential Expressions.

$= a^{6/12 + 8/12 - 3/12}$

$= a^{11/12}$ • Simplify.

HOW TO • 4 Simplify: $(x^6y^4)^{3/2}$

$(x^6y^4)^{3/2} = x^{6(3/2)}y^{4(3/2)}$ • Use the Rule for Simplifying Powers of Products.

$= x^9y^6$ • Simplify.

HOW TO • 5 Simplify: $3x^{-3/4}(2x^{11/4} - x^{-1/4})$

$3x^{-3/4}(2x^{11/4} - x^{-1/4}) = 3x^{-3/4}(2x^{11/4}) - 3x^{-3/4}(x^{-1/4})$ • Use the Distributive Property to remove parentheses.

$= 6x^{-3/4 + 11/4} - 3x^{-3/4 + (-1/4)}$ • Use the Rule for Multiplying Exponential Expressions.

$= 6x^{8/4} - 3x^{-4/4} = 6x^2 - 3x^{-1}$ • Simplify.

$= 6x^2 - \frac{3}{x}$

Tips for Success

Remember that a HOW TO indicates a worked-out example. Using paper and pencil, work through the example. See *AIM for Success* in the Preface.

HOW TO • 6 Simplify: $\left(\frac{8a^3b^{-4}}{64a^{-9}b^2}\right)^{2/3}$

$\left(\frac{8a^3b^{-4}}{64a^{-9}b^2}\right)^{2/3} = \left(\frac{2^3a^3b^{-4}}{2^6a^{-9}b^2}\right)^{2/3}$ • Rewrite 8 as 2^3 and 64 as 2^6.

$= (2^{-3}a^{12}b^{-6})^{2/3}$ • Use the Rule for Dividing Exponential Expressions.

$= 2^{-2}a^8b^{-4}$ • Use the Rule for Simplifying Powers of Products.

$= \frac{a^8}{2^2b^4} = \frac{a^8}{4b^4}$ • Use the Rule of Negative Exponents and simplify.

EXAMPLE • 1

Simplify: $64^{-2/3}$

Solution $64^{-2/3} = (2^6)^{-2/3} = 2^{-4}$ • $64 = 2^6$

$= \frac{1}{2^4} = \frac{1}{16}$

YOU TRY IT • 1

Simplify: $16^{-3/4}$

Your solution

EXAMPLE • 2

Simplify: $(-49)^{3/2}$

Solution
The base of the exponential expression is a negative number, while the denominator of the exponent is a positive even number.
Therefore, $(-49)^{3/2}$ is not a real number.

YOU TRY IT • 2

Simplify: $(-81)^{3/4}$

Your solution

EXAMPLE • 3

Simplify: $(x^{1/2}y^{-3/2}z^{1/4})^{-3/2}$

Solution $(x^{1/2}y^{-3/2}z^{1/4})^{-3/2}$

$= x^{-3/4}y^{9/4}z^{-3/8}$

$= \frac{y^{9/4}}{x^{3/4}z^{3/8}}$

• Use the Rule for Simplifying Powers of Products.

YOU TRY IT • 3

Simplify: $(x^{3/4}y^{1/2}z^{-2/3})^{-4/3}$

Your solution

EXAMPLE • 4

Simplify: $\frac{x^{1/2}y^{-5/4}}{x^{-4/3}y^{1/3}}$

Solution $\frac{x^{1/2}y^{-5/4}}{x^{-4/3}y^{1/3}}$

$= x^{3/6-(-8/6)}y^{-15/12-4/12}$

$= x^{11/6}y^{-19/12} = \frac{x^{11/6}}{y^{19/12}}$

• Use the Rule for Dividing Exponential Expressions.

YOU TRY IT • 4

Simplify: $\left(\frac{16a^{-2}b^{4/3}}{9a^4b^{-2/3}}\right)^{-1/2}$

Your solution

Solutions on pp. S26–S27

OBJECTIVE B **To write exponential expressions as radical expressions and to write radical expressions as exponential expressions** 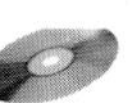

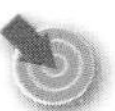

Point of Interest

The radical sign was introduced in 1525 by Christoff Rudolff in a book called *Coss.* He modified the symbol to indicate square roots, cube roots, and fourth roots. The idea of using an index, as we do in our modern notation, did not occur until some years later.

Recall that $a^{1/n}$ is the *n*th root of *a*. The expression $\sqrt[n]{a}$ is another symbol for the *n*th root of *a*.

If *a* is a real number, then $a^{1/n} = \sqrt[n]{a}$.

In the expression $\sqrt[n]{a}$, the symbol $\sqrt{}$ is called a **radical,** *n* is the **index** of the radical, and *a* is the **radicand.** When $n = 2$, the radical expression represents a square root and the index 2 is usually not written.

An exponential expression with a rational exponent can be written as a radical expression.

Writing Exponential Expressions as Radical Expressions

If $a^{1/n}$ is a real number, then $a^{1/n} = \sqrt[n]{a}$ and $a^{m/n} = a^{m \cdot 1/n} = (a^m)^{1/n} = \sqrt[n]{a^m}$.

The expression $a^{m/n}$ can also be written $a^{m/n} = (a^{1/n})^m = (\sqrt[n]{a})^m$.

The exponential expression at the right has been written as a radical expression.

$$y^{2/3} = (y^2)^{1/3}$$
$$= \sqrt[3]{y^2}$$

The radical expressions at the right have been written as exponential expressions.

$$\sqrt[5]{x^6} = (x^6)^{1/5} = x^{6/5}$$
$$\sqrt{17} = 17^{1/2}$$

HOW TO • 7 Write $(5x)^{2/5}$ as a radical expression.

$(5x)^{2/5} = \sqrt[5]{(5x)^2}$ • The denominator of the rational exponent is the index of the radical. The numerator is the power of the radicand.

$= \sqrt[5]{25x^2}$ • Simplify.

HOW TO • 8 Write $\sqrt[3]{x^4}$ as an exponential expression with a rational exponent.

$\sqrt[3]{x^4} = (x^4)^{1/3}$ • The index of the radical is the denominator of the rational exponent. The power of the radicand is the numerator of the rational exponent.

$= x^{4/3}$ • Simplify.

HOW TO • 9 Write $\sqrt[3]{a^3 + b^3}$ as an exponential expression with a rational exponent.

$$\sqrt[3]{a^3 + b^3} = (a^3 + b^3)^{1/3}$$

Note that $(a^3 + b^3)^{1/3} \neq a + b$.

EXAMPLE • 5

Write $(3x)^{5/4}$ as a radical expression.

Solution $(3x)^{5/4} = \sqrt[4]{(3x)^5} = \sqrt[4]{243x^5}$

YOU TRY IT • 5

Write $(2x^3)^{3/4}$ as a radical expression.

Your solution

EXAMPLE • 6

Write $-2x^{2/3}$ as a radical expression.

Solution $-2x^{2/3} = -2(x^2)^{1/3} = -2\sqrt[3]{x^2}$

YOU TRY IT • 6

Write $-5a^{5/6}$ as a radical expression.

Your solution

EXAMPLE • 7

Write $\sqrt[4]{3a}$ as an exponential expression.

Solution $\sqrt[4]{3a} = (3a)^{1/4}$

YOU TRY IT • 7

Write $\sqrt[3]{3ab}$ as an exponential expression.

Your solution

EXAMPLE • 8

Write $\sqrt{a^2 - b^2}$ as an exponential expression.

Solution $\sqrt{a^2 - b^2} = (a^2 - b^2)^{1/2}$

YOU TRY IT • 8

Write $\sqrt[4]{x^4 + y^4}$ as an exponential expression.

Your solution

Solutions on p. S27

OBJECTIVE C To simplify radical expressions that are roots of perfect powers 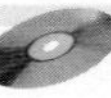

Every positive number has two square roots, one a positive number and one a negative number. For example, because $(5)^2 = 25$ and $(-5)^2 = 25$, there are two square roots of 25: 5 and -5.

The symbol $\sqrt{}$ is used to indicate the positive square root, or **principal square root.** To indicate the negative square root of a number, a negative sign is placed in front of the radical.

$\sqrt{25} = 5$

$-\sqrt{25} = -5$

The square root of zero is zero.

$\sqrt{0} = 0$

The square root of a negative number is not a real number because the square of a real number must be positive.

$\sqrt{-25}$ is not a real number.

Note that

$\sqrt{(-5)^2} = \sqrt{25} = 5$ and $\sqrt{5^2} = \sqrt{25} = 5$

This is true for all real numbers and is stated as the following result.

For any real number a, $\sqrt{a^2} = |a|$ and $-\sqrt{a^2} = -|a|$. If a is a positive real number, then $\sqrt{a^2} = a$ and $(\sqrt{a})^2 = a$.

Integrating Technology

See the Keystroke Guide: *Radical Expressions* for instructions on using a graphing calculator to evaluate a numerical radical expression.

Besides square roots, we can also determine cube roots, fourth roots, and so on.

$\sqrt[3]{8} = 2$, because $2^3 = 8$. • The cube root of a positive number is positive.

$\sqrt[3]{-8} = -2$, because $(-2)^3 = -8$. • The cube root of a negative number is negative.

$\sqrt[4]{625} = 5$, because $5^4 = 625$.

$\sqrt[5]{243} = 3$, because $3^5 = 243$.

The following properties hold true for finding the nth root of a real number.

If n is an even integer, then $\sqrt[n]{a^n} = |a|$ and $-\sqrt[n]{a^n} = -|a|$. If n is an odd integer, then $\sqrt[n]{a^n} = a$.

For example,

$$\sqrt[6]{y^6} = |y| \qquad -\sqrt[12]{x^{12}} = -|x| \qquad \sqrt[5]{b^5} = b$$

Take Note

Note that when the index is an even natural number, the nth root requires absolute value symbols.

$\sqrt[6]{y^6} = |y|$ but $\sqrt[5]{y^5} = y$

Because we stated that variables within radicals represent *positive* numbers, we will omit the absolute value symbols when writing an answer.

For the remainder of this chapter, we will assume that variable expressions inside a radical represent positive numbers. Therefore, it is not necessary to use the absolute value signs.

HOW TO • 10 Simplify: $\sqrt[4]{x^4y^8}$

$\sqrt[4]{x^4y^8} = (x^4y^8)^{1/4}$ • The radicand is a perfect fourth power because the exponents on the variables are divisible by 4. Write the radical expression as an exponential expression.

$= xy^2$ • Use the Rule for Simplifying Powers of Products.

HOW TO • 11 Simplify: $\sqrt[3]{125c^9d^6}$

$\sqrt[3]{125c^9d^6} = (5^3c^9d^6)^{1/3}$ • The radicand is a perfect cube because 125 is a perfect cube ($125 = \mathbf{5^3}$) and all the exponents on the variables are divisible by 3.

$= 5c^3d^2$ • Use the Rule for Simplifying Powers of Products.

Note that a variable expression is a perfect power if the exponents on the factors are evenly divisible by the index of the radical.

The chart below shows roots of perfect powers. Knowledge of these roots is very helpful when simplifying radical expressions.

Square Roots		*Cube Roots*	*Fourth Roots*	*Fifth Roots*
$\sqrt{1} = 1$	$\sqrt{36} = 6$	$\sqrt[3]{1} = 1$	$\sqrt[4]{1} = 1$	$\sqrt[5]{1} = 1$
$\sqrt{4} = 2$	$\sqrt{49} = 7$	$\sqrt[3]{8} = 2$	$\sqrt[4]{16} = 2$	$\sqrt[5]{32} = 2$
$\sqrt{9} = 3$	$\sqrt{64} = 8$	$\sqrt[3]{27} = 3$	$\sqrt[4]{81} = 3$	$\sqrt[5]{243} = 3$
$\sqrt{16} = 4$	$\sqrt{81} = 9$	$\sqrt[3]{64} = 4$	$\sqrt[4]{256} = 4$	
$\sqrt{25} = 5$	$\sqrt{100} = 10$	$\sqrt[3]{125} = 5$	$\sqrt[4]{625} = 5$	

Take Note
From the chart, $\sqrt[5]{243} = 3$, which means that $3^5 = 243$. From this we know that $(-3)^5 = -243$, which means $\sqrt[5]{-243} = -3$.

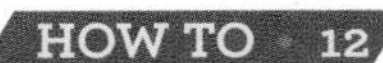

HOW TO • 12 Simplify: $\sqrt[5]{-243x^5y^{15}}$

$\sqrt[5]{-243x^5y^{15}} = -3xy^3$ • From the chart, 243 is a perfect fifth power, and each exponent is divisible by 5. Therefore, the radicand is a perfect fifth power.

EXAMPLE • 9

Simplify: $\sqrt[3]{-125a^6b^9}$

Solution
The radicand is a perfect cube.

$\sqrt[3]{-125a^6b^9} = -5a^2b^3$ • Divide each exponent by 3.

YOU TRY IT • 9

Simplify: $\sqrt[3]{-8x^{12}y^3}$

Your solution

EXAMPLE • 10

Simplify: $-\sqrt[4]{16a^4b^8}$

Solution
The radicand is a perfect fourth power.

$-\sqrt[4]{16a^4b^8} = -2ab^2$ • Divide each exponent by 4.

YOU TRY IT • 10

Simplify: $-\sqrt[4]{81x^{12}y^8}$

Your solution

Solutions on p. S27

9.1 EXERCISES

OBJECTIVE A To simplify expressions with rational exponents

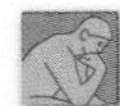

1. Which of the following are not real numbers?
(i) $(-7)^{1/2}$ (ii) $(-7)^{1/3}$ (iii) $(-7)^{1/4}$ (iv) $(-7)^{1/5}$

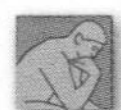

2. If $x^{1/n} = a$, what is a^n?

For Exercises 3 to 74, simplify.

3. $8^{1/3}$
4. $16^{1/2}$
5. $9^{3/2}$
6. $25^{3/2}$
7. $27^{-2/3}$
8. $64^{-1/3}$
9. $32^{2/5}$
10. $16^{3/4}$
11. $(-25)^{5/2}$
12. $(-36)^{1/4}$
13. $\left(\frac{25}{49}\right)^{-3/2}$
14. $\left(\frac{8}{27}\right)^{-2/3}$
15. $x^{1/2}x^{1/2}$
16. $a^{1/3}a^{5/3}$
17. $y^{-1/4}y^{3/4}$
18. $x^{2/5} \cdot x^{-4/5}$
19. $x^{-2/3} \cdot x^{3/4}$
20. $x \cdot x^{-1/2}$
21. $a^{1/3} \cdot a^{3/4} \cdot a^{-1/2}$
22. $y^{-1/6} \cdot y^{2/3} \cdot y^{1/2}$
23. $\frac{a^{1/2}}{a^{3/2}}$
24. $\frac{b^{1/3}}{b^{4/3}}$
25. $\frac{y^{-3/4}}{y^{1/4}}$
26. $\frac{x^{-3/5}}{x^{1/5}}$
27. $\frac{y^{2/3}}{y^{-5/6}}$
28. $\frac{b^{3/4}}{b^{-3/2}}$
29. $(x^2)^{-1/2}$
30. $(a^8)^{-3/4}$
31. $(x^{-2/3})^6$
32. $(y^{-5/6})^{12}$
33. $(a^{-1/2})^{-2}$
34. $(b^{-2/3})^{-6}$
35. $(x^{-3/8})^{-4/5}$
36. $(y^{-3/2})^{-2/9}$
37. $(a^{1/2} \cdot a)^2$
38. $(b^{2/3} \cdot b^{1/6})^6$
39. $(x^{-1/2} \cdot x^{3/4})^{-2}$
40. $(a^{1/2} \cdot a^{-2})^3$
41. $(y^{-1/2} \cdot y^{2/3})^{2/3}$
42. $(b^{-2/3} \cdot b^{1/4})^{-4/3}$

43. $(x^{-3}y^{6})^{-1/3}$

44. $(a^{2}b^{-6})^{-1/2}$

45. $(x^{-2}y^{1/3})^{-3/4}$

46. $(a^{-2/3}b^{2/3})^{3/2}$

47. $\left(\dfrac{x^{1/2}}{y^{2}}\right)^{4}$

48. $\left(\dfrac{b^{-3/4}}{a^{-1/2}}\right)^{8}$

49. $\dfrac{x^{1/4} \cdot x^{-1/2}}{x^{2/3}}$

50. $\dfrac{b^{1/2} \cdot b^{-3/4}}{b^{1/4}}$

51. $\left(\dfrac{y^{2/3} \cdot y^{-5/6}}{y^{1/9}}\right)^{9}$

52. $\left(\dfrac{a^{1/3} \cdot a^{-2/3}}{a^{1/2}}\right)^{4}$

53. $\left(\dfrac{b^{2} \cdot b^{-3/4}}{b^{-1/2}}\right)^{-1/2}$

54. $\dfrac{(x^{-5/6} \cdot x^{3})^{-2/3}}{x^{4/3}}$

55. $(a^{2/3}b^{2})^{6}(a^{3}b^{3})^{1/3}$

56. $(x^{3}y^{-1/2})^{-2}(x^{-3}y^{2})^{1/6}$

57. $(16x^{-2}y^{4})^{-1/2}(xy^{1/2})$

58. $(27s^{3}t^{-6})^{1/3}(s^{-1/3}t^{5/6})^{6}$

59. $(x^{-2/3}y^{-3})^{3}(27x^{-3}y^{6})^{-1/3}$

60. $(9x^{-2/3}y^{4/3})^{-1/2}(x^{2/3}y^{-8/3})^{1/2}$

61. $\dfrac{(4a^{4/3}b^{-2})^{-1/2}}{(a^{1/6}b^{-3/2})^{2}}$

62. $\dfrac{(4x^{-2}y^{4})^{1/2}}{(8x^{6}y^{-3/2})^{-2/3}}$

63. $\left(\dfrac{x^{1/2}y^{-3/4}}{y^{2/3}}\right)^{-6}$

64. $\left(\dfrac{x^{1/2}y^{-5/4}}{y^{-3/4}}\right)^{-4}$

65. $\left(\dfrac{b^{-3}}{64a^{-1/2}}\right)^{-2/3}$

66. $\left(\dfrac{49c^{5/3}}{a^{-1/4}b^{5/6}}\right)^{-3/2}$

67. $y^{3/2}(y^{1/2} - y^{-1/2})$

68. $y^{3/5}(y^{2/5} + y^{-3/5})$

69. $a^{-1/4}(a^{5/4} - a^{9/4})$

70. $x^{4/3}(x^{2/3} + x^{-1/3})$

71. $x^{n} \cdot x^{n/2}$

72. $a^{n/2} \cdot a^{-n/3}$

73. $\dfrac{y^{n/2}}{y^{-n}}$

74. $\dfrac{b^{m/3}}{b^{m}}$

OBJECTIVE B **To write exponential expressions as radical expressions and to write radical expressions as exponential expressions**

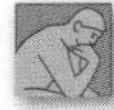
75. True or false? $8x^{1/3} = 2\sqrt[3]{x}$

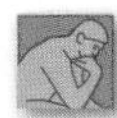
76. True or false? $(\sqrt[3]{x})^5 = (x^5)^{1/3}$

For Exercises 77 to 92, rewrite the exponential expression as a radical expression.

77. $3^{1/4}$

78. $5^{1/2}$

79. $a^{3/2}$

80. $b^{4/3}$

81. $(2t)^{5/2}$

82. $(3x)^{2/3}$

83. $-2x^{2/3}$

84. $-3a^{2/5}$

85. $(a^2b)^{2/3}$

86. $(x^2y^3)^{3/4}$

87. $(a^2b^4)^{3/5}$

88. $(a^3b^7)^{3/2}$

89. $(4x - 3)^{3/4}$

90. $(3x - 2)^{1/3}$

91. $x^{-2/3}$

92. $b^{-3/4}$

For Exercises 93 to 108, rewrite the radical expression as an exponential expression.

93. $\sqrt{14}$

94. $\sqrt{7}$

95. $\sqrt[3]{x}$

96. $\sqrt[4]{y}$

97. $\sqrt[3]{x^4}$

98. $\sqrt[4]{a^3}$

99. $\sqrt[5]{b^3}$

100. $\sqrt[4]{b^5}$

101. $\sqrt[3]{2x^2}$

102. $\sqrt[5]{4y^7}$

103. $-\sqrt{3x^5}$

104. $-\sqrt[4]{4x^5}$

105. $3x\sqrt[3]{y^2}$

106. $2y\sqrt{x^3}$

107. $\sqrt{a^2 - 2}$

108. $\sqrt{3 - y^2}$

OBJECTIVE C To simplify radical expressions that are roots of perfect powers

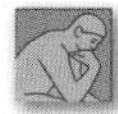

For Exercises 109 to 112, assume that x is a negative real number. State whether the expression simplifies to a positive number, a negative number, or a number that is not a real number.

109. $-\sqrt[3]{8x^{15}}$ **110.** $-\sqrt{9x^{8}}$ **111.** $\sqrt{-4x^{12}}$ **112.** $\sqrt[3]{-27x^{9}}$

For Exercises 113 to 136, simplify.

113. $\sqrt{x^{16}}$ **114.** $\sqrt{y^{14}}$ **115.** $-\sqrt{x^{8}}$ **116.** $-\sqrt{a^{6}}$

117. $\sqrt[3]{x^{3}y^{9}}$ **118.** $\sqrt[3]{a^{6}b^{12}}$ **119.** $-\sqrt[3]{x^{15}y^{3}}$ **120.** $-\sqrt[3]{a^{9}b^{9}}$

121. $\sqrt{16a^{4}b^{12}}$ **122.** $\sqrt{25x^{8}y^{2}}$ **123.** $\sqrt{-16x^{4}y^{2}}$ **124.** $\sqrt{-9a^{4}b^{8}}$

125. $\sqrt[3]{27x^{9}}$ **126.** $\sqrt[3]{8a^{21}b^{6}}$ **127.** $\sqrt[3]{-64x^{9}y^{12}}$ **128.** $\sqrt[3]{-27a^{3}b^{15}}$

129. $-\sqrt[4]{x^{8}y^{12}}$ **130.** $-\sqrt[4]{a^{16}b^{4}}$ **131.** $\sqrt[5]{x^{20}y^{10}}$ **132.** $\sqrt[5]{a^{5}b^{25}}$

133. $\sqrt[4]{81x^{4}y^{20}}$ **134.** $\sqrt[4]{16a^{8}b^{20}}$ **135.** $\sqrt[5]{32a^{5}b^{10}}$ **136.** $\sqrt[5]{-32x^{15}y^{20}}$

Applying the Concepts

137. Determine whether the following statements are true or false. If the statement is false, correct the right-hand side of the equation.

a. $\sqrt{(-2)^2} = -2$ **b.** $\sqrt[3]{(-3)^3} = -3$ **c.** $\sqrt[n]{a} = a^{1/n}$

d. $\sqrt[n]{a^n + b^n} = a + b$ **e.** $(a^{1/2} + b^{1/2})^2 = a + b$ **f.** $\sqrt[m]{a^n} = a^{m/n}$

138. Simplify.

a. $\sqrt[3]{\sqrt{x^{6}}}$ **b.** $\sqrt[4]{\sqrt{a^{8}}}$ **c.** $\sqrt{\sqrt{81y^{8}}}$

d. $\sqrt{\sqrt[n]{a^{4n}}}$ **e.** $\sqrt[n]{\sqrt{b^{6n}}}$ **f.** $\sqrt{\sqrt[3]{x^{12}y^{24}}}$

139. If x is any real number, is $\sqrt{x^2} = x$ always true? Show why or why not.

SECTION

9.2 Operations on Radical Expressions

OBJECTIVE A To simplify radical expressions

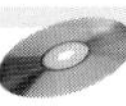

Point of Interest

The Latin expression for irrational numbers was *numerus surdus,* which literally means "inaudible number." A prominent 16th-century mathematician wrote of irrational numbers, ". . . just as an infinite number is not a number, so an irrational number is not a true number, but lies hidden in some sort of cloud of infinity." In 1872, Richard Dedekind wrote a paper that established the first logical treatment of irrational numbers.

If a number is not a perfect power, its root can only be approximated; examples include $\sqrt{5}$ and $\sqrt[3]{3}$. These numbers are irrational numbers. Their decimal representations never terminate or repeat.

$$\sqrt{5} = 2.2360679\ldots \qquad \sqrt[3]{3} = 1.4422495\ldots$$

A radical expression is in simplest form when the radicand contains no factor that is a perfect power. The Product Property of Radicals is used to simplify radical expressions whose radicands are not perfect powers.

The Product Property of Radicals

If $\sqrt[n]{a}$ and $\sqrt[n]{b}$ are positive real numbers, then $\sqrt[n]{ab} = \sqrt[n]{a} \cdot \sqrt[n]{b}$ and $\sqrt[n]{a} \cdot \sqrt[n]{b} = \sqrt[n]{ab}$.

HOW TO • 1 Simplify: $\sqrt{48}$

$\sqrt{48} = \sqrt{16 \cdot 3}$ • Write the radicand as the product of a perfect square and a factor that does not contain a perfect square.

$= \sqrt{16}\sqrt{3}$ • Use the Product Property of Radicals to write the expression as a product.

$= 4\sqrt{3}$ • Simplify $\sqrt{16}$.

Note that 48 must be written as the product of a perfect square and *a factor that does not contain a perfect square.* Therefore, it would not be correct to rewrite $\sqrt{48}$ as $\sqrt{4 \cdot 12}$ and simplify the expression as shown at the right. Although 4 is a perfect square factor of 48, 12 contains a perfect square ($12 = 4 \cdot 3$) and thus $\sqrt{12}$ can be simplified further. Remember to find the *largest* perfect power that is a factor of the radicand.

$$\sqrt{48} = \sqrt{4 \cdot 12}$$
$$= \sqrt{4}\sqrt{12}$$
$$= 2\sqrt{12}$$

Not in simplest form

HOW TO • 2 Simplify: $\sqrt{18x^2y^3}$

$\sqrt{18x^2y^3} = \sqrt{9x^2y^2 \cdot 2y}$ • Write the radicand as the product of a perfect square and factors that do not contain a perfect square.

$= \sqrt{9x^2y^2}\sqrt{2y}$ • Use the Product Property of Radicals to write the expression as a product.

$= 3xy\sqrt{2y}$ • Simplify.

HOW TO • 3 Simplify: $\sqrt[3]{x^7}$

$\sqrt[3]{x^7} = \sqrt[3]{x^6 \cdot x}$ • Write the radicand as the product of a perfect cube and a factor that does not contain a perfect cube.

$= \sqrt[3]{x^6}\sqrt[3]{x}$ • Use the Product Property of Radicals to write the expression as a product.

$= x^2\sqrt[3]{x}$ • Simplify.

HOW TO • 4 Simplify: $\sqrt[4]{32x^7}$

$\sqrt[4]{32x^7} = \sqrt[4]{16x^4(2x^3)}$ • Write the radicand as the product of a perfect fourth power and factors that do not contain a perfect fourth power.

$= \sqrt[4]{16x^4}\sqrt[4]{2x^3}$ • Use the Product Property of Radicals to write the expression as a product.

$= 2x\sqrt[4]{2x^3}$ • Simplify.

EXAMPLE • 1

Simplify: $\sqrt[4]{x^9}$

Solution

$\sqrt[4]{x^9} = \sqrt[4]{x^8 \cdot x}$ • x^8 is a perfect fourth power.

$= \sqrt[4]{x^8}\sqrt[4]{x}$

$= x^2\sqrt[4]{x}$

YOU TRY IT • 1

Simplify: $\sqrt[5]{x^7}$

Your solution

EXAMPLE • 2

Simplify: $\sqrt[3]{-27a^5b^{12}}$

Solution

$\sqrt[3]{-27a^5b^{12}}$

$= \sqrt[3]{-27a^3b^{12}(a^2)}$ • $-27a^3b^{12}$ is a perfect cube.

$= \sqrt[3]{-27a^3b^{12}}\sqrt[3]{a^2}$

$= -3ab^4\sqrt[3]{a^2}$

YOU TRY IT • 2

Simplify: $\sqrt[3]{-64x^8y^{18}}$

Your solution

Solutions on p. S27

OBJECTIVE B To add or subtract radical expressions 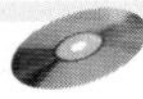

The Distributive Property is used to simplify the sum or difference of radical expressions that have the same radicand and the same index. For example,

$$3\sqrt{5} + 8\sqrt{5} = (3 + 8)\sqrt{5} = 11\sqrt{5}$$
$$2\sqrt[3]{3x} - 9\sqrt[3]{3x} = (2 - 9)\sqrt[3]{3x} = -7\sqrt[3]{3x}$$

Radical expressions that are in simplest form and have unlike radicands or different indices cannot be simplified by the Distributive Property. The expressions below cannot be simplified by the Distributive Property.

$$3\sqrt[4]{2} - 6\sqrt[4]{3} \qquad 2\sqrt[4]{4x} + 3\sqrt[3]{4x}$$

HOW TO • 5 Simplify: $3\sqrt{32x^2} - 2x\sqrt{2} + \sqrt{128x^2}$

$3\sqrt{32x^2} - 2x\sqrt{2} + \sqrt{128x^2}$
$= 3\sqrt{16x^2}\sqrt{2} - 2x\sqrt{2} + \sqrt{64x^2}\sqrt{2}$
$= 3 \cdot 4x\sqrt{2} - 2x\sqrt{2} + 8x\sqrt{2}$
$= 12x\sqrt{2} - 2x\sqrt{2} + 8x\sqrt{2}$
$= 18x\sqrt{2}$

• First simplify each term. Then combine like terms by using the Distributive Property.

EXAMPLE • 3

Simplify: $5b\sqrt[4]{32a^7b^5} - 2a\sqrt[4]{162a^3b^9}$

Solution

$5b\sqrt[4]{32a^7b^5} - 2a\sqrt[4]{162a^3b^9}$
$= 5b\sqrt[4]{16a^4b^4 \cdot 2a^3b} - 2a\sqrt[4]{81b^8 \cdot 2a^3b}$
$= 5b \cdot 2ab\sqrt[4]{2a^3b} - 2a \cdot 3b^2\sqrt[4]{2a^3b}$
$= 10ab^2\sqrt[4]{2a^3b} - 6ab^2\sqrt[4]{2a^3b}$
$= 4ab^2\sqrt[4]{2a^3b}$

YOU TRY IT • 3

Simplify: $3xy\sqrt[3]{81x^5y} - \sqrt[3]{192x^8y^4}$

Your solution

Solution on p. S27

OBJECTIVE C To multiply radical expressions 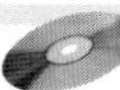

The Product Property of Radicals is used to multiply radical expressions with the same index.

$\sqrt{3x} \cdot \sqrt{5y} = \sqrt{3x \cdot 5y} = \sqrt{15xy}$

HOW TO • 6 Multiply: $\sqrt[3]{2a^5b}\sqrt[3]{16a^2b^2}$

$\sqrt[3]{2a^5b}\sqrt[3]{16a^2b^2} = \sqrt[3]{32a^7b^3}$ • Use the Product Property of Radicals to multiply the radicands.

$= \sqrt[3]{8a^6b^3 \cdot 4a}$ • Simplify.

$= 2a^2b\sqrt[3]{4a}$

HOW TO • 7 Multiply: $\sqrt{2x}(\sqrt{8x} - \sqrt{3})$

$\sqrt{2x}(\sqrt{8x} - \sqrt{3}) = \sqrt{2x}(\sqrt{8x}) - \sqrt{2x}(\sqrt{3})$ • Use the Distributive Property.

$= \sqrt{16x^2} - \sqrt{6x}$ • Simplify.

$= 4x - \sqrt{6x}$

HOW TO • 8 Multiply: $(2\sqrt{5} - 3)(3\sqrt{5} + 4)$

$(2\sqrt{5} - 3)(3\sqrt{5} + 4) = 6(\sqrt{5})^2 + 8\sqrt{5} - 9\sqrt{5} - 12$ • Use the FOIL method to multiply the numbers.

$= 30 + 8\sqrt{5} - 9\sqrt{5} - 12$

$= 18 - \sqrt{5}$ • Combine like terms.

Take Note

The concept of conjugate is used in a number of different instances. Make sure you understand this idea.

The conjugate of $\sqrt{3} - 4$ is $\sqrt{3} + 4$.

The conjugate of $\sqrt{3} + 4$ is $\sqrt{3} - 4$.

The conjugate of $\sqrt{5a} + \sqrt{b}$ is $\sqrt{5a} - \sqrt{b}$.

The expressions $a + b$ and $a - b$ are **conjugates** of each other—that is, binomial expressions that differ only in the sign of a term. Recall that $(a + b)(a - b) = a^2 - b^2$. This identity is used to simplify conjugate radical expressions.

HOW TO • 9 Multiply: $(\sqrt{11} - 3)(\sqrt{11} + 3)$

$$(\sqrt{11} - 3)(\sqrt{11} + 3) = (\sqrt{11})^2 - 3^2 = 11 - 9$$
$$= 2$$

• The radical expressions are conjugates.

EXAMPLE • 4

Multiply: $\sqrt{3}(\sqrt{15} - \sqrt{21})$

Solution

$$\sqrt{3}(\sqrt{15} - \sqrt{21}) = \sqrt{45} - \sqrt{63}$$
$$= 3\sqrt{5} - 3\sqrt{7}$$

YOU TRY IT • 4

Multiply: $5\sqrt{2}(\sqrt{6} + \sqrt{24})$

Your solution

EXAMPLE • 5

Multiply: $(2 + 3\sqrt{5})(3 - \sqrt{5})$

Solution

$$(2 + 3\sqrt{5})(3 - \sqrt{5}) = 6 - 2\sqrt{5} + 9\sqrt{5} - 3(\sqrt{5})^2$$
$$= 6 + 7\sqrt{5} - 3 \cdot 5$$
$$= 6 + 7\sqrt{5} - 15$$
$$= -9 + 7\sqrt{5}$$

YOU TRY IT • 5

Multiply: $(4 - 2\sqrt{7})(1 + 3\sqrt{7})$

Your solution

EXAMPLE • 6

Expand: $(3 - \sqrt{x + 1})^2$

Solution

$$(3 - \sqrt{x + 1})^2 = (3 - \sqrt{x + 1})(3 - \sqrt{x + 1})$$
$$= 9 - 3\sqrt{x + 1} - 3\sqrt{x + 1} + (\sqrt{x + 1})^2$$
$$= 9 - 6\sqrt{x + 1} + (x + 1)$$
$$= x - 6\sqrt{x + 1} + 10$$

YOU TRY IT • 6

Expand: $(4 - \sqrt{2x})^2$

Your solution

Solutions on p. S27

OBJECTIVE D To divide radical expressions 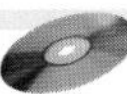

The Quotient Property of Radicals is used to divide radical expressions with the same index.

The Quotient Property of Radicals

If $\sqrt[n]{a}$ and $\sqrt[n]{b}$ are real numbers and $b \neq 0$, then

$$\sqrt[n]{\frac{a}{b}} = \frac{\sqrt[n]{a}}{\sqrt[n]{b}} \quad \text{and} \quad \frac{\sqrt[n]{a}}{\sqrt[n]{b}} = \sqrt[n]{\frac{a}{b}}$$

HOW TO 10 Divide: $\dfrac{\sqrt{5a^4b^7c^2}}{\sqrt{ab^3c}}$

$$\frac{\sqrt{5a^4b^7c^2}}{\sqrt{ab^3c}} = \sqrt{\frac{5a^4b^7c^2}{ab^3c}}$$

• Use the Quotient Property of Radicals.

$$= \sqrt{5a^3b^4c}$$

• Simplify the radicand.

$$= \sqrt{a^2b^4 \cdot 5ac} = ab^2\sqrt{5ac}$$

A radical expression is in simplest form when no radical exists in the denominator of an expression. The radical expressions at the right are *not* in simplest form.

$$\frac{3}{\sqrt{7}} \qquad \frac{6}{\sqrt[3]{2x}} \qquad \frac{6}{3-\sqrt{5}}$$

Not in simplest form

The procedure used to remove a radical expression from the denominator is called **rationalizing the denominator.** The idea is to multiply the numerator and denominator by an expression that will result in a denominator that is a perfect root of the index.

HOW TO 11 Simplify: $\dfrac{3}{\sqrt{7}}$

$$\frac{3}{\sqrt{7}} = \frac{3}{\sqrt{7}} \cdot \boxed{\frac{\sqrt{7}}{\sqrt{7}}}$$

• Multiply the numerator and denominator by $\sqrt{7}$.

$$= \frac{3\sqrt{7}}{\sqrt{49}} = \frac{3\sqrt{7}}{7}$$

• $\sqrt{7} \cdot \sqrt{7} = \sqrt{49} = 7$. Because 49 is a perfect square, the denominator can now be written without a radical.

Take Note

$\dfrac{\sqrt{7}}{\sqrt{7}} = 1$. Therefore, we are multiplying by 1 and not changing the value of the expression.

Take Note

Because the index of the radical is , we must multiply by a factor that will produce a perfect third power. Ask, "What must $4x$ be multiplied by to produce a perfect third power?"

$4x \cdot ? = 8x^3$

$4x \cdot 2x^2 = 8x^3$

We must multiply the numerator and denominator by $\sqrt[3]{2x^2}$.

HOW TO 12 Simplify: $\dfrac{6}{\sqrt[3]{4x}}$

$$\frac{6}{\sqrt[3]{4x}} = \frac{6}{\sqrt[3]{4x}} \cdot \frac{\sqrt[3]{2x^2}}{\sqrt[3]{2x^2}}$$

• Multiply the numerator and denominator by $\sqrt[3]{2x^2}$. See the Take Note at the left.

$$= \frac{6\sqrt[3]{2x^2}}{\sqrt[3]{8x^3}} = \frac{6\sqrt[3]{2x^2}}{2x}$$

• $\sqrt[3]{4x} \cdot \sqrt[3]{2x^2} = \sqrt[3]{8x^3} = 2x$. Because $8x^3$ is a perfect cube, the denominator can now be written without a radical.

$$= \frac{3\sqrt[3]{2x^2}}{x}$$

• Divide by the common factor.

Recall that the expressions $a + b$ and $a - b$ are conjugates and that $(a + b)(a - b) = a^2 - b^2$. This result can be used to simplify a square root radical expression that has two terms in the denominator.

HOW TO • 13 Simplify: $\dfrac{6}{3 - \sqrt{5}}$

$$\frac{6}{3 - \sqrt{5}} = \frac{6}{3 - \sqrt{5}} \cdot \frac{3 + \sqrt{5}}{3 + \sqrt{5}} = \frac{6(3 + \sqrt{5})}{3^2 - (\sqrt{5})^2}$$

• Multiply the numerator and denominator by $3 + \sqrt{5}$, the conjugate of $3 - \sqrt{5}$.

$$= \frac{6(3 + \sqrt{5})}{9 - 5} = \frac{6(3 + \sqrt{5})}{4}$$

$$= \frac{3(3 + \sqrt{5})}{2} = \frac{9 + 3\sqrt{5}}{2}$$

EXAMPLE • 7

Simplify: $\sqrt{\dfrac{3}{2x}}$

Solution

$$\sqrt{\frac{3}{2x}} = \frac{\sqrt{3}}{\sqrt{2x}}$$

• Quotient Property of Radicals

$$= \frac{\sqrt{3}}{\sqrt{2x}} \cdot \frac{\sqrt{2x}}{\sqrt{2x}}$$

• Rationalize the denominator.

$$= \frac{\sqrt{6x}}{\sqrt{4x^2}} = \frac{\sqrt{6x}}{2x}$$

YOU TRY IT • 7

Simplify: $\sqrt{\dfrac{5}{6}}$

Your solution

EXAMPLE • 8

Simplify: $\dfrac{6}{\sqrt[4]{9x}}$

Solution

Ask "What must $9x$ be multiplied by to produce a perfect fourth power?"

$9x \cdot ? = 81x^4$
$9x \cdot 9x^3 = 81x^4$

Multiply the numerator and denominator by $\sqrt[4]{9x^3}$.

$$\frac{6}{\sqrt[4]{9x}} = \frac{6\sqrt[4]{9x^3}}{\sqrt[4]{9x} \cdot \sqrt[4]{9x^3}} = \frac{6\sqrt[4]{9x^3}}{\sqrt[4]{81x^4}} = \frac{6\sqrt[4]{9x^3}}{3x} = \frac{2\sqrt[4]{9x^3}}{x}$$

YOU TRY IT • 8

Simplify: $\dfrac{3x}{\sqrt[3]{3x^2}}$

Your solution

EXAMPLE • 9

Simplify: $\dfrac{1 - \sqrt{5}}{2 + 3\sqrt{5}}$

Solution

$$\frac{1 - \sqrt{5}}{2 + 3\sqrt{5}} = \frac{1 - \sqrt{5}}{2 + 3\sqrt{5}} \cdot \frac{2 - 3\sqrt{5}}{2 - 3\sqrt{5}}$$

• The conjugate of $2 + 3\sqrt{5}$ is $2 - 3\sqrt{5}$.

$$= \frac{2 - 3\sqrt{5} - 2\sqrt{5} + 3\sqrt{25}}{2^2 - (3\sqrt{5})^2}$$

$$= \frac{2 - 5\sqrt{5} + 3 \cdot 5}{4 - 9 \cdot 5} = \frac{17 - 5\sqrt{5}}{-41} = \frac{-17 + 5\sqrt{5}}{41}$$

YOU TRY IT • 9

Simplify: $\dfrac{3}{2 - \sqrt{x}}$

Your solution

Solutions on p. S27

9.2 EXERCISES

OBJECTIVE A To simplify radical expressions

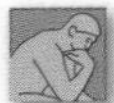

For Exercises 1 to 4, state whether the given expression is in simplest form.

1. $\sqrt{108}$
2. $\sqrt[3]{y^7}$
3. $\sqrt[4]{8ab^2c^3}$
4. $2b\sqrt[4]{8b^3}$

For Exercises 5 to 20, simplify.

5. $\sqrt{x^4y^3z^5}$
6. $\sqrt{x^3y^6z^9}$
7. $\sqrt{8a^3b^8}$
8. $\sqrt{24a^9b^6}$
9. $\sqrt{45x^2y^3z^5}$
10. $\sqrt{60xy^7z^{12}}$
11. $\sqrt[4]{48x^4y^5z^6}$
12. $\sqrt[4]{162x^9y^8z^2}$
13. $\sqrt[3]{a^{16}b^8}$
14. $\sqrt[3]{a^5b^8}$
15. $\sqrt[3]{-125x^2y^4}$
16. $\sqrt[3]{-216x^5y^9}$
17. $\sqrt[3]{a^4b^5c^6}$
18. $\sqrt[3]{a^8b^{11}c^{15}}$
19. $\sqrt[4]{16x^9y^5}$
20. $\sqrt[4]{64x^8y^{10}}$

OBJECTIVE B To add or subtract radical expressions

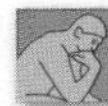

21. True or false? $\sqrt{9+a}+\sqrt{a+9}=2\sqrt{a+9}$

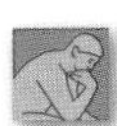

22. True or false? $\sqrt{9+a}=3+\sqrt{a}$

For Exercises 23 to 48, simplify.

23. $2\sqrt{x}-8\sqrt{x}$
24. $3\sqrt{y}+12\sqrt{y}$
25. $\sqrt{8x}-\sqrt{32x}$
26. $\sqrt{27a}-\sqrt{8a}$
27. $\sqrt{18b}+\sqrt{75b}$
28. $2\sqrt{2x^3}+4x\sqrt{8x}$
29. $3\sqrt{8x^2y^3}-2x\sqrt{32y^3}$
30. $2\sqrt{32x^2y^3}-xy\sqrt{98y}$
31. $2a\sqrt{27ab^5}+3b\sqrt{3a^3b}$
32. $\sqrt[3]{128}+\sqrt[3]{250}$
33. $\sqrt[3]{16}-\sqrt[3]{54}$
34. $2\sqrt[3]{3a^4}-3a\sqrt[3]{81a}$
35. $2b\sqrt[3]{16b^2}+\sqrt[3]{128b^5}$
36. $3\sqrt[3]{x^5y^7}-8xy\sqrt[3]{x^2y^4}$

37. $3\sqrt[4]{32a^5} - a\sqrt[4]{162a}$

38. $2a\sqrt[4]{16ab^5} + 3b\sqrt[4]{256a^5b}$

39. $2\sqrt{50} - 3\sqrt{125} + \sqrt{98}$

40. $3\sqrt{108} - 2\sqrt{18} - 3\sqrt{48}$

41. $\sqrt{9b^3} - \sqrt{25b^3} + \sqrt{49b^3}$

42. $\sqrt{4x^7y^5} + 9x^2\sqrt{x^3y^5} - 5xy\sqrt{x^5y^3}$

43. $2x\sqrt{8xy^2} - 3y\sqrt{32x^3} + \sqrt{4x^3y^3}$

44. $5a\sqrt{3a^3b} + 2a^2\sqrt{27ab} - 4\sqrt{75a^5b}$

45. $\sqrt[3]{54xy^3} - 5\sqrt[3]{2xy^3} + y\sqrt[3]{128x}$

46. $2\sqrt[3]{24x^3y^4} + 4x\sqrt[3]{81y^4} - 3y\sqrt[3]{24x^3y}$

47. $2a\sqrt[4]{32b^5} - 3b\sqrt[4]{162a^4b} + \sqrt[4]{2a^4b^5}$

48. $6y\sqrt[4]{48x^5} - 2x\sqrt[4]{243xy^4} - 4\sqrt[4]{3x^5y^4}$

OBJECTIVE C To multiply radical expressions

For Exercises 49 to 83, simplify.

49. $\sqrt{8}\sqrt{32}$

50. $\sqrt{14}\sqrt{35}$

51. $\sqrt[3]{4}\sqrt[3]{8}$

52. $\sqrt[3]{6}\sqrt[3]{36}$

53. $\sqrt{x^2y^5}\sqrt{xy}$

54. $\sqrt{a^3b}\sqrt{ab^4}$

55. $\sqrt{2x^2y}\sqrt{32xy}$

56. $\sqrt{5x^3y}\sqrt{10x^3y^4}$

57. $\sqrt[3]{x^2y}\sqrt[3]{16x^4y^2}$

58. $\sqrt[3]{4a^2b^3}\sqrt[3]{8ab^5}$

59. $\sqrt[4]{12ab^3}\sqrt[4]{4a^5b^2}$

60. $\sqrt[4]{36a^2b^4}\sqrt[4]{12a^5b^3}$

61. $\sqrt{3}(\sqrt{27} - \sqrt{3})$

62. $\sqrt{10}(\sqrt{10} - \sqrt{5})$

63. $\sqrt{x}(\sqrt{x} - \sqrt{2})$

64. $\sqrt{y}(\sqrt{y} - \sqrt{5})$

65. $\sqrt{2x}(\sqrt{8x} - \sqrt{32})$

66. $\sqrt{3a}(\sqrt{27a^2} - \sqrt{a})$

67. $(3 - 2\sqrt{5})(2 + \sqrt{5})$

68. $(6 + 5\sqrt{2})(4 - 2\sqrt{2})$

69. $(-2 + \sqrt{7})(3 + 5\sqrt{7})$

70. $(5 - 2\sqrt{5})(7 - 3\sqrt{5})$

71. $(6 + 3\sqrt{2})(4 - 2\sqrt{2})$

72. $(10 - 3\sqrt{5})(3 + 2\sqrt{5})$

73. $(5 - 2\sqrt{7})(5 + 2\sqrt{7})$

74. $(4 + 2\sqrt{3})(4 - 2\sqrt{3})$

75. $(3 - \sqrt{2x})(1 + 5\sqrt{2x})$

76. $(7 + \sqrt{x})(4 - 2\sqrt{x})$

77. $(2 + 2\sqrt{x})(1 + 5\sqrt{x})$

78. $(3 - 2\sqrt{x})^2$

79. $(2 + \sqrt{x})^2$

80. $(5 - \sqrt{x + 2})^2$

81. $(4 - \sqrt{2x + 1})^2$

82. $(\sqrt{5} - 2\sqrt{7})(3\sqrt{5} + 2\sqrt{7})$

83. $(\sqrt{6} - 5\sqrt{3})(3\sqrt{6} + 4\sqrt{3})$

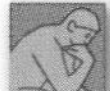

84. True or false? If $a > 0$, then $(\sqrt{a} - 1)(\sqrt{a} + 1) > a$.

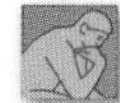

85. True or false? If $a > 0$, then $(\sqrt{a} + 1)^2 > a + 1$.

OBJECTIVE D To divide radical expressions

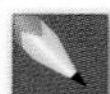

86. When is a radical expression in simplest form?

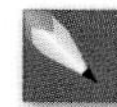

87. Explain what it means to rationalize the denominator of a radical expression and how to do so.

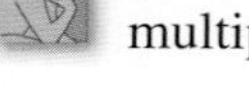

For Exercises 88 to 91, by what expression should the numerator and denominator be multiplied in order to rationalize the denominator?

88. $\dfrac{1}{\sqrt{6}}$

89. $\dfrac{7}{\sqrt[3]{2x^5}}$

90. $\dfrac{8x}{\sqrt[4]{27x}}$

91. $\dfrac{4}{\sqrt{3} - x}$

For Exercises 92 to 128, simplify.

92. $\dfrac{\sqrt{32x^2}}{\sqrt{2x}}$

93. $\dfrac{\sqrt{60y^4}}{\sqrt{12y}}$

94. $\dfrac{\sqrt{42a^3b^5}}{\sqrt{14a^2b}}$

95. $\dfrac{\sqrt{65ab^4}}{\sqrt{5ab}}$

96. $\dfrac{1}{\sqrt{5}}$

97. $\dfrac{1}{\sqrt{2}}$

98. $\dfrac{1}{\sqrt{2x}}$

99. $\dfrac{2}{\sqrt{3y}}$

100. $\dfrac{5}{\sqrt{5x}}$

101. $\dfrac{9}{\sqrt{3a}}$

102. $\sqrt{\dfrac{x}{5}}$

103. $\sqrt{\dfrac{y}{2}}$

104. $\dfrac{3}{\sqrt[3]{2}}$

105. $\dfrac{5}{\sqrt[3]{9}}$

106. $\dfrac{3}{\sqrt[3]{4x^2}}$

107. $\dfrac{5}{\sqrt[3]{3y}}$

108. $\dfrac{\sqrt{40x^3y^2}}{\sqrt{80x^2y^3}}$

109. $\dfrac{\sqrt{15a^2b^5}}{\sqrt{30a^5b^3}}$

110. $\dfrac{\sqrt{24a^2b}}{\sqrt{18ab^4}}$

111. $\dfrac{\sqrt{12x^3y}}{\sqrt{20x^4y}}$

112. $\dfrac{5}{\sqrt{3}-2}$

113. $\dfrac{-2}{1-\sqrt{2}}$

114. $\dfrac{-3}{2-\sqrt{3}}$

115. $\dfrac{-4}{3-\sqrt{2}}$

116. $\dfrac{2}{\sqrt{5}+2}$

117. $\dfrac{5}{2-\sqrt{7}}$

118. $\dfrac{3}{\sqrt{y}-2}$

119. $\dfrac{-7}{\sqrt{x}-3}$

120. $\dfrac{\sqrt{2}-\sqrt{3}}{\sqrt{2}+\sqrt{3}}$

121. $\dfrac{\sqrt{3}+\sqrt{4}}{\sqrt{2}+\sqrt{3}}$

122. $\dfrac{2+3\sqrt{7}}{5-2\sqrt{7}}$

123. $\dfrac{2+3\sqrt{5}}{1-\sqrt{5}}$

124. $\dfrac{2\sqrt{3}-1}{3\sqrt{3}+2}$

125. $\dfrac{2\sqrt{a}-\sqrt{b}}{4\sqrt{a}+3\sqrt{b}}$

126. $\dfrac{2\sqrt{x}-4}{\sqrt{x}+2}$

127. $\dfrac{3\sqrt{y}-y}{\sqrt{y}+2y}$

128. $\dfrac{3\sqrt{x}-4\sqrt{y}}{3\sqrt{x}-2\sqrt{y}}$

Applying the Concepts

129. Determine whether the following statements are true or false. If the statement is false, correct the right-hand side of the equation.

a. $\sqrt[2]{3} \cdot \sqrt[3]{4} = \sqrt[5]{12}$

b. $\sqrt{3} \cdot \sqrt{3} = 3$

c. $\sqrt[3]{x} \cdot \sqrt[3]{x} = x$

d. $\sqrt{x} + \sqrt{y} = \sqrt{x+y}$

e. $\sqrt[2]{2} + \sqrt[3]{3} = \sqrt[5]{2+3}$

f. $8\sqrt[5]{a} - 2\sqrt[5]{a} = 6\sqrt[5]{a}$

130. Multiply: $(\sqrt[3]{a}+\sqrt[3]{b})(\sqrt[3]{a^2}-\sqrt[3]{ab}+\sqrt[3]{b^2})$

131. Rewrite $\dfrac{\sqrt[4]{(a+b)^3}}{\sqrt{a+b}}$ as an expression with a single radical.

SECTION

9.3 Solving Equations Containing Radical Expressions

OBJECTIVE A To solve a radical equation 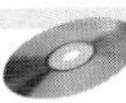

An equation that contains a variable expression in a radicand is called a **radical equation.**

$$\left.\begin{aligned}\sqrt[3]{2x-5}+x&=7\\ \sqrt{x+1}-\sqrt{x}&=4\end{aligned}\right\}\text{ Radical equations}$$

The following property is used to solve a radical equation.

> **The Property of Raising Each Side of an Equation to a Power**
>
> If two numbers are equal, then the same powers of the numbers are equal.
>
> If $a = b$, then $a^n = b^n$.

HOW TO • 1 Solve: $\sqrt{x-2} - 6 = 0$

$\sqrt{x-2} - 6 = 0$

$\sqrt{x-2} = 6$ • Isolate the radical by adding 6 to each side of the equation.

$(\sqrt{x-2})^2 = 6^2$ • Square each side of the equation.

$x - 2 = 36$ • Simplify and solve for x.

$x = 38$

Check:

$$\begin{array}{r|l} \sqrt{x-2} - 6 & 0 \\ \hline \sqrt{38-2} - 6 & 0 \\ \sqrt{36} - 6 & 0 \\ 6 - 6 & 0 \\ \end{array}$$

$0 = 0$

38 checks as a solution. The solution is 38.

HOW TO • 2 Solve: $\sqrt[3]{x+2} = -3$

$\sqrt[3]{x+2} = -3$

$(\sqrt[3]{x+2})^3 = (-3)^3$ • Cube each side of the equation.

$x + 2 = -27$ • Solve the resulting equation.

$x = -29$

Check:

$$\begin{array}{r|l} \sqrt[3]{x+2} & -3 \\ \hline \sqrt[3]{-29+2} & -3 \\ \sqrt[3]{-27} & -3 \\ \end{array}$$

$-3 = -3$

-29 checks as a solution. The solution is -29.

Raising each side of an equation to an *even* power may result in an equation that has extraneous solutions. Therefore, you *must* check proposed solutions of an equation if one of the steps in solving the equation was to raise each side to an even power.

Take Note

Note that

$(2-\sqrt{x})^2$
$= (2-\sqrt{x})(2-\sqrt{x})$
$= 4 - 4\sqrt{x} + x$

HOW TO • 3 Solve: $\sqrt{2x-1} + \sqrt{x} = 2$

$\sqrt{2x-1} + \sqrt{x} = 2$

$\sqrt{2x-1} = 2 - \sqrt{x}$ • Solve for one of the radical expressions.

$(\sqrt{2x-1})^2 = (2-\sqrt{x})^2$ • Square each side. Recall that $(a-b)^2 = a^2 - 2ab + b^2$.

$2x - 1 = 4 - 4\sqrt{x} + x$

$x - 5 = -4\sqrt{x}$

$(x-5)^2 = (-4\sqrt{x})^2$ • Square each side again.

$x^2 - 10x + 25 = 16x$

$x^2 - 26x + 25 = 0$ • Solve the quadratic equation by factoring.

$(x-25)(x-1) = 0$

$x = 25$ or $x = 1$

Check:

$\sqrt{2x-1} + \sqrt{x}$	$= 2$
$\sqrt{2(25)-1} + \sqrt{25}$	2
$7 + 5$	2
$12 \neq 2$	

$\sqrt{2x-1} + \sqrt{x}$	$= 2$
$\sqrt{2(1)-1} + \sqrt{1}$	2
$1 + 1$	2
$2 = 2$	

25 does not check as a solution. 1 checks as a solution. The solution is 1. Here 25 is an extraneous solution.

Take Note

The proposed solutions of the equation were 1 and 25. However, 25 did not check as a solution. Therefore, it is an extraneous solution.

EXAMPLE • 1

Solve: $\sqrt{x-1} + \sqrt{x+4} = 5$

Solution

$\sqrt{x-1} + \sqrt{x+4} = 5$

$\sqrt{x+4} = 5 - \sqrt{x-1}$ • Subtract $\sqrt{x-1}$.

$(\sqrt{x+4})^2 = (5-\sqrt{x-1})^2$ • Square each side.

$x + 4 = 25 - 10\sqrt{x-1} + x - 1$

$-20 = -10\sqrt{x-1}$

$2 = \sqrt{x-1}$

$2^2 = (\sqrt{x-1})^2$ • Square each side.

$4 = x - 1$

$5 = x$

The solution checks. The solution is 5.

YOU TRY IT • 1

Solve: $\sqrt{x} - \sqrt{x+5} = 1$

Your solution

EXAMPLE • 2

Solve: $\sqrt[3]{3x-1} = -4$

Solution

$\sqrt[3]{3x-1} = -4$

$(\sqrt[3]{3x-1})^3 = (-4)^3$ • Cube each side.

$3x - 1 = -64$

$3x = -63$

$x = -21$

The solution checks. The solution is -21.

YOU TRY IT • 2

Solve: $\sqrt[4]{x-8} = 3$

Your solution

Solutions on p. S27

OBJECTIVE B To solve application problems 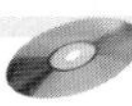

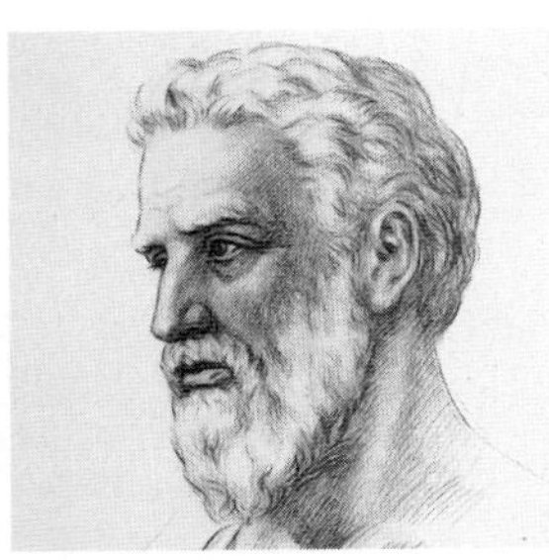

Pythagoras

A right triangle contains one 90° angle. The side opposite the 90° angle is called the hypotenuse. The other two sides are called legs.

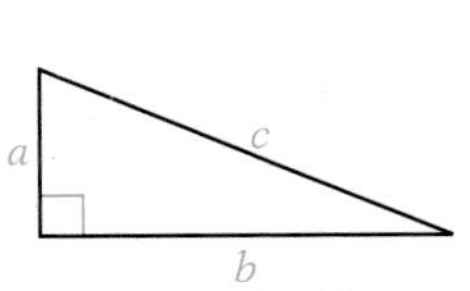

Pythagoras, a Greek mathematician, discovered that the square of the hypotenuse of a right triangle is equal to the sum of the squares of the two legs. Recall that this is called the Pythagorean Theorem.

a c b

$c^2 = a^2 + b^2$

EXAMPLE • 3

A ladder 20 ft long is leaning against a building. How high on the building will the ladder reach when the bottom of the ladder is 8 ft from the building? Round to the nearest tenth.

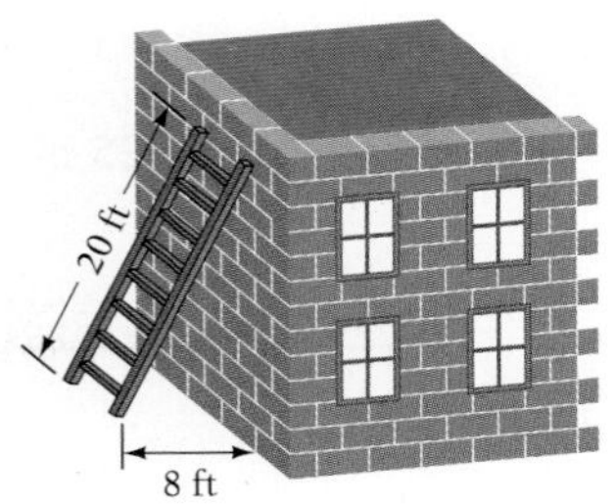

Strategy

To find the distance, use the Pythagorean Theorem. The hypotenuse is the length of the ladder. One leg is the distance from the bottom of the ladder to the base of the building. The distance along the building from the ground to the top of the ladder is the unknown leg.

Solution

$c^2 = a^2 + b^2$ • Pythagorean Theorem

$20^2 = 8^2 + b^2$ • Replace c by **20** and a by 8.

$400 = 64 + b^2$ • Solve for b.

$336 = b^2$

$\sqrt{336} = \sqrt{b^2}$ • Take the square root of each side.

$\sqrt{336} = b$

$18.3 \approx b$

The distance is approximately 18.3 ft.

YOU TRY IT • 3

Find the diagonal of a rectangle that is 6 cm long and 3 cm wide. Round to the nearest tenth.

Your strategy

Your solution

Solution on pp. S27–S28

EXAMPLE • 4

An object is dropped from a high building. Find the distance the object has fallen when its speed reaches 96 ft/s. Use the equation $v = \sqrt{64d}$, where v is the speed of the object in feet per second and d is the distance in feet.

Strategy

To find the distance the object has fallen, replace v in the equation with the given value and solve for d.

Solution

$$v = \sqrt{64d}$$

$$96 = \sqrt{64d} \quad \text{• Replace } v \text{ by 96.}$$

$$(96)^2 = (\sqrt{64d})^2 \quad \text{• Square each side.}$$

$$9216 = 64d$$

$$144 = d$$

The object has fallen 144 ft.

YOU TRY IT • 4

How far would a submarine periscope have to be above the water to locate a ship 5.5 mi away? The equation for the distance in miles that the lookout can see is $d = \sqrt{1.5h}$, where h is the height in feet above the surface of the water. Round to the nearest hundredth.

Your strategy

Your solution

EXAMPLE • 5

Find the length of a pendulum that makes one swing in 1.5 s. The equation for the time of one swing is given by $T = 2\pi\sqrt{\frac{L}{32}}$, where T is the time in seconds and L is the length in feet. Use 3.14 for π. Round to the nearest hundredth.

Strategy

To find the length of the pendulum, replace T in the equation with the given value and solve for L.

Solution

$$T = 2\pi\sqrt{\frac{L}{32}}$$

$$1.5 = 2(3.14)\sqrt{\frac{L}{32}} \quad \text{• Replace } T \text{ by 1.5 and } \pi \text{ by 3.14.}$$

$$\frac{1.5}{2(3.14)} = \sqrt{\frac{L}{32}} \quad \text{• Divide each side by 2(3.14).}$$

$$\left[\frac{1.5}{2(3.14)}\right]^2 = \left(\sqrt{\frac{L}{32}}\right)^2 \quad \text{• Square each side.}$$

$$\left(\frac{1.5}{6.28}\right)^2 = \frac{L}{32} \quad \text{• Solve for } L. \text{ Multiply each side by 32.}$$

$$1.83 \approx L$$

The length of the pendulum is 1.83 ft.

YOU TRY IT • 5

Find the distance required for a car to reach a velocity of 88 ft/s when the acceleration is 22 ft/s^2. Use the equation $v = \sqrt{2as}$, where v is the velocity in feet per second, a is the acceleration, and s is the distance in feet.

Your strategy

Your solution

Solutions on p. S28

9.3 EXERCISES

OBJECTIVE A To solve a radical equation

For Exercises 1 to 21, solve for x.

1. $\sqrt[3]{4x} = -2$

2. $\sqrt[3]{6x} = -3$

3. $\sqrt{3x - 2} = 5$

4. $\sqrt{3x + 9} - 12 = 0$

5. $\sqrt{4x - 3} + 9 = 4$

6. $\sqrt{2x - 5} + 4 = 1$

7. $\sqrt[3]{2x - 6} = 4$

8. $\sqrt[3]{x - 2} = 3$

9. $\sqrt[4]{3x} + 2 = 5$

10. $\sqrt[4]{4x + 1} = 2$

11. $\sqrt[3]{2x - 3} + 5 = 2$

12. $\sqrt[3]{x - 4} + 7 = 5$

13. $\sqrt{x} + 2 = x - 4$

14. $\sqrt{x + 1} + 5 = x + 4$

15. $\sqrt{x} + \sqrt{x - 5} = 5$

16. $\sqrt{x + 3} + \sqrt{x - 1} = 2$

17. $\sqrt{2x + 5} - \sqrt{2x} = 1$

18. $\sqrt{3x} - \sqrt{3x - 5} = 1$

19. $\sqrt{2x} - \sqrt{x - 1} = 1$

20. $\sqrt{2x - 5} + \sqrt{x + 1} = 3$

21. $\sqrt{2x + 2} + \sqrt{x} = 3$

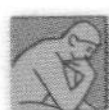

22. The equation $\sqrt{x} = -4$ has no solution. Why?

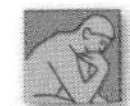

23. Without attempting to solve the equation, explain why $\sqrt{x} - \sqrt{x + 5} = 1$ has no solution. *Hint:* See Exercise 22.

OBJECTIVE B To solve application problems

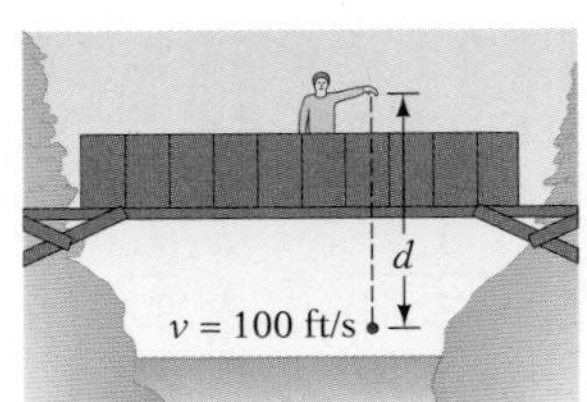

24. Physics An object is dropped from a bridge. Find the distance the object has fallen when its speed reaches 100 ft/s. Use the equation $v = \sqrt{64d}$, where v is the speed in feet per second and d is the distance in feet.

25. Physics The time it takes for an object to fall a certain distance is given by the equation $t = \sqrt{\frac{2d}{g}}$, where t is the time in seconds, d is the distance in feet, and g is the acceleration due to gravity. If an astronaut above the moon's surface drops an object, how far will it have fallen in 3 s? The acceleration on the moon's surface is 5.5 feet per second per second.

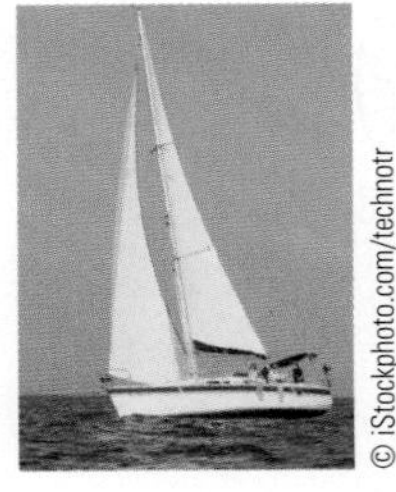
© iStockphoto.com/technotr

26. Sailing The total recommended area A, in square feet, of the sails for a certain sailboat is given by $A = 16\sqrt[3]{d^2}$, where d is the displacement of the hull in cubic feet. If a sailboat has a total of 400 ft^2 of sail, what is the displacement of the hull of the sailboat?

27. Water Tanks A 6-foot-high conical water tank is filled to the top. When a valve at the bottom of the tank is opened, the height h, in feet, of the water in the tank is given by $h = (88.18 - 3.18t)^{2/5}$, where t is the time in seconds after the valve is opened.

a. Find the height of the water 10 s after the valve is opened. Round to the nearest tenth.

b. How long will it take to empty the tank? Round to the nearest tenth.

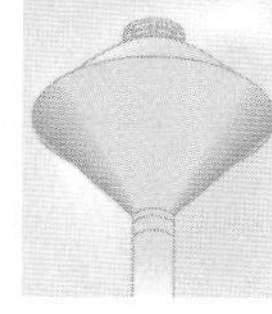

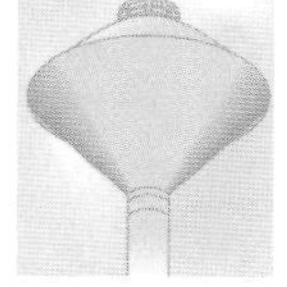

28. Water Tanks The velocity v, in feet per second, of the water pouring out of a small hole in the bottom of a cylindrical tank is given by $v = \sqrt{64h + 10}$, where h is the height, in feet, of the water in the tank. What is the height of the water in the tank when the velocity of the water leaving the tank is 14 ft/s? Round to the nearest tenth.

29. Pendulums Find the length of a pendulum that makes one swing in 3 s. The equation for the time of one swing of a pendulum is $T = 2\pi\sqrt{\frac{L}{32}}$, where T is the time in seconds and L is the length in feet. Round to the nearest hundredth.

30. Satellites Read the article at the right. At what height above Earth's surface is the A-Train in orbit? Use the equation $v = \sqrt{\frac{4 \times 10^{14}}{h + 6.4 \times 10^{6}}}$, where v is the speed of the satellite in meters per second and h is the height, in meters, above Earth's surface. Round to the nearest ten thousand.

> **In the News**
>
> **NASA to Add OCO to A-Train**
>
> The 2009 launch of the Orbiting Carbon Observatory adds a sixth satellite to the "A-Train" of satellites that orbit Earth at about 7500 m/s, providing scientists with a wealth of data that can be used to study climate change.
>
> *Source:* Jet Propulsion Laboratory

31. Television High definition television (HDTV) gives consumers a wider viewing area, more like a film screen in a theater. A regular television with a 27-inch diagonal measurement has a screen 16.2 in. tall. An HDTV screen with the same 16.2-inch height would have a diagonal measuring 33 in. How many inches wider is the HDTV screen? Round to the nearest hundredth.

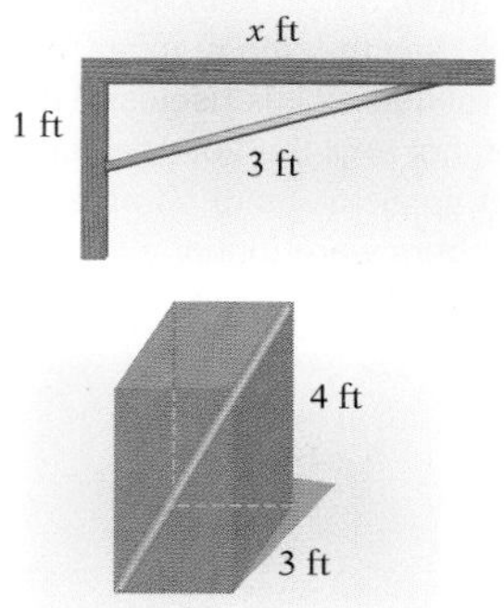

32. Construction A carpenter is inserting a 3-foot brace between two beams as shown in the figure. How far from the vertical beam will the brace reach? Round to the nearest tenth.

33. Moving Boxes A moving box has a base that measures 2 ft by 3 ft, and the box is 4 ft tall. Find the length of the longest pole that could be placed in the box. Round to the nearest tenth.

34. A 10-foot ladder is resting against a wall, with the bottom of the ladder 6 ft from the wall. The top of the ladder begins sliding down the wall at a constant rate of 2 ft/s. Is the bottom of the ladder sliding away from the wall at the same rate?

Applying the Concepts

35. Solve: $\sqrt{3x - 2} = \sqrt{2x - 3} + \sqrt{x - 1}$

36. Geometry Find the length of the side labeled x.

SECTION
9.4 Complex Numbers

OBJECTIVE A To simplify a complex number 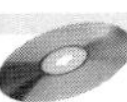

The radical expression $\sqrt{-4}$ is not a real number because there is no real number whose square is -4. However, the solution of an algebraic equation is sometimes the square root of a negative number.

For example, the equation $x^2 + 1 = 0$ does not have a real number solution because there is no real number whose square is -1.

$$x^2 + 1 = 0$$
$$x^2 = -1$$

Around the 17th century, a new number, called an **imaginary number,** was defined so that a negative number would have a square root. The letter i was chosen to represent the number whose square is -1.

$$i^2 = -1$$

An imaginary number is defined in terms of i.

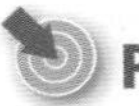

Point of Interest

The first written occurrence of an imaginary number was in a book published in 1545 by Hieronimo Cardan, where he wrote (in our modern notation) $5 + \sqrt{-15}$. He went on to say that the number "is as refined as it is useless." It was not until the 20th century that applications of complex numbers were found.

Definition of $\sqrt{-a}$

If a is a positive real number, then the principal square root of negative a is the imaginary number $i\sqrt{a}$.

$$\sqrt{-a} = i\sqrt{a}$$

Here are some examples.

$$\sqrt{-16} = i\sqrt{16} = 4i$$
$$\sqrt{-12} = i\sqrt{12} = 2i\sqrt{3}$$
$$\sqrt{-21} = i\sqrt{21}$$
$$\sqrt{-1} = i\sqrt{1} = i$$

It is customary to write i in front of a radical to avoid confusing $\sqrt{a}\,i$ with $\sqrt{ai}$.

The real numbers and imaginary numbers make up the complex numbers.

Complex Number

A **complex number** is a number of the form $a + bi$, where a and b are real numbers and $i = \sqrt{-1}$. The number a is the **real part** of $a + bi$, and the number b is the **imaginary part.**

Take Note

The *imaginary part* of a complex number is a real number. As another example, the imaginary part of $6 - 8i$ is -8.

Examples of complex numbers are shown at the right.

Real Part	*Imaginary Part*
a	$+ bi$
3	$+ 2i$
8	$- 10i$

Complex numbers $a + bi$	Real Numbers $a + 0i$		A *real number* is a complex number in which $b = 0$.
	Imaginary Numbers $0 + bi$		An *imaginary number* is a complex number in which $a = 0$.

Tips for Success

Be sure you understand how to simplify expressions such as those in Example 1 and You Try It 1, as it is a prerequisite for solving quadratic equations in Chapter 10.

HOW TO • 1 Simplify: $7 - \sqrt{-50}$

$7 - \sqrt{-50} = 7 - i\sqrt{50}$ • Use the definition of $\sqrt{-a}$ to write $\sqrt{-50} = i\sqrt{50}$.

$= 7 - i\sqrt{25 \cdot 2}$ • Simplify the radical expression.

$= 7 - 5i\sqrt{2}$

EXAMPLE • 1

Simplify: $\sqrt{-80}$

Solution

$\sqrt{-80} = i\sqrt{80} = i\sqrt{16 \cdot 5} = 4i\sqrt{5}$

YOU TRY IT • 1

Simplify: $\sqrt{-45}$

Your solution

EXAMPLE • 2

Evaluate $-b + \sqrt{b^2 - 4ac}$ when $a = 2$, $b = -2$, and $c = 3$. Write the result as a complex number.

Solution

$-b + \sqrt{b^2 - 4ac}$

$-(-2) + \sqrt{(-2)^2 - 4(2)(3)} = 2 + \sqrt{4 - 24}$

$= 2 + \sqrt{-20} = 2 + i\sqrt{20}$

$= 2 + i\sqrt{4 \cdot 5} = 2 + 2i\sqrt{5}$

YOU TRY IT • 2

Evaluate $-b + \sqrt{b^2 - 4ac}$ when $a = 1$, $b = 6$, and $c = 25$. Write the result as a complex number.

Your solution

Solutions on p. S28

OBJECTIVE B To add or subtract complex numbers 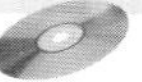

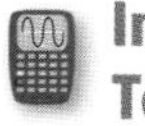

Integrating Technology

See the Keystroke Guide: *Complex Numbers* for instructions on using a graphing calculator to perform operations on complex numbers.

Addition and Subtraction of Complex Numbers

To add two complex numbers, add the real parts and add the imaginary parts. To subtract two complex numbers, subtract the real parts and subtract the imaginary parts.

$$(a + bi) + (c + di) = (a + c) + (b + d)i$$
$$(a + bi) - (c + di) = (a - c) + (b - d)i$$

HOW TO • 2 Add: $(3 - 5i) + (2 + 3i)$

$(3 - 5i) + (2 + 3i) = (3 + 2) + (-5 + 3)i$ • Add the real parts and add the imaginary parts of the complex number.
$= 5 - 2i$

HOW TO • 3 Subtract: $(5 + 6i) - (7 - 3i)$

$(5 + 6i) - (7 - 3i) = (5 - 7) + [6 - (-3)]i$ • Subtract the real parts and subtract the imaginary parts of the complex number.
$= -2 + 9i$

The additive inverse of the complex number $a + bi$ is $-a - bi$. The sum of these two numbers is 0.

$$(a + bi) + (-a - bi) = (a - a) + (b - b)i = 0 + 0i = 0$$

EXAMPLE • 3

Simplify: $(3 + 2i) + (6 - 5i)$

Solution

$(3 + 2i) + (6 - 5i) = 9 - 3i$

YOU TRY IT • 3

Simplify: $(-4 + 2i) - (6 - 8i)$

Your solution

EXAMPLE • 4

Simplify: $(9 - \sqrt{-8}) - (5 + \sqrt{-32})$

Solution

$(9 - \sqrt{-8}) - (5 + \sqrt{-32})$
$= (9 - i\sqrt{8}) - (5 + i\sqrt{32})$
$= (9 - i\sqrt{4 \cdot 2}) - (5 + i\sqrt{16 \cdot 2})$
$= (9 - 2i\sqrt{2}) - (5 + 4i\sqrt{2})$
$= 4 - 6i\sqrt{2}$

YOU TRY IT • 4

Simplify: $(16 - \sqrt{-45}) - (3 + \sqrt{-20})$

Your solution

Solutions on p. S28

OBJECTIVE C To multiply complex numbers 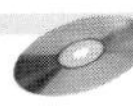

When multiplying complex numbers, we often find that the term i^2 is a part of the product. Recall that $i^2 = -1$.

HOW TO • 4 Simplify: $2i \cdot 3i$

$2i \cdot 3i = 6i^2$ • Multiply the imaginary numbers.
$= 6(-1)$ • Replace i^2 by -1.
$= -6$ • Simplify.

Take Note

This example illustrates an important point. When working with an expression that contains a square root of a negative number, always rewrite the number as the product of a real number and i before continuing.

HOW TO • 5 Simplify: $\sqrt{-6} \cdot \sqrt{-24}$

$\sqrt{-6} \cdot \sqrt{-24} = i\sqrt{6} \cdot i\sqrt{24}$ • Write each radical as the product of a real number and i.

$= i^2\sqrt{144}$ • Multiply the imaginary numbers.

$= -\sqrt{144}$ • Replace i^2 by -1.

$= -12$ • Simplify the radical expression.

Note from the last example that it would have been incorrect to multiply the radicands of the two radical expressions. To illustrate,

$$\sqrt{-6} \cdot \sqrt{-24} = \sqrt{(-6)(-24)} = \sqrt{144} = 12, \textit{ not } -12$$

The Product Property of Radicals does *not* hold true when both radicands are negative and the index is an even number.

HOW TO • 6 Simplify: $4i(3 - 2i)$

$4i(3 - 2i) = 12i - 8i^2$ • Use the Distributive Property to remove parentheses.

$= 12i - 8(-1)$ • Replace i^2 by -1.

$= 8 + 12i$ • Write the answer in the form $a + bi$.

The product of two complex numbers is defined as follows.

Multiplication of Complex Numbers

$(a + bi)(c + di) = (ac - bd) + (ad + bc)i$

One way to remember this rule is to think of the FOIL method.

HOW TO • 7 Simplify: $(2 + 4i)(3 - 5i)$

$(2 + 4i)(3 - 5i) = 6 - 10i + 12i - 20i^2$ • Use the FOIL method to find the product.

$= 6 + 2i - 20i^2$

$= 6 + 2i - 20(-1)$ • Replace i^2 by -1.

$= 26 + 2i$ • Write the answer in the form $a + bi$.

The conjugate of $a + bi$ is $a - bi$.

The product of complex conjugates, $(a + bi)(a - bi)$, is the real number $a^2 + b^2$.

$$\begin{aligned}(a + bi)(a - bi) &= a^2 - b^2i^2 \\ &= a^2 - b^2(-1) \\ &= a^2 + b^2\end{aligned}$$

HOW TO • 8 Simplify: $(2 + 3i)(2 - 3i)$

$(2 + 3i)(2 - 3i) = 2^2 + 3^2$ • $a = 2, b = 3$ The product of the conjugates is $2^2 + 3^2$.

$= 4 + 9$

$= 13$

Note that the product of a complex number and its conjugate is a real number.

EXAMPLE 5

Simplify: $(2i)(-5i)$

Solution

$(2i)(-5i) = -10i^2 = (-10)(-1) = 10$

YOU TRY IT 5

Simplify: $(-3i)(-10i)$

Your solution

EXAMPLE 6

Simplify: $\sqrt{-10} \cdot \sqrt{-5}$

Solution

$$\begin{aligned}\sqrt{-10} \cdot \sqrt{-5} &= i\sqrt{10} \cdot i\sqrt{5} \\ &= i^2\sqrt{50} = -\sqrt{25 \cdot 2} = -5\sqrt{2}\end{aligned}$$

YOU TRY IT 6

Simplify: $-\sqrt{-8} \cdot \sqrt{-5}$

Your solution

EXAMPLE 7

Simplify: $3i(2 - 4i)$

Solution

$$\begin{aligned}3i(2 - 4i) &= 6i - 12i^2 \quad \text{• The Distributive Property} \\ &= 6i - 12(-1) \\ &= 12 + 6i\end{aligned}$$

YOU TRY IT 7

Simplify: $-6i(3 + 4i)$

Your solution

EXAMPLE 8

Simplify: $(3 - 4i)(2 + 5i)$

Solution

$$\begin{aligned}(3 - 4i)(2 + 5i) &= 6 + 15i - 8i - 20i^2 \quad \text{• FOIL} \\ &= 6 + 7i - 20i^2 \\ &= 6 + 7i - 20(-1) = 26 + 7i\end{aligned}$$

YOU TRY IT 8

Simplify: $(4 - 3i)(2 - i)$

Your solution

EXAMPLE 9

Expand: $(3 + 4i)^2$

Solution

$$\begin{aligned}(3 + 4i)^2 &= (3 + 4i)(3 + 4i) \\ &= 9 + 12i + 12i + 16i^2 \\ &= 9 + 24i + 16(-1) \\ &= 9 + 24i - 16 \\ &= -7 + 24i\end{aligned}$$

YOU TRY IT 9

Expand: $(5 - 3i)^2$

Your solution

EXAMPLE 10

Simplify: $(4 + 5i)(4 - 5i)$

Solution

$$\begin{aligned}(4 + 5i)(4 - 5i) &= 4^2 + 5^2 \quad \text{• Conjugates} \\ &= 16 + 25 = 41\end{aligned}$$

YOU TRY IT 10

Simplify: $(3 + 6i)(3 - 6i)$

Your solution

Solutions on p. S28

OBJECTIVE D To divide complex numbers

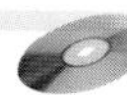

A rational expression containing one or more complex numbers is in simplest form when no imaginary number remains in the denominator.

HOW TO • 9 Simplify: $\frac{2-3i}{2i}$

$$\frac{2-3i}{2i} = \frac{2-3i}{2i} \cdot \frac{i}{i}$$
• Multiply the numerator and denominator by $\frac{i}{i}$.

$$= \frac{2i-3i^2}{2i^2}$$

$$= \frac{2i-3(-1)}{2(-1)}$$
• Replace i^2 by -1.

$$= \frac{3+2i}{-2} = -\frac{3}{2} - i$$
• Simplify. Write the answer in the form $a + bi$.

HOW TO • 10 Simplify: $\frac{3+2i}{1+i}$

$$\frac{3+2i}{1+i} = \frac{3+2i}{1+i} \cdot \frac{1-i}{1-i}$$
• Multiply the numerator and denominator by the conjugate of $1 + i$.

$$= \frac{3-3i+2i-2i^2}{1^2+1^2}$$

$$= \frac{3-i-2(-1)}{1+1}$$
• Replace i^2 by -1 and simplify.

$$= \frac{5-i}{2} = \frac{5}{2} - \frac{1}{2}i$$
• Write the answer in the form $a + bi$.

EXAMPLE • 11

Simplify: $\frac{5+4i}{3i}$

Solution

$$\frac{5+4i}{3i} = \frac{5+4i}{3i} \cdot \frac{i}{i} = \frac{5i+4i^2}{3i^2}$$

$$= \frac{5i+4(-1)}{3(-1)} = \frac{-4+5i}{-3} = \frac{4}{3} - \frac{5}{3}i$$

YOU TRY IT • 11

Simplify: $\frac{2-3i}{4i}$

Your solution

EXAMPLE • 12

Simplify: $\frac{5-3i}{4+2i}$

Solution

$$\frac{5-3i}{4+2i} = \frac{5-3i}{4+2i} \cdot \frac{4-2i}{4-2i}$$

$$= \frac{20-10i-12i+6i^2}{4^2+2^2}$$

$$= \frac{20-22i+6(-1)}{16+4}$$

$$= \frac{14-22i}{20} = \frac{7-11i}{10} = \frac{7}{10} - \frac{11}{10}i$$

YOU TRY IT • 12

Simplify: $\frac{2+5i}{3-2i}$

Your solution

Solutions on p. S28

9.4 EXERCISES

OBJECTIVE A To simplify a complex number

1. What is an imaginary number? What is a complex number?

2. Are all real numbers also complex numbers? Are all complex numbers also real numbers?

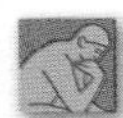

3. Fill in the blank with $\sqrt{a}$ or $i\sqrt{a}$.
If $a > 0$, then $\sqrt{-a} =$ __?__.

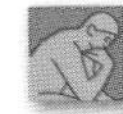

4. Does $\sqrt{a^2} = (\sqrt{a})^2$ for all values of a?

For Exercises 5 to 12, simplify.

5. $\sqrt{-25}$
6. $\sqrt{-64}$
7. $\sqrt{-98}$
8. $\sqrt{-72}$
9. $3 + \sqrt{-45}$
10. $7 - \sqrt{-63}$
11. $-6 - \sqrt{-100}$
12. $4 + \sqrt{-49}$

For Exercises 13 to 24, evaluate $-b + \sqrt{b^2 - 4ac}$ for the given values of a, b, and c. Write the result as a complex number.

13. $a = 1, b = 4, c = 5$
14. $a = 1, b = -6, c = 13$
15. $a = 2, b = -4, c = 10$
16. $a = 4, b = -12, c = 45$
17. $a = 3, b = -8, c = 6$
18. $a = 3, b = -2, c = 9$
19. $a = 4, b = 2, c = 7$
20. $a = 4, b = 5, c = 10$
21. $a = -2, b = 5, c = -6$
22. $a = -1, b = 4, c = -29$
23. $a = -3, b = 4, c = -6$
24. $a = -5, b = 1, c = -5$

OBJECTIVE B To add or subtract complex numbers

For Exercises 25 to 34, simplify.

25. $(2 + 4i) + (6 - 5i)$
26. $(6 - 9i) + (4 + 2i)$
27. $(-2 - 4i) - (6 - 8i)$
28. $(3 - 5i) + (8 - 2i)$
29. $(8 - 2i) - (2 + 4i)$
30. $(5 - 5i) - (11 - 6i)$
31. $5 + (6 - 4i)$
32. $-7 + (3 + 5i)$
33. $3i - (6 + 5i)$
34. $(7 + 3i) - 8i$

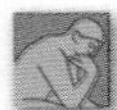

35. If the sum of two complex numbers is an imaginary number, what must be true of the complex numbers?

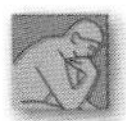

36. If the sum of two complex numbers is a real number, what must be true of the complex numbers?

OBJECTIVE C To multiply complex numbers

For Exercises 37 to 54, simplify.

37. $(7i)(-9i)$

38. $(-6i)(-4i)$

39. $\sqrt{-2}\sqrt{-8}$

40. $\sqrt{-5}\sqrt{-45}$

41. $(5 + 2i)(5 - 2i)$

42. $(3 + 8i)(3 - 8i)$

43. $2i(6 + 2i)$

44. $-3i(4 - 5i)$

45. $-i(4 - 3i)$

46. $i(6 + 2i)$

47. $(5 - 2i)(3 + i)$

48. $(2 - 4i)(2 - i)$

49. $(6 + 5i)(3 + 2i)$

50. $(4 - 7i)(2 + 3i)$

51. $(2 + 5i)^2$

52. $(3 - 4i)^2$

53. $\left(\frac{6}{5} + \frac{3}{5}i\right)\left(\frac{2}{3} - \frac{1}{3}i\right)$

54. $(2 - i)\left(\frac{2}{5} + \frac{1}{5}i\right)$

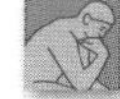

55. True or false? For all real numbers a and b, the product $(a + bi)(a - bi)$ is a positive real number.

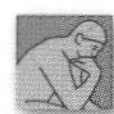

56. Given that $\left(\frac{\sqrt{2}}{2} + \frac{\sqrt{2}}{2}i\right)^2 = i$ (try to verify this statement), what is a square root of i?

OBJECTIVE D To divide complex numbers

For Exercises 57 to 71, simplify.

57. $\frac{3}{i}$

58. $\frac{4}{5i}$

59. $\frac{2 - 3i}{-4i}$

60. $\frac{16 + 5i}{-3i}$

61. $\frac{4}{5 + i}$

62. $\frac{6}{5 + 2i}$

63. $\frac{2}{2 - i}$

64. $\frac{5}{4 - i}$

65. $\frac{1 - 3i}{3 + i}$

66. $\frac{2 + 12i}{5 + i}$

67. $\frac{3i}{1 + 4i}$

68. $\frac{-2i}{2 - 3i}$

69. $\frac{2 - 3i}{3 + i}$

70. $\frac{3 + 5i}{1 - i}$

71. $\frac{5 + 3i}{3 - i}$

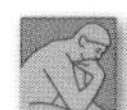

72. True or false? The quotient of two imaginary numbers is an imaginary number.

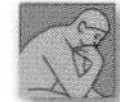

73. True or false? The reciprocal of an imaginary number is an imaginary number.

Applying the Concepts

74. Evaluate i^n for $n = 0, 1, 2, 3, 4, 5, 6,$ and 7. Make a conjecture about the value of i^n for any natural number. Using your conjecture, evaluate i^{76}.

75. **a.** Is $3i$ a solution of $2x^2 + 18 = 0$?
b. Is $3 + i$ a solution of $x^2 - 6x + 10 = 0$?

FOCUS ON PROBLEM SOLVING

Another Look at Polya's Four-Step Process

Polya's four general steps to follow when attempting to solve a problem are to understand the problem, devise a plan, carry out the plan, and review the solution. (See the Focus on Problem Solving in Chapter 4.) In the process of devising a plan (Step 2), it may be appropriate to write a mathematical expression or an equation. We will illustrate this with the following problem.

A = 1
B = 2
C = 3
D = 4
E = 5
F = 6
G = 7
H = 8
I = 9
J = 10
K = 11
L = 12
M = 13
N = 14
O = 15
P = 16
Q = 17
R = 18
S = 19
T = 20
U = 21
V = 22
W = 23
X = 24
Y = 25
Z = 26

Number the letters of the alphabet in sequence from 1 to 26. (See the list at the left.) Find a word for which the product of the numerical values of the letters of the word equals 1,000,000. We will agree that a "word" is any sequence of letters that contains at least one vowel but is not necessarily in the dictionary.

Understand the Problem

Consider REZB. The product of the values of the letters is $18 \cdot 5 \cdot 26 \cdot 2 = 4680$. This "word" is a sequence of letters with at least one vowel. However, the product of the numerical values of the letters is not 1,000,000. Thus this word does not solve our problem.

Devise a Plan

Actually, we should have known that the product of the values of the letters in REZB could not equal 1,000,000. The letter R has a factor of 9, and the letter Z has a factor of 13. Neither of these two numbers is a factor of 1,000,000. Consequently, R and Z cannot be letters in the word we are trying to find. This observation leads to an important consideration: The value of each of the letters that make up our word must be a factor of 1,000,000. To find these letters, consider the prime factorization of 1,000,000.

$$1{,}000{,}000 = 2^6 \cdot 5^6$$

Looking at the prime factorization, we note that only letters whose numerical values contain 2 or 5 as factors are possible candidates. These letters are B, D, E, H, J, P, T, and Y. One additional point: Because 1 times any number is the number, the letter A can be part of any word we construct.

Our task is now to construct a word from these letters such that the product of the numerical values is 1,000,000. From the prime factorization above, we must have 2 as a factor six times and 5 as a factor six times.

Carry Out the Plan

We must construct a word with the characteristics described in our plan. Here is a possibility:

THEBEYE

Review the Solution

You should multiply the values of all the letters and verify that the product is 1,000,000. To ensure that you have an understanding of the problem, find other "words" that satisfy the conditions of the problem.

PROJECTS AND GROUP ACTIVITIES

Solving Radical Equations with a Graphing Calculator

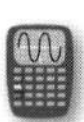

The radical equation $\sqrt{x-2} - 6 = 0$ was solved algebraically at the beginning of Section 9.3. To solve $\sqrt{x-2} - 6 = 0$ with a graphing calculator, use the left side of the equation to write an equation in two variables.

$$y = \sqrt{x-2} - 6$$

The graph of $y = \sqrt{x-2} - 6$ is shown below. The solution set of the equation $y = \sqrt{x-2} - 6$ is the set of ordered pairs (x, y) whose coordinates make the equation a true statement. The x-coordinate at which $y = 0$ is the solution of the equation $\sqrt{x-2} - 6 = 0$. The solution is the x-intercept of the curve given by the equation $y = \sqrt{x-2} - 6$. The solution is 38.

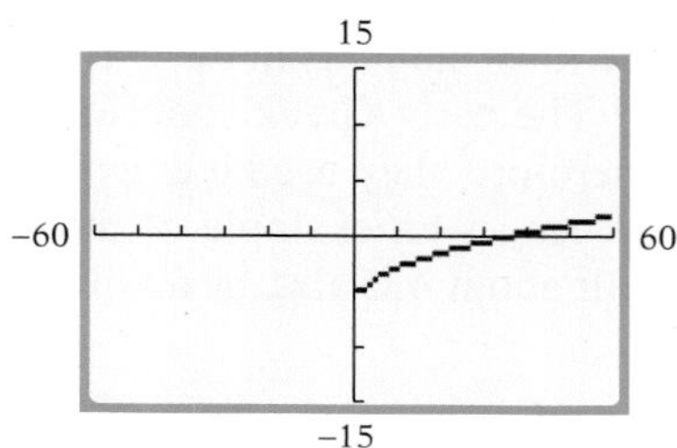

The solution of the radical equation $\sqrt[3]{x-2} = -0.8$ is 1.488. A graphing calculator can be used to find any rational or irrational solution to a predetermined degree of accuracy. The graph of $y = \sqrt[3]{x-2} + 0.8$ is shown below. The graph intersects the x-axis at $x = 1.488$.

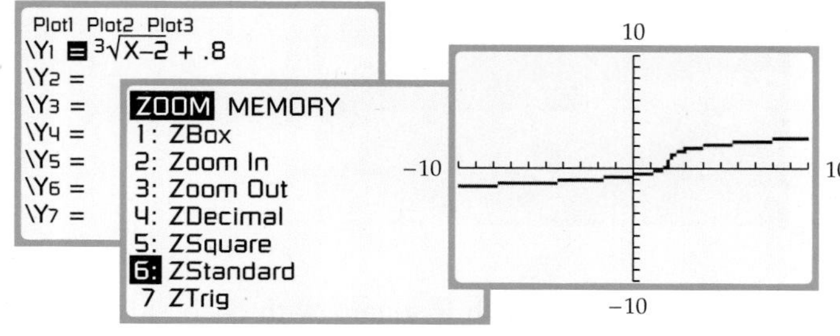

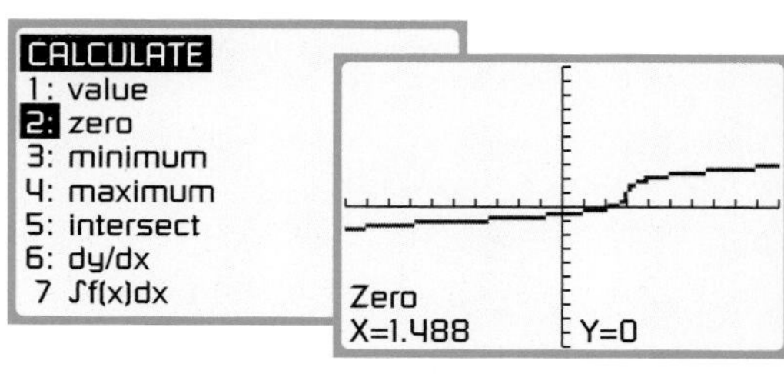

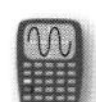

Integrating Technology

See the Keystroke Guide: *Graph* for instructions on using a graphing calculator to graph a function, and *Zero* for instructions on finding the x-intercepts of a function.

To graph the equation $y = \sqrt[3]{x-2} + 0.8$ on a TI-84, enter

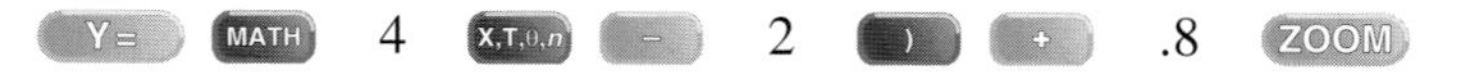

Enter 2ND CALC 2 to find the x-intercept of the equation.

You were to solve Exercises 1 to 21 in Section 9.3 by algebraic means. Try some of those exercises again, this time using a graphing calculator.

Use a graphing calculator to find the solutions of the equations in Exercises 1 to 3. Round to the nearest thousandth.

1. $\sqrt{x + 0.3} = 1.3$ **2.** $\sqrt[3]{x + 1.2} = -1.1$ **3.** $\sqrt[4]{3x - 1.5} = 1.4$

The Golden Rectangle

Take the time to look closely at each of the four rectangles shown at the left. Which rectangle seems to be the most pleasant to look at? Record the responses of all the students in your class. Which rectangle was named most frequently as the most appealing?

Rectangle A

The length of rectangle A is three times the width. The ratio of the length to the width is 3 to 1.

Rectangle B

Rectangle B is a square. Its length is equal to its width. The ratio of the length to the width is 1 to 1.

Rectangle C

The ratio of the length of rectangle C to its width is 1.2 to 1.

Rectangle D

The shape of rectangle D is referred to as the golden rectangle. In a golden rectangle, the ratio of the length to the width is approximately 1.6 to 1.

Did most students in the class select the golden rectangle as the one most pleasant to look at? The early Greeks considered a golden rectangle to be the most pleasing to the eye. Therefore, they used it extensively in their art and architecture. The Parthenon in Athens, Greece, is an example of the use of the golden rectangle in Greek architecture. It was built about 435 B.C. as a temple to the goddess Athena.

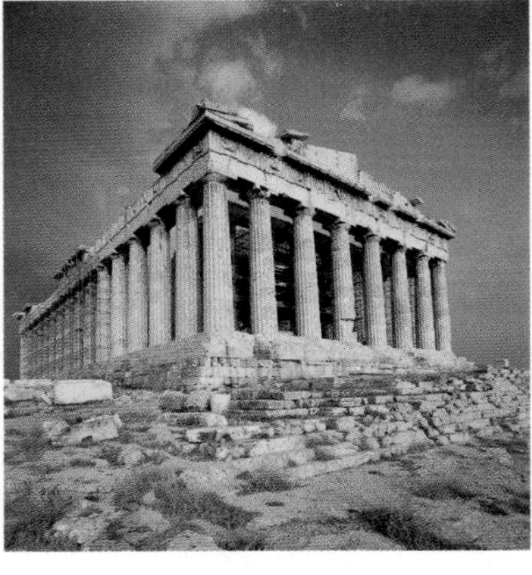

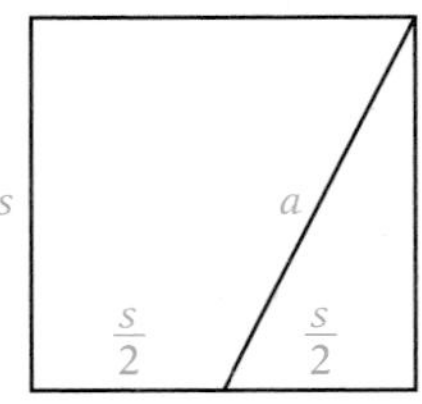

1. Construct a golden rectangle. Beginning with a square with each side of length s, draw a line segment from the midpoint of the base of the square to a vertex, as shown. Call this distance a. Make a rectangle whose width is that of the square and whose length is $\frac{s}{2} + a$.

2. Find the area of the rectangle in terms of s.

3. Find the exact ratio of the length to the width.

4. Find some examples of the golden rectangle in art and architecture. (*Note:* The Internet is a wonderful source of information on the golden rectangle. Simply enter “golden rectangle” in a search engine. You will be directed to hundreds of websites where you can learn about this rectangle’s appearance in art, architecture, music, nature, and mathematics.)

CHAPTER 9
SUMMARY

KEY WORDS	EXAMPLES
$a^{1/n}$ is the *nth root of a.* [9.1A, p. 522]	$16^{1/2} = 4$ because $4^2 = 16$.
The expression $\sqrt[n]{a}$ is another symbol for the *n*th root of *a*. In the expression $\sqrt[n]{a}$, the symbol $\sqrt{\ }$ is called a *radical, n* is the *index* of the radical, and *a* is the *radicand.* [9.1B, p. 524]	$125^{1/3} = \sqrt[3]{125} = 5$ The index is 3, and the radicand is 125.
The symbol $\sqrt{\ }$ is used to indicate the positive square root, or *principal square root*, of a number. [9.1C, p. 526]	$\sqrt{16} = 4$ $-\sqrt{16} = -4$
The expressions $a + b$ and $a - b$ are called *conjugates* of each other. The product of conjugates of the form $(a + b)(a - b)$ is $a^2 - b^2$. [9.2C, p. 535]	$(x + 3)(x - 3) = x^2 - 3^2 = x^2 - 9$
The procedure used to remove a radical from the denominator of a radical expression is called *rationalizing the denominator.* [9.2D, p. 536]	$\frac{2}{1-\sqrt{3}} = \frac{2}{1-\sqrt{3}} \cdot \frac{1+\sqrt{3}}{1+\sqrt{3}} = \frac{2(1+\sqrt{3})}{(1-\sqrt{3})(1+\sqrt{3})} = \frac{2+2\sqrt{3}}{1-3} = -1-\sqrt{3}$
A *radical equation* is an equation that contains a variable expression in a radicand. [9.3A, p. 542]	$\sqrt{x-2} - 3 = 6$ is a radical equation.
A *complex number* is a number of the form $a + bi$, where a and b are real numbers and $i = \sqrt{-1}$. For the complex number $a + bi$, a is the *real part* of the complex number and b is the *imaginary part* of the complex number. [9.4A, p. 548]	$3 + 2i$ is a complex number. 3 is the real part and 2 is the imaginary part of the complex number.

ESSENTIAL RULES AND PROCEDURES	EXAMPLES
Rule for Rational Exponents [9.1A, p. 522] If m and n are positive integers and $a^{1/n}$ is a real number, then $a^{m/n} = (a^{1/n})^m$.	$8^{2/3} = (8^{1/3})^2 = 2^2 = 4$

Definition of *n*th Root of *a* [9.1B, p. 524]
If a is a real number, then $a^{1/n} = \sqrt[n]{a}$.

$x^{1/3} = \sqrt[3]{x}$

Writing Exponential Expressions as Radical Expressions [9.1B, p. 524]
If $a^{1/n}$ is a real number, then $a^{m/n} = a^{m \cdot 1/n} = (a^m)^{1/n} = \sqrt[n]{a^m}$.
The expression $a^{m/n}$ can also be written $(\sqrt[n]{a})^m$.

$b^{3/4} = \sqrt[4]{b^3}$
$8^{2/3} = (\sqrt[3]{8})^2 = 2^2 = 4$

Product Property of Radicals [9.2A, p. 532]
If $\sqrt[n]{a}$ and $\sqrt[n]{b}$ are positive real numbers, then $\sqrt[n]{ab} = \sqrt[n]{a} \cdot \sqrt[n]{b}$ and $\sqrt[n]{a} \cdot \sqrt[n]{b} = \sqrt[n]{ab}$.

$\sqrt{9 \cdot 7} = \sqrt{9} \cdot \sqrt{7} = 3\sqrt{7}$

Quotient Property of Radicals [9.2D, p. 536]
If $\sqrt[n]{a}$ and $\sqrt[n]{b}$ are positive real numbers and $b \neq 0$, then
$\sqrt[n]{\dfrac{a}{b}} = \dfrac{\sqrt[n]{a}}{\sqrt[n]{b}}$ and $\dfrac{\sqrt[n]{a}}{\sqrt[n]{b}} = \sqrt[n]{\dfrac{a}{b}}$.

$\sqrt[3]{\dfrac{5}{27}} = \dfrac{\sqrt[3]{5}}{\sqrt[3]{27}} = \dfrac{\sqrt[3]{5}}{3}$

Property of Raising Each Side of an Equation to a Power [9.3A, p. 542]
If $a = b$, then $a^n = b^n$.

If $x = 4$, then $x^2 = 16$.

Pythagorean Theorem [9.3B, p. 544]
The square of the hypotenuse of a right triangle is equal to the sum of the squares of the two legs.
$c^2 = a^2 + b^2$

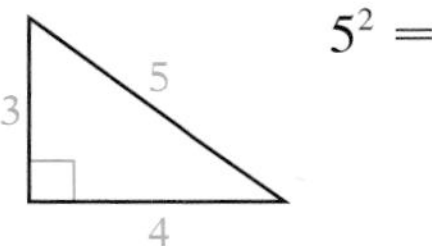

$5^2 = 3^2 + 4^2$

Definition of $\sqrt{-a}$ [9.4A, p. 548]
If a is a positive real number, then the principal square root of negative a is the imaginary number $i\sqrt{a}$: $\sqrt{-a} = i\sqrt{a}$.

$\sqrt{-8} = i\sqrt{8} = 2i\sqrt{2}$

Addition and Subtraction of Complex Numbers [9.4B, p. 549]
$(a + bi) + (c + di) = (a + c) + (b + d)i$
$(a + bi) - (c + di) = (a - c) + (b - d)i$

$(2 + 4i) + (3 + 6i)$
$= (2 + 3) + (4 + 6)i$
$= 5 + 10i$
$(4 + 3i) - (7 + 4i)$
$= (4 - 7) + (3 - 4)i$
$= -3 - i$

Multiplication of Complex Numbers [9.4C, p. 551]
$(a + bi)(c + di) = (ac - bd) + (ad + bc)i$
One way to remember this rule is to think of the FOIL method.

$(2 - 3i)(5 + 4i)$
$= 10 + 8i - 15i - 12i^2$
$= 10 - 7i - 12(-1)$
$= 22 - 7i$

CHAPTER 9

CONCEPT REVIEW

Test your knowledge of the concepts presented in this chapter. Answer each question. Then check your answers against the ones provided in the Answer Section.

1. How do you write an expression that contains a rational exponent as a radical expression?

2. In Chapter 9, why is it assumed that all variables represent positive numbers?

3. How can you tell when a radical expression is in simplest form?

4. How do you rationalize the denominator of a radical expression that has two terms in the denominator?

5. When can you add two radical expressions?

6. Why do you have to isolate the radical when solving an equation containing a radical expression?

7. How can you tell if a solution is an extraneous solution?

8. How do you add two complex numbers?

9. How can $\sqrt{36} + \sqrt{-36}$ be written as a complex number?

10. How do you find the product of two variable expressions with the same base and fractional exponents?

CHAPTER 9

REVIEW EXERCISES

1. Simplify: $(16x^{-4}y^{12})^{1/4}(100x^6y^{-2})^{1/2}$

2. Solve: $\sqrt[4]{3x-5}=2$

3. Multiply: $(6-5i)(4+3i)$

4. Rewrite $7y\sqrt[3]{x^2}$ as an exponential expression.

5. Multiply: $(\sqrt{3}+8)(\sqrt{3}-2)$

6. Solve: $\sqrt{4x+9}+10=11$

7. Divide: $\dfrac{x^{-3/2}}{x^{7/2}}$

8. Simplify: $\dfrac{8}{\sqrt{3y}}$

9. Simplify: $\sqrt[3]{-8a^6b^{12}}$

10. Simplify: $\sqrt{50a^4b^3}-ab\sqrt{18a^2b}$

11. Simplify: $\dfrac{14}{4-\sqrt{2}}$

12. Simplify: $\dfrac{5+2i}{3i}$

13. Simplify: $\sqrt{18a^3b^6}$

14. Subtract: $(17+8i)-(15-4i)$

15. Simplify: $3x\sqrt[3]{54x^8y^{10}}-2x^2y\sqrt[3]{16x^5y^7}$

16. Multiply: $\sqrt[3]{16x^4y}\sqrt[3]{4xy^5}$

17. Multiply: $i(3-7i)$

18. Simplify: $\dfrac{(4a^{-2/3}b^4)^{-1/2}}{(a^{-1/6}b^{3/2})^2}$

19. Simplify: $\sqrt[5]{-64a^8b^{12}}$

20. Divide: $\dfrac{5 + 9i}{1 - i}$

21. Multiply: $\sqrt{-12}\sqrt{-6}$

22. Solve: $\sqrt{x - 5} + \sqrt{x + 6} = 11$

23. Simplify: $\sqrt[4]{81a^8b^{12}}$

24. Simplify: $\dfrac{9}{\sqrt[3]{3x}}$

25. Subtract: $(-8 + 3i) - (4 - 7i)$

26. Expand: $(2 + \sqrt{2x - 1})^2$

27. Simplify: $4x\sqrt{12x^2y} + \sqrt{3x^4y} - x^2\sqrt{27y}$

28. Simplify: $81^{-1/4}$

29. Simplify: $(a^{16})^{-5/8}$

30. Simplify: $-\sqrt{49x^6y^{16}}$

31. Rewrite $4a^{2/3}$ as a radical expression.

32. Simplify: $(9x^2y^4)^{-1/2}(x^6y^6)^{1/3}$

33. Simplify: $\sqrt[4]{x^6y^8z^{10}}$

34. Simplify: $\sqrt{54} + \sqrt{24}$

35. Simplify: $\sqrt{48x^5y} - x\sqrt{80x^2y}$

36. Simplify: $\sqrt{32}\sqrt{50}$

37. Simplify: $\sqrt{3x}(3 + \sqrt{3x})$

38. Simplify: $\dfrac{\sqrt{125x^6}}{\sqrt{5x^3}}$

39. Simplify: $\dfrac{2 - 3\sqrt{7}}{6 - \sqrt{7}}$

40. Simplify: $\sqrt{-36}$

41. Evaluate $-b + \sqrt{b^2 - 4ac}$ when $a = 1$, $b = -8$, and $c = 25$.

42. Evaluate $-b + \sqrt{b^2 - 4ac}$ when $a = 1$, $b = 2$, and $c = 9$.

43. Add: $(5 + 2i) + (4 - 3i)$

44. Simplify: $(3 + 2\sqrt{5})(3 - 2\sqrt{5})$

45. Simplify: $(3 - 9i) - 7$

46. Expand: $(4 - i)^2$

47. Simplify: $\dfrac{-6}{i}$

48. Divide: $\dfrac{7}{2 - i}$

49. Solve: $\sqrt{2x - 7} + 2 = 5$

50. Solve: $\sqrt[3]{9x} = -6$

51. **Geometry** Find the width of a rectangle that has a diagonal of 13 in. and a length of 12 in.

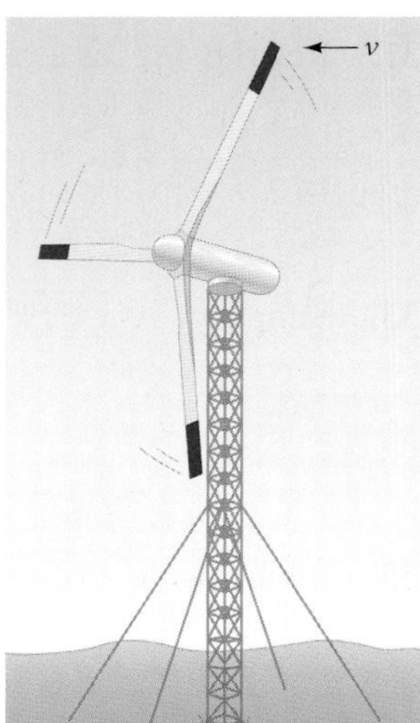

52. **Energy** The velocity of the wind determines the amount of power generated by a windmill. A typical equation for this relationship is $v = 4.05\sqrt[3]{P}$, where v is the velocity in miles per hour and P is the power in watts. Find the amount of power generated by a 20-mph wind. Round to the nearest whole number.

53. **Automotive Technology** Find the distance required for a car to reach a velocity of 88 ft/s when the acceleration is 16 ft/s^2. Use the equation $v = \sqrt{2as}$, where v is the velocity in feet per second, a is the acceleration, and s is the distance in feet.

54. **Home Maintenance** A 12-foot ladder is leaning against a house in preparation for washing the windows. How far from the house is the bottom of the ladder when the top of the ladder touches the house 10 ft above the ground? Round to the nearest hundredth.

CHAPTER 9

TEST

1. Write $\frac{1}{2}\sqrt[4]{x^3}$ as an exponential expression.

2. Simplify: $\sqrt[3]{54x^7y^3} - x\sqrt[3]{128x^4y^3} - x^2\sqrt[3]{2xy^3}$

3. Write $3y^{2/5}$ as a radical expression.

4. Multiply: $(2 + 5i)(4 - 2i)$

5. Expand: $(3 - 2\sqrt{x})^2$

6. Simplify: $\dfrac{r^{2/3}r^{-1}}{r^{-1/2}}$

7. Solve: $\sqrt{x + 12} - \sqrt{x} = 2$

8. Multiply: $\sqrt[4]{4a^5b^3}\sqrt[4]{8a^3b^7}$

9. Multiply: $\sqrt{3x}(\sqrt{x} - \sqrt{25x})$

10. Subtract: $(5 - 2i) - (8 - 4i)$

11. Simplify: $\sqrt{32x^4y^7}$

12. Multiply: $(2\sqrt{3} + 4)(3\sqrt{3} - 1)$

13. Simplify: $\sqrt{-5}\sqrt{-20}$

14. Simplify: $\dfrac{4 - 2\sqrt{5}}{2 - \sqrt{5}}$

15. Add: $\sqrt{18a^3} + a\sqrt{50a}$

16. Multiply: $(\sqrt{a} - 3\sqrt{b})(2\sqrt{a} + 5\sqrt{b})$

17. Simplify: $\dfrac{(2x^{1/3}y^{-2/3})^6}{(x^{-4}y^8)^{1/4}}$

18. Simplify: $\dfrac{10x}{\sqrt[3]{5x^2}}$

19. Simplify: $\dfrac{2 + 3i}{1 - 2i}$

20. Solve: $\sqrt[3]{2x - 2} + 4 = 2$

21. Simplify: $\left(\dfrac{4a^4}{b^2}\right)^{-3/2}$

22. Simplify: $\sqrt[3]{27a^4b^3c^7}$

23. Divide: $\dfrac{\sqrt{32x^5y}}{\sqrt{2xy^3}}$

24. Simplify: $\dfrac{5x}{\sqrt{5x}}$

25. **Physics** An object is dropped from a high building. Find the distance the object has fallen when its speed reaches 192 ft/s. Use the equation $v = \sqrt{64d}$, where v is the speed of the object in feet per second and d is the distance in feet.

CUMULATIVE REVIEW EXERCISES

1. Simplify: $2^3 \cdot 3 - 4(3 - 4 \cdot 5)$

2. Evaluate $4a^2b - a^3$ when $a = -2$ and $b = 3$.

3. Simplify: $-3(4x - 1) - 2(1 - x)$

4. Solve: $5 - \frac{2}{3}x = 4$

5. Solve: $2[4 - 2(3 - 2x)] = 4(1 - x)$

6. Solve: $6x - 3(2x + 2) > 3 - 3(x + 2)$

7. Solve: $2 + |4 - 3x| = 5$

8. Solve: $|2x + 3| \leq 9$

9. Find the area of the triangle shown in the figure below.

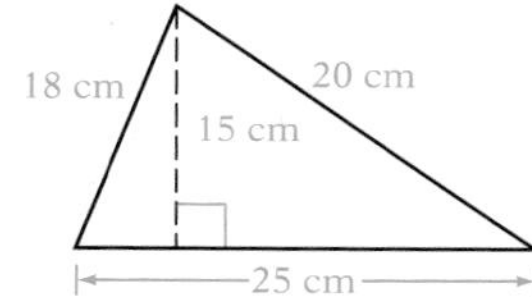

10. Find the volume of a rectangular solid with a length of 3.5 ft, a width of 2 ft, and a height of 2 ft.

11. Graph $3x - 2y = -6$. State the slope and y-intercept.

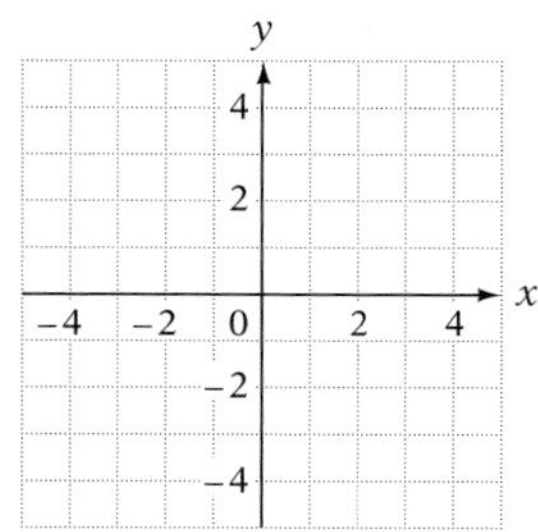

12. Graph the solution set of $3x + 2y \leq 4$.

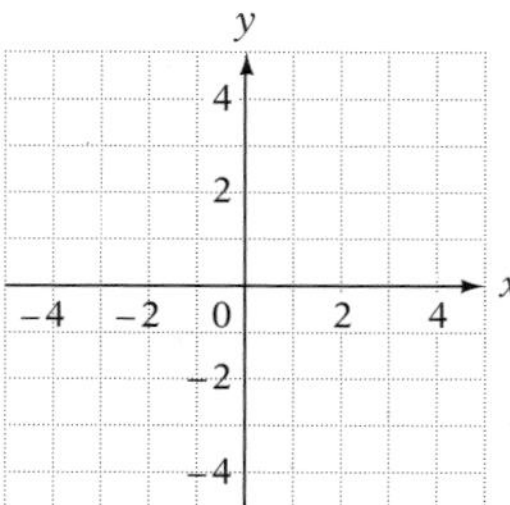

13. Find the equation of the line that passes through the points (2, 3) and (−1, 2).

14. Solve by the addition method:

$$2x - y = 4$$
$$-2x + 3y = 0$$

15. Simplify: $(2^{-1}x^2y^{-6})(2^{-1}y^{-4})^{-2}$

16. Factor: $81x^2 - y^2$

17. Factor: $x^5 + 2x^3 - 3x$

18. Solve $P = \dfrac{R - C}{n}$ for C.

19. Simplify: $\left(\dfrac{x^{-2/3}y^{1/2}}{y^{-1/3}}\right)^6$

20. Subtract: $\sqrt{40x^3} - x\sqrt{90x}$

21. Multiply: $(\sqrt{3} - 2)(\sqrt{3} - 5)$

22. Simplify: $\dfrac{4}{\sqrt{6} - \sqrt{2}}$

23. Simplify: $\dfrac{2i}{3 - i}$

24. Solve: $\sqrt[3]{3x - 4} + 5 = 1$

25. The two triangles are similar triangles. Find the length of side DE.

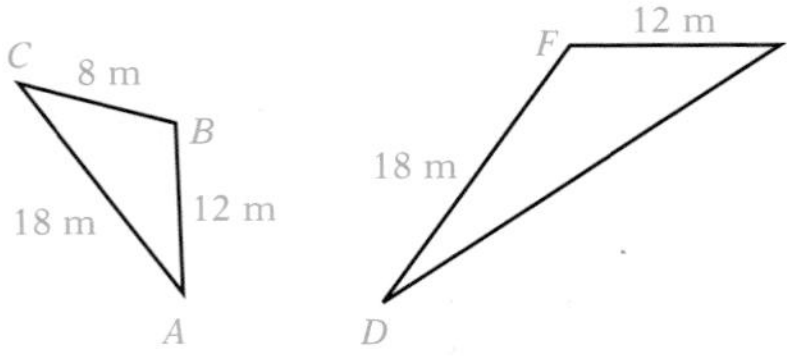

26. **Investments** An investor has a total of \$10,000 deposited in two simple interest accounts. On one account, the annual simple interest rate is 3.5%. On the second account, the annual simple interest rate is 4.5%. How much is invested in the 3.5% account if the total annual interest earned is \$425?

27. **Travel** A sales executive traveled 25 mi by car and then an additional 625 mi by plane. The rate of the plane was five times the rate of the car. The total time of the trip was 3 h. Find the rate of the plane.

28. **Astronomy** How long does it take light to travel to Earth from the moon when the moon is 232,500 mi from Earth? Light travels 1.86×10^5 mi/s.

29. **Investments** The graph shows the amount invested and the annual interest earned on the investment. Find the slope of the line between the two points shown on the graph. Then write a sentence that states the meaning of the slope.

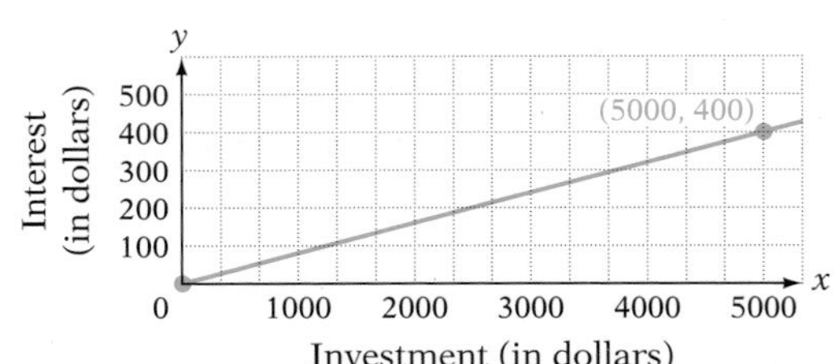

30. **Oceanography** How far would a submarine periscope have to be above the water for the lookout to locate a ship 7 mi away? The equation for the distance in miles that the lookout can see is $d = \sqrt{1.5h}$, where h is the height in feet above the surface of the water. Round to the nearest tenth of a foot.

CHAPTER 10

Quadratic Equations

digitalvision/First Light

OBJECTIVES

SECTION 10.1

A To solve a quadratic equation by factoring

B To write a quadratic equation given its solutions

C To solve a quadratic equation by taking square roots

SECTION 10.2

A To solve a quadratic equation by completing the square

SECTION 10.3

A To solve a quadratic equation by using the quadratic formula

SECTION 10.4

A To solve an equation that is quadratic in form

B To solve a radical equation that is reducible to a quadratic equation

C To solve a rational equation that is reducible to a quadratic equation

SECTION 10.5

A To solve a nonlinear inequality

SECTION 10.6

A To solve application problems

ARE YOU READY?

Take the Chapter 10 Prep Test to find out if you are ready to learn to:

- Solve a quadratic equation by factoring, by taking square roots, by completing the square, and by using the quadratic formula
- Solve an equation that is reducible to a quadratic equation
- Solve a nonlinear inequality

PREP TEST

Do these exercises to prepare for Chapter 10.

1. Simplify: $\sqrt{18}$

2. Simplify: $\sqrt{-9}$

3. Simplify: $\dfrac{3x-2}{x-1} - 1$

4. Evaluate $b^2 - 4ac$ when $a = 2$, $b = -4$, and $c = 1$.

5. Is $4x^2 + 28x + 49$ a perfect square trinomial?

6. Factor: $4x^2 - 4x + 1$

7. Factor: $9x^2 - 4$

8. Graph $\{x|x < -1\} \cap \{x|x < 4\}$.

−5 −4 −3 −2 −1 0 1 2 3 4 5

9. Solve: $x(x-1) = x + 15$

10. Solve: $\dfrac{4}{x-3} = \dfrac{16}{x}$

SECTION

10.1 Solving Quadratic Equations by Factoring or by Taking Square Roots

OBJECTIVE A To solve a quadratic equation by factoring

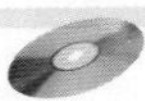

Recall that a *quadratic equation* is an equation of the form $ax^2 + bx + c = 0$, where a and b are coefficients, c is a constant, and $a \neq 0$.

$$\text{Quadratic Equations} \begin{cases} 3x^2 - x + 2 = 0, & a = 3, \quad b = -1, \quad c = 2 \\ -x^2 + 4 = 0, & a = -1, \quad b = 0, \quad c = 4 \\ 6x^2 - 5x = 0, & a = 6, \quad b = -5, \quad c = 0 \end{cases}$$

A quadratic equation is in *standard form* when the polynomial is in descending order and equal to zero. Because the degree of the polynomial $ax^2 + bx + c$ is 2, a quadratic equation is also called a **second-degree equation.**

As we discussed earlier, quadratic equations sometimes can be solved by using the Principle of Zero Products. This method is reviewed here.

The Principle of Zero Products

If a and b are real numbers and $ab = 0$, then $a = 0$ or $b = 0$.

The Principle of Zero Products states that if the product of two factors is zero, then at least one of the factors must be zero.

Take Note

Recall that the steps involved in solving a quadratic equation by factoring are

1. Write the equation in standard form.
2. Factor.
3. Use the Principle of Zero Products to set each factor equal to 0.
4. Solve each equation.
5. Check the solutions.

HOW TO • 1 Solve by factoring: $3x^2 = 2 - 5x$

$$3x^2 = 2 - 5x$$

$$3x^2 + 5x - 2 = 0$$ • Write the equation in standard form.

$$(3x - 1)(x + 2) = 0$$ • Factor.

$$3x - 1 = 0 \qquad x + 2 = 0$$ • Use the Principle of Zero Products to write two equations.

$$3x = 1 \qquad x = -2$$ • Solve each equation.

$$x = \frac{1}{3}$$

$\frac{1}{3}$ and -2 check as solutions. The solutions are $\frac{1}{3}$ and -2.

HOW TO • 2 Solve by factoring: $x^2 - 6x = -9$

$$x^2 - 6x = -9$$

$$x^2 - 6x + 9 = 0$$ • Write the equation in standard form.

$$(x - 3)(x - 3) = 0$$ • Factor.

$$x - 3 = 0 \qquad x - 3 = 0$$ • Use the Principle of Zero Products.

$$x = 3 \qquad x = 3$$ • Solve each equation.

3 checks as a solution. The solution is 3.

Take Note

When a quadratic equation has two solutions that are the same number, the solution is called a **double root** of the equation. The number 3 is a double root of $x^2 - 6x = -9$.

EXAMPLE 1

Solve by factoring: $2x(x - 3) = x + 4$

Solution

$$2x(x - 3) = x + 4$$
$$2x^2 - 6x = x + 4$$
$$2x^2 - 7x - 4 = 0 \quad \text{• Write in standard form.}$$
$$(2x + 1)(x - 4) = 0 \quad \text{• Solve by factoring.}$$

$$2x + 1 = 0 \qquad x - 4 = 0$$
$$2x = -1 \qquad x = 4$$
$$x = -\frac{1}{2}$$

The solutions are $-\frac{1}{2}$ and 4.

YOU TRY IT 1

Solve by factoring: $2x^2 = 7x - 3$

Your solution

EXAMPLE 2

Solve for x by factoring: $x^2 - 4ax - 5a^2 = 0$

Solution

This is a literal equation. Solve for x in terms of a.

$$x^2 - 4ax - 5a^2 = 0$$
$$(x + a)(x - 5a) = 0$$
$$x + a = 0 \qquad x - 5a = 0$$
$$x = -a \qquad x = 5a$$

The solutions are $-a$ and $5a$.

YOU TRY IT 2

Solve for x by factoring: $x^2 - 3ax - 4a^2 = 0$

Your solution

Solutions on p. S28

OBJECTIVE B To write a quadratic equation given its solutions 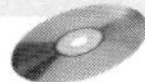

As shown below, the solutions of the equation $(x - r_1)(x - r_2) = 0$ are r_1 and r_2.

$$(x - r_1)(x - r_2) = 0$$
$$x - r_1 = 0 \qquad x - r_2 = 0$$
$$x = r_1 \qquad x = r_2$$

Check:

$$\begin{array}{c|c} (x - r_1)(x - r_2) & 0 \\ \hline (r_1 - r_1)(r_1 - r_2) & 0 \\ 0 \cdot (r_1 - r_2) & 0 \\ 0 = 0 & \end{array} \qquad \begin{array}{c|c} (x - r_1)(x - r_2) & 0 \\ \hline (r_2 - r_1)(r_2 - r_2) & 0 \\ (r_2 - r_1) \cdot 0 & 0 \\ 0 = 0 & \end{array}$$

Using the equation $(x - r_1)(x - r_2) = 0$ and the fact that r_1 and r_2 are solutions of this equation, it is possible to write a quadratic equation given its solutions.

HOW TO 3 Write a quadratic equation that has solutions 4 and -5.

$$(x - r_1)(x - r_2) = 0$$
$$(x - 4)[x - (-5)] = 0 \quad \text{• Replace } r_1 \text{ by 4 and } r_2 \text{ by } -5.$$
$$(x - 4)(x + 5) = 0 \quad \text{• Simplify.}$$
$$x^2 + x - 20 = 0 \quad \text{• Multiply.}$$

HOW TO • 4 Write a quadratic equation with integer coefficients and solutions $\frac{2}{3}$ and $\frac{1}{2}$.

$$(x - r_1)(x - r_2) = 0$$

$$\left(x - \frac{2}{3}\right)\left(x - \frac{1}{2}\right) = 0$$ • Replace r_1 by $\frac{2}{3}$ and r_2 by $\frac{1}{2}$.

$$x^2 - \frac{7}{6}x + \frac{1}{3} = 0$$ • Multiply.

$$6\left(x^2 - \frac{7}{6}x + \frac{1}{3}\right) = 6 \cdot 0$$ • Multiply each side of the equation by the LCM of the denominators.

$$6x^2 - 7x + 2 = 0$$

EXAMPLE • 3

Write a quadratic equation with integer coefficients and solutions $\frac{1}{2}$ and -4.

Solution

$$(x - r_1)(x - r_2) = 0$$

$$\left(x - \frac{1}{2}\right)[x - (-4)] = 0$$ • $r_1 = \frac{1}{2}, r_2 = -4$

$$\left(x - \frac{1}{2}\right)(x + 4) = 0$$

$$x^2 + \frac{7}{2}x - 2 = 0$$

$$2\left(x^2 + \frac{7}{2}x - 2\right) = 2 \cdot 0$$

$$2x^2 + 7x - 4 = 0$$

YOU TRY IT • 3

Write a quadratic equation with integer coefficients and solutions 3 and $-\frac{1}{2}$.

Your solution

Solution on p. S29

OBJECTIVE C To solve a quadratic equation by taking square roots

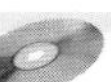

The solution of the quadratic equation $x^2 = 16$ is shown at the right.

$$x^2 = 16$$
$$x^2 - 16 = 0$$
$$(x - 4)(x + 4) = 0$$
$$x - 4 = 0 \qquad x + 4 = 0$$
$$x = 4 \qquad x = -4$$

The solution can also be found by taking the square root of each side of the equation and writing the positive and negative square roots of the radicand. The notation ± 4 means $x = 4$ or $x = -4$.

$$x^2 = 16$$
$$\sqrt{x^2} = \sqrt{16}$$
$$x = \pm 4$$

The solutions are 4 and -4.

HOW TO • 5 Solve by taking square roots: $3x^2 = 54$

$$3x^2 = 54$$

$$x^2 = 18$$ • Solve for x^2.

$$\sqrt{x^2} = \sqrt{18}$$ • Take the square root of each side of the equation.

$$x = \pm\sqrt{18}$$ • Simplify.

$$x = \pm 3\sqrt{2}$$

The solutions are $3\sqrt{2}$ and $-3\sqrt{2}$. • $3\sqrt{2}$ and $-3\sqrt{2}$ check as solutions.

Solving a quadratic equation by taking the square root of each side of the equation can lead to solutions that are complex numbers.

HOW TO • 6 Solve by taking square roots: $2x^2 + 18 = 0$

$2x^2 + 18 = 0$

$2x^2 = -18$

$x^2 = -9$ • Solve for x^2.

$\sqrt{x^2} = \sqrt{-9}$ • Take the square root of each side of the equation.

$x = \pm\sqrt{-9}$ • Simplify.

$x = \pm 3i$

Check:

$2x^2 + 18$	0
$2(3i)^2 + 18$	0
$2(-9) + 18$	0
$-18 + 18$	0
$0 = 0$	

$2x^2 + 18$	0
$2(-3i)^2 + 18$	0
$2(-9) + 18$	0
$-18 + 18$	0
$0 = 0$	

The solutions are $3i$ and $-3i$.

Tips for Success
Always check the solution of an equation.

An equation containing the square of a binomial can be solved by taking square roots.

HOW TO • 7 Solve by taking square roots: $(x + 2)^2 - 24 = 0$

$(x + 2)^2 - 24 = 0$

$(x + 2)^2 = 24$ • Solve for $(x + 2)^2$.

$\sqrt{(x + 2)^2} = \sqrt{24}$ • Take the square root of each side of the equation. Then simplify.

$x + 2 = \pm\sqrt{24}$

$x + 2 = \pm 2\sqrt{6}$

$x + 2 = 2\sqrt{6}$ $\quad$ $x + 2 = -2\sqrt{6}$ • Solve for x.

$x = -2 + 2\sqrt{6}$ $\quad$ $x = -2 - 2\sqrt{6}$

The solutions are $-2 + 2\sqrt{6}$ and $-2 - 2\sqrt{6}$.

EXAMPLE • 4

Solve by taking square roots: $3(x - 2)^2 + 12 = 0$

Solution

$3(x - 2)^2 + 12 = 0$

$3(x - 2)^2 = -12$

$(x - 2)^2 = -4$ • Solve for $(x - 2)^2$.

$\sqrt{(x - 2)^2} = \sqrt{-4}$ • Take the square root of each side of the equation.

$x - 2 = \pm\sqrt{-4}$

$x - 2 = \pm 2i$

$x - 2 = 2i$ $\quad$ $x - 2 = -2i$ • Solve for x.

$x = 2 + 2i$ $\quad$ $x = 2 - 2i$

The solutions are $2 + 2i$ and $2 - 2i$.

YOU TRY IT • 4

Solve by taking square roots: $2(x + 1)^2 - 24 = 0$

Your solution

Solution on p. S29

10.1 EXERCISES

OBJECTIVE A To solve a quadratic equation by factoring

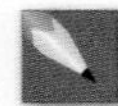

1. Explain why the restriction $a \neq 0$ is necessary in the definition of a quadratic equation.

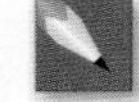

2. What does the Principle of Zero Products state? How is it used to solve a quadratic equation?

For Exercises 3 to 6, write the quadratic equation in standard form with the coefficient of x^2 positive. Name the values of a, b, and c.

3. $2x^2 - 4x = 5$
4. $x^2 = 3x + 1$
5. $5x = 4x^2 + 6$
6. $3x^2 = 7$

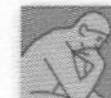

For Exercises 7 to 10, solve for x.

7. $(x - 3)(x - 5) = 0$
8. $(x + 2)(x - 4) = 0$
9. $(x + 7)(x + 1) = 0$
10. $x(x + 3) = 0$

For Exercises 11 to 41, solve by factoring.

11. $x^2 - 4x = 0$
12. $y^2 + 6y = 0$
13. $t^2 - 25 = 0$
14. $p^2 - 81 = 0$
15. $s^2 - s - 6 = 0$
16. $v^2 + 4v - 5 = 0$
17. $y^2 - 6y + 9 = 0$
18. $x^2 + 10x + 25 = 0$
19. $9z^2 - 18z = 0$
20. $4y^2 + 20y = 0$
21. $r^2 - 3r = 10$
22. $p^2 + 5p = 6$
23. $v^2 + 10 = 7v$
24. $t^2 - 16 = 15t$
25. $2x^2 - 9x - 18 = 0$
26. $3y^2 - 4y - 4 = 0$
27. $4z^2 - 9z + 2 = 0$
28. $2s^2 - 9s + 9 = 0$
29. $3w^2 + 11w = 4$
30. $2r^2 + r = 6$
31. $6x^2 = 23x + 18$

32. $6x^2 = 7x - 2$

33. $4 - 15u - 4u^2 = 0$

34. $3 - 2y - 8y^2 = 0$

35. $x + 18 = x(x - 6)$

36. $t + 24 = t(t + 6)$

37. $4s(s + 3) = s - 6$

38. $3v(v - 2) = 11v + 6$

39. $u^2 - 2u + 4 = (2u - 3)(u + 2)$

40. $(3v - 2)(2v + 1) = 3v^2 - 11v - 10$

41. $(3x - 4)(x + 4) = x^2 - 3x - 28$

For Exercises 42 to 50, solve for x by factoring.

42. $x^2 + 14ax + 48a^2 = 0$

43. $x^2 - 9bx + 14b^2 = 0$

44. $x^2 + 9xy - 36y^2 = 0$

45. $x^2 - 6cx - 7c^2 = 0$

46. $x^2 - ax - 20a^2 = 0$

47. $2x^2 + 3bx + b^2 = 0$

48. $3x^2 - 4cx + c^2 = 0$

49. $3x^2 - 14ax + 8a^2 = 0$

50. $3x^2 - 11xy + 6y^2 = 0$

OBJECTIVE B To write a quadratic equation given its solutions

For Exercises 51 to 80, write a quadratic equation that has integer coefficients and has as solutions the given pair of numbers.

51. 2 and 5

52. 3 and 1

53. -2 and -4

54. -1 and -3

55. 6 and -1

56. -2 and 5

57. 3 and -3

58. 5 and -5

59. 4 and 4

60. 2 and 2

61. 0 and 5

62. 0 and -2

63. 0 and 3

64. 0 and -1

65. 3 and $\frac{1}{2}$

66. 2 and $\frac{2}{3}$

67. $-\frac{3}{4}$ and 2

68. $-\frac{1}{2}$ and 5

69. $-\frac{5}{3}$ and -2

70. $-\frac{3}{2}$ and -1

71. $-\frac{2}{3}$ and $\frac{2}{3}$

72. $-\frac{1}{2}$ and $\frac{1}{2}$

73. $\frac{1}{2}$ and $\frac{1}{3}$

74. $\frac{3}{4}$ and $\frac{2}{3}$

75. $\frac{6}{5}$ and $-\frac{1}{2}$

76. $\frac{3}{4}$ and $-\frac{3}{2}$

77. $-\frac{1}{4}$ and $-\frac{1}{2}$

78. $-\frac{5}{6}$ and $-\frac{2}{3}$

79. $\frac{3}{5}$ and $-\frac{1}{10}$

80. $\frac{7}{2}$ and $-\frac{1}{4}$

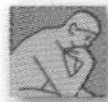

81. If 0 is a solution of $ax^2 + bx + c = 0$, what is the value of c?

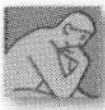

82. If u and v are solutions of $x^2 + bx + c = 0$, what are the values of b and c?

OBJECTIVE C **To solve a quadratic equation by taking square roots**

For Exercises 83 to 112, solve by taking square roots.

83. $y^2 = 49$

84. $x^2 = 64$

85. $z^2 = -4$

86. $v^2 = -16$

87. $s^2 - 4 = 0$

88. $r^2 - 36 = 0$

89. $4x^2 - 81 = 0$

90. $9x^2 - 16 = 0$

91. $y^2 + 49 = 0$

92. $z^2 + 16 = 0$

93. $v^2 - 48 = 0$

94. $s^2 - 32 = 0$

95. $r^2 - 75 = 0$

96. $u^2 - 54 = 0$

97. $z^2 + 18 = 0$

98. $t^2 + 27 = 0$

99. $(x - 1)^2 = 36$

100. $(x + 2)^2 = 25$

101. $3(y + 3)^2 = 27$

102. $4(s - 2)^2 = 36$

103. $5(z + 2)^2 = 125$

104. $2(y - 3)^2 = 18$

105. $\left(v - \frac{1}{2}\right)^2 = \frac{1}{4}$

106. $\left(r + \frac{2}{3}\right)^2 = \frac{1}{9}$

107. $(x + 5)^2 - 6 = 0$

108. $(t - 1)^2 - 15 = 0$

109. $(v - 3)^2 + 45 = 0$

110. $(x + 5)^2 + 32 = 0$

111. $\left(u + \frac{2}{3}\right)^2 - 18 = 0$

112. $\left(z - \frac{1}{2}\right)^2 - 20 = 0$

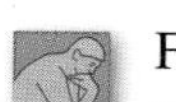

For Exercises 113 to 116, assume that a and b are positive real numbers. In each case, state how many real or complex number solutions the equation has.

113. $(x - a)^2 + b = 0$

114. $x^2 + a = 0$

115. $(x - a)^2 = 0$

116. $(x - a)^2 - b = 0$

Applying the Concepts

For Exercises 117 to 124, write a quadratic equation that has as solutions the given pair of numbers.

117. $\sqrt{2}$ and $-\sqrt{2}$

118. $\sqrt{5}$ and $-\sqrt{5}$

119. i and $-i$

120. $2i$ and $-2i$

121. $2\sqrt{2}$ and $-2\sqrt{2}$

122. $3\sqrt{2}$ and $-3\sqrt{2}$

123. $i\sqrt{2}$ and $-i\sqrt{2}$

124. $2i\sqrt{3}$ and $-2i\sqrt{3}$

For Exercises 125 to 128, solve for x.

125. $(2x - 1)^2 = (2x + 3)^2$

126. $(x - 4)^2 = (x + 2)^2$

127. $(2x + 1)^2 = (x - 7)^2$

128. $(x + 3)^2 = (2x - 3)^2$

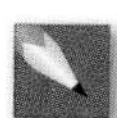

129. Use the Principle of Zero Products to explain why an equation with solutions -2, 3, and 5 could not be a quadratic equation.

SECTION

10.2 Solving Quadratic Equations by Completing the Square

OBJECTIVE A To solve a quadratic equation by completing the square

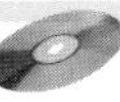

Recall that a perfect-square trinomial is the square of a binomial.

Perfect-Square Trinomial		*Square of a Binomial*
$x^2 + 8x + 16$	=	$(x + 4)^2$
$x^2 - 10x + 25$	=	$(x - 5)^2$
$x^2 + 2ax + a^2$	=	$(x + a)^2$

For each perfect-square trinomial, the square of $\frac{1}{2}$ of the coefficient of x equals the constant term.

$$\left(\frac{1}{2}\textbf{ coefficient of } x\right)^2 = \textbf{constant term}$$

$x^2 + 8x + 16$, $\left(\frac{1}{2} \cdot 8\right)^2 = 16$

$x^2 - 10x + 25$, $\left[\frac{1}{2}(-10)\right]^2 = 25$

$x^2 + 2ax + a^2$, $\left(\frac{1}{2} \cdot 2a\right)^2 = a^2$

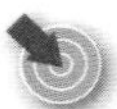

Point of Interest

Early attempts to solve quadratic equations were primarily geometric. The Persian mathematician al-Khowarizmi (c. A.D. 800) essentially completed a square of $x^2 + 12x$ as follows.

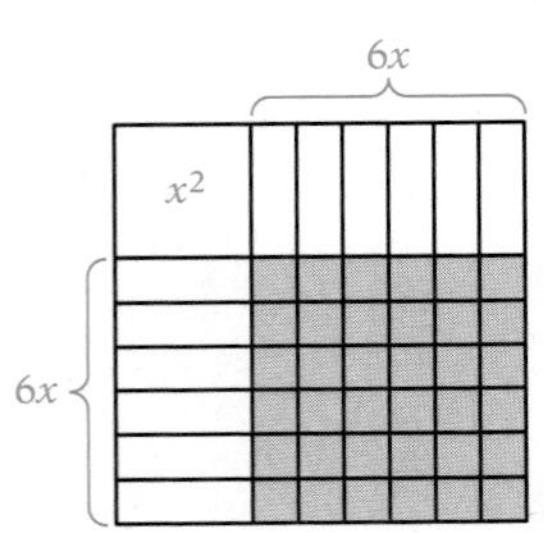

This relationship can be used to write the constant term for a perfect-square trinomial. Adding to a binomial the constant term that makes it a perfect-square trinomial is called **completing the square.**

HOW TO • 1 Complete the square on $x^2 + 12x$. Write the resulting perfect-square trinomial as the square of a binomial.

$\left[\frac{1}{2}(12)\right]^2 = (6)^2 = 36$ • Find the constant term.

$x^2 + 12x + 36$ • Complete the square on $x^2 + 12x$ by adding the constant term.

$x^2 + 12x + 36 = (x + 6)^2$ • Write the resulting perfect-square trinomial as the square of a binomial.

HOW TO • 2 Complete the square on $z^2 - 3z$. Write the resulting perfect-square trinomial as the square of a binomial.

$\left[\frac{1}{2} \cdot (-3)\right]^2 = \left(-\frac{3}{2}\right)^2 = \frac{9}{4}$ • Find the constant term.

$z^2 - 3z + \frac{9}{4}$ • Complete the square on $z^2 - 3z$ by adding the constant term.

$z^2 - 3z + \frac{9}{4} = \left(z - \frac{3}{2}\right)^2$ • Write the resulting perfect-square trinomial as the square of a binomial.

Any quadratic equation can be solved by completing the square. Add to each side of the equation the term that completes the square. Rewrite the equation in the form $(x + a)^2 = b$. Then take the square root of each side of the equation.

Point of Interest

Mathematicians have studied quadratic equations for centuries. Many of the initial equations were a result of trying to solve a geometry problem. One of the most famous, which dates from around 500 B.C., is "squaring the circle." The question was "Is it possible to construct a square whose area is that of a given circle?" For these early mathematicians, to *construct* meant to draw with only a straightedge and a compass. It was approximately 2300 years later that mathematicians were able to prove that such a construction is impossible.

HOW TO • 3 Solve by completing the square: $x^2 - 6x - 15 = 0$

$x^2 - 6x - 15 = 0$

$x^2 - 6x = 15$ • Add 15 to each side of the equation.

$x^2 - 6x + 9 = 15 + 9$ • Complete the square. Add $\left[\frac{1}{2}(-6)\right]^2 = (-3)^2 = 9$ to each side of the equation.

$(x - 3)^2 = 24$ • Factor the perfect-square trinomial.

$\sqrt{(x - 3)^2} = \sqrt{24}$ • Take the square root of each side of the equation.

$x - 3 = \pm\sqrt{24}$ • Solve for x.

$x - 3 = \pm 2\sqrt{6}$

$x - 3 = 2\sqrt{6}$ $\qquad$ $x - 3 = -2\sqrt{6}$

$x = 3 + 2\sqrt{6}$ $\qquad$ $x = 3 - 2\sqrt{6}$

Check:

$x^2 - 6x - 15 = 0$	
$(3 + 2\sqrt{6})^2 - 6(3 + 2\sqrt{6}) - 15$	0
$9 + 12\sqrt{6} + 24 - 18 - 12\sqrt{6} - 15$	0
$0 = 0$	

$x^2 - 6x - 15 = 0$	
$(3 - 2\sqrt{6})^2 - 6(3 - 2\sqrt{6}) - 15$	0
$9 - 12\sqrt{6} + 24 - 18 + 12\sqrt{6} - 15$	0
$0 = 0$	

The solutions are $3 + 2\sqrt{6}$ and $3 - 2\sqrt{6}$.

Take Note

The exact solutions of the equation $x^2 - 6x - 15 = 0$ are $3 + 2\sqrt{6}$ and $3 - 2\sqrt{6}$. Approximate solutions of the equation are 7.899 and −1.899.

In the example above, the solutions of the equation $x^2 - 6x - 15 = 0$ are $3 + 2\sqrt{6}$ and $3 - 2\sqrt{6}$. These are the exact solutions. However, in some situations it may be preferable to have decimal approximations of the solutions of a quadratic equation. Approximate solutions can be found by using a calculator and then rounding to the desired degree of accuracy.

$$3 + 2\sqrt{6} \approx 7.899 \text{ and } 3 - 2\sqrt{6} \approx -1.899$$

To the nearest thousandth, the approximate solutions of the equation $x^2 - 6x - 15 = 0$ are 7.899 and −1.899.

HOW TO • 4 Solve $2x^2 - x - 2 = 0$ by completing the square. Find the exact solutions, and approximate the solutions to the nearest thousandth.

In order for us to complete the square on an expression, the coefficient of the squared term must be 1. After adding the constant term to each side of the equation, multiply each side of the equation by $\frac{1}{2}$.

$2x^2 - x - 2 = 0$

$2x^2 - x = 2$ • Add 2 to each side of the equation.

$\frac{1}{2}(2x^2 - x) = \frac{1}{2} \cdot 2$ • Multiply each side of the equation by $\frac{1}{2}$.

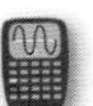

Integrating Technology

A graphing calculator can be used to check a proposed solution of a quadratic equation. For the equation at the right, enter the expression $2x^2 - x - 2$ in Y_1. Then evaluate the function for $\frac{1+\sqrt{17}}{4}$ and $\frac{1-\sqrt{17}}{4}$. The value of each should be zero. See the Keystroke Guide: *Evaluating Functions*.

$$x^2 - \frac{1}{2}x = 1$$

• The coefficient of x^2 is now 1.

$$x^2 - \frac{1}{2}x + \frac{1}{16} = 1 + \frac{1}{16}$$

• Complete the square. Add $\left[\frac{1}{2}\left(-\frac{1}{2}\right)\right]^2 = \left(-\frac{1}{4}\right)^2 = \frac{1}{16}$ to each side of the equation.

$$\left(x - \frac{1}{4}\right)^2 = \frac{17}{16}$$

• Factor the perfect-square trinomial.

$$\sqrt{\left(x - \frac{1}{4}\right)^2} = \sqrt{\frac{17}{16}}$$

• Take the square root of each side of the equation.

$$x - \frac{1}{4} = \pm\frac{\sqrt{17}}{4}$$

• Solve for x.

$$x - \frac{1}{4} = \frac{\sqrt{17}}{4} \qquad x - \frac{1}{4} = -\frac{\sqrt{17}}{4}$$

$$x = \frac{1}{4} + \frac{\sqrt{17}}{4} \qquad x = \frac{1}{4} - \frac{\sqrt{17}}{4}$$

The exact solutions are $\frac{1+\sqrt{17}}{4}$ and $\frac{1-\sqrt{17}}{4}$.

$$\frac{1+\sqrt{17}}{4} \approx 1.281 \qquad \frac{1-\sqrt{17}}{4} \approx -0.781$$

To the nearest thousandth, the solutions are 1.281 and −0.781.

Check of exact solutions:

$2x^2 - x - 2$	$= 0$
$2\left(\frac{1+\sqrt{17}}{4}\right)^2 - \frac{1+\sqrt{17}}{4} - 2$	0
$2\left(\frac{18+2\sqrt{17}}{16}\right) - \frac{1+\sqrt{17}}{4} - 2$	0
$\frac{18+2\sqrt{17}}{8} - \frac{2+2\sqrt{17}}{8} - 2$	0
$\frac{16}{8} - 2$	0
$0 = 0$	

$2x^2 - x - 2$	$= 0$
$2\left(\frac{1-\sqrt{17}}{4}\right)^2 - \frac{1-\sqrt{17}}{4} - 2$	0
$2\left(\frac{18-2\sqrt{17}}{16}\right) - \frac{1-\sqrt{17}}{4} - 2$	0
$\frac{18-2\sqrt{17}}{8} - \frac{2-2\sqrt{17}}{8} - 2$	0
$\frac{16}{8} - 2$	0
$0 = 0$	

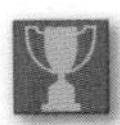

Tips for Success

This is a new skill and one that is difficult for many students. Be sure to do all you need to do in order to be successful at solving equations by completing the square: Read through the introductory material, work through the HOW TO examples, study the paired Examples, do the You Try Its, and check your solutions against the ones in the back of the book. See *AIM for Success* in the Preface.

The above example illustrates all the steps required in solving a quadratic equation by completing the square.

Procedure for Solving a Quadratic Equation by Completing the Square

1. Write the equation in the form $ax^2 + bx = -c$.
2. Multiply both sides of the equation by $\frac{1}{a}$.
3. Complete the square on $x^2 + \frac{b}{a}x$. Add the number that completes the square to both sides of the equation.
4. Factor the perfect-square trinomial.
5. Take the square root of each side of the equation.
6. Solve the resulting equation for x.
7. Check the solutions.

EXAMPLE 1

Solve by completing the square: $4x^2 - 8x + 1 = 0$

Solution

$4x^2 - 8x + 1 = 0$

$4x^2 - 8x = -1$ • Write in the form $ax^2 + bx = -c$.

$\frac{1}{4}(4x^2 - 8x) = \frac{1}{4}(-1)$ • Multiply both sides by $\frac{1}{a}$.

$x^2 - 2x = -\frac{1}{4}$

$x^2 - 2x + 1 = -\frac{1}{4} + 1$ • Complete the square.

$(x - 1)^2 = \frac{3}{4}$ • Factor.

$\sqrt{(x - 1)^2} = \sqrt{\frac{3}{4}}$ • Take square roots.

$x - 1 = \pm\frac{\sqrt{3}}{2}$

$x - 1 = \frac{\sqrt{3}}{2}$ $\quad x - 1 = -\frac{\sqrt{3}}{2}$ • Solve for x.

$x = 1 + \frac{\sqrt{3}}{2}$ $\quad x = 1 - \frac{\sqrt{3}}{2}$

$= \frac{2 + \sqrt{3}}{2}$ $\quad = \frac{2 - \sqrt{3}}{2}$

The solutions are $\frac{2 + \sqrt{3}}{2}$ and $\frac{2 - \sqrt{3}}{2}$.

YOU TRY IT 1

Solve by completing the square: $4x^2 - 4x - 1 = 0$

Your solution

EXAMPLE 2

Solve by completing the square: $x^2 + 4x + 5 = 0$

Solution

$x^2 + 4x + 5 = 0$

$x^2 + 4x = -5$

$x^2 + 4x + 4 = -5 + 4$ • Complete the square.

$(x + 2)^2 = -1$ • Factor.

$\sqrt{(x + 2)^2} = \sqrt{-1}$ • Take square roots.

$x + 2 = \pm i$

$x + 2 = i$ $\quad x + 2 = -i$ • Solve for x.

$x = -2 + i$ $\quad x = -2 - i$

The solutions are $-2 + i$ and $-2 - i$.

YOU TRY IT 2

Solve by completing the square: $x^2 + 4x + 8 = 0$

Your solution

Solutions on p. S29

10.2 EXERCISES

OBJECTIVE A To solve a quadratic equation by completing the square

For Exercises 1 to 48, solve by completing the square.

1. $x^2 - 4x - 5 = 0$

2. $y^2 + 6y + 5 = 0$

3. $v^2 + 8v - 9 = 0$

4. $w^2 - 2w - 24 = 0$

5. $z^2 - 6z + 9 = 0$

6. $u^2 + 10u + 25 = 0$

7. $r^2 + 4r - 7 = 0$

8. $s^2 + 6s - 1 = 0$

9. $x^2 - 6x + 7 = 0$

10. $y^2 + 8y + 13 = 0$

11. $z^2 - 2z + 2 = 0$

12. $t^2 - 4t + 8 = 0$

13. $s^2 - 5s - 24 = 0$

14. $v^2 + 7v - 44 = 0$

15. $x^2 + 5x - 36 = 0$

16. $y^2 - 9y + 20 = 0$

17. $p^2 - 3p + 1 = 0$

18. $r^2 - 5r - 2 = 0$

19. $t^2 - t - 1 = 0$

20. $u^2 - u - 7 = 0$

21. $y^2 - 6y = 4$

22. $w^2 + 4w = 2$

23. $x^2 = 8x - 15$

24. $z^2 = 4z - 3$

25. $v^2 = 4v - 13$

26. $x^2 = 2x - 17$

27. $p^2 + 6p = -13$

28. $x^2 + 4x = -20$

29. $y^2 - 2y = 17$

30. $x^2 + 10x = 7$

31. $z^2 = z + 4$

32. $r^2 = 3r - 1$

33. $x^2 + 13 = 2x$

34. $6v^2 - 7v = 3$

35. $4x^2 - 4x + 5 = 0$

36. $4t^2 - 4t + 17 = 0$

37. $9x^2 - 6x + 2 = 0$

38. $9y^2 - 12y + 13 = 0$

39. $2s^2 = 4s + 5$

40. $3u^2 = 6u + 1$

41. $2r^2 = 3 - r$

42. $2x^2 = 12 - 5x$

43. $y - 2 = (y - 3)(y + 2)$

44. $8s - 11 = (s - 4)(s - 2)$

45. $6t - 2 = (2t - 3)(t - 1)$

46. $2z + 9 = (2z + 3)(z + 2)$

47. $(x - 4)(x + 1) = x - 3$

48. $(y - 3)^2 = 2y + 10$

For Exercises 49 to 54, solve by completing the square. Approximate the solutions to the nearest thousandth.

49. $z^2 + 2z = 4$

50. $t^2 - 4t = 7$

51. $2x^2 = 4x - 1$

52. $3y^2 = 5y - 1$

53. $4z^2 + 2z = 1$

54. $4w^2 - 8w = 3$

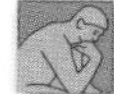

55. For what values of c does the equation $x^2 + 4x + c = 0$ have real number solutions?

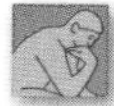

56. For what values of c does the equation $x^2 - 6x + c = 0$ have complex number solutions?

Applying the Concepts

For Exercises 57 to 59, solve for x by completing the square.

57. $x^2 - ax - 2a^2 = 0$

58. $x^2 + 3ax - 4a^2 = 0$

59. $x^2 + 3ax - 10a^2 = 0$

60. **Sports** The height h (in feet) of a baseball above the ground t seconds after it is hit can be approximated by the equation $h = -16t^2 + 70t + 4$. Using this equation, determine when the ball will hit the ground. Round to the nearest hundredth. (*Hint:* The ball hits the ground when $h = 0$.)

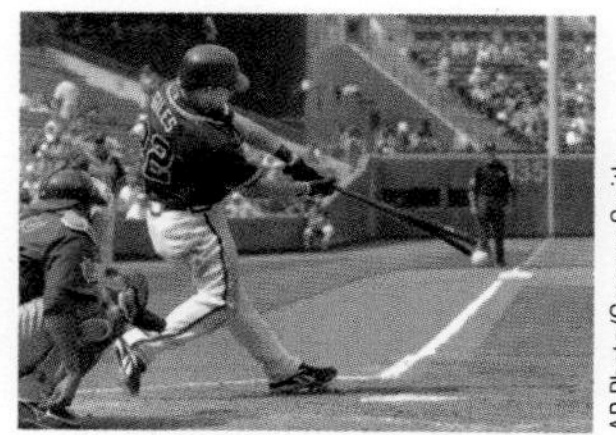

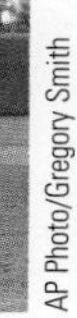

AP Photo/Gregory Smith

61. **Sports** After a baseball is hit, there are two equations that can be considered. One gives the height h (in feet) the ball is above the ground t seconds after it is hit. The second is the distance s (in feet) the ball is from home plate t seconds after it is hit. A model of this situation is given by $h = -16t^2 + 70t + 4$ and $s = 44.5t$. Using this model, determine whether the ball will clear a fence 325 ft from home plate.

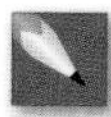

62. Explain how to complete the square on $x^2 + bx$.

SECTION

10.3 Solving Quadratic Equations by Using the Quadratic Formula

OBJECTIVE A To solve a quadratic equation by using the quadratic formula 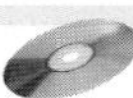

A general formula known as the **quadratic formula** can be derived by applying the method of completing the square to the standard form of a quadratic equation. This formula can be used to solve any quadratic equation. The equation $ax^2 + bx + c = 0$ is solved by completing the square as follows.

$$ax^2 + bx + c = 0$$

Add the opposite of the constant term to each side of the equation.

$$ax^2 + bx + c + (-c) = 0 + (-c)$$

$$ax^2 + bx = -c$$

Multiply each side of the equation by the reciprocal of a, the coefficient of x^2.

$$\frac{1}{a}(ax^2 + bx) = \frac{1}{a}(-c)$$

$$x^2 + \frac{b}{a}x = -\frac{c}{a}$$

Complete the square by adding $\left(\frac{1}{2} \cdot \frac{b}{a}\right)^2 = \frac{b^2}{4a^2}$ to each side of the equation.

$$x^2 + \frac{b}{a}x + \frac{b^2}{4a^2} = \frac{b^2}{4a^2} - \frac{c}{a}$$

Simplify the right side of the equation.

$$x^2 + \frac{b}{a}x + \frac{b^2}{4a^2} = \frac{b^2}{4a^2} - \left(\frac{c}{a} \cdot \frac{4a}{4a}\right)$$

$$x^2 + \frac{b}{a}x + \frac{b^2}{4a^2} = \frac{b^2}{4a^2} - \frac{4ac}{4a^2}$$

$$x^2 + \frac{b}{a}x + \frac{b^2}{4a^2} = \frac{b^2 - 4ac}{4a^2}$$

Factor the perfect-square trinomial on the left side of the equation.

$$\left(x + \frac{b}{2a}\right)^2 = \frac{b^2 - 4ac}{4a^2}$$

Take the square root of each side of the equation.

$$\sqrt{\left(x + \frac{b}{2a}\right)^2} = \sqrt{\frac{b^2 - 4ac}{4a^2}}$$

$$x + \frac{b}{2a} = \pm\frac{\sqrt{b^2 - 4ac}}{2a}$$

Solve for x.

$$x + \frac{b}{2a} = \frac{\sqrt{b^2 - 4ac}}{2a} \qquad x + \frac{b}{2a} = -\frac{\sqrt{b^2 - 4ac}}{2a}$$

$$x = -\frac{b}{2a} + \frac{\sqrt{b^2 - 4ac}}{2a} \qquad x = -\frac{b}{2a} - \frac{\sqrt{b^2 - 4ac}}{2a}$$

$$= \frac{-b + \sqrt{b^2 - 4ac}}{2a} \qquad = \frac{-b - \sqrt{b^2 - 4ac}}{2a}$$

Point of Interest

Although mathematicians have studied quadratic equations since around 500 B.C., it was not until the 18th century that the formula was written as it is today. Of further note, the word *quadratic* has the same Latin root as the word *square*.

The Quadratic Formula

The solutions of $ax^2 + bx + c = 0$, $a \neq 0$, are

$$\frac{-b + \sqrt{b^2 - 4ac}}{2a} \quad \text{and} \quad \frac{-b - \sqrt{b^2 - 4ac}}{2a}$$

The quadratic formula is frequently written as $x = \dfrac{-b \pm \sqrt{b^2 - 4ac}}{2a}$.

HOW TO • 1 Solve by using the quadratic formula: $2x^2 + 5x + 3 = 0$

$$x = \frac{-b \pm \sqrt{b^2 - 4ac}}{2a}$$

- The equation $2x^2 + 5x + 3 = 0$ is in standard form. $a = 2$, $b = 5$, $c = 3$
- Replace a, b, and c in the quadratic formula with these values.

$$= \frac{-(5) \pm \sqrt{(5)^2 - 4(2)(3)}}{2(2)}$$

$$= \frac{-5 \pm \sqrt{25 - 24}}{4}$$

$$= \frac{-5 \pm \sqrt{1}}{4} = \frac{-5 \pm 1}{4}$$

$$x = \frac{-5 + 1}{4} = \frac{-4}{4} = -1 \qquad x = \frac{-5 - 1}{4} = \frac{-6}{4} = -\frac{3}{2}$$

The solutions are -1 and $-\frac{3}{2}$.

✓ Take Note

The solutions of this quadratic equation are rational numbers. When this happens, the equation could have been solved by factoring and using the Principle of Zero Products. This may be easier than applying the quadratic formula.

HOW TO • 2 Solve $3x^2 = 4x + 6$ by using the quadratic formula. Find the exact solutions, and approximate the solutions to the nearest thousandth.

$$3x^2 = 4x + 6$$

$$3x^2 - 4x - 6 = 0$$

$$x = \frac{-b \pm \sqrt{b^2 - 4ac}}{2a}$$

- Write the equation in standard form. Subtract $4x$ and 6 from each side of the equation. $a = 3$, $b = -4$, $c = -6$

$$= \frac{-(-4) \pm \sqrt{(-4)^2 - 4(3)(-6)}}{2(3)}$$

- Replace a, b, and c in the quadratic formula with these values.

$$= \frac{4 \pm \sqrt{16 - (-72)}}{6}$$

$$= \frac{4 \pm \sqrt{88}}{6} = \frac{4 \pm 2\sqrt{22}}{6}$$

$$= \frac{\cancel{2}(2 \pm \sqrt{22})}{\cancel{2} \cdot 3} = \frac{2 \pm \sqrt{22}}{3}$$

Check:

$3x^2$	$= 4x + 6$
$3\left(\frac{2 + \sqrt{22}}{3}\right)^2$	$4\left(\frac{2 + \sqrt{22}}{3}\right) + 6$
$3\left(\frac{4 + 4\sqrt{22} + 22}{9}\right)$	$\frac{8}{3} + \frac{4\sqrt{22}}{3} + \frac{18}{3}$
$3\left(\frac{26 + 4\sqrt{22}}{9}\right)$	$\frac{26}{3} + \frac{4\sqrt{22}}{3}$
$\frac{26 + 4\sqrt{22}}{3}$	$= \frac{26 + 4\sqrt{22}}{3}$

$3x^2$	$= 4x + 6$
$3\left(\frac{2 - \sqrt{22}}{3}\right)^2$	$4\left(\frac{2 - \sqrt{22}}{3}\right) + 6$
$3\left(\frac{4 - 4\sqrt{22} + 22}{9}\right)$	$\frac{8}{3} - \frac{4\sqrt{22}}{3} + \frac{18}{3}$
$3\left(\frac{26 - 4\sqrt{22}}{9}\right)$	$\frac{26}{3} - \frac{4\sqrt{22}}{3}$
$\frac{26 - 4\sqrt{22}}{3}$	$= \frac{26 - 4\sqrt{22}}{3}$

The exact solutions are $\frac{2 + \sqrt{22}}{3}$ and $\frac{2 - \sqrt{22}}{3}$.

$$\frac{2 + \sqrt{22}}{3} \approx 2.230 \qquad \frac{2 - \sqrt{22}}{3} \approx -0.897$$

To the nearest thousandth, the solutions are 2.230 and -0.897.

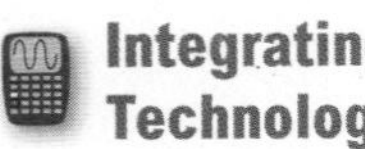

Integrating Technology

See the Projects and Group Activities at the end of this chapter for instructions on using a graphing calculator to solve a quadratic equation. Instructions are also provided in the Keystroke Guide: *Zero*.

HOW TO 3 Solve by using the quadratic formula: $4x^2 = 8x - 13$

$$4x^2 = 8x - 13$$
$$4x^2 - 8x + 13 = 0 \quad \text{• Write the equation in standard form.}$$
$$x = \frac{-b \pm \sqrt{b^2 - 4ac}}{2a} \quad \text{• Use the quadratic formula.}$$
$$= \frac{-(-8) \pm \sqrt{(-8)^2 - 4 \cdot 4 \cdot 13}}{2 \cdot 4} \quad \text{• } a = 4, b = -8, c = 13$$
$$= \frac{8 \pm \sqrt{64 - 208}}{8} = \frac{8 \pm \sqrt{-144}}{8}$$
$$= \frac{8 \pm 12i}{8} = \frac{2 \pm 3i}{2}$$

The solutions are $1 + \frac{3}{2}i$ and $1 - \frac{3}{2}i$.

Of the three preceding examples, the first two had real number solutions; the last one had complex number solutions.

In the quadratic formula, the quantity $b^2 - 4ac$ is called the **discriminant.** When a, b, and c are real numbers, the discriminant determines whether a quadratic equation will have a double root, two real number solutions that are not equal, or two complex number solutions.

The Effect of the Discriminant on the Solutions of a Quadratic Equation

1. If $b^2 - 4ac = 0$, the equation has two equal real number solutions, a double root.
2. If $b^2 - 4ac > 0$, the equation has two unequal real number solutions.
3. If $b^2 - 4ac < 0$, the equation has two complex number solutions.

HOW TO 4 Use the discriminant to determine whether $x^2 - 4x - 5 = 0$ has two equal real number solutions, two unequal real number solutions, or two complex number solutions.

$b^2 - 4ac$ • Evaluate the discriminant.

$(-4)^2 - 4(1)(-5) = 16 + 20 = 36$ $\quad a = 1, b = -4, c = -5$

$36 > 0$

Because $b^2 - 4ac > 0$, the equation has two unequal real number solutions.

EXAMPLE 1

Solve by using the quadratic formula:
$2x^2 - x + 5 = 0$

Solution

$$2x^2 - x + 5 = 0$$
$$a = 2, b = -1, c = 5$$
$$x = \frac{-b \pm \sqrt{b^2 - 4ac}}{2a}$$
$$= \frac{-(-1) \pm \sqrt{(-1)^2 - 4(2)(5)}}{2 \cdot 2}$$
$$= \frac{1 \pm \sqrt{1 - 40}}{4} = \frac{1 \pm \sqrt{-39}}{4}$$
$$= \frac{1 \pm i\sqrt{39}}{4}$$

The solutions are $\frac{1}{4} + \frac{\sqrt{39}}{4}i$ and $\frac{1}{4} - \frac{\sqrt{39}}{4}i$.

YOU TRY IT 1

Solve by using the quadratic formula:
$x^2 - 2x + 10 = 0$

Your solution

Solution on p. S29

EXAMPLE 2

Solve by using the quadratic formula:
$2x^2 = (x - 2)(x - 3)$

Solution

$2x^2 = (x - 2)(x - 3)$

$2x^2 = x^2 - 5x + 6$ • Write in standard form.

$x^2 + 5x - 6 = 0$

$a = 1, b = 5, c = -6$

$$x = \frac{-b \pm \sqrt{b^2 - 4ac}}{2a}$$

$$= \frac{-5 \pm \sqrt{5^2 - 4(1)(-6)}}{2 \cdot 1}$$

$$= \frac{-5 \pm \sqrt{25 + 24}}{2} = \frac{-5 \pm \sqrt{49}}{2}$$

$$= \frac{-5 \pm 7}{2}$$

$$x = \frac{-5 + 7}{2} \qquad x = \frac{-5 - 7}{2}$$

$$= \frac{2}{2} = 1 \qquad = \frac{-12}{2} = -6$$

The solutions are 1 and -6.

YOU TRY IT 2

Solve by using the quadratic formula:
$4x^2 = 4x - 1$

Your solution

EXAMPLE 3

Use the discriminant to determine whether $4x^2 - 2x + 5 = 0$ has two equal real number solutions, two unequal real number solutions, or two complex number solutions.

Solution

$a = 4, b = -2, c = 5$

$$b^2 - 4ac = (-2)^2 - 4(4)(5)$$
$$= 4 - 80$$
$$= -76$$

$-76 < 0$

Because the discriminant is less than zero, the equation has two complex number solutions.

YOU TRY IT 3

Use the discriminant to determine whether $3x^2 - x - 1 = 0$ has two equal real number solutions, two unequal real number solutions, or two complex number solutions.

Your solution

Solutions on p. S29

10.3 EXERCISES

OBJECTIVE A **To solve a quadratic equation by using the quadratic formula**

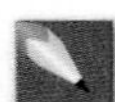

1. Write the quadratic formula. What does each variable in the formula represent?

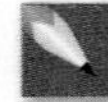

2. Write the expression that appears under the radical symbol in the quadratic formula. What is this quantity called? What can it be used to determine?

For Exercises 3 to 29, solve by using the quadratic formula.

3. $x^2 - 3x - 10 = 0$

4. $z^2 - 4z - 8 = 0$

5. $y^2 + 5y - 36 = 0$

6. $z^2 - 3z - 40 = 0$

7. $w^2 = 8w + 72$

8. $t^2 = 2t + 35$

9. $v^2 = 24 - 5v$

10. $x^2 = 18 - 7x$

11. $2y^2 + 5y - 1 = 0$

12. $4p^2 - 7p + 1 = 0$

13. $8s^2 = 10s + 3$

14. $12t^2 = 5t + 2$

15. $x^2 = 14x - 4$

16. $v^2 = 12v - 24$

17. $2z^2 - 2z - 1 = 0$

18. $6w^2 = 9w - 1$

19. $z^2 + 2z + 2 = 0$

20. $p^2 - 4p + 5 = 0$

21. $y^2 - 2y + 5 = 0$

22. $x^2 + 6x + 13 = 0$

23. $s^2 - 4s + 13 = 0$

24. $t^2 - 6t + 10 = 0$

25. $2w^2 - 2w - 5 = 0$

26. $2v^2 + 8v + 3 = 0$

27. $2x^2 + 6x + 5 = 0$

28. $2y^2 + 2y + 13 = 0$

29. $4t^2 - 6t + 9 = 0$

For Exercises 30 to 35, solve by using the quadratic formula. Approximate the solutions to the nearest thousandth.

30. $x^2 - 6x - 6 = 0$

31. $p^2 - 8p + 3 = 0$

32. $r^2 - 2r = 4$

33. $w^2 + 4w = 1$

34. $3t^2 = 7t + 1$

35. $2y^2 = y + 5$

For Exercises 36 to 41, use the discriminant to determine whether the quadratic equation has two equal real number solutions, two unequal real number solutions, or two complex number solutions.

36. $2z^2 - z + 5 = 0$

37. $3y^2 + y + 1 = 0$

38. $9x^2 - 12x + 4 = 0$

39. $4x^2 + 20x + 25 = 0$

40. $2v^2 - 3v - 1 = 0$

41. $3w^2 + 3w - 2 = 0$

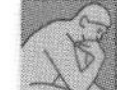

42. Suppose $a > 0$ and $c < 0$. Does the equation $ax^2 + bx + c = 0$ (i) always have real number solutions, (ii) never have real number solutions, or (iii) sometimes have real number solutions, depending on the value of b?

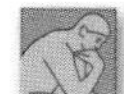

43. If $a > 0$ and $c > 0$, what is the smallest value of b in $ax^2 + bx + c = 0$ that will guarantee that the equation will have real number solutions?

Applying the Concepts

44. Sports The height h, in feet, of a baseball above the ground t seconds after it is hit by a Little Leaguer can be approximated by the equation $h = -0.01t^2 + 2t + 3.5$. Does the ball reach a height of 100 ft?

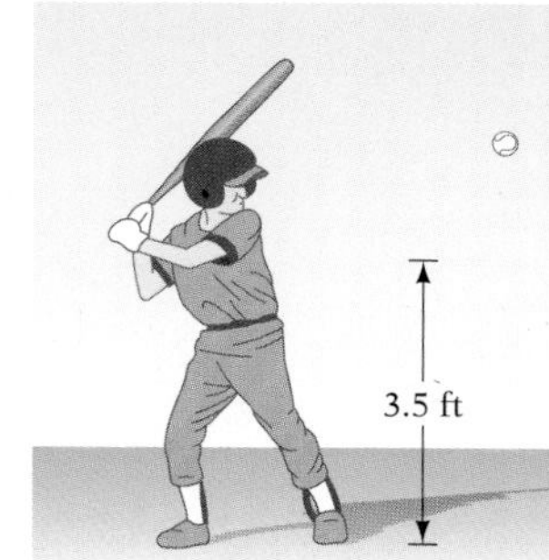

45. Sports The height h, in feet, of an arrow shot upward can be approximated by the equation $h = 128t - 16t^2$, where t is the time in seconds. Does the arrow reach a height of 275 ft?

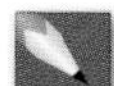

46. Name the three methods of solving a quadratic equation that have been discussed. What are the advantages and disadvantages of each?

For what values of p do the quadratic equations in Exercises 47 and 48 have two unequal real number solutions? Write the answer in set-builder notation.

47. $x^2 - 6x + p = 0$

48. $x^2 + 10x + p = 0$

For what values of p do the quadratic equations in Exercises 49 and 50 have two complex number solutions? Write the answer in interval notation.

49. $x^2 - 2x + p = 0$

50. $x^2 + 4x + p = 0$

51. Find all values of x that satisfy the equation $x^2 + ix + 2 = 0$.

52. Show that the equation $x^2 + bx - 1 = 0$, $b \in$ real numbers, always has real number solutions regardless of the value of b.

53. Show that the equation $2x^2 + bx - 2 = 0$, $b \in$ real numbers, always has real number solutions regardless of the value of b.

SECTION

10.4 Solving Equations That Are Reducible to Quadratic Equations

OBJECTIVE A To solve an equation that is quadratic in form 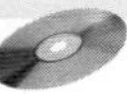

Certain equations that are not quadratic can be expressed in quadratic form by making suitable substitutions. An equation is **quadratic in form** if it can be written as $au^2 + bu + c = 0$.

The equation $x^4 - 4x^2 - 5 = 0$ is quadratic in form.

$$x^4 - 4x^2 - 5 = 0$$
$$(x^2)^2 - 4(x^2) - 5 = 0$$
$$u^2 - 4u - 5 = 0 \quad \text{• Let } x^2 = u.$$

The equation $y - y^{1/2} - 6 = 0$ is quadratic in form.

$$y - y^{1/2} - 6 = 0$$
$$(y^{1/2})^2 - (y^{1/2}) - 6 = 0$$
$$u^2 - u - 6 = 0 \quad \text{• Let } y^{1/2} = u.$$

The key to recognizing equations that are quadratic in form is as follows. When the equation is written in standard form, the exponent on one variable term is $\frac{1}{2}$ the exponent on the other variable term.

HOW TO • 1 Solve: $z + 7z^{1/2} - 18 = 0$

$$z + 7z^{1/2} - 18 = 0 \quad \text{• The equation is quadratic in form.}$$
$$(z^{1/2})^2 + 7(z^{1/2}) - 18 = 0$$
$$u^2 + 7u - 18 = 0 \quad \text{• Let } z^{1/2} = u.$$
$$(u - 2)(u + 9) = 0 \quad \text{• Solve by factoring.}$$

$$u - 2 = 0 \qquad u + 9 = 0$$
$$u = 2 \qquad u = -9$$

$$z^{1/2} = 2 \qquad z^{1/2} = -9 \quad \text{• Replace } u \text{ by } z^{1/2}.$$
$$\sqrt{z} = 2 \qquad \sqrt{z} = -9$$
$$(\sqrt{z})^2 = 2^2 \qquad (\sqrt{z})^2 = (-9)^2 \quad \text{• Solve for } z.$$
$$z = 4 \qquad z = 81$$

Check each solution.

Check:

$z + 7z^{1/2} - 18$	0
$4 + 7(4)^{1/2} - 18$	0
$4 + 7 \cdot 2 - 18$	0
$4 + 14 - 18$	0
$0 = 0$	

$z + 7z^{1/2} - 18$	0
$81 + 7(81)^{1/2} - 18$	0
$81 + 7 \cdot 9 - 18$	0
$81 + 63 - 18$	0
$126 \neq 0$	

4 checks as a solution, but 81 does not check as a solution.

The solution is 4.

✓ Take Note

When each side of an equation is squared, the resulting equation may have a solution that is not a solution of the original equation.

EXAMPLE 1

Solve: $x^4 + x^2 - 12 = 0$

Solution

$$x^4 + x^2 - 12 = 0$$
$$(x^2)^2 + (x^2) - 12 = 0$$
$$u^2 + u - 12 = 0$$
$$(u - 3)(u + 4) = 0$$

$u - 3 = 0$ $\quad$ $u + 4 = 0$

$u = 3$ $\quad$ $u = -4$

Replace u by x^2.

$x^2 = 3$ $\quad$ $x^2 = -4$

$\sqrt{x^2} = \sqrt{3}$ $\quad$ $\sqrt{x^2} = \sqrt{-4}$

$x = \pm\sqrt{3}$ $\quad$ $x = \pm 2i$

The solutions are $\sqrt{3}$, $-\sqrt{3}$, $2i$, and $-2i$.

YOU TRY IT 1

Solve: $x - 5x^{1/2} + 6 = 0$

Your solution

Solution on p. S30

OBJECTIVE B **To solve a radical equation that is reducible to a quadratic equation**

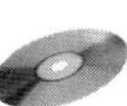

Certain equations containing radicals can be expressed as quadratic equations.

HOW TO 2 Solve: $\sqrt{x + 2} + 4 = x$

$\sqrt{x + 2} + 4 = x$

$\sqrt{x + 2} = x - 4$ • Solve for the radical expression.

$(\sqrt{x + 2})^2 = (x - 4)^2$ • Square each side of the equation.

$x + 2 = x^2 - 8x + 16$ • Simplify.

$0 = x^2 - 9x + 14$ • Write the equation in standard form.

$0 = (x - 7)(x - 2)$ • Solve for x.

$x - 7 = 0$ $\quad$ $x - 2 = 0$

$x = 7$ $\quad$ $x = 2$

Check each solution.

Check:

$\sqrt{x + 2} + 4 = x$	
$\sqrt{7 + 2} + 4$	7
$\sqrt{9} + 4$	7
$3 + 4$	7
$7 = 7$	

$\sqrt{x + 2} + 4 = x$	
$\sqrt{2 + 2} + 4$	2
$\sqrt{4} + 4$	2
$2 + 4$	2
$6 \neq 2$	

7 checks as a solution, but 2 does not check as a solution.

The solution is 7.

EXAMPLE • 2

Solve: $\sqrt{7y-3}+3=2y$

Solution

$\sqrt{7y-3}+3=2y$ • Solve for the radical.

$\sqrt{7y-3}=2y-3$

$(\sqrt{7y-3})^2=(2y-3)^2$ • Square each side.

$7y-3=4y^2-12y+9$ • Write in standard form.

$0=4y^2-19y+12$

$0=(4y-3)(y-4)$ • Solve by factoring.

$4y-3=0$ $\quad$ $y-4=0$

$4y=3$ $\quad$ $y=4$

$y=\frac{3}{4}$

4 checks as a solution.

$\frac{3}{4}$ does not check as a solution.

The solution is 4.

YOU TRY IT • 2

Solve: $\sqrt{2x+1}+x=7$

Your solution

EXAMPLE • 3

Solve: $\sqrt{2y+1}-\sqrt{y}=1$

Solution

$\sqrt{2y+1}-\sqrt{y}=1$

Solve for one of the radical expressions.

$\sqrt{2y+1}=\sqrt{y}+1$

$(\sqrt{2y+1})^2=(\sqrt{y}+1)^2$ • Square each side.

$2y+1=y+2\sqrt{y}+1$

$y=2\sqrt{y}$

$y^2=(2\sqrt{y})^2$ • Square each side.

$y^2=4y$

$y^2-4y=0$

$y(y-4)=0$

$y=0$ $\quad$ $y-4=0$

$y=4$

0 and 4 check as solutions.

The solutions are 0 and 4.

YOU TRY IT • 3

Solve: $\sqrt{2x-1}+\sqrt{x}=2$

Your solution

Solutions on p. S30

OBJECTIVE C **To solve a rational equation that is reducible to a quadratic equation** 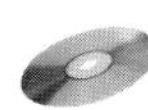

After each side of a rational equation has been multiplied by the LCM of the denominators, the resulting equation may be a quadratic equation.

HOW TO • 3 Solve: $\frac{1}{r} + \frac{1}{r+1} = \frac{3}{2}$

$$\frac{1}{r} + \frac{1}{r+1} = \frac{3}{2}$$

$$2r(r+1)\left(\frac{1}{r} + \frac{1}{r+1}\right) = 2r(r+1) \cdot \frac{3}{2}$$

• Multiply each side of the equation by the LCM of the denominators.

$$2(r+1) + 2r = r(r+1) \cdot 3$$
$$2r + 2 + 2r = 3r(r+1)$$
$$4r + 2 = 3r^2 + 3r$$
$$0 = 3r^2 - r - 2$$
$$0 = (3r+2)(r-1)$$

• Write the equation in standard form.
• Solve for r by factoring.

$$3r + 2 = 0 \qquad r - 1 = 0$$
$$3r = -2 \qquad r = 1$$
$$r = -\frac{2}{3}$$

$-\frac{2}{3}$ and 1 check as solutions.

The solutions are $-\frac{2}{3}$ and 1.

EXAMPLE • 4

Solve: $\frac{9}{x-3} = 2x + 1$

Solution

$$\frac{9}{x-3} = 2x + 1$$

$$(x-3)\frac{9}{x-3} = (x-3)(2x+1)$$

• Clear denominators.

$$9 = 2x^2 - 5x - 3$$

• Write in standard form.

$$0 = 2x^2 - 5x - 12$$
$$0 = (2x+3)(x-4)$$

• Solve by factoring.

$$2x + 3 = 0 \qquad x - 4 = 0$$
$$2x = -3 \qquad x = 4$$
$$x = -\frac{3}{2}$$

The solutions are $-\frac{3}{2}$ and 4.

YOU TRY IT • 4

Solve: $3y + \frac{25}{3y-2} = -8$

Your solution

Solution on p. S30

10.4 EXERCISES

OBJECTIVE A To solve an equation that is quadratic in form

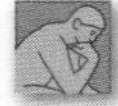

For Exercises 1 to 4, state whether the equation could be solved by writing it as a quadratic equation of the form $u^2 - 8u - 20 = 0$.

1. $x^{10} - 8x^5 - 20 = 0$ **2.** $x^{16} - 8x^4 - 20 = 0$ **3.** $x^{1/10} - 8x^{1/5} - 20 = 0$ **4.** $x^{2/5} - 8x^{1/5} - 20 = 0$

For Exercises 5 to 22, solve.

5. $x^4 - 13x^2 + 36 = 0$ **6.** $y^4 - 5y^2 + 4 = 0$ **7.** $z^4 - 6z^2 + 8 = 0$

8. $t^4 - 12t^2 + 27 = 0$ **9.** $p - 3p^{1/2} + 2 = 0$ **10.** $v - 7v^{1/2} + 12 = 0$

11. $x - x^{1/2} - 12 = 0$ **12.** $w - 2w^{1/2} - 15 = 0$ **13.** $z^4 + 3z^2 - 4 = 0$

14. $y^4 + 5y^2 - 36 = 0$ **15.** $x^4 + 12x^2 - 64 = 0$ **16.** $x^4 - 81 = 0$

17. $p + 2p^{1/2} - 24 = 0$ **18.** $v + 3v^{1/2} - 4 = 0$ **19.** $y^{2/3} - 9y^{1/3} + 8 = 0$

20. $z^{2/3} - z^{1/3} - 6 = 0$ **21.** $9w^4 - 13w^2 + 4 = 0$ **22.** $4y^4 - 7y^2 - 36 = 0$

OBJECTIVE B To solve a radical equation that is reducible to a quadratic equation

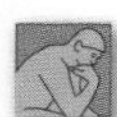

23. For which of Exercises 25 to 42 will the first step in solving the equation be to square each side of the equation?

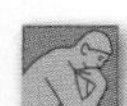

24. For which of Exercises 25 to 42 will it be necessary to square each side of the equation twice?

For Exercises 25 to 42, solve.

25. $\sqrt{x+1} + x = 5$ **26.** $\sqrt{x-4} + x = 6$ **27.** $x = \sqrt{x} + 6$

28. $\sqrt{2y-1} = y - 2$ **29.** $\sqrt{3w+3} = w + 1$ **30.** $\sqrt{2s+1} = s - 1$

31. $\sqrt{4y+1} - y = 1$ **32.** $\sqrt{3s+4} + 2s = 12$ **33.** $\sqrt{10x+5} - 2x = 1$

34. $\sqrt{t+8} = 2t + 1$

35. $\sqrt{p+11} = 1 - p$

36. $x - 7 = \sqrt{x-5}$

37. $\sqrt{x-1} - \sqrt{x} = -1$

38. $\sqrt{y} + 1 = \sqrt{y+5}$

39. $\sqrt{2x-1} = 1 - \sqrt{x-1}$

40. $\sqrt{x+6} + \sqrt{x+2} = 2$

41. $\sqrt{t+3} + \sqrt{2t+7} = 1$

42. $\sqrt{5-2x} = \sqrt{2-x} + 1$

OBJECTIVE C To solve a rational equation that is reducible to a quadratic equation

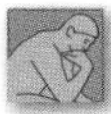

43. To solve Exercise 47, the first step will be to multiply each side of the equation by what expression?

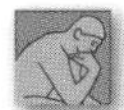

44. To solve Exercise 52, the first step will be to multiply each side of the equation by what expression?

For Exercises 45 to 56, solve.

45. $x = \dfrac{10}{x-9}$

46. $z = \dfrac{5}{z-4}$

47. $\dfrac{y-1}{y+2} + y = 1$

48. $\dfrac{2p-1}{p-2} + p = 8$

49. $\dfrac{3r+2}{r+2} - 2r = 1$

50. $\dfrac{2v+3}{v+4} + 3v = 4$

51. $\dfrac{2}{2x+1} + \dfrac{1}{x} = 3$

52. $\dfrac{3}{s} - \dfrac{2}{2s-1} = 1$

53. $\dfrac{16}{z-2} + \dfrac{16}{z+2} = 6$

54. $\dfrac{2}{y+1} + \dfrac{1}{y-1} = 1$

55. $\dfrac{t}{t-2} + \dfrac{2}{t-1} = 4$

56. $\dfrac{4t+1}{t+4} + \dfrac{3t-1}{t+1} = 2$

Applying the Concepts

57. Solve: $(x^2 - 7)^{1/2} = (x - 1)^{1/2}$

58. Solve: $(\sqrt{x} - 2)^2 - 5\sqrt{x} + 14 = 0$ (*Hint:* Let $u = \sqrt{x} - 2$.)

59. Solve: $(\sqrt{x} + 3)^2 - 4\sqrt{x} - 17 = 0$ (*Hint:* Let $u = \sqrt{x} + 3$.)

60. **Mathematics** The fourth power of a number is twenty-five less than ten times the square of the number. Find the number.

SECTION

10.5 Quadratic Inequalities and Rational Inequalities

OBJECTIVE A To solve a nonlinear inequality 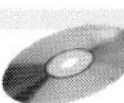

A **quadratic inequality** is one that can be written in the form $ax^2 + bx + c < 0$ or $ax^2 + bx + c > 0$, where $a \neq 0$. The symbols $\leq$ and $\geq$ can also be used. The solution set of a quadratic inequality can be found by solving a compound inequality.

To solve $x^2 - 3x - 10 > 0$, first factor the trinomial.

$$x^2 - 3x - 10 > 0$$
$$(x + 2)(x - 5) > 0$$

There are two cases for which the product of the factors will be positive: (1) both factors are positive, or (2) both factors are negative.

(1) $x + 2 > 0$ and $x - 5 > 0$
(2) $x + 2 < 0$ and $x - 5 < 0$

Solve each pair of compound inequalities.

(1) $x + 2 > 0$ and $x - 5 > 0$
$x > -2$ $\qquad$ $x > 5$
$\{x \mid x > -2\} \cap \{x \mid x > 5\} = \{x \mid x > 5\}$

(2) $x + 2 < 0$ and $x - 5 < 0$
$x < -2$ $\qquad$ $x < 5$
$\{x \mid x < -2\} \cap \{x \mid x < 5\} = \{x \mid x < -2\}$

Because the two cases for which the product will be positive are connected by *or*, the solution set is the union of the solution sets of the individual inequalities.

$$\{x \mid x > 5\} \cup \{x \mid x < -2\} = \{x \mid x > 5 \text{ or } x < -2\}$$

Although the solution set of any quadratic inequality can be found by using the method outlined above, a graphical method is often easier to use.

HOW TO • 1 Solve and graph the solution set of $x^2 - x - 6 < 0$.

$$x^2 - x - 6 < 0$$

Factor the trinomial.

$$(x - 3)(x + 2) < 0$$

On a number line, draw vertical lines indicating the numbers that make each factor equal to zero.

−5 −4 −3 −2 −1 0 1 2 3 4 5

$x - 3 = 0$ $\qquad$ $x + 2 = 0$
$x = 3$ $\qquad$ $x = -2$

For each factor, place plus signs above the number line for those regions where the factor is positive and minus signs where the factor is negative.

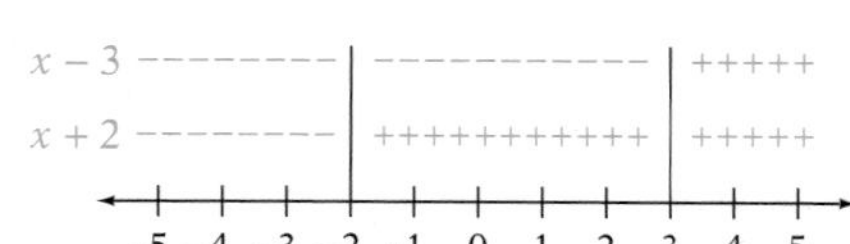

Because $x^2 - x - 6 < 0$, the solution set will be the regions where one factor is positive and the other factor is negative.

Write the solution set.

$$\{x \mid -2 < x < 3\}$$

The graph of the solution set of $x^2 - x - 6 < 0$ is shown at the right.

−5 −4 −3 −2 −1 0 1 2 3 4 5

Take Note

For each factor, choose a number in each region. For example: When $x = -4$, $x - 3$ is negative; when $x = 1$, $x - 3$ is negative; and when $x = 4$, $x - 3$ is positive. When $x = -4$, $x + 2$ is negative; when $x = 1$, $x + 2$ is positive; and when $x = 4$, $x + 2$ is positive.

HOW TO • 2 Solve and graph the solution set of $(x - 2)(x + 1)(x - 4) > 0$.

On a number line, identify for each factor the regions where the factor is positive and those where the factor is negative.

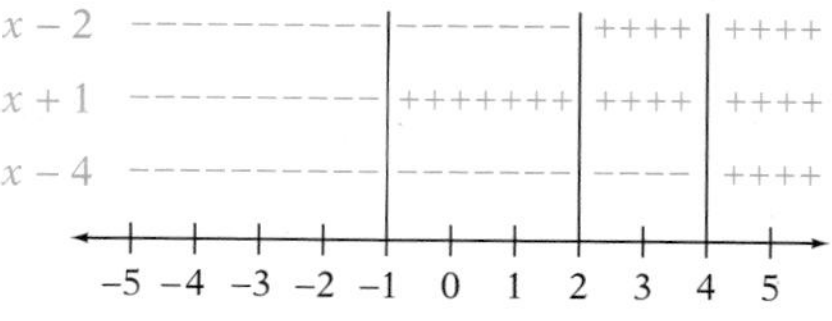

There are two regions where the product of the three factors is positive.

Write the solution set. $\{x \mid -1 < x < 2 \text{ or } x > 4\}$

The graph of the solution set of $(x - 2)(x + 1)(x - 4) > 0$ is shown at the right.

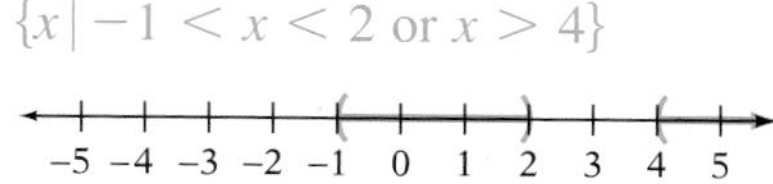

HOW TO • 3 Solve: $\dfrac{2x - 5}{x - 4} \le 1$

$$\frac{2x - 5}{x - 4} \le 1$$

Rewrite the inequality so that zero appears on the right side of the inequality.

$$\frac{2x - 5}{x - 4} - 1 \le 0$$

Simplify.

$$\frac{2x - 5}{x - 4} - \frac{x - 4}{x - 4} \le 0$$

$$\frac{x - 1}{x - 4} \le 0$$

On a number line, identify for each factor of the numerator and each factor of the denominator the regions where the factor is positive and those where the factor is negative.

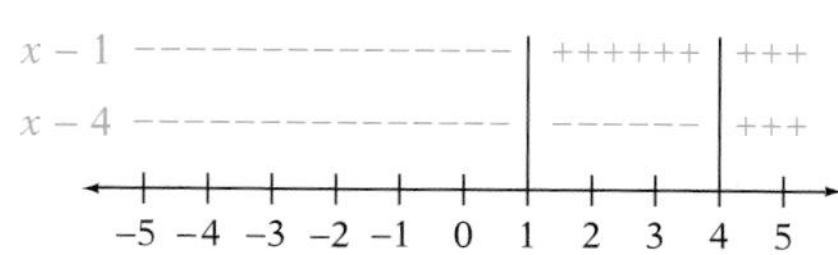

The region where the quotient of the two factors is negative is between 1 and 4.

Write the solution set. $\{x \mid 1 \le x < 4\}$

Note that 1 is part of the solution set but 4 is not because the denominator of the rational expression is zero when $x = 4$.

EXAMPLE • 1

Solve and graph the solution set of $2x^2 - x - 3 \ge 0$.

Solution

$$2x^2 - x - 3 \ge 0$$
$$(2x - 3)(x + 1) \ge 0$$

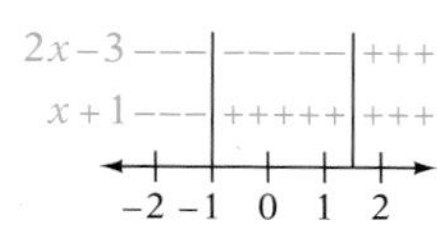

$$\left\{x \mid x \le -1 \text{ or } x \ge \frac{3}{2}\right\}$$

YOU TRY IT • 1

Solve and graph the solution set of $2x^2 - x - 10 \le 0$.

Your solution

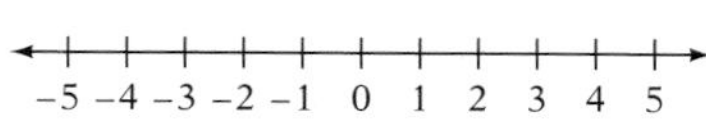

Solution on p. S30

10.5 EXERCISES

OBJECTIVE A To solve a nonlinear inequality

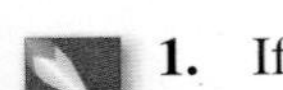

1. If $(x - 3)(x - 5) > 0$, what must be true of the values of $x - 3$ and $x - 5$?

2. For the inequality $\frac{x - 2}{x - 3} \le 1$, which of the values 1, 2, and 3 is not a possible element of the solution set? Why?

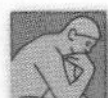

For Exercises 3 to 6, for the given values of x, state whether the inequality is true or false.

3. $(x - 3)(x + 2)(x - 4) > 0$ **a.** $x = 2$ **b.** $x = -2$ **c.** $x = -3$

4. $\frac{(x + 1)(x - 5)}{x - 3} \ge 0$ **a.** $x = 2$ **b.** $x = 3$ **c.** $x = -1$

5. $(x + 5)(x - 6) \le 0$ **a.** $x < 6$ **b.** $-5 \le x \le 6$

6. $\frac{x - 4}{x + 3} > 0$ **a.** $x < -3$ **b.** $x > 4$

For Exercises 7 to 22, solve and graph the solution set.

7. $(x - 4)(x + 2) > 0$

−5 −4 −3 −2 −1 0 1 2 3 4 5

8. $(x + 1)(x - 3) > 0$

−5 −4 −3 −2 −1 0 1 2 3 4 5

9. $x^2 - 3x + 2 \ge 0$

−5 −4 −3 −2 −1 0 1 2 3 4 5

10. $x^2 + 5x + 6 > 0$

−5 −4 −3 −2 −1 0 1 2 3 4 5

11. $x^2 - x - 12 < 0$

−5 −4 −3 −2 −1 0 1 2 3 4 5

12. $x^2 + x - 20 < 0$

−5 −4 −3 −2 −1 0 1 2 3 4 5

13. $(x - 1)(x + 2)(x - 3) < 0$

−5 −4 −3 −2 −1 0 1 2 3 4 5

14. $(x + 4)(x - 2)(x + 1) > 0$

−5 −4 −3 −2 −1 0 1 2 3 4 5

15. $(x + 4)(x - 2)(x - 1) \ge 0$

−5 −4 −3 −2 −1 0 1 2 3 4 5

16. $(x - 1)(x + 5)(x - 2) \le 0$

−5 −4 −3 −2 −1 0 1 2 3 4 5

17. $\frac{x - 4}{x + 2} > 0$

−5 −4 −3 −2 −1 0 1 2 3 4 5

18. $\frac{x + 2}{x - 3} > 0$

−5 −4 −3 −2 −1 0 1 2 3 4 5

19. $\dfrac{x-3}{x+1} \le 0$

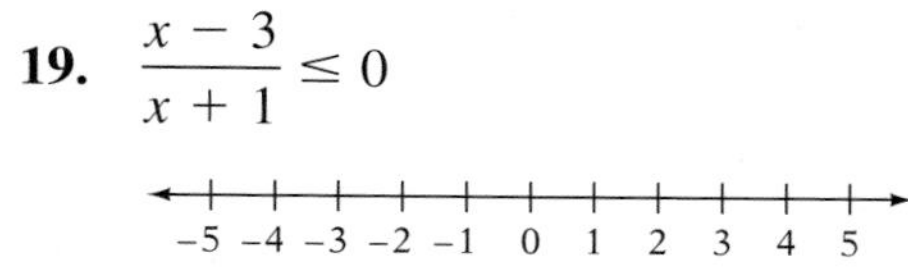

20. $\dfrac{x-1}{x} > 0$

21. $\dfrac{(x-1)(x+2)}{x-3} \le 0$

22. $\dfrac{(x+3)(x-1)}{x-2} \ge 0$

For Exercises 23 to 37, solve.

23. $x^2 - 16 > 0$

24. $x^2 - 4 \ge 0$

25. $x^2 - 9x \le 36$

26. $x^2 + 4x > 21$

27. $4x^2 - 8x + 3 < 0$

28. $2x^2 + 11x + 12 \ge 0$

29. $\dfrac{3}{x-1} < 2$

30. $\dfrac{x}{(x-1)(x+2)} \ge 0$

31. $\dfrac{x-2}{(x+1)(x-1)} \le 0$

32. $\dfrac{1}{x} < 2$

33. $\dfrac{x}{2x-1} \ge 1$

34. $\dfrac{x}{2x-3} \le 1$

35. $\dfrac{x}{2-x} \le -3$

36. $\dfrac{3}{x-2} > \dfrac{2}{x+2}$

37. $\dfrac{3}{x-5} > \dfrac{1}{x+1}$

Applying the Concepts

For Exercises 38 to 43, graph the solution set.

38. $(x+2)(x-3)(x+1)(x+4) > 0$

39. $(x-1)(x+3)(x-2)(x-4) \ge 0$

40. $(x^2 + 2x - 8)(x^2 - 2x - 3) < 0$

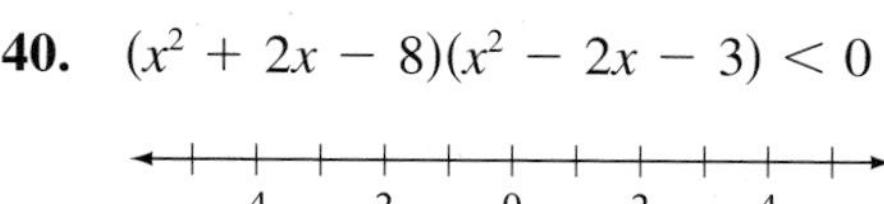

41. $(x^2 + 2x - 3)(x^2 + 3x + 2) \ge 0$

42. $(x^2 + 1)(x^2 - 3x + 2) > 0$

43. $\dfrac{x^2(3-x)(2x+1)}{(x+4)(x+2)} \ge 0$

SECTION 10.6 Applications of Quadratic Equations

OBJECTIVE A To solve application problems 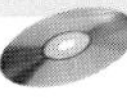

There are various applications of quadratic equations.

HOW TO • 1 Assuming no air resistance, the height h, in feet, of a soccer ball x feet from where it is kicked at an angle of 45° to the ground can be given by $h = -\frac{32}{v^2}x^2 + x$, where v is the initial speed of the soccer ball in feet per second. If a soccer ball is kicked with an initial speed of 50 ft/s, how far from where the ball was kicked is the height of the ball 15 ft? Round to the nearest tenth.

Strategy

To find where the soccer ball will be 15 ft above the ground, use the equation $h = -\frac{32}{v^2}x^2 + x$. Substitute 15 for h and 50 for v, and then solve for x.

Solution

$$h = -\frac{32}{v^2}x^2 + x$$

$$15 = -\frac{32}{50^2}x^2 + x$$ • Replace h by **15** and v by **50**.

$$0 = -0.0128x^2 + x - 15$$ • Write the equation in standard form.

$$x = \frac{-b \pm \sqrt{b^2 - 4ac}}{2a}$$ • Solve by using the quadratic formula.

$$x = \frac{-1 \pm \sqrt{1^2 - 4(-0.0128)(-15)}}{2(-0.0128)}$$ • $a = -0.0128, b = 1, c = -15$

$$= \frac{-1 \pm \sqrt{0.232}}{-0.0256} \approx \frac{-1 \pm 0.48166}{-0.0256}$$

$$x = \frac{-1 + 0.48166}{-0.0256} \quad \text{or} \quad x = \frac{-1 - 0.48166}{-0.0256}$$

$$= \frac{-0.51834}{-0.0256} \qquad = \frac{-1.48166}{-0.0256}$$

$$\approx 20.2 \qquad \approx 57.9$$

The ball is 15 ft above the ground when it is 20.2 ft and 57.9 ft from where it was kicked.

A drawing of the flight of the ball is shown below. Note that the ball is 15 ft high at two locations, 20.2 ft from the kicker and 57.9 ft from the kicker.

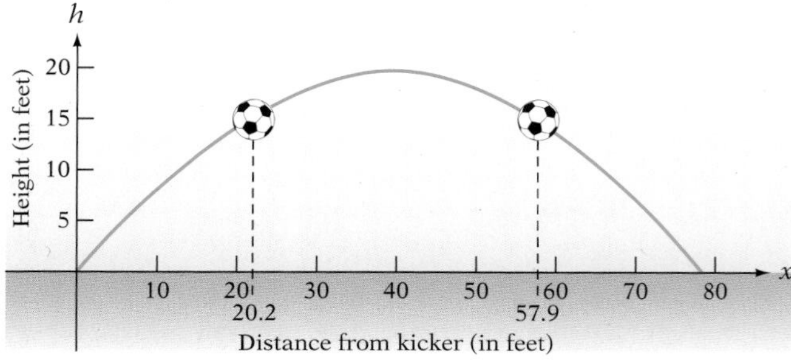

EXAMPLE • 1

A small pipe takes 2 h longer to empty a tank than does a larger pipe. After working together for 1 h, the larger pipe quits. It takes the smaller pipe 1 more hour to empty the tank. How long would it take each pipe, working alone, to empty the tank?

Strategy

- This is a work problem.
- The unknown time for the larger pipe to empty the tank working alone: t
- The unknown time for the smaller pipe to empty the tank working alone: $t + 2$
- The larger pipe operates for 1 h. The smaller pipe operates for 2 h.

	Rate	·	*Time*	=	*Part*
Larger pipe	$\frac{1}{t}$	·	1	=	$\frac{1}{t}$
Smaller pipe	$\frac{1}{t+2}$	·	2	=	$\frac{2}{t+2}$

- The sum of the part of the task completed by the larger pipe and the part completed by the smaller pipe equals 1.

Solution

$$\frac{1}{t} + \frac{2}{t+2} = 1$$

$$t(t+2)\left(\frac{1}{t} + \frac{2}{t+2}\right) = t(t+2) \cdot 1$$

$$(t+2) + 2t = t^2 + 2t$$

$$0 = t^2 - t - 2$$

$$0 = (t+1)(t-2)$$

$$t + 1 = 0 \qquad t - 2 = 0$$

$$t = -1 \qquad t = 2$$

Because time cannot be negative, $t = -1$ is not possible. It takes the larger pipe, working alone, 2 h to empty the tank.

The time for the smaller pipe to empty the tank working alone is $t + 2$.

$t + 2$

$2 + 2 = 4$ • **Replace t by 2.**

It takes the smaller pipe, working alone, 4 h to empty the tank.

YOU TRY IT • 1

It takes William 3 h longer than it does Olivia to detail a car. Working together, they can detail the car in 2 h. How long would it take William, working alone, to detail the car?

Your strategy

Your solution

Solution on pp. S30–S31

EXAMPLE • 2

In 8 h, two campers rowed 15 mi down a river and then rowed back to their campsite. The rate of the river's current was 1 mph. Find the rate at which the campers row in still water.

Strategy

- Unknown rowing rate of the campers: r

	Distance	÷	*Rate*	=	*Time*
Down river	15	÷	$r + 1$	=	$\frac{15}{r + 1}$
Up river	15	÷	$r - 1$	=	$\frac{15}{r - 1}$

- The total time of the trip was 8 h.

Solution

$$\frac{15}{r + 1} + \frac{15}{r - 1} = 8$$

$$(r + 1)(r - 1)\left(\frac{15}{r + 1} + \frac{15}{r - 1}\right) = (r + 1)(r - 1) \cdot 8$$

$$15(r - 1) + 15(r + 1) = (r^2 - 1)8$$

$$30r = 8r^2 - 8$$

$$0 = 8r^2 - 30r - 8$$

$$0 = 2(4r + 1)(r - 4)$$

$4r + 1 = 0 \qquad r - 4 = 0$

$r = -\frac{1}{4} \qquad r = 4$

The rate cannot be negative, so the solution $-\frac{1}{4}$ is not possible. The rowing rate was 4 mph.

YOU TRY IT • 2

The rate of a jet in calm air is 250 mph. Flying with the wind, the jet can fly 1200 mi in 2 h less time than is required to make the return trip against the wind. Find the rate of the wind.

Your strategy

Your solution

EXAMPLE • 3

The length of a rectangle is 5 in. more than the width. The area is 36 in^2. Find the width of the rectangle.

Strategy

- Width of the rectangle: w
- Length of the rectangle: $w + 5$
- Use the formula $A = lw$.

Solution

$$A = lw$$

$$36 = (w + 5)w$$

$$36 = w^2 + 5w$$

$$0 = w^2 + 5w - 36$$

$$0 = (w + 9)(w - 4)$$

$w + 9 = 0 \qquad w - 4 = 0$

$w = -9 \qquad w = 4$

The width cannot be negative. The width of the rectangle is 4 in.

YOU TRY IT • 3

The base of a triangle is 4 in. more than twice the height. The area is 35 in^2. Find the height of the triangle.

Your strategy

Your solution

Solutions on p. S31

10.6 EXERCISES

OBJECTIVE A To solve application problems

1. **Safety** A car with good tire tread can stop in less distance than a car with poor tread. The formula for the stopping distance d, in feet, of a car with good tread on dry cement is approximated by $d = 0.04v^2 + 0.5v$, where v is the speed of the car. If the driver must be able to stop within 60 ft, what is the maximum safe speed, to the nearest mile per hour, of the car?

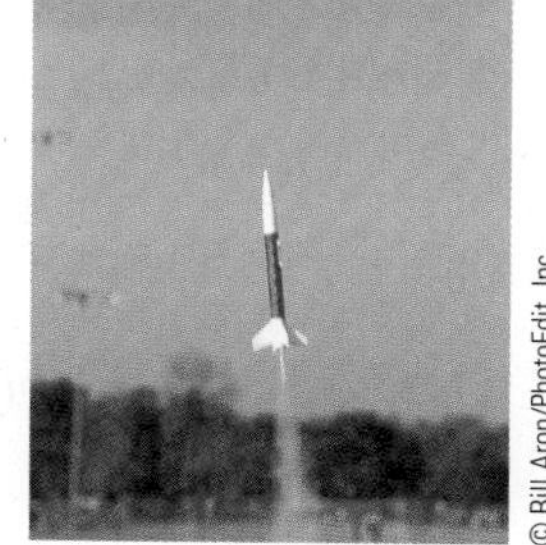
© Bill Aron/PhotoEdit, Inc.

2. **Rockets** A model rocket is launched with an initial velocity of 200 ft/s. The height h, in feet, of the rocket t seconds after the launch is given by $h = -16t^2 + 200t$. How many seconds after the launch will the rocket be 300 ft above the ground? Round to the nearest hundredth of a second.

3. **Physics** The height of a projectile fired upward is given by the formula $s = v_0 t - 16t^2$, where s is the height in feet, v_0 is the initial velocity, and t is the time in seconds. Find the time for a projectile to return to Earth if it has an initial velocity of 200 ft/s.

4. **Physics** The height of a projectile fired upward is given by the formula $s = v_0 t - 16t^2$, where s is the height in feet, v_0 is the initial velocity, and t is the time in seconds. Find the time for a projectile to reach a height of 64 ft if it has an initial velocity of 128 ft/s. Round to the nearest hundredth of a second.

5. **Construction** The height h, in feet, of an arch is given by the equation $h(x) = -\frac{3}{64}x^2 + 27$, where $|x|$ is the distance in feet from the center of the arch.

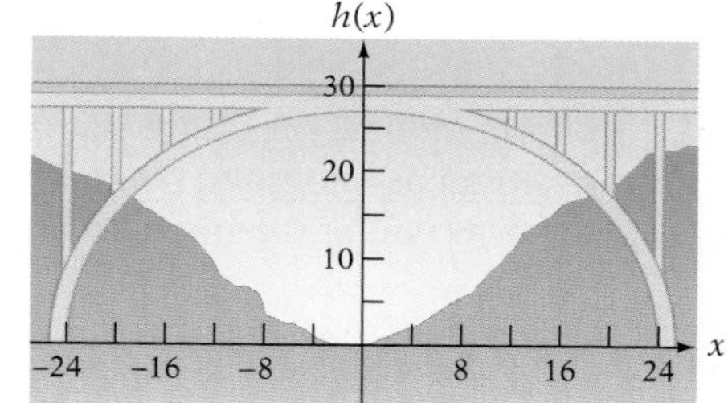

   a. What is the maximum height of the arch?
   b. What is the height of the arch 8 ft to the right of the center?
   c. How far from the center is the arch 8 ft tall? Round to the nearest hundredth.

6. **Physics** Using Torricelli's Principle, it can be shown that the depth d of a liquid in a bottle with a hole of area 0.5 cm^2 in its side can be approximated by $d = 0.0034t^2 - 0.52518t + 20$, where t is the time since a stopper was removed from the hole. When will the depth be 10 cm? Round to the nearest tenth of a second.

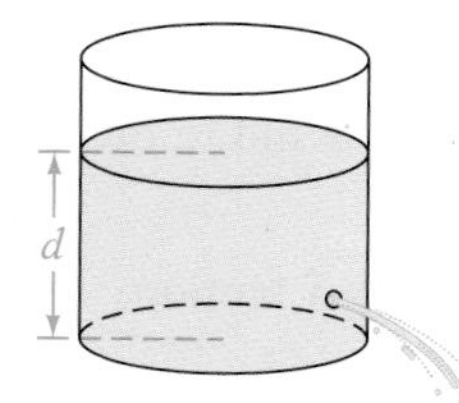

7. **Sports** The Water Cube was built in Beijing, China, to house the National Swimming Center for the 2008 Olympics. Although not actually a cube (its height is not equal to its length and width), the Water Cube is designed to look like a "cube" of water molecules. The volume of the 31-meter-high Water Cube is 971,199 m^3. Find the length of a side of its square base. Recall that $V = LWH$.

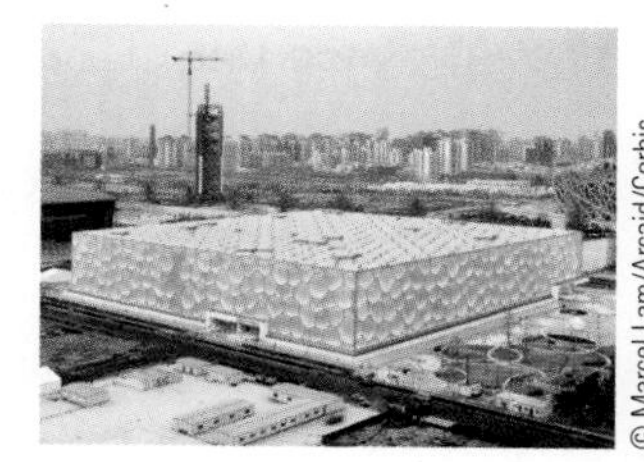
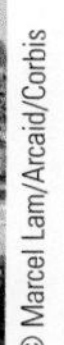
© Marcel Lam/Arcaid/Corbis

8. Suppose a large pipe and a smaller pipe, working together, can empty a swimming pool in 4 h. Is it possible for the larger pipe, working alone, to empty the tank in less than 4 h?

9. **Tanks** A small pipe can fill a tank in 6 min more time than it takes a larger pipe to fill the same tank. Working together, the pipes can fill the tank in 4 min. How long would it take each pipe, working alone, to fill the tank?

10. **Landscaping** It takes a small sprinkler 16 min longer to soak a lawn than it takes a larger sprinkler. Working together, the sprinklers can soak the lawn in 6 min. How long would it take each sprinkler, working alone, to soak the lawn?

11. **Parallel Processing** Parallel processing is the simultaneous use of more than one computer to run a program. Suppose one computer, working alone, takes 4 h longer than a second computer to run a program that can determine whether a large number is prime. After both computers work together for 1 h, the faster computer crashes. The slower computer continues working for another 2 h before completing the program. How long would it take the faster computer, working alone, to run the program?

12. **Payroll** It takes one printer, working alone, 6 h longer to print a weekly payroll than it takes a second printer. Working together, the two printers can print the weekly payroll in 4 h. How long would it take each printer, working alone, to print the payroll?

13. **Wood Flooring** It takes an apprentice carpenter 2 h longer to install a small section of wood floor than it does a more experienced carpenter. After the carpenters work together for 2 h, the experienced carpenter leaves for another job. It then takes the apprentice carpenter 2 h longer to complete the installation. How long would it take the apprentice carpenter, working alone, to install the floor? Round to the nearest tenth.

14. **Uniform Motion** A cruise ship made a trip of 100 mi in 8 h. The ship traveled the first 40 mi at a constant rate before increasing its speed by 5 mph. Then it traveled another 60 mi at the increased speed. Find the rate of the cruise ship for the first 40 mi.

15. **Uniform Motion** The Concorde's speed in calm air was 1320 mph. Flying with the wind, the Concorde could fly from New York to London, a distance of approximately 4000 mi, in 0.5 h less time than is required to make the return trip. Find the rate of the wind to the nearest mile per hour.

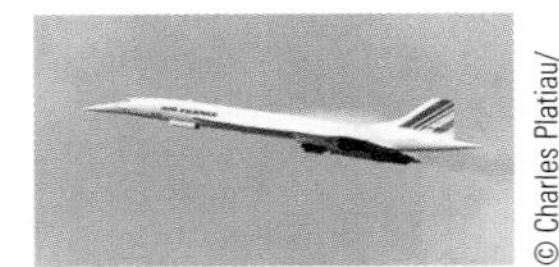

© Charles Platiau/Reuters/Corbis

16. **Uniform Motion** A car travels 120 mi. A second car, traveling 10 mph faster than the first car, makes the same trip in 1 h less time. Find the speed of each car.

17. **Air Force One** The Air Force uses the designation VC-25 for the plane on which the president of the United States flies. When the president is on the plane, its call sign is Air Force One. The plane's speed in calm air is 630 mph. Flying with the jet stream, the plane can fly from Washington, D.C., to London, a distance of approximately 3660 mi, in 1.75 h less time than is required to make the return trip. Find the rate of the jet stream to the nearest mile per hour.

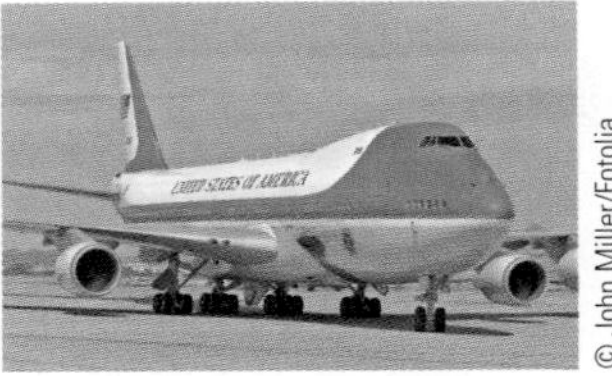

© John Miller/Fotolia

18. **Uniform Motion** For a portion of the Green River in Utah, the rate of the river's current is 4 mph. A tour guide can row 5 mi down this river and back in 3 h. Find the rowing rate of the guide in calm water.

19. **Geography** The state of Colorado is almost perfectly rectangular, with its north border 111 mi longer than its west border. If the state encompasses 104,000 mi^2, estimate the dimensions of Colorado. Round to the nearest mile.

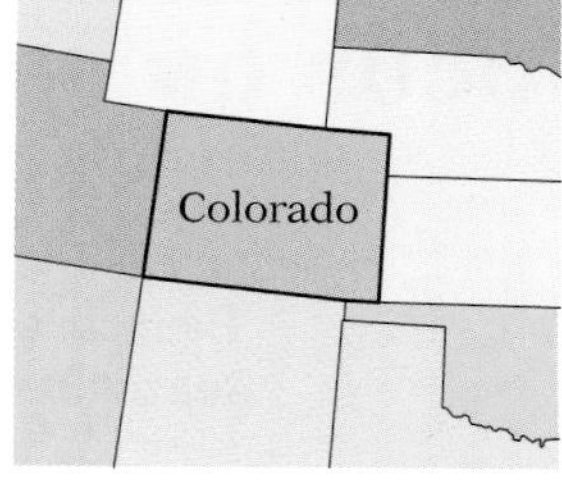

20. **Geometry** The length of a rectangle is 2 ft less than three times the width of the rectangle. The area of the rectangle is 65 ft^2. Find the length and width of the rectangle.

21. **Geometry** The length of the base of a triangle is 1 cm less than five times the height of the triangle. The area of the triangle is 21 cm^2. Find the height of the triangle and the length of the base of the triangle.

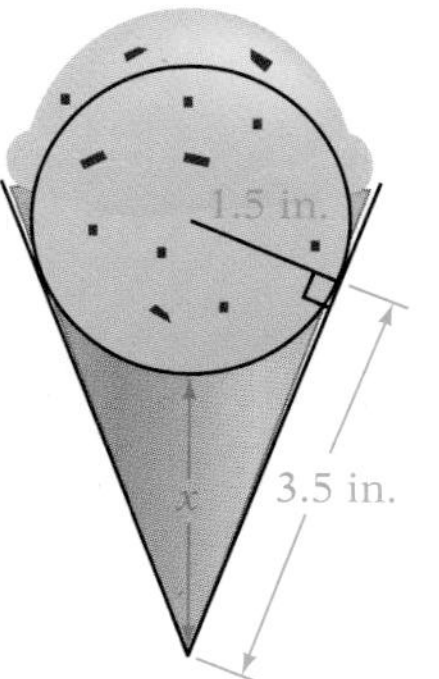

22. **Ice Cream** A perfectly spherical scoop of mint chocolate chip ice cream is placed in a cone, as shown in the figure. How far is the bottom of the scoop of ice cream from the bottom of the cone? Round to the nearest tenth. (*Hint:* A line segment from the center of the ice cream to the point at which the ice cream touches the cone is perpendicular to the edge of the cone.)

23. **Geometry** A square piece of cardboard is formed into a box by cutting 10-centimeter squares from each of the four corners and then folding up the sides, as shown in the figure. If the volume V of the box is to be 49,000 cm^3, what size square piece of cardboard is needed? Recall that $V = LWH$.

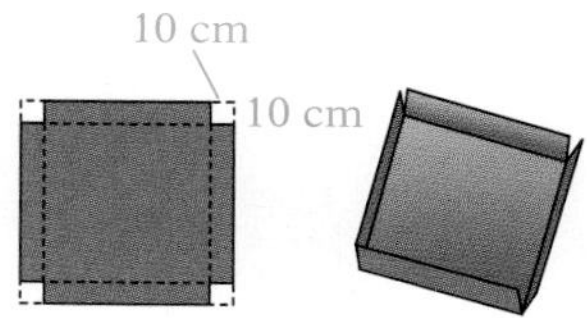

24. **Construction** A homeowner hires a mason to lay a brick border around a cement patio that measures 8 ft by 10 ft. If the total area of the patio and border is 168 ft^2, what is the width of the border?

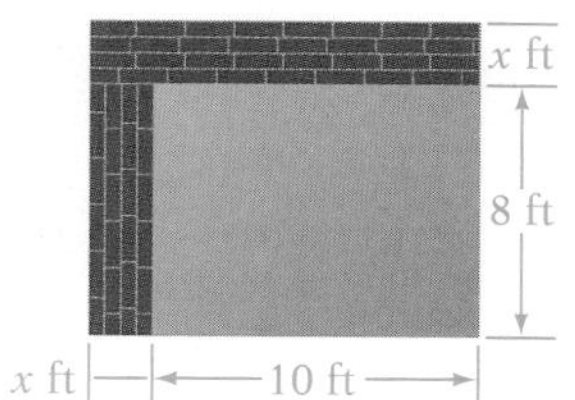

25. **Fencing** A dog trainer has 80 ft of fencing with which to create a rectangular work area for dogs. If the trainer wants to enclose an area of 300 ft^2, what will be the dimensions of the work area?

26. **Sports** Read the article at the right.
   a. The screen behind the east end zone is a rectangle with length 13 ft less than three times its width. Find the length and width of the east end zone screen.

   b. The screen behind the west end zone is a rectangle with length 1 ft less than twice its width. Find the length and width of the west end zone screen.

In the News

Dolphins on the Big Screen

The Miami Dolphins' stadium, Landshark Stadium, is home to one of the world's largest digital displays—a 6850-square-foot screen installed behind the east end zone. A smaller screen, of area 4950 ft^2, sits behind the west end zone.

Source: Business Wire

Applying the Concepts

27. **Geometry** The surface area of the ice cream cone shown at the right is given by $A = \pi r^2 + \pi rs$, where r is the radius of the circular top of the cone and s is the slant height of the cone. If the surface area of the cone is 11.25π in^2 and the slant height is 6 in., find the radius of the cone.

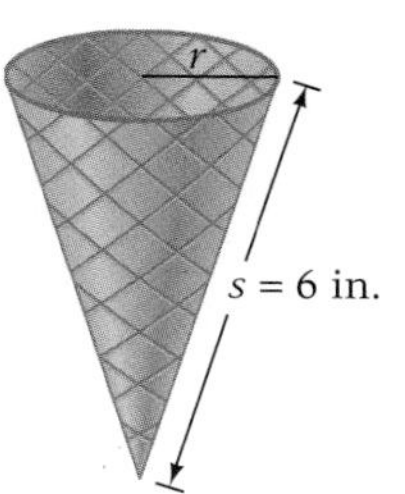

FOCUS ON PROBLEM SOLVING

Inductive and Deductive Reasoning

Consider the following sums of odd positive integers.

$1 + 3 = 4 = 2^2$	Sum of the first two odd numbers is 2^2.
$1 + 3 + 5 = 9 = 3^2$	Sum of the first three odd numbers is 3^2.
$1 + 3 + 5 + 7 = 16 = 4^2$	Sum of the first four odd numbers is 4^2.
$1 + 3 + 5 + 7 + 9 = 25 = 5^2$	Sum of the first five odd numbers is 5^2.

1. Make a conjecture about the value of the sum

$$1 + 3 + 5 + 7 + 9 + 11 + 13 + 15$$

without adding the numbers.

If the pattern continues, a possible conjecture is that the sum of the first eight odd positive integers is $8^2 = 64$. By adding the numbers, you can verify that this is correct. Inferring that the pattern established by a few cases is valid for all cases is an example of **inductive reasoning.**

2. Use inductive reasoning to find the next figure in the sequence below.

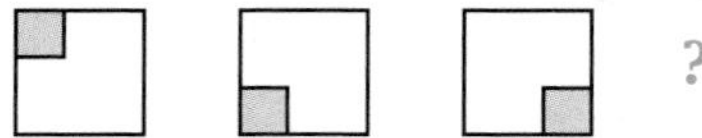

?

The fact that a pattern *appears* to be true does not *prove* it is true for all cases. For example, consider the polynomial $n^2 - n + 41$. If we begin substituting positive integer values for n and evaluating the expression, we produce the following table.

n	$n^2 - n + 41$	
1	$1^2 - 1 + 41 = 41$	41 is a prime number.
2	$2^2 - 2 + 41 = 43$	43 is a prime number.
3	$3^2 - 3 + 41 = 47$	47 is a prime number.
4	$4^2 - 4 + 41 = 53$	53 is a prime number.

Even if we try a number like 30, we have $30^2 - 30 + 41 = 911$, and 911 is a prime number. Thus it appears that the value of this polynomial is a prime number for any value of n. However, the conjecture is not true. For instance, if $n = 41$, we have $41^2 - 41 + 41 = 41^2 = 1681$, which is not a prime number because it is divisible by 41. This illustrates that inductive reasoning may lead to incorrect conclusions and that an inductive proof must be available to prove conjectures. There is such a proof, called *mathematical induction,* that you may study in a future math course.

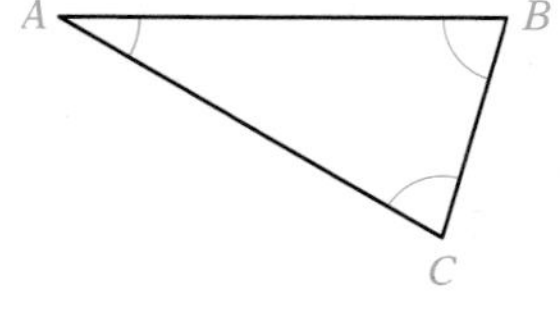

$\angle A + \angle B + \angle C = 180°$

Now consider the true statement that the sum of the measures of the interior angles of a triangle is 180°. The figure at the left is a triangle; therefore, the sum of the measures of the interior angles must be 180°. This is an example of *deductive reasoning.* **Deductive reasoning** uses a rule or statement of fact to reach a conclusion.

3. Use deductive reasoning to complete the following sentence. All even numbers are divisible by 2. Because 14,386 is an even number, . . .

4. a. Use the pattern given below to find the three missing terms in __ __ __ 64.

A 2 3 4 B 6 7 8 C 10 11 12 D 14 15 16. . .

b. Did you use inductive or deductive reasoning?

5. a. Draw the next figure in the sequence:

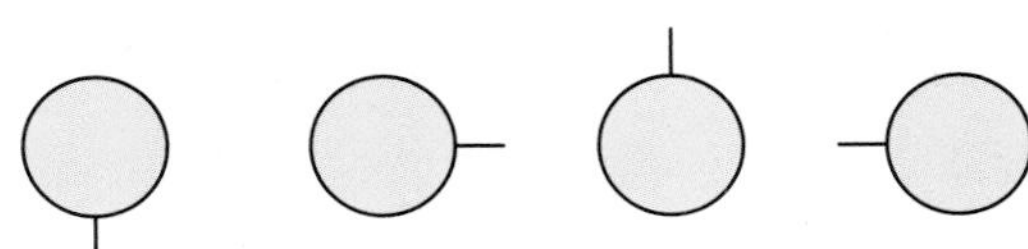

b. Did you use inductive or deductive reasoning?

6. a. If ◇◇◇◇◇ = ‡‡‡ and ‡‡‡ = ∇∇∇∇, how many ◇'s equal ∇∇∇∇∇∇∇∇?

b. Did you use inductive or deductive reasoning?

7. a. If ♠♠ = ♦♦♦♦♦♦, and ♦♦♦ = ♣♣, and ♣♣♣♣ = ♥, how many ♥'s equal ♠♠♠♠♠♠?

b. Did you use inductive or deductive reasoning?

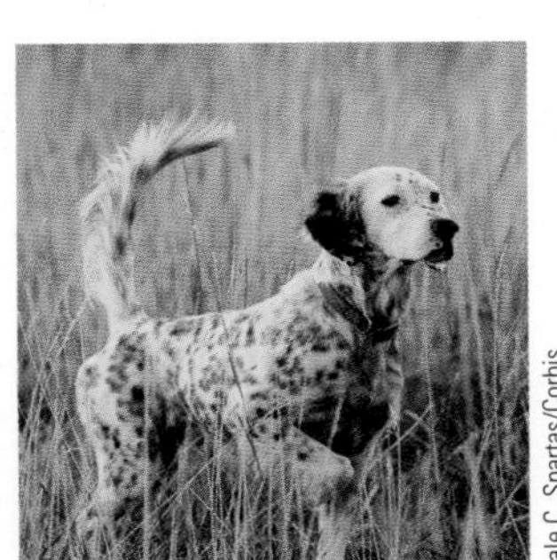
Dale C. Spartas/Corbis

For Exercises 8 to 10, determine whether inductive or deductive reasoning is being used.

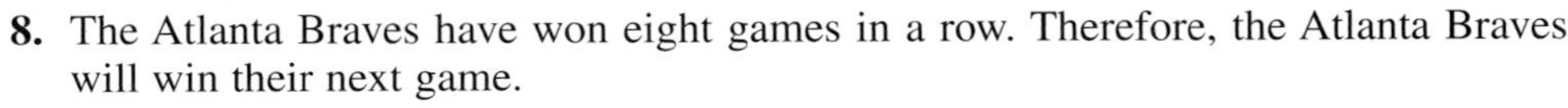

8. The Atlanta Braves have won eight games in a row. Therefore, the Atlanta Braves will win their next game.

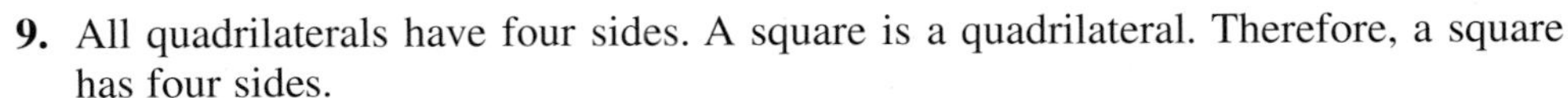

9. All quadrilaterals have four sides. A square is a quadrilateral. Therefore, a square has four sides.

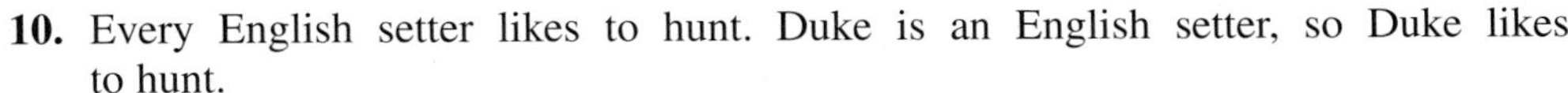

10. Every English setter likes to hunt. Duke is an English setter, so Duke likes to hunt.

PROJECTS AND GROUP ACTIVITIES

Using a Graphing Calculator to Solve a Quadratic Equation

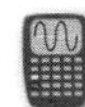

Recall that an x-intercept of the graph of an equation is a point at which the graph crosses the x-axis. For the graph in Figure 1, the x-intercepts are $(-2, 0)$ and $(3, 0)$.

Recall also that to find the x-intercept of a graph, set $y = 0$ and then solve for x. For the equation in Figure 1, if we set $y = 0$, the resulting equation is $0 = x^2 - x - 6$, which is a quadratic equation. Solving this equation by factoring, we have

$$0 = x^2 - x - 6$$
$$0 = (x + 2)(x - 3)$$

$$x + 2 = 0 \qquad x - 3 = 0$$
$$x = -2 \qquad x = 3$$

9; −5; 5; −9; (−2, 0); (3, 0); $Y_1 = x^2 - x - 6$

FIGURE 1

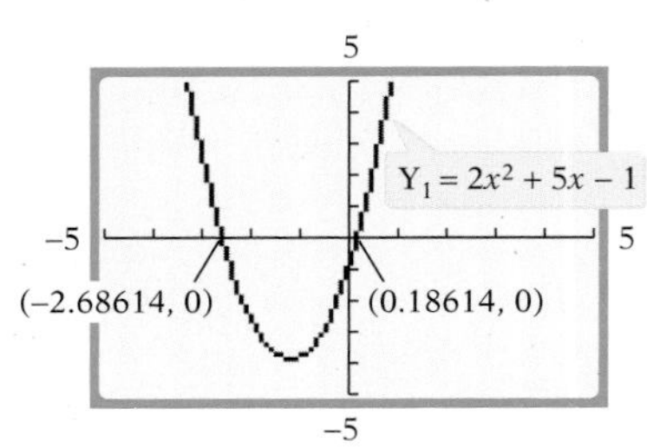

FIGURE 2

Thus the solutions of the equation are the x-coordinates of the x-intercepts of the graph.

This connection between the solutions of an equation and the x-intercepts of its graph enables us to find approximations of the real number solutions of an equation graphically. For example, to use a TI-84 to approximate the solutions of $2x^2 + 5x - 1 = 0$ graphically, use the following keystrokes to graph $y = 2x^2 + 5x - 1$ and find the decimal approximations of the x-coordinates of the x-intercepts. Use the window shown in the graph in Figure 2.

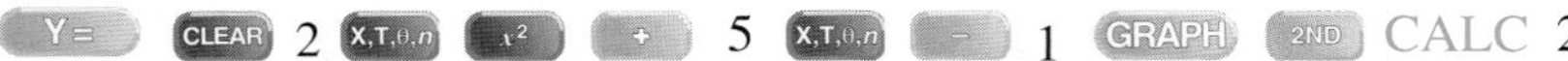

Use the arrow keys to move to the left of the leftmost x-intercept. Press ENTER. Use the arrow keys to move to a point just to the right of the leftmost x-intercept. Press ENTER twice. The x-coordinate at the bottom of the screen is the approximation of one solution of the equation. To find the other solution, use the same procedure but move the cursor first to the left and then to the right of the rightmost x-intercept.

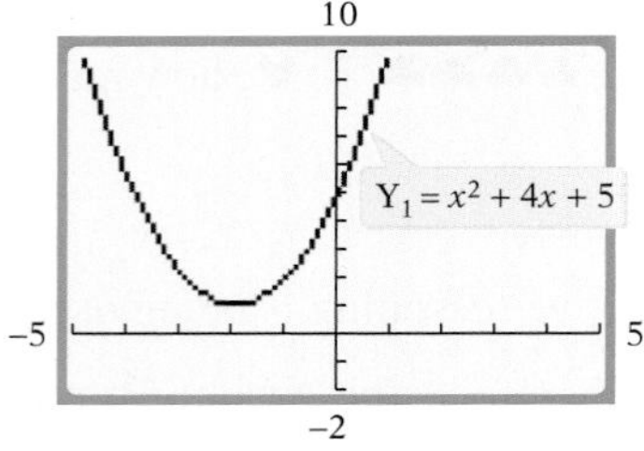

FIGURE 3

Attempting to find the solutions of an equation graphically will not necessarily yield all the solutions. Because the x-coordinates of the x-intercepts of a graph are *real* numbers, only real number solutions can be found. For instance, consider the equation $x^2 + 4x + 5 = 0$. The graph of $y = x^2 + 4x + 5$ is shown in Figure 3. Note that the graph has no x-intercepts and that consequently it has no real number solutions. However, $x^2 + 4x + 5 = 0$ does have complex number solutions that can be obtained by using the quadratic formula. They are $-2 + i$ and $-2 - i$.

Use a graphing calculator to approximate the solutions of the equations in Exercises 1 to 4.

1. $x^2 + 6x + 1 = 0$

2. $-x^2 + x + 3 = 0$

3. $2x^2 + 4x - 1 = 0$

4. $x^2 + 4x + 5 = 0$

CHAPTER 10

SUMMARY

KEY WORDS	EXAMPLES
A *quadratic equation* is an equation of the form $ax^2 + bx + c = 0$, where $a \neq 0$. A quadratic equation is also called a *second-degree equation.* A quadratic equation is in *standard form* when the polynomial is in descending order and equal to zero. [10.1A, p. 570]	$3x^2 + 4x - 7 = 0$, $x^2 - 1 = 0$, and $4x^2 + 8x = 0$ are quadratic equations. They are all written in standard form.
When a quadratic equation has two solutions that are the same number, the solution is called a *double root* of the equation. [10.1A, p. 570]	$x^2 - 4x + 4 = 0$ $(x - 2)(x - 2) = 0$ $x - 2 = 0 \quad x - 2 = 0$ $x = 2 \quad x = 2$ 2 is a double root.

For an equation of the form $ax^2 + bx + c = 0$, the quantity $b^2 - 4ac$ is called the *discriminant.* [10.3A, p. 586]

$$2x^2 - 3x + 5 = 0$$
$$a = 2, b = -3, c = 5$$
$$b^2 - 4ac = (-3)^2 - 4(2)(5)$$
$$= 9 - 40 = -31$$

An equation is *quadratic in form* if it can be written as $au^2 + bu + c = 0$. [10.4A, p. 590]

$$6x^4 - 5x^2 - 4 = 0$$
$$6(x^2)^2 - 5(x^2) - 4 = 0$$
$$6u^2 - 5u - 4 = 0$$

The equation is quadratic in form.

A *quadratic inequality* is one that can be written in the form $ax^2 + bx + c < 0$ or $ax^2 + bx + c > 0$, where $a \neq 0$. The symbols $\leq$ and $\geq$ can also be used. [10.5A, p. 596]

$3x^2 + 5x - 8 \leq 0$ is a quadratic inequality.

ESSENTIAL RULES AND PROCEDURES — **EXAMPLES**

Principle of Zero Products [10.1A, p. 570]
If a and b are real numbers and $ab = 0$, then $a = 0$ or $b = 0$.

$$(x - 3)(x + 4) = 0$$
$$x - 3 = 0 \qquad x + 4 = 0$$
$$x = 3 \qquad x = -4$$

To Solve a Quadratic Equation by Factoring [10.1A, p. 570]

1. Write the equation in standard form.
2. Factor the polynomial.
3. Use the Principle of Zero Products to set each factor equal to 0.
4. Solve each equation.
5. Check the solutions.

$$x^2 + 2x = 35$$
$$x^2 + 2x - 35 = 0$$
$$(x - 5)(x + 7) = 0$$
$$x - 5 = 0 \qquad x + 7 = 0$$
$$x = 5 \qquad x = -7$$

To Write a Quadratic Equation Given Its Solutions [10.1B, p. 571]

Use the equation $(x - r_1)(x - r_2) = 0$. Replace r_1 with one solution and r_2 with the other solution. Then multiply the two factors.

Write a quadratic equation that has solutions -3 and 6.

$$(x - r_1)(x - r_2) = 0$$
$$[x - (-3)](x - 6) = 0$$
$$(x + 3)(x - 6) = 0$$
$$x^2 - 3x - 18 = 0$$

To Solve a Quadratic Equation by Taking Square Roots [10.1C, pp. 572, 573]

1. Solve for x^2 or for $(x + a)^2$.
2. Take the square root of each side of the equation.
3. Simplify.
4. Check the solutions.

$$(x + 2)^2 - 9 = 0$$
$$(x + 2)^2 = 9$$
$$\sqrt{(x + 2)^2} = \sqrt{9}$$
$$x + 2 = \pm\sqrt{9}$$
$$x + 2 = \pm 3$$
$$x + 2 = 3 \qquad x + 2 = -3$$
$$x = 1 \qquad x = -5$$

To Complete the Square [10.2A, p. 578]

Add to a binomial of the form $x^2 + bx$ the square of $\frac{1}{2}$ of the coefficient of x, making it a perfect-square trinomial.

To complete the square on $x^2 - 8x$, add $\left[\frac{1}{2}(-8)\right]^2 = 16$: $x^2 - 8x + 16$.

To Solve a Quadratic Equation by Completing the Square [10.2A, p. 580]

1. Write the equation in the form $ax^2 + bx = -c$.
2. Multiply both sides of the equation by $\frac{1}{a}$.
3. Complete the square on $x^2 + \frac{b}{a}x$. Add the number that completes the square to both sides of the equation.
4. Factor the perfect-square trinomial.
5. Take the square root of each side of the equation.
6. Solve the resulting equation for x.
7. Check the solutions.

$$x^2 + 4x - 1 = 0$$
$$x^2 + 4x = 1$$
$$x^2 + 4x + 4 = 1 + 4$$
$$(x + 2)^2 = 5$$
$$\sqrt{(x + 2)^2} = \sqrt{5}$$
$$x + 2 = \pm\sqrt{5}$$
$$x + 2 = \sqrt{5} \qquad x + 2 = -\sqrt{5}$$
$$x = -2 + \sqrt{5} \qquad x = -2 - \sqrt{5}$$

Quadratic Formula [10.3A, p. 584]

The solutions of $ax^2 + bx + c = 0$, $a \neq 0$, are $x = \frac{-b \pm \sqrt{b^2 - 4ac}}{2a}$.

$$2x^2 - 3x + 4 = 0$$
$$a = 2,\ b = -3,\ c = 4$$
$$x = \frac{-(-3) \pm \sqrt{(-3)^2 - 4(2)(4)}}{2(2)}$$
$$= \frac{3 \pm \sqrt{-23}}{4} = \frac{3 \pm i\sqrt{23}}{4}$$
$$= \frac{3}{4} \pm \frac{\sqrt{23}}{4}i$$

The Effect of the Discriminant on the Solutions of a Quadratic Equation [10.3A, p. 586]

1. If $b^2 - 4ac = 0$, the equation has two equal real number solutions, a double root.
2. If $b^2 - 4ac > 0$, the equation has two unequal real number solutions.
3. If $b^2 - 4ac < 0$, the equation has two complex number solutions.

$x^2 + 8x + 16 = 0$ has a double root because $b^2 - 4ac = 8^2 - 4(1)(16) = 0$.

$2x^2 + 3x - 5 = 0$ has two unequal real number solutions because $b^2 - 4ac = 3^2 - 4(2)(-5) = 49$.

$3x^2 + 2x + 4 = 0$ has two complex number solutions because $b^2 - 4ac = 2^2 - 4(3)(4) = -44$.

CHAPTER 10

CONCEPT REVIEW

Test your knowledge of the concepts presented in this chapter. Answer each question. Then check your answers against the ones provided in the Answer Section.

1. How do you write a quadratic equation if you are given the solutions?

2. What does the $\pm$ symbol mean?

3. Complete the square on the expression $x^2 - 18x$.

4. Given exact solutions $4 + 3\sqrt{2}$ and $4 - 3\sqrt{2}$, what are the approximate solutions? Round to the nearest ten-thousandth.

5. What is the procedure for solving a quadratic equation by completing the square?

6. What is the quadratic formula?

7. What does the discriminant tell you about the solutions to a quadratic equation?

8. Why is it especially important to check the solutions to radical equations?

9. What is a quadratic inequality?

10. In a quadratic inequality, how does the inequality symbol indicate whether or not to include the endpoints of the solution set?

11. In a rational inequality, why is it important to check the elements of the solution set?

CHAPTER 10

REVIEW EXERCISES

1. Solve by factoring: $2x^2 - 3x = 0$

2. Solve for x by factoring: $6x^2 + 9cx = 6c^2$

3. Solve by taking square roots: $x^2 = 48$

4. Solve by taking square roots: $\left(x + \frac{1}{2}\right)^2 + 4 = 0$

5. Solve by completing the square: $x^2 + 4x + 3 = 0$

6. Solve by completing the square: $7x^2 - 14x + 3 = 0$

7. Solve by using the quadratic formula: $12x^2 - 25x + 12 = 0$

8. Solve by using the quadratic formula: $x^2 - x + 8 = 0$

9. Write a quadratic equation that has integer coefficients and has solutions 0 and -3.

10. Write a quadratic equation that has integer coefficients and has solutions $\frac{3}{4}$ and $-\frac{2}{3}$.

11. Solve by completing the square: $x^2 - 2x + 8 = 0$

12. Solve by completing the square: $(x - 2)(x + 3) = x - 10$

13. Solve by using the quadratic formula: $3x(x - 3) = 2x - 4$

14. Use the discriminant to determine whether $3x^2 - 5x + 3 = 0$ has two equal real number solutions, two unequal real number solutions, or two complex number solutions.

15. Solve: $(x + 3)(2x - 5) < 0$

16. Solve: $(x - 2)(x + 4)(2x + 3) \leq 0$

17. Solve: $x^{2/3} + x^{1/3} - 12 = 0$

18. Solve: $2(x - 1) + 3\sqrt{x - 1} - 2 = 0$

19. Solve: $3x = \dfrac{9}{x - 2}$

20. Solve: $\dfrac{3x + 7}{x + 2} + x = 3$

21. Solve and graph the solution set:
$\dfrac{x - 2}{2x - 3} \geq 0$

−5 −4 −3 −2 −1 0 1 2 3 4 5

22. Solve and graph the solution set:
$\dfrac{(2x - 1)(x + 3)}{x - 4} \leq 0$

−5 −4 −3 −2 −1 0 1 2 3 4 5

23. Solve: $x = \sqrt{x} + 2$

24. Solve: $2x = \sqrt{5x + 24} + 3$

25. Solve: $\dfrac{x - 2}{2x + 3} - \dfrac{x - 4}{x} = 2$

26. Solve: $1 - \dfrac{x + 4}{2 - x} = \dfrac{x - 3}{x + 2}$

27. Write a quadratic equation that has integer coefficients and has solutions $\frac{1}{3}$ and -3.

28. Solve by factoring: $2x^2 + 9x = 5$

29. Solve: $2(x + 1)^2 - 36 = 0$

30. Solve by using the quadratic formula:
$x^2 + 6x + 10 = 0$

31. Solve: $\dfrac{2}{x - 4} + 3 = \dfrac{x}{2x - 3}$

32. Solve: $x^4 - 28x^2 + 75 = 0$

33. Solve: $\sqrt{2x - 1} + \sqrt{2x} = 3$

34. Solve: $2x^{2/3} + 3x^{1/3} - 2 = 0$

35. Solve: $\sqrt{3x - 2} + 4 = 3x$

36. Solve by completing the square: $x^2 - 10x + 7 = 0$

37. Solve: $\dfrac{2x}{x - 4} + \dfrac{6}{x + 1} = 11$

38. Solve by using the quadratic formula: $9x^2 - 3x = 1$

39. Solve: $2x = 4 - 3\sqrt{x - 1}$

40. Solve: $1 - \dfrac{x + 3}{3 - x} = \dfrac{x - 4}{x + 3}$

41. Use the discriminant to determine whether $2x^2 - 5x = 6$ has two equal real number solutions, two unequal real number solutions, or two complex number solutions.

42. Solve: $x^2 - 3x \leq 10$

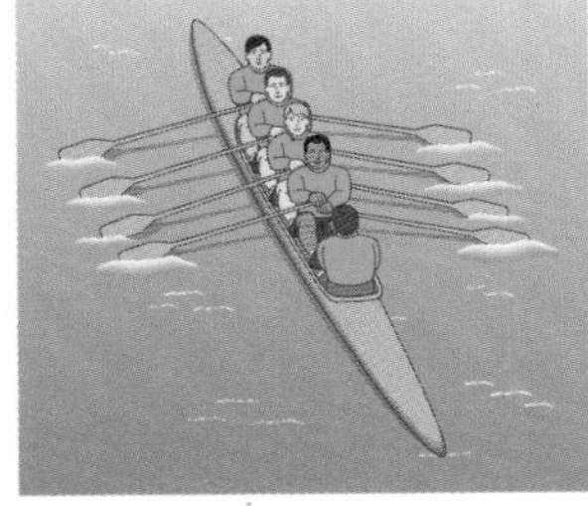

43. **Sports** To prepare for an upcoming race, a sculling crew rowed 16 mi down a river and back in 6 h. If the rate of the river's current is 2 mph, find the sculling crew's rate of rowing in calm water.

44. **Geometry** The length of a rectangle is 2 cm more than twice the width. The area of the rectangle is 60 cm^2. Find the length and width of the rectangle.

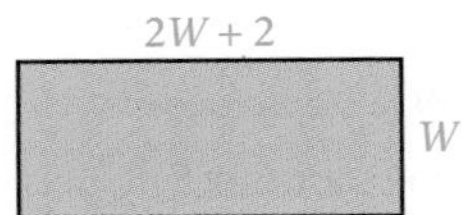

45. **Integers** The sum of the squares of three consecutive even integers is fifty-six. Find the three integers.

46. **Computers** An older computer requires 12 min longer to print the payroll than does a newer computer. Together the computers can print the payroll in 8 min. Find the time required for the new computer, working alone, to print the payroll.

47. **Uniform Motion** A car travels 200 mi. A second car, making the same trip, travels 10 mph faster than the first car and makes the trip in 1 h less time. Find the speed of each car.

CHAPTER 10

TEST

1. Solve by factoring: $3x^2 + 10x = 8$

2. Solve by factoring: $6x^2 - 5x - 6 = 0$

3. Write a quadratic equation that has integer coefficients and has solutions 3 and -3.

4. Write a quadratic equation that has integer coefficients and has solutions $\frac{1}{2}$ and -4.

5. Solve by taking square roots: $3(x - 2)^2 - 24 = 0$

6. Solve by completing the square: $x^2 - 6x - 2 = 0$

7. Solve by completing the square: $3x^2 - 6x = 2$

8. Solve by using the quadratic formula: $2x^2 - 2x = 1$

9. Solve by using the quadratic formula: $x^2 + 4x + 12 = 0$

10. Solve: $2x + 7x^{1/2} - 4 = 0$

11. Solve: $x^4 - 4x^2 + 3 = 0$

12. Solve: $\sqrt{2x + 1} + 5 = 2x$

13. Solve: $\sqrt{x-2} = \sqrt{x} - 2$

14. Solve: $\dfrac{2x}{x-3} + \dfrac{5}{x-1} = 1$

15. Solve and graph the solution set of $(x-2)(x+4)(x-4) < 0$.

16. Solve and graph the solution set of $\dfrac{2x-3}{x+4} \le 0$.

17. Use the discriminant to determine whether $9x^2 + 24x = -16$ has two equal real number solutions, two unequal real number solutions, or two complex number solutions.

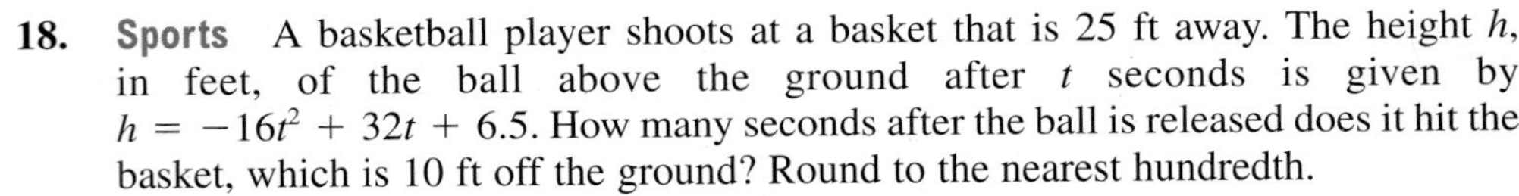

18. **Sports** A basketball player shoots at a basket that is 25 ft away. The height h, in feet, of the ball above the ground after t seconds is given by $h = -16t^2 + 32t + 6.5$. How many seconds after the ball is released does it hit the basket, which is 10 ft off the ground? Round to the nearest hundredth.

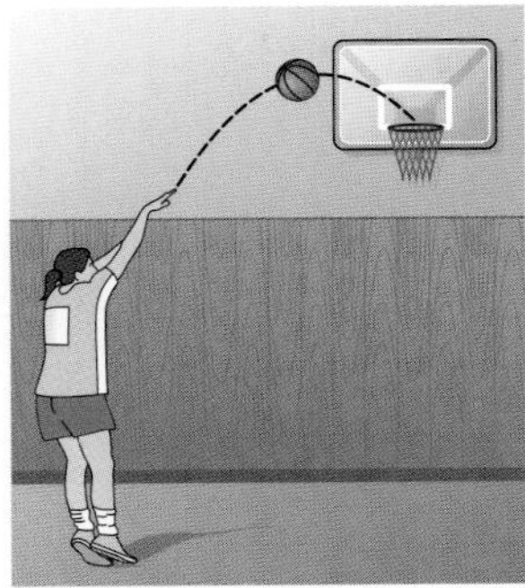

19. **Woodworking** It takes Clive 6 h longer to stain a bookcase than it does Cora. Working together, they can stain the bookcase in 4 h. Working alone, how long would it take Cora to stain the bookcase?

20. **Uniform Motion** The rate of a river's current is 2 mph. A canoe was rowed 6 mi down the river and back in 4 h. Find the rowing rate in calm water.

CUMULATIVE REVIEW EXERCISES

1. Evaluate $2a^2 - b^2 \div c^2$ when $a = 3$, $b = -4$, and $c = -2$.

2. Solve: $|3x - 2| < 8$

3. Find the volume of a cylinder with a height of 6 m and a radius of 3 m. Give the exact measure.

4. Given $f(x) = \frac{2x - 3}{x^2 - 1}$, find $f(-2)$.

5. Find the slope of the line containing the points $(3, -4)$ and $(-1, 2)$.

6. Find the x- and y-intercepts of the graph of $6x - 5y = 15$.

7. Find the equation of the line that contains the point $(1, 2)$ and is parallel to the line $x - y = 1$.

8. Solve the system of equations.

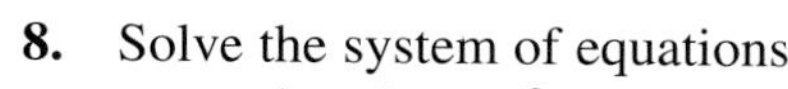

$$\begin{aligned} x + y + z &= 2 \\ -x + 2y - 3z &= -9 \\ x - 2y - 2z &= -1 \end{aligned}$$

9. Graph the solution set:

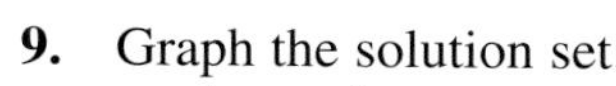

$$\begin{aligned} x + y &\le 3 \\ 2x - y &< 4 \end{aligned}$$

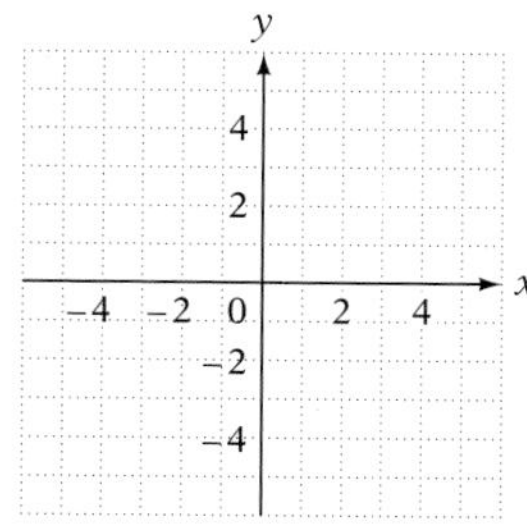

10. Triangles ABC and DEF are similar. Find the height of triangle DEF.

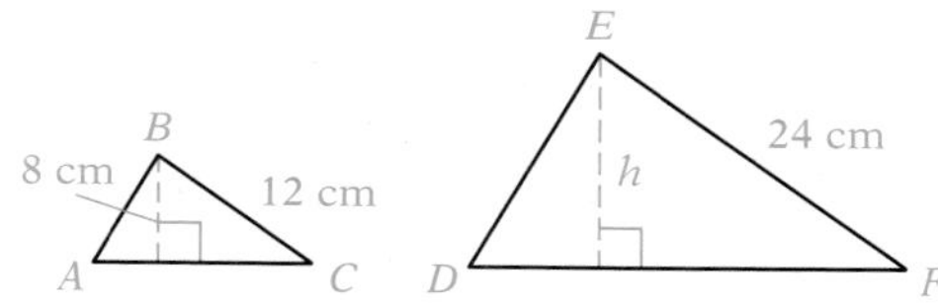

11. Divide: $(3x^3 - 13x^2 + 10) \div (3x - 4)$

12. Factor: $-3x^3y + 6x^2y^2 - 9xy^3$

13. Factor: $6x^2 - 7x - 20$

14. Multiply: $\frac{x^2 + 2x + 1}{8x^2 + 8x} \cdot \frac{4x^3 - 4x^2}{x^2 - 1}$

15. Solve: $\frac{x}{x+2} - \frac{4x}{x+3} = 1$

16. Solve $S = \frac{n}{2}(a + b)$ for b.

17. Multiply: $a^{-1/2}(a^{1/2} - a^{3/2})$

18. Multiply: $-2i(7 - 4i)$

19. Solve: $\sqrt{3x + 1} - 1 = x$

20. Solve: $x^4 - 6x^2 + 8 = 0$

21. **Mechanics** A piston rod for an automobile is $9\frac{3}{8}$ in. with a tolerance of $\frac{1}{64}$ in. Find the lower and upper limits of the length of the piston rod.

22. **Geometry** The base of a triangle is $(x + 8)$ feet. The height is $(2x - 4)$ feet. Find the area of the triangle in terms of the variable x.

23. **Depreciation** The graph shows the relationship between the value of a building and the time, in years, since depreciation began. Find the slope of the line between the two points shown on the graph. Write a sentence that states the meaning of the slope.

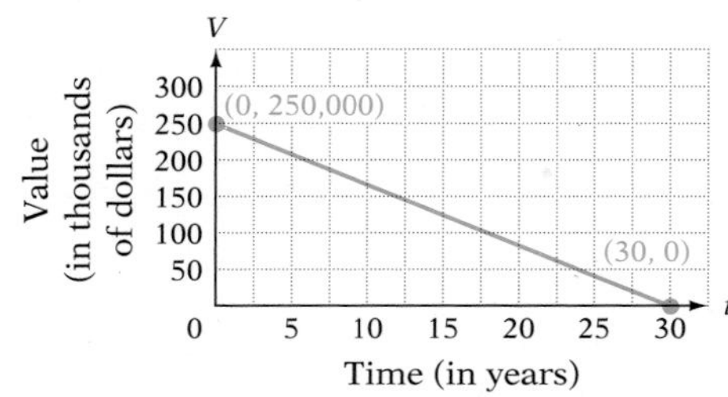

24. **Home Maintenance** How high on a house will a 17-foot ladder reach when the bottom of the ladder is 8 ft from the house?

25. Use the discriminant to determine whether $2x^2 + 4x + 3 = 0$ has one real number solution, two real number solutions, or two complex number solutions.

CHAPTER

11

Functions and Relations

Don Smith/The Image Bank/Getty Images

OBJECTIVES

SECTION 11.1

A To graph a quadratic function

B To find the x-intercepts of a parabola

C To find the minimum or maximum of a quadratic function

D To solve application problems

SECTION 11.2

A To graph functions

SECTION 11.3

A To perform operations on functions

B To find the composition of two functions

SECTION 11.4

A To determine whether a function is one-to-one

B To find the inverse of a function

ARE YOU READY?

Take the Chapter 11 Prep Test to find out if you are ready to learn to:

- Graph functions
- Find the minimum and maximum values of a quadratic function
- Perform operations on functions
- Find the composition of two functions
- Determine whether a function is one-to-one
- Find the inverse of a function

PREP TEST

Do these exercises to prepare for Chapter 11.

1. Evaluate $-\frac{b}{2a}$ for $b = -4$ and $a = 2$.

2. Given $y = -x^2 + 2x + 1$, find the value of y when $x = -2$.

3. Given $f(x) = x^2 - 3x + 2$, find $f(-4)$.

4. Evaluate $p(r) = r^2 - 5$ when $r = 2 + h$.

5. Solve: $0 = 3x^2 - 7x - 6$

6. Solve by using the quadratic formula: $0 = x^2 - 4x + 1$

7. Solve $x = 2y + 4$ for y.

8. Find the domain and range of the relation $\{(-2, 4), (3, 5), (4, 6), (6, 5)\}$. Is the relation a function?

9. Graph: $x = -2$

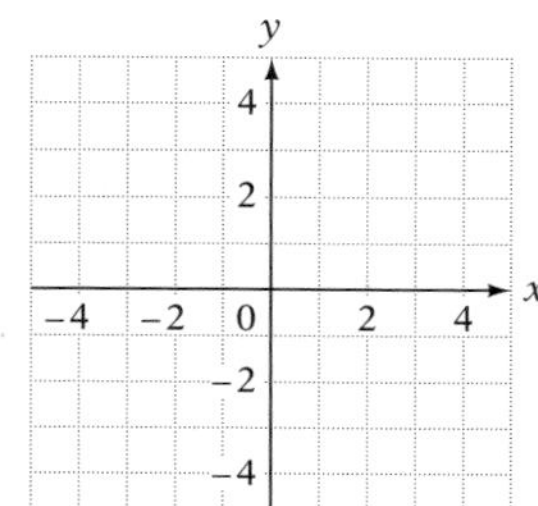

SECTION

11.1 Properties of Quadratic Functions

OBJECTIVE A To graph a quadratic function 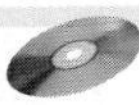

Recall that a linear function is one that can be expressed by the equation $f(x) = mx + b$. The graph of a linear function has certain characteristics. It is a straight line with slope m and y-intercept $(0, b)$. A **quadratic function** is one that can be expressed by the equation $f(x) = ax^2 + bx + c,\ a \neq 0$. The graph of this function, called a **parabola,** also has certain characteristics. The graph of a quadratic function can be drawn by finding ordered pairs that belong to the function.

Take Note

Sometimes the value of the independent variable is called the **input** because it is *put in* place of the independent variable. The result of evaluating the function is called the **output.**

An **input/output table** shows the results of evaluating a function for various values of the independent variable. An input/output table for $f(x) = x^2 - 2x - 3$ is shown at the right.

HOW TO • 1 Graph $f(x) = x^2 - 2x - 3$.

By evaluating the function for various values of x, find enough ordered pairs to determine the shape of the graph.

x	$f(x) = x^2 - 2x - 3$	$f(x)$	(x, y)
-2	$f(-2) = (-2)^2 - 2(-2) - 3$	5	$(-2, 5)$
-1	$f(-1) = (-1)^2 - 2(-1) - 3$	0	$(-1, 0)$
0	$f(0) = (0)^2 - 2(0) - 3$	-3	$(0, -3)$
1	$f(1) = (1)^2 - 2(1) - 3$	-4	$(1, -4)$
2	$f(2) = (2)^2 - 2(2) - 3$	-3	$(2, -3)$
3	$f(3) = (3)^2 - 2(3) - 3$	0	$(3, 0)$
4	$f(4) = (4)^2 - 2(4) - 3$	5	$(4, 5)$

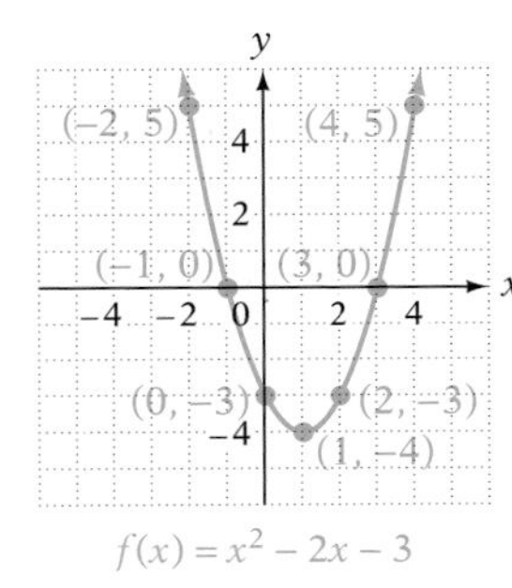

$f(x) = x^2 - 2x - 3$

Because the value of $f(x) = x^2 - 2x - 3$ is a real number for all values of x, the domain of f is all real numbers. From the graph, it appears that no value of y is less than -4. Thus the range is $\{y \mid y \geq -4\}$. The range can also be determined algebraically, as shown below, by completing the square.

Take Note

In completing the square, 1 is both added and subtracted. Because $1 - 1 = 0$, the expression $x^2 - 2x - 3$ is not changed. Note that

$$\begin{aligned}&(x - 1)^2 - 4\\ &\quad = (x^2 - 2x + 1) - 4\\ &\quad = x^2 - 2x - 3\end{aligned}$$

which is the original expression.

$$\begin{aligned} f(x) &= x^2 - 2x - 3 \\ &= (x^2 - 2x) - 3 \\ &= (x^2 - 2x + 1) - 1 - 3 \\ &= (x - 1)^2 - 4 \end{aligned}$$

- Group the variable terms.
- Complete the square on $x^2 - 2x$. Add and subtract $\left[\frac{1}{2}(-2)\right]^2 = 1$ to and from $x^2 - 2x$.
- Factor and combine like terms.

Because the square of a real number is always nonnegative, we have

$$(x - 1)^2 \geq 0$$

$$(x - 1)^2 - 4 \geq -4$$ • Subtract 4 from each side of the inequality.

$$f(x) \geq -4$$ • $f(x) = (x - 1)^2 - 4$

$$y \geq -4$$

From the last inequality, the range is $\{y \mid y \geq -4\}$.

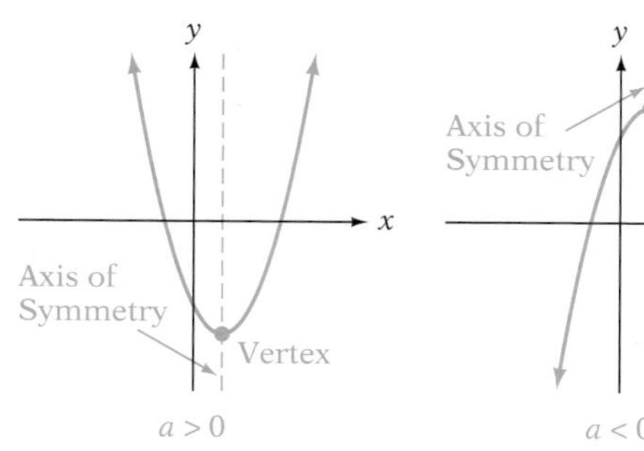

In general, the graph of $f(x) = ax^2 + bx + c,\ a \neq 0$, resembles a "cup" shape as shown at the left. When $a > 0$, the parabola opens up and the **vertex of the parabola** is the point with the smallest y-coordinate. When $a < 0$, the parabola opens down and the vertex is the point with the largest y-coordinate. The **axis of symmetry of a parabola** is the vertical line that passes through the vertex of the parabola and is parallel to the y-axis. To understand the axis of symmetry, think of folding the graph along that vertical line. The two portions of the graph will match up.

The vertex and axis of symmetry of a parabola can be found by completing the square.

HOW TO • 2 Find the coordinates of the vertex and the equation of the axis of symmetry for the graph of $F(x) = x^2 + 4x + 3$.

To find the coordinates of the vertex, complete the square.

$F(x) = x^2 + 4x + 3$

$= (x^2 + 4x) + 3$ • Group the variable terms.

$= (x^2 + 4x + 4) - 4 + 3$ • Complete the square on $x^2 + 4x$. Add and subtract $\left[\frac{1}{2}(4)\right]^2 = 4$ to and from $x^2 + 4x$.

$= (x + 2)^2 - 1$ • Factor and combine like terms.

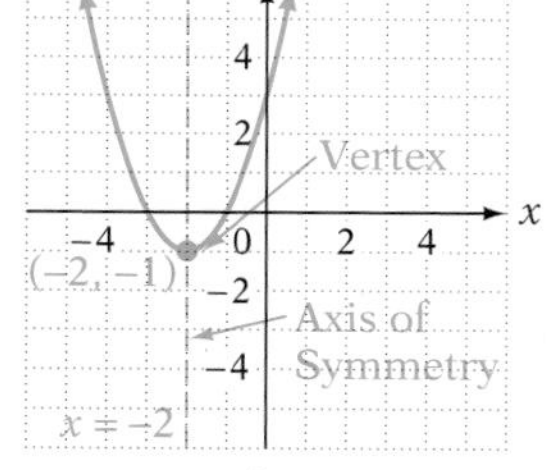

$F(x) = x^2 + 4x + 3$

Because a, the coefficient of x^2, is positive ($a = 1$), the parabola opens up and the vertex is the point with the smallest y-coordinate. Because $(x + 2)^2 \geq 0$ for all values of x, the smallest y-coordinate occurs when $(x + 2)^2 = 0$. The quantity $(x + 2)^2$ is equal to zero when $x = -2$. Therefore, the x-coordinate of the vertex is -2.

To find the y-coordinate of the vertex, evaluate the function at $x = -2$.

$F(x) = (x + 2)^2 - 1$

$F(-2) = (-2 + 2)^2 - 1 = -1$

The y-coordinate of the vertex is -1.

From the results above, the coordinates of the vertex are $(-2, -1)$.

The axis of symmetry is the vertical line that passes through the vertex. The equation of the vertical line that passes through the point $(-2, -1)$ is $x = -2$. The equation of the axis of symmetry is $x = -2$.

By following the process illustrated in the preceding example and completing the square on $f(x) = ax^2 + bx + c$, we can find a formula for the coordinates of the vertex and the equation of the axis of symmetry of a parabola.

Vertex and Axis of Symmetry of a Parabola

Let $f(x) = ax^2 + bx + c$ be the equation of a parabola. The coordinates of the vertex are $\left(-\frac{b}{2a}, f\left(-\frac{b}{2a}\right)\right)$. The equation of the axis of symmetry is $x = -\frac{b}{2a}$.

HOW TO • 3 Find the coordinates of the vertex of the parabola whose equation is $g(x) = -2x^2 + 3x + 1$. Then graph the equation.

x-coordinate of the vertex: $-\dfrac{b}{2a} = -\dfrac{3}{2(-2)} = \dfrac{3}{4}$

• From the equation $g(x) = -2x^2 + 3x + 1$, $a = -2$, $b = 3$.

y-coordinate of the vertex: $g(x) = -2x^2 + 3x + 1$

$$g\left(\frac{3}{4}\right) = -2\left(\frac{3}{4}\right)^2 + 3\left(\frac{3}{4}\right) + 1$$

$$g\left(\frac{3}{4}\right) = \frac{17}{8}$$

• Evaluate the function at the value of the x-coordinate of the vertex.

The vertex is $\left(\frac{3}{4}, \frac{17}{8}\right)$.

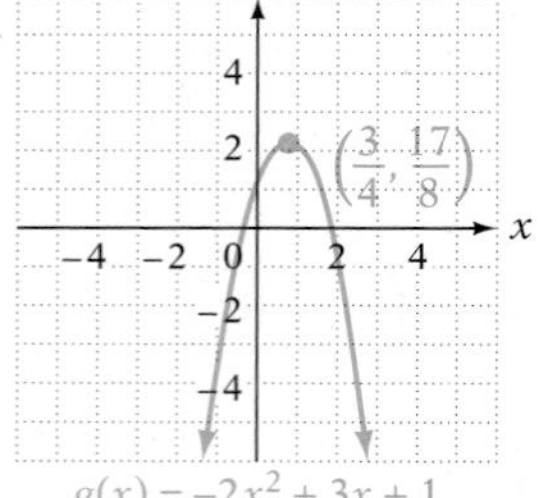

$g(x) = -2x^2 + 3x + 1$

Because a is negative, the graph opens down. Find a few ordered pairs that belong to the function and then sketch the graph, as shown at the left.

Take Note
Once the coordinates of the vertex are found, the range of the quadratic function can be determined.

Once the y-coordinate of the vertex is known, the range of the function can be determined. Here, the graph of g opens down, so the y-coordinate of the vertex is the largest value of y. Therefore, the range of g is $\left\{y \mid y \le \frac{17}{8}\right\}$. The domain is $\{x \mid x \in \text{real numbers}\}$.

EXAMPLE • 1

Find the coordinates of the vertex and the equation of the axis of symmetry for the parabola whose equation is $y = -x^2 + 4x + 1$. Then graph the equation.

Solution

x-coordinate of vertex:

$-\dfrac{b}{2a} = -\dfrac{4}{2(-1)} = 2$

y-coordinate of vertex:

$y = -x^2 + 4x + 1$
$= -(2)^2 + 4(2) + 1$
$= 5$

Vertex: $(2, 5)$
Axis of symmetry: $x = 2$

YOU TRY IT • 1

Find the coordinates of the vertex and the equation of the axis of symmetry for the parabola whose equation is $y = 4x^2 + 4x + 1$. Then graph the equation.

Your solution

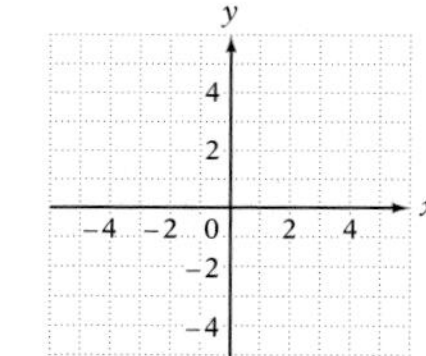

EXAMPLE • 2

Find the domain and range of $f(x) = 0.5x^2 - 3$. Then graph the equation.

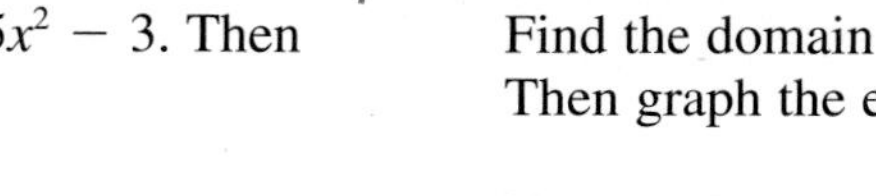

Solution

x-coordinate of vertex:

$-\dfrac{b}{2a} = -\dfrac{0}{2(0.5)} = 0$

y-coordinate of vertex:

$f(x) = 0.5x^2 - 3$
$f(0) = 0.5(0)^2 - 3$
$= -3$

Vertex: $(0, -3)$
The domain is $\{x \mid x \in \text{real numbers}\}$.
The range is $\{y \mid y \ge -3\}$.

YOU TRY IT • 2

Find the domain and range of $f(x) = -x^2 - 2x - 1$. Then graph the equation.

Your solution

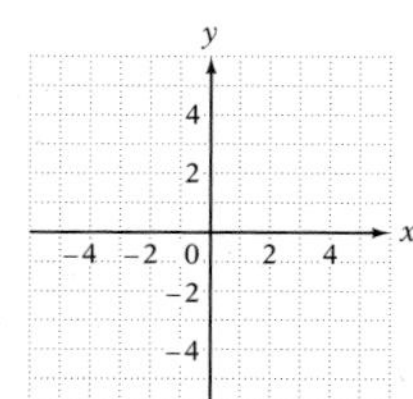

Solutions on p. S31

OBJECTIVE B To find the x-intercepts of a parabola

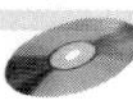

Recall that a point at which a graph crosses the x- or y-axis is called an *intercept* of the graph. The x-intercepts of the graph of an equation occur when $y = 0$; the y-intercepts occur when $x = 0$.

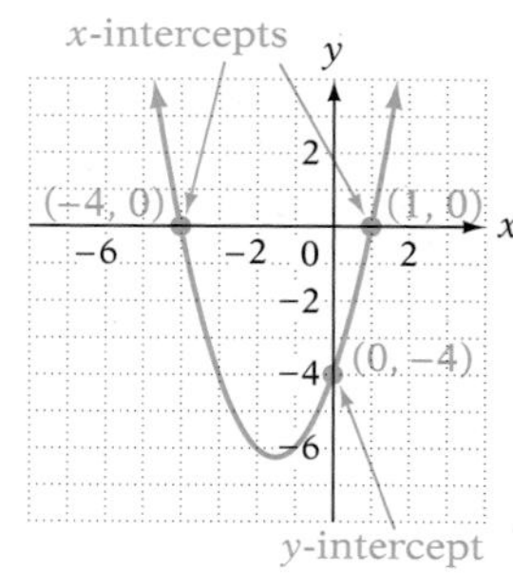

The graph of $y = x^2 + 3x - 4$ is shown at the right. The points whose coordinates are $(-4, 0)$ and $(1, 0)$ are the x-intercepts of the graph. The point whose coordinates are $(0, -4)$ is the y-intercept of the graph.

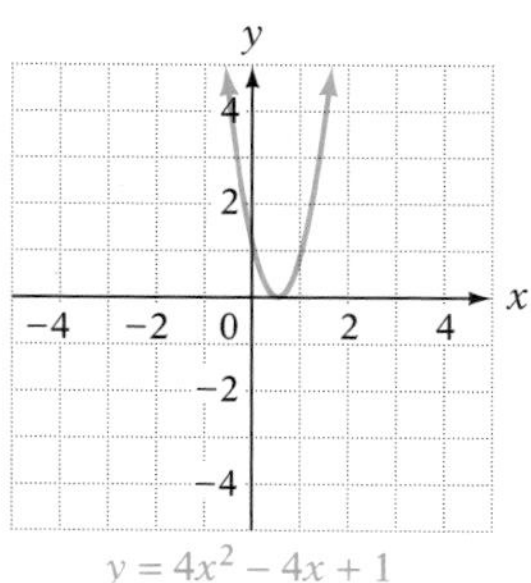

$y = 4x^2 - 4x + 1$

HOW TO 4 Find the x-intercepts of the graph of $y = 4x^2 - 4x + 1$.

To find the x-intercepts, let $y = 0$ and then solve for x.

$$y = 4x^2 - 4x + 1$$
$$0 = 4x^2 - 4x + 1 \quad \text{• Let } y = 0.$$
$$0 = (2x - 1)(2x - 1) \quad \text{• Solve for } x \text{ by factoring and using the Principle of Zero Products.}$$

$$2x - 1 = 0 \qquad 2x - 1 = 0$$
$$2x = 1 \qquad 2x = 1$$
$$x = \frac{1}{2} \qquad x = \frac{1}{2}$$

The x-intercept is $\left(\frac{1}{2}, 0\right)$.

Integrating Technology

See the Keystroke Guide: *Zero* for instructions on using a graphing calculator to find the x-intercepts of the graph of a parabola.

In the preceding example, the parabola has only one x-intercept. In this case, the parabola is said to be *tangent* to the x-axis at $x = \frac{1}{2}$.

HOW TO 5 Find the x-intercepts of the graph of $y = 2x^2 - x - 6$.

To find the x-intercepts, let $y = 0$ and solve for x.

$$y = 2x^2 - x - 6$$
$$0 = 2x^2 - x - 6 \quad \text{• Let } y = 0.$$
$$0 = (2x + 3)(x - 2) \quad \text{• Solve for } x \text{ by factoring and using the Principle of Zero Products.}$$

$$2x + 3 = 0 \qquad x - 2 = 0$$
$$x = -\frac{3}{2} \qquad x = 2$$

The x-intercepts are $\left(-\frac{3}{2}, 0\right)$ and $(2, 0)$.

Take Note

A zero of a function is the x-coordinate of an x-intercept of the graph of the function. Because the x-intercepts of the graph of $f(x) = 2x^2 - x - 6$ are $\left(-\frac{3}{2}, 0\right)$ and $(2, 0)$, the zeros are $-\frac{3}{2}$ and 2.

If the equation above, $y = 2x^2 - x - 6$, is written in function notation as $f(x) = 2x^2 - x - 6$, then to find the x-intercepts you would let $f(x) = 0$ and solve for x. A value of x for which $f(x) = 0$ is called a **zero of a function.** Thus $-\frac{3}{2}$ and 2 are *zeros* of $f(x) = 2x^2 - x - 6$.

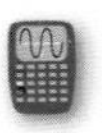

Integrating Technology

See the Keystroke Guide: *Zero* for instructions on using a graphing calculator to find the zeros of a function.

Take Note

The graph of $f(x) = x^2 - 2x - 1$ is shown at the right. Note that the zeros are the x-intercepts of the graph of f.

HOW TO 6 Find the zeros of $f(x) = x^2 - 2x - 1$.

To find the zeros, let $f(x) = 0$ and solve for x.

$$f(x) = x^2 - 2x - 1$$
$$0 = x^2 - 2x - 1$$
$$x = \frac{-b \pm \sqrt{b^2 - 4ac}}{2a}$$
• Because $x^2 - 2x - 1$ does not easily factor, use the quadratic formula to solve for x.
$$= \frac{-(-2) \pm \sqrt{(-2)^2 - 4(1)(-1)}}{2(1)}$$
• $a = 1, b = -2, c = -1$
$$= \frac{2 \pm \sqrt{4 + 4}}{2} = \frac{2 \pm \sqrt{8}}{2}$$
$$= \frac{2 \pm 2\sqrt{2}}{2}$$
$$= 1 \pm \sqrt{2}$$

The zeros of the function are $1 - \sqrt{2}$ and $1 + \sqrt{2}$.

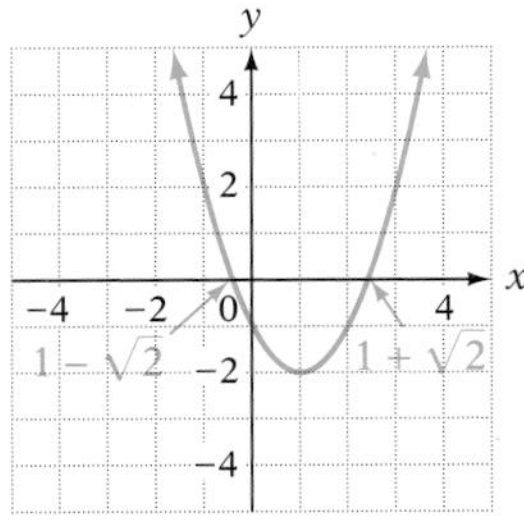

The preceding examples suggest that there is a relationship among the x-intercepts of the graph of a function, the zeros of a function, and the solutions of an equation. In fact, those three concepts are different ways of discussing the same number. The choice depends on the focus of the discussion. If we are discussing graphing, then the intercept is our focus; if we are discussing functions, then the zero of the function is our focus; and if we are discussing equations, the solution of the equation is our focus.

The x-axis is a real number line. Therefore, if a graph crosses the x-axis, it must have a real number zero. It is also true that if a graph does not cross the x-axis, it does not have a real number zero. The zeros in this case are complex numbers.

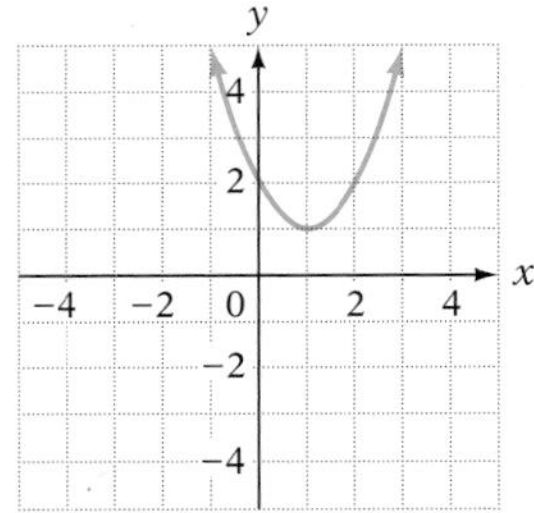

The graph of $f(x) = x^2 - 2x + 2$ is shown at the right. The graph has no x-intercepts. Thus f has no real zeros.

Using the quadratic formula, the complex number solutions of $x^2 - 2x + 2 = 0$ are $1 - i$ and $1 + i$. Thus the zeros of $f(x) = x^2 - 2x + 2$ are the complex numbers $1 - i$ and $1 + i$.

Tips for Success

The paragraph at the right begins with the word "Recall." This signals that the content refers to material presented earlier in the text. The ideas presented here will be more meaningful if you return to the discussion of the word *discriminant* on page 586 and review the concepts presented there.

Recall that the *discriminant* of $ax^2 + bx + c$ is the expression $b^2 - 4ac$ and that this expression can be used to determine whether $ax^2 + bx + c = 0$ has zero, one, or two real number solutions. Because there is a connection between the solutions of $ax^2 + bx + c = 0$ and the x-intercepts of the graph of $y = ax^2 + bx + c$, the discriminant can be used to determine the number of x-intercepts of a parabola.

The Effect of the Discriminant on the Number of *x*-Intercepts of the Graph of a Parabola

1. If $b^2 - 4ac = 0$, the parabola has one x-intercept.
2. If $b^2 - 4ac > 0$, the parabola has two x-intercepts.
3. If $b^2 - 4ac < 0$, the parabola has no x-intercepts.

HOW TO • 7 Use the discriminant to determine the number of x-intercepts of the graph of the parabola whose equation is $y = 2x^2 - x + 2$.

$b^2 - 4ac$ • Evaluate the discriminant.
$(-1)^2 - 4(2)(2) = 1 - 16 = -15$ • $a = 2, b = -1, c = 2$
$-15 < 0$

The discriminant is less than zero. Therefore, the parabola has no x-intercepts.

EXAMPLE • 3

Find the x-intercepts of the graph of $y = 2x^2 - 5x + 2$.

Solution

$y = 2x^2 - 5x + 2$
$0 = 2x^2 - 5x + 2$ • Let $y = 0$.
$0 = (2x - 1)(x - 2)$ • Solve for x by factoring.

$2x - 1 = 0 \qquad x - 2 = 0$
$2x = 1 \qquad x = 2$
$x = \frac{1}{2}$

The x-intercepts are $\left(\frac{1}{2}, 0\right)$ and $(2, 0)$.

YOU TRY IT • 3

Find the x-intercepts of the graph of $y = x^2 + 3x + 4$.

Your solution

EXAMPLE • 4

Find the zeros of $f(x) = x^2 + 4x + 5$.

Solution

$f(x) = x^2 + 4x + 5$
$0 = x^2 + 4x + 5$ • Let $f(x) = 0$.

$x = \frac{-b \pm \sqrt{b^2 - 4ac}}{2a}$ • Use the quadratic formula.

$= \frac{-4 \pm \sqrt{4^2 - 4(1)(5)}}{2(1)}$ • $a = 1, b = 4, c = 5$

$= \frac{-4 \pm \sqrt{16 - 20}}{2} = \frac{-4 \pm \sqrt{-4}}{2}$

$= \frac{-4 \pm 2i}{2} = -2 \pm i$

The zeros of the function are $-2 + i$ and $-2 - i$.

YOU TRY IT • 4

Find the zeros of $g(x) = x^2 - x + 6$.

Your solution

EXAMPLE • 5

Use the discriminant to determine the number of x-intercepts of the graph of $y = x^2 - 6x + 9$.

Solution

$b^2 - 4ac$ • Evaluate the discriminant.
$(-6)^2 - 4(1)(9)$ • $a = 1, b = -6, c = 9$
$= 36 - 36 = 0$

The discriminant is equal to zero.
The parabola has one x-intercept.

YOU TRY IT • 5

Use the discriminant to determine the number of x-intercepts of the graph of $y = x^2 - x - 6$.

Your solution

Solutions on p. S32

OBJECTIVE C To find the minimum or maximum of a quadratic function 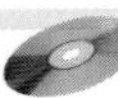

Tips for Success
After studying this objective, you should be able to describe in words what the minimum or maximum value of a quadratic function is, draw a graph of a quadratic function with either a maximum or a minimum value, determine from a quadratic equation whether it has a minimum or a maximum value, and calculate algebraically the minimum or maximum value of a quadratic function.

The graph of $f(x) = x^2 - 2x + 3$ is shown at the right. Because a is positive, the parabola opens up. The vertex of the parabola is the lowest point on the parabola. It is the point that has the minimum y-coordinate. This point represents the **minimum value of the function.**

y
Vertex (1, 2)
Minimum y-coordinate = 2
x

The graph of $f(x) = -x^2 + 2x + 1$ is shown at the right. Because a is negative, the parabola opens down. The vertex of the parabola is the highest point on the parabola. It is the point that has the maximum y-coordinate. This point represents the **maximum value of the function.**

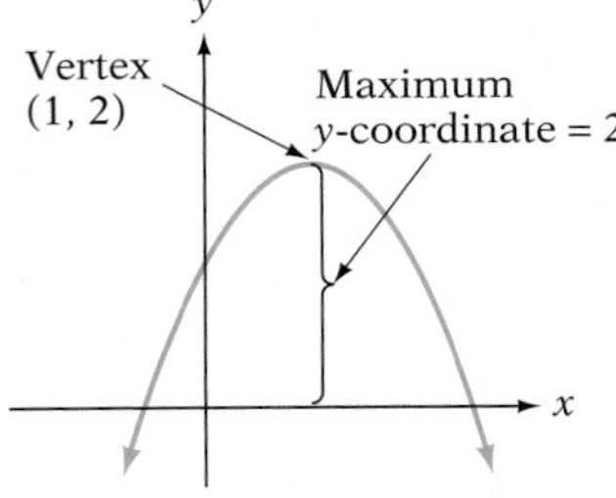

To find the minimum or maximum value of a quadratic function, first find the x-coordinate of the vertex. Then evaluate the function at that value.

HOW TO • 8 Find the maximum or minimum value of $f(x) = -2x^2 + 4x + 3$.

$$x = -\frac{b}{2a} = -\frac{4}{2(-2)} = 1$$

• Find the x-coordinate of the vertex. $a = -2, b = 4$

$$\begin{aligned} f(x) &= -2x^2 + 4x + 3 \\ f(1) &= -2(1)^2 + 4(1) + 3 \\ &= -2 + 4 + 3 \\ &= 5 \end{aligned}$$

• Evaluate the function at $x = 1$.

The maximum value of the function is 5.

• Because $a < 0$, the graph of f opens down. Therefore, the function has a maximum value.

Integrating Technology
See the Projects and Group Activities at the end of this chapter for instructions on using a graphing calculator to find the minimum or maximum value of a function.

EXAMPLE • 6

Find the maximum or minimum value of $f(x) = 2x^2 - 3x + 1$.

Solution

$$x = -\frac{b}{2a} = -\frac{-3}{2(2)} = \frac{3}{4}$$

• The x-coordinate of the vertex

$$\begin{aligned} f(x) &= 2x^2 - 3x + 1 \\ f\left(\frac{3}{4}\right) &= 2\left(\frac{3}{4}\right)^2 - 3\left(\frac{3}{4}\right) + 1 \\ &= \frac{9}{8} - \frac{9}{4} + 1 = -\frac{1}{8} \end{aligned}$$

• $x = \frac{3}{4}$

Because a is positive, the graph opens up. The function has a minimum value.
The minimum value of the function is $-\frac{1}{8}$.

YOU TRY IT • 6

Find the maximum or minimum value of $f(x) = -3x^2 + 4x - 1$.

Your solution

Solution on p. S32

OBJECTIVE D To solve application problems

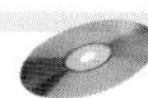

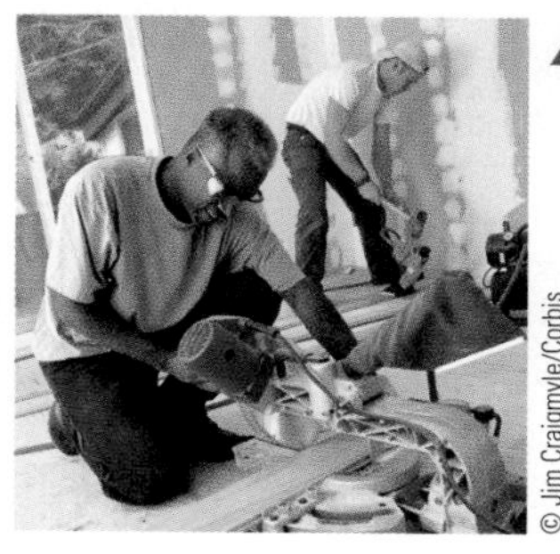

© Jim Craigmyle/Corbis

Point of Interest

Calculus is a branch of mathematics that demonstrates, among other things, how to find the maximum or minimum values of functions other than quadratic functions. These are very important problems in applied mathematics. For instance, an automotive engineer wants to design a car whose shape will *minimize* the effects of air resistance. The same engineer tries to *maximize* the efficiency of the car's engine. Similarly, an economist may try to determine what business practices will *minimize* cost and *maximize* profit.

HOW TO 9 A carpenter is forming a rectangular floor for a storage shed. The perimeter of the rectangle is 44 ft. What dimensions of the rectangle would give the floor a maximum area? What is the maximum area?

We are given the perimeter of the rectangle, and we want to find the dimensions of the rectangle that will yield the maximum area for the floor. Use the equation for the perimeter of a rectangle.

$$P = 2L + 2W$$
$$44 = 2L + 2W$$ • $P = 44$
$$22 = L + W$$ • Divide both sides of the equation by 2.
$$22 - L = W$$ • Solve the equation for W.

Now use the equation for the area of a rectangle. Use substitution to express the area in terms of L.

$$A = LW$$
$$A = L(22 - L)$$ • From the equation above, $W = 22 - L$. Substitute $22 - L$ for W.
$$A = 22L - L^2$$ • The area of the rectangle is $22L - L^2$.

To find the length of the rectangle, find the L-coordinate of the vertex of the function $f(L) = -L^2 + 22L$.

$$L = -\frac{b}{2a} = -\frac{22}{2(-1)} = 11$$ • For the equation $f(L) = -L^2 + 22L$, $a = -1$ and $b = 22$.

The length of the rectangle is 11 ft.

To find the width, replace L in $22 - L$ by the L-coordinate of the vertex and evaluate.

$$W = 22 - L$$
$$W = 22 - 11 = 11$$ • Replace L by 11 and evaluate.

The width of the rectangle is 11 ft.

The dimensions of the rectangle that would give the floor a maximum area are 11 ft by 11 ft.

To find the maximum area of the floor, evaluate $f(L) = -L^2 + 22L$ at the L-coordinate of the vertex.

$$f(L) = -L^2 + 22L$$
$$f(11) = -(11)^2 + 22(11)$$ • Evaluate the function at 11.
$$= -121 + 242 = 121$$

The maximum area of the floor is 121 ft².

The graph of $f(L) = -L^2 + 22L$ is shown at the right. Note that the vertex of the parabola is (11, 121). For any value of L less than 11, the area of the floor will be less than 121 ft². For any value of L greater than 11, the area of the floor will be less than 121 ft². The maximum value of the function is 121, and the maximum value occurs when $L = 11$.

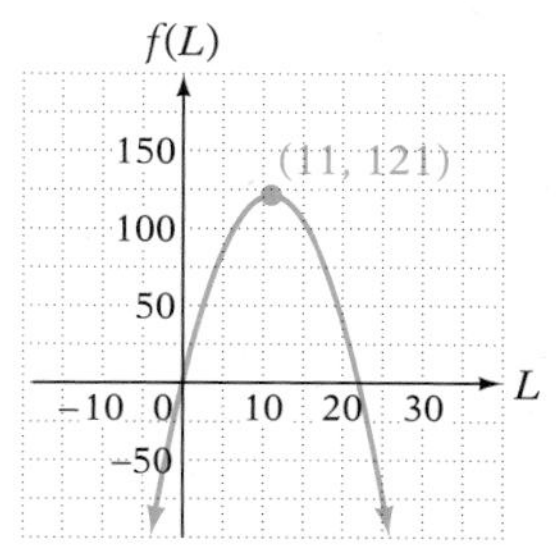

EXAMPLE • 7

A mining company has determined that the cost c, in dollars per ton, of mining a mineral is given by $c(x) = 0.2x^2 - 2x + 12$, where x is the number of tons of the mineral mined. Find the number of tons of the mineral that should be mined to minimize the cost. What is the minimum cost per ton?

Strategy

- To find the number of tons of the mineral that should be mined to minimize the cost, find the x-coordinate of the vertex.
- To find the minimum cost, evaluate $c(x)$ at the x-coordinate of the vertex.

Solution

$c(x) = 0.2x^2 - 2x + 12$ • $a = 0.2, b = -2, c = 12$

$x = -\frac{b}{2a} = -\frac{(-2)}{2(0.2)} = 5$ • The x-coordinate of the vertex

To minimize the cost, 5 tons of the mineral should be mined.

$c(x) = 0.2x^2 - 2x + 12$

$c(5) = 0.2(5)^2 - 2(5) + 12$ • $x = 5$

$= 5 - 10 + 12$

$= 7$

The minimum cost is $7 per ton.

Note: The graph of $c(x) = 0.2x^2 - 2x + 12$ is shown below. The vertex of the parabola is (5, 7). For any value of x less than 5, the cost per ton is greater than $7. For any value of x greater than 5, the cost per ton is greater than $7. Seven is the minimum value of the function, and the minimum value occurs when $x = 5$.

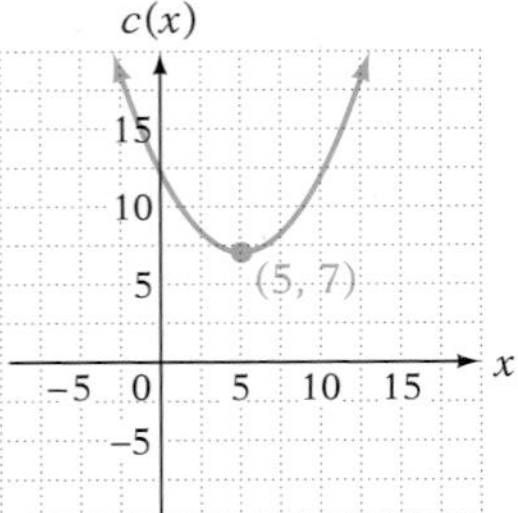

YOU TRY IT • 7

The height s, in feet, of a ball thrown straight up is given by $s(t) = -16t^2 + 64t$, where t is the time in seconds. Find the time it takes the ball to reach its maximum height. What is the maximum height?

Your strategy

Your solution

Solution on p. S32

EXAMPLE 8

Find two numbers whose difference is 10 and whose product is a minimum. What is the minimum product of the two numbers?

Strategy

- Let x represent one number. Because the difference between the two numbers is 10,

$$x + 10$$

represents the other number.
[*Note:* $(x + 10) - (x) = 10$]
Then their product is represented by

$$x(x + 10) = x^2 + 10x$$

- To find one of the two numbers, find the x-coordinate of the vertex of $f(x) = x^2 + 10x$.
- To find the other number, replace x in $x + 10$ by the x-coordinate of the vertex and evaluate.
- To find the minimum product, evaluate the function at the x-coordinate of the vertex.

Solution

$f(x) = x^2 + 10x$ • $a = 1,\ b = 10,\ c = 0$

$x = -\dfrac{b}{2a} = -\dfrac{10}{2(1)} = -5$ • One number is -5.

$x + 10$
$-5 + 10 = 5$ • The other number is 5.

The numbers are -5 and 5.

$$\begin{aligned} f(x) &= x^2 + 10x \\ f(-5) &= (-5)^2 + 10(-5) \\ &= 25 - 50 \\ &= -25 \end{aligned}$$

The minimum product of the two numbers is -25.

YOU TRY IT 8

A rectangular fence is being constructed along a stream to enclose a picnic area. If there are 100 ft of fence available, what dimensions of the rectangle will produce the maximum area for picnicking?

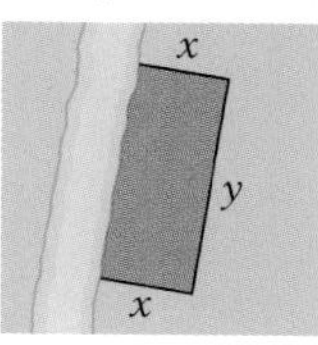

Your strategy

Your solution

Solution on p. S32

11.1 EXERCISES

OBJECTIVE A To graph a quadratic function

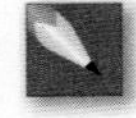
1. What is a quadratic function?

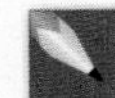
2. Describe the graph of a parabola.

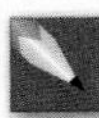
3. What is the vertex of a parabola?

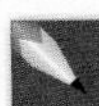
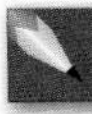
4. What is the axis of symmetry of the graph of a parabola?

5. The equation of the axis of symmetry of a parabola is $x = -5$. What is the x-coordinate of the vertex of the parabola?

6. The equation of the axis of symmetry of a parabola is $x = 8$. What is the x-coordinate of the vertex of the parabola?

7. The coordinates of the vertex of a parabola are $(7, -9)$. What is the equation of the axis of symmetry of the parabola?

8. The coordinates of the vertex of a parabola are $(-4, 10)$. What is the equation of the axis of symmetry of the parabola?

For Exercises 9 to 23, find the coordinates of the vertex and the equation of the axis of symmetry for the parabola given by the equation. Then graph the equation.

9. $y = x^2 - 2x - 4$

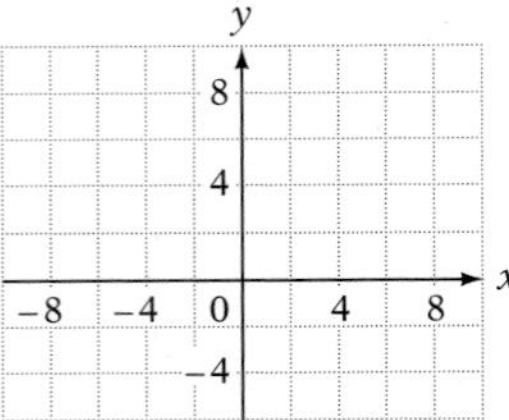

10. $y = x^2 + 4x - 4$

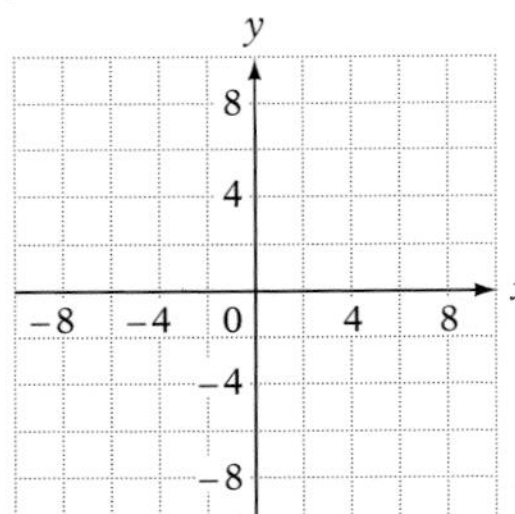

11. $y = -x^2 + 2x - 3$

12. $y = -x^2 + 4x - 5$

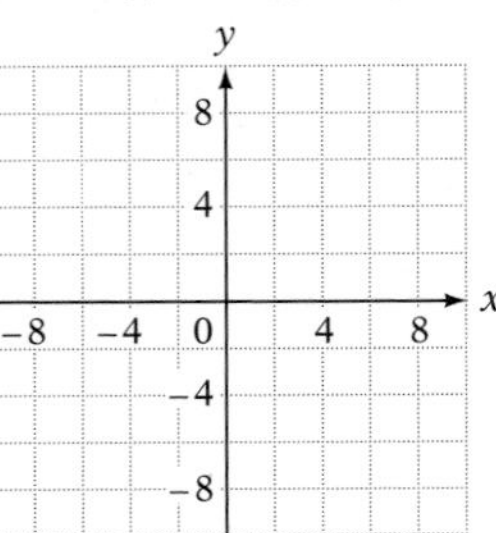

13. $f(x) = x^2 - x - 6$

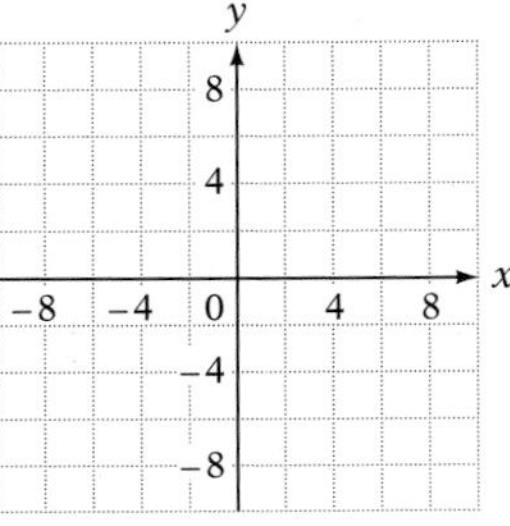

14. $G(x) = x^2 - x - 2$

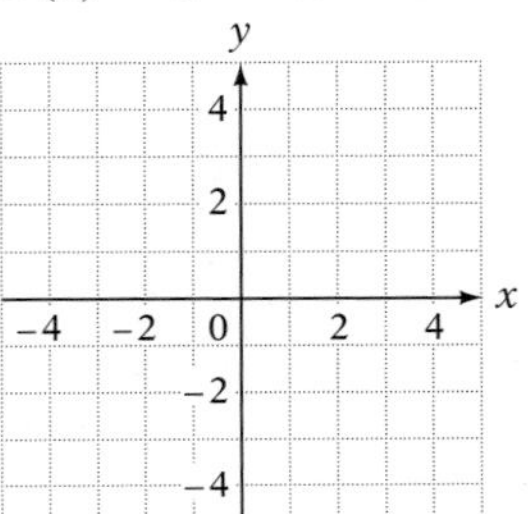

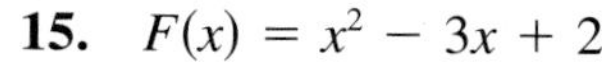

15. $F(x) = x^2 - 3x + 2$

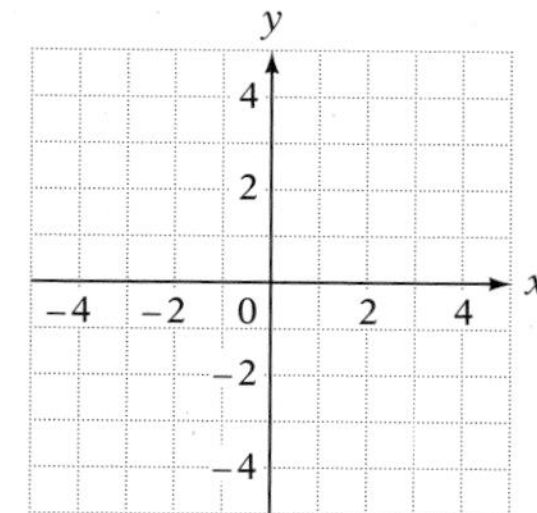

16. $y = 2x^2 - 4x + 1$

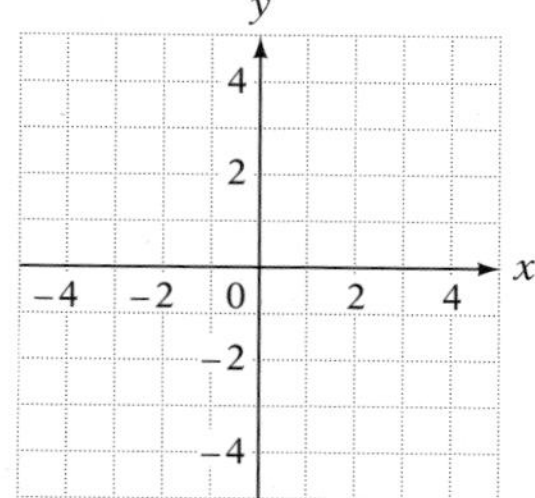

17. $y = -2x^2 + 6x$

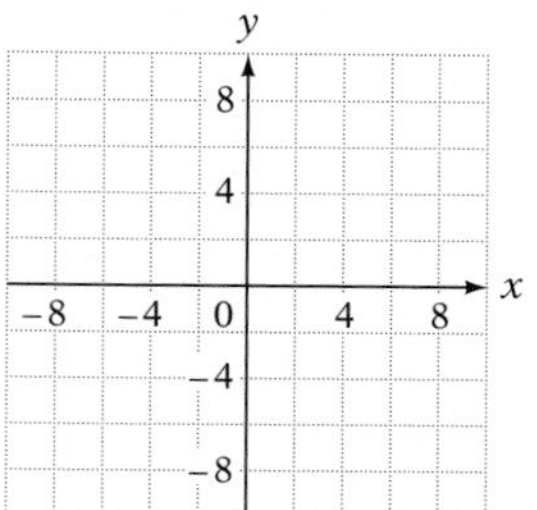

18. $y = \frac{1}{2}x^2 + 4$

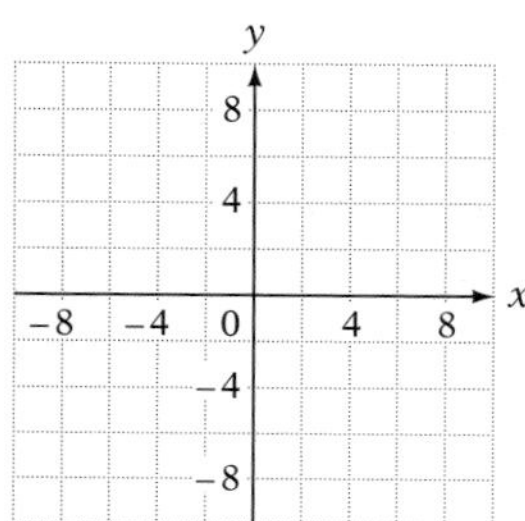

19. $y = -\frac{1}{4}x^2 - 1$

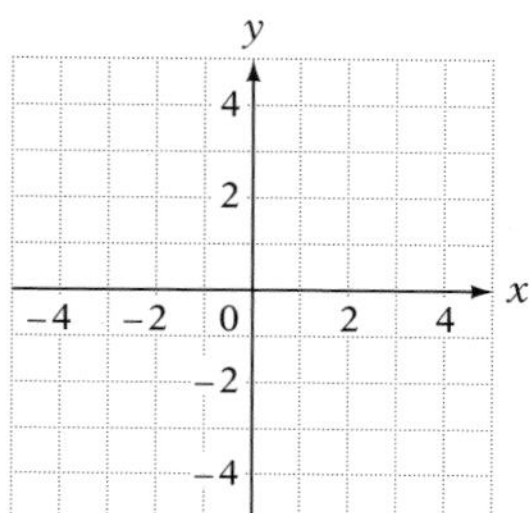

20. $h(x) = \frac{1}{2}x^2 - x + 1$

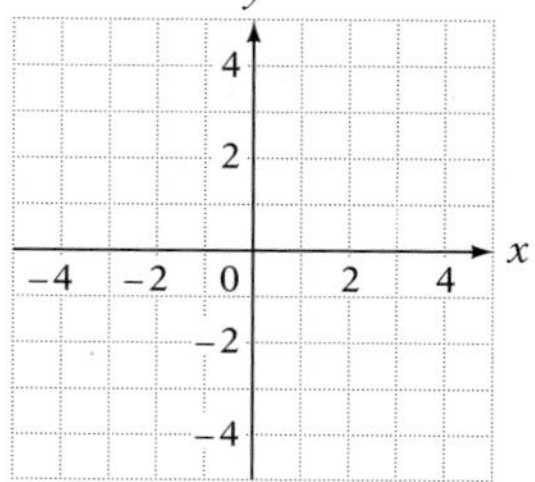

21. $P(x) = -\frac{1}{2}x^2 + 2x - 3$

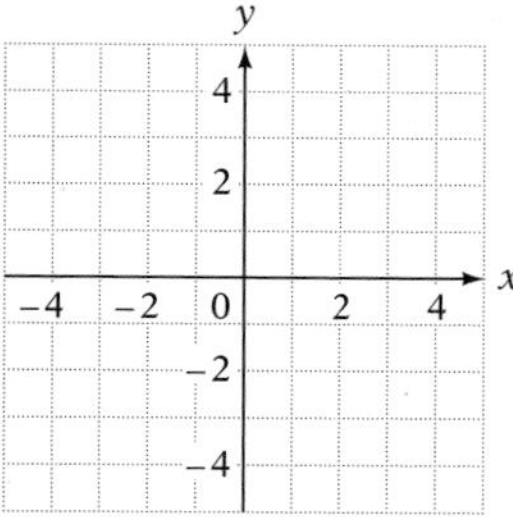

22. $y = \frac{1}{2}x^2 + 2x - 6$

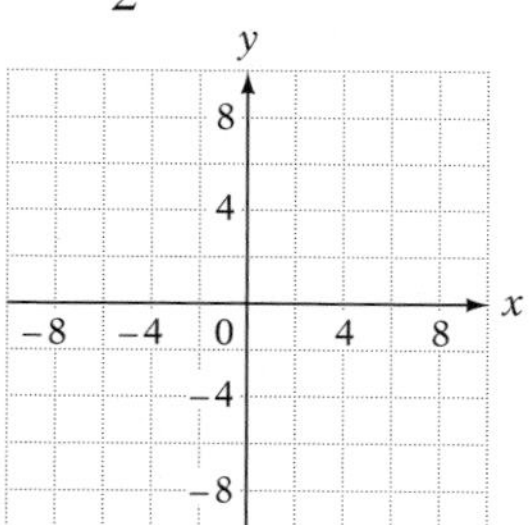

23. $y = -\frac{1}{2}x^2 + x - 3$

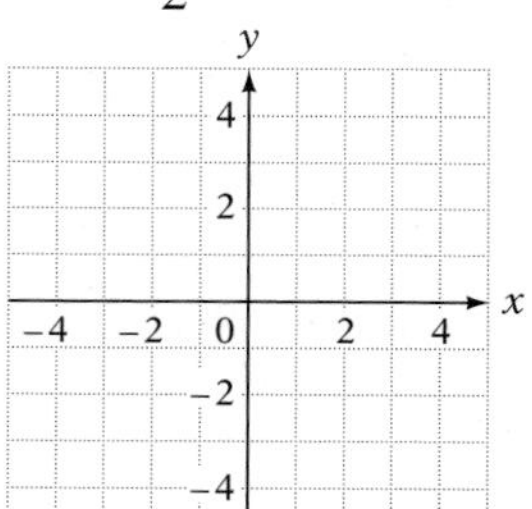

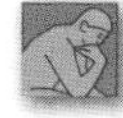

24. Is 3 in the range of the function given by $f(x) = (x + 2)^2 + 4$?

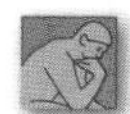

25. What are the domain and range of the function given by $f(x) = (x - 1)^2 + 2$?

For Exercises 26 to 31, state the domain and range of the function.

26. $f(x) = 2x^2 - 4x - 5$

27. $f(x) = 2x^2 + 8x + 3$

28. $f(x) = -2x^2 - 3x + 2$

29. $f(x) = -x^2 + 6x - 9$

30. $f(x) = -x^2 - 4x - 5$

31. $f(x) = x^2 + 4x - 3$

OBJECTIVE B To find the x-intercepts of a parabola

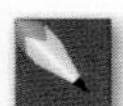

32. **a.** What is an x-intercept of the graph of a parabola?
b. How many x-intercepts can the graph of a parabola have?

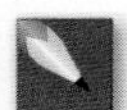

33. **a.** What is the y-intercept of the graph of a parabola?
b. How many y-intercepts can the graph of a parabola have?

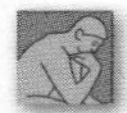

34. The zeros of $f(x) = x^2 - 2x - 3$ are -1 and 3. **a.** What are the x-intercepts of the graph of f? **b.** What are the solutions of $x^2 - 2x - 3 = 0$?

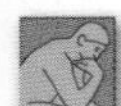

35. The x-intercepts of the graph of $f(x) = x^2 + 3x - 4$ are $(-4, 0)$ and $(1, 0)$. **a.** What are the zeros of f? **b.** What are the solutions of $x^2 + 3x - 4 = 0$?

For Exercises 36 to 47, find the x-intercepts of the graph of the parabola given by each equation.

36. $y = x^2 - 4$

37. $y = x^2 - 9$

38. $y = 2x^2 - 4x$

39. $y = 3x^2 + 6x$

40. $y = x^2 - x - 2$

41. $y = x^2 - 2x - 8$

42. $y = 2x^2 - x - 1$

43. $y = 2x^2 - 5x - 3$

44. $y = x^2 + 2x - 1$

45. $y = x^2 + 4x - 3$

46. $y = x^2 + 6x + 10$

47. $y = -x^2 - 4x - 5$

For Exercises 48 to 65, find the zeros of f.

48. $f(x) = x^2 + 3x + 2$

49. $f(x) = x^2 - 4x - 5$

50. $f(x) = 2x^2 - x - 3$

51. $f(x) = 3x^2 - 2x - 8$

52. $f(x) = x^2 - 6x + 9$

53. $h(x) = 4x^2 - 4x + 1$

54. $f(x) = 2x^2 - 3x$

55. $f(x) = -3x^2 + 4x$

56. $f(x) = 2x^2 - 32$

57. $f(x) = -3x^2 + 12$

58. $f(x) = 2x^2 - 4$

59. $f(x) = 2x^2 - 54$

60. $f(x) = x^2 - 4x + 1$

61. $f(x) = x^2 - 2x - 17$

62. $f(x) = x^2 - 6x + 4$

63. $f(x) = x^2 + 4x + 5$

64. $f(x) = x^2 - 8x + 20$

65. $f(x) = x^2 + 4x + 13$

For Exercises 66 to 77, use the discriminant to determine the number of x-intercepts of the graph.

66. $y = 2x^2 + 2x - 1$

67. $y = -x^2 - x + 3$

68. $y = x^2 - 8x + 16$

69. $y = x^2 - 10x + 25$

70. $y = -3x^2 - x - 2$

71. $y = -2x^2 + x - 1$

72. $y = -2x^2 + x + 1$

73. $y = 4x^2 - x - 2$

74. $y = 2x^2 + x + 1$

75. $y = 2x^2 + x + 4$

76. $y = -3x^2 + 2x - 8$

77. $y = 4x^2 + 2x - 5$

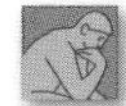

78. Let $f(x) = (x - 3)^2 + a$. For what values of a will the graph of f have **a.** two x-intercepts, **b.** one x-intercept, and **c.** no x-intercepts?

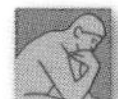

79. Let $f(x) = -(x + 1)^2 + a$. For what values of a will the graph of f have **a.** two x-intercepts, **b.** one x-intercept, and **c.** no x-intercepts?

OBJECTIVE C To find the minimum or maximum of a quadratic function

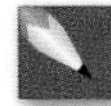

80. What is the minimum or maximum value of a quadratic function?

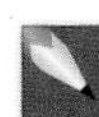

81. Describe how to find the minimum or maximum value of a quadratic function.

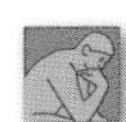

82. Does the function have a minimum or a maximum value?
a. $f(x) = -x^2 + 6x - 1$ **b.** $f(x) = 2x^2 - 4$ **c.** $f(x) = -5x^2 + x$

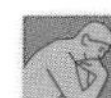

83. Does the function have a minimum or a maximum value?
a. $f(x) = 3x^2 - 2x + 4$ **b.** $f(x) = -x^2 + 9$ **c.** $f(x) = 6x^2 - 3x$

For Exercises 84 to 95, find the minimum or maximum value of the quadratic function.

84. $f(x) = x^2 - 2x + 3$

85. $f(x) = 2x^2 + 4x$

86. $f(x) = -2x^2 + 4x - 3$

87. $f(x) = -2x^2 + 4x - 5$

88. $f(x) = -2x^2 - 3x + 4$

89. $f(x) = -2x^2 - 3x$

90. $f(x) = 2x^2 + 3x - 8$

91. $f(x) = 3x^2 + 3x - 2$

92. $f(x) = -3x^2 + x - 6$

93. $f(x) = -x^2 - x + 2$

94. $f(x) = x^2 - 5x + 3$

95. $f(x) = 3x^2 + 5x + 2$

OBJECTIVE D To solve application problems

96. **Physics** The height s, in feet, of a rock thrown upward at an initial speed of 64 ft/s from a cliff 50 ft above an ocean beach is given by the function $s(t) = -16t^2 + 64t + 50$, where t is the time in seconds. Find the maximum height above the beach that the rock will attain.

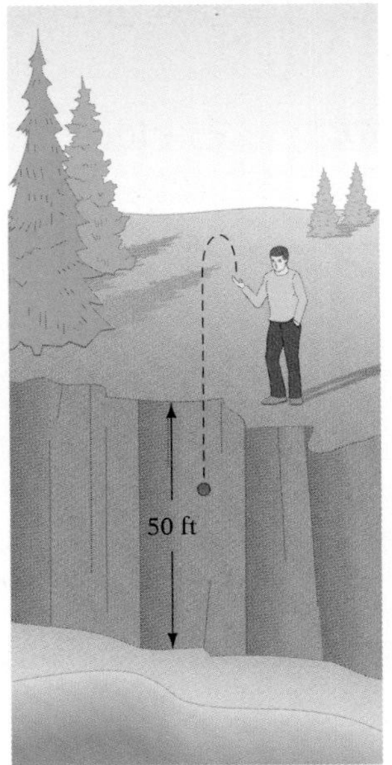

97. **Business** A manufacturer of microwave ovens believes that the revenue R, in dollars, the company receives is related to the price P, in dollars, of an oven by the function $R(P) = 125P - 0.25P^2$. What price will give the maximum revenue?

98. **Business** A tour operator believes that the profit P, in dollars, from selling x tickets is given by $P(x) = 40x - 0.25x^2$. Using this model, what is the maximum profit the tour operator can expect?

99. **Weightlessness** Read the article at the right. Suppose the height h, in meters, of the airplane is modeled by the equation $h(t) = -1.42t^2 + 119t + 6000$, where t is the number of seconds elapsed after the plane enters its parabolic path. Show that the maximum height of the airplane is approximately the value given in the article. Round values to the nearest unit.

In the News

Zero-G Painting Sells Sky High

A painting completed while the artist floated weightless in zero gravity sold at auction for over \$300,000. The British artist flew from the Star City cosmonaut training center in an airplane that produces a weightless enviroment by flying in a series of parabolic paths. During its flight, the plane may reach a maximum height of about 8500 m above Earth.

Sources: BBC News, www.msnbc.msn.com, www.space-travellers.com

100. **Weightlessness** The airplane described in Exercise 99 is used to prepare Russian cosmonauts for their work in the weightless environment of space travel. NASA uses a similar technique to train American astronauts. Suppose the height h, in feet, of NASA's airplane is modeled by the equation $h(t) = -6.63t^2 + 431t + 25{,}000$, where t is the number of seconds elapsed after the plane enters its parabolic path. Find the maximum height of the airplane. Round to the nearest thousand feet.

101. **Construction** The suspension cable that supports a small footbridge hangs in the shape of a parabola. The height h, in feet, of the cable above the bridge is given by the function $h(x) = 0.25x^2 - 0.8x + 25$, where x is the distance in feet from one end of the bridge. What is the minimum height of the cable above the bridge?

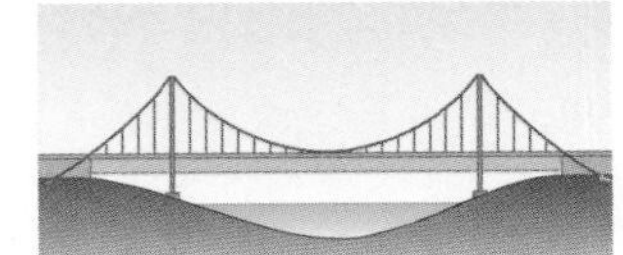

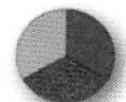

102. Telescopes An equation that models the thickness h, in inches, of the mirror in the telescope at the Palomar Mountain Observatory is given by $h(x) = 0.000379x^2 - 0.0758x + 24$, where x is measured in inches from the edge of the mirror. Find the minimum thickness of the mirror.

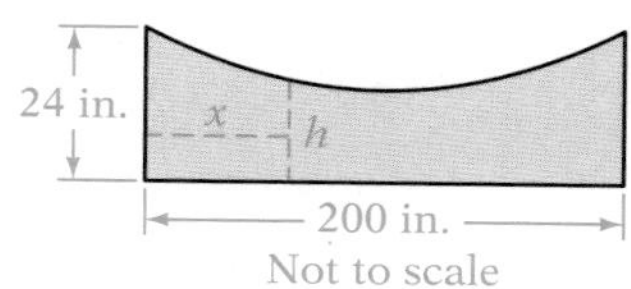

103. Mathematics Find two numbers whose sum is 20 and whose product is a maximum.

104. Mathematics Find two numbers whose product is a maximum and whose sum is 1.

105. Mathematics Find two numbers whose difference is 14 and whose product is a minimum.

106. Mathematics Find two numbers whose product is a maximum and whose difference is 1.

107. Ranching A rancher has 200 ft of fencing with which to build a rectangular corral alongside an existing fence. Determine the dimensions of the corral that will maximize the enclosed area.

108. Construction A courtyard at the corner of two buildings is to be enclosed using 100 ft of redwood fencing. Find the dimensions of the courtyard that will maximize the area.

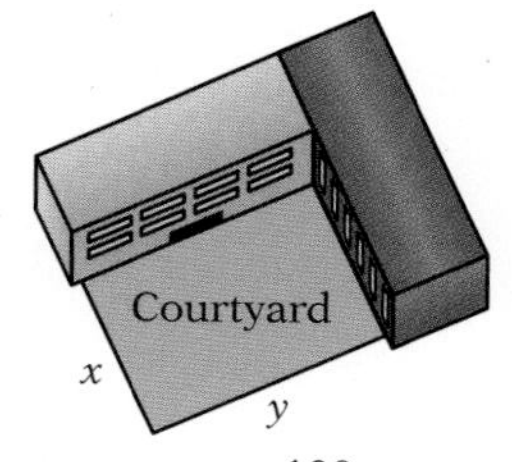

109. Recreation A large lot in a park is going to be split into two softball fields, and each field will be enclosed with a fence. The parks and recreation department has 2100 ft of fencing to enclose the fields. What dimensions will enclose the greatest area?

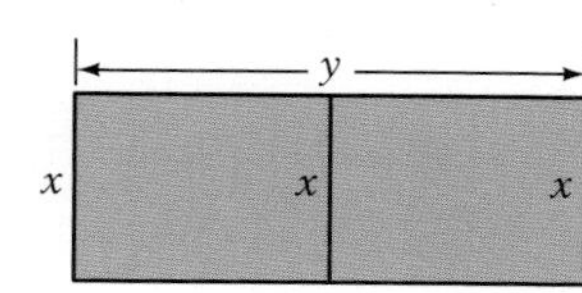

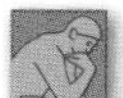

110. If the height h, in feet, of a ball t seconds after it has been tossed directly upward is given by $h(t) = -16(t - 2)^2 + 40$, what is the maximum height the ball will attain?

Applying the Concepts

111. One root of the quadratic equation $2x^2 - 5x + k = 0$ is 4. What is the other root?

112. What is the value of k if the vertex of the graph of $y = x^2 - 8x + k$ is a point on the x-axis?

113. The zeros of $f(x) = mx^2 + nx + 1$ are -2 and 3. What are the zeros of $g(x) = nx^2 + mx - 1$?

SECTION

11.2 Graphs of Functions

OBJECTIVE A To graph functions 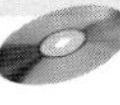

The graphs of the polynomial functions $L(x) = mx + b$ (a straight line) and $Q(x) = ax^2 + bx + c$, $a \neq 0$ (a parabola) were discussed in previous objectives. The graphs of other functions can also be drawn by finding ordered pairs that belong to the function, plotting the points that correspond to the ordered pairs, and then drawing a curve through the points.

Take Note

The units along the x-axis are different from those along the y-axis. If the units along the x-axis were the same as those along the y-axis, the graph would appear narrower than the one shown.

HOW TO 1 Graph: $F(x) = x^3$

Select several values of x and evaluate the function.

x	$F(x) = x^3$	$F(x)$	(x, y)
-2	$(-2)^3$	-8	$(-2, -8)$
-1	$(-1)^3$	-1	$(-1, -1)$
0	0^3	0	$(0, 0)$
1	1^3	1	$(1, 1)$
2	2^3	8	$(2, 8)$

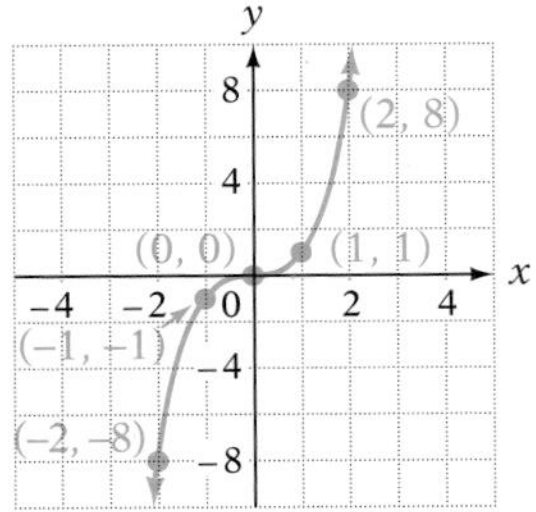

Plot the ordered pairs and draw a graph through the points.

HOW TO 2 Graph: $g(x) = x^3 - 4x + 5$

Select several values of x and evaluate the function.

x	$g(x) = x^3 - 4x + 5$	$g(x)$	(x, y)
-3	$(-3)^3 - 4(-3) + 5$	-10	$(-3, -10)$
-2	$(-2)^3 - 4(-2) + 5$	5	$(-2, 5)$
-1	$(-1)^3 - 4(-1) + 5$	8	$(-1, 8)$
0	$(0)^3 - 4(0) + 5$	5	$(0, 5)$
1	$(1)^3 - 4(1) + 5$	2	$(1, 2)$
2	$(2)^3 - 4(2) + 5$	5	$(2, 5)$

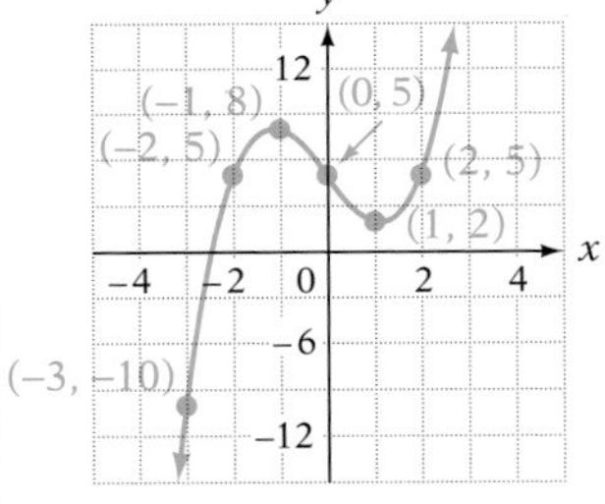

Plot the ordered pairs and draw a graph through the points.

Note from the graphs of the two cubic functions above that the shapes of the graphs can be quite different. The following graphs of typical cubic polynomial functions show the general shape of a cubic polynomial.

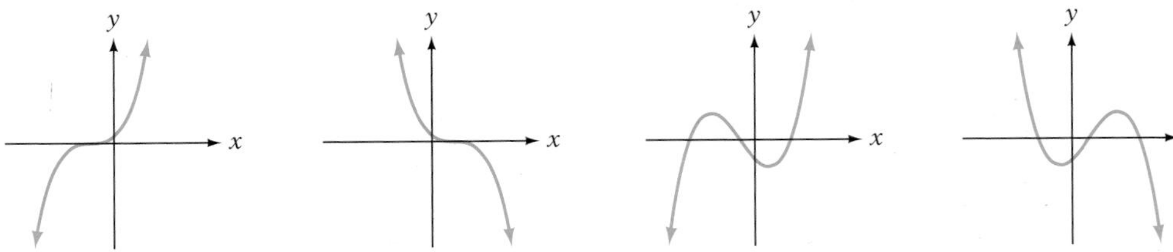

As the degree of a polynomial increases, the graph of the polynomial function can change significantly. In such cases, it may be necessary to plot many points before an accurate graph can be drawn. Only polynomials of degree 3 are considered here.

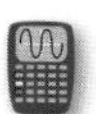

Integrating Technology

See the Keystroke Guide: *Graph* for instructions on using a graphing calculator to graph a function.

HOW TO • 3 Graph: $f(x) = |x + 2|$

This is an **absolute-value function.**

x	$f(x) = \|x + 2\|$	$f(x)$	(x, y)
-3	$\|-3 + 2\|$	1	$(-3, 1)$
-2	$\|-2 + 2\|$	0	$(-2, 0)$
-1	$\|-1 + 2\|$	1	$(-1, 1)$
0	$\|0 + 2\|$	2	$(0, 2)$
1	$\|1 + 2\|$	3	$(1, 3)$
2	$\|2 + 2\|$	4	$(2, 4)$

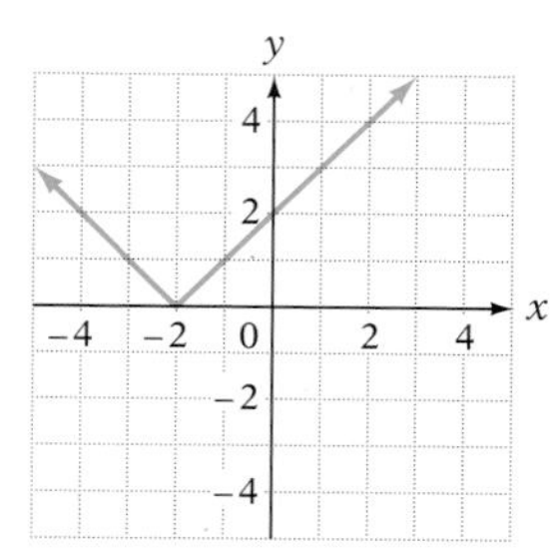

In general, the graph of the absolute value of a linear polynomial is V-shaped.

HOW TO • 4 Graph: $R(x) = \sqrt{2x - 4}$

This is a **radical function.** Because the square root of a negative number is not a real number, the domain of this function requires that $2x - 4 \geq 0$. Solve this inequality for x.

$$2x - 4 \geq 0$$
$$2x \geq 4$$
$$x \geq 2$$

The domain is $\{x \mid x \geq 2\}$. This means that only values of x that are greater than or equal to 2 can be chosen as values at which to evaluate the function. In this case, some of the y-coordinates must be approximated.

x	$R(x) = \sqrt{2x - 4}$	$R(x)$	(x, y)
2	$\sqrt{2(2) - 4}$	0	$(2, 0)$
3	$\sqrt{2(3) - 4}$	1.41	$(3, 1.41)$
4	$\sqrt{2(4) - 4}$	2	$(4, 2)$
5	$\sqrt{2(5) - 4}$	2.45	$(5, 2.45)$
6	$\sqrt{2(6) - 4}$	2.83	$(6, 2.83)$

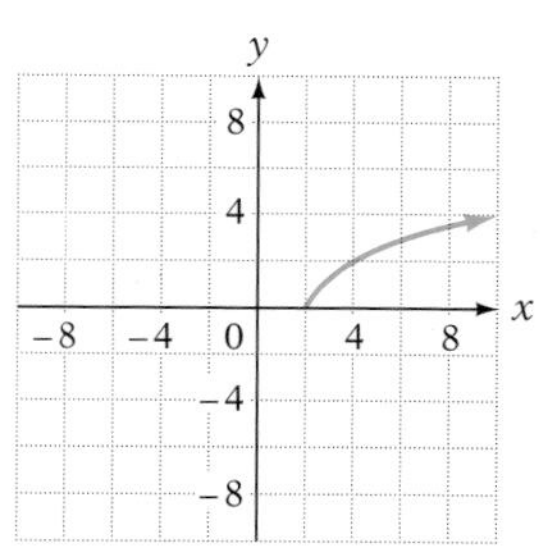

Recall that a function is a special type of relation in which no two ordered pairs have the same first coordinate. Graphically, this means that the graph of a function cannot pass through two points that have the same x-coordinate and different y-coordinates. For instance, the graph at the right is not the graph of a function because there are ordered pairs with the same x-coordinate and different y-coordinates.

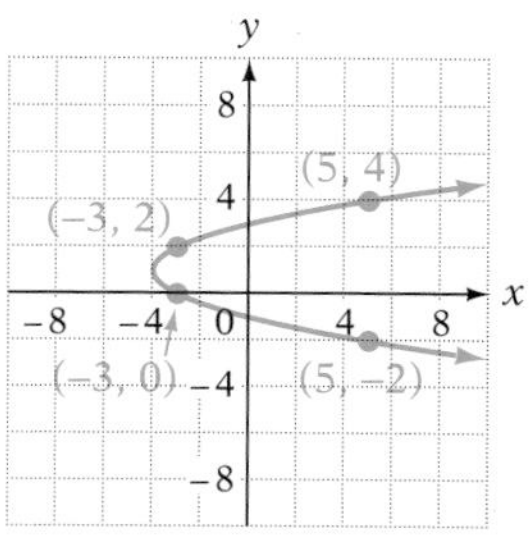

The last graph illustrates a general statement that can be made about whether a graph defines a function. It is called the **vertical-line test.**

Vertical-Line Test

A graph defines a function if any vertical line intersects the graph at no more than one point.

For example, the graph of a nonvertical straight line is the graph of a function. Any vertical line intersects the graph no more than once. The graph of a circle, however, is not the graph of a function. There are vertical lines that intersect the graph at more than one point.

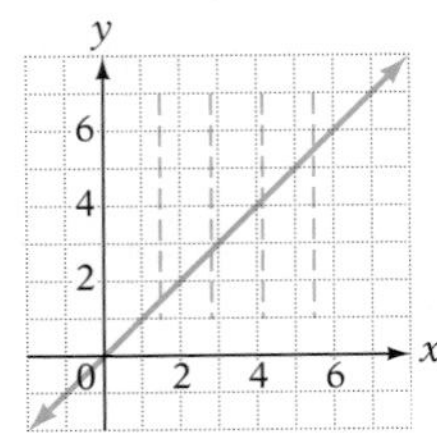

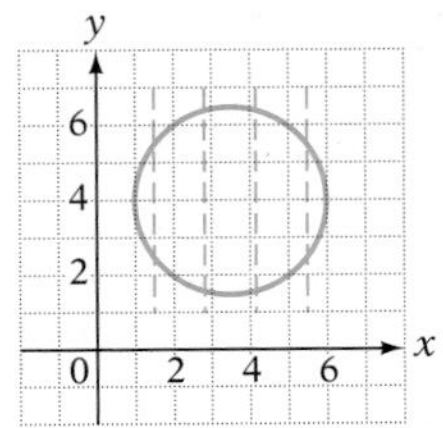

Some practical applications can be modeled by a graph that is not the graph of a function. Below is an example of such an application.

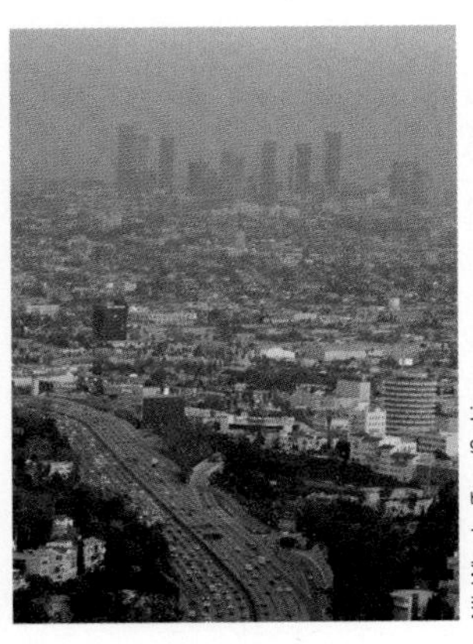
Nik Wheeler/Terra/Corbis

One of the causes of smog is an inversion layer of air in which temperatures at higher altitudes are warmer than those at lower altitudes. The graph at the right shows the altitudes at which various temperatures were recorded. As shown by the dashed lines in the graph, there are two altitudes at which the temperature was 25°C. This means that there are two ordered pairs (shown in the graph) with the same first coordinate but different second coordinates. The graph does not define a function.

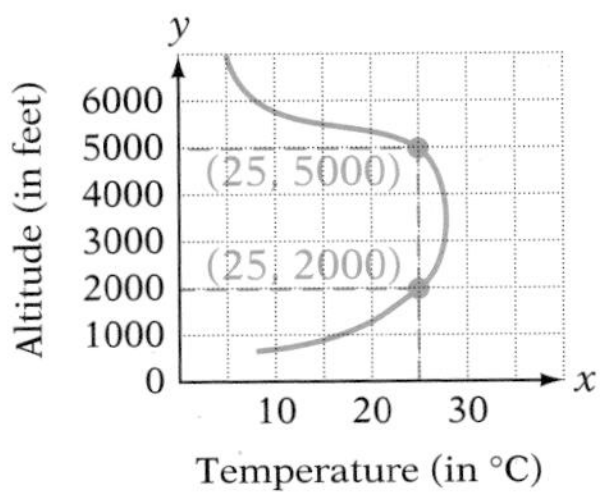

When a graph does define a function, the domain and range can be estimated from the graph.

✓ Take Note
To determine the domain, think of collapsing the graph onto the x-axis and determining the interval on the x-axis that the graph covers. To determine the range, think of collapsing the graph onto the y-axis and determining the interval on the y-axis that the graph covers.

HOW TO • 5 Determine the domain and range of the function given by the graph at the right. Write the answer in set-builder notation.

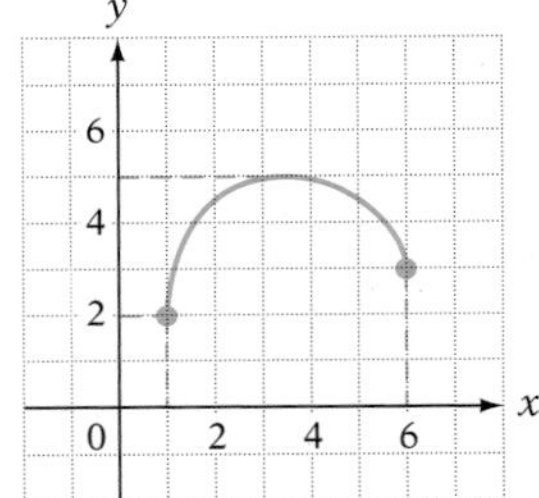

The solid dots on the graph indicate its beginning and ending points.

The domain is the set of x-coordinates.
Domain: $\{x \mid 1 \le x \le 6\}$

The range is the set of y-coordinates.
Range: $\{y \mid 2 \le y \le 5\}$

✓ Take Note
In HOW TO 5, the domain and range are given in set-builder notation. In HOW TO 6, the domain and range are given in interval notation. Either method may be used.

HOW TO • 6 Determine the domain and range of the function given by the graph at the right. Write the answer in interval notation.

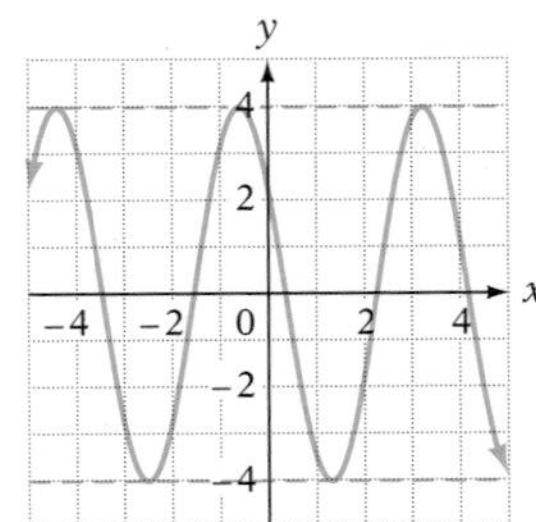

The arrows on the graph indicate that the graph continues in the same manner.

The domain is the set of x-coordinates.
Domain: $(-\infty, \infty)$

The range is the set of y-coordinates.
Range: $[-4, 4]$

EXAMPLE • 1

Use the vertical-line test to determine whether the graph shown is the graph of a function.

Solution

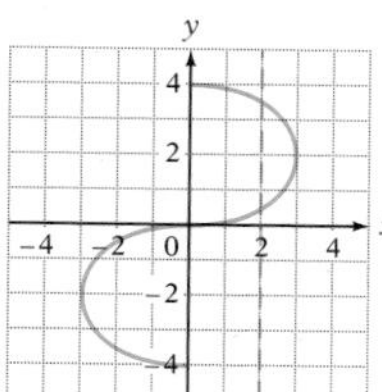

A vertical line intersects the graph more than once. The graph is not the graph of a function.

YOU TRY IT • 1

Use the vertical-line test to determine whether the graph shown is the graph of a function.

Your solution

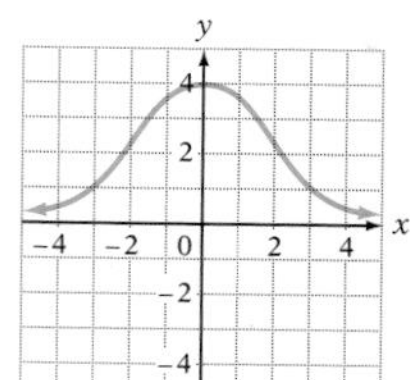

EXAMPLE • 2

Graph and write in set-builder notation the domain and range of the function defined by $f(x) = x^3 - 3x$.

Solution

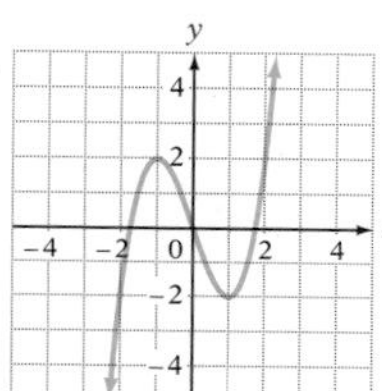

Domain: $\{x|x \in \text{real numbers}\}$
Range: $\{y|y \in \text{real numbers}\}$

YOU TRY IT • 2

Graph and write in set-builder notation the domain and range of the function defined by $f(x) = -\frac{1}{2}x^3 + 2x$.

Your solution

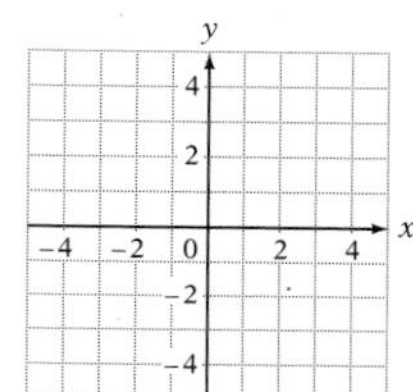

EXAMPLE • 3

Graph and write in interval notation the domain and range of the function defined by $f(x) = |x| + 2$.

Solution

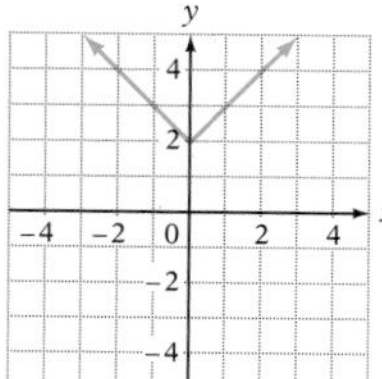

Domain: $(-\infty, \infty)$
Range: $[2, \infty)$

YOU TRY IT • 3

Graph and write in interval notation the domain and range of the function defined by $f(x) = |x - 3|$.

Your solution

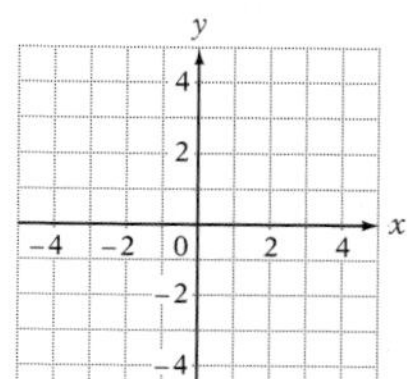

EXAMPLE • 4

Graph and write in set-builder notation the domain and range of the function defined by $f(x) = \sqrt{2 - x}$.

Solution

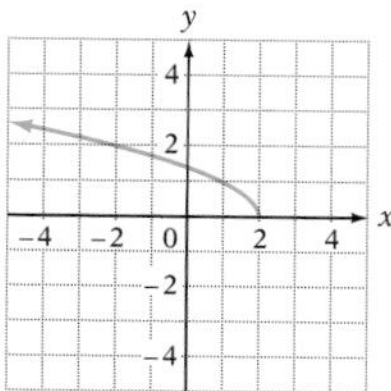

Domain: $\{x|x \le 2\}$
Range: $\{y|y \ge 0\}$

YOU TRY IT • 4

Graph and write in set-builder notation the domain and range of the function defined by $f(x) = -\sqrt{x - 1}$.

Your solution

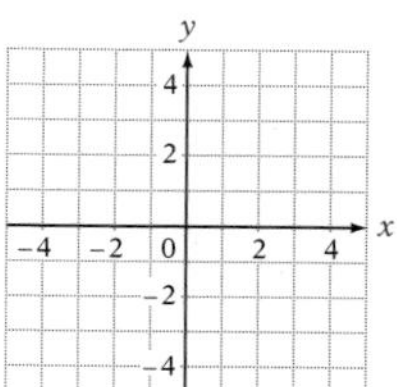

Solutions on pp. S32–S33

11.2 EXERCISES

OBJECTIVE A To graph functions

For Exercises 1 to 6, use the vertical-line test to determine whether the graph is the graph of a function.

1.

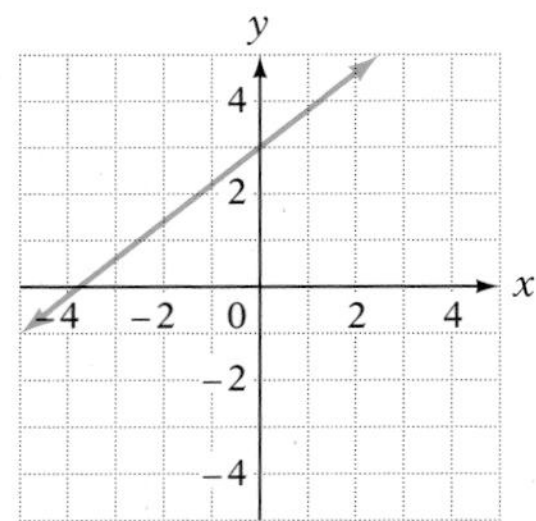

2.

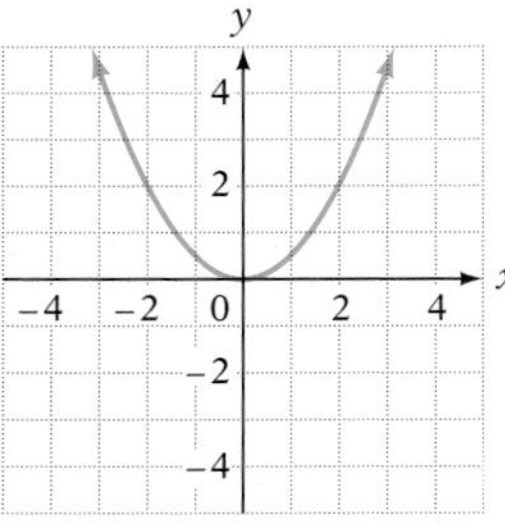

3.

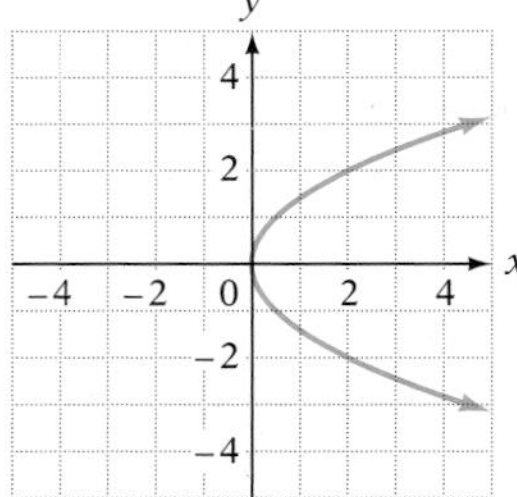

4.

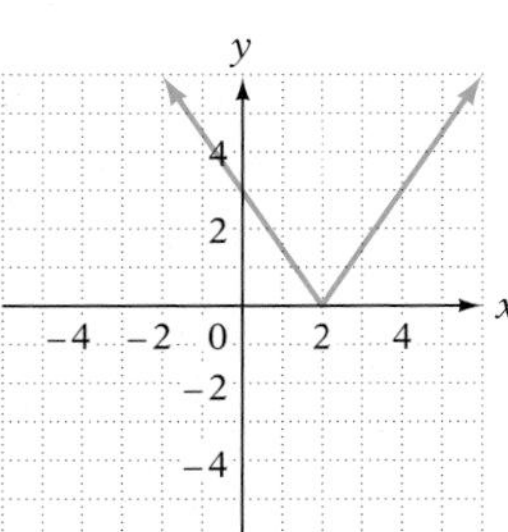

5.

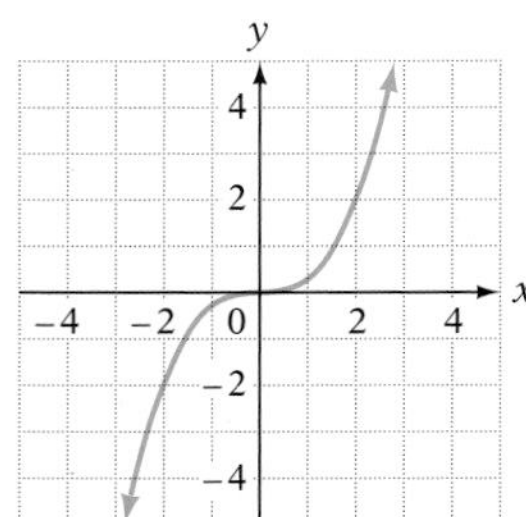

6.

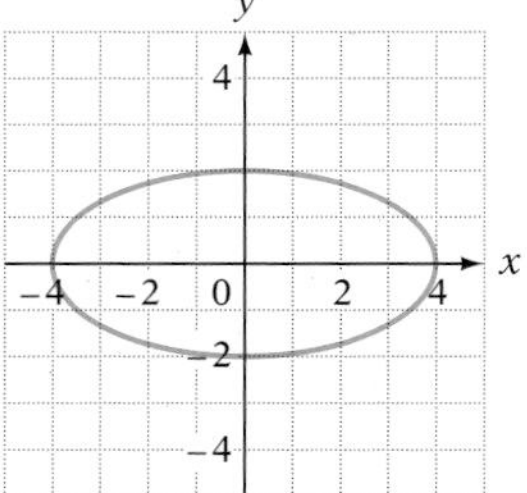

7. Is 4 in the domain of $f(x) = \sqrt{3 - x}$?

8. Is 4 in the range of $f(x) = |2x + 5| + 1$?

For Exercises 9 to 20, graph the function and state its domain and range in set-builder notation.

9. $f(x) = 1 - x^3$

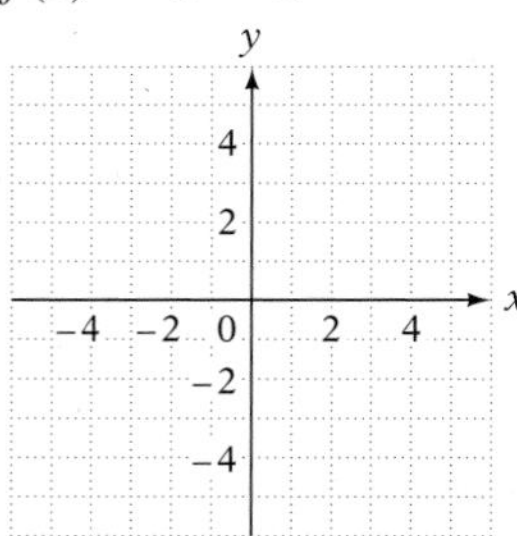

10. $f(x) = x^3 - 1$

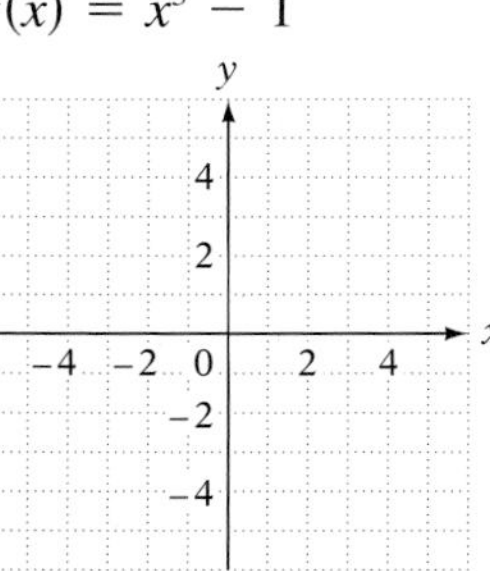

11. $f(x) = 3|2 - x|$

12. $f(x) = |2x - 1|$

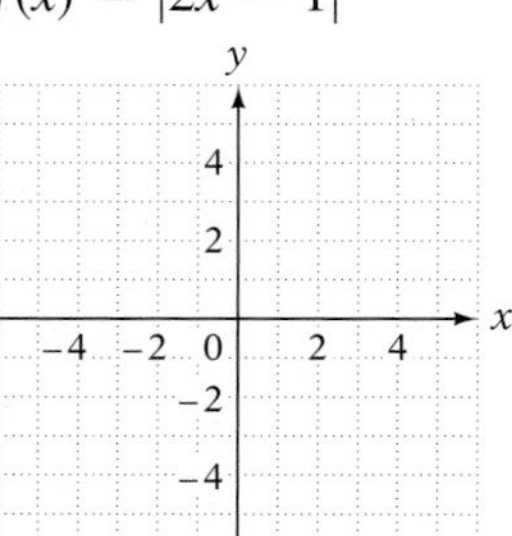

13. $f(x) = \sqrt{4 - x}$

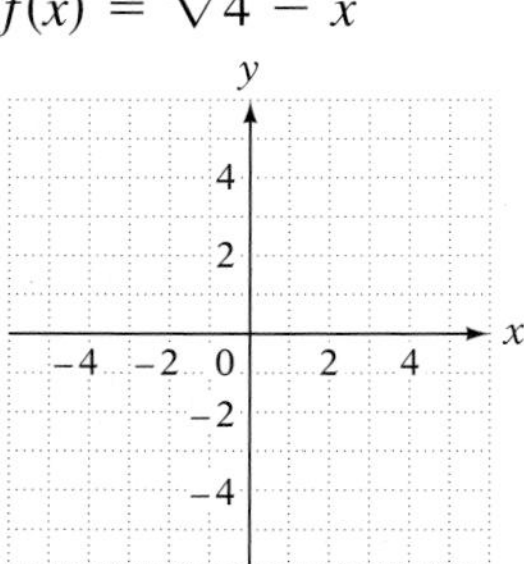

14. $f(x) = \sqrt{1 + x}$

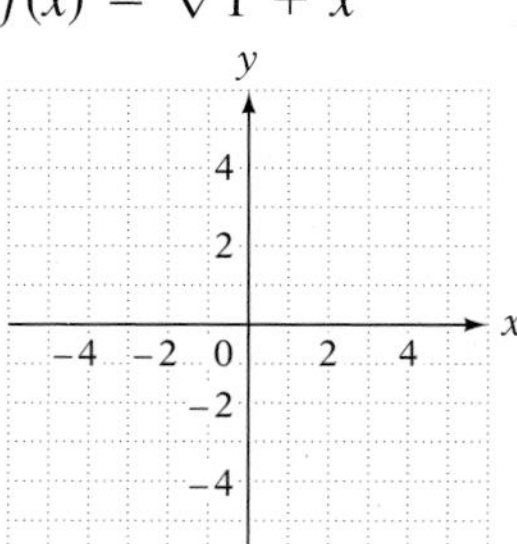

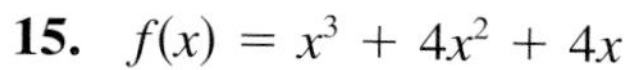

15. $f(x) = x^3 + 4x^2 + 4x$

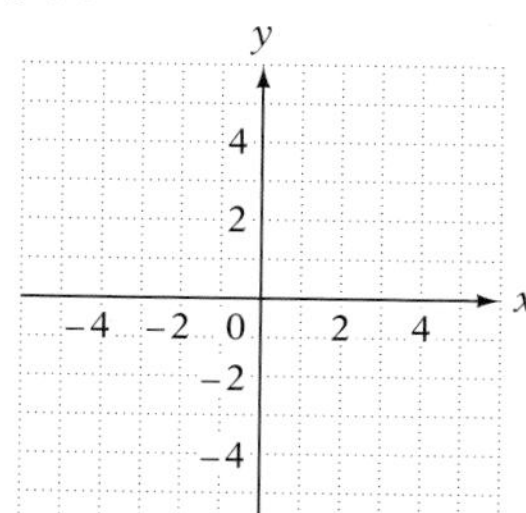

16. $f(x) = x^3 - x^2 - x + 1$

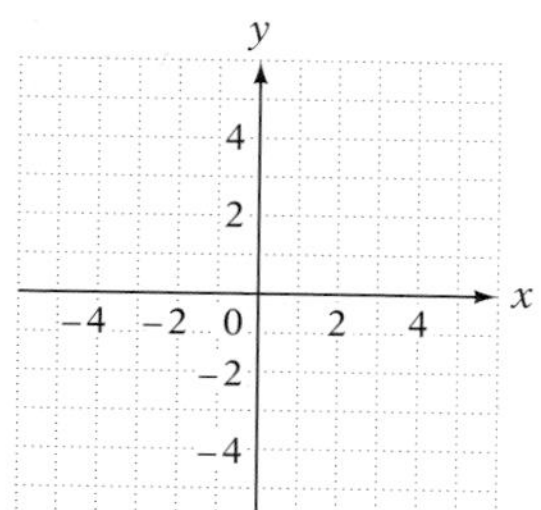

17. $f(x) = |2x + 2|$

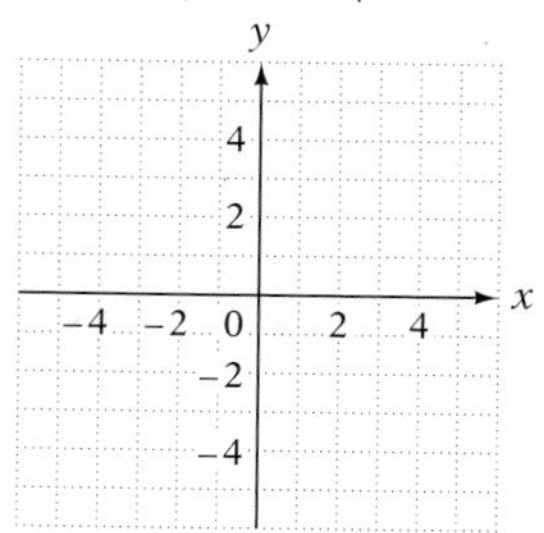

18. $f(x) = 2|x + 1|$

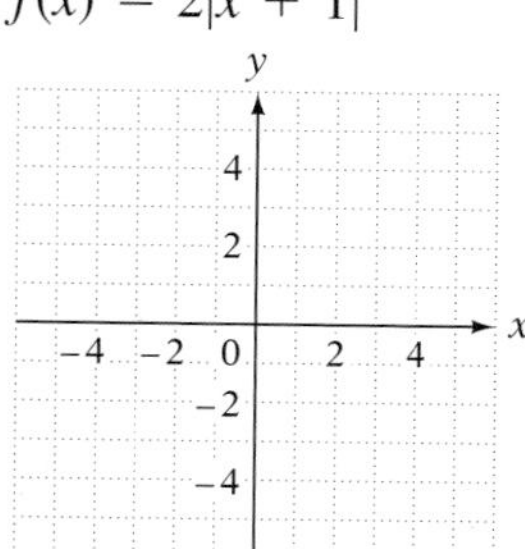

19. $f(x) = -\sqrt{x + 2}$

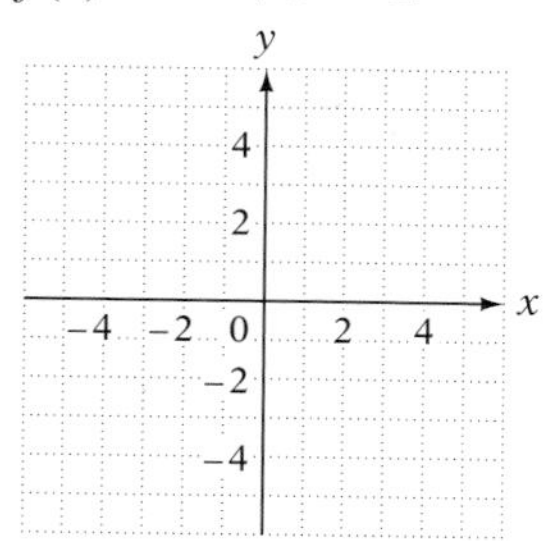

20. $f(x) = -\sqrt{x - 3}$

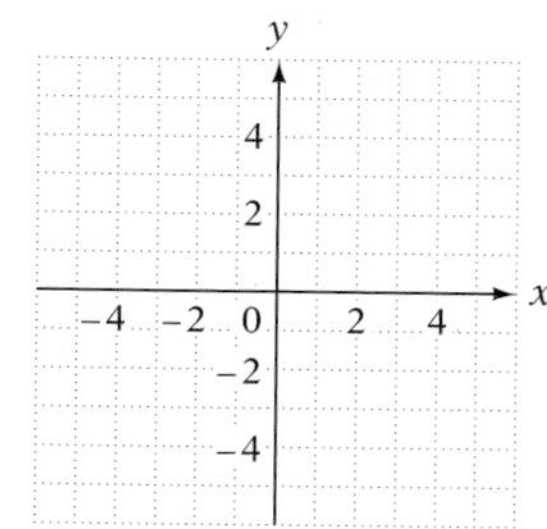

Applying the Concepts

21. If $f(x) = \sqrt{x - 2}$ and $f(a) = 4$, find a.

22. If $f(x) = \sqrt{x + 5}$ and $f(a) = 3$, find a.

23. $f(a, b) =$ the sum of a and b
$g(a, b) =$ the product of a and b
Find $f(2, 5) + g(2, 5)$.

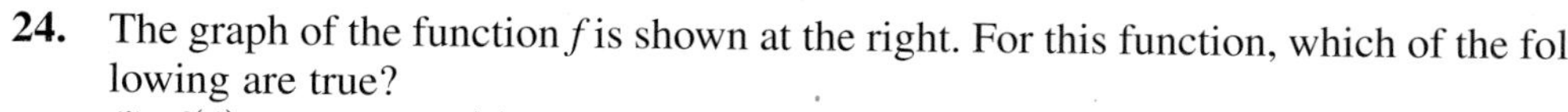

24. The graph of the function f is shown at the right. For this function, which of the following are true?
(i) $f(4) = 1$ (ii) $f(0) = 3$ (iii) $f(-3) = 2$

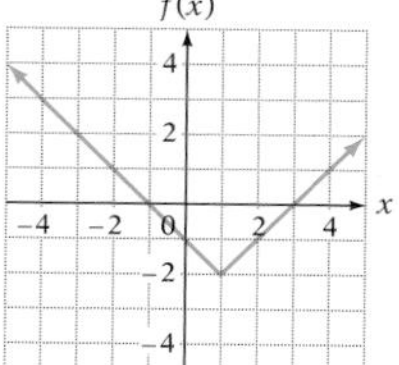

25. Let $f(x)$ be the digit in the xth decimal place of the repeating decimal $0.\overline{387}$. For example, $f(3) = 7$ because 7 is the digit in the third decimal place. Find $f(14)$.

26. Given $f(x) = (x + 1)(x - 1)$, for what values of x is $f(x)$ negative? Write your answer in set-builder notation.

27. Given $f(x) = (x + 2)(x - 2)$, for what values of x is $f(x)$ negative? Write your answer in set-builder notation.

28. Given $f(x) = -|x + 3|$, for what value of x is $f(x)$ greatest?

29. Given $f(x) = |2x - 2|$, for what value of x is $f(x)$ smallest?

SECTION

11.3 Algebra of Functions

OBJECTIVE A To perform operations on functions 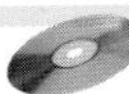

The operations of addition, subtraction, multiplication, and division of functions are defined as follows.

Operations on Functions

If f and g are functions and x is an element of the domain of each function, then

$$(f+g)(x) = f(x) + g(x) \qquad (f \cdot g)(x) = f(x) \cdot g(x)$$

$$(f-g)(x) = f(x) - g(x) \qquad \left(\frac{f}{g}\right)(x) = \frac{f(x)}{g(x)},\ g(x) \neq 0$$

HOW TO • 1 Given $f(x) = x^2 + 1$ and $g(x) = 3x - 2$, find $(f+g)(3)$ and $(f \cdot g)(-1)$.

$$\begin{aligned}(f+g)(3) &= f(3) + g(3)\\ &= [(3)^2 + 1] + [3(3) - 2]\\ &= 10 + 7 = 17\end{aligned}$$

$$\begin{aligned}(f \cdot g)(-1) &= f(-1) \cdot g(-1)\\ &= [(-1)^2 + 1] \cdot [3(-1) - 2]\\ &= 2 \cdot (-5) = -10\end{aligned}$$

Using $f(x) = x^2 + 1$ and $g(x) = 3x - 2$ from the last example, let $S(x)$ be the sum of the two functions. Then

$$\begin{aligned}S(x) &= (f+g)(x) = f(x) + g(x) && \text{• The definition of addition of functions}\\ &= [x^2 + 1] + [3x - 2] && \text{• } f(x) = x^2 + 1, g(x) = 3x - 2\\ S(x) &= x^2 + 3x - 1\end{aligned}$$

Now evaluate $S(3)$.

$$\begin{aligned}S(x) &= x^2 + 3x - 1\\ S(3) &= (3)^2 + 3(3) - 1\\ &= 9 + 9 - 1\\ &= 17 = (f+g)(3)\end{aligned}$$

Note that $S(3) = 17$ and $(f+g)(3) = 17$. This shows that adding $f(x) + g(x)$ and then evaluating is the same as evaluating $f(x)$ and $g(x)$ and then adding. The same is true for the other operations on functions. For instance, let $P(x)$ be the product of the functions $f(x) = x^2 + 1$ and $g(x) = 3x - 2$. Then

$$\begin{aligned}P(x) &= (f \cdot g)(x) = f(x) \cdot g(x)\\ &= (x^2 + 1)(3x - 2)\\ &= 3x^3 - 2x^2 + 3x - 2\\ P(-1) &= 3(-1)^3 - 2(-1)^2 + 3(-1) - 2\\ &= -3 - 2 - 3 - 2\\ &= -10 && \text{• Note that } P(-1) = -10 \text{ and } (f \cdot g)(-1) = -10.\end{aligned}$$

HOW TO • 2 Given $f(x) = 2x^2 - 5x + 3$ and $g(x) = x^2 - 1$, find $\left(\frac{f}{g}\right)(1)$.

$$\left(\frac{f}{g}\right)(1) = \frac{f(1)}{g(1)}$$

$$= \frac{2(1)^2 - 5(1) + 3}{(1)^2 - 1} = \frac{0}{0}$$ • Not a real number

Because $\frac{0}{0}$ is not defined, the expression $\left(\frac{f}{g}\right)(1)$ cannot be evaluated.

EXAMPLE • 1

Given $f(x) = x^2 - x + 1$ and $g(x) = x^3 - 4$, find $(f - g)(3)$.

Solution

$$\begin{aligned}(f - g)(3) &= f(3) - g(3)\\ &= (3^2 - 3 + 1) - (3^3 - 4)\\ &= 7 - 23\\ &= -16\end{aligned}$$

$(f - g)(3) = -16$

YOU TRY IT • 1

Given $f(x) = x^2 + 2x$ and $g(x) = 5x - 2$, find $(f + g)(-2)$.

Your solution

EXAMPLE • 2

Given $f(x) = x^2 + 2$ and $g(x) = 2x + 3$, find $(f \cdot g)(-2)$.

Solution

$$\begin{aligned}(f \cdot g)(-2) &= f(-2) \cdot g(-2)\\ &= [(-2)^2 + 2] \cdot [2(-2) + 3]\\ &= 6(-1)\\ &= -6\end{aligned}$$

$(f \cdot g)(-2) = -6$

YOU TRY IT • 2

Given $f(x) = 4 - x^2$ and $g(x) = 3x - 4$, find $(f \cdot g)(3)$.

Your solution

EXAMPLE • 3

Given $f(x) = x^2 + 4x + 4$ and $g(x) = x^3 - 2$, find $\left(\frac{f}{g}\right)(3)$.

Solution

$$\begin{aligned}\left(\frac{f}{g}\right)(3) &= \frac{f(3)}{g(3)}\\ &= \frac{3^2 + 4(3) + 4}{3^3 - 2}\\ &= \frac{25}{25}\\ &= 1\end{aligned}$$

$\left(\frac{f}{g}\right)(3) = 1$

YOU TRY IT • 3

Given $f(x) = x^2 - 4$ and $g(x) = x^2 + 2x + 1$, find $\left(\frac{f}{g}\right)(4)$.

Your solution

Solutions on p. S33

OBJECTIVE B To find the composition of two functions 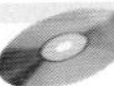

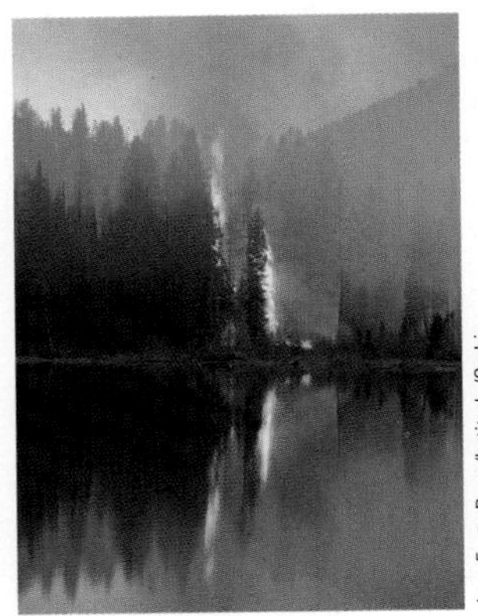
Jose Fuste Raga/Latitude/Corbis

Composition of functions is another way in which functions can be combined. This method of combining functions uses the output of one function as the input for a second function.

Suppose a forest fire is started by lightning striking a tree, and the spread of the fire can be approximated by a circle whose radius r, in feet, is given by $r(t) = 24\sqrt{t}$, where t is the number of hours after the tree is struck by the lightning. The area of the fire is the area of a circle and is given by the formula $A(r) = \pi r^2$. Because the area of the fire depends on the radius of the circle and the radius depends on the time since the tree was struck, there is a relationship between the area of the fire and time. This relationship can be found by evaluating the formula for the area of a circle using $r(t) = 24\sqrt{t}$.

$$\begin{aligned} A(r) &= \pi r^2 \\ A[r(t)] &= \pi[r(t)]^2 && \text{• Replace } r \text{ by } r(t). \\ &= \pi[24\sqrt{t}]^2 && \text{• } r(t) = 24\sqrt{t} \\ &= 576\pi t && \text{• Simplify.} \end{aligned}$$

The result is the function $A(t) = 576\pi t$, which gives the area of the fire in terms of the time since the lightning struck. For instance, when $t = 3$, we have

$$\begin{aligned} A(t) &= 576\pi t \\ A(3) &= 576\pi(3) \\ &\approx 5429 \end{aligned}$$

Three hours after the lightning strikes, the area of the fire is approximately 5429 ft^2.

The function produced above, in which one function was evaluated at another function, is referred to as the *composition* of A with r. The notation $A \circ r$ is used to denote this composition of functions. That is,

$$(A \circ r)(t) = 576\pi t$$

Definition of the Composition of Two Functions

Let f and g be two functions such that $g(x)$ is in the domain of f for all x in the domain of g. Then the **composition of the two functions,** denoted by $f \circ g$, is the function whose value at x is given by $(f \circ g)(x) = f[g(x)]$.

The function defined by $(f \circ g)(x)$ is also called the **composite** of f and g and represents a **composite function.** We read $(f \circ g)(x)$ or $f[g(x)]$ as "f of g of x."

The function machine at the right illustrates the composition of the two functions $g(x) = x^2$ and $f(x) = 2x$. Note that a composite function combines two functions. First, one function pairs an input with an output. Then that output is used as the input for a second function, which in turn produces a final output.

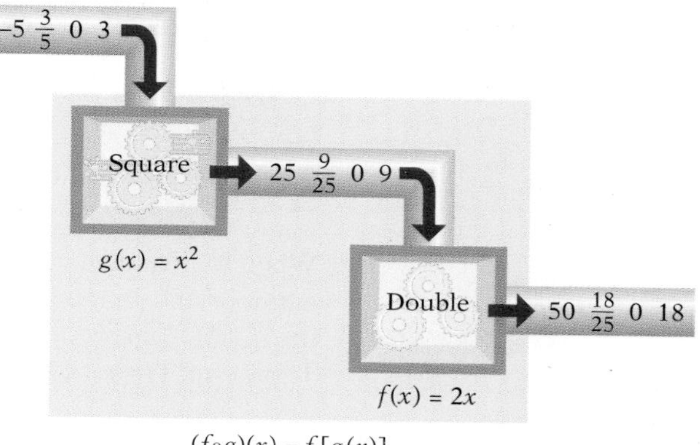

$(f \circ g)(x) = f[g(x)]$

Consider $f(x) = 3x - 1$ and $g(x) = x^2 + 1$. The expression $(f \circ g)(-2)$ or, equivalently, $f[g(-2)]$ means to evaluate the function f at $g(-2)$.

$$g(x) = x^2 + 1$$
$$g(-2) = (-2)^2 + 1 \quad \text{• Evaluate } g \text{ at } -2.$$
$$g(-2) = 5$$
$$f(x) = 3x - 1$$
$$f(5) = 3(5) - 1 = 14 \quad \text{• Evaluate } f \text{ at } g(-2) = 5.$$

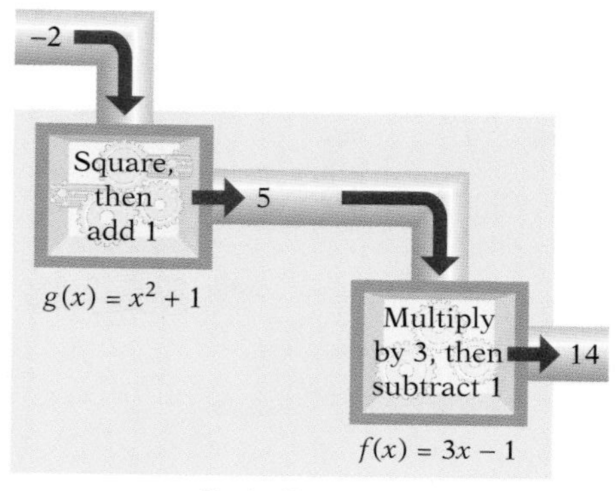

$f[g(-2)] = 14$

If we apply our function machine analogy, the composition of functions looks something like the figure at the left.

We can find a general expression for $f[g(x)]$ by evaluating f at $g(x)$. For instance, using $f(x) = 3x - 1$ and $g(x) = x^2 + 1$, we have

$$f(x) = 3x - 1$$
$$f[g(x)] = 3[g(x)] - 1 \quad \text{• Replace } x \text{ by } g(x).$$
$$= 3[x^2 + 1] - 1 \quad \text{• Replace } g(x) \text{ by } x^2 + 1.$$
$$= 3x^2 + 2 \quad \text{• Simplify.}$$

The requirement in the definition of the composition of two functions that $g(x)$ be in the domain of f for all x in the domain of g is important. For instance, let

$$f(x) = \frac{1}{x - 1} \qquad \text{and} \qquad g(x) = 3x - 5$$

When $x = 2$,

$$g(2) = 3(2) - 5 = 1$$
$$f[g(2)] = f(1) = \frac{1}{1 - 1} = \frac{1}{0} \quad \text{• This is not a real number.}$$

In this case, $g(2)$ is not in the domain of f. Thus the composition is not defined at 2.

HOW TO • 3 Given $f(x) = x^3 - x + 1$ and $g(x) = 2x^2 - 10$, evaluate $(g \circ f)(2)$.

$$f(2) = (2)^3 - (2) + 1 = 7$$
$$(g \circ f)(2) = g[f(2)]$$
$$= g(7)$$
$$= 2(7)^2 - 10 = 88$$

HOW TO • 4 Given $f(x) = 3x - 2$ and $g(x) = x^2 - 2x$, find $(f \circ g)(x)$.

$$\begin{aligned}(f \circ g)(x) &= f[g(x)] \\ &= 3(x^2 - 2x) - 2 \\ &= 3x^2 - 6x - 2\end{aligned}$$

When we evaluate compositions of functions, the order in which the functions are applied is important. In the two diagrams below, the order of the *square* function, $g(x) = x^2$, and the *double* function, $f(x) = 2x$, is interchanged. Note that the final outputs are different. Therefore, $(f \circ g)(x) \neq (g \circ f)(x)$.

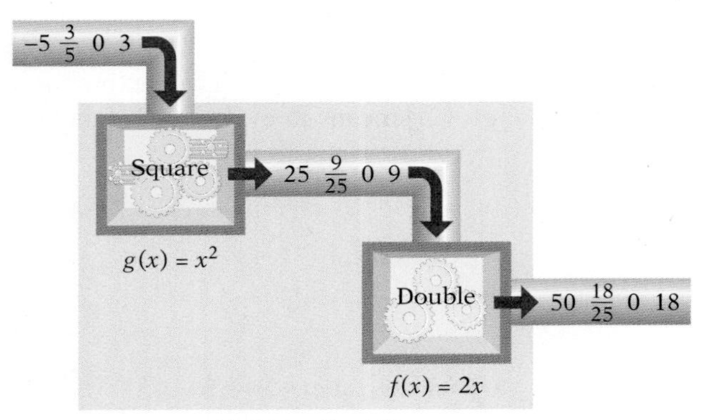

$(f \circ g)(x) = f[g(x)]$

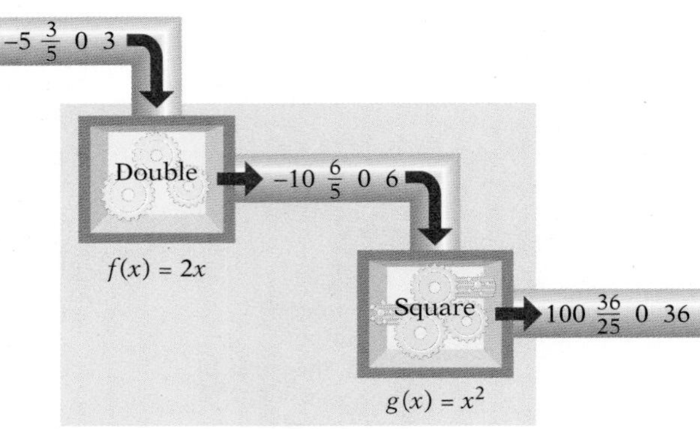

$(g \circ f)(x) = g[f(x)]$

EXAMPLE • 4

Given $f(x) = x^2 - x$ and $g(x) = 3x - 2$, find $f[g(3)]$.

Solution

$g(x) = 3x - 2$

$g(3) = 3(3) - 2 = 9 - 2 = 7$ • Evaluate g at 3.

$f(x) = x^2 - x$ • Evaluate f at $g(3) = 7$.

$f[g(3)] = f(7) = 7^2 - 7 = 42$

YOU TRY IT • 4

Given $f(x) = 1 - 2x$ and $g(x) = x^2$, find $f[g(-1)]$.

Your solution

EXAMPLE • 5

Given $s(t) = t^2 + 3t - 1$ and $v(t) = 2t + 1$, determine $s[v(t)]$.

Solution

$s(t) = t^2 + 3t - 1$ • Evaluate s at $v(t)$.

$$\begin{aligned}s[v(t)] &= (2t + 1)^2 + 3(2t + 1) - 1 \\ &= (4t^2 + 4t + 1) + 6t + 3 - 1 \\ &= 4t^2 + 10t + 3\end{aligned}$$

YOU TRY IT • 5

Given $L(s) = s + 1$ and $M(s) = s^3 + 1$, determine $M[L(s)]$.

Your solution

Solutions on p. S33

11.3 EXERCISES

OBJECTIVE A To perform operations on functions

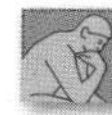

1. Let $f(x) = x^2 + 4$ and $g(x) = \sqrt{x + 4}$. For the given value of x, is it possible to find $(f + g)(x)$?

a. 0 **b.** -4 **c.** 1 **d.** -5

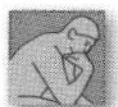

2. Let $f(1) = -5$ and $g(1) = 3$. Find each of the following.

a. $(f + g)(1)$ **b.** $(f - g)(1)$ **c.** $(f \cdot g)(1)$ **d.** $\left(\frac{f}{g}\right)(1)$

For Exercises 3 to 10, let $f(x) = 2x^2 - 3$ and $g(x) = -2x + 4$. Find:

3. $f(2) - g(2)$ **4.** $f(3) - g(3)$ **5.** $f(0) + g(0)$ **6.** $f(1) + g(1)$

7. $(f \cdot g)(2)$ **8.** $(f \cdot g)(-1)$ **9.** $\left(\frac{f}{g}\right)(4)$ **10.** $\left(\frac{g}{f}\right)(-3)$

For Exercises 11 to 18, let $f(x) = 2x^2 + 3x - 1$ and $g(x) = 2x - 4$. Find:

11. $f(1) + g(1)$ **12.** $f(-3) + g(-3)$ **13.** $f(4) - g(4)$

14. $f(-2) - g(-2)$ **15.** $(f \cdot g)(1)$ **16.** $(f \cdot g)(-2)$

17. $\left(\frac{f}{g}\right)(2)$ **18.** $(f \cdot g)\left(\frac{1}{2}\right)$

For Exercises 19 to 21, let $f(x) = x^2 + 3x - 5$ and $g(x) = x^3 - 2x + 3$. Find:

19. $f(2) - g(2)$ **20.** $(f \cdot g)(-3)$ **21.** $\left(\frac{f}{g}\right)(-2)$

OBJECTIVE B To find the composition of two functions

22. Explain the meaning of the notation $f[g(3)]$.

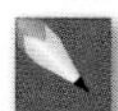

23. Explain the meaning of the expression $(f \circ g)(-2)$.

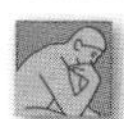

24. Let $f(x) = x^2 + 1$ and $g(x) = \frac{x}{x^2 - 4}$.

a. Is 2 in the domain of $f \circ g$? **b.** Is 2 in the domain of $g \circ f$?

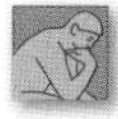

25. Suppose $g(4) = -3$ and $(f \circ g)(4) = 5$. What is the value of $f(-3)$?

Given $f(x) = 2x - 3$ and $g(x) = 4x - 1$, evaluate the composite functions in Exercises 26 to 31.

26. $f[g(0)]$ **27.** $g[f(0)]$ **28.** $f[g(2)]$ **29.** $g[f(-2)]$ **30.** $f[g(x)]$ **31.** $g[f(x)]$

Given $h(x) = 2x + 4$ and $f(x) = \frac{1}{2}x + 2$, evaluate the composite functions in Exercises 32 to 37.

32. $h[f(0)]$ **33.** $f[h(0)]$ **34.** $h[f(2)]$ **35.** $f[h(-1)]$ **36.** $h[f(x)]$ **37.** $f[h(x)]$

Given $g(x) = x^2 + 3$ and $h(x) = x - 2$, evaluate the composite functions in Exercises 38 to 43.

38. $g[h(0)]$ **39.** $h[g(0)]$ **40.** $g[h(4)]$ **41.** $h[g(-2)]$ **42.** $g[h(x)]$ **43.** $h[g(x)]$

Given $f(x) = x^2 + x + 1$ and $h(x) = 3x + 2$, evaluate the composite functions in Exercises 44 to 49.

44. $f[h(0)]$ **45.** $h[f(0)]$ **46.** $f[h(-1)]$ **47.** $h[f(-2)]$ **48.** $f[h(x)]$ **49.** $h[f(x)]$

Given $f(x) = x - 2$ and $g(x) = x^3$, evaluate the composite functions in Exercises 50 to 55.

50. $f[g(2)]$ **51.** $f[g(-1)]$ **52.** $g[f(2)]$ **53.** $g[f(-1)]$ **54.** $f[g(x)]$ **55.** $g[f(x)]$

56. Oil Spills Suppose the spread of an oil leak from a tanker can be approximated by a circle with the tanker at its center and radius r, in feet. The radius of the spill t hours after the beginning of the leak is given by $r(t) = 45t$.
a. Find the area of the spill as a function of time.
b. What is the area of the spill after 3 h? Round to the nearest whole number.

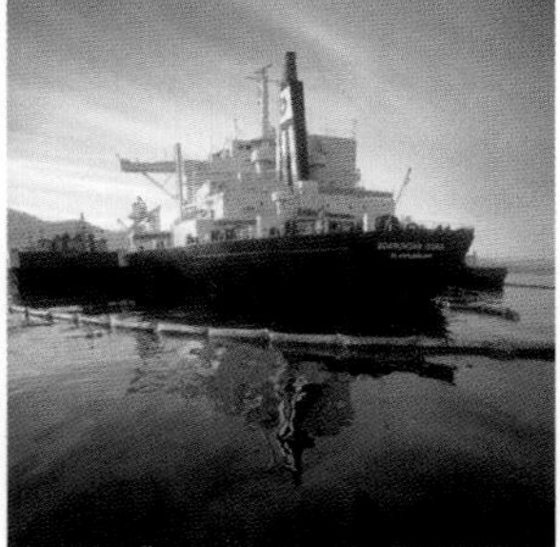
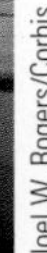
© Joel W. Rogers/Corbis

57. Manufacturing Suppose the manufacturing cost, in dollars, per digital camera is given by the function $M(x) = \frac{50x + 10{,}000}{x}$. A camera store will sell the cameras by marking up the manufacturing cost per camera, $M(x)$, by 60%.
a. Express the selling price of a camera as a function of the number of cameras to be manufactured. That is, find $S \circ M$.
b. Find $(S \circ M)(5000)$.
c. Write a sentence that explains the meaning of the answer to part **b.**

58. Manufacturing The number of electric scooters e that a factory can produce per day is a function of the number of hours h it operates and is given by $e(h) = 250h$ for $0 \le h \le 10$. The daily cost c to manufacture e electric scooters is given by the function $c(e) = 0.05e^2 + 60e + 1000$.
a. Find $(c \circ e)(h)$.
b. Evaluate $(c \circ e)(10)$.
c. Write a sentence that explains the meaning of the answer to part **b.**

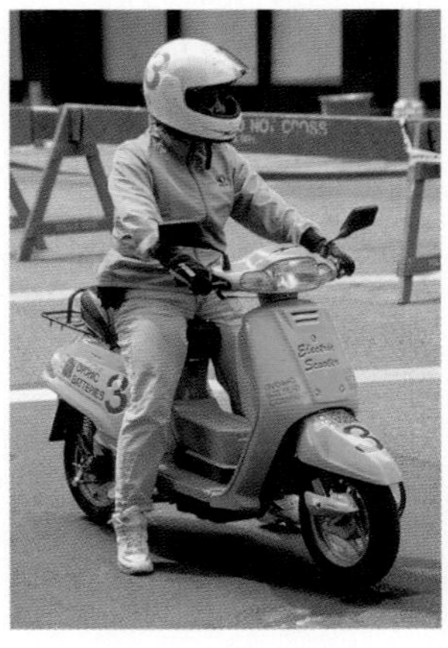
Robert Brenner/PhotoEdit, Inc.

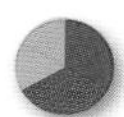

59. Electric Cars Read the article at the right. The garage's income for the conversion of vehicles is given by $I(n) = 12{,}500n$, where n is the number of vehicles converted. The number of vehicles the garage has converted is given by $n(m) = 4m$, where m is the number of months.

a. Find $(I \circ n)(m)$.

b. Evalute $(I \circ n)(3)$.

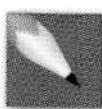

c. Write a sentence that explains the meaning of the answer to part **b.**

60. Electric Cars A company in Santa Cruz, California, sells kits for converting cars to electric power. The company's income, in dollars, from the sale of conversion kits is given by $I(n) = 10{,}000n$, where n is the number of conversion kits sold. The company sells an average of 9 kits per week.

a. Find the income the company receives from conversion kits as a function of the number of weeks w.

b. What is the company's income from the sale of conversion kits after 1 year?

In the News

Going Electric

If you don't want to wait for car manufacturers to come out with a reasonably priced electric car, you can take your gasoline-powered vehicle to a garage right here in Walton, Kansas, that charges $12,500 per vehicle to convert cars and trucks to electric power. The garage is able to convert four vehicles per month.

Source: Associated Press

61. Automobile Rebates A car dealership offers a $1500 rebate and a 10% discount off the price of a new car. Let p be the sticker price of a new car on the dealer's lot, r the price after the rebate, and d the discounted price. Then $r(p) = p - 1500$ and $d(p) = 0.90p$.

a. Write a composite function for the dealer taking the rebate first and then the discount.

b. Write a composite function for the dealer taking the discount first and then the rebate.

c. Which composite function would you prefer the dealer use when you buy a new car?

Applying the Concepts

The graphs of f and g are shown at the right. Use the graphs to determine the values of the composite functions in Exercises 62 to 67.

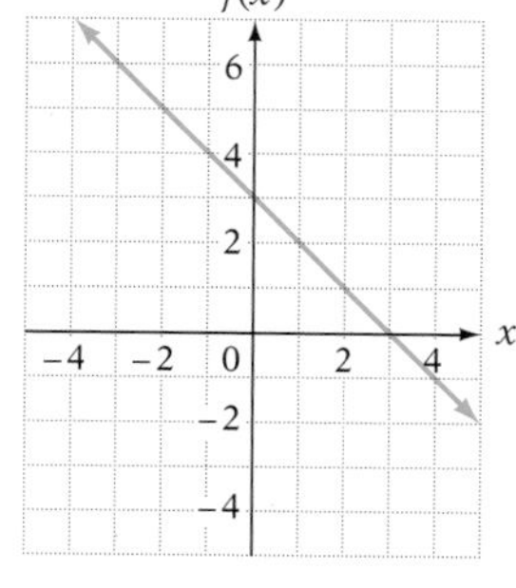

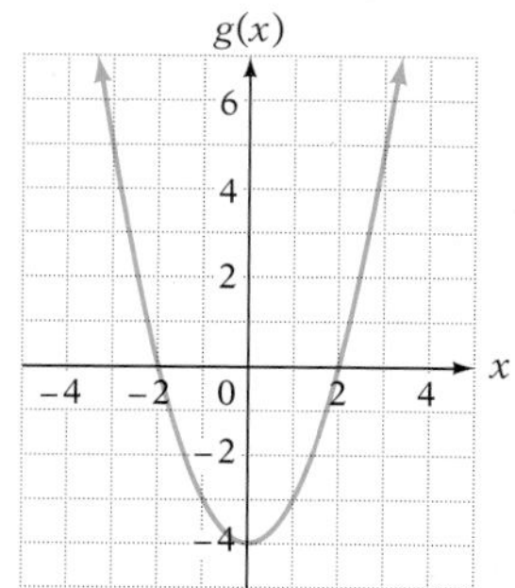

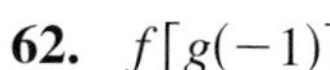

62. $f[g(-1)]$ **63.** $g[f(1)]$ **64.** $(g \circ f)(2)$

65. $(f \circ g)(3)$ **66.** $g[f(3)]$ **67.** $f[g(0)]$

For Exercises 68 to 73, let $g(x) = x^2 - 1$. Find:

68. $g(2 + h)$ **69.** $g(3 + h) - g(3)$ **70.** $g(-1 + h) - g(-1)$

71. $\dfrac{g(1 + h) - g(1)}{h}$ **72.** $\dfrac{g(-2 + h) - g(-2)}{h}$ **73.** $\dfrac{g(a + h) - g(a)}{h}$

For Exercises 74 to 79, let $f(x) = 2x$, $g(x) = 3x - 1$, and $h(x) = x - 2$. Find:

74. $f(g[h(2)])$ **75.** $g(h[f(1)])$ **76.** $h(g[f(-1)])$

77. $f(h[g(0)])$ **78.** $f(g[h(x)])$ **79.** $g(f[h(x)])$

SECTION

11.4 One-to-One and Inverse Functions

OBJECTIVE A To determine whether a function is one-to-one 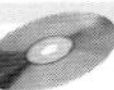

Recall that a function is a set of ordered pairs in which no two ordered pairs that have the same first coordinate have different second coordinates. This means that given any x, there is only one y that can be paired with that x. A **one-to-one function** satisfies the additional condition that given any y, there is only one x that can be paired with that y. One-to-one functions are commonly written as 1–1.

> **One-to-One Function**
>
> A function f is a 1–1 function if for any a and b in the domain of f, $f(a) = f(b)$ implies $a = b$.

This definition states that if the y-coordinates of an ordered pair are equal, $f(a) = f(b)$, then the x-coordinates must be equal, $a = b$.

The function defined by $f(x) = 2x + 1$ is a 1–1 function. To show this, determine $f(a)$ and $f(b)$. Then form the equation $f(a) = f(b)$.

$f(a) = 2a + 1 \qquad f(b) = 2b + 1$

$$f(a) = f(b)$$
$$2a + 1 = 2b + 1$$
$$2a = 2b \quad \text{• Subtract 1 from each side of the equation.}$$
$$a = b \quad \text{• Divide each side of the equation by 2.}$$

Because $f(a) = f(b)$ implies $a = b$, the function is a 1–1 function.

Consider the function defined by $g(x) = x^2 - x$. Evaluate g at -2 and 3.

$$g(-2) = (-2)^2 - (-2) = 6 \qquad g(3) = 3^2 - 3 = 6$$

Note that $g(-2) = 6$ and $g(3) = 6$, but $-2 \neq 3$. Thus g is not a 1–1 function.

The graphs of $f(x) = 2x + 1$ and $g(x) = x^2 - x$ are shown at the right. Note that a horizontal line intersects the graph of f at no more than one point. However, a horizontal line intersects the graph of g at more than one point.

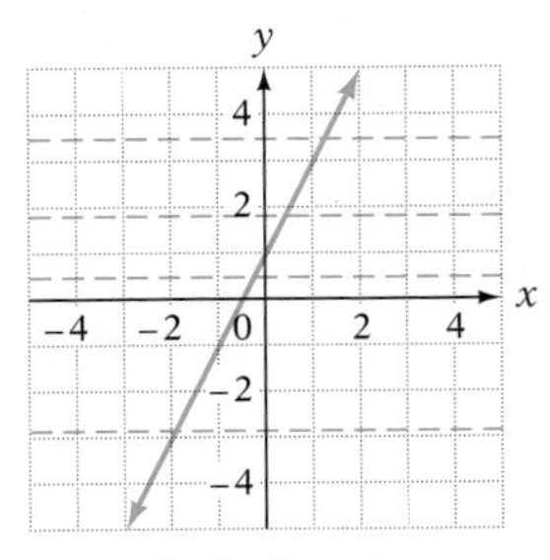

$f(x) = 2x + 1$

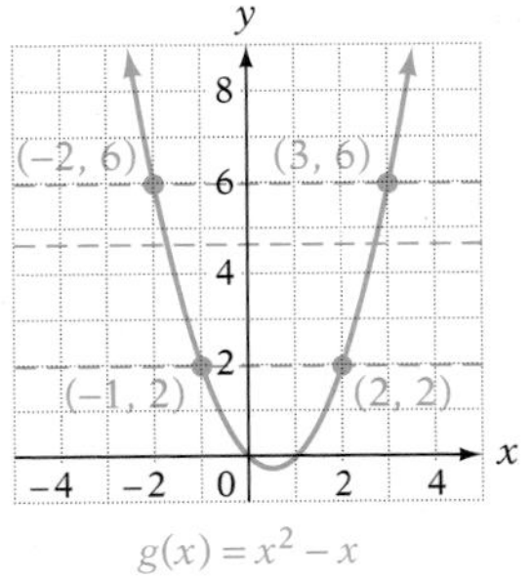

$g(x) = x^2 - x$

Looking at the graph of f, note that for each y-coordinate there is only one x-coordinate. Thus f is a 1–1 function. From the graph of g, however, there are *two* x-coordinates for a given y-coordinate. For instance, $(-2, 6)$ and $(3, 6)$ are the coordinates of two points on the graph for which the y-coordinates are the same and the x-coordinates are different. Therefore, g is not a 1–1 function.

> **Horizontal-Line Test**
>
> The graph of a function represents the graph of a 1–1 function if any horizontal line intersects the graph at no more than one point.

EXAMPLE • 1

Determine whether the graph is the graph of a 1–1 function.

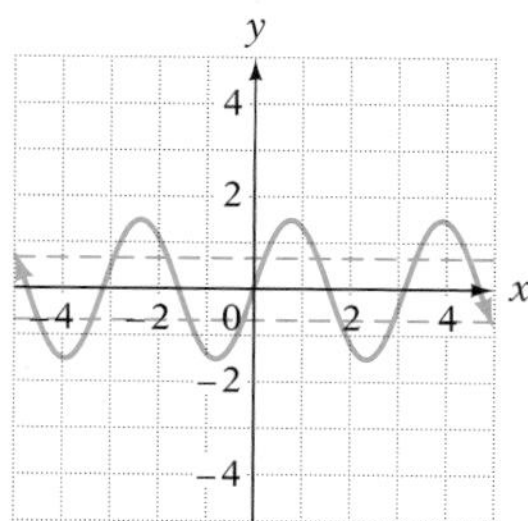

Solution

Because a horizontal line intersects the graph more than once, the graph is not the graph of a 1–1 function.

YOU TRY IT • 1

Determine whether the graph is the graph of a 1–1 function.

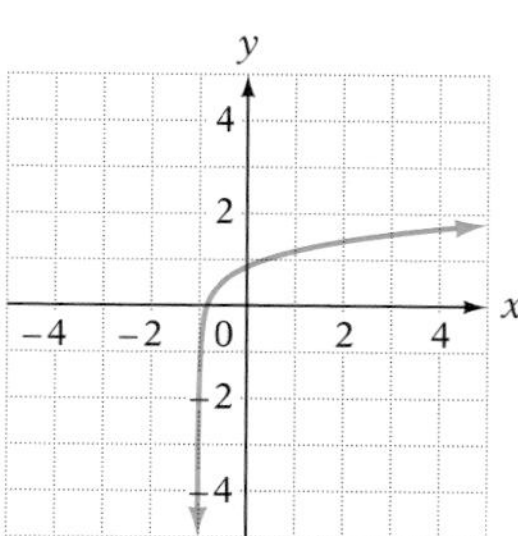

Your solution

Solution on p. S33

OBJECTIVE B To find the inverse of a function

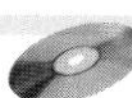

The **inverse** of a function f is the set of ordered pairs formed by reversing the coordinates of each ordered pair of f.

For instance, let $f(x) = 3x$ with domain $\{-2, -1, 0, 1, 2\}$. The ordered pairs of f are $\{(-2, -6), (-1, -3), (0, 0), (1, 3), (2, 6)\}$. Reversing the coordinates of the ordered pairs of f gives $\{(-6, -2), (-3, -1), (0, 0), (3, 1), (6, 2)\}$. This set of ordered pairs is a function and is called the *inverse function* of f. The inverse function of f is denoted by f^{-1}.

From the ordered pairs of f and f^{-1}, we have

Domain of f: $\{-2, -1, 0, 1, 2\}$ Range of f: $\{-6, -3, 0, 3, 6\}$

Domain of f^{-1}: $\{-6, -3, 0, 3, 6\}$ Range of f^{-1}: $\{-2, -1, 0, 1, 2\}$

The domain of f^{-1} is the range of f, and the range of f^{-1} is the domain of f.

Because the function f multiplies an element of the domain by 3, a reasonable guess would be that the *inverse* function would *divide* an element by 3. To test this, we can write $f^{-1}(x) = \frac{x}{3}$, where the symbol $f^{-1}(x)$ is read "f inverse of x." Using $\{-6, -3, 0, 3, 6\}$ as the domain of f^{-1}, the ordered pairs of f^{-1} are $\{(-6, -2), (-3, -1), (0, 0), (3, 1), (6, 2)\}$. This is the same result we obtained by interchanging the coordinates of the ordered pairs of f.

Take Note

It is important to note that $f^{-1}(x)$ is *not* the reciprocal of $f(x)$. $f^{-1}(x) = \frac{x}{3}$; the *reciprocal* of f is $\frac{1}{f(x)} = \frac{1}{3x}$.

Note that $f^{-1}(6) = 2$, whereas evaluating the reciprocal of f at $x = 6$ gives $\frac{1}{18}$.

Now let $g(x) = x^2$ with domain $\{-2, -1, 0, 1, 2\}$. The ordered pairs of g are $\{(-2, 4), (-1, 1), (0, 0), (1, 1), (2, 4)\}$. Reversing the coordinates of the ordered pairs of g gives $\{(4, -2), (1, -1), (0, 0), (1, 1), (4, 2)\}$. This set of ordered pairs *is not* a function. There are ordered pairs with the same first coordinate and different second coordinates. This illustrates that not every function has an inverse function.

The graphs of $f(x) = 3x$ and $g(x) = x^2$, with the set of real numbers as the domain, are shown below.

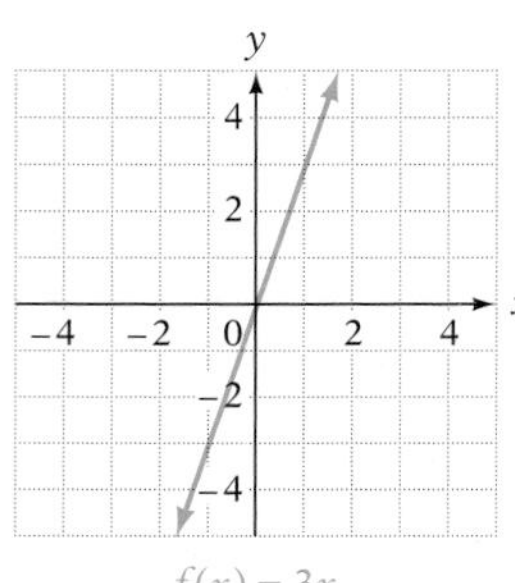

$f(x) = 3x$

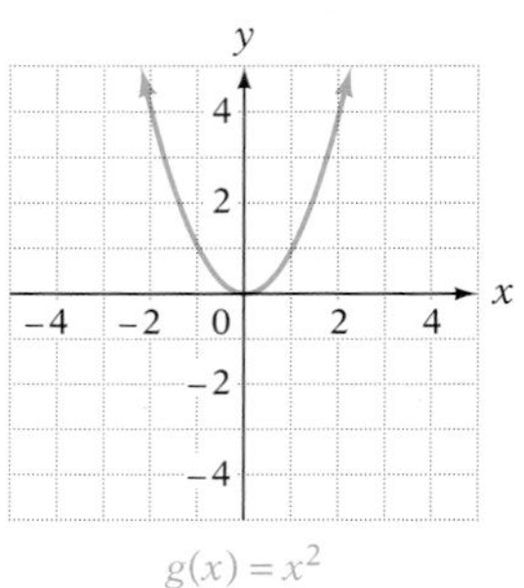

$g(x) = x^2$

By the horizontal-line test, f is a 1–1 function but g is not a 1–1 function.

Condition for an Inverse Function

A function f has an inverse function f^{-1} if and only if f is a 1–1 function.

HOW TO • 1 Find the inverse of the function $\{(-2, -8), (-1, -1), (0, 0), (1, 1), (2, 8)\}$.

Reverse the coordinates of the ordered pairs of the function.

$$\{(-8, -2), (-1, -1), (0, 0), (1, 1), (8, 2)\}$$

Because no two ordered pairs have the same first coordinate, this is a function. The inverse function is $\{(-8, -2), (-1, -1), (0, 0), (1, 1), (8, 2)\}$.

We can use the fact that the inverse of a function is found by interchanging x and y to determine the inverse of a function that is given by an equation.

HOW TO • 2 Find the inverse of $f(x) = 3x + 6$.

$f(x) = 3x + 6$

$y = 3x + 6$ • Replace $f(x)$ by y.

$x = 3y + 6$ • Interchange x and y.

$x - 6 = 3y$ • Solve for y.

$\frac{1}{3}x - 2 = y$

$f^{-1}(x) = \frac{1}{3}x - 2$ • Replace y by $f^{-1}(x)$.

The inverse of the function $f(x) = 3x + 6$ is $f^{-1}(x) = \frac{1}{3}x - 2$.

The composition of a function and its inverse has a special property.

✓ Take Note

Inverse functions can be likened to, for instance, multiplying a number by 5 and then dividing the result by 5; you will be back to the original number. Evaluating f at a number and then evaluating f^{-1} at the output of f returns you to the original number.

Composition of Inverse Functions Property

If f is a function and f^{-1} is its inverse function, then $f^{-1}[f(x)] = x$ and $f[f^{-1}(x)] = x$.

In words, $f^{-1}[f(x)] = x$ states that if $f(a) = b$, then $f^{-1}(b) = a$, the original number used to evaluate f. A similar statement can be made for $f[f^{-1}(x)] = x$.

For instance, consider $f(x) = 3x + 6$ and $f^{-1}(x) = \frac{1}{3}x - 2$ from the previous example. Choose *any* value of x in the domain of f, say, $x = 4$. Then $f(4) = 3 \cdot 4 + 6 = 18$. Now evaluate f^{-1} at 18. $f^{-1}(18) = \frac{1}{3}(18) - 2 = 4$, the original value of x that was chosen.

The Composition of Inverse Functions Property can be used to determine whether two functions are inverses of each other.

HOW TO • 3 Are $f(x) = 2x - 4$ and $g(x) = \frac{1}{2}x + 2$ inverses of each other?

To determine whether the functions are inverses, use the Composition of Inverse Functions Property.

$$f[g(x)] = 2\left(\frac{1}{2}x + 2\right) - 4 = x + 4 - 4 = x$$

$$g[f(x)] = \frac{1}{2}(2x - 4) + 2 = x - 2 + 2 = x$$

Because $f[g(x)] = x$ and $g[f(x)] = x$, the functions are inverses of each other.

Inverse functions occur naturally in many real-world situations.

HOW TO • 4 The equation $F(x) = \frac{9}{5}x + 32$ converts a Celsius temperature x to a Fahrenheit temperature $F(x)$. Find the inverse of this function, and write a sentence that explains its meaning.

$$F(x) = \frac{9}{5}x + 32$$

$$y = \frac{9}{5}x + 32$$ • Replace $F(x)$ by y.

$$x = \frac{9}{5}y + 32$$ • Interchange x and y.

$$x - 32 = \frac{9}{5}y$$ • Solve for y.

$$\frac{5}{9}(x - 32) = y$$

$$F^{-1}(x) = \frac{5}{9}(x - 32)$$ • Replace y by $F^{-1}(x)$.

The inverse is $F^{-1}(x) = \frac{5}{9}(x - 32)$. This equation gives a formula for converting a Fahrenheit temperature to a Celsius temperature.

EXAMPLE • 2

Find the inverse of $f(x) = 2x - 3$.

Solution

$$f(x) = 2x - 3$$

$$y = 2x - 3 \quad \text{• Replace } f(x) \text{ by } y.$$

$$x = 2y - 3 \quad \text{• Interchange } x \text{ and } y.$$

$$x + 3 = 2y \quad \text{• Solve for } y.$$

$$\frac{x}{2} + \frac{3}{2} = y$$

$$f^{-1}(x) = \frac{x}{2} + \frac{3}{2} \quad \text{• Replace } y \text{ by } f^{-1}(x).$$

The inverse of the function is given by $f^{-1}(x) = \frac{x}{2} + \frac{3}{2}$.

YOU TRY IT • 2

Find the inverse of $f(x) = \frac{1}{2}x + 4$.

Your solution

EXAMPLE • 3

Are $f(x) = 3x - 6$ and $g(x) = \frac{1}{3}x + 2$ inverses of each other?

Solution

$$f[g(x)] = 3\left(\frac{1}{3}x + 2\right) - 6 = x + 6 - 6 = x$$

$$g[f(x)] = \frac{1}{3}(3x - 6) + 2 = x - 2 + 2 = x$$

Yes, the functions are inverses of each other.

YOU TRY IT • 3

Are $f(x) = 2x - 6$ and $g(x) = \frac{1}{2}x - 3$ inverses of each other?

Your solution

EXAMPLE • 4

To convert feet to inches, we can use the formula $f(x) = 12x$, where x is the number of feet and $f(x)$ is the number of inches. Find $f^{-1}(x)$, and write a sentence that explains its meaning.

Solution

$$f(x) = 12x$$

$$y = 12x \quad \text{• Replace } f(x) \text{ by } y.$$

$$x = 12y \quad \text{• Interchange } x \text{ and } y.$$

$$\frac{x}{12} = y \quad \text{• Solve for } y.$$

$$\frac{x}{12} = f^{-1}(x) \quad \text{• Replace } y \text{ by } f^{-1}(x).$$

The inverse is $f^{-1}(x) = \frac{x}{12}$. This equation gives a formula for converting inches to feet.

YOU TRY IT • 4

The speed of sound s, in feet per second, depends on the temperature of the air and can be approximated by $s(x) = \frac{3}{5}x + 1125$, where x is the air temperature in degrees Celsius. Find $s^{-1}(x)$, and write a sentence that explains its meaning.

Your solution

Solutions on p. S33

11.4 EXERCISES

OBJECTIVE A **To determine whether a function is one-to-one**

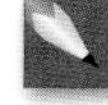
1. What is a 1–1 function?

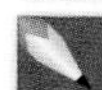
2. What is the horizontal-line test?

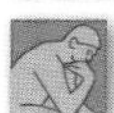
3. **a.** Suppose f is a function and $f(3) = 5$. Is it possible that there exists another value c for which $f(c) = 5$?
b. Suppose f is a 1–1 function and $f(3) = 5$. Is it possible that there exists another value c for which $f(c) = 5$?

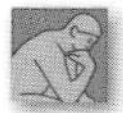
4. The domain of a 1–1 function f has n elements. How many elements are in the range of f?

For Exercises 5 to 16, determine whether the graph represents the graph of a 1–1 function.

5.
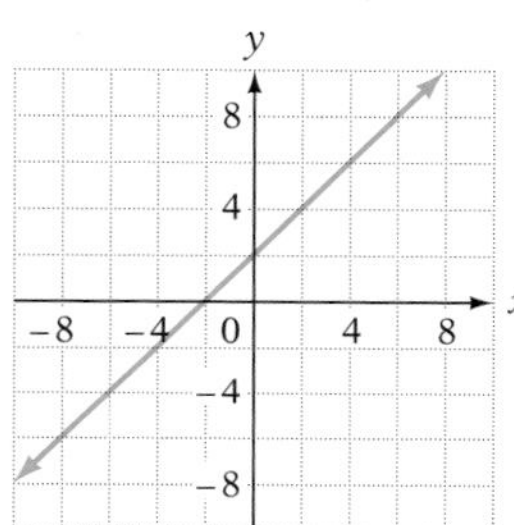

6.
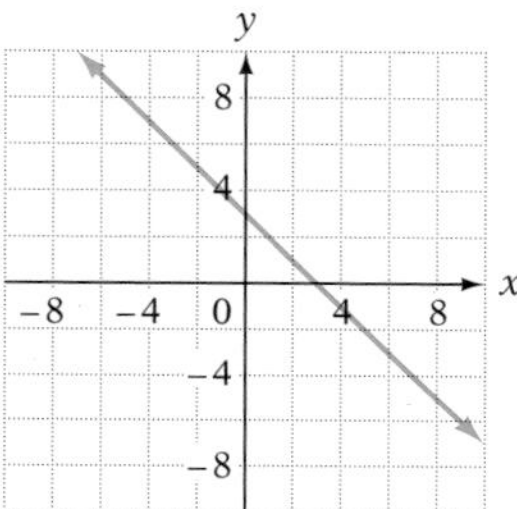

7.
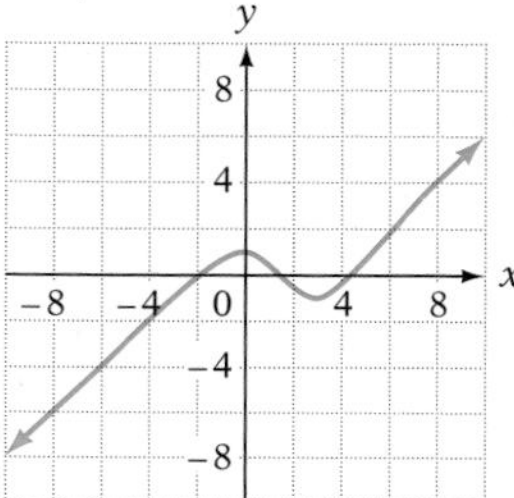

8.
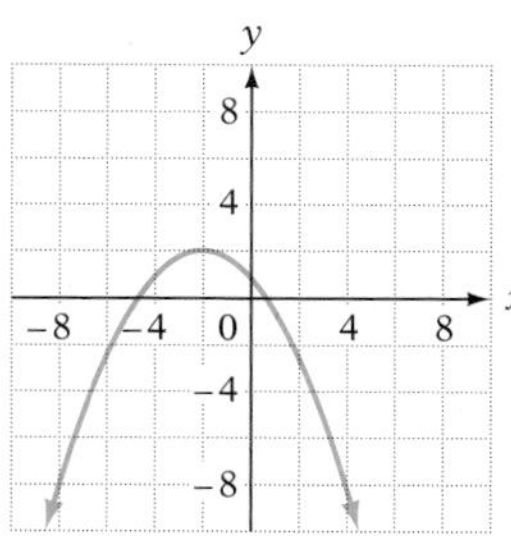

9.
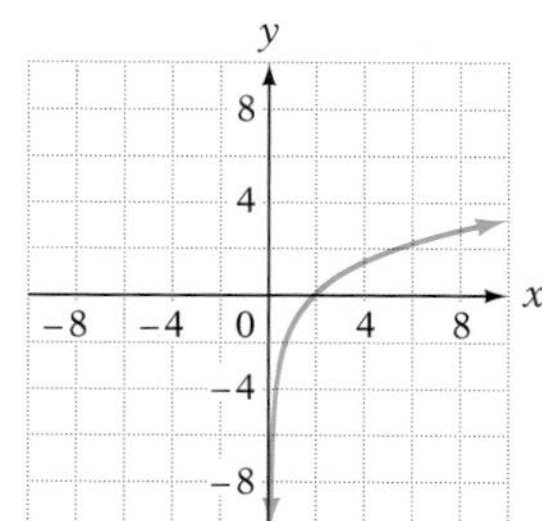

10.
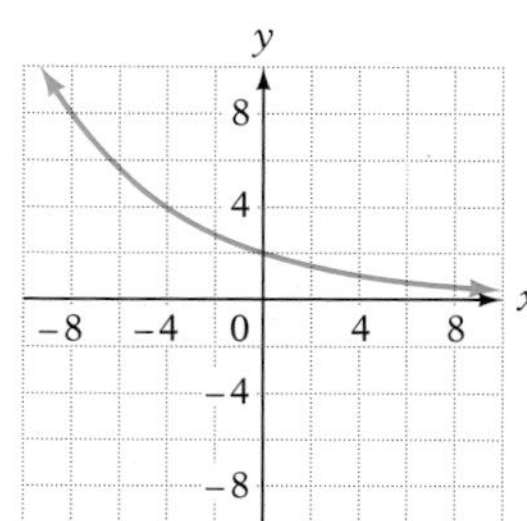

11.
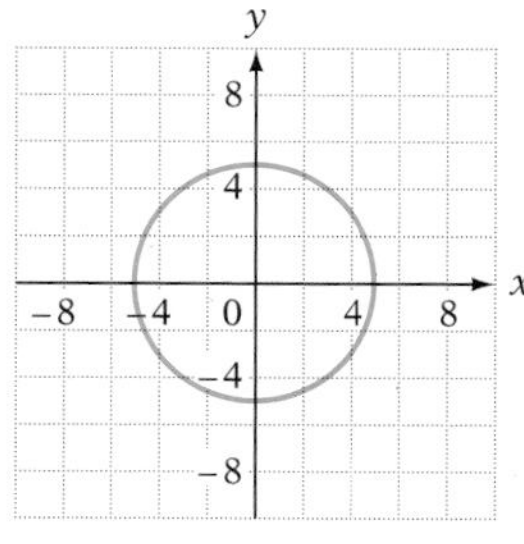

12.
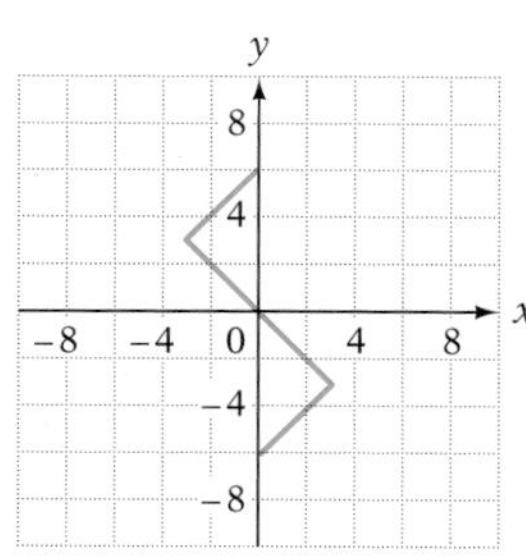

13.
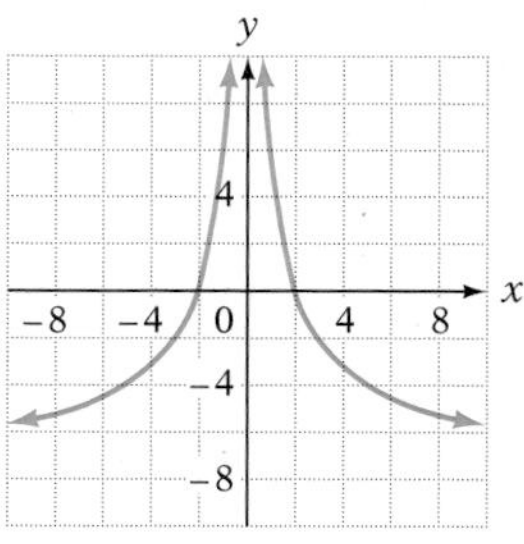

14.
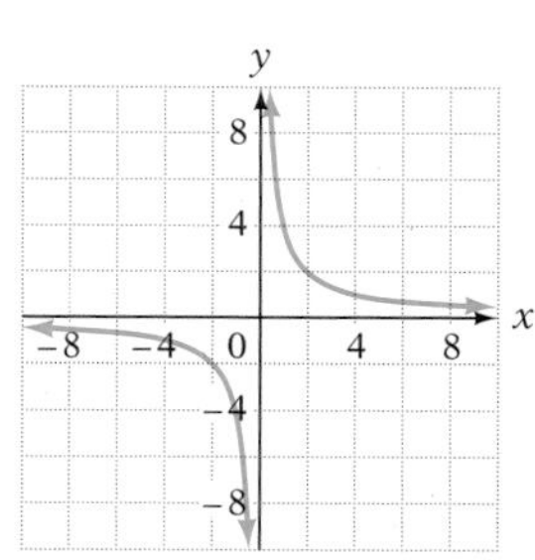

15.
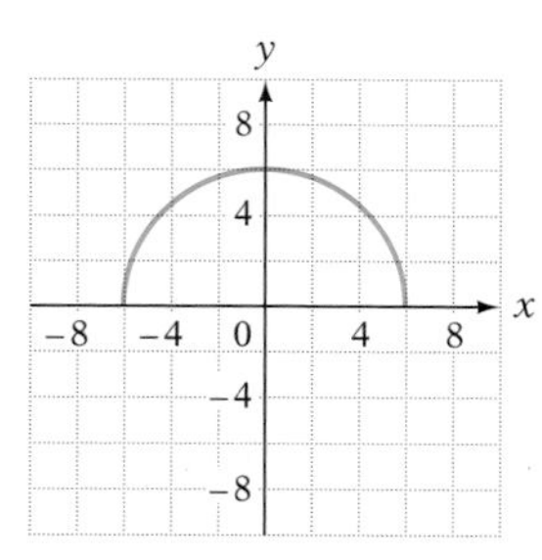

16.
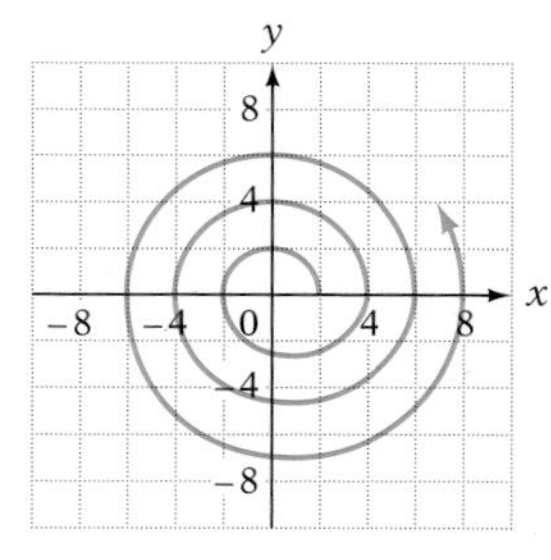

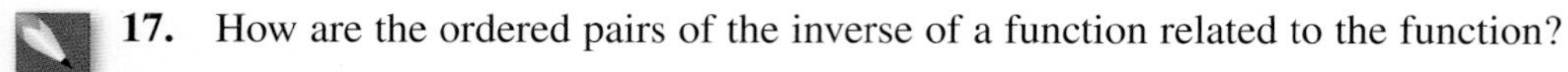

OBJECTIVE B To find the inverse of a function

17. How are the ordered pairs of the inverse of a function related to the function?

18. Why is it that not all functions have an inverse function?

19. Suppose the domain of f is $\{1, 2, 3, 4, 5\}$, and the range of f is $\{3, 7, 11\}$. Does f have an inverse function?

20. Suppose that f is a 1–1 function and that $f(3) = -2$. Which of the following is (are) possible? (i) $f^{-1}(-2) = -2$ (ii) $f^{-1}(-2) = 3$ (iii) $f^{-1}(-2) = 6$

21. Let f be a 1–1 function with $f(3) = 4$, $f(4) = 5$, and $f(6) = 7$. What is the value of **a.** $f^{-1}(5)$, **b.** $f^{-1}(4)$, and **c.** $f^{-1}(7)$?

22. Let f be a 1–1 function with $f(-3) = 3$, $f(-4) = 7$, and $f(0) = 8$. What is the value of **a.** $f^{-1}(7)$, **b.** $f^{-1}(3)$, and **c.** $f^{-1}(8)$?

For Exercises 23 to 30, find the inverse of the function. If the function does not have an inverse function, write "no inverse."

23. $\{(1, 0), (2, 3), (3, 8), (4, 15)\}$

24. $\{(1, 0), (2, 1), (-1, 0), (-2, 0)\}$

25. $\{(3, 5), (-3, -5), (2, 5), (-2, -5)\}$

26. $\{(-5, -5), (-3, -1), (-1, 3), (1, 7)\}$

27. $\{(0, -2), (-1, 5), (3, 3), (-4, 6)\}$

28. $\{(-2, -2), (0, 0), (2, 2), (4, 4)\}$

29. $\{(-2, -3), (-1, 3), (0, 3), (1, 3)\}$

30. $\{(2, 0), (1, 0), (3, 0), (4, 0)\}$

For Exercises 31 to 48, find $f^{-1}(x)$.

31. $f(x) = 4x - 8$

32. $f(x) = 3x + 6$

33. $f(x) = 2x + 4$

34. $f(x) = x - 5$

35. $f(x) = \frac{1}{2}x - 1$

36. $f(x) = \frac{1}{3}x + 2$

37. $f(x) = -2x + 2$

38. $f(x) = -3x - 9$

39. $f(x) = \frac{2}{3}x + 4$

40. $f(x) = \frac{3}{4}x - 4$

41. $f(x) = -\frac{1}{3}x + 1$

42. $f(x) = -\frac{1}{2}x + 2$

43. $f(x) = 2x - 5$

44. $f(x) = 3x + 4$

45. $f(x) = 5x - 2$

46. $f(x) = 4x - 2$

47. $f(x) = 6x - 3$

48. $f(x) = -8x + 4$

For Exercises 49 to 51, let $f(x) = 3x - 5$. Find:

49. $f^{-1}(0)$

50. $f^{-1}(2)$

51. $f^{-1}(4)$

For Exercises 52 to 54, state whether the graph is the graph of a function. If it is the graph of a function, does the function have an inverse?

52.

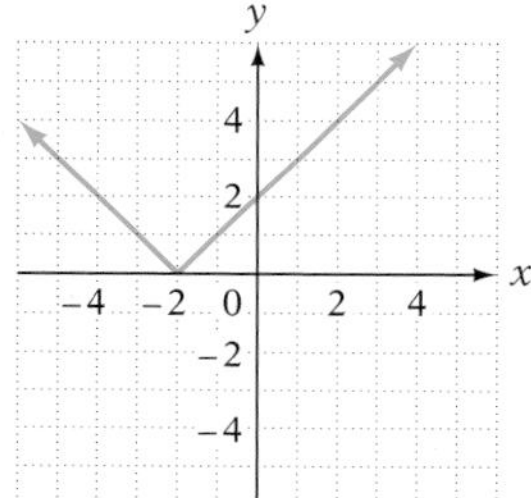

53.

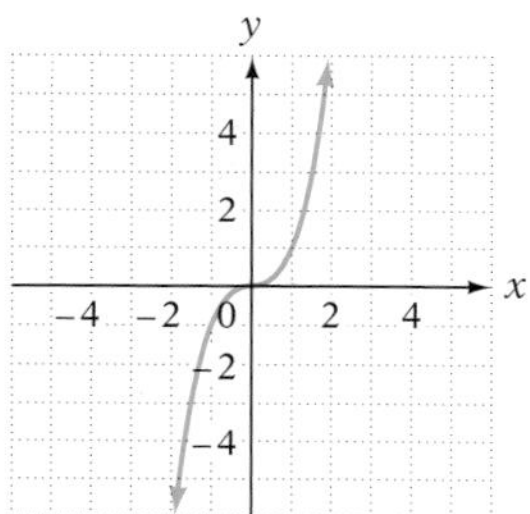

54.

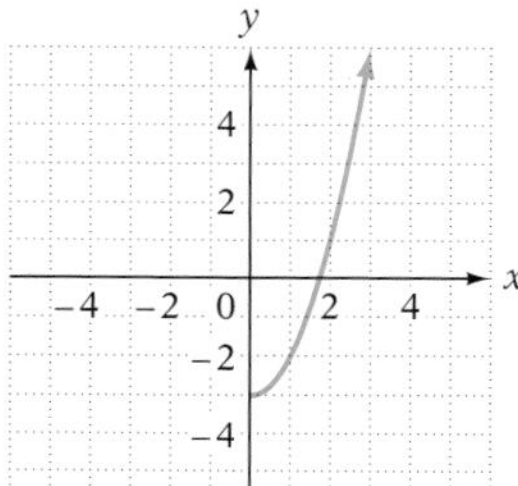

For Exercises 55 to 62, use the Composition of Inverse Functions Property to determine whether the functions are inverses.

55. $f(x) = 4x;\ g(x) = \dfrac{x}{4}$

56. $g(x) = x + 5;\ h(x) = x - 5$

57. $f(x) = 3x;\ h(x) = \dfrac{1}{3x}$

58. $h(x) = x + 2;\ g(x) = 2 - x$

59. $g(x) = 3x + 2;\ f(x) = \dfrac{1}{3}x - \dfrac{2}{3}$

60. $h(x) = 4x - 1;\ f(x) = \dfrac{1}{4}x + \dfrac{1}{4}$

61. $f(x) = \dfrac{1}{2}x - \dfrac{3}{2};\ g(x) = 2x + 3$

62. $g(x) = -\dfrac{1}{2}x - \dfrac{1}{2};\ h(x) = -2x + 1$

63. To convert ounces to pounds, we can use the formula $f(x) = \frac{x}{16}$, where x is the number of ounces and $f(x)$ is the number of pounds. Find $f^{-1}(x)$, and write a sentence that explains its meaning.

64. To convert pounds to kilograms, we can use the formula $f(x) = \frac{x}{2.2}$, where x is the number of pounds and $f(x)$ is the number of kilograms. Find $f^{-1}(x)$, and write a sentence that explains its meaning.

65. The function given by $f(x) = x + 30$ converts a dress size x in the United States to a dress size $f(x)$ in France. Find the inverse of this function, and write a sentence that explains its meaning.

66. The function given by $f(x) = x - 31$ converts a shoe size x in France to a shoe size $f(x)$ in the United States. Find the inverse of this function, and write a sentence that explains its meaning.

67. The target heart rate $f(x)$ for a certain athlete is given by $f(x) = 90x + 65$, where x is the training intensity percent. Find the inverse of this function, and write a sentence that explains its meaning.

68. A data messaging service charges \$5 per month plus \$.10 per message. The function given by $f(x) = 0.1x + 5$ gives the monthly cost $f(x)$ for sending x messages. Find the inverse of this function, and write a sentence that explains its meaning.

Applying the Concepts

Recall that interchanging the x- and y-coordinates of each ordered pair of a 1–1 function forms the inverse function. Given the graphs in Exercises 69 to 74, interchange the x- and y-coordinates of the given ordered pairs and then plot the new ordered pairs. Draw a smooth graph through the ordered pairs. The graph is the graph of the inverse function.

69.

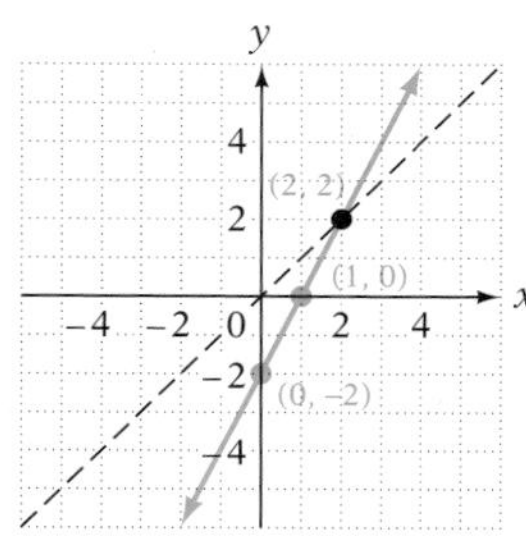

70.

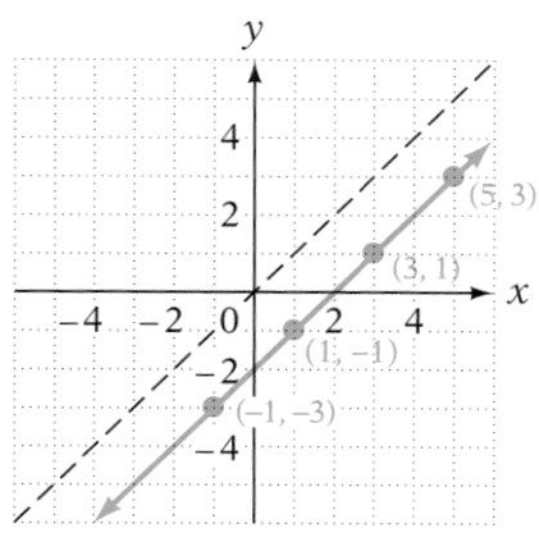

71.

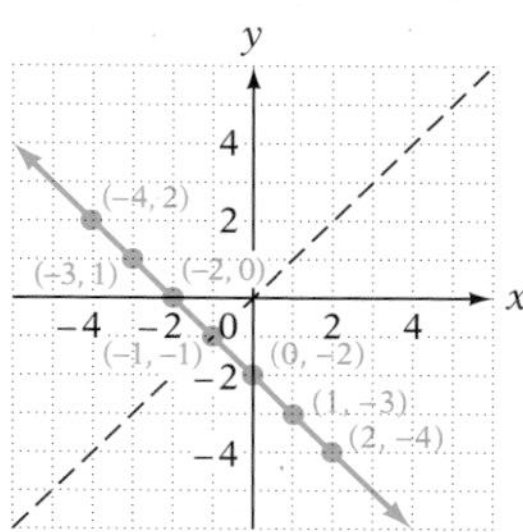

72.

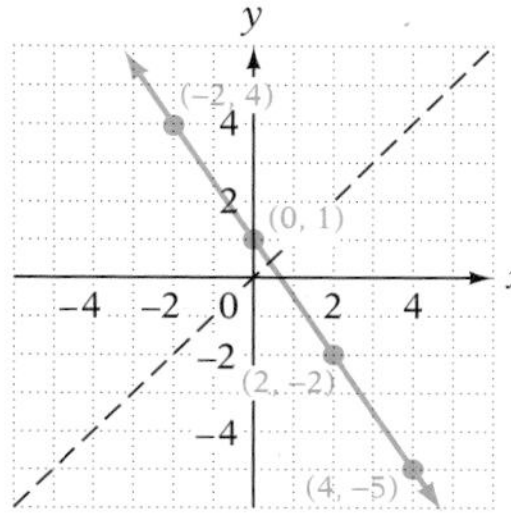

73.

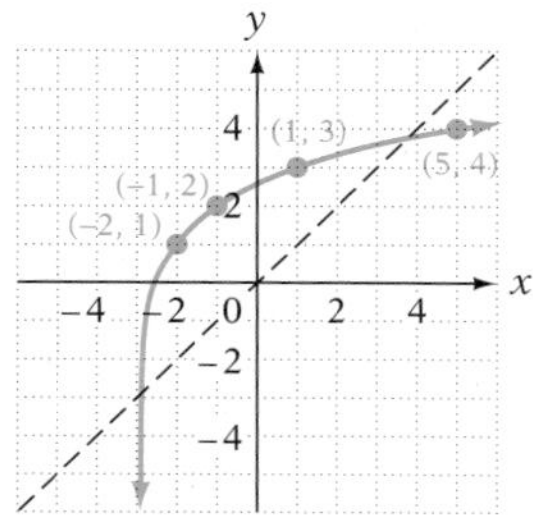

74.

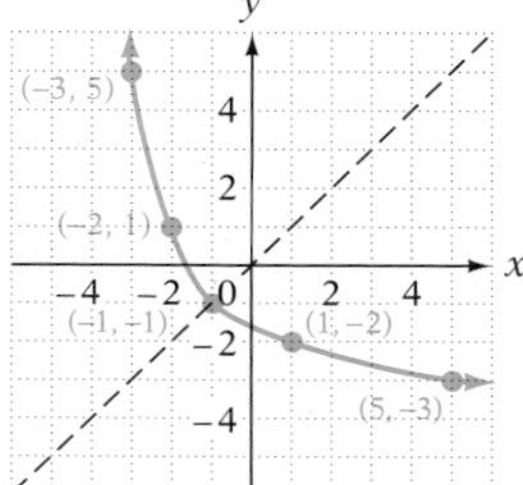

Each of the tables in Exercises 75 and 76 defines a function. Is the inverse of the function a function? Explain your answer.

75. Grading Scale

Score	*Grade*
90–100	A
80–89	B
70–79	C
60–69	D
0–59	F

76. First-Class Postage Rates

Weight	*Cost*
$0 < w < 1$	\$.42
$1 < w \le 2$	\$.59
$2 < w \le 3$	\$.76
$3 < w \le 3.5$	\$.93

77. Is the inverse of a constant function a function? Explain your answer.

78. The graphs of all functions given by $f(x) = mx + b$, $m \neq 0$, are straight lines. Are all of these functions 1–1 functions? If so, explain why. If not, give an example of a linear function that is not 1–1.

FOCUS ON PROBLEM SOLVING

Algebraic Manipulation and Graphing Techniques

Problem solving is often easier when we have both algebraic manipulation and graphing techniques at our disposal. Solving quadratic equations and graphing quadratic equations in two variables are the techniques used here to solve problems involving profit.

A company's revenue, R, is the total amount of money the company earned by selling its products. The cost, C, is the total amount of money the company spent to manufacture and sell its products. A company's profit, P, is the difference between the revenue and the cost: $P = R - C$. A company's revenue and cost may be represented by equations.

A company manufactures and sells woodstoves. The total monthly cost, in dollars, to produce n woodstoves is $C = 30n + 2000$. Write a variable expression for the company's monthly profit if the revenue, in dollars, obtained from selling all n woodstoves is $R = 150n - 0.4n^2$.

$$P = R - C$$
$$P = 150n - 0.4n^2 - (30n + 2000) \quad \text{• Replace } R \text{ by } 150n - 0.4n^2 \text{ and } C \text{ by } 30n + 2000. \text{ Then simplify.}$$
$$P = -0.4n^2 + 120n - 2000$$

How many woodstoves must the company manufacture and sell in order to make a profit of $6000 a month?

$$P = -0.4n^2 + 120n - 2000$$
$$6000 = -0.4n^2 + 120n - 2000 \quad \text{• Substitute 6000 for } P.$$
$$0 = -0.4n^2 + 120n - 8000 \quad \text{• Write the equation in standard form.}$$
$$0 = n^2 - 300n + 20{,}000 \quad \text{• Divide each side of the equation by } -0.4.$$
$$0 = (n - 100)(n - 200) \quad \text{• Factor.}$$
$$n - 100 = 0 \qquad n - 200 = 0 \quad \text{• Solve for } n.$$
$$n = 100 \qquad n = 200$$

The company will make a monthly profit of $6000 if either 100 or 200 woodstoves are manufactured and sold.

The graph of $P = -0.4n^2 + 120n - 2000$ is shown at the right. Note that when $P = 6000$, the values of n are 100 and 200.

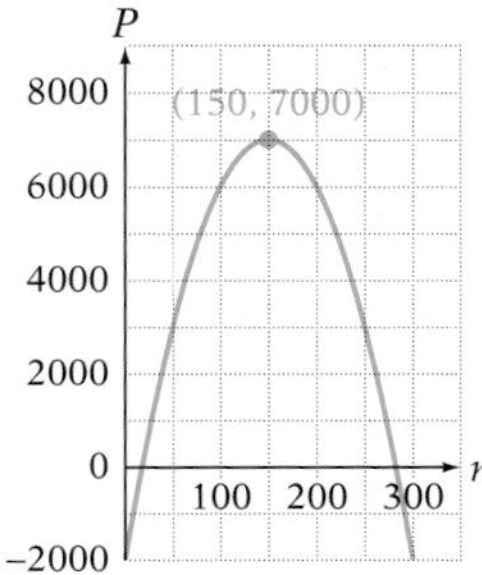

Also note that the coordinates of the highest point on the graph are (150, 7000). This means that the company makes a maximum profit of $7000 per month when 150 woodstoves are manufactured and sold.

1. The total cost, in dollars, for a company to produce and sell n guitars per month is $C = 240n + 1200$. The company's revenue, in dollars, from selling all n guitars is $R = 400n - 2n^2$.

   a. How many guitars must the company produce and sell each month in order to make a monthly profit of $1200?

   b. Graph the profit equation. What is the maximum monthly profit that the company can make?

PROJECTS AND GROUP ACTIVITIES

Finding the Maximum or Minimum of a Function Using a Graphing Calculator

Recall that for a quadratic function, the maximum or minimum value of the function is the y-coordinate of the vertex of the graph of the function. We determined algebraically that the coordinates of the vertex are $\left(-\frac{b}{2a}, f\left(-\frac{b}{2a}\right)\right)$. It is also possible to use a graphing calculator to approximate a maximum or minimum value. The method used to find the value depends on the calculator model. However, it is generally true for most calculators that the point at which the function has a maximum or minimum must be shown on the screen. If it is not, you must select a different viewing window.

For a TI-84 graphing calculator, press 2ND CALC 3 to determine a minimum value or 2ND CALC 4 to determine a maximum value. Move the cursor to a point on the curve that is to the left of the minimum (or maximum). Press ENTER. Move the cursor to a point on the curve that is to the right of the minimum (or maximum). Press ENTER twice. The minimum (or maximum) value is shown as the y-coordinate at the bottom of the screen.

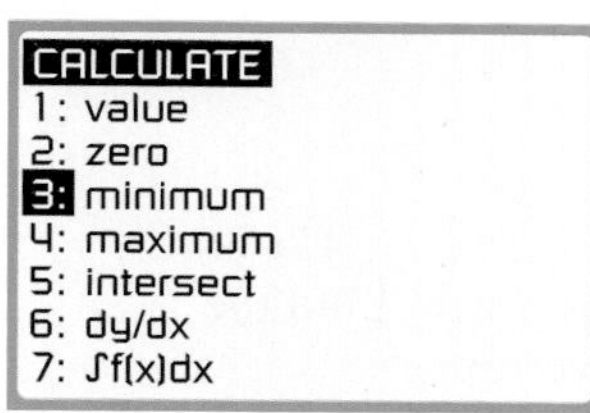

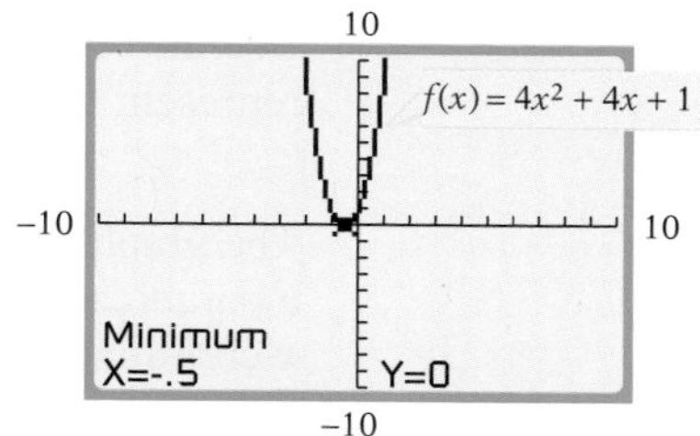

1. Find the maximum or minimum value of $f(x) = -x^2 - 3x + 2$ by using a graphing calculator.
2. Find the maximum or minimum value of $f(x) = x^2 - 4x - 2$ by using a graphing calculator.
3. Find the maximum or minimum value of $f(x) = -\sqrt{2}x^2 - \pi x + 2$ by using a graphing calculator.
4. Find the maximum or minimum value of $f(x) = \frac{x^2}{\sqrt{5}} - \sqrt{7}x - \sqrt{11}$ by using a graphing calculator.
5. **Business** A manufacturer of camera lenses estimates that the average monthly cost C of production is given by the function $C(x) = 0.1x^2 - 20x + 2000$, where x is the number of lenses produced each month. Find the number of lenses the company should produce in order to minimize the average cost.
6. **Compensation** The net annual income I, in dollars, of a family physician can be modeled by the equation $I(x) = -290(x - 48)^2 + 148{,}000$, where x is the age of the physician and $27 \leq x \leq 70$.
   a. Find the age at which the physician's income will be a maximum.
   b. Find the maximum income.

Business Applications of Maximum and Minimum Values of Quadratic Functions

A company's revenue R is the total amount of money the company earns by selling its products. The cost C is the total amount of money the company spends to manufacture and sell its products. A company's profit P is the difference between the revenue and the cost.

$$P = R - C$$

A company's revenue and cost may be represented by equations.

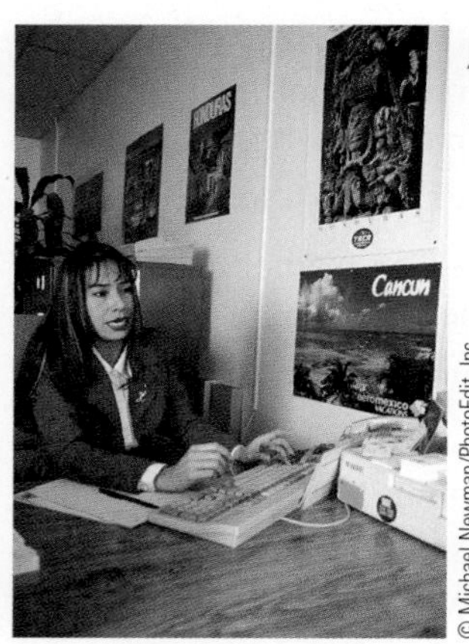

© Michael Newman/PhotoEdit, Inc.

HOW TO • 1 The owners of a travel agency are selling tickets for a tour. The travel agency must pay a cost per person of \$10. A maximum of 180 people can be accommodated on the tour. For every \$.25 increase in the price of a ticket over the \$10 cost, the owners estimate that they will sell one ticket less. Therefore, the price per ticket can be represented by

$$\begin{aligned} p &= 10 + 0.25(180 - x) \\ &= 55 - 0.25x \end{aligned}$$

where x represents the number of tickets sold. Determine the maximum profit and the price per ticket that yields the maximum profit.

Solution

The cost is represented by $10x$ and the revenue is represented by $x(55 - 0.25x)$, where x represents the number of tickets sold.

Use the equation $P = R - C$ to represent the profit.

$$P = R - C$$
$$P = x(55 - 0.25x) - 10x$$
$$P = -0.25x^2 + 45x$$

The graph of $P = -0.25x^2 + 45x$ is a parabola that opens down. Find the x-coordinate of the vertex.

$$x = -\frac{b}{2a} = -\frac{45}{2(-0.25)} = 90$$

To find the maximum profit, evaluate $P = -0.25x^2 + 45x$ at 90.

$$P = -0.25x^2 + 45x$$
$$P = -0.25(90^2) + 45(90)$$
$$P = 2025$$

To find the price per ticket, evaluate $p = 55 - 0.25x$ at 90.

$$p = 55 - 0.25x$$
$$p = 55 - 0.25(90)$$
$$p = 32.50$$

The travel agency can expect a maximum profit of \$2025 when 90 people take the tour at a ticket price of \$32.50 per person.

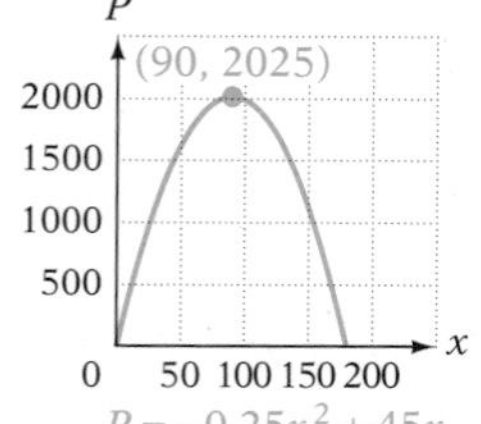

$P = -0.25x^2 + 45x$

The graph of the profit function is shown at the left.

1. **Business** A company's total monthly cost, in dollars, for producing and selling n DVDs per month is $C = 25n + 4000$. The company's revenue, in dollars, from selling all n DVDs is $R = 275n - 0.2n^2$.
   a. How many DVDs must the company produce and sell each month in order to make the maximum monthly profit?
   b. What is the maximum monthly profit?

2. **Business** A shipping company has determined that its cost to deliver x packages per week is $1000 + 6x$. The price per package that the company charges when it sends x packages per week is $24 - 0.02x$. Determine the company's maximum weekly profit and the price per package that yields the maximum profit.

CHAPTER 11

SUMMARY

KEY WORDS	EXAMPLES
A *quadratic function* is one that can be expressed by the equation $f(x) = ax^2 + bx + c, a \neq 0$. [11.1A, p. 620]	$f(x) = 4x^2 + 3x - 5$ is a quadratic function.

The value of the independent variable is sometimes called the *input.* The result of evaluating the function is called the *output.* An *input/output table* shows the results of evaluating a function for various values of the independent variable. [11.1A, p. 620]

$f(x) = x^2 + 4x - 6$

x	y
−2	−10
−1	−9
0	−6
1	−1
2	6

The graph of a quadratic function is a *parabola.* When $a > 0$, the parabola opens up and the *vertex* of the parabola is the point with the smallest y-coordinate. When $a < 0$, the parabola opens down and the *vertex* of the parabola is the point with the largest y-coordinate. The *axis of symmetry* is the vertical line that passes through the vertex of the parabola and is parallel to the y-axis. [11.1A, pp. 620, 621]

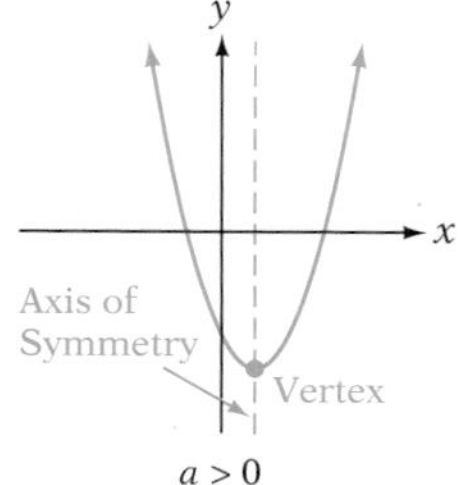

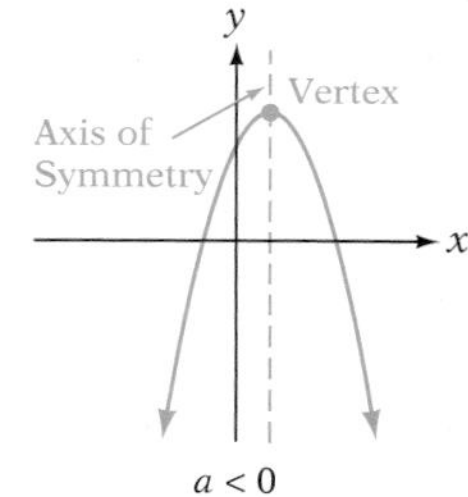

A point at which a graph crosses the x-axis is called an *x-intercept* of the graph. The x-intercepts occur when $y = 0$. [11.1B, p. 623]

$$x^2 - 5x + 6 = 0$$
$$(x - 2)(x - 3) = 0$$
$$x - 2 = 0 \quad x - 3 = 0$$
$$x = 2 \quad x = 3$$

The x-intercepts are (2, 0) and (3, 0).

A value of x for which $f(x) = 0$ is called a *zero* of the function. [11.1B, p. 623]

$$f(x) = x^2 - 7x + 10$$
$$0 = x^2 - 7x + 10$$
$$0 = (x - 2)(x - 5)$$
$$x - 2 = 0 \quad x - 5 = 0$$
$$x = 2 \quad x = 5$$

2 and 5 are zeros of the function.

The graph of a quadratic function has a *minimum value* if $a > 0$ and a *maximum value* if $a < 0$. [11.1C, p. 626]

The graph of $f(x) = 3x^2 - x + 4$ has a minimum value.
The graph of $f(x) = -x^2 - 6x + 5$ has a maximum value.

A function is a *one-to-one function* if, for any a and b in the domain of f, $f(a) = f(b)$ implies $a = b$. This means that given any y, there is only one x that can be paired with that y. One-to-one functions are commonly written as 1–1. [11.4A, p. 650]

A non-constant linear function is a 1–1 function.
A quadratic function is not a 1–1 function.

The *inverse of a function* is the set of ordered pairs formed by reversing the coordinates of each ordered pair of the function. [11.4B, p. 651]

The inverse of the function $\{(1, 2), (2, 4), (3, 6), (4, 8), (5, 10)\}$ is $\{(2, 1), (4, 2), (6, 3), (8, 4), (10, 5)\}$.

ESSENTIAL RULES AND PROCEDURES — EXAMPLES

Vertex and Axis of Symmetry of a Parabola [11.1A, p. 621]
Let $f(x) = ax^2 + bx + c$ be the equation of a parabola. The coordinates of the vertex are $\left(-\frac{b}{2a}, f\left(-\frac{b}{2a}\right)\right)$. The equation of the axis of symmetry is $x = -\frac{b}{2a}$.

$f(x) = x^2 - 2x - 4$
$a = 1, b = -2$
$$-\frac{b}{2a} = -\frac{-2}{2(1)} = 1$$
$f(1) = 1^2 - 2(1) - 4 = -5$
The coordinates of the vertex are $(1, -5)$.
The axis of symmetry is $x = 1$.

Effect of the Discriminant on the Number of x-Intercepts of a Parabola [11.1B, p. 624]

1. If $b^2 - 4ac = 0$, the parabola has one x-intercept.
2. If $b^2 - 4ac > 0$, the parabola has two x-intercepts.
3. If $b^2 - 4ac < 0$, the parabola has no x-intercepts.

$x^2 + 8x + 16 = 0$ has one x-intercept because $b^2 - 4ac = 8^2 - 4(1)(16) = 0$.
$2x^2 + 3x - 5 = 0$ has two x-intercepts because $b^2 - 4ac = 3^2 - 4(2)(-5) = 49$.
$3x^2 + 2x + 4 = 0$ has no x-intercepts because $b^2 - 4ac = 2^2 - 4(3)(4) = -44$.

To Find the Minimum or Maximum Value of a Quadratic Function [11.1C, p. 626]
First find the x-coordinate of the vertex. Then evaluate the function at that value.

$f(x) = x^2 - 2x - 4$
$a = 1, b = -2$
$$-\frac{b}{2a} = -\frac{-2}{2(1)} = 1$$
$f(1) = 1^2 - 2(1) - 4 = -5$
$a > 0$. The minimum value of the function is -5.

Vertical-Line Test [11.2A, p. 637]
A graph defines a function if any vertical line intersects the graph at no more than one point.

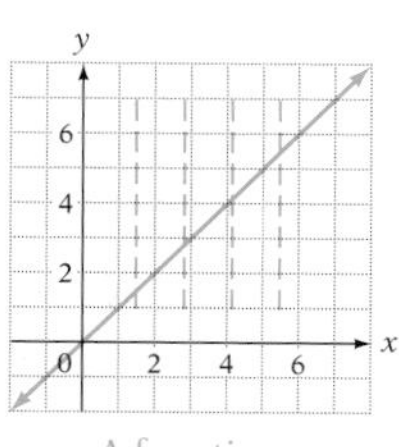
A function

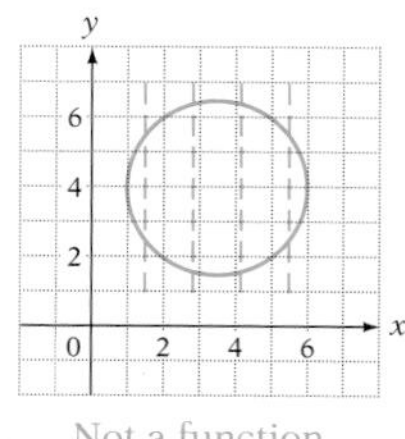
Not a function

Operations on Functions [11.3A, p. 642]
If f and g are functions and x is an element of the domain of each function, then

$(f + g)(x) = f(x) + g(x)$
$(f - g)(x) = f(x) - g(x)$
$(f \cdot g)(x) = f(x) \cdot g(x)$
$\left(\frac{f}{g}\right)(x) = \frac{f(x)}{g(x)}, g(x) \neq 0$

Given $f(x) = x + 2$ and $g(x) = 2x$, then

$$(f + g)(4) = f(4) + g(4) = (4 + 2) + 2(4) = 6 + 8 = 14$$

$$(f - g)(4) = f(4) - g(4) = (4 + 2) - 2(4) = 6 - 8 = -2$$

$$(f \cdot g)(4) = f(4) \cdot g(4) = (4 + 2) \cdot 2(4) = 6 \cdot 8 = 48$$

$$\left(\frac{f}{g}\right)(4) = \frac{f(4)}{g(4)} = \frac{4 + 2}{2(4)} = \frac{6}{8} = \frac{3}{4}$$

Definition of the Composition of Two Functions [11.3B, p. 644]
The composition of two functions f and g, symbolized by $f \circ g$, is the function whose value at x is given by $(f \circ g)(x) = f[g(x)]$.

Given $f(x) = x - 4$ and $g(x) = 4x$, then

$$(f \circ g)(2) = f[g(2)] = f(8) \quad \text{because } g(2) = 8$$
$$= 8 - 4 = 4$$

Horizontal-Line Test [11.4A, p. 651]
The graph of a function represents the graph of a 1–1 function if any horizontal line intersects the graph at no more than one point.

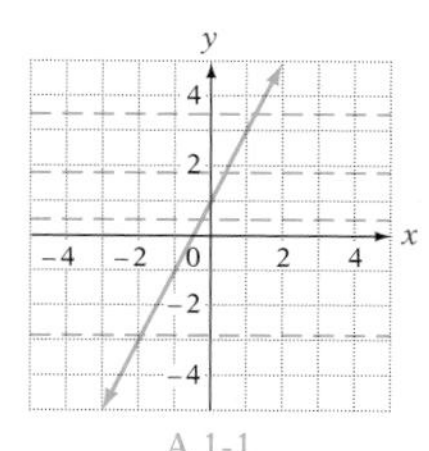
A 1-1 function

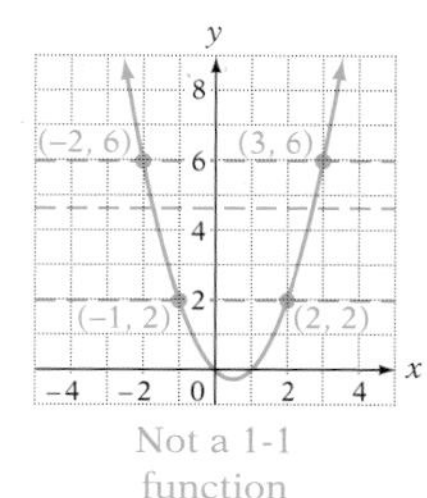

Not a 1-1 function

Condition for an Inverse Function [11.4B, p. 652]
A function f has an inverse function if and only if f is a 1–1 function.

The function $f(x) = x^2$ does not have an inverse function. When $y = 4$, $x = 2$ or -2; therefore, the function $f(x)$ is not a 1–1 function.

Composition of Inverse Functions Property [11.4B, p. 653]
$f^{-1}[f(x)] = x$ and $f[f^{-1}(x)] = x$.

$$f(x) = 2x - 3 \qquad f^{-1}(x) = \frac{1}{2}x + \frac{3}{2}$$

$$f^{-1}[f(x)] = \frac{1}{2}(2x - 3) + \frac{3}{2} = x$$

$$f[f^{-1}(x)] = 2\left(\frac{1}{2}x + \frac{3}{2}\right) - 3 = x$$

CHAPTER 11

CONCEPT REVIEW

Test your knowledge of the concepts presented in this chapter. Answer each question. Then check your answers against the ones provided in the Answer Section.

1. Given a quadratic function, how do you find the vertex and axis of symmetry of its graph?

2. Given its vertex and axis of symmetry, can you accurately graph a parabola?

3. Suppose you know the vertex of a parabola and whether it opens up or down. How can you determine the range of the function?

4. What does the discriminant tell you about the x-intercepts of a parabola?

5. For the graph of $f(x) = ax^2 + bx + c$, $a \neq 0$, how does the coefficient of the x^2 term change the shape of the parabola?

6. What is the vertical-line test?

7. What is the shape of the graph of the absolute value of a linear polynomial?

8. Describe the four basic operations on functions.

9. What does the notation $(f \circ g)(x)$ represent?

10. How does the horizontal-line test determine a 1–1 function?

11. What condition must be met for a function to have an inverse?

CHAPTER 11

REVIEW EXERCISES

1. Is the graph below the graph of a function?

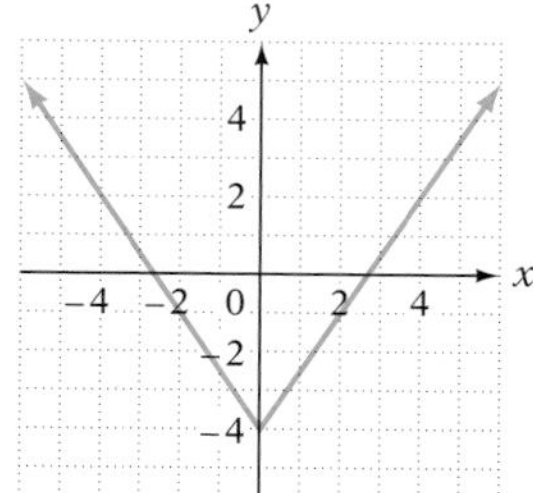

2. Is the graph below the graph of a 1–1 function?

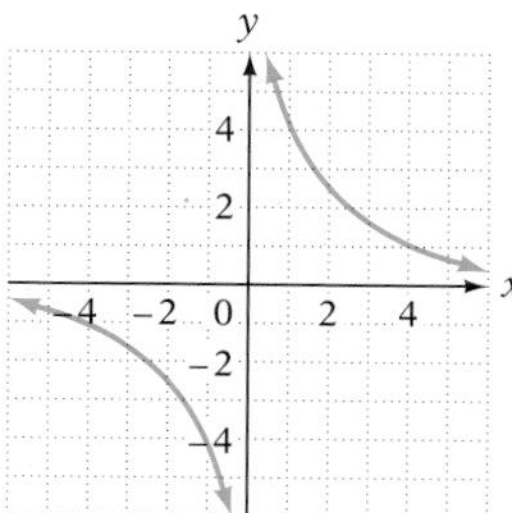

3. Graph $f(x) = 3x^3 - 2$. State the domain and range.

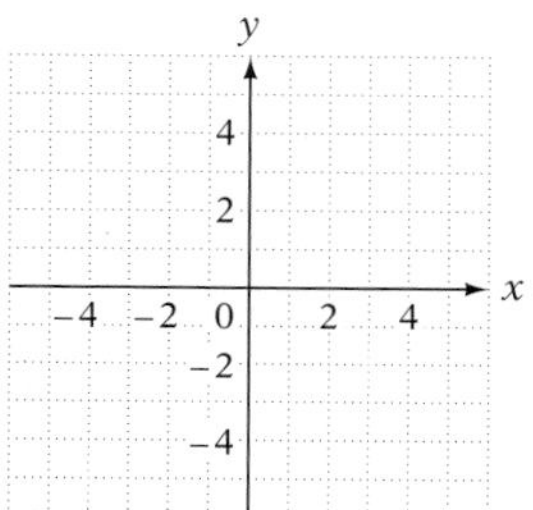

4. Graph $f(x) = \sqrt{x + 4}$. State the domain and range.

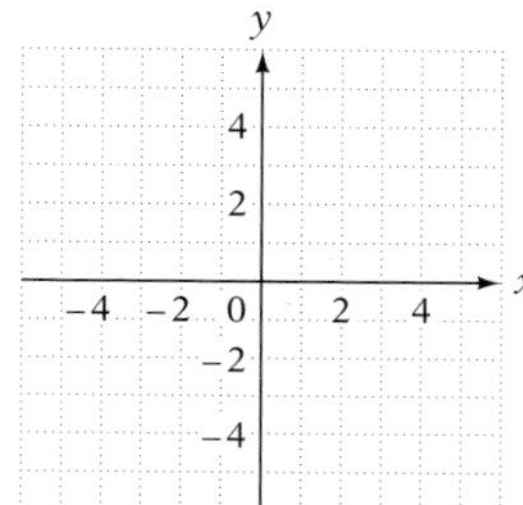

5. Graph $f(x) = |x| - 3$. State the domain and range.

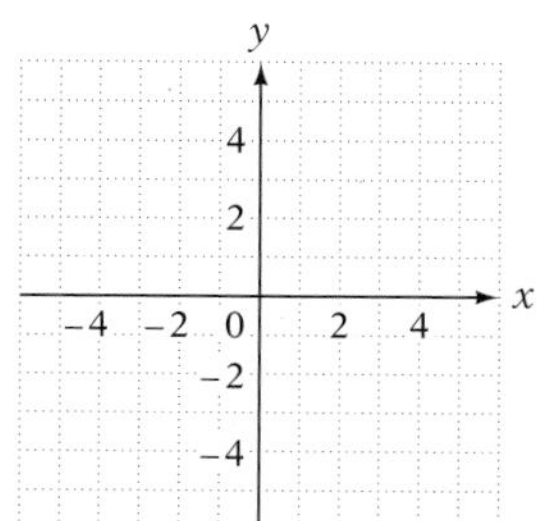

6. Find the coordinates of the vertex and the equation of the axis of symmetry for the parabola whose equation is $y = x^2 - 2x + 3$. Then graph the equation.

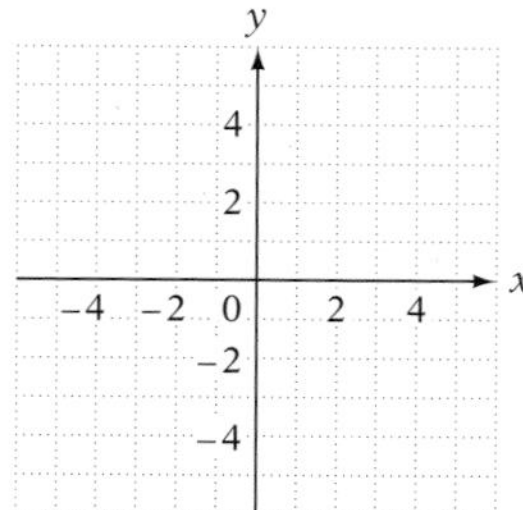

7. Use the discriminant to determine the number of x-intercepts of the graph of $y = -3x^2 + 4x + 6$.

8. Use the discriminant to determine the number of x-intercepts of the graph of $f(x) = 2x^2 + x + 5$.

9. Find the x-intercepts of the graph of $y = 3x^2 + 9x$.

10. Find the x-intercepts of the graph of $f(x) = x^2 - 6x + 7$.

11. Find the zeros of $f(x) = 2x^2 - 7x - 15$.

12. Find the zeros of $f(x) = x^2 - 2x + 10$.

13. Find the maximum value of $f(x) = -2x^2 + 4x + 1$.

14. Find the minimum value of $f(x) = x^2 - 7x + 8$.

15. Given $f(x) = x^2 + 4$ and $g(x) = 4x - 1$, find $f[g(0)]$.

16. Given $f(x) = 6x + 8$ and $g(x) = 4x + 2$, find $g[f(-1)]$.

17. Given $f(x) = 3x^2 - 4$ and $g(x) = 2x + 1$, find $f[g(x)]$.

18. Given $f(x) = 2x^2 + x - 5$ and $g(x) = 3x - 1$, find $g[f(x)]$.

For Exercises 19 to 22, use $f(x) = x^2 + 2x - 3$ and $g(x) = x^2 - 2$.

19. Evaluate: $(f + g)(2)$

20. Evaluate: $(f - g)(-4)$

21. Evaluate: $(f \cdot g)(-4)$

22. Evaluate: $\left(\frac{f}{g}\right)(3)$

23. Find the inverse of $f(x) = -6x + 4$.

24. Find the inverse of $f(x) = \frac{2}{3}x - 12$.

25. Use the Composition of Inverse Functions Property to determine whether $f(x) = -\frac{1}{4}x + \frac{5}{4}$ and $g(x) = -4x + 5$ are inverses of each other.

26. Use the Composition of Inverse Functions Property to determine whether $f(x) = \frac{1}{2}x$ and $g(x) = 2x + 1$ are inverses of each other.

27. **Deep-sea Diving** The pressure p, in pounds per square inch, on a scuba diver x feet below the surface of the water can be approximated by $p(x) = 0.4x + 15$. Find $p^{-1}(x)$, and write a sentence that explains its meaning.

28. **Business** The monthly profit P, in dollars, for a company from the sale of x youth baseball gloves is given by $P(x) = -x^2 + 100x + 2500$. How many gloves should be made each month to maximize profit? What is the maximum monthly profit?

29. **Geometry** The perimeter of a rectangle is 28 ft. What dimensions would give the rectangle a maximum area?

CHAPTER 11

TEST

1. Find the coordinates of the vertex and the equation of the axis of symmetry for the parabola whose equation is $y = x^2 - 6x + 4$. Then graph the equation.

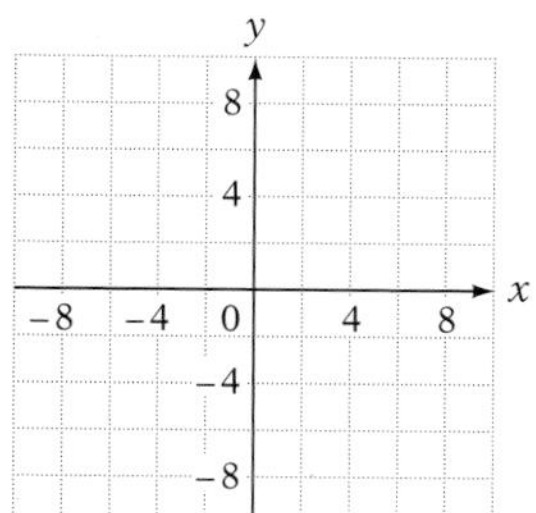

2. Graph $f(x) = \left|\frac{1}{2}x\right| - 2$. State the domain and range.

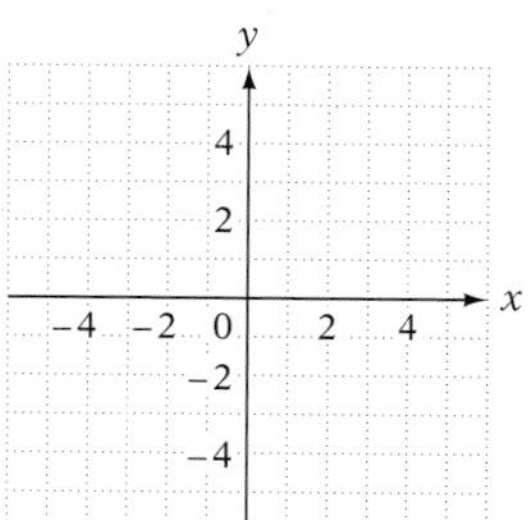

3. Graph $f(x) = -\sqrt{3 - x}$. State the domain and range.

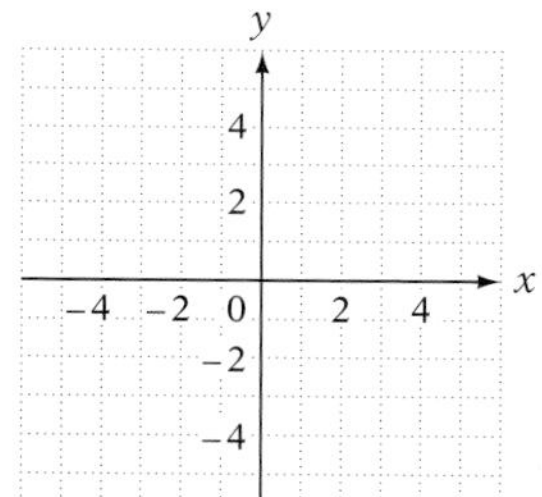

4. Graph $f(x) = x^3 - 3x + 2$. State the domain and range.

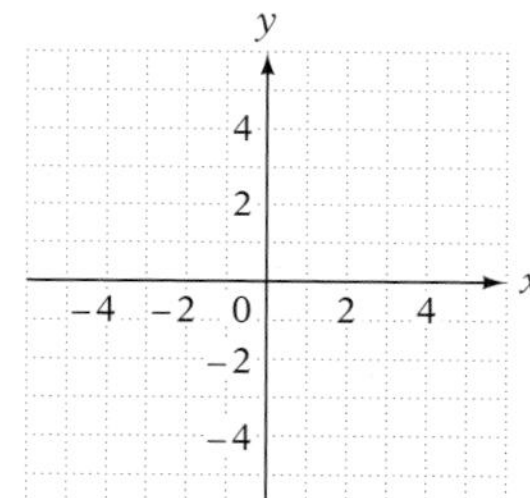

5. Use the discriminant to determine the number of x-intercepts of the graph of $f(x) = 3x^2 - 4x + 6$.

6. Find the coordinates of the x-intercepts of the graph of $y = 3x^2 - 7x - 6$.

7. Find the maximum value of $f(x) = -x^2 + 8x - 7$.

8. Find the domain and range of $f(x) = 2x^2 + 4x - 5$.

9. Given $f(x) = x^2 + 2x - 3$ and $g(x) = x^3 - 1$, find $(f - g)(2)$.

10. Given $f(x) = x^3 + 1$ and $g(x) = 2x - 3$, find $(f \cdot g)(-3)$.

11. Given $f(x) = 4x - 5$ and $g(x) = x^2 + 3x + 4$, find $\left(\frac{f}{g}\right)(-2)$.

12. Given $f(x) = x^2 + 4$ and $g(x) = 2x^2 + 2x + 1$, find $(f - g)(-4)$.

13. Given $f(x) = 2x - 7$ and $g(x) = x^2 - 2x - 5$, find $f[g(2)]$.

14. Given $f(x) = x^2 + 1$ and $g(x) = x^2 + x + 1$, find $g[f(-2)]$.

15. Given $f(x) = x^2 - 1$ and $g(x) = 3x + 2$, find $g[f(x)]$.

16. Given $f(x) = 2x^2 - 7$ and $g(x) = x - 1$, find $f[g(x)]$.

17. Does the function $\{(1, 4), (2, 5), (3, 6), (4, 5), (5, 4)\}$ have an inverse function?

18. Find the inverse of the function $\{(2, 6), (3, 5), (4, 4), (5, 3)\}$.

19. Find the inverse of $f(x) = 4x - 2$.

20. Find the inverse of $f(x) = \frac{1}{4}x - 4$.

21. Use the Composition of Inverse Functions Property to determine whether $f(x) = \frac{1}{2}x + 2$ and $g(x) = 2x - 4$ are inverses of each other.

22. Use the Composition of Inverse Functions Property to determine whether $f(x) = \frac{2}{3}x - 2$ and $g(x) = \frac{3}{2}x + 3$ are inverses of each other.

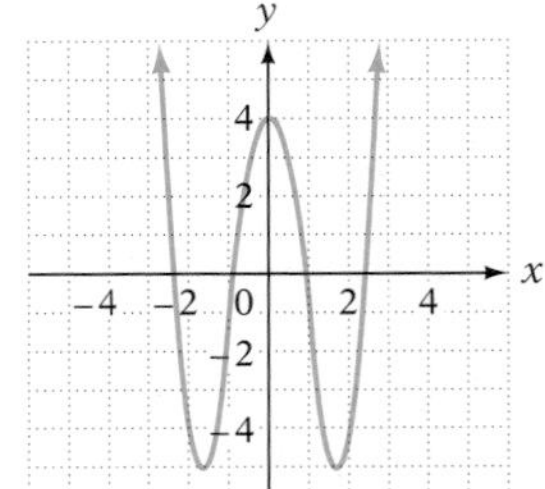

23. Determine whether the graph at the right is the graph of a 1–1 function.

24. Delivery Service A company that delivers documents in a city via bicycle messenger charges \$5 per message plus \$1.25 per mile for the service. The total cost $C(x)$ to deliver a document to a location x miles away is given by $C(x) = 1.25x + 5$. Find $C^{-1}(x)$, and write a sentence that explains its meaning.

25. Business The daily production cost $C(x)$, in dollars, to manufacture small speakers is given by $C(x) = x^2 - 50x + 675$, where x is the number of speakers produced per day. How many speakers should be made per day to minimize the daily production cost? What is the minimum daily production cost?

26. Mathematics Find two numbers whose sum is 28 and whose product is a maximum.

27. Geometry The perimeter of a rectangle is 200 cm. What dimensions would give the rectangle a maximum area? What is the maximum area?

CUMULATIVE REVIEW EXERCISES

1. Evaluate $-3a + \left|\frac{3b - ab}{3b - c}\right|$ when $a = 2$, $b = 2$, and $c = -2$.

2. Graph $\{x \mid x < -3\} \cap \{x \mid x > -4\}$.

   −5 −4 −3 −2 −1 0 1 2 3 4 5

3. Solve: $\dfrac{3x - 1}{6} - \dfrac{5 - x}{4} = \dfrac{5}{6}$

4. Solve: $4x - 2 < -10$ or $3x - 1 > 8$
   Write the solution set in set-builder notation.

5. Solve: $|8 - 2x| \geq 0$

6. Simplify: $\left(\dfrac{3a^3b}{2a}\right)^2\left(\dfrac{a^2}{-3b^2}\right)^3$

7. Multiply: $(x - 4)(2x^2 + 4x - 1)$

8. Solve by using the addition method:
   $6x - 2y = -3$
   $4x + y = 5$

9. Factor: $x^3y + x^2y^2 - 6xy^3$

10. Solve: $(b + 2)(b - 5) = 2b + 14$

11. Solve: $x^2 - 2x > 15$

12. Subtract: $\dfrac{x^2 + 4x - 5}{2x^2 - 3x + 1} - \dfrac{x}{2x - 1}$

13. Solve: $\dfrac{5}{x^2 + 7x + 12} = \dfrac{9}{x + 4} - \dfrac{2}{x + 3}$

14. Divide: $\dfrac{4 - 6i}{2i}$

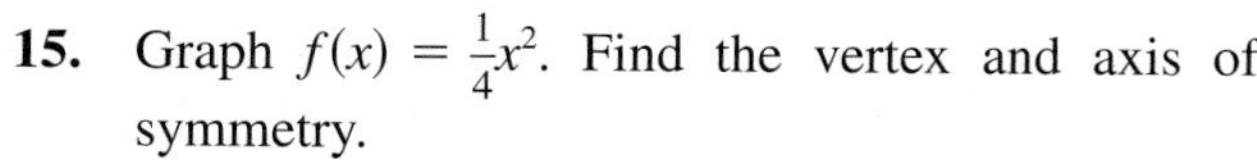

15. Graph $f(x) = \frac{1}{4}x^2$. Find the vertex and axis of symmetry.

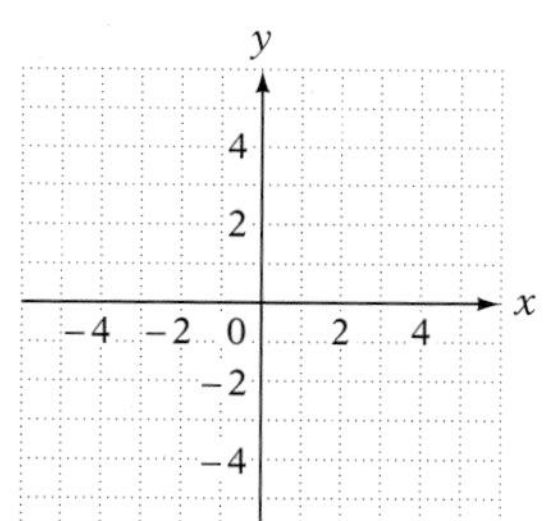

16. Graph the solution set of $3x - 4y \geq 8$.

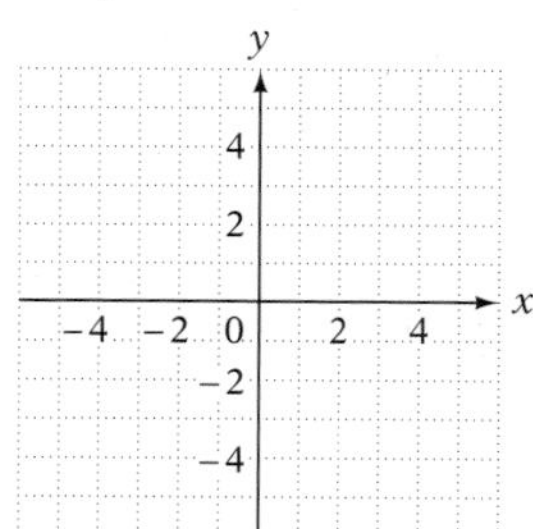

17. Find the equation of the line containing the points whose coordinates are $(-3, 4)$ and $(2, -6)$.

18. Find the equation of the line that contains the point whose coordinates are $(-3, 1)$ and is perpendicular to the line $2x - 3y = 6$.

19. Solve: $3x^2 = 3x - 1$

20. Solve: $\sqrt{8x + 1} = 2x - 1$

21. Find the minimum value of $f(x) = 2x^2 - 3$.

22. Find the range of $f(x) = |3x - 4|$ if the domain is $\{0, 1, 2, 3\}$.

23. Is this set of ordered pairs a function? $\{(-3, 0), (-2, 0), (-1, 1), (0, 1)\}$

24. Solve: $\sqrt[3]{5x - 2} = 2$

25. Given $g(x) = 3x - 5$ and $h(x) = \frac{1}{2}x + 4$, find $g[h(2)]$.

26. Find the inverse of $f(x) = -3x + 9$.

27. **Mixtures** Find the cost per pound of a tea mixture made from 30 lb of tea costing \$4.50 per pound and 45 lb of tea costing \$3.60 per pound.

28. **Mixtures** How many pounds of an 80% copper alloy must be mixed with 50 lb of a 20% copper alloy to make an alloy that is 40% copper?

29. **Mixtures** Six ounces of an insecticide are mixed with 16 gal of water to make a spray for spraying an orange grove. How much additional insecticide is required if it is to be mixed with 28 gal of water?

30. **Tanks** A large pipe can fill a tank in 8 min less time than it takes a smaller pipe to fill the same tank. Working together, both pipes can fill the tank in 3 min. How long would it take the larger pipe, working alone, to fill the tank?

31. **Physics** The distance d that a spring stretches varies directly as the force f used to stretch the spring. If a force of 50 lb can stretch a spring 30 in., how far can a force of 40 lb stretch the spring?

32. **Music** The frequency of vibration f in a pipe in an open pipe organ varies inversely as the length L of the pipe. If the air in a pipe 2 m long vibrates 60 times per minute, find the frequency in a pipe that is 1.5 m long.

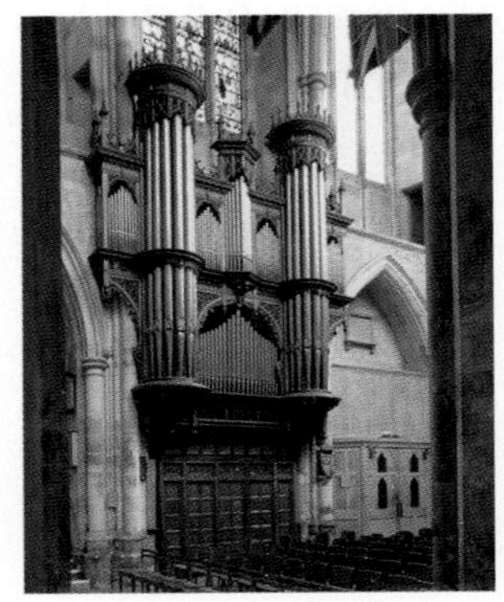

CHAPTER 12

Exponential and Logarithmic Functions

OBJECTIVES

SECTION 12.1

A To evaluate an exponential function
B To graph an exponential function

SECTION 12.2

A To find the logarithm of a number
B To use the Properties of Logarithms to simplify expressions containing logarithms
C To use the Change-of-Base Formula

SECTION 12.3

A To graph a logarithmic function

SECTION 12.4

A To solve an exponential equation
B To solve a logarithmic equation

SECTION 12.5

A To solve application problems

ARE YOU READY?

Take the Chapter 12 Prep Test to find out if you are ready to learn to:

- Evaluate an exponential function
- Graph an exponential function and a logarithmic function
- Find the logarithm of a number
- Use the Properties of Logarithms
- Use the Change-of-Base Formula
- Solve an exponential equation and a logarithmic equation

PREP TEST

Do these exercises to prepare for Chapter 12.

1. Simplify: 3^{-2}

2. Simplify: $\left(\frac{1}{2}\right)^{-4}$

3. Complete: $\frac{1}{8} = 2^{?}$

4. Evaluate $f(x) = x^4 + x^3$ for $x = -1$ and $x = 3$.

5. Solve: $3x + 7 = x - 5$

6. Solve: $16 = x^2 - 6x$

7. Evaluate $A(1 + i)^n$ for $A = 5000$, $i = 0.04$, and $n = 6$. Round to the nearest hundredth.

8. Graph: $f(x) = x^2 - 1$

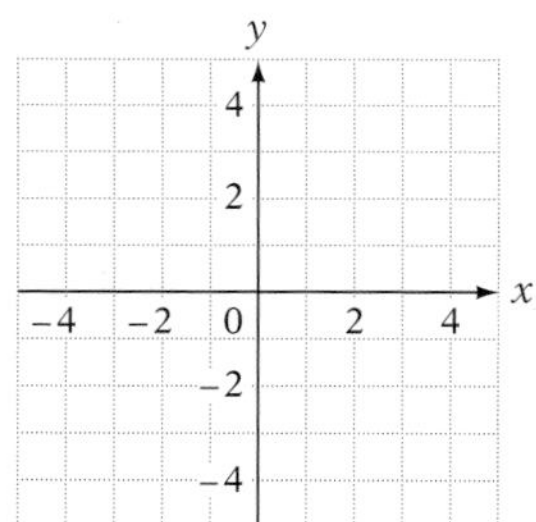

SECTION

12.1 Exponential Functions

OBJECTIVE A To evaluate an exponential function

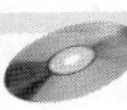

The growth of a \$500 savings account that earns 5% annual interest compounded daily is shown in the graph at the right. In approximately 14 years, the savings account contains approximately \$1000, twice the initial amount. The growth of this savings account is an example of an exponential function.

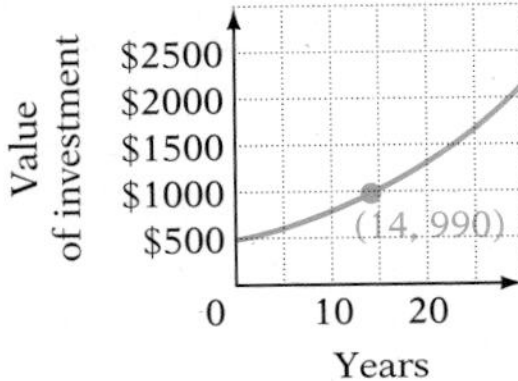

The pressure of the atmosphere at a certain altitude is shown in the graph at the right. This is another example of an exponential function. From the graph, we read that the air pressure is approximately 6.5 lb/in^2 at an altitude of 20,000 ft.

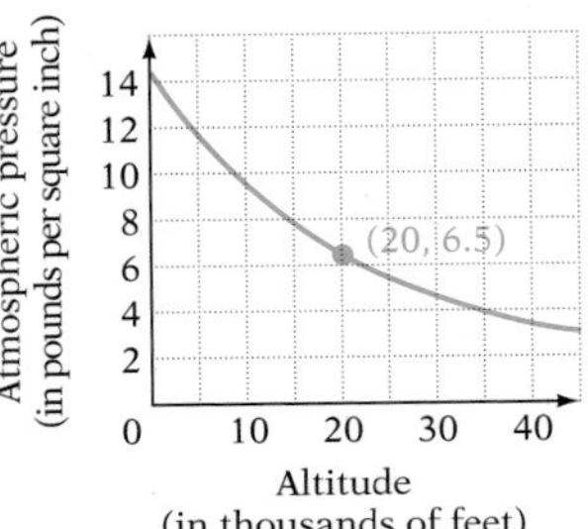

Definition of an Exponential Function

The **exponential function** with base b is defined by

$$f(x) = b^x$$

where $b > 0$, $b \neq 1$, and x is any real number.

In the definition of an exponential function, b, the base, is required to be positive. If the base were a negative number, the value of the function would be a complex number for some values of x. For instance, the value of $f(x) = (-4)^x$ when $x = \frac{1}{2}$ is $f\left(\frac{1}{2}\right) = (-4)^{1/2} = \sqrt{-4} = 2i$. To avoid complex number values of a function, the base of the exponential function is always a positive number.

Integrating Technology

A graphing calculator can be used to evaluate an exponential expression at an irrational number.

```
4^√(2)
                7.102993301
```

HOW TO • 1 Evaluate $f(x) = 2^x$ at $x = 3$ and $x = -2$.

$f(3) = 2^3 = 8$ • Substitute 3 for x and simplify.

$f(-2) = 2^{-2} = \frac{1}{2^2} = \frac{1}{4}$ • Substitute −2 for x and simplify.

To evaluate an exponential expression at an irrational number such as $\sqrt{2}$, we obtain an approximation to the value of the function by approximating the irrational number. For instance, the value of $f(x) = 4^x$ at $x = \sqrt{2}$ can be approximated by using an approximation of $\sqrt{2}$.

$$f(\sqrt{2}) = 4^{\sqrt{2}} \approx 4^{1.4142} \approx 7.1029$$

✓ Take Note

The natural exponential function is an extremely important function. It is used extensively in applied problems in virtually all disciplines, from archaeology to zoology. Leonhard Euler (1707–1783) was the first to use the letter e as the base of the natural exponential function.

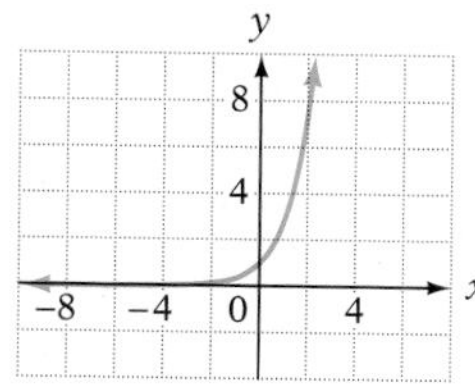

Because $f(x) = b^x$ $(b > 0, b \neq 1)$ can be evaluated at both rational and irrational numbers, the domain of f is all real numbers. And because $b^x > 0$ for all values of x, the range of f is the positive real numbers.

A frequently used base in applications of exponential functions is an irrational number designated by e. **The number e is approximately 2.71828183.** It is an irrational number, so it has a nonterminating, nonrepeating decimal representation.

Natural Exponential Function

The function defined by $f(x) = e^x$ is called the **natural exponential function.**

The e^x key on a calculator can be used to evaluate the natural exponential function. The graph of $y = e^x$ is shown at the left.

EXAMPLE 1

Evaluate $f(x) = \left(\frac{1}{2}\right)^x$ at $x = 2$ and $x = -3$.

Solution

$$f(x) = \left(\frac{1}{2}\right)^x$$

$$f(2) = \left(\frac{1}{2}\right)^2 = \frac{1}{4} \qquad \bullet\ x = 2$$

$$f(-3) = \left(\frac{1}{2}\right)^{-3} = 2^3 = 8 \qquad \bullet\ x = -3$$

YOU TRY IT 1

Evaluate $f(x) = \left(\frac{2}{3}\right)^x$ at $x = 3$ and $x = -2$.

Your solution

EXAMPLE 2

Evaluate $f(x) = 2^{3x-1}$ at $x = 1$ and $x = -1$.

Solution

$$f(x) = 2^{3x-1}$$

$$f(1) = 2^{3(1)-1} = 2^2 = 4 \qquad \bullet\ x = 1$$

$$f(-1) = 2^{3(-1)-1} = 2^{-4} = \frac{1}{2^4} = \frac{1}{16} \qquad \bullet\ x = -1$$

YOU TRY IT 2

Evaluate $f(x) = 2^{2x+1}$ at $x = 0$ and $x = -2$.

Your solution

EXAMPLE 3

Evaluate $f(x) = e^{2x}$ at $x = 1$ and $x = -1$. Round to the nearest ten-thousandth.

Solution

$$f(x) = e^{2x}$$

$$f(1) = e^{2\cdot 1} = e^2 \approx 7.3891 \qquad \bullet\ x = 1$$

$$f(-1) = e^{2(-1)} = e^{-2} \approx 0.1353 \qquad \bullet\ x = -1$$

YOU TRY IT 3

Evaluate $f(x) = e^{2x-1}$ at $x = 2$ and $x = -2$. Round to the nearest ten-thousandth.

Your solution

Solutions on p. S34

OBJECTIVE B To graph an exponential function 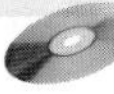

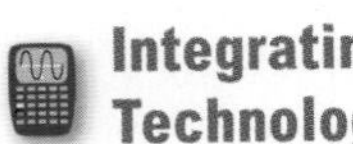

Integrating Technology

See the Keystroke Guide: *Graph* for instructions on using a graphing calculator to graph functions.

Some properties of an exponential function can be seen from its graph.

HOW TO • 2 Graph $f(x) = 2^x$.

Think of this as the equation $y = 2^x$.

Choose values of x and find the corresponding values of y. The results can be recorded in a table.

Graph the ordered pairs on a rectangular coordinate system.

Connect the points with a smooth curve.

x	$f(x) = y$
-2	$2^{-2} = \frac{1}{4}$
-1	$2^{-1} = \frac{1}{2}$
0	$2^0 = 1$
1	$2^1 = 2$
2	$2^2 = 4$
3	$2^3 = 8$

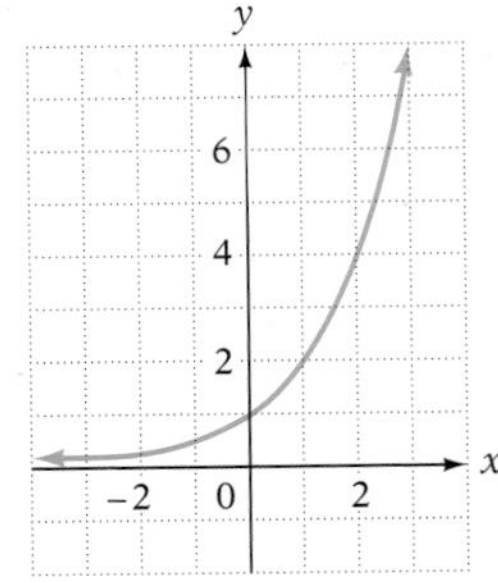

Note that any vertical line would intersect the graph at only one point. Therefore, by the vertical-line test, the graph of $f(x) = 2^x$ is the graph of a function. Also note that any horizontal line would intersect the graph at only one point. Therefore, the graph of $f(x) = 2^x$ is the graph of a one-to-one function.

HOW TO • 3 Graph $f(x) = \left(\frac{1}{2}\right)^x$.

Think of this as the equation $y = \left(\frac{1}{2}\right)^x$.

Choose values of x and find the corresponding values of y.

Graph the ordered pairs on a rectangular coordinate system.

Connect the points with a smooth curve.

x	$f(x) = y$
-3	$\left(\frac{1}{2}\right)^{-3} = 8$
-2	$\left(\frac{1}{2}\right)^{-2} = 4$
-1	$\left(\frac{1}{2}\right)^{-1} = 2$
0	$\left(\frac{1}{2}\right)^{0} = 1$
1	$\left(\frac{1}{2}\right)^{1} = \frac{1}{2}$
2	$\left(\frac{1}{2}\right)^{2} = \frac{1}{4}$

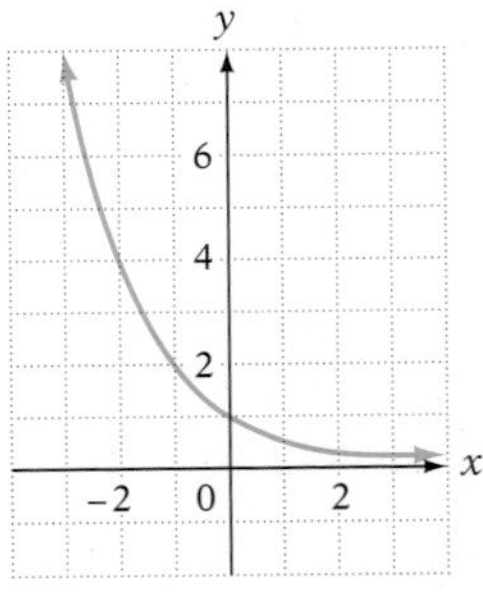

Applying the vertical-line and horizontal-line tests reveals that the graph of $f(x) = \left(\frac{1}{2}\right)^x$ is also the graph of a one-to-one function.

HOW TO • 4 Graph $f(x) = 2^{-x}$.

Think of this as the equation $y = 2^{-x}$.

Choose values of x and find the corresponding values of y.

Graph the ordered pairs on a rectangular coordinate system.

Connect the points with a smooth curve.

x	y
-3	8
-2	4
-1	2
0	1
1	$\frac{1}{2}$
2	$\frac{1}{4}$

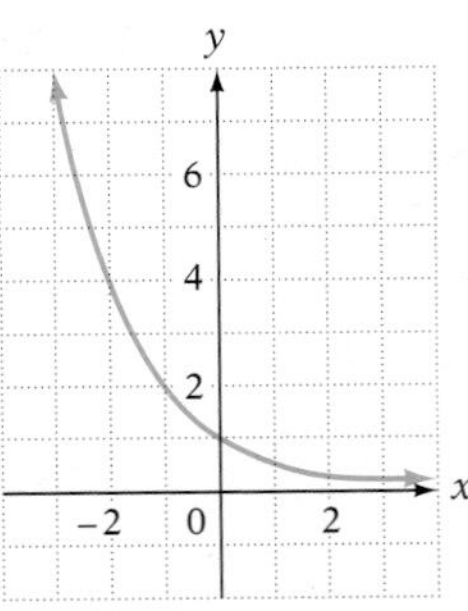

✓ Take Note

Note that because $2^{-x} = (2^{-1})^x = \left(\frac{1}{2}\right)^x$, the graphs of $f(x) = 2^{-x}$ and $f(x) = \left(\frac{1}{2}\right)^x$ are the same.

EXAMPLE • 4

Graph: $f(x) = 3^{\frac{1}{2}x-1}$

Solution

x	y
-2	$\frac{1}{9}$
0	$\frac{1}{3}$
2	1
4	3

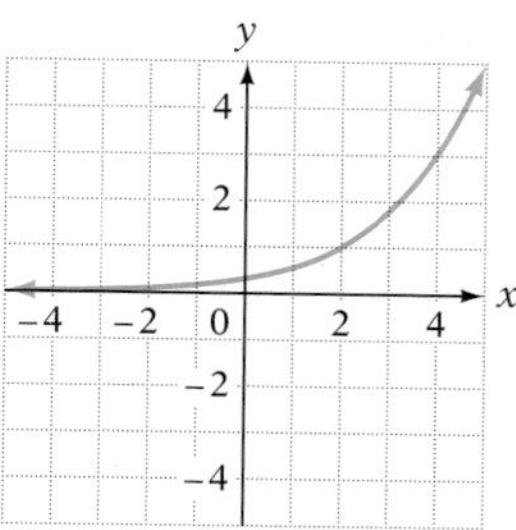

YOU TRY IT • 4

Graph: $f(x) = 2^{-\frac{1}{2}x}$

Your solution

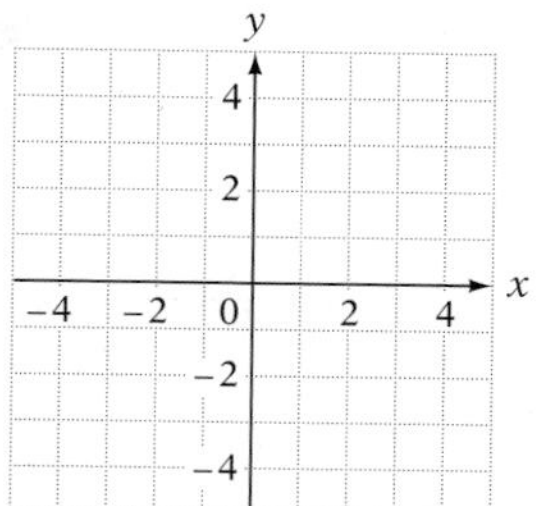

EXAMPLE • 5

Graph: $f(x) = 2^x - 1$

Solution

x	y
-2	$-\frac{3}{4}$
-1	$-\frac{1}{2}$
0	0
1	1
2	3
3	7

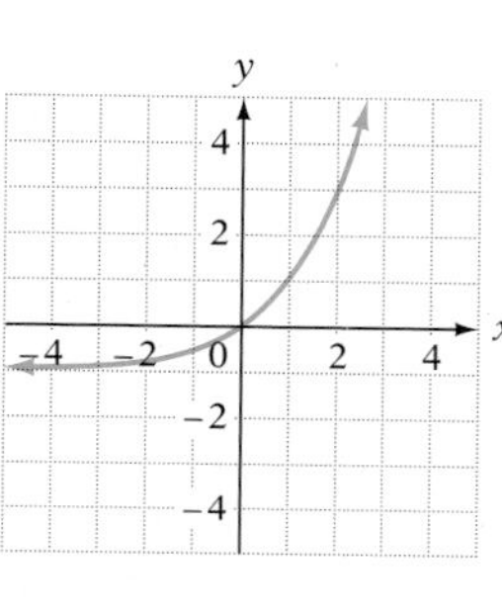

YOU TRY IT • 5

Graph: $f(x) = 2^x + 1$

Your solution

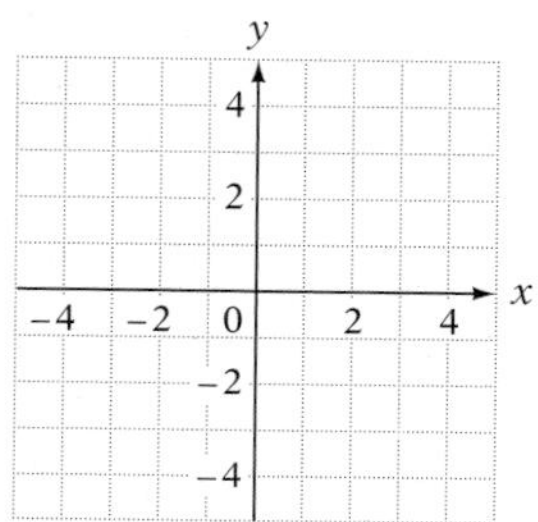

EXAMPLE • 6

Graph: $f(x) = \left(\frac{1}{3}\right)^x - 2$

Solution

x	y
-2	7
-1	1
0	-1
1	$-\frac{5}{3}$
2	$-\frac{17}{9}$

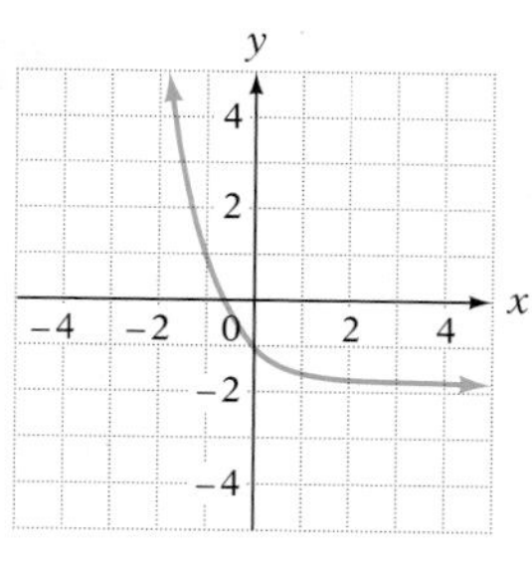

YOU TRY IT • 6

Graph: $f(x) = 2^{-x} + 2$

Your solution

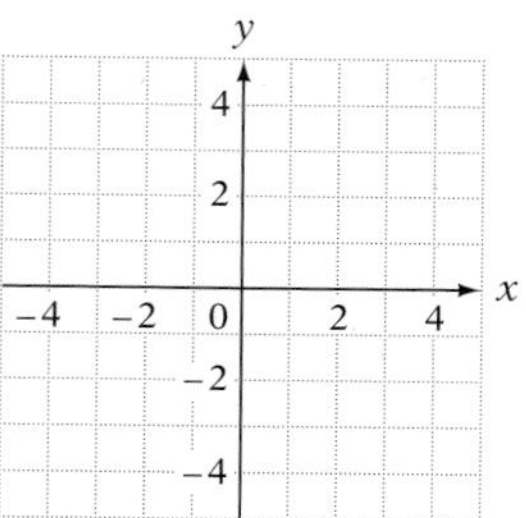

EXAMPLE • 7

Graph: $f(x) = 2^{-\frac{1}{2}x} - 1$

Solution

x	y
-6	7
-4	3
-2	1
0	0
2	$-\frac{1}{2}$
4	$-\frac{3}{4}$

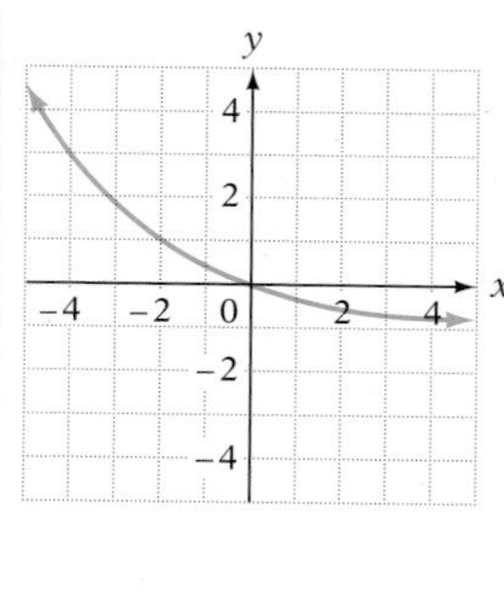

YOU TRY IT • 7

Graph: $f(x) = \left(\frac{1}{2}\right)^{-\frac{1}{2}x} + 2$

Your solution

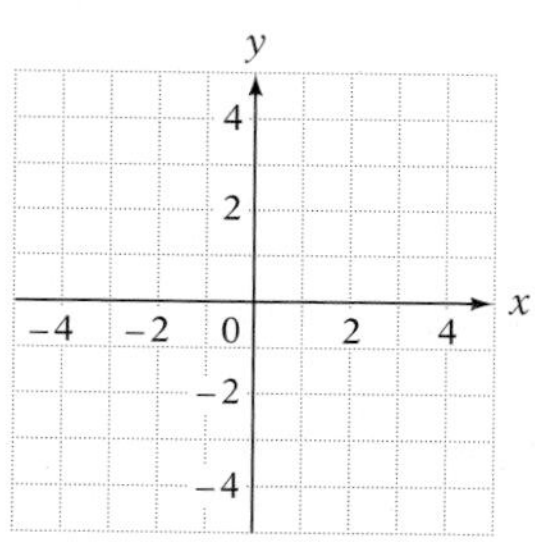

Solutions on p. S34

12.1 EXERCISES

OBJECTIVE A To evaluate an exponential function

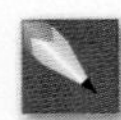

1. What is an exponential function?

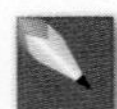

2. What is the natural exponential function?

3. Which of the following cannot be the base of an exponential function?

   (i) 7 **(ii)** $\frac{1}{4}$ **(iii)** -5 **(iv)** 0.01

4. Which of the following cannot be the base of an exponential function?

   (i) 0.9 **(ii)** 476 **(iii)** 8 **(iv)** $-\frac{1}{2}$

5. Given $f(x) = 3^x$, evaluate the following.
   a. $f(2)$ **b.** $f(0)$ **c.** $f(-2)$

6. Given $H(x) = 2^x$, evaluate the following.
   a. $H(-3)$ **b.** $H(0)$ **c.** $H(2)$

7. Given $g(x) = 2^{x+1}$, evaluate the following.
   a. $g(3)$ **b.** $g(1)$ **c.** $g(-3)$

8. Given $F(x) = 3^{x-2}$, evaluate the following.
   a. $F(-4)$ **b.** $F(-1)$ **c.** $F(0)$

9. Given $P(x) = \left(\frac{1}{2}\right)^{2x}$, evaluate the following.

   a. $P(0)$ **b.** $P\left(\frac{3}{2}\right)$ **c.** $P(-2)$

10. Given $R(t) = \left(\frac{1}{3}\right)^{3t}$, evaluate the following.

    a. $R\left(-\frac{1}{3}\right)$ **b.** $R(1)$ **c.** $R(-2)$

11. Given $G(x) = e^{x/2}$, evaluate the following. Round to the nearest ten-thousandth.

    a. $G(4)$ **b.** $G(-2)$ **c.** $G\left(\frac{1}{2}\right)$

12. Given $f(x) = e^{2x}$, evaluate the following. Round to the nearest ten-thousandth.

    a. $f(-2)$ **b.** $f\left(-\frac{2}{3}\right)$ **c.** $f(2)$

13. Given $H(r) = e^{-r+3}$, evaluate the following. Round to the nearest ten-thousandth.
    a. $H(-1)$ **b.** $H(3)$ **c.** $H(5)$

14. Given $P(t) = e^{-\frac{1}{2}t}$, evaluate the following. Round to the nearest ten-thousandth.

    a. $P(-3)$ **b.** $P(4)$ **c.** $P\left(\frac{1}{2}\right)$

15. Given $F(x) = 2^{x^2}$, evaluate the following.

a. $F(2)$ **b.** $F(-2)$ **c.** $F\left(\frac{3}{4}\right)$

16. Given $Q(x) = 2^{-x^2}$, evaluate the following.

a. $Q(3)$ **b.** $Q(-1)$ **c.** $Q(-2)$

17. Given $f(x) = e^{-x^2/2}$, evaluate the following. Round to the nearest ten-thousandth.

a. $f(-2)$ **b.** $f(2)$ **c.** $f(-3)$

18. Given $f(x) = e^{-2x} + 1$, evaluate the following. Round to the nearest ten-thousandth.

a. $f(-1)$ **b.** $f(3)$ **c.** $f(-2)$

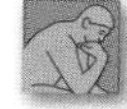

19. Suppose a and b are real numbers with $a < b$. If $f(x) = 2^{-x}$, then is $f(a) < f(b)$ or is $f(a) > f(b)$?

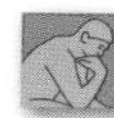

20. Suppose that u and v are real numbers with $u < v$ and that $f(x) = b^x$ $(b > 0, b \neq 1)$. If $f(u) < f(v)$, then is $0 < b < 1$ or is $b > 1$?

OBJECTIVE B To graph an exponential function

Graph the functions in Exercises 21 to 32.

21. $f(x) = 3^x$

22. $f(x) = 3^{-x}$

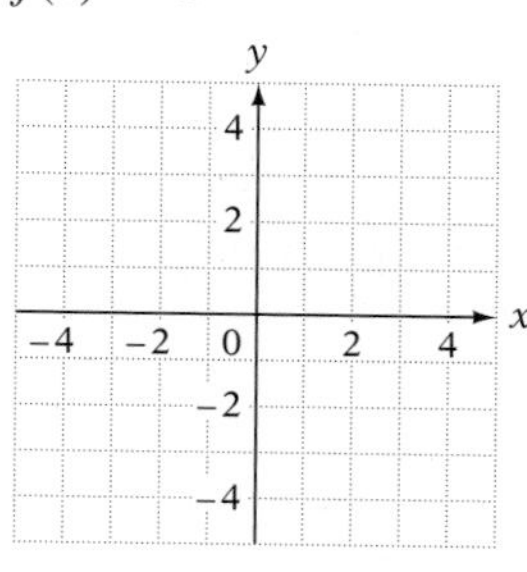

23. $f(x) = 2^{x+1}$

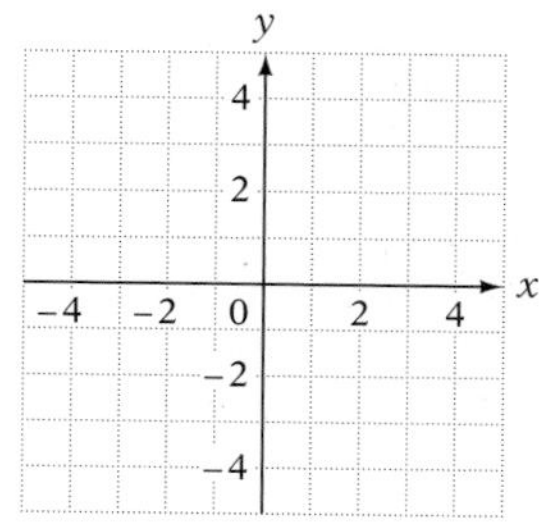

24. $f(x) = 2^{x-1}$

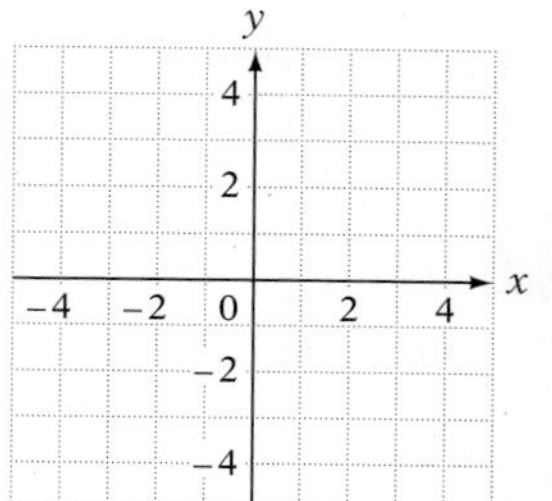

25. $f(x) = \left(\frac{1}{3}\right)^x$

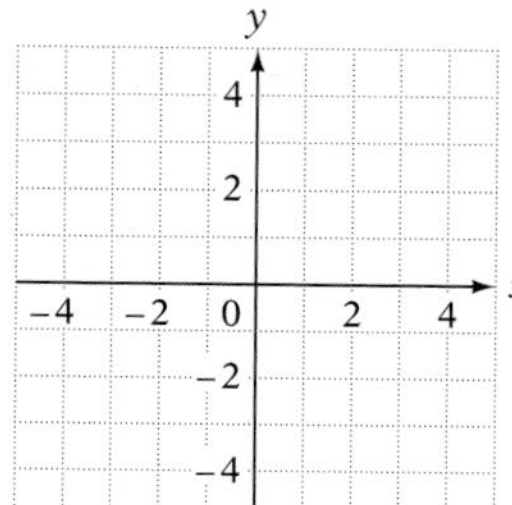

26. $f(x) = \left(\frac{2}{3}\right)^x$

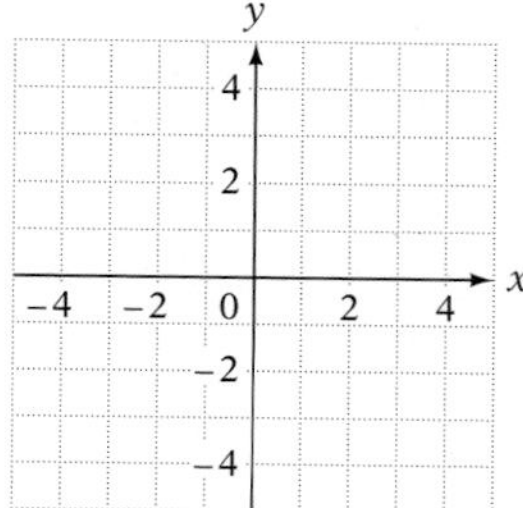

27. $f(x) = 2^{-x} + 1$

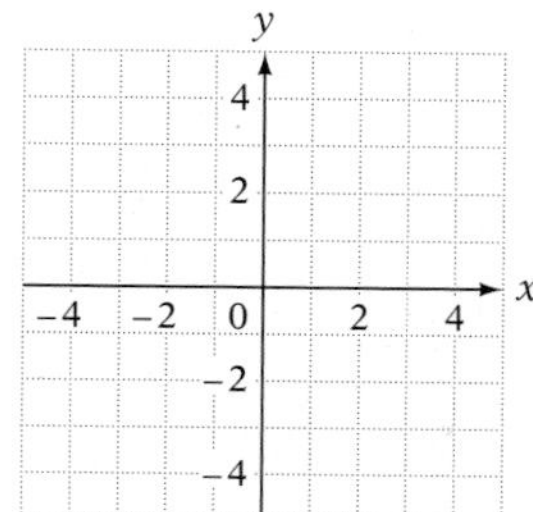

28. $f(x) = 2^x - 3$

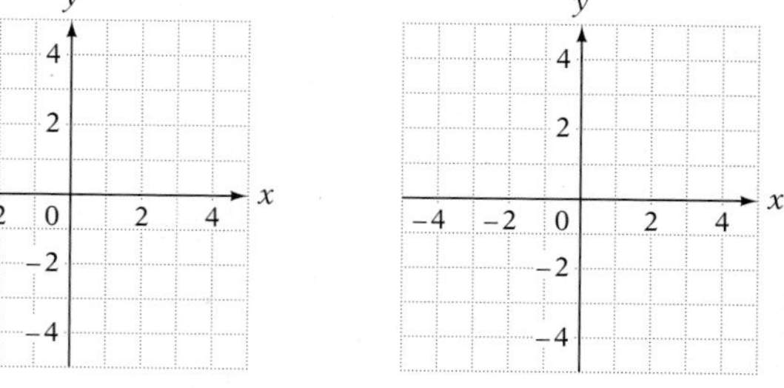

29. $f(x) = \left(\frac{1}{3}\right)^{-x}$

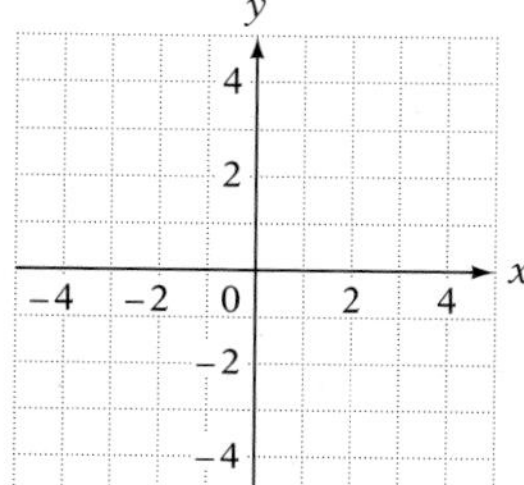

30. $f(x) = \left(\frac{3}{2}\right)^{-x}$

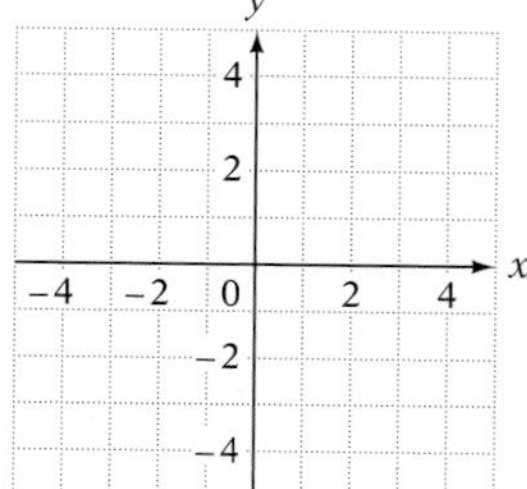

31. $f(x) = \left(\frac{1}{2}\right)^{-x} + 2$

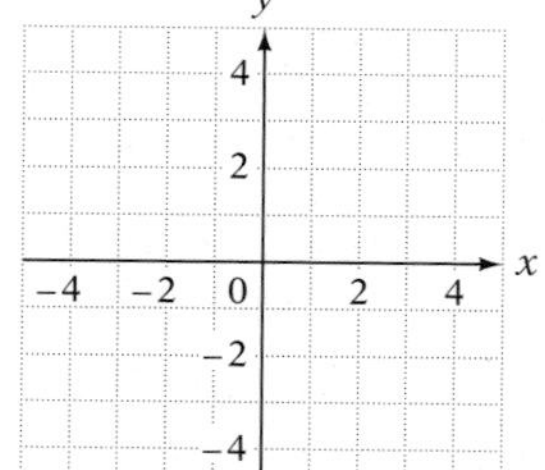

32. $f(x) = \left(\frac{1}{2}\right)^x - 1$

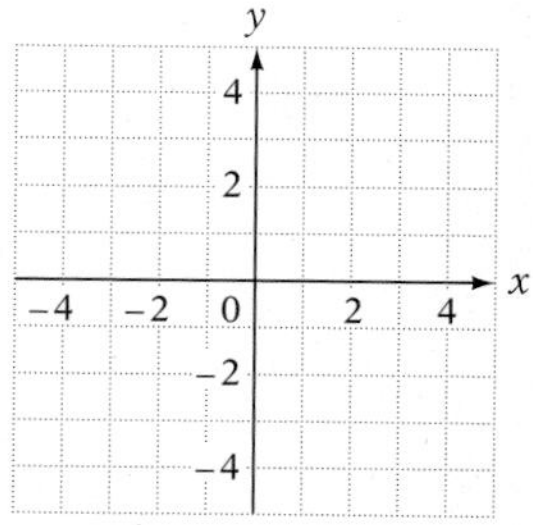

33. Graph $f(x) = 3^x$ and $f(x) = 3^{-x}$, and find the point of intersection of the two graphs.

34. Graph $f(x) = 2^{x+1}$ and $f(x) = 2^{-x+1}$, and find the point of intersection of the two graphs.

35. Graph $f(x) = \left(\frac{1}{3}\right)^x$. What are the x- and y-intercepts of the graph of the function?

36. Graph $f(x) = \left(\frac{1}{3}\right)^{-x}$. What are the x- and y-intercepts of the graph of the function?

37. Which of the following have the same graph?

(i) $f(x) = 3^x$ **(ii)** $f(x) = \left(\frac{1}{3}\right)^x$ **(iii)** $f(x) = \left(\frac{1}{3}\right)^{-x}$ **(iv)** $f(x) = 3^{-x}$

38. Which of the following have the same graph?

(i) $f(x) = \left(\frac{1}{4}\right)^{-x}$ **(ii)** $f(x) = 4^{-x}$ **(iii)** $f(x) = 4^x$ **(iv)** $f(x) = \left(\frac{1}{4}\right)^x$

Applying the Concepts

Use a graphing calculator to graph the functions in Exercises 39 to 41.

39. $P(x) = (\sqrt{3})^x$

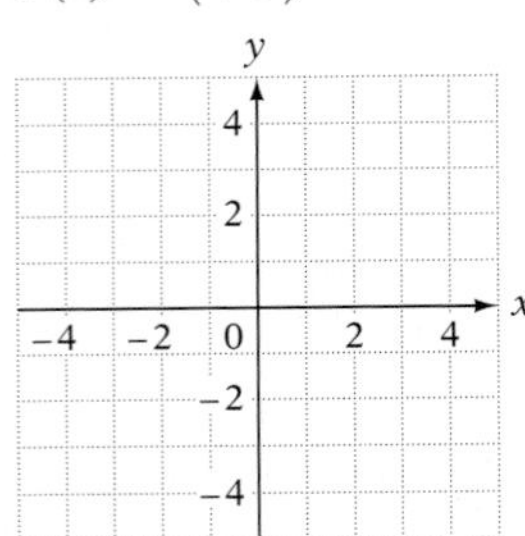

40. $Q(x) = (\sqrt{3})^{-x}$

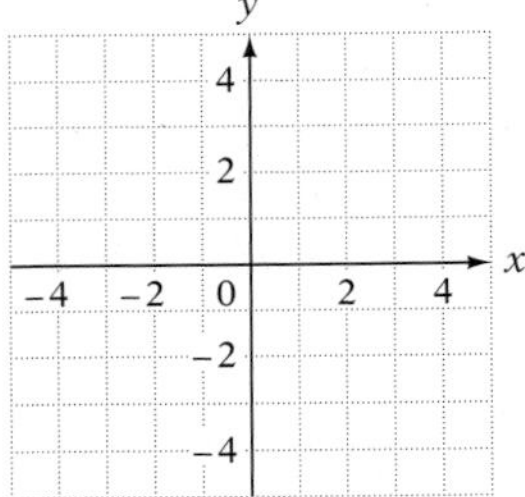

41. $f(x) = \pi^x$

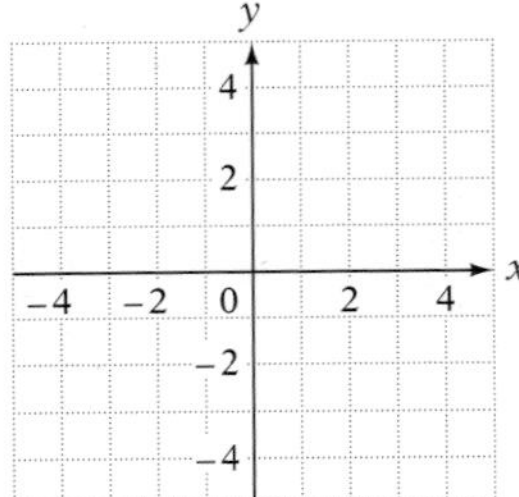

42. Evaluate $\left(1 + \frac{1}{n}\right)^n$ for $n = 100$, 1000, 10,000, and 100,000, and compare the results with the value of e, the base of the natural exponential function. On the basis of your evaluation, complete the following sentence:

As n increases, $\left(1 + \frac{1}{n}\right)^n$ becomes closer to ________.

43. **Physics** If air resistance is ignored, the speed v, in feet per second, of an object t seconds after it has been dropped is given by $v = 32t$. However, if air resistance is considered, then the speed depends on the mass (and on other things). For a certain mass, the speed t seconds after an object has been dropped is given by $v = 32(1 - e^{-t})$.

a. Graph this equation. *Suggestion:* Use Xmin = 0, Xmax = 5.5, Ymin = 0, Ymax = 40, and Yscl = 5.

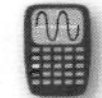

b. The point whose approximate coordinates are (2, 27.7) is on this graph. Write a sentence that explains the meaning of these coordinates.

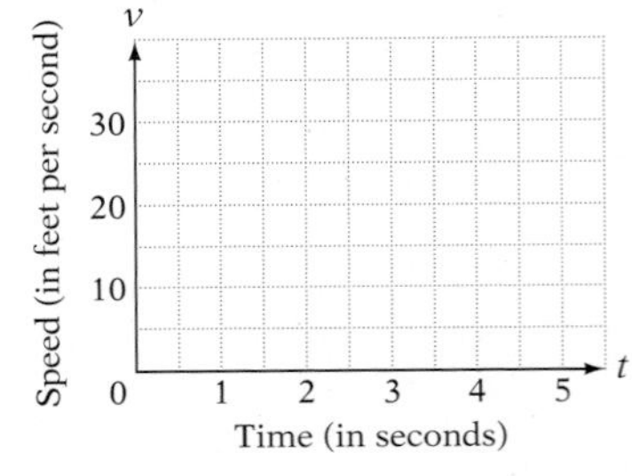

SECTION

12.2 Introduction to Logarithms

OBJECTIVE A To find the logarithm of a number 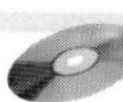

Suppose a bacteria colony that originally contained 1000 bacteria doubled in size every hour. Then the table at the right would show the number of bacteria in that colony after 1 h, 2 h, and 3 h.

Time (in hours)	*Number of Bacteria*
0	1000
1	2000
2	4000
3	8000

The exponential function $A = 1000(2^t)$, where A is the number of bacteria in the colony at time t, is a model of the growth of the colony. For instance, when $t = 3$ h, we have

$$A = 1000(2^t)$$

$$A = 1000(2^3) \quad \text{• Replace } t \text{ by } 3.$$

$$A = 1000(8) = 8000$$

After 3 h, there are 8000 bacteria in the colony.

Now we ask, "How long will it take for there to be 32,000 bacteria in the colony?" To answer the question, we must solve the *exponential equation* $32{,}000 = 1000(2^t)$. By trial and error, we find that when $t = 5$,

$$A = 1000(2^t)$$

$$A = 1000(2^5) \quad \text{• Replace } t \text{ by } 5.$$

$$A = 1000(32) = 32{,}000$$

After 5 h, there will be 32,000 bacteria in the colony.

Now suppose we want to know how long it will take before the colony reaches 50,000 bacteria. To answer that question, we must find t so that $50{,}000 = 1000(2^t)$. Using trial and error again, we find that

$$1000(2^5) = 32{,}000 \quad \text{and} \quad 1000(2^6) = 64{,}000$$

Because 50,000 is between 32,000 and 64,000, we conclude that t is between 5 h and 6 h. If we try $t = 5.5$ (halfway between 5 and 6), then

$$A = 1000(2^t)$$

$$A = 1000(2^{5.5}) \quad \text{• Replace } t \text{ by } 5.5.$$

$$A \approx 1000(45.25) = 45{,}250$$

In 5.5 h, there are approximately 45,250 bacteria in the colony. Because this is less than 50,000, the value of t must be a little greater than 5.5.

We could continue to use trial and error to find the correct value of t, but it would be more efficient if we could just solve the exponential equation $50{,}000 = 1000(2^t)$ for t. If we follow the procedures for solving equations that were discussed earlier in the text, we have

Integrating Technology

Using a calculator, we can verify that $2^{5.644} \approx 50$. On a graphing calculator, press 2 ^ 5.644.

$$50{,}000 = 1000(2^t)$$

$$50 = 2^t \quad \text{• Divide each side of the equation by 1000.}$$

To proceed to the next step, it would be helpful to have a function that would find the power of 2 that produces 50.

Take Note

Note that when we tried $t = 5.5$ earlier, we stated that the actual value of t must be greater than 5.5 and that 5.644 is a little greater than 5.5.

Around the mid-sixteenth century, mathematicians created such a function, which we now call a *logarithmic function*. We write the solution of $50 = 2^t$ as $t = \log_2 50$. This is read "t equals the logarithm base 2 of 50" and it means "t equals the power of 2 that produces 50." When logarithms were first introduced, tables were used to find a numerical value of t. Today, a calculator is used. Using a calculator, we can approximate the value of t as 5.644. This means that $2^{5.644} \approx 50$.

The equivalence of the expressions $50 = 2^t$ and $t = \log_2 50$ are described in the following definition of **logarithm.**

Take Note

Read $\log_b x$ as "the logarithm of x, base b" or "log base b of x."

Tips for Success

Be sure you can rewrite an exponential equation as a logarithmic equation and a logarithmic equation as an exponential equation. This relationship is very important.

Definition of Logarithm

For $x > 0$, $b > 0$, $b \neq 1$, $y = \log_b x$ is equivalent to $x = b^y$.

The table at the right shows equivalent statements written in both exponential and logarithmic form.

Exponential Form	*Logarithmic Form*
$2^4 = 16$	$\log_2 16 = 4$
$\left(\frac{2}{3}\right)^2 = \frac{4}{9}$	$\log_{2/3}\left(\frac{4}{9}\right) = 2$
$10^{-1} = 0.1$	$\log_{10}(0.1) = -1$

HOW TO • 1 Write $\log_3 81 = 4$ in exponential form.

$\log_3 81 = 4$ is equivalent to $3^4 = 81$.

HOW TO • 2 Write $10^{-2} = 0.01$ in logarithmic form.

$10^{-2} = 0.01$ is equivalent to $\log_{10}(0.01) = -2$.

It is important to note that the exponential function is a 1–1 function and thus has an inverse function. The inverse function of the exponential function is called a logarithm.

The 1–1 property of exponential functions can be used to evaluate some logarithms.

1–1 Property of Exponential Functions

For $b > 0$, $b \neq 1$, if $b^u = b^v$, then $u = v$.

HOW TO • 3 Evaluate $\log_2 8$.

$\log_2 8 = x$ • Write an equation.

$8 = 2^x$ • Write the equation in its equivalent exponential form.

$2^3 = 2^x$ • Write 8 as 2^3.

$3 = x$ • Use the 1–1 Property of Exponential Functions.

$\log_2 8 = 3$

HOW TO • 4 Solve $\log_4 x = -2$ for x.

$\log_4 x = -2$

$4^{-2} = x$ • Write the equation in its equivalent exponential form.

$\frac{1}{16} = x$ • Simplify the negative exponent.

The solution is $\frac{1}{16}$.

Integrating Technology

The logarithms of most numbers are irrational numbers. Therefore, the value displayed by a calculator is an approximation.

Logarithms to the base 10 are called **common logarithms.** Usually the base, 10, is omitted when writing the common logarithm of a number. Therefore, $\log_{10} x$ is written $\log x$. To find the common logarithm of most numbers, a calculator is necessary. A calculator was used to find the value of log 384, shown below.

$$\log 384 \approx 2.5843312$$

When e (the base of the natural exponential function) is used as the base of a logarithm, the logarithm is referred to as the **natural logarithm** and is abbreviated $\ln x$. This is read "el en x." Use a calculator to approximate natural logarithms.

$$\ln 23 \approx 3.135494216$$

EXAMPLE • 1

Evaluate: $\log_3 \frac{1}{9}$

Solution

$\log_3 \frac{1}{9} = x$ • Write an equation.

$\frac{1}{9} = 3^x$ • Write the equivalent exponential form.

$3^{-2} = 3^x$ • $\frac{1}{9} = 3^{-2}$

$-2 = x$ • The bases are the same. The exponents are equal.

$\log_3 \frac{1}{9} = -2$

YOU TRY IT • 1

Evaluate: $\log_4 64$

Your solution

Solution on p. S34

EXAMPLE • 2

Solve for x: $\log_5 x = 2$

Solution

$$\log_5 x = 2$$
$$5^2 = x \quad \text{• Write the equivalent exponential form.}$$
$$25 = x$$

The solution is 25.

YOU TRY IT • 2

Solve for x: $\log_2 x = -4$

Your solution

EXAMPLE • 3

Solve $\log x = -1.5$ for x. Round to the nearest ten-thousandth.

Solution

$$\log x = -1.5 \quad \text{• Write the equivalent exponential form.}$$
$$10^{-1.5} = x$$
$$0.0316 \approx x \quad \text{• Use a calculator.}$$

YOU TRY IT • 3

Solve $\ln x = 3$ for x. Round to the nearest ten-thousandth.

Your solution

Solutions on p. S34

OBJECTIVE B To use the Properties of Logarithms to simplify expressions containing logarithms

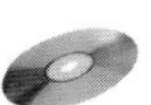

Take Note

The properties stated here can be proved using the definition of logarithm and the properties of exponents. For instance, to prove the Logarithm Property of 1, let $\log_b 1 = x$. Then, by the definition of logarithm, $b^x = 1$. Because $b^0 = 1$, we have $x = 0$.

From the definition of a logarithm, a number of properties of logarithms can be stated.

Logarithm Property of 1

If $b > 0$, $b \neq 1$, then $\log_b 1 = 0$.

Here are some examples.

$$\log_9 1 = 0 \qquad \log_2 1 = 0 \qquad \log_{3/4} 1 = 0$$

Logarithm of the Base Property

If $b > 0$, $b \neq 1$, then $\log_b b = 1$.

Here are some examples.

$$\log_7 7 = 1 \qquad \log 10 = 1 \qquad \log_{1/2}\left(\frac{1}{2}\right) = 1$$

1–1 Property of Logarithms

Suppose that $b > 0$, $b \neq 1$, and x and y are positive real numbers. If $\log_b x = \log_b y$, then $x = y$.

For instance, if $\log_b 7 = \log_b x$, then $x = 7$.

The next property is a direct result of the fact that logarithmic functions and exponential functions are inverses of one another.

Take Note

By the Inverse Property of Logarithms, $9^{\log_9 5x} = 5x$.

Inverse Property of Logarithms

If $b > 0$, $b \neq 1$, and $x > 0$, then $\log_b b^x = x$ and $b^{\log_b x} = x$.

Because a logarithm is related to an exponential expression, logarithms have properties that are similar to those of exponential expressions. The proofs of the next three properties can be found in the Appendix.

The following property relates to the logarithm of the product of two numbers.

Take Note

Pay close attention to this property. Note, for instance, that this property states that $\log_3(4p) = \log_3 4 + \log_3 p$. It also states that $\log_5 9 + \log_5 z = \log_5(9z)$. It does *not* state any relationship regarding the expression $\log_b(x + y)$. **This expression cannot be simplified.**

The Logarithm Property of the Product of Two Numbers

For any positive real numbers x, y, and b, $b \neq 1$, $\log_b(xy) = \log_b x + \log_b y$.

In words, this property states that the logarithm of the product of two numbers equals the sum of the logarithms of the two numbers.

HOW TO • 5 Write $\log_b(6z)$ in expanded form.

$\log_b(6z) = \log_b 6 + \log_b z$ • Use the Logarithm Property of Products.

HOW TO • 6 Write $\log_b 12 + \log_b r$ as a single logarithm.

$\log_b 12 + \log_b r = \log_b(12r)$ • Use the Logarithm Property of Products.

The Logarithm Property of Products can be extended to include the logarithm of the product of more than two factors. For instance,

$$\log_b(7rt) = \log_b 7 + \log_b r + \log_b t$$

The next property relates to the logarithm of the quotient of two numbers.

Take Note

This property is used to rewrite expressions such as $\log_5\left(\frac{m}{8}\right) = \log_5 m - \log_5 8$. It does *not* state any relationship regarding the expression $\frac{\log_b x}{\log_b y}$. **This expression cannot be simplified.**

The Logarithm Property of the Quotient of Two Numbers

For any positive real numbers x, y, and b, $b \neq 1$, $\log_b \frac{x}{y} = \log_b x - \log_b y$.

In words, this property states that the logarithm of the quotient of two numbers equals the difference of the logarithms of the two numbers.

HOW TO • 7 Write $\log_b \frac{p}{8}$ in expanded form.

$\log_b \frac{p}{8} = \log_b p - \log_b 8$ • Use the Logarithm Property of Quotients.

HOW TO • 8 Write $\log_b y - \log_b v$ as a single logarithm.

$\log_b y - \log_b v = \log_b \frac{y}{v}$ • Use the Logarithm Property of Quotients.

Another property of logarithms is used to simplify powers of a number.

The Logarithm Property of the Power of a Number

For any positive real numbers x and b, $b \neq 1$, and for any real number r, $\log_b x^r = r \log_b x$.

In words, this property states that the logarithm of the power of a number equals the power times the logarithm of the number.

HOW TO • 9 Rewrite $\log_b x^3$ in terms of $\log_b x$.

$\log_b x^3 = 3 \log_b x$ • Use the Logarithm Property of Powers.

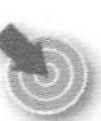

Point of Interest

Logarithms were developed independently by Jobst Burgi (1552–1632) and John Napier (1550–1617) as a means of simplifying the calculations of astronomers. The idea was to devise a method by which two numbers could be multiplied by performing additions. Napier is usually given credit for logarithms because he published his results first.

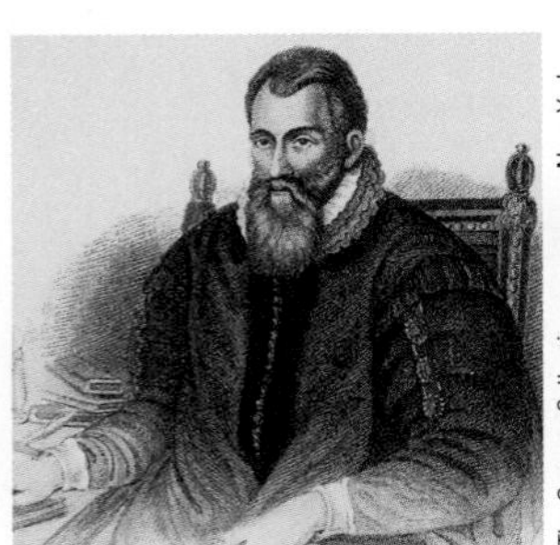

John Napier

The Granger Collection, New York

HOW TO • 10 Rewrite $\frac{2}{3}\log_b x$ with a coefficient of 1.

$\frac{2}{3}\log_b x = \log_b x^{2/3}$ • Use the Logarithm Property of Powers.

Here are some additional examples that combine some of the properties of logarithms we have discussed.

HOW TO • 11 Write $\log_b \frac{x^3}{y^2 z}$ in expanded form.

$\log_b \frac{x^3}{y^2 z} = \log_b x^3 - \log_b y^2 z$ • Use the Logarithm Property of Quotients.

$= \log_b x^3 - (\log_b y^2 + \log_b z)$ • Use the Logarithm Property of Products.

$= 3\log_b x - (2\log_b y + \log_b z)$ • Use the Logarithm Property of Powers.

$= 3\log_b x - 2\log_b y - \log_b z$ • Write in simplest form.

HOW TO • 12 Write $2\log_b x + 4\log_b y - \log_b z$ as a single logarithm with a coefficient of 1.

$2\log_b x + 4\log_b y - \log_b z = \log_b x^2 + \log_b y^4 - \log_b z$ • Use the Logarithm Property of Powers.

$= \log_b (x^2y^4) - \log_b z$ • Use the Logarithm Property of Products.

$= \log_b \frac{x^2y^4}{z}$ • Use the Logarithm Property of Quotients.

EXAMPLE • 4

Write $\log \sqrt{x^3 y}$ in expanded form.

Solution

$\log \sqrt{x^3y} = \log (x^3y)^{1/2} = \frac{1}{2}\log (x^3y)$ • Power Property

$= \frac{1}{2}(\log x^3 + \log y)$ • Product Property

$= \frac{1}{2}(3\log x + \log y)$ • Power Property

$= \frac{3}{2}\log x + \frac{1}{2}\log y$ • Distributive Property

YOU TRY IT • 4

Write $\log_8 \sqrt[3]{xy^2}$ in expanded form.

Your solution

EXAMPLE • 5

Write $\frac{1}{2}(\log_3 x - 3\log_3 y + \log_3 z)$ as a single logarithm with a coefficient of 1.

Solution

$\frac{1}{2}(\log_3 x - 3\log_3 y + \log_3 z)$

$= \frac{1}{2}(\log_3 x - \log_3 y^3 + \log_3 z)$

$= \frac{1}{2}\left(\log_3 \frac{x}{y^3} + \log_3 z\right)$

$= \frac{1}{2}\left(\log_3 \frac{xz}{y^3}\right) = \log_3\left(\frac{xz}{y^3}\right)^{1/2} = \log_3 \sqrt{\frac{xz}{y^3}}$

YOU TRY IT • 5

Write $\frac{1}{3}(\log_4 x - 2\log_4 y + \log_4 z)$ as a single logarithm with a coefficient of 1.

Your solution

Solutions on p. S34

OBJECTIVE C To use the Change-of-Base Formula 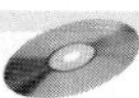

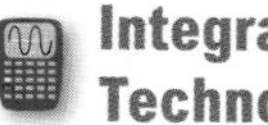

Integrating Technology

To evaluate $\frac{\log 22}{\log 5}$ using a graphing calculator, use the keystrokes LOG 22) ÷ LOG 5) ENTER. The display should read 1.92057266.

If natural logarithms are used, use the LN key rather than the LOG key. The result will be the same.

Although only common logarithms and natural logarithms are programmed into calculators, the logarithms for other positive bases can be found.

HOW TO • 13 Evaluate $\log_5 22$.

$\log_5 22 = x$ • Write an equation.

$5^x = 22$ • Write the equation in its equivalent exponential form.

$\log 5^x = \log 22$ • Apply the common logarithm to each side of the equation.

$x \log 5 = \log 22$ • Use the Power Property of Logarithms.

$x = \frac{\log 22}{\log 5}$ • Exact answer

$x \approx 1.92057$ • Approximate answer

$\log_5 22 \approx 1.92057$

In the third step above, the natural logarithm, instead of the common logarithm, could have been applied to each side of the equation. The same result would have been obtained.

Using a procedure similar to the one used to evaluate $\log_5 22$, we can derive a formula for changing bases.

Change-of-Base Formula

$$\log_a N = \frac{\log_b N}{\log_b a}$$

HOW TO • 14 Evaluate $\log_2 14$.

$\log_2 14 = \frac{\log 14}{\log 2}$ • Use the Change-of-Base Formula with $N = 14, a = 2, b = 10$.

≈ 3.80735

In the last example, common logarithms were used. Here is the same example using natural logarithms. Note that the answers are the same.

$$\log_2 14 = \frac{\ln 14}{\ln 2} \approx 3.80735$$

EXAMPLE • 6

Evaluate $\log_8 0.137$ by using natural logarithms.

Solution

$\log_8 0.137 = \frac{\ln 0.137}{\ln 8} \approx -0.95592$

YOU TRY IT • 6

Evaluate $\log_3 0.834$ by using natural logarithms.

Your solution

EXAMPLE • 7

Evaluate $\log_2 90.813$ by using common logarithms.

Solution

$\log_2 90.813 = \frac{\log 90.813}{\log 2} \approx 6.50483$

YOU TRY IT • 7

Evaluate $\log_7 6.45$ by using common logarithms.

Your solution

Solutions on p. S34

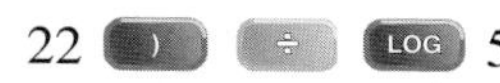

12.2 EXERCISES

OBJECTIVE A To find the logarithm of a number

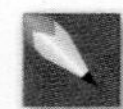

1. **a.** What is a common logarithm?
b. How is the common logarithm of $4z$ written?

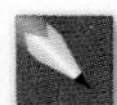

2. **a.** What is a natural logarithm?
b. How is the natural logarithm of $3x$ written?

For Exercises 3 to 10, write the exponential expression in logarithmic form.

3. $5^2 = 25$

4. $10^3 = 1000$

5. $4^{-2} = \frac{1}{16}$

6. $3^{-3} = \frac{1}{27}$

7. $10^y = x$

8. $e^y = x$

9. $a^x = w$

10. $b^y = c$

For Exercises 11 to 18, write the logarithmic expression in exponential form.

11. $\log_3 9 = 2$

12. $\log_2 32 = 5$

13. $\log 0.01 = -2$

14. $\log_5 \frac{1}{5} = -1$

15. $\ln x = y$

16. $\log x = y$

17. $\log_b u = v$

18. $\log_c x = y$

Evaluate the expressions in Exercises 19 to 30.

19. $\log_3 81$

20. $\log_7 49$

21. $\log_2 128$

22. $\log_5 125$

23. $\log 100$

24. $\log 0.001$

25. $\ln e^3$

26. $\ln e^2$

27. $\log_8 1$

28. $\log_3 243$

29. $\log_5 625$

30. $\log_2 64$

For Exercises 31 to 38, solve for x.

31. $\log_3 x = 2$

32. $\log_5 x = 1$

33. $\log_4 x = 3$

34. $\log_2 x = 6$

35. $\log_7 x = -1$

36. $\log_8 x = -2$

37. $\log_6 x = 0$

38. $\log_4 x = 0$

For Exercises 39 to 46, solve for x. Round to the nearest hundredth.

39. $\log x = 2.5$

40. $\log x = 3.2$

41. $\log x = -1.75$

42. $\log x = -2.1$

43. $\ln x = 2$

44. $\ln x = 1.4$

45. $\ln x = -\frac{1}{2}$

46. $\ln x = -1.7$

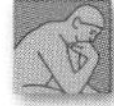

47. If $\log_3 x > 0$, then $x >$ ____.

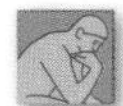

48. Why is there no real number value for the expression $\log_5 (-2)$?

OBJECTIVE B To use the Properties of Logarithms to simplify expressions containing logarithms

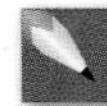

49. What is the Product Property of Logarithms?

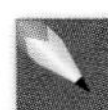

50. What is the Quotient Property of Logarithms?

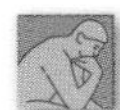

51. True or false? Using the Product Property of Logarithms, $\log_5 (-2) + \log_5 (-3) = \log_5 [(-2)(-3)] = \log_5 6$

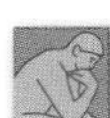

52. For what values of x is the equation $\log_b x^2 = 2 \log_b x$ true?

For Exercises 53 to 64, evaluate the expression.

53. $\log_{12} 1$

54. $\ln 1$

55. $\ln e$

56. $\log_{10} 10$

57. $\log_3 3^x$

58. $8^{\log_8 p}$

59. $e^{\ln v}$

60. $\ln e^{3x}$

61. $2^{\log_2 (x^2 + 1)}$

62. $\log_4 4^{3x + 1}$

63. $\log_5 5^{x^2 - x - 1}$

64. $8^{\log_8 (3x - 7)}$

For Exercises 65 to 88, write the logarithm in expanded form.

65. $\log_8 (xz)$

66. $\log_7 (rt)$

67. $\log_3 x^5$

68. $\log_2 y^7$

69. $\log_b \frac{r}{s}$

70. $\log_c \frac{z}{4}$

71. $\log_3 (x^2 y^6)$

72. $\log_4 (t^4 u^2)$

73. $\log_7 \frac{u^3}{v^4}$

74. $\log_{10} \frac{s^5}{t^2}$

75. $\log_2 (rs)^2$

76. $\log_3 (x^2 y)^3$

77. $\ln (x^2 yz)$

78. $\ln (xy^2 z^3)$

79. $\log_5 \frac{xy^2}{z^4}$

80. $\log_b \frac{r^2 s}{t^3}$

81. $\log_8 \frac{x^2}{yz^2}$

82. $\log_9 \frac{x}{y^2 z^3}$

83. $\log_4 \sqrt{x^3 y}$

84. $\log_3 \sqrt{x^5 y^3}$

85. $\log_7 \sqrt{\frac{x^3}{y}}$

86. $\log_b \sqrt[3]{\frac{r^2}{t}}$

87. $\log_3 \frac{t}{\sqrt{x}}$

88. $\log_4 \frac{x}{\sqrt{y^2 z}}$

For Exercises 89 to 112, express as a single logarithm with a coefficient of 1.

89. $\log_3 x^3 + \log_3 y^2$

90. $\log_7 x + \log_7 z^2$

91. $\ln x^4 - \ln y^2$

92. $\ln x^2 - \ln y$

93. $3 \log_7 x$

94. $4 \log_8 y$

95. $3 \ln x + 4 \ln y$

96. $2 \ln x - 5 \ln y$

97. $2(\log_4 x + \log_4 y)$

98. $3(\log_5 r + \log_5 t)$

99. $2 \log_3 x - \log_3 y + 2 \log_3 z$

100. $4 \log_5 r - 3 \log_5 s + \log_5 t$

101. $\ln x - (2 \ln y + \ln z)$

102. $2 \log_b x - 3(\log_b y + \log_b z)$

103. $\frac{1}{2}(\log_6 x - \log_6 y)$

104. $\frac{1}{3}(\log_8 x - \log_8 y)$

105. $2(\log_4 s - 2 \log_4 t + \log_4 r)$

106. $3(\log_9 x + 2 \log_9 y - 2 \log_9 z)$

107. $\ln x - 2(\ln y + \ln z)$

108. $\ln t - 3(\ln u + \ln v)$

109. $\frac{1}{2}(3 \log_4 x - 2 \log_4 y + \log_4 z)$

110. $\frac{1}{3}(4 \log_5 t - 5 \log_5 u - 7 \log_5 v)$

111. $\frac{1}{2}\log_2 x - \frac{2}{3}\log_2 y + \frac{1}{2}\log_2 z$

112. $\frac{2}{3}\log_3 x + \frac{1}{3}\log_3 y - \frac{1}{2}\log_3 z$

OBJECTIVE C To use the Change-of-Base Formula

Evaluate the expressions in Exercises 113 to 128. Round to the nearest ten-thousandth.

113. $\log_8 6$

114. $\log_4 8$

115. $\log_5 30$

116. $\log_6 28$

117. $\log_3 0.5$

118. $\log_5 0.6$

119. $\log_7 1.7$

120. $\log_6 3.2$

121. $\log_5 15$

122. $\log_3 25$

123. $\log_{12} 120$

124. $\log_9 90$

125. $\log_4 2.55$

126. $\log_8 6.42$

127. $\log_5 67$

128. $\log_8 35$

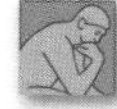

129. Use the Change-of-Base Formula to write $\log_5 x$ in terms of $\log x$.

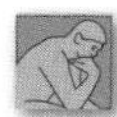

130. Use the Change-of-Base Formula to write $\log_5 x$ in terms of $\ln x$.

Applying the Concepts

131. For each of the following, answer True or False. Assume all variables represent positive numbers.

a. $\log_3(-9) = -2$

b. $x^y = z$ and $\log_x z = y$ are equivalent equations.

c. $\log(x^{-1}) = \dfrac{1}{\log x}$

d. $\log \dfrac{x}{y} = \log x - \log y$

e. $\log(x \cdot y) = \log x \cdot \log y$

f. If $\log x = \log y$, then $x = y$.

132. Complete each statement. For part a, use the equation $\log_a b = c$.

a. $a^c =$ ________

b. $\log_a(a^c) =$ ________

SECTION

12.3 Graphs of Logarithmic Functions

OBJECTIVE A To graph a logarithmic function 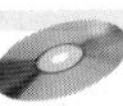

The graph of a logarithmic function can be drawn by using the relationship between the exponential and logarithmic functions.

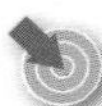

Point of Interest

Although logarithms were originally developed to assist with computations, logarithmic functions have a much broader use today. These functions occur in geology, acoustics, chemistry, and economics, for example.

HOW TO • 1 Graph: $f(x) = \log_2 x$

Think of $f(x) = \log_2 x$ as the equation $y = \log_2 x$.

$$f(x) = \log_2 x$$
$$y = \log_2 x$$

Write the equivalent exponential equation.

$$x = 2^y$$

Because the equation is solved for x in terms of y, it is easier to choose values of y and find the corresponding values of x. The results can be recorded in a table.

Graph the ordered pairs on a rectangular coordinate system.

Connect the points with a smooth curve.

x	y
$\frac{1}{4}$	-2
$\frac{1}{2}$	-1
1	0
2	1
4	2

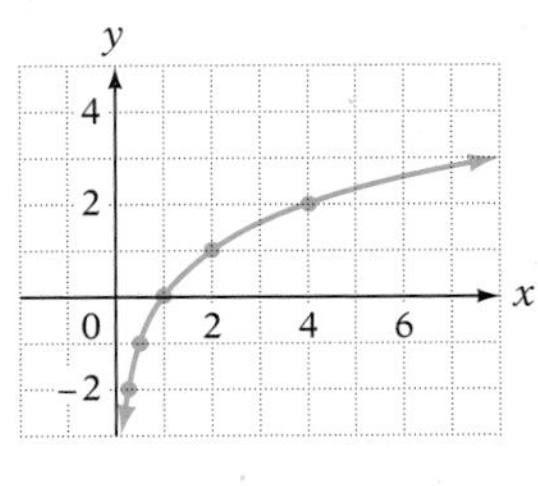

Applying the vertical-line and horizontal-line tests reveals that $f(x) = \log_2 x$ is a one-to-one function.

HOW TO • 2 Graph: $f(x) = \log_2 x + 1$

Think of $f(x) = \log_2 x + 1$ as the equation $y = \log_2 x + 1$.

$$f(x) = \log_2 x + 1$$
$$y = \log_2 x + 1$$

Solve for $\log_2 x$.

$$y - 1 = \log_2 x$$

Write the equivalent exponential equation.

$$2^{y-1} = x$$

Choose values of y and find the corresponding values of x.

Graph the ordered pairs on a rectangular coordinate system.

Connect the points with a smooth curve.

x	y
$\frac{1}{4}$	-1
$\frac{1}{2}$	0
1	1
2	2
4	3

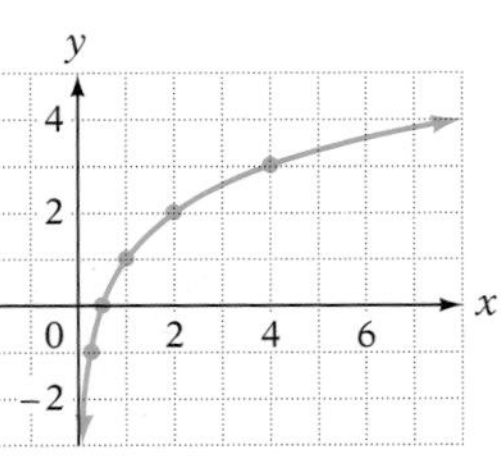

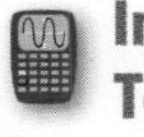

Integrating Technology

See the Projects and Group Activities at the end of this chapter for suggestions on graphing logarithmic functions using a graphing calculator.

EXAMPLE • 1

Graph: $f(x) = \log_3 x$

Solution

$$f(x) = \log_3 x$$
$$y = \log_3 x \quad \bullet\ f(x) = y$$
$$3^y = x \quad \bullet\ \text{Write the equivalent exponential equation.}$$

x	y
$\frac{1}{9}$	-2
$\frac{1}{3}$	-1
1	0
3	1

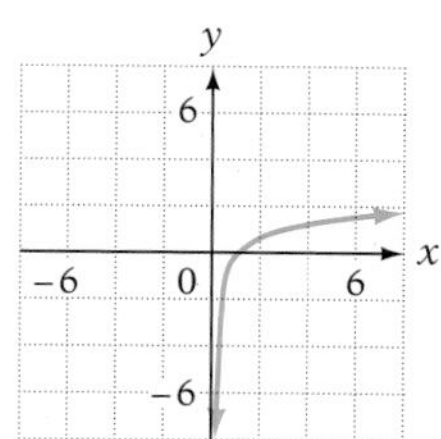

YOU TRY IT • 1

Graph: $f(x) = \log_2(x - 1)$

Your solution

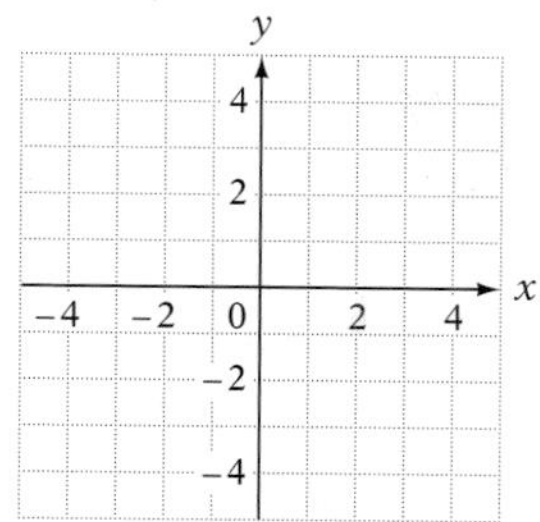

EXAMPLE • 2

Graph: $f(x) = 2 \log_3 x$

Solution

$$f(x) = 2 \log_3 x$$
$$y = 2 \log_3 x \quad \bullet\ f(x) = y$$
$$\frac{y}{2} = \log_3 x \quad \bullet\ \text{Divide both sides by 2.}$$
$$3^{y/2} = x \quad \bullet\ \text{Write the equivalent exponential equation.}$$

x	y
$\frac{1}{9}$	-4
$\frac{1}{3}$	-2
1	0
3	2

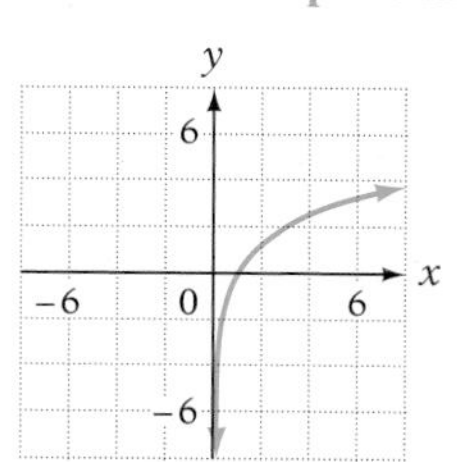

YOU TRY IT • 2

Graph: $f(x) = \log_3(2x)$

Your solution

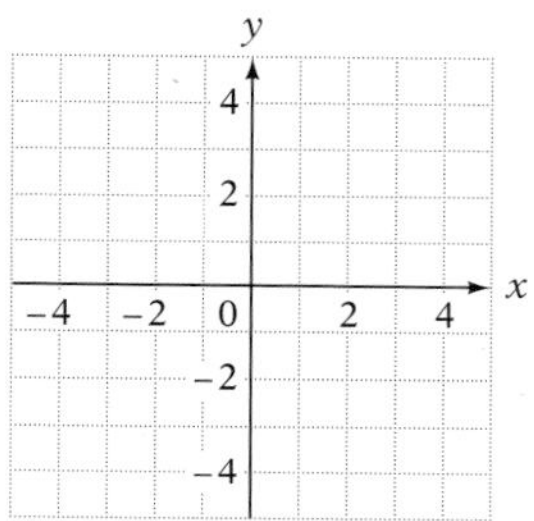

EXAMPLE • 3

Graph: $f(x) = -\log_2(x - 2)$

Solution

$$f(x) = -\log_2(x - 2)$$
$$y = -\log_2(x - 2) \quad \bullet\ f(x) = y$$
$$-y = \log_2(x - 2) \quad \bullet\ \text{Multiply both sides by } -1.$$
$$2^{-y} = x - 2 \quad \bullet\ \text{Write the equivalent exponential equation.}$$
$$2^{-y} + 2 = x$$

x	y
6	-2
4	-1
3	0
$\frac{5}{2}$	1
$\frac{9}{4}$	2
$\frac{17}{8}$	3

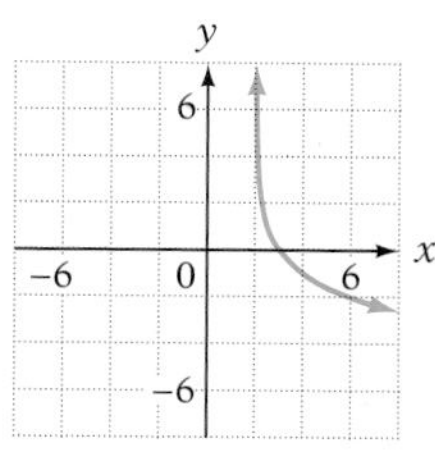

YOU TRY IT • 3

Graph: $f(x) = -\log_3(x + 1)$

Your solution

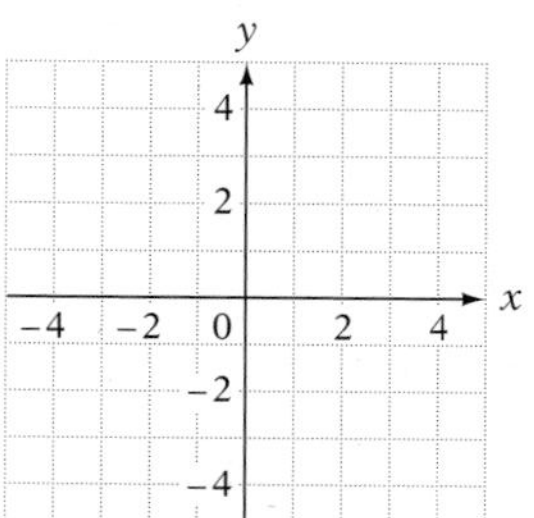

Solutions on p. S35

12.3 EXERCISES

OBJECTIVE A To graph a logarithmic function

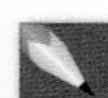

1. What is the relationship between the graphs of $x = 3^y$ and $y = \log_3 x$?

2. What is the relationship between the graphs of $y = 3^x$ and $y = \log_3 x$?

For Exercises 3 to 14, graph the function.

3. $f(x) = \log_4 x$

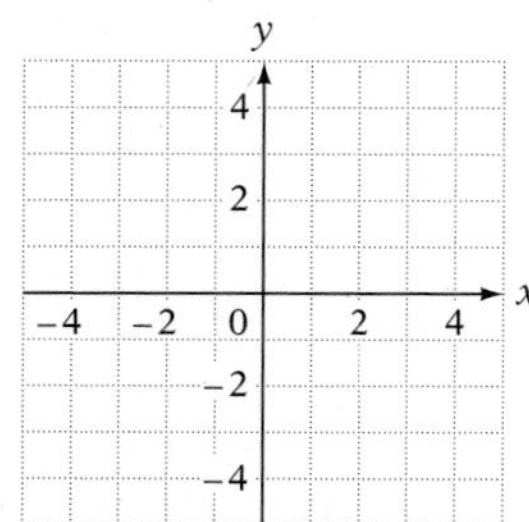

4. $f(x) = \log_2 (x + 1)$

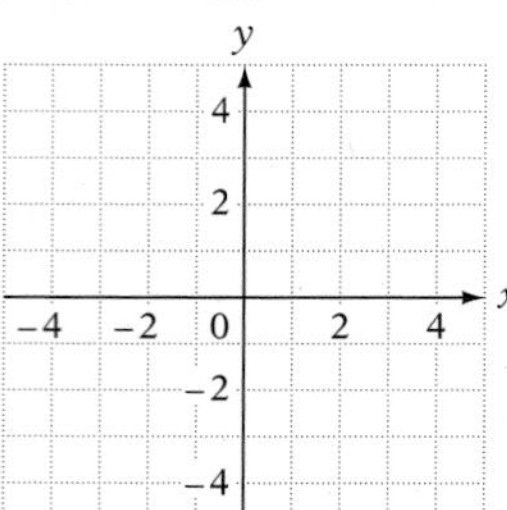

5. $f(x) = \log_3 (2x - 1)$

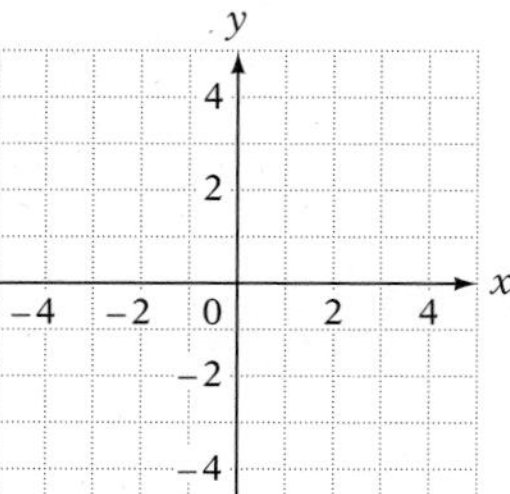

6. $f(x) = \log_2 \left(\frac{1}{2}x\right)$

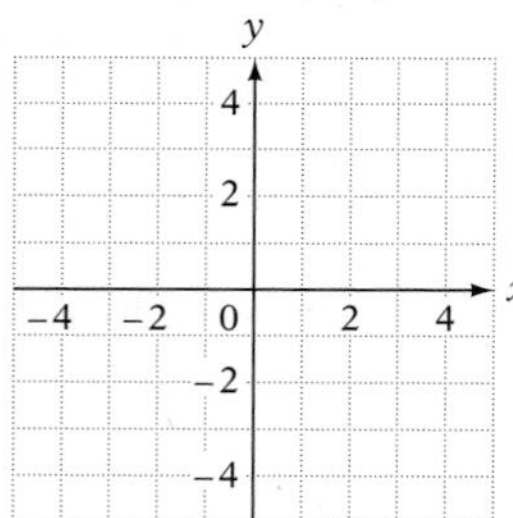

7. $f(x) = 3 \log_2 x$

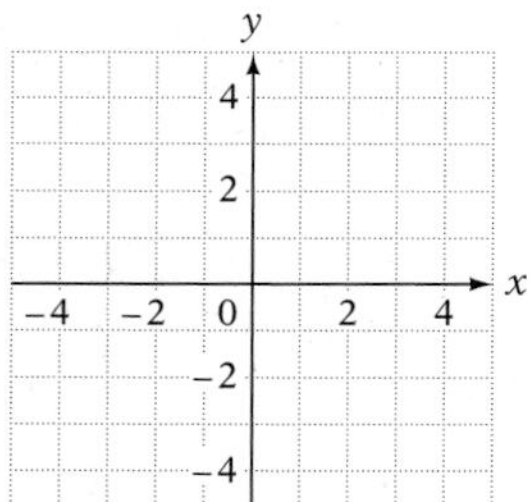

8. $f(x) = \frac{1}{2}\log_2 x$

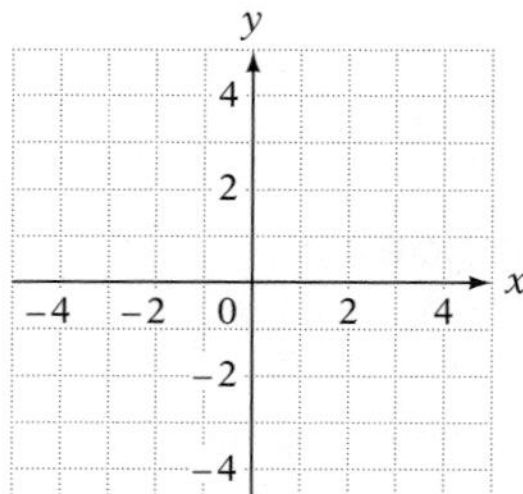

9. $f(x) = -\log_2 x$

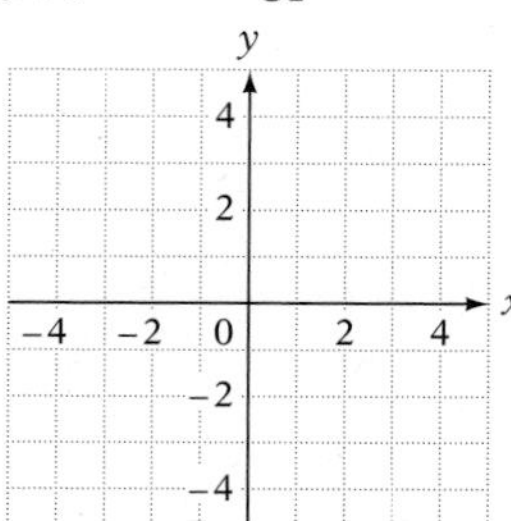

10. $f(x) = -\log_3 x$

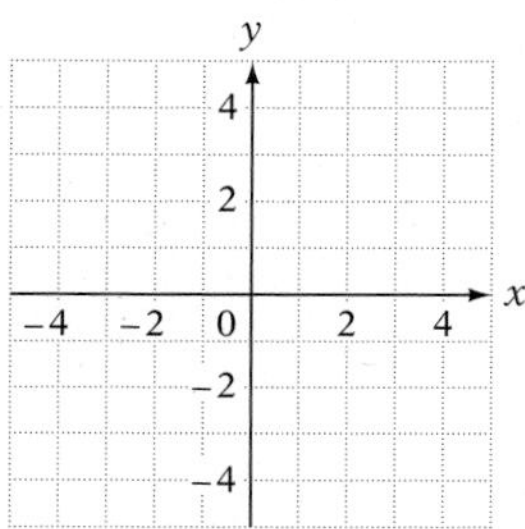

11. $f(x) = \log_2 (x - 1)$

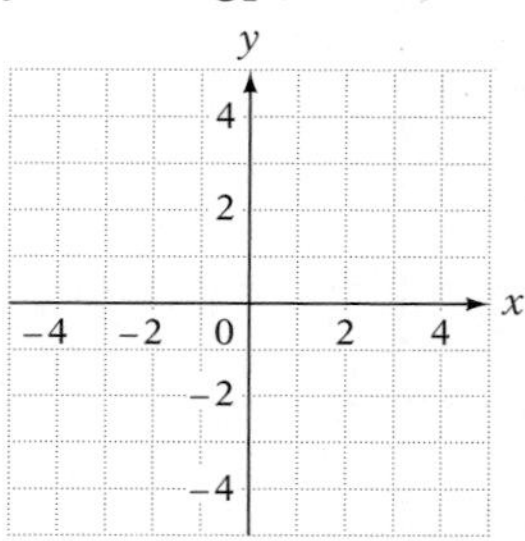

12. $f(x) = \log_3(2 - x)$

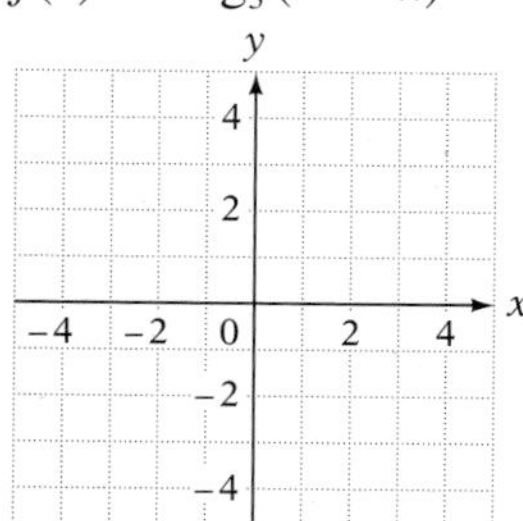

13. $f(x) = -\log_2(x - 1)$

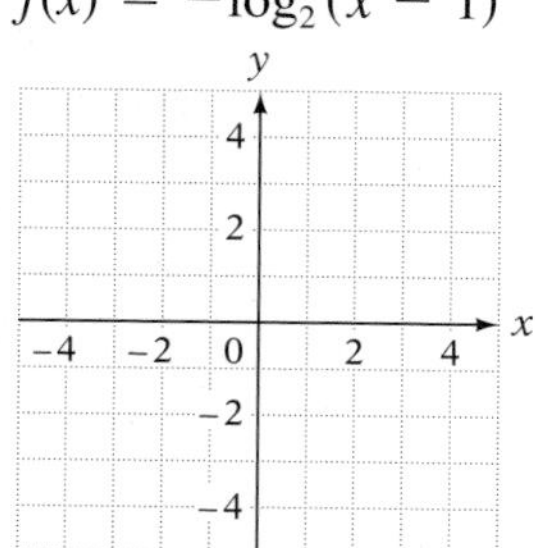

14. $f(x) = -\log_2(1 - x)$

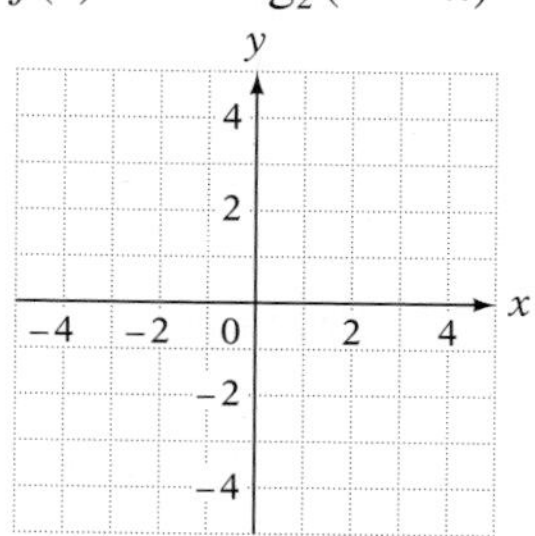

15. If $a \neq b$, at what point do the graphs of $y = \log_b x$ and $y = \log_a x$ intersect?

16. If $a \neq b$, what is the difference between the y-coordinates of the graphs of $y = \log_b b$ and $y = \log_a a$?

Applying the Concepts

Use a graphing calculator to graph the functions in Exercises 17 to 22.

17. $f(x) = x - \log_2(1 - x)$

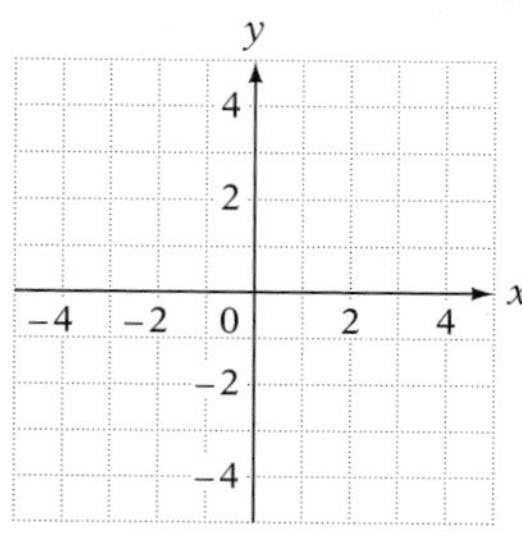

18. $f(x) = -\frac{1}{2}\log_2 x - 1$

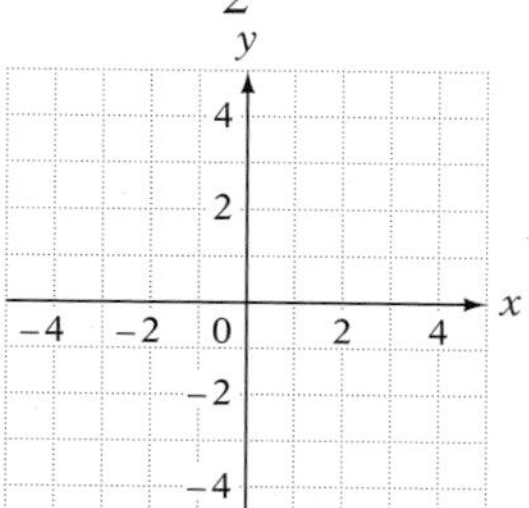

19. $f(x) = \frac{x}{2} - 2\log_2(x + 1)$

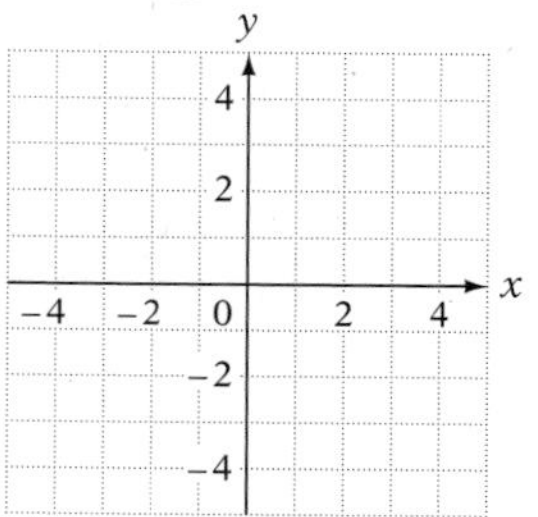

20. $f(x) = x + \log_3(2 - x)$

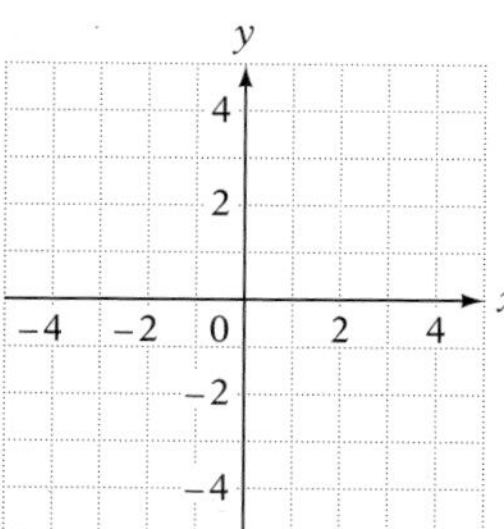

21. $f(x) = x^2 - 10\ln(x - 1)$

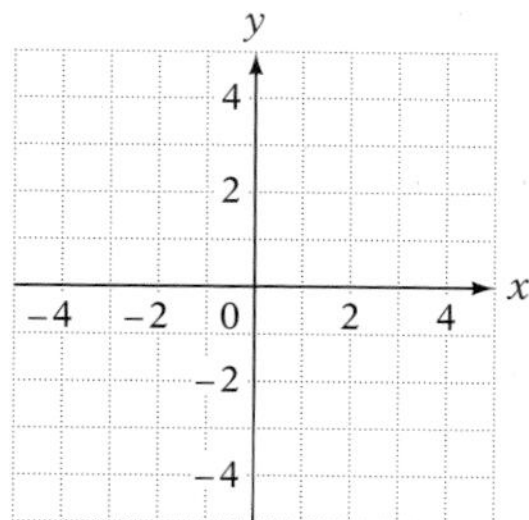

22. $f(x) = \frac{x}{3} - 3\log_2(x + 3)$

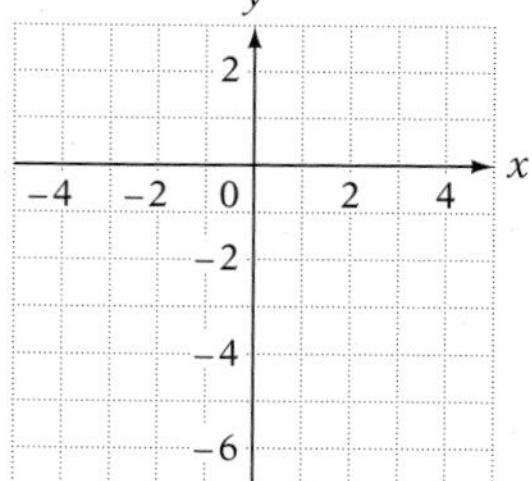

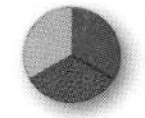

23. **Astronomy** Astronomers use the *distance modulus* of a star as a method of determining the star's distance from Earth. The formula is $M = 5\log s - 5$, where M is the distance modulus and s is the star's distance from Earth in parsecs. (One parsec $\approx 1.9 \times 10^{13}$ mi)

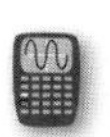

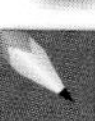

a. Graph the equation.

b. The point with coordinates (25.1, 2) is on the graph. Write a sentence that describes the meaning of this ordered pair.

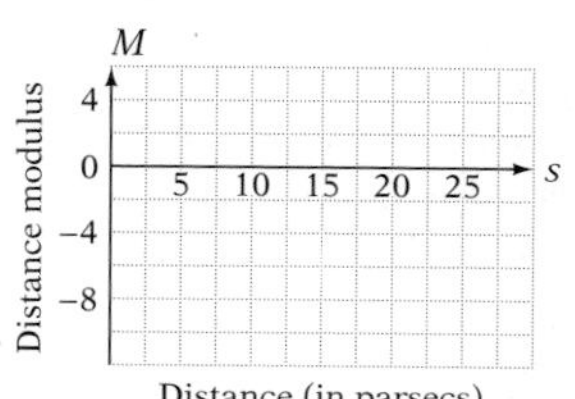

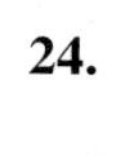

24. **Typing** Without practice, the proficiency of a typist decreases. The equation $s = 60 - 7\ln(t + 1)$, where s is the typing speed in words per minute and t is the number of months without typing, approximates this decrease.

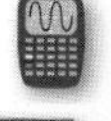

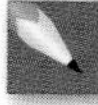

a. Graph the equation.

b. The point with coordinates (4, 49) is on the graph. Write a sentence that describes the meaning of this ordered pair.

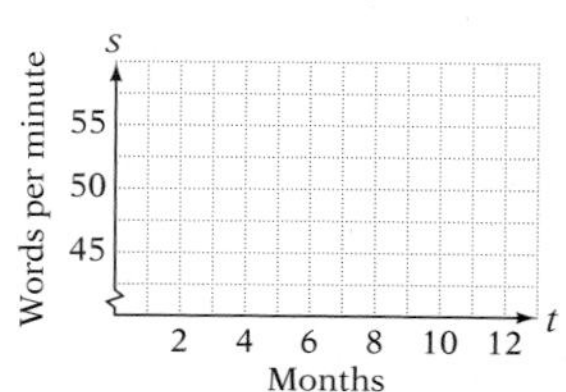

SECTION

12.4 Solving Exponential and Logarithmic Equations

OBJECTIVE A To solve an exponential equation 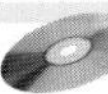

An **exponential equation** is one in which a variable occurs in an exponent. The equations at the right are exponential equations.

$$6^{2x+1} = 6^{3x-2}$$
$$4^x = 3$$
$$2^{x+1} = 7$$

An exponential equation in which each side of the equation can be expressed in terms of the same base can be solved by using the 1–1 Property of Exponential Functions. Recall that the 1–1 Property of Exponential Functions states that for $b > 0, b \neq 1$,

$$\text{if } b^x = b^y, \text{ then } x = y.$$

In the two examples below, this property is used in solving exponential equations.

HOW TO 1 Solve: $10^{3x+5} = 10^{x-3}$

$$10^{3x+5} = 10^{x-3}$$
$$3x + 5 = x - 3$$

• Use the 1–1 Property of Exponential Functions to equate the exponents.

$$2x + 5 = -3$$

• Solve the resulting equation.

$$2x = -8$$
$$x = -4$$

Check:

$$\begin{array}{c|c} \multicolumn{2}{c}{10^{3x+5} = 10^{x-3}} \\ \hline 10^{3(-4)+5} & 10^{-4-3} \\ 10^{-12+5} & 10^{-7} \end{array}$$
$$10^{-7} = 10^{-7}$$

The solution is −4.

 **Take Note**

The 1–1 Property of Exponential Functions requires that the bases be equal. For the problem at the right, we can write $9 = 3^2$ and $27 = 3^3$. After simplifying, the bases of the exponential expressions are equal.

HOW TO 2 Solve: $9^{x+1} = 27^{x-1}$

$$9^{x+1} = 27^{x-1}$$
$$(3^2)^{x+1} = (3^3)^{x-1}$$

• $3^2 = 9$; $3^3 = 27$

$$3^{2x+2} = 3^{3x-3}$$
$$2x + 2 = 3x - 3$$

• Use the 1–1 Property of Exponential Functions to equate the exponents.

$$2 = x - 3$$

• Solve for x.

$$5 = x$$

Check:

$$\begin{array}{c|c} \multicolumn{2}{c}{9^{x+1} = 27^{x-1}} \\ \hline 9^{5+1} & 27^{5-1} \\ 9^6 & 27^4 \end{array}$$
$$531{,}441 = 531{,}441$$

The solution is 5.

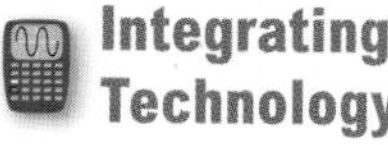

Integrating Technology

To evaluate $\frac{\log 7}{\log 4}$ on a scientific calculator, use the keystrokes

7 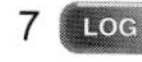  4 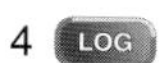

The display should read 1.4036775.

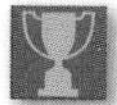

Tips for Success

Always check the solution of an equation, even when the solution is an approximation. For the equation at the right:

$$3^{x+1} = 5$$

$3^{0.4650+1}$	5
$3^{1.4650}$	5
$5.000145 \approx 5$	

When both sides of an exponential equation cannot easily be expressed in terms of the same base, logarithms are used to solve the exponential equation.

HOW TO • 3 Solve: $4^x = 7$

$4^x = 7$

$\log 4^x = \log 7$ • Take the common logarithm of each side of the equation.

$x \log 4 = \log 7$ • Rewrite the equation using the Properties of Logarithms.

$x = \frac{\log 7}{\log 4} \approx 1.4037$ • Solve for x.

The solution is 1.4037. • Note that $\frac{\log 7}{\log 4} \neq \log 7 - \log 4$.

HOW TO • 4 Solve: $3^{x+1} = 5$

$3^{x+1} = 5$

$\log 3^{x+1} = \log 5$ • Take the common logarithm of each side of the equation.

$(x+1)\log 3 = \log 5$ • Rewrite the equation using the Properties of Logarithms.

$x + 1 = \frac{\log 5}{\log 3}$

$x + 1 \approx 1.4650$ • Solve for x.

$x \approx 0.4650$

The solution is 0.4650.

EXAMPLE • 1

Solve for n: $(1.1)^n = 2$

Solution

$(1.1)^n = 2$

$\log (1.1)^n = \log 2$ • Take the log of each side.

$n \log 1.1 = \log 2$ • Power Property

$n = \frac{\log 2}{\log 1.1}$ • Divide both sides by log 1.1.

$n \approx 7.2725$

The solution is 7.2725.

YOU TRY IT • 1

Solve for n: $(1.06)^n = 1.5$

Your solution

EXAMPLE • 2

Solve for x: $3^{2x} = 4$

Solution

$3^{2x} = 4$

$\log 3^{2x} = \log 4$ • Take the log of each side.

$2x \log 3 = \log 4$ • Power Property

$2x = \frac{\log 4}{\log 3}$ • Divide both sides by log 3.

$2x \approx 1.2619$

$x \approx 0.6310$ • Divide both sides by 2.

The solution is 0.6310.

YOU TRY IT • 2

Solve for x: $4^{3x} = 25$

Your solution

Solutions on p. S35

OBJECTIVE B To solve a logarithmic equation 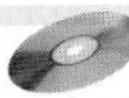

The 1–1 Property of Logarithms from Section 12.2 can be used to solve some logarithmic equations.

✓ Take Note
It is important to check a proposed solution of a logarithmic equation. Consider the equation $\log(2x - 3) = \log(x - 2)$.

$$\log(2x - 3) = \log(x - 2)$$
$$2x - 3 = x - 2$$
$$x = 1$$

Substituting 1 into the original equation gives $\log(-1) = \log(-1)$. However, logarithms of negative numbers are not defined. Therefore, the equation $\log(2x - 3) = \log(x - 2)$ has no solution.

HOW TO • 5 Solve: $\log_5(3x - 1) = \log_5(7 - x)$

$\log_5(3x - 1) = \log_5(7 - x)$

$3x - 1 = 7 - x$ • Use the 1–1 Property of Logarithms.

$4x = 8$ • Solve for x.

$x = 2$

Because logarithms are defined only for positive numbers, we must check the solution.

Check: $\log_5(3x - 1) = \log_5(7 - x)$

$\log_5[3(2) - 1] \mid \log_5(7 - 2)$ • Replace x by **2**.

$\log_5(5) = \log_5(5)$

The solution checks. The solution is 2.

Solving some logarithmic equations may require using several of the properties of logarithms.

HOW TO • 6 Solve: $\log_3 6 - \log_3(2x + 3) = \log_3(x + 1)$

$\log_3 6 - \log_3(2x + 3) = \log_3(x + 1)$

$\log_3 \frac{6}{2x + 3} = \log_3(x + 1)$ • Use the Quotient Property of Logarithms.

$\frac{6}{2x + 3} = x + 1$ • Use the 1–1 Property of Logarithms.

$6 = (2x + 3)(x + 1)$

$6 = 2x^2 + 5x + 3$

$0 = 2x^2 + 5x - 3$ • Write in standard form.

$0 = (2x - 1)(x + 3)$ • Factor and use the Principle of Zero Products.

$2x - 1 = 0 \qquad x + 3 = 0$

$x = \frac{1}{2} \qquad x = -3$

-3 does not check as a solution. The solution is $\frac{1}{2}$.

HOW TO • 7 Solve: $\log_9 x + \log_9(x - 8) = 1$

$\log_9 x + \log_9(x - 8) = 1$

$\log_9[x(x - 8)] = 1$ • Use the Product Property of Logarithms.

$9^1 = x(x - 8)$ • Write the equation in exponential form.

$9 = x^2 - 8x$

$0 = x^2 - 8x - 9$ • Write in standard form.

$0 = (x - 9)(x + 1)$ • Factor and use the Principle of Zero Products.

$x - 9 = 0 \qquad x + 1 = 0$

$x = 9 \qquad x = -1$

-1 does not check as a solution. The solution is 9.

EXAMPLE • 3

Solve for x: $\log_3(2x - 1) = 2$

Solution

$$\log_3(2x - 1) = 2$$
$$3^2 = 2x - 1 \quad \text{• Write in exponential form.}$$
$$9 = 2x - 1$$
$$10 = 2x$$
$$5 = x$$

The solution is 5.

YOU TRY IT • 3

Solve for x: $\log_4(x^2 - 3x) = 1$

Your solution

EXAMPLE • 4

Solve for x:
$\log_2(3x + 8) = \log_2(2x + 2) + \log_2(x - 2)$

Solution

$$\log_2(3x + 8) = \log_2(2x + 2) + \log_2(x - 2)$$
$$\log_2(3x + 8) = \log_2[(2x + 2)(x - 2)]$$
$$\log_2(3x + 8) = \log_2(2x^2 - 2x - 4)$$
$$3x + 8 = 2x^2 - 2x - 4 \quad \text{• 1–1 Property of Logarithms}$$
$$0 = 2x^2 - 5x - 12$$
$$0 = (2x + 3)(x - 4) \quad \text{• Solve by factoring.}$$
$$2x + 3 = 0 \qquad x - 4 = 0$$
$$x = -\frac{3}{2} \qquad x = 4$$

$-\frac{3}{2}$ does not check as a solution. The solution is 4.

YOU TRY IT • 4

Solve for x: $\log_3 x + \log_3(x + 3) = \log_3 4$

Your solution

EXAMPLE • 5

Solve for x: $\log_3(5x + 4) - \log_3(2x - 1) = 2$

Solution

$$\log_3(5x + 4) - \log_3(2x - 1) = 2$$
$$\log_3 \frac{5x + 4}{2x - 1} = 2 \quad \text{• Quotient Property}$$
$$3^2 = \frac{5x + 4}{2x - 1} \quad \text{• Definition of logarithm}$$
$$9 = \frac{5x + 4}{2x - 1}$$
$$(2x - 1)9 = (2x - 1)\left(\frac{5x + 4}{2x - 1}\right)$$
$$18x - 9 = 5x + 4$$
$$13x = 13$$
$$x = 1$$

1 checks as a solution. The solution is 1.

YOU TRY IT • 5

Solve for x: $\log_3 x + \log_3(x + 6) = 3$

Your solution

Solutions on p. S35

12.4 EXERCISES

OBJECTIVE A To solve an exponential equation

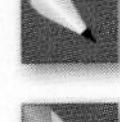

1. What is an exponential equation?

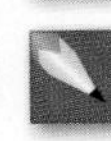

2. **a.** What does the 1–1 Property of Exponential Functions state?
   b. Provide an example of when you would use this property.

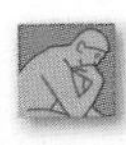

3. Let $2^x = \frac{1}{3}$. Without solving the equation, is $x < 0$ or is $x > 0$?

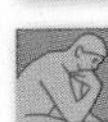

4. Without solving, determine which of the following equations have no solution.
   (i) $5^{-x} = 6$ (ii) $5^x = -6$ (iii) $5^{-x} = -6$ (iv) $5^x = 6$

For Exercises 5 to 36, solve for x. If necessary, round to the nearest ten-thousandth.

5. $5^{4x-1} = 5^{x-2}$
6. $7^{4x-3} = 7^{2x+1}$
7. $8^{x-4} = 8^{5x+8}$
8. $10^{4x-5} = 10^{x+4}$
9. $9^x = 3^{x+1}$
10. $2^{x-1} = 4^x$
11. $8^{x+2} = 16^x$
12. $9^{3x} = 81^{x-4}$
13. $16^{2-x} = 32^{2x}$
14. $27^{2x-3} = 81^{4-x}$
15. $25^{3-x} = 125^{2x-1}$
16. $8^{4x-7} = 64^{x-3}$
17. $5^x = 6$
18. $7^x = 10$
19. $8^{x/4} = 0.4$
20. $5^{x/2} = 0.5$
21. $2^{3x} = 5$
22. $3^{6x} = 0.5$
23. $2^{-x} = 7$
24. $3^{-x} = 14$
25. $2^{x-1} = 6$
26. $4^{x+1} = 9$
27. $3^{2x-1} = 4$
28. $4^{-x+2} = 12$
29. $\left(\frac{1}{2}\right)^{x+1} = 3$
30. $\left(\frac{3}{5}\right)^{-2x} = 2$
31. $3 \cdot 2^x = 7$
32. $5 \cdot 3^{2-x} = 4$
33. $7 = 10\left(\frac{1}{2}\right)^{x/8}$
34. $8 = 15\left(\frac{1}{2}\right)^{x/22}$
35. $15 = 12e^{0.05x}$
36. $7 = 42e^{-3x}$

OBJECTIVE B To solve a logarithmic equation

For Exercises 37 to 54, solve for x. If necessary, round to the nearest ten-thousandth.

37. $\log x = \log (1 - x)$
38. $\ln (3x - 2) = \ln (x + 1)$
39. $\ln (3x + 2) = \ln (5x + 4)$
40. $\log_3 (x - 2) = \log_3 (2x)$
41. $\log_2 (8x) - \log_2 (x^2 - 1) = \log_2 3$
42. $\log_5 (3x) - \log_5 (x^2 - 1) = \log_5 2$
43. $\log_9 x + \log_9 (2x - 3) = \log_9 2$
44. $\log_6 x + \log_6 (3x - 5) = \log_6 2$
45. $\log_2 (2x - 3) = 3$

46. $\log_4(3x + 1) = 2$

47. $\ln(3x + 2) = 4$

48. $\ln(2x + 3) = -1$

49. $\log_2(x + 1) + \log_2(x + 3) = 3$

50. $\log_4(3x - 4) + \log_4(x + 6) = 2$

51. $\log_5(2x) - \log_5(x - 1) = 1$

52. $\log_3(3x) - \log_3(2x - 1) = 2$

53. $\log_8(6x) = \log_8 2 + \log_8(x - 4)$

54. $\log_7(5x) = \log_7 3 + \log_7(2x + 1)$

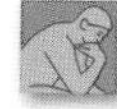

55. If u and v are positive numbers and $u < v$, then $\log u < \log v$. Use this fact to explain why the equation $\log(x - 2) - \log x = 3$ has no solution. Do not solve the equation.

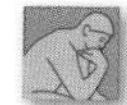

56. Sometimes it is easy to represent one number as a power of another. For instance, 81 represented as a power of 3 is $3^4 = 81$. Although it is not as easy, show that we could represent 81 as a power of 7.

Applying the Concepts

For Exercises 57 to 60, solve for x. Round to the nearest ten-thousandth.

57. $3^{x+1} = 2^{x-2}$

58. $2^{2x} = 5^{x+1}$

59. $7^{2x-1} = 3^{2x+3}$

60. $4^{3x+2} = 3^{2x-5}$

61. **Physics** A model for the distance s, in feet, that an object experiencing air resistance will fall in t seconds is given by $s = 312.5 \ln \dfrac{e^{0.32t} + e^{-0.32t}}{2}$.

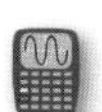

a. Graph this equation. *Suggestion:* Use Xmin = 0, Xmax = 4.5, Ymin = 0, Ymax = 140, and Yscl = 20.

b. Determine, to the nearest hundredth of a second, the time it takes for the object to fall 100 ft.

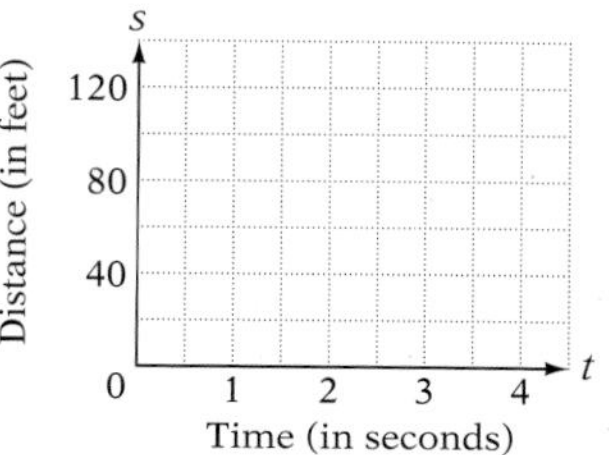

62. **Physics** A model for the distance s, in feet, that an object experiencing air resistance will fall in t seconds is given by $s = 78 \ln \dfrac{e^{0.8t} + e^{-0.8t}}{2}$.

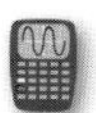

a. Graph this equation. *Suggestion:* Use Xmin = 0, Xmax = 4.5, Ymin = 0, Ymax = 140, and Yscl = 20.

b. Determine, to the nearest hundredth of a second, the time it takes for the object to fall 125 ft.

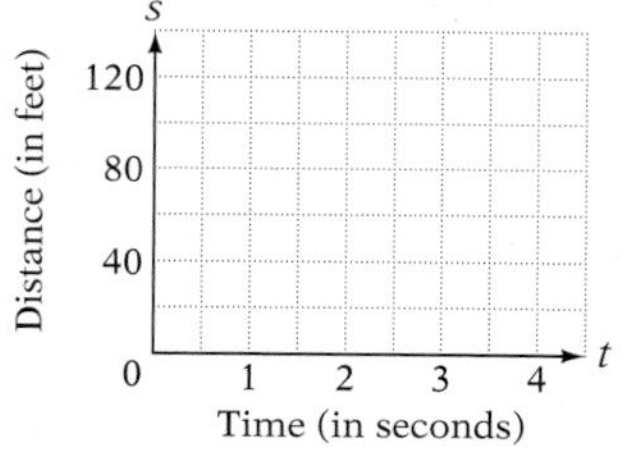

SECTION

12.5 Applications of Exponential and Logarithmic Functions

OBJECTIVE A To solve application problems

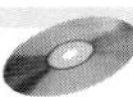

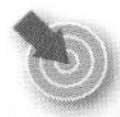

Point of Interest

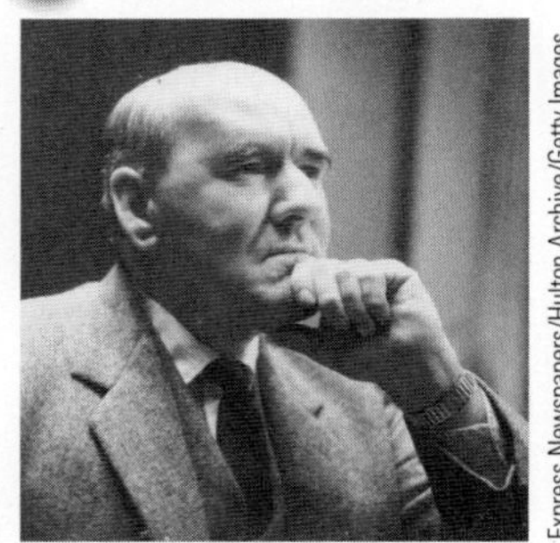

Express Newspapers/Hulton Archive/Getty Images

C. Northcote Parkinson

Parkinson's Law, named after C. Northcote Parkinson, is sometimes stated as "A job will expand to fill the time allotted for the job." However, Parkinson actually said that in any new government administration, administrative employees will be added at the rate of about 5% to 6% per year. This is an example of exponential growth and means that a staff of 500 will grow to approximately 630 by the end of a 4-year term.

A biologist places one single-celled bacterium in a culture, and each hour that particular species of bacteria divides into two bacteria. After 1 h, there will be two bacteria. After 2 h, each of those two bacteria will divide and there will be four bacteria. After 3 h, each of the four bacteria will divide and there will be eight bacteria.

The table at the right shows the number of bacteria in the culture after various intervals of time t, in hours. Values in this table could also be found by using the exponential equation $N = 2^t$.

Time, t	*Number of Bacteria, N*
0	1
1	2
2	4
3	8
4	16

The equation $N = 2^t$ is an example of an **exponential growth equation.** In general, any equation that can be written in the form $A = A_0b^{kt}$, where A is the size at time t, A_0 is the initial size, $b > 1$, and k is a positive real number, is an exponential growth equation. These equations are important not only in population growth studies but also in physics, chemistry, psychology, and economics.

Recall that interest is the amount of money that one pays (or receives) when borrowing (or investing) money. **Compound interest** is interest that is computed not only on the original principal but also on the interest already earned. The compound interest formula is an exponential equation.

The **compound interest formula** is $A = P(1 + i)^n$, where P is the original value of an investment, i is the interest rate per compounding period, n is the total number of compounding periods, and A is the value of the investment after n periods.

HOW TO • 1 An investment broker deposits $1000 into an account that earns 8% annual interest compounded quarterly. What is the value of the investment after 3 years?

$i = \frac{8\%}{4} = \frac{0.08}{4} = 0.02$ • Find i, the interest rate per quarter. The quarterly rate is the annual rate divided by 4, the number of quarters in 1 year.

$n = 4 \cdot 3 = 12$ • Find n, the number of compounding periods. The investment is compounded quarterly, 4 times a year, for 3 years.

$A = P(1 + i)^n$ • Use the compound interest formula.

$A = 1000(1 + 0.02)^{12}$ • Replace P, i, and n by their values.

$A \approx 1268$ • Solve for A.

The value of the investment after 3 years is approximately $1268.

Exponential decay offers another example of an exponential equation. A common illustration of exponential decay is the decay of a radioactive element. One form of the equation that is used to model radioactive decay is $A = A_0(0.5)^{t/k}$, where A is the amount of the substance remaining after a time period t, A_0 is the initial amount of the radioactive material, and k is the *half-life* of the material. The **half-life** of a radioactive substance is the time it takes for one-half of the material to disintegrate.

The table at the right indicates the amount of an initial 10-microgram sample of tritium that remains after various intervals of time t in years. Note from the table that after each 12-year period of time, the amount of tritium is reduced by one-half. The half-life of tritium is 12 years. The equation that models the decay of the 10-microgram sample is $A = 10(0.5)^{t/12}$, where t is in years and A is in micrograms.

Time, t	*Amount, A*
0	10
12	5
24	2.5
36	1.25

HOW TO • 2 Yttrium-90 is a radioactive isotope that is used to treat some cancers. A sample that originally contained 5 mg of yttrium-90 was measured again after 20 h and found to have 4 mg of yttrium-90. What is the half-life of yttrium-90? Round to the nearest whole number.

$$A = A_0(0.5)^{t/k}$$

$$4 = 5(0.5)^{20/k}$$ • Replace A by **4**, A_0 by **5**, and t by **20**.

$$\frac{4}{5} = (0.5)^{20/k}$$ • Divide each side by 5 to isolate the exponential expression.

$$\log(0.8) = \log(0.5)^{20/k}$$ • Take the common logarithm of each side.

$$\log(0.8) = \frac{20}{k}\log(0.5)$$ • Use the Power Property of Logarithms.

$$k\log(0.8) = 20\log(0.5)$$ • Solve for k.

$$k = \frac{20\log(0.5)}{\log(0.8)}$$ • Exact answer

$$k \approx 62.13$$ • Approximate answer

The half-life of yttrium-90 is approximately 62 h.

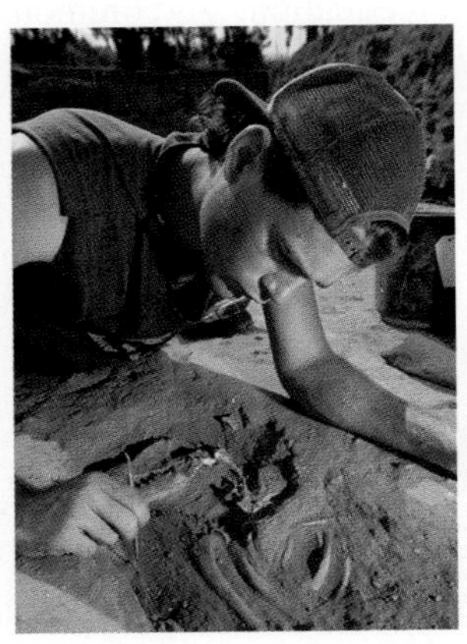
© Richard T. Nowitz/Corbis

A method by which an archaeologist can measure the age of a bone is called *carbon dating*. Carbon dating is based on a radioactive isotope of carbon called carbon-14, which has a half-life of approximately 5570 years. The exponential decay equation is given by $A = A_0(0.5)^{t/5570}$, where A_0 is the original amount of carbon-14 present in the bone, t is the age of the bone in years, and A is the amount of carbon-14 present after t years.

HOW TO • 3 A bone that originally contained 100 mg of carbon-14 now has 70 mg of carbon-14. What is the approximate age of the bone?

$A = A_0(0.5)^{t/5570}$ • Use the exponential decay equation.

$70 = 100(0.5)^{t/5570}$ • Replace A by **70** and A_0 by **100**, and solve for t.

$0.7 = (0.5)^{t/5570}$ • Divide each side by 100.

$\log 0.7 = \log(0.5)^{t/5570}$ • Take the common logarithm of each side of the equation.

$\log 0.7 = \frac{t}{5570}\log 0.5$ • Power Property

$\frac{5570 \log 0.7}{\log 0.5} = t$ • Multiply by 5570, and divide by log 0.5.

$2866 \approx t$

The bone is approximately 2866 years old.

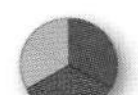

Logarithmic functions are used to scale very large or very small numbers into numbers that are easier to comprehend. For instance, the *Richter scale magnitude* of an earthquake uses a logarithmic function to convert the intensity of shock waves I into a number M, which for most earthquakes is in the range of 0 to 10. The intensity I of an earthquake is often given in terms of the constant I_0, where I_0 is the intensity of the smallest earthquake, called a *zero-level earthquake,* that can be measured on a seismograph near the earthquake's epicenter. An earthquake with an intensity I has a Richter scale magnitude of $M = \log \frac{I}{I_0}$, where I_0 is the measure of the intensity of a zero-level earthquake.

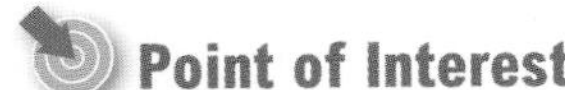
Point of Interest

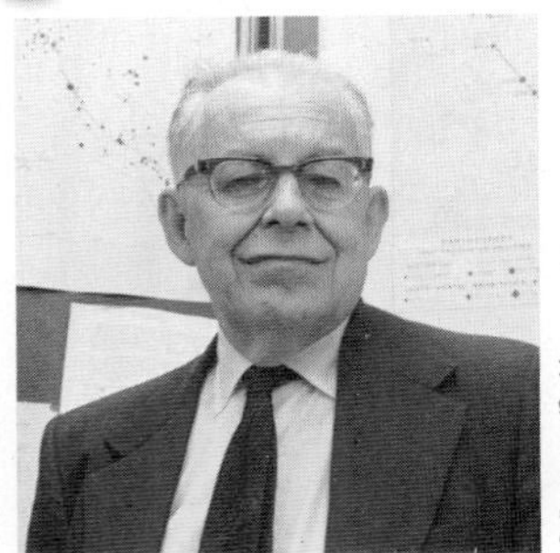
© Bettmann/Corbis

Charles F. Richter

The Richter scale was created by seismologist Charles F. Richter in 1935. Note that a tenfold increase in the intensity level of an earthquake increases the Richter scale magnitude of the earthquake by only 1.

HOW TO • 4 The Richter magnitude of an earthquake that occurred on May, 2008, in Sichuan, China, was 7.9. Find the intensity of the earthquake in terms of I_0. Round to the nearest thousand.

$M = \log \frac{I}{I_0}$

$7.9 = \log \frac{I}{I_0}$ • Replace M by **7.9**.

$\frac{I}{I_0} = 10^{7.9}$ • Write in exponential form.

$I = 10^{7.9}I_0$ • Solve for I.

$I \approx 79{,}432{,}823I_0$

The Sichuan earthquake had an intensity that was 79,433,000 times the intensity of a zero-level earthquake.

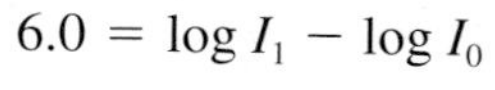

HOW TO • 5 In April 2008, there was an earthquake near Wells, Nevada, that measured 6.0 on the Richter scale. In July 2008, an earthquake of magnitude 5.4 occurred near Los Angeles, California. How many times stronger was the Wells earthquake than the Los Angeles earthquake? Round to the nearest whole number.

Let I_1 be the intensity of the Wells earthquake, and let I_2 be the intensity of the Los Angeles earthquake. Use the Quotient Property of Logarithms to write $M = \log \frac{I}{I_0}$ as $M = \log I - \log I_0$. For the Wells earthquake, $6.0 = \log I_1 - \log I_0$. For the Los Angeles earthquake, $5.4 = \log I_2 - \log I_0$. Using these equations, we have

$$6.0 = \log I_1 - \log I_0$$
$$5.4 = \log I_2 - \log I_0$$

$$0.6 = \log I_1 - \log I_2 \qquad \text{• Subtract the equations.}$$
$$0.6 = \log \frac{I_1}{I_2} \qquad \text{• Use the Quotient Property of Logarithms.}$$
$$\frac{I_1}{I_2} = 10^{0.6} \qquad \text{• Write in exponential form.}$$
$$I_1 = 10^{0.6}\, I_2 \qquad \text{• Solve for } I_1.$$
$$I_1 \approx 3.981 I_2$$

The Wells earthquake was approximately 4 times stronger than the Los Angeles earthquake.

Point of Interest

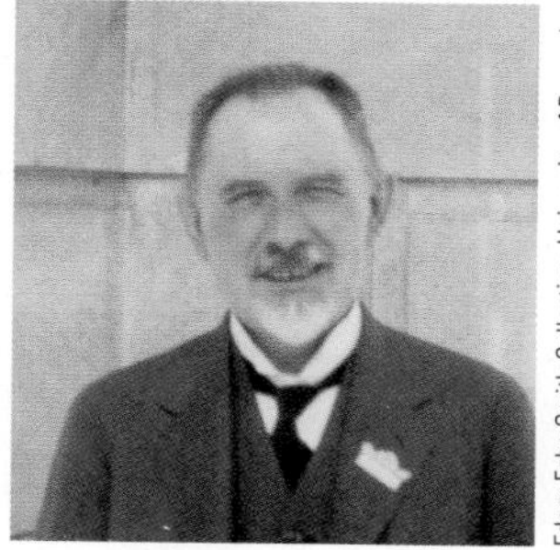

Søren Sørensen

The pH scale was created by the Danish biochemist Søren Sørensen in 1909 to measure the acidity of water used in the brewing of beer. pH is an abbreviation for *pondus hydrogenii*, which translates as "potential hydrogen."

A chemist measures the acidity or alkalinity of a solution by measuring the concentration of hydrogen ions, H^+, in the solution using the formula $pH = -\log(H^+)$. A neutral solution such as distilled water has a pH of 7, acids have a pH less than 7, and alkaline solutions (also called basic solutions) have a pH greater than 7.

HOW TO • 6 Find the pH of orange juice that has a hydrogen ion concentration, H^+, of 2.9×10^{-4}. Round to the nearest tenth.

$$pH = -\log(H^+)$$
$$= -\log(2.9 \times 10^{-4}) \qquad \text{• } H^+ = \mathbf{2.9 \times 10^{-4}}$$
$$\approx 3.5376$$

The pH of the orange juice is approximately 3.5.

EXAMPLE • 1

An investment of $3000 is placed into an account that earns 12% annual interest compounded monthly. In approximately how many years will the investment be worth twice the original amount?

Strategy

To find the time, solve the compound interest formula for n. Use $A = 6000$, $P = 3000$, and $i = \frac{12\%}{12} = \frac{0.12}{12} = 0.01$.

Solution

$$A = P(1 + i)^n$$
$$6000 = 3000(1 + 0.01)^n$$
$$6000 = 3000(1.01)^n$$
$$2 = (1.01)^n \qquad \text{• Divide by 3000.}$$
$$\log 2 = \log(1.01)^n \qquad \text{• Take the log of each side.}$$
$$\log 2 = n \log 1.01 \qquad \text{• Power Property}$$
$$\frac{\log 2}{\log 1.01} = n \qquad \text{• Divide both sides by log 1.01.}$$
$$70 \approx n$$

70 months ÷ 12 ≈ 5.8 years

In approximately 6 years, the investment will be worth $6000.

YOU TRY IT • 1

Find the hydrogen ion concentration, H^+, of vinegar that has a pH of 2.9.

Your strategy

Your solution

EXAMPLE • 2

Rhenium-186 is a radioactive isotope with a half-life of approximately 3.78 days. Rhenium-186 is sometimes used for pain management. If a patient receives a dose of 5 micrograms of rhenium-186, how long (in days) will it take for the rhenium-186 level to reach 2 micrograms?

Strategy

- Use the equation $A = A_0(0.5)^{t/k}$.
- Replace A with 2, A_0 with 5, and k with 3.78. Then solve for t.

Solution

$$A = A_0(0.5)^{t/k}$$
$$2 = 5(0.5)^{t/3.78}$$
$$\frac{2}{5} = (0.5)^{t/3.78}$$
$$\log(0.4) = \frac{t}{3.78}\log(0.5)$$
$$\frac{3.78\log(0.4)}{\log(0.5)} = t$$
$$4.997 \approx t$$

In approximately 5 days, there will be 2 micrograms of rhenium-186 remaining in the patient's body.

YOU TRY IT • 2

The percent of light p, as a decimal, that passes through a substance of thickness d, in meters, is given by $\log p = -kd$. The value of k for some opaque glass is 6. How thick a piece of this glass is necessary so that 20% of the light passes through the glass?

Your strategy

Your solution

Solutions on p. S36

12.5 EXERCISES

OBJECTIVE A To solve application problems

Compound Interest For Exercises 1 to 4, use the compound interest formula $A = P(1 + i)^n$, where P is the original value of an investment, i is the interest rate per compounding period, n is the total number of compounding periods, and A is the value of the investment after n periods.

1. An investment broker deposits $1000 into an account that earns 8% annual interest compounded quarterly. What is the value of the investment after 2 years? Round to the nearest dollar.

2. A financial advisor recommends that a client deposit $2500 into a fund that earns 7.5% annual interest compounded monthly. What will be the value of the investment after 3 years? Round to the nearest cent.

3. To save for college tuition, the parents of a preschooler invest $5000 in a bond fund that earns 6% annual interest compounded monthly. In approximately how many years will the investment be worth $15,000?

4. A hospital administrator deposits $10,000 into an account that earns 6% annual interest compounded monthly. In approximately how many years will the investment be worth $15,000?

Radioactivity For Exercises 5 to 8, use the exponential decay equation $A = A_0(0.5)^{t/k}$, where A is the amount of a radioactive material present after time t, k is the half-life of the radioactive substance, and A_0 is the original amount of the radioactive substance. Round to the nearest tenth.

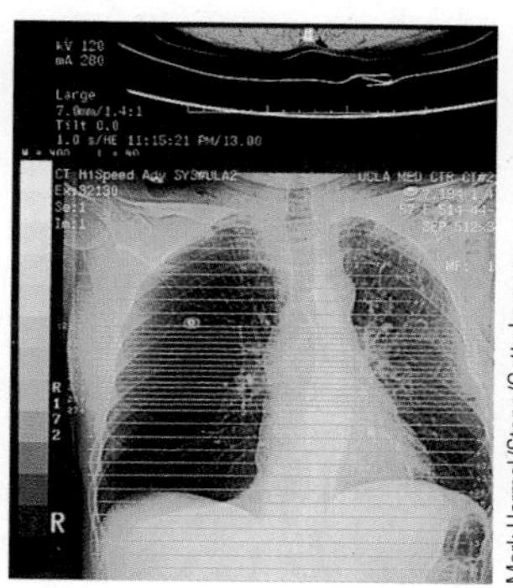

Mark Harmel/Stone/Getty Images

5. An isotope of technetium is used to prepare images of internal body organs. This isotope has a half-life of approximately 6 h. A patient is injected with 30 mg of this isotope.
a. What is the technetium level in the patient after 3 h?
b. How long (in hours) will it take for the technetium level to reach 20 mg?

6. Iodine-131 is an isotope that is used to study the functioning of the thyroid gland. This isotope has a half-life of approximately 8 days. A patient is given an injection that contains 8 micrograms of iodine-131.
a. What is the iodine level in the patient after 5 days?
b. How long (in days) will it take for the iodine level to reach 5 micrograms?

7. A sample of promethium-147 (used in some luminous paints) weighs 25 mg. One year later, the sample weighs 18.95 mg. What is the half-life of promethium-147, in years?

8. Francium-223 is a very rare radioactive isotope discovered in 1939 by Marguerite Percy. A 3-microgram sample of francium-223 decays to 2.54 micrograms in 5 min. What is the half-life of francium-223, in minutes?

Seismology For Exercises 9 to 12, use the Richter scale equation $M = \log \frac{I}{I_0}$, where M is the magnitude of an earthquake, I is the intensity of the shock waves, and I_0 is the measure of the intensity of a zero-level earthquake.

9. An earthquake in Japan on March 2, 1933, measured 8.9 on the Richter scale. Find the intensity of the earthquake in terms of I_0. Round to the nearest whole number.

10. An earthquake that occurred in China in 1978 measured 8.2 on the Richter scale. Find the intensity of the earthquake in terms of I_0. Round to the nearest whole number.

11. In 2008, a 6.9-magnitude earthquake occurred off the coast of Honshu, Japan. In the same year, a 6.4-magnitude earthquake occurred near Quetta, Pakistan. How many times stronger was the Honshu earthquake than the Quetta earthquake? Round to the nearest tenth.

12. In 2008, a 6.4-magnitude earthquake occurred off the coast of Oregon. In the same year, a 7.2-magnitude earthquake occurred near Xinjiang, China. How many times stronger was the Xinjiang earthquake than the one off the coast of Oregon? Round to the nearest tenth.

Seismology Shown at the right is a *seismogram,* which is used to measure the magnitude of an earthquake. The magnitude is determined by the amplitude A of a shock wave and the difference in time t, in seconds, between the occurrences of two types of waves called *primary waves* and *secondary waves.* As you can see on the graph, a primary wave is abbreviated *p-wave* and a secondary wave is abbreviated *s-wave*. The amplitude A of a wave is one-half the difference between its highest and lowest points. For this graph, A is 23 mm. The equation is $M = \log A + 3 \log 8t - 2.92$. Use this information for Exercises 13 to 15.

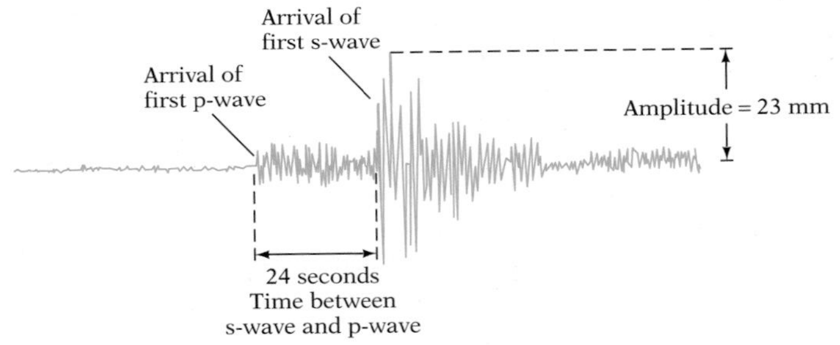

13. Determine the magnitude of the earthquake for the seismogram given in the figure. Round to the nearest tenth.

14. Find the magnitude of an earthquake that has a seismogram with an amplitude of 30 mm and for which t is 21 s.

15. Find the magnitude of an earthquake that has a seismogram with an amplitude of 28 mm and for which t is 28 s.

Chemistry For Exercises 16 to 19, use the equation $pH = -\log(H^+)$, where H^+ is the hydrogen ion concentration of a solution.

16. Find the pH of milk, which has a hydrogen ion concentration of 3.97×10^{-7}. Round to the nearest tenth.

17. Find the pH of a baking soda solution for which the hydrogen ion concentration is 3.98×10^{-9}. Round to the nearest tenth.

18. The pH of pure water is 7. What is the hydrogen ion concentration of pure water?

19. Peanuts grow best in soils that have a pH between 5.3 and 6.6. What is the range of hydrogen ion concentration for these soils?

Sound For Exercises 20 to 23, use the equation $D = 10(\log I + 16)$, where D is the number of decibels of a sound and I is the intensity of the sound in watts per square centimeter.

20. Find the number of decibels of normal conversation. The intensity of the sound of normal conversation is approximately 3.2×10^{-10} watts/cm^2. Round to the nearest whole number.

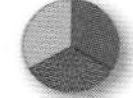

21. The loudest sound made by any animal is made by the blue whale and can be heard from more than 500 mi away. The intensity of the sound is 630 watts/cm^2. Find the number of decibels of sound emitted by the blue whale. Round to the nearest whole number.

22. Although pain thresholds for sound vary in humans, a decibel level of 125 decibels will produce pain (and even hearing loss) for most people. What is the intensity, in watts per square centimeter, of 125 decibels?

23. The purr of a cat is approximately 25 decibels. What is the intensity, in watts per square centimeter, of 25 decibels?

Light For Exercises 24 and 25, use the equation $\log P = -kd$, which gives the percent P, as a decimal, of light passing through a substance of thickness d, in meters.

24. The value of k for a swimming pool is approximately 0.05. At what depth, in meters, will the percent of light be 75% of the light at the surface of the pool?

25. The constant k for a piece of blue stained glass is 20. What percent of light will pass through a piece of this glass that is 0.005 m thick?

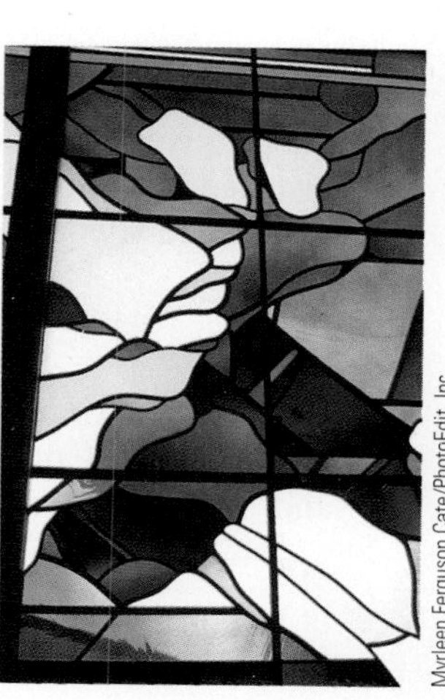

26. Earth Science Earth's atmospheric pressure changes as you rise above its surface. At an altitude of h kilometers, where $0 < h < 80$, the pressure P in newtons per square centimeter is approximately modeled by the equation $P(h) = 10.13e^{-0.116h}$.
a. What is the approximate pressure at 40 km above Earth's surface?

b. What is the approximate pressure on Earth's surface?

c. Does atmospheric pressure increase or decrease as you rise above Earth's surface?

27. Chemistry The intensity I of an x-ray after it has passed through a material that is x centimeters thick is given by $I = I_0e^{-kx}$, where I_0 is the initial intensity of the x-ray and k is a number that depends on the material. The constant k for copper is 3.2. Find the thickness of copper necessary so that the intensity of an x-ray after passing through the copper is 25% of the original intensity. Round to the nearest tenth.

28. Disk Storage The cost per megabyte of disk storage in 1981 was \$700. In 2010, the cost per megabyte of disk storage was about \$.00002. Assuming that the cost per megabyte is decreasing exponentially, find an exponential model of the form $C(t) = Ae^{kt}$ that gives the cost per megabyte C in year t. Use 1981 for $t = 0$, and round k to the nearest thousandth. According to the model, in what year was the cost per megabyte of storage \$.10?

29. Steroid Use When air resistance is considered, the height f, in feet, of a baseball x feet from home plate and hit at a certain angle can be approximated by $f(x) = \left(\frac{0.5774v + 155.3}{v}\right)x + 565.3 \ln\left(\frac{v - 0.2747x}{v}\right) + 3.5$, where v is the speed of the ball in feet per second when it leaves the bat.
a. If $v = 160$ ft/s, show that a baseball will hit near the bottom of a 15-foot-high fence 375 ft from home plate.

b. If the speed in part (a) is increased by 4%, show that the baseball will clear the fence by approximately 9 ft.

In the News

Steroids Increase Number of Home Runs

Steroids can help batters hit 50 percent more home runs by boosting their muscle mass by just 10 percent, a U.S. physicist said on Thursday. Calculations show that by putting on 10 percent more muscle mass, a batter can swing about 5 percent faster, increasing the ball's speed by 4 percent as it leaves the bat. According to Roger Tobin of Tufts University in Boston, depending on the ball's trajectory, this added speed could take it into home run territory 50 percent more often.

Source: www.reuters.com

30. Steroid Use In Exercise 29, a 4% increase in the speed of the ball as it leaves the bat gives $v = 166.4$ ft/s. Using f from Exercise 29, determine the greatest distance from home plate the 15-foot fence could be placed so that a ball hit with speed v would clear the fence.

31. Ball Flight Air resistance plays a large role in the flight of a ball. If air resistance is ignored, then the height h, in feet, of a baseball x feet from home plate and hit at the same angle as in Exercise 29 can be approximated by $h(x) = \frac{-21.33}{v^2}x^2 + 0.5774x + 3.5$. If the speed of a ball as it leaves the bat is 160 ft/s, how much farther would the ball travel before hitting the ground if air resistance were ignored? Round to the nearest foot.

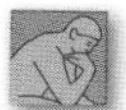

32. Assuming that an exponential model is an appropriate model for each of the following, would the situation be modeled using exponential growth or exponential decay?

a. The amount of a medication in a patient's system t hours after taking the medicine

b. The atmospheric pressure x meters above the surface of Earth

c. The temperature of a roast t minutes after it is put in a hot oven

d. The temperature of a hot cup of tea that is put in a refrigerator

e. The spread of a contagious disease

Applying the Concepts

33. Uranium Dating Read the article at the right. The carbon-14 dating method does not work on rocks, so methods based on different radioactive elements are used. One such method uses uranium-235, which has a half-life of approximately 713 million years. Use the uranium-235 dating equation $A = A_0\left(\frac{1}{2}\right)^{t/713{,}000{,}000}$ to estimate the ratio of the current amount of uranium-235 in the Hudson Bay rock to the original amount of uranium-235 in the rock. (*Hint:* You are finding $\frac{A}{A_0}$ when $t = 4.28$ billion years.) Round to the nearest thousandth.

In the News

Scientists Discover Ancient Rocks

Scientists have found that a pinkish tract of bedrock on the eastern shore of Canada's Hudson Bay contains the oldest known rocks on Earth, formed 4.28 billion years ago, not long after Earth itself was formed.

Source: news.yahoo.com

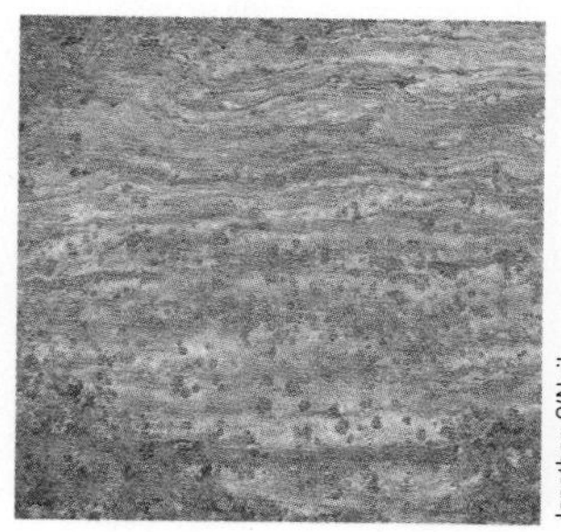

Jonathan O'Neil

34. Investments The value of an investment in an account that earns an annual interest rate of 7% compounded daily grows according to the equation $A = A_0\left(1 + \frac{0.07}{365}\right)^{365t}$, where A_0 is the original value of an investment and t is the time in years. Find the time for the investment to double in value. Round to the nearest year.

35. Continuous Compounding In some scenarios, economists will use continuous compounding of an investment. The equation for continuous compounding is $A = A_0e^{rt}$, where A is the present value of an initial investment of A_0 at an annual interest rate r (as a decimal), and t is the time in years.

a. Find the value of an investment of \$5000 after 3 years if interest is compounded continuously at an annual rate of 6%.

b. If an investor wanted to grow a continuously compounded investment of \$1000 to \$1250 in 2 years, what interest rate must the investor receive? Round to the nearest tenth.

FOCUS ON PROBLEM SOLVING

Proof by Contradiction The four-step plan for solving problems that we have used before is restated here.

1. Understand the problem.
2. Devise a plan.
3. Carry out the plan.
4. Review the solution.

One of the techniques that can be used in the second step is a method called *proof by contradiction.* In this method, you assume that the conditions of the problem you are trying to solve can be met and then show that your assumption leads to a condition you already know is not true.

To illustrate this method, suppose we try to prove that $\sqrt{2}$ is a rational number. We begin by recalling that a rational number is one that can be written as the quotient of integers. Therefore, let a and b be two integers with no common factors. If $\sqrt{2}$ is a rational number, then

$$\sqrt{2} = \frac{a}{b}$$

$$(\sqrt{2})^2 = \left(\frac{a}{b}\right)^2 \quad \text{• Square each side.}$$

$$2 = \frac{a^2}{b^2}$$

$$2b^2 = a^2 \quad \text{• Multiply each side by } b^2.$$

From the last equation, a^2 is an even number. Because a^2 is an even number, a is an even number. Now divide each side of the last equation by 2.

$$2b^2 = a^2$$

$$2b^2 = a \cdot a$$

$$b^2 = a \cdot x \quad \text{• } x = \frac{a}{2}$$

Because a is an even number, $a \cdot x$ is an even number. Because $a \cdot x$ is an even number, b^2 is an even number, and this in turn means that b is an even number. This result, however, contradicts the assumption that a and b are two integers with no common factors. Because this assumption is now known not to be true, we conclude that our original assumption, that $\sqrt{2}$ is a rational number, is false. This proves that $\sqrt{2}$ is an irrational number.

Try a proof by contradiction for the following problem: "Is it possible to write numbers using each of the digits 0, 1, 2, 3, 4, 5, 6, 7, 8, and 9 exactly once such that the sum of the numbers is exactly 100?"* Here are some suggestions. First note that the sum of the 10 digits is 45. This means that some of the digits used must be tens digits. Let x be the sum of those digits.

a. What is the sum of the remaining units digits?
b. Express "the sum of the units digits and the tens digits equals 100" as an equation.
c. Solve the equation for x.
d. Explain why this result means that it is impossible to satisfy both conditions of the problem.

*Problem adapted from G. Polya, *How to Solve It: A New Aspect of Mathematical Method. Copyright* © 1947 by G. Polya. Reprinted by permission of Princeton University Press.

PROJECTS AND GROUP ACTIVITIES

Solving Exponential and Logarithmic Equations Using a Graphing Calculator

A graphing calculator can be used to draw the graphs of logarithmic functions. Note that there are two logarithmic keys, LOG and LN, on a graphing calculator. The first key gives the values of common logarithms (base 10), and the second gives the values of natural logarithms (base e).

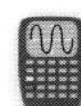

To graph $y = \ln x$, press the Y= key. Clear any equations already entered.

Press LN X,T,θ,n) GRAPH. The graph is shown at the left with a viewing window of Xmin = −1, Xmax = 8, Ymin = −5, and Ymax = 5.

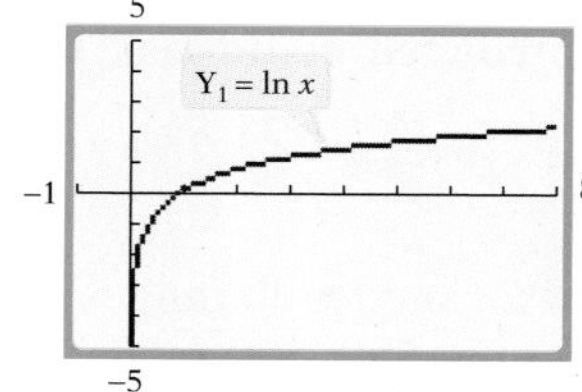

Some exponential and logarithmic equations cannot be solved algebraically. In these cases, a graphical approach may be appropriate. Here is an example.

HOW TO 1 Solve $\ln(2x + 4) = x^2$ for x. Round to the nearest hundredth.

Rewrite the equation by subtracting x^2 from each side.

$$\ln(2x + 4) = x^2$$
$$\ln(2x + 4) - x^2 = 0$$

The zeros of $f(x) = \ln(2x + 4) - x^2$ are the solutions of $\ln(2x + 4) - x^2 = 0$.

Graph f and use the zero feature of the graphing calculator to estimate the solutions to the nearest hundredth.

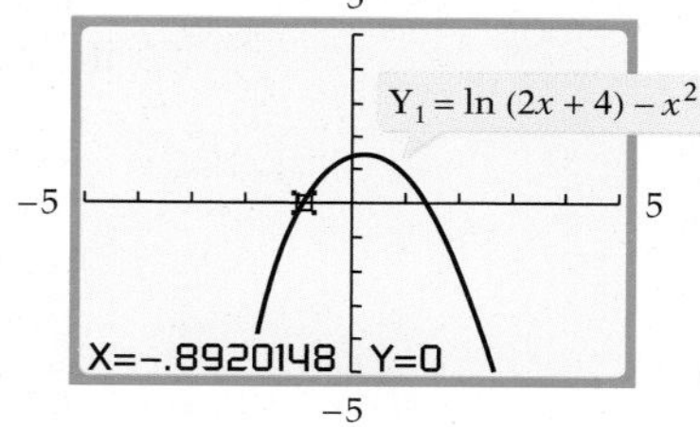

The zeros are approximately −0.89 and 1.38.

The solutions are −0.89 and 1.38.

For Exercises 1 to 8, solve for x by graphing. Round to the nearest hundredth.

1. $2^x = 2x + 4$

2. $3^x = -x - 1$

3. $e^x = -2x - 2$

4. $e^x = 3x + 4$

5. $\log(2x - 1) = -x + 3$

6. $\log(x + 4) = -2x + 1$

7. $\ln(x + 2) = x^2 - 3$

8. $\ln x = -x^2 + 1$

To graph a logarithmic function with a base other than base 10 or base e, the Change-of-Base Formula is used. Here is the Change-of-Base Formula, discussed earlier:

$$\log_a N = \frac{\log_b N}{\log_b a}$$

To graph $y = \log_2 x$, first change from base 2 logarithms to the equivalent base 10 logarithms or natural logarithms. Base 10 is shown here.

$$y = \log_2 x = \frac{\log_{10} x}{\log_{10} 2} \qquad \bullet\ a = 2,\ b = 10,\ N = x$$

Then graph the equivalent equation $y = \frac{\log x}{\log 2}$. The graph is shown at the left with a viewing window of Xmin = −1, Xmax = 8, Ymin = −5, and Ymax = 5.

Credit Reports and FICO® Scores[1]

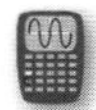
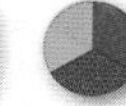

When a consumer applies for a loan, the lender generally wants to know the consumer's credit history. For this, the lender turns to a credit reporting agency. These agencies maintain files on millions of borrowers. They provide the lender with a credit report, which lists information such as the types of credit the consumer uses, the length of time the consumer's credit accounts have been open, the amounts owed by the consumer, and whether the consumer has paid his or her bills on time. Along with the credit report, the lender can buy a credit score based on the information in the report. The credit score gives the lender a quick measure of the consumer's credit risk, or how likely the consumer is to repay a debt. It answers the lender's question: "If I lend this person money (or give this person a credit card), what is the probability that I will be paid back in a timely manner?"

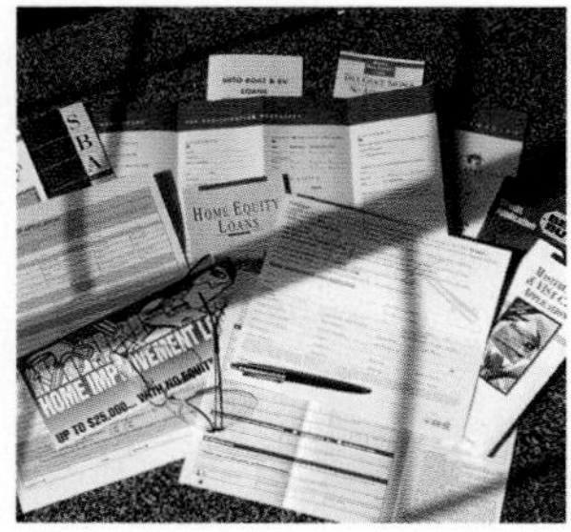

The most widely used credit bureau score is the FICO® score, so named because scores are produced from software developed by Fair Isaac Corporation (FICO). FICO scores range from 300 to 850. The higher the score, the lower the predicted credit risk for lenders.

The following graph shows the delinquency rate, or credit risk, associated with ranges of FICO scores. For example, the delinquency rate of consumers in the 550–599 range of scores is 51%. This means that within the next 2 years, for every 100 borrowers in this range, approximately 51 will default on a loan, file for bankruptcy, or fail to pay a credit card bill within 90 days of the due date.

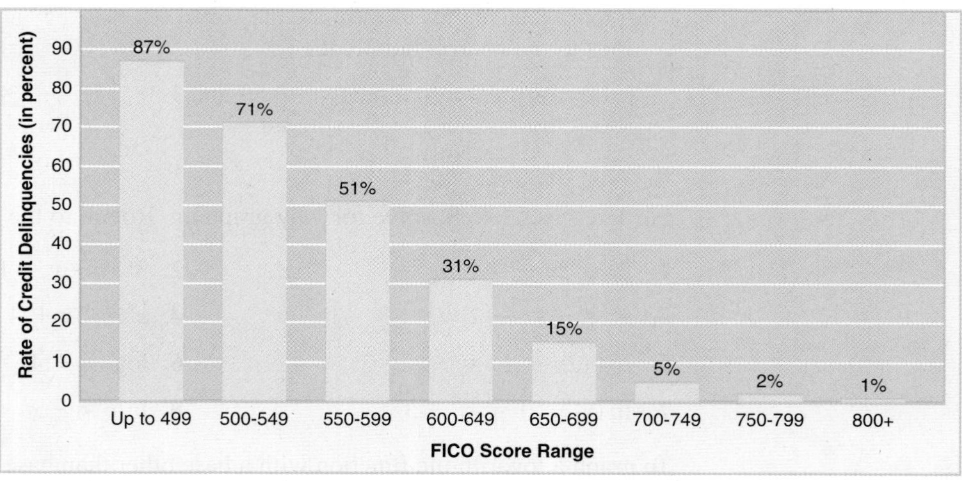

Because the delinquency rate depends on the FICO score, we can let the independent variable x represent the FICO score and the dependent variable y represent the delinquency rate. We will use the middle of each range for the values of x. The resulting ordered pairs are recorded in the table at the right.

x	y
400	87
525	71
575	51
625	31
675	15
725	5
775	2
825	1

[1]Source: www.myfico.com

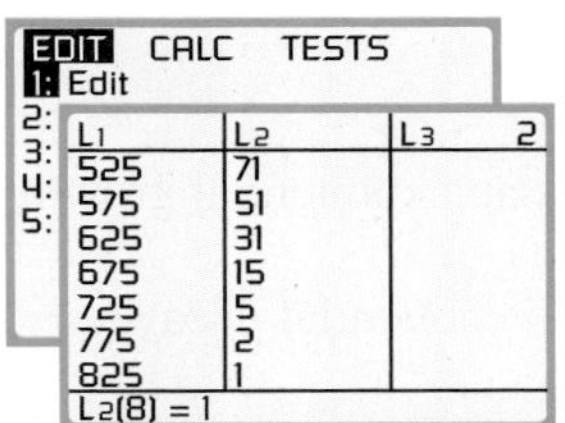

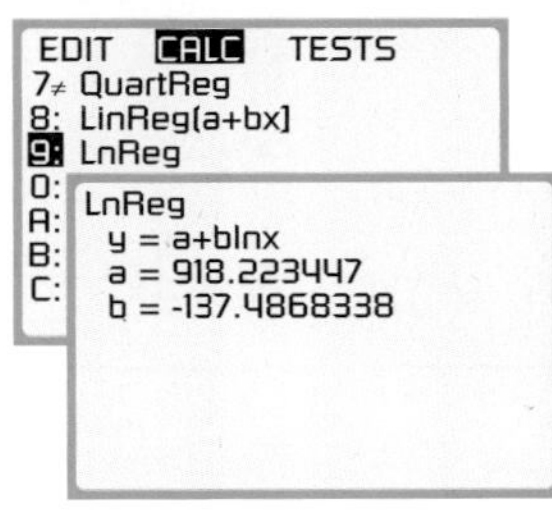

1. Use your graphing calculator to find a logarithmic equation that approximates these data. Here are instructions for the TI-84:

   Press STAT ENTER. This will bring up a table into which you can enter data. Enter the values of x in L1. Enter the values of y in L2.

   Press STAT again, press the right arrow key to highlight CALC, and arrow down to LnReg (ln for the natural logarithm). Press ENTER twice. The values for a and b in the equation $y = a + b \ln x$ will appear on the screen.

2. Use your equation to predict the delinquency rate of a consumer with a score of 500. Round to the nearest whole number.
3. The equation pairs a delinquency rate of 36% with what score?
4. Use the graph on page 714 to determine the highest FICO score that will result in a delinquency rate of 71% or more.
5. Use the graph on page 714 to determine the lowest FICO score that will result in a delinquency rate of 5% or lower.
6. Use the Internet to find the name of at least one of the major credit reporting agencies.

CHAPTER 12 SUMMARY

KEY WORDS	EXAMPLES
A function of the form $f(x) = b^x$, where b is a positive real number not equal to 1, is an *exponential function*. The number b is the *base* of the exponential function. [12.1A, p. 674]	$f(x) = 3^x$ is an exponential function. 3 is the base of the function.
The function defined by $f(x) = e^x$ is called the *natural exponential function*. [12.1A, p. 675]	$f(x) = 2e^{x-1}$ is a natural exponential function. e is an irrational number approximately equal to 2.71828183.
Because the exponential function is a 1–1 function, it has an inverse function that is called a *logarithm*. The definition of logarithm is: For $x > 0$, $b > 0$, $b \neq 1$, $y = \log_b x$ is equivalent to $x = b^y$. [12.2A, p. 682]	$\log_2 8 = 3$ is equivalent to $8 = 2^3$.
Logarithms with base 10 are called *common logarithms*. We usually omit the base, 10, when writing the common logarithm of a number. [12.2A, p. 683]	$\log_{10} 100 = 2$ is usually written $\log 100 = 2$.
When e (the base of the natural exponential function) is used as a base of a logarithm, the logarithm is referred to as the *natural logarithm* and is abbreviated $\ln x$. [12.2A, p. 683]	$\log_e 100 \approx 4.61$ is usually written $\ln 100 \approx 4.61$.

An *exponential equation* is one in which a variable occurs in an exponent. [12.4A, p. 696]	$2^x = 12$ is an exponential equation.
An *exponential growth equation* is an equation that can be written in the form $A = A_0 b^{kt}$, where A is the size at time t, A_0 is the initial size, $b > 1$, and k is a positive real number. In an *exponential decay equation*, the base is between 0 and 1. [12.5A, pp. 702–703]	$P = 1000(1.03)^n$ is an exponential growth equation. $A = 10(0.5)^x$ is an exponential decay equation.

ESSSENTIAL RULES AND PROCEDURES	EXAMPLES
1–1 Property of Exponential Functions [12.2A, p. 683] For $b > 0$, $b \neq 1$, if $b^u = b^v$, then $u = v$.	If $b^x = b^5$, then $x = 5$.
Logarithm Property of 1 [12.2B, p. 684] For any positive real numbers x, y, and b, $b \neq 1$, $\log_b 1 = 0$.	$\log_6 1 = 0$
Inverse Property of Logarithms [12.2B, p. 684] For any positive real numbers x, y, and b, $b \neq 1$, $b^{\log_b x} = x$ and $\log_b b^x = x$.	$\log_3 3^4 = 4$
1–1 Property of Logarithms [12.2B, p. 684] For any positive real numbers x, y, and b, $b \neq 1$, if $\log_b x = \log_b y$, then $x = y$.	If $\log_5 (x - 2) = \log_5 3$, then $x - 2 = 3$.
Logarithm Property of the Product of Two Numbers [12.2B, p. 685] For any positive real numbers x, y, and b, $b \neq 1$, $\log_b (xy) = \log_b x + \log_b y$.	$\log_b (3x) = \log_b 3 + \log_b x$
Logarithm Property of the Quotient of Two Numbers [12.2B, p. 685] For any positive real numbers x, y, and b, $b \neq 1$, $\log_b \frac{x}{y} = \log_b x - \log_b y$.	$\log_b \frac{x}{20} = \log_b x - \log_b 20$
Logarithm Property of the Power of a Number [12.2B, p. 685] For any positive real numbers x and b, $b \neq 1$, and for any real number r, $\log_b x^r = r \log_b x$.	$\log_b x^5 = 5 \log_b x$
Change-of-Base Formula [12.2C, p. 687] $\log_a N = \frac{\log_b N}{\log_b a}$	$\log_3 12 = \frac{\log 12}{\log 3}$ $\quad \log_6 16 = \frac{\ln 16}{\ln 6}$

CHAPTER 12

CONCEPT REVIEW

Test your knowledge of the concepts presented in this chapter. Answer each question. Then check your answers against the ones provided in the Answer Section.

1. How do you know that a function is an exponential function?

2. Why are there limitations on the values of the base of an exponential function?

3. What kind of number is *e*, and what is its approximate value?

4. Why are there limitations on the values used in a logarithmic function?

5. What is the relationship between logarithmic functions and exponential functions?

6. Why is $\log_b(xy) = (\log_b x)(\log_b y)$ a false statement?

7. Why is $\log_b\left(\dfrac{x}{y}\right) = \dfrac{\log_b x}{\log_b y}$ a false statement?

8. What practical purpose does the Change-of-Base Formula have?

9. What is the domain of a logarithmic function?

10. What is the compound interest formula?

11. What is the relationship between earthquakes and logarithms?

CHAPTER 12

REVIEW EXERCISES

1. Evaluate $f(x) = e^{x-2}$ at $x = 2$.

2. Write $\log_5 25 = 2$ in exponential form.

3. Graph: $f(x) = 3^{-x} + 2$

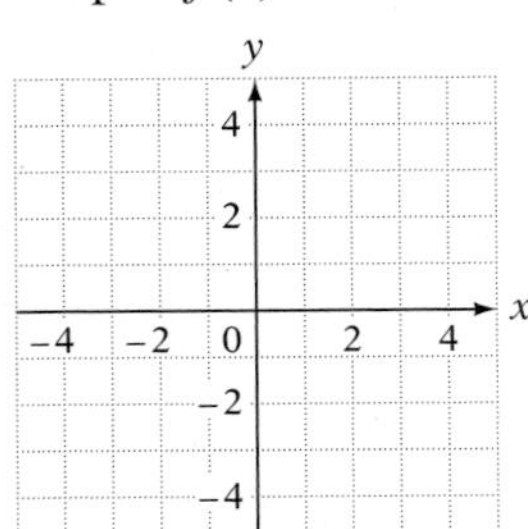

4. Graph: $f(x) = \log_3(x - 1)$

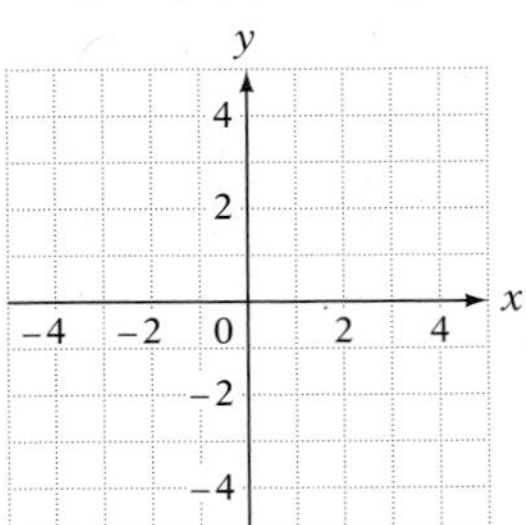

5. Write $\log_3 \sqrt[5]{x^2y^4}$ in expanded form.

6. Write $2 \log_3 x - 5 \log_3 y$ as a single logarithm with a coefficient of 1.

7. Solve: $27^{2x+4} = 81^{x-3}$

8. Solve: $\log_5 \dfrac{7x + 2}{3x} = 1$

9. Find $\log_6 22$. Round to the nearest ten-thousandth.

10. Solve: $\log_2 x = 5$

11. Solve: $\log_3 (x + 2) = 4$

12. Solve: $\log_{10} x = 3$

13. Write $\frac{1}{3}(\log_7 x + 4 \log_7 y)$ as a single logarithm with a coefficient of 1.

14. Write $\log_8 \sqrt{\dfrac{x^5}{y^3}}$ in expanded form.

15. Write $2^5 = 32$ in logarithmic form.

16. Find $\log_3 1.6$. Round to the nearest ten-thousandth.

17. Solve $3^{x+2} = 5$. Round to the nearest thousandth.

18. Evaluate $f(x) = \left(\frac{2}{3}\right)^{x+2}$ at $x = -3$.

19. Solve: $\log_2(x + 3) - \log_2(x - 1) = 3$

20. Solve: $\log_3(2x + 3) + \log_3(x - 2) = 2$

21. Graph: $f(x) = \left(\frac{2}{3}\right)^{x+1}$

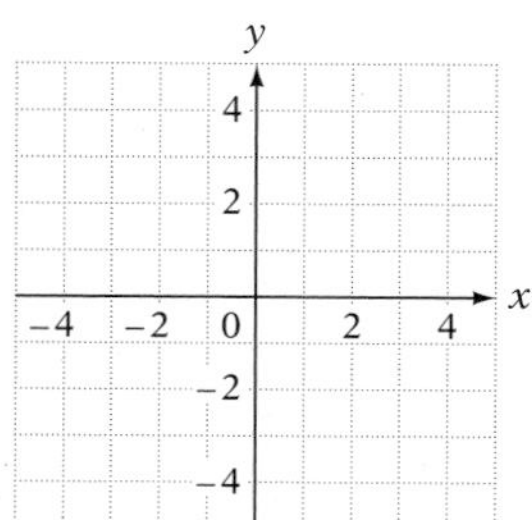

22. Graph: $f(x) = \log_2(2x - 1)$

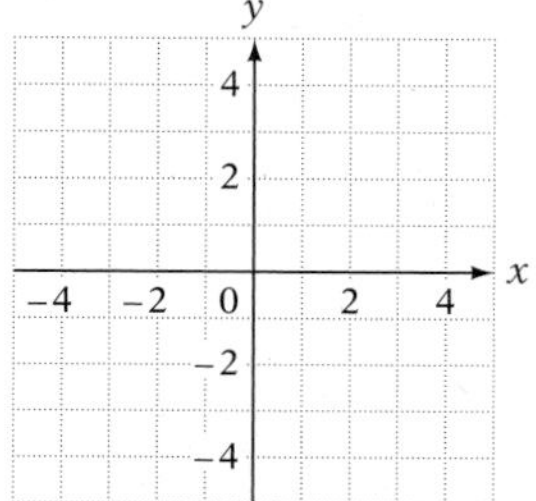

23. Evaluate: $\log_6 36$

24. Write $\frac{1}{3}(\log_2 x - \log_2 y)$ as a single logarithm with a coefficient of 1.

25. Solve for x: $9^{2x} = 3^{x+3}$

26. Solve for x: $5 \cdot 3^{x/2} = 12$. Round to the nearest ten-thousandth.

27. Solve for x: $\log_5 x = -1$

28. Write $3^4 = 81$ in logarithmic form.

29. Solve for x: $\log x + \log(x - 2) = \log 15$

30. Write $\log_5 \sqrt[3]{x^2 y}$ in expanded form.

31. Solve for x: $6e^{-2x} = 17$. Round to the nearest ten-thousandth.

32. Evaluate $f(x) = 7^{x+2}$ at $x = -3$.

33. Evaluate: $\log_2 16$

34. Solve for x: $\log_6 x = \log_6 2 + \log_6 (2x - 3)$

35. Evaluate $\log_2 5$. Round to the nearest ten-thousandth.

36. Solve for x: $4^x = 8^{x-1}$

37. Solve for x: $\log_5 x = 4$

38. Write $3 \log_b x - 7 \log_b y$ as a single logarithm with a coefficient of 1.

39. Evaluate $f(x) = 5^{-x-1}$ at $x = -2$.

40. Solve $5^{x-2} = 7$ for x. Round to the nearest ten-thousandth.

41. Investments Use the compound interest formula $P = A(1 + i)^n$, where A is the original value of an investment, i is the interest rate per compounding period, n is the number of compounding periods, and P is the value of the investment after n periods, to find the value of an investment after 2 years. The amount of the investment is \$4000, and it is invested at 8% compounded monthly. Round to the nearest dollar.

42. Seismology An earthquake in Japan in September, 2003, had an intensity of $I = 199{,}526{,}232I_0$. Find the Richter scale magnitude of the earthquake. Use the Richter scale equation $M = \log \frac{I}{I_0}$, where M is the magnitude of an earthquake, I is the intensity of the shock waves, and I_0 is the measure of the intensity of a zero-level earthquake. Round to the nearest tenth.

43. Radioactivity Use the exponential decay equation $A = A_0 (0.5)^{t/k}$, where A is the amount of a radioactive material present after time t, k is the half-life of the radioactive material, and A_0 is the original amount of radioactive material, to find the half-life of a material that decays from 25 mg to 15 mg in 20 days. Round to the nearest whole number.

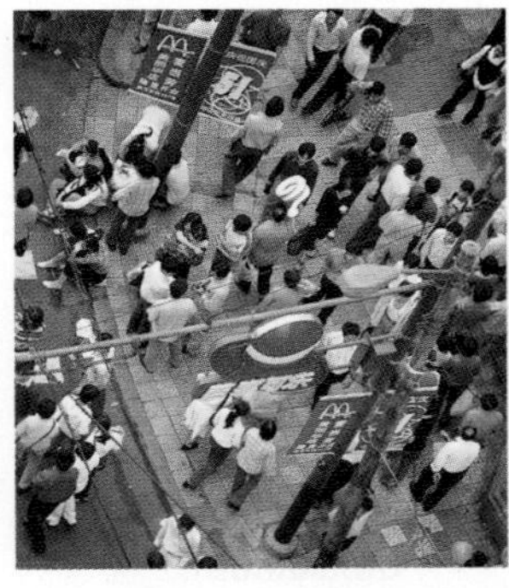

44. Sound The number of decibels D of a sound can be given by the equation $D = 10(\log I + 16)$, where I is the intensity of the sound measured in watts per square centimeter. Find the number of decibels of sound emitted from a busy street corner for which the intensity of the sound is 5×10^{-6} watts per square centimeter.

CHAPTER 12

TEST

1. Evaluate $f(x) = \left(\frac{2}{3}\right)^x$ at $x = 0$.

2. Evaluate $f(x) = 3^{x+1}$ at $x = -2$.

3. Graph: $f(x) = 2^x - 3$

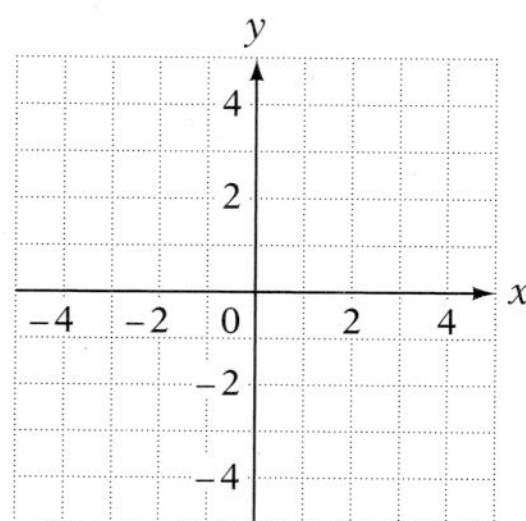

4. Graph: $f(x) = 2^x + 2$

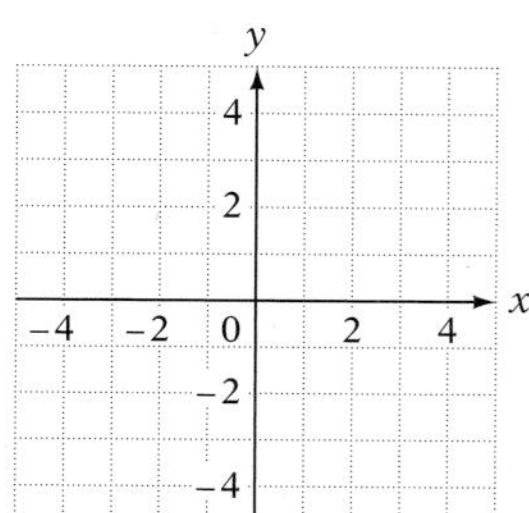

5. Evaluate: $\log_4 16$

6. Solve for x: $\log_3 x = -2$

7. Graph: $f(x) = \log_2(2x)$

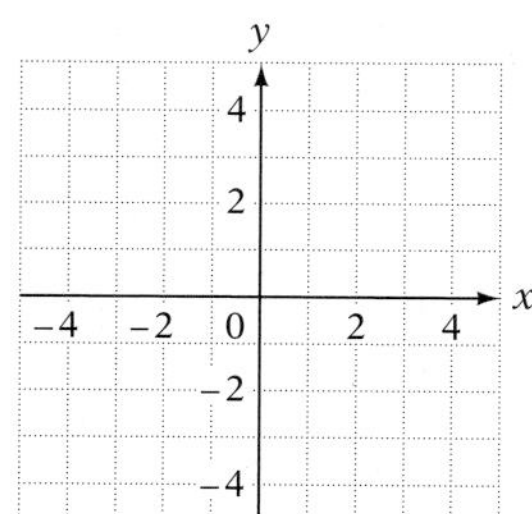

8. Graph: $f(x) = \log_3(x + 1)$

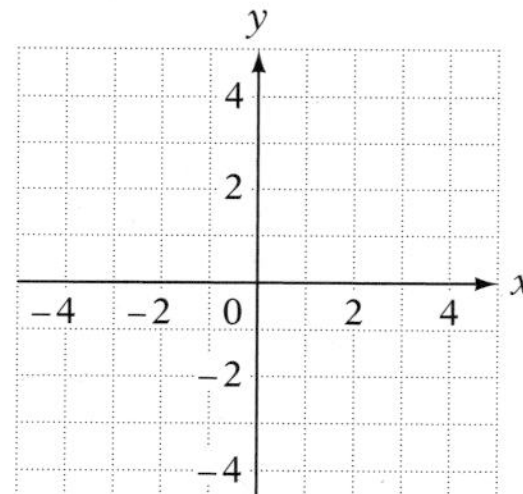

9. Write $\log_6 \sqrt{xy^3}$ in expanded form.

10. Write $\frac{1}{2}(\log_3 x - \log_3 y)$ as a single logarithm with a coefficient of 1.

11. Write $\ln \frac{x}{\sqrt{z}}$ in expanded form.

12. Write $3 \ln x - \ln y - \frac{1}{2} \ln z$ as a single logarithm with a coefficient of 1.

13. Solve for x: $3^{7x+1} = 3^{4x-5}$

14. Solve for x: $8^x = 2^{x-6}$

15. Solve for x: $3^x = 17$

16. Solve for x: $\log x + \log(x - 4) = \log 12$

17. Solve for x: $\log_6 x + \log_6(x - 1) = 1$

18. Find $\log_5 9$.

19. Find $\log_3 19$.

20. Solve for x: $5^{2x-5} = 9$

21. Solve for x: $2e^{x/4} = 9$

22. Solve for x: $\log_5(30x) - \log_5(x + 1) = 2$

23. **Carbon Dating** A shard from a vase originally contained 250 mg of carbon-14 and now contains 170 mg of carbon-14. Use the equation $A = A_0(0.5)^{t/5570}$, where A_0 is the original amount of carbon-14 in the shard and A is the amount of carbon-14 in the shard t years later, to find the approximate age of the shard. Round to the nearest whole number.

24. **Sound** Use the decibel equation $D = 10(\log I + 16)$, where D is the decibel level and I is the intensity of a sound in watts per square centimeter, to find the intensity of a 75-decibel dial tone.

25. **Radioactivity** Use the exponential decay equation $A = A_0(0.5)^{t/k}$, where A is the amount of a radioactive material present after time t, k is the half-life of the material, and A_0 is the original amount of radioactive material, to find the half-life of a material that decays from 10 mg to 9 mg in 5 h. Round to the nearest whole number.

CUMULATIVE REVIEW EXERCISES

1. Solve: $4 - 2[x - 3(2 - 3x) - 4x] = 2x$

2. Find the equation of the line that contains the point $(2, -2)$ and is parallel to the line $2x - y = 5$.

3. Factor: $4x^{2n} + 7x^n + 3$

4. Simplify: $\dfrac{1 - \frac{5}{x} + \frac{6}{x^2}}{1 + \frac{1}{x} - \frac{6}{x^2}}$

5. Simplify: $\dfrac{\sqrt{xy}}{\sqrt{x} - \sqrt{y}}$

6. Solve by completing the square: $x^2 - 4x - 6 = 0$

7. Write a quadratic equation that has integer coefficients and has solutions $\frac{1}{3}$ and -3.

8. Graph the solution set: $2x - y < 3$
$x + y < 1$

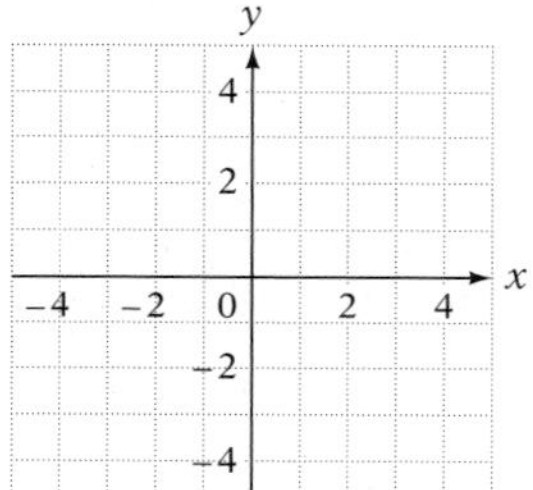

9. Solve by the addition method:
$$3x - y + z = 3$$
$$x + y + 4z = 7$$
$$3x - 2y + 3z = 8$$

10. Subtract: $\dfrac{x - 4}{2 - x} - \dfrac{1 - 6x}{2x^2 - 7x + 6}$

11. Solve: $x^2 + 4x - 5 \le 0$
Write the solution set in set-builder notation.

12. Solve: $|2x - 5| \le 3$

13. Graph: $f(x) = \left(\frac{1}{2}\right)^x + 1$

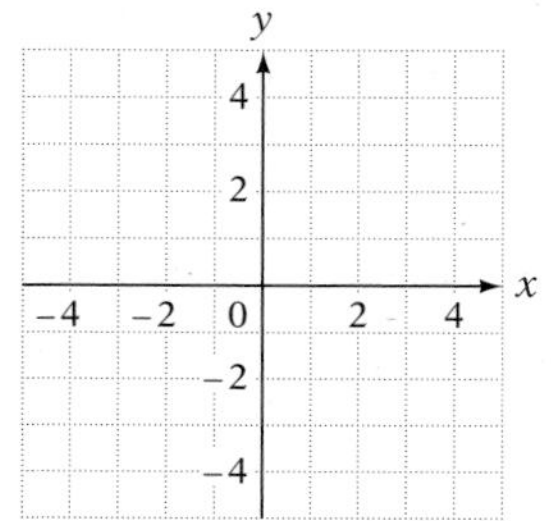

14. Graph: $f(x) = \log_2 x - 1$

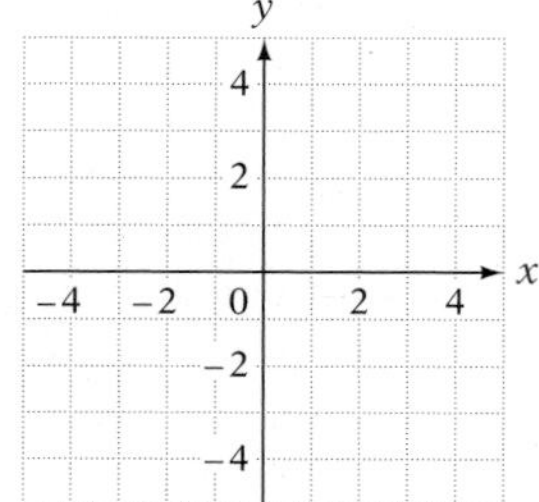

15. Evaluate $f(x) = 2^{-x-1}$ at $x = -3$.

16. Solve for x: $\log_5 x = 3$

17. Write $3 \log_b x - 5 \log_b y$ as a single logarithm with a coefficient of 1.

18. Find $\log_3 7$.

19. Solve for x: $4^{5x-2} = 4^{3x+2}$

20. Solve for x: $\log x + \log(2x + 3) = \log 2$

21. **Banking** A bank offers two types of business checking accounts. One account has a charge of \$5 per month plus 2 cents per check. The second account has a charge of \$2 per month plus 8 cents per check. How many checks can a customer who has the second type of account write if it is to cost the customer less than the first type of checking account?

22. **Mixtures** Find the cost per pound of a mixture made from 16 lb of chocolate that costs \$4.00 per pound and 24 lb of chocolate that costs \$2.50 per pound.

23. **Uniform Motion** A plane can fly at a rate of 225 mph in calm air. Traveling with the wind, the plane flew 1000 mi in the same amount of time it took to fly 800 mi against the wind. Find the rate of the wind.

24. **Physics** The distance d that a spring stretches varies directly as the force f used to stretch the spring. If a force of 20 lb stretches a spring 6 in., how far will a force of 34 lb stretch the spring?

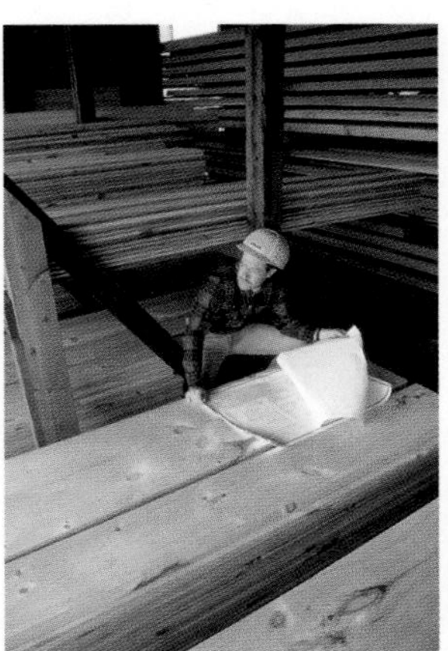

25. **Carpentry** A carpenter purchased 80 ft of redwood and 140 ft of fir for a total cost of \$67. A second purchase, at the same prices, included 140 ft of redwood and 100 ft of fir for a total cost of \$81. Find the cost of redwood and of fir.

26. **Investments** The compound interest formula is $A = P(1 + i)^n$, where P is the original value of an investment, i is the interest rate per compounding period, n is the total number of compounding periods, and A is the value of the investment after n periods. Use the compound interest formula to find how many years it will take for an investment of \$5000 to double in value. The investment earns 7% annual interest and is compounded semiannually.

FINAL EXAM

1. Simplify:
$12 - 8[3 - (-2)]^2 \div 5 - 3$

2. Evaluate $\dfrac{a^2 - b^2}{a - b}$ when $a = 3$ and $b = -4$.

3. Simplify: $5 - 2[3x - 7(2 - x) - 5x]$

4. Solve: $\dfrac{3}{4}x - 2 = 4$

5. Solve: $8 - |5 - 3x| = 1$

6. Find the volume of a sphere with a diameter of 8 ft. Round to the nearest tenth.

7. Graph $2x - 3y = 9$ using the x- and y-intercepts.

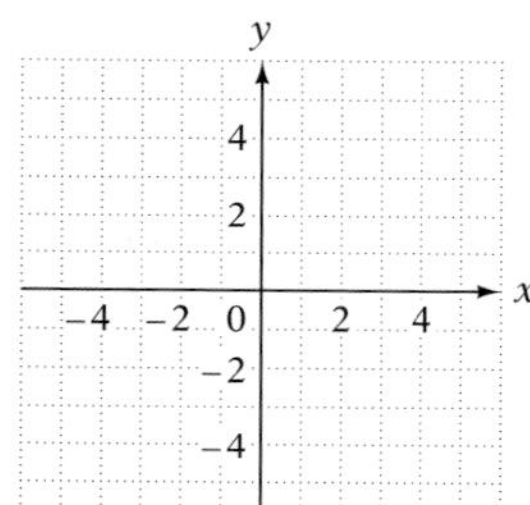

8. Find the equation of the line containing the points $(3, -2)$ and $(1, 4)$.

9. Find the equation of the line that contains the point $(-2, 1)$ and is perpendicular to the line $3x - 2y = 6$.

10. Simplify: $2a[5 - a(2 - 3a) - 2a] + 3a^2$

11. Factor: $8 - x^3y^3$

12. Factor: $x - y - x^3 + x^2y$

13. Divide: $(2x^3 - 7x^2 + 4) \div (2x - 3)$

14. Divide: $\dfrac{x^2 - 3x}{2x^2 - 3x - 5} \div \dfrac{4x - 12}{4x^2 - 4}$

15. Subtract: $\dfrac{x - 2}{x + 2} - \dfrac{x + 3}{x - 3}$

16. Simplify: $\dfrac{\dfrac{3}{x} + \dfrac{1}{x + 4}}{\dfrac{1}{x} + \dfrac{3}{x + 4}}$

17. Solve: $\dfrac{5}{x-2} - \dfrac{5}{x^2-4} = \dfrac{1}{x+2}$

18. Solve $a_n = a_1 + (n-1)d$ for d.

19. Simplify: $\left(\dfrac{4x^2y^{-1}}{3x^{-1}y}\right)^{-2}\left(\dfrac{2x^{-1}y^2}{9x^{-2}y^2}\right)^3$

20. Simplify: $\left(\dfrac{3x^{2/3}y^{1/2}}{6x^2y^{4/3}}\right)^6$

21. Subtract: $x\sqrt{18x^2y^3} - y\sqrt{50x^4y}$

22. Simplify: $\dfrac{\sqrt{16x^5y^4}}{\sqrt{32xy^7}}$

23. Simplify: $\dfrac{3}{2+i}$

24. Write a quadratic equation that has integer coefficients and has solutions $-\frac{1}{2}$ and 2.

25. Solve by using the quadratic formula:
$2x^2 - 3x - 1 = 0$

26. Solve: $x^{2/3} - x^{1/3} - 6 = 0$

27. Graph: $f(x) = -x^2 + 4$

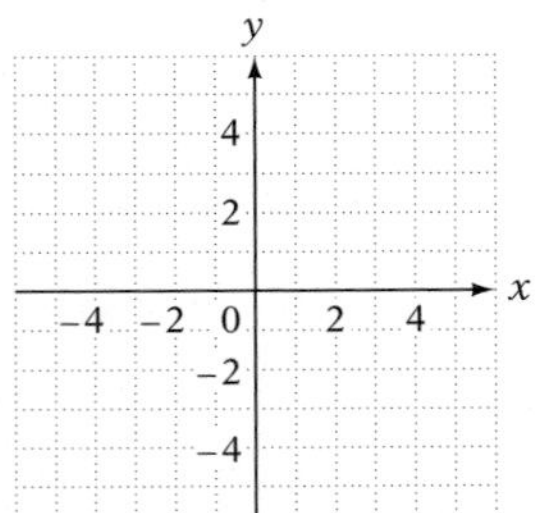

28. Graph: $f(x) = -\dfrac{1}{2}x - 3$

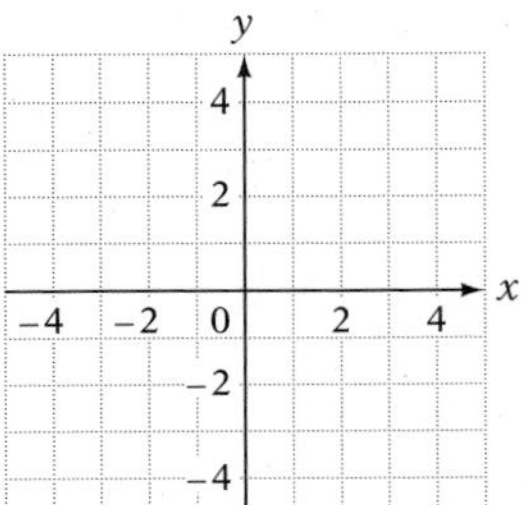

29. Solve: $\dfrac{2}{x} - \dfrac{2}{2x+3} = 1$

30. Find the inverse of the function $f(x) = \dfrac{2}{3}x - 4$.

31. Solve by the addition method:
$3x - 2y = 1$
$5x - 3y = 3$

32. Simplify: $\sqrt{49x^6}$

33. Solve: $2 - 3x < 6$ and $2x + 1 > 4$

34. Solve: $|2x + 5| < 3$

35. Graph the solution set: $3x + 2y > 6$

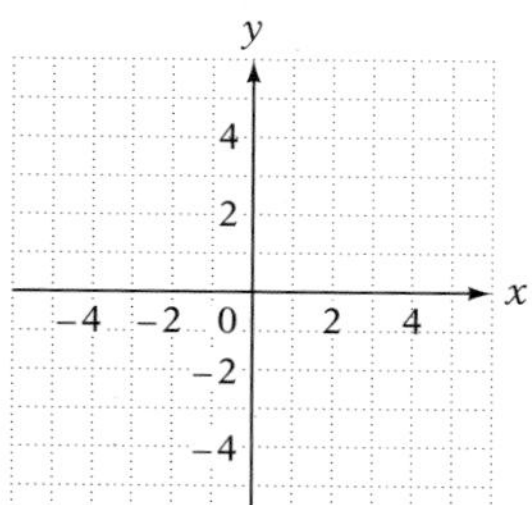

36. Graph: $f(x) = 3^{-x} - 2$

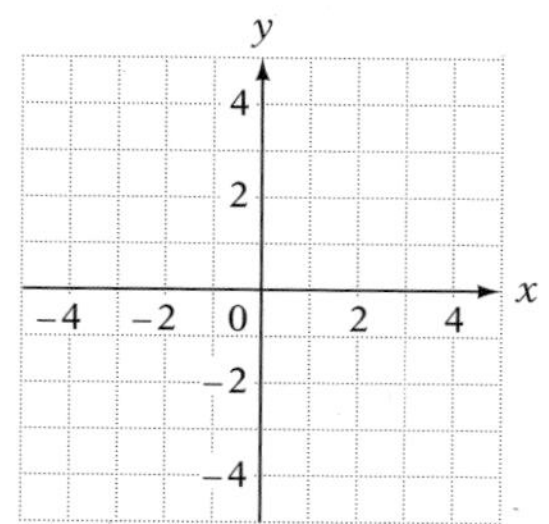

37. Graph: $f(x) = \log_2 (x + 1)$

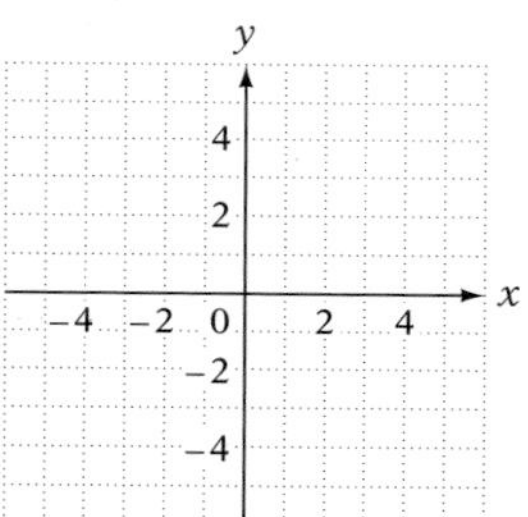

38. Write $2(\log_2 a - \log_2 b)$ as a single logarithm with a coefficient of 1.

39. Solve for x: $\log_3 x - \log_3 (x - 3) = \log_3 2$

40. **Education** An average score of 70–79 in a history class receives a C grade. A student has grades of 64, 58, 82, and 77 on four history tests. Find the range of scores on the fifth test that will give the student a C grade for the course.

41. **Uniform Motion** A jogger and a cyclist set out at 8 A.M. from the same point headed in the same direction. The average speed of the cyclist is two and a half times the average speed of the jogger. In 2 h, the cyclist is 24 mi ahead of the jogger. How far did the cyclist ride in that time?

42. **Investments** You have a total of $12,000 invested in two simple interest accounts. On one account, a money market fund, the annual simple interest rate is 8.5%. On the other account, a tax-free bond fund, the annual simple interest rate is 6.4%. The total annual interest earned by the two accounts is $936. How much do you have invested in each account?

43. **Geometry** The length of a rectangle is 1 ft less than three times the width. The area of the rectangle is 140 ft^2. Find the length and width of the rectangle.

44. **The Stock Market** Three hundred shares of a utility stock earn a yearly dividend of $486. How many additional shares of the utility stock would give a total dividend income of $810?

45. **Travel** An account executive traveled 45 mi by car and then an additional 1050 mi by plane. The rate of the plane was seven times the rate of the car. The total time for the trip was $3\frac{1}{4}$ h. Find the rate of the plane.

46. **Physics** An object is dropped from the top of a building. Find the distance the object has fallen when the speed reaches 75 ft/s. Use the equation $v = \sqrt{64d}$, where v is the speed of the object and d is the distance. Round to the nearest whole number.

47. **Travel** A small plane made a trip of 660 mi in 5 h. The plane traveled the first 360 mi at a constant rate before increasing its speed by 30 mph. Then it traveled another 300 mi at the increased speed. Find the rate of the plane for the first 360 mi.

48. **Light** The intensity (L) of a light source is inversely proportional to the square of the distance (d) from the source. If the intensity is 8 foot-candles at a distance of 20 ft, what is the intensity when the distance is 4 ft?

49. **Travel** A motorboat traveling with the current can go 30 mi in 2 h. Against the current, it takes 3 h to go the same distance. Find the rate of the motorboat in calm water and the rate of the current.

50. **Investments** An investor deposits $4000 into an account that earns 9% annual interest compounded monthly. Use the compound interest formula $P = A(1 + i)^n$, where A is the original value of the investment, i is the interest rate per compounding period, n is the total number of compounding periods, and P is the value of the investment after n periods, to find the value of the investment after 2 years. Round to the nearest cent.

CHAPTER

R

Review of Introductory Algebra Topics

Digital Vision/Getty Images

OBJECTIVES

SECTION R.1

A To evaluate a variable expression
B To simplify a variable expression

SECTION R.2

A To solve a first-degree equation in one variable
B To solve an inequality in one variable

SECTION R.3

A To graph points in a rectangular coordinate system
B To graph a linear equation in two variables
C To evaluate a function
D To find the equation of a line

SECTION R.4

A To multiply and divide monomials
B To add and subtract polynomials
C To multiply polynomials
D To divide polynomials
E To factor polynomials of the form $ax^2 + bx + c$

SECTION

R.1 Variable Expressions

OBJECTIVE A To evaluate a variable expression 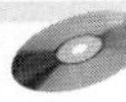

Whenever an expression contains more than one operation, the operations must be performed in a specified order, as listed below in the Order of Operations Agreement.

1.3A* Order of Operations Agreement

The Order of Operations Agreement

Step 1 Perform operations inside grouping symbols. Grouping symbols include parentheses ( ), brackets [ ], braces { }, the absolute value symbol | |, and fraction bars.

Step 2 Simplify exponential expressions.

Step 3 Do multiplication and division as they occur from left to right.

Step 4 Do addition and subtraction as they occur from left to right.

EXAMPLE • 1

Evaluate: $-2(7 - 3)^2 + 4 - 2(5 - 2)$

Solution

$-2(7 - 3)^2 + 4 - 2(5 - 2)$

$= -2(4)^2 + 4 - 2(3)$ • Perform operations inside parentheses.

$= -2(16) + 4 - 2(3)$ • Simplify the exponential expression.

$= -32 + 4 - 2(3)$ • Do the multiplication and division from left to right.

$= -32 + 4 - 6$

$= -28 - 6$ • Do the addition and subtraction from left to right.

$= -28 + (-6) = -34$

YOU TRY IT • 1

Evaluate: $(-4)(6 - 8)^2 - (-12 \div 4)$

Your solution

Solution on p. S36

1.4A Evaluate variable expressions

A **variable** is a letter that represents a quantity that is unknown or that can change or vary. An expression that contains one or more variables is a **variable expression.** $3x - 4y + 7z$ is a variable expression. It contains the variables x, y, and z.

Replacing a variable in a variable expression by a number and then simplifying the resulting numerical expression is called **evaluating the variable expression.** The number substituted for the variable is called the **value of the variable.** The result is called the **value of the variable expression.**

*Review this objective for more detailed coverage of this topic.

EXAMPLE 2

Evaluate $5ab^3 + 2a^2b^2 - 4$ when $a = 3$ and $b = -2$.

Solution

$5ab^3 + 2a^2b^2 - 4$

$5(3)(-2)^3 + 2(3)^2(-2)^2 - 4$ • Replace *a* by **3** and *b* by **−2**.

$= 5(3)(-8) + 2(9)(4) - 4$ • Use the Order of Operations Agreement to simplify the numerical expression.

$= -120 + 72 - 4$

$= -48 - 4$

$= -48 + (-4)$

$= -52$

YOU TRY IT 2

Evaluate $3xy^2 - 3x^2y$ when $x = -2$ and $y = 5$.

Your solution

Solution on p. S36

OBJECTIVE B To simplify a variable expression 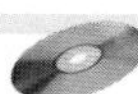

1.4B 1.4C Simplify variable expressions using the Properties of Addition and Multiplication

The Properties of Real Numbers are used to simplify variable expressions.

EXAMPLE 3

Simplify: $-\frac{1}{3}(-3y)$

Solution

$-\frac{1}{3}(-3y)$

$= \left[-\frac{1}{3}(-3)\right]y$ • Use the Associative Property of Multiplication to regroup factors.

$= 1y$ • Use the Inverse Property of Multiplication.

$= y$ • Use the Multiplication Property of One.

YOU TRY IT 3

Simplify: $-5(-3a)$

Your solution

Solution on p. S36

A variable expression is shown at the right. The expression can be rewritten by writing subtraction as addition of the opposite. Note that the expression has four addends. The **terms** of a variable expression are the addends of the expression. The expression has four terms.

$3x^2 - 4xy + 5z - 2$

$3x^2 + (-4xy) + 5z + (-2)$

Four terms

$3x^2 - 4xy + 5z - 2$

Variable terms — Constant term

The terms $3x^2$, $-4xy$, and $5z$ are **variable terms.** The term -2 is a **constant term,** or simply a **constant.**

Like terms of a variable expression are terms that have the same variable part. The terms $3x$ and $-7x$ are like terms. Constant terms are also like terms. Thus -6 and 9 are like terms.

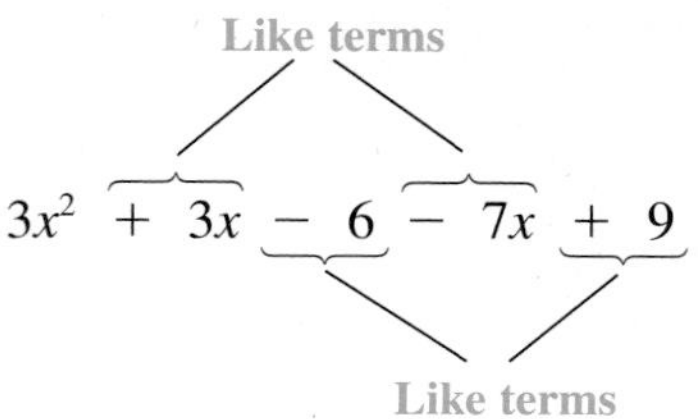

By using the Commutative Property of Multiplication, we can rewrite the Distributive Property as $ba + ca = (b + c)a$. This form of the Distributive Property is used to **combine like terms** of a variable expression by adding their coefficients. For instance,

$$7x + 9x = (7 + 9)x$$
$$= 16x$$

• Use the Distributive Property: $ba + ca = (b + c)a$.

EXAMPLE 4

Simplify: $4x^2 + 5x - 6x^2 - 7x$

Solution

$4x^2 + 5x - 6x^2 - 7x$

$= 4x^2 - 6x^2 + 5x - 7x$

$= (4x^2 - 6x^2) + (5x - 7x)$

• Use the Associative and Commutative Properties of Addition to rearrange and group like terms.

• Use the Distributive Property to combine like terms.

$= -2x^2 + (-2x)$

$= -2x^2 - 2x$

YOU TRY IT 4

Simplify: $2z^2 - 5z - 3z^2 + 6z$

Your solution

Solution on p. S36

1.4D Simplify variable expressions using the Distributive Property

The Distributive Property also is used to remove parentheses from a variable expression. Here is an example.

$$4(2x + 5z) = 4(2x) + 4(5z)$$

• Use the Distributive Property: $a(b + c) = ab + ac$.

$$= (4 \cdot 2)x + (4 \cdot 5)z$$

• Use the Associative Property of Multiplication to regroup factors.

$$= 8x + 20z$$

• Multiply $4 \cdot 2$ and $4 \cdot 5$.

The Distributive Property can be extended to expressions containing more than two terms. For instance,

$$4(2x + 3y + 5z) = 4(2x) + 4(3y) + 4(5z)$$
$$= 8x + 12y + 20z$$

EXAMPLE • 5

Simplify.

a. $-3(2x + 4)$

b. $6(3x - 4y + z)$

Solution

a. $-3(2x + 4)$

$= -3(2x) + (-3)(4)$ • Use the Distributive Property.

$= -6x - 12$

b. $6(3x - 4y + z)$

$= 6(3x) - 6(4y) + 6(z)$ • Use the Distributive Property.

$= 18x - 24y + 6z$

YOU TRY IT • 5

Simplify.

a. $-3(5y - 2)$

b. $-2(4x + 2y - 6z)$

Your solution

Solution on p. S36

Take Note

Recall that the Distributive Property states that if a, b, and c are real numbers, then

$$a(b + c) = ab + ac$$

To simplify the expression $5 + 3(4x - 2)$, use the Distributive Property to remove the parentheses.

$5 + 3(4x - 2) = 5 + 3(4x) - 3(2)$ • Use the Distributive Property.

$= 5 + 12x - 6$

$= 12x - 1$ • Add the like terms 5 and -6.

EXAMPLE • 6

Simplify.

a. $3(2x - 4) - 5(3x + 2)$

b. $3a - 2[7a - 2(2a + 1)]$

Solution

a. $3(2x - 4) - 5(3x + 2)$

$= 6x - 12 - 15x - 10$ • Use the Distributive Property.

$= -9x - 22$ • Combine like terms.

b. $3a - 2[7a - 2(2a + 1)]$

$= 3a - 2[7a - 4a - 2]$ • Use the Distributive Property.

$= 3a - 2[3a - 2]$ • Combine like terms.

$= 3a - 6a + 4$ • Use the Distributive Property.

$= -3a + 4$ • Combine like terms.

YOU TRY IT • 6

Simplify.

a. $7(-3x - 4y) - 3(3x + y)$

b. $2y - 3[5 - 3(3 + 2y)]$

Your solution

Solution on p. S36

R.1 EXERCISES

OBJECTIVE A To evaluate a variable expression

Evaluate the variable expression when $a = 2$, $b = 3$, and $c = -4$.

1. $a - 2c$

2. $-3a + 4b$

3. $3b - 3c$

4. $-3c + 4$

5. $16 \div (2c)$

6. $6b \div (-a)$

7. $3b - (a + c)^2$

8. $(a - b)^2 + 2c$

9. $(b - 3a)^2 + bc$

Evaluate the variable expression when $a = -1$, $b = 3$, $c = -2$, and $d = 4$.

10. $\dfrac{b - a}{d}$

11. $\dfrac{d - b}{a}$

12. $\dfrac{2d + b}{-a}$

13. $\dfrac{b - d}{c - a}$

14. $2(b + c) - 2a$

15. $3(b - a) - bc$

16. $\dfrac{-4bc}{2a + c}$

17. $\dfrac{abc}{b - d}$

18. $(d - a)^2 - (b - c)^2$

19. $(-b + d)^2 + (-a + c)^2$

20. $4ab + (2c)^2$

21. $3cd - (4a)^2$

Evaluate the variable expression when $a = 2.7$, $b = -1.6$, and $c = -0.8$.

22. $c^2 - ab$

23. $(a + b)^2 - c$

24. $\dfrac{b^3}{c} - 4a$

OBJECTIVE B To simplify a variable expression

Simplify each of the following.

25. $x + 7x$

26. $12y + 9y$

27. $8b - 5b$

28. $4y - 11y$

29. $-12a + 17a$

30. $-15xy + 7xy$

31. $4x + 5x + 2x$

32. $-5x^2 - 10x^2 + x^2$

33. $6x - 2y + 9x$

34. $3x - 7y - 6x + 4x$

35. $5a + 6a - 2a$

36. $2a - 5a + 3a$

37. $12y^2 + 10y^2$

38. $2z^2 - 9z^2$

39. $\frac{3}{4}x - \frac{1}{4}x$

40. $\frac{2}{5}y - \frac{3}{5}y$

41. $-4(5x)$

42. $-2(-8y)$

43. $(6a)(-4)$

44. $-5(7x^2)$

45. $\frac{1}{4}(4x)$

46. $\frac{12x}{5}\left(\frac{5}{12}\right)$

47. $\frac{1}{3}(21x)$

48. $-\frac{5}{8}(24a^2)$

49. $(36y)\left(\frac{1}{12}\right)$

50. $-(z + 4)$

51. $-3(a + 5)$

52. $(4 - 3b)9$

53. $(-2x - 6)8$

54. $3(5x^2 + 2x)$

55. $-5(2y^2 - 1)$

56. $4(x^2 - 3x + 5)$

57. $6(3x^2 - 2xy - y^2)$

58. $5a - (4a + 6)$

59. $3 - (10 + 8y)$

60. $12(y - 2) + 3(7 - 4y)$

61. $-5[2x + 3(5 - x)]$

62. $-3[2x - (x + 7)]$

63. $-5a - 2[2a - 4(a + 7)]$

SECTION R.2 Equations and Inequalities

OBJECTIVE A To solve a first-degree equation in one variable 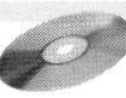

An **equation** expresses the equality of two mathematical expressions. Each of the equations below is a **first-degree equation in one variable.** *First degree* means that the variable has an exponent of 1.

$$x + 11 = 14$$
$$3a + 5 = 8a$$
$$2(6y - 1) = 3$$
$$4 - 3(2n - 1) = 6n - 5$$

A **solution** of an equation is a number that, when substituted for the variable, results in a true equation.

3 is a solution of the equation $x + 4 = 7$ because $3 + 4 = 7$.
9 is not a solution of the equation $x + 4 = 7$ because $9 + 4 \neq 7$.

To **solve an equation** means to find a solution of the equation. In solving an equation, the goal is to rewrite the given equation with the variable alone on one side of the equation and a constant term on the other side of the equation.

variable = *constant* or *constant* = *variable*

The constant is the solution of the equation.

The following properties of equations are used to rewrite equations in this form.

2.1B 2.1C Solving equations using the Addition and Multiplication Properties of Equations

Properties of Equations

Addition Property of Equations

The same number can be added to each side of an equation without changing the solution of the equation. In symbols, the equation $a = b$ has the same solution as the equation $a + c = b + c$.

Multiplication Property of Equations

Each side of an equation can be multiplied by the same nonzero number without changing the solution of the equation. In symbols, if $c \neq 0$, then the equation $a = b$ has the same solution as the equation $ac = bc$.

Take Note

Subtraction is defined as addition of the opposite.
$a - b = a + (-b)$

The Addition Property of Equations is used to remove a term from one side of the equation by adding the opposite of that term to each side of the equation. Because subtraction is defined in terms of addition, the Addition Property of Equations also makes it possible to subtract the same number from each side of an equation without changing the solution of the equation.

For example, to solve the equation $t + 9 = -4$, subtract the constant term (9) from each side of the equation.

$$\begin{aligned} t + 9 &= -4 \\ t + 9 - 9 &= -4 - 9 \\ t &= -13 \end{aligned}$$

Now the variable is alone on one side of the equation and a constant term (-13) is on the other side. The solution is the constant. The solution is -13.

To solve $7 = y - 8$, add 8 to each side of the equation.

$$\begin{aligned} 7 &= y - 8 \\ 7 + 8 &= y - 8 + 8 \\ 15 &= y \end{aligned}$$

Take Note

Division is defined as multiplication by the reciprocal.

$$a \div b = a \cdot \frac{1}{b}$$

The equation is in the form *constant* = *variable*. The solution is the constant. The solution is 15.

The Multiplication Property of Equations is used to remove a coefficient by multiplying each side of the equation by the reciprocal of the coefficient. Because division is defined in terms of multiplication, each side of an equation can be divided by the same nonzero number without changing the solution of the equation.

Take Note

When using the Multiplication Property of Equations, multiply each side of the equation by the reciprocal of the coefficient when the coefficient is a fraction. Divide each side of the equation by the coefficient when the coefficient is an integer or a decimal.

For example, to solve the equation $-5q = 120$, divide each side of the equation by the coefficient -5.

$$\begin{aligned} -5q &= 120 \\ \frac{-5q}{-5} &= \frac{120}{-5} \\ q &= -24 \end{aligned}$$

Now the variable is alone on one side of the equation and a constant (-24) is on the other side. The solution is the constant. The solution is -24.

2.2B 2.2C Solve general equations

In solving more complicated first-degree equations in one variable, use the following sequence of steps.

Steps for Solving a First-Degree Equation in One Variable

1. Use the Distributive Property to remove parentheses.
2. Combine any like terms on the right side of the equation and any like terms on the left side of the equation.
3. Use the Addition Property to rewrite the equation with only one variable term.
4. Use the Addition Property to rewrite the equation with only one constant term.
5. Use the Multiplication Property to rewrite the equation with the variable alone on one side of the equation and a constant on the other side of the equation.

If one of these steps is not needed to solve a given equation, proceed to the next step.

EXAMPLE • 1

Solve.

a. $5x + 9 = 23 - 2x$

b. $8x - 3(4x - 5) = -2x + 6$

Solution

a.

$$\begin{aligned} 5x + 9 &= 23 - 2x \\ 5x + 2x + 9 &= 23 - 2x + 2x && \text{• Step 3} \\ 7x + 9 &= 23 \\ 7x + 9 - 9 &= 23 - 9 && \text{• Step 4} \\ 7x &= 14 \\ \frac{7x}{7} &= \frac{14}{7} && \text{• Step 5} \\ x &= 2 \end{aligned}$$

The solution is 2.

b.

$$\begin{aligned} 8x - 3(4x - 5) &= -2x + 6 \\ 8x - 12x + 15 &= -2x + 6 && \text{• Step 1} \\ -4x + 15 &= -2x + 6 && \text{• Step 2} \\ -4x + 2x + 15 &= -2x + 2x + 6 && \text{• Step 3} \\ -2x + 15 &= 6 \\ -2x + 15 - 15 &= 6 - 15 && \text{• Step 4} \\ -2x &= -9 \\ \frac{-2x}{-2} &= \frac{-9}{-2} && \text{• Step 5} \\ x &= \frac{9}{2} \end{aligned}$$

The solution is $\frac{9}{2}$.

YOU TRY IT • 1

Solve.

a. $4x + 3 = 7x + 9$

b. $4 - (5x - 8) = 4x + 3$

Your solution

Solution on p. S36

OBJECTIVE B To solve an inequality in one variable 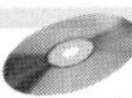

An **inequality** contains the symbol $>$, $<$, $\geq$, or $\leq$. An inequality expresses the relative order of two mathematical expressions. Here are some examples of inequalities in one variable.

$$\left.\begin{aligned} 4x &\geq 12 \\ 2x + 7 &\leq 9 \\ x^2 + 1 &> 3x \end{aligned}\right\} \text{Inequalities in one variable}$$

2.5A Solve an inequality in one variable

A solution of an inequality in one variable is a number that, when substituted for the variable, results in a true inequality. For the inequality $x < 4$ shown below, 3, 0, and -5 are solutions of the inequality because replacing the variable by these numbers results in a true inequality.

$$x < 4 \qquad x < 4 \qquad x < 4$$
$$3 < 4 \quad \text{True} \qquad 0 < 4 \quad \text{True} \qquad -5 < 4 \quad \text{True}$$

The number 7 is not a solution of the inequality $x < 4$ because $7 < 4$ is a false inequality. Besides the numbers 3, 0, and -5, there is an infinite number of other solutions of the inequality $x < 4$. Any number less than 4 is a solution; for instance, -5.2, $\frac{5}{2}$, π, and 1 are also solutions of the inequality. The set of all the solutions of an inequality is called the **solution set of the inequality.** The solution set of the inequality $x < 4$ is written in set-builder notation as $\{x|x < 4\}$. This is read, "the set of all x such that x is less than 4."

The graph of the solution set of $x < 4$ is shown at the right.

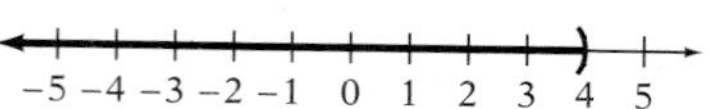

In solving an inequality, the goal is to rewrite the given inequality in the form

variable < *constant* or *variable* > *constant*

The Addition Property of Inequalities is used to rewrite an inequality in this form.

Take Note

The Addition Property of Inequalities states that the same number can be added to each side of an inequality without changing the solution set of the inequality.

The Addition Property of Inequalities

If $a > b$ and c is a real number, then the inequalities $a > b$ and $a + c > b + c$ have the same solution set.

If $a < b$ and c is a real number, then the inequalities $a < b$ and $a + c < b + c$ have the same solution set.

The Addition Property of Inequalities is also true for the symbols $\leq$ and $\geq$.

The Addition Property of Inequalities is used to remove a term from one side of an inequality by adding the additive inverse of that term to each side of the inequality. Because subtraction is defined in terms of addition, the same number can be subtracted from each side of an inequality without changing the solution set of the inequality.

As shown in the example below, the Addition Property of Inequalities applies to variable terms as well as to constants.

HOW TO • 1 Solve $4x - 5 \leq 3x - 2$. Write the solution set in set-builder notation.

$$4x - 5 \leq 3x - 2$$
$$4x - 3x - 5 \leq 3x - 3x - 2$$ • Subtract $3x$ from each side of the inequality.
$$x - 5 \leq -2$$ • Simplify.
$$x - 5 + 5 \leq -2 + 5$$ • Add 5 to each side of the inequality.
$$x \leq 3$$ • Simplify.

The solution set is $\{x|x \leq 3\}$.

When multiplying or dividing an inequality by a number, the inequality symbol may be reversed, depending on whether the number is positive or negative. Look at the following two examples.

$3 < 5$		$3 < 5$	
$2(3) < 2(5)$	• Multiply by positive 2. The inequality symbol remains the same.	$-2(3) > -2(5)$	• Multiply by negative 2. The inequality symbol is reversed in order to make the inequality a true statement.
$6 < 10$	• $6 < 10$ is a true statement.	$-6 > -10$	

This is summarized in the Multiplication Property of Inequalities.

The Multiplication Property of Inequalities

Rule 1

If $a > b$ and $c > 0$, then $ac > bc$.

If $a < b$ and $c > 0$, then $ac < bc$.

Rule 2

If $a > b$ and $c < 0$, then $ac < bc$.

If $a < b$ and $c < 0$, then $ac > bc$.

Rule 1 states that when each side of an inequality is multiplied by a positive number, the inequality symbol remains the same. Rule 2 states that when each side of an inequality is multiplied by a negative number, the inequality symbol must be reversed.

Here are a few more examples of this property.

Rule 1		Rule 2	
$-4 < -2$	$5 > -3$	$7 < 9$	$-2 > -6$
$-4(2) < -2(2)$	$5(3) > -3(3)$	$7(-2) > 9(-2)$	$-2(-3) < -6(-3)$
$-8 < -4$	$15 > -9$	$-14 > -18$	$6 < 18$

The Multiplication Property of Inequalities is also true for the symbols $\leq$ and $\geq$.

Use the Multiplication Property of Inequalities to remove a coefficient other than 1 from one side of an inequality so that the inequality can be rewritten with the variable alone on one side of the inequality and a constant term on the other side.

Because division is defined in terms of multiplication, when each side of an inequality is divided by a positive number, the inequality symbol remains the same. When each side of an inequality is divided by a negative number, the inequality symbol must be reversed.

Take Note

Solving inequalities in one variable is similar to solving equations in one variable *except* that when you multiply or divide by a negative number, you must reverse the inequality symbol.

HOW TO • 2 Solve $-3x < 9$. Write the solution set in set-builder notation.

$-3x < 9$

$\frac{-3x}{-3} > \frac{9}{-3}$ • Divide each side of the inequality by the coefficient -3 and reverse the inequality symbol.

$x > -3$ • Simplify.

$\{x|x > -3\}$ • Write the answer in set-builder notation.

EXAMPLE 2

Solve $x + 3 > 4x + 6$. Write the solution set in set-builder notation.

Solution

$$x + 3 > 4x + 6$$
$$x - 4x + 3 > 4x - 4x + 6 \quad \text{• Subtract } 4x \text{ from each side.}$$
$$-3x + 3 > 6$$
$$-3x + 3 - 3 > 6 - 3 \quad \text{• Subtract 3 from each side.}$$
$$-3x > 3$$
$$\frac{-3x}{-3} < \frac{3}{-3} \quad \text{• Divide each side by } -3\text{. Reverse the inequality symbol.}$$
$$x < -1$$

$\{x|x < -1\}$

YOU TRY IT 2

Solve $3x - 1 \leq 5x - 7$. Write the solution set in set-builder notation.

Your solution

Solution on p. S36

When an inequality contains parentheses, often the first step in solving the inequality is to use the Distributive Property to remove the parentheses.

EXAMPLE 3

Solve $-2(x - 7) > 3 - 4(2x - 3)$. Write the solution set in set-builder notation.

Solution

$$-2(x - 7) > 3 - 4(2x - 3)$$
$$-2x + 14 > 3 - 8x + 12 \quad \text{• Use the Distributive Property.}$$
$$-2x + 14 > 15 - 8x \quad \text{• Combine like terms.}$$
$$-2x + 8x + 14 > 15 - 8x + 8x \quad \text{• Add } 8x \text{ to each side.}$$
$$6x + 14 > 15$$
$$6x + 14 - 14 > 15 - 14 \quad \text{• Subtract 14 from each side.}$$
$$6x > 1$$
$$\frac{6x}{6} > \frac{1}{6} \quad \text{• Divide each side by 6.}$$
$$x > \frac{1}{6}$$

$\left\{x\middle|x > \frac{1}{6}\right\}$

YOU TRY IT 3

Solve $3 - 2(3x + 1) < 7 - 2x$. Write the solution set in set-builder notation.

Your solution

Solution on p. S36

R.2 EXERCISES

OBJECTIVE A To solve a first-degree equation in one variable

Solve.

1. $x + 7 = -5$
2. $9 + b = 21$
3. $-9 = z - 8$
4. $b - 11 = 11$
5. $-48 = 6z$
6. $-9a = -108$
7. $-\frac{3}{4}x = 15$
8. $\frac{5}{2}x = -10$
9. $-\frac{x}{4} = -2$
10. $\frac{2x}{5} = -8$
11. $4 - 2b = 2 - 4b$
12. $4y - 10 = 6 + 2y$
13. $5x - 3 = 9x - 7$
14. $3m + 5 = 2 - 6m$
15. $6a - 1 = 2 + 2a$
16. $5x + 7 = 8x + 5$
17. $2 - 6y = 5 - 7y$
18. $4b + 15 = 3 - 2b$
19. $2(x + 1) + 5x = 23$
20. $9n - 15 = 3(2n - 1)$
21. $7a - (3a - 4) = 12$
22. $5(3 - 2y) = 3 - 4y$
23. $9 - 7x = 4(1 - 3x)$
24. $2(3b + 5) - 1 = 10b + 1$
25. $2z - 2 = 5 - (9 - 6z)$
26. $4a + 3 = 7 - (5 - 8a)$
27. $5(6 - 2x) = 2(5 - 3x)$
28. $4(3y + 1) = 2(y - 8)$
29. $2(3b - 5) = 4(6b - 2)$
30. $3(x - 4) = 1 - (2x - 7)$

OBJECTIVE B To solve an inequality in one variable

Solve. Write the answer in set-builder notation.

31. $x - 5 > -2$
32. $5 + n \geq 4$
33. $-2 + n \geq 0$

34. $x - 3 < 2$

35. $8x \leq -24$

36. $-4x < 8$

37. $3n > 0$

38. $-2n \leq -8$

39. $2x - 1 > 7$

40. $5x - 2 \leq 8$

41. $4 - 3x < 10$

42. $7 - 2x \geq 1$

43. $3x - 1 > 2x + 2$

44. $6x + 4 \leq 8 + 5x$

45. $8x + 1 \geq 2x + 13$

46. $6x + 3 > 4x - 1$

47. $-3 - 4x > -11$

48. $4x - 2 < x - 11$

49. $4x - 2 > 3x + 1$

50. $7x + 5 \leq 9 + 6x$

51. $9x + 2 \geq 3x + 14$

52. $8x + 1 > 6x - 3$

53. $-5 - 2x > -13$

54. $5x - 3 < x - 11$

55. $4(2x - 1) > 3x - 2(3x - 5)$

56. $2 - 5(x + 1) \geq 3(x - 1) - 8$

57. $3(4x + 3) \leq 7 - 4(x - 2)$

58. $3 + 2(x + 5) \geq x + 5(x + 1) + 1$

59. $3 - 4(x + 2) \leq 6 + 4(2x + 1)$

60. $12 - 2(3x - 2) \geq 5x - 2(5 - x)$

SECTION

R.3 Linear Equations in Two Variables

OBJECTIVE A To graph points in a rectangular coordinate system

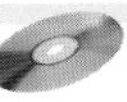

A **rectangular coordinate system** is formed by two number lines, one horizontal and one vertical. The point of intersection is called the **origin.** The two axes are called the **coordinate axes** or simply the **axes.** Generally, the horizontal axis is labeled the x-axis, and the vertical axis is labeled the y-axis. In this case, the axes form what is called the ***xy*-plane.**

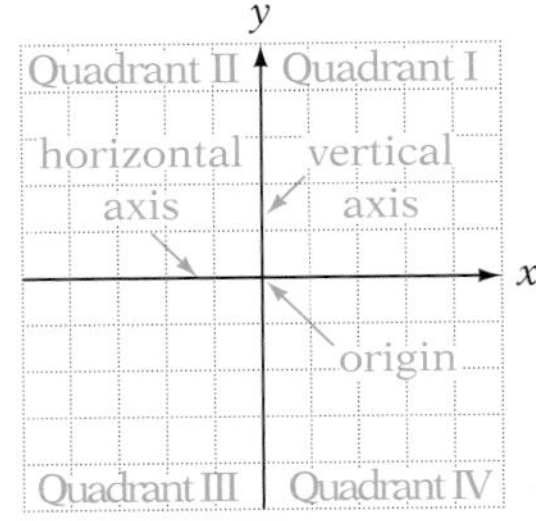

4.1A Points in the plane

Each point in the plane can be identified by a pair of numbers called an **ordered pair.** The first number of the ordered pair measures a horizontal change from the y-axis and is called the **abscissa** or ***x*-coordinate.** The second number of the ordered pair measures a vertical change from the x-axis and is called the **ordinate** or ***y*-coordinate.** The ordered pair (x, y) associated with a point is also called the **coordinates** of the point.

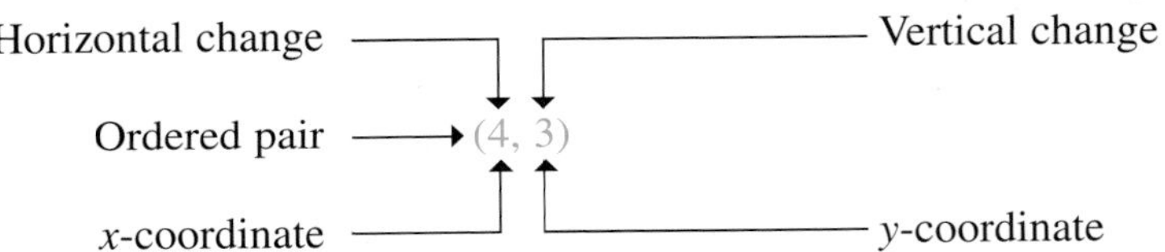

To **graph**, or **plot**, a point means to place a dot at the coordinates of the point. For example, to graph the ordered pair $(4, 3)$, start at the origin. Move 4 units to the right and then 3 units up. Draw a dot. To graph $(-3, -4)$, start at the origin. Move 3 units to the left and then 4 units down. Draw a dot.

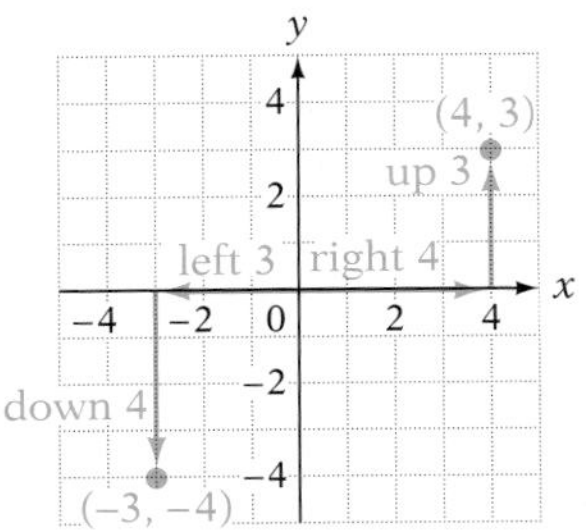

The **graph of an ordered pair** is the dot drawn at the coordinates of the point in the plane. The graphs of the ordered pairs $(4, 3)$ and $(-3, -4)$ are shown above.

The graphs of the points $(2, 3)$ and $(3, 2)$ are different points. The *order* in which the numbers in an *ordered* pair are listed is important.

EXAMPLE • 1

Graph the ordered pairs $(-4, 2)$, $(3, 4)$, $(0, -1)$, $(2, 0)$, and $(-1, -3)$.

Solution

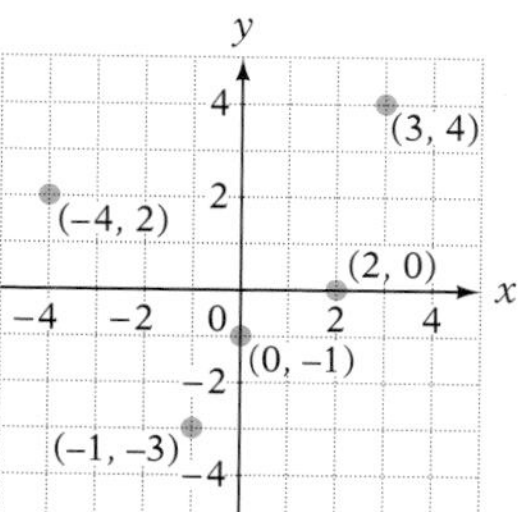

YOU TRY IT • 1

Graph the ordered pairs $(-2, 4)$, $(4, 0)$, $(0, 3)$, $(-4, -3)$, and $(5, -1)$.

Your solution

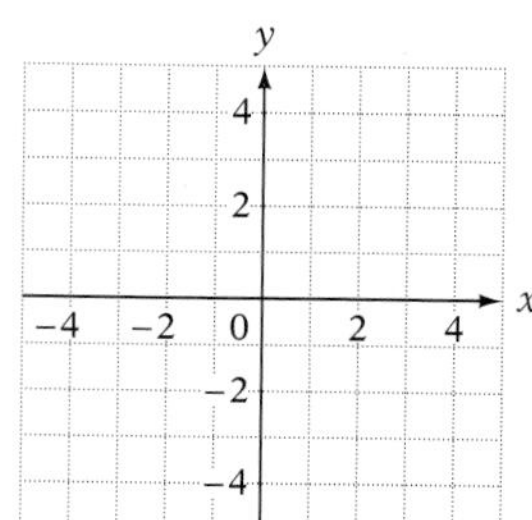

Solution on p. S37

OBJECTIVE B To graph a linear equation in two variables 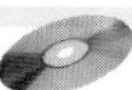

The equations below are examples of equations in two variables.

$$y = 3x - 4$$
$$2x - y = 7$$
$$y = x^2 + 1$$

4.1B Solutions of equations in two variables

A **solution of an equation in two variables** is an ordered pair (x, y) whose coordinates make the equation a true statement.

HOW TO • 1 Is the ordered pair $(-4, 9)$ a solution of the equation $y = -2x + 1$?

y	$= -2x + 1$	
9	$-2(-4) + 1$	• Replace x by -4 and y by 9.
9	$8 + 1$	• Simplify the right side.
$9 = 9$		• Compare the results. If the resulting equation is true, the ordered pair is a solution of the equation. If it is not true, the ordered pair is not a solution of the equation.

Yes, the ordered pair $(-4, 9)$ is a solution of the equation $y = -2x + 1$.

Besides $(-4, 9)$, there are many other ordered pairs that are solutions of the equation $y = -2x + 1$. For example, $(0, 1)$, $(3, -5)$, and $\left(-\frac{3}{2}, 4\right)$ are also solutions. In general, an equation in two variables has an infinite number of solutions. By choosing any value of x and substituting that value into the equation, we can calculate a corresponding value of y.

HOW TO • 2 Find the ordered-pair solution of $y = \frac{2}{5}x - 4$ that corresponds to $x = 5$.

$y = \frac{2}{5}x - 4$

$y = \frac{2}{5}(5) - 4$ • We are given that $x = 5$. Replace x by 5 in the equation.

$y = 2 - 4$ • Simplify the right side.

$y = -2$ • When $x = 5, y = -2$.

The ordered-pair solution of $y = \frac{2}{5}x - 4$ for $x = 5$ is $(5, -2)$.

4.3A Graph an equation of the form $y = mx + b$

Solutions of an equation in two variables can be graphed in the rectangular coordinate system.

HOW TO • 3 Graph the ordered-pair solutions of $y = -2x + 1$ for $x = -2, -1, 0, 1$, and 2.

x	$y = -2x + 1$	y	(x, y)
-2	$-2(-2) + 1$	5	$(-2, 5)$
-1	$-2(-1) + 1$	3	$(-1, 3)$
0	$-2(0) + 1$	1	$(0, 1)$
1	$-2(1) + 1$	-1	$(1, -1)$
2	$-2(2) + 1$	-3	$(2, -3)$

• Use the given values of x to determine ordered-pair solutions of the equation. It is convenient to record these in a table.

The ordered-pair solutions of $y = -2x + 1$ for $x = -2, -1, 0, 1$, and 2 are $(-2, 5)$, $(-1, 3)$, $(0, 1)$, $(1, -1)$, and $(2, -3)$. These are graphed at the right.

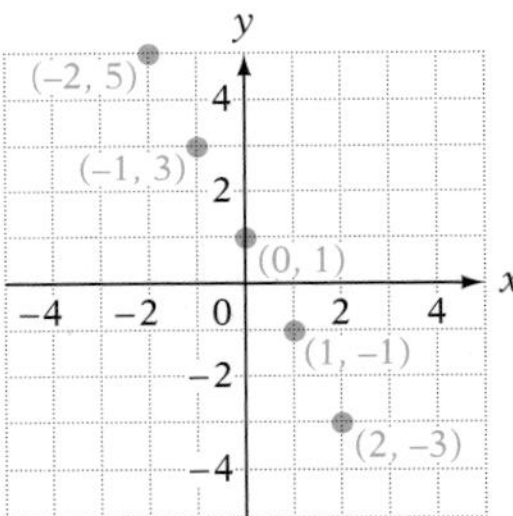

The **graph of an equation in two variables** is a graph of all the ordered-pair solutions of the equation. Consider the equation $y = -2x + 1$ above. The ordered-pair solutions $(-2, 5)$, $(-1, 3)$, $(0, 1)$, $(1, -1)$, and $(2, -3)$ are graphed in the figure above. We can choose values of x that are not integers to produce more ordered pairs to graph, such as $\left(\frac{5}{2}, -4\right)$ and $\left(-\frac{1}{2}, 2\right)$. Choosing still other values of x would result in more and more ordered pairs being graphed. The result would be so many dots that the graph would appear as a straight line, as shown below. This is the graph of $y = -2x + 1$.

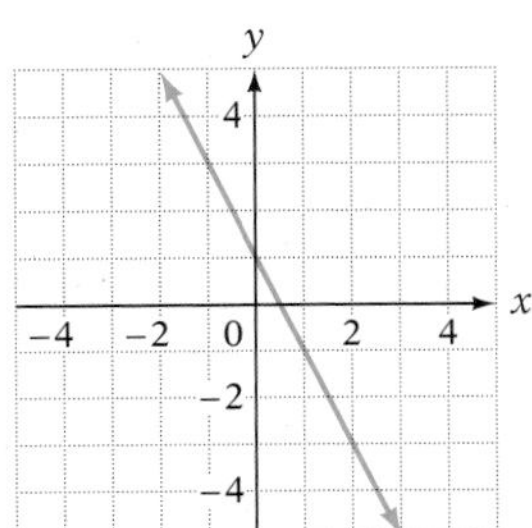

The equation $y = -2x + 1$ is an example of a linear equation because its graph is a straight line. It is also called a first-degree equation in two variables because the exponent on each variable is 1.

✓ Take Note

The equation $y = x^2 + 2x - 4$ is not a linear equation in two variables because there is a term with a variable squared. The equation $y = \frac{5}{x-3}$ is not a linear equation because a variable occurs in the denominator of a fraction.

Linear Equation in Two Variables

An equation of the form $y = mx + b$, where m is the coefficient of x and b is a constant, is a linear equation in two variables. The graph of a linear equation in two variables is a straight line.

Examples of linear equations are shown at the right.

$$y = 5x + 3 \qquad (m = 5, b = 3)$$
$$y = x - 4 \qquad (m = 1, b = -4)$$
$$y = -\frac{3}{4}x \qquad \left(m = -\frac{3}{4}, b = 0\right)$$

To graph a linear equation, find ordered-pair solutions of the equation. Do this by choosing any value of x and finding the corresponding value of y. Repeat this procedure, choosing different values for x, until you have found the number of solutions desired. Because the graph of a linear equation in two variables is a straight line, and a straight line is determined by two points, it is necessary to find only two solutions. However, it is recommended that at least three points be used to ensure accuracy.

HOW TO • 4 Graph: $y = 2x - 3$

x	$y = 2x - 3$	y
0	2(0) − 3	−3
2	2(2) − 3	1
−1	2(−1) − 3	−5

• Choose any values of x. Then find the corresponding values of y. The numbers 0, 2, and −1 were chosen arbitrarily for x.

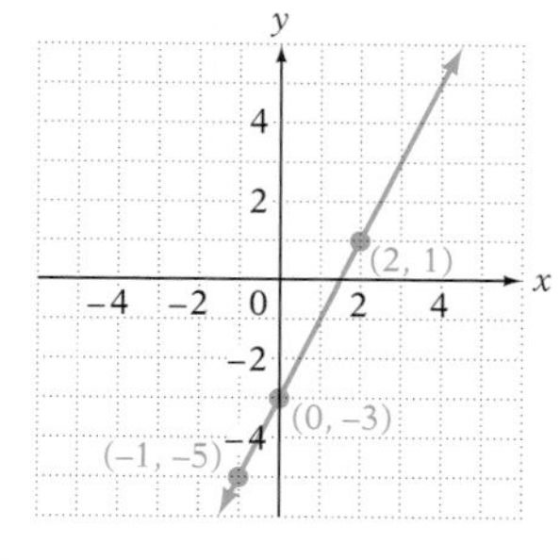

• Graph the ordered-pair solutions (0, −3), (2, 1), and (−1, −5). Draw a straight line through the points.

Remember that a graph is a drawing of the ordered-pair solutions of an equation. Therefore, every point on the graph is a solution of the equation, and every solution of the equation is a point on the graph.

When graphing an equation of the form $y = mx + b$, if m is a fraction, choose values of x that will simplify the evaluation. This is illustrated in Example 2. Note that the values of x chosen are multiples of the denominator, 2.

EXAMPLE • 2

Graph $y = -\frac{3}{2}x - 3$.

Solution

x	y
0	−3
−2	0
−4	3

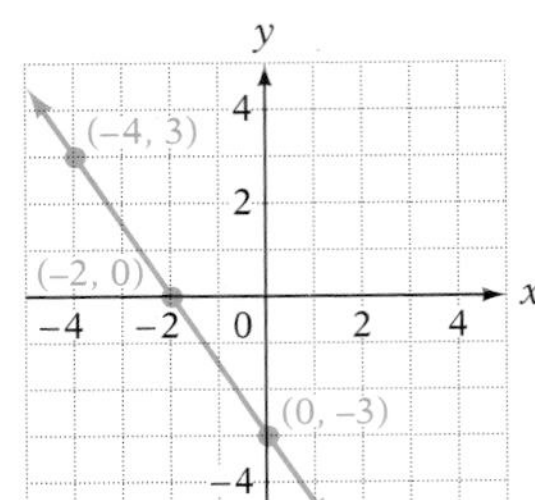

YOU TRY IT • 2

Graph $y = \frac{3}{5}x - 4$.

Your solution

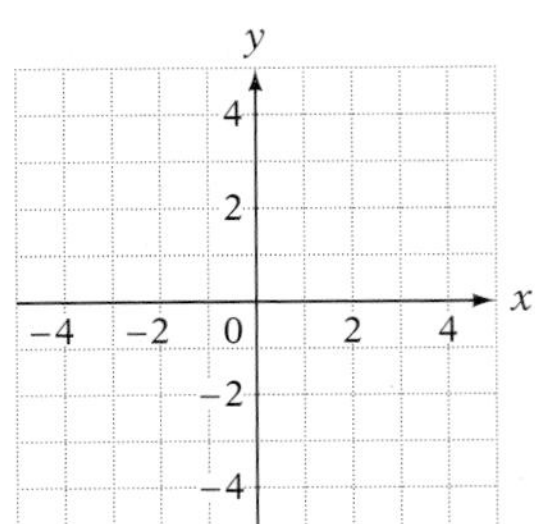

Solution on p. S37

4.3B Graph an equation of the form $Ax + By = C$

An equation of the form $Ax + By = C$ is also a linear equation in two variables. Examples of these equations are shown below.

$$3x + 4y = 12 \quad (A = 3, B = 4, C = 12)$$
$$x - 5y = -10 \quad (A = 1, B = -5, C = -10)$$
$$2x - y = 0 \quad (A = 2, B = -1, C = 0)$$

One method of graphing an equation of the form $Ax + By = C$ involves first solving the equation for y and then following the same procedure used for graphing an equation of the form $y = mx + b$. To solve the equation for y means to rewrite the equation so that y is alone on one side of the equation and the term containing x and the constant are on the other side of the equation. The Addition and Multiplication Properties of Equations are used to rewrite an equation of the form $Ax + By = C$ in the form $y = mx + b$.

HOW TO • 5 Graph $3x + 2y = 6$.

$3x + 2y = 6$ • The equation is in the form $Ax + By = C$.

$2y = -3x + 6$ • Solve the equation for y. Subtract $3x$ from each side of the equation.

$y = -\frac{3}{2}x + 3$ • Divide each side of the equation by 2. Note that each term on the right side is divided by 2.

• Find at least three solutions.

x	y
0	3
2	0
4	−3

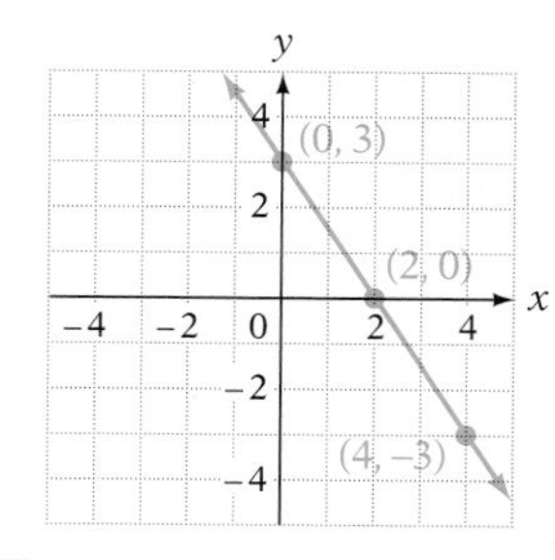

• Graph the ordered pairs (0, 3), (2, 0), and (4, −3). Draw a straight line through the points.

EXAMPLE • 3

Graph $2x + 3y = 9$.

Solution

$$2x + 3y = 9$$
$$3y = -2x + 9$$
$$y = -\frac{2}{3}x + 3$$

x	y
−3	5
0	3
3	1

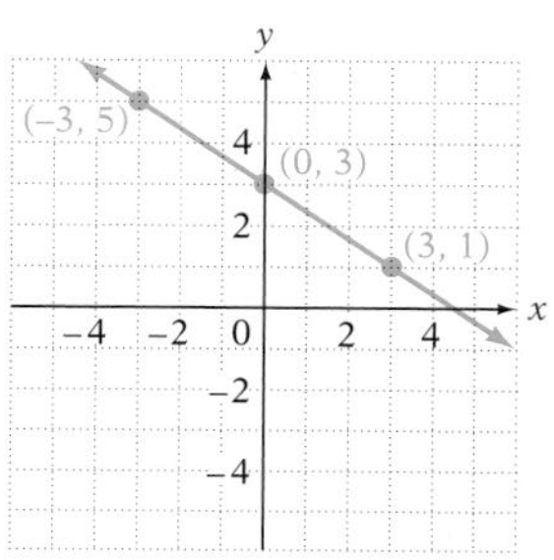

YOU TRY IT • 3

Graph $-3x + 2y = 4$.

Your solution

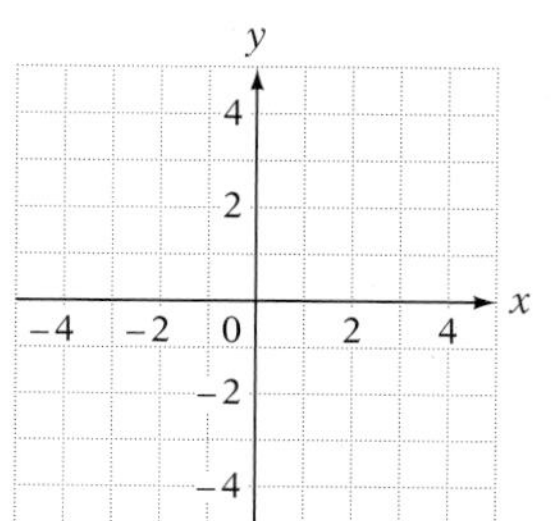

Solution on p. S37

4.4A Slope

The **slope** of a line is a measure of the slant of the line. The symbol for slope is m.

For an equation of the form $y = mx + b$, m is the slope. Here are a few examples:

The slope of the line $y = -2x + 5$ is -2.
The slope of the line $y = 8x$ is 8.
The slope of the line $y = \frac{3}{4}x - 1$ is $\frac{3}{4}$.

The slope of a line containing two points is the ratio of the change in the y values between the two points to the change in the x values.

Take Note

$\text{Slope} = m = \frac{\text{change in } y}{\text{change in } x}$

Slope Formula

The slope of the line containing the two points $P_1(x_1, y_1)$ and $P_2(x_2, y_2)$ is given by

$$m = \frac{y_2 - y_1}{x_2 - x_1}, \; x_1 \neq x_2$$

EXAMPLE • 4

Find the slope of the line containing the points $(-4, -3)$ and $(-1, 1)$.

Solution

Let $(x_1, y_1) = (-4, -3)$ and $(x_2, y_2) = (-1, 1)$.

$$m = \frac{y_2 - y_1}{x_2 - x_1} = \frac{1 - (-3)}{-1 - (-4)} = \frac{4}{3}$$

The slope is $\frac{4}{3}$.

YOU TRY IT • 4

Find the slope of the line containing the points $(-2, 3)$ and $(1, -3)$.

Your solution

Solution on p. S37

4.4B Slope-intercept form of a straight line

One important characteristic of the graph of a linear equation is its *intercepts*. An ***x*-intercept** is a point at which the graph crosses the x-axis. A ***y*-intercept** is a point at which the graph crosses the *y*-axis.

The graph of the equation $y = \frac{1}{2}x - 2$ is shown at the right. The *x*-intercept of the graph is $(4, 0)$. The *y*-intercept of the graph is $(0, -2)$.

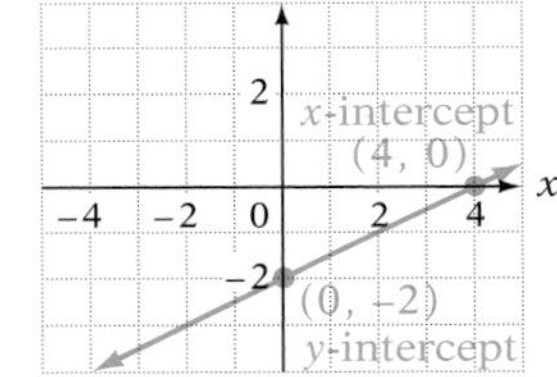

Note that at an *x*-intercept, the *y*-coordinate is 0. At a *y*-intercept, the *x*-coordinate is 0.

We can find the *y*-intercept of a linear equation by letting $x = 0$.

HOW TO • 6 Find the *y*-intercept of the graph of the equation $y = 3x + 4$.

$y = 3x + 4$

$y = 3(0) + 4$ • **To find the *y*-intercept, let $x = 0$.**

$y = 0 + 4$

$y = 4$

The *y*-intercept is $(0, 4)$.

Note that the constant term of $y = mx + b$ is b, and the *y*-intercept is $(0, b)$.

In general, **for any equation of the form $y = mx + b$, the *y*-intercept is $(0, b)$.**

Because the slope and the *y*-intercept can be determined directly from the equation $y = mx + b$, this equation is called the slope-intercept form of a straight line.

Slope-Intercept Form of a Straight Line

The equation $y = mx + b$ is called the **slope-intercept form of a straight line.** The slope of the line is m, the coefficient of x. The y-intercept is $(0, b)$.

The following equations are written in slope-intercept form.

$y = -4x + 3$	Slope $= -4$, y-intercept $= (0, 3)$
$y = \frac{2}{5}x - 1$	Slope $= \frac{2}{5}$, y-intercept $= (0, -1)$
$y = -x$	Slope $= -1$, y-intercept $= (0, 0)$

When an equation is in slope-intercept form, it is possible to quickly draw a graph of the function.

HOW TO • 7 Graph $x + 2y = 4$ by using the slope and y-intercept.

Solve the equation for y.

$$x + 2y = 4$$
$$2y = -x + 4$$
$$y = -\frac{1}{2}x + 2$$

From the equation $y = -\frac{1}{2}x + 2$, the slope is $-\frac{1}{2}$ and the y-intercept is $(0, 2)$.

Rewrite the slope $-\frac{1}{2}$ as $\frac{-1}{2}$.

Beginning at the y-intercept, move right 2 units (change in x) and then down 1 unit (change in y).

The point whose coordinates are (2, 1) is a second point on the graph. Draw a straight line through the points (0, 2) and (2, 1).

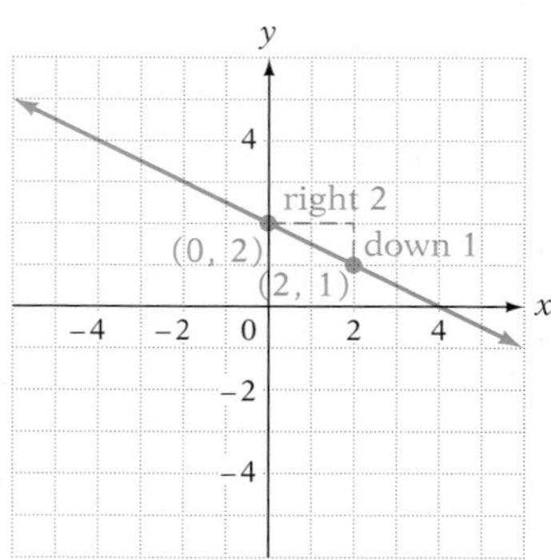

Take Note

Recall that slope $= m = \frac{\text{change in } y}{\text{change in } x}$. For the example at the right, $m = -\frac{1}{2} = \frac{-1}{2} = \frac{\text{change in } y}{\text{change in } x}$. Therefore, the change in y is -1 and the change in x is 2.

EXAMPLE • 5

Graph $y = -\frac{3}{2}x + 4$ by using the slope and the y-intercept.

YOU TRY IT • 5

Graph $y = -\frac{2}{3}x + 2$ by using the slope and the y-intercept.

Solution

From the equation $y = -\frac{3}{2}x + 4$, the slope is $-\frac{3}{2}$ and the y-intercept is (0, 4).

Rewrite the slope $-\frac{3}{2}$ as $\frac{-3}{2}$.

Place a dot at the y-intercept.

Starting at the y-intercept, move right 2 units (the change in x) and down 3 units (the change in y). Place a dot at that location.

Draw a line through the two points.

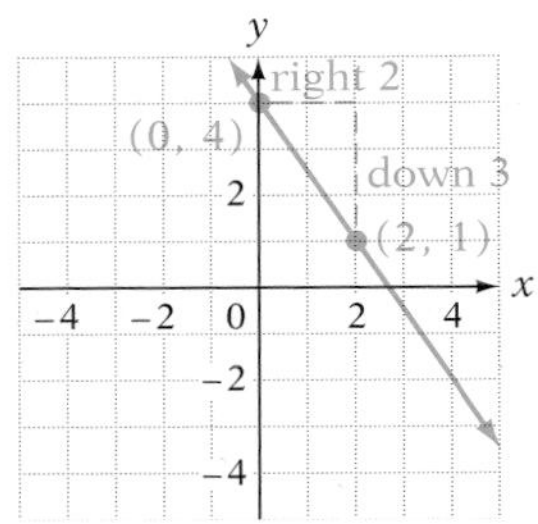

Your solution

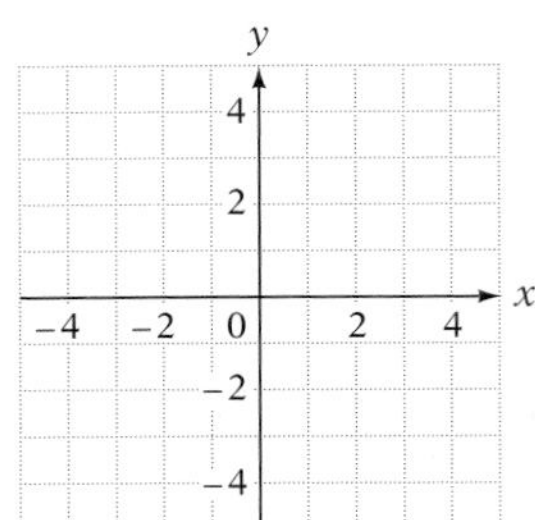

Solution on p. S37

OBJECTIVE C To evaluate a function 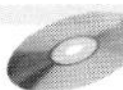

4.2A Evaluate a function

A **relation** is a set of ordered pairs. A **function** is a relation in which no two ordered pairs have the same first coordinate and different second coordinates.

The relation {(0, 4), (1, 5), (1, 6), (2, 7)} is not a function because the ordered pairs (1, 5), and (1, 6) have the same first coordinate and different second coordinates.

The relation {(0, 4), (1, 5), (2, 6), (3, 7)} is a function.

The phrase "y is a function of x," or a similar phrase with different variables, is used to describe those equations in two variables that define functions. To emphasize that the equation represents a function, function notation is used. For example, the square function is written in function notation as follows:

Take Note
The symbol $f(x)$ is read "the value of f at x" or "f of x."

$$f(x) = x^2$$

The process of determining $f(x)$ for a given value of x is called **evaluating the function.** For instance, to evaluate $f(x) = x^2$ when $x = 3$, replace x by 3 and simplify.

$$f(x) = x^2$$
$$f(3) = 3^2 = 9$$

The value of the function is 9 when $x = 3$. An ordered pair of the function is (3, 9).

EXAMPLE • 6

Evaluate $f(x) = 2x - 4$ when $x = 3$. Use your answer to write an ordered pair of the function.

Solution

$f(x) = 2x - 4$
$f(3) = 2(3) - 4$ • $x = 3$
$f(3) = 6 - 4$
$f(3) = 2$

An ordered pair of the function is (3, 2).

YOU TRY IT • 6

Evaluate the function $f(x) = 4 - 2x$ at $x = -3$. Use your answer to write an ordered pair of the function.

Your solution

Solution on p. S37

OBJECTIVE D To find the equation of a line

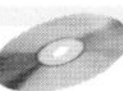

4.5A Find the equation of a line given a point and the slope

When the slope of a line and a point on the line are known, the equation of the line can be determined by using the point-slope formula.

Point-Slope Formula

Let m be the slope of a line, and let (x_1, y_1) be the coordinates of a point on the line. The equation of the line can be found using the point-slope formula:

$$y - y_1 = m(x - x_1)$$

EXAMPLE • 7

Find the equation of the line that contains the point $(1, -3)$ and has slope -2.

Solution

$y - y_1 = m(x - x_1)$
$y - (-3) = -2(x - 1)$ • $x_1 = 1, y_1 = -3, m = -2$
$y + 3 = -2x + 2$
$y = -2x - 1$

The equation of the line is $y = -2x - 1$.

YOU TRY IT • 7

Find the equation of the line that contains the point $(-2, 2)$ and has slope $-\frac{1}{2}$.

Your solution

Solution on p. S37

R.3 EXERCISES

OBJECTIVE A To graph points in a rectangular coordinate system

Graph the ordered pairs.

1. (2, 3), (4, 0), (−4, 1), (−2, −2)

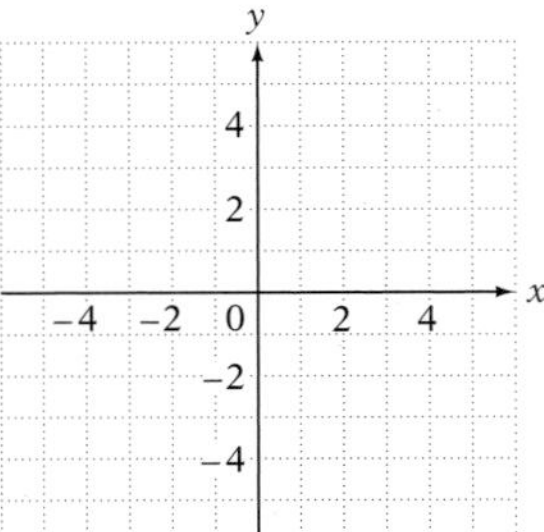

2. (0, 2), (−4, −1), (2, 0), (1, −3)

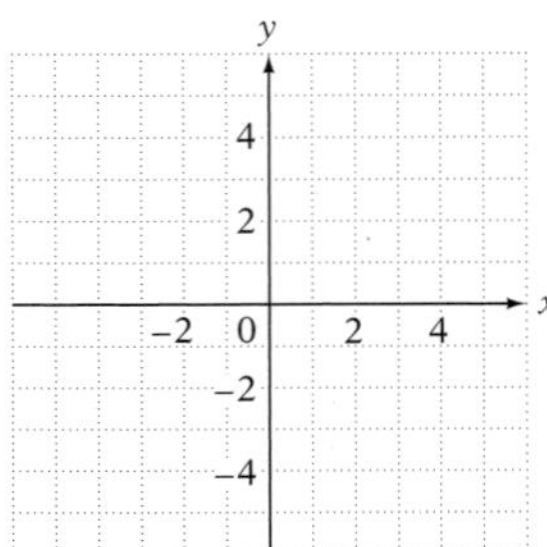

3. (−2, 5), (3, 4), (0, 0), (−3, −2)

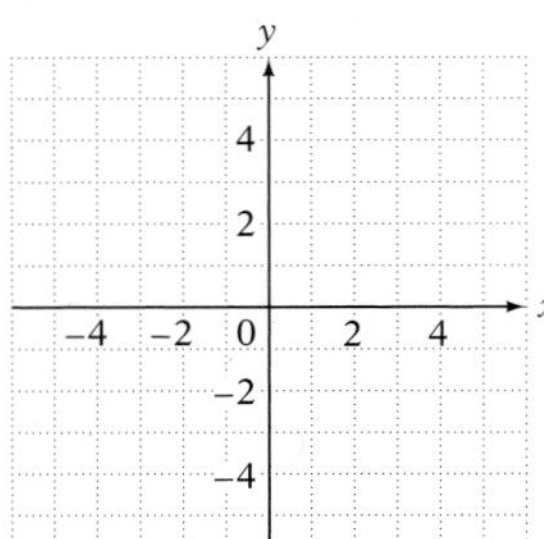

4. (1, −4), (−2, 0), (−1, −5), (0, 4)

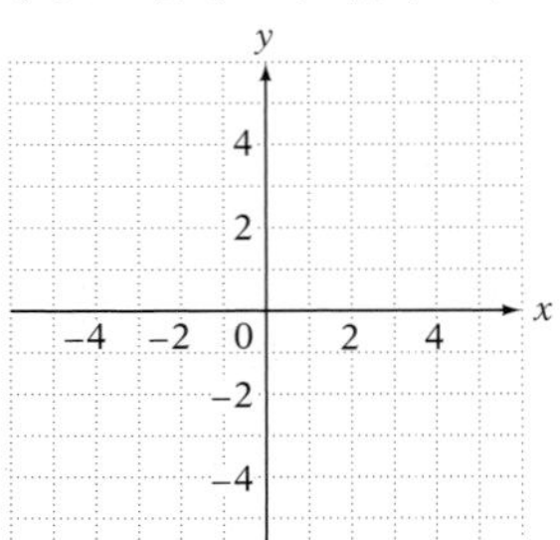

OBJECTIVE B To graph a linear equation in two variables

Graph.

5. $y = 2x + 1$

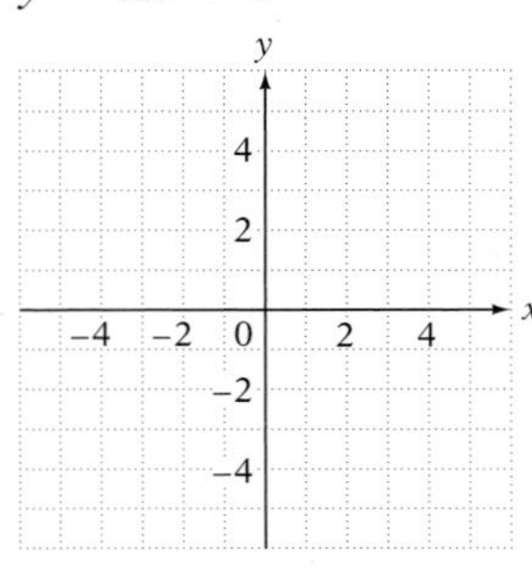

6. $y = 2x - 4$

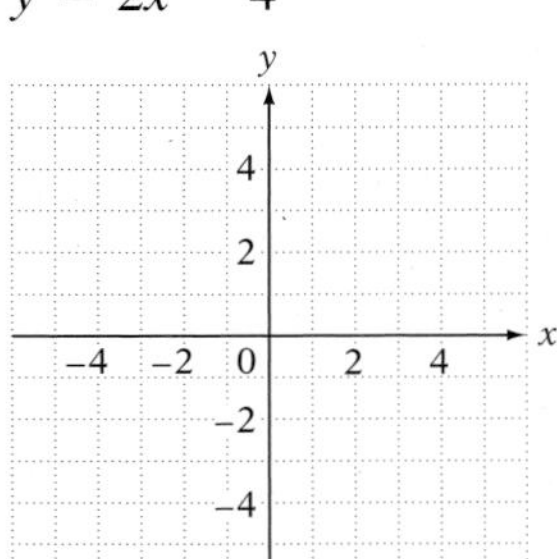

7. $y = -3x + 4$

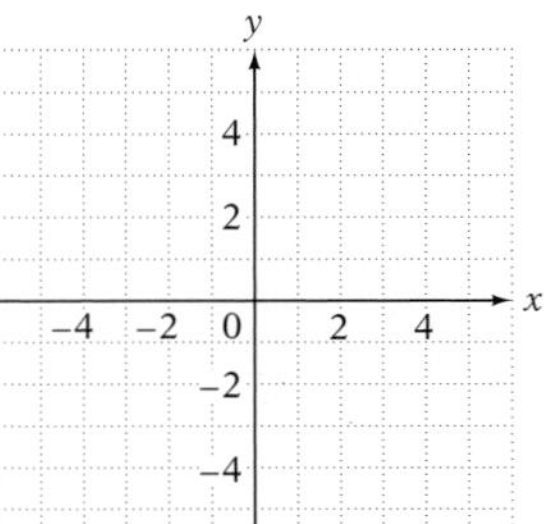

8. $y = -4x + 1$

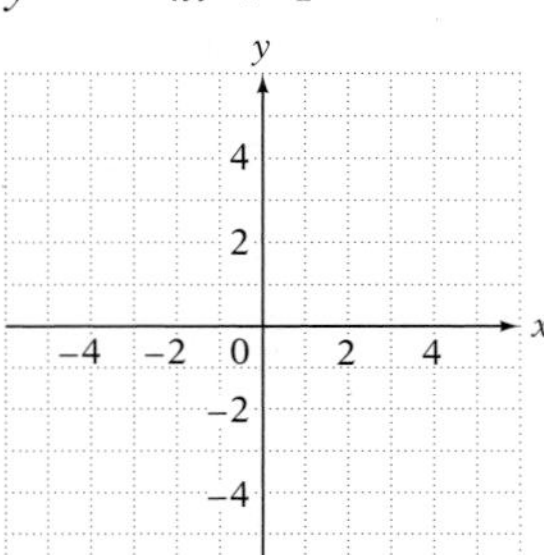

9. $y = 3x$

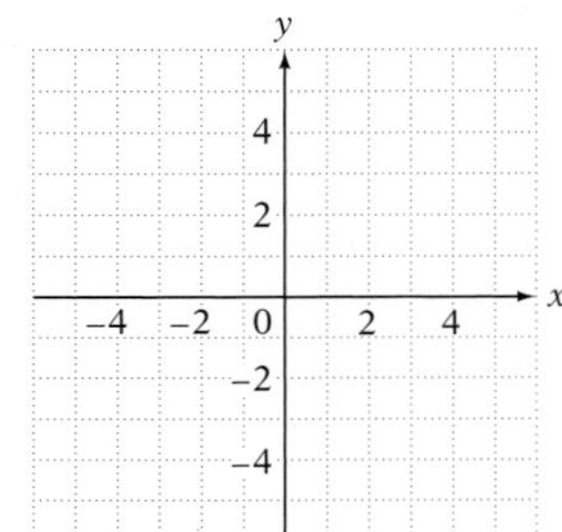

10. $y = 2x$

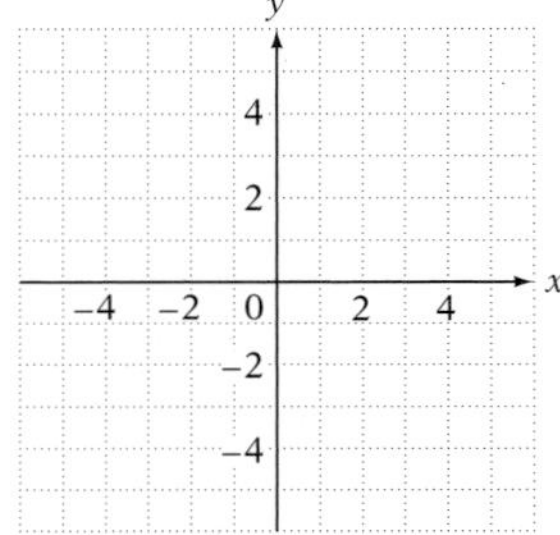

11. $y = -\frac{4}{3}x$

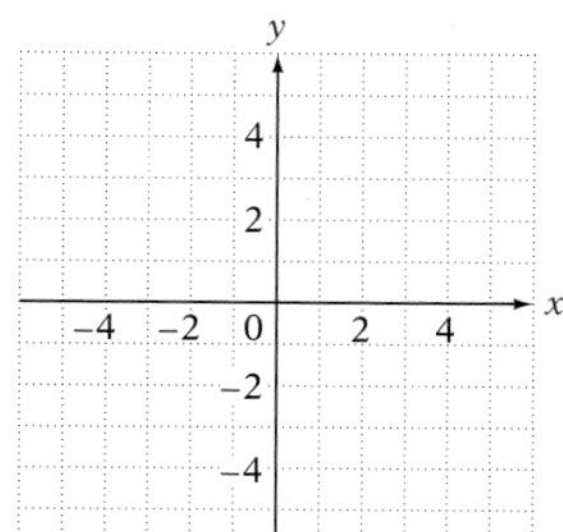

12. $y = -\frac{5}{2}x$

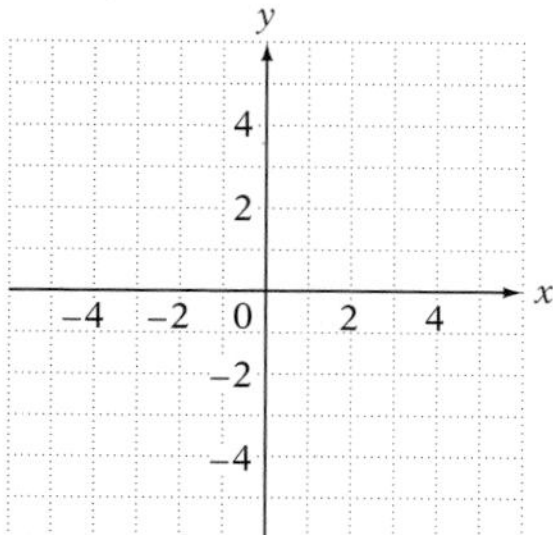

13. $y = \frac{3}{2}x - 1$

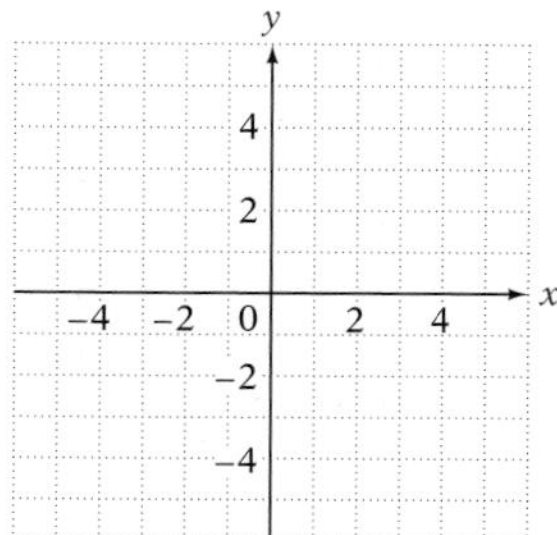

14. $y = \frac{2}{3}x + 1$

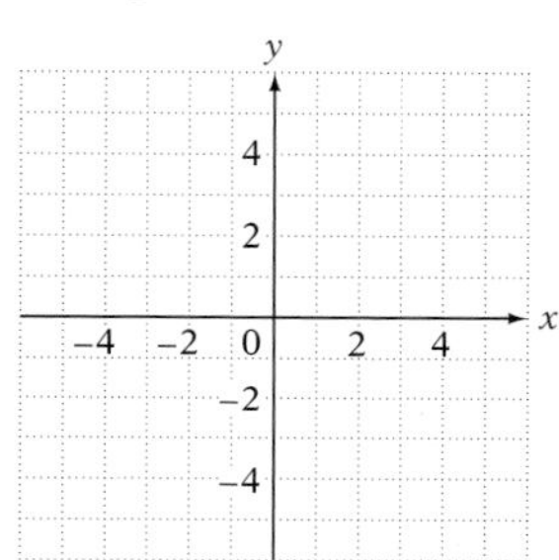

15. $y = -\frac{2}{3}x + 1$

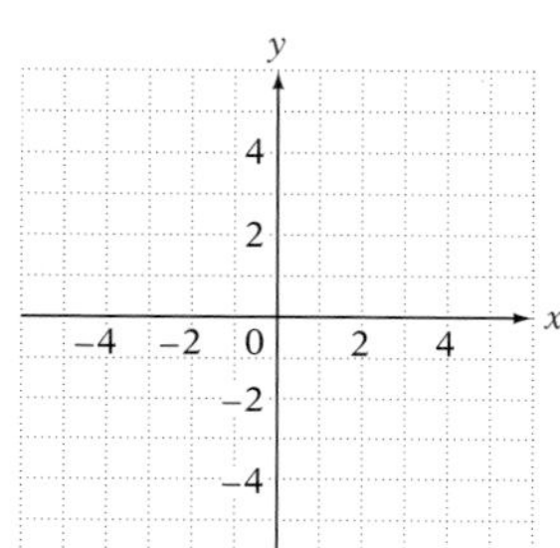

16. $y = -\frac{1}{2}x + 3$

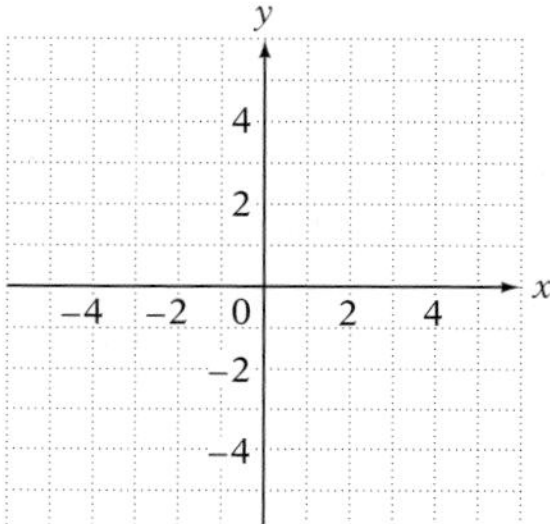

17. $2x + y = -3$

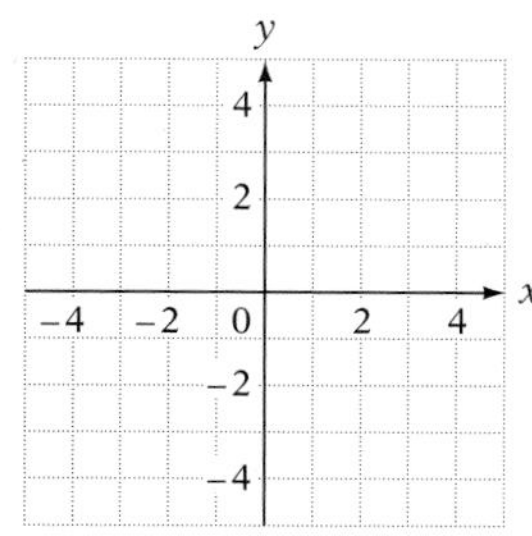

18. $2x - y = 3$

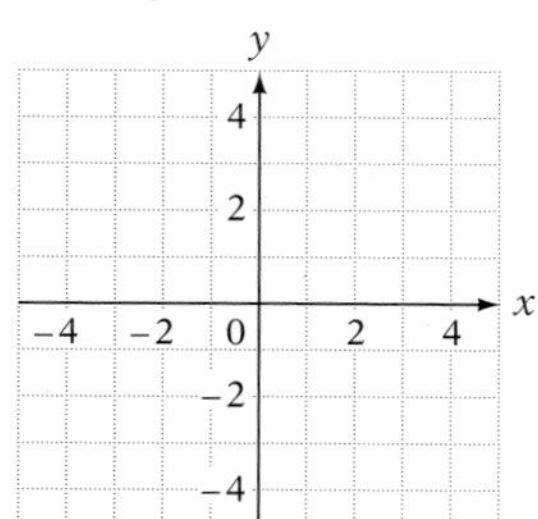

19. $x - 4y = 8$

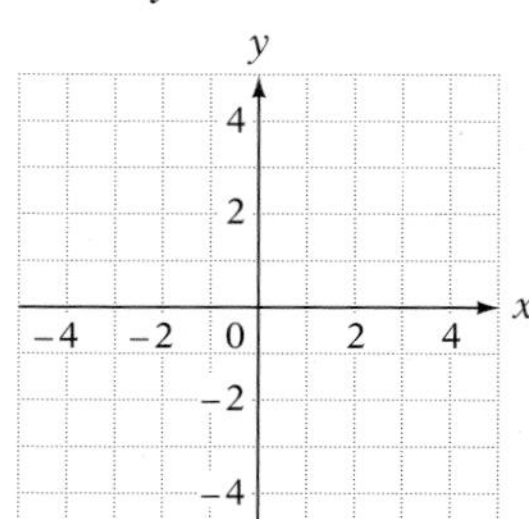

20. $2x + 5y = 10$

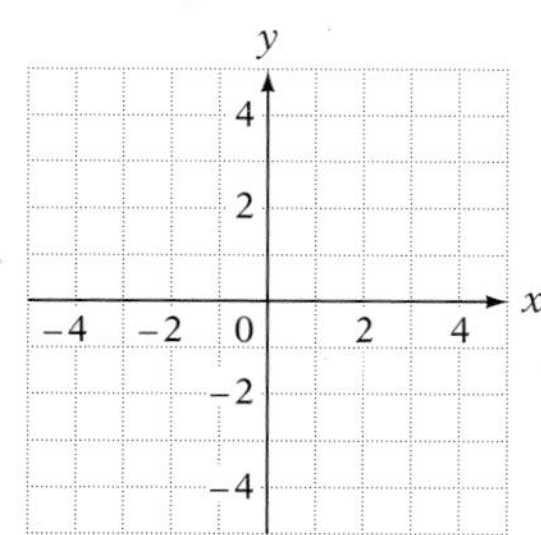

21. $3x - 2y = 8$

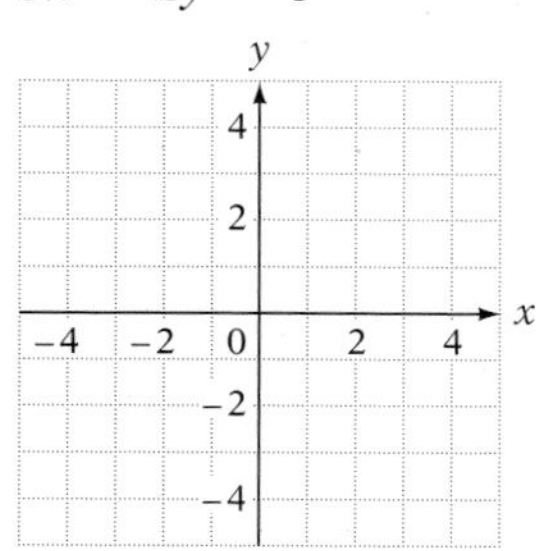

22. $3x - y = -2$

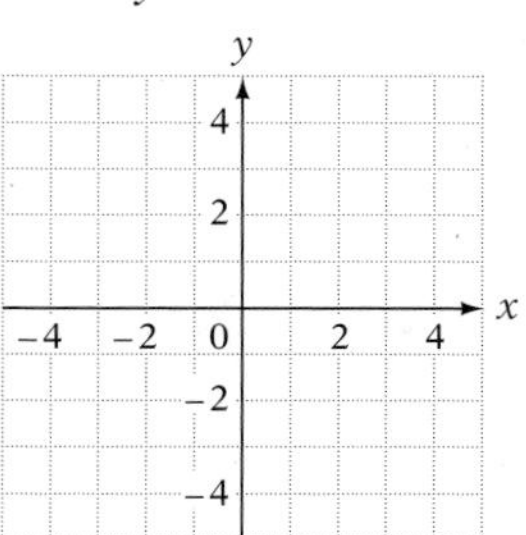

Find the slope of the line containing the given points.

23. $P_1(2, 1), P_2(3, 4)$

24. $P_1(4, 2), P_2(3, 4)$

25. $P_1(-2, 1), P_2(2, 2)$

26. $P_1(-1, 3), P_2(2, 4)$

27. $P_1(1, 3), P_2(5, -3)$

28. $P_1(2, 4), P_2(4, -1)$

29. $P_1(-1, 2), P_2(-1, 3)$

30. $P_1(3, -4), P_2(3, 5)$

31. $P_1(5, 1), P_2(-2, 1)$

32. $P_1(4, -2), P_2(3, -2)$

33. $P_1(3, 0), P_2(2, -1)$

34. $P_1(0, -1), P_2(3, -2)$

Give the slope and the y-intercept of the graph of the equation.

35. $y = \frac{5}{2}x - 4$

36. $y = -3x + 7$

37. $y = x$

Graph by using the slope and the y-intercept.

38. $y = \frac{2}{3}x - 3$

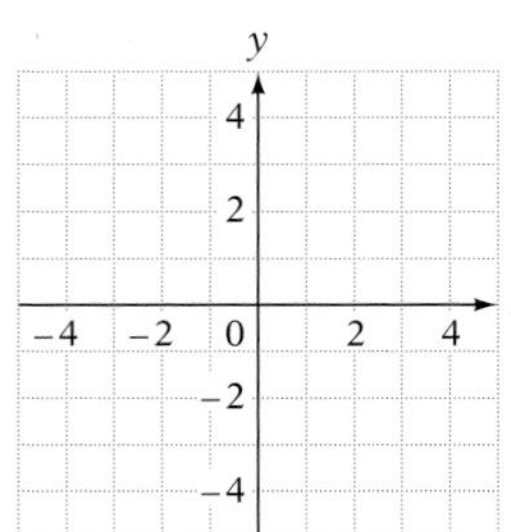

39. $y = \frac{1}{2}x + 2$

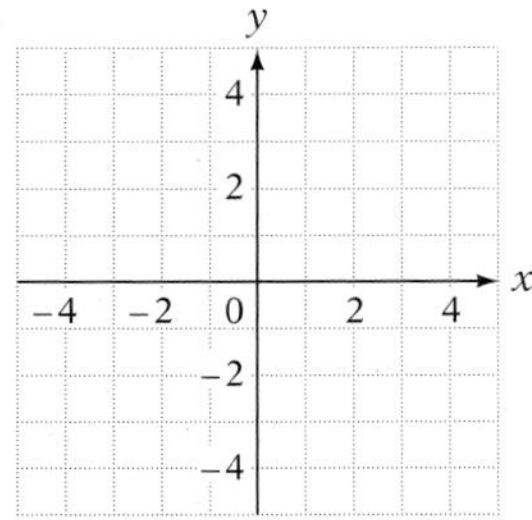

40. $y = \frac{3}{4}x$

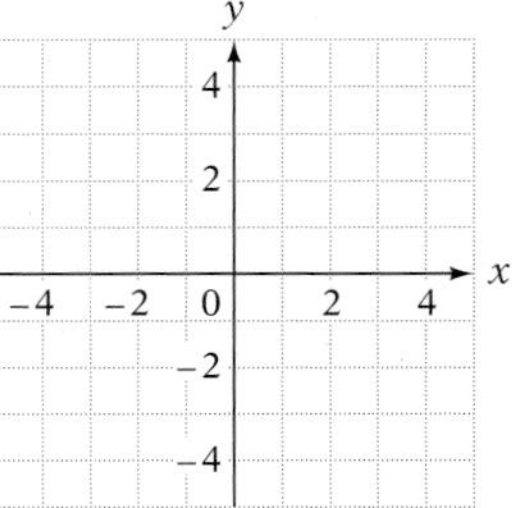

41. $y = -\frac{3}{2}x$

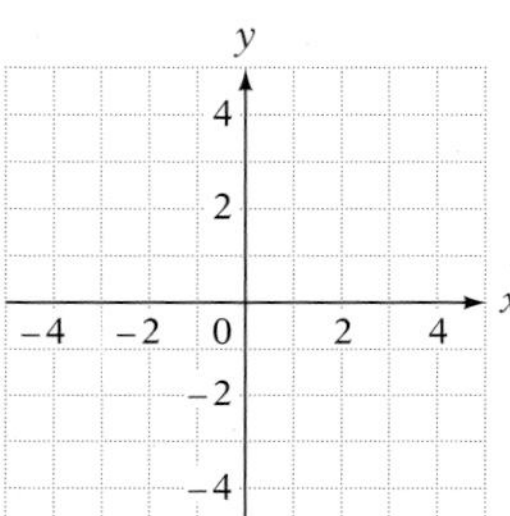

42. $y = \frac{2}{3}x - 1$

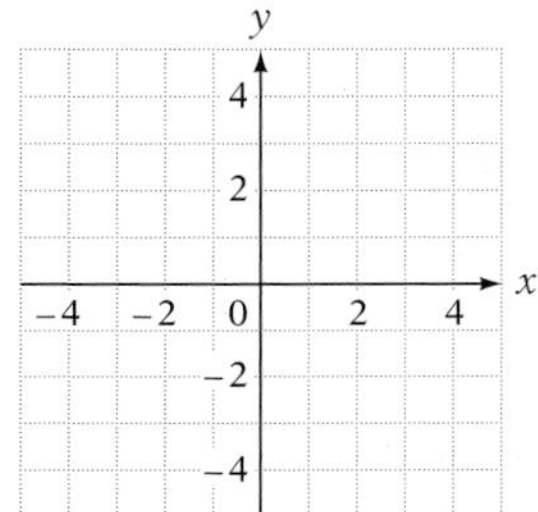

43. $y = -\frac{1}{2}x + 2$

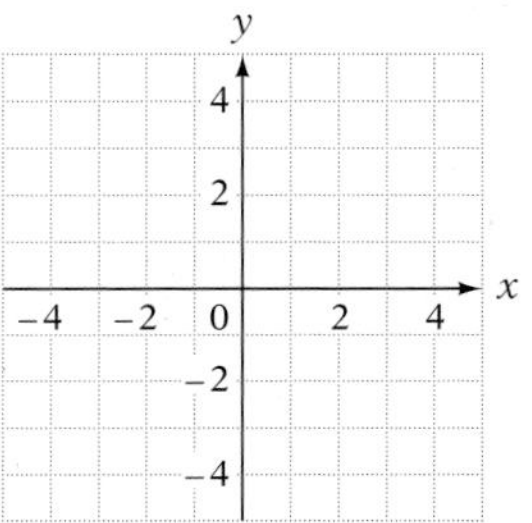

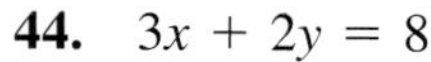

44. $3x + 2y = 8$

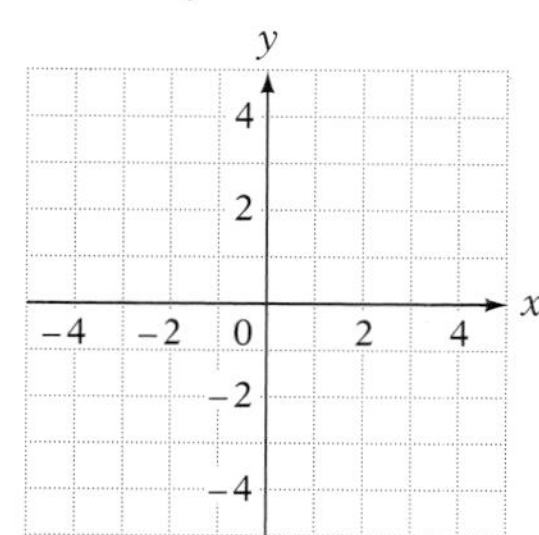

45. $x - 3y = 3$

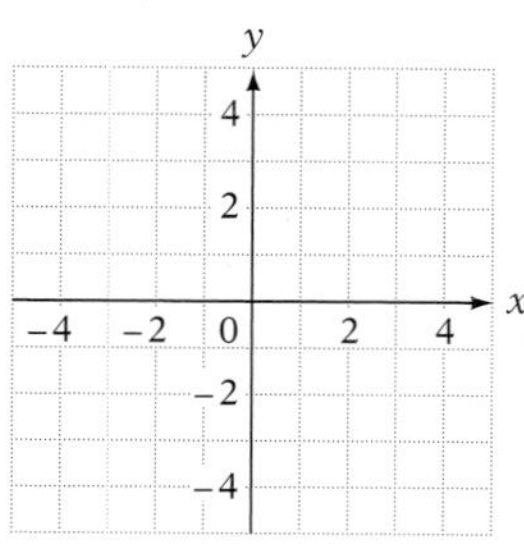

46. $4x + y = 2$

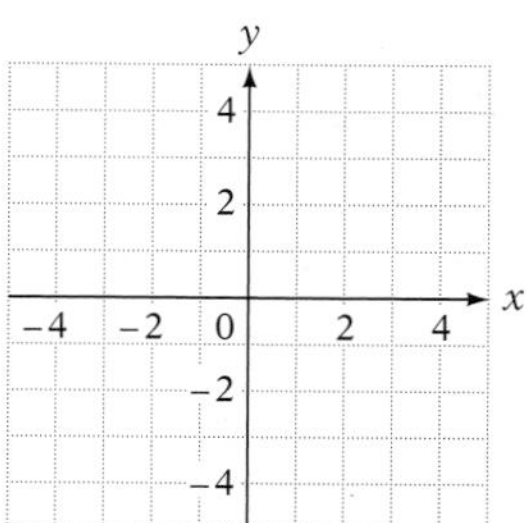

OBJECTIVE C To evaluate a function

Evaluate the function for the given value of the variable. Use your answer to write an ordered pair of the function.

47. $g(x) = -3x + 1; x = -4$

48. $f(x) = 2x + 7; x = -2$

49. $p(x) = 6 - 8x; x = -1$

50. $h(x) = 4x - 2; x = 3$

51. $f(t) = t^2 - t - 3; t = 2$

52. $t(x) = 5 - 7x; x = 0$

53. $h(x) = -3x^2 + x - 1; x = -2$

54. $p(n) = n^2 - 4n - 7; n = -3$

55. $g(t) = 4t^3 - 2t; t = -1$

OBJECTIVE D To find the equation of a line

Find the equation of the line that contains the given point and has the given slope.

56. Point $(2, -3)$; $m = 3$

57. Point $(-1, 2)$; $m = -3$

58. Point $(-3, 5)$; $m = 3$

59. Point $(4, -5)$; $m = -2$

60. Point $(3, 1)$; $m = \frac{1}{3}$

61. Point $(5, -3)$; $m = -\frac{3}{5}$

62. Point $(4, -2)$; $m = \frac{3}{4}$

63. Point $(-3, -2)$; $m = -\frac{2}{3}$

64. Point $(2, 3)$; $m = -\frac{1}{2}$

SECTION

R.4 Polynomials

OBJECTIVE A To multiply and divide monomials

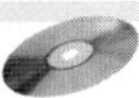

6.1A 6.1B Multiply and divide monomials

A **monomial** is a number, a variable, or a product of a number and variables. The following rules and definitions are used to multiply and divide monomials and to write monomials in simplest form.

Rule for Multiplying Exponential Expressions

If m and n are integers, then $x^m \cdot x^n = x^{m+n}$.

Rule for Simplifying Powers of Exponential Expressions

If m and n are integers, then $(x^m)^n = x^{mn}$.

Rule for Simplifying Powers of Products

If m, n, and p are integers, then $(x^m y^n)^p = x^{mp}y^{np}$.

Rule for Dividing Exponential Expressions

If m and n are integers and $x \neq 0$, then $\frac{x^m}{x^n} = x^{m-n}$.

Rule for Simplifying Powers of Quotients

If m, n, and p are integers and $y \neq 0$, then $\left(\frac{x^m}{y^n}\right)^p = \frac{x^{mp}}{y^{np}}$.

Definition of Zero as an Exponent

If $x \neq 0$, then $x^0 = 1$. The expression 0^0 is undefined.

Definition of Negative Exponents

If n is a positive integer and $x \neq 0$, then $x^{-n} = \frac{1}{x^n}$ and $\frac{1}{x^{-n}} = x^n$.

HOW TO • 1 Simplify: $(3x^4)^2(4x^3)$

$$\begin{aligned}(3x^4)^2(4x^3) &= (3^{1\cdot 2}x^{4\cdot 2})(4x^3) \\ &= (3^2x^8)(4x^3) \\ &= (9x^8)(4x^3) \\ &= (9\cdot 4)(x^8\cdot x^3) \\ &= 36x^{8+3} \\ &= 36x^{11}\end{aligned}$$

• Use the Rule for Simplifying Powers of Products to simplify $(3x^4)^2$.

• Use the Rule for Multiplying Exponential Expressions.

An exponential expression is in simplest form when there are no negative exponents in the expression. For example, the expression y^{-7} is not in simplest form; use the Definition of Negative Exponents to rewrite the expression with a positive exponent: $y^{-7} = \frac{1}{y^7}$. The expression $\frac{1}{c^{-4}}$ is not in simplest form; use the Definition of Negative Exponents to rewrite the expression with a positive exponent: $\frac{1}{c^{-4}} = c^4$.

HOW TO • 2 Simplify: $\frac{6x^2}{8x^9}$

$$\frac{6x^2}{8x^9} = \frac{3x^2}{4x^9} = \frac{3x^{2-9}}{4}$$

- Divide the coefficients by their common factors. Use the Rule for Dividing Exponential Expressions.

$$= \frac{3x^{-7}}{4} = \frac{3}{4} \cdot \frac{x^{-7}}{1} = \frac{3}{4} \cdot \frac{1}{x^7}$$

- Rewrite the expression with only positive exponents.

$$= \frac{3}{4x^7}$$

HOW TO • 3 Simplify: $\left(\frac{a^4}{b^3}\right)^{-2}$

$$\left(\frac{a^4}{b^3}\right)^{-2} = \frac{a^{4(-2)}}{b^{3(-2)}} = \frac{a^{-8}}{b^{-6}}$$

- Use the Rule for Simplifying Powers of Quotients.

$$= \frac{b^6}{a^8}$$

- Rewrite the expression with positive exponents.

EXAMPLE • 1

Simplify.

a. $(-2x)(3x^{-2})^{-3}$

b. $\left(\frac{3a^2b^{-1}}{27a^{-3}b^{-4}}\right)^{-2}$

Solution

a. $(-2x)(3x^{-2})^{-3}$

$= (-2x)(3^{-3}x^6)$

- Use the Rule for Simplifying Powers of Products.

$= \frac{-2x \cdot x^6}{3^3}$

- Write the expression with positive exponents.

$= -\frac{2x^7}{27}$

- Use the Rule for Multiplying Exponential Expressions. Simplify 3^3.

YOU TRY IT • 1

Simplify.

a. $(-2ab)(2a^3b^{-2})^{-3}$

b. $\left(\frac{2x^2y^{-4}}{4x^{-2}y^{-5}}\right)^{-3}$

Your solution

b. Use the Rule for Simplifying Powers of Quotients. Then simplify the expression and write it with positive exponents.

$$\left(\frac{3a^2b^{-1}}{27a^{-3}b^{-4}}\right)^{-2}$$

$$= \left(\frac{a^2b^{-1}}{9a^{-3}b^{-4}}\right)^{-2}$$

$$= \frac{a^{2(-2)}b^{(-1)(-2)}}{9^{1(-2)}a^{(-3)(-2)}b^{(-4)(-2)}}$$

$$= \frac{a^{-4}b^2}{9^{-2}a^6b^8}$$

$$= 9^2a^{-4-6}b^{2-8}$$

$$= 81a^{-10}b^{-6} = \frac{81}{a^{10}b^6}$$

Solution on p. S37

OBJECTIVE B To add and subtract polynomials 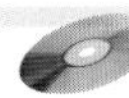

A **polynomial** is a variable expression in which the terms are monomials. The polynomial $15t^2 - 2t + 3$ has three terms: $15t^2$, $-2t$, and 3. Note that each of these three terms is a monomial.

A polynomial of *one* term is a **monomial.** $-7x^2$ is a monomial.
A polynomial of *two* terms is a **binomial.** $4y + 3$ is a binomial.
A polynomial of *three* terms is a **trinomial.** $6b^2 + 5b - 8$ is a trinomial.

6.2B Add polynomials

Polynomials can be added by combining like terms. This is illustrated in Example 2 below.

EXAMPLE • 2

Add: $(8x^2 - 4x - 9) + (2x^2 + 9x - 9)$

Solution

$(8x^2 - 4x - 9) + (2x^2 + 9x - 9)$
$= (8x^2 + 2x^2) + (-4x + 9x) + (-9 - 9)$
$= 10x^2 + 5x - 18$

YOU TRY IT • 2

Add: $(-4x^3 + 2x^2 - 8) + (4x^3 + 6x^2 - 7x + 5)$

Your solution

Solution on p. S37

6.2B Subtract polynomials

The additive inverse of the polynomial $(3x^2 - 7x + 8)$ is $-(3x^2 - 7x + 8)$.

To find the additive inverse of a polynomial, change the sign of each term inside the parentheses.

$$-(3x^2 - 7x + 8) = -3x^2 + 7x - 8$$

To subtract two polynomials, add the additive inverse of the second polynomial to the first.

HOW TO • 4 Simplify: $(5a^2 - a + 2) - (-2a^3 + 3a - 3)$

$(5a^2 - a + 2) - (-2a^3 + 3a - 3)$

$= (5a^2 - a + 2) + (2a^3 - 3a + 3)$ • Rewrite subtraction as addition of the additive inverse.

$= 2a^3 + 5a^2 - 4a + 5$ • Combine like terms.

EXAMPLE • 3

Subtract: $(7c^2 - 9c - 12) - (9c^2 + 5c - 8)$

Solution

$(7c^2 - 9c - 12) - (9c^2 + 5c - 8)$

$= (7c^2 - 9c - 12) + (-9c^2 - 5c + 8)$

$= -2c^2 - 14c - 4$

YOU TRY IT • 3

Subtract:

$(-4w^3 + 8w - 8) - (3w^3 - 4w^2 + 2w - 1)$

Your solution

Solution on p. S37

OBJECTIVE C To multiply polynomials 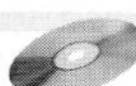

6.3A Multiply a polynomial by a monomial

The Distributive Property is used to multiply a polynomial by a monomial. Each term of the polynomial is multiplied by the monomial.

HOW TO • 5 Multiply: $3x^3(4x^4 - 2x + 5)$

$3x^3(4x^4 - 2x + 5)$

$= 3x^3(4x^4) - 3x^3(2x) + 3x^3(5)$ • Use the Distributive Property. Multiply each term of the polynomial by $3x^3$.

$= 12x^7 - 6x^4 + 15x^3$ • Use the Rule for Multiplying Exponential Expressions.

EXAMPLE • 4

Multiply: $2xy(3x^2 - xy + 2y^2)$

Solution

$2xy(3x^2 - xy + 2y^2)$

$= 2xy(3x^2) - 2xy(xy) + 2xy(2y^2)$

$= 6x^3y - 2x^2y^2 + 4xy^3$

YOU TRY IT • 4

Multiply: $3mn^2(2m^2 - 3mn - 1)$

Your solution

Solution on p. S37

6.3B Multiply polynomials

A vertical format similar to that used for multiplication of whole numbers is used to multiply two polynomials. The product $(2y - 3)(y^2 + 2y + 5)$ is shown below.

$$\begin{array}{rrrrl} & & y^2 + & 2y + 5 & \\ & & & 2y - 3 & \\ \hline & -3y^2 - & 6y - & 15 & \text{This is } -3(y^2 + 2y + 5). \\ 2y^3 + & 4y^2 + & 10y & & \text{This is } 2y(y^2 + 2y + 5). \text{ Like terms are placed in the same column.} \\ \hline 2y^3 + & y^2 + & 4y - & 15 & \text{Add the terms in each column.} \end{array}$$

EXAMPLE 5

Multiply: $(2a^2 + 4a - 5)(3a + 5)$

Solution

$$\begin{array}{rrrr} & 2a^2 + & 4a - & 5 \\ & & 3a + & 5 \\ \hline & 10a^2 + & 20a - & 25 \\ 6a^3 + & 12a^2 - & 15a & \\ \hline 6a^3 + & 22a^2 + & 5a - & 25 \end{array}$$

• Align like terms in the same column.

YOU TRY IT 5

Multiply: $(3c^2 - 4c + 5)(2c - 3)$

Your solution

Solution on p. S38

6.3B Multiply two binomials

It is frequently necessary to multiply two binomials. The product is computed by using a method called FOIL. The letters of FOIL stand for **F**irst, **O**uter, **I**nner, and **L**ast. The FOIL method is based on the Distributive Property and involves adding the products of the first terms, the outer terms, the inner terms, and the last terms.

The product $(2x + 3)(3x + 4)$ is shown below using FOIL.

First, Last, Inner, Outer

First terms, Outer terms, Inner terms, Last terms

$$(2x + 3) \cdot (3x + 4) = (2x)(3x) + (2x)(4) + (3)(3x) + (3)(4)$$

$$= 6x^2 + 8x + 9x + 12$$

$$= 6x^2 + 17x + 12$$

EXAMPLE 6

Multiply: $(4x - 3)(2x + 5)$

Solution

$(4x - 3)(2x + 5)$

$= (4x)(2x) + (4x)(5) + (-3)(2x) + (-3)(5)$

$= 8x^2 + 20x + (-6x) + (-15)$

$= 8x^2 + 14x - 15$

YOU TRY IT 6

Multiply: $(4y - 7)(3y - 5)$

Your solution

Solution on p. S38

OBJECTIVE D To divide polynomials 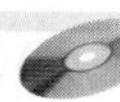

6.4B Divide polynomials

To divide two polynomials, use a method similar to that used for division of whole numbers.

To divide $(x^2 - 5x + 8) \div (x - 3)$:

Step 1

$$\begin{array}{r} x \phantom{{}-5x+8} \\ x-3\overline{)x^2 - 5x + 8} \\ \underline{x^2 - 3x} \downarrow \\ -2x + 8 \end{array}$$

Think: $x\overline{)x^2} = \dfrac{x^2}{x} = x$

Multiply: $x(x - 3) = x^2 - 3x$

Subtract: $(x^2 - 5x) - (x^2 - 3x) = -2x$

Bring down the $+8$.

Step 2

$$\begin{array}{r} x - 2 \\ x-3\overline{)x^2 - 5x + 8} \\ \underline{x^2 - 3x} \phantom{{}+8} \\ -2x + 8 \\ \underline{-2x + 6} \\ 2 \end{array}$$

Think: $x\overline{)-2x} = \dfrac{-2x}{x} = -2$

Multiply: $-2(x - 3) = -2x + 6$

Subtract: $(-2x + 8) - (-2x + 6) = 2$

The remainder is 2.

Check: (Quotient × Divisor) + Remainder = Dividend

$(x - 2)(x - 3) + 2 = x^2 - 3x - 2x + 6 + 2 = x^2 - 5x + 8$

$(x^2 - 5x + 8) \div (x - 3) = x - 2 + \dfrac{2}{x - 3}$

EXAMPLE • 7

Divide: $(6x + 2x^3 + 26) \div (x + 2)$

Solution

Arrange the terms of the dividend in descending order. There is no x^2 term in $2x^3 + 6x + 26$. Insert $0x^2$ for the missing term so that like terms will be in the same columns.

$$\begin{array}{r} 2x^2 - 4x + 14 \\ x+2\overline{)2x^3 + 0x^2 + 6x + 26} \\ \underline{2x^3 + 4x^2} \phantom{{}+6x+26} \\ -4x^2 + 6x \phantom{{}+26} \\ \underline{-4x^2 - 8x} \phantom{{}+26} \\ 14x + 26 \\ \underline{14x + 28} \\ -2 \end{array}$$

Check: $(2x^2 - 4x + 14)(x + 2) - 2$
$= 2x^3 + 6x + 28 - 2 = 2x^3 + 6x + 26$

$(6x + 2x^3 + 26) \div (x + 2) = 2x^2 - 4x + 14 - \dfrac{2}{x + 2}$

YOU TRY IT • 7

Divide: $(x^3 - 7 - 2x) \div (x - 2)$

Your solution

Solution on p. S38

OBJECTIVE E To factor polynomials of the form $ax^2 + bx + c$ 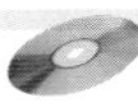

7.1A Factor a monomial from a polynomial

A polynomial is in factored form when it is written as a product of other polynomials. Factoring can be thought of as the reverse of multiplication.

Polynomial	Factored Form
$2x^3 + 6x^2 - 10x$	$= 2x(x^2 + 3x - 5)$
$x^2 - 3x - 28$	$= (x + 4)(x - 7)$

To factor out a common monomial from the terms of a polynomial, first find the greatest common factor (GCF) of the terms.

The GCF of two or more monomials is the product of the GCF of the coefficients and the common variable factors.

$$10a^3b = \mathbf{2} \cdot 5 \cdot \boldsymbol{a} \cdot \boldsymbol{a} \cdot a \cdot \boldsymbol{b}$$
$$4a^2b^2 = \mathbf{2} \cdot 2 \cdot \boldsymbol{a} \cdot \boldsymbol{a} \cdot \boldsymbol{b} \cdot b$$
$$\text{GCF} = \mathbf{2} \cdot \boldsymbol{a} \cdot \boldsymbol{a} \cdot \boldsymbol{b} = 2a^2b$$

Note that the exponent on each variable in the GCF is the same as the smallest exponent on that variable in either of the monomials.

The GCF of $10a^3b$ and $4a^2b^2$ is $2a^2b$.

HOW TO • 6 Factor: $5x^3 - 35x^2 + 10x$

The GCF is $5x$.
• **Find the GCF of the terms $5x^3$, $-35x^2$, and $10x$.**

$$\frac{5x^3 - 35x^2 + 10x}{5x} = x^2 - 7x + 2$$
• **Divide each term of the polynomial by the GCF.**

$$5x^3 - 35x^2 + 10x = 5x(x^2 - 7x + 2)$$
• **Write the polynomial as the product of the GCF and the quotient found above.**

$$5x(x^2 - 7x + 2) = 5x^3 - 35x^2 + 10x$$
• **Check the factorization by multiplying.**

EXAMPLE • 8

Factor: $16x^4y^5 + 8x^4y^2 - 12x^3y$

Solution

The GCF is $4x^3y$.

$$16x^4y^5 + 8x^4y^2 - 12x^3y = 4x^3y(4xy^4 + 2xy - 3)$$

YOU TRY IT • 8

Factor: $6x^4y^2 - 9x^3y^2 + 12x^2y^4$

Your solution

Solution on p. S38

A **quadratic trinomial** is a trinomial of the form $ax^2 + bx + c$, where a and b are coefficients and c is a constant. Examples of quadratic trinomials are shown below.

$$\begin{array}{ll} x^2 + 9x + 14 & a = 1, b = 9, c = 14 \\ x^2 - 2x - 15 & a = 1, b = -2, c = -15 \\ 3x^2 - x + 4 & a = 3, b = -1, c = 4 \end{array}$$

To **factor a quadratic trinomial** means to express the trinomial as the product of two binomials. For example,

Trinomial		**Factored Form**
$2x^2 - x - 1$	=	$(2x + 1)(x - 1)$
$y^2 - 3y + 2$	=	$(y - 1)(y - 2)$

7.2A Factor a trinomial of the form $x^2 + bx + c$

We will begin by factoring trinomials of the form $x^2 + bx + c$, where the coefficient of x^2 is 1.

The method by which factors of a trinomial are found is based on FOIL. Consider the following binomial products, noting the relationship between the constant terms of the binomials and the terms of the trinomial.

Sum of the binomial constants

Product of the binomial constants

$$\begin{array}{ll} (x + 4)(x + 5) = x \cdot x + 5x + 4x + 4 \cdot 5 & = x^2 + 9x + 20 \\ (x - 6)(x + 8) = x \cdot x + 8x - 6x + (-6)(8) & = x^2 + 2x - 48 \\ (x - 3)(x - 2) = x \cdot x - 2x - 3x + (-3)(-2) & = x^2 - 5x + 6 \end{array}$$

Take Note

Once the correct factors are found, it is not necessary to try the remaining factors.

Take Note

Always check the proposed factorization by multiplying the factors.

HOW TO • 7 Factor: $x^2 - 7x + 10$

Find two integers whose product is 10 and whose sum is -7.

Negative Factors of 10	**Sum**	
$-1, -10$	-11	
$\mathbf{-2, -5}$	$\mathbf{-7}$	• These are the correct factors.

$x^2 - 7x + 10 = (x - 2)(x - 5)$ • Write the trinomial as a product of factors.

Check:

$$\begin{aligned} (x - 2)(x - 5) &= x^2 - 5x - 2x + 10 \\ &= x^2 - 7x + 10 \end{aligned}$$

EXAMPLE • 9

Factor: $x^2 + 6x - 27$

Solution

Two factors of -27 whose sum is 6 are -3 and 9.

$x^2 + 6x - 27 = (x - 3)(x + 9)$

YOU TRY IT • 9

Factor: $x^2 - 8x + 15$

Your solution

Solution on p. S38

7.3A Factor a trinomial of the form $ax^2 + bx + c$ by using trial factors

To use the trial factor method to factor a trinomial of the form $ax^2 + bx + c$, where $a \neq 1$, use the factors of a and the factors of c to write all of the possible binomial factors of the trinomial. Then use FOIL to determine the correct factorization. To reduce the number of trial factors that must be considered, remember the following guidelines.

Use the signs of the constant term and the coefficient of x in the trinomial to determine the signs of the binomial factors. If the constant term is positive, the signs of the binomial factors will be the same as the sign of the coefficient of x in the trinomial. If the sign of the constant term is negative, the constant terms in the binomials will have different signs.

HOW TO 8 Factor: $2x^2 - 7x + 3$

Because the constant term is positive (+3) and the coefficient of x is negative (−7), the binomial constants will be negative.

Positive Factors of 2 (a = 2)	**Negative Factors of 3 (c = 3)**
1, 2	−1, −3

Write trial factors. Use the **O**uter and **I**nner products of FOIL to determine the middle term, $-7x$, of the trinomial.

Trial Factors	**Middle Term**	
$(x - 1)(2x - 3)$	$-3x - 2x = -5x$	
$\mathbf{(x - 3)(2x - 1)}$	$\mathbf{-x - 6x = -7x}$	• $-7x$ is the middle term.

$2x^2 - 7x + 3 = (x - 3)(2x - 1)$

Check: $(x - 3)(2x - 1) = 2x^2 - x - 6x + 3 = 2x^2 - 7x + 3$

HOW TO 9 Factor: $5x^2 + 22x - 15$

The constant term is negative (−15). The binomial constants will have different signs.

Positive Factors of 5 (a = 5)	**Factors of −15 (c = −15)**
1, 5	−1, 15
	1, −15
	−3, 5
	3, −5

Write trial factors. Use the **O**uter and **I**nner products of FOIL to determine the middle term, $22x$, of the trinomial.

Trial Factors	**Middle Term**	
$(x - 1)(5x + 15)$	common factor	
$(x + 15)(5x - 1)$	$-x + 75x = 74x$	
$(x + 1)(5x - 15)$	common factor	
$(x - 15)(5x + 1)$	$x - 75x = -74x$	
$(x - 3)(5x + 5)$	common factor	
$\mathbf{(x + 5)(5x - 3)}$	$\mathbf{-3x + 25x = 22x}$	• $22x$ is the middle term.
$(x + 3)(5x - 5)$	common factor	
$(x - 5)(5x + 3)$	$3x - 25x = -22x$	

$5x^2 + 22x - 15 = (x + 5)(5x - 3)$

Check: $(x + 5)(5x - 3) = 5x^2 - 3x + 25x - 15 = 5x^2 + 22x - 15$

Take Note

It is not necessary to test trial factors that have a common factor. If the trinomial does not have a common factor, then its factors cannot have a common factor.

EXAMPLE 10

Factor: $2x^2 + 3x - 5$

Solution

Positive Factors of 2	Factors of −5
1, 2	1, −5
	−1, 5

Trial Factors	Middle Term
$(x + 1)(2x - 5)$	$-5x + 2x = -3x$
$(x - 5)(2x + 1)$	$x - 10x = -9x$
$(x - 1)(2x + 5)$	$5x - 2x = 3x$
$(x + 5)(2x - 1)$	$-x + 10x = 9x$

$2x^2 + 3x - 5 = (x - 1)(2x + 5)$

Check: $(x - 1)(2x + 5) = 2x^2 + 5x - 2x - 5$
$= 2x^2 + 3x - 5$

YOU TRY IT 10

Factor: $3x^2 - x - 2$

Your solution

Solution on p. S38

7.4D Factor completely

A polynomial is factored completely when it is written as a product of factors that are nonfactorable over the integers.

The first step in any factoring problem is to determine whether the terms of the polynomial have a common factor. If they do, factor it out first.

EXAMPLE 11

Factor: $5x^2y + 60xy + 100y$

Solution

There is a common factor, $5y$.
Factor out the GCF.

$5x^2y + 60xy + 100y = 5y(x^2 + 12x + 20)$

Factor $x^2 + 12x + 20$. The two factors of 20 whose sum is 12 are 2 and 10.

$5y(x^2 + 12x + 20) = 5y(x + 2)(x + 10)$

$5x^2y + 60xy + 100y = 5y(x + 2)(x + 10)$

Check:
$5y(x + 2)(x + 10) = (5xy + 10y)(x + 10)$
$= 5x^2y + 50xy + 10xy + 100y$
$= 5x^2y + 60xy + 100y$

YOU TRY IT 11

Factor: $4a^3 - 4a^2 - 24a$

Your solution

Solution on p. S38

R.4 EXERCISES

OBJECTIVE A **To multiply and divide monomials**

Simplify.

1. $z^3 \cdot z \cdot z^4$

2. $b \cdot b^2 \cdot b^6$

3. $(x^3)^5$

4. $(b^2)^4$

5. $(x^2y^3)^6$

6. $(m^4n^2)^3$

7. $\dfrac{a^8}{a^2}$

8. $\dfrac{c^{12}}{c^5}$

9. $(-m^3n)(m^6n^2)$

10. $(-r^4t^3)(r^2t^9)$

11. $(-2a^3bc^2)^3$

12. $(-4xy^3z^2)^2$

13. $\dfrac{m^4n^7}{m^3n^5}$

14. $\dfrac{a^5b^6}{a^3b^2}$

15. $\dfrac{-16a^7}{24a^6}$

16. $\dfrac{18b^5}{-45b^4}$

17. $(9mn^4p)(-3mp^2)$

18. $(-3v^2wz)(-4vz^4)$

19. $(-2n^2)(-3n^4)^3$

20. $(-3m^3n)(-2m^2n^3)^3$

21. $\dfrac{14x^4y^6z^2}{16x^3y^9z}$

22. $\dfrac{25x^4y^7z^2}{20x^5y^9z^{11}}$

23. $(-2x^3y^2)^3(-xy^2)^4$

24. $(-m^4n^2)^5(-2m^3n^3)^3$

25. $4x^{-7}$

26. $-6y^{-1}$

27. $d^{-4}d^{-6}$

28. $x^{-3}x^{-5}$

29. $\dfrac{x^{-3}}{x^2}$

30. $\dfrac{x^4}{x^{-5}}$

31. $\dfrac{1}{3x^{-2}}$

32. $\dfrac{2}{5c^{-6}}$

33. $(x^2y^{-4})^3$

34. $(x^3y^5)^{-4}$

35. $(3x^{-1}y^{-2})^2$

36. $(5xy^{-3})^{-2}$

37. $(2x^{-1})(x^{-3})$

38. $(-2x^{-5})(x^7)$

39. $\dfrac{3x^{-2}y^2}{6xy^2}$

40. $\dfrac{2x^{-2}y}{8xy}$

41. $\dfrac{2x^{-1}y^{-4}}{4xy^2}$

42. $\dfrac{3a^{-2}b}{ab}$

43. $(x^{-2}y)^2(xy)^{-2}$

44. $(x^{-1}y^2)^{-3}(x^2y^{-4})^{-3}$

45. $\left(\dfrac{x^2y^{-1}}{xy}\right)^{-4}$

46. $\left(\dfrac{x^{-2}y^{-4}}{x^{-2}y}\right)^{-2}$

47. $\left(\dfrac{4a^{-2}b}{8a^3b^{-4}}\right)^2$

48. $\left(\dfrac{6ab^{-2}}{3a^{-2}b}\right)^{-2}$

OBJECTIVE B **To add and subtract polynomials**

Add or subtract.

49. $(4b^2 - 5b) + (3b^2 + 6b - 4)$

50. $(2c^2 - 4) + (6c^2 - 2c + 4)$

51. $(2a^2 - 7a + 10) + (a^2 + 4a + 7)$

52. $(-6x^2 + 7x + 3) + (3x^2 + x + 3)$

53. $(x^2 - 2x + 1) - (x^2 + 5x + 8)$

54. $(3x^2 + 2x - 2) - (5x^2 - 5x + 6)$

55. $(-2x^3 + x - 1) - (-x^2 + x - 3)$

56. $(2x^2 + 5x - 3) - (3x^3 + 2x - 5)$

57. $(x^3 - 7x + 4) + (2x^2 + x - 10)$

58. $(3y^3 + y^2 + 1) + (-4y^3 - 6y - 3)$

59. $(5x^3 + 7x - 7) + (10x^2 - 8x + 3)$

60. $(3y^3 + 4y + 9) + (2y^2 + 4y - 21)$

61. $(2y^3 + 6y - 2) - (y^3 + y^2 + 4)$

62. $(-2x^2 - x + 4) - (-x^3 + 3x - 2)$

63. $(4y^3 - y - 1) - (2y^2 - 3y + 3)$

64. $(3x^2 - 2x - 3) - (2x^3 - 2x^2 + 4)$

OBJECTIVE C To multiply polynomials

Multiply.

65. $4b(3b^3 - 12b^2 - 6)$

66. $-2a^2(3a^2 - 2a + 3)$

67. $3b(3b^4 - 3b^2 + 8)$

68. $-2x^2(2x^2 - 3x - 7)$

69. $-2x^2y(x^2 - 3xy + 2y^2)$

70. $3ab^2(3a^2 - 2ab + 4b^2)$

71. $(x^2 + 3x + 2)(x + 1)$

72. $(x^2 - 2x + 7)(x - 2)$

73. $(a - 3)(a^2 - 3a + 4)$

74. $(2x - 3)(x^2 - 3x + 5)$

75. $(-2b^2 - 3b + 4)(b - 5)$

76. $(-a^2 + 3a - 2)(2a - 1)$

77. $(x^3 - 3x + 2)(x - 4)$

78. $(y^3 + 4y^2 - 8)(2y - 1)$

79. $(y + 2)(y^3 + 2y^2 - 3y + 1)$

80. $(2a - 3)(2a^3 - 3a^2 + 2a - 1)$

81. $(a - 3)(a + 4)$

82. $(b - 6)(b + 3)$

83. $(y - 7)(y - 3)$

84. $(a - 8)(a - 9)$

85. $(2x + 1)(x + 7)$

86. $(y + 2)(5y + 1)$

87. $(3x - 1)(x + 4)$

88. $(7x - 2)(x + 4)$

89. $(4x - 3)(x - 7)$

90. $(2x - 3)(4x - 7)$

91. $(3y - 8)(y + 2)$

92. $(5y - 9)(y + 5)$

93. $(7a - 16)(3a - 5)$

94. $(5a - 12)(3a - 7)$

95. $(x + y)(2x + y)$

96. $(2a + b)(a + 3b)$

97. $(3x - 4y)(x - 2y)$

98. $(2a - b)(3a + 2b)$

99. $(5a - 3b)(2a + 4b)$

100. $(2x + 3)(2x - 3)$

101. $(4x - 7)(4x + 7)$

OBJECTIVE D To divide polynomials

Divide.

102. $(b^2 - 14b + 49) \div (b - 7)$

103. $(x^2 - x - 6) \div (x - 3)$

104. $(2x^2 + 5x + 2) \div (x + 2)$

105. $(2y^2 - 13y + 21) \div (y - 3)$

106. $(x^2 + 1) \div (x - 1)$

107. $(x^2 + 4) \div (x + 2)$

108. $(6x^2 - 7x) \div (3x - 2)$

109. $(6y^2 + 2y) \div (2y + 4)$

110. $(a^2 + 5a + 10) \div (a + 2)$

111. $(b^2 - 8b - 9) \div (b - 3)$

112. $(2y^2 - 9y + 8) \div (2y + 3)$

113. $(3x^2 + 5x - 4) \div (x - 4)$

114. $(8x + 3 + 4x^2) \div (2x - 1)$

115. $(10 + 21y + 10y^2) \div (2y + 3)$

116. $(x^3 + 3x^2 + 5x + 3) \div (x + 1)$

117. $(x^3 - 6x^2 + 7x - 2) \div (x - 1)$

118. $(x^4 - x^2 - 6) \div (x^2 + 2)$

119. $(x^4 + 3x^2 - 10) \div (x^2 - 2)$

OBJECTIVE E To factor polynomials of the form $ax^2 + bx + c$

Factor.

120. $8x + 12$

121. $12y^2 - 5y$

122. $10x^4 - 12x^2$

123. $10x^2yz^2 + 15xy^3z$

124. $x^3 - 3x^2 - x$

125. $5x^2 - 15x + 35$

126. $3x^3 + 6x^2 + 9x$

127. $3y^4 - 9y^3 - 6y^2$

128. $2x^3 + 6x^2 - 14x$

129. $x^4y^4 - 3x^3y^3 + 6x^2y^2$

130. $4x^5y^5 - 8x^4y^4 + x^3y^3$

131. $16x^2y - 8x^3y^4 - 48x^2y^2$

132. $x^2 + 5x + 6$

133. $x^2 + x - 2$

134. $x^2 + x - 6$

135. $a^2 + a - 12$

136. $a^2 - 2a - 35$

137. $a^2 - 3a + 2$

138. $a^2 - 5a + 4$

139. $b^2 + 7b - 8$

140. $y^2 + 6y - 55$

141. $z^2 - 4z - 45$

142. $y^2 - 8y + 15$

143. $z^2 - 14z + 45$

144. $p^2 + 12p + 27$

145. $b^2 + 9b + 20$

146. $y^2 - 8y + 32$

147. $y^2 - 9y + 81$

148. $p^2 + 24p + 63$

149. $x^2 - 15x + 56$

150. $5x^2 + 6x + 1$

151. $2y^2 + 7y + 3$

152. $2a^2 - 3a + 1$

153. $3a^2 - 4a + 1$

154. $4x^2 - 3x - 1$

155. $2x^2 - 5x - 3$

156. $6t^2 - 11t + 4$

157. $10t^2 + 11t + 3$

158. $8x^2 + 33x + 4$

159. $10z^2 + 3z - 4$

160. $3x^2 + 14x - 5$

161. $3z^2 + 95z + 10$

162. $8z^2 - 36z + 1$

163. $2t^2 - t - 10$

164. $2t^2 + 5t - 12$

165. $12y^2 + 19y + 5$

166. $5y^2 - 22y + 8$

167. $11a^2 - 54a - 5$

168. $4z^2 + 11z + 6$

169. $6b^2 - 13b + 6$

170. $6x^2 + 35x - 6$

171. $3x^2 + 15x + 18$

172. $3a^2 + 3a - 18$

173. $ab^2 + 7ab - 8a$

174. $3y^3 - 15y^2 + 18y$

175. $2y^4 - 26y^3 - 96y^2$

176. $3y^4 + 54y^3 + 135y^2$

177. $2x^3 - 11x^2 + 5x$

178. $2x^3 + 3x^2 - 5x$

179. $10t^2 - 5t - 50$

180. $16t^2 + 40t - 96$

181. $6p^3 + 5p^2 + p$

182. $12x^2y - 36xy + 27y$

Appendix A

Keystroke Guide for the TI-84 Plus

Basic Operations

Numerical calculations are performed on the **home screen.** You can always return to the home screen by pressing [2ND] QUIT. Pressing [CLEAR] erases the home screen.

To evaluate the expression $-2(3 + 5) - 8 \div 4$, use the following keystrokes.

[(-)] 2 [(] 3 [+] 5 [)] [−] 8 [÷] 4 [ENTER]

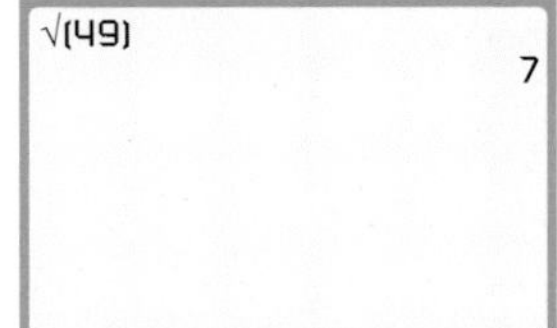

Note: There is a difference between the key to enter a negative number, [(-)], and the key for subtraction, [−]. You cannot use these keys interchangeably.

✓ Take Note

The descriptions in the margins (for example, Basic Operations and Complex Numbers) are the same as those used in the text and are arranged alphabetically.

The [2ND] key is used to access the commands in gold writing above a key. For instance, to evaluate $\sqrt{49}$, press [2ND] $\sqrt{\ }$ 49 [)] [ENTER].

The [ALPHA] key is used to place a letter on the screen. One reason to do this is to store a value of a variable. The following keystrokes give A the value of 5.

5 [STO▸] [ALPHA] A [ENTER]

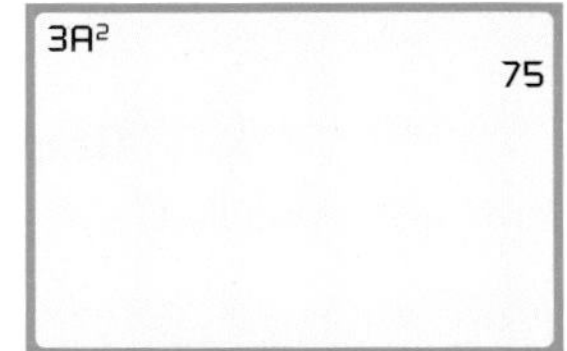

This value is now available in calculations. For instance, we can find the value of $3a^2$ by using the following keystrokes: 3 [ALPHA] A [x^2]. To display the value of the variable on the screen, press [2ND] RCL [ALPHA] A.

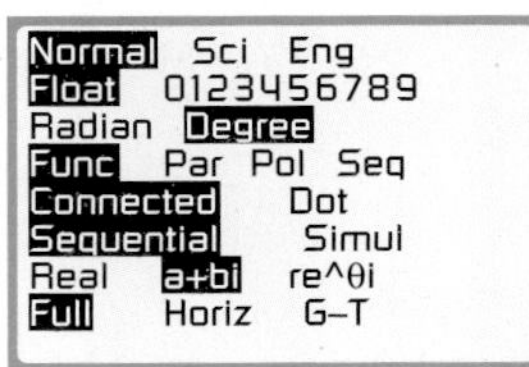

Note: When you use the [ALPHA] key, only capital letters are available on the TI-83 calculator.

Complex Numbers

To perform operations on complex numbers, first press [MODE] and then use the arrow keys to select a+bi. Then press [ENTER] [2ND] QUIT.

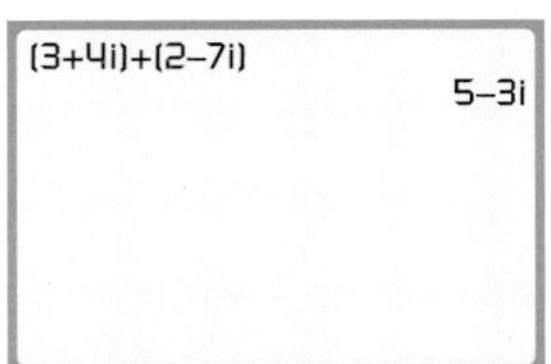

Addition of complex numbers To add $(3 + 4i) + (2 - 7i)$, use the keystrokes

[(] 3 [+] 4 [2ND] i [)] [+]
[(] 2 [−] 7 [2ND] i [)] [ENTER].

Division of complex numbers To divide $\frac{26 + 2i}{2 + 4i}$, use the keystrokes [(] 26 [+] 2 [2ND] i [)] [÷] [(] 2 [+] 4 [2ND] i [)] [ENTER].

(26+2i)/(2+4i)
3–5i

Note: Operations for subtraction and multiplication are similar.

Additional operations on complex numbers can be found by selecting CPX under the [MATH] key.

To find the absolute value of $2 - 5i$, press [MATH] (scroll to CPX) (scroll to abs) [ENTER] [(] 2 [−] 5 [2ND] i [)] [ENTER].

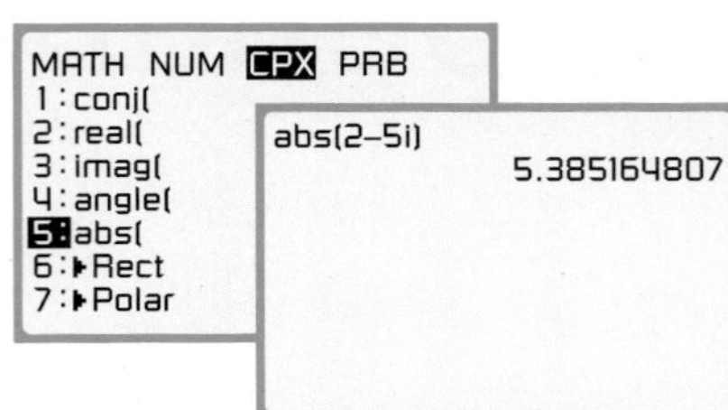

Evaluating Functions

There are various methods of evaluating a function but all methods require that the expression be entered as one of the ten functions Y_1 to Y_0. To evaluate $f(x) = \frac{x^2}{x - 1}$ when $x = -3$, enter the expression into, for instance, Y_1, and then press [VARS] [▸] 1 1 [(] [(−)] 3 [)] [ENTER].

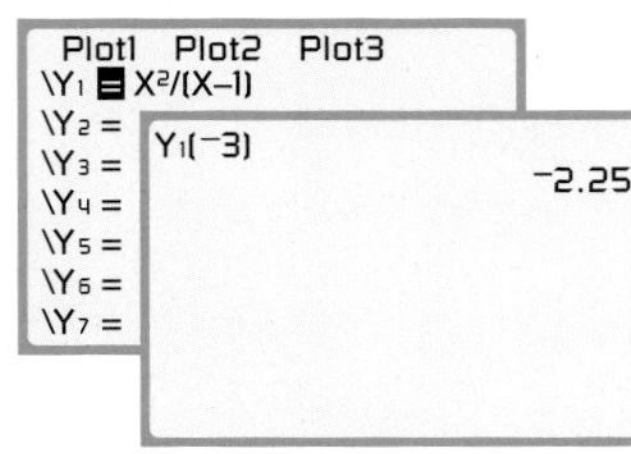

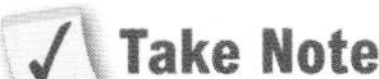

Use the down arrow key to scroll past Y_7 to see Y_8, Y_9, and Y_0.

Note: If you try to evaluate a function at a number that is not in the domain of the function, you will get an error message. For instance, 1 is not in the domain of $f(x) = \frac{x^2}{x - 1}$. If we try to evaluate the function at 1, the error screen at the right appears.

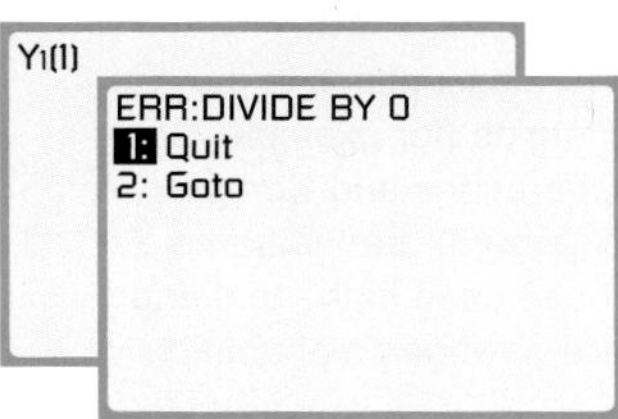

Evaluating Variable Expressions

To evaluate a variable expression, first store the values of each variable. Then enter the variable expression. For instance, to evaluate $s^2 + 2sl$ when $s = 4$ and $l = 5$, use the following keystrokes.

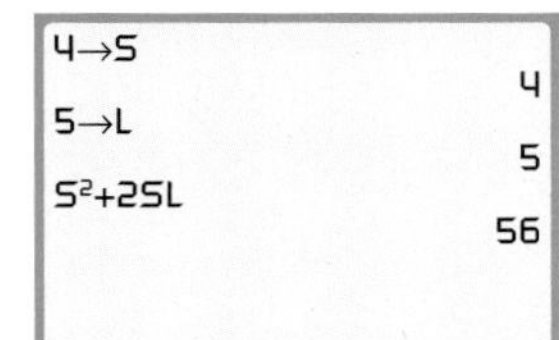

4 [STO▸] [ALPHA] S [ENTER] 5 [STO▸] [ALPHA] L [ENTER] [ALPHA] S [x²] [+] 2 [ALPHA] S [ALPHA] L [ENTER]

Graph

To graph a function, use the [Y=] key to enter the expression for the function, select a suitable viewing window, and then press [GRAPH]. For instance, to graph $f(x) = 0.1x^3 - 2x - 1$ in the standard viewing window, use the following keystrokes.

[Y=] 0.1 [X,T,θ,n] [^] 3 [−] 2 [X,T,θ,n] [−] 1 [ZOOM] (scroll to 6) [ENTER]

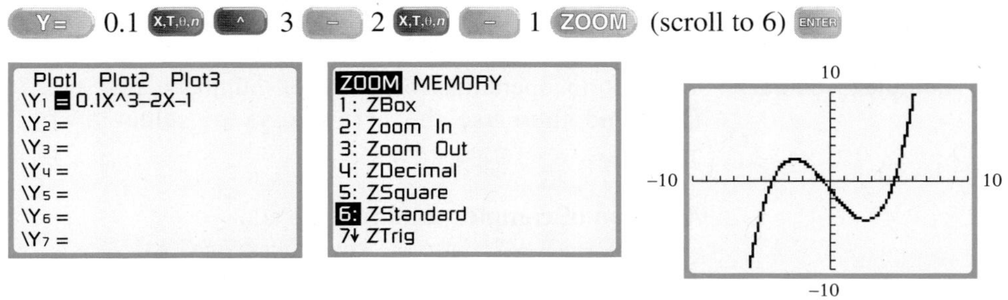

Note: For the keystrokes above, you do not have to scroll to 6. Alternatively, use [ZOOM] 6. This will select the standard viewing window and automatically start the graph. Use the [WINDOW] key to create a custom window for a graph.

Graphing Inequalities

To illustrate this feature, we will graph $y \le 2x - 1$. Enter $2x - 1$ into Y_1. Because $y \le 2x - 1$, we want to shade below the graph. Move the cursor to the left of Y_1 and press [ENTER] three times. Press [GRAPH].

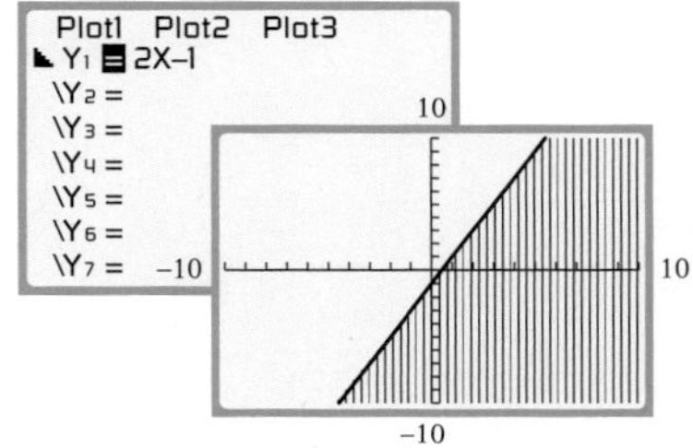

Note: To shade above the graph, move the cursor to the left of Y1 and press ENTER two times. An inequality with the symbol $\leq$ or $\geq$ should be graphed with a solid line, and an inequality with the symbol $<$ or $>$ should be graphed with a dashed line. However, a graphing calculator does not distinguish between a solid line and a dashed line.

To graph the solution set of a system of inequalities, solve each inequality for y and graph each inequality. The solution set is the intersection of the two inequalities. The solution set of $\begin{matrix} 3x + 2y > 10 \\ 4x - 3y \leq 5 \end{matrix}$ is shown at the right.

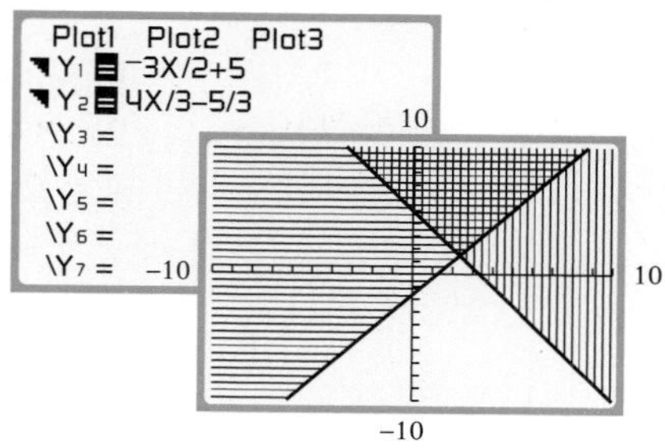

Intersect The INTERSECT feature is used to solve a system of equations. To illustrate this feature, we will use the system of equations $\begin{matrix} 2x - 3y = 13 \\ 3x + 4y = -6 \end{matrix}$.

Note: Some equations can be solved by this method. See the section "Solve an equation" on the next page. Also, this method is used to find a number in the domain of a function for a given number in the range. See the section "Find a domain element" on the next page.

Solve each of the equations in the system of equations for y. In this case, we have $y = \frac{2}{3}x - \frac{13}{3}$ and $y = -\frac{3}{4}x - \frac{3}{2}$.

Use the Y-editor to enter $\frac{2}{3}x - \frac{13}{3}$ into Y1 and $-\frac{3}{4}x - \frac{3}{2}$ into Y2. Graph the two functions in the standard viewing window. (If the window does not show the point of intersection of the two graphs, adjust the window until you can see the point of intersection.)

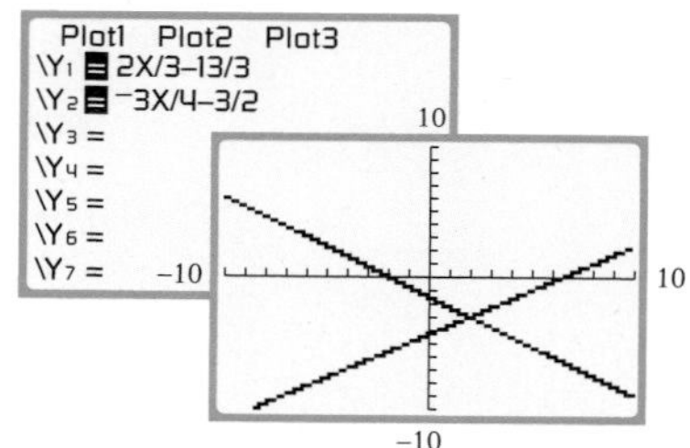

Press 2ND CALC (scroll to 5, intersect) ENTER.

Alternatively, you can just press 2ND CALC 5.

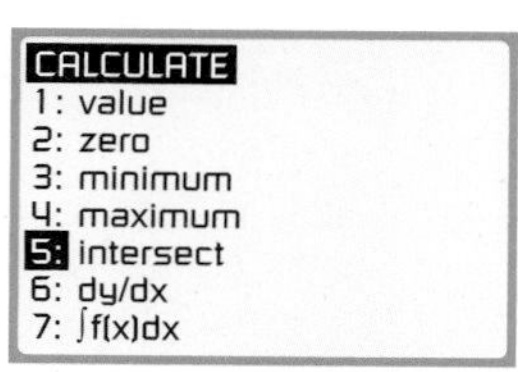

First curve? is shown at the bottom of the screen and identifies one of the two graphs on the screen. Press ENTER.

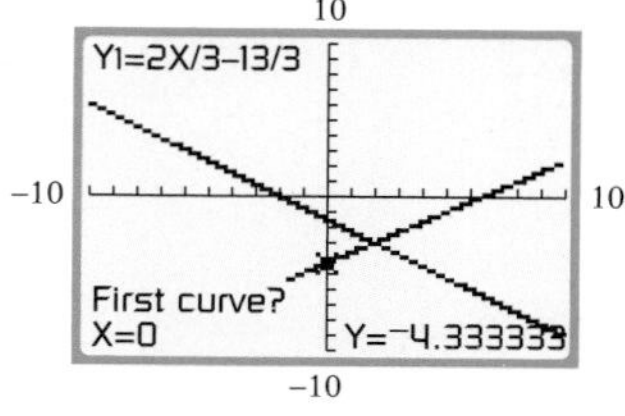

Second curve? is shown at the bottom of the screen and identifies the second of the two graphs on the screen. Press ENTER.

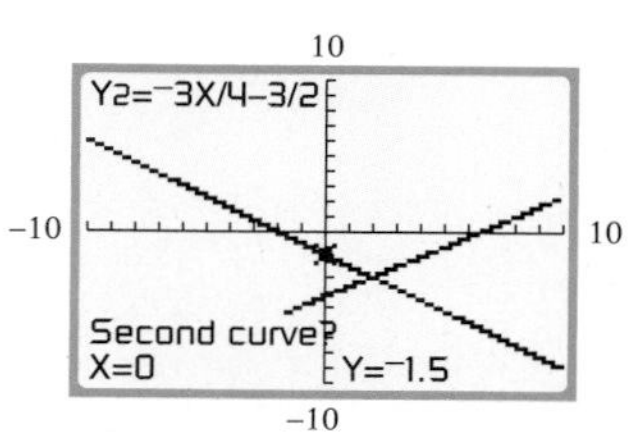

Guess? shown at the bottom of the screen asks you to use the left or right arrow key to move the cursor to the *approximate* location of the point of intersection. (If there are two or more points of intersection, it does not matter which one you choose first.) Press ENTER.

The solution of the system of equations is (2, −3).

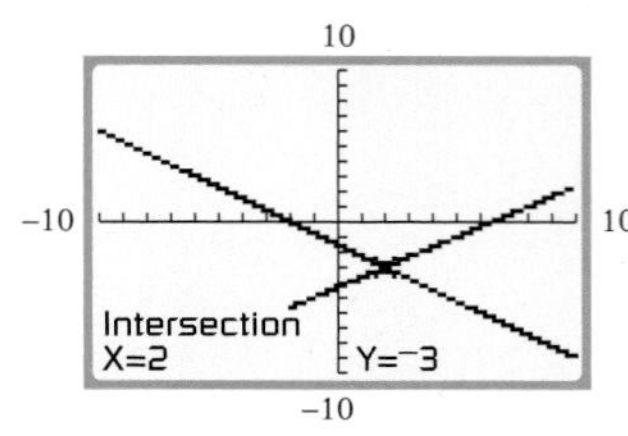

Solve an equation To illustrate the steps involved, we will solve the equation $2x + 4 = -3x - 1$. The idea is to write the equation as the system of equations $y = 2x + 4$, $y = -3x - 1$ and then use the steps for solving a system of equations.

Use the Y-editor to enter the left and right sides of the equation into Y_1 and Y_2. Graph the two functions and then follow the steps for Intersect.

The solution is −1, the x-coordinate of the point of intersection.

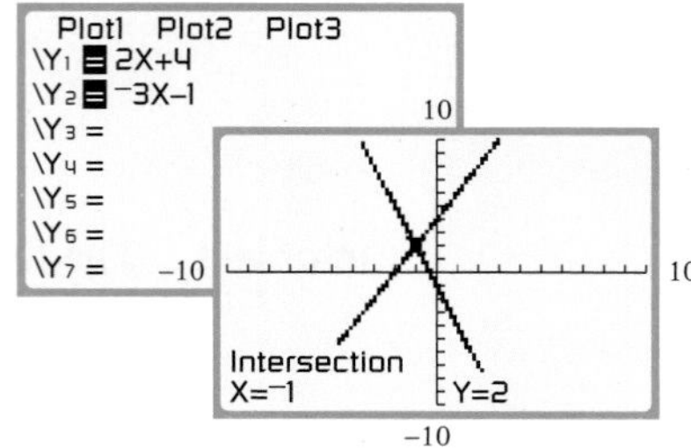

Find a domain element For this example, we will find a number in the domain of $f(x) = -\frac{2}{3}x + 2$ that corresponds to 4 in the range of the function. This is like solving the system of equations $y = -\frac{2}{3}x + 2$ and $y = 4$.

Use the Y = editor to enter the expression for the function in Y_1 and the desired output, 4, in Y_2. Graph the two functions and then follow the steps for Intersect.

The point of intersection is (−3, 4). The number −3 in the domain of f produces an output of 4 in the range of f.

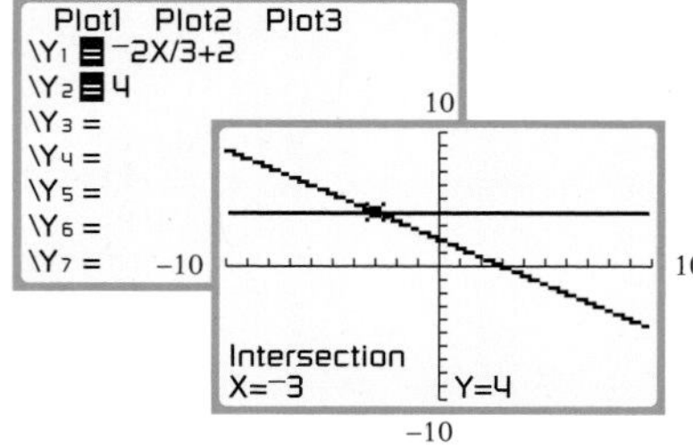

Math Pressing MATH gives you access to many built-in functions. The following keystrokes will convert 0.125 to a fraction: .125 MATH 1 ENTER.

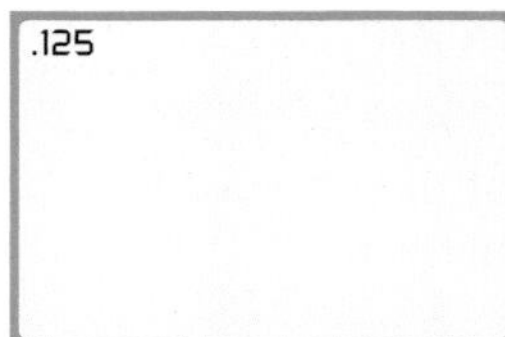

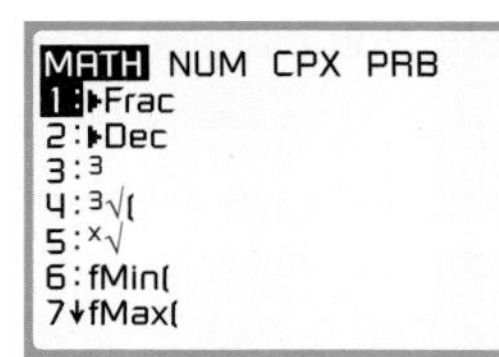

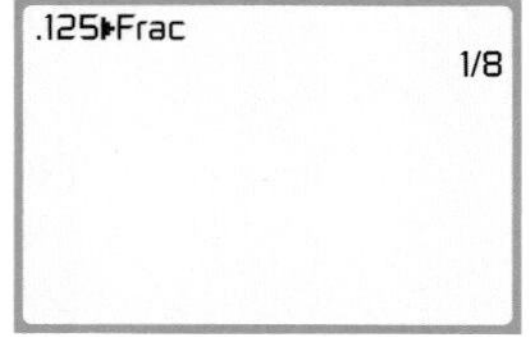

Additional built-in functions under MATH can be found by pressing MATH ▸. For instance, to evaluate $-|-25|$, press (−) MATH ▸ 1 (−) 25) ENTER.

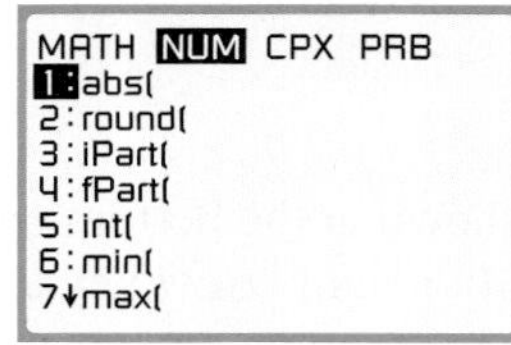

See your owner's manual for assistance with other functions under the MATH key.

Min and Max The local minimum and the local maximum values of a function are calculated by accessing the CALC menu. For this demonstration, we will find the minimum value and the maximum value of $f(x) = 0.2x^3 + 0.3x^2 - 3.6x + 2$.

Enter the function into Y_1. Press 2ND CALC (scroll to 3 for minimum of the function) ENTER.

Alternatively, you can just press 2ND CALC 3.

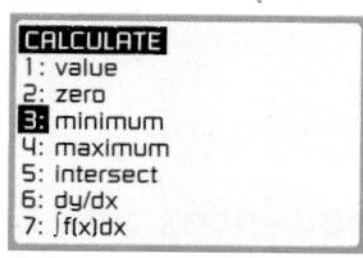

Left Bound? shown at the bottom of the screen asks you to use the left or right arrow key to move the cursor to the *left* of the minimum. Press ENTER.

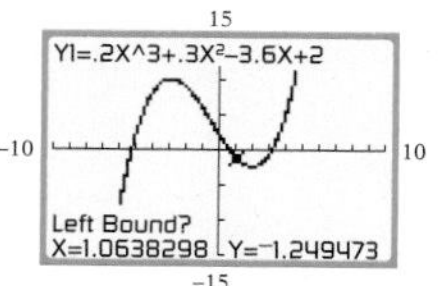

Right Bound? shown at the bottom of the screen asks you to use the left or right arrow key to move the cursor to the *right* of the minimum. Press ENTER.

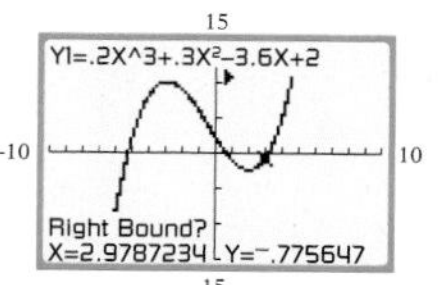

Guess? shown at the bottom of the screen asks you to use the left or right arrow key to move the cursor to the *approximate* location of the minimum. Press ENTER.

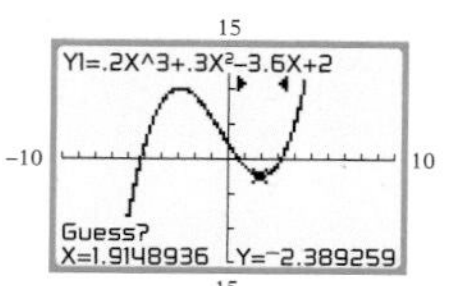

The minimum value of the function is the y-coordinate. For this example, the minimum value of the function is -2.4.

The x-coordinate for the minimum is 2. However, because of rounding errors in the calculation, it is shown as a number close to 2.

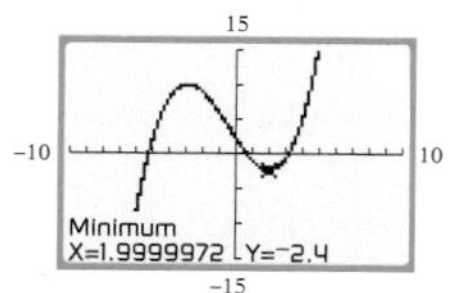

To find the maximum value of the function, follow the same steps as above except select maximum under the CALC menu. The screens for this calculation are shown below.

CALCULATE
1: value
2: zero
3: minimum
4: maximum
5: intersect
6: dy/dx
7: ∫f(x)dx

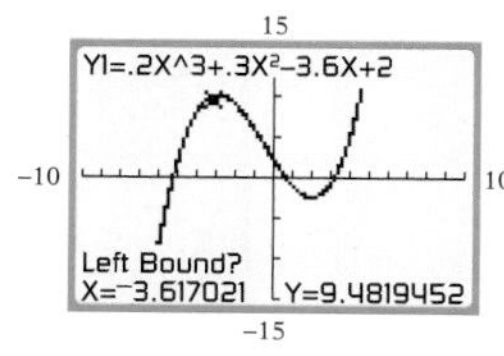

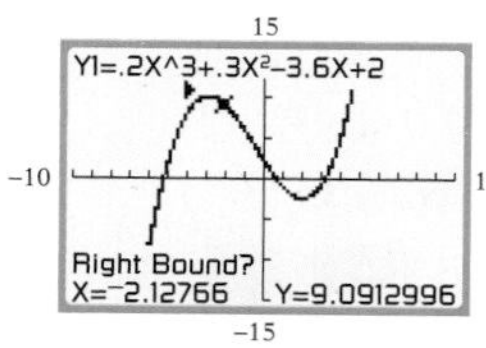

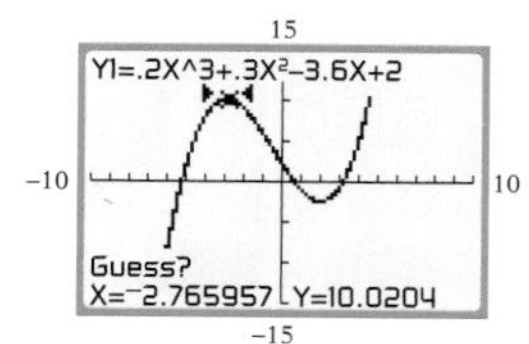

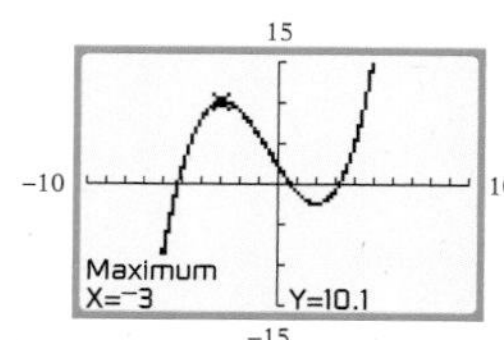

The maximum value of the function is 10.1.

Radical Expressions To evaluate a square-root expression, press 2ND √.

For instance, to evaluate $0.15\sqrt{p^2 + 4p + 10}$ when $p = 100{,}000$, first store 100,000 in P. Then press 0.15

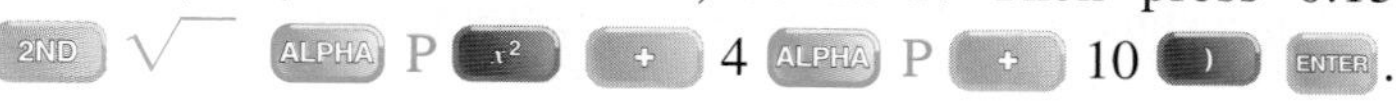

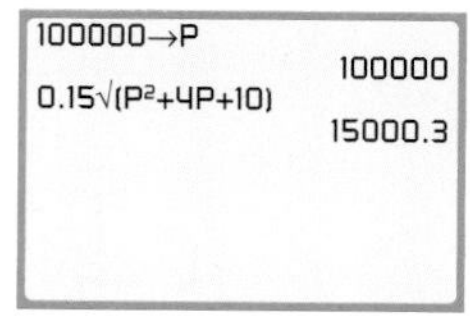

To evaluate a radical expression other than a square root, access $\sqrt[x]{}$ by pressing MATH. For instance, to evaluate $\sqrt[4]{67}$, press 4 (the index of the radical) MATH (scroll to 5) ENTER 67 ENTER.

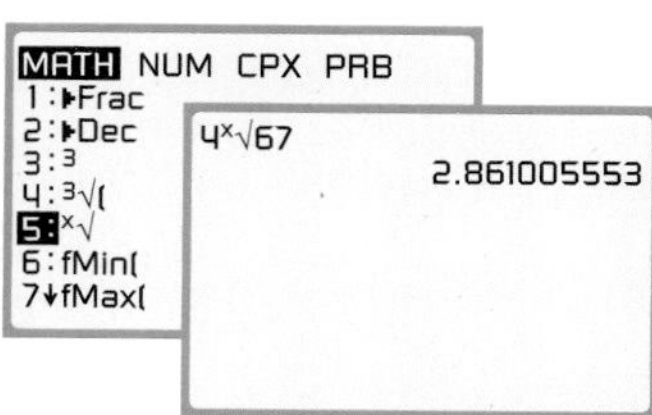

Scientific Notation To enter a number in scientific notation, use 2ND EE. For instance, to find $\frac{3.45 \times 10^{-12}}{1.5 \times 10^{25}}$, press 3.45 2ND EE (–) 12 ÷ 1.5 2ND EE 25 ENTER. The answer is 2.3×10^{-37}.

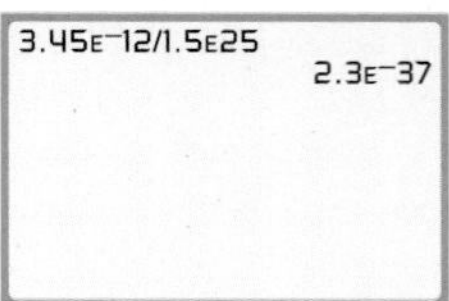

Sequences and Series The terms of a sequence and the sum of a series can be calculated by using the 2ND LIST feature.

Store a sequence A sequence is stored in one of the lists L1 through L6. For instance, to store the sequence 1, 3, 5, 7, 9 in L1, use the following keystrokes.

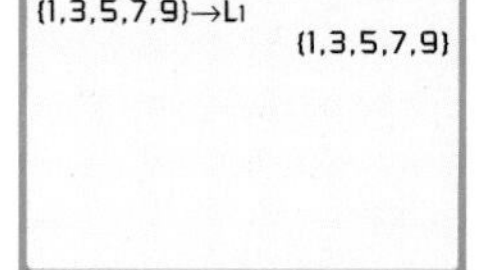

2ND { 1 , 3 , 5 , 7 ,
9 2ND } STO▸ 2ND L1 ENTER

Display the terms of a sequence The terms of a sequence are displayed by using the function seq(expression, variable, begin, end, increment). For instance, to display the 3rd through 8th terms of the sequence given by $a_n = n^2 + 6$, enter the following keystrokes.

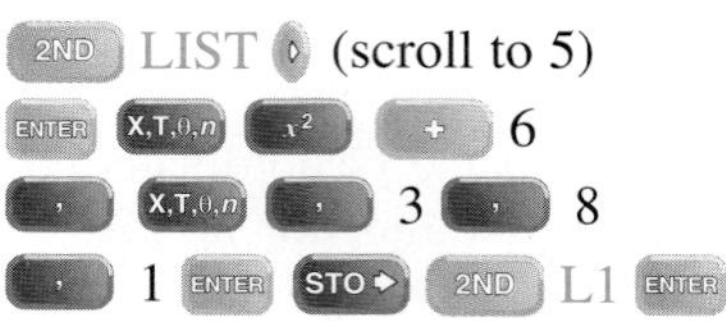

NAMES OPS MATH
1:SortA(
2:SortD(
3:dim(
4:Fill(
5:seq(
6:cumSum(
7↓ΔList(

seq(X²+6,X,3,8,1)
{15 22 31 42 55...

The keystrokes STO▸ 2ND L1 ENTER store the terms of the sequence in L1. This is not necessary but is sometimes helpful if additional work will be done with that sequence.

Find a sequence of partial sums To find a sequence of partial sums, use the cumSum(function. For instance, to find the sequence of partial sums for 2, 4, 6, 8, 10, use the following keystrokes.

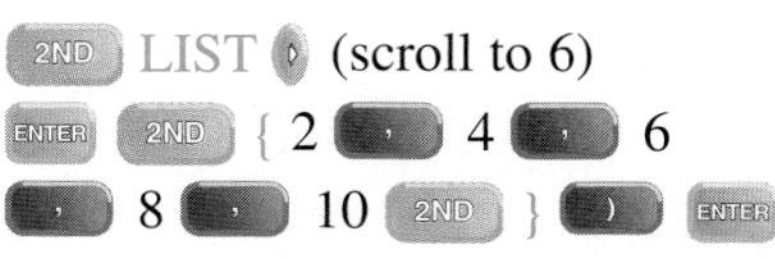

NAMES OPS MATH
1:SortA(
2:SortD(
3:dim(
4:Fill(
5:seq(
6:cumSum(
7↓ΔList(

cumSum({2,4,6,8,10})
{2 6 12 20 30}

If a sequence is stored as a list in L1, then the sequence of partial sums can be calculated by pressing 2ND LIST ▸ (scroll to 6 [or press 6]) ENTER 2ND L1) ENTER.

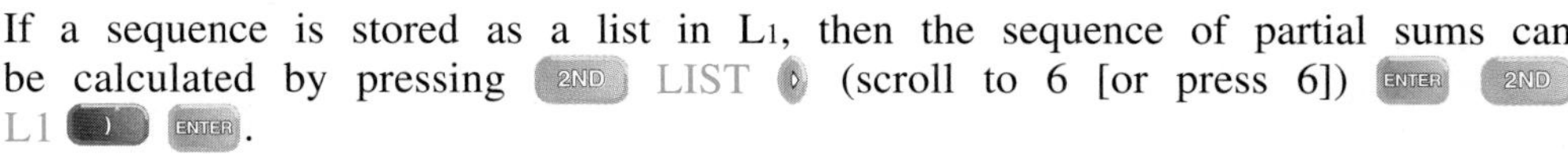

Find the sum of a series The sum of a series is calculated using sum<list, start, end>. For instance, to find $\sum_{n=3}^{6} (n^2 + 2)$, enter the following keystrokes.

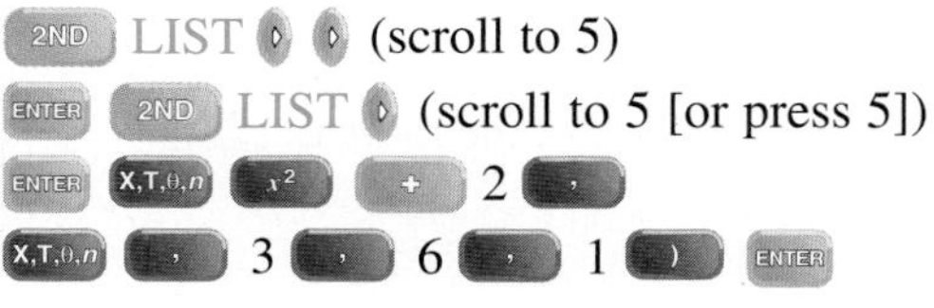

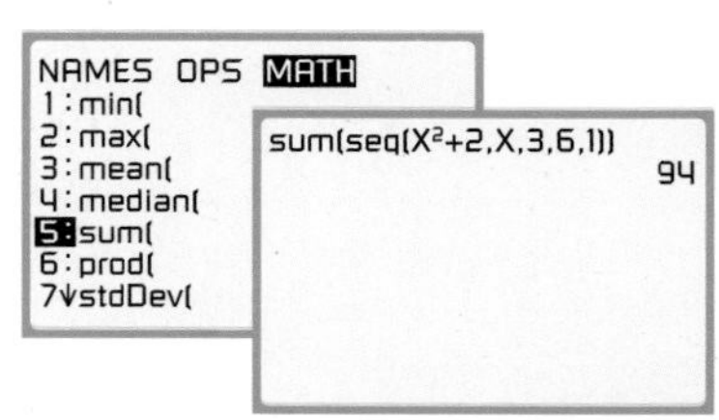

Table There are three steps in creating an input/output table for a function. First use the Y= editor to input the function. The second step is setting up the table, and the third step is displaying the table.

To set up the table, press 2ND TBLSET. **TblStart** is the first value of the independent variable in the input/output table. Δ**Tbl** is the difference between successive values. Setting this to 1 means that, for this table, the input values are $-2, -1, 0, 1, 2, \ldots$ If $\Delta\text{Tbl} = 0.5$, then the input values are $-2, -1.5, -1, -0.5, 0, 0.5, \ldots$

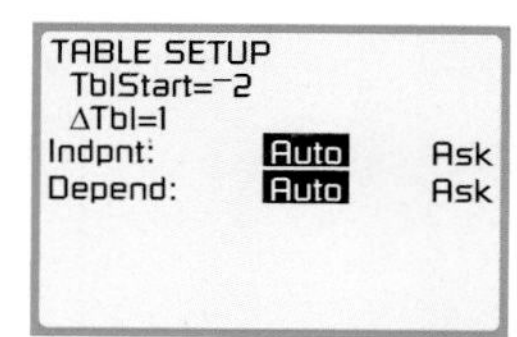

Indpnt is the independent variable. When this is set to **Auto**, values of the independent variable are automatically entered into the table. **Depend** is the dependent variable. When this is set to **Auto**, values of the dependent variable are automatically entered into the table.

To display the table, press 2ND TABLE. An input/output table for $f(x) = x^2 - 1$ is shown at the right.

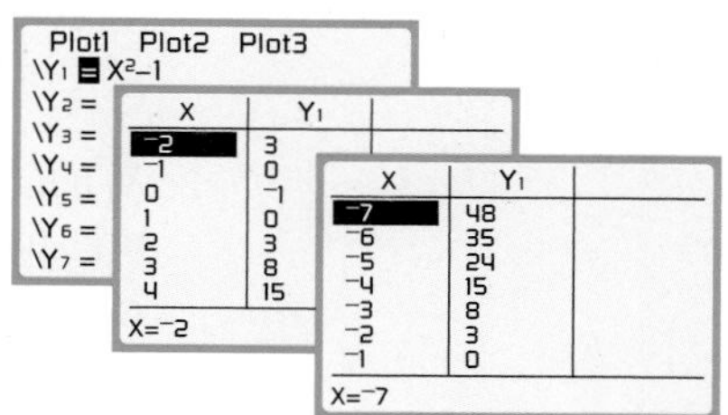

Once the table is on the screen, the up and down arrow keys can be used to display more values in the table. For the table at the right, we used the up arrow key to move to $x = -7$.

An input/output table for any given input can be created by selecting **Ask** for the independent variable. The table at the right shows an input/output table for $f(x) = \frac{4x}{x-2}$ for selected values of x. Note the word **ERROR** when 2 was entered. This occurred because f is not defined when $x = 2$.

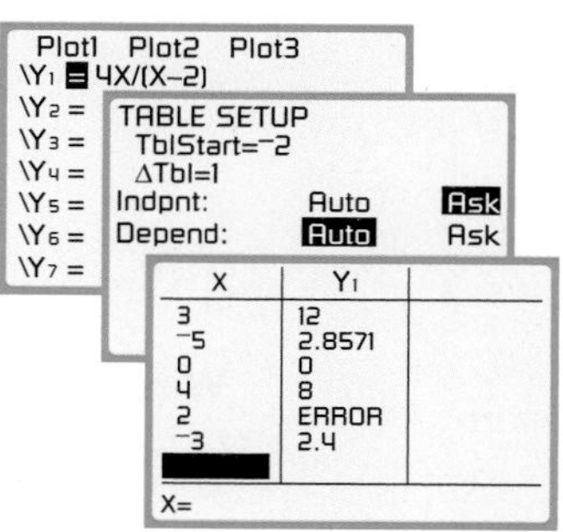

Note: Using the table feature in **Ask** mode is the same as evaluating a function for given values of the independent variable. For instance, from the table at the right, we have $f(4) = 8$.

Test The TEST feature has many uses, one of which is to graph the solution set of a linear inequality in one variable. To illustrate this feature, we will graph the solution set of $x - 1 < 4$. Press Y= X,T,θ,n − 1 2ND TEST (scroll to 5) ENTER 4 GRAPH.

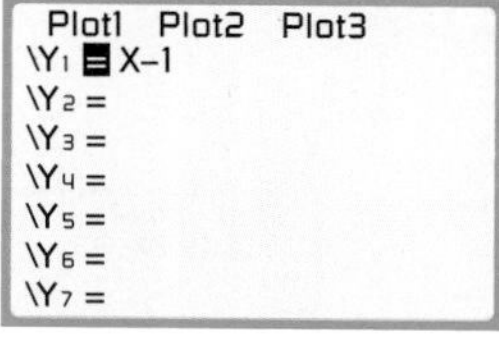

Trace Once a graph is drawn, pressing TRACE will place a cursor on the screen, and the coordinates of the point below the cursor are shown at the bottom of the screen. Use the left and right arrow keys to move the cursor along the graph. For the graph at the right, we have $f(4.8) = 3.4592$, where $f(x) = 0.1x^3 - 2x + 2$ is shown at the top left of the screen.

In TRACE mode, you can evaluate a function at any value of the independent variable that is within Xmin and Xmax. To do this, first graph the function. Now press TRACE (the value of x) ENTER. For the graph at the left below, we used $x = -3.5$. If a value of x is chosen outside the window, an error message is displayed.

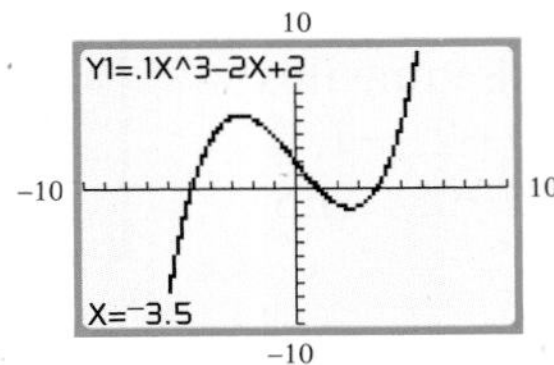

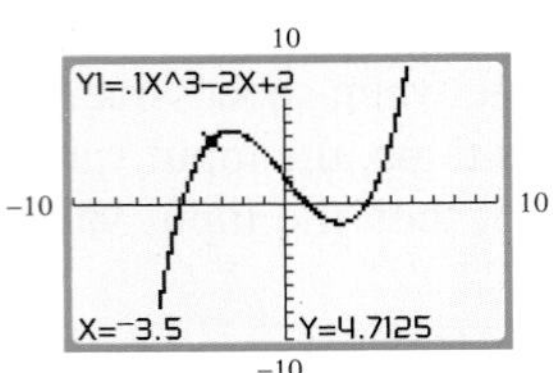

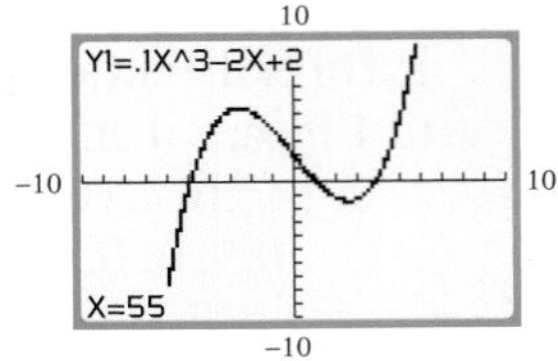

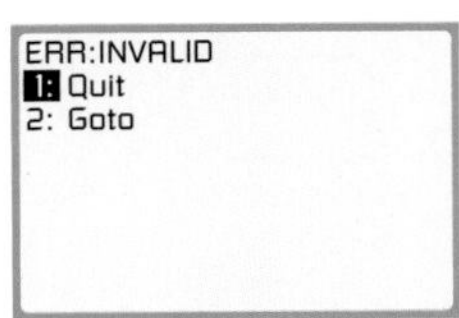

In the example above in which we entered -3.5 for x, the value of the function was calculated as 4.7125. This means that $f(-3.5) = 4.7125$. The keystrokes 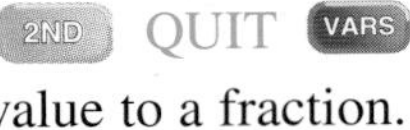 2ND QUIT VARS ▷ 11 MATH 1 ENTER will convert the decimal value to a fraction.

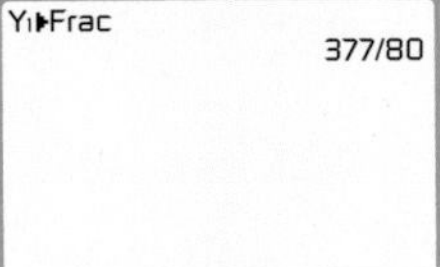

When the TRACE feature is used with two or more graphs, the up and down arrow keys are used to move between the graphs. The graphs below are for the functions $f(x) = 0.1x^3 - 2x + 2$ and $g(x) = 2x - 3$. By using the up and down arrows, we can place the cursor on either graph. The right and left arrows are used to move along the graph.

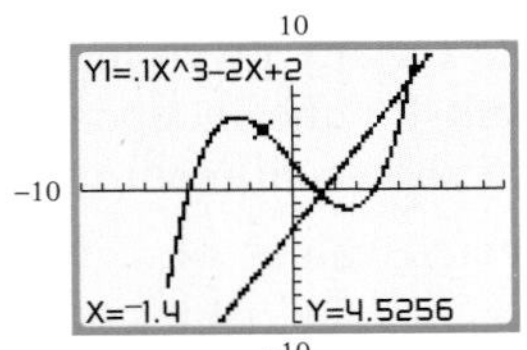

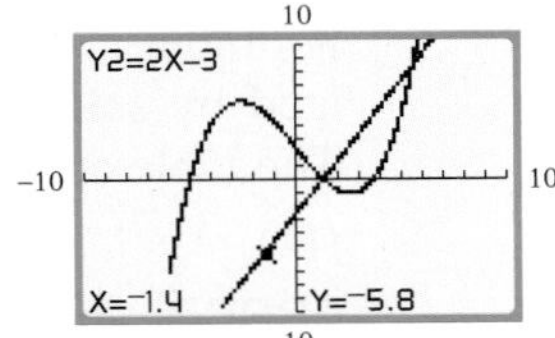

Window The viewing window for a graph is controlled by pressing WINDOW. Xmin and Xmax are the minimum value and maximum value, respectively, of the independent variable shown on the graph. Xscl is the distance between tic marks on the x-axis. Ymin and Ymax are the minimum value and maximum value, respectively, of the dependent variable shown on the graph. Yscl is the distance between tic marks on the y-axis. Leave Xres as 1.

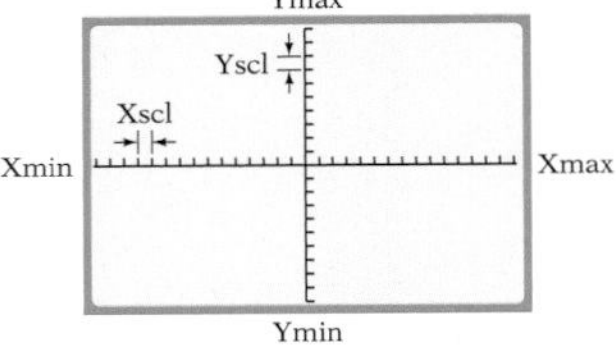

Note: In the standard viewing window, the distance between tic marks on the x-axis is different from the distance between tic marks on the y-axis. This will distort a graph. A more accurate picture of a graph can be created by using a square viewing window. See ZOOM.

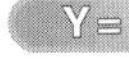

The Y= editor is used to enter the expression for a function. There are ten possible functions, labeled Y_1 to Y_0, that can be active at any one time. For instance, to enter $f(x) = x^2 + 3x - 2$ as Y_1, use the following keystrokes.

Y= X,T,θ,n x^2 + 3 X,T,θ,n − 2

Plot1 Plot2 Plot3
\Y1 = X²+3X−2
\Y2 =
\Y3 =
\Y4 =
\Y5 =
\Y6 =
\Y7 =

Note: If an expression is already entered for Y_1, place the cursor anywhere on that expression and press CLEAR.

To enter $s = \frac{2v - 1}{v^3 - 3}$ into Y_2, place the cursor to the right of the equals sign for Y_2. Then press (2 X,T,θ,n − 1) ÷ (X,T,θ,n ^ 3 − 3).

```
Plot1 Plot2 Plot3
\Y1 ■ X²+3X−2
\Y2 ■ (2X−1)/(X^3−3)
\Y3 =
\Y4 =
\Y5 =
\Y6 =
\Y7 =
```

Note: When we enter an equation, the independent variable, v in the expression above, is entered using X,T,θ,n. The dependent variable, s in the expression above, is one of Y_1 to Y_0. Also note the use of parentheses to ensure the correct order of operations.

Observe the black rectangle that covers the equals sign for the two examples we have shown. This rectangle means that the function is "active." If we were to press GRAPH, then the graph of both functions would appear. You can make a function inactive by using the arrow keys to move the cursor over the equals sign of that function and then pressing ENTER. This will remove the black rectangle. We have done that for Y_2, as shown at the right. Now if GRAPH is pressed, only Y_1 will be graphed.

```
Plot1 Plot2 Plot3
\Y1 ■ X²+3X−2
\Y2 = (2X−1)/(X^3−3)
\Y3 =
\Y4 =
\Y5 =
\Y6 =
\Y7 =
```

It is also possible to control the appearance of the graph by moving the cursor on the Y= screen to the left of any Y. With the cursor in this position, pressing ENTER will change the appearance of the graph. The options are shown at the right.

```
Plot1 Plot2 Plot3
\Y1 = Default graph line
\Y2 = Bold graph line
◥Y3 = Shade above graph
◣Y4 = Shade below graph
-oY5 = Draw path of graph
oY6 = Travel path of graph
\Y7 = Dashed graph line
```

Zero The ZERO feature of a graphing calculator is used for various calculations: to find the x-intercepts of a function, to solve some equations, and to find the zero of a function.

***x*-intercepts** To illustrate the procedure for finding x-intercepts, we will use $f(x) = x^2 + x - 2$.

First, use the Y-editor to enter the expression for the function and then graph the function in the standard viewing window. (It may be necessary to adjust this window so that the intercepts are visible.) Once the graph is displayed, use the keystrokes below to find the x-intercepts of the graph of the function.

Press 2ND CALC (scroll to 2 for **zero** of the function) ENTER.

Alternatively, you can just press 2ND CALC 2.

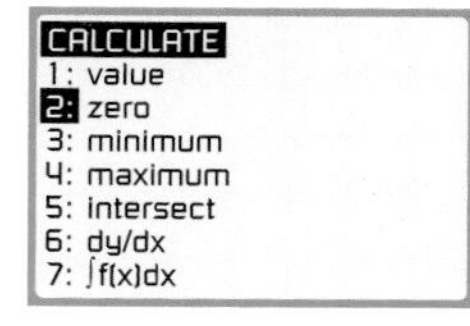

Left Bound? shown at the bottom of the screen asks you to use the left or right arrow key to move the cursor to the *left* of the desired x-intercept. Press ENTER.

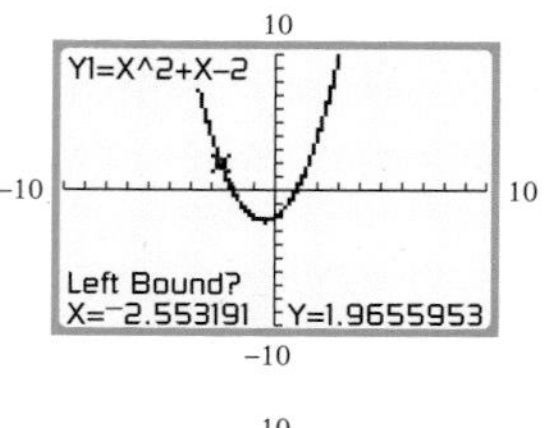

Right Bound? shown at the bottom of the screen asks you to use the left or right arrow key to move the cursor to the *right* of the desired x-intercept. Press ENTER.

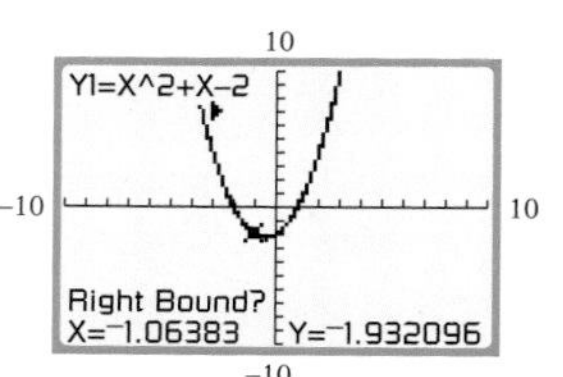

Guess? shown at the bottom of the screen asks you to use the left or right arrow key to move the cursor to the *approximate* location of the desired x-intercept. Press ENTER.

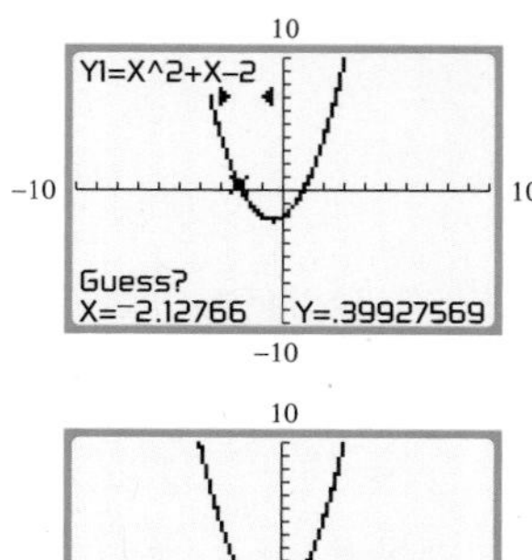

The x-coordinate of an x-intercept is -2. Therefore, an x-intercept is $(-2, 0)$.

To find the other x-intercept, follow the same steps as above. The screens for this calculation are shown below.

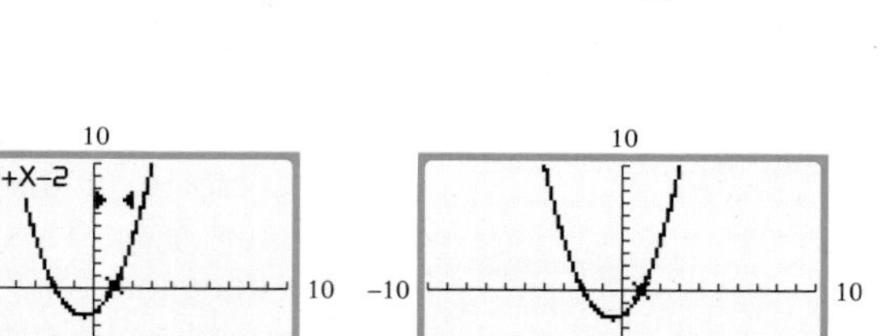

A second x-intercept is $(1, 0)$.

Solve an equation To use the ZERO feature to solve an equation, first rewrite the equation with all terms on one side. For instance, one way to solve the equation $x^3 - x + 1 = -2x + 3$ is first to rewrite it as $x^3 + x - 2 = 0$. Enter $x^3 + x - 2$ into Y_1 and then follow the steps for finding x-intercepts.

Find the real zeros of a function To find the real zeros of a function, follow the steps for finding x-intercepts.

Zoom Pressing ZOOM allows you to select some preset viewing windows. This key also gives you access to ZBox, Zoom In, and Zoom Out. These functions enable you to redraw a selected portion of a graph in a new window. Some windows used frequently in this text are shown below.

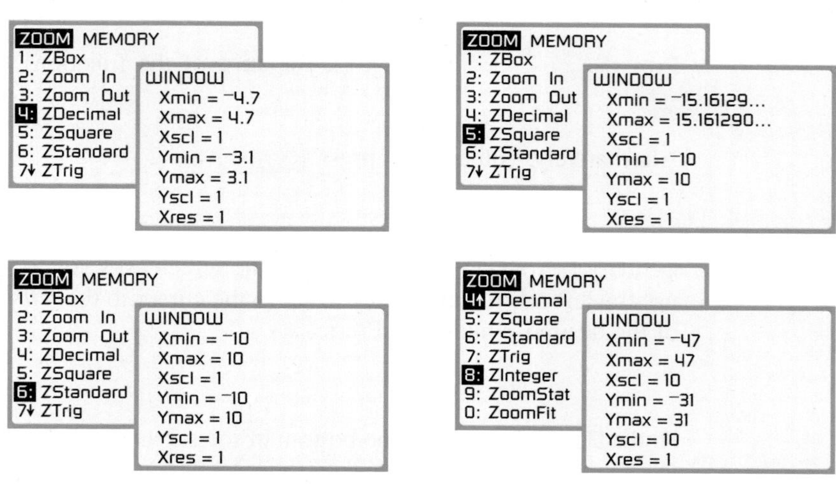

Appendix B

Proofs and Tables

Proofs of Logarithmic Properties

In each of the following proofs of logarithmic properties, it is assumed that the Properties of Exponents are true for all real number exponents.

The Logarithm Property of the Product of Two Numbers

For any positive real numbers x, y, and b, $b \neq 1$, $\log_b xy = \log_b x + \log_b y$.

Proof: Let $\log_b x = m$ and $\log_b y = n$.

Write each equation in its equivalent exponential form. $\quad x = b^m \quad y = b^n$

Use substitution and the Properties of Exponents. $\quad xy = b^m b^n$

$xy = b^{m+n}$

Write the equation in its equivalent logarithmic form. $\quad \log_b xy = m + n$

Substitute $\log_b x$ for m and $\log_b y$ for n. $\quad \log_b xy = \log_b x + \log_b y$

The Logarithm Property of the Quotient of Two Numbers

For any positive real numbers x, y, and b, $b \neq 1$, $\log_b \frac{x}{y} = \log_b x - \log_b y$.

Proof: Let $\log_b x = m$ and $\log_b y = n$.

Write each equation in its equivalent exponential form. $\quad x = b^m \quad y = b^n$

Use substitution and the Properties of Exponents. $\quad \frac{x}{y} = \frac{b^m}{b^n}$

$\frac{x}{y} = b^{m-n}$

Write the equation in its equivalent logarithmic form. $\quad \log_b \frac{x}{y} = m - n$

Substitute $\log_b x$ for m and $\log_b y$ for n. $\quad \log_b \frac{x}{y} = \log_b x - \log_b y$

The Logarithm Property of the Power of a Number

For any real numbers x, r, and b, $b \neq 1$, $\log_b x^r = r \log_b x$.

Proof: Let $\log_b x = m$.

Write the equation in its equivalent exponential form. $\quad x = b^m$

Raise both sides to the r power. $\quad x^r = (b^m)^r$

$x^r = b^{mr}$

Write the equation in its equivalent logarithmic form. $\quad \log_b x^r = mr$

Substitute $\log_b x$ for m. $\quad \log_b x^r = r \log_b x$

Table of Symbols

Symbol	Meaning
$+$	add
$-$	subtract
$\cdot, \times, (a)(b)$	multiply
$\frac{a}{b}, \div$	divide
$(\ )$	parentheses, a grouping symbol
$[\ ]$	brackets, a grouping symbol
π	pi, a number approximately equal to $\frac{22}{7}$ or 3.14
$-a$	the opposite, or additive inverse, of a
$\frac{1}{a}$	the reciprocal, or multiplicative inverse, of a
$=$	is equal to
$\approx$	is approximately equal to
$\neq$	is not equal to
$<$	is less than
$\leq$	is less than or equal to
$>$	is greater than
$\geq$	is greater than or equal to
(a, b)	an ordered pair whose first component is a and whose second component is b
$^\circ$	degree (for angles)
$\sqrt{a}$	the principal square root of a
$\varnothing, \{\ \}$	the empty set
$\lvert a \rvert$	the absolute value of a
$\cup$	union of two sets
$\cap$	intersection of two sets
$\in$	is an element of (for sets)
$\notin$	is not an element of (for sets)

Table of Measurement Abbreviations

U.S. Customary System

Length		**Capacity**		**Weight**		**Area**	
in.	inches	oz	fluid ounces	oz	ounces	in^2	square inches
ft	feet	c	cups	lb	pounds	ft^2	square feet
yd	yards	qt	quarts				
mi	miles	gal	gallons				

Metric System

Length		**Capacity**		**Weight/Mass**		**Area**	
mm	millimeter (0.001 m)	ml	milliliter (0.001 L)	mg	milligram (0.001 g)	cm^2	square centimeters
cm	centimeter (0.01 m)	cl	centiliter (0.01 L)	cg	centigram (0.01 g)	m^2	square meters
dm	decimeter (0.1 m)	dl	deciliter (0.1 L)	dg	decigram (0.1 g)		
m	meter	L	liter	g	gram		
dam	decameter (10 m)	dal	decaliter (10 L)	dag	decagram (10 g)		
hm	hectometer (100 m)	hl	hectoliter (100 L)	hg	hectogram (100 g)		
km	kilometer (1000 m)	kl	kiloliter (1000 L)	kg	kilogram (1000 g)		

Time

h	hours	min	minutes	s	seconds

Table of Properties

Properties of Real Numbers

The Associative Property of Addition
If a, b, and c are real numbers, then
$(a + b) + c = a + (b + c)$.

The Associative Property of Multiplication
If a, b, and c are real numbers, then
$(a \cdot b) \cdot c = a \cdot (b \cdot c)$.

The Commutative Property of Addition
If a and b are real numbers, then
$a + b = b + a$.

The Commutative Property of Multiplication
If a and b are real numbers, then
$a \cdot b = b \cdot a$.

The Addition Property of Zero
If a is a real number, then
$a + 0 = 0 + a = a$.

The Multiplication Property of One
If a is a real number, then
$a \cdot 1 = 1 \cdot a = a$.

The Multiplication Property of Zero
If a is a real number, then
$a \cdot 0 = 0 \cdot a = 0$.

The Inverse Property of Multiplication
If a is a real number and $a \neq 0$, then
$a \cdot \frac{1}{a} = \frac{1}{a} \cdot a = 1$.

The Inverse Property of Addition
If a is a real number, then
$a + (-a) = (-a) + a = 0$.

Distributive Property
If a, b, and c are real numbers, then
$a(b + c) = ab + ac$.

Properties of Equations

Addition Property of Equations
If $a = b$, then $a + c = b + c$.

Multiplication Property of Equations
If $a = b$ and $c \neq 0$, then $a \cdot c = b \cdot c$.

Properties of Inequalities

Addition Property of Inequalities
If $a > b$, then $a + c > b + c$.
If $a < b$, then $a + c < b + c$.

Multiplication Property of Inequalities
If $a > b$ and $c > 0$, then $ac > bc$.
If $a < b$ and $c > 0$, then $ac < bc$.
If $a > b$ and $c < 0$, then $ac < bc$.
If $a < b$ and $c < 0$, then $ac > bc$.

Properties of Exponents

If m and n are integers, then $x^m \cdot x^n = x^{m+n}$.
If m and n are integers, then $(x^m)^n = x^{mn}$.

If $x \neq 0$, then $x^0 = 1$.

If m and n are integers and $x \neq 0$, then $\frac{x^m}{x^n} = x^{m-n}$.

If m, n, and p are integers, then $(x^m \cdot y^n)^p = x^{mp}y^{np}$.
If n is a positive integer and $x \neq 0$, then
$x^{-n} = \frac{1}{x^n}$ and $\frac{1}{x^{-n}} = x^n$.

If m, n, and p are integers and $y \neq 0$, then $\left(\frac{x^m}{y^n}\right)^p = \frac{x^{mp}}{y^{np}}$.

Principle of Zero Products

If $a \cdot b = 0$, then $a = 0$ or $b = 0$.

Properties of Radical Expressions

If a and b are positive real numbers, then $\sqrt{ab} = \sqrt{a}\sqrt{b}$. If a and b are positive real numbers, then $\sqrt{\frac{a}{b}} = \frac{\sqrt{a}}{\sqrt{b}}$.

Property of Squaring Both Sides of an Equation

If a and b are real numbers and $a = b$, then $a^2 = b^2$.

Properties of Logarithms

If x, y, and b are positive real numbers and $b \neq 1$, then
$\log_b(xy) = \log_b x + \log_b y$.
If x, y, and b are positive real numbers and $b \neq 1$, then
$\log_b \frac{x}{y} = \log_b x - \log_b y$.

If x and b are positive real numbers, $b \neq 1$, and r is any real number, then $\log_b x^r = r \log_b x$.
If x and b are positive real numbers and $b \neq 1$, then
$\log_b b^x = x$.

Table of Algebraic and Geometric Formulas

Slope of a Line

$$m = \frac{y_2 - y_1}{x_2 - x_1}, x_1 \neq x_2$$

Point-slope Formula for a Line

$$y - y_1 = m(x - x_1)$$

Quadratic Formula

$$x = \frac{-b \pm \sqrt{b^2 - 4ac}}{2a}$$

discriminant $= b^2 - 4ac$

Perimeter and Area of a Triangle, and Sum of the Measures of the Angles

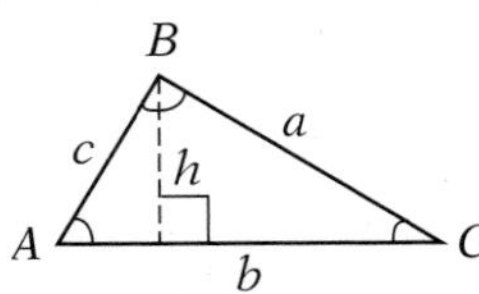

$$P = a + b + c$$

$$A = \frac{1}{2}bh$$

$$A + B + C = 180°$$

Pythagorean Theorem

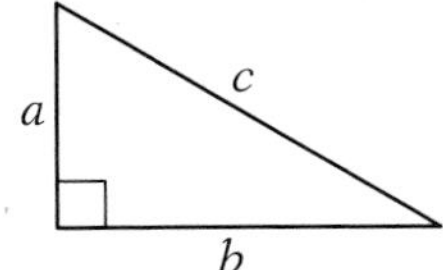

$$a^2 + b^2 = c^2$$

Perimeter and Area of a Rectangle

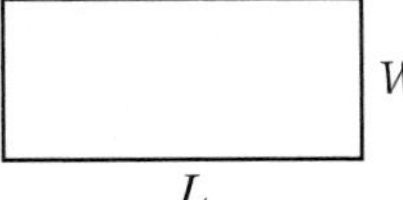

$$P = 2L + 2W$$
$$A = LW$$

Perimeter and Area of a Square

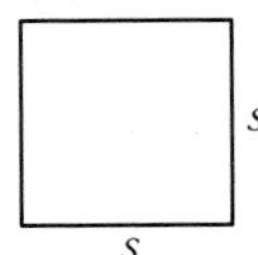

$$P = 4s$$
$$A = s^2$$

Area of a Trapezoid

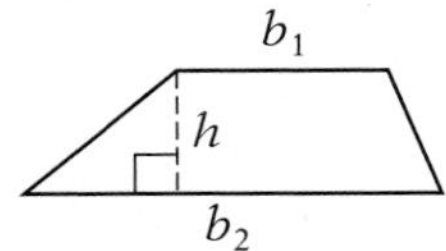

$$A = \frac{1}{2}h(b_1 + b_2)$$

Circumference and Area of a Circle

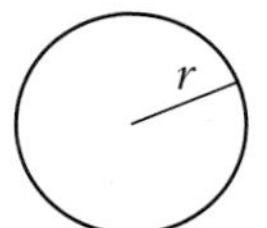

$$C = 2\pi r$$
$$A = \pi r^2$$

Volume and Surface Area of a Rectangular Solid

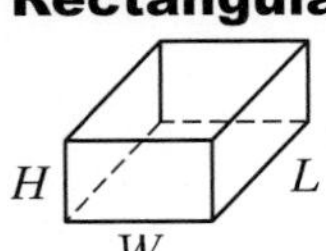

$$V = LWH$$
$$SA = 2LW + 2LH + 2WH$$

Volume and Surface Area of a Sphere

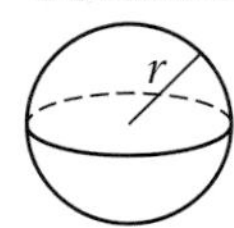

$$V = \frac{4}{3}\pi r^3$$

$$SA = 4\pi r^2$$

Volume and Surface Area of a Right Circular Cylinder

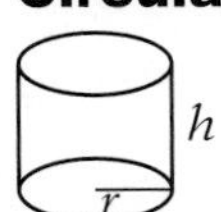

$$V = \pi r^2 h$$
$$SA = 2\pi r^2 + 2\pi rh$$

Volume and Surface Area of a Right Circular Cone

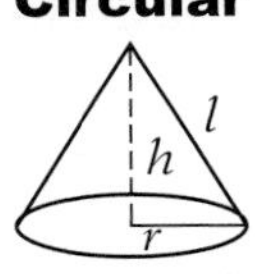

$$V = \frac{1}{3}\pi r^2 h$$

$$SA = \pi r^2 + \pi rl$$

Solutions to "You Try It"

SOLUTIONS TO CHAPTER 1 "YOU TRY IT"

SECTION 1.1

You Try It 1 Replace z by each of the elements of the set and determine whether the inequality is true.

$z > -5$

$-10 > -5$ False

$-5 > -5$ False

$6 > -5$ True

The inequality is true for 6.

You Try It 2

a. Replace d in $-d$ by each element of the set and determine the value of the expression.

$-d$

$-(-11) = 11$

$-(0) = 0$

$-(8) = -8$

b. Replace d in $|d|$ by each element of the set and determine the value of the expression.

$|d|$

$|-11| = 11$

$|0| = 0$

$|8| = 8$

You Try It 3 $100 + (-43) = 57$

You Try It 4

$$\begin{aligned}&-51 + 42 + 17 + (-102)\\ &= -9 + 17 + (-102)\\ &= 8 + (-102)\\ &= -94\end{aligned}$$

You Try It 5 $-8 + 7 = -1$

You Try It 6

$$\begin{aligned}&-9 - (-12) - 17 - 4\\ &= -9 + 12 + (-17) + (-4)\\ &= 3 + (-17) + (-4)\\ &= -14 + (-4)\\ &= -18\end{aligned}$$

You Try It 7 $-11 - (-12) = -11 + 12 = 1$

You Try It 8

$$\begin{aligned}8(-9)10 &= -72(10)\\ &= -720\end{aligned}$$

You Try It 9

$$\begin{aligned}(-2)3(-8)7 &= -6(-8)7\\ &= 48(7)\\ &= 336\end{aligned}$$

You Try It 10 $-9(34) = -306$

You Try It 11 $(-135) \div (-9) = 15$

You Try It 12 $\dfrac{-72}{4} = -18$

You Try It 13

$$\begin{aligned}-\frac{36}{-12} &= -(-3)\\ &= 3\end{aligned}$$

You Try It 14 $\dfrac{-72}{-8} = 9$

You Try It 15

Strategy

To find the average daily low temperature:

- Add the seven temperature readings.
- Divide the sum by 7.

Solution

$-6 + (-7) + 0 + (-5) + (-8) + (-1) + (-1) = -28$

$-28 \div 7 = -4$

The average daily low temperature was $-4°C$.

SECTION 1.2

You Try It 1

$$\begin{array}{r} 0.444\\ 9\overline{)4.000}\\ \underline{-3\,6}\\ 40\\ \underline{-36}\\ 40\\ \underline{-36}\\ 4\end{array}$$

$\dfrac{4}{9} = 0.\overline{4}$

You Try It 2

$125\% = 125\left(\dfrac{1}{100}\right) = \dfrac{125}{100} = \dfrac{5}{4}$

$125\% = 125(0.01) = 1.25$

You Try It 3

$$\frac{1}{3} = \frac{1}{3}(100\%) = \frac{100}{3}\% = 33\frac{1}{3}\%$$

You Try It 4

$$0.043 = 0.043(100\%) = 4.3\%$$

You Try It 5

The LCM of 8, 6, and 4 is 24.

$$-\frac{7}{8} - \frac{5}{6} + \frac{3}{4} = -\frac{21}{24} - \frac{20}{24} + \frac{18}{24} = \frac{-21}{24} + \frac{-20}{24} + \frac{18}{24} = \frac{-21 - 20 + 18}{24} = \frac{-23}{24} = -\frac{23}{24}$$

You Try It 6

$$16.127 - 67.91 = 16.127 + (-67.91) = -51.783$$

You Try It 7

The quotient is positive.

$$-\frac{3}{8} \div \left(-\frac{5}{12}\right) = \frac{3}{8} \div \frac{5}{12} = \frac{3}{8} \cdot \frac{12}{5} = \frac{3 \cdot 12}{8 \cdot 5} = \frac{3 \cdot \overset{1}{\cancel{2}} \cdot \overset{1}{\cancel{2}} \cdot 3}{2 \cdot \underset{1}{\cancel{2}} \cdot \underset{1}{\cancel{2}} \cdot 5} = \frac{9}{10}$$

You Try It 8

The product is negative.

```
  5.44
×  3.8
  4352
 1632
20.672
```

$$-5.44(3.8) = -20.672$$

You Try It 9 $-6^3 = -(6 \cdot 6 \cdot 6) = -216$

You Try It 10 $(-3)^4 = (-3)(-3)(-3)(-3) = 81$

You Try It 11

$$(3^3) \cdot (-2)^3 = (3)(3)(3) \cdot (-2)(-2)(-2) = 27(-8) = -216$$

You Try It 12

$$\left(-\frac{2}{5}\right)^2 = \left(-\frac{2}{5}\right)\left(-\frac{2}{5}\right) = \frac{4}{25}$$

You Try It 13

$$-3(0.3)^3 = -3(0.3)(0.3)(0.3) = -0.9(0.3)(0.3) = -0.27(0.3) = -0.081$$

You Try It 14

$$-5\sqrt{32} = -5\sqrt{16 \cdot 2} = -5\sqrt{16}\sqrt{2} = -5 \cdot 4\sqrt{2} = -20\sqrt{2}$$

You Try It 15

$$\sqrt{216} = \sqrt{36 \cdot 6} = \sqrt{36}\sqrt{6} = 6\sqrt{6}$$

You Try It 16

Strategy To find the difference:

- Read the numbers from the graph that correspond to 2001 (7.2, 9.3, 8.2, 6.0, 20.8).
- Add the numbers.
- Subtract the sum from the total number of barrels produced each day in 2007, calculated in Example 16 (55.1 million).

Solution

$7.2 + 9.3 + 8.2 + 6.0 + 20.8 = 51.5$

$55.1 - 51.5 = 3.6$

The difference is 3.6 million barrels of oil per day.

SECTION 1.3

You Try It 1

$18 - 5[8 - 2(2 - 5)] \div 10$

$= 18 - 5[8 - 2(-3)] \div 10$ • Perform operations inside grouping symbols.

$= 18 - 5[8 + 6] \div 10$

$= 18 - 5[14] \div 10$

$= 18 - 70 \div 10$ • Do multiplication and division from left to right.

$= 18 - 7$

$= 11$ • Do addition and subtraction from left to right.

You Try It 2

$$(6.97 - 4.72)^2 \cdot 4.5 \div 0.05 = (2.25)^2 \cdot 4.5 \div 0.05 = 5.0625 \cdot 4.5 \div 0.05 = 22.78125 \div 0.05 = 455.625$$

SECTION 1.4

You Try It 1

$$\frac{a^2 + b^2}{a + b}$$

$$\frac{5^2 + (-3)^2}{5 + (-3)} = \frac{25 + 9}{5 + (-3)} = \frac{34}{2} = 17$$

You Try It 2

$$x^3 - 2(x + y) + z^2$$

$$(2)^3 - 2[2 + (-4)] + (-3)^2 = 8 - 2(-2) + 9 = 8 + 4 + 9 = 12 + 9 = 21$$

You Try It 3 $3a - 2b - 5a + 6b = -2a + 4b$

You Try It 4 $-3y^2 + 7 + 8y^2 - 14 = 5y^2 - 7$

You Try It 5 $-5(4y^2) = -20y^2$

You Try It 6 $-7(-2a) = 14a$

You Try It 7 $-\frac{3}{5}\left(-\frac{7}{9}a\right) = \frac{7}{15}a$

You Try It 8 $(3a - 1)5 = 15a - 5$

You Try It 9 $-8(-2a + 7b) = 16a - 56b$

You Try It 10 $3(12x^2 - x + 8) = 36x^2 - 3x + 24$

You Try It 11
$$\begin{aligned} 3y - 2(y - 7x) &= 3y - 2y + 14x \\ &= y + 14x \end{aligned}$$

You Try It 12
$$\begin{aligned} -2(x - 2y) - (-x + 3y) &= -2x + 4y + x - 3y \\ &= -x + y \end{aligned}$$

You Try It 13
$$\begin{aligned} 3y - 2[x - 4(2 - 3y)] &= 3y - 2[x - 8 + 12y] \\ &= 3y - 2x + 16 - 24y \\ &= -2x - 21y + 16 \end{aligned}$$

You Try It 14 the speed of the older model: s
the speed of the new jet plane is twice the speed of the older model: $2s$

You Try It 15 the length of the longer piece: y
the length of the shorter piece: $6 - y$

SECTION 1.5

You Try It 1 $A = \{-9, -7, -5, -3, -1\}$

You Try It 2 $A = \{1, 3, 5, \ldots\}$

You Try It 3 $A \cup B = \{-2, -1, 0, 1, 2, 3, 4\}$

You Try It 4 $C \cap D = \{10, 16\}$

You Try It 5 $A \cap B = \varnothing$

You Try It 6 $\{x|x \le 3\}$ is the set of real numbers less than or equal to 3. This set extends forever in the negative direction. In interval notation, this set is written $(-\infty, 3]$.

You Try It 7 $\{x|-5 \le x \le -3\}$ is the set of real numbers between -5 and -3, including -5 and -3. In interval notation, this set is written $[-5, -3]$.

You Try It 8 The interval $(-3, \infty)$ is the set of real numbers greater than -3. In set-builder notation, this set is written $\{x|x > -3\}$.

You Try It 9 The interval $[0, 4)$ is the set of real numbers between 0 and 4, including 0 and excluding 4. In set-builder notation, this set is written $\{x|0 \le x < 4\}$.

You Try It 10 The graph is the set of real numbers between -4 and 4, including -4 and 4. Use brackets at -4 and 4.

−5 −4 −3 −2 −1 0 1 2 3 4 5

You Try It 11 The graph is the set of real numbers greater than -3. Use a parenthesis at -3.

−5 −4 −3 −2 −1 0 1 2 3 4 5

You Try It 12 The graph is the set of real numbers greater than -2. Use a parenthesis at -2.

−5 −4 −3 −2 −1 0 1 2 3 4 5

You Try It 13 The graph is the set of real numbers between 2 and 5, including 2 and 5. Use brackets at 2 and 5.

−5 −4 −3 −2 −1 0 1 2 3 4 5

SOLUTIONS TO CHAPTER 2 "YOU TRY IT"

SECTION 2.1

You Try It 1
$$\begin{array}{r|l} \multicolumn{2}{c}{5 - 4x = 8x + 2} \\ \hline 5 - 4\left(\frac{1}{4}\right) & 8\left(\frac{1}{4}\right) + 2 \\ 5 - 1 & 2 + 2 \\ \multicolumn{2}{c}{4 = 4} \end{array}$$

Yes, $\frac{1}{4}$ is a solution.

You Try It 2
$$\begin{array}{r|l} \multicolumn{2}{c}{10x - x^2 = 3x - 10} \\ \hline 10(5) - (5)^2 & 3(5) - 10 \\ 50 - 25 & 15 - 10 \\ \multicolumn{2}{c}{25 \ne 5} \end{array}$$

No, 5 is not a solution.

You Try It 3
$$\begin{aligned} \frac{5}{6} &= y - \frac{3}{8} \\ \frac{5}{6} + \frac{3}{8} &= y - \frac{3}{8} + \frac{3}{8} \\ \frac{29}{24} &= y \end{aligned}$$

The solution is $\frac{29}{24}$.

You Try It 4
$$\begin{aligned} -\frac{2x}{5} &= 6 \\ \left(-\frac{5}{2}\right)\left(-\frac{2}{5}x\right) &= \left(-\frac{5}{2}\right)(6) \\ x &= -15 \end{aligned}$$

The solution is -15.

You Try It 5

$$4x - 8x = 16$$
$$-4x = 16$$
$$\frac{-4x}{-4} = \frac{16}{-4}$$
$$x = -4$$

The solution is -4.

You Try It 6

$$P \cdot B = A$$
$$\frac{1}{6}B = 18 \qquad \bullet\ 16\frac{2}{3}\% = \frac{1}{6}$$
$$6 \cdot \frac{1}{6}B = 6 \cdot 18$$
$$B = 108$$

18 is $16\frac{2}{3}\%$ of 108.

You Try It 7

Strategy Use the basic percent equation. $B = 300$, the U.S. population; $A = 97.5$, the number of people who watched the game; P is the unknown percent.

Solution

$$P \cdot B = A$$
$$P \cdot 300 = 97.5 \qquad \bullet\ B = 300, A = 97.5$$
$$\frac{P \cdot 300}{300} = \frac{97.5}{300}$$
$$P = 0.325$$

32.5% of the U.S. population watched Super Bowl XLII.

You Try It 8

Strategy To find how much Clarissa must deposit into the account:

- Find the amount of interest earned on the municipal bond by solving $I = Prt$ for I using $P = 1000$, $r = 6.4\% = 0.064$, and $t = 1$.
- Solve $I = Prt$ for P using the amount of interest earned on the municipal bond as I. $r = 8\% = 0.08$, and $t = 1$.

Solution

$$I = Prt$$
$$= 1000(0.064)(1) = 64$$

The interest earned on the municipal bond was \$64.

$$I = Prt$$
$$64 = P(0.08)(1) \qquad \bullet\ I = 64, r = 0.08, t = 1$$
$$64 = 0.08P$$
$$\frac{64}{0.08} = \frac{0.08P}{0.08}$$
$$800 = P$$

Clarissa must invest \$800 in the account.

You Try It 9

Strategy To find the number of ounces of cereal in the bowl, solve $Q = Ar$ for A using $Q = 2$ and $r = 25\% = 0.25$.

Solution

$$Q = Ar$$
$$2 = A(0.25) \qquad \bullet\ Q = 2, r = 0.25$$
$$\frac{2}{0.25} = \frac{A(0.25)}{0.25}$$
$$8 = A$$

The bowl contains 8 oz of cereal.

You Try It 10

Strategy To find the distance, solve the equation $d = rt$ for d. The time is 3 h. Therefore, $t = 3$. The plane is moving against the wind, which means the headwind is slowing the actual speed of the plane. 250 mph $-$ 25 mph $=$ 225 mph. Thus $r = 225$.

Solution

$$d = rt$$
$$d = 225(3) \qquad \bullet\ r = 225, t = 3$$
$$= 675$$

The plane travels 675 mi in 3 h.

SECTION 2.2

You Try It 1

$$5x + 7 = 10$$
$$5x + 7 - 7 = 10 - 7 \qquad \bullet\ \text{Subtract 7.}$$
$$5x = 3$$
$$\frac{5x}{5} = \frac{3}{5} \qquad \bullet\ \text{Divide by 5.}$$
$$x = \frac{3}{5}$$

The solution is $\frac{3}{5}$.

You Try It 2

$$2 = 11 + 3x$$
$$2 - 11 = 11 - 11 + 3x \qquad \bullet\ \text{Subtract 11.}$$
$$-9 = 3x$$
$$\frac{-9}{3} = \frac{3x}{3} \qquad \bullet\ \text{Divide by 3.}$$
$$-3 = x$$

The solution is -3.

You Try It 3

$$\frac{5}{8} - \frac{2x}{3} = \frac{5}{4}$$
$$\frac{5}{8} - \frac{5}{8} - \frac{2}{3}x = \frac{5}{4} - \frac{5}{8}$$ • Recall that $\frac{2x}{3} = \frac{2}{3}x$.
$$-\frac{2}{3}x = \frac{5}{8}$$
$$-\frac{3}{2}\left(-\frac{2}{3}x\right) = -\frac{3}{2}\left(\frac{5}{8}\right)$$ • Multiply by $-\frac{3}{2}$.
$$x = -\frac{15}{16}$$

The solution is $-\frac{15}{16}$.

You Try It 4

$$\frac{2}{3}x + 3 = \frac{7}{2}$$
$$6\left(\frac{2}{3}x + 3\right) = 6\left(\frac{7}{2}\right)$$
$$6\left(\frac{2}{3}x\right) + 6(3) = 6\left(\frac{7}{2}\right)$$ • Distributive Property
$$4x + 18 = 21$$
$$4x + 18 - 18 = 21 - 18$$ • Subtract 18.
$$4x = 3$$
$$\frac{4x}{4} = \frac{3}{4}$$ • Divide by 4.
$$x = \frac{3}{4}$$

The solution is $\frac{3}{4}$.

You Try It 5

$$x - 5 + 4x = 25$$
$$5x - 5 = 25$$
$$5x - 5 + 5 = 25 + 5$$
$$5x = 30$$
$$\frac{5x}{5} = \frac{30}{5}$$
$$x = 6$$

The solution is 6.

You Try It 6

$$5x + 4 = 6 + 10x$$
$$5x - 10x + 4 = 6 + 10x - 10x$$ • Subtract $10x$.
$$-5x + 4 = 6$$
$$-5x + 4 - 4 = 6 - 4$$ • Subtract 4.
$$-5x = 2$$
$$\frac{-5x}{-5} = \frac{2}{-5}$$ • Divide by -5.
$$x = -\frac{2}{5}$$

The solution is $-\frac{2}{5}$.

You Try It 7

$$5x - 10 - 3x = 6 - 4x$$
$$2x - 10 = 6 - 4x$$ • Combine like terms.
$$2x + 4x - 10 = 6 - 4x + 4x$$ • Add $4x$.
$$6x - 10 = 6$$
$$6x - 10 + 10 = 6 + 10$$ • Add 10.
$$6x = 16$$
$$\frac{6x}{6} = \frac{16}{6}$$ • Divide by 6.
$$x = \frac{8}{3}$$

The solution is $\frac{8}{3}$.

You Try It 8

$$5x - 4(3 - 2x) = 2(3x - 2) + 6$$
$$5x - 12 + 8x = 6x - 4 + 6$$ • Distributive Property
$$13x - 12 = 6x + 2$$
$$13x - 6x - 12 = 6x - 6x + 2$$ • Subtract $6x$.
$$7x - 12 = 2$$
$$7x - 12 + 12 = 2 + 12$$ • Add 12.
$$7x = 14$$
$$\frac{7x}{7} = \frac{14}{7}$$ • Divide by 7.
$$x = 2$$

The solution is 2.

You Try It 9

$$-2[3x - 5(2x - 3)] = 3x - 8$$
$$-2[3x - 10x + 15] = 3x - 8$$ • Distributive Property
$$-2[-7x + 15] = 3x - 8$$
$$14x - 30 = 3x - 8$$
$$14x - 3x - 30 = 3x - 3x - 8$$ • Subtract $3x$.
$$11x - 30 = -8$$
$$11x - 30 + 30 = -8 + 30$$ • Add 30.
$$11x = 22$$
$$\frac{11x}{11} = \frac{22}{11}$$ • Divide by 11.
$$x = 2$$

The solution is 2.

You Try It 10

Strategy Given: $F_1 = 45$
$F_2 = 80$
$d = 25$

Unknown: x

Solution

$$F_1x = F_2(d - x)$$
$$45x = 80(25 - x)$$
$$45x = 2000 - 80x$$
$$45x + 80x = 2000 - 80x + 80x$$
$$125x = 2000$$
$$\frac{125x}{125} = \frac{2000}{125}$$
$$x = 16$$

The fulcrum is 16 ft from the 45-pound force.

SECTION 2.3

You Try It 1

The smaller number: n
The larger number: $12 - n$

The total of three times the smaller number and six	amounts to	seven less than the product of four and the larger number

$$\begin{aligned} 3n + 6 &= 4(12 - n) - 7 \\ 3n + 6 &= 48 - 4n - 7 \\ 3n + 6 &= 41 - 4n \\ 3n + 4n + 6 &= 41 - 4n + 4n \\ 7n + 6 &= 41 \\ 7n + 6 - 6 &= 41 - 6 \\ 7n &= 35 \\ \frac{7n}{7} &= \frac{35}{7} \\ n &= 5 \end{aligned}$$

$12 - n = 12 - 5 = 7$

The smaller number is 5.

The larger number is 7.

You Try It 2

Strategy

- First integer: n
  Second integer: $n + 1$
  Third integer: $n + 2$
- The sum of the three integers is -6.

Solution

$$\begin{aligned} n + (n + 1) + (n + 2) &= -6 \\ 3n + 3 &= -6 \\ 3n &= -9 \\ n &= -3 \end{aligned}$$

$n + 1 = -3 + 1 = -2$
$n + 2 = -3 + 2 = -1$

The three consecutive integers are -3, -2, and -1.

You Try It 3

Strategy

To find the number of tickets that you are purchasing, write and solve an equation using x to represent the number of tickets purchased.

Solution

\$3.50 plus \$17.50 for each ticket	is	\$161

$$\begin{aligned} 3.50 + 17.50x &= 161 \\ 3.50 - 3.50 + 17.50x &= 161 - 3.50 \\ 17.50x &= 157.50 \\ \frac{17.50x}{17.50} &= \frac{157.50}{17.50} \\ x &= 9 \end{aligned}$$

You are purchasing 9 tickets.

You Try It 4

Strategy

To find the length, write and solve an equation using x to represent the length of the shorter piece and $22 - x$ to represent the length of the longer piece.

Solution

The length of the longer piece	is	4 in. more than twice the length of the shorter piece

$$\begin{aligned} 22 - x &= 2x + 4 \\ 22 - x - 2x &= 2x - 2x + 4 \\ 22 - 3x &= 4 \\ 22 - 22 - 3x &= 4 - 22 \\ -3x &= -18 \\ \frac{-3x}{-3} &= \frac{-18}{-3} \\ x &= 6 \end{aligned}$$

$22 - x = 22 - 6 = 16$

The length of the shorter piece is 6 in.

The length of the longer piece is 16 in.

SECTION 2.4

You Try It 1

Strategy

- Pounds of \$.55 fertilizer: x

	Amount	Cost	Value
\$.80 fertilizer	20	0.80	0.80(20)
\$.55 fertilizer	x	0.55	$0.55x$
\$.75 fertilizer	$20 + x$	0.75	$0.75(20 + x)$

- The sum of the values before mixing equals the value after mixing.

Solution

$$\begin{aligned} 0.80(20) + 0.55x &= 0.75(20 + x) \\ 16 + 0.55x &= 15 + 0.75x \\ 16 - 0.20x &= 15 \\ -0.20x &= -1 \\ x &= 5 \end{aligned}$$

5 lb of the \$.55 fertilizer must be added.

You Try It 2

Strategy

- Liters of 6% solution: x

	Amount	*Percent*	*Quantity*
6% solution	x	0.06	$0.06x$
12% solution	5	0.12	5(0.12)
8% solution	$x + 5$	0.08	$0.08(x + 5)$

- The sum of the quantities before mixing equals the quantity after mixing.

Solution

$$\begin{aligned} 0.06x + 5(0.12) &= 0.08(x + 5) \\ 0.06x + 0.60 &= 0.08x + 0.40 \\ -0.02x + 0.60 &= 0.40 \\ -0.02x &= -0.20 \\ x &= 10 \end{aligned}$$

The pharmacist adds 10 L of the 6% solution to the 12% solution to get an 8% solution.

You Try It 3

Strategy

- Rate of the first train: r
  Rate of the second train: $2r$

	Rate	*Time*	*Distance*
1st train	r	3	$3r$
2nd train	$2r$	3	$3(2r)$

- The sum of the distances traveled by the two trains equals 288 mi.

Solution

$$\begin{aligned} 3r + 3(2r) &= 288 \\ 3r + 6r &= 288 \\ 9r &= 288 \\ r &= 32 \end{aligned}$$

$$2r = 2(32) = 64$$

The first train is traveling at 32 mph. The second train is traveling at 64 mph.

You Try It 4

Strategy

- Time spent flying out: t
  Time spent flying back: $5 - t$

	Rate	*Time*	*Distance*
Out	150	t	$150t$
Back	100	$5 - t$	$100(5 - t)$

- The distance out equals the distance back.

Solution

$$\begin{aligned} 150t &= 100(5 - t) \\ 150t &= 500 - 100t \\ 250t &= 500 \\ t &= 2 \quad \text{(The time out was 2 h.)} \end{aligned}$$

$$\text{The distance out} = 150t = 150(2) = 300 \text{ mi}$$

The parcel of land was 300 mi away.

SECTION 2.5

You Try It 1

$$\begin{aligned} 2x - 1 &< 6x + 7 \\ -4x - 1 &< 7 && \text{• Subtract } 6x \text{ from each side.} \\ -4x &< 8 && \text{• Add 1 to each side.} \\ \frac{-4x}{-4} &> \frac{8}{-4} && \text{• Divide each side by } -4. \\ x &> -2 \end{aligned}$$

$\{x \mid x > -2\}$

−5 −4 −3 −2 −1 0 1 2 3 4 5

You Try It 2

$$\begin{aligned} 5x - 2 &\le 4 - 3(x - 2) \\ 5x - 2 &\le 4 - 3x + 6 \\ 5x - 2 &\le 10 - 3x \\ 8x - 2 &\le 10 \\ 8x &\le 12 \\ \frac{8x}{8} &\le \frac{12}{8} \\ x &\le \frac{3}{2} \end{aligned}$$

$\left(-\infty, \frac{3}{2}\right]$

You Try It 3

$$\begin{aligned} -2 &\le 5x + 3 \le 13 \\ -2 - 3 &\le 5x + 3 - 3 \le 13 - 3 && \text{• Subtract 3 from each of the three parts.} \\ -5 &\le 5x \le 10 \\ \frac{-5}{5} &\le \frac{5x}{5} \le \frac{10}{5} && \text{• Divide each of the three parts by 5.} \\ -1 &\le x \le 2 \end{aligned}$$

$[-1, 2]$

You Try It 4

$$\begin{aligned} 2 - 3x &> 11 & \text{or} \qquad 5 + 2x &> 7 \\ -3x &> 9 & 2x &> 2 \\ x &< -3 & x &> 1 \\ \{x \mid x &< -3\} & \{x \mid x &> 1\} \end{aligned}$$

$$\{x \mid x < -3\} \cup \{x \mid x > 1\} = \{x \mid x < -3 \text{ or } x > 1\}$$

You Try It 5

Strategy To find the maximum height, substitute the given values in the inequality $\frac{1}{2}bh < A$ and solve.

Solution

$$\frac{1}{2}bh < A$$
$$\frac{1}{2}(12)(x+2) < 50$$
$$6(x+2) < 50$$
$$6x + 12 < 50$$
$$6x < 38$$
$$x < \frac{19}{3}$$

The largest integer less than $\frac{19}{3}$ is 6.

$x + 2 = 6 + 2 = 8$

The maximum height of the triangle is 8 in.

You Try It 6

Strategy To find the range of scores, write and solve an inequality using N to represent the score on the last test.

Solution

$$80 \le \frac{72 + 94 + 83 + 70 + N}{5} \le 89$$
$$80 \le \frac{319 + N}{5} \le 89$$
$$5 \cdot 80 \le 5\left(\frac{319 + N}{5}\right) \le 5 \cdot 89$$
$$400 \le 319 + N \le 445$$
$$400 - 319 \le 319 + N - 319 \le 445 - 319$$
$$81 \le N \le 126$$

Because 100 is the maximum score, the range of scores to receive a B grade is $81 \le N \le 100$.

SECTION 2.6

You Try It 1

$|2x - 3| = 5$

$$2x - 3 = 5 \qquad 2x - 3 = -5$$
$$2x = 8 \qquad 2x = -2 \qquad \text{• Add 3.}$$
$$x = 4 \qquad x = -1 \qquad \text{• Divide by 2.}$$

The solutions are 4 and -1.

You Try It 2 $|x - 3| = -2$

There is no solution to this equation because the absolute value of a number must be nonnegative.

You Try It 3

$$5 - |3x + 5| = 3$$
$$-|3x + 5| = -2 \qquad \text{• Subtract 5.}$$
$$|3x + 5| = 2 \qquad \text{• Multiply by } -1.$$

$$3x + 5 = 2 \qquad 3x + 5 = -2$$
$$3x = -3 \qquad 3x = -7$$
$$x = -1 \qquad x = -\frac{7}{3}$$

The solutions are -1 and $-\frac{7}{3}$.

You Try It 4

$$|3x + 2| < 8$$
$$-8 < 3x + 2 < 8$$
$$-8 - 2 < 3x + 2 - 2 < 8 - 2$$
$$-10 < 3x < 6$$
$$\frac{-10}{3} < \frac{3x}{3} < \frac{6}{3}$$
$$-\frac{10}{3} < x < 2$$
$$\left\{x \,\middle|\, -\frac{10}{3} < x < 2\right\}$$

You Try It 5 $|3x - 7| < 0$

The absolute value of a number must be nonnegative.

The solution set is the empty set.

$\varnothing$

You Try It 6 $|2x + 7| \ge -1$

The absolute value of a number is nonnegative.

The solution set is the set of real numbers.

You Try It 7

$|5x + 3| > 8$

$$5x + 3 < -8 \quad \text{or} \quad 5x + 3 > 8$$
$$5x < -11 \qquad 5x > 5$$
$$x < -\frac{11}{5} \qquad x > 1$$
$$\left\{x \middle| x < -\frac{11}{5}\right\} \qquad \{x|x > 1\}$$
$$\left\{x \middle| x < -\frac{11}{5}\right\} \cup \{x|x > 1\}$$
$$= \left\{x \middle| x < -\frac{11}{5} \text{ or } x > 1\right\}$$

You Try It 8

Strategy

Let b represent the diameter of the bushing, T the tolerance, and d the lower and upper limits of the diameter. Solve the absolute value inequality $|d - b| \le T$ for d.

Solution

$$|d - b| \le T$$
$$|d - 2.55| \le 0.003$$
$$-0.003 \le d - 2.55 \le 0.003$$
$$-0.003 + 2.55 \le d - 2.55 + 2.55 \le 0.003 + 2.55$$
$$2.547 \le d \le 2.553$$

The lower and upper limits of the diameter of the bushing are 2.547 in. and 2.553 in.

SOLUTIONS TO CHAPTER 3 "YOU TRY IT"

SECTION 3.1

You Try It 1

$$QR + RS + ST = QT$$
$$24 + RS + 17 = 62$$
$$41 + RS = 62$$
$$RS = 21$$

$RS = 21$ cm

You Try It 2

$$AC = AB + BC$$
$$AC = \frac{1}{4}(BC) + BC$$
$$AC = \frac{1}{4}(16) + 16$$
$$AC = 4 + 16$$
$$AC = 20$$

$AC = 20$ ft

You Try It 3

Strategy Supplementary angles are two angles whose sum is 180°. To find the supplement, let x represent the supplement of a 129° angle. Write an equation and solve for x.

Solution

$$x + 129° = 180°$$
$$x = 51°$$

The supplement of a 129° angle is a 51° angle.

You Try It 4

Strategy To find the measure of $\angle a$, write an equation using the fact that the sum of the measure of $\angle a$ and 68° is 118°. Solve for $\angle a$.

Solution

$$\angle a + 68° = 118°$$
$$\angle a = 50°$$

The measure of $\angle a$ is 50°.

You Try It 5

Strategy The angles labeled are adjacent angles of intersecting lines and are, therefore, supplementary angles. To find x, write an equation and solve for x.

Solution

$$(x + 16°) + 3x = 180°$$
$$4x + 16° = 180°$$
$$4x = 164°$$
$$x = 41°$$

You Try It 6

Strategy $3x = y$ because corresponding angles have the same measure. $y + (x + 40°) = 180°$ because adjacent angles of intersecting lines are supplementary angles. Substitute $3x$ for y and solve for x.

Solution

$$3x + (x + 40°) = 180°$$
$$4x + 40° = 180°$$
$$4x = 140°$$
$$x = 35°$$

You Try It 7

Strategy

- To find the measure of angle b, use the fact that $\angle b$ and $\angle x$ are supplementary angles.
- To find the measure of angle c, use the fact that the sum of the measures of the interior angles of a triangle is 180°.
- To find the measure of angle y, use the fact that $\angle c$ and $\angle y$ are vertical angles.

Solution

$$\angle b + \angle x = 180°$$
$$\angle b + 100° = 180°$$
$$\angle b = 80°$$

$$\angle a + \angle b + \angle c = 180°$$
$$45° + 80° + \angle c = 180°$$
$$125° + \angle c = 180°$$
$$\angle c = 55°$$

$$\angle y = \angle c = 55°$$

You Try It 8

Strategy To find the measure of the third angle, use the facts that the measure of a right angle is 90° and the sum of the measures of the interior angles of a triangle is 180°. Write an equation using x to represent the measure of the third angle. Solve the equation for x.

Solution

$$x + 90^\circ + 34^\circ = 180^\circ$$
$$x + 124^\circ = 180^\circ$$
$$x = 56^\circ$$

The measure of the third angle is 56°.

SECTION 3.2

You Try It 1

Strategy To find the perimeter, use the formula for the perimeter of a square. Substitute 60 ft for s and solve for P.

Solution

$$P = 4s$$
$$P = 4(60 \text{ ft})$$
$$P = 240 \text{ ft}$$

The perimeter of the infield is 240 ft.

You Try It 2

Strategy To find the perimeter, use the formula for the perimeter of a rectangle.

Substitute 11 in. for L and $8\frac{1}{2}$ in. for W and solve for P.

Solution

$$P = 2L + 2W$$
$$P = 2(11 \text{ in.}) + 2\left(8\frac{1}{2} \text{ in.}\right)$$
$$P = 2(11 \text{ in.}) + 2\left(\frac{17}{2} \text{ in.}\right)$$
$$P = 22 \text{ in.} + 17 \text{ in.}$$
$$P = 39 \text{ in.}$$

The perimeter of a standard piece of typing paper is 39 in.

You Try It 3

Strategy To find the circumference, use the circumference formula that involves the diameter. Leave the answer in terms of π.

Solution

$$C = \pi d$$
$$C = \pi(9 \text{ in.})$$
$$C = 9\pi \text{ in.}$$

The circumference is 9π in.

You Try It 4

Strategy To find the number of rolls of wallpaper to be purchased:

- Use the formula for the area of a rectangle to find the area of one wall.
- Multiply the area of one wall by the number of walls to be covered (2).
- Divide the area of wall to be covered by the area that one roll of wallpaper will cover (30).

Solution

$A = LW$

$A = (12 \text{ ft})(8 \text{ ft}) = 96 \text{ ft}^2$ The area of one wall is 96 ft^2.

$2(96 \text{ ft}^2) = 192 \text{ ft}^2$ The area of the two walls is 192 ft^2.

$192 \div 30 = 6.4$

Because a portion of a seventh roll is needed, 7 rolls of wallpaper should be purchased.

You Try It 5

Strategy To find the area, use the formula for the area of a circle. An approximation is asked for; use the π key on a calculator. $r = 11$ cm.

Solution

$$A = \pi r^2$$
$$A = \pi(11 \text{ cm})^2$$
$$A = 121\pi \text{ cm}^2$$
$$A \approx 380.13 \text{ cm}^2$$

The area is approximately 380.13 cm^2.

SECTION 3.3

You Try It 1

Strategy To find the volume, use the formula for the volume of a cube. $s = 2.5$ m.

Solution

$$V = s^3$$
$$V = (2.5 \text{ m})^3 = 15.625 \text{ m}^3$$

The volume of the cube is 15.625 m^3.

You Try It 2

Strategy To find the volume:

- Find the radius of the base of the cylinder. $d = 8$ ft.
- Use the formula for the volume of a cylinder. Leave the answer in terms of π.

Solution

$$r = \frac{1}{2}d = \frac{1}{2}(8 \text{ ft}) = 4 \text{ ft}$$
$$V = \pi r^2 h = \pi(4 \text{ ft})^2(22 \text{ ft})$$
$$= \pi(16 \text{ ft}^2)(22 \text{ ft}) = 352\pi \text{ ft}^3$$

The volume of the cylinder is 352π ft^3.

You Try It 3

Strategy To find the surface area of the cylinder:

- Find the radius of the base of the cylinder. $d = 6$ ft.
- Use the formula for the surface area of a cylinder. An approximation is asked for; use the π key on a calculator.

Solution

$r = \frac{1}{2}d = \frac{1}{2}(6 \text{ ft}) = 3 \text{ ft}$

$SA = 2\pi r^2 + 2\pi rh$
$SA = 2\pi(3 \text{ ft})^2 + 2\pi(3 \text{ ft})(8 \text{ ft})$
$SA = 2\pi(9 \text{ ft}^2) + 2\pi(3 \text{ ft})(8 \text{ ft})$
$SA = 18\pi \text{ ft}^2 + 48\pi \text{ ft}^2$
$SA = 66\pi \text{ ft}^2$
$SA \approx 207.35 \text{ ft}^2$

The surface area of the cylinder is approximately 207.35 ft^2.

You Try It 4

Strategy

To find which solid has the larger surface area:

- Use the formula for the surface area of a cube to find the surface area of the cube. $s = 10$ cm.
- Find the radius of the sphere. $d = 8$ cm.
- Use the formula for the surface area of a sphere to find the surface area of the sphere. Because this number is to be compared to another number, use the π key on a calculator to approximate the surface area.
- Compare the two numbers.

Solution

$SA = 6s^2$
$SA = 6(10 \text{ cm})^2 = 6(100 \text{ cm}^2) = 600 \text{ cm}^2$
The surface area of the cube is 600 cm^2.

$r = \frac{1}{2}d = \frac{1}{2}(8 \text{ cm}) = 4 \text{ cm}$

$SA = 4\pi r^2$

$SA = 4\pi(4 \text{ cm})^2 = 4\pi(16 \text{ cm}^2) = 64\pi \text{ cm}^2 \approx 201.06 \text{ cm}^2$
The surface area of the sphere is approximately 201.06 cm^2.

$600 > 201.06$

The cube has a larger surface area than the sphere.

SOLUTIONS TO CHAPTER 4 "YOU TRY IT"

SECTION 4.1

You Try It 1

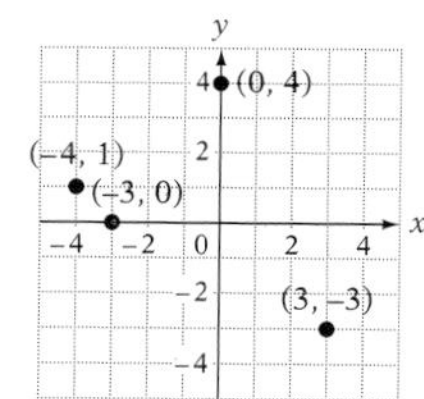

You Try It 2 $A(4, -2), B(-2, 4)$
The abscissa of D is 0.
The ordinate of C is 0.

You Try It 3

$x - 3y = -14$	
$-2 - 3(4)$	-14
$-2 - 12$	-14
$-14 =$	-14

Yes, $(-2, 4)$ is a solution of $x - 3y = -14$.

You Try It 4 Replace x by -2 and solve for y.

$y = \frac{3x}{x+1} = \frac{3(-2)}{-2+1} = \frac{-6}{-1} = 6$

The ordered-pair solution is $(-2, 6)$.

You Try It 5 $y = -\frac{1}{2}x + 2$

x	y
−4	4
−2	3
0	2
2	1

You Try It 6

Strategy To draw a scatter diagram:

- Draw a coordinate grid with the horizontal axis representing the time in seconds and the vertical axis representing the temperature in °F. Because the temperatures start at 72°, it is more convenient to start labeling the vertical axis at 70 (any number less than 72 could be used). The jagged line on the vertical axis indicates that the graph does not start at zero.
- Graph the ordered pairs (0, 72), (10, 76), (20, 88), (30, 106), (40, 126), (50, 140), and (60, 160).

Solution

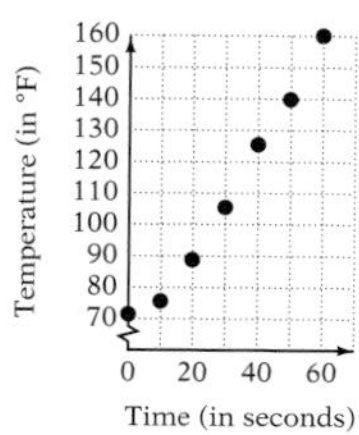

SECTION 4.2

You Try It 1 Domain: $\{-1, 3, 4, 6\}$ • The domain is the set of first coordinates.
Range: $\{5\}$

You Try It 2

$$G(x) = \frac{3x}{x+2}$$
$$G(-4) = \frac{3(-4)}{-4+2} = \frac{-12}{-2} = 6$$

You Try It 3

$$f(x) = x^2 - 11$$
$$f(3h) = (3h)^2 - 11$$
$$= 9h^2 - 11$$

You Try It 4

$$h(z) = 3z + 1$$
$$h(0) = 3(0) + 1 = 1$$
$$h\left(\frac{1}{3}\right) = 3\left(\frac{1}{3}\right) + 1 = 2$$
$$h\left(\frac{2}{3}\right) = 3\left(\frac{2}{3}\right) + 1 = 3$$
$$h(1) = 3(1) + 1 = 4$$

The range is {1, 2, 3, 4}.

You Try It 5

$$f(x) = \frac{2}{x-5}$$

For $x = 5, f(5) = \frac{2}{5-5} = \frac{2}{0}$, which is not a real number.

5 is excluded from the domain of the function.

SECTION 4.3

You Try It 1

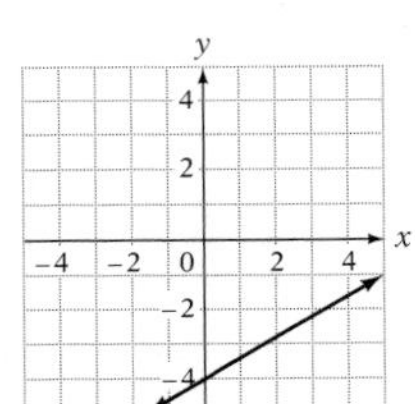

You Try It 2

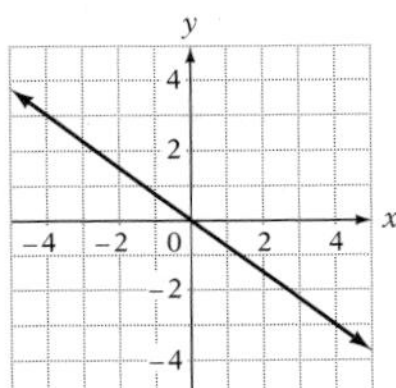

You Try It 3

$$-3x + 2y = 4$$
$$2y = 3x + 4$$
$$y = \frac{3}{2}x + 2$$

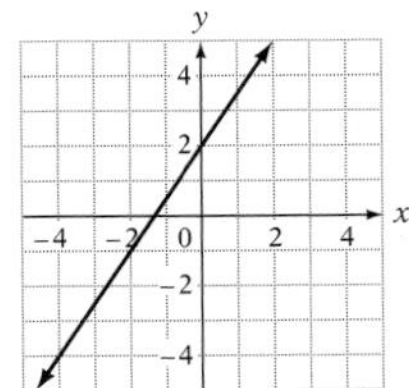

You Try It 4

$$y - 3 = 0$$
$$y = 3$$

• The graph of $y = 3$ goes through the point (0, 3).

You Try It 5

x-intercept:

$$3x - y = 2$$
$$3x - 0 = 2 \quad \text{• Let } y = 0.$$
$$3x = 2$$
$$x = \frac{2}{3}$$

x-intercept: $\left(\frac{2}{3}, 0\right)$

y-intercept:

$$3x - y = 2$$
$$3(0) - y = 2 \quad \text{• Let } x = 0.$$
$$-y = 2$$
$$y = -2$$

y-intercept: $(0, -2)$

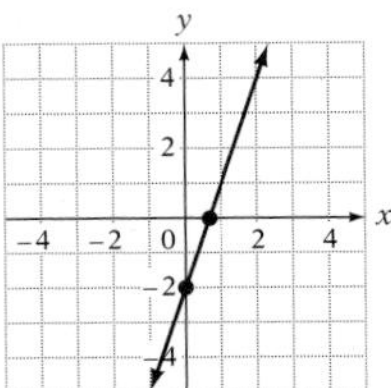

You Try It 6

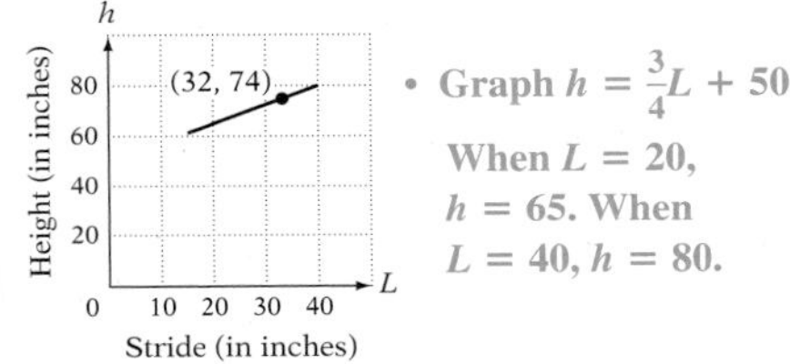

• Graph $h = \frac{3}{4}L + 50$. When $L = 20$, $h = 65$. When $L = 40$, $h = 80$.

The ordered pair (32, 74) means that a person with a stride of 32 in. is 74 in. tall.

SECTION 4.4

You Try It 1

Choose $P_1(4, -3)$ and $P_2(2, 7)$.

$$m = \frac{y_2 - y_1}{x_2 - x_1} = \frac{7 - (-3)}{2 - 4} = \frac{10}{-2} = -5$$

The slope is -5.

You Try It 2

Choose $P_1(6, -1)$ and $P_2(6, 7)$.

$$m = \frac{y_2 - y_1}{x_2 - x_1} = \frac{7 - (-1)}{6 - 6} = \frac{8}{0}$$

Division by zero is not defined.

The slope of the line is undefined.

You Try It 3 Choose $P_1(5, 25{,}000)$ and $P_2(2, 55{,}000)$.

$$m = \frac{y_2 - y_1}{x_2 - x_1}$$
$$= \frac{55{,}000 - 25{,}000}{2 - 5} \quad \bullet\ (x_1, y_1) = (5, 25{,}000),\ (x_2, y_2) = (2, 55{,}000)$$
$$= \frac{30{,}000}{-3}$$
$$= -10{,}000$$

A slope of $-10{,}000$ means that the value of the recycling truck is decreasing by \$10,000 per year.

You Try It 4

$$2x + 3y = 6$$
$$3y = -2x + 6$$
$$y = -\frac{2}{3}x + 2$$
$$m = -\frac{2}{3} = \frac{-2}{3}$$

y-intercept $= (0, 2)$

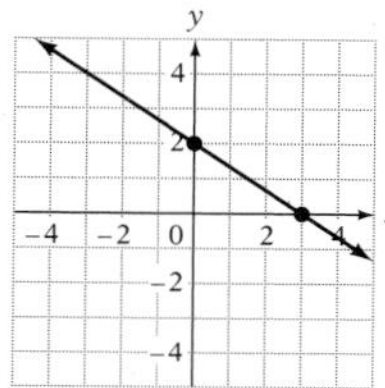

You Try It 5 Locate $(-3, -2)$.

$$m = 3 = \frac{3}{1}$$

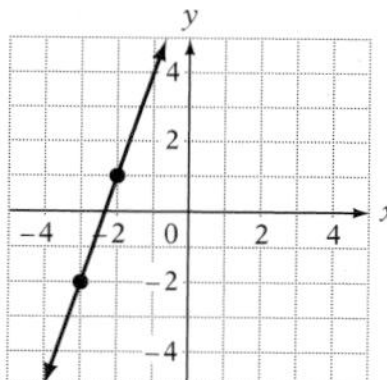

SECTION 4.5

You Try It 1 $m = -\frac{1}{3}$ $\quad (x_1, y_1) = (-3, -2)$

$$y - y_1 = m(x - x_1)$$
$$y - (-2) = -\frac{1}{3}[x - (-3)]$$
$$y + 2 = -\frac{1}{3}(x + 3)$$
$$y + 2 = -\frac{1}{3}x - 1$$
$$y = -\frac{1}{3}x - 3$$

The equation of the line is $y = -\frac{1}{3}x - 3$.

You Try It 2 $m = -3 \quad (x_1, y_1) = (4, -3)$

$$y - y_1 = m(x - x_1)$$
$$y - (-3) = -3(x - 4)$$
$$y + 3 = -3x + 12$$
$$y = -3x + 9$$

The equation of the line is $y = -3x + 9$.

You Try It 3 Choose $P_1(2, 0)$ and $P_2(5, 3)$.

$$m = \frac{y_2 - y_1}{x_2 - x_1} = \frac{3 - 0}{5 - 2} = \frac{3}{3} = 1$$
$$y - y_1 = m(x - x_1)$$
$$y - 0 = 1(x - 2)$$
$$y = 1(x - 2)$$
$$y = x - 2$$

The equation of the line is $y = x - 2$.

You Try It 4 Choose $P_1(2, 3)$ and $P_2(-5, 3)$.

$$m = \frac{y_2 - y_1}{x_2 - x_1} = \frac{3 - 3}{-5 - 2} = \frac{0}{-7} = 0$$

The line has zero slope, so the line is a horizontal line.
All points on the line have an ordinate of 3.
The equation of the line is $y = 3$.

You Try It 5

Strategy

- Select the independent and dependent variables. The function is to predict the Celsius temperature, so that quantity is the dependent variable, y. The Fahrenheit temperature is the independent variable, x.
- From the given data, two ordered pairs are (212, 100) and (32, 0). Use these ordered pairs to determine the linear function.

Solution Choose $P_1(32, 0)$ and $P_2(212, 100)$.

$$m = \frac{y_2 - y_1}{x_2 - x_1} = \frac{100 - 0}{212 - 32} = \frac{100}{180} = \frac{5}{9}$$
$$y - y_1 = m(x - x_1)$$
$$y - 0 = \frac{5}{9}(x - 32)$$
$$y = \frac{5}{9}(x - 32), \text{ or } C = \frac{5}{9}(F - 32)$$

The linear function is $f(F) = \frac{5}{9}(F - 32)$.

SECTION 4.6

You Try It 1

$$m_1 = \frac{1 - (-3)}{7 - (-2)} = \frac{4}{9}$$

• $(x_1, y_1) = (-2, -3)$, $(x_2, y_2) = (7, 1)$

$$m_2 = \frac{-5 - 1}{6 - 4} = \frac{-6}{2} = -3$$

• $(x_1, y_1) = (4, 1)$, $(x_2, y_2) = (6, -5)$

$$m_1 \cdot m_2 = \frac{4}{9}(-3) = -\frac{4}{3}$$

No, the lines are not perpendicular.

You Try It 2

$$\begin{aligned} 5x + 2y &= 2 \\ 2y &= -5x + 2 \\ y &= -\frac{5}{2}x + 1 \end{aligned}$$

$$m_1 = -\frac{5}{2}$$

$$\begin{aligned} 5x + 2y &= -6 \\ 2y &= -5x - 6 \\ y &= -\frac{5}{2}x - 3 \end{aligned}$$

$$m_2 = -\frac{5}{2}$$

$$m_1 = m_2 = -\frac{5}{2}$$

Yes, the lines are parallel.

You Try It 3

$$\begin{aligned} x - 4y &= 3 \\ -4y &= -x + 3 \\ y &= \frac{1}{4}x - \frac{3}{4} \end{aligned}$$

$$m_1 = \frac{1}{4}$$

$$\begin{aligned} m_1 \cdot m_2 &= -1 \\ \frac{1}{4} \cdot m_2 &= -1 \\ m_2 &= -4 \end{aligned}$$

$$\begin{aligned} y - y_1 &= m(x - x_1) \\ y - 2 &= -4[x - (-2)] \\ y - 2 &= -4(x + 2) \\ y - 2 &= -4x - 8 \\ y &= -4x - 6 \end{aligned}$$

• $(x_1, y_1) = (-2, 2)$

The equation of the line is $y = -4x - 6$.

SECTION 4.7

You Try It 1

$$\begin{aligned} x + 3y &> 6 \\ 3y &> -x + 6 \\ y &> -\frac{1}{3}x + 2 \end{aligned}$$

You Try It 2 $y < 2$

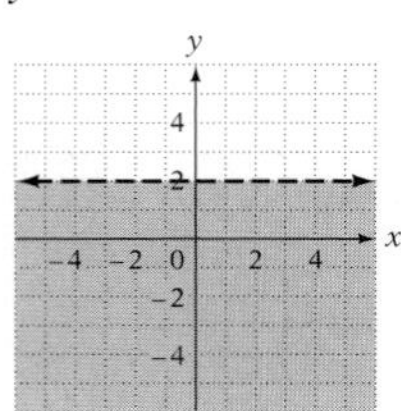

SOLUTIONS TO CHAPTER 5 "YOU TRY IT"

SECTION 5.1

You Try It 1

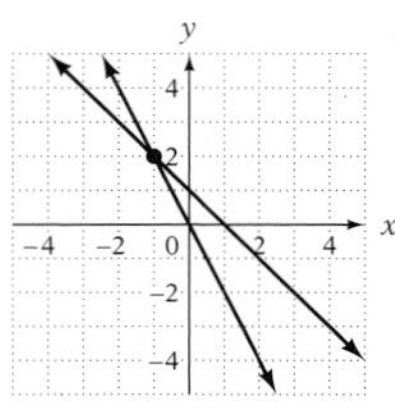

• Find the point of intersection of the graphs of the equations.

The solution is $(-1, 2)$.

You Try It 2

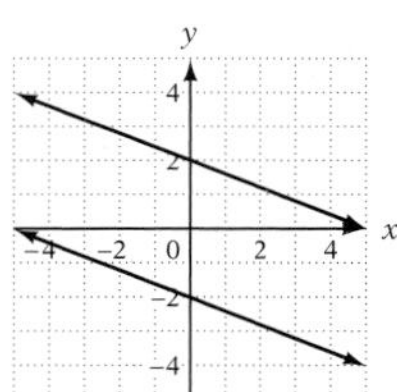

• Graph the two equations.

The lines are parallel and therefore do not intersect. The system of equations has no solution. The system of equations is inconsistent.

You Try It 3

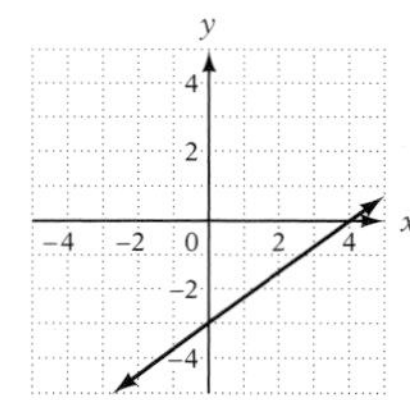

• Graph the two equations.

The two equations represent the same line. The system of equations is dependent. The solutions are the ordered pairs $\left(x, \frac{3}{4}x - 3\right)$.

You Try It 4

(1) $3x - y = 3$

(2) $6x + 3y = -4$

Solve Equation (1) for y.

$$\begin{aligned} 3x - y &= 3 \\ -y &= -3x + 3 \\ y &= 3x - 3 \end{aligned}$$

Substitute into Equation (2).

$$\begin{aligned} 6x + 3y &= -4 \\ 6x + 3(3x - 3) &= -4 \\ 6x + 9x - 9 &= -4 \\ 15x - 9 &= -4 \\ 15x &= 5 \\ x &= \frac{5}{15} = \frac{1}{3} \end{aligned}$$

Substitute the value of x into Equation (1).

$$\begin{aligned} 3x - y &= 3 \\ 3\left(\frac{1}{3}\right) - y &= 3 \\ 1 - y &= 3 \\ -y &= 2 \\ y &= -2 \end{aligned}$$

The solution is $\left(\frac{1}{3}, -2\right)$.

You Try It 5

(1) $y = 2x - 3$

(2) $3x - 2y = 6$

$$\begin{aligned} 3x - 2y &= 6 \\ 3x - 2(2x - 3) &= 6 \\ 3x - 4x + 6 &= 6 \\ -x + 6 &= 6 \\ -x &= 0 \\ x &= 0 \end{aligned}$$

Substitute the value of x into Equation (1).

$$\begin{aligned} y &= 2x - 3 \\ y &= 2(0) - 3 \\ y &= 0 - 3 \\ y &= -3 \end{aligned}$$

The solution is $(0, -3)$.

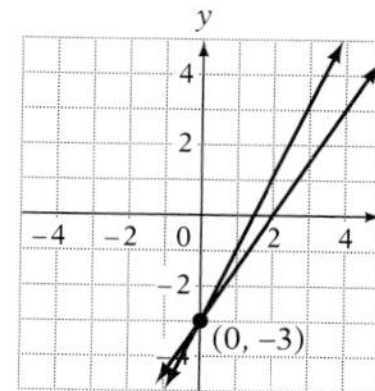

• Graph the two equations.

You Try It 6

(1) $6x - 3y = 6$

(2) $2x - y = 2$

Solve Equation (2) for y.

$$\begin{aligned} 2x - y &= 2 \\ -y &= -2x + 2 \\ y &= 2x - 2 \end{aligned}$$

Substitute into Equation (1).

$$\begin{aligned} 6x - 3y &= 6 \\ 6x - 3(2x - 2) &= 6 \\ 6x - 6x + 6 &= 6 \\ 6 &= 6 \end{aligned}$$

The system of equations is dependent. The solutions are the ordered pairs $(x, 2x - 2)$.

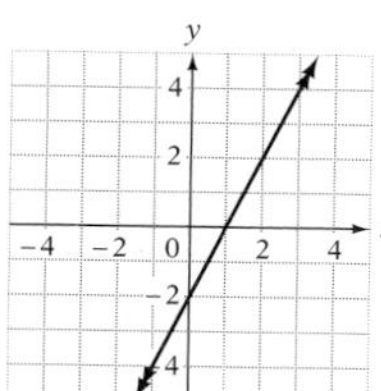

• Graph the two equations.

You Try It 7

Strategy

• Amount invested at 4.2%: x
Amount invested at 6%: y

	Principal	*Rate*	*Interest*
Amount at 4.2%	x	0.042	$0.042x$
Amount at 6%	y	0.06	$0.06y$

• The total investment is \$13,600.
The two accounts earn the same interest.

Solution

$$\begin{aligned} x + y &= 13{,}600 \\ 0.042x &= 0.06y \\ x &= \frac{10}{7}y \end{aligned}$$

Substitute $\frac{10}{7}y$ for x in $x + y = 13{,}600$ and solve for y.

$$\begin{aligned} \frac{10}{7}y + y &= 13{,}600 \\ \frac{17}{7}y &= 13{,}600 \\ y &= 5600 \end{aligned}$$

$$\begin{aligned} x + 5600 &= 13{,}600 \\ x &= 8000 \end{aligned}$$

\$8000 must be invested at 4.2%, and \$5600 must be invested at 6%.

SECTION 5.2

You Try It 1 (1) $2x + 5y = 6$
(2) $3x - 2y = 6x + 2$

Write Equation (2) in the form $Ax + By = C$.

$$3x - 2y = 6x + 2$$
$$-3x - 2y = 2$$

Solve the system: $2x + 5y = 6$, $-3x - 2y = 2$

Eliminate y.

$$2(2x + 5y) = 2(6)$$
$$5(-3x - 2y) = 5(2)$$

$$4x + 10y = 12$$
$$-15x - 10y = 10$$
$$-11x = 22 \quad \text{• Add the equations.}$$
$$x = -2 \quad \text{• Solve for } x.$$

Replace x in Equation (1).

$$2x + 5y = 6$$
$$2(-2) + 5y = 6$$
$$-4 + 5y = 6$$
$$5y = 10$$
$$y = 2$$

The solution is $(-2, 2)$.

You Try It 2 $2x + y = 5$
$4x + 2y = 6$

Eliminate y.

$$-2(2x + y) = -2(5)$$
$$4x + 2y = 6$$

$$-4x - 2y = -10$$
$$4x + 2y = 6$$
$$0x + 0y = -4 \quad \text{• Add the equations.}$$
$$0 = -4$$

This is not a true equation.
The system is inconsistent and therefore has no solution.

You Try It 3 (1) $x - y + z = 6$
(2) $2x + 3y - z = 1$
(3) $x + 2y + 2z = 5$

Eliminate z. Add Equations (1) and (2).

$$x - y + z = 6$$
$$2x + 3y - z = 1$$
$$3x + 2y = 7 \quad \text{• Equation (4)}$$

Multiply Equation (2) by 2 and add to Equation (3).

$$4x + 6y - 2z = 2$$
$$x + 2y + 2z = 5$$
$$5x + 8y = 7 \quad \text{• Equation (5)}$$

Solve the system of two equations.

(4) $3x + 2y = 7$
(5) $5x + 8y = 7$

Multiply Equation (4) by -4 and add to Equation (5).

$$-12x - 8y = -28$$
$$5x + 8y = 7$$
$$-7x = -21$$
$$x = 3$$

Replace x by 3 in Equation (4).

$$3x + 2y = 7$$
$$3(3) + 2y = 7$$
$$9 + 2y = 7$$
$$2y = -2$$
$$y = -1$$

Replace x by 3 and y by -1 in Equation (1).

$$x - y + z = 6$$
$$3 - (-1) + z = 6$$
$$4 + z = 6$$
$$z = 2$$

The solution is $(3, -1, 2)$.

SECTION 5.3

You Try It 1

Strategy

- Rate of the rowing team in calm water: t
  Rate of the current: c

	Rate	Time	Distance
With current	$t + c$	2	$2(t + c)$
Against current	$t - c$	2	$2(t - c)$

- The distance traveled with the current is 18 mi.
  The distance traveled against the current is 10 mi.

$$2(t + c) = 18$$
$$2(t - c) = 10$$

Solution

$$2(t + c) = 18 \qquad \frac{1}{2} \cdot 2(t + c) = \frac{1}{2} \cdot 18$$
$$2(t - c) = 10 \qquad \frac{1}{2} \cdot 2(t - c) = \frac{1}{2} \cdot 10$$
$$t + c = 9$$
$$t - c = 5$$
$$2t = 14$$
$$t = 7$$

$$t + c = 9$$
$$7 + c = 9 \quad \text{• Substitute 7 for } t.$$
$$c = 2$$

The rate of the rowing team in calm water is 7 mph.
The rate of the current is 2 mph.

You Try It 2

Strategy

- Number of dimes: d
  Number of nickels: n
  Number of quarters: q
- There are 19 coins in a bank that contains only nickels, dimes, and quarters. $(n + d + q = 19)$
  The value of the coins is \$2.
  $(5n + 10d + 25q = 200)$
  There are twice as many nickels as dimes. $(n = 2d)$

Solution

$$(1) \quad n + d + q = 19$$
$$(2) \quad n = 2d$$
$$(3) \quad 5n + 10d + 25q = 200$$

Solve the system of equations. Substitute $2d$ for n in Equation (1) and Equation (3).

$$2d + d + q = 19$$
$$5(2d) + 10d + 25q = 200$$

$$(4) \quad 3d + q = 19$$
$$(5) \quad 20d + 25q = 200$$

Solve the system of equations in two variables by multiplying Equation (4) by -25 and then adding it to Equation (5).

$$\begin{aligned} -75d - 25q &= -475 \\ 20d + 25q &= 200 \\ \hline -55d &= -275 \\ d &= 5 \end{aligned}$$

Substituting the value of d into Equation (2), we have $n = 10$. Substituting the values of d and n into Equation (1), we have $q = 4$.

There are 10 nickels, 5 dimes, and 4 quarters in the bank.

SECTION 5.4

You Try It 1 Shade above the solid line $y = 2x - 3$.
Shade above the dashed line $y = -3x$.

The solution set of the system is the intersection of the solution sets of the individual inequalities.

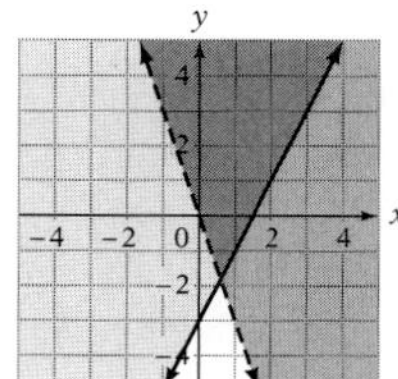

You Try It 2

$$\begin{aligned} 3x + 4y &> 12 \\ 4y &> -3x + 12 \\ y &> -\frac{3}{4}x + 3 \end{aligned}$$

Shade above the dashed line $y = -\frac{3}{4}x + 3$.

Shade below the dashed line $y = \frac{3}{4}x - 1$.

The solution set of the system is the intersection of the solution sets of the individual inequalities.

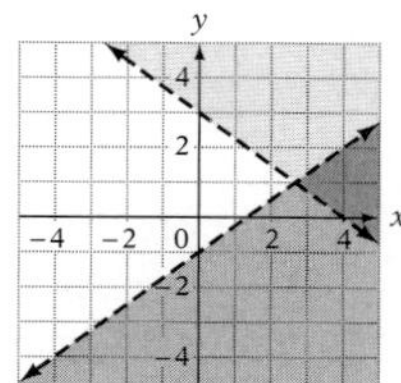

SOLUTIONS TO CHAPTER 6 "YOU TRY IT"

SECTION 6.1

You Try It 1

$$\begin{aligned} (-3a^2b^4)(-2ab^3)^4 &= (-3a^2b^4)[(-2)^4a^4b^{12}] \\ &= (-3a^2b^4)(16a^4b^{12}) \\ &= -48a^6b^{16} \end{aligned}$$

You Try It 2

$$\begin{aligned} (-4ab^4)^2(2a^4b^2)^4 &= [(-4)^2a^2b^8][2^4a^{16}b^8] \\ &= [16a^2b^8][16a^{16}b^8] \\ &= 256a^{18}b^{16} \end{aligned}$$

You Try It 3

$$\begin{aligned} (y^{n-3})^2 &= y^{(n-3)2} \quad \text{• Multiply the exponents.} \\ &= y^{2n-6} \end{aligned}$$

You Try It 4

$$\begin{aligned} \frac{20r^{-2}t^{-5}}{-16r^{-3}s^{-2}} &= -\frac{4 \cdot 5r^{-2-(-3)}s^2t^{-5}}{4 \cdot 4} \\ &= -\frac{5rs^2}{4t^5} \end{aligned}$$

You Try It 5

$$\begin{aligned} \frac{(9u^{-6}v^4)^{-1}}{(6u^{-3}v^{-2})^{-2}} &= \frac{9^{-1}u^6v^{-4}}{6^{-2}u^6v^4} \\ &= 9^{-1} \cdot 6^2u^0v^{-8} \\ &= \frac{36}{9v^8} \\ &= \frac{4}{v^8} \end{aligned}$$

You Try It 6

$$\begin{aligned} \frac{a^{2n+1}}{a^{n+3}} &= a^{2n+1-(n+3)} \quad \text{• Subtract the exponents.} \\ &= a^{2n+1-n-3} \\ &= a^{n-2} \end{aligned}$$

You Try It 7 $942{,}000{,}000 = 9.42 \times 10^8$

You Try It 8 $2.7 \times 10^{-5} = 0.000027$

You Try It 9

$$\frac{5{,}600{,}000 \times 0.000000081}{900 \times 0.000000028} = \frac{5.6 \times 10^6 \times 8.1 \times 10^{-8}}{9 \times 10^2 \times 2.8 \times 10^{-8}} = \frac{(5.6)(8.1) \times 10^{6+(-8)-2-(-8)}}{(9)(2.8)} = 1.8 \times 10^4 = 18{,}000$$

You Try It 10

Strategy To find the number of arithmetic operations:

- Find the reciprocal of 9.74×10^{-16}, which is the number of operations performed in 1 s.
- Write the number of seconds in 1 min (60) in scientific notation.
- Multiply the number of arithmetic operations per second by the number of seconds in 1 min.

Solution

$$\frac{1}{9.74 \times 10^{-16}} \approx 1.03 \times 10^{15}$$

$60 = 6.0 \times 10$

$(1.03 \times 10^{15})(6.0 \times 10) = 6.18 \times 10^{16}$

The computer can perform 6.18×10^{16} operations in 1 min.

SECTION 6.2

You Try It 1

$R(x) = -2x^4 - 5x^3 + 2x - 8$

$R(2) = -2(2)^4 - 5(2)^3 + 2(2) - 8$ • Replace x by 2. Simplify.

$= -2(16) - 5(8) + 4 - 8$

$= -32 - 40 + 4 - 8$

$= -76$

You Try It 2 The leading coefficient is -3, the constant term is -12, and the degree is 4.

You Try It 3

a. Yes, this is a polynomial function.

b. No, this is not a polynomial function. A polynomial function does not have a variable expression raised to a negative power.

c. No, this is not a polynomial function. A polynomial function does not have a variable expression within a radical.

You Try It 4

x	y
-4	5
-3	0
-2	-3
-1	-4
0	-3
1	0
2	5

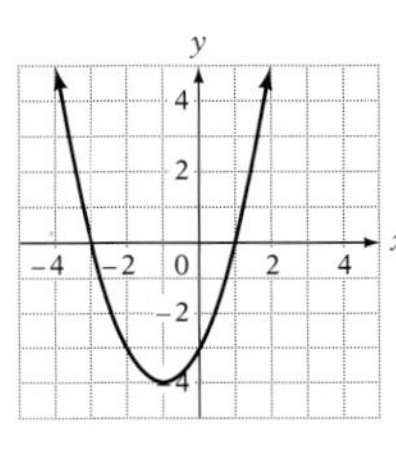

You Try It 5

x	y
-3	28
-2	9
-1	2
0	1
1	0
2	-7
3	-26

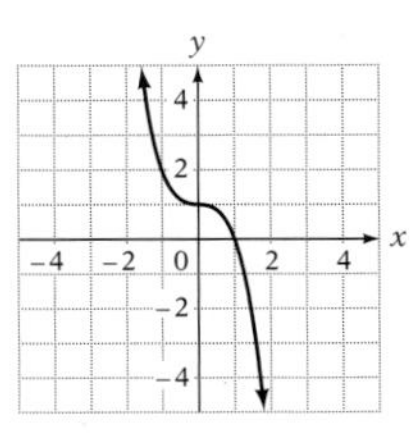

You Try It 6

$$\begin{array}{r} -3x^2 - 4x + 9 \\ -5x^2 - 7x + 1 \\ \hline -8x^2 - 11x + 10 \end{array}$$

You Try It 7 Add the additive inverse of $6x^2 + 3x - 7$ to $-5x^2 + 2x - 3$.

$$\begin{array}{r} -5x^2 + 2x - 3 \\ -6x^2 - 3x + 7 \\ \hline -11x^2 - x + 4 \end{array}$$

You Try It 8

$S(x) = (4x^3 - 3x^2 + 2) + (-2x^2 + 2x - 3)$

$= 4x^3 - 5x^2 + 2x - 1$

$S(-1) = 4(-1)^3 - 5(-1)^2 + 2(-1) - 1$

$= 4(-1) - 5(1) - 2 - 1$

$= -4 - 5 - 2 - 1$

$= -12$

You Try It 9

$D(x) = P(x) - R(x)$

$D(x) = (5x^{2n} - 3x^n - 7) - (-2x^{2n} - 5x^n + 8)$

$= (5x^{2n} - 3x^n - 7) + (2x^{2n} + 5x^n - 8)$

$= 7x^{2n} + 2x^n - 15$

SECTION 6.3

You Try It 1

$(2b^2 - 7b - 8)(-5b)$

$= 2b^2(-5b) - 7b(-5b) - 8(-5b)$ • Use the Distributive Property.

$= -10b^3 + 35b^2 + 40b$

You Try It 2 $x^2 - 2x[x - x(4x - 5) + x^2]$

$= x^2 - 2x[x - 4x^2 + 5x + x^2]$

$= x^2 - 2x[6x - 3x^2]$

$= x^2 - 12x^2 + 6x^3$

$= 6x^3 - 11x^2$

You Try It 3

$$\begin{array}{r} -2b^2 + 5b - 4 \\ -3b + 2 \\ \hline -4b^2 + 10b - 8 \\ 6b^3 - 15b^2 + 12b \\ \hline 6b^3 - 19b^2 + 22b - 8 \end{array}$$

• $2(-2b^2 + 5b - 4)$
• $-3b(-2b^2 + 5b - 4)$

You Try It 4

$(3x - 4)(2x - 3) = 6x^2 - 9x - 8x + 12$ • FOIL

$= 6x^2 - 17x + 12$

You Try It 5

$(3ab + 4)(5ab - 3) = 15a^2b^2 - 9ab + 20ab - 12$

$= 15a^2b^2 + 11ab - 12$

You Try It 6 $(3x - 7)(3x + 7)$
$= 9x^2 - 49$
• The sum and difference of two terms

You Try It 7 $(2ab + 7)(2ab - 7)$
$= 4a^2b^2 - 49$
• The sum and difference of two terms

You Try It 8 $(3x - 4y)^2$
$= 9x^2 - 24xy + 16y^2$
• The square of a binomial

You Try It 9 $(5xy + 4)^2$
$= 25x^2y^2 + 40xy + 16$
• The square of a binomial

You Try It 10

Strategy To find the area, replace the variables b and h in the equation $A = \frac{1}{2}bh$ by the given values, and solve for A.

Solution

$$A = \frac{1}{2}bh$$
$$A = \frac{1}{2}(2x + 6)(x - 4)$$
$$A = (x + 3)(x - 4)$$
$$A = x^2 - 4x + 3x - 12 \quad \text{• FOIL}$$
$$A = x^2 - x - 12$$

The area is $(x^2 - x - 12)$ ft^2.

You Try It 11

Strategy To find the volume, subtract the volume of the small rectangular solid from the volume of the large rectangular solid.

Large rectangular solid:
Length $= L_1 = 12x$
Width $= W_1 = 7x + 2$
Height $= H_1 = 5x - 4$
Small rectangular solid:
Length $= L_2 = 12x$
Width $= W_2 = x$
Height $= H_2 = 2x$

Solution

$V =$ Volume of large rectangular solid $-$ volume of small rectangular solid

$$V = (L_1 \cdot W_1 \cdot H_1) - (L_2 \cdot W_2 \cdot H_2)$$
$$V = (12x)(7x + 2)(5x - 4) - (12x)(x)(2x)$$
$$= (84x^2 + 24x)(5x - 4) - (12x^2)(2x)$$
$$= (420x^3 - 336x^2 + 120x^2 - 96x) - 24x^3$$
$$= 396x^3 - 216x^2 - 96x$$

The volume is $(396x^3 - 216x^2 - 96x)$ ft^3.

You Try It 12

Strategy To find the area, replace the variable r in the equation $A = \pi r^2$ by the given value, and solve for A.

Solution

$$A = \pi r^2$$
$$A \approx 3.14(2x + 3)^2$$
$$= 3.14(4x^2 + 12x + 9)$$
$$= 12.56x^2 + 37.68x + 28.26$$

The area is $(12.56x^2 + 37.68x + 28.26)$ cm^2.

SECTION 6.4

You Try It 1

$$\frac{4x^3y + 8x^2y^2 - 4xy^3}{2xy} = \frac{4x^3y}{2xy} + \frac{8x^2y^2}{2xy} - \frac{4xy^3}{2xy}$$
$$= 2x^2 + 4xy - 2y^2$$

Check: $2xy(2x^2 + 4xy - 2y^2) = 4x^3y + 8x^2y^2 - 4xy^3$

You Try It 2

$$\begin{array}{r} 5x - 1 \\ 3x + 4\overline{)15x^2 + 17x - 20} \\ \underline{15x^2 + 20x} \\ -3x - 20 \\ \underline{-3x - 4} \\ -16 \end{array}$$

$$\frac{15x^2 + 17x - 20}{3x + 4} = 5x - 1 - \frac{16}{3x + 4}$$

You Try It 3

$$\begin{array}{r} x^2 + 3x - 1 \\ 3x - 1\overline{)3x^3 + 8x^2 - 6x + 2} \\ \underline{3x^3 - x^2} \\ 9x^2 - 6x \\ \underline{9x^2 - 3x} \\ -3x + 2 \\ \underline{-3x + 1} \\ 1 \end{array}$$

$$\frac{3x^3 + 8x^2 - 6x + 2}{3x - 1} = x^2 + 3x - 1 + \frac{1}{3x - 1}$$

You Try It 4

$$\begin{array}{r} 3x^2 - 2x + 4 \\ x^2 - 3x + 2\overline{)3x^4 - 11x^3 + 16x^2 - 16x + 8} \\ \underline{3x^4 - 9x^3 + 6x^2} \\ -2x^3 + 10x^2 - 16x \\ \underline{-2x^3 + 6x^2 - 4x} \\ 4x^2 - 12x + 8 \\ \underline{4x^2 - 12x + 8} \\ 0 \end{array}$$

$$\frac{3x^4 - 11x^3 + 16x^2 - 16x + 8}{x^2 - 3x + 2} = 3x^2 - 2x + 4$$

You Try It 5

$$\begin{array}{r|rrr} -2 & 6 & 8 & -5 \\ & & -12 & 8 \\ \hline & 6 & -4 & 3 \end{array}$$

$$(6x^2 + 8x - 5) \div (x + 2)$$
$$= 6x - 4 + \frac{3}{x + 2}$$

You Try It 6

2	5	−12	−8	16
		10	−4	−24
	5	−2	−12	−8

$(5x^3 - 12x^2 - 8x + 16) \div (x - 2)$

$= 5x^2 - 2x - 12 - \dfrac{8}{x-2}$

You Try It 7

3	2	−3	−8	0	−2
		6	9	3	9
	2	3	1	3	7

$(2x^4 - 3x^3 - 8x^2 - 2) \div (x - 3)$

$= 2x^3 + 3x^2 + x + 3 + \dfrac{7}{x-3}$

You Try It 8

2	2	−3	−5
		4	2
	2	1	−3

$P(2) = -3$

You Try It 9

−3	2	−5	0	7
		−6	33	−99
	2	−11	33	−92

$P(-3) = -92$

SOLUTIONS TO CHAPTER 7 "YOU TRY IT"

SECTION 7.1

You Try It 1

The GCF is $7a^2$.

$14a^2 - 21a^4b = 7a^2(2) + 7a^2(-3a^2b)$
$= 7a^2(2 - 3a^2b)$

You Try It 2

The GCF is 9.

$27b^2 + 18b + 9$
$= 9(3b^2) + 9(2b) + 9(1)$
$= 9(3b^2 + 2b + 1)$

You Try It 3

The GCF is $3x^2y^2$.

$6x^4y^2 - 9x^3y^2 + 12x^2y^4$
$= 3x^2y^2(2x^2) + 3x^2y^2(-3x) + 3x^2y^2(4y^2)$
$= 3x^2y^2(2x^2 - 3x + 4y^2)$

You Try It 4

$2y(5x - 2) - 3(2 - 5x)$
$= 2y(5x - 2) + 3(5x - 2)$ • $5x - 2$ is the common factor.
$= (5x - 2)(2y + 3)$

You Try It 5

$a^2 - 3a + 2ab - 6b$
$= (a^2 - 3a) + (2ab - 6b)$
$= a(a - 3) + 2b(a - 3)$ • $a - 3$ is the common factor.
$= (a - 3)(a + 2b)$

You Try It 6

$2mn^2 - n + 8mn - 4$
$= (2mn^2 - n) + (8mn - 4)$
$= n(2mn - 1) + 4(2mn - 1)$ • $2mn - 1$ is the common factor.
$= (2mn - 1)(n + 4)$

You Try It 7

$3xy - 9y - 12 + 4x$
$= (3xy - 9y) - (12 - 4x)$ • $-12 + 4x = -(12 - 4x)$
$= 3y(x - 3) - 4(3 - x)$ • $-(3 - x) = (x - 3)$
$= 3y(x - 3) + 4(x - 3)$ • $x - 3$ is the common factor.
$= (x - 3)(3y + 4)$

SECTION 7.2

You Try It 1

Find the positive factors of 20 whose sum is 9.

Factors	*Sum*
1, 20	21
2, 10	12
4, 5	9

$x^2 + 9x + 20 = (x + 4)(x + 5)$

You Try It 2

Find the factors of −18 whose sum is 7.

Factors	*Sum*
+1, −18	−17
−1, +18	17
+2, −9	−7
−2, +9	7
+3, −6	−3
−3, +6	3

$x^2 + 7x - 18 = (x + 9)(x - 2)$

You Try It 3

The GCF is $-2x$.

$-2x^3 + 14x^2 - 12x = -2x(x^2 - 7x + 6)$

Factor the trinomial $x^2 - 7x + 6$. Find two negative factors of 6 whose sum is −7.

Factors	*Sum*
−1, −6	−7
−2, −3	−5

$-2x^3 + 14x^2 - 12x = -2x(x - 6)(x - 1)$

You Try It 4

The GCF is 3.

$3x^2 - 9xy - 12y^2 = 3(x^2 - 3xy - 4y^2)$

Factor the trinomial.

Find the factors of -4 whose sum is -3.

Factors	Sum
$+1, -4$	-3
$-1, +4$	3
$+2, -2$	0

$3x^2 - 9xy - 12y^2 = 3(x + y)(x - 4y)$

SECTION 7.3

You Try It 1

Factor the trinomial $2x^2 - x - 3$.

Positive factors of 2: 1, 2

Factors of -3: $+1, -3$; $-1, +3$

Trial Factors	Middle Term
$(x + 1)(2x - 3)$	$-3x + 2x = -x$
$(x - 3)(2x + 1)$	$x - 6x = -5x$
$(x - 1)(2x + 3)$	$3x - 2x = x$
$(x + 3)(2x - 1)$	$-x + 6x = 5x$

$2x^2 - x - 3 = (x + 1)(2x - 3)$

You Try It 2

The GCF is $-3y$.

$-45y^3 + 12y^2 + 12y = -3y(15y^2 - 4y - 4)$

Factor the trinomial $15y^2 - 4y - 4$.

Positive factors of 15: 1, 15; 3, 5

Factors of -4: $1, -4$; $-1, 4$; $2, -2$

Trial Factors	Middle Term
$(y + 1)(15y - 4)$	$-4y + 15y = 11y$
$(y - 4)(15y + 1)$	$y - 60y = -59y$
$(y - 1)(15y + 4)$	$4y - 15y = -11y$
$(y + 4)(15y - 1)$	$-y + 60y = 59y$
$(y + 2)(15y - 2)$	$-2y + 30y = 28y$
$(y - 2)(15y + 2)$	$2y - 30y = -28y$
$(3y + 1)(5y - 4)$	$-12y + 5y = -7y$
$(3y - 4)(5y + 1)$	$3y - 20y = -17y$
$(3y - 1)(5y + 4)$	$12y - 5y = 7y$
$(3y + 4)(5y - 1)$	$-3y + 20y = 17y$
$(3y + 2)(5y - 2)$	$-6y + 10y = 4y$
$(3y - 2)(5y + 2)$	$6y - 10y = -4y$

$-45y^3 + 12y^2 + 12y = -3y(3y - 2)(5y + 2)$

You Try It 3

Factors of -14 $[2(-7)]$	Sum
$+1, -14$	-13
$-1, +14$	13
$+2, -7$	-5
$-2, +7$	5

$$\begin{aligned}2a^2 + 13a - 7 &= 2a^2 - a + 14a - 7\\ &= (2a^2 - a) + (14a - 7)\\ &= a(2a - 1) + 7(2a - 1)\\ &= (2a - 1)(a + 7)\end{aligned}$$

$2a^2 + 13a - 7 = (2a - 1)(a + 7)$

You Try It 4

The GCF is $5x$.

$15x^3 + 40x^2 - 80x = 5x(3x^2 + 8x - 16)$

Factors of -48 $[3(-16)]$	Sum
$+1, -48$	-47
$-1, +48$	47
$+2, -24$	-22
$-2, +24$	22
$+3, -16$	-13
$-3, +16$	13
$+4, -12$	-8
$-4, +12$	8

$$\begin{aligned}3x^2 + 8x - 16 &= 3x^2 - 4x + 12x - 16\\ &= (3x^2 - 4x) + (12x - 16)\\ &= x(3x - 4) + 4(3x - 4)\\ &= (3x - 4)(x + 4)\end{aligned}$$

$$\begin{aligned}15x^3 + 40x^2 - 80x &= 5x(3x^2 + 8x - 16)\\ &= 5x(3x - 4)(x + 4)\end{aligned}$$

SECTION 7.4

You Try It 1

$$\begin{aligned}x^2 - 36y^4 &= x^2 - (6y^2)^2\\ &= (x + 6y^2)(x - 6y^2)\end{aligned}$$

• Difference of two squares

You Try It 2

$9x^2 + 12x + 4 = (3x + 2)^2$

• Perfect-square trinomial

You Try It 3

$$\begin{aligned}&(a + b)^2 - (a - b)^2\\ &= [(a + b) + (a - b)][(a + b) - (a - b)]\\ &= (a + b + a - b)(a + b - a + b)\\ &= (2a)(2b) = 4ab\end{aligned}$$

• Difference of two squares

You Try It 4

$$\begin{aligned}a^3b^3 - 27 &= (ab)^3 - 3^3\\ &= (ab - 3)(a^2b^2 + 3ab + 9)\end{aligned}$$

• Difference of two cubes

You Try It 5

$8x^3 + y^3z^3 = (2x)^3 + (yz)^3$ • Sum of two cubes
$= (2x + yz)(4x^2 - 2xyz + y^2z^2)$

You Try It 6

$(x - y)^3 + (x + y)^3$ • Sum of two cubes
$= [(x - y) + (x + y)]$
$\times [(x - y)^2 - (x - y)(x + y) + (x + y)^2]$
$= 2x[x^2 - 2xy + y^2 - (x^2 - y^2) + x^2 + 2xy + y^2]$
$= 2x(x^2 - 2xy + y^2 - x^2 + y^2 + x^2 + 2xy + y^2)$
$= 2x(x^2 + 3y^2)$

You Try It 7 Let $u = x^2$.

$3x^4 + 4x^2 - 4 = 3u^2 + 4u - 4$
$= (u + 2)(3u - 2)$
$= (x^2 + 2)(3x^2 - 2)$

You Try It 8

$18x^3 - 6x^2 - 60x = 6x(3x^2 - x - 10)$ • GCF
$= 6x(3x + 5)(x - 2)$

You Try It 9 $4x - 4y - x^3 + x^2y$ • Factor by grouping.
$= (4x - 4y) - (x^3 - x^2y)$
$= 4(x - y) - x^2(x - y)$
$= (x - y)(4 - x^2)$
$= (x - y)(2 + x)(2 - x)$

You Try It 10

$x^{4n} - x^{2n}y^{2n} = x^{2n+2n} - x^{2n}y^{2n}$
$= x^{2n}(x^{2n} - y^{2n})$ • GCF
$= x^{2n}[(x^n)^2 - (y^n)^2]$ • Difference of two squares
$= x^{2n}(x^n + y^n)(x^n - y^n)$

SECTION 7.5

You Try It 1

$2x(x + 7) = 0$
$2x = 0 \quad x + 7 = 0$ • Principle of Zero Products
$x = 0 \quad x = -7$

The solutions are 0 and −7.

You Try It 2

$4x^2 - 9 = 0$ • Difference of two squares
$(2x - 3)(2x + 3) = 0$
$2x - 3 = 0 \quad 2x + 3 = 0$ • Principle of Zero Products
$2x = 3 \quad 2x = -3$
$x = \frac{3}{2} \quad x = -\frac{3}{2}$

The solutions are $\frac{3}{2}$ and $-\frac{3}{2}$.

You Try It 3

$(x + 2)(x - 7) = 52$
$x^2 - 5x - 14 = 52$
$x^2 - 5x - 66 = 0$
$(x + 6)(x - 11) = 0$
$x + 6 = 0 \quad x - 11 = 0$ • Principle of Zero Products
$x = -6 \quad x = 11$

The solutions are −6 and 11.

You Try It 4

Strategy First consecutive positive integer: n

Second consecutive positive integer: $n + 1$

The sum of the squares of the two consecutive positive integers is 61.

Solution

$n^2 + (n + 1)^2 = 61$
$n^2 + n^2 + 2n + 1 = 61$
$2n^2 + 2n + 1 = 61$
$2n^2 + 2n - 60 = 0$
$2(n^2 + n - 30) = 0$
$2(n - 5)(n + 6) = 0$

$n - 5 = 0 \quad n + 6 = 0$ • Principle of Zero Products
$n = 5 \quad n = -6$

Because −6 is not a positive integer, it is not a solution.

$n = 5$
$n + 1 = 5 + 1 = 6$

The two integers are 5 and 6.

You Try It 5

Strategy Width $= x$
Length $= 2x + 4$

The area of the rectangle is 96 in^2. Use the equation $A = L \cdot W$.

Solution

$A = L \cdot W$
$96 = (2x + 4)x$
$96 = 2x^2 + 4x$
$0 = 2x^2 + 4x - 96$
$0 = 2(x^2 + 2x - 48)$
$0 = 2(x + 8)(x - 6)$
$x + 8 = 0 \quad x - 6 = 0$ • Principle of Zero Products
$x = -8 \quad x = 6$

Because the width cannot be a negative number, −8 is not a solution.

$x = 6$
$2x + 4 = 2(6) + 4 = 12 + 4 = 16$

The length is 16 in. The width is 6 in.

SOLUTIONS TO CHAPTER 8 "YOU TRY IT"

SECTION 8.1

You Try It 1

$$\frac{6x^5y}{12x^2y^3} = \frac{\overset{1}{\cancel{2}} \cdot \overset{1}{\cancel{3}} \cdot x^5y}{\underset{1}{\cancel{2}} \cdot 2 \cdot \underset{1}{\cancel{3}} \cdot x^2y^3} = \frac{x^3}{2y^2}$$

You Try It 2

$$\frac{x^2 + 4x - 12}{x^2 - 3x + 2} = \frac{\overset{1}{\cancel{(x - 2)}}(x + 6)}{(x - 1)\underset{1}{\cancel{(x - 2)}}} = \frac{x + 6}{x - 1}$$

You Try It 3

$$\frac{x^2+2x-24}{16-x^2} = \frac{\overset{-1}{\cancel{(x-4)}}(x+6)}{\underset{1}{\cancel{(4-x)}}(4+x)} = -\frac{x+6}{x+4}$$

• $\frac{x-4}{4-x} = \frac{x-4}{-1(x-4)} = -1$

You Try It 4

$$\frac{12x^2+3x}{10x-15}\cdot\frac{8x-12}{9x+18} = \frac{3x(4x+1)}{5(2x-3)}\cdot\frac{4(2x-3)}{9(x+2)} = \frac{\overset{1}{\cancel{3}}x(4x+1)\cdot 2\cdot 2\overset{1}{\cancel{(2x-3)}}}{5\underset{1}{\cancel{(2x-3)}}\cdot \underset{1}{\cancel{3}}\cdot 3(x+2)} = \frac{4x(4x+1)}{15(x+2)}$$

You Try It 5

$$\frac{x^2+2x-15}{9-x^2}\cdot\frac{x^2-3x-18}{x^2-7x+6}$$

$$= \frac{(x-3)(x+5)}{(3-x)(3+x)}\cdot\frac{(x+3)(x-6)}{(x-1)(x-6)}$$ • Factor.

$$= \frac{\overset{-1}{\cancel{(x-3)}}(x+5)\cdot\overset{1}{\cancel{(x+3)}}\overset{1}{\cancel{(x-6)}}}{\underset{1}{\cancel{(3-x)}}\underset{1}{\cancel{(3+x)}}\cdot(x-1)\underset{1}{\cancel{(x-6)}}} = -\frac{x+5}{x-1}$$

You Try It 6

$$\frac{a^2}{4bc^2-2b^2c} \div \frac{a}{6bc-3b^2}$$

$$= \frac{a^2}{4bc^2-2b^2c}\cdot\frac{6bc-3b^2}{a}$$ • Multiply by the reciprocal.

$$= \frac{a^2\cdot 3\overset{1}{\cancel{b}}\overset{1}{\cancel{(2c-b)}}}{2\underset{1}{\cancel{b}}c\underset{1}{\cancel{(2c-b)}}\cdot a} = \frac{3a}{2c}$$

You Try It 7

$$\frac{3x^2+26x+16}{3x^2-7x-6} \div \frac{2x^2+9x-5}{x^2+2x-15}$$

$$= \frac{3x^2+26x+16}{3x^2-7x-6}\cdot\frac{x^2+2x-15}{2x^2+9x-5}$$ • Multiply by the reciprocal.

$$= \frac{\overset{1}{\cancel{(3x+2)}}(x+8)\cdot\overset{1}{\cancel{(x+5)}}\overset{1}{\cancel{(x-3)}}}{\underset{1}{\cancel{(3x+2)}}\underset{1}{\cancel{(x-3)}}\cdot(2x-1)\underset{1}{\cancel{(x+5)}}} = \frac{x+8}{2x-1}$$

SECTION 8.2

You Try It 1

The LCM is $(2x-5)(x+4)$.

$$\frac{2x}{2x-5} = \frac{2x}{2x-5}\cdot\frac{x+4}{x+4} = \frac{2x^2+8x}{(2x-5)(x+4)}$$

$$\frac{3}{x+4} = \frac{3}{x+4}\cdot\frac{2x-5}{2x-5} = \frac{6x-15}{(2x-5)(x+4)}$$

You Try It 2

$2x^2-11x+15 = (x-3)(2x-5)$; $x^2-3x = x(x-3)$

The LCM is $x(x-3)(2x-5)$.

$$\frac{3x}{2x^2-11x+15} = \frac{3x}{(x-3)(2x-5)}\cdot\frac{x}{x} = \frac{3x^2}{x(x-3)(2x-5)}$$

$$\frac{x-2}{x^2-3x} = \frac{x-2}{x(x-3)}\cdot\frac{2x-5}{2x-5} = \frac{2x^2-9x+10}{x(x-3)(2x-5)}$$

You Try It 3

$2x-x^2 = x(2-x) = -x(x-2)$;

$3x^2-5x-2 = (x-2)(3x+1)$

The LCM is $x(x-2)(3x+1)$.

$$\frac{2x-7}{2x-x^2} = -\frac{2x-7}{x(x-2)}\cdot\frac{3x+1}{3x+1} = -\frac{6x^2-19x-7}{x(x-2)(3x+1)}$$

$$\frac{3x-2}{3x^2-5x-2} = \frac{3x-2}{(x-2)(3x+1)}\cdot\frac{x}{x} = \frac{3x^2-2x}{x(x-2)(3x+1)}$$

You Try It 4

The LCM is ab.

$$\frac{2}{b}-\frac{1}{a}+\frac{4}{ab} = \frac{2}{b}\cdot\frac{a}{a}-\frac{1}{a}\cdot\frac{b}{b}+\frac{4}{ab} = \frac{2a}{ab}-\frac{b}{ab}+\frac{4}{ab} = \frac{2a-b+4}{ab}$$

You Try It 5

The LCM is $a(a-5)(a+5)$.

$$\frac{a-3}{a^2-5a}+\frac{a-9}{a^2-25}$$

$$= \frac{a-3}{a(a-5)}\cdot\frac{a+5}{a+5}+\frac{a-9}{(a-5)(a+5)}\cdot\frac{a}{a}$$

$$= \frac{(a-3)(a+5)+a(a-9)}{a(a-5)(a+5)}$$

$$= \frac{(a^2+2a-15)+(a^2-9a)}{a(a-5)(a+5)}$$

$$= \frac{a^2+2a-15+a^2-9a}{a(a-5)(a+5)}$$

$$= \frac{2a^2-7a-15}{a(a-5)(a+5)} = \frac{(2a+3)(a-5)}{a(a-5)(a+5)}$$

$$= \frac{(2a+3)\overset{1}{\cancel{(a-5)}}}{a\underset{1}{\cancel{(a-5)}}(a+5)} = \frac{2a+3}{a(a+5)}$$

You Try It 6

The LCM is $(x-4)(x+1)$.

$$\frac{2x}{x-4}-\frac{x-1}{x+1}+\frac{2}{x^2-3x-4}$$

$$= \frac{2x}{x-4}\cdot\frac{x+1}{x+1}-\frac{x-1}{x+1}\cdot\frac{x-4}{x-4}+\frac{2}{(x-4)(x+1)}$$

$$= \frac{2x(x+1)-(x-1)(x-4)+2}{(x-4)(x+1)}$$

$$= \frac{(2x^2+2x)-(x^2-5x+4)+2}{(x-4)(x+1)}$$

$$= \frac{x^2+7x-2}{(x-4)(x+1)}$$

SECTION 8.3

You Try It 1

The LCM of the denominators in the numerator and denominator is $x - 3$.

$$\frac{2x + 5 + \frac{14}{x-3}}{4x + 16 + \frac{49}{x-3}} = \frac{2x + 5 + \frac{14}{x-3}}{4x + 16 + \frac{49}{x-3}} \cdot \frac{x-3}{x-3}$$

$$= \frac{(2x+5)(x-3) + \frac{14}{x-3}(x-3)}{(4x+16)(x-3) + \frac{49}{x-3}(x-3)}$$

$$= \frac{2x^2 - x - 15 + 14}{4x^2 + 4x - 48 + 49} = \frac{2x^2 - x - 1}{4x^2 + 4x + 1}$$

$$= \frac{(2x+1)(x-1)}{(2x+1)(2x+1)} = \frac{\overset{1}{\cancel{(2x+1)}}(x-1)}{\underset{1}{\cancel{(2x+1)}}(2x+1)} = \frac{x-1}{2x+1}$$

You Try It 2

The LCM of 2 and $\frac{1}{x}$ is x.

$$2 - \frac{1}{2 - \frac{1}{x}} = 2 - \frac{1}{2 - \frac{1}{x}} \cdot \frac{x}{x}$$

$$= 2 - \frac{1 \cdot x}{2 \cdot x - \frac{1}{x} \cdot x} = 2 - \frac{x}{2x-1}$$

The LCM of the denominators is $2x - 1$.

$$2 - \frac{x}{2x-1} = 2 \cdot \frac{2x-1}{2x-1} - \frac{x}{2x-1}$$

$$= \frac{4x-2}{2x-1} - \frac{x}{2x-1}$$

$$= \frac{4x - 2 - x}{2x-1} = \frac{3x-2}{2x-1}$$

SECTION 8.4

You Try It 1

$$\frac{x}{x+6} = \frac{3}{x}$$ • The LCM is $x(x + 6)$.

$$\frac{x\overset{1}{\cancel{(x+6)}}}{1} \cdot \frac{x}{\underset{1}{\cancel{x+6}}} = \frac{x(x+6)}{1} \cdot \frac{3}{x}$$ • Multiply by the LCM.

$$x^2 = (x+6)3$$ • Simplify.

$$x^2 = 3x + 18$$

$$x^2 - 3x - 18 = 0$$

$$(x+3)(x-6) = 0$$ • Factor.

$$x + 3 = 0 \qquad x - 6 = 0$$ • Principle of Zero Products

$$x = -3 \qquad x = 6$$

Both -3 and 6 check as solutions.

The solutions are -3 and 6.

You Try It 2

$$\frac{5x}{x+2} = 3 - \frac{10}{x+2}$$ • The LCM is $x + 2$.

$$\frac{(x+2)}{1} \cdot \frac{5x}{x+2} = \frac{(x+2)}{1}\left(3 - \frac{10}{x+2}\right)$$ • Clear denominators.

$$\frac{\overset{1}{\cancel{x+2}}}{1} \cdot \frac{5x}{\underset{1}{\cancel{x+2}}} = \frac{x+2}{1} \cdot 3 - \frac{\overset{1}{\cancel{x+2}}}{1} \cdot \frac{10}{\underset{1}{\cancel{x+2}}}$$

$$5x = (x+2)3 - 10$$ • Solve for x.

$$5x = 3x + 6 - 10$$

$$5x = 3x - 4$$

$$2x = -4$$

$$x = -2$$

-2 does not check as a solution.

The equation has no solution.

SECTION 8.5

You Try It 1

$$\frac{2}{x+3} = \frac{6}{5x+5}$$

$$\frac{(x+3)(5x+5)}{1} \cdot \frac{2}{x+3} = \frac{(x+3)(5x+5)}{1} \cdot \frac{6}{5x+5}$$

$$\frac{\overset{1}{\cancel{(x+3)}}(5x+5)}{1} \cdot \frac{2}{\underset{1}{\cancel{x+3}}} = \frac{(x+3)\overset{1}{\cancel{(5x+5)}}}{1} \cdot \frac{6}{\underset{1}{\cancel{5x+5}}}$$

$$(5x+5)2 = (x+3)6$$ • Solve for x.

$$10x + 10 = 6x + 18$$

$$4x + 10 = 18$$

$$4x = 8$$

$$x = 2$$

The solution is 2.

You Try It 2

Strategy To find the total area that 256 ceramic tiles will cover, write and solve a proportion using x to represent the number of square feet that 256 tiles will cover.

Solution

$$\frac{9}{16} = \frac{x}{256}$$ • Write a proportion.

$$256\left(\frac{9}{16}\right) = 256\left(\frac{x}{256}\right)$$ • Clear denominators.

$$144 = x$$

An area of 144 ft^2 can be tiled using 256 ceramic tiles.

You Try It 3

Strategy To find the area of triangle AOB:
- Solve a proportion to find the length of AO (the height of triangle AOB).
- Use the formula for the area of a triangle. AB is the base and AO is the height.

Solution

$$\frac{CD}{AB} = \frac{DO}{AO} \quad \text{• Write a proportion.}$$

$$\frac{4}{10} = \frac{3}{AO} \quad \text{• Substitute.}$$

$$10 \cdot AO \cdot \frac{4}{10} = 10 \cdot AO \cdot \frac{3}{AO}$$

$$4(AO) = 30$$

$$AO = 7.5$$

$$A = \frac{1}{2}bh \quad \text{• Area of a triangle}$$

$$= \frac{1}{2}(10)(7.5) \quad \text{• Substitute.}$$

$$= 37.5$$

The area of triangle AOB is 37.5 cm^2.

SECTION 8.6

You Try It 1

$$5x - 2y = 10$$

$$5x - 5x - 2y = -5x + 10 \quad \text{• Subtract } 5x.$$

$$-2y = -5x + 10$$

$$\frac{-2y}{-2} = \frac{-5x + 10}{-2} \quad \text{• Divide by } -2.$$

$$y = \frac{5}{2}x - 5$$

You Try It 2

$$s = \frac{A + L}{2}$$

$$2 \cdot s = 2\left(\frac{A + L}{2}\right) \quad \text{• Multiply by 2.}$$

$$2s = A + L$$

$$2s - A = A - A + L \quad \text{• Subtract } A.$$

$$2s - A = L$$

You Try It 3

$$S = a + (n - 1)d$$

$$S = a + nd - d$$

$$S - a = a - a + nd - d \quad \text{• Subtract } a.$$

$$S - a = nd - d$$

$$S - a + d = nd - d + d \quad \text{• Add } d.$$

$$S - a + d = nd$$

$$\frac{S - a + d}{d} = \frac{nd}{d} \quad \text{• Divide by } d.$$

$$\frac{S - a + d}{d} = n$$

You Try It 4

$$S = rS + C$$

$$S - rS = rS - rS + C \quad \text{• Subtract } rS.$$

$$S - rS = C$$

$$(1 - r)S = C \quad \text{• Factor.}$$

$$\frac{(1 - r)S}{1 - r} = \frac{C}{1 - r} \quad \text{• Divide by } 1 - r.$$

$$S = \frac{C}{1 - r}$$

SECTION 8.7

You Try It 1

Strategy
- Time for one printer to complete the job: t

	Rate	*Time*	*Part*
1st printer	$\frac{1}{t}$	2	$\frac{2}{t}$
2nd printer	$\frac{1}{t}$	5	$\frac{5}{t}$

- The sum of the parts of the task completed must equal 1.

Solution

$$\frac{2}{t} + \frac{5}{t} = 1$$

$$t\left(\frac{2}{t} + \frac{5}{t}\right) = t \cdot 1$$

$$2 + 5 = t$$

$$7 = t$$

Working alone, one printer takes 7 h to print the payroll.

You Try It 2

Strategy

- Rate sailing across the lake: r
  Rate sailing back: $3r$

	Distance	Rate	Time
Across	6	r	$\frac{6}{r}$
Back	6	$3r$	$\frac{6}{3r}$

- The total time for the trip was 2 h.

Solution

$$\frac{6}{r} + \frac{6}{3r} = 2$$

$$3r\left(\frac{6}{r} + \frac{6}{3r}\right) = 3r(2) \quad \text{• Multiply by the LCM, } 3r.$$

$$3r \cdot \frac{6}{r} + 3r \cdot \frac{6}{3r} = 6r$$

$$18 + 6 = 6r \quad \text{• Solve for } r.$$

$$24 = 6r$$

$$4 = r$$

The rate sailing across the lake was 4 km/h.

SECTION 8.8

You Try It 1

Strategy

To find the distance:

- Write the basic direct variation equation, replace the variables by the given values, and solve for k.
- Write the direct variation equation, replacing k by its value. Substitute 5 for t and solve for s.

Solution

$s = kt^2$ • Direct variation equation

$64 = k(2)^2$ • Replace s by 64 and t by 2.

$64 = k \cdot 4$

$16 = k$

$s = 16t^2 = 16(5)^2 = 400$ • $k = 16, t = 5$

The object will fall 400 ft in 5 s.

You Try It 2

Strategy

To find the resistance:

- Write the basic inverse variation equation, replace the variables by the given values, and solve for k.
- Write the inverse variation equation, replacing k by its value. Substitute 0.02 for d and solve for R.

Solution

$$R = \frac{k}{d^2} \quad \text{• Inverse variation equation}$$

$$0.5 = \frac{k}{(0.01)^2} \quad \text{• Replace } R \text{ by 0.5 and } d \text{ by 0.01.}$$

$$0.5 = \frac{k}{0.0001}$$

$$0.00005 = k$$

$$R = \frac{0.00005}{d^2} = \frac{0.00005}{(0.02)^2} = 0.125 \quad \text{• } k = 0.00005, d = 0.02$$

The resistance is 0.125 ohm.

You Try It 3

Strategy

- Write the basic combined variation equation, replace the variables by the given values, and solve for k.
- Write the combined variation equation, replacing k by its value and substituting 4 for W, 8 for d, and 16 for L. Solve for s.

Solution

$$s = \frac{kWd^2}{L} \quad \text{• Combined variation equation}$$

$$1200 = \frac{k(2)(12)^2}{12} \quad \text{• Replace } s \text{ by 1200, } W \text{ by 2, } d \text{ by 12, and } L \text{ by 12.}$$

$$1200 = 24k$$

$$50 = k$$

$$s = \frac{50Wd^2}{L} \quad \text{• Replace } k \text{ by 50 in the combined variation equation.}$$

$$= \frac{50(4)8^2}{16} \quad \text{• Replace } W \text{ by 4, } d \text{ by 8, and } L \text{ by 16.}$$

$$= 800$$

The strength of the beam is 800 lb.

SOLUTIONS TO CHAPTER 9 "YOU TRY IT"

SECTION 9.1

You Try It 1

$$16^{-3/4} = (2^4)^{-3/4} \quad \text{• } 16 = 2^4$$

$$= 2^{-3}$$

$$= \frac{1}{2^3} = \frac{1}{8}$$

You Try It 2

$(-81)^{3/4}$

The base of the exponential expression is negative, and the denominator of the exponent is a positive even number.

Therefore, $(-81)^{3/4}$ is not a real number.

You Try It 3

$$(x^{3/4}y^{1/2}z^{-2/3})^{-4/3} = x^{-1}y^{-2/3}z^{8/9}$$

$$= \frac{z^{8/9}}{xy^{2/3}}$$

You Try It 4

$$\left(\frac{16a^{-2}b^{4/3}}{9a^4b^{-2/3}}\right)^{-1/2} = \left(\frac{2^4a^{-6}b^2}{3^2}\right)^{-1/2}$$

• Use the Rule for Dividing Exponential Expressions.

$$= \frac{2^{-2}a^3b^{-1}}{3^{-1}}$$

• Use the Rule for Simplifying Powers of Products.

$$= \frac{3a^3}{2^2b} = \frac{3a^3}{4b}$$

You Try It 5 $(2x^3)^{3/4} = \sqrt[4]{(2x^3)^3}$
$= \sqrt[4]{8x^9}$

You Try It 6 $-5a^{5/6} = -5(a^5)^{1/6}$
$= -5\sqrt[6]{a^5}$

You Try It 7 $\sqrt[3]{3ab} = (3ab)^{1/3}$

You Try It 8 $\sqrt[4]{x^4 + y^4} = (x^4 + y^4)^{1/4}$

You Try It 9 $\sqrt[3]{-8x^{12}y^3} = -2x^4y$

You Try It 10 $-\sqrt[4]{81x^{12}y^8} = -3x^3y^2$

SECTION 9.2

You Try It 1 $\sqrt[5]{x^7} = \sqrt[5]{x^5 \cdot x^2}$
$= \sqrt[5]{x^5}\sqrt[5]{x^2}$ • x^5 is a perfect fifth power.
$= x\sqrt[5]{x^2}$

You Try It 2

$\sqrt[3]{-64x^8y^{18}} = \sqrt[3]{-64x^6y^{18}(x^2)}$
$= \sqrt[3]{-64x^6y^{18}}\sqrt[3]{x^2}$ • $-64x^6y^{18}$ is a perfect third power.
$= -4x^2y^6\sqrt[3]{x^2}$

You Try It 3

$3xy\sqrt[3]{81x^5y} - \sqrt[3]{192x^8y^4}$
$= 3xy\sqrt[3]{27x^3 \cdot 3x^2y} - \sqrt[3]{64x^6y^3 \cdot 3x^2y}$
$= 3xy\sqrt[3]{27x^3}\sqrt[3]{3x^2y} - \sqrt[3]{64x^6y^3}\sqrt[3]{3x^2y}$
$= 3xy \cdot 3x\sqrt[3]{3x^2y} - 4x^2y\sqrt[3]{3x^2y}$
$= 9x^2y\sqrt[3]{3x^2y} - 4x^2y\sqrt[3]{3x^2y} = 5x^2y\sqrt[3]{3x^2y}$

You Try It 4

$5\sqrt{2}(\sqrt{6} + \sqrt{24})$
$= 5\sqrt{12} + 5\sqrt{48}$ • The Distributive Property
$= 5\sqrt{4 \cdot 3} + 5\sqrt{16 \cdot 3}$ • Simplify each radical.
$= 5 \cdot 2\sqrt{3} + 5 \cdot 4\sqrt{3}$
$= 10\sqrt{3} + 20\sqrt{3} = 30\sqrt{3}$

You Try It 5

$(4 - 2\sqrt{7})(1 + 3\sqrt{7})$
$= 4 + 12\sqrt{7} - 2\sqrt{7} - 6(\sqrt{7})^2$ • The FOIL method
$= 4 + 10\sqrt{7} - 6 \cdot 7$
$= 4 + 10\sqrt{7} - 42 = -38 + 10\sqrt{7}$

You Try It 6

$(4 - \sqrt{2x})^2$
$= (4 - \sqrt{2x})(4 - \sqrt{2x})$
$= 16 - 4\sqrt{2x} - 4\sqrt{2x} + (\sqrt{2x})^2$ • The FOIL method
$= 16 - 8\sqrt{2x} + 2x$
$= 2x - 8\sqrt{2x} + 16$

You Try It 7

$$\sqrt{\frac{5}{6}} = \frac{\sqrt{5}}{\sqrt{6}}$$

$$= \frac{\sqrt{5}}{\sqrt{6}} \cdot \frac{\sqrt{6}}{\sqrt{6}} = \frac{\sqrt{30}}{6}$$

• Rationalize the denominator.

You Try It 8

$$\frac{3x}{\sqrt[3]{3x^2}} = \frac{3x}{\sqrt[3]{3x^2}} \cdot \frac{\sqrt[3]{9x}}{\sqrt[3]{9x}} = \frac{3x\sqrt[3]{9x}}{\sqrt[3]{27x^3}}$$

• Rationalize the denominator.

$$= \frac{3x\sqrt[3]{9x}}{3x} = \sqrt[3]{9x}$$

You Try It 9

$$\frac{3}{2 - \sqrt{x}} = \frac{3}{2 - \sqrt{x}} \cdot \frac{2 + \sqrt{x}}{2 + \sqrt{x}}$$

• Rationalize the denominator.

$$= \frac{3(2 + \sqrt{x})}{2^2 - (\sqrt{x})^2} = \frac{6 + 3\sqrt{x}}{4 - x}$$

SECTION 9.3

You Try It 1

$\sqrt{x} - \sqrt{x + 5} = 1$
$\sqrt{x} = 1 + \sqrt{x + 5}$ • Add $\sqrt{x + 5}$ to each side.
$(\sqrt{x})^2 = (1 + \sqrt{x + 5})^2$ • Square each side.
$x = 1 + 2\sqrt{x + 5} + x + 5$
$-6 = 2\sqrt{x + 5}$
$-3 = \sqrt{x + 5}$
$(-3)^2 = (\sqrt{x + 5})^2$ • Square each side.
$9 = x + 5$
$4 = x$

4 does not check as a solution. The equation has no solution.

You Try It 2 $\sqrt[4]{x - 8} = 3$
$(\sqrt[4]{x - 8})^4 = 3^4$ • Raise each side to the fourth power.
$x - 8 = 81$
$x = 89$

Check:

$\sqrt[4]{x - 8} = 3$	
$\sqrt[4]{89 - 8}$	3
$\sqrt[4]{81}$	3
$3 = 3$	

The solution is 89.

You Try It 3

Strategy To find the diagonal, use the Pythagorean Theorem. One leg is the length of the rectangle. The second leg is the width of the rectangle. The hypotenuse is the diagonal of the rectangle.

Solution

$c^2 = a^2 + b^2$ • Pythagorean Theorem
$c^2 = (6)^2 + (3)^2$ • Replace *a* by 6 and *b* by 3.
$c^2 = 36 + 9$ • Solve for *c*.
$c^2 = 45$
$(c^2)^{1/2} = (45)^{1/2}$ • Raise each side to the $\frac{1}{2}$ power.
$c = \sqrt{45}$ • $a^{1/2} = \sqrt{a}$
$c \approx 6.7$

The diagonal is 6.7 cm.

You Try It 4

Strategy To find the height, replace *d* in the equation with the given value and solve for *h*.

Solution

$d = \sqrt{1.5h}$
$5.5 = \sqrt{1.5h}$ • Replace *d* by 5.5.
$(5.5)^2 = (\sqrt{1.5h})^2$ • Square each side.
$30.25 = 1.5h$
$20.17 \approx h$

The periscope must be approximately 20.17 ft above the water.

You Try It 5

Strategy To find the distance, replace the variables *v* and *a* in the equation by their given values and solve for *s*.

Solution

$v = \sqrt{2as}$
$88 = \sqrt{2 \cdot 22s}$ • Replace *v* by 88 and *a* by 22.
$88 = \sqrt{44s}$
$(88)^2 = (\sqrt{44s})^2$ • Square each side.
$7744 = 44s$
$176 = s$

The distance required is 176 ft.

SECTION 9.4

You Try It 1 $\sqrt{-45} = i\sqrt{45} = i\sqrt{9 \cdot 5} = 3i\sqrt{5}$

You Try It 2

$-b + \sqrt{b^2 - 4ac}$
$-(6) + \sqrt{(6)^2 - 4(1)(25)} = -6 + \sqrt{36 - 100}$
$= -6 + \sqrt{-64} = -6 + i\sqrt{64}$
$= -6 + 8i$

You Try It 3 $(-4 + 2i) - (6 - 8i) = -10 + 10i$

You Try It 4 $(16 - \sqrt{-45}) - (3 + \sqrt{-20})$
$= (16 - i\sqrt{45}) - (3 + i\sqrt{20})$
$= (16 - i\sqrt{9 \cdot 5}) - (3 + i\sqrt{4 \cdot 5})$
$= (16 - 3i\sqrt{5}) - (3 + 2i\sqrt{5})$
$= 13 - 5i\sqrt{5}$

You Try It 5 $(-3i)(-10i) = 30i^2 = 30(-1) = -30$

You Try It 6

$-\sqrt{-8} \cdot \sqrt{-5} = -i\sqrt{8} \cdot i\sqrt{5} = -i^2\sqrt{40}$
$= -(-1)\sqrt{40} = \sqrt{4 \cdot 10} = 2\sqrt{10}$

You Try It 7

$-6i(3 + 4i) = -18i - 24i^2$ • The Distributive Property
$= -18i - 24(-1)$
$= 24 - 18i$

You Try It 8

$(4 - 3i)(2 - i) = 8 - 4i - 6i + 3i^2$ • FOIL
$= 8 - 10i + 3i^2$
$= 8 - 10i + 3(-1)$
$= 5 - 10i$

You Try It 9 $(5 - 3i)^2 = (5 - 3i)(5 - 3i)$
$= 25 - 15i - 15i + 9i^2$
$= 25 - 30i + 9(-1)$
$= 25 - 30i - 9 = 16 - 30i$

You Try It 10 $(3 + 6i)(3 - 6i) = 3^2 + 6^2$ • Conjugates
$= 9 + 36$
$= 45$

You Try It 11 $\frac{2 - 3i}{4i} = \frac{2 - 3i}{4i} \cdot \frac{i}{i}$
$= \frac{2i - 3i^2}{4i^2}$
$= \frac{2i - 3(-1)}{4(-1)}$
$= \frac{3 + 2i}{-4} = -\frac{3}{4} - \frac{1}{2}i$

You Try It 12

$\frac{2 + 5i}{3 - 2i} = \frac{2 + 5i}{3 - 2i} \cdot \frac{3 + 2i}{3 + 2i} = \frac{6 + 4i + 15i + 10i^2}{3^2 + 2^2}$
$= \frac{6 + 19i + 10(-1)}{9 + 4} = \frac{-4 + 19i}{13}$
$= -\frac{4}{13} + \frac{19}{13}i$

SOLUTIONS TO CHAPTER 10 "YOU TRY IT"

SECTION 10.1

You Try It 1

$2x^2 = 7x - 3$
$2x^2 - 7x + 3 = 0$ • Write in standard form.
$(2x - 1)(x - 3) = 0$ • Solve by factoring.

$2x - 1 = 0$ $x - 3 = 0$
$2x = 1$ $x = 3$
$x = \frac{1}{2}$

The solutions are $\frac{1}{2}$ and 3.

You Try It 2 $x^2 - 3ax - 4a^2 = 0$
$(x + a)(x - 4a) = 0$

$x + a = 0$ $x - 4a = 0$
$x = -a$ $x = 4a$

The solutions are $-a$ and $4a$.

You Try It 3

$$(x - r_1)(x - r_2) = 0$$

$$(x - 3)\left[x - \left(-\frac{1}{2}\right)\right] = 0 \qquad \bullet\ r_1 = 3, r_2 = -\frac{1}{2}$$

$$(x - 3)\left(x + \frac{1}{2}\right) = 0$$

$$x^2 - \frac{5}{2}x - \frac{3}{2} = 0$$

$$2\left(x^2 - \frac{5}{2}x - \frac{3}{2}\right) = 2 \cdot 0$$

$$2x^2 - 5x - 3 = 0$$

You Try It 4

$$2(x + 1)^2 - 24 = 0$$

$$2(x + 1)^2 = 24$$

$$(x + 1)^2 = 12 \qquad \bullet \text{ Solve for } (x + 1)^2.$$

$$\sqrt{(x + 1)^2} = \sqrt{12} \qquad \bullet \text{ Take the square root of each side of the equation.}$$

$$x + 1 = \pm\sqrt{12}$$

$$x + 1 = \pm 2\sqrt{3}$$

$$x + 1 = 2\sqrt{3} \qquad x + 1 = -2\sqrt{3} \qquad \bullet \text{ Solve for } x.$$

$$x = -1 + 2\sqrt{3} \qquad x = -1 - 2\sqrt{3}$$

The solutions are $-1 + 2\sqrt{3}$ and $-1 - 2\sqrt{3}$.

SECTION 10.2

You Try It 1

$$4x^2 - 4x - 1 = 0 \qquad \bullet \text{ Write in the form } ax^2 + bx = -c.$$

$$4x^2 - 4x = 1$$

$$\frac{1}{4}(4x^2 - 4x) = \frac{1}{4} \cdot 1 \qquad \bullet \text{ Multiply both sides by } \frac{1}{a}.$$

$$x^2 - x = \frac{1}{4}$$

$$x^2 - x + \frac{1}{4} = \frac{1}{4} + \frac{1}{4} \qquad \bullet \text{ Complete the square.}$$

$$\left(x - \frac{1}{2}\right)^2 = \frac{2}{4} \qquad \bullet \text{ Factor.}$$

$$\sqrt{\left(x - \frac{1}{2}\right)^2} = \sqrt{\frac{2}{4}} \qquad \bullet \text{ Take square roots.}$$

$$x - \frac{1}{2} = \pm\frac{\sqrt{2}}{2}$$

$$x - \frac{1}{2} = \frac{\sqrt{2}}{2} \qquad x - \frac{1}{2} = -\frac{\sqrt{2}}{2} \qquad \bullet \text{ Solve for } x.$$

$$x = \frac{1}{2} + \frac{\sqrt{2}}{2} \qquad x = \frac{1}{2} - \frac{\sqrt{2}}{2}$$

The solutions are $\frac{1 + \sqrt{2}}{2}$ and $\frac{1 - \sqrt{2}}{2}$.

You Try It 2

$$x^2 + 4x + 8 = 0$$

$$x^2 + 4x = -8$$

$$x^2 + 4x + 4 = -8 + 4 \qquad \bullet \text{ Complete the square.}$$

$$(x + 2)^2 = -4 \qquad \bullet \text{ Factor.}$$

$$\sqrt{(x + 2)^2} = \sqrt{-4} \qquad \bullet \text{ Take square roots.}$$

$$x + 2 = \pm 2i$$

$$x + 2 = 2i \qquad x + 2 = -2i \qquad \bullet \text{ Solve for } x.$$

$$x = -2 + 2i \qquad x = -2 - 2i$$

The solutions are $-2 + 2i$ and $-2 - 2i$.

SECTION 10.3

You Try It 1 $x^2 - 2x + 10 = 0$

$$a = 1, b = -2, c = 10$$

$$x = \frac{-b \pm \sqrt{b^2 - 4ac}}{2a}$$

$$= \frac{-(-2) \pm \sqrt{(-2)^2 - 4(1)(10)}}{2 \cdot 1}$$

$$= \frac{2 \pm \sqrt{4 - 40}}{2} = \frac{2 \pm \sqrt{-36}}{2}$$

$$= \frac{2 \pm 6i}{2} = 1 \pm 3i$$

The solutions are $1 + 3i$ and $1 - 3i$.

You Try It 2

$$4x^2 = 4x - 1$$

$$4x^2 - 4x + 1 = 0 \qquad \bullet \text{ Write in standard form.}$$

$$a = 4, b = -4, c = 1$$

$$x = \frac{-b \pm \sqrt{b^2 - 4ac}}{2a}$$

$$= \frac{-(-4) \pm \sqrt{(-4)^2 - 4(4)(1)}}{2 \cdot 4}$$

$$= \frac{4 \pm \sqrt{16 - 16}}{8} = \frac{4 \pm \sqrt{0}}{8}$$

$$= \frac{4}{8} = \frac{1}{2}$$

The solution is $\frac{1}{2}$.

You Try It 3 $3x^2 - x - 1 = 0$

$$a = 3, b = -1, c = -1$$

$$b^2 - 4ac =$$

$$(-1)^2 - 4(3)(-1) = 1 + 12 = 13$$

$$13 > 0$$

Because the discriminant is greater than zero, the equation has two unequal real number solutions.

SECTION 10.4

You Try It 1

$$\begin{aligned} x - 5x^{1/2} + 6 &= 0 \\ (x^{1/2})^2 - 5(x^{1/2}) + 6 &= 0 \\ u^2 - 5u + 6 &= 0 \\ (u-2)(u-3) &= 0 \end{aligned}$$

$$u - 2 = 0 \qquad u - 3 = 0$$
$$u = 2 \qquad u = 3$$

Replace u by $x^{1/2}$.

$$x^{1/2} = 2 \qquad x^{1/2} = 3$$
$$\sqrt{x} = 2 \qquad \sqrt{x} = 3$$
$$(\sqrt{x})^2 = 2^2 \qquad (\sqrt{x})^2 = 3^2$$
$$x = 4 \qquad x = 9$$

The solutions are 4 and 9.

You Try It 2

$$\begin{aligned} \sqrt{2x+1} + x &= 7 \\ \sqrt{2x+1} &= 7 - x && \text{• Solve for the radical.} \\ (\sqrt{2x+1})^2 &= (7-x)^2 && \text{• Square each side.} \\ 2x + 1 &= 49 - 14x + x^2 && \text{• Write in standard form.} \\ 0 &= x^2 - 16x + 48 \\ 0 &= (x-4)(x-12) && \text{• Solve by factoring.} \end{aligned}$$

$$x - 4 = 0 \qquad x - 12 = 0$$
$$x = 4 \qquad x = 12$$

4 checks as a solution.
12 does not check as a solution.

The solution is 4.

You Try It 3

$$\sqrt{2x-1} + \sqrt{x} = 2$$

Solve for one of the radical expressions.

$$\begin{aligned} \sqrt{2x-1} &= 2 - \sqrt{x} \\ (\sqrt{2x-1})^2 &= (2 - \sqrt{x})^2 && \text{• Square each side.} \\ 2x - 1 &= 4 - 4\sqrt{x} + x \\ x - 5 &= -4\sqrt{x} \\ (x-5)^2 &= (-4\sqrt{x})^2 && \text{• Square each side.} \\ x^2 - 10x + 25 &= 16x \\ x^2 - 26x + 25 &= 0 \\ (x-1)(x-25) &= 0 \end{aligned}$$

$$x - 1 = 0 \qquad x - 25 = 0$$
$$x = 1 \qquad x = 25$$

1 checks as a solution.
25 does not check as a solution.

The solution is 1.

You Try It 4

$$3y + \frac{25}{3y-2} = -8$$

$$(3y-2)\left(3y + \frac{25}{3y-2}\right) = (3y-2)(-8) \qquad \text{• Clear denominators.}$$

$$(3y-2)(3y) + (3y-2)\left(\frac{25}{3y-2}\right) = (3y-2)(-8)$$

$$\begin{aligned} 9y^2 - 6y + 25 &= -24y + 16 \\ 9y^2 + 18y + 9 &= 0 && \text{• Write in standard form.} \\ 9(y^2 + 2y + 1) &= 0 && \text{• Solve by factoring.} \\ 9(y+1)(y+1) &= 0 \end{aligned}$$

$$y + 1 = 0 \qquad y + 1 = 0$$
$$y = -1 \qquad y = -1$$

The solution is -1.

SECTION 10.5

You Try It 1

$$2x^2 - x - 10 \le 0$$
$$(2x-5)(x+2) \le 0$$

2x − 5 --- | ---------- | +++
x + 2 --- | ++++++++++ | +++

(number line: −3 −2 −1 0 1 2 3)

(number line: −5 −4 −3 −2 −1 0 1 2 3 4 5, with the interval from −2 to 5/2 shaded, closed at both ends)

$$\left\{x \,\middle|\, -2 \le x \le \frac{5}{2}\right\}$$

SECTION 10.6

You Try It 1

Strategy

- The unknown time for Olivia working alone: t
- The unknown time for William working alone: $t + 3$
- The time working together: 2 h

	Rate	·	*Time*	=	*Part*
Olivia	$\frac{1}{t}$	·	2	=	$\frac{2}{t}$
William	$\frac{1}{t+3}$	·	2	=	$\frac{2}{t+3}$

- The sum of the parts of the task completed by Olivia and William equals 1.

Solution

$$\frac{2}{t} + \frac{2}{t+3} = 1$$

$$t(t+3)\left(\frac{2}{t} + \frac{2}{t+3}\right) = t(t+3) \cdot 1$$

$$2(t+3) + 2t = t^2 + 3t$$

$$2t + 6 + 2t = t^2 + 3t$$

$$0 = t^2 - t - 6$$

$$0 = (t+2)(t-3)$$

$t + 2 = 0 \qquad t - 3 = 0$

$t = -2 \qquad t = 3$

The time cannot be negative. It takes Olivia 3 h, working alone, to detail the car. The time for William to detail the car is $t + 3$.

$t + 3$

$3 + 3 = 6$

It takes William 6 h, working alone, to detail the car.

You Try It 2

Strategy

- Unknown rate of the wind: r

	Distance	÷	*Rate*	=	*Time*
With the wind	1200	÷	$250 + r$	=	$\frac{1200}{250 + r}$
Against the wind	1200	÷	$250 - r$	=	$\frac{1200}{250 - r}$

- The time with the wind was 2 h less than the time against the wind.

Solution

$$\frac{1200}{250 + r} = \frac{1200}{250 - r} - 2$$

$$(250 + r)(250 - r)\left(\frac{1200}{250 + r}\right) = (250 + r)(250 - r)\left(\frac{1200}{250 - r} - 2\right)$$

$$(250 - r)1200 = (250 + r)1200 - 2(250 + r)(250 - r)$$

$$300{,}000 - 1200r = 300{,}000 + 1200r - 2(62{,}500 - r^2)$$

$$300{,}000 - 1200r = 300{,}000 + 1200r - 125{,}000 + 2r^2$$

$$0 = 2r^2 + 2400r - 125{,}000$$

$$0 = 2(r^2 + 1200 - 62{,}500)$$

$$0 = 2(r + 1250)(r - 50)$$

$r + 1250 = 0 \qquad r - 50 = 0$

$r = -1250 \qquad r = 50$

The rate cannot be negative. The rate of the wind was 50 mph.

You Try It 3

Strategy

- Height of the triangle: h
- Base of the triangle: $2h + 4$
- Use the formula $A = \frac{1}{2}bh$.

Solution

$$A = \frac{1}{2}bh$$

$$35 = \frac{1}{2}(2h + 4)h$$

$$35 = h^2 + 2h$$

$$0 = h^2 + 2h - 35$$

$$0 = (h + 7)(h - 5)$$

$h + 7 = 0 \qquad h - 5 = 0$

$h = -7 \qquad h = 5$

The height cannot be negative.
The height of the triangle is 5 in.

SOLUTIONS TO CHAPTER 11 "YOU TRY IT"

SECTION 11.1

You Try It 1

x-coordinate of vertex:

$$-\frac{b}{2a} = -\frac{4}{2(4)} = -\frac{1}{2}$$

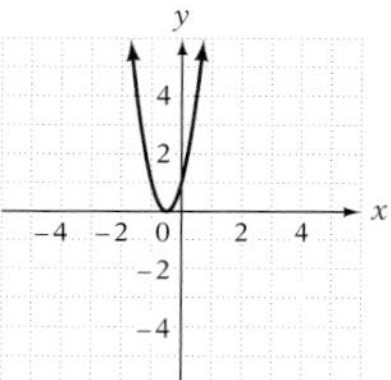

y-coordinate of vertex:

$$y = 4x^2 + 4x + 1$$

$$= 4\left(-\frac{1}{2}\right)^2 + 4\left(-\frac{1}{2}\right) + 1$$

$$= 1 - 2 + 1$$

$$= 0$$

Vertex: $\left(-\frac{1}{2}, 0\right)$

Axis of symmetry: $x = -\frac{1}{2}$

You Try It 2

x-coordinate of vertex:

$$-\frac{b}{2a} = -\frac{-2}{2(-1)} = -1$$

y-coordinate of vertex:

$$f(x) = -x^2 - 2x - 1$$

$$f(-1) = -(-1)^2 - 2(-1) - 1$$

$$= -1 + 2 - 1$$

$$= 0$$

Vertex: $(-1, 0)$

The domain is $\{x \mid x \in \text{real numbers}\}$. The range is $\{y \mid y \le 0\}$.

You Try It 3

$y = x^2 + 3x + 4$
$0 = x^2 + 3x + 4$

$$x = \frac{-b \pm \sqrt{b^2 - 4ac}}{2a}$$

$$= \frac{-3 \pm \sqrt{3^2 - 4(1)(4)}}{2 \cdot 1} \quad \bullet\ a = 1, b = 3, c = 4$$

$$= \frac{-3 \pm \sqrt{-7}}{2}$$

$$= \frac{-3 \pm i\sqrt{7}}{2}$$

The equation has no real number solutions. There are no x-intercepts.

You Try It 4

$g(x) = x^2 - x + 6$
$0 = x^2 - x + 6$

$$x = \frac{-b \pm \sqrt{b^2 - 4ac}}{2a}$$

$$= \frac{-(-1) \pm \sqrt{(-1)^2 - 4(1)(6)}}{2(1)} \quad \bullet\ a = 1, b = -1, c = 6$$

$$= \frac{1 \pm \sqrt{1 - 24}}{2}$$

$$= \frac{1 \pm \sqrt{-23}}{2} = \frac{1 \pm i\sqrt{23}}{2}$$

The zeros of the function are $\frac{1}{2} + \frac{\sqrt{23}}{2}i$ and $\frac{1}{2} - \frac{\sqrt{23}}{2}i$.

You Try It 5

$y = x^2 - x - 6$
$a = 1, b = -1, c = -6$

$b^2 - 4ac$

$(-1)^2 - 4(1)(-6) = 1 + 24 = 25$

Because the discriminant is greater than zero, the parabola has two x-intercepts.

You Try It 6

$f(x) = -3x^2 + 4x - 1$

$$x = -\frac{b}{2a} = -\frac{4}{2(-3)} = \frac{2}{3} \quad \bullet\ \text{The } x\text{-coordinate of the vertex}$$

$f(x) = -3x^2 + 4x - 1$

$$f\left(\frac{2}{3}\right) = -3\left(\frac{2}{3}\right)^2 + 4\left(\frac{2}{3}\right) - 1 \quad \bullet\ x = \frac{2}{3}$$

$$= -\frac{4}{3} + \frac{8}{3} - 1 = \frac{1}{3}$$

Because a is negative, the function has a maximum value.

The maximum value of the function is $\frac{1}{3}$.

You Try It 7

Strategy

- To find the time it takes the ball to reach its maximum height, find the t-coordinate of the vertex.
- To find the maximum height, evaluate the function at the t-coordinate of the vertex.

Solution

$$t = -\frac{b}{2a} = -\frac{64}{2(-16)} = 2 \quad \bullet\ \text{The } t\text{-coordinate of the vertex}$$

The ball reaches its maximum height in 2 s.

$s(t) = -16t^2 + 64t$
$s(2) = -16(2)^2 + 64(2) = -64 + 128 = 64 \quad \bullet\ t = 2$

The maximum height is 64 ft.

You Try It 8

Strategy

$P = 2x + y$
$100 = 2x + y \quad \bullet\ P = 100$
$100 - 2x = y \quad \bullet$ Solve for y.

Express the area of the rectangle in terms of x.

$A = xy$
$A = x(100 - 2x) \quad \bullet\ y = 100 - 2x$
$A = -2x^2 + 100x$

- To find the width, find the x-coordinate of the vertex of $f(x) = -2x^2 + 100x$.
- To find the length, replace x in $y = 100 - 2x$ by the x-coordinate of the vertex.

Solution

$$x = -\frac{b}{2a} = -\frac{100}{2(-2)} = 25$$

The width is 25 ft.

$100 - 2x = 100 - 2(25)$
$= 100 - 50 = 50$

The length is 50 ft.

SECTION 11.2

You Try It 1

Any vertical line intersects the graph at most once. The graph is the graph of a function.

You Try It 2

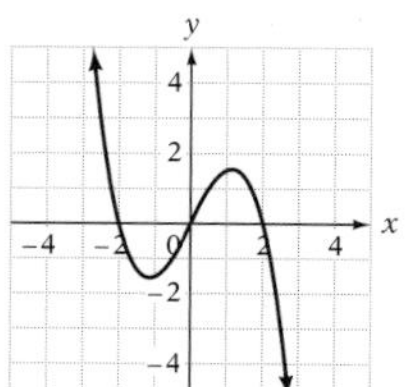

Domain: $\{x \mid x \in \text{real numbers}\}$
Range: $\{y \mid y \in \text{real numbers}\}$

You Try It 3

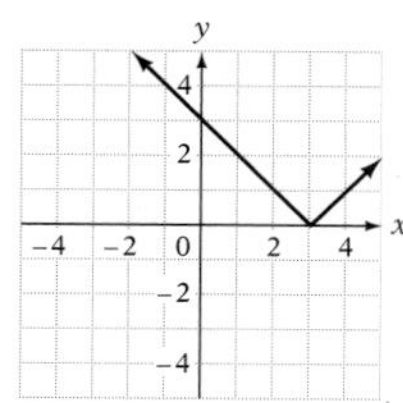

Domain: $(-\infty, \infty)$
Range: $[0, \infty)$

You Try It 4

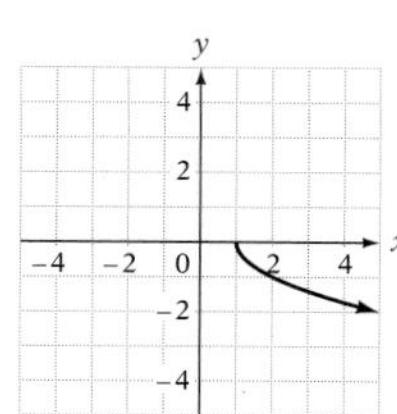

Domain: $\{x \mid x \geq 1\}$
Range: $\{y \mid y \leq 0\}$

SECTION 11.3

You Try It 1

$$\begin{aligned}(f+g)(-2) &= f(-2) + g(-2)\\ &= [(-2)^2 + 2(-2)] + [5(-2) - 2]\\ &= (4-4) + (-10-2)\\ &= -12\end{aligned}$$
$$(f+g)(-2) = -12$$

You Try It 2

$$\begin{aligned}(f \cdot g)(3) &= f(3) \cdot g(3)\\ &= (4 - 3^2) \cdot [3(3) - 4]\\ &= (4-9) \cdot (9-4)\\ &= (-5)(5)\\ &= -25\end{aligned}$$
$$(f \cdot g)(3) = -25$$

You Try It 3

$$\begin{aligned}\left(\frac{f}{g}\right)(4) &= \frac{f(4)}{g(4)}\\ &= \frac{4^2 - 4}{4^2 + 2 \cdot 4 + 1}\\ &= \frac{16-4}{16+8+1}\\ &= \frac{12}{25}\end{aligned}$$
$$\left(\frac{f}{g}\right)(4) = \frac{12}{25}$$

You Try It 4

$g(x) = x^2$
$g(-1) = (-1)^2 = 1$ • Evaluate g at -1.
$f(x) = 1 - 2x$
$f[g(-1)] = f(1) = 1 - 2(1) = -1$ • Evaluate f at $g(-1) = 1$.
$f[g(-1)] = -1$

You Try It 5

$$\begin{aligned}M(s) &= s^3 + 1\\ M[L(s)] &= (s+1)^3 + 1 \quad \text{• Evaluate } M \text{ at } L(s).\\ &= s^3 + 3s^2 + 3s + 1 + 1\\ &= s^3 + 3s^2 + 3s + 2\\ M[L(s)] &= s^3 + 3s^2 + 3s + 2\end{aligned}$$

SECTION 11.4

You Try It 1 Because any horizontal line intersects the graph at most once, the graph is the graph of a 1–1 function.

You Try It 2

$$\begin{aligned}f(x) &= \frac{1}{2}x + 4\\ y &= \frac{1}{2}x + 4\\ x &= \frac{1}{2}y + 4\\ x - 4 &= \frac{1}{2}y\\ 2x - 8 &= y\\ f^{-1}(x) &= 2x - 8\end{aligned}$$

The inverse of the function is given by $f^{-1}(x) = 2x - 8$.

You Try It 3

$$\begin{aligned}f[g(x)] &= 2\left(\frac{1}{2}x - 3\right) - 6\\ &= x - 6 - 6 = x - 12\end{aligned}$$

No, $g(x)$ is not the inverse of $f(x)$.

You Try It 4

$s(x) = \frac{3}{5}x + 1125$

$y = \frac{3}{5}x + 1125$ • Replace $s(x)$ by y.

$x = \frac{3}{5}y + 1125$ • Interchange x and y.

$x - 1125 = \frac{3}{5}y$ • Solve for y.

$\frac{5}{3}(x - 1125) = y$

$\frac{5}{3}x - 1875 = y$

$\frac{5}{3}x - 1875 = s^{-1}(x)$ • Replace y by $s^{-1}(x)$.

$s^{-1}(x) = \frac{5}{3}x - 1875$. The equation gives the temperature of the air given the speed of sound.

SOLUTIONS TO CHAPTER 12 "YOU TRY IT"

SECTION 12.1

You Try It 1

$f(x) = \left(\frac{2}{3}\right)^x$

$f(3) = \left(\frac{2}{3}\right)^3 = \frac{8}{27}$ • $x = 3$

$f(-2) = \left(\frac{2}{3}\right)^{-2} = \left(\frac{3}{2}\right)^2 = \frac{9}{4}$ • $x = -2$

You Try It 2

$f(x) = 2^{2x+1}$

$f(0) = 2^{2(0)+1} = 2^1 = 2$ • $x = 0$

$f(-2) = 2^{2(-2)+1} = 2^{-3} = \frac{1}{2^3} = \frac{1}{8}$ • $x = -2$

You Try It 3

$f(x) = e^{2x-1}$

$f(2) = e^{2\cdot2-1} = e^3 \approx 20.0855$ • $x = 2$

$f(-2) = e^{2(-2)-1} = e^{-5} \approx 0.0067$ • $x = -2$

You Try It 4

x	y
-4	4
-2	2
0	1
2	$\frac{1}{2}$
4	$\frac{1}{4}$

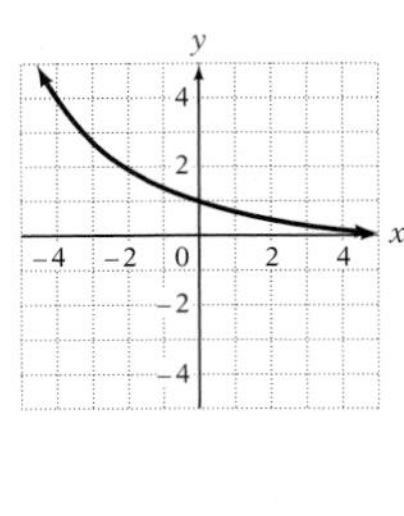

You Try It 5

x	y
-2	$\frac{5}{4}$
-1	$\frac{3}{2}$
0	2
1	3
2	5

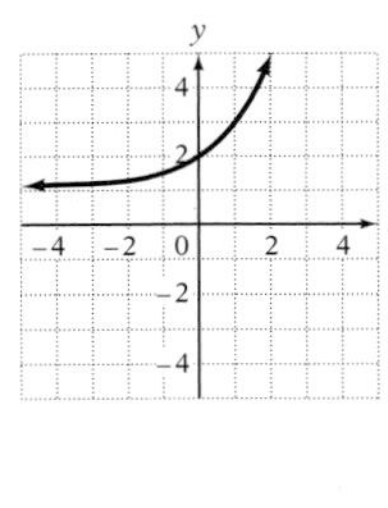

You Try It 6

x	y
-2	6
-1	4
0	3
1	$\frac{5}{2}$
2	$\frac{9}{4}$

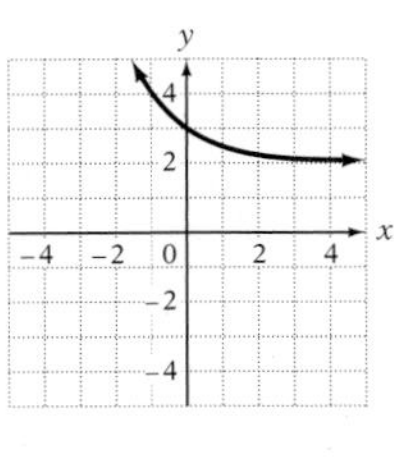

You Try It 7

x	y
-4	$\frac{9}{4}$
-2	$\frac{5}{2}$
0	3
2	4
4	6

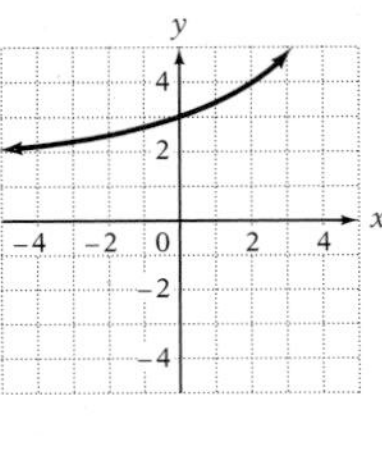

SECTION 12.2

You Try It 1

$\log_4 64 = x$ • Write the equivalent exponential form.

$64 = 4^x$

$4^3 = 4^x$ • $64 = 4^3$

$3 = x$ • The bases are the same. The exponents are equal.

$\log_4 64 = 3$

You Try It 2

$\log_2 x = -4$

$2^{-4} = x$ • Write the equivalent exponential form.

$\frac{1}{2^4} = x$

$\frac{1}{16} = x$

The solution is $\frac{1}{16}$.

You Try It 3

$\ln x = 3$ • Write the equivalent exponential form.

$e^3 = x$

$20.0855 \approx x$

You Try It 4

$\log_8 \sqrt[3]{xy^2} = \log_8 (xy^2)^{1/3} = \frac{1}{3}\log_8 (xy^2)$ • Power Property

$= \frac{1}{3}(\log_8 x + \log_8 y^2)$ • Product Property

$= \frac{1}{3}(\log_8 x + 2\log_8 y)$ • Power Property

$= \frac{1}{3}\log_8 x + \frac{2}{3}\log_8 y$ • Distributive Property

You Try It 5

$\frac{1}{3}(\log_4 x - 2\log_4 y + \log_4 z)$

$= \frac{1}{3}(\log_4 x - \log_4 y^2 + \log_4 z)$

$= \frac{1}{3}\left(\log_4 \frac{x}{y^2} + \log_4 z\right) = \frac{1}{3}\left(\log_4 \frac{xz}{y^2}\right)$

$= \log_4 \left(\frac{xz}{y^2}\right)^{1/3} = \log_4 \sqrt[3]{\frac{xz}{y^2}}$

You Try It 6 $\log_3 0.834 = \frac{\ln 0.834}{\ln 3} \approx -0.16523$

You Try It 7 $\log_7 6.45 = \frac{\log 6.45}{\log 7} \approx 0.95795$

SECTION 12.3

You Try It 1

$$\begin{aligned} f(x) &= \log_2(x-1) \\ y &= \log_2(x-1) \quad &\text{• } f(x) = y \\ 2^y &= x-1 \quad &\text{• Write the equivalent exponential equation.} \\ 2^y + 1 &= x \end{aligned}$$

x	y
$\frac{5}{4}$	-2
$\frac{3}{2}$	-1
2	0
3	1
5	2

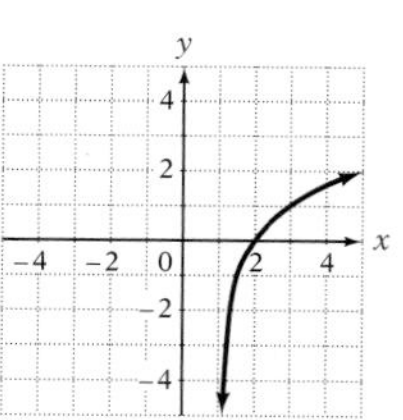

You Try It 2

$$\begin{aligned} f(x) &= \log_3(2x) \\ y &= \log_3(2x) \quad &\text{• } f(x) = y \\ 3^y &= 2x \quad &\text{• Write the equivalent exponential equation.} \\ \frac{3^y}{2} &= x \end{aligned}$$

x	y
$\frac{1}{18}$	-2
$\frac{1}{6}$	-1
$\frac{1}{2}$	0
$\frac{3}{2}$	1
$\frac{9}{2}$	2

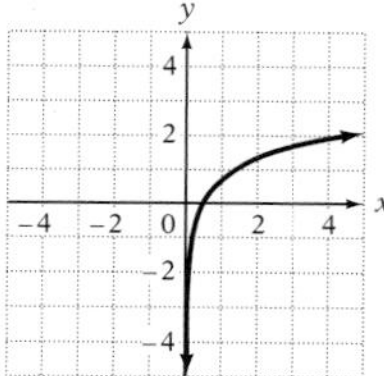

You Try It 3

$$\begin{aligned} f(x) &= -\log_3(x+1) \\ y &= -\log_3(x+1) \quad &\text{• } f(x) = y \\ -y &= \log_3(x+1) \quad &\text{• Multiply both sides by } -1. \\ 3^{-y} &= x+1 \quad &\text{• Write the equivalent exponential equation.} \\ 3^{-y} - 1 &= x \end{aligned}$$

x	y
8	-2
2	-1
0	0
$-\frac{2}{3}$	1
$-\frac{8}{9}$	2

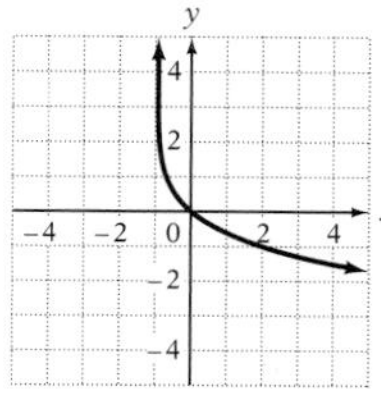

SECTION 12.4

You Try It 1

$$\begin{aligned} (1.06)^n &= 1.5 \\ \log(1.06)^n &= \log 1.5 \quad &\text{• Take the log of each side.} \\ n \log 1.06 &= \log 1.5 \quad &\text{• Power Property} \\ n &= \frac{\log 1.5}{\log 1.06} \quad &\text{• Divide both sides by log 1.06.} \\ n &\approx 6.9585 \end{aligned}$$

The solution is 6.9585.

You Try It 2

$$\begin{aligned} 4^{3x} &= 25 \\ \log 4^{3x} &= \log 25 \quad &\text{• Take the log of each side.} \\ 3x \log 4 &= \log 25 \quad &\text{• Power Property} \\ 3x &= \frac{\log 25}{\log 4} \quad &\text{• Divide both sides by log 4.} \\ 3x &\approx 2.3219 \\ x &\approx 0.7740 \quad &\text{• Divide both sides by 3.} \end{aligned}$$

The solution is 0.7740.

You Try It 3

$\log_4(x^2 - 3x) = 1$

$$\begin{aligned} 4^1 &= x^2 - 3x \quad &\text{• Write in exponential form.} \\ 4 &= x^2 - 3x \\ 0 &= x^2 - 3x - 4 \\ 0 &= (x+1)(x-4) \end{aligned}$$

$x + 1 = 0 \qquad x - 4 = 0$

$x = -1 \qquad x = 4$

The solutions are -1 and 4.

You Try It 4

$$\begin{aligned} \log_3 x + \log_3(x+3) &= \log_3 4 \\ \log_3[x(x+3)] &= \log_3 4 \quad &\text{• Product Property} \\ x(x+3) &= 4 \quad &\text{• 1–1 Property of Logarithms} \\ x^2 + 3x &= 4 \\ x^2 + 3x - 4 &= 0 \\ (x+4)(x-1) &= 0 \quad &\text{• Solve by factoring.} \end{aligned}$$

$x + 4 = 0 \qquad x - 1 = 0$

$x = -4 \qquad x = 1$

-4 does not check as a solution. The solution is 1.

You Try It 5

$$\begin{aligned} \log_3 x + \log_3(x+6) &= 3 \\ \log_3[x(x+6)] &= 3 \quad &\text{• Product Property} \\ x(x+6) &= 3^3 \quad &\text{• Write the equivalent exponential equation.} \\ x^2 + 6x &= 27 \\ x^2 + 6x - 27 &= 0 \\ (x+9)(x-3) &= 0 \quad &\text{• Solve by factoring.} \end{aligned}$$

$x + 9 = 0 \qquad x - 3 = 0$

$x = -9 \qquad x = 3$

-9 does not check as a solution. The solution is 3.

SECTION 12.5

You Try It 1

Strategy To find the hydrogen ion concentration, replace pH by 2.9 in the equation $\text{pH} = -\log(\text{H}^+)$ and solve for H^+.

Solution

$$\begin{aligned} \text{pH} &= -\log(\text{H}^+) \\ 2.9 &= -\log(\text{H}^+) \\ -2.9 &= \log(\text{H}^+) && \text{• Multiply by } -1. \\ 10^{-2.9} &= \text{H}^+ && \text{• Write the equivalent exponential equation.} \\ 0.00126 &\approx \text{H}^+ \end{aligned}$$

The hydrogen ion concentration is approximately 0.00126.

You Try It 2

Strategy Substitute 0.20 for p and 6 for k into the equation $\log p = -kd$ and solve for d.

Solution

$$\begin{aligned} \log p &= -kd \\ \log 0.20 &= -6d \\ \frac{\log 0.20}{-6} &= d \\ 0.116 &\approx d \end{aligned}$$

The opaque glass must be 0.116 m thick.

SOLUTIONS TO CHAPTER R "YOU TRY IT"

SECTION R.1

You Try It 1

$$\begin{aligned} (-4)(6-8)^2 - (-12 \div 4) &= -4(-2)^2 - (-3) \\ &= -4(4) - (-3) \\ &= -16 - (-3) \\ &= -16 + 3 \\ &= -13 \end{aligned}$$

You Try It 2

$3xy^2 - 3x^2y$

$$\begin{aligned} 3(-2)(5)^2 - 3(-2)^2(5) &= 3(-2)(25) - 3(4)(5) \\ &= -6(25) - 3(4)(5) \\ &= -150 - 3(4)(5) \\ &= -150 - 12(5) \\ &= -150 - 60 \\ &= -150 + (-60) \\ &= -210 \end{aligned}$$

You Try It 3

$$\begin{aligned} -5(-3a) &= [-5(-3)]a \\ &= 15a \end{aligned}$$

You Try It 4

$$\begin{aligned} 2z^2 - 5z - 3z^2 + 6z &= 2z^2 - 3z^2 - 5z + 6z \\ &= (2z^2 - 3z^2) + (-5z + 6z) \\ &= -1z^2 + z \\ &= -z^2 + z \end{aligned}$$

You Try It 5

a. $$\begin{aligned} -3(5y-2) &= -3(5y) - (-3)(2) \\ &= -15y + 6 \end{aligned}$$

b. $$\begin{aligned} -2(4x + 2y - 6z) &= -2(4x) + (-2)(2y) - (-2)(6z) \\ &= -8x - 4y + 12z \end{aligned}$$

You Try It 6

a. $$\begin{aligned} 7(-3x - 4y) - 3(3x + y) &= -21x - 28y - 9x - 3y \\ &= -30x - 31y \end{aligned}$$

b. $$\begin{aligned} 2y - 3[5 - 3(3 + 2y)] &= 2y - 3[5 - 9 - 6y] \\ &= 2y - 3[-4 - 6y] \\ &= 2y + 12 + 18y \\ &= 20y + 12 \end{aligned}$$

SECTION R.2

You Try It 1

a. $$\begin{aligned} 4x + 3 &= 7x + 9 \\ 4x - 7x + 3 &= 7x - 7x + 9 \\ -3x + 3 &= 9 \\ -3x + 3 - 3 &= 9 - 3 \\ -3x &= 6 \\ \frac{-3x}{-3} &= \frac{6}{-3} \\ x &= -2 \end{aligned}$$

The solution is -2.

b. $$\begin{aligned} 4 - (5x - 8) &= 4x + 3 \\ 4 - 5x + 8 &= 4x + 3 \\ -5x + 12 &= 4x + 3 \\ -5x - 4x + 12 &= 4x - 4x + 3 \\ -9x + 12 &= 3 \\ -9x + 12 - 12 &= 3 - 12 \\ -9x &= -9 \\ \frac{-9x}{-9} &= \frac{-9}{-9} \\ x &= 1 \end{aligned}$$

The solution is 1.

You Try It 2

$$\begin{aligned} 3x - 1 &\le 5x - 7 \\ 3x - 5x - 1 &\le 5x - 5x - 7 \\ -2x - 1 &\le -7 \\ -2x - 1 + 1 &\le -7 + 1 \\ -2x &\le -6 \\ \frac{-2x}{-2} &\ge \frac{-6}{-2} \\ x &\ge 3 \end{aligned}$$

$\{x | x \ge 3\}$

You Try It 3

$$\begin{aligned} 3 - 2(3x + 1) &< 7 - 2x \\ 3 - 6x - 2 &< 7 - 2x \\ 1 - 6x &< 7 - 2x \\ 1 - 6x + 2x &< 7 - 2x + 2x \\ 1 - 4x &< 7 \\ 1 - 1 - 4x &< 7 - 1 \\ -4x &< 6 \\ \frac{-4x}{-4} &> \frac{6}{-4} \\ x &> -\frac{3}{2} \end{aligned}$$

$\left\{x \middle| x > -\frac{3}{2}\right\}$

SECTION R.3

You Try It 1

You Try It 2

$$y = \frac{3}{5}x - 4$$

x	y
5	−1
0	−4
−5	−7

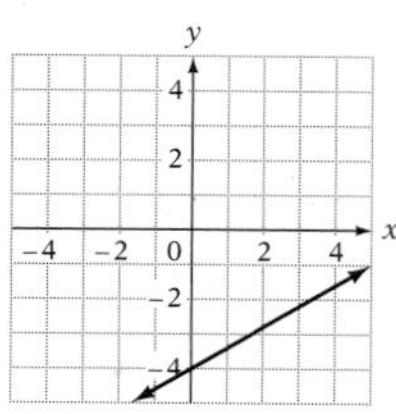

You Try It 3

$$\begin{aligned} -3x + 2y &= 4 \\ 2y &= 3x + 4 \\ y &= \frac{3}{2}x + 2 \end{aligned}$$

x	y
2	5
0	2
−2	−1

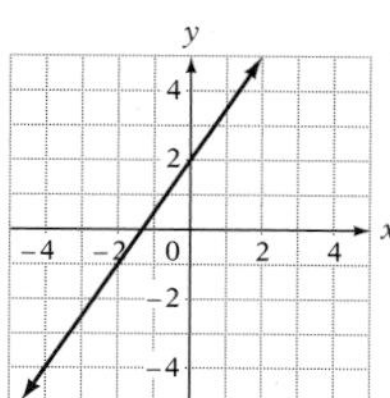

You Try It 4

Let $(x_1, y_1) = (-2, 3)$ and $(x_2, y_2) = (1, -3)$.

$$m = \frac{y_2 - y_1}{x_2 - x_1} = \frac{-3 - 3}{1 - (-2)} = \frac{-6}{3} = -2$$

The slope is −2.

You Try It 5

$$y = -\frac{2}{3}x + 2$$

$$m = -\frac{2}{3} = \frac{-2}{3}$$

y-intercept = (0, 2)

Place a dot at the y-intercept.
Starting at the y-intercept, move to the right 3 units (the change in x) and down 2 units (the change in y).
Place a dot at that location.
Draw a line through the two points.

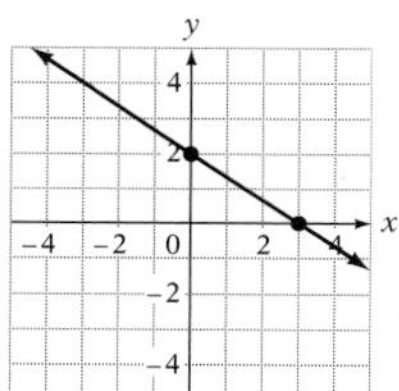

You Try It 6

$$\begin{aligned} f(x) &= 4 - 2x \\ f(-3) &= 4 - 2(-3) \\ f(-3) &= 4 + 6 \\ f(-3) &= 10 \end{aligned}$$

An ordered pair of the function is (−3, 10).

You Try It 7

$$(x_1, y_1) = (-2, 2), m = -\frac{1}{2}$$

$$\begin{aligned} y - y_1 &= m(x - x_1) \\ y - 2 &= -\frac{1}{2}[x - (-2)] \\ y - 2 &= -\frac{1}{2}(x + 2) \\ y - 2 &= -\frac{1}{2}x - 1 \\ y &= -\frac{1}{2}x + 1 \end{aligned}$$

SECTION R.4

You Try It 1

a. $$\begin{aligned} (-2ab)(2a^3b^{-2})^{-3} &= (-2ab)(2^{-3}a^{-9}b^6) \\ &= (-2^{-2})a^{-8}b^7 \\ &= -\frac{b^7}{2^2a^8} \\ &= -\frac{b^7}{4a^8} \end{aligned}$$

b. $$\begin{aligned} \left(\frac{2x^2y^{-4}}{4x^{-2}y^{-5}}\right)^{-3} &= \left(\frac{x^2y^{-4}}{2x^{-2}y^{-5}}\right)^{-3} \\ &= \frac{x^{2(-3)}y^{(-4)(-3)}}{2^{1(-3)}x^{(-2)(-3)}y^{(-5)(-3)}} \\ &= \frac{x^{-6}y^{12}}{2^{-3}x^6y^{15}} \\ &= 2^3x^{-6-6}y^{12-15} \\ &= 2^3x^{-12}y^{-3} \\ &= \frac{8}{x^{12}y^3} \end{aligned}$$

You Try It 2

$$\begin{aligned} &(-4x^3 + 2x^2 - 8) + (4x^3 + 6x^2 - 7x + 5) \\ &\quad = (-4x^3 + 4x^3) + (2x^2 + 6x^2) + (-7x) + (-8 + 5) \\ &\quad = 8x^2 - 7x - 3 \end{aligned}$$

You Try It 3

$$\begin{aligned} &(-4w^3 + 8w - 8) - (3w^3 - 4w^2 + 2w - 1) \\ &\quad = (-4w^3 + 8w - 8) + (-3w^3 + 4w^2 - 2w + 1) \\ &\quad = (-4w^3 - 3w^3) + 4w^2 + (8w - 2w) + (-8 + 1) \\ &\quad = -7w^3 + 4w^2 + 6w - 7 \end{aligned}$$

You Try It 4

$$\begin{aligned} &3mn^2(2m^2 - 3mn - 1) \\ &\quad = 3mn^2(2m^2) - 3mn^2(3mn) - 3mn^2(1) \\ &\quad = 6m^3n^2 - 9m^2n^3 - 3mn^2 \end{aligned}$$

You Try It 5

$$\begin{array}{r} 3c^2 - 4c + 5 \\ 2c - 3 \\ \hline -9c^2 + 12c - 15 \\ 6c^3 - 8c^2 + 10c \\ \hline 6c^3 - 17c^2 + 22c - 15 \end{array}$$

You Try It 6

$(4y - 7)(3y - 5)$

$= (4y)(3y) + (4y)(-5) + (-7)(3y) + (-7)(-5)$

$= 12y^2 - 20y - 21y + 35$

$= 12y^2 - 41y + 35$

You Try It 7

$(x^3 - 7 - 2x) \div (x - 2)$

$$\begin{array}{r} x^2 + 2x + 2 \\ x - 2\overline{)x^3 + 0x^2 - 2x - 7} \\ \underline{x^3 - 2x^2} \\ 2x^2 - 2x \\ \underline{2x^2 - 4x} \\ 2x - 7 \\ \underline{2x - 4} \\ -3 \end{array}$$

Check: $(x^2 + 2x + 2)(x - 2) - 3$
$= x^3 - 2x - 4 - 3$
$= x^3 - 2x - 7$

$(x^3 - 7 - 2x) \div (x - 2)$
$= x^2 + 2x + 2 - \dfrac{3}{x - 2}$

You Try It 8 The GCF is $3x^2y^2$.

$6x^4y^2 - 9x^3y^2 + 12x^2y^4$
$= 3x^2y^2(2x^2 - 3x + 4y^2)$

You Try It 9 Two factors of 15 whose sum is -8 are -3 and -5.

$x^2 - 8x + 15 = (x - 3)(x - 5)$

You Try It 10

$3x^2 - x - 2$

Positive Factors of 3	Factors of −2
1, 3	1, −2
	−1, 2

Trial Factors	Middle Term
$(x + 1)(3x - 2)$	$-2x + 3x = x$
$(x - 2)(3x + 1)$	$x - 6x = -5x$
$(x - 1)(3x + 2)$	$2x - 3x = -x$
$(x + 2)(3x - 1)$	$-x + 6x = 5x$

$3x^2 - x - 2 = (x - 1)(3x + 2)$

Check: $(x - 1)(3x + 2) = 3x^2 + 2x - 3x - 2$
$= 3x^2 - x - 2$

You Try It 11

$4a^3 - 4a^2 - 24a$

There is a common factor, $4a$. Factor out the GCF.

$4a^3 - 4a^2 - 24a = 4a(a^2 - a - 6)$

Factor $a^2 - a - 6$. The two factors of -6 whose sum is -1 are 2 and -3.

$4a(a^2 - a - 6) = 4a(a + 2)(a - 3)$

$4a^3 - 4a^2 - 24a = 4a(a + 2)(a - 3)$

Check: $4a(a + 2)(a - 3) = (4a^2 + 8a)(a - 3)$
$= 4a^3 - 12a^2 + 8a^2 - 24a$
$= 4a^3 - 4a^2 - 24a$

Answers to Selected Exercises

ANSWERS TO CHAPTER 1 SELECTED EXERCISES

PREP TEST

1. 127.16 **2.** 46,514 **3.** 4517 **4.** 11,396 **5.** 508 **6.** 24 **7.** 4 **8.** $3 \cdot 7$ **9.** $\frac{2}{5}$

SECTION 1.1

1. $8 > -6$ **3.** $-12 < 1$ **5.** $42 > 19$ **7.** $0 > -31$ **9.** $53 > -46$ **11.** False **13.** True **15.** False **17.** True **19.** False **21.** −23, −18 **23.** 21, 37 **25.** (i) **27.** −4 **29.** 9 **31.** 28 **33.** 14 **35.** −77 **37.** 0 **39.** 74 **41.** −82 **43.** −81 **45.** $|-83| > |58|$ **47.** $|43| < |-52|$ **49.** $|-68| > |-42|$ **51.** $|-45| < |-61|$ **53.** 19, 0, −28 **55.** −45, 0, −17 **57.** True **59.** −11 **61.** −5 **63.** −83 **65.** −46 **67.** 0 **69.** −5 **71.** 9 **73.** 1 **75.** 8 **77.** −7 **79.** −9 **81.** 9 **83.** −3 **85.** 18 **87.** −10 **89.** −41 **91.** −12 **93.** 0 **95.** −9 **97.** 11 **99.** −18 **101.** 0 **103.** 2 **105.** −138 **107.** −8 **109.** −12 **111.** Negative **113.** Positive **115.** 42 **117.** −28 **119.** 60 **121.** −253 **123.** −238 **125.** −114 **127.** −2 **129.** 8 **131.** −7 **133.** −12 **135.** −6 **137.** −7 **139.** 11 **141.** −14 **143.** 15 **145.** −16 **147.** 0 **149.** −29 **151.** Undefined **153.** −11 **155.** Undefined **157.** −105 **159.** 252 **161.** −240 **163.** 96 **165.** −216 **167.** −315 **169.** 420 **171.** 0 **173.** Negative **175.** The difference is 153°F. **177.** After a rise of 5°C, the temperature is −14°C. **179.** The difference in height is 20,370 m. **181. a.** −5, −7, −6 **b.** Woods' score for the first 3 days was −18. **c.** His final score was −19. **183.** The average daily high temperature for the week was −4°F. **185.** False **187.** The average score of the 10 golfers was −2. **189.** The student scored 71 points. **191.** 22; 2 **193. a.** True **b.** True

SECTION 1.2

1. 0.125 **3.** $0.\overline{2}$ **5.** $0.1\overline{6}$ **7.** 0.5625 **9.** $0.58\overline{3}$ **11.** 0.24 **13.** 0.225 **15.** $0.\overline{45}$ **19.** $\frac{2}{5}$, 0.40 **21.** $\frac{22}{25}$, 0.88 **23.** $\frac{8}{5}$, 1.6 **25.** $\frac{87}{100}$, 0.87 **27.** $\frac{9}{2}$, 4.50 **29.** $\frac{3}{70}$ **31.** $\frac{3}{8}$ **33.** $\frac{1}{400}$ **35.** $\frac{1}{16}$ **37.** $\frac{23}{400}$ **39.** 0.091 **41.** 0.167 **43.** 0.009 **45.** 0.0915 **47.** 0.1823 **49.** 37% **51.** 2% **53.** 12.5% **55.** 136% **57.** 0.4% **59.** 83% **61.** $37\frac{1}{2}\%$ **63.** $44\frac{4}{9}\%$ **65.** 45% **67.** 250% **69.** Greater than 1% **71.** $\frac{5}{26}$ **73.** $\frac{11}{8}$ **75.** $\frac{1}{12}$ **77.** $\frac{7}{24}$ **79.** 0 **81.** $\frac{3}{8}$ **83.** −1.06 **85.** −23.845 **87.** −10.7893 **89.** −37.19 **91.** −17.5 **93.** 19.61 **95.** 2 **97.** 1 **99.** $\frac{2}{9}$ **101.** $-\frac{2}{9}$ **103.** $-\frac{2}{27}$ **105.** $\frac{3}{2}$ **107.** $-\frac{8}{9}$ **109.** $\frac{2}{3}$ **111.** 4.164 **113.** 4.347 **115.** −2.22 **117.** 0.75 **119.** 2.32 **121.** 3.83 **123. a.** Less than 1 **b.** Greater than 1 **125.** 2401 **127.** −64 **129.** −8 **131.** −125 **133.** $-\frac{27}{64}$ **135.** 3.375 **137.** −1 **139.** −8 **141.** −6750 **143.** Negative **145.** Positive **147.** 4 **149.** 7 **151.** $4\sqrt{2}$ **153.** $2\sqrt{2}$ **155.** $18\sqrt{2}$ **157.** $10\sqrt{10}$ **159.** $\sqrt{15}$ **161.** $\sqrt{29}$ **163.** $-54\sqrt{2}$ **165.** $3\sqrt{5}$ **167.** 0 **169.** $48\sqrt{2}$ **171.** 15.492 **173.** 16.971 **175.** 16 **177.** −11 and −10 **179.** 2 and 3 **181.** The difference between the temperatures was 55.57°C. **183.** The difference between the record high and low temperatures is 146.3°F. **185.** The temperature began below 0°F. **187.** The difference is 35.438°C.

189. **a.** The closing price on the previous day was \$13.42. **b.** The closing price on the previous day was \$36.83.
191. Answers will vary. For example, **a.** 0.15, **b.** 1.05, **c.** 0.001.

SECTION 1.3

3. 0 **5.** −11 **7.** 20 **9.** −10 **11.** 20 **13.** 29 **15.** 11 **17.** 7 **19.** −11 **21.** 6 **23.** 15 **25.** 4 **27.** 5 **29.** −1 **31.** 4 **33.** 0.51 **35.** 1.7 **37.** Answers will vary. For example, $\frac{17}{24}$ and $\frac{33}{48}$. **39.** Answers will vary. For example, **a.** $\frac{1}{2}$, **b.** 1, **c.** 2.

SECTION 1.4

1. −9 **3.** 41 **5.** −7 **7.** 13 **9.** −15 **11.** 41 **13.** 1 **15.** 5 **17.** 1 **19.** 57 **21.** 5 **23.** 8 **25.** −3 **27.** −2 **29.** −4 **31.** Positive **33.** Negative **35.** $14x$ **37.** $5a$ **39.** $-6y$ **41.** $7 - 3b$ **43.** $5a$ **45.** $-2ab$ **47.** $5xy$ **49.** 0 **51.** $-\frac{5}{6}x$ **53.** $6.5x$ **55.** $0.45x$ **57.** $7a$ **59.** $-14x^2$ **61.** $-\frac{11}{24}x$ **63.** $17x - 3y$ **65.** $-2a - 6b$ **67.** $-3x - 8y$ **69.** $-4x^2 - 2x$ **71.** (iv) and (v) **73.** $60x$ **75.** $-10a$ **77.** $30y$ **79.** $72x$ **81.** $-28a$ **83.** $108b$ **85.** $-56x^2$ **87.** x^2 **89.** x **91.** a **93.** b **95.** x **97.** n **99.** $2x$ **101.** $-2x$ **103.** $-15a^2$ **105.** $6y$ **107.** $3y$ **109.** $-2x$ **111.** $-9y$ **113.** $8x - 6$ **115.** $-2a - 14$ **117.** $-6y + 24$ **119.** $-x - 2$ **121.** $35 - 21b$ **123.** $2 - 5y$ **125.** $15x^2 + 6x$ **127.** $2y - 18$ **129.** $-15x - 30$ **131.** $-6x^2 - 28$ **133.** $-6y^2 + 21$ **135.** $3x^2 - 3y^2$ **137.** $-4x + 12y$ **139.** $-6a^2 + 7b^2$ **141.** $4x^2 - 12x + 20$ **143.** $\frac{3}{2}x - \frac{9}{2}y + 6$ **145.** $-12a^2 - 20a + 28$ **147.** $12x^2 - 9x + 12$ **149.** $10x^2 - 20xy - 5y^2$ **151.** $-8b^2 + 6b - 9$ **153.** (iii) **155.** $a - 7$ **157.** $-11x + 13$ **159.** $-4y - 4$ **161.** $-2x - 16$ **163.** $14y - 45$ **165.** $a + 7b$ **167.** $6x + 28$ **169.** $5x - 75$ **171.** $4x - 4$ **173.** $2x - 9$ **175.** $1.24x + 0.36$ **177.** $-0.01x + 40$ **179.** $\frac{x}{18}$ **181.** $x + 20$ **183.** $11x - 8$ **185.** $40 - \frac{x}{20}$ **187.** $x^2 + 2x$ **189.** $10(x - 50)$; $10x - 500$ **191.** $x - (x + 3)$; -3 **193.** $(2x - 4) + x$; $3x - 4$ **195.** $x + 3x$; $4x$ **197.** $(x + 6) + 5$; $x + 11$ **199.** $x - (x + 10)$; -10

201. Number of visitors to the Metropolitan Museum of Art: M; number of visitors to the Louvre: $M + 3{,}800{,}000$
203. Number of visitors to Google websites: G; number of visitors to Microsoft websites: $G - 63{,}000{,}000$
205. Length of one piece: S; length of second piece: $12 - S$ **207.** Distance traveled by the faster car: x; distance traveled by the slower car: $200 - x$ **209.** Number of bones in your body: N; number of bones in your foot: $\frac{1}{4}N$ **211.** Number of people surveyed: N; number of people who would pay down their debt: $0.43N$ **213.** $\frac{1}{4}x$

SECTION 1.5

3. $A = \{16, 17, 18, 19, 20, 21\}$ **5.** $A = \{9, 11, 13, 15, 17\}$ **7.** $A = \{b, c\}$ **9.** $A \cup B = \{3, 4, 5, 6\}$
11. $A \cup B = \{-10, -9, -8, 8, 9, 10\}$ **13.** $A \cup B = \{a, b, c, d, e, f\}$ **15.** $A \cup B = \{1, 3, 7, 9, 11, 13\}$
17. $A \cap B = \{4, 5\}$ **19.** $A \cap B = \varnothing$ **21.** $A \cap B = \{c, d, e\}$ **23.** Answers may vary. For example, $A = \{1, 2, 3\}$ and $B = \{1, 2, 4, 5\}$. **25.** $\{x \mid x > -5, x \in \text{negative integers}\}$ **27.** $\{x \mid x > 30, x \in \text{integers}\}$
29. $\{x \mid x > 8, x \in \text{real numbers}\}$ **31.** $(1, 2)$ **33.** $(3, \infty)$ **35.** $[-4, 5)$ **37.** $(-\infty, 2]$ **39.** $[-3, 1]$
41. $\{x \mid -5 < x < -3\}$ **43.** $\{x \mid x \le -2\}$ **45.** $\{x \mid -3 \le x \le -2\}$ **47.** $\{x \mid x \le 6\}$

49. −5 −4 −3 −2 −1 0 1 2 3 4 5 **51.** −5 −4 −3 −2 −1 0 1 2 3 4 5 **53.** −5 −4 −3 −2 −1 0 1 2 3 4 5

55. −5 −4 −3 −2 −1 0 1 2 3 4 5 **57.** −5 −4 −3 −2 −1 0 1 2 3 4 5 **59.** −5 −4 −3 −2 −1 0 1 2 3 4 5

61. −5 −4 −3 −2 −1 0 1 2 3 4 5 **63.** −5 −4 −3 −2 −1 0 1 2 3 4 5 **65.** None **67.** $m \ge 250$

CHAPTER 1 CONCEPT REVIEW*

1. The opposite of a number is the number with its sign changed. If a number is positive, its opposite is negative. If a number is negative, its opposite is positive. The absolute value of a number is its distance from zero on the number line. Because distance is never negative, the absolute value of a number is never negative. The absolute value of a positive number is positive. The absolute value of a negative number is positive. [1.1B]
2. A minus sign indicates the operation of subtraction. A negative sign indicates the sign of a number. [1.1C]
3. The product of two nonzero numbers with the same sign is positive. The product of two nonzero numbers with different signs is negative. [1.1D]
4. The steps in the Order of Operations Agreement are:
   1. Perform operations inside grouping symbols. Grouping symbols include parentheses ( ), brackets [ ], braces { }, the absolute value symbol | |, and the fraction bar.
   2. Simplify exponential expressions.
   3. Do multiplication and division as they occur from left to right.
   4. Do addition and subtraction as they occur from left to right. [1.3A]
5. When adding fractions, you have to convert to equivalent fractions with a common denominator. One way to explain this is that you can combine like things, but you cannot combine unlike things. You can combine 4 apples and 5 apples and get 9 apples. You cannot combine 4 apples and 5 oranges and get one type of fruit. In adding whole numbers, you add like things: ones, tens, hundreds, and so on. In adding fractions, you can combine 2 *ninths* and 5 *ninths* and get 7 *ninths,* but you cannot add 2 *ninths* and 3 *fifths.* [1.2C]
6. No, you do not have to find a common denominator when multiplying two fractions. The product of two fractions is the product of the numerators over the product of the denominators. [1.2D]
7. To find the reciprocal of a fraction, interchange the numerator and denominator. [1.4C]
8. In a term, the numerical coefficient is the number. The variable part consists of the variables and their exponents. [1.4A]
9. Like terms of a variable expression are terms with the same variable part. Constant terms are also considered like terms. [1.4A]
10. The empty set or null set is represented by $\varnothing$ or { }. [1.5A]
11. $A \cup B$ is the union of two sets A and B. $A \cap B$ is the intersection of two sets A and B. [1.5A]
12. The roster method encloses a list of the elements of a set in braces; for example, {1, 2, 3, 4, 5}. Set-builder notation uses a rule to describe the elements of the set; for example, $\{x \mid x > 4, x \in \text{real numbers}\}$. [1.5A, 1.5B]

CHAPTER 1 REVIEW EXERCISES

1. −4, 0 [1.1A] **2.** 4 [1.1B] **3.** −5 [1.1B] **4.** −13 [1.1C] **5.** 1 [1.1C] **6.** −42 [1.1D]
7. −20 [1.1D] **8.** 0.28 [1.2A] **9.** 0.062 [1.2B] **10.** 62.5% [1.2B] **11.** $\frac{7}{12}$ [1.2C] **12.** −1.068 [1.2C]
13. $-\frac{72}{85}$ [1.2D] **14.** −4.6224 [1.2D] **15.** $\frac{16}{81}$ [1.2E] **16.** 12 [1.2F] **17.** $-6\sqrt{30}$ [1.2F]
18. 31 [1.3A] **19.** 29 [1.4A] **20.** $8a - 4b$ [1.4B] **21.** $36y$ [1.4C] **22.** $10x - 35$ [1.4D]
23. $7x + 46$ [1.4D] **24.** $-90x + 25$ [1.4D] **25.** {1, 3, 5, 7} [1.5A] **26.** $A \cap B = \{1, 5, 9\}$ [1.5A]
27. [number line: −5 −4 −3 −2 −1 0 1 2 3 4 5] [1.5B] **28.** [number line: −5 −4 −3 −2 −1 0 1 2 3 4 5] [1.5B] **29.** $(-4, \infty)$ [1.5B]
30. The student's score is 98. [1.1E] **31.** 50.8% of the candy consumed was chocolate. [1.2G]
32. $2x - \frac{1}{2}x$; $\frac{3}{2}x$ [1.4E] **33.** Let A be the number of American League players' cards; $5A$ [1.4E]

CHAPTER 1 TEST

1. $-2 > -40$ [1.1A, Example 1] **2.** 4 [1.1B, Example 2] **3.** −4 [1.1B, How To 1] **4.** −14 [1.1C, How To 6]
5. −16 [1.1C, Example 4] **6.** 4 [1.1C, How To 7] **7.** 17 [1.1D, Example 11] **8.** $0.\overline{7}$ [1.2A, How To 1]
9. $\frac{9}{20}$, 0.45 [1.2B, Example 2] **10.** $\frac{1}{15}$ [1.2C, How To 3] **11.** −5.3578 [1.2D, How To 8] **12.** $-\frac{1}{2}$ [1.2D, How To 7]
13. 12 [1.2E, Example 13] **14.** $-6\sqrt{5}$ [1.2F, You Try It 14] **15.** 17 [1.3A, How To 3] **16.** 22 [1.4A, How To 1]

Note: The numbers in brackets following the answers in the Concept Review are a reference to the objective that corresponds to that problem. For example, the reference [1.2A] stands for Section 1.2, Objective A. This notation will be used for all Prep Tests, Concept Reviews, Chapter Reviews, Chapter Tests, and Cumulative Reviews throughout the text.

17. $5x$ [1.4B, How To 2] **18.** $2x$ [1.4C, How To 6] **19.** $-6x^2 + 21y^2$ [1.4D, Example 10]
20. $-x + 6$ [1.4D, Example 11] **21.** $-7x + 33$ [1.4D, Example 13] **22.** $\{-2, -1, 0, 1, 2, 3\}$ [1.5A, Example 1]
23. $\{x \mid x < -3, x \in \text{real numbers}\}$ [1.5B, How To 6] **24.** $A \cup B = \{1, 2, 3, 4, 5, 6, 7, 8\}$ [1.5A, Example 3]
25. −5 −4 −3 −2 −1 0 1 2 3 4 5 [1.5B, Example 11] **26.** −5 −4 −3 −2 −1 0 1 2 3 4 5 [1.5B, Example 13]
27. $10(x - 3)$; $10x - 30$ [1.4E, How To 16] **28.** Let s be the speed of the catcher's return throw; $2s$ [1.4E, You Try It 14]
29. a. The balance of trade increased from the previous year in 1981, 1988, 1989, 1990, 1991, and 1995.
b. The difference was \$288.6 billion. **c.** The difference was greatest between 1999 and 2000.
d. The trade balance was approximately 4 times greater in 1990 than in 1980. **e.** The average trade balance per quarter for the year 2000 was −\$92.425 billion. [1.2G, Example 16] **30.** The difference is 215.4°F. [1.2G, You Try It 16]

ANSWERS TO CHAPTER 2 SELECTED EXERCISES

PREP TEST

1. 0.09 [1.2A] **2.** 75% [1.2B] **3.** 63 [1.4A] **4.** $0.65R$ [1.4B] **5.** $\frac{7}{6}x$ [1.4B] **6.** $9x - 18$ [1.4D]
7. $1.66x + 1.32$ [1.4D] **8.** $5 - 2n$ [1.4E] **9.** Speed of the old card: s; speed of the new card: $5s$ [1.4E]
10. $5 - x$ [1.4E]

SECTION 2.1

1. Yes **3.** No **5.** No **7.** Yes **9.** No **11.** Yes **13.** No **15.** Yes **17.** No **19.** Negative
23. 2 **25.** 15 **27.** 6 **29.** 3 **31.** 0 **33.** −7 **35.** −7 **37.** −12 **39.** −5 **41.** 15 **43.** 9
45. 14 **47.** −1 **49.** 1 **51.** $-\frac{1}{2}$ **53.** $-\frac{3}{4}$ **55.** $\frac{1}{12}$ **57.** $-\frac{7}{12}$ **59.** 0.6529 **61.** −0.283
63. 9.257 **65.** −3 **67.** 0 **69.** −2 **71.** 9 **73.** 80 **75.** 0 **77.** −7 **79.** 12 **81.** −18
83. 15 **85.** −20 **87.** 0 **89.** $\frac{8}{3}$ **91.** $\frac{1}{3}$ **93.** $-\frac{1}{2}$ **95.** $-\frac{3}{2}$ **97.** $\frac{15}{7}$ **99.** 4 **101.** 3
103. 4.745 **105.** 2.06 **107.** −2.13 **109.** Positive **111.** Negative **113.** Equal to **115.** 28 **117.** 0.72
119. 64 **121.** 24% **123.** 7.2 **125.** 400 **127.** 9 **129.** 25% **131.** 200% **133.** 400 **135.** 7.7
137. 200 **139.** 400 **141.** 30 **143.** 98.1% of the runners who started the course finished the race. **145.** You need to know the number of people over 3 years old in the U.S. that are enrolled in school. **147. a.** The approximate teen population in 2007 was 240,000 teens. **b.** 42,000 teens would have been smokers in 2007. **149.** Andrea must invest \$1875.
151. Octavia will earn the greater amount of interest. **153.** \$1500 was invested at 8%. **155.** There are 1.8 g of platinum in the necklace. **157.** There are 131.25 lb of wool in the carpet. **159.** The percent concentration of sugar is 50%. **161.** The percent concentration of the resulting mixture is 6%. **163. a.** The distance walked by Joe is greater than the distance walked by John. **b.** Joe's time spent walking is equal to John's time spent walking. **c.** The total distance traveled is 2 mi. **165.** The runner will travel 3 mi. **167.** Marcella's average rate of speed is 36 mph. **169.** It would take Palmer 2.5 h to walk the course.
171. The two joggers will meet 40 min after they start. **173.** It will take them 0.5 h. **175.** Negative

SECTION 2.2

1. 3 **3.** 6 **5.** −1 **7.** −3 **9.** 2 **11.** 2 **13.** 5 **15.** −3 **17.** 6 **19.** 3 **21.** 1 **23.** 6
25. −7 **27.** 0 **29.** $\frac{3}{4}$ **31.** $\frac{4}{9}$ **33.** $\frac{1}{3}$ **35.** $-\frac{1}{2}$ **37.** $-\frac{3}{4}$ **39.** $\frac{1}{3}$ **41.** $-\frac{1}{6}$ **43.** 1 **45.** 1
47. 0 **49.** 0.15 **51.** $\frac{2}{5}$ **53.** $-\frac{4}{3}$ **55.** $-\frac{3}{2}$ **57.** 18 **59.** 8 **61.** −16 **63.** 25 **65.** $\frac{3}{4}$
67. $\frac{3}{8}$ **69.** $\frac{16}{9}$ **71.** $\frac{1}{18}$ **73.** $\frac{15}{2}$ **75.** $-\frac{18}{5}$ **77.** 2 **79.** 3 **81.** Negative **83.** Negative
85. Subtract $2x$ from each side. **87.** 3 **89.** −2 **91.** −3 **93.** 2 **95.** −2 **97.** −0.2 **99.** 0
101. −2 **103.** −2 **105.** −2 **107.** 4 **109.** $\frac{3}{4}$ **111.** $\frac{3}{2}$ **113.** −14 **115.** 7 **117.** (ii) **119.** 1

121. 4 **123.** −1 **125.** −1 **127.** 24 **129.** 495 **131.** $\frac{1}{2}$ **133.** $-\frac{1}{3}$ **135.** $\frac{10}{3}$ **137.** $-\frac{1}{4}$ **139.** 0
141. The score given by the third judge was 8.5. **143.** The score given by the third judge was 8. **145. a.** The fulcrum is 5 ft from the other person. **b.** The person who is 3 ft from the fulcrum is heavier. **c.** No, the seesaw will not balance.
147. The fulcrum is 10 ft from the child. **149.** The fulcrum must be placed 4.8 ft from the 90-pound child.
151. The force on the lip of the can is 1770 lb. **153.** The break-even point is 260 barbecues. **155.** The break-even point is 520 recorders. **157.** The oxygen consumption is 54.8 ml/min.

SECTION 2.3

1. $x - 15 = 7$; 22 **3.** $9 - x = 7$; 2 **5.** $5 - 2x = 1$; 2 **7.** $2x + 5 = 15$; 5 **9.** $4x - 6 = 22$; 7
11. $3(4x - 7) = 15$; 3 **13.** $3x = 2(20 - x)$; 8, 12 **15.** $2x - (14 - x) = 1$; 5, 9 **17.** 15, 17, 19 **19.** −1, 1, 3
21. 4, 6 **23.** (iii) **25.** The yearly costs for a robot are \$600. **27.** The lengths of the sides are 20 m, 20 m, and 6 m. **29.** The customer used the service for 11 min. **31.** The executive used the phone for 162 min.
33. The customer pays \$.15 per text message over 300 messages. **35.** The length is 13 m; the width is 8 m.
37. The shorter piece is 3 ft; the longer piece is 9 ft. **39.** The larger scholarship is \$5000.

SECTION 2.4

1. (iii) and (v) **3.** The amount of \$.80 meal used is 300 lb. **5.** The cost per ounce is \$3. **7.** 25 lb of sunflower seeds are needed. **9.** The amount of diet supplement is 2 lb; the amount of vitamin supplement is 3 lb. **11.** The cost per pound of the trail mix is \$2.90. **13.** The goldsmith used 56 oz of the \$4.30 alloy and 144 oz of the \$1.80 alloy.
15. The cost per pound of the coffee mixture is \$6.98. **17.** 1016 adult tickets were sold. **19.** False
21. The percent concentration of tomato juice is $33\frac{1}{3}\%$. **23.** The amount of chicken feed that is 50% corn is 80 lb.
25. The amount of the darker green paint is $1\frac{2}{3}$ gal. **27.** The chemist should use 20 ml of the 13% solution and 30 ml of the 18% solution. **29.** The percent concentration of the resulting alloy is 27%. **31.** The amount of 60% rye grass used is 30 lb.
33. The tea blend is made from 3 lb of 20% jasmine and 2 lb of 15% jasmine. **35.** The amount of pure bran flakes needed is 10 oz. **37.** (i) **39.** The first plane is traveling at a rate of 105 mph; the second plane is traveling at a rate of 130 mph.
41. The planes will be 3000 km apart at 11 A.M. **43.** After 2 h, the cabin cruiser will be alongside the motorboat.
45. The corporate offices are 120 mi from the airport. **47.** The rate of the car is 68 mph. **49.** The distance between the airports is 300 mi. **51.** The planes will pass each other 2.5 h after the plane leaves Seattle. **53.** The cyclists will meet after 1.5 h. **55.** The car overtakes the cyclist 48 mi from the starting point. **57.** 10 oz of water must be evaporated.
59. It is impossible to average 60 mph.

SECTION 2.5

3. (i), (iii) **5.** $\{x \mid x < 5\}$ **7.** $\{x \mid x \le 2\}$
9. $\{x \mid x < -4\}$ **11.** $\{x \mid x > 3\}$ **13.** $\{x \mid x > 4\}$ **15.** $\{x \mid x \le 2\}$ **17.** $\{x \mid x > -2\}$
19. $\{x \mid x \ge 2\}$ **21.** $\{x \mid x > -2\}$ **23.** $\{x \mid x \le 3\}$ **25.** $\{x \mid x < 2\}$ **27.** $\{x \mid x < -3\}$ **29.** $\{x \mid x \le 5\}$
31. $\left\{x \mid x \ge -\frac{1}{2}\right\}$ **33.** Both positive and negative numbers **35.** Both positive and negative numbers
37. $\left(-\infty, \frac{23}{16}\right)$ **39.** $\left(-\infty, \frac{8}{3}\right)$ **41.** $(1, \infty)$ **43.** $\left(\frac{14}{11}, \infty\right)$ **45.** $(-\infty, 1]$ **47.** $\left(-\infty, \frac{7}{4}\right]$
49. $\left[-\frac{5}{4}, \infty\right)$ **51.** $(-\infty, 2]$ **55.** $[-2, 4]$ **57.** $(-\infty, 3) \cup (5, \infty)$ **59.** $(-4, 2)$ **61.** $(-\infty, -4) \cup (6, \infty)$
63. $(-\infty, -3)$ **65.** $\varnothing$ **67.** $(-2, 1)$ **69.** All real numbers **71.** Two intervals of real numbers
73. $\{x \mid x < -2 \text{ or } x > 2\}$ **75.** $\{x \mid 2 < x < 6\}$ **77.** $\{x \mid -3 < x < -2\}$ **79.** $\left\{x \mid x > 5 \text{ or } x < -\frac{5}{3}\right\}$ **81.** $\varnothing$

83. The set of real numbers **85.** $\left\{x\middle|\frac{17}{7} \le x \le \frac{45}{7}\right\}$ **87.** $\left\{x\middle|-5 < x < \frac{17}{3}\right\}$ **89.** The set of real numbers

91. $\varnothing$ **93.** $t \le 42$ **95.** $t \le 42$ **97.** The smallest number is -12. **99.** The maximum width of the rectangle is 11 cm. **101.** The advertisement can run for a maximum of 104 days. **103.** The homeowner can pay a maximum of \$19 per gallon of paint. **105.** $32° < F < 86°$ **107.** George's amount of sales must be \$44,000 or more. **109.** The first account is less expensive if more than 200 checks are written. **111.** $58 \le n \le 100$ **113.** The three even integers are 10, 12, and 14; or 12, 14, and 16; or 14, 16, and 18. **115. a.** Always true **b.** Sometimes true **c.** Sometimes true **d.** Sometimes true **e.** Always true

SECTION 2.6

1. Yes **3.** Yes **5.** 7 and -7 **7.** 4 and -4 **9.** 6 and -6 **11.** 7 and -7 **13.** No solution **15.** No solution **17.** 1 and -5 **19.** 8 and 2 **21.** 2 **23.** No solution **25.** $-\frac{3}{2}$ and 3 **27.** $\frac{3}{2}$ **29.** No solution **31.** 7 and -3 **33.** 2 and $-\frac{10}{3}$ **35.** 1 and 3 **37.** $\frac{3}{2}$ **39.** No solution **41.** $\frac{11}{6}$ and $-\frac{1}{6}$ **43.** $-\frac{1}{3}$ and -1 **45.** No solution **47.** 3 and 0 **49.** No solution **51.** 1 and $\frac{13}{3}$ **53.** No solution **55.** $\frac{7}{3}$ and $\frac{1}{3}$ **57.** $-\frac{1}{2}$ **59.** $-\frac{1}{2}$ and $-\frac{7}{2}$ **61.** $-\frac{8}{3}$ and $\frac{10}{3}$ **63.** No solution **65.** Two positive solutions **67.** Two negative solutions **69.** $\{x|x > 3 \text{ or } x < -3\}$ **71.** $\{x|x > 1 \text{ or } x < -3\}$ **73.** $\{x|4 \le x \le 6\}$ **75.** $\{x|x \ge 5 \text{ or } x \le -1\}$ **77.** $\{x|-3 < x < 2\}$ **79.** $\left\{x\middle|x > 2 \text{ or } x < -\frac{14}{5}\right\}$ **81.** $\varnothing$ **83.** The set of real numbers **85.** $\left\{x\middle|x \le -\frac{1}{3} \text{ or } x \ge 3\right\}$ **87.** $\left\{x\middle|-2 \le x \le \frac{9}{2}\right\}$ **89.** $\{x|x = 2\}$ **91.** $\left\{x\middle|x < -2 \text{ or } x > \frac{22}{9}\right\}$ **93.** $\left\{x\middle|-\frac{3}{2} < x < \frac{9}{2}\right\}$ **95.** $\left\{x\middle|x < 0 \text{ or } x > \frac{4}{5}\right\}$ **97.** $\{x|x > 5 \text{ or } x < 0\}$ **99.** All negative solutions **101.** The desired dosage is 3 ml. The tolerance is 0.2 ml. **103.** The lower and upper limits of the diameter of the bushing are 1.742 in. and 1.758 in. **105.** The lower and upper limits of the percent of voters who felt the economy was the most important election issue are 38% and 44%. **107.** The lower and upper limits of the length of the piston rod are $9\frac{19}{32}$ in. and $9\frac{21}{32}$ in. **109.** The lower and upper limits of the resistor are 28,420 ohms and 29,580 ohms. **111. a.** $\{x|x \ge -3\}$ **b.** $\{a|a \le 4\}$ **113. a.** $\le$ **b.** $\ge$ **c.** $\ge$ **d.** $=$

CHAPTER 2 CONCEPT REVIEW*

1. The Addition Property of Equations is used to remove a term from one side of an equation by adding the opposite of that term to each side of the equation. [2.1B]
2. To check the solution of an equation, substitute the proposed solution for the variable in the original equation. Simplify the resulting numerical expressions. If the left and right sides are equal, the proposed answer is a solution of the equation. If the left and right sides are not equal, the proposed solution is not a solution of the equation. [2.1B]
3. To solve an equation containing parentheses, first use the Distributive Property to remove the parentheses. Combine like terms on each side of the equation. Use the Addition Property of Equations to rewrite the equation with only one variable term. Then use the Addition Property of Equations to rewrite the equation with only one constant term. Use the Multiplication Property of Equations to rewrite the equation in the form *variable* = *constant*. The constant is the solution. [2.2C]
4. Consecutive integers differ by 1; for example, 8 and 9 are consecutive integers. Consecutive even integers are even numbers that differ by 2; for example, 8 and 10 are consecutive even integers. [2.3A]
5. When mixing a 15% solution with a 20% solution, the percent concentration of the resulting solution must be between 15% and 20%. [2.4B]
6. In solving a percent mixture problem, use the formula $Ar = Q$, where A is the amount of solution or alloy, r is the percent of concentration, and Q is the quantity of a substance in the solution or alloy. [2.4B]

Note: The numbers in brackets following the answers in the Concept Review are a reference to the objective that corresponds to that problem. For example, the reference [1.2A] stands for Section 1.2, Objective A. This notation will be used for all Prep Tests, Concept Reviews, Chapter Reviews, Chapter Tests, and Cumulative Reviews throughout the text.

7. In solving a uniform motion problem, use the formula $rt = d$, where r is the rate of travel, t is the time spent traveling, and d is the distance traveled. [2.4C]

8. The Multiplication Property of Equations states that both sides of an equation can be multiplied by the same nonzero number without changing the solution of the equation. The Multiplication Property of Inequalities consists of two rules: (1) Each side of an inequality can be multiplied by the same positive number without changing the solution set; (2) If each side of an inequality is multiplied by the same negative number, the inequality symbol must be reversed in order to keep the solution set of the inequality unchanged. [2.5A]

9. The compound inequality $-c < ax + b < c$ is equivalent to an absolute value inequality of the form $|ax + b| < c$, $c > 0$. [2.6B]

10. The compound inequality $ax + b < -c$ or $ax + b > c$ is equivalent to an absolute value inequality of the form $|ax + b| > c$. [2.6B]

11. To check the solution to an absolute value equation, substitute the proposed solution for the variable in the original equation. Simplify the resulting numerical expressions. If the left and right sides are equal, the proposed solution is a solution of the equation. If the left and right sides are not equal, the proposed solution is not a solution of the equation. [2.6A]

CHAPTER 2 REVIEW EXERCISES

1. 6 [2.2B] **2.** $\left\{x \middle| x > \frac{5}{3}\right\}$ [2.5A] **3.** No [2.1A] **4.** -9 [2.1B] **5.** $\left\{x \middle| -3 < x < \frac{4}{3}\right\}$ [2.5B] **6.** $-\frac{40}{7}$ [2.2B] **7.** $-\frac{2}{3}$ [2.1C] **8.** -1 and 9 [2.6A] **9.** $\{x \mid 1 < x < 4\}$ [2.6B] **10.** $-\frac{9}{19}$ [2.2C] **11.** $\frac{26}{17}$ [2.2C] **12.** $\{x \mid x > 2 \text{ or } x < -2\}$ [2.5B] **13.** $\left\{x \middle| x \geq 2 \text{ or } x \leq \frac{1}{2}\right\}$ [2.6B] **14.** 250% [2.1D] **15.** $-\frac{17}{2}$ [2.2B] **16.** No solution [2.6A] **17.** The set of real numbers [2.5B] **18.** $\frac{5}{2}$ [2.2C] **19.** -15 [2.1B] **20.** $-\frac{1}{12}$ [2.1B] **21.** 7 [2.1C] **22.** $\frac{2}{3}$ [2.1C] **23.** $\frac{8}{5}$ [2.2B] **24.** 3 [2.2B] **25.** 9 [2.2C] **26.** 8 [2.2C] **27.** $(-\infty, 1)$ [2.5A] **28.** $[-4, \infty)$ [2.5A] **29.** $\left\{x \middle| x \leq -\frac{39}{2}\right\}$ [2.5A] **30.** $[1, \infty)$ [2.5A] **31.** $\{x \mid -1 < x < 2\}$ [2.5B] **32.** $-\frac{5}{2}, \frac{11}{2}$ [2.6A] **33.** $-\frac{8}{5}$ [2.6A] **34.** $\varnothing$ [2.6B] **35.** The island is 16 mi from the dock. [2.4C] **36.** The mixture must contain 52 gal of apple juice. [2.4A] **37.** The executive's amount of sales must be $55,000 or more. [2.5C] **38.** $5x - 4 = 16$; $x = 4$ [2.3A] **39.** The lower and upper limits of the diameter of the bushing are 2.747 in. and 2.753 in. [2.6C] **40.** The integers are 6 and 14. [2.3A] **41.** $82 \leq x \leq 100$ [2.5C] **42.** The speed of one plane is 520 mph. The speed of the other plane is 440 mph. [2.4C] **43.** 375 lb of the 30% tin alloy and 125 lb of the 70% tin alloy were used. [2.4B] **44.** The lower and upper limits of the length of the piston are $10\frac{11}{32}$ in. and $10\frac{13}{32}$ in. [2.6C]

CHAPTER 2 TEST

1. -2 [2.1B, How To 2] **2.** $-\frac{1}{8}$ [2.1B, How To 3] **3.** $\frac{5}{6}$ [2.1C, How To 4] **4.** 4 [2.2A, Example 1] **5.** $\frac{32}{3}$ [2.2A, How To 1] **6.** $-\frac{1}{5}$ [2.2B, Example 7] **7.** 1 [2.2C, Example 9] **8.** No [2.1A, How To 1] **9.** $\frac{12}{7}$ [2.2C, Example 6] **10.** $\{x \mid x \leq -3\}$ [2.5A, How To 5] **11.** 0.04 [2.1D, How To 7] **12.** $\{x \mid x > -2\}$ [2.5B, How To 10] **13.** $\varnothing$ [2.5B, Example 4] **14.** $-\frac{9}{5}$ and 3 [2.6A, Example 1] **15.** 7 and -2 [2.6A, Example 3] **16.** $\left\{x \middle| \frac{1}{3} \leq x \leq 3\right\}$ [2.6B, Example 4] **17.** $\left\{x \middle| x < -\frac{1}{2} \text{ or } x > 2\right\}$ [2.6B, How To 4] **18.** It costs less to rent from Gambelli Agency if the car is driven less than 120 mi. [2.5C, Example 5] **19.** The lower and upper limits of the diameter of the bushing are 2.648 in. and 2.652 in. [2.6C, You Try It 8] **20.** The integers are 4 and 11. [2.3A, Example 1] **21.** 1.25 gal are needed. [2.4B, How To 2] **22.** The price of the hamburger mixture is $2.70 per pound. [2.4A, Example 1] **23.** The jogger ran a total distance of 12 mi. [2.4C, Example 4] **24.** The rate of the slower train is 60 mph. The rate of the faster train is 65 mph. [2.4C, You Try It 3] **25.** It is necessary to add 100 oz of pure water. [2.4B, How To 2]

CUMULATIVE REVIEW EXERCISES

1. 6 [1.1C] **2.** −48 [1.1D] **3.** $-\frac{19}{48}$ [1.2C] **4.** 54 [1.2E] **5.** $\frac{49}{40}$ [1.3A] **6.** 6 [1.4A] **7.** $-17x$ [1.4B] **8.** $-5a - 4b$ [1.4B] **9.** $2x$ [1.4C] **10.** $36y$ [1.4C] **11.** $2x^2 + 6x - 4$ [1.4D] **12.** $-4x + 14$ [1.4D] **13.** $6x - 34$ [1.4D] **14.** $A \cap B = \{-4, 0\}$ [1.5A] **15.** −5 −4 −3 −2 −1 0 1 2 3 4 5 [1.5B] **16.** Yes [2.1A] **17.** −25 [2.1C] **18.** −3 [2.2A] **19.** 3 [2.2A] **20.** −3 [2.2B] **21.** $\frac{1}{2}$ [2.2B] **22.** 13 [2.2C] **23.** $\{x|x \leq 1\}$ [2.5A] **24.** $\{x|-4 \leq x \leq 1\}$ [2.5B] **25.** −1 and 4 [2.6A] **26.** $\left\{x|x > 2 \text{ or } x < -\frac{4}{3}\right\}$ [2.6B] **27.** $\frac{11}{20}$ [1.2B] **28.** 103% [1.2B] **29.** 25% of 120 is 30. [2.1D] **30.** $6x + 13 = 3x - 5; x = -6$ [2.3A] **31.** 20 lb of oat flour must be used. [2.4A] **32.** 25 g of pure gold must be added. [2.4B] **33.** The length of the track is 120 m. [2.4C]

ANSWERS TO CHAPTER 3 SELECTED EXERCISES

PREP TEST

1. 43 [2.1B] **2.** 51 [2.1B] **3.** 56 [1.3A] **4.** 56.52 [1.4A] **5.** 113.04 [1.4A] **6.** 120 [1.4A]

SECTION 3.1

1. 40°; acute **3.** 115°; obtuse **5.** 90°; right **7.** The complement is 28°. **9.** The supplement is 18°. **11.** BC is 14 cm. **13.** QS is 28 ft. **15.** EG is 30 m. **17.** The measure of $\angle MON$ is 86°. **19.** 71° **21.** 30° **23.** 36° **25.** 127° **27.** 116° **29.** 20° **31.** 20° **33.** 20° **35.** 141° **37.** An acute angle **39.** An obtuse angle **41.** 106° **43.** 11° **45.** $\angle a = 38°, \angle b = 142°$ **47.** $\angle a = 47°, \angle b = 133°$ **49.** 20° **51.** 47° **53.** True **55.** $\angle x = 155°, \angle y = 70°$ **57.** $\angle a = 45°, \angle b = 135°$ **59.** $90° - x$ **61.** The measure of the third angle is 60°. **63.** The measure of the third angle is 35°. **65.** The measure of the third angle is 102°. **67.** False

SECTION 3.2

1. Hexagon **3.** Pentagon **5.** Scalene **7.** Equilateral **9.** Obtuse **11.** Acute **13.** 56 in. **15.** 14 ft **17.** 47 mi **19.** 8π cm or approximately 25.13 cm **21.** 11π mi or approximately 34.56 mi **23.** 17π ft or approximately 53.41 ft **25.** The perimeter is 17.4 cm. **27.** The perimeter is 8 cm. **29.** The perimeter is 24 m. **31.** The perimeter is 48.8 cm. **33.** The perimeter is 17.5 in. **35.** The length of a diameter is 8.4 cm. **37.** The circumference is 1.5π in. **39.** The circumference is 226.19 cm. **41.** 60 ft of fencing should be purchased. **43.** The carpet must be nailed down along 44 ft. **45.** The length is 120 ft. **47.** The length of the third side is 10 in. **49.** The length of each side is 12 in. **51.** The length of a diameter is 2.55 cm. **53.** The length is 13.19 ft. **55.** The bicycle travels 50.27 ft. **57.** The circumference is 39,935.93 km. **59.** The perimeter of the square is greater. **61.** 60 ft^2 **63.** 20.25 in^2 **65.** 546 ft^2 **67.** 16π cm^2 or approximately 50.27 cm^2 **69.** 30.25π mi^2 or approximately 95.03 mi^2 **71.** 72.25π ft^2 or approximately 226.98 ft^2 **73.** The area is 156.25 cm^2. **75.** The area is 570 in^2. **77.** The area is 192 in^2. **79.** The area is 13.5 ft^2. **81.** The area is 330 cm^2. **83.** The area is 25π in^2. **85.** The area is 9.08 ft^2. **87.** The area is $10{,}000\pi$ in^2. **89.** The area is 126 ft^2. **91.** 7500 yd^2 must be purchased. **93.** The width is 10 in. **95.** The length of the base is 20 m. **97.** You should buy 2 qt. **99.** It will cost $98. **101.** The increase in area is 113.10 in^2. **103.** The area is 216 m^2. **105.** The cost will be $878. **107.** The area of the first figure is equal to the area of the second figure. **109.** The area of the resulting rectangle is 4 times larger.

SECTION 3.3

1. 840 in^3 **3.** 15 ft^3 **5.** 4.5π cm^3 or approximately 14.14 cm^3 **7.** The volume is 34 m^3. **9.** The volume is 42.875 in^3. **11.** The volume is 36π ft^3. **13.** The volume is 8143.01 cm^3. **15.** The volume is 75π in^3. **17.** The volume is 120 in^3. **19.** The width is 2.5 ft. **21.** The height is 6.40 in. **23.** The volume is 172,800 ft^3. **25.** The sphere has the greater volume. **27.** 94 m^2 **29.** 56 m^2 **31.** 96π in^2 or approximately 301.59 in^2 **33.** The surface area is 184 ft^2. **35.** The surface area is 69.36 m^2. **37.** The surface area is 225π cm^2. **39.** The surface area is 402.12 in^2. **41.** The surface area is 6π ft^2. **43.** The surface area is 297 in^2. **45.** The width is 3 cm. **47.** 11 cans of paint should be purchased. **49.** 456 in^2 of glass are needed. **51.** The surface area of the pyramid is 22.53 cm^2 larger. **53. a.** Always true **b.** Never true **c.** Sometimes true

CHAPTER 3 CONCEPT REVIEW*

1. Perpendicular lines are intersecting lines that form right angles. [3.1A]
2. Two angles are complementary when the sum of their measures is 90°. [3.1A]
3. The obtuse angles $\angle a$, $\angle c$, $\angle w$, and $\angle y$ have the same measure. [3.1B]
4. The sum of the measures of the three angles in a triangle is 180°. [3.1C]
5. To find the radius of a circle when you know the diameter, multiply the diameter by $\frac{1}{2}$. [3.2A]
6. The formula for the perimeter of a rectangle is $P = 2L + 2W$, where P is the perimeter, L is the length, and W is the width. [3.2A]
7. To find the circumference of a circle, multiply π times the diameter or multiply 2π times the radius. [3.2A]
8. The formula for the area of a triangle is $A = \frac{1}{2}bh$, where A is the area, b is the base, and h is the height of the triangle. [3.2B]
9. To find the volume of a rectangular solid, you need to know the length, the width, and the height. [3.3A]
10. The perimeter of a plane geometric figure is a measure of the distance around the figure. Perimeter is measured in linear units. Area is a measure of the amount of space in a region. Area is measured in square units. [3.2A, 3.2B]
11. The surface area of a rectangular solid is the sum of the areas of six rectangles. [3.3B]

CHAPTER 3 REVIEW EXERCISES

1. $\angle x = 22°$, $\angle y = 158°$ [3.1C] **2.** $\angle x = 68°$ [3.1B] **3.** AC is 44 cm. [3.1A] **4.** 19° [3.1A] **5.** The volume is 96 cm^3. [3.3A] **6.** $\angle a = 138°$, $\angle b = 42°$ [3.1B] **7.** The surface area is 220 ft^2. [3.3B] **8.** The supplement is 148°. [3.1A] **9.** The area is 78 cm^2. [3.2B] **10.** The area is 63 m^2. [3.2B] **11.** The volume is 39 ft^3. [3.3A] **12.** The measure of the third angle is 95°. [3.1C] **13.** The length of the base is 8 cm. [3.2B] **14.** The volume is 288π mm^3. [3.3A] **15.** The volume is $\frac{784\pi}{3}$ cm^3. [3.3A] **16.** Each side measures 21.5 cm. [3.2A] **17.** 4 cans of paint should be purchased. [3.3B] **18.** 208 yd of fencing are needed. [3.2A] **19. a.** The area of the floor is 45 ft^2. [3.2B] **b.** The volume of a cell is 315 ft^3. [3.3A] **20.** The area is 276 m^2. [3.2B]

CHAPTER 3 TEST

1. The radius is 0.75 m. [3.3A, How To 1] **2.** The circumference is 31.42 cm. [3.2A, Example 3] **3.** The perimeter is 26 ft. [3.2A, How To 2] **4.** $BC = 3$ [3.1A, Example 1] **5.** The volume is 268.08 ft^3. [3.3A, How To 1] **6.** The area is 63.62 cm^2. [3.2B, Example 5] **7.** $\angle a = 100°$, $\angle b = 80°$ [3.1B, Example 6] **8.** 75° [3.1A, You Try It 3] **9.** $\angle a = 135°$, $\angle b = 45°$ [3.1B, Example 6] **10.** The area is 55 m^2. [3.2B, How To 5] **11.** The volume is 169.65 m^3. [3.3A, You Try It 2] **12.** The perimeter is 6.8 m. [3.2A, How To 2] **13.** 58° [3.1A, Example 3] **14.** The surface area is 164.93 ft^2. [3.3B, You Try It 3] **15.** There are 113.10 in^2 more in the larger pizza. [3.2B, You Try It 5] **16.** The measures of the other two angles are 58° and 90°. [3.1C, You Try It 8] **17.** The bicycle travels 73.3 ft. [3.2A, How To 4] **18.** The area of the room is 28 yd^2. [3.2B, How To 5] **19.** The volume of the silo is 1145.11 ft^3. [3.3A, You Try It 2] **20.** The area is 11 m^2. [3.2B, How To 8]

**Note:* The numbers in brackets following the answers in the Concept Review are a reference to the objective that corresponds to that problem. For example, the reference [1.2A] stands for Section 1.2, Objective A. This notation will be used for all Prep Tests, Concept Reviews, Chapter Reviews, Chapter Tests, and Cumulative Reviews throughout the text.

CUMULATIVE REVIEW EXERCISES

1. $-3, 0,$ and 1 [1.1A] **2.** 0.089 [1.2B] **3.** 35% [1.2B] **4.** $-\frac{2}{3}$ [1.2D] **5.** -24.51 [1.2D] **6.** $-5\sqrt{5}$ [1.2F] **7.** -28 [1.3A] **8.** -8 [1.4A] **9.** $-3m + 3n$ [1.4B] **10.** $21y$ [1.4C] **11.** $7x + 9$ [1.4D] **12.** $\{-2, -1\}$ [1.5A] **13.** $C \cup D = \{-10, 0, 10, 20, 30\}$ [1.5A] **14.** [number line: −5 −4 −3 −2 −1 0 1 2 3 4 5] [1.5B] **15.** 5 [2.2B] **16.** $\frac{1}{2}$ [2.2C] **17.** $(-\infty, -4)$ [2.5A] **18.** $\{x | x \geq 2\}$ [2.5A] **19.** $\{x | x < -3 \text{ or } x > 4\}$ [2.5B] **20.** $\{x | 2 \leq x \leq 6\}$ [2.5B] **21.** $1, -\frac{1}{3}$ [2.6A] **22.** $\{x | 6 \leq x \leq 10\}$ [2.6B] **23.** $\angle x = 131°$ [3.1B] **24.** $4x - 10 = 2; x = 3$ [2.3B] **25.** The third angle measures 122°. [3.1C] **26.** Michael will earn $312.50 from the two accounts in one year. [2.1D] **27.** The third side measures 4.5 m. [3.2A] **28.** The women's median annual earnings are 70.8% of the men's median annual earnings. [1.2G] **29.** The area is 20.25π cm². [3.2B] **30.** The height of the box is 3 ft. [3.3A]

ANSWERS TO CHAPTER 4 SELECTED EXERCISES

PREP TEST

1. $-4x + 12$ [1.4D] **2.** 10 [1.2F] **3.** -2 [1.3A] **4.** 11 [1.4A] **5.** 2.5 [1.4A] **6.** 5 [1.4A] **7.** 1 [1.4A] **8.** 4 [2.2A]

SECTION 4.1

1.

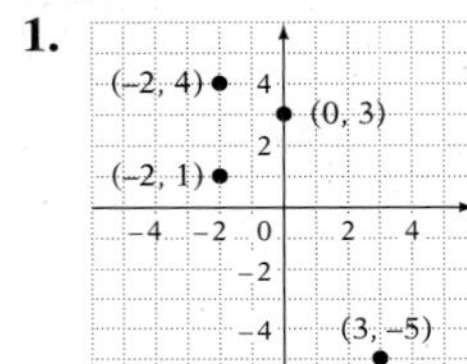

3. (−3, 0), (0, 2), (0, 0), (0, −5) **5.** (−1, 4), (0, 2), (4, 0), (−2, −3) **7.** $A(2, 3), B(4, 0), C(-4, 1), D(-2, -2)$

9. $A(-2, 5), B(3, 4), C(0, 0), D(-3, -2)$ **11. a.** $2, -4$ **b.** $1, -3$ **13. a.** y-axis **b.** x-axis **15.** Yes **17.** No **19.** No **21.** Negative **23.** (3, 7) **25.** (6, 3) **27.** (0, 1) **29.** (−5, 0)

31. (2, 4), (0, 0), (−1, −2), (−2, −4) **33.** (3, 3), (0, 1), (−3, −1) **35.**

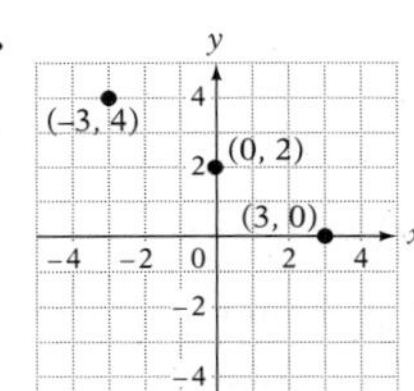

37. a. 25 g of cerium selenate will dissolve at 50°C. **b.** 5 g of cerium selenate will dissolve at 80°C.

39. Rainfall in previous hour (in inches) vs. Hour (11 A.M., 2 P.M., 5 P.M.)

41.

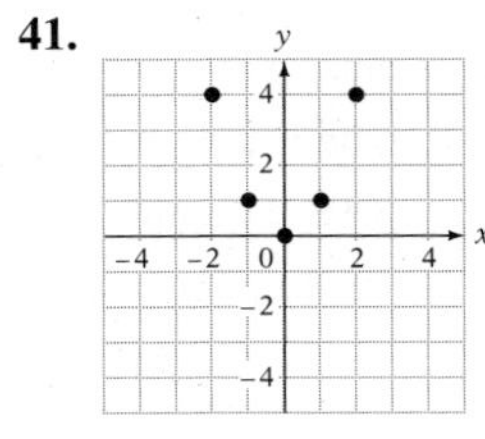

45.

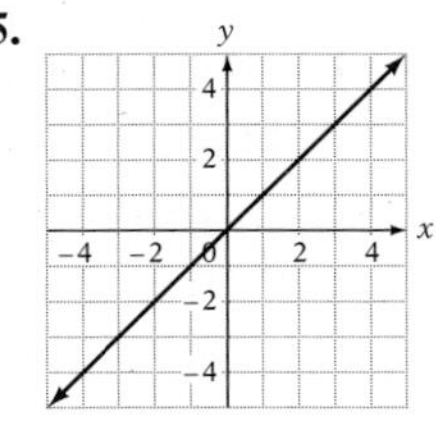

SECTION 4.2

3. Yes **5.** Yes **7.** No **9.** Function **11.** Function **13.** Function **15.** Not a function **17. a.** Yes **b.** $y = \$125$ **19.** True **21.** $f(3) = 11$ **23.** $f(0) = -4$ **25.** $G(0) = 4$ **27.** $G(-2) = 10$ **29.** $q(3) = 5$ **31.** $q(-2) = 0$ **33.** $F(4) = 24$ **35.** $F(-3) = -4$ **37.** $H(1) = 1$ **39.** $H(t) = \frac{3t}{t + 2}$ **41.** $s(-1) = 6$ **43.** $s(a) = a^3 - 3a + 4$ **45.** $4h$ **47.** The annual cost savings is $114.29. **49. a.** $4.75 per game **b.** $4.00 per game

51. Domain = {1, 2, 3, 4, 5}
Range = {1, 4, 7, 10, 13}
53. Domain = {0, 2, 4, 6}
Range = {1, 2, 3, 4}
55. Domain = {1, 3, 5, 7, 9}
Range = {0}
57. Domain = {−2, −1, 0, 1, 2}
Range = {0, 1, 2}
59. Domain = {−2, −1, 0, 1, 2}
Range = {−3, 3, 6, 7, 9}
61. 1 **63.** −8 **65.** None **67.** None **69.** 0 **71.** None **73.** None
75. −2 **77.** None **79.** Range = {−3, 1, 5, 9} **81.** Range = {−23, −13, −8, −3, 7} **83.** Range = {0, 1, 4}
85. Range = {2, 14, 26, 42} **87.** Range = $\left\{-5, \frac{5}{3}, 5\right\}$ **89.** Range = $\left\{-1, -\frac{1}{2}, -\frac{1}{3}, 1\right\}$ **91.** Range = {−38, −8, 2}
95. a. {(−2, −8), (−1, −1), (0, 0), (1, 1), (2, 8)} **b.** Yes, the set of ordered pairs defines a function because each member of the domain is assigned to exactly one member of the range. **97.** The power produced will be 50.625 watts.
99. a. The speed of the paratrooper 11.5 s after beginning the jump is 36.3 ft/s. **b.** The paratrooper is falling at a speed of approximately 30 ft/s. **101. a.** 275,000 malware attacks **b.** 700,000 malware attacks

SECTION 4.3

3.
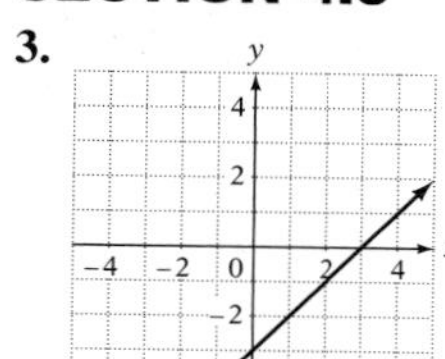

5.
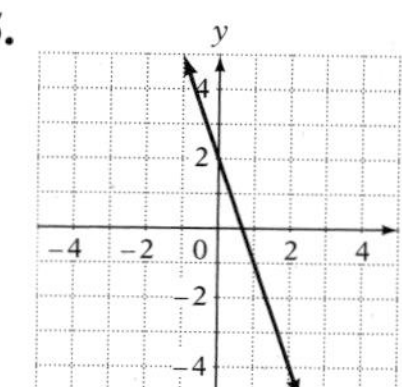

7.
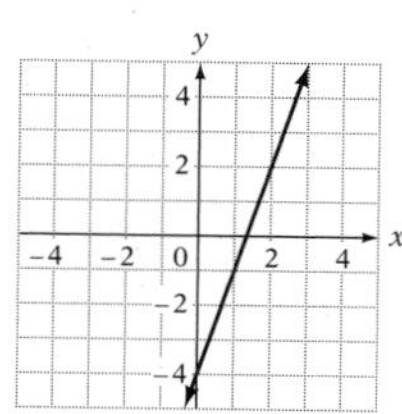

9.
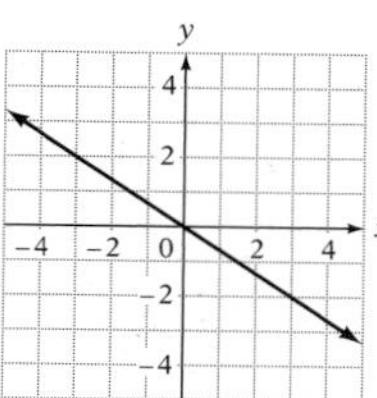

11.
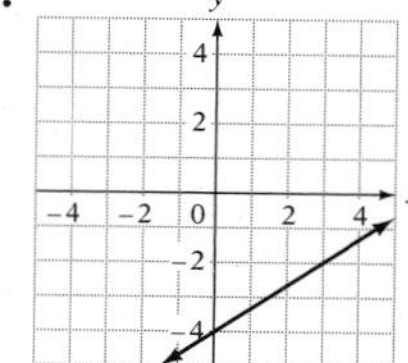

13.
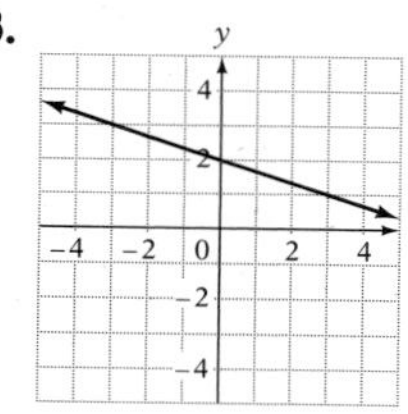

15.
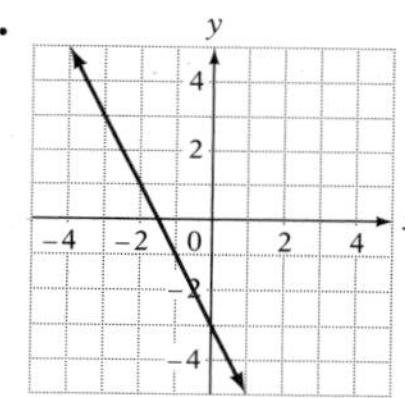

17.
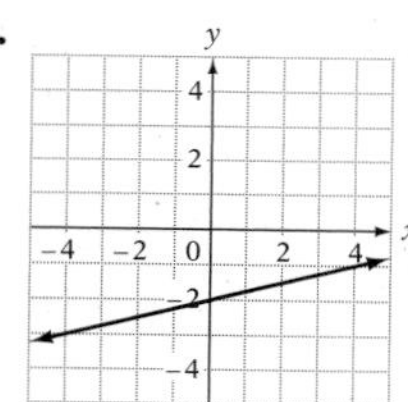

19.
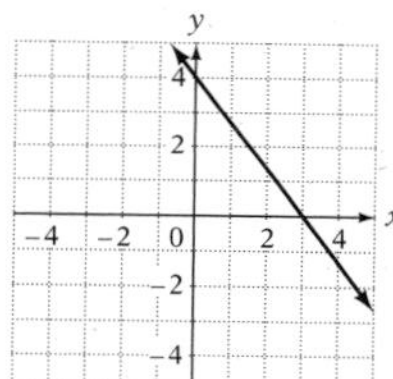

21.
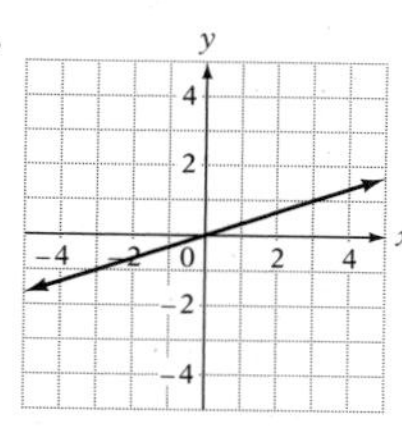

23.
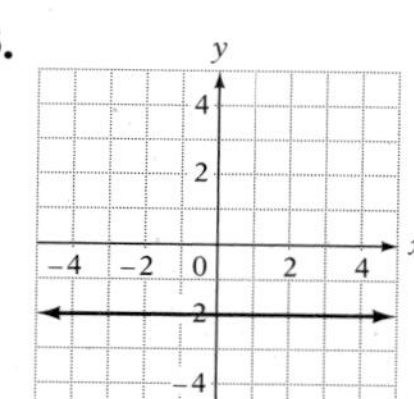

25.
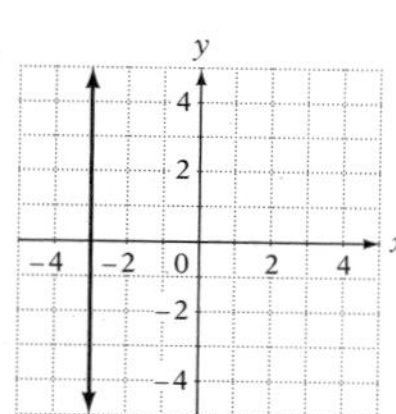

27.
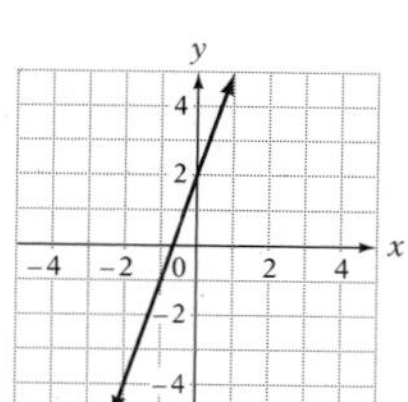

29.
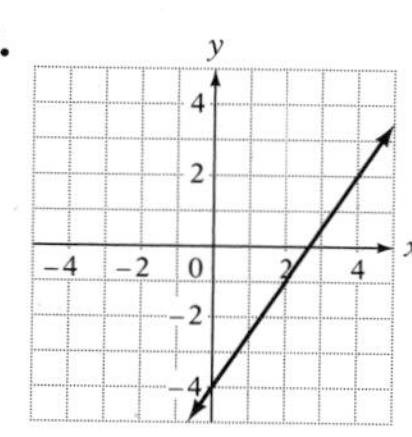

31. No. If $B = 0$, it is not possible to solve $Ax + By = C$ for y.

33. x-intercept: (−4, 0); y-intercept: (0, 2)

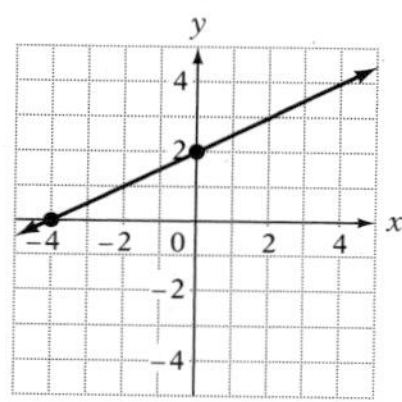

35. x-intercept: $\left(\frac{9}{2}, 0\right)$; y-intercept: (0, −3)

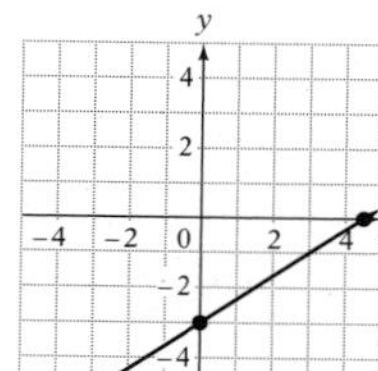

37. x-intercept: $\left(\frac{3}{2}, 0\right)$; y-intercept: $(0, 3)$ **39.** x-intercept: $\left(\frac{4}{3}, 0\right)$; y-intercept: $(0, 2)$

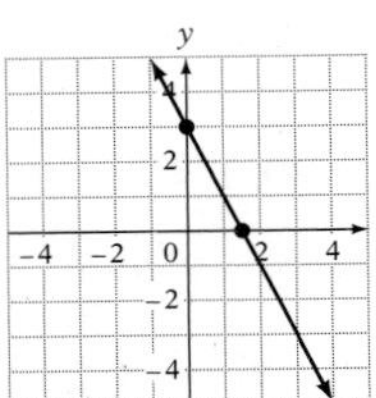

41. No. The graph of the equation $x = a$, $a \neq 0$, is a vertical line and has no y-intercept. **43.** The heart of the hummingbird will beat 8400 times in 7 min of flight. **45.** Marlys receives \$165 for tutoring 15 h.

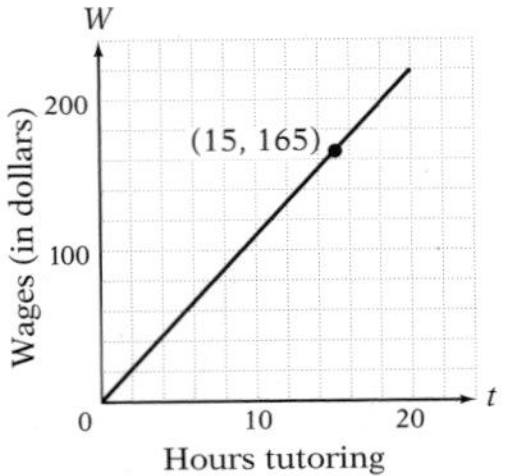

47. The cost of manufacturing 50 pairs of skis is \$9000.

49. a.

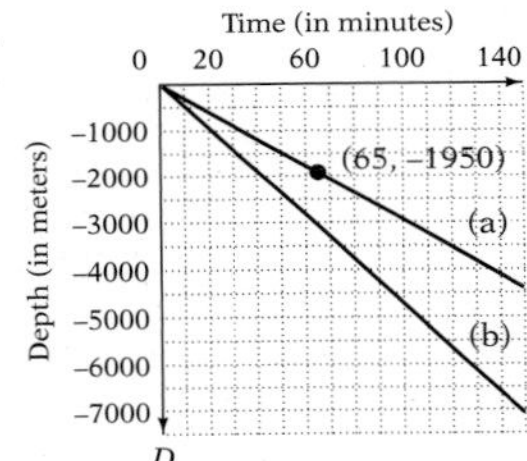
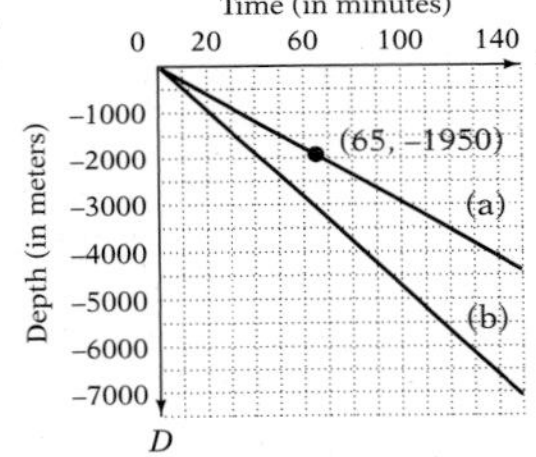

After 65 min, *Alvin* is 1950 m below sea level. **b.** Below

51. Yes. The points with y-coordinate less than -6500. **57.**

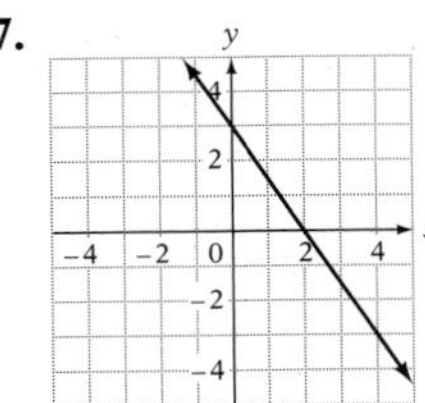

SECTION 4.4

1. -1 **3.** $\frac{1}{3}$ **5.** $-\frac{2}{3}$ **7.** $-\frac{3}{4}$ **9.** Undefined **11.** $\frac{7}{5}$ **13.** 0 **15.** $-\frac{1}{2}$ **17.** Undefined **19.** a and c

21. $m = 40$. The average speed of the motorist is 40 mph. **23.** $m = -5$. The temperature of the oven decreases 5°/min. **25.** $m = -0.05$. For each mile the car is driven, approximately 0.05 gal of fuel is used. **27.** $m \approx 352.4$. The average speed of the runner was 352.4 m/min. **29. a.** No **b.** Yes **31.** $-3, 5, -3, (0, 5)$ **33.** $4, 0, 4, (0, 0)$

35.

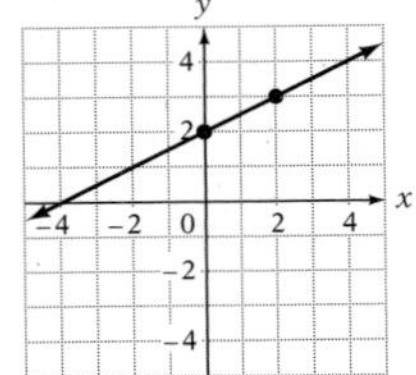

37.

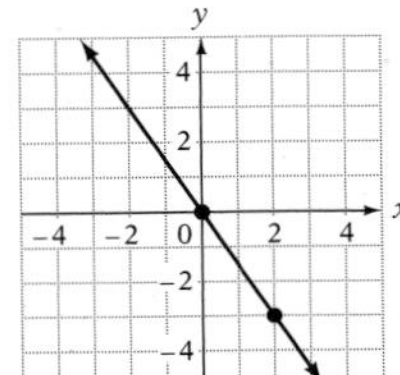

39.

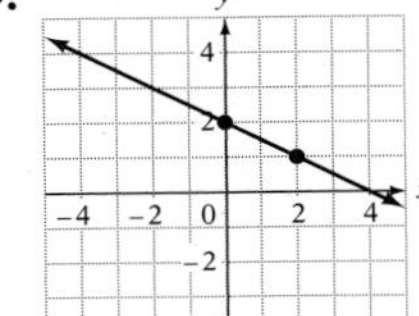

41.

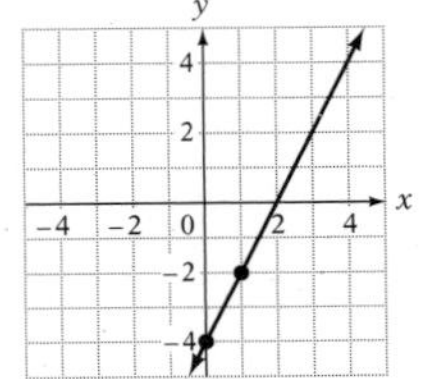

43.

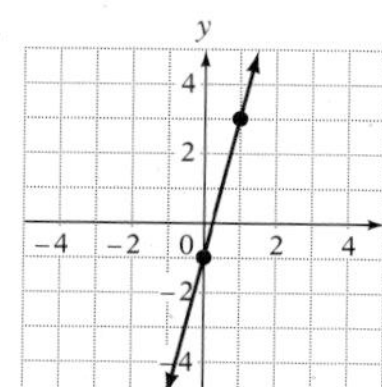

45.

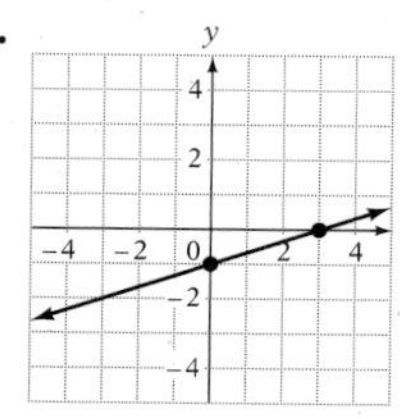

47.

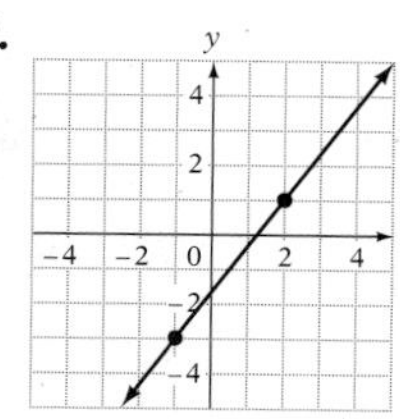

49.

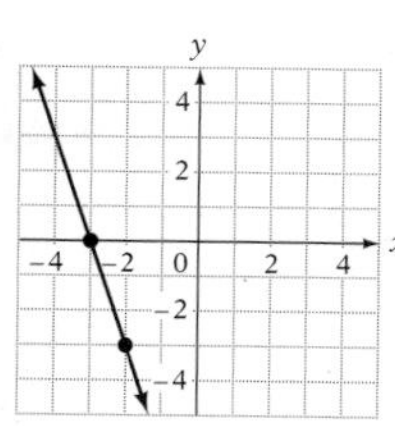

51.

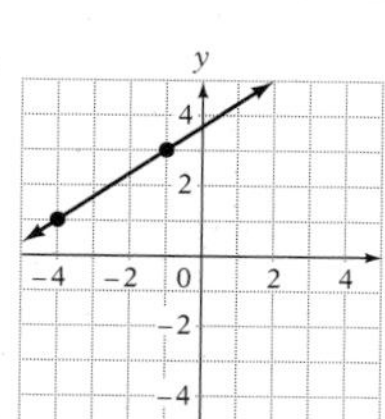

53. a. Below **b.** Negative **55.** Increases by 2 **57.** Increases by 2 **59.** Decreases by $\frac{2}{3}$

61. i. D; ii. C; iii. B; iv. F; v. E; vi. A **63.** $k = 10$ **65.** $k = -4$

SECTION 4.5

3. Check that the coefficient of x is the given slope. Check that the coordinates of the given point are a solution of your equation. **5.** $y = 2x + 5$ **7.** $y = \frac{1}{2}x + 2$ **9.** $y = \frac{5}{4}x + \frac{21}{4}$ **11.** $y = -\frac{5}{3}x + 5$ **13.** $y = -3x + 9$ **15.** $y = -3x + 4$ **17.** $y = \frac{2}{3}x - \frac{7}{3}$ **19.** $y = \frac{1}{2}x$ **21.** $y = 3x - 9$ **23.** $y = -\frac{2}{3}x + 7$ **25.** $y = -x - 3$ **27.** $y = \frac{7}{5}x - \frac{27}{5}$ **29.** $y = -\frac{2}{5}x + \frac{3}{5}$ **31.** $x = 3$ **33.** $y = -\frac{5}{4}x - \frac{15}{2}$ **35.** $y = -3$ **37.** $y = -2x + 3$ **39.** $x = -5$ **41.** Check that the coordinates of each given point are a solution of your equation. **43.** $y = x + 2$ **45.** $y = -2x - 3$ **47.** $y = \frac{2}{3}x + \frac{5}{3}$ **49.** $y = \frac{1}{3}x + \frac{10}{3}$ **51.** $y = \frac{3}{2}x - \frac{1}{2}$ **53.** $y = -\frac{3}{2}x + 3$ **55.** $y = -1$ **57.** $y = x - 1$ **59.** $y = -x + 1$ **61.** $y = -\frac{8}{3}x + \frac{25}{3}$ **63.** $y = \frac{1}{2}x - 1$ **65.** $y = -4$ **67.** $y = \frac{3}{4}x$ **69.** $y = -\frac{4}{3}x + \frac{5}{3}$ **71.** $x = -2$ **73.** $y = x - 1$ **75.** $y = \frac{4}{3}x + \frac{7}{3}$ **77.** $y = -x + 3$ **79. a.** $f(x) = 1200x, 0 \le x \le 26\frac{2}{3}$ **b.** The height of the plane 11 min after takeoff is 13,200 ft. **81. a.** $f(x) = 0.625x - 1170.5$ **b.** In 2012, 87% of trees at 2600 ft will be hardwoods. **83. a.** $f(x) = -0.032x + 16, 0 \le x \le 500$ **b.** After 150 mi are driven, 11.2 gal are left in the tank. **85. a.** $f(x) = -20x + 230{,}000$ **b.** 60,000 motorcycles would be sold at \$8500 each. **87. a.** $f(x) = 63x$ **b.** There are 315 Calories in a 5-ounce serving. **89.** Substitute 15,000 for $f(x)$ and solve for x. **91.** $f(x) = x + 3$ **93.** 0 **95. a.** -10 **b.** 6 **99.** Answers will vary. Possible answers include (0, 4), (3, 2), and (9, −2). **101.** $y = -2x + 1$

SECTION 4.6

3. No. If two nonvertical lines are perpendicular, the product of their slopes is −1. Therefore, one line must have a positive slope and one line must have a negative slope. **5.** −5 **7.** $-\frac{1}{4}$ **9.** Yes **11.** No **13.** No **15.** Yes **17.** Yes **19.** Yes **21.** No **23.** Yes **25.** $y = 2x - 8$ **27.** $y = \frac{3}{2}x + 2$ **29.** $y = \frac{2}{3}x - \frac{8}{3}$ **31.** $y = \frac{1}{3}x - \frac{1}{3}$ **33.** $y = -\frac{5}{3}x - \frac{14}{3}$ **35.** $y = -2x + 15$ **37.** $\frac{A_1}{B_1} = \frac{A_2}{B_2}$

SECTION 4.7

3. Yes **5.** No **7.**

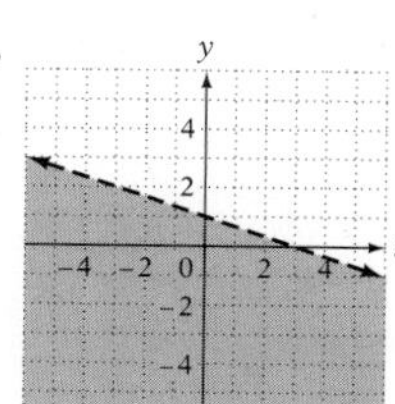

9.

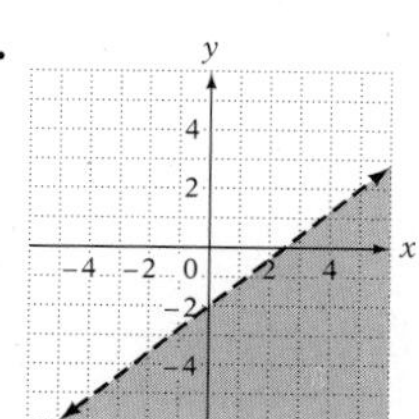

11.

13.

15.

17.

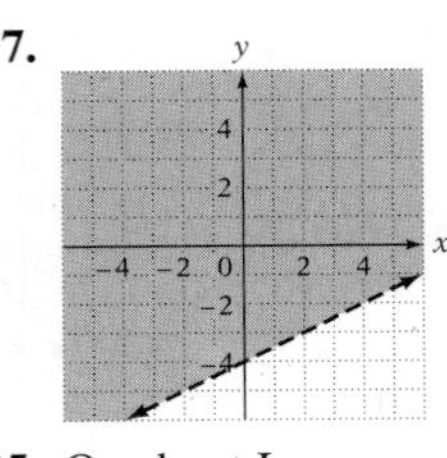

19.

21.

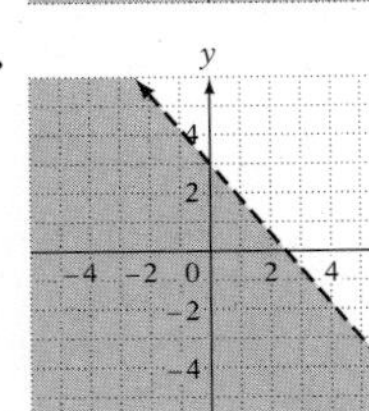

23.

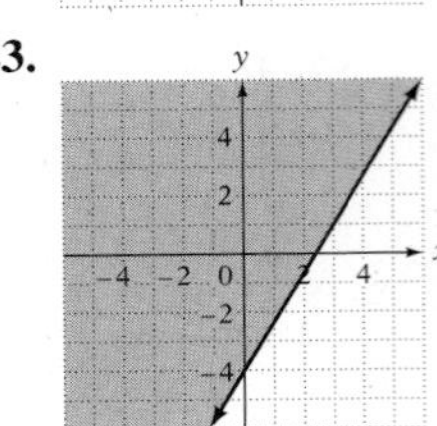

25. Quadrant I

CHAPTER 4 CONCEPT REVIEW*

1. The ordinate is the second number in an ordered pair. The abscissa is the first number in an ordered pair. [4.1A]

2. A linear equation in two variables has an infinite number of solutions. [4.1B]

3. A relation is a function when no two ordered pairs of the relation have the same first coordinate. [4.2A]

4. The value of a dependent variable y depends on the value of the independent variable x. The value of the independent variable x is not dependent on the value of any other variable. [4.2A]

5. In the general equation $y = mx + b$, m represents the slope and b is the ordinate of the y-intercept. [4.4B]

6. A straight line is determined by two points. However, to ensure the accuracy of a graph, find three ordered-pair solutions. If the three solutions do not lie on a straight line, there has been an error in calculating an ordered-pair solution or in plotting the points. [4.3A]

7. The equation of a vertical line is of the form $x = a$. The equation of a horizontal line is of the form $y = b$. [4.3B]

8. In the ordered pair for an x-intercept, the y coordinate is 0. In the ordered pair for a y-intercept, the x coordinate is 0. [4.3C]

9. A line that has undefined slope is a vertical line. It is the graph of an equation of the form $x = a$. [4.4A]

10. Given two ordered pairs on a line, use the slope formula to find the slope of the line. The slope formula is $m = \frac{y_2 - y_1}{x_2 - x_1}$, where m is the slope, and (x_1, y_1) and (x_2, y_2) are the two points on the line. [4.4A]

11. Parallel lines never meet; the distance between them is always the same. In a rectangular coordinate system, parallel lines have the same slope. Perpendicular lines are two lines that intersect at right angles. In a rectangular coordinate system, two lines are perpendicular if the product of their slopes is -1. [4.6A]

12. The point-slope formula states that if (x_1, y_1) is a point on a line with slope m, then $y - y_1 = m(x - x_1)$. [4.5A]

CHAPTER 4 REVIEW EXERCISES

1. (4, 2) [4.1B] **2.** $P(-2) = -2$; $P(a) = 3a + 4$ [4.2A] **3.** [4.1B] **4.** [4.1A]

5. Range = $\{-1, 1, 5\}$ [4.2A] **6.** Domain = $\{-1, 0, 1, 5\}$ [4.2A]
Range = $\{0, 2, 4\}$ **7.** Yes [4.6A]

**Note:* The numbers in brackets following the answers in the Concept Review are a reference to the objective that corresponds to that problem. For example, the reference [1.2A] stands for Section 1.2, Objective A. This notation will be used for all Prep Tests, Concept Reviews, Chapter Reviews, Chapter Tests, and Cumulative Reviews throughout the text.

8. -4 [4.2A] **9.** x-intercept: $(-3, 0)$ y-intercept: $(0, -2)$

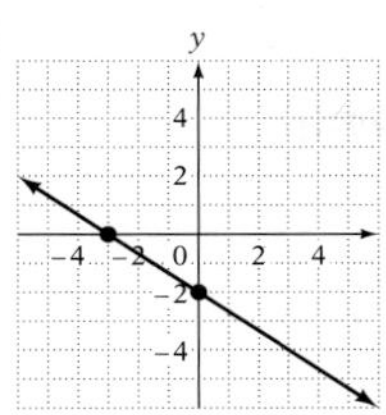

[4.3C]

10. x-intercept: $(-2, 0)$ y-intercept: $(0, -3)$

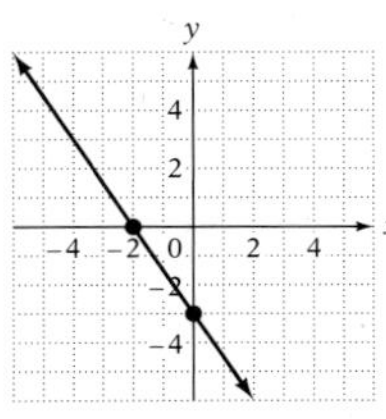

[4.3C] **11.**

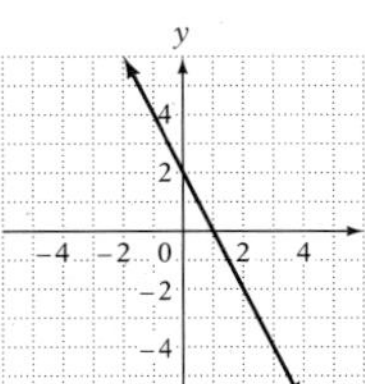

[4.3A] **12.**

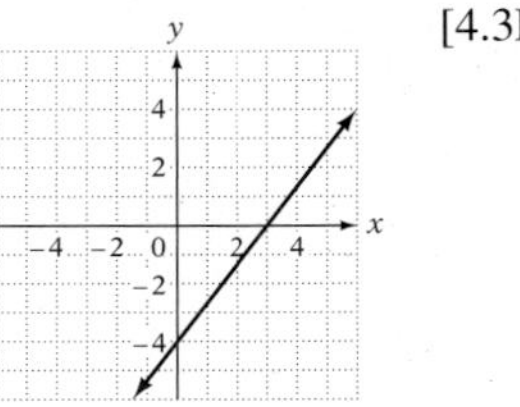

[4.3B]

13. -1 [4.4A] **14.** $y = \frac{5}{2}x + \frac{23}{2}$ [4.5A] **15.**

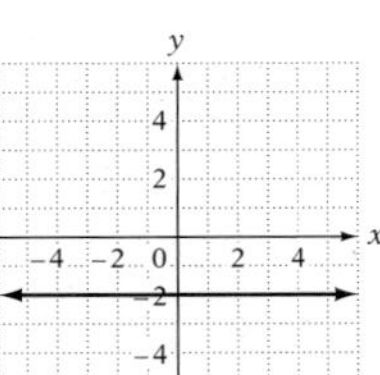

[4.1A] **16.**

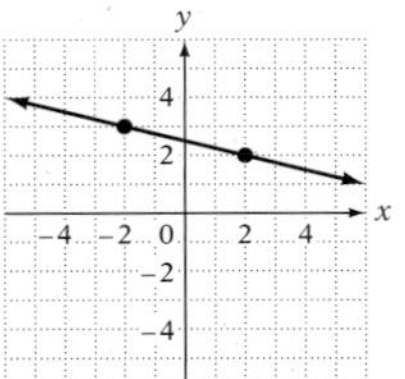

[4.4B]

17. $\{-2, -1, 2\}$ [4.2A] **18. a.** $y = -x + 295, 0 \le x \le 295$ **b.** When the rate is \$120, 175 rooms will be occupied. [4.5C]

19. $y = -4x - 5$ [4.6A] **20.** $y = \frac{5}{2}x + 8$ [4.6A] **21.**

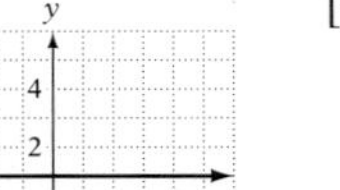

[4.3B]

22.

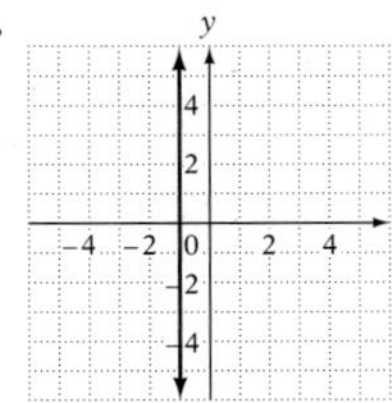

[4.3B] **23.** $y = -\frac{2}{3}x + 1$ [4.5A] **24.** $y = \frac{1}{4}x + 4$ [4.5B] **25.** $(-4, 0), (0, 6)$ [4.3C]

26. $(-1, -7)$ [4.1B] **27.**

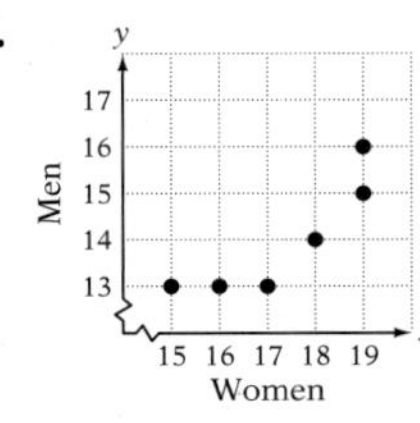

[4.1C] **28.**

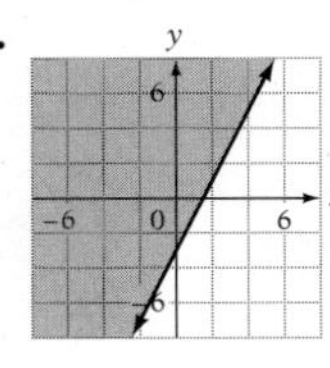
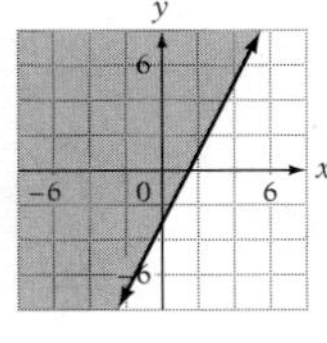

[4.7A] **29.**

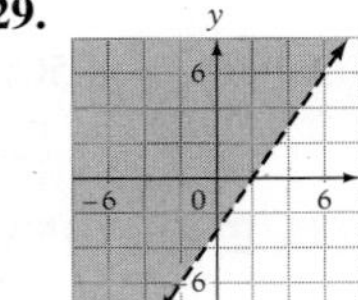
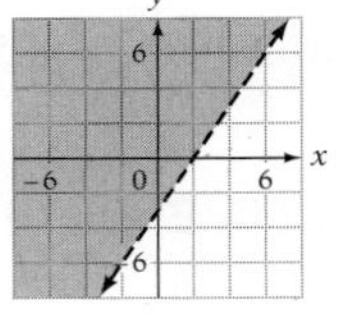

[4.7A]

30. $y = -\frac{7}{6}x + \frac{5}{3}$ [4.5B] **31.** $y = 2x$ [4.6A] **32.** $y = -3x + 7$ [4.6A] **33.** $y = \frac{3}{2}x + 2$ [4.6A]

34.

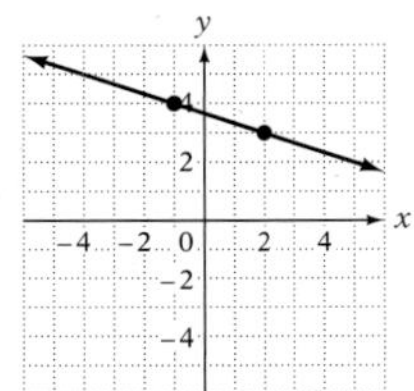

[4.4B] **35.** After 4 h, the car has traveled 220 mi.

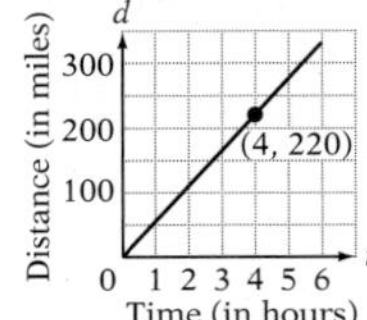

[4.3D]

36. The slope is 20. The manufacturing cost is \$20 per calculator. [4.4A] **37. a.** $y = 80x + 25{,}000$ **b.** The house will cost \$185,000 to build. [4.5C]

CHAPTER 4 TEST

1. [4.1B, How To 3] **2.** $(-3, 0)$ [4.1B, Example 4] **3.**

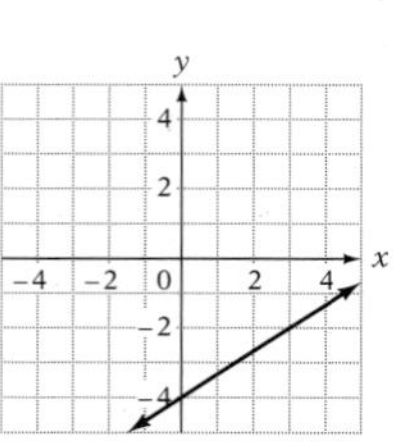

[4.3A, How To 1]

4. [4.3B, Example 3] **5.** $x = -2$ [4.5A, How To 3] **6.** No [4.6A, How To 4] **7.** $-\frac{1}{6}$ [4.4A, How To 2]

8. $P(2) = 9$ [4.2A, How To 2] **9.** [4.3C, How To 6] **10.** [4.4B, How To 7]

11. $y = \frac{2}{5}x + 4$ [4.5A, How To 2] **12.** 0 [4.2A, You Try It 5] **13.** $y = -\frac{7}{5}x + \frac{1}{5}$ [4.5B, How To 4]

14. $y = -3$ [4.5A, How To 3] **15.** Domain $= \{-4, -2, 0, 3\}$ [4.2A, How To 1] Range $= \{0, 2, 5\}$ **16.** $y = -\frac{3}{2}x + \frac{7}{2}$ [4.6A, How To 2]

17. $y = 2x + 1$ [4.6A, How To 6] **18.** [4.7A, How To 1] **19.** The slope is $-10{,}000$. The value of the house decreases by \$10,000 each year. [4.4A, You Try It 3] **20. a.** $y = -\frac{3}{10}x + 175$ **b.** When the tuition is \$300, 85 students will enroll. [4.5C, Example 5]

CUMULATIVE REVIEW EXERCISES

1. $-5, -3$ [1.1A] **2.** 0.85 [1.2B] **3.** $9\sqrt{5}$ [1.2F] **4.** -12 [1.3A] **5.** $-\frac{5}{8}$ [1.4A]

6. $-4d - 9$ [1.4B] **7.** $-32z$ [1.4C] **8.** $-13x + 7y$ [1.4D] **9.** [1.5B]

10. $\frac{3}{2}$ [2.2A] **11.** 1 [2.2C] **12.** $\{x|x > -1\}$ [2.5A] **13.** $\varnothing$ [2.5B] **14.** $\left\{x|0 < x < \frac{10}{3}\right\}$ [2.6B]

15. $f(2) = 6$ [4.2A] **16.** The slope is 2. [4.4A] **17.**

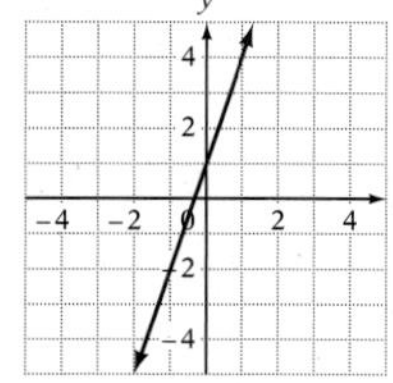

[4.3A]

18. 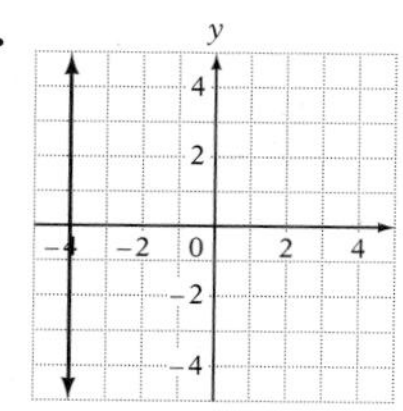 [4.3B] **19.** 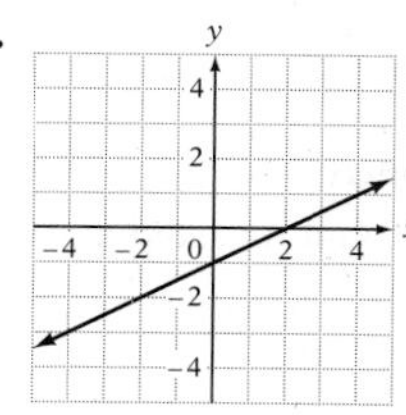 [4.4B] **20.** 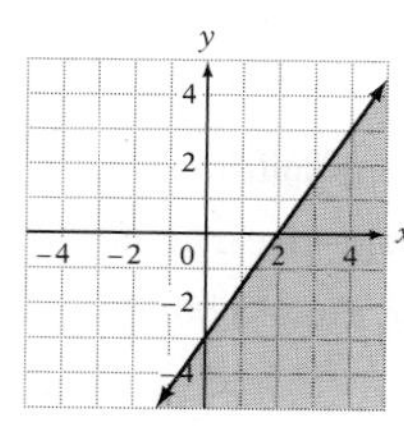 [4.7A]

21. $y = -\frac{1}{3}x - 2$ [4.5B] **22.** $y = -\frac{3}{2}x + 7$ [4.6A] **23.** The first plane is traveling at 400 mph. The second plane is traveling at 200 mph. [2.4C] **24.** 20 lb of \$6 coffee and 40 lb of \$9 coffee should be used. [2.4A]

25. The slope is $-\frac{10{,}000}{3}$. The value of the backhoe decreases by \$3333.33 each year. [4.4A]

ANSWERS TO CHAPTER 5 SELECTED EXERCISES

PREP TEST

1. $6x + 5y$ [1.4D] **2.** 7 [1.4A] **3.** 0 [2.2A] **4.** -3 [2.2C] **5.** 1000 [2.2C]

6. 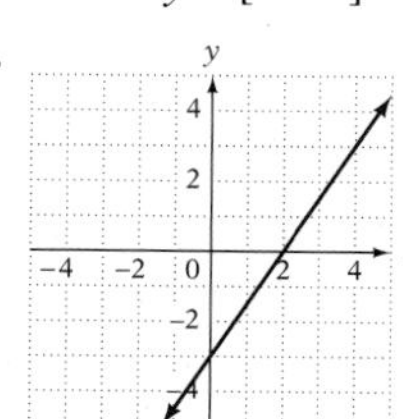 [4.3B] **7.** 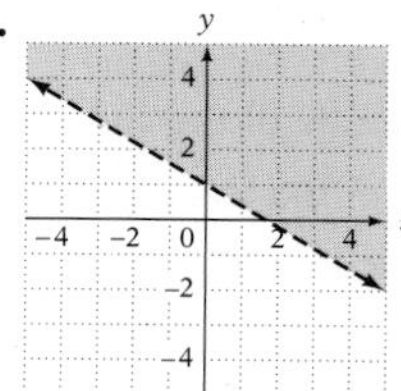 [4.7A]

SECTION 5.1

1. No **3.** Yes **5.** Independent **7.** Inconsistent **9.** Independent **11.**

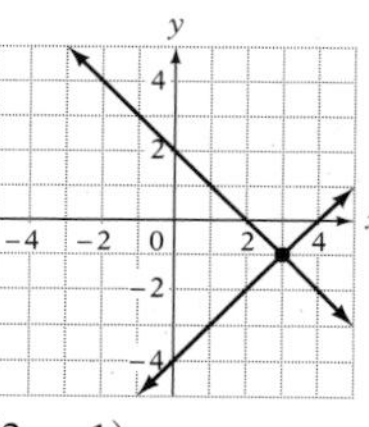

(3, −1)

13.

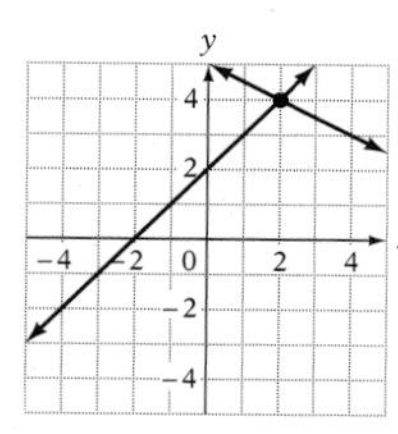

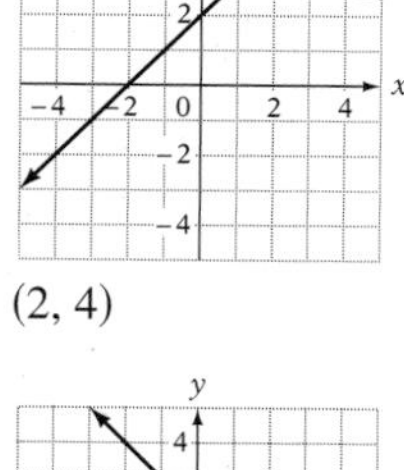

(2, 4)

15.

(4, 3)

17.

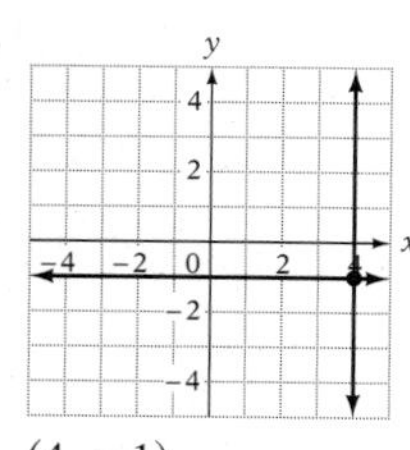

(4, −1)

19.

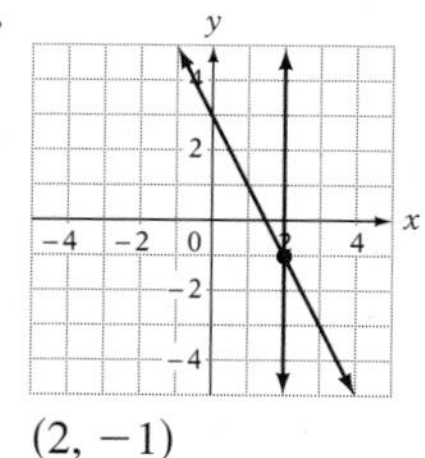

(2, −1)

21.

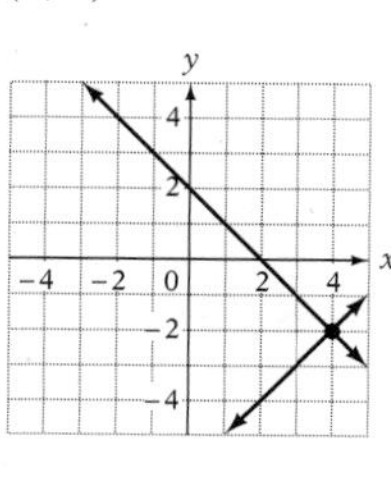

(4, −2)

23.

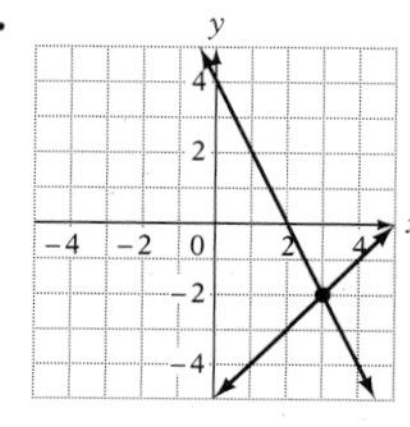

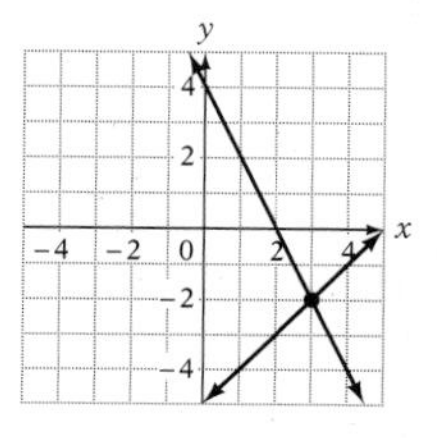

(3, −2)

25.

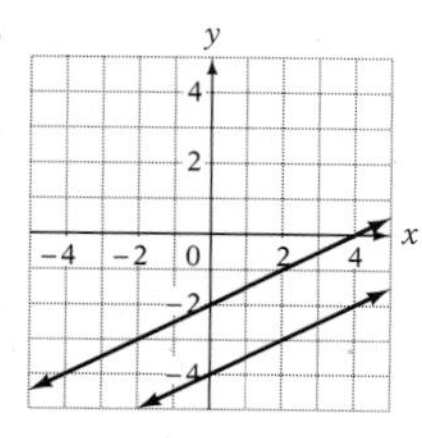

No solution

27.

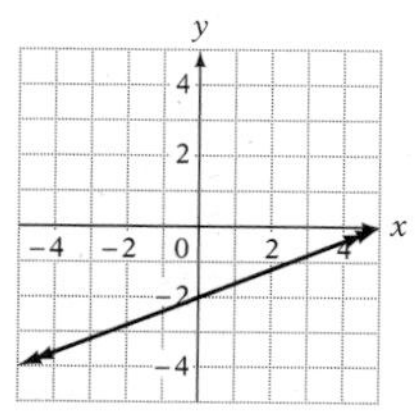

$\left(x, \frac{2}{5}x - 2\right)$

29. (2, −1) **31.** (1, 2) **33.** (3, −2) **35.** $\left(-\frac{5}{2}, -4\right)$ **37.** (−1, 4) **39.** (−2, 5) **41.** (1, 5) **43.** (2, 0)

45. $\left(\frac{2}{3}, -1\right)$ **47.** No solution **49.** $(-2, 5)$ **51.** $(0, 0)$ **53.** $\left(x, \frac{1}{2}x - 4\right)$ **55.** $(1, 5)$ **57.** $\left(\frac{2}{3}, 3\right)$ **59.** $\frac{2}{3}$ **61.** 5.5% and 7.2% **63.** The amount invested in the insured bond fund is \$5500. **65.** The amount invested at 6.5% must be \$4000. **67.** The amount invested at 3.5% is \$23,625. The amount invested at 4.5% is \$18,375. **69.** There is \$6000 invested in the mutual bond fund. **71.** $(1.20, 1.40)$ **73.** $(0.54, 1.03)$

SECTION 5.2

1. Answers may vary. Possible answers are 6 and -5. **3.** $(6, 1)$ **5.** $(1, 1)$ **7.** $(2, 1)$ **9.** $(-2, 1)$ **11.** $(x, 3x - 4)$ **13.** $\left(-\frac{1}{2}, 2\right)$ **15.** No solution **17.** $(-1, -2)$ **19.** $(-5, 4)$ **21.** $(2, 5)$ **23.** $\left(\frac{1}{2}, \frac{3}{4}\right)$ **25.** $(0, 0)$ **27.** $(-1, 3)$ **29.** $\left(\frac{2}{3}, -\frac{2}{3}\right)$ **31.** $(-2, 3)$ **33.** $(2, -1)$ **35.** $(10, -5)$ **37.** $\left(-\frac{1}{2}, \frac{2}{3}\right)$ **39.** $\left(\frac{5}{3}, \frac{1}{3}\right)$ **41.** No solution **43.** $(1, -1)$ **45.** $(2, 1, 3)$ **47.** $(1, -1, 2)$ **49.** $(1, 2, 4)$ **51.** $(2, -1, -2)$ **53.** $(-2, -1, 3)$ **55.** No solution **57.** $(1, 4, 1)$ **59.** $(1, 3, 2)$ **61.** $(1, -1, 3)$ **63.** $(0, 2, 0)$ **65.** $(1, 5, 2)$ **67.** $(-2, 1, 1)$ **69. a.** (iii) **b.** (ii) **c.** (i) **71.** $(1, -1)$ **73.** $(1, -1)$

SECTION 5.3

1. Less than **3.** The rate of the motorboat in calm water is 15 mph. The rate of the current is 3 mph. **5.** The rate of the plane in calm air is 502.5 mph. The rate of the wind is 47.5 mph. **7.** The rate of the team in calm water is 8 km/h. The rate of the current is 2 km/h. **9.** The rate of the plane in calm air is 180 mph. The rate of the wind is 20 mph. **11.** The rate of the plane in calm air is 110 mph. The rate of the wind is 10 mph. **13.** Greater than **15.** There are 30 nickels in the bank. **17.** The cost of the pine is \$1.10/ft. The cost of the redwood is \$3.30/ft. **19.** The cost of the wool carpet is \$52/yd. **21.** The company plans to manufacture 25 mountain bikes during the week. **23.** The owner drove 322 mi in the city and 72 mi on the highway. **25.** The chemist should use 480 g of the first alloy and 40 g of the second alloy. **27.** The Model VI computer costs \$4000. **29.** The investor placed \$10,400 in the 8% account, \$5200 in the 6% account, and \$9400 in the 4% account. **31.** The measures of the two angles are 35° and 145°. **33.** The age of the oil painting is 85 years, and the age of the watercolor is 50 years.

SECTION 5.4

1.

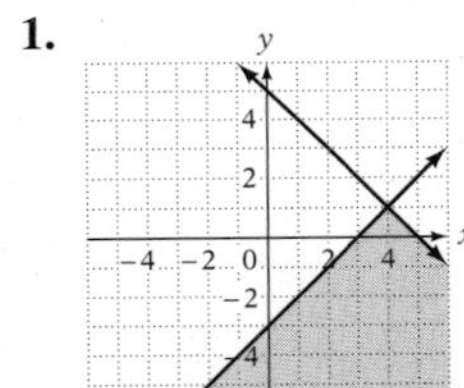

3.

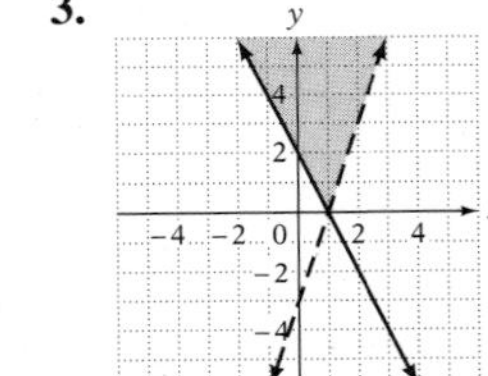

5.

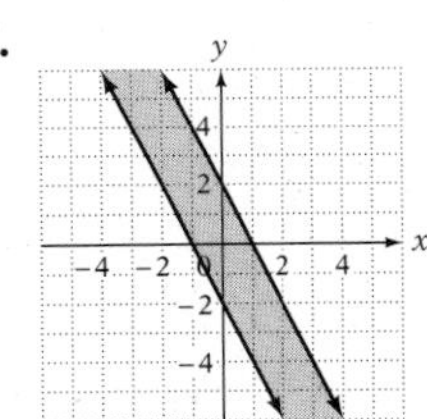

7.

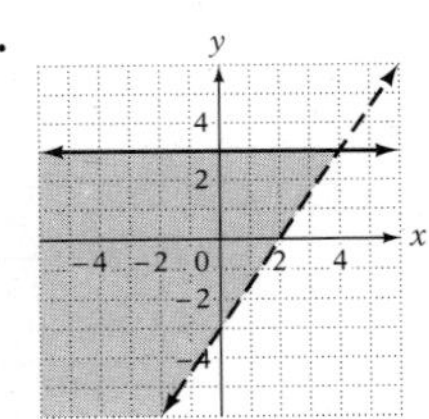

9.

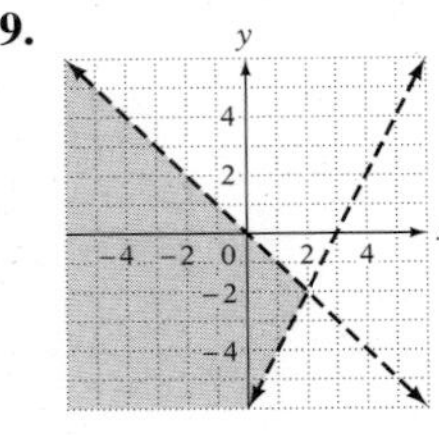

11.

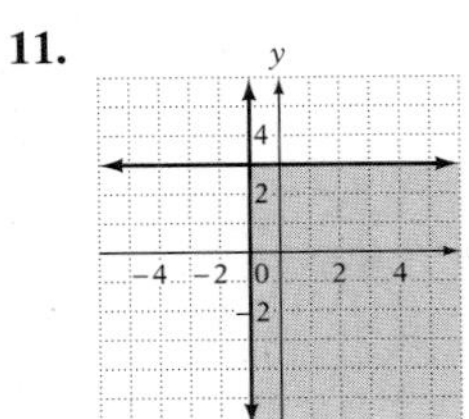

13.

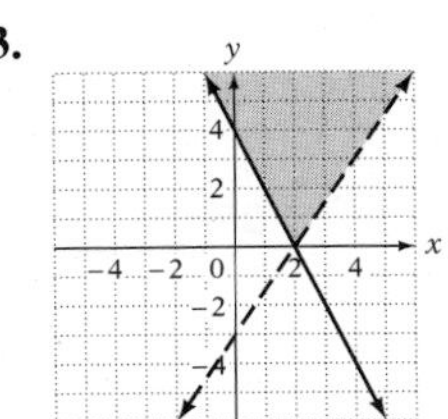

15.

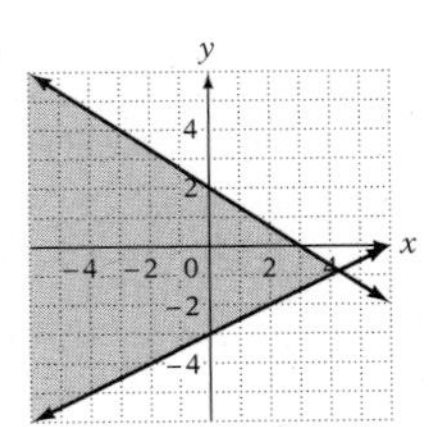

17.

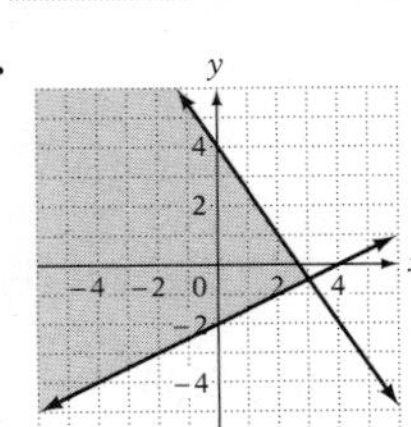

19. The points below the line $x + y = b$ **21.** The region between the parallel lines $x + y = a$ and $x + y = b$

23. 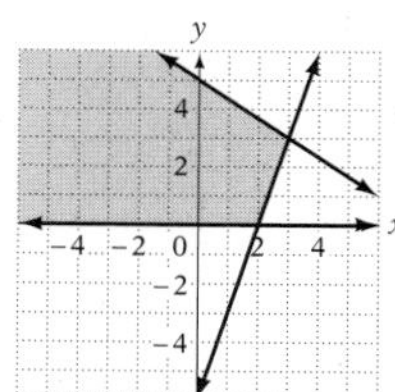

25. 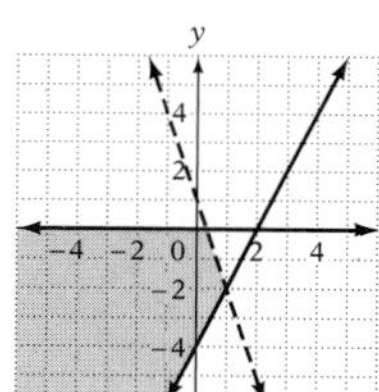

CHAPTER 5 CONCEPT REVIEW*

1. To determine the solution after graphing a system of linear equations, find the point of intersection of the lines. The point of intersection is the ordered-pair solution. [5.1A]
2. An independent system of equations has exactly one solution; the graphs of the equations intersect at one point. A dependent system has an infinite number of solutions; the graphs of the equations are the same line. [5.1A]
3. The graph of an inconsistent system of equations is the graph of two parallel lines. [5.1A]
4. The graph of a dependent system of equations looks like the graph of one line; because the graphs of the equations are the same line, one graph lies directly on top of the other. [5.1A]
5. To solve a system of two linear equations by the substitution method, solve one of the equations in the system for one of its variables. Suppose you have solved one equation for *y*. Substitute the expression for *y* into the other equation. You now have an equation with only one variable in it. Let's suppose the variable is *x*. Solve this equation for *x*. This is the first coordinate of the ordered-pair solution of the system of equations. Substitute the value of *x* into the equation that has been solved for *y*. Evaluate the numerical expression. This is the second coordinate of the ordered-pair solution of the system of equations. [5.1B]
6. To solve a simple interest problem, use the formula $Pr = I$, where P is the principal, r is the simple interest rate, and I is the simple interest. [5.1C]
7. To solve a system of two linear equations by the addition method, rewrite one or both of the equations in the system so that the coefficients of one of the variables are opposites. Suppose that the coefficients of the *y* terms are opposites. Add the two equations. The result is an equation with no *y* term; solve it for *x*. The solution of this equation is the first coordinate of the ordered-pair solution of the system of equations. Substitute the value of *x* into either of the original equations in the system, and solve the resulting equation for *y*. This is the second coordinate of the ordered-pair solution of the system of equations. [5.2A]
8. When using the addition method, after adding the two equations in the system of two equations, a true equation that contains no variable, such as $0 = 0$, tells you that the system of equations is dependent. [5.2A]
9. In a rate-of-wind problem, the expression $p + w$ represents the rate of the plane flying with the wind. The expression $p - w$ represents the rate of the plane flying against the wind. [5.3A]
10. If we have only one equation in two variables, we cannot solve that equation for the values of both variables. In application problems with two variables, we need to write two equations. When we have two equations—a system of equations—we can solve the system of equations for the value of each of the two variables. [5.3B]
11. To check the solution of a system of three equations in three variables, substitute the values of the three variables into each of the equations of the system. In each case, simplify the resulting numerical expression. If the left and right sides are equal in each case, the ordered triple is the solution of the system. If the left and right sides are not equal in all three cases, the proposed solution is not the solution of the system. [5.2B]
12. To solve a system of linear inequalities, first graph the solution set of each linear inequality. The solution set of the system of linear inequalities is the region of the plane represented by the intersection of the shaded areas. [5.4A]

CHAPTER 5 REVIEW EXERCISES

1. $\left(6, -\frac{1}{2}\right)$ [5.1B] **2.** $(-4, 7)$ [5.2A] **3.**

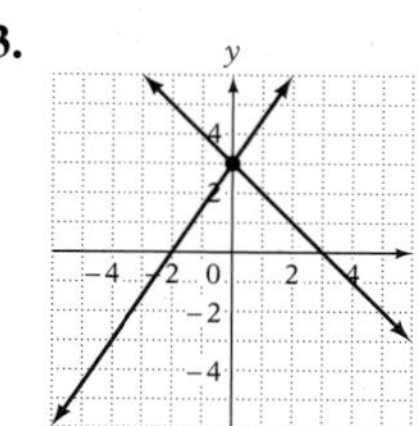

(0, 3) [5.1A]

4.

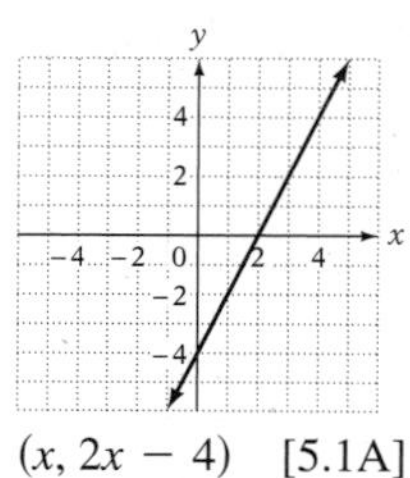

$(x, 2x - 4)$ [5.1A]

5. $\left(x, -\frac{1}{4}x + \frac{3}{2}\right)$ [5.1B] **6.** $\left(x, \frac{1}{3}x - 2\right)$ [5.2A] **7.** $(3, -1, -2)$ [5.2B] **8.** $(5, -2, 3)$ [5.2B] **9.** Yes [5.1A]

**Note:* The numbers in brackets following the answers in the Concept Review are a reference to the objective that corresponds to that problem. For example, the reference [1.2A] stands for Section 1.2, Objective A. This notation will be used for all Prep Tests, Concept Reviews, Chapter Reviews, Chapter Tests, and Cumulative Reviews throughout the text.

10. $\left(\frac{1}{2}, -\frac{5}{2}\right)$ [5.1B] **11.** $(3, -1)$ [5.1B] **12.** $(3, 2)$ [5.2A] **13.** $(-1, -3, 4)$ [5.2B] **14.** $(2, 3, -5)$ [5.2B]

15.

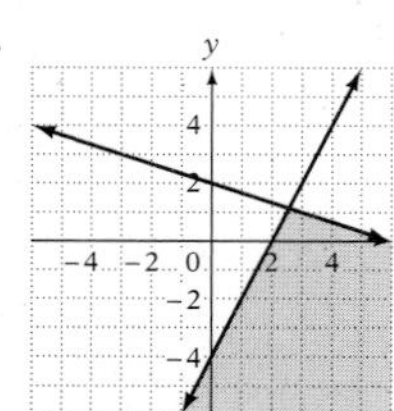

[5.4A] **16.** [5.4A]

17. The rate of the cabin cruiser in calm water is 16 mph. The rate of the current is 4 mph. [5.3A] **18.** The rate of the plane in calm air is 175 mph. The rate of the wind is 25 mph. [5.3A] **19.** On Friday, 100 children attended. [5.3B] **20.** The amount invested at 3% is $5000. The amount invested at 7% is $15,000. [5.1C]

CHAPTER 5 TEST

1.

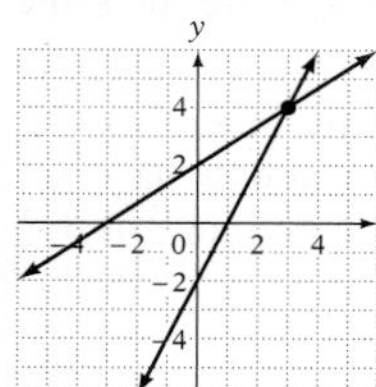

$(3, 4)$ [5.1A, Example 1] **2.** $(-5, 0)$ [5.1A, Example 1]

3.

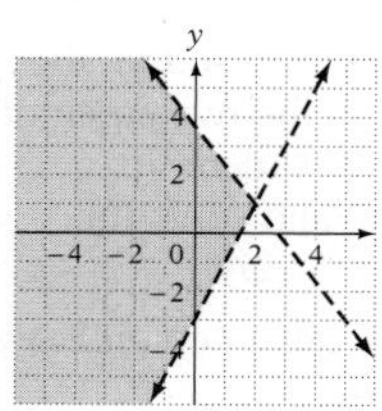

[5.4A, How To 1] **4.** [5.4A, How To 1] **5.** $\left(\frac{3}{4}, \frac{7}{8}\right)$ [5.1B, How To 3]

6. $(-3, -4)$ [5.1B, How To 3] **7.** $(2, -1)$ [5.1B, How To 3] **8.** $(-2, 1)$ [5.2A, How To 1] **9.** No solution [5.2A, You Try It 2] **10.** $(1, 1)$ [5.2A, Example 1] **11.** No solution [5.2B, How To 5] **12.** $(2, -1, -2)$ [5.2B, How To 6] **13.** $\left(-\frac{1}{3}, -\frac{10}{3}\right)$ [5.1B, Example 4] **14.** Yes [5.1A, How To 1] **15.** $\left(\frac{1}{5}, -\frac{6}{5}, \frac{3}{5}\right)$ [5.2B, How To 4] **16.** The rate of the plane in calm air is 150 mph. The rate of the wind is 25 mph. [5.3A, Example 1] **17.** The cost of cotton is $9/yd. The cost of wool is $14/yd. [5.3B, How To 2] **18.** The amount invested at 2.7% is $9000. The amount invested at 5.1% is $6000. [5.1C, How To 5]

CUMULATIVE REVIEW EXERCISES

1. $-6\sqrt{10}$ [1.2F] **2.** 22 [2.2C] **3.** $3x - 24$ [1.4D] **4.** -4 [1.4A] **5.** $\{x|x < 6\}$ [2.5B] **6.** $\{x| -4 < x < 8\}$ [2.6B] **7.** $\{x|x > 4 \text{ or } x < -1\}$ [2.6B] **8.** $F(2) = 1$ [4.2A]

9. (number line: −5 −4 −3 −2 −1 0 1 2 3 4 5) [1.5C] **10.** $y = -\frac{2}{3}x + \frac{5}{3}$ [4.5A] **11.** $y = 5x - 11$ [4.5B]

12. $y = -\frac{3}{2}x - 1$ [4.6A] **13.**

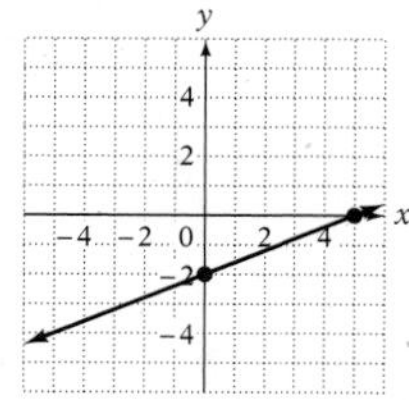

[4.3C]

14.

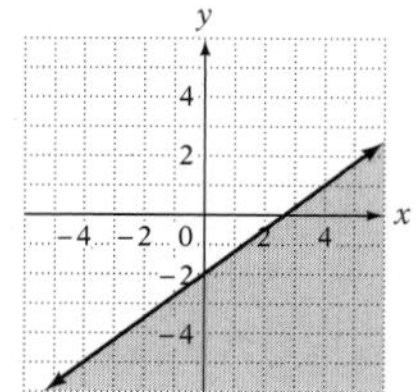

[4.7A]

15.

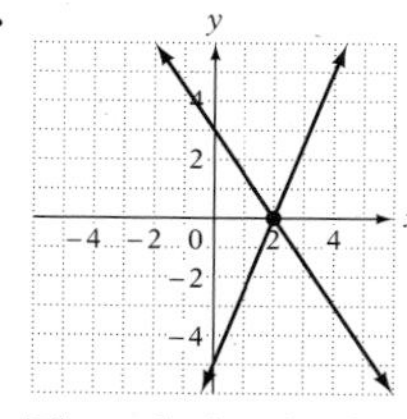

[5.1A]

The solution is (2, 0).

16.

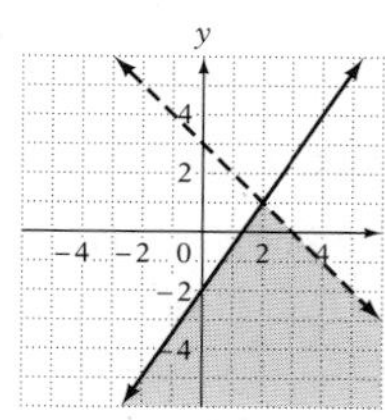

[5.4A]

17. (1, 0, −1) [5.2B] **18.** (2, 1, −1) [5.2B]

19. (2, −3) [5.2A] **20.** (−5, −11) [5.1B] **21.** 60 ml of pure water must be used. [2.4A] **22.** The rate of the wind is 12.5 mph. [5.3A] **23.** One pound of steak costs \$10. [5.3B] **24.** The lower limit is 10,200 ohms. The upper limit is 13,800 ohms. [2.6C] **25.** The slope is 40. The monthly income increases by \$40 for every increase of \$1000 in sales. [4.4A]

ANSWERS TO CHAPTER 6 SELECTED EXERCISES

PREP TEST

1. 1 [1.1C] **2.** −18 [1.1D] **3.** $\frac{2}{3}$ [1.1D] **4.** 48 [1.4A] **5.** 0 [1.1D] **6.** No [1.4B] **7.** $5x^2 - 9x - 6$ [1.4B] **8.** 0 [1.4B] **9.** $-6x + 24$ [1.4D] **10.** $-7xy + 10y$ [1.4D]

SECTION 6.1

1. a^4b^4 **3.** $-18x^3y^4$ **5.** x^8y^{16} **7.** $81x^8y^{12}$ **9.** $729a^{10}b^6$ **11.** $64a^{24}b^{18}$ **13.** x^5y^{11} **15.** $-45a^7b^5$ **17.** $-192x^{14}y$ **19.** $42a^6b^2c^5$ **21.** $54a^{13}b^{17}$ **23.** $-432a^7b^{11}$ **25.** $-6x^4y^4z^5$ **27.** $-6x^5y^5z^4$ **29.** $-12a^2b^9c^2$ **31.** x^{2n+1} **33.** y^{6n-2} **35.** a^{2n^2-6n} **37.** x^{15n+10} **39.** 33 **41.** 243 **43.** y^3 **45.** $\frac{a^3b^2}{4}$ **47.** $\frac{x}{y^4}$ **49.** $\frac{1}{2}$ **51.** $-\frac{1}{9}$ **53.** $\frac{1}{x^6y^{10}}$ **55.** $\frac{15}{a^6b}$ **57.** $\frac{4z^2}{9y^9}$ **59.** $\frac{x^{13}}{4y^{18}}$ **61.** $\frac{3}{4x^3}$ **63.** $-\frac{1}{2x^2}$ **65.** $\frac{1}{y^8}$ **67.** x^5y^5 **69.** $\frac{a^8}{b^9}$ **71.** $-\frac{1}{243a^5b^{10}}$ **73.** $\frac{3a^4}{4b^3}$ **75.** $\frac{16x^2}{y^6}$ **77.** $\frac{ab}{72}$ **79.** $-\frac{32x^{19}}{y^{14}}$ **81.** $\frac{1}{b^{4n}}$ **83.** $-\frac{1}{y^{6n}}$ **85.** y^{n-2} **87.** $\frac{x^{n-5}}{y^6}$ **89.** $\frac{8b^{15}}{3a^{18}}$ **91.** 0 **93.** 4.67×10^{-6} **95.** 1.7×10^{-10} **97.** 2×10^{11} **99.** 0.000000123 **101.** 8,200,000,000,000,000 **103.** 0.039 **105.** 150,000 **107.** 20,800,000 **109.** 0.000000015 **111.** 0.0000000000178 **113.** 11,000,000 **115.** 0.000008 **117.** Greater than **119.** It would take a spaceship 2.6×10^{11} years to cross the galaxy. **121.** The mass of a proton is 1.83664508×10^3 times larger than the mass of an electron. **123.** The mass of the sun is 3.3898305×10^5 times larger than the mass of Earth. **125.** The signals from Earth to Mars traveled $1.0\overline{81} \times 10^7$ mi/min. **127.** The lodgepole pine trees released $1.\overline{6} \times 10^2$ seeds for each surviving seedling.

SECTION 6.2

1. $P(3) = 13$ **3.** $R(2) = 10$ **5.** $f(-1) = -11$ **7.** Polynomial: **a.** −1 **b.** 8 **c.** 2 **9.** Not a polynomial **11.** Not a polynomial **13.** Polynomial: **a.** 3 **b.** π **c.** 5 **15.** Polynomial: **a.** −5 **b.** 2 **c.** 3 **17.** Polynomial: **a.** 14 **b.** 14 **c.** 0 **19.**

21.

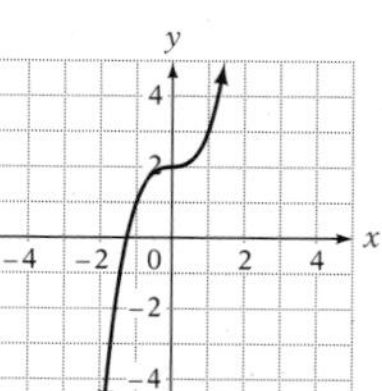

23.

25. $f(c) - g(c) > 0$; $f(c) - g(c) > 0$ **27.** $6x^2 - 6x + 5$ **29.** $-x^2 + 1$ **31.** $5y^2 - 15y + 2$ **33.** $7a^2 - a + 2$ **35.** $3x^2 - 3xy - 2y^2$ **37.** $8x^2 - 2xy + 5y^2$ **39.** $S(x) = 3x^4 - 8x^2 + 2x$; $S(2) = 20$ **41. a.** $k = 8$ **b.** $k = -4$

SECTION 6.3

1. Distributive Property **3.** $2x^2 - 6x$ **5.** $6x^4 - 3x^3$ **7.** $6x^2y - 9xy^2$ **9.** $x^{n+1} + x^n$ **11.** $x^{2n} + x^ny^n$ **13.** $-4b^2 + 10b$ **15.** $-6a^4 + 4a^3 - 6a^2$ **17.** $-3y^4 - 4y^3 + 2y^2$ **19.** $-20x^2 + 15x^3 - 15x^4 - 20x^5$

21. $-2x^4y + 6x^3y^2 - 4x^2y^3$ **23.** $-3x^3 - 12x^2 + 28x$ **25.** $5y^2 - 11y$ **27.** $6y^2 - 31y$ **29.** $9b^5 - 9b^3 + 24b$ **31.** $x^2 + 5x - 14$ **33.** $8y^2 + 2y - 21$ **35.** $4a^2 - 7ac - 15c^2$ **37.** $25x^2 - 70x + 49$ **39.** $8x^2 + 8xy - 30y^2$ **41.** $x^2y^2 + xy - 12$ **43.** $2x^4 - 15x^2 + 25$ **45.** $10x^4 - 15x^2y + 5y^2$ **47.** $x^3 + 2x^2 - 11x + 20$ **49.** $10a^3 - 27a^2b + 26ab^2 - 12b^3$ **51.** $4x^4 - 13x^2 + 16x - 15$ **53.** $4x^5 - 16x^4 + 15x^3 - 4x^2 + 28x - 45$ **55.** $x^4 - 3x^3 - 6x^2 + 29x - 21$ **57.** $2a^3 + 7a^2 - 43a + 42$ **59.** $6x^3 + 5x^2 - 61x - 60$ **61.** $2y^5 - 10y^4 - y^3 - y^2 + 3$ **63.** 5 **65.** $9x^2 - 4$ **67.** $36 - x^2$ **69.** $4a^2 - 9b^2$ **71.** $9a^2b^2 - 16$ **73.** $x^4 - 1$ **75.** $x^2 - 10x + 25$ **77.** $9a^2 + 30ab + 25b^2$ **79.** $x^4 - 6x^2 + 9$ **81.** $4x^4 - 12x^2y^2 + 9y^4$ **83.** $9m^2n^2 - 30mn + 25$ **85.** $-x^2 + 2xy$ **87.** $-4xy$ **89.** False **91.** ft^2 **93.** $(3x^2 + 10x - 8)\text{ ft}^2$ **95.** $(x^2 + 3x)\text{ m}^2$ **97.** $(x^3 + 9x^2 + 27x + 27)\text{ cm}^3$ **99.** $(2x^3)\text{ in}^3$ **101.** $(78.5x^2 + 125.6x + 50.24)\text{ in}^2$ **103. a.** $a^3 - b^3$ **b.** $x^3 + y^3$ **105. a.** $k = 5$ **b.** $k = 1$ **107.** $a^2 + b^2$

SECTION 6.4

1. $x - 2$ **3.** $-x + 2$ **5.** $xy + 2$ **7.** $x^2 + 3x - 5$ **9.** $3b^3 + 4b^2 + 2b$ **11.** $a^4 - 6a^2 + 1$ **13.** $6x^3 + 21x^2 - 15x$ **15.** $x + 8$ **17.** $x^2 + 3x + 6 + \frac{20}{x - 3}$ **19.** $3x + 5 + \frac{3}{2x + 1}$ **21.** $5x + 7 + \frac{2}{2x - 1}$ **23.** $4x^2 + 6x + 9 + \frac{18}{2x - 3}$ **25.** $3x^2 + 1 + \frac{1}{2x^2 - 5}$ **27.** $x^2 - 3x - 10$ **29.** $x^2 - 2x + 1 - \frac{1}{x - 3}$ **31.** $2x^3 - 3x^2 + x - 4$ **33.** $2x + \frac{x + 2}{x^2 + 2x - 1}$ **35.** $x^2 + 4x + 6 + \frac{10x + 8}{x^2 - 2x - 1}$ **37.** $x^2 + 4 + \frac{3}{2x + 1}$ **39.** False **41.** 3 **43.** $2x - 8$ **45.** $3x - 8$ **47.** $3x + 3 - \frac{1}{x - 1}$ **49.** $2x^2 - 3x + 9$ **51.** $4x^2 + 8x + 15 + \frac{12}{x - 2}$ **53.** $2x^2 - 3x + 7 - \frac{8}{x + 4}$ **55.** $3x^3 + 2x^2 + 12x + 19 + \frac{33}{x - 2}$ **57.** $3x^3 - x + 4 - \frac{2}{x + 1}$ **59.** $3x + 1 + \frac{8}{x - 2}$ **61.** $x - 3$ **63.** $P(3) = 8$ **65.** $R(4) = 43$ **67.** $P(-2) = -39$ **69.** $Z(-3) = -60$ **71.** $Q(2) = 31$ **73.** $F(-3) = 178$ **75.** $P(5) = 122$ **77.** $R(-3) = 302$ **79.** $Q(2) = 0$ **81. a.** $a^2 - ab + b^2$ **b.** $x^4 - x^3y + x^2y^2 - xy^3 + y^4$ **c.** $x^5 - x^4y + x^3y^2 - x^2y^3 + xy^4 - y^5$

CHAPTER 6 CONCEPT REVIEW*

1. Writing the terms in descending order before adding helps us to add the like terms of the polynomials. [6.2B]
2. When multiplying the terms $4p^3$ and $7p^6$, we add the exponents 3 and 6 to get $28p^9$. [6.1A]
3. To multiply two binomials, use the FOIL method: Add the products of the First terms, the Outer terms, the Inner terms, and the Last terms. [6.3B]
4. To square a binomial, multiply it times itself. [6.3C]
5. A number written in scientific notation is of the form $a \times 10^n$, where a is a number between 1 and 10 and n is an integer. To write a large number in scientific notation, move the decimal point to the right of the first digit. The exponent on 10 is positive and equal to the number of places the decimal point has been moved. For example, $35{,}000{,}000{,}000{,}000 = 3.5 \times 10^{13}$. To write a very small number in scientific notation, move the decimal point to the right of the first nonzero digit. This is the value of a. The exponent on 10 is negative. The absolute value of the exponent is equal to the number of places the decimal point has been moved. [6.1C]
6. The equation used to check polynomial division is (Quotient × divisor) + remainder = dividend. [6.4B]
7. A binomial is a factor of a polynomial if, when the polynomial is divided by the binomial, there is no remainder. [6.4B]
8. To use synthetic division for division of a polynomial, the divisor must be of the form $x - a$. [6.4C]
9. When using synthetic division with a polynomial that has a missing term, insert a zero for the missing term in the polynomial. [6.4B]
10. To evaluate a polynomial function at a given value of the variable, use synthetic division to divide the polynomial by the value at which the polynomial is being evaluated. The remainder is the value of the polynomial function. [6.4D]

CHAPTER 6 REVIEW EXERCISES

1. $21y^2 + 4y - 1$ [6.2B] **2.** $5x + 4 + \frac{6}{3x - 2}$ [6.4B] **3.** $144x^2y^{10}z^{14}$ [6.1B] **4.** $25y^2 - 70y + 49$ [6.3C] **5.** $\frac{b^6}{a^4}$ [6.1B] **6.** 1 [6.4D] **7.** $4x^2 - 8xy + 5y^2$ [6.2B] **8.** $4b^4 + 12b^2 - 1$ [6.4A] **9.** $-\frac{1}{2a}$ [6.1B]

**Note:* The numbers in brackets following the answers in the Concept Review are a reference to the objective that corresponds to that problem. For example, the reference [1.2A] stands for Section 1.2, Objective A. This notation will be used for all Prep Tests, Concept Reviews, Chapter Reviews, Chapter Tests, and Cumulative Reviews throughout the text.

10. $-6a^3b^6$ [6.1A] **11.** $\frac{2x^3}{3}$ [6.1B] **12.** $4x^2 + 3x - 8 + \frac{50}{x+6}$ [6.4B/6.4C] **13.** -7 [6.2A] **14.** $13y^3 - 12y^2 - 5y - 1$ [6.2B] **15.** $b^2 + 5b + 2 + \frac{7}{b-7}$ [6.4B/6.4C] **16.** $12x^5y^3 + 8x^3y^2 - 28x^2y^4$ [6.3A] **17.** $2ax - 4ay - bx + 2by$ [6.3B] **18.** $8b^2 - 2b - 15$ [6.3B] **19.** $33x^2 + 24x$ [6.3A] **20.** $x^4y^8z^4$ [6.1A] **21.** $16x^2 - 24xy + 9y^2$ [6.3C] **22.** $x^3 + 4x^2 + 16x + 64 + \frac{252}{x-4}$ [6.4B/6.4C] **23.** $2x^2 - 5x - 2$ [6.2B] **24.** $70xy^2z^6$ [6.1B] **25.** $-\frac{x^3}{4y^2z^3}$ [6.1B] **26.** 9.48×10^8 [6.1C] **27.** 2×10^{-6} [6.1C] **28.** 68 [6.4D] **29.** $4x^4 - 2x^2 + 5$ [6.4A] **30.** $2x - 3 - \frac{4}{6x+1}$ [6.4B] **31.** $a^{3n+3} - 5a^{2n+4} + 2a^{2n+3}$ [6.3A] **32.** $x^4 + 3x^3 - 23x^2 - 29x + 6$ [6.3B] **33.** $-8x^3 - 14x^2 + 18x$ [6.3A] **34.** $6y^3 + 17y^2 - 2y - 21$ [6.3B] **35.** $16u^{12}v^{16}$ [6.1A] **36.** $2x^3 + 9x^2 - 3x - 12$ [6.2B] **37.** $2x^2 + 3x - 8$ [6.2B] **38.** $a^2 - 49$ [6.3C] **39.** $100a^{15}b^{13}$ [6.1A] **40.** 14,600,000 [6.1C] **41.** $-108x^{18}$ [6.1A] **42.** $2y - 9$ [6.4B] **43.** $-\frac{1}{16}$ [6.1B] **44.** $10a^2 + 31a - 63$ [6.3B] **45.** $-x + 2 + \frac{1}{x+3}$ [6.4B/6.4C] **46.** 1.27×10^{-7} [6.1C] **47.** $-4y + 8$ [6.4A] **48.** $\frac{c^{10}}{2b^{17}}$ [6.1B] **49.** $6x^3 - 29x^2 + 14x + 24$ [6.3B] **50.** $\frac{x^4y^6}{9}$ [6.1B] **51.** $-54a^{13}b^5c^7$ [6.1A] **52.** $25a^2 - 4b^2$ [6.3C] **53.** 0.00254 [6.1C] **54.** $8a^3b^3 - 4a^2b^4 + 6ab^5$ [6.3A] **55.**

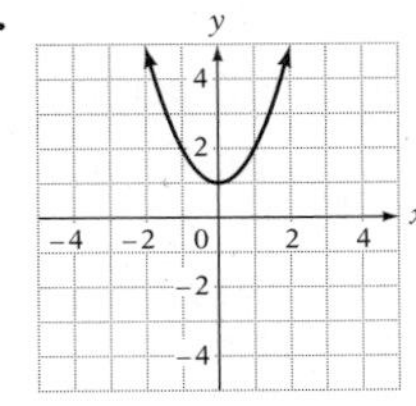

[6.2A]

56. **a.** 3 **b.** 8 **c.** 5 [6.2A] **57.** The mass of the moon is 8.103×10^{19} tons. [6.1D] **58.** The area is $(9x^2 - 12x + 4)$ in^2. [6.3D] **59.** The Great Galaxy of Andromeda is 1.291224×10^{19} mi from Earth. [6.1D] **60.** The area is $(10x^2 - 29x - 21)$ cm^2. [6.3D]

CHAPTER 6 TEST

1. $4x^3 - 6x^2$ [6.3A, How To 1] **2.** -8 [6.4D, How To 6] **3.** $-\frac{4}{x^6}$ [6.1B, Example 4] **4.** $-6x^3y^6$ [6.1A, How To 1] **5.** $x - 1 + \frac{2}{x+1}$ [6.4B, Example 3] **6.** $x^3 - 7x^2 + 17x - 15$ [6.3B, How To 5] **7.** $-8a^6b^3$ [6.1A, How To 3] **8.** $\frac{9y^{10}}{x^{10}}$ [6.1B, Example 5] **9.** $a^2 + 3ab - 10b^2$ [6.3B, Example 4] **10.** -3 [6.2A, How To 1] **11.** $x + 7$ [6.4C, How To 5] **12.** $6y^4 - 9y^3 + 18y^2$ [6.3A, How To 1] **13.** $-4x^4 + 8x^3 - 3x^2 - 14x + 21$ [6.3B, How To 5] **14.** $16y^2 - 9$ [6.3C, How To 8] **15.** $6x^3 + 3x^2 - 2x$ [6.4A, How To 1] **16.** $8ab^4$ [6.1B, Example 4] **17.** $\frac{2b^7}{a^{10}}$ [6.1B, Example 5] **18.** $-5a^3 + 3a^2 - 4a + 3$ [6.2B, How To 6] **19.** $4x^2 - 20x + 25$ [6.3C, How To 9] **20.** $x^2 - 5x + 10 - \frac{23}{x+3}$ [6.4C, Example 6] **21.** $10x^2 - 43xy + 28y^2$ [6.3B, Example 4] **22.** $3x^3 + 6x^2 - 8x + 3$ [6.2B, How To 3] **23.** 3.02×10^{-9} [6.1C, Example 7] **24.** There are 6.048×10^6 s in 10 weeks. [6.1D, Example 10] **25.** The area of the circle is $(\pi x^2 - 10\pi x + 25\pi)$ m^2. [6.3D, Example 10]

CUMULATIVE REVIEW EXERCISES

1. -3 and 3 [1.1A] **2.** -83 [1.1B] **3.** 6 [1.3A] **4.** $-\frac{5}{4}$ [1.4A] **5.** $-50\sqrt{3}$ [1.2F] **6.** The Inverse Property of Addition [1.4B] **7.** $-186x + 8$ [1.4D] **8.** $-\frac{1}{6}$ [2.2A] **9.** $-\frac{11}{4}$ [2.2B] **10.** -1 and $\frac{7}{3}$ [2.6A] **11.** 18 [4.2A] **12.** Yes [4.2A] **13.** $-\frac{1}{6}$ [4.4A] **14.** $y = -\frac{3}{2}x + \frac{1}{2}$ [4.5A] **15.** $y = \frac{2}{3}x + \frac{16}{3}$ [4.6A] **16.** $(-1, -2)$ [5.1B] **17.** $\left(-\frac{9}{7}, \frac{2}{7}, \frac{11}{7}\right)$ [5.2B] **18.** $5x - 3xy$ [1.4B]

19. $4x^3 - 7x + 3$ [6.3B] **20.** 5.01×10^{-6} [6.1C] **21.**

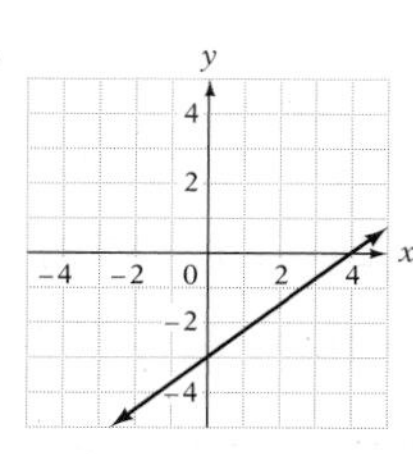

[4.3B] **22.**

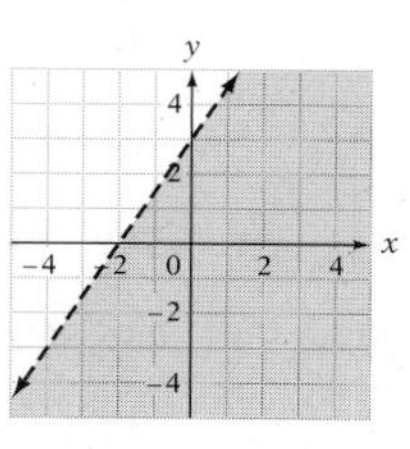

[4.7A]

23.

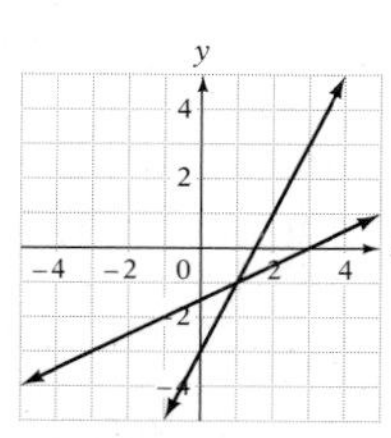

[5.1A] **24.**

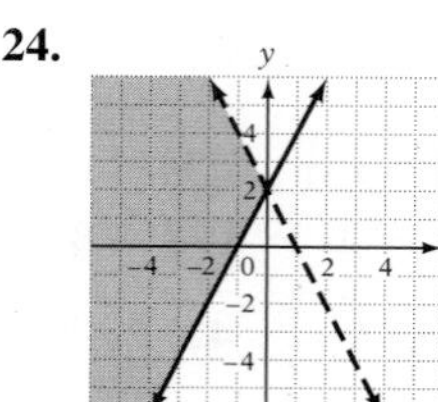

[5.4A] **25.** $\frac{b^5}{a^8}$ [6.1B] **26.** $\frac{y^2}{25x^6}$ [6.1B]

The solution is $(1, -1)$.

27. The two integers are 9 and 15. [2.3A] **28.** 40 oz of pure gold must be used. [2.4A] **29.** The cyclists are traveling at 5 mph and 7.5 mph. [2.4C] **30.** $2000 is invested in the 4% account. [5.1C] **31.** The slope is 50. The slope represents the average speed in miles per hour. [4.4A] **32.** The length is 15 m. The width is 6 m. [3.2A] **33.** The area is $(4x^2 + 12x + 9)$ m². [6.3D]

ANSWERS TO CHAPTER 7 SELECTED EXERCISES

PREP TEST

1. $2 \cdot 3 \cdot 5$ [1.2D] **2.** $-12y + 15$ [1.4D] **3.** $-a + b$ [1.4D] **4.** $-3a + 3b$ [1.4D] **5.** 0 [2.1C] **6.** $-\frac{1}{2}$ [2.2A] **7.** $x^2 - 2x - 24$ [6.3B] **8.** $6x^2 - 11x - 10$ [6.3B] **9.** x^3 [6.1B] **10.** $3x^3y$ [6.1B]

SECTION 7.1

3. $5(a + 1)$ **5.** $8(2 - a^2)$ **7.** $4(2x + 3)$ **9.** $6(5a - 1)$ **11.** $x(7x - 3)$ **13.** $a^2(3 + 5a^3)$ **15.** $y(14y + 11)$ **17.** $2x(x^3 - 2)$ **19.** $2x^2(5x^2 - 6)$ **21.** $4a^5(2a^3 - 1)$ **23.** $xy(xy - 1)$ **25.** $3xy(xy^3 - 2)$ **27.** $xy(x - y^2)$ **29.** $5y(y^2 - 4y + 1)$ **31.** $3y^2(y^2 - 3y - 2)$ **33.** $3y(y^2 - 3y + 8)$ **35.** $a^2(6a^3 - 3a - 2)$ **37.** $ab(2a - 5ab + 7b)$ **39.** $2b(2b^4 + 3b^2 - 6)$ **41.** $x^2(8y^2 - 4y + 1)$ **43. a.** (i), (ii), and (iii) **b.** (iii) **45.** $(a + z)(y + 7)$ **47.** $(a - b)(3r + s)$ **49.** $(m - 7)(t - 7)$ **51.** $(4a + b)(2y - 1)$ **53.** $(x + 2)(x + 2y)$ **55.** $(p - 2)(p - 3r)$ **57.** $(a + 6)(b - 4)$ **59.** $(2z - 1)(z + y)$ **61.** $(2v - 3y)(4v + 7)$ **63.** $(2x - 5)(x - 3y)$ **65.** $(y - 2)(3y - a)$ **67.** $(3x - y)(y + 1)$ **69.** $(3s + t)(t - 2)$ **71.** P doubles.

SECTION 7.2

1. $(x + 1)(x + 2)$ **3.** $(x + 1)(x - 2)$ **5.** $(a + 4)(a - 3)$ **7.** $(a - 1)(a - 2)$ **9.** $(a + 2)(a - 1)$ **11.** $(b - 3)(b - 3)$ **13.** $(b + 8)(b - 1)$ **15.** $(y + 11)(y - 5)$ **17.** $(y - 2)(y - 3)$ **19.** $(z - 5)(z - 9)$ **21.** $(z + 8)(z - 20)$ **23.** $(p + 3)(p + 9)$ **25.** $(x + 10)(x + 10)$ **27.** $(b + 4)(b + 5)$ **29.** $(x + 3)(x - 14)$ **31.** $(b + 4)(b - 5)$ **33.** $(y + 3)(y - 17)$ **35.** $(p + 3)(p - 7)$ **37.** Nonfactorable over the integers **39.** $(x - 5)(x - 15)$ **41.** $(p + 3)(p + 21)$ **43.** $(x + 2)(x + 19)$ **45.** $(x + 9)(x - 4)$ **47.** $(a + 4)(a - 11)$ **49.** $(a - 3)(a - 18)$ **51.** $(z + 21)(z - 7)$ **53.** $(c + 12)(c - 15)$ **55.** $(p + 9)(p + 15)$ **57.** $(c + 2)(c + 9)$ **59.** $(x + 15)(x - 5)$ **61.** $(x + 25)(x - 4)$ **63.** $(b - 4)(b - 18)$ **65.** $(a + 45)(a - 3)$ **67.** $(b - 7)(b - 18)$ **69.** $(z + 12)(z + 12)$ **71.** $(x - 4)(x - 25)$ **73.** $(x + 16)(x - 7)$ **75.** Positive **77.** $3(x + 2)(x + 3)$ **79.** $-(x - 2)(x + 6)$ **81.** $a(b + 8)(b - 1)$ **83.** $x(y + 3)(y + 5)$ **85.** $-2a(a + 1)(a + 2)$ **87.** $4y(y + 6)(y - 3)$ **89.** $2x(x^2 - x + 2)$ **91.** $6(z + 5)(z - 3)$ **93.** $3a(a + 3)(a - 6)$ **95.** $(x + 7y)(x - 3y)$ **97.** $(a - 5b)(a - 10b)$ **99.** $(s + 8t)(s - 6t)$ **101.** Nonfactorable over the integers **103.** $z^2(z + 10)(z - 8)$ **105.** $b^2(b + 2)(b - 5)$ **107.** $3y^2(y + 3)(y + 15)$ **109.** $-x^2(x + 1)(x - 12)$ **111.** $3y(x + 3)(x - 5)$ **113.** $-3x(x - 3)(x - 9)$ **115.** $(x - 3y)(x - 5y)$ **117.** $(a - 6b)(a - 7b)$ **119.** $(y + z)(y + 7z)$

121. $3y(x + 21)(x - 1)$ **123.** $3x(x + 4)(x - 3)$ **125.** $2(t - 5s)(t - 7s)$ **127.** $3(a + 3b)(a - 11b)$ **129.** $5x(x + 2y)(x + 4y)$ **131. a.** Yes **b.** No **133.** $-19, -11, -9, 9, 11, 19$ **135.** 3, 4 **137.** 5, 8, 9 **139.** 2 **141.** An infinite number

SECTION 7.3

1. $(x + 1)(2x + 1)$ **3.** $(y + 3)(2y + 1)$ **5.** $(a - 1)(2a - 1)$ **7.** $(b - 5)(2b - 1)$ **9.** $(x + 1)(2x - 1)$ **11.** $(x - 3)(2x + 1)$ **13.** $(t + 2)(2t - 5)$ **15.** $(p - 5)(3p - 1)$ **17.** $(3y - 1)(4y - 1)$ **19.** Nonfactorable over the integers **21.** $(2t - 1)(3t - 4)$ **23.** $(x + 4)(8x + 1)$ **25.** Nonfactorable over the integers **27.** $(3y + 1)(4y + 5)$ **29.** $(a + 7)(7a - 2)$ **31.** $(b - 4)(3b - 4)$ **33.** $(z - 14)(2z + 1)$ **35.** $(p + 8)(3p - 2)$ **37.** $2(x + 1)(2x + 1)$ **39.** $5(y - 1)(3y - 7)$ **41.** $x(x - 5)(2x - 1)$ **43.** $b(a - 4)(3a - 4)$ **45.** Nonfactorable over the integers **47.** $-3x(x + 4)(x - 3)$ **49.** $4(4y - 1)(5y - 1)$ **51.** $z(2z + 3)(4z + 1)$ **53.** $y(2x - 5)(3x + 2)$ **55.** $5(t + 2)(2t - 5)$ **57.** $p(p - 5)(3p - 1)$ **59.** $2(z + 4)(13z - 3)$ **61.** $2y(y - 4)(5y - 2)$ **63.** $yz(z + 2)(4z - 3)$ **65.** $3a(2a + 3)(7a - 3)$ **67.** $y(3x - 5y)(3x - 5y)$ **69.** $xy(3x - 4y)(3x - 4y)$ **71.** Odd **73.** $(2x - 3)(3x - 4)$ **75.** $(b + 7)(5b - 2)$ **77.** $(3a + 8)(2a - 3)$ **79.** $(z + 2)(4z + 3)$ **81.** $(2p + 5)(11p - 2)$ **83.** $(y + 1)(8y + 9)$ **85.** $(6t - 5)(3t + 1)$ **87.** $(b + 12)(6b - 1)$ **89.** $(3x + 2)(3x + 2)$ **91.** $(2b - 3)(3b - 2)$ **93.** $(3b + 5)(11b - 7)$ **95.** $(3y - 4)(6y - 5)$ **97.** $(3a + 7)(5a - 3)$ **99.** $(2y - 5)(4y - 3)$ **101.** $(2z + 3)(4z - 5)$ **103.** Nonfactorable over the integers **105.** $(2z - 5)(5z - 2)$ **107.** $(6z + 5)(6z + 7)$ **109.** $(x + y)(3x - 2y)$ **111.** $(a + 2b)(3a - b)$ **113.** $(y - 2z)(4y - 3z)$ **115.** $-(z - 7)(z + 4)$ **117.** $-(x - 1)(x + 8)$ **119.** $3(x + 5)(3x - 4)$ **121.** $4(2x - 3)(3x - 2)$ **123.** $a^2(5a + 2)(7a - 1)$ **125.** $5(b - 7)(3b - 2)$ **127.** $(x - 7y)(3x - 5y)$ **129.** $3(8y - 1)(9y + 1)$ **131.** $-(x - 1)(x + 21)$ **133.** Two positive **135.** Two negative **139.** $x(x - 1)$ **141.** $(2y + 1)(y + 3)$ **143.** $(4y - 3)(y - 3)$

SECTION 7.4

1. 4; $25x^6$; $100x^4y^4$ **3.** $4z^4$ **5.** $9a^2b^3$ **7.** (iv) **9.** $(x + 4)(x - 4)$ **11.** $(2x + 1)(2x - 1)$ **13.** $(4x + 11)(4x - 11)$ **15.** $(1 + 3a)(1 - 3a)$ **17.** $(xy + 10)(xy - 10)$ **19.** Nonfactorable over the integers **21.** $(5 + ab)(5 - ab)$ **23.** $(a^n + 1)(a^n - 1)$ **25.** $(x - 6)^2$ **27.** $(b - 1)^2$ **29.** $(4x - 5)^2$ **31.** Nonfactorable over the integers **33.** Nonfactorable over the integers **35.** $(x + 3y)^2$ **37.** $(5a - 4b)^2$ **39.** $(x^n + 3)^2$ **41.** $(x - 7)(x - 1)$ **43.** $(x - y - a - b)(x - y + a + b)$ **45.** 8; x^9; $27c^{15}d^{18}$ **47.** $2x^3$ **49.** $4a^2b^6$ **51.** Yes **53.** Yes **55.** $(x - 3)(x^2 + 3x + 9)$ **57.** $(2x - 1)(4x^2 + 2x + 1)$ **59.** $(x - y)(x^2 + xy + y^2)$ **61.** $(m + n)(m^2 - mn + n^2)$ **63.** $(4x + 1)(16x^2 - 4x + 1)$ **65.** $(3x - 2y)(9x^2 + 6xy + 4y^2)$ **67.** $(xy + 4)(x^2y^2 - 4xy + 16)$ **69.** Nonfactorable over the integers **71.** Nonfactorable over the integers **73.** $(a - 2b)(a^2 - ab + b^2)$ **75.** $(x^{2n} + y^n)(x^{4n} - x^{2n}y^n + y^{2n})$ **77.** $(x^n + 2)(x^{2n} - 2x^n + 4)$ **79.** No, polynomials cannot have square roots as variable terms. **81.** $(xy - 5)(xy - 3)$ **83.** $(xy - 12)(xy - 5)$ **85.** $(x^2 - 3)(x^2 - 6)$ **87.** $(b^2 + 5)(b^2 - 18)$ **89.** $(x^2y^2 - 2)(x^2y^2 - 6)$ **91.** $(x^n + 1)(x^n + 2)$ **93.** $(3xy - 5)(xy - 3)$ **95.** $(2ab - 3)(3ab - 7)$ **97.** $(x^2 + 1)(2x^2 - 15)$ **99.** $(2x^n - 1)(x^n - 3)$ **101.** $(2a^n + 5)(3a^n + 2)$ **103.** $3(2x - 3)^2$ **105.** $a(3a - 1)(9a^2 + 3a + 1)$ **107.** $5(2x + 1)(2x - 1)$ **109.** $y^3(y + 11)(y - 5)$ **111.** $(4x^2 + 9)(2x + 3)(2x - 3)$ **113.** $2a(2 - a)(4 + 2a + a^2)$ **115.** $b^3(ab - 1)(a^2b^2 + ab + 1)$ **117.** $2x^2(2x - 5)^2$ **119.** $(x^2 + y^2)(x + y)(x - y)$ **121.** $(x^2 + y^2)(x^4 - x^2y^2 + y^4)$ **123.** Nonfactorable over the integers **125.** $2a(2a - 1)(4a^2 + 2a + 1)$ **127.** $a^2b^2(a + 4b)(a - 12b)$ **129.** $(x - 2)^2(x + 2)$ **131.** $(x - 4)(x + 4)(2x + 1)$ **133.** $(2x + 1)(2x - 1)(x + y)(x - y)$ **135.** $b^n(b + 2)(3b - 2)$ **137.** 8

SECTION 7.5

3. $-3, -2$ **5.** 7, 3 **7.** 0, 5 **9.** 0, 9 **11.** $0, -\frac{3}{2}$ **13.** $0, \frac{2}{3}$ **15.** $-2, 5$ **17.** $-9, 9$ **19.** $-\frac{7}{2}, \frac{7}{2}$ **21.** $-\frac{1}{3}, \frac{1}{3}$ **23.** $-2, -4$ **25.** $-7, 2$ **27.** $-\frac{1}{2}, 5$ **29.** $-\frac{1}{3}, -\frac{1}{2}$ **31.** 0, 3 **33.** 0, 7 **35.** $-1, -4$ **37.** 2, 3 **39.** $\frac{1}{2}, -4$ **41.** $\frac{1}{3}, 4$ **43.** 3, 9 **45.** $-2, 9$ **47.** $-1, -2$ **49.** $-9, 5$ **51.** $-7, 4$ **53.** $-2, -3$ **55.** $-8, 9$ **57.** 1, 4 **59.** $-5, 2$ **61.** Less than **63.** The number is 6. **65.** The numbers are 2 and 4. **67.** (ii) **69.** The numbers are 4 and 5. **71.** The numbers are 3 and 7. **73.** There will be 12 consecutive numbers. **75.** There are 6 teams in the league. **77.** The object will hit the ground 3 s later. **79.** The golf ball will return to the ground 3.75 s later. **81.** The length is 15 in. The width is 5 in. **83.** The height of the triangle is 14 m. **85.** The width of the lane is 16 ft. **87.** The dimensions of the type area are 4 in. by 7 in. **89.** The width of the iris is 3 mm. **91.** 2, 32 **93.** −9, −1 **95.** 1, 18

CHAPTER 7 CONCEPT REVIEW*

1. In factoring a polynomial, we always first check to see if the terms of the polynomial have a common factor. If they do, we factor out the GCF of the terms. [7.4D]
2. When factoring, the terms of a polynomial do not have to be like terms. If they were like terms, we would combine them, and the result would be a monomial. [7.1A]
3. To check the answer to a factorization, multiply the factors. The product must be the original polynomial. [7.2A]
4. When factoring by grouping, after we group the first two terms and group the last two terms, we factor the GCF from each group. [7.1B]
5. A polynomial of the form $x^2 + bx + c$ or $ax^2 + bx + c$ is nonfactorable over the integers when it does not factor into the product of two binomials that have integer coefficients and constants. [7.2A]
6. To factor a polynomial completely means to write the polynomial as a product of factors that are nonfactorable over the integers. [7.2B]
7. When factoring a polynomial of the form $x^2 + bx + c$, we begin by finding the possible factors of c because we are looking for two numbers whose product is c and whose sum is b. [7.2A]
8. Trial factors can be used when factoring a trinomial of the form $ax^2 + bx + c$. We use the factors of a and the factors of c to write all the possible binomial pairs that, when multiplied, have ax^2 and c in their product. We test each pair of trial factors to find which one has bx as the middle term of the product when the factors are multiplied. [7.3A]
9. The middle term of a trinomial of the form $x^2 + bx + c$ or $ax^2 + bx + c$ is bx. [7.3A]
10. The binomial factors of the difference of two squares $a^2 - b^2$ are $a + b$ and $a - b$. [7.4A]
11. The square of a binomial is a perfect-square trinomial. For example, $(2x - 5)^2$ is the square of a binomial. $(2x - 5)(2x - 5) = 4x^2 - 20x + 25$, so $4x^2 - 20x + 25$ is a perfect-square trinomial. [7.4A]
12. To solve an equation by factoring, the equation must be set equal to zero in order to use the Principle of Zero Products, which states that if the product of two factors is zero, then at least one of the factors must be zero. [7.5A]

CHAPTER 7 REVIEW EXERCISES

1. $(b - 10)(b - 3)$ [7.2A] **2.** $(x - 3)(4x + 5)$ [7.1B] **3.** Nonfactorable over the integers [7.3A] **4.** $(7x^2y^2 + 3)(3x^2y^2 + 2)$ [7.4C] **5.** $7y^3(2y^6 - 7y^3 + 1)$ [7.1A] **6.** $(y + 9)(y - 4)$ [7.2A] **7.** $(2x - 7)(3x - 4)$ [7.3A] **8.** $3ab(4a + b)$ [7.1A] **9.** $(a^3 + 10)(a^3 - 10)$ [7.4A] **10.** $n^2(n - 3)(n + 1)$ [7.2B] **11.** $(6y - 1)(2y + 3)$ [7.3A] **12.** $2b(2b - 7)(3b - 4)$ [7.4D] **13.** $(3y^2 + 5z)(3y^2 - 5z)$ [7.4A] **14.** $(c + 6)(c + 2)$ [7.2A] **15.** $(6a - 5)(3a + 2)$ [7.3B] **16.** $\frac{1}{4}, -7$ [7.5A] **17.** $4x(x - 6)(x + 1)$ [7.2B] **18.** $(4a - 3b)(16a^2 + 12ab + 9b^2)$ [7.4B] **19.** $(a - 12)(2a + 5)$ [7.3B] **20.** 7, −3 [7.5A] **21.** $(3a - 5b)(7x + 2y)$ [7.1B] **22.** $(6x^4 - 1)(6x^4 - 5)$ [7.4C] **23.** $(2x + 5)(5x + 2y)$ [7.1B] **24.** $5(x - 3)(x + 2)$ [7.2B] **25.** $3(x + 6)^2$ [7.4D] **26.** $(x - 5)(3x - 2)$ [7.3B] **27.** The length is 100 yd. The width is 60 yd. [7.5B] **28.** The distance is 20 ft. [7.5B] **29.** The width of the frame is 1.5 in. [7.5B] **30.** A side of the original garden plot was 20 ft. [7.5B]

CHAPTER 7 TEST

1. $(b + 6)(a - 3)$ [7.1B, How To 5] **2.** $2y^2(y - 8)(y + 1)$ [7.2B, Example 3] **3.** $4(x + 4)(2x - 3)$ [7.3B, Example 4] **4.** $(2x + 1)(3x + 8)$ [7.3A, How To 1] **5.** $(a - 16)(a - 3)$ [7.2A, How To 1] **6.** $2x(3x^2 - 4x + 5)$ [7.1A, How To 2] **7.** $(x + 5)(x - 3)$ [7.2A, Example 2] **8.** $-\frac{1}{2}, \frac{1}{2}$ [7.5A, You Try It 2] **9.** $5(x^2 - 9x - 3)$ [7.1A, How To 2] **10.** $(p + 6)^2$ [7.4A, How To 2] **11.** 3, 5 [7.5A, Example 3] **12.** $3(x + 2y)^2$ [7.4D, Example 8]

**Note:* The numbers in brackets following the answers in the Concept Review are a reference to the objective that corresponds to that problem. For example, the reference [1.2A] stands for Section 1.2, Objective A. This notation will be used for all Prep Tests, Concept Reviews, Chapter Reviews, Chapter Tests, and Cumulative Reviews throughout the text.

13. $(b + 4)(b - 4)$ [7.4A, Example 1] **14.** $3y^2(2x + 1)(x + 1)$ [7.3B, Example 4]
15. $(3x - 2)(9x^2 + 6x + 4)$ [7.4B, How To 5] **16.** $(2a^2 - 5)(3a^2 + 1)$ [7.4C, You Try It 7]
17. $(p + 1)(x - 1)$ [7.1B, How To 4] **18.** $3(a + 5)(a - 5)$ [7.4D, Example 8]
19. Nonfactorable over the integers [7.3B, How To 6] **20.** $(x - 12)(x + 3)$ [7.2A, How To 2]
21. $(2a - 3b)^2$ [7.4A, How To 2] **22.** $(2x + 7y)(2x - 7y)$ [7.4A, How To 1] **23.** $\frac{3}{2}, -7$ [7.5A, How To 1]
24. The two numbers are 7 and 3. [7.5B, Example 4] **25.** The length is 15 cm. The width is 6 cm. [7.5B, You Try It 5]

CUMULATIVE REVIEW EXERCISES

1. 7 [1.1C] **2.** 4 [1.3A] **3.** -7 [1.4A] **4.** $15x^2$ [1.4C] **5.** 12 [1.4D] **6.** $\frac{2}{3}$ [2.1C] **7.** $\frac{7}{4}$ [2.2B]
8. 3 [2.2C] **9.** 45 [2.1D] **10.** 1 [4.2A] **11.** [4.3A] **12.**

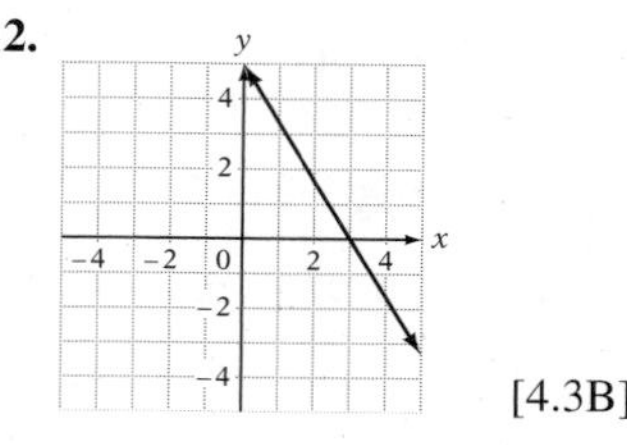

[4.3B]

13. $y = \frac{2}{3}x + 6$ [4.5A] **14.** $(1, 6)$ [5.1B] **15.** $(-1, -2)$ [5.2A] **16.** $9a^6b^4$ [6.1A]
17. $x^3 - 3x^2 - 6x + 8$ [6.3B] **18.** $4x + 8 + \frac{21}{2x - 3}$ [6.4B] **19.** $\frac{y^6}{x^8}$ [6.1B] **20.** $(a - b)(3 - x)$ [7.1B]
21. $5xy^2(3 - 4y^2)$ [7.1A] **22.** $(x - 7y)(x + 2y)$ [7.2B] **23.** $\frac{2}{3}$ and -7 [7.5A] **24.** $\frac{5}{2}$ and 4 [7.5A]
25. The third angle measures 59°. [3.1C] **26.** The width is 15 ft. [3.2A] **27.** The pieces are 4 ft long and 6 ft long. [2.3B]
28. \$8000 is invested at 4%. [5.1C] **29.** The distance to the resort is 168 mi. [2.4C] **30.** The length of the base of the triangle is 12 in. [7.5B]

ANSWERS TO CHAPTER 8 SELECTED EXERCISES

PREP TEST

1. 50 [1.2C] **2.** $-\frac{1}{6}$ [1.2D] **3.** $-\frac{3}{2}$ [1.2D] **4.** $\frac{1}{24}$ [1.2C] **5.** $\frac{5}{24}$ [1.2C] **6.** $\frac{1}{3}$ [1.4A] **7.** -2 [2.2C]
8. $\frac{10}{7}$ [2.2C] **9.** The rate of the first plane is 110 mph, and the rate of the second plane is 130 mph. [2.4C]

SECTION 8.1

3. $\frac{3}{4x}$ **5.** $\frac{1}{x + 3}$ **7.** -1 **9.** $\frac{2}{3y}$ **11.** $-\frac{3}{4x}$ **13.** $\frac{a}{b}$ **15.** $-\frac{2}{x}$ **17.** $\frac{y - 2}{y - 3}$ **19.** $\frac{x + 5}{x + 4}$ **21.** $\frac{x + 4}{x - 3}$
23. $-\frac{x + 2}{x + 5}$ **25.** $\frac{2(x + 2)}{x + 3}$ **27.** $\frac{2x - 1}{2x + 3}$ **29.** $-\frac{x + 7}{x + 6}$ **31.** $\frac{2}{3xy}$ **33.** $\frac{8xy^2ab}{3}$ **35.** $\frac{2}{9}$ **37.** $\frac{y^2}{x}$ **39.** $\frac{y(x + 4)}{x(x + 1)}$
41. $\frac{x^3(x - 7)}{y^2(x - 4)}$ **43.** $-\frac{y}{x}$ **45.** $\frac{x + 3}{x + 1}$ **47.** $\frac{x - 5}{x + 3}$ **49.** $-\frac{x + 3}{x + 5}$ **51.** $-\frac{x + 3}{x - 12}$ **53.** $\frac{x + 2}{x + 4}$ **55.** 1 **57.** 1
61. $\frac{7a^3y^2}{40bx}$ **63.** $\frac{4}{3}$ **65.** $\frac{3a}{2}$ **67.** $\frac{x^2(x + 4)}{y^2(x + 2)}$ **69.** $\frac{x(x - 2)}{y(x - 6)}$ **71.** $-\frac{3by}{ax}$ **73.** $\frac{(x + 6)(x - 3)}{(x + 7)(x - 6)}$ **75.** 1 **77.** $-\frac{x + 8}{x - 4}$
79. $\frac{2n + 1}{2n - 3}$ **81.** Yes **83.** No **85.** $\frac{4}{25}$

SECTION 8.2

1. a. Six **b.** Four **3.** $\frac{9y^3}{12x^2y^4}, \frac{17x}{12x^2y^4}$ **5.** $\frac{2x^2 - 4x}{6x^2(x - 2)}, \frac{3x - 6}{6x^2(x - 2)}$ **7.** $\frac{3x - 1}{2x(x - 5)}, -\frac{6x^3 - 30x^2}{2x(x - 5)}$ **9.** $\frac{6x^2 + 9x}{(2x + 3)(2x - 3)}, \frac{10x^2 - 15x}{(2x + 3)(2x - 3)}$
11. $\frac{2x}{(x + 3)(x - 3)}, \frac{x^2 + 4x + 3}{(x + 3)(x - 3)}$ **13.** $\frac{6}{6(x + 2y)(x - 2y)}, \frac{5x + 10y}{6(x + 2y)(x - 2y)}$ **15.** $\frac{3x^2 - 3x}{(x + 1)(x - 1)^2}, \frac{5x^2 + 5x}{(x + 1)(x - 1)^2}$
17. $-\frac{x - 3}{(x - 2)(x^2 + 2x + 4)}, \frac{2x - 4}{(x - 2)(x^2 + 2x + 4)}$ **19.** $\frac{2x^2 + 6x}{(x - 1)(x + 3)^2}, -\frac{x^2 - x}{(x - 1)(x + 3)^2}$
21. $-\frac{12x^2 - 8x}{(2x - 3)(2x - 5)(3x - 2)}, \frac{6x^2 - 9x}{(2x - 3)(2x - 5)(3x - 2)}$ **23.** $\frac{5}{(3x - 4)(2x - 3)}, -\frac{4x^2 - 6x}{(3x - 4)(2x - 3)}, \frac{3x^2 - x - 4}{(3x - 4)(2x - 3)}$
25. $\frac{2x^2 + 10x}{(x + 5)(x - 3)}, -\frac{2x - 6}{(x + 5)(x - 3)}, -\frac{x - 1}{(x + 5)(x - 3)}$ **27.** True **29.** $\frac{1}{2x^2}$ **31.** $\frac{1}{x + 2}$ **33.** $\frac{12ab - 9b + 8a}{30a^2b^2}$ **35.** $\frac{5 - 16b + 12a}{40ab}$

37. $\frac{7}{12x}$ **39.** $\frac{2xy - 8x + 3y}{10x^2y^2}$ **41.** $-\frac{a(2a - 13)}{(a + 1)(a - 2)}$ **43.** $\frac{5x^2 - 6x + 10}{(5x - 2)(2x - 5)}$ **45.** $\frac{a}{b(a - b)}$ **47.** $\frac{a^2 + 18a - 9}{a(a - 3)}$ **49.** $\frac{17x^2 + 20x - 25}{x(6x - 5)}$ **51.** $\frac{6}{(x - 3)^2(x + 3)}$ **53.** $-\frac{2(x - 1)}{(x + 2)^2}$ **55.** $-\frac{5x^2 - 17x + 8}{(x + 4)(x - 2)}$ **57.** $\frac{3x^n + 2}{(x^n + 1)(x^n - 1)}$ **59.** 1 **61.** $\frac{x^2 - 52x + 160}{4(x + 3)(x - 3)}$ **63.** $\frac{3x - 1}{4x + 1}$ **65.** $\frac{2(5x - 3)}{(x + 4)(x - 3)(x + 3)}$ **67.** $\frac{x - 2}{x + 3}$ **69.** 1 **71.** $\frac{x + 1}{2x - 1}$ **73.** $\frac{1}{2x - 1}$ **75.** $\frac{1}{x^2 + 4}$ **77.** $\frac{3 - a}{3a}$ **79.** $-\frac{2x^2 + 5x - 2}{(x + 2)(x + 1)}$ **81.** $\frac{b - a}{b + 2a}$ **83.** $\frac{2}{x + 2}$ **85. a.** $f(4) = \frac{2}{3}$; $g(4) = 4$; $S(4) = 4\frac{2}{3}$ **b.** Yes

SECTION 8.3

3. $\frac{5}{23}$ **5.** $\frac{2}{5}$ **7.** $\frac{x}{x - 1}$ **9.** $-\frac{a}{a + 2}$ **11.** $-\frac{1}{x(x + h)}$ **13.** $\frac{(x - 1)(x + 1)}{x^2 + 1}$ **15.** $-\frac{a - 1}{a + 1}$ **17.** $\frac{2}{5}$ **19.** $-\frac{1}{2}$ **21.** $\frac{x + 1}{x - 4}$ **23.** $\frac{x - 2}{x - 3}$ **25.** $\frac{x - 3}{x + 4}$ **27.** $\frac{x + 2}{x + 5}$ **29.** $\frac{x + 2}{x - 1}$ **31.** $-\frac{x + 4}{2x + 3}$ **33.** $\frac{x + y}{x - y}$ **35.** $\frac{x^2 - x - 1}{x^2 + x + 1}$ **37.** $-\frac{2(a + 1)}{7a - 4}$ **39.** $-\frac{2x}{x^2 + 1}$ **41.** $\frac{a(a^2 + a + 1)}{a^2 + 1}$ **43.** $-\frac{a^2}{1 - 2a}$ **45.** $-\frac{3x + 2}{x - 2}$ **47.** $\frac{a - 1}{a}$ **49.** $\frac{3}{10}$ **51.** $\frac{2}{x + 2}$

SECTION 8.4

1. $-1, 2$ **3.** 0, 9 **5.** 1 **7.** 3 **9.** 2 **11.** 2 **13.** $\frac{2}{3}$ **15.** 4 **17.** -3 **19.** $\frac{3}{4}$ **21.** 7 **23.** -7 **25.** -1 **27.** -1 **29.** No solution **31.** $2, -6$ **33.** $-\frac{2}{3}, 5$ **35.** $-1, 6$ **39.** 1 **41.** $0, -\frac{5}{2}$ **43.** -3

SECTION 8.5

1. 9 **3.** 12 **5.** 7 **7.** 6 **9.** 1 **11.** -6 **13.** 4 **15.** $-\frac{2}{3}$ **17.** Fifty million Americans do not have health insurance. **19.** 20,000 voters voted in favor of the amendment. **21.** The amount of sugar required is 6 c. **23.** There are approximately 800 fish in the lake. **25.** The person is 64 in. tall. **27.** She will be able to deduct $1111 on her tax return. **29.** The length of a claw is approximately 1.28 in. **31.** The length of side *AC* is 6.7 cm. **33.** The height is 2.9 m. **35.** The perimeter is 22.5 ft. **37.** The area is 48 m^2. **39.** True **41.** The length of *BC* is 6.25 cm. **43.** The length of *QO* is 10 ft. **45.** The distance across the river is 35 m. **47.** The first person won $1.25 million.

SECTION 8.6

1. $y = -3x + 10$ **3.** $y = 4x - 3$ **5.** $y = -\frac{3}{2}x + 3$ **7.** $y = \frac{2}{5}x - 2$ **9.** $y = -\frac{2}{7}x + 2$ **11.** $y = -\frac{1}{3}x + 2$ **13.** $y = 3x + 8$ **15.** $y = -\frac{2}{3}x - 3$ **17.** $x = -6y + 10$ **19.** $x = \frac{1}{2}y + 3$ **21.** $x = -\frac{3}{4}y + 3$ **23.** $x = 4y + 3$ **25.** $t = \frac{d}{r}$ **27.** $T = \frac{PV}{nR}$ **29.** $l = \frac{P - 2w}{2}$ **31.** $b_1 = \frac{2A - hb_2}{h}$ **33.** $h = \frac{3V}{A}$ **35.** $S = C - Rt$ **37.** $P = \frac{A}{1 + rt}$ **39.** $w = \frac{A}{S + 1}$ **41. a.** $S = \frac{F + BV}{B}$ **b.** The required selling price is $180. **c.** The required selling price is $75.

SECTION 8.7

3. *t* is less than *k*. **5.** With both skiploaders working together, it would take 3 h to remove the earth. **7.** It would take the new machine 12 h to complete the task. **9.** It would take 30 min to print the first edition with both presses operating. **11.** Working alone, the apprentice could construct the wall in 15 h. **13.** It will take the second technician 3 h to complete the wiring. **15.** Working alone, it would have taken one of the welders 40 h to complete the welds. **17.** It would have taken one machine 28 h to fill the boxes. **19.** Zachary picked $\frac{m}{n}$ of the row of peas. Eli picked $\frac{n - m}{n}$ of the row of peas. **21.** Time $= \frac{\text{distance}}{\text{rate}}$; Rate $= \frac{\text{distance}}{\text{time}}$ **23.** The rate of the jet plane was 600 mph. **25.** The rate of the freight train is 30 mph. The rate of the express train is 50 mph. **27.** Camille's walking speed is 4 mph. **29.** The rate of the car is 48 mph. **31.** The hiker walked the first 9 mi at a rate of 3 mph. **33.** The rate of the jet stream is 50 mph. **35.** The rate of the current is 5 mph. **39.** The bus usually travels 60 mph.

SECTION 8.8

1. The profit is \$80,000. **3.** The pressure is 6.75 lb/in^2. **5.** In 10 s, the object will fall 1600 ft. **7.** It takes five computers 100 s to solve the problem. **9.** The load on the sail is 360 lb. **11.** The agency would receive a commission of \$13,800. **13.** The current is 7.5 amps. **15.** The intensity is 48 foot-candles when the distance is 5 ft. **17.** y is doubled. **19.** inversely **21.** inversely

CHAPTER 8 CONCEPT REVIEW*

1. A rational expression is in simplest form when the numerator and denominator have no common factors. [8.1A]

2. To divide two rational expressions, change the division to a multiplication and change the divisor to its reciprocal. Then multiply. [8.1C]

3. To find the LCM of two polynomials, first factor each polynomial completely. The LCM is the product of each factor the greatest number of times it occurs in any one factorization. [8.2A]

4. When subtracting two rational expressions, both expressions must have the same denominator before subtraction can take place. [8.2B]

5. To add rational expressions:

1. Find the LCM of the denominators.

2. Write each fraction as an equivalent fraction using the LCM as the denominator.

3. Add the numerators and place the result over the common denominator.

4. Write the answer in simplest form. [8.2B]

6. To simplify a complex fraction by Method 1:

1. Determine the LCM of the denominators of the fractions in the numerator and denominator of the complex fraction.

2. Multiply the numerator and denominator of the complex fraction by the LCM.

3. Simplify.

To simplify a complex fraction by Method 2:

1. Simplify the numerator to a single fraction and simplify the denominator to a single fraction.

2. Using the definition for dividing fractions, multiply the numerator by the reciprocal of the denominator.

3. Simplify. [8.3A]

7. When solving an equation that contains fractions, we first clear the denominators in order to rewrite the equation without any fractions. [8.4A]

8. If the units in a comparison are different, the comparison is a rate. [8.5A]

9. The ratios of corresponding sides of similar triangles are equal. Therefore, we can write a proportion by setting one ratio of corresponding sides equal to the ratio of another pair of corresponding sides. If there is one unknown side in the proportion, we can solve the proportion for the unknown side. [8.5C]

10. When solving a literal equation for a particular variable, the goal is to rewrite the equation so that the variable being solved for is on one side of the equation and all numbers and other variables are on the other side. [8.6A]

11. If a job is completed in x hours, the rate of work is $\frac{1}{x}$ of the job each hour. [8.7A]

CHAPTER 8 REVIEW EXERCISES

1. $\frac{b^3y}{10ax}$ [8.1C] **2.** $\frac{7x+22}{60x}$ [8.2B] **3.** $\frac{x-2}{3x-10}$ [8.3A] **4.** $-\frac{x+6}{x+3}$ [8.1A] **5.** $\frac{2x^4}{3y^7}$ [8.1A] **6.** 62 [8.5A] **7.** $\frac{(3y-2)^2}{(y-1)(y-2)}$ [8.1C] **8.** $\frac{by^3}{6ax^2}$ [8.1B] **9.** $\frac{x}{x-7}$ [8.3A] **10.** $\frac{3x^2-x}{(2x+3)(6x-1)(3x-1)}, \frac{24x^3-4x^2}{(2x+3)(6x-1)(3x-1)}$ [8.2A] **11.** $a = \frac{T-2bc}{2b+2c}$ [8.6A] **12.** 2 [8.4A] **13.** $c = \frac{100m}{i}$ [8.6A] **14.** The equation has no solution. [8.4A] **15.** $\frac{1}{x^2}$ [8.1C] **16.** $\frac{2y-3}{5y-7}$ [8.2B] **17.** $\frac{1}{x+3}$ [8.2B] **18.** $(5x-3)(2x-1)(4x-1)$ [8.2A] **19.** $y = -\frac{4}{9}x + 2$ [8.6A] **20.** $\frac{8x+5}{3x-4}$ [8.1B] **21.** 5 [8.4A] **22.** $\frac{3x-1}{x-5}$ [8.2B] **23.** 10 [8.5A]

*Note: The numbers in brackets following the answers in the Concept Review are a reference to the objective that corresponds to that problem. For example, the reference [1.2A] stands for Section 1.2, Objective A. This notation will be used for all Prep Tests, Concept Reviews, Chapter Reviews, Chapter Tests, and Cumulative Reviews throughout the text.

24. 12 [8.5A] **25.** The perimeter of triangle *ABC* is 24 in. [8.5C] **26.** It would take 6 h to fill the pool using both hoses. [8.7A] **27.** The rate of the car is 45 mph. [8.7B] **28.** The rate of the wind is 20 mph. [8.7B] **29.** The pitcher's ERA is 1.35. [8.5B] **30.** The current is 2 amps. [8.8A]

CHAPTER 8 TEST

1. $\frac{2x^3}{3y^3}$ [8.1A, Example 1] **2.** $-\frac{x+5}{x+1}$ [8.1A, Example 3] **3.** $\frac{x+1}{x^3(x-2)}$ [8.1B, How To 3] **4.** $\frac{(x-5)(2x-1)}{(x+3)(2x+5)}$ [8.1B, Example 5] **5.** $\frac{x+5}{x+4}$ [8.1C, Example 7] **6.** $3(2x-1)(x+1)$ [8.2A, How To 1] **7.** $\frac{3x+6}{x(x+2)(x-2)}; \frac{x^2}{x(x+2)(x-2)}$ [8.2A, How To 3] **8.** $\frac{2}{x+5}$ [8.2B, How To 5] **9.** $\frac{5}{(2x-1)(3x+1)}$ [8.2B, Example 6] **10.** $\frac{x^2-4x+5}{(x-2)(x+3)}$ [8.2B, Example 6] **11.** $\frac{x-3}{x-2}$ [8.3A, How To 1] **12.** 2 [8.4A, Example 2] **13.** The equation has no solution. [8.4A, Example 2] **14.** -1 [8.5A, You Try It 1] **15.** The area is 64.8 m^2. [8.5C, How To 3] **16.** $t = \frac{d-s}{r}$ [8.6A, How To 1] **17.** 14 rolls of wallpaper are needed. [8.5B, You Try It 2] **18.** It would take 10 min with both landscapers working. [8.7A, How To 1] **19.** The rate of the cyclist is 10 mph. [8.7B, Example 2] **20.** The resistance is 0.4 ohm. [8.8A, Example 3]

CUMULATIVE REVIEW EXERCISES

1. $\frac{31}{30}$ [1.3A] **2.** 21 [1.4A] **3.** $5x - 2y$ [1.4B] **4.** $-8x + 26$ [1.4D] **5.** $-\frac{9}{2}$ [2.2A] **6.** -12 [2.2C] **7.** 10 [2.1D] **8.** $\{x|x \le 8\}$ [2.5A] **9.** The volume is 200 ft^3. [3.3A] **10.**

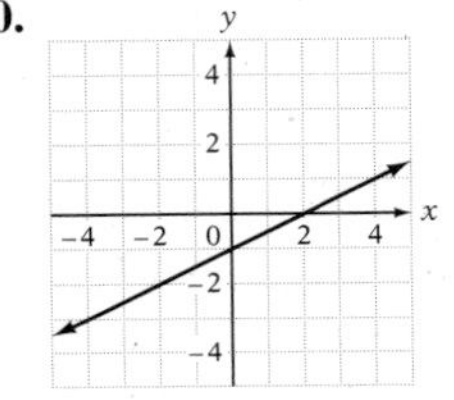

[4.3B]

11. $\frac{3}{7}$ [4.2A] **12.** $y = \frac{3}{2}x + 2$ [4.6A] **13.** $(2, 1, -1)$ [5.2B] **14.** a^3b^7 [6.1A] **15.** $\frac{a^5}{b^6}$ [6.1B] **16.** 3.5×10^{-8} [6.1C] **17.** $-4a^4 + 6a^3 - 2a^2$ [6.3A] **18.** $a^2 + ab - 12b^2$ [6.3B] **19.** $x^2 + 2x + 4$ [6.4B] **20.** $(y-6)(y-1)$ [7.2A] **21.** $(4x+1)(3x-1)$ [7.3A/7.3B] **22.** $a(2a-3)(a+5)$ [7.3A/7.3B] **23.** $4(b+5)(b-5)$ [7.4D] **24.** -3 and $\frac{5}{2}$ [7.5A] **25.** $\frac{2x^3}{3y^5}$ [8.1A] **26.** $-\frac{x-2}{x+5}$ [8.1A] **27.** 1 [8.1C] **28.** $\frac{3}{(2x-1)(x+1)}$ [8.2B] **29.** $\frac{x+3}{x+5}$ [8.3A] **30.** 4 [8.4A] **31.** The alloy contains 70% silver. [2.4B] **32.** It would cost \$160. [8.5B] **33.** It would take 6 min to fill the tank. [8.7A]

ANSWERS TO CHAPTER 9 SELECTED EXERCISES

PREP TEST

1. 16 [1.1D] **2.** 32 [1.2E] **3.** 9 [1.2D] **4.** $\frac{1}{12}$ [1.2C] **5.** $-5x - 1$ [1.4D] **6.** $\frac{xy^5}{4}$ [6.1B] **7.** $9x^2 - 12x + 4$ [6.3C] **8.** $-12x^2 + 14x + 10$ [6.3B] **9.** $36x^2 - 1$ [6.3C] **10.** -1 and 15 [7.5A]

SECTION 9.1

1. (i), (iii) **3.** 2 **5.** 27 **7.** $\frac{1}{9}$ **9.** 4 **11.** Not a real number **13.** $\frac{343}{125}$ **15.** x **17.** $y^{1/2}$ **19.** $x^{1/12}$ **21.** $a^{7/12}$ **23.** $\frac{1}{a}$ **25.** $\frac{1}{y}$ **27.** $y^{3/2}$ **29.** $\frac{1}{x}$ **31.** $\frac{1}{x^4}$ **33.** a **35.** $x^{3/10}$ **37.** a^3 **39.** $\frac{1}{x^{1/2}}$ **41.** $y^{1/9}$ **43.** $\frac{x}{y^2}$ **45.** $\frac{x^{3/2}}{y^{1/4}}$ **47.** $\frac{x^2}{y^8}$ **49.** $\frac{1}{x^{11/12}}$ **51.** $\frac{1}{y^{5/2}}$ **53.** $\frac{1}{b^{7/8}}$ **55.** a^5b^{13} **57.** $\frac{x^2}{4y^{3/2}}$ **59.** $\frac{1}{3xy^{11}}$ **61.** $\frac{b^4}{2a}$ **63.** $\frac{y^{17/2}}{x^3}$ **65.** $\frac{16b^2}{a^{1/3}}$ **67.** $y^2 - y$

69. $a - a^2$ **71.** $x^{3n/2}$ **73.** $y^{3n/2}$ **75.** False **77.** $\sqrt[4]{3}$ **79.** $\sqrt{a^3}$ **81.** $\sqrt{32t^5}$ **83.** $-2\sqrt[3]{x^2}$ **85.** $\sqrt[3]{a^4b^2}$
87. $\sqrt[5]{a^6b^{12}}$ **89.** $\sqrt[4]{(4x-3)^3}$ **91.** $\frac{1}{\sqrt[3]{x^2}}$ **93.** $14^{1/2}$ **95.** $x^{1/3}$ **97.** $x^{4/3}$ **99.** $b^{3/5}$ **101.** $(2x^2)^{1/3}$ **103.** $-(3x^5)^{1/2}$
105. $3xy^{2/3}$ **107.** $(a^2-2)^{1/2}$ **109.** Positive **111.** Not a real number **113.** x^8 **115.** $-x^4$ **117.** xy^3 **119.** $-x^5y$
121. $4a^2b^6$ **123.** Not a real number **125.** $3x^3$ **127.** $-4x^3y^4$ **129.** $-x^2y^3$ **131.** x^4y^2 **133.** $3xy^5$ **135.** $2ab^2$
137. a. False; 2 **b.** True **c.** True **d.** False; $(a^n + b^n)^{1/n}$ **e.** False; $a + 2a^{1/2}b^{1/2} + b$ **f.** False; $a^{n/m}$

SECTION 9.2

1. No **3.** Yes **5.** $x^2yz^2\sqrt{yz}$ **7.** $2ab^4\sqrt{2a}$ **9.** $3xyz^2\sqrt{5yz}$ **11.** $2xyz\sqrt[4]{3yz^2}$ **13.** $a^5b^2\sqrt[3]{ab^2}$ **15.** $-5y\sqrt[3]{x^2y}$
17. $abc^2\sqrt[3]{ab^2}$ **19.** $2x^2y\sqrt[4]{xy}$ **21.** True **23.** $-6\sqrt{x}$ **25.** $-2\sqrt{2x}$ **27.** $3\sqrt{2b} + 5\sqrt{3b}$ **29.** $-2xy\sqrt{2y}$
31. $6ab^2\sqrt{3ab} + 3ab\sqrt{3ab}$ **33.** $-\sqrt[3]{2}$ **35.** $8b\sqrt[3]{2b^2}$ **37.** $3a\sqrt[4]{2a}$ **39.** $17\sqrt{2} - 15\sqrt{5}$ **41.** $5b\sqrt{b}$
43. $-8xy\sqrt{2x} + 2xy\sqrt{xy}$ **45.** $2y\sqrt[3]{2x}$ **47.** $-4ab\sqrt[4]{2b}$ **49.** 16 **51.** $2\sqrt[3]{4}$ **53.** $xy^3\sqrt{x}$ **55.** $8xy\sqrt{x}$
57. $2x^2y\sqrt[3]{2}$ **59.** $2ab\sqrt[4]{3a^2b}$ **61.** 6 **63.** $x - \sqrt{2x}$ **65.** $4x - 8\sqrt{x}$ **67.** $-4 - \sqrt{5}$ **69.** $29 - 7\sqrt{7}$ **71.** 12
73. -3 **75.** $-10x + 14\sqrt{2x} + 3$ **77.** $10x + 12\sqrt{x} + 2$ **79.** $x + 4\sqrt{x} + 4$ **81.** $2x - 8\sqrt{2x+1} + 17$
83. $-42 - 33\sqrt{2}$ **85.** True **89.** $\sqrt[3]{4x}$ **91.** $\sqrt{3} + x$ **93.** $y\sqrt{5y}$ **95.** $b\sqrt{13b}$ **97.** $\frac{\sqrt{2}}{2}$ **99.** $\frac{2\sqrt{3y}}{3y}$ **101.** $\frac{3\sqrt{3a}}{a}$
103. $\frac{\sqrt{2y}}{2}$ **105.** $\frac{5\sqrt[3]{3}}{3}$ **107.** $\frac{5\sqrt[3]{9y^2}}{3y}$ **109.** $\frac{b\sqrt{2a}}{2a^2}$ **111.** $\frac{\sqrt{15x}}{5x}$ **113.** $2 + 2\sqrt{2}$ **115.** $-\frac{12 + 4\sqrt{2}}{7}$ **117.** $-\frac{10 + 5\sqrt{7}}{3}$
119. $-\frac{7\sqrt{x} + 21}{x - 9}$ **121.** $-\sqrt{6} + 3 - 2\sqrt{2} + 2\sqrt{3}$ **123.** $-\frac{17 + 5\sqrt{5}}{4}$ **125.** $\frac{8a - 10\sqrt{ab} + 3b}{16a - 9b}$ **127.** $\frac{3 - 7\sqrt{y} + 2y}{1 - 4y}$
129. a. False; $\sqrt[6]{432}$ **b.** True **c.** False; $x^{2/3}$ **d.** False; $\sqrt{x} + \sqrt{y}$ is in simplest form. **e.** False; $\sqrt[2]{2} + \sqrt[3]{3}$ is in simplest form. **f.** True **131.** $\sqrt[4]{a+b}$

SECTION 9.3

1. -2 **3.** 9 **5.** No solution **7.** 35 **9.** 27 **11.** -12 **13.** 9 **15.** 9 **17.** 2 **19.** 2 **21.** 1
23. $\sqrt{x} < \sqrt{x+5}$. Therefore, $\sqrt{x} - \sqrt{x+5} < 0$ and cannot be equal to a positive number. **25.** On the moon, an object will fall 24.75 ft in 3 s. **27. a.** The height of the water is 5.0 ft after the valve has been opened for 10 s. **b.** It takes 27.7 s to empty the tank. **29.** The length of the pendulum is 7.30 ft. **31.** The HDTV screen is approximately 7.15 in. wider.
33. The longest pole that can be placed in the box is 5.4 ft. **35.** 2

SECTION 9.4

3. $i\sqrt{a}$ **5.** $5i$ **7.** $7i\sqrt{2}$ **9.** $3 + 3i\sqrt{5}$ **11.** $-6 - 10i$ **13.** $-4 + 2i$ **15.** $4 + 8i$ **17.** $8 + 2i\sqrt{2}$
19. $-2 + 6i\sqrt{3}$ **21.** $-5 + i\sqrt{23}$ **23.** $-4 + 2i\sqrt{14}$ **25.** $8 - i$ **27.** $-8 + 4i$ **29.** $6 - 6i$ **31.** $11 - 4i$
33. $-6 - 2i$ **35.** The real parts of the complex numbers are additive inverses. **37.** 63 **39.** -4 **41.** 29
43. $-4 + 12i$ **45.** $-3 - 4i$ **47.** $17 - i$ **49.** $8 + 27i$ **51.** $-21 + 20i$ **53.** 1 **55.** True **57.** $-3i$
59. $\frac{3}{4} + \frac{1}{2}i$ **61.** $\frac{10}{13} - \frac{2}{13}i$ **63.** $\frac{4}{5} + \frac{2}{5}i$ **65.** $-i$ **67.** $\frac{12}{17} + \frac{3}{17}i$ **69.** $\frac{3}{10} - \frac{11}{10}i$ **71.** $\frac{6}{5} + \frac{7}{5}i$ **73.** True
75. a. Yes **b.** Yes

CHAPTER 9 CONCEPT REVIEW*

1. When writing an expression containing a rational exponent as a radical expression, the denominator of the rational exponent is the index of the radical. The numerator of the rational exponent becomes the power of the radicand. [9.1B]
2. Expressions that contain rational exponents do not always represent real numbers when the base of the exponential expression is a negative number. For this reason, all variables in this chapter represent positive numbers. [9.1C]
3. A radical expression is in simplest form when the radicand contains no factor that is a perfect power. [9.2A]
4. To rationalize the denominator of a radical expression that has two terms in the denominator, multiply the numerator and denominator by the conjugate of the denominator. [9.2D]
5. You can add two radical expressions that have the same radicand and the same index. [9.2B]
6. When solving an equation containing a radical expression, you isolate the radical so that when you raise each side of the equation to the appropriate power, the resulting equation does not contain a radical expression. [9.3A]

**Note:* The numbers in brackets following the answers in the Concept Review are a reference to the objective that corresponds to that problem. For example, the reference [1.2A] stands for Section 1.2, Objective A. This notation will be used for all Prep Tests, Concept Reviews, Chapter Reviews, Chapter Tests, and Cumulative Reviews throughout the text.

7. A solution is an extraneous solution if, when that value is substituted into the original equation, the result is an equation that is not true. [9.3A]

8. To add two complex numbers, add the real parts and add the imaginary parts. [9.4B]

9. To rewrite $\sqrt{36} + \sqrt{-36}$ as a complex number, first rewrite the expression in the form $a + bi$: $\sqrt{36} + i\sqrt{36}$. Then simplify each radical: $6 + 6i$. [9.4A]

10. To find the product of two variable expressions with the same base and fractional exponents, add the exponents. The base remains unchanged. [9.1A]

CHAPTER 9 REVIEW EXERCISES

1. $20x^2y^2$ [9.1A] **2.** 7 [9.3A] **3.** $39 - 2i$ [9.4C] **4.** $7x^{2/3}y$ [9.1B] **5.** $6\sqrt{3} - 13$ [9.2C] **6.** -2 [9.3A] **7.** $\frac{1}{x^5}$ [9.1A] **8.** $\frac{8\sqrt{3y}}{3y}$ [9.2D] **9.** $-2a^2b^4$ [9.1C] **10.** $2a^2b\sqrt{2b}$ [9.2B] **11.** $4 + \sqrt{2}$ [9.2D] **12.** $\frac{2}{3} - \frac{5}{3}i$ [9.4D] **13.** $3ab^3\sqrt{2a}$ [9.2A] **14.** $2 + 12i$ [9.4B] **15.** $5x^3y^3\sqrt[3]{2x^2y}$ [9.2B] **16.** $4xy^2\sqrt[3]{x^2}$ [9.2C] **17.** $7 + 3i$ [9.4C] **18.** $\frac{a^{2/3}}{2b^5}$ [9.1A] **19.** $-2ab^2\sqrt[5]{2a^3b^2}$ [9.2A] **20.** $-2 + 7i$ [9.4D] **21.** $-6\sqrt{2}$ [9.4C] **22.** 30 [9.3A] **23.** $3a^2b^3$ [9.1C] **24.** $\frac{3\sqrt[3]{9x^2}}{x}$ [9.2D] **25.** $-12 + 10i$ [9.4B] **26.** $2x + 4\sqrt{2x-1} + 3$ [9.2C] **27.** $6x^2\sqrt{3y}$ [9.2B] **28.** $\frac{1}{3}$ [9.1A] **29.** $\frac{1}{a^{10}}$ [9.1A] **30.** $-7x^3y^8$ [9.1C] **31.** $4\sqrt[3]{a^2}$ [9.1B] **32.** $\frac{x}{3}$ [9.1A] **33.** $xy^2z^2\sqrt[4]{x^2z^2}$ [9.2A] **34.** $5\sqrt{6}$ [9.2B] **35.** $4x^2\sqrt{3xy} - 4x^2\sqrt{5y}$ [9.2B] **36.** 40 [9.2C] **37.** $3x + 3\sqrt{3x}$ [9.2C] **38.** $5x\sqrt{x}$ [9.2D] **39.** $\frac{-9 - 16\sqrt{7}}{29}$ [9.2D] **40.** $6i$ [9.4A] **41.** $8 + 6i$ [9.4A] **42.** $-2 + 4i\sqrt{2}$ [9.4A] **43.** $9 - i$ [9.4B] **44.** -11 [9.2C] **45.** $-4 - 9i$ [9.4B] **46.** $15 - 8i$ [9.4C] **47.** $6i$ [9.4D] **48.** $\frac{14}{5} + \frac{7}{5}i$ [9.4D] **49.** 8 [9.3A] **50.** -24 [9.3A] **51.** The width is 5 in. [9.3B] **52.** The amount of power is 120 watts. [9.3B] **53.** The distance required is 242 ft. [9.3B] **54.** The distance is 6.63 ft. [9.3B]

CHAPTER 9 TEST

1. $\frac{1}{2}x^{3/4}$ [9.1B, How To 8] **2.** $-2x^2y\sqrt[3]{2x}$ [9.2B, Example 3] **3.** $3\sqrt[5]{y^2}$ [9.1B, Example 6] **4.** $18 + 16i$ [9.4C, How To 7] **5.** $4x + 12\sqrt{x} + 9$ [9.2C, Example 6] **6.** $r^{1/6}$ [9.1A, Example 4] **7.** 4 [9.3A, How To 3] **8.** $2a^2b^2\sqrt[4]{2b^2}$ [9.2C, How To 6] **9.** $-4x\sqrt{3}$ [9.2C, How To 7] **10.** $-3 + 2i$ [9.4B, How To 3] **11.** $4x^2y^3\sqrt{2y}$ [9.2A, How To 2] **12.** $14 + 10\sqrt{3}$ [9.2C, How To 8] **13.** -10 [9.4C, How To 5] **14.** 2 [9.2D, Example 9] **15.** $8a\sqrt{2a}$ [9.2B, Example 3] **16.** $2a - \sqrt{ab} - 15b$ [9.2C, How To 8] **17.** $\frac{64x^3}{y^6}$ [9.1A, How To 6] **18.** $2\sqrt[3]{25x}$ [9.2D, You Try It 8] **19.** $-\frac{4}{5} + \frac{7}{5}i$ [9.4D, How To 10] **20.** -3 [9.3A, How To 2] **21.** $\frac{b^3}{8a^6}$ [9.1A, How To 6] **22.** $3abc^2\sqrt[3]{ac}$ [9.2A, Example 2] **23.** $\frac{4x^2}{y}$ [9.2D, How To 10] **24.** $\sqrt{5x}$ [9.2D, How To 11] **25.** The distance is 576 ft. [9.3B, Example 4]

CUMULATIVE REVIEW EXERCISES

1. 92 [1.3A] **2.** 56 [1.4A] **3.** $-10x + 1$ [1.4D] **4.** $\frac{3}{2}$ [2.2A] **5.** $\frac{2}{3}$ [2.2C] **6.** $\{x|x > 1\}$ [2.5A] **7.** $\frac{1}{3}, \frac{7}{3}$ [2.6A] **8.** $\{x|-6 \le x \le 3\}$ [2.6B] **9.** The area is 187.5 cm². [3.2B] **10.** The volume is 14 ft³. [3.3A]

11.

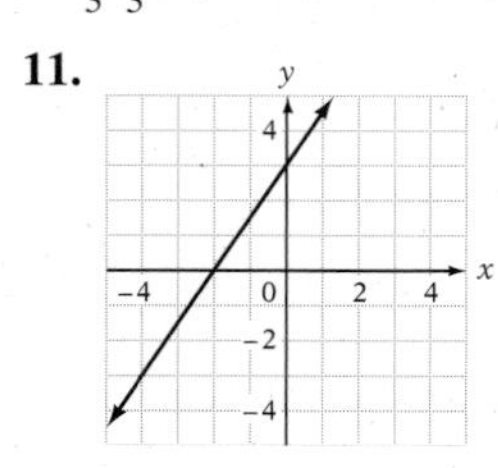

$m = \frac{3}{2}$

$b = 3$

[4.3B]

12. [4.7A]

13. $y = \frac{1}{3}x + \frac{7}{3}$ [4.5B]

14. (3, 2) [5.2A] **15.** $2x^2y^2$ [6.1B] **16.** $(9x + y)(9x - y)$ [7.4A] **17.** $x(x^2 + 3)(x + 1)(x - 1)$ [7.4D]
18. $C = R - nP$ [8.6A] **19.** $\frac{y^5}{x^4}$ [9.1A] **20.** $-x\sqrt{10x}$ [9.2B] **21.** $13 - 7\sqrt{3}$ [9.2C] **22.** $\sqrt{6} + \sqrt{2}$ [9.2D]
23. $-\frac{1}{5} + \frac{3}{5}i$ [9.4D] **24.** -20 [9.3A] **25.** The length of side DE is 27 m. [8.5C] **26.** $2500 is invested at 3.5%. [5.1C]
27. The rate of the plane was 250 mph. [8.7B] **28.** It takes 1.25 s for light to travel to Earth from the moon. [6.1D]
29. The slope is 0.08. The slope represents the simple interest rate on the investment. The interest rate is 8%. [4.4A]
30. The periscope must be approximately 32.7 ft above the water. [9.3B]

ANSWERS TO CHAPTER 10 SELECTED EXERCISES

PREP TEST

1. $3\sqrt{2}$ [1.2F] **2.** $3i$ [9.4A] **3.** $\frac{2x - 1}{x - 1}$ [8.2B] **4.** 8 [1.4A] **5.** Yes [7.4A] **6.** $(2x - 1)^2$ [7.4A]
7. $(3x + 2)(3x - 2)$ [7.4A] **8.** (number line from −5 to 5 with parenthesis at −1, shaded to the left) [1.5B] **9.** -3 and 5 [7.5A] **10.** 4 [8.5A]

SECTION 10.1

3. $2x^2 - 4x - 5 = 0$; $a = 2, b = -4, c = -5$ **5.** $4x^2 - 5x + 6 = 0$; $a = 4, b = -5, c = 6$ **7.** 3 and 5
9. -7 and -1 **11.** 0 and 4 **13.** -5 and 5 **15.** -2 and 3 **17.** 3 **19.** 0 and 2 **21.** -2 and 5
23. 2 and 5 **25.** $-\frac{3}{2}$ and 6 **27.** $\frac{1}{4}$ and 2 **29.** -4 and $\frac{1}{3}$ **31.** $-\frac{2}{3}$ and $\frac{9}{2}$ **33.** -4 and $\frac{1}{4}$ **35.** -2 and 9
37. -2 and $-\frac{3}{4}$ **39.** -5 and 2 **41.** -4 and $-\frac{3}{2}$ **43.** $2b$ and $7b$ **45.** $-c$ and $7c$ **47.** $-b$ and $-\frac{b}{2}$
49. $\frac{2a}{3}$ and $4a$ **51.** $x^2 - 7x + 10 = 0$ **53.** $x^2 + 6x + 8 = 0$ **55.** $x^2 - 5x - 6 = 0$ **57.** $x^2 - 9 = 0$
59. $x^2 - 8x + 16 = 0$ **61.** $x^2 - 5x = 0$ **63.** $x^2 - 3x = 0$ **65.** $2x^2 - 7x + 3 = 0$ **67.** $4x^2 - 5x - 6 = 0$
69. $3x^2 + 11x + 10 = 0$ **71.** $9x^2 - 4 = 0$ **73.** $6x^2 - 5x + 1 = 0$ **75.** $10x^2 - 7x - 6 = 0$
77. $8x^2 + 6x + 1 = 0$ **79.** $50x^2 - 25x - 3 = 0$ **81.** 0 **83.** -7 and 7 **85.** $-2i$ and $2i$ **87.** -2 and 2
89. $-\frac{9}{2}$ and $\frac{9}{2}$ **91.** $-7i$ and $7i$ **93.** $-4\sqrt{3}$ and $4\sqrt{3}$ **95.** $-5\sqrt{3}$ and $5\sqrt{3}$ **97.** $-3i\sqrt{2}$ and $3i\sqrt{2}$
99. -5 and 7 **101.** -6 and 0 **103.** -7 and 3 **105.** 0 and 1 **107.** $-5 - \sqrt{6}$ and $-5 + \sqrt{6}$
109. $3 - 3i\sqrt{5}$ and $3 + 3i\sqrt{5}$ **111.** $-\frac{2 - 9\sqrt{2}}{3}$ and $-\frac{2 + 9\sqrt{2}}{3}$ **113.** Two complex solutions
115. Two equal real solutions **117.** $x^2 - 2 = 0$ **119.** $x^2 + 1 = 0$ **121.** $x^2 - 8 = 0$ **123.** $x^2 + 2 = 0$
125. $-\frac{1}{2}$ **127.** -8 and 2

SECTION 10.2

1. -1 and 5 **3.** -9 and 1 **5.** 3 **7.** $-2 - \sqrt{11}$ and $-2 + \sqrt{11}$ **9.** $3 - \sqrt{2}$ and $3 + \sqrt{2}$ **11.** $1 - i$ and $1 + i$
13. -3 and 8 **15.** -9 and 4 **17.** $\frac{3 - \sqrt{5}}{2}$ and $\frac{3 + \sqrt{5}}{2}$ **19.** $\frac{1 - \sqrt{5}}{2}$ and $\frac{1 + \sqrt{5}}{2}$ **21.** $3 - \sqrt{13}$ and $3 + \sqrt{13}$
23. 3 and 5 **25.** $2 - 3i$ and $2 + 3i$ **27.** $-3 - 2i$ and $-3 + 2i$ **29.** $1 - 3\sqrt{2}$ and $1 + 3\sqrt{2}$
31. $\frac{1 - \sqrt{17}}{2}$ and $\frac{1 + \sqrt{17}}{2}$ **33.** $1 - 2i\sqrt{3}$ and $1 + 2i\sqrt{3}$ **35.** $\frac{1}{2} - i$ and $\frac{1}{2} + i$ **37.** $\frac{1}{3} - \frac{1}{3}i$ and $\frac{1}{3} + \frac{1}{3}i$
39. $\frac{2 - \sqrt{14}}{2}$ and $\frac{2 + \sqrt{14}}{2}$ **41.** $-\frac{3}{2}$ and 1 **43.** $1 - \sqrt{5}$ and $1 + \sqrt{5}$ **45.** $\frac{1}{2}$ and 5 **47.** $2 - \sqrt{5}$ and $2 + \sqrt{5}$
49. -3.236 and 1.236 **51.** 0.293 and 1.707 **53.** -0.809 and 0.309 **55.** $c \le 4$ **57.** $-a$ and $2a$ **59.** $-5a$ and $2a$
61. No, the ball will have gone only 197.2 ft when it hits the ground.

SECTION 10.3

3. -2 and 5 **5.** -9 and 4 **7.** $4 - 2\sqrt{22}$ and $4 + 2\sqrt{22}$ **9.** -8 and 3 **11.** $\frac{-5 - \sqrt{33}}{4}$ and $\frac{-5 + \sqrt{33}}{4}$

13. $-\frac{1}{4}$ and $\frac{3}{2}$ **15.** $7 - 3\sqrt{5}$ and $7 + 3\sqrt{5}$ **17.** $\frac{1 - \sqrt{3}}{2}$ and $\frac{1 + \sqrt{3}}{2}$ **19.** $-1 - i$ and $-1 + i$

21. $1 - 2i$ and $1 + 2i$ **23.** $2 - 3i$ and $2 + 3i$ **25.** $\frac{1 - \sqrt{11}}{2}$ and $\frac{1 + \sqrt{11}}{2}$ **27.** $-\frac{3}{2} - \frac{1}{2}i$ and $-\frac{3}{2} + \frac{1}{2}i$

29. $\frac{3}{4} - \frac{3\sqrt{3}}{4}i$ and $\frac{3}{4} + \frac{3\sqrt{3}}{4}i$ **31.** 0.394 and 7.606 **33.** -4.236 and 0.236 **35.** -1.351 and 1.851

37. Two complex number solutions **39.** Two equal real number solutions **41.** Two unequal real number solutions

43. $\sqrt{4ac}$ **45.** No. The arrow does not reach a height of 275 ft. (The discriminant is less than zero.) **47.** $\{p \mid p < 9\}$

49. $(1, \infty)$ **51.** $-2i$ and i

SECTION 10.4

1. Yes **3.** No **5.** $-2, 2, -3,$ and 3 **7.** $-2, 2, -\sqrt{2},$ and $\sqrt{2}$ **9.** 1 and 4 **11.** 16 **13.** $-2i, 2i, -1,$ and 1

15. $-4i, 4i, -2,$ and 2 **17.** 16 **19.** 1 and 512 **21.** $-\frac{2}{3}, \frac{2}{3}, -1,$ and 1 **23.** Exercises 28, 29, 30, 34, 35, 36, 38, 39, 42

25. 3 **27.** 9 **29.** -1 and 2 **31.** 0 and 2 **33.** $-\frac{1}{2}$ and 2 **35.** -2 **37.** 1 **39.** 1 **41.** -3

43. $y + 2$ **45.** -1 and 10 **47.** -3 and 1 **49.** -1 and 0 **51.** $-\frac{1}{3}$ and $\frac{1}{2}$ **53.** $-\frac{2}{3}$ and 6 **55.** $\frac{4}{3}$ and 3

57. -2 and 3 **59.** 4

SECTION 10.5

3. a. True **b.** False **c.** False **5. a.** False **b.** True **7.** [number line: −5 −4 −3 −2 −1 0 1 2 3 4 5] $\{x \mid x < -2 \text{ or } x > 4\}$

9. [number line: −5 −4 −3 −2 −1 0 1 2 3 4 5] $\{x \mid x \le 1 \text{ or } x \ge 2\}$ **11.** [number line: −5 −4 −3 −2 −1 0 1 2 3 4 5] $\{x \mid -3 < x < 4\}$

13. [number line: −5 −4 −3 −2 −1 0 1 2 3 4 5] $\{x \mid x < -2 \text{ or } 1 < x < 3\}$ **15.** [number line: −5 −4 −3 −2 −1 0 1 2 3 4 5] $\{x \mid -4 \le x \le 1 \text{ or } x \ge 2\}$

17. [number line: −5 −4 −3 −2 −1 0 1 2 3 4 5] $\{x \mid x < -2 \text{ or } x > 4\}$ **19.** [number line: −5 −4 −3 −2 −1 0 1 2 3 4 5] $\{x \mid -1 < x \le 3\}$

21. [number line: −5 −4 −3 −2 −1 0 1 2 3 4 5] $\{x \mid x \le -2 \text{ or } 1 \le x < 3\}$ **23.** $\{x \mid x > 4 \text{ or } x < -4\}$ **25.** $\{x \mid -3 \le x \le 12\}$

27. $\left\{x \,\middle|\, \frac{1}{2} < x < \frac{3}{2}\right\}$ **29.** $\left\{x \,\middle|\, x < 1 \text{ or } x > \frac{5}{2}\right\}$ **31.** $\{x \mid x < -1 \text{ or } 1 < x \le 2\}$ **33.** $\left\{x \,\middle|\, \frac{1}{2} < x \le 1\right\}$

35. $\{x \mid 2 < x \le 3\}$ **37.** $\{x \mid x > 5 \text{ or } -4 < x < -1\}$ **39.** [number line: −4 −2 0 2 4]

41. [number line: −4 −2 0 2 4] **43.** [number line: −4 −2 0 2 4]

SECTION 10.6

1. The maximum speed is 33 mph. **3.** The rocket takes 12.5 s to return to Earth. **5. a.** The maximum height is 27 ft. **b.** The height of the arch is 24 ft. **c.** The arch is 8 ft tall 20.13 ft from the center. **7.** The length of a side of the square base is 177 m. **9.** It would take the larger pipe 6 min to fill the tank. It would take the smaller pipe 12 min to fill the tank. **11.** It would take the faster computer 2 h to run the program. **13.** It would take the apprentice carpenter 6.8 h to install the floor. **15.** The rate of the wind was approximately 108 mph. **17.** The rate of the jet stream is 93 mph. **19.** The dimensions of Colorado are 272 mi by 383 mi. **21.** The height is 3 cm. The base is 14 cm. **23.** The piece of cardboard needed is 90 cm by 90 cm. **25.** The dimensions of the work area are 10 ft by 30 ft. **27.** The radius of the cone is 1.5 in.

CHAPTER 10 CONCEPT REVIEW*

1. Given the solutions of a quadratic equation, write the quadratic equation by using the equation $(x - r_1)(x - r_2) = 0$, where r_1 and r_2 are the given solutions. [10.1B]

2. The symbol ± means plus or minus. For example, ±6 means plus 6 or minus 6. [10.1C]

**Note:* The numbers in brackets following the answers in the Concept Review are a reference to the objective that corresponds to that problem. For example, the reference [1.2A] stands for Section 1.2, Objective A. This notation will be used for all Prep Tests, Concept Reviews, Chapter Reviews, Chapter Tests, and Cumulative Reviews throughout the text.

3. To complete the square on $x^2 - 18x$, square the product of $\frac{1}{2}$ and the coefficient of x: $\left[\frac{1}{2}(-18)\right]^2 = 81$. Add 81 to the binomial: $x^2 - 18x + 81$. [10.2A]

4. To approximate $4 + 3\sqrt{2}$, find the approximation of $\sqrt{2}$: 1.4142. Substitute the approximate value of $\sqrt{2}$ in $4 + 3\sqrt{2}$ and simplify: $4 + 3(1.4142) = 8.2426$. To approximate $4 - 3\sqrt{2}$, substitute the approximate value of $\sqrt{2}$ in $4 - 3\sqrt{2}$ and simplify: $4 - 3(1.4142) = -0.2426$. [10.2A]

5. To solve a quadratic equation by completing the square:

1. Write the equation in the form $ax^2 + bx = -c$.
2. Multiply both sides of the equation by $\frac{1}{a}$.
3. Complete the square on $x^2 + \frac{b}{a}x$. Add the number that completes the square to both sides of the equation.
4. Factor the perfect-square trinomial.
5. Take the square root of each side of the equation.
6. Solve the resulting equation for x.
7. Check the solutions. [10.2A]

6. The quadratic formula states that the quadratic equation $ax^2 + bx + c = 0$, $a \neq 0$, has solutions $x = \frac{-b \pm \sqrt{b^2 - 4ac}}{2a}$. [10.3A]

7. In the quadratic formula, the discriminant is the quantity $b^2 - 4ac$. If $b^2 - 4ac = 0$, the equation has two equal real number solutions, a double root. If $b^2 - 4ac > 0$, the equation has two unequal real number solutions. If $b^2 - 4ac < 0$, the equation has two complex number solutions. [10.3A]

8. It is important to check the solutions to a radical equation because when both sides of an equation are squared, the resulting equation may have a solution that is not a solution of the original equation. [10.4A]

9. A quadratic inequality is one that can be written in the form $ax^2 + bx + c < 0$ or $ax^2 + bx + c > 0$, where $a \neq 0$. The symbols $\leq$ and $\geq$ can also be used. [10.5A]

10. In a quadratic inequality, the symbol $<$ or $>$ indicates that the endpoints are not included in the solution set. The symbol $\leq$ or $\geq$ indicates that the endpoints are included in the solution set. [10.5A]

11. In a rational inequality, we must check that no element of the solution results in a denominator of zero when substituted for the variable in the original inequality. [10.5A]

CHAPTER 10 REVIEW EXERCISES

1. 0 and $\frac{3}{2}$ [10.1A] **2.** $-2c$ and $\frac{c}{2}$ [10.1A] **3.** $-4\sqrt{3}$ and $4\sqrt{3}$ [10.1C] **4.** $-\frac{1}{2} - 2i$ and $-\frac{1}{2} + 2i$ [10.1C] **5.** -3 and -1 [10.2A] **6.** $\frac{7 - 2\sqrt{7}}{7}$ and $\frac{7 + 2\sqrt{7}}{7}$ [10.2A] **7.** $\frac{3}{4}$ and $\frac{4}{3}$ [10.3A] **8.** $\frac{1}{2} - \frac{\sqrt{31}}{2}i$ and $\frac{1}{2} + \frac{\sqrt{31}}{2}i$ [10.3A] **9.** $x^2 + 3x = 0$ [10.1B] **10.** $12x^2 - x - 6 = 0$ [10.1B] **11.** $1 - i\sqrt{7}$ and $1 + i\sqrt{7}$ [10.2A] **12.** $-2i$ and $2i$ [10.2A] **13.** $\frac{11 - \sqrt{73}}{6}$ and $\frac{11 + \sqrt{73}}{6}$ [10.3A] **14.** Two complex number solutions [10.3A] **15.** $\left\{x \mid -3 < x < \frac{5}{2}\right\}$ [10.5A] **16.** $\left\{x \mid x \leq -4 \text{ or } -\frac{3}{2} \leq x \leq 2\right\}$ [10.5A] **17.** -64 and 27 [10.4A] **18.** $\frac{5}{4}$ [10.4B] **19.** -1 and 3 [10.4C] **20.** -1 [10.4C] **21.** [number line: −5 −4 −3 −2 −1 0 1 2 3 4 5] $\left\{x \mid x < \frac{3}{2} \text{ or } x \geq 2\right\}$ [10.5A]

22. [number line: −5 −4 −3 −2 −1 0 1 2 3 4 5] $\left\{x \mid x \leq -3 \text{ or } \frac{1}{2} \leq x < 4\right\}$ [10.5A] **23.** 4 [10.4B] **24.** 5 [10.4B] **25.** $\frac{-3 - \sqrt{249}}{10}$ and $\frac{-3 + \sqrt{249}}{10}$ [10.4C] **26.** $\frac{-11 - \sqrt{129}}{2}$ and $\frac{-11 + \sqrt{129}}{2}$ [10.4C] **27.** $3x^2 + 8x - 3 = 0$ [10.1B] **28.** -5 and $\frac{1}{2}$ [10.1A] **29.** $-1 + 3\sqrt{2}$ and $-1 - 3\sqrt{2}$ [10.1C] **30.** $-3 + i$ and $-3 - i$ [10.3A] **31.** 2 and 3 [10.4C] **32.** $-5, 5, -\sqrt{3}$, and $\sqrt{3}$ [10.4A] **33.** $\frac{25}{18}$ [10.4B] **34.** -8 and $\frac{1}{8}$ [10.4A] **35.** 2 [10.4B] **36.** $5 + 3\sqrt{2}$ and $5 - 3\sqrt{2}$ [10.2A] **37.** 5 and $-\frac{4}{9}$ [10.4C] **38.** $\frac{1 + \sqrt{5}}{6}$ and $\frac{1 - \sqrt{5}}{6}$ [10.3A] **39.** $\frac{5}{4}$ [10.4B] **40.** $\frac{-13 + \sqrt{217}}{2}$ and $\frac{-13 - \sqrt{217}}{2}$ [10.4C] **41.** Two unequal real number solutions [10.3A] **42.** $\{x \mid -2 \leq x \leq 5\}$ [10.5A] **43.** The sculling crew's rate of rowing in calm water is 6 mph. [10.6A] **44.** The width of the rectangle is 5 cm. The length of the rectangle is 12 cm. [10.6A] **45.** The integers are 2, 4, and 6 or -6, -4, and -2. [10.6A] **46.** Working alone, the new computer can print the payroll in 12 min. [10.6A] **47.** The rate of the first car is 40 mph. The rate of the second car is 50 mph. [10.6A]

CHAPTER 10 TEST

1. -4 and $\frac{2}{3}$ [10.1A, How To 1] **2.** $-\frac{2}{3}$ and $\frac{3}{2}$ [10.1A, How To 1] **3.** $x^2 - 9 = 0$ [10.1B, How To 3] **4.** $2x^2 + 7x - 4 = 0$ [10.1B, Example 3] **5.** $2 - 2\sqrt{2}$ and $2 + 2\sqrt{2}$ [10.1C, Example 4] **6.** $3 - \sqrt{11}$ and $3 + \sqrt{11}$ [10.2A, How To 3] **7.** $\frac{3 - \sqrt{15}}{3}$ and $\frac{3 + \sqrt{15}}{3}$ [10.2A, How To 4] **8.** $\frac{1 - \sqrt{3}}{2}$ and $\frac{1 + \sqrt{3}}{2}$ [10.3A, How To 2] **9.** $-2 - 2i\sqrt{2}$ and $-2 + 2i\sqrt{2}$ [10.3A, How To 3] **10.** $\frac{1}{4}$ [10.4A, How To 1] **11.** $\sqrt{3}, -\sqrt{3}, 1$, and -1 [10.4A, Example 1] **12.** 4 [10.4B, How To 2] **13.** No solution [10.4B, Example 3] **14.** -9 and 2 [10.4C, How To 3] **15.** −5 −4 −3 −2 −1 0 1 2 3 4 5 $\{x \mid x < -4 \text{ or } 2 < x < 4\}$ [10.5A, How To 2] **16.** −5 −4 −3 −2 −1 0 1 2 3 4 5 $\left\{x \mid -4 < x \le \frac{3}{2}\right\}$ [10.5A, How To 3] **17.** Two equal real number solutions [10.3A, How To 4] **18.** The ball hits the basket 1.88 s after it is released. [10.6A, How To 1] **19.** It would take Cora 6 h to stain the bookcase. [10.6A, Example 1] **20.** The rate of the canoe in calm water is 4 mph. [10.6A, Example 2]

CUMULATIVE REVIEW EXERCISES

1. 14 [1.4A] **2.** $\left\{x \mid -2 < x < \frac{10}{3}\right\}$ [2.6B] **3.** The volume is 54π m³. [3.3A] **4.** $-\frac{7}{3}$ [4.2A] **5.** $-\frac{3}{2}$ [4.4A] **6.** $\left(\frac{5}{2}, 0\right), (0, -3)$ [4.3C] **7.** $y = x + 1$ [4.6A] **8.** $(1, -1, 2)$ [5.2B] **9.**

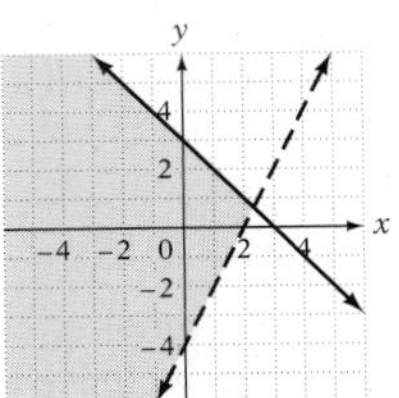

[5.4A]

10. The height of triangle *DEF* is 16 cm. [8.5C] **11.** $x^2 - 3x - 4 - \frac{6}{3x - 4}$ [6.4B] **12.** $-3xy(x^2 - 2xy + 3y^2)$ [7.1A] **13.** $(2x - 5)(3x + 4)$ [7.3A/7.3B] **14.** $\frac{x}{2}$ [8.1B] **15.** $-\frac{3}{2}$ and -1 [8.4A] **16.** $b = \frac{2S - an}{n}$ [8.6A] **17.** $1 - a$ [9.1A] **18.** $-8 - 14i$ [9.4C] **19.** 0, 1 [9.3A] **20.** $2, -2, \sqrt{2}, -\sqrt{2}$ [10.4A] **21.** The lower limit is $9\frac{23}{64}$ in. The upper limit is $9\frac{25}{64}$ in. [2.6C] **22.** The area is $(x^2 + 6x - 16)$ square feet. [6.3D] **23.** The slope is $-\frac{25{,}000}{3}$. The building decreases \$8333.33 in value each year. [4.4A] **24.** The ladder will reach 15 feet up on the house. [9.3B] **25.** There are two complex number solutions. [10.3A]

ANSWERS TO CHAPTER 11 SELECTED EXERCISES

PREP TEST

1. 1 [1.4A] **2.** -7 [1.4A] **3.** 30 [4.2A] **4.** $h^2 + 4h - 1$ [4.2A] **5.** $-\frac{2}{3}$ and 3 [10.1A] **6.** $2 - \sqrt{3}$ and $2 + \sqrt{3}$ [10.3A] **7.** $y = \frac{1}{2}x - 2$ [8.6A] **8.** Domain: $\{-2, 3, 4, 6\}$; Range: $\{4, 5, 6\}$ The relation is a function. [4.2A] **9.**

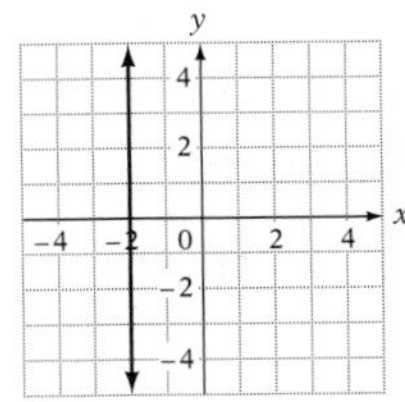

[4.3B]

SECTION 11.1

5. -5 **7.** $x = 7$

9.

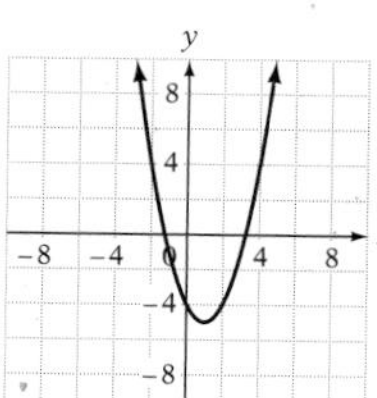

Vertex: $(1, -5)$

Axis of symmetry: $x = 1$

11.

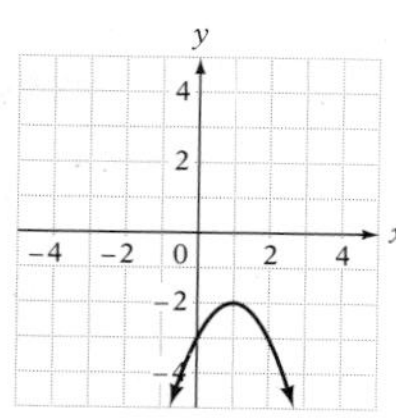

Vertex: $(1, -2)$

Axis of symmetry: $x = 1$

13.

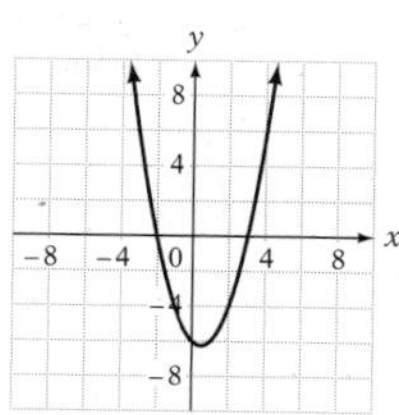

Vertex: $\left(\frac{1}{2}, -\frac{25}{4}\right)$

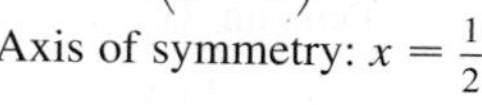

Axis of symmetry: $x = \frac{1}{2}$

15.

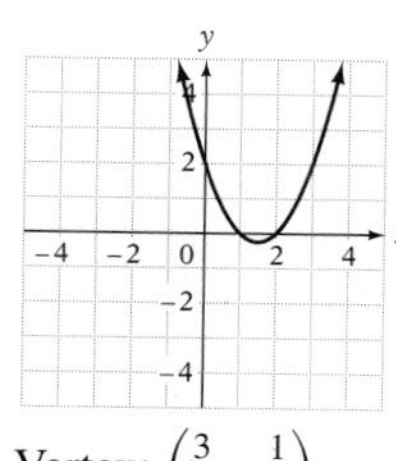

Vertex: $\left(\frac{3}{2}, -\frac{1}{4}\right)$

Axis of symmetry: $x = \frac{3}{2}$

17.

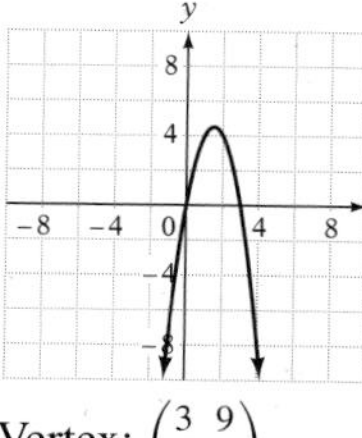

Vertex: $\left(\frac{3}{2}, \frac{9}{2}\right)$

Axis of symmetry: $x = \frac{3}{2}$

19.

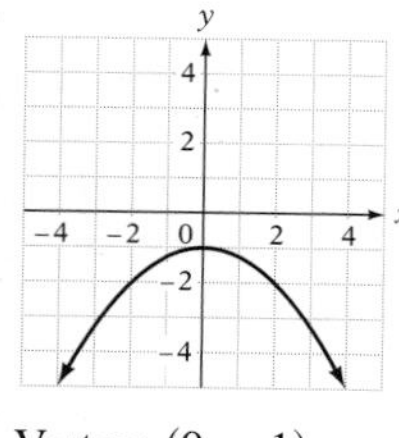

Vertex: $(0, -1)$

Axis of symmetry: $x = 0$

21.

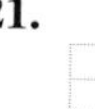

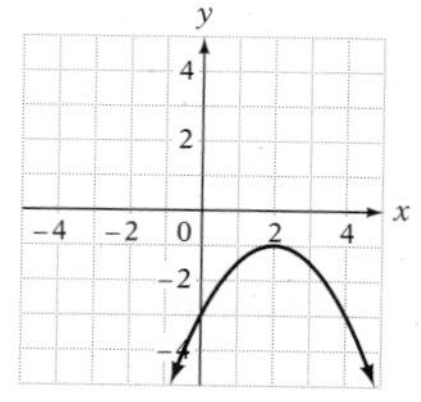

Vertex: $(2, -1)$

Axis of symmetry: $x = 2$

23.

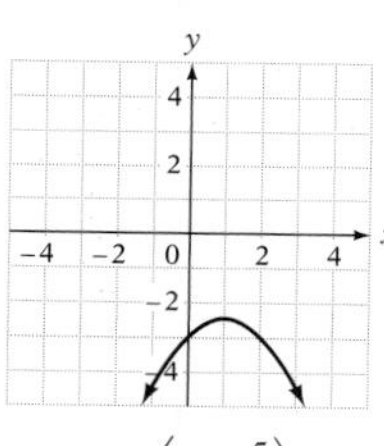

Vertex: $\left(1, -\frac{5}{2}\right)$

Axis of symmetry: $x = 1$

25. Domain: $\{x \mid x \in \text{real numbers}\}$ Range: $\{y \mid y \geq 2\}$ **27.** Domain: $\{x \mid x \in \text{real numbers}\}$ Range: $\{y \mid y \geq -5\}$

29. Domain: $\{x \mid x \in \text{real numbers}\}$ Range: $\{y \mid y \leq 0\}$ **31.** Domain: $\{x \mid x \in \text{real numbers}\}$ Range: $\{y \mid y \geq -7\}$ **35. a.** -4 and 1 **b.** -4 and 1

37. $(3, 0)$ and $(-3, 0)$ **39.** $(0, 0)$ and $(-2, 0)$ **41.** $(4, 0)$ and $(-2, 0)$ **43.** $\left(-\frac{1}{2}, 0\right)$ and $(3, 0)$

45. $(-2 + \sqrt{7}, 0)$ and $(-2 - \sqrt{7}, 0)$ **47.** No x-intercepts **49.** -1 and 5 **51.** $-\frac{4}{3}$ and 2 **53.** $\frac{1}{2}$

55. 0 and $\frac{4}{3}$ **57.** -2 and 2 **59.** $-3\sqrt{3}$ and $3\sqrt{3}$ **61.** $1 - 3\sqrt{2}$ and $1 + 3\sqrt{2}$ **63.** $-2 - i$ and $-2 + i$

65. $-2 - 3i$ and $-2 + 3i$ **67.** Two **69.** One **71.** No x-intercepts **73.** Two **75.** No x-intercepts **77.** Two

79. a. $a > 0$ **b.** $a = 0$ **c.** $a < 0$ **83. a.** Minimum **b.** Maximum **c.** Minimum **85.** Minimum value: -2

87. Maximum value: -3 **89.** Maximum value: $\frac{9}{8}$ **91.** Minimum value: $-\frac{11}{4}$ **93.** Maximum value: $\frac{9}{4}$

95. Minimum value: $-\frac{1}{12}$ **97.** A price of \$250 will give the maximum revenue. **99.** The vertex of the parabola is approximately (42, 8500), so the maximum height of the plane is about 8500 m. **101.** The minimum height is 24.36 ft.

103. The numbers are 10 and 10. **105.** The numbers are -7 and 7. **107.** The length is 100 ft and the width is 50 ft.

109. The dimensions are 350 ft by 525 ft. **111.** $-\frac{3}{2}$ **113.** -2 and 3

SECTION 11.2

1. Yes **3.** No **5.** Yes **7.** No

9.

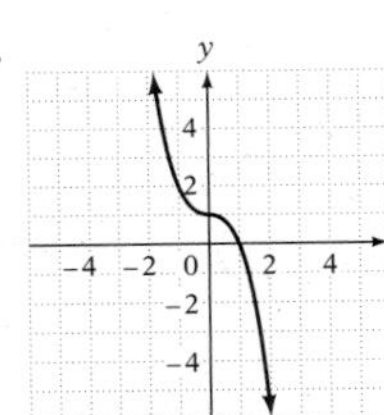

Domain: $\{x|x \in \text{real numbers}\}$
Range: $\{y|y \in \text{real numbers}\}$

11.

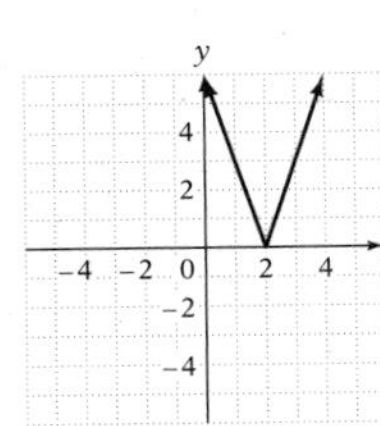

Domain: $\{x|x \in \text{real numbers}\}$
Range: $\{y|y \geq 0\}$

13.

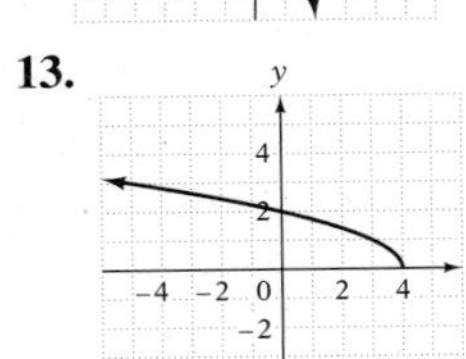

Domain: $\{x|x \leq 4\}$
Range: $\{y|y \geq 0\}$

15.

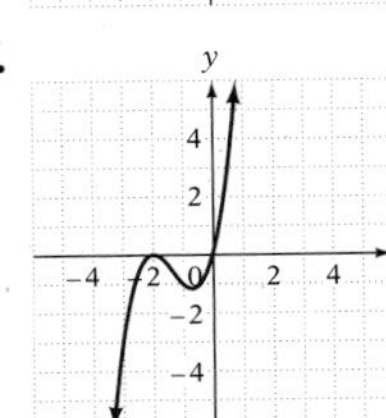

Domain: $\{x|x \in \text{real numbers}\}$
Range: $\{y|y \in \text{real numbers}\}$

17.

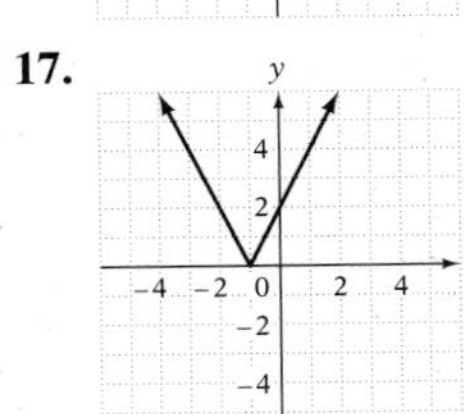

Domain: $\{x|x \in \text{real numbers}\}$
Range: $\{y|y \geq 0\}$

19.

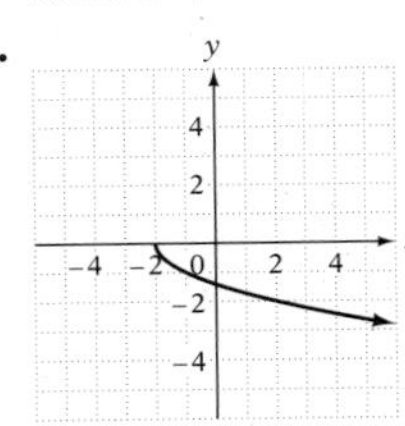

Domain: $\{x|x \geq -2\}$
Range: $\{y|y \leq 0\}$

21. $a = 18$ **23.** 17 **25.** $f(14) = 8$ **27.** $\{x|-2 < x < 2\}$ **29.** $x = 1$

SECTION 11.3

1. a. Yes **b.** Yes **c.** Yes **d.** No **3.** 5 **5.** 1 **7.** 0 **9.** $-\frac{29}{4}$ **11.** 2 **13.** 39 **15.** −8 **17.** Undefined **19.** −2 **21.** 7 **25.** 5 **27.** −13 **29.** −29 **31.** $8x - 13$ **33.** 4 **35.** 3 **37.** $x + 4$ **39.** 1 **41.** 5 **43.** $x^2 + 1$ **45.** 5 **47.** 11 **49.** $3x^2 + 3x + 5$ **51.** −3 **53.** −27 **55.** $x^3 - 6x^2 + 12x - 8$ **57. a.** $S[M(x)] = 80 + \frac{16,000}{x}$ **b.** \$83.20 **c.** When 5000 digital cameras are manufactured, the camera store sells each camera for \$83.20. **59. a.** $I[n(m)] = 50,000m$ **b.** \$150,000 **c.** The garage's income from conversions during 3 months is \$150,000. **61. a.** $d[r(p)] = 0.90p - 1350$ **b.** $r[d(p)] = 0.90p - 1500$ **c.** $r[d(p)]$ (The cost is less.) **63.** 0 **65.** −2 **67.** 7 **69.** $h^2 + 6h$ **71.** $2 + h$ **73.** $2a + h$ **75.** −1 **77.** −6 **79.** $6x - 13$

SECTION 11.4

3. a. Yes **b.** No **5.** Yes **7.** No **9.** Yes **11.** No **13.** No **15.** No **19.** No **21. a.** 4 **b.** 3 **c.** 6 **23.** $\{(0, 1), (3, 2), (8, 3), (15, 4)\}$ **25.** No inverse **27.** $\{(-2, 0), (5, -1), (3, 3), (6, -4)\}$ **29.** No inverse **31.** $f^{-1}(x) = \frac{1}{4}x + 2$ **33.** $f^{-1}(x) = \frac{1}{2}x - 2$ **35.** $f^{-1}(x) = 2x + 2$ **37.** $f^{-1}(x) = -\frac{1}{2}x + 1$ **39.** $f^{-1}(x) = \frac{3}{2}x - 6$ **41.** $f^{-1}(x) = -3x + 3$ **43.** $f^{-1}(x) = \frac{1}{2}x + \frac{5}{2}$ **45.** $f^{-1}(x) = \frac{1}{5}x + \frac{2}{5}$ **47.** $f^{-1}(x) = \frac{1}{6}x + \frac{1}{2}$ **49.** $\frac{5}{3}$ **51.** 3 **53.** Yes; Yes **55.** Yes **57.** No **59.** Yes **61.** Yes **63.** $f^{-1}(x) = 16x$; The inverse function converts pounds to ounces. **65.** $f^{-1}(x) = x - 30$; The inverse function converts a dress size in France to a dress size in the United States. **67.** $f^{-1}(x) = \frac{1}{90}x - \frac{13}{18}$; The inverse function gives the intensity of training percent for a given target heart rate for the athlete.

69.

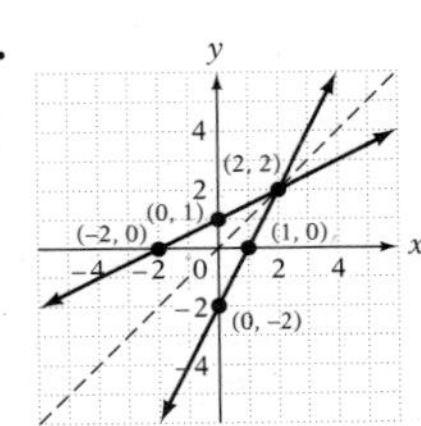

71.

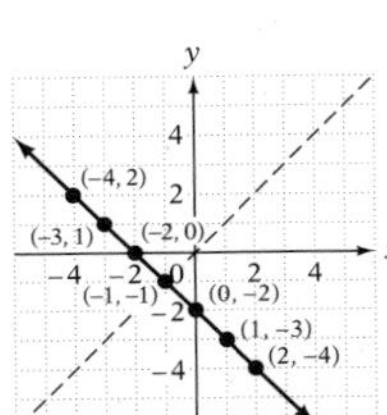

73.

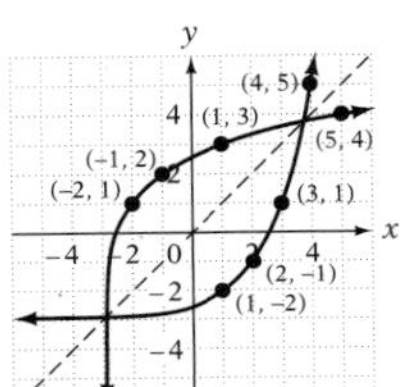

CHAPTER 11 CONCEPT REVIEW*

1. Given a quadratic equation $y = ax^2 + bx + c$, the axis of symmetry is $x = -\frac{b}{2a}$. The coordinates of the vertex are $\left(-\frac{b}{2a}, f\left(-\frac{b}{2a}\right)\right)$. [11.1A]

2. Given its vertex and axis of symmetry, you cannot accurately graph a parabola because the vertex and axis of symmetry are not enough to determine how wide or narrow the parabola is. [11.1A]

3. If you are given the vertex of a parabola, you can determine the range of the function. If the parabola opens up, the range is all y values greater than or equal to the y-coordinate of the vertex. If the parabola opens down, the range is all y values less than or equal to the y-coordinate of the vertex. [11.1A]

4. The following describes the effect of the discriminant on the x-intercepts of a parabola:
1. If $b^2 - 4ac = 0$, the parabola has one x-intercept.
2. If $b^2 - 4ac > 0$, the parabola has two x-intercepts.
3. If $b^2 - 4ac < 0$, the parabola has no x-intercepts. [11.1B]

5. For the graph of $f(x) = ax^2 + bx + c$, $a \neq 0$, the coefficient of the x^2 term determines how wide or narrow the parabola is. The larger the absolute value of the coefficient of the x^2 term, the narrower the parabola; the smaller the absolute value of the coefficient of the x^2 term, the wider the parabola. [11.1A]

6. The vertical-line test states that a graph defines a function if any vertical line intersects the graph at no more than one point. [11.2A]

7. The graph of the absolute value of a linear polynomial is V-shaped. [11.2A]

8. The four basic operations on functions are:
1. $(f + g)(x) = f(x) + g(x)$
2. $(f - g)(x) = f(x) - g(x)$
3. $(f \cdot g)(x) = f(x) \cdot g(x)$
4. $\left(\frac{f}{g}\right)(x) = \frac{f(x)}{g(x)}, g(x) \neq 0$ [11.3A]

9. The notation $(f \circ g)(x)$ refers to the composite function $f[g(x)]$. [11.3B]

10. The horizontal-line test states that the graph of a function represents a 1–1 function if any horizontal line intersects the graph at no more than one point. [11.4A]

11. For a function to have an inverse, it must be a 1–1 function. [11.4B]

CHAPTER 11 REVIEW EXERCISES

1. Yes [11.2A] **2.** Yes [11.4A] **3.**

Domain: $\{x \mid x \in \text{real numbers}\}$
Range: $\{y \mid y \in \text{real numbers}\}$ [11.2A]

4.

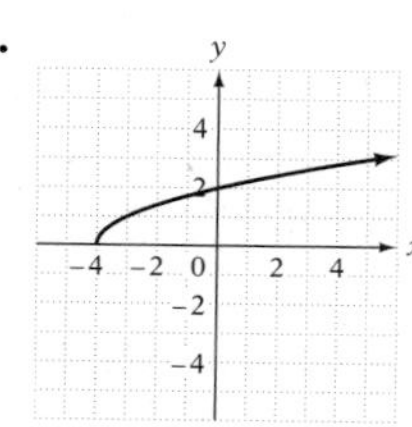

Domain: $\{x \mid x \geq -4\}$
Range: $\{y \mid y \geq 0\}$ [11.2A]

5.

Domain: $\{x \mid x \in \text{real numbers}\}$
Range: $\{y \mid y \geq -3\}$ [11.2A]

6.

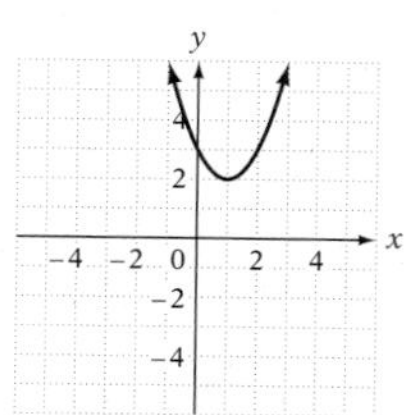

Vertex: $(1, 2)$
Axis of symmetry: $x = 1$ [11.1A]

7. Two x-intercepts [11.1B] **8.** No x-intercepts [11.1B]

9. $(0, 0)$ and $(-3, 0)$ [11.1B] **10.** $(3 - \sqrt{2}, 0)$ and $(3 + \sqrt{2}, 0)$ [11.1B] **11.** $-\frac{3}{2}$ and 5 [11.1B]

12. $1 - 3i$ and $1 + 3i$ [11.1B] **13.** 3 [11.1C] **14.** $-\frac{17}{4}$ [11.1C] **15.** 5 [11.3B] **16.** 10 [11.3B]

**Note:* The numbers in brackets following the answers in the Concept Review are a reference to the objective that corresponds to that problem. For example, the reference [1.2A] stands for Section 1.2, Objective A. This notation will be used for all Prep Tests, Concept Reviews, Chapter Reviews, Chapter Tests, and Cumulative Reviews throughout the text.

17. $12x^2 + 12x - 1$ [11.3B] **18.** $6x^2 + 3x - 16$ [11.3B] **19.** 7 [11.3A] **20.** -9 [11.3A] **21.** 70 [11.3A]

22. $\frac{12}{7}$ [11.3A] **23.** $f^{-1}(x) = -\frac{1}{6}x + \frac{2}{3}$ [11.4B] **24.** $f^{-1}(x) = \frac{3}{2}x + 18$ [11.4B] **25.** Yes [11.4B] **26.** No [11.4B]

27. $p^{-1}(x) = 2.5x - 37.5$; The inverse function gives a diver's depth below the surface of the water for a given pressure on the diver. [11.4B] **28.** Fifty gloves should be made each month. The maximum profit for a month is \$5000. [11.1D] **29.** The dimensions are 7 ft by 7 ft. [11.1D]

CHAPTER 11 TEST

1.

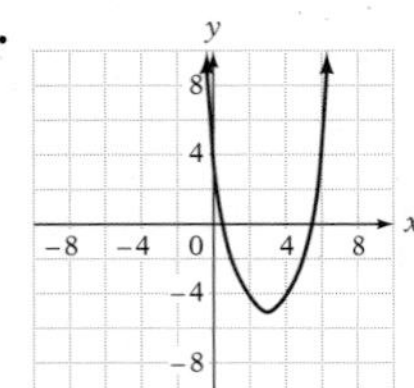

Vertex: $(3, -5)$
Axis of symmetry: $x = 3$
[11.1A, How To 2]

2.

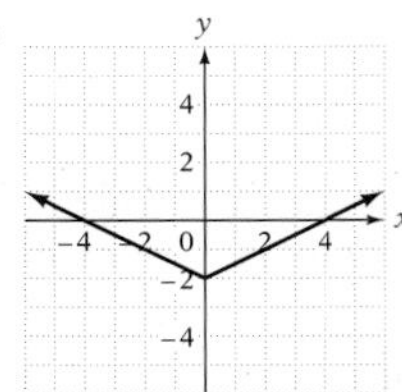

Domain: $\{x|x \in \text{real numbers}\}$
Range: $\{y|y \geq -2\}$
[11.2A, How To 3]

3.

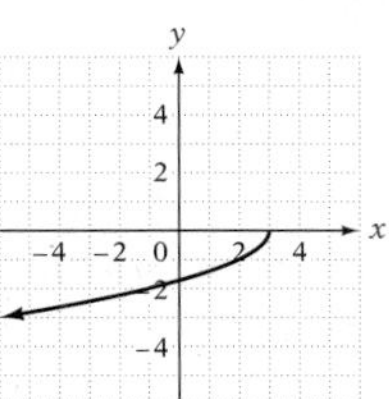

Domain: $\{x|x \leq 3\}$
Range: $\{y|y \leq 0\}$
[11.2A, How To 4]

4.

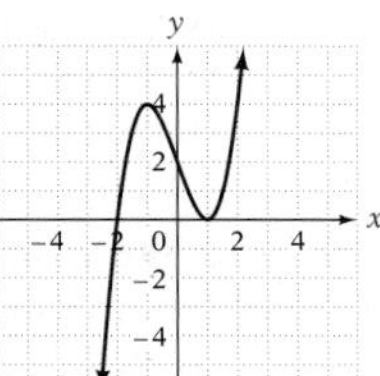

Domain: $\{x|x \in \text{real numbers}\}$
Range: $\{y|y \in \text{real numbers}\}$
[11.2A, How To 2]

5. No x-intercepts [11.1B, How To 7] **6.** $\left(-\frac{2}{3}, 0\right)$ and $(3, 0)$ [11.1B, How To 4] **7.** 9 [11.1C, How To 8]

8. Domain: $\{x|x \in \text{real numbers}\}$; Range: $\{y|y \geq -7\}$ [11.1A, You Try It 2] **9.** -2 [11.3A, Example 1]

10. 234 [11.3A, Example 2] **11.** $-\frac{13}{2}$ [11.3A, Example 3] **12.** -5 [11.3A, Example 1] **13.** -17 [11.3B, Example 4]

14. 31 [11.3B, Example 4] **15.** $3x^2 - 1$ [11.3B, How To 4] **16.** $f[g(x)] = 2x^2 - 4x - 5$ [11.3B, How To 4]

17. No [11.4B, How To 1] **18.** $\{(6, 2), (5, 3), (4, 4), (3, 5)\}$ [11.4B, How To 1] **19.** $f^{1}(x) = \frac{1}{4}x + \frac{1}{2}$ [11.4B, How To 2]

20. $f^{1}(x) = 4x + 16$ [11.4B, You Try It 2] **21.** Yes [11.4B, Example 3] **22.** Yes [11.4B, Example 3] **23.** No [11.4A, Example 1] **24.** $C^{-1}(x) = 0.8x - 4$; The inverse function gives the number of miles to the location for the given cost. [11.4B, Example 4] **25.** Twenty-five speakers should be made. The minimum daily production cost is \$50. [11.1D, Example 7] **26.** 14 and 14 [11.1D, Example 8] **27.** Dimensions of 50 cm by 50 cm would give a maximum area of 2500 cm². [11.1D, How To 9]

CUMULATIVE REVIEW EXERCISES

1. $-\frac{23}{4}$ [1.4A] **2.** −5 −4 −3 −2 −1 0 1 2 3 4 5 [1.5B] **3.** 3 [2.2C] **4.** $\{x|x < -2 \text{ or } x > 3\}$ [2.5B]

5. The set of all real numbers [2.6B] **6.** $-\frac{a^{10}}{12b^4}$ [6.1B] **7.** $2x^3 - 4x^2 - 17x + 4$ [6.3B] **8.** $\left(\frac{1}{2}, 3\right)$ [5.2A]

9. $xy(x + 3y)(x - 2y)$ [7.2B] **10.** -3 and 8 [7.5A] **11.** $\{x|x < -3 \text{ or } x > 5\}$ [10.5A]

12. $\frac{5}{2x - 1}$ [8.2B] **13.** -2 [8.4A] **14.** $-3 - 2i$ [9.4D] **15.** Vertex: $(0, 0)$
Axis of symmetry: $x = 0$ [11.1A]

16. 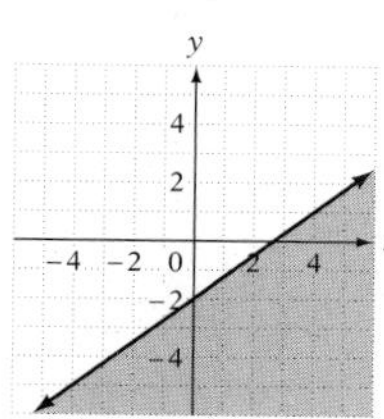 [4.7A] **17.** $y = -2x - 2$ [4.5B] **18.** $y = -\frac{3}{2}x - \frac{7}{2}$ [4.6A]

19. $\frac{1}{2} + \frac{\sqrt{3}}{6}i$ and $\frac{1}{2} - \frac{\sqrt{3}}{6}i$ [10.3A] **20.** 3 [10.4B] **21.** -3 [11.1C] **22.** $\{1, 2, 4, 5\}$ [4.2A] **23.** Yes [4.2A]
24. 2 [9.3A] **25.** 10 [11.3B] **26.** $f^{-1}(x) = -\frac{1}{3}x + 3$ [11.4B] **27.** The cost per pound of the mixture is \$3.96. [2.4A]
28. 25 lb of the 80% copper alloy must be used. [2.4B] **29.** An additional 4.5 oz of insecticide are required. [8.5B]
30. It would take the larger pipe 4 min to fill the tank. [10.6A] **31.** A force of 40 lb will stretch the spring 24 in. [8.8A]
32. The frequency is 80 vibrations/min. [8.8A]

ANSWERS TO CHAPTER 12 SELECTED EXERCISES

PREP TEST

1. $\frac{1}{9}$ [6.1B] **2.** 16 [6.1B] **3.** -3 [6.1B] **4.** 0; 108 [4.2A/6.2A] **5.** -6 [2.2B] **6.** -2 and 8 [10.1A]
7. 6326.60 [1.4A] **8.** 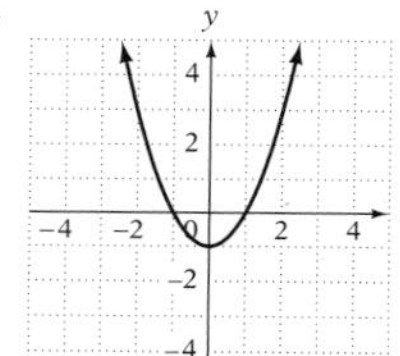 [11.1A]

SECTION 12.1

3. (iii) **5. a.** 9 **b.** 1 **c.** $\frac{1}{9}$ **7. a.** 16 **b.** 4 **c.** $\frac{1}{4}$ **9. a.** 1 **b.** $\frac{1}{8}$ **c.** 16 **11. a.** 7.3891 **b.** 0.3679
c. 1.2840 **13. a.** 54.5982 **b.** 1 **c.** 0.1353 **15. a.** 16 **b.** 16 **c.** 1.4768 **17. a.** 0.1353 **b.** 0.1353 **c.** 0.0111
19. $f(a) > f(b)$ **21.** 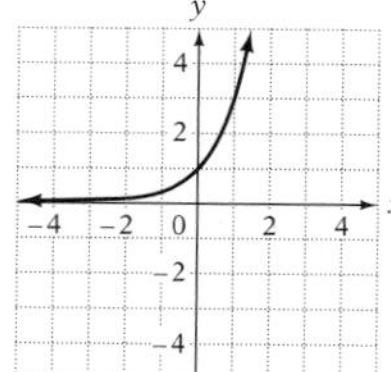 **23.** 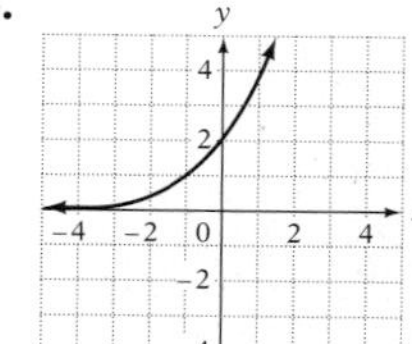 **25.** 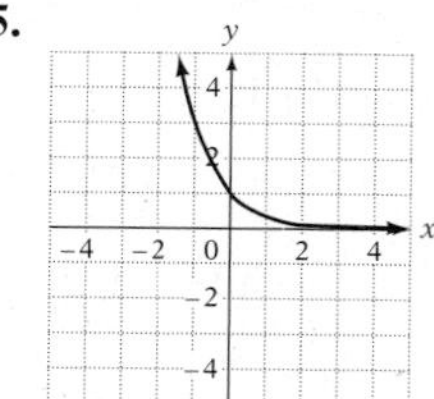 **27.** 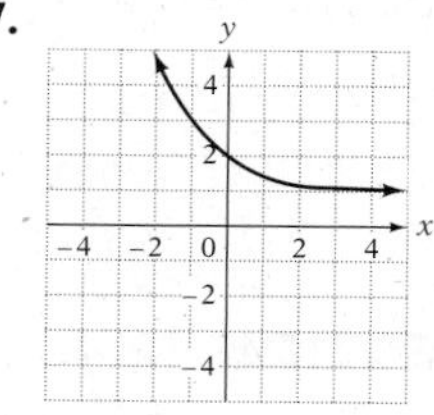

29. 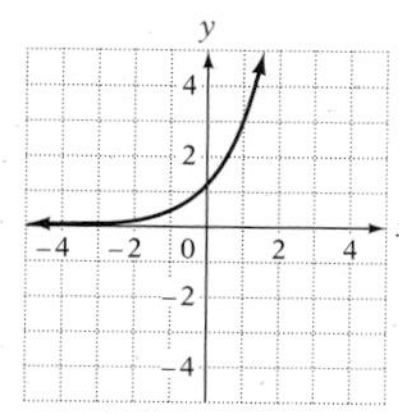 **31.** 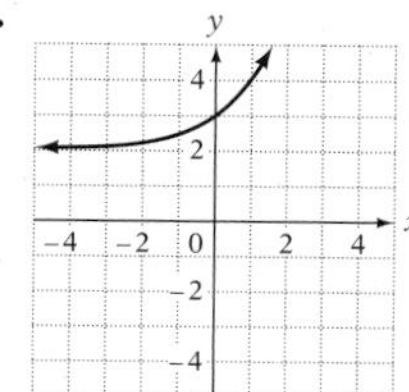 **33.** (0, 1) **35.** No x-intercept; (0, 1) **37.** (i) and (iii), (ii) and (iv)

39. 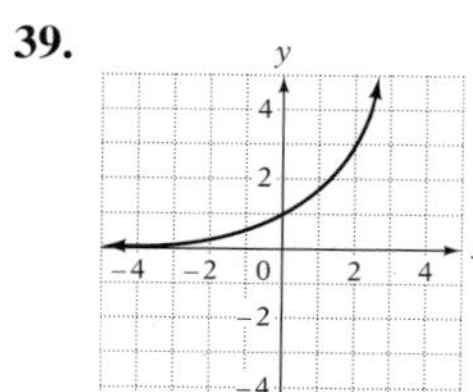 **41.**  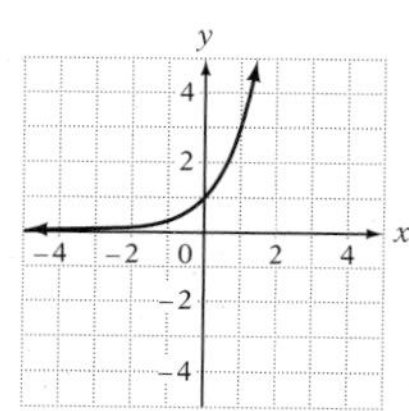 **43. a.** 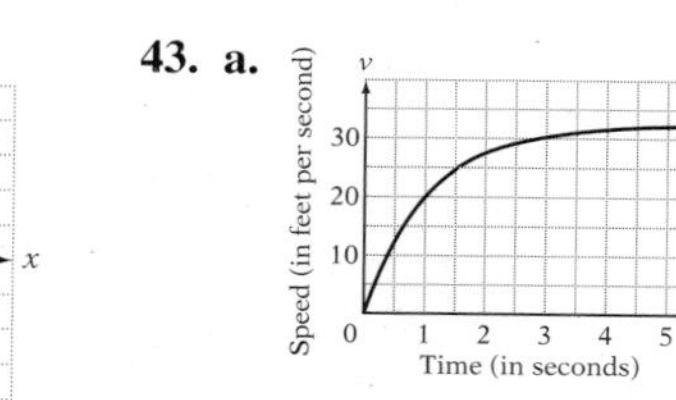

b. The point (2, 27.7) means that after 2 s, the object is falling at a speed of 27.7 ft/s.

SECTION 12.2

3. $\log_5 25 = 2$ **5.** $\log_4 \frac{1}{16} = -2$ **7.** $\log_{10} x = y$ **9.** $\log_a w = x$ **11.** $3^2 = 9$ **13.** $10^{-2} = 0.01$ **15.** $e^y = x$ **17.** $b^v = u$ **19.** 4 **21.** 7 **23.** 2 **25.** 3 **27.** 0 **29.** 4 **31.** 9 **33.** 64 **35.** $\frac{1}{7}$ **37.** 1 **39.** 316.23 **41.** 0.02 **43.** 7.39 **45.** 0.61 **47.** 1 **51.** False **53.** 0 **55.** 1 **57.** x **59.** v **61.** $x^2 + 1$ **63.** $x^2 - x - 1$ **65.** $\log_8 x + \log_8 z$ **67.** $5 \log_3 x$ **69.** $\log_b r - \log_b s$ **71.** $2 \log_3 x + 6 \log_3 y$ **73.** $3 \log_7 u - 4 \log_7 v$ **75.** $2 \log_2 r + 2 \log_2 s$ **77.** $2 \ln x + \ln y + \ln z$ **79.** $\log_5 x + 2 \log_5 y - 4 \log_5 z$ **81.** $2 \log_8 x - \log_8 y - 2 \log_8 z$ **83.** $\frac{3}{2} \log_4 x + \frac{1}{2} \log_4 y$ **85.** $\frac{3}{2} \log_7 x - \frac{1}{2} \log_7 y$ **87.** $\log_3 t - \frac{1}{2} \log_3 x$ **89.** $\log_3 (x^3y^2)$ **91.** $\ln \frac{x^4}{y^2}$ **93.** $\log_7 x^3$ **95.** $\ln(x^3y^4)$ **97.** $\log_4 (x^2y^2)$ **99.** $\log_3 \frac{x^2z^2}{y}$ **101.** $\ln \frac{x}{y^2z}$ **103.** $\log_6 \sqrt{\frac{x}{y}}$ **105.** $\log_4 \frac{s^2r^2}{t^4}$ **107.** $\ln \frac{x}{y^2z^2}$ **109.** $\log_4 \sqrt{\frac{x^3z}{y^2}}$ **111.** $\log_2 \frac{\sqrt{xz}}{\sqrt[3]{y^2}}$ **113.** 0.8617 **115.** 2.1133 **117.** −0.6309 **119.** 0.2727 **121.** 1.6826 **123.** 1.9266 **125.** 0.6752 **127.** 2.6125 **129.** $\log_5 x = \frac{\log x}{\log 5}$ **131. a.** False **b.** True **c.** False **d.** True **e.** False **f.** True

SECTION 12.3

3.

5.

7.

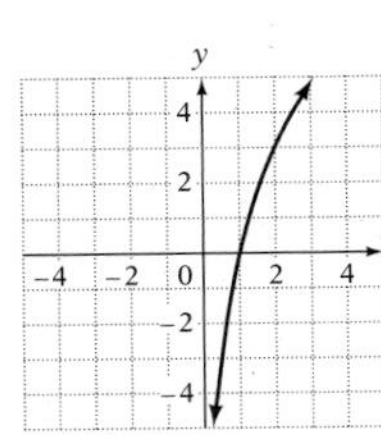

9.

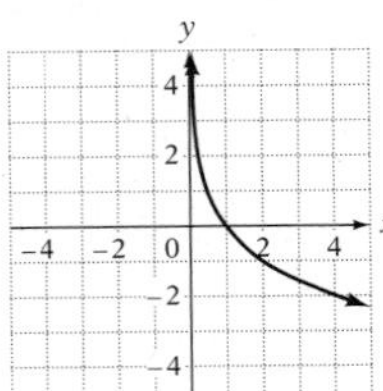

11.

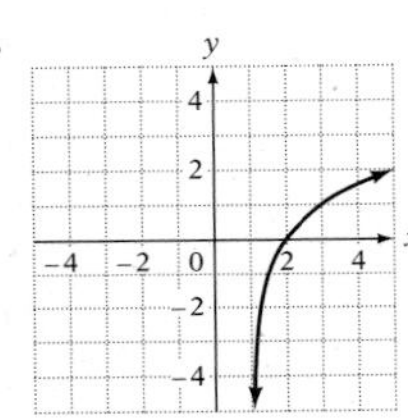

13.

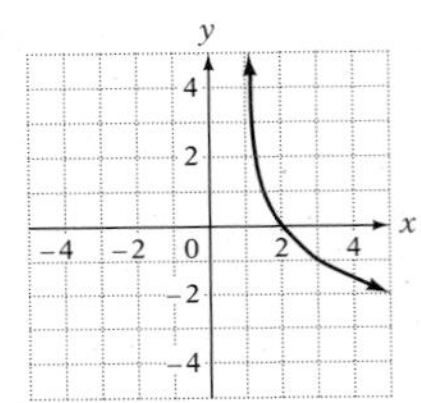

15. (1, 0) **17.**

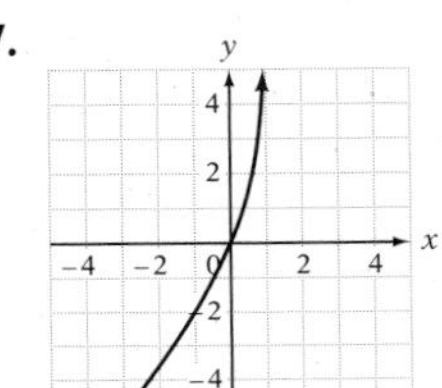

19.

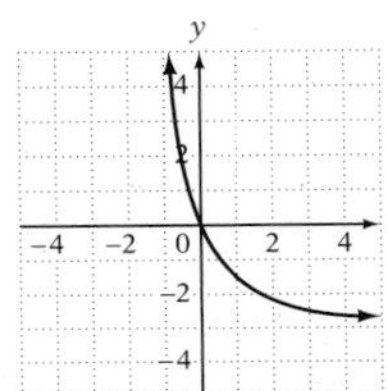

21.

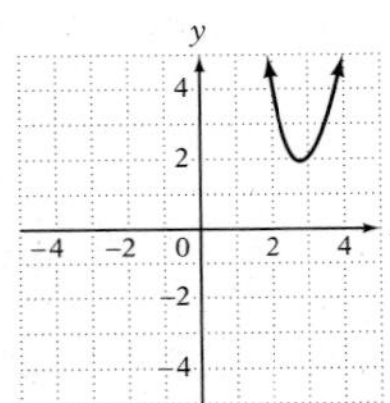

23. a.

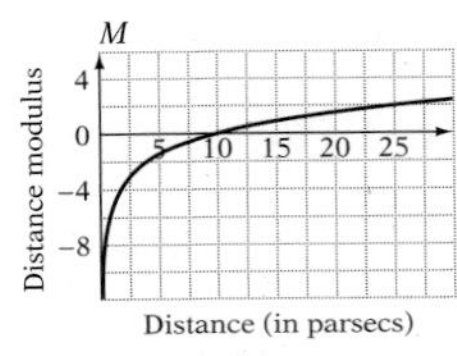

b. The point (25.1, 2) means that a star that is 25.1 parsecs from Earth has a distance modulus of 2.

SECTION 12.4

3. $x < 0$ **5.** $-\frac{1}{3}$ **7.** −3 **9.** 1 **11.** 6 **13.** $\frac{4}{7}$ **15.** $\frac{9}{8}$ **17.** 1.1133 **19.** −1.7626 **21.** 0.7740 **23.** −2.8074 **25.** 3.5850 **27.** 1.1309 **29.** −2.5850 **31.** 1.2224 **33.** 4.1166 **35.** 4.4629 **37.** $\frac{1}{2}$ **39.** No solution **41.** 3 **43.** 2 **45.** $\frac{11}{2}$ **47.** 17.5327 **49.** 1 **51.** $\frac{5}{3}$ **53.** No solution **55.** $x - 2 < x$, and therefore $\log(x - 2) < \log x$. This means that $\log(x - 2) - \log x < 0$ and cannot equal the positive number 3. **57.** −6.1285 **59.** 3.0932

61. a.

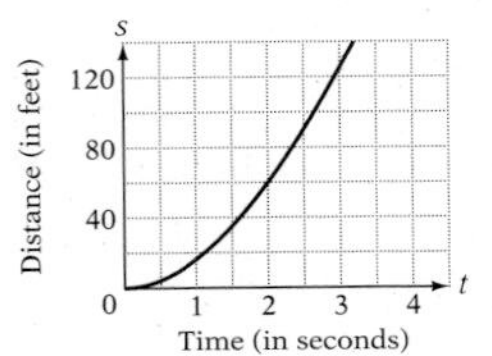

b. It will take 2.64 s for the object to fall 100 ft.

SECTION 12.5

1. The value of the investment after 2 years is $1172. **3.** The investment will be worth $15,000 in approximately 18 years. **5. a.** After 3 h, the technetium level will be 21.2 mg. **b.** The technetium level will reach 20 mg after 3.5 h. **7.** The half-life is 2.5 years. **9.** The intensity of the earthquake was $794{,}328{,}235I_0$. **11.** The Honshu earthquake was approximately 3.2 times stronger than the Quetta earthquake. **13.** The magnitude of the earthquake for the seismogram given is 5.3. **15.** The magnitude of the earthquake for the seismogram given is 5.6. **17.** The pH of the baking soda solution is 8.4. **19.** The range of hydrogen ion concentration is 2.5×10^{-7} to 5.0×10^{-6}. **21.** A blue whale emits 188 decibels of sound. **23.** The sound intensity of a cat's purr is 3.16×10^{-14} watts/cm^2. **25.** 79.4% of the light will pass through the glass. **27.** The thickness of the copper is 0.4 cm. **29. a.** When $v = 160$, $f(375) = 0.43$ ft. **b.** When $v = 166.4$, $f(375) = 24.4$ ft. **31.** The ball would travel 324 ft farther if air resistance were ignored. **33.** The ratio of the current amount of uranium-235 to the original amount of uranium-235 is 0.016. **35. a.** The value of the investment after 3 years is $5986.09. **b.** The investor needs an annual interest rate of 11.2%.

CHAPTER 12 CONCEPT REVIEW*

1. You know that a function is an exponential function if the base is a constant and the exponent contains a variable. [12.1A]

2. There are limitations on the value of the base of an exponential function; it is required to be positive. If the base were a negative number, the value of the function would be a complex number for some values of the variable. To avoid complex number values of a function, the base of the exponential function is always a positive number. Also, $b \neq 1$ because if $b = 1$ in the expression b^x, then the expression is always equal to 1. [12.1A]

3. The number e is an irrational number that is approximately equal to 2.71828183. [12.1A]

4. Because a logarithm is the inverse function of the exponential function, it has the same restrictions on the base b: $b > 0$, $b \neq 1$. See the answer to question 2. [12.2A]

5. The relationship between logarithmic functions and exponential functions is that $y = \log_b x$ is equivalent to $x = b^y$. [12.2A]

6. Remember that a logarithm is an exponent. When we multiply monomials, we add the exponents. Therefore, $\log_b (xy) = \log_b x + \log_b y$. [12.2B]

7. Remember that a logarithm is an exponent. When we divide monomials, we subtract the exponents. Therefore, $\log_b \left(\frac{x}{y}\right) = \log_b x - \log_b y$. [12.2B]

8. Using the Change-of-Base Formula, we can rewrite a logarithm in any base as a common logarithm or a natural logarithm. [12.2C]

9. The domain of a logarithmic function is $\{x \mid x > 0\}$. [12.3A]

10. The compound interest formula is $A = P(1 + i)^n$, where P is the original value of an investment, i is the interest rate per compounding period, n is the total number of compounding periods, and A is the value of the investment after n periods. [12.5A]

11. The Richter scale magnitude of an earthquake uses a logarithmic function to convert the intensity of shock waves I into a number M, which for most earthquakes is in the range of 0 to 10. The intensity I of an earthquake is often given in terms of the constant I_0, where I_0 is the intensity of the smallest earthquake, called a zero-level earthquake, that can be measured on a seismograph near the earthquake's epicenter. An earthquake with an intensity I has a Richter scale magnitude of $M = \log \frac{I}{I_0}$, where I_0 is the measure of the intensity of a zero-level earthquake. [12.5A]

CHAPTER 12 REVIEW EXERCISES

1. 1 [12.1A] **2.** $5^2 = 25$ [12.2A] **3.** [12.1B] **4.** [12.3A]

5. $\frac{2}{5} \log_3 x + \frac{4}{5} \log_3 y$ [12.2B] **6.** $\log_3 \frac{x^2}{y^5}$ [12.2B] **7.** -12 [12.4A] **8.** $\frac{1}{4}$ [12.4B] **9.** 1.7251 [12.2C] **10.** 32 [12.2A] **11.** 79 [12.4B] **12.** 1000 [12.2A] **13.** $\log_7 \sqrt[3]{xy^4}$ [12.2B] **14.** $\frac{5}{2} \log_8 x - 3 \log_8 y$ [12.2B]

*Note: The numbers in brackets following the answers in the Concept Review are a reference to the objective that corresponds to that problem. For example, the reference [1.2A] stands for Section 1.2, Objective A. This notation will be used for all Prep Tests, Concept Reviews, Chapter Reviews, Chapter Tests, and Cumulative Reviews throughout the text.

15. $\log_2 32 = 5$ [12.2A] **16.** 0.4278 [12.2C] **17.** -0.535 [12.4A] **18.** $\frac{3}{2}$ [12.1A] **19.** $\frac{11}{7}$ [12.4B]

20. 3 [12.4B] **21.** [12.1B] **22.** [12.3A] **23.** 2 [12.2A]

24. $\log_2 \sqrt[3]{\frac{x}{y}}$ [12.2B] **25.** 1 [12.4A] **26.** 1.5938 [12.4A] **27.** $\frac{1}{5}$ [12.2A] **28.** $\log_3 81 = 4$ [12.2A]

29. 5 [12.4B] **30.** $\frac{2}{3}\log_5 x + \frac{1}{3}\log_5 y$ [12.2B] **31.** -0.5207 [12.4A] **32.** $\frac{1}{7}$ [12.1A] **33.** 4 [12.2A]

34. 2 [12.4B] **35.** 2.3219 [12.2C] **36.** 3 [12.4A] **37.** 625 [12.2A] **38.** $\log_b \frac{x^3}{y^7}$ [12.2B]

39. 5 [12.1A] **40.** 3.2091 [12.4A] **41.** The value of the investment in 2 years is $4692. [12.5A]

42. The Richter scale magnitude of the earthquake is 8.3. [12.5A] **43.** The half-life is 27 days. [12.5A]

44. The sound emitted from the busy street corner is 107 decibels. [12.5A]

CHAPTER 12 TEST

1. $f(0) = 1$ [12.1A, Example 1] **2.** $f(-2) = \frac{1}{3}$ [12.1A, Example 2] **3.**

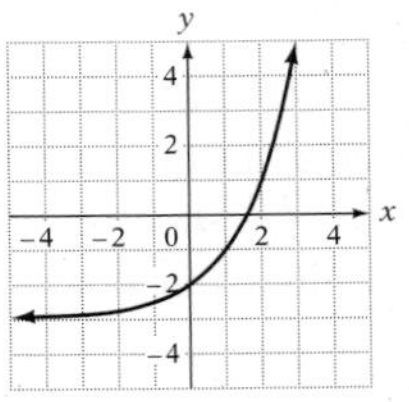

[12.1B, Example 5]

4.

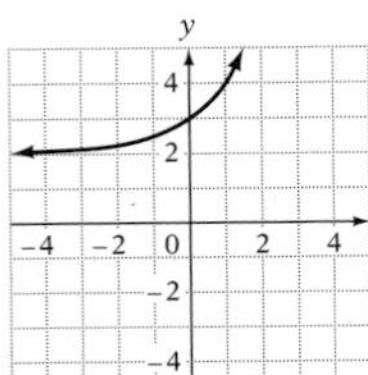

[12.1B, You Try It 5] **5.** 2 [12.2A, You Try It 1] **6.** $\frac{1}{9}$ [12.2A, How To 4]

7.

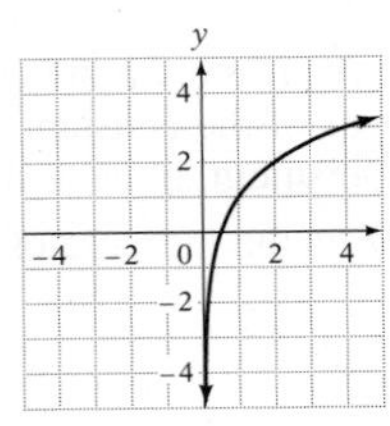

[12.3A, You Try It 2] **8.** [12.3A, Example 3]

9. $\frac{1}{2}\log_6 x + \frac{3}{2}\log_6 y$ [12.2B, You Try It 5] **10.** $\log_3 \sqrt{\frac{x}{y}}$ [12.2B, Example 6] **11.** $\ln x - \frac{1}{2}\ln z$ [12.2B, How To 13]

12. $\ln \frac{x^3}{y\sqrt{z}}$ [12.2B, How To 14] **13.** -2 [12.4A, How To 1] **14.** -3 [12.4A, How To 2]

15. 2.5789 [12.4A, Example 2] **16.** 6 [12.4B, How To 7] **17.** 3 [12.4B, How To 7]

18. 1.3652 [12.2C, How To 16] **19.** 2.6801 [12.2C, How To 16] **20.** 3.1826 [12.4A, How To 4]

21. 6.0163 [12.4A, How To 4] **22.** 5 [12.4B, How To 6] **23.** The shard is 3099 years old. [12.5A, How To 2]

24. The intensity of the sound is 3.16×10^{-9} watts/cm^2. [12.5A, Example 1] **25.** The half-life is 33 h. [12.5A, How To 2]

CUMULATIVE REVIEW EXERCISES

1. $\frac{8}{7}$ [2.2C] **2.** $y = 2x - 6$ [4.6A] **3.** $(4x^n + 3)(x^n + 1)$ [7.3A/7.3B] **4.** $\frac{x-3}{x+3}$ [8.3A] **5.** $\frac{x\sqrt{y} + y\sqrt{x}}{x - y}$ [9.2D]
6. $2 + \sqrt{10}$ and $2 - \sqrt{10}$ [10.2A] **7.** $3x^2 + 8x - 3 = 0$ [10.1B] **8.**

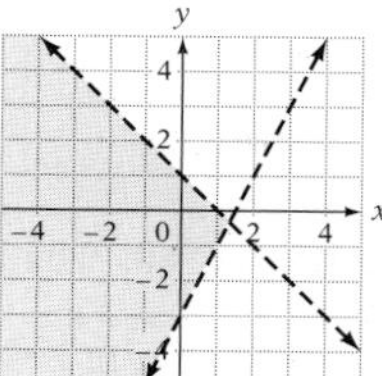

[5.4A] **9.** $(0, -1, 2)$ [5.2B]

10. $-\frac{2x^2 - 17x + 13}{(x-2)(2x-3)}$ [8.2B] **11.** $\{x \mid -5 \le x \le 1\}$ [10.5A] **12.** $\{x \mid 1 \le x \le 4\}$ [2.6B]

13.

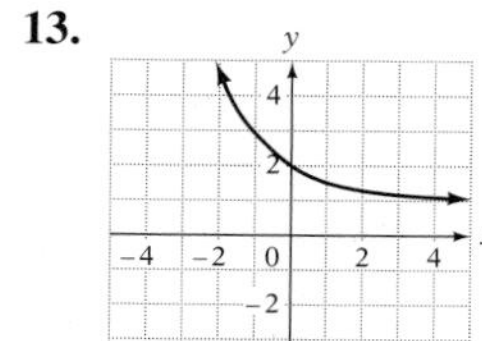

[12.1B] **14.**

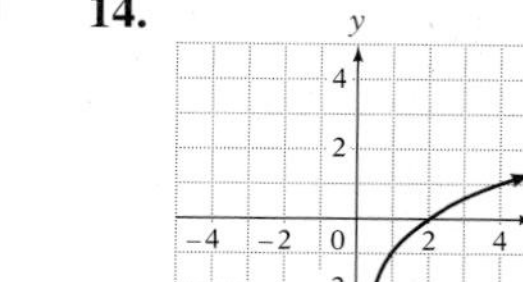

[12.3A] **15.** 4 [12.1A]

16. 125 [12.2A] **17.** $\log_b \frac{x^3}{y^5}$ [12.2B] **18.** 1.7712 [12.2C] **19.** 2 [12.4A] **20.** $\frac{1}{2}$ [12.4B]
21. The customer can write at most 49 checks. [2.5C] **22.** The cost per pound of the mixture is \$3.10. [2.4A]
23. The rate of the wind is 25 mph. [8.7B] **24.** The spring will stretch 10.2 in. [8.8A] **25.** The cost of redwood is \$.40 per foot. The cost of fir is \$.25 per foot. [5.3B] **26.** In approximately 10 years, the investment will be worth \$10,000. [12.5A]

FINAL EXAM

1. -31 [1.3A] **2.** -1 [1.4A] **3.** $-10x + 33$ [1.4D] **4.** 8 [2.2A] **5.** $4, -\frac{2}{3}$ [2.6A] **6.** The volume is 268.1 ft³. [3.3A] **7.**

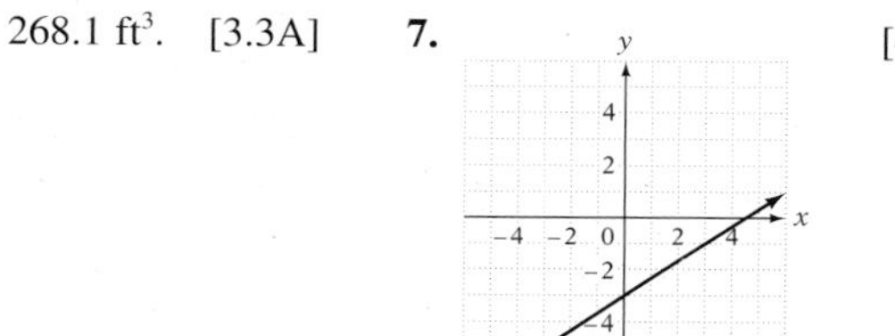

[4.3C] **8.** $y = -3x + 7$ [4.5B] **9.** $y = -\frac{2}{3}x - \frac{1}{3}$ [4.6A]

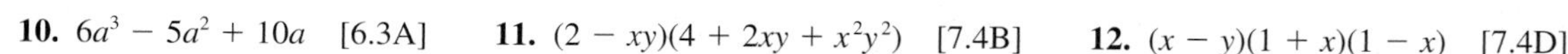

10. $6a^3 - 5a^2 + 10a$ [6.3A] **11.** $(2 - xy)(4 + 2xy + x^2y^2)$ [7.4B] **12.** $(x - y)(1 + x)(1 - x)$ [7.4D]
13. $x^2 - 2x - 3 - \frac{5}{2x - 3}$ [6.4B] **14.** $\frac{x(x-1)}{2x-5}$ [8.1C] **15.** $\frac{-10x}{(x+2)(x-3)}$ [8.2B] **16.** $\frac{x+3}{x+1}$ [8.3A]
17. $-\frac{7}{4}$ [8.4A] **18.** $d = \frac{a_n - a_1}{n - 1}$ [8.6A] **19.** $\frac{y^4}{162x^3}$ [6.1B] **20.** $\frac{1}{64x^8y^5}$ [9.1A] **21.** $-2x^2y\sqrt{2y}$ [9.2B]
22. $\frac{x^2\sqrt{2y}}{2y^2}$ [9.2D] **23.** $\frac{6}{5} - \frac{3}{5}i$ [9.4D] **24.** $2x^2 - 3x - 2 = 0$ [10.1B] **25.** $\frac{3+\sqrt{17}}{4}, \frac{3-\sqrt{17}}{4}$ [10.3A]

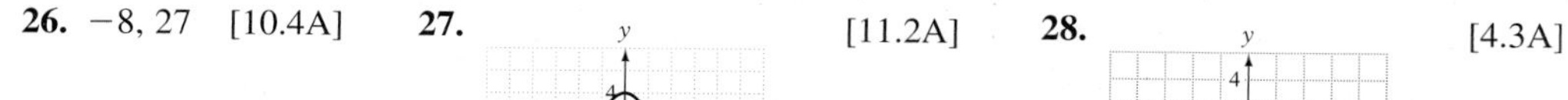

26. $-8, 27$ [10.4A] **27.**

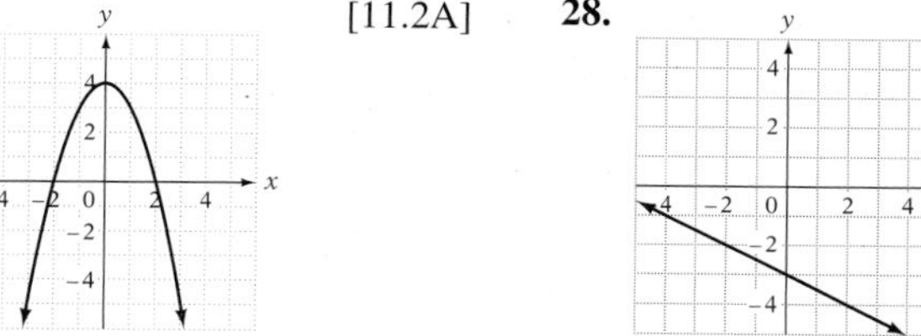

[11.2A] **28.** [4.3A]

29. $-2, \frac{3}{2}$ [10.4C] **30.** $f^{-1}(x) = \frac{3}{2}x + 6$ [11.4B] **31.** $(3, 4)$ [5.2A] **32.** $7x^3$ [9.1C]
33. $\left\{x \mid x > \frac{3}{2}\right\}$ [2.5B] **34.** $\{x \mid -4 < x < -1\}$ [2.6B]

35. 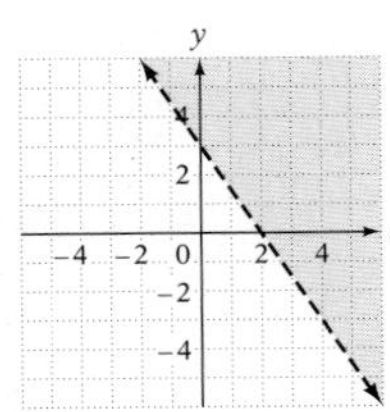 [4.7A] **36.** 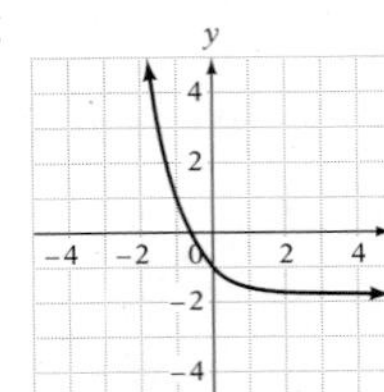 [12.1B] **37.** 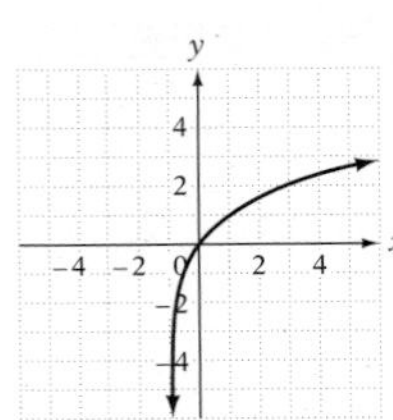 [12.3A]

38. $\log_2 \frac{a^2}{b^2}$ [12.2B] **39.** 6 [12.4B] **40.** The range of scores is $69 \le x \le 100$. [2.5C]

41. The cyclist rode 40 mi. [2.4C] **42.** There is \$8000 invested at 8.5% and \$4000 invested at 6.4%. [5.1C] **43.** The length is 20 ft. The width is 7 ft. [10.6A] **44.** An additional 200 shares are needed. [8.5B] **45.** The rate of the plane was 420 mph. [8.7B] **46.** The object has fallen 88 ft when the speed reaches 75 ft/s. [9.3B] **47.** The rate of the plane for the first 360 mi was 120 mph. [8.7B] **48.** The intensity is 200 foot-candles. [8.8A] **49.** The rate of the boat in calm water is 12.5 mph. The rate of the current is 2.5 mph. [5.3A] **50.** The value of the investment after 2 years is \$4785.65. [12.5A]

ANSWERS TO CHAPTER R SELECTED EXERCISES

SECTION R.1

1. 10 **3.** 21 **5.** -2 **7.** 5 **9.** -3 **11.** -1 **13.** 1 **15.** 18 **17.** -6 **19.** 2 **21.** -40 **23.** 2.01 **25.** $8x$ **27.** $3b$ **29.** $5a$ **31.** $11x$ **33.** $15x - 2y$ **35.** $9a$ **37.** $22y^2$ **39.** $\frac{1}{2}x$ **41.** $-20x$ **43.** $-24a$ **45.** x **47.** $7x$ **49.** $3y$ **51.** $-3a - 15$ **53.** $-16x - 48$ **55.** $-10y^2 + 5$ **57.** $18x^2 - 12xy - 6y^2$ **59.** $-8y - 7$ **61.** $5x - 75$ **63.** $-a + 56$

SECTION R.2

1. -12 **3.** -1 **5.** -8 **7.** -20 **9.** 8 **11.** -1 **13.** 1 **15.** $\frac{3}{4}$ **17.** 3 **19.** 3 **21.** 2 **23.** -1 **25.** $\frac{1}{2}$ **27.** 5 **29.** $-\frac{1}{9}$ **31.** $\{x \mid x > 3\}$ **33.** $\{n \mid n \ge 2\}$ **35.** $\{x \mid x \le -3\}$ **37.** $\{n \mid n > 0\}$ **39.** $\{x \mid x > 4\}$ **41.** $\{x \mid x > -2\}$ **43.** $\{x \mid x > 3\}$ **45.** $\{x \mid x \ge 2\}$ **47.** $\{x \mid x < 2\}$ **49.** $\{x \mid x > 3\}$ **51.** $\{x \mid x \ge 2\}$ **53.** $\{x \mid x < 4\}$ **55.** $\left\{x \mid x > \frac{14}{11}\right\}$ **57.** $\left\{x \mid x \le \frac{3}{8}\right\}$ **59.** $\left\{x \mid x \ge -\frac{5}{4}\right\}$

SECTION R.3

1. 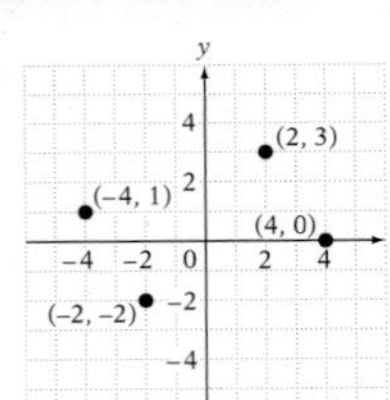

3. 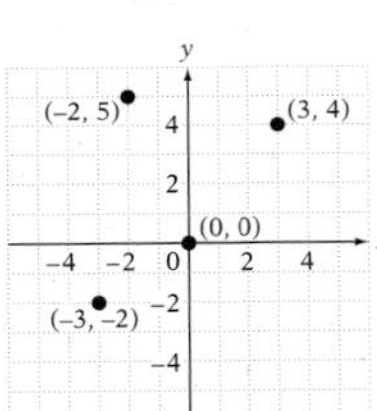

5. 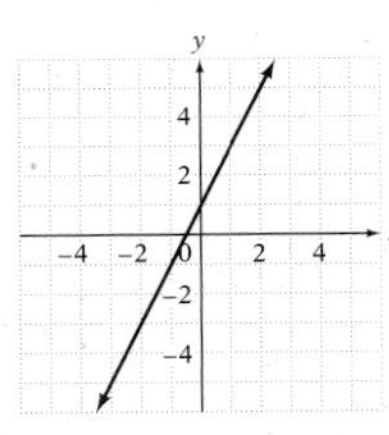

7. 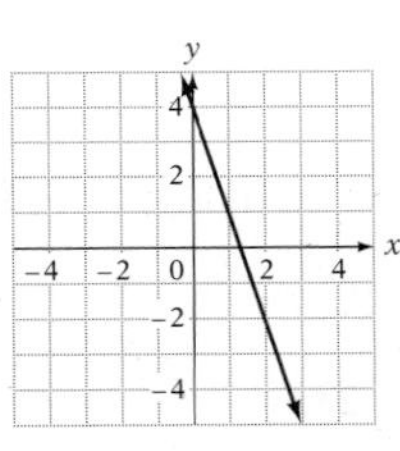

9. 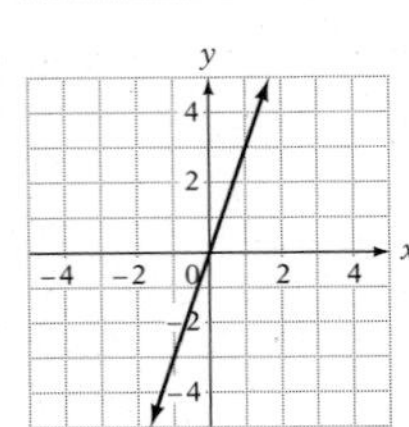

11. 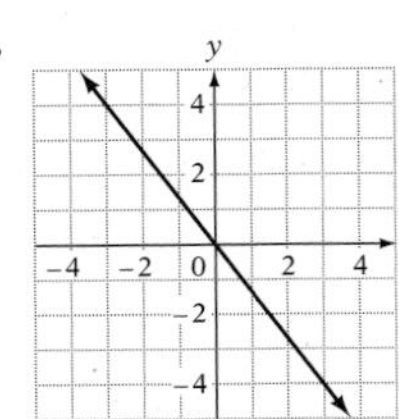

13. 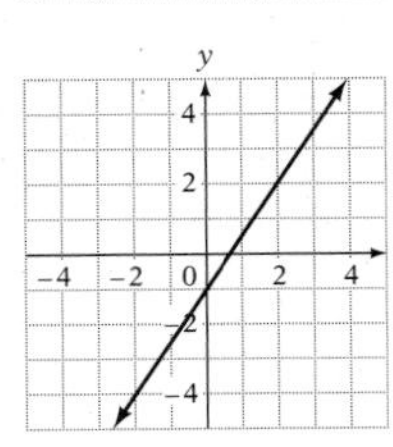

15. 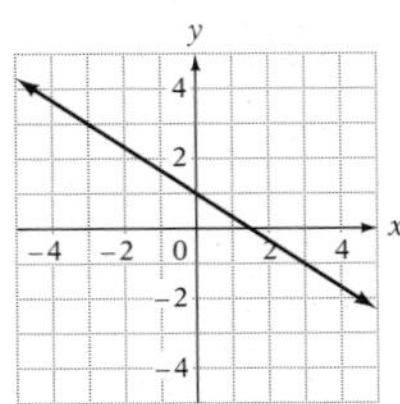

17.

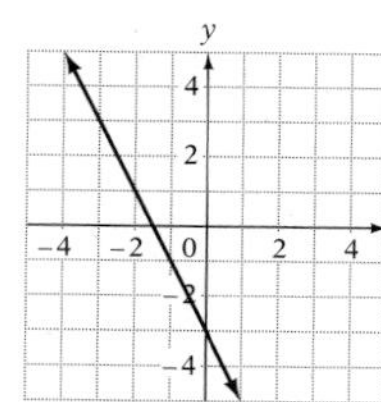

19.

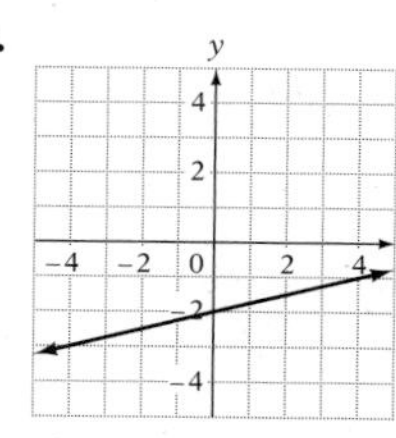

21.

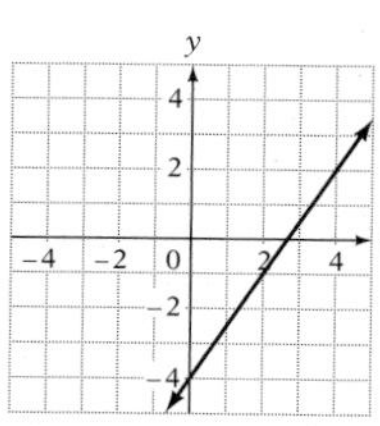

23. 3 **25.** $\frac{1}{4}$ **27.** $-\frac{3}{2}$

29. Undefined **31.** 0 **33.** 1 **35.** $m = \frac{5}{2}$, $b = (0, -4)$ **37.** $m = 1$, $b = (0, 0)$ **39.**

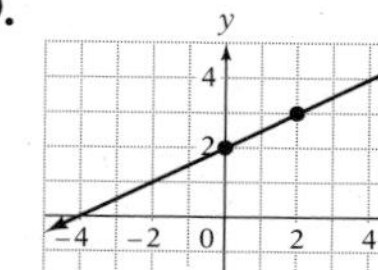

41.

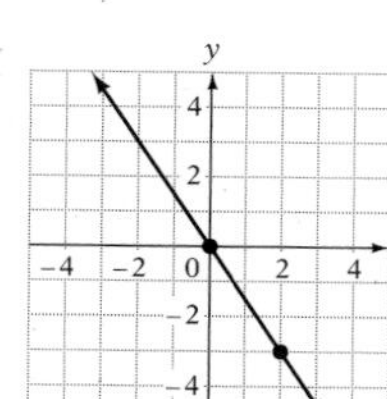

43.

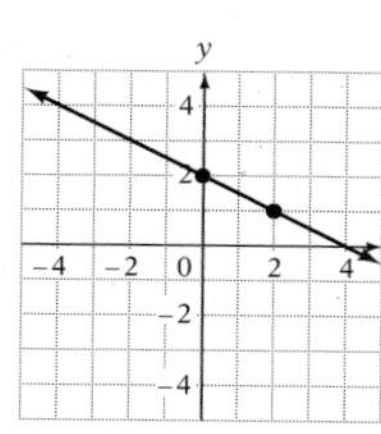

45.

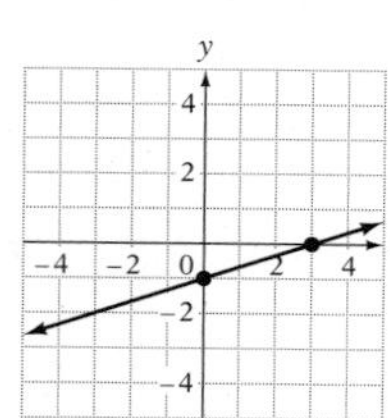

47. 13; $(-4, 13)$ **49.** 14; $(-1, 14)$

51. -1; $(2, -1)$ **53.** -15; $(-2, -15)$ **55.** -2; $(-1, -2)$ **57.** $y = -3x - 1$ **59.** $y = -2x + 3$ **61.** $y = -\frac{3}{5}x$
63. $y = -\frac{2}{3}x - 4$

SECTION R.4

1. z^8 **3.** x^{15} **5.** $x^{12}y^{18}$ **7.** a^6 **9.** $-m^9n^3$ **11.** $-8a^9b^3c^6$ **13.** mn^2 **15.** $-\frac{2a}{3}$ **17.** $-27m^2n^4p^3$
19. $54n^{14}$ **21.** $\frac{7xz}{8y^3}$ **23.** $-8x^{13}y^{14}$ **25.** $\frac{4}{x^7}$ **27.** $\frac{1}{d^{10}}$ **29.** $\frac{1}{x^5}$ **31.** $\frac{x^2}{3}$ **33.** $\frac{x^6}{y^{12}}$ **35.** $\frac{9}{x^2y^4}$ **37.** $\frac{2}{x^4}$
39. $\frac{1}{2x^3}$ **41.** $\frac{1}{2x^2y^6}$ **43.** $\frac{1}{x^6}$ **45.** $\frac{y^8}{x^4}$ **47.** $\frac{b^{10}}{4a^{10}}$ **49.** $7b^2 + b - 4$ **51.** $3a^2 - 3a + 17$ **53.** $-7x - 7$
55. $-2x^3 + x^2 + 2$ **57.** $x^3 + 2x^2 - 6x - 6$ **59.** $5x^3 + 10x^2 - x - 4$ **61.** $y^3 - y^2 + 6y - 6$
63. $4y^3 - 2y^2 + 2y - 4$ **65.** $12b^4 - 48b^3 - 24b$ **67.** $9b^5 - 9b^3 + 24b$ **69.** $-2x^4y + 6x^3y^2 - 4x^2y^3$
71. $x^3 + 4x^2 + 5x + 2$ **73.** $a^3 - 6a^2 + 13a - 12$ **75.** $-2b^3 + 7b^2 + 19b - 20$ **77.** $x^4 - 4x^3 - 3x^2 + 14x - 8$
79. $y^4 + 4y^3 + y^2 - 5y + 2$ **81.** $a^2 + a - 12$ **83.** $y^2 - 10y + 21$ **85.** $2x^2 + 15x + 7$ **87.** $3x^2 + 11x - 4$
89. $4x^2 - 31x + 21$ **91.** $3y^2 - 2y - 16$ **93.** $21a^2 - 83a + 80$ **95.** $2x^2 + 3xy + y^2$ **97.** $3x^2 - 10xy + 8y^2$
99. $10a^2 + 14ab - 12b^2$ **101.** $16x^2 - 49$ **103.** $x + 2$ **105.** $2y - 7$ **107.** $x - 2 + \frac{8}{x + 2}$
109. $3y - 5 + \frac{20}{2y + 4}$ **111.** $b - 5 - \frac{24}{b - 3}$ **113.** $3x + 17 + \frac{64}{x - 4}$ **115.** $5y + 3 + \frac{1}{2y + 3}$ **117.** $x^2 - 5x + 2$
119. $x^2 + 5$ **121.** $y(12y - 5)$ **123.** $5xyz(2xz + 3y^2)$ **125.** $5(x^2 - 3x + 7)$ **127.** $3y^2(y^2 - 3y - 2)$
129. $x^2y^2(x^2y^2 - 3xy + 6)$ **131.** $8x^2y(2 - xy^3 - 6y)$ **133.** $(x + 2)(x - 1)$ **135.** $(a + 4)(a - 3)$
137. $(a - 1)(a - 2)$ **139.** $(b + 8)(b - 1)$ **141.** $(z + 5)(z - 9)$ **143.** $(z - 5)(z - 9)$ **145.** $(b + 4)(b + 5)$
147. Nonfactorable over the integers **149.** $(x - 7)(x - 8)$ **151.** $(y + 3)(2y + 1)$ **153.** $(a - 1)(3a - 1)$
155. $(x - 3)(2x + 1)$ **157.** $(2t + 1)(5t + 3)$ **159.** $(2z - 1)(5z + 4)$ **161.** Nonfactorable over the integers
163. $(t + 2)(2t - 5)$ **165.** $(3y + 1)(4y + 5)$ **167.** $(a - 5)(11a + 1)$ **169.** $(2b - 3)(3b - 2)$
171. $3(x + 2)(x + 3)$ **173.** $a(b + 8)(b - 1)$ **175.** $2y^2(y + 3)(y - 16)$ **177.** $x(x - 5)(2x - 1)$
179. $5(t + 2)(2t - 5)$ **181.** $p(2p + 1)(3p + 1)$

Glossary

abscissa The first number of an ordered pair; it measures a horizontal distance and is also called the first coordinate of an ordered pair. [4.1]

absolute value of a number The distance of the number from zero on the number line. [1.1, 2.6]

absolute value inequality An inequality that contains the absolute-value symbol. [2.6]

absolute value equation An equation that contains the absolute-value symbol. [2.6]

acute angle An angle whose measure is between 0° and 90°. [3.1]

acute triangle A triangle that has three acute angles. [3.2]

addend In addition, a number being added. [1.1]

addition method An algebraic method of finding an exact solution of a system of linear equations. [5.2]

additive inverse of a polynomial The polynomial with the sign of every term changed. [6.2]

additive inverses Numbers that are the same distance from zero on the number line but lie on different sides of zero; also called opposites. [1.1, 1.4]

adjacent angles Two angles that share a common side. [3.1]

alternate exterior angles Two nonadjacent angles that are on opposite sides of the transversal and lie outside the parallel lines. [3.1]

alternate interior angles Two nonadjacent angles that are on opposite sides of the transversal and between the parallel lines. [3.1]

analytic geometry Geometry in which a coordinate system is used to study relationships between variables. [4.1]

angle Figure formed when two rays start from the same point. [3.1]

area A measure of the amount of surface in a region. [3.2]

asymptotes The two straight lines that a hyperbola "approaches." [11.5 – online]

axes The two number lines that form a rectangular coordinate system; also called coordinate axes. [4.1]

axis of symmetry of a parabola A line of symmetry that passes through the vertex of the parabola and is parallel to the y-axis for an equation of the form $y = ax^2 + bx + c$ or parallel to the x-axis for an equation of the form $x = ay^2 + by + c$. [11.2, 11.5 – online]

base In an exponential expression, the number that is taken as a factor as many times as indicated by the exponent. [1.2]

basic percent equation Percent times base equals amount. [2.1]

binomial A polynomial of two terms. [6.2]

center of a circle The central point that is equidistant from all the points that make up a circle. [3.2, 11.5 – online]

center of an ellipse The intersection of the two axes of symmetry of the ellipse. [11.5 – online]

circle Plane figure in which all points are the same distance from its center. [3.2, 11.5 – online]

circumference The perimeter of a circle. [3.2]

clearing denominators Removing denominators from an equation that contains fractions by multiplying each side of the equation by the LCM of the denominators. [2.2]

coefficient The number part of a variable term. [1.4]

cofactor of an element of a matrix $(-1)^{i+j}$ times the minor of that element, where i is the row number of the element and j is its column number. [5.5 – online]

combined variation A variation in which two or more types of variation occur at the same time. [8.8]

combining like terms Using the Distributive Property to add the coefficients of like variable terms; adding like terms of a variable expression. [1.4]

common logarithms Logarithms to the base 10. [12.2]

common monomial factor A monomial factor that is a factor of the terms in a polynomial. [7.1]

complementary angles Two angles whose measures have the sum 90°. [3.1]

completing the square Adding to a binomial the constant term that makes it a perfect-square trinomial. [10.2]

complex fraction A fraction whose numerator or denominator contains one or more fractions. [8.3]

complex number A number of the form $a + bi$, where a and b are real numbers and $i = \sqrt{-1}$. [9.4]

composition of two functions The operation on two functions f and g denoted by $f \circ g$. The value of the composition of f and g is given by $(f \circ g)(x) = f[g(x)]$. [11.3]

compound inequality Two inequalities joined with a connective word such as *and* or *or*. [2.5]

compound interest Interest that is computed not only on the original principal but also on the interest already earned. [12.5]

conic section A curve that can be constructed from the intersection of a plane and a right circular cone. The four conic sections are the parabola, hyperbola, ellipse, and circle. [11.5 – online]

conjugates Binomial expressions that differ only in the sign of a term. The expressions $a + b$ and $a - b$ are conjugates. [9.2]

consecutive even integers Even integers that follow one another in order. [2.3]

consecutive integers Integers that follow one another in order. [2.3]

consecutive odd integers Odd integers that follow one another in order. [2.3]

constant function A function given by $f(x) = b$, where b is a constant. Its graph is a horizontal line passing through $(0, b)$. [4.3]

constant of proportionality k in a variation equation; also called the constant of variation. [8.8]

constant of variation k in a variation equation; also called the constant of proportionality. [8.8]

constant term A term that includes no variable part; also called a constant. [1.4]

coordinate axes The two number lines that form a rectangular coordinate system; also called axes. [4.1]

coordinates of a point The numbers in the ordered pair that is associated with the point. [4.1]

corresponding angles Two angles that are on the same side of the transversal and are both acute angles or are both obtuse angles. [3.1]

cube A rectangular solid in which all six faces are squares. [3.3]

cube root of a perfect cube One of the three equal factors of the perfect cube. [7.4]

cubic function A third-degree polynomial function. [6.2]

decimal notation Notation in which a number consists of a whole-number part, a decimal point, and a decimal part. [1.2]

degree A unit used to measure angles. [3.1]

degree of a monomial The sum of the exponents of the variables. [6.1]

degree of a polynomial The greatest of the degrees of any of the polynomial's terms. [6.2]

dependent system of equations A system of equations whose graphs coincide. [5.1]

dependent variable In a function, the variable whose value depends on the value of another variable known as the independent variable. [4.2]

descending order The terms of a polynomial in one variable are arranged in descending order when the exponents of the variable decrease from left to right. [6.2]

determinant A number associated with a square matrix. [5.5 – online]

diameter of a circle A line segment with endpoints on the circle and going through the center. [3.2]

diameter of a sphere A line segment with endpoints on the sphere and going through the center. [3.3]

difference of two perfect squares A polynomial in the form $a^2 - b^2$. [7.4]

difference of two perfect cubes A polynomial in the form $a^3 - b^3$. [7.4]

direct variation A special function that can be expressed as the equation $y = kx$, where k is a constant called the constant of variation or the constant of proportionality. [8.8]

discriminant For an equation of the form $ax^2 + bx + c = 0$, the quantity $b^2 - 4ac$ is called the discriminant. [10.3]

domain The set of the first coordinates of all the ordered pairs of a relation. [4.2]

double root When a quadratic equation has two solutions that are the same number, the solution is called a double root of the equation. [10.1]

element of a matrix A number in a matrix. [5.5 – online]

elements of a set The objects in the set. [1.1, 1.5]

ellipse An oval shape that is one of the conic sections. [11.5 – online]

empty set The set that contains no elements; also called the null set. [1.5]

equation A statement of the equality of two mathematical expressions. [2.1]

equilateral triangle A triangle in which all three sides are of equal length. [3.2]

equivalent equations Equations that have the same solution. [2.1]

evaluating a function Replacing x in $f(x)$ with some value and then simplifying the numerical expression that results. [4.2]

evaluating a variable expression Replacing each variable by its value and then simplifying the resulting numerical expression. [1.4]

even integer An integer that is divisible by 2. [2.3]

expanding by cofactors A technique for finding the value of a 3×3 or larger determinant. [5.5 – online]

exponent In an exponential expression, the raised number that indicates how many times the factor, or base, occurs in the multiplication. [1.2]

exponential equation An equation in which the variable occurs in the exponent. [12.4]

exponential form The expression 2^6 is in exponential form. Compare *factored form.* [1.2]

exponential function The exponential function with base b is defined by $f(x) = b^x$, where b is a positive real number not equal to 1. [12.1]

exterior angle An angle adjacent to an interior angle of a triangle. [3.1]

extraneous solution When each side of an equation is raised to an even power, the resulting equation may have a solution that is not a solution of the original equation. Such a solution is called an extraneous solution. [9.3]

factor In multiplication, a number being multiplied. [1.1]

factored form The multiplication $2 \cdot 2 \cdot 2 \cdot 2 \cdot 2 \cdot 2$ is in factored form. Compare *exponential form.* [1.2]

factoring a polynomial Writing the polynomial as a product of other polynomials. [7.1]

factoring a quadratic trinomial Expressing the trinomial as the product of two binomials. [7.2]

FOIL A method of finding the product of two binomials. The letters stand for First, Outer, Inner, and Last. [6.3]

formula A literal equation that states a rule about measurement. [8.6]

function A relation in which no two ordered pairs that have the same first coordinate have different second coordinates. [4.2]

function notation A function designated by $f(x)$, which is the value of the function at x. [4.2]

graph of an integer A heavy dot directly above the number on the number line. [1.1]

graph of a function A graph of the ordered pairs that belong to the function. [4.3]

graph of an ordered pair The dot drawn at the coordinates of the point in the plane. [4.1]

graphing a point in the plane Placing a dot at the location given by the ordered pair; also called plotting a point in the plane. [4.1]

greater than A number that lies to the right of another number on the number line is said to be greater than that number. [1.1]

greatest common factor (GCF) The greatest common factor of two or more integers is the greatest integer that is a factor of all the integers. [7.1]

greatest common factor (GCF) of two or more monomials The greatest common factor of two or more monomials is the product of the GCF of the coefficients and the common variable factors. [7.1]

grouping symbols Parentheses ( ), brackets [ ], braces { }, the absolute value symbol, and the fraction bar. [1.3]

half-plane The solution set of a linear inequality in two variables. [4.7]

horizontal-line test A graph of a function represents the graph of a one-to-one function if any horizontal line intersects the graph at no more than one point. [11.4]

hyperbola A conic section formed by the intersection of a cone and a plane perpendicular to the base of the cone. [11.5 – online]

hypotenuse In a right triangle, the side opposite the 90° angle. [9.3]

imaginary number A number of the form ai, where a is a real number and $i = \sqrt{-1}$. [9.4]

imaginary part of a complex number For the complex number $a + bi$, b is the imaginary part. [9.4]

inconsistent system of equations A system of equations that has no solution. [5.1]

independent system of equations A system of equations whose graphs intersect at only one point. [5.1]

independent variable In a function, the variable that varies independently and whose value determines the value of the dependent variable. [4.2]

index In the expression $\sqrt[n]{a}$, n is the index of the radical. [9.1]

inequality An expression that contains the symbol $>$, $<$, $\geq$ (is greater than or equal to), or $\leq$ (is less than or equal to). [1.5]

integers The numbers . . . , −3, −2, −1, 0, 1, 2, 3, ... [1.1]

interior angle of a triangle Angle within the region enclosed by a triangle. [3.1]

intersecting lines Lines that cross at a point in the plane. [3.1]

intersection of two sets The set that contains all elements that are common to both of the sets. [1.5]

inverse of a function The set of ordered pairs formed by reversing the coordinates of each ordered pair of the function. [11.4]

inverse variation A function that can be expressed as the equation $y = \frac{k}{x}$, where k is a constant. [8.8]

irrational number The decimal representation of an irrational number never terminates or repeats and can only be approximated. [1.2, 9.2]

isosceles triangle A triangle that has two sides of equal length; the angles opposite the equal sides are of equal measure. [3.2]

joint variation A variation in which a variable varies directly as the product of two or more variables. A joint variation can be expressed as the equation $z = kxy$, where k is a constant. [8.8]

leading coefficient In a polynomial, the coefficient of the variable with the largest exponent. [6.2]

least common denominator The smallest number that is a multiple of each denominator in question. [1.2]

least common multiple (LCM) The LCM of two or more numbers is the smallest number that is a multiple of each of those numbers. [1.2]

least common multiple (LCM) of two or more polynomials The simplest polynomial of least degree that contains the factors of each polynomial. [8.2]

leg In a right triangle, one of the two sides that are not opposite the 90° angle. [9.3]

less than A number that lies to the left of another number on the number line is said to be less than that number. [1.1]

like terms Terms of a variable expression that have the same variable part. Having no variable part, constant terms are like terms. [1.4]

line Having no width, it extends indefinitely in two directions in a plane. [3.1]

linear equation in three variables An equation of the form $Ax + By + Cz = D$, where A, B, and C are coefficients of the variables and D is a constant. [5.2]

linear equation in two variables An equation of the form $y = mx + b$, where m is the coefficient of x and b is a constant; also called a linear function. [4.3]

linear function A function that can be expressed in the form $y = mx + b$. Its graph is a straight line. [4.3, 6.2]

linear inequality in two variables An inequality of the form $y > mx + b$ or $Ax + By > C$. The symbol $>$ could be replaced by $\geq$, $<$, or $\leq$. [4.7]

line segment Part of a line; it has two endpoints. [3.1]

literal equation An equation that contains more than one variable. [8.6]

logarithm For b greater than zero and not equal to 1, the statement $y = \log_b x$ (the logarithm of x to the base b) is equivalent to $x = b^y$. [12.2]

lower limit In a tolerance, the lowest acceptable value. [2.6]

main fraction bar The fraction bar that is placed between the numerator and denominator of a complex fraction. [8.3]

matrix A rectangular array of numbers. [5.5 – online]

maximum value of a function The greatest value that the function can take on. [11.1]

minimum value of a function The least value that the function can take on. [11.1]

minor of an element The minor of an element in a 3×3 determinant is the 2×2 determinant obtained by eliminating the row and column that contain that element. [5.5 – online]

monomial A number, a variable, or a product of a number and variables; a polynomial of one term. [6.1, 6.2]

multiplication The process of finding the product of two numbers. [1.1]

multiplicative inverse The multiplicative inverse of a nonzero real number a is $\frac{1}{a}$; also called the reciprocal. [1.4]

natural exponential function The function defined by $f(x) = e^x$, where $e \approx 2.71828$. [12.1]

natural logarithm When e (the base of the natural exponential function) is used as the base of a logarithm, the logarithm is referred to as the natural logarithm and is abbreviated $\ln x$. [12.2]

natural numbers The numbers 1, 2, 3 . . . ; also called the positive integers. [1.1]

negative integers The numbers . . . , −3, −2, −1. [1.1]

negative reciprocal The negative reciprocal of a nonzero real number a is $-\frac{1}{a}$. [11.1]

negative slope The slope of a line that slants downward to the right. [4.4]

nonfactorable over the integers A polynomial is nonfactorable over the integers if it does not factor using only integers. [7.2]

n*th root of *a A number b such that $b^n = a$. The nth root of a can be written $a^{1/n}$ or $\sqrt[n]{a}$. [9.1]

null set The set that contains no elements; also called the empty set. [1.5]

numerical coefficient The number part of a variable term. When the numerical coefficient is 1 or -1, the 1 is usually not written. [1.4]

obtuse angle An angle whose measure is between 90° and 180°. [3.1]

obtuse triangle A triangle that has one obtuse angle. [3.2]

odd integer An integer that is not divisible by 2. [2.3]

one-to-one function In a one-to-one function, given any y, there is only one x that can be paired with the given y. [11.4]

opposites Numbers that are the same distance from zero on the number line but lie on different sides of zero; also called additive inverses. [1.1]

order $m \times n$ A matrix of m rows and n columns is of order $m \times n$. [5.5 – online]

Order of Operations Agreement A set of rules that tells us in what order to perform the operations that occur in a numerical expression. [1.3]

ordered pair A pair of numbers expressed in the form (a, b) and used to locate a point in the plane determined by a rectangular coordinate system. [4.1]

ordered triple Three numbers expressed in the form (x, y, z) and used to locate a point in the xyz-coordinate system. [5.2]

ordinate The second number of an ordered pair; it measures a vertical distance and is also called the second coordinate of an ordered pair. [4.1]

origin The point of intersection of the two number lines that form a rectangular coordinate system. [4.1]

parabola The graph of a quadratic function is called a parabola. [11.1]

parallel lines Lines that never meet; the distance between them is always the same. In a rectangular coordinate system, parallel lines have the same slope and thus do not intersect. [3.1, 4.6]

parallelogram Four-sided plane figure with opposite sides parallel. [3.2]

percent Parts of 100. [1.2]

perfect cube The product of the same three factors. [7.4]

perfect square The product of a term and itself. [1.2, 7.4]

perfect-square trinomial The square of a binomial. [7.4]

perimeter The distance around a plane geometric figure. [3.2]

perpendicular lines Intersecting lines that form right angles. The slopes of perpendicular lines are negative reciprocals of each other. [3.1, 4.6]

plane A flat surface that extends indefinitely. [3.1]

plane figure A figure that lies entirely in a plane. [3.1]

plotting a point in the plane Placing a dot at the location given by the ordered pair; also called graphing a point in the plane. [4.1]

point-slope formula The equation $y - y_1 = m(x - x_1)$, where m is the slope of a line and (x_1, y_1) is a point on the line. [4.5]

polygon A closed figure determined by three or more line segments that lie in a plane. [3.2]

polynomial A variable expression in which the terms are monomials. [6.2]

positive integers The numbers 1, 2, 3 . . . ; also called the natural numbers. [1.1]

positive slope The slope of a line that slants upward to the right. [4.4]

prime polynomial A polynomial that is nonfactorable over the integers. [7.2]

principal square root The positive square root of a number. [1.2, 9.1]

product In multiplication, the result of multiplying two numbers. [1.1]

product of the sum and difference of two terms A polynomial that can be expressed in the form $(a + b)(a - b)$. [7.4]

proportion An equation that states the equality of two ratios or rates. [8.4]

Pythagorean Theorem The square of the hypotenuse of a right triangle is equal to the sum of the squares of the two legs. [9.3]

quadrant One of the four regions into which a rectangular coordinate system divides the plane. [4.1]

quadratic equation An equation of the form $ax^2 + bx + c = 0$, where a and b are coefficients, c is a constant, and $a \neq 0$; also called a second-degree equation. [7.5, 10.1]

quadratic equation in standard form A quadratic equation written in descending order and set equal to zero. [7.5]

quadratic formula A general formula, derived by applying the method of completing the square to the standard form of a quadratic equation, used to solve quadratic equations. [10.3]

quadratic function A function that can be expressed by the equation $f(x) = ax^2 + bx + c$, where a is not equal to zero. [6.2, 11.1]

quadratic inequality An inequality that can be written in the form $ax^2 + bx + c < 0$ or $ax^2 + bx + c > 0$, where a is not equal to zero. The symbols $\le$ and $\ge$ can also be used. [10.5]

quadratic trinomial A trinomial of the form $ax^2 + bx + c$, where a and b are nonzero coefficients and c is a nonzero constant. [7.4]

quadrilateral A four-sided closed figure. [3.2]

quotient In division, the result of dividing the divisor into the dividend. [1.1]

radical equation An equation that contains a variable expression in a radicand. [9.3]

radical sign The symbol $\sqrt{\ }$, which is used to indicate the positive, or principal, square root of a number. [1.2, 9.1]

radicand In a radical expression, the expression under the radical sign. [1.2, 9.1]

radius of a circle Line segment from the center of the circle to a point on the circle. [3.2, 11.5 – online]

radius of a sphere A line segment going from the center to a point on the sphere. [3.3]

range The set of the second coordinates of all the ordered pairs of a relation. [4.2]

rate The quotient of two quantities that have different units. [8.5]

rate of work That part of a task that is completed in one unit of time. [8.7]

ratio The quotient of two quantities that have the same unit. [8.5]

rational expression A fraction in which the numerator or denominator is a polynomial. [8.1]

rational number A number of the form $\frac{a}{b}$, where a and b are integers and b is not equal to zero. [1.2]

rationalizing the denominator The procedure used to remove a radical from the denominator of a fraction. [9.2]

ray Line that starts at a point and extends indefinitely in one direction. [3.1]

real numbers The rational numbers and the irrational numbers taken together. [1.2]

real part of a complex number For the complex number $a + bi$, a is the real part. [9.4]

reciprocal The reciprocal of a nonzero real number a is $\frac{1}{a}$; also called the multiplicative inverse. [1.2, 1.4]

reciprocal of a rational expression The rational expression with the numerator and denominator interchanged. [8.1]

rectangle A parallelogram that has four right angles. [3.2]

rectangular coordinate system A coordinate system formed by two number lines, one horizontal and one vertical, that intersect at the zero point of each line. [4.1]

rectangular solid A solid in which all six faces are rectangles. [3.3]

regular polygon A polygon in which each side has the same length and each angle has the same measure. [3.2]

relation A set of ordered pairs. [4.2]

repeating decimal A decimal that is formed when division of the numerator of its fractional counterpart by the denominator results in a decimal part wherein a block of digits repeats infinitely. [1.2]

right angle An angle whose measure is 90°. [3.1]

right triangle A triangle that contains a 90° angle. [3.1]

roster method A method of designating a set by enclosing a list of its elements in braces. [1.5]

scalene triangle A triangle that has no sides of equal length; no two of its angles are of equal measure. [3.2]

scatter diagram A graph of collected data as points in a coordinate system. [4.1]

scientific notation Notation in which a number is expressed as the product of a number between 1 and 10 and a power of 10. [6.1]

second-degree equation An equation of the form $ax^2 + bx + c = 0$, where a and b are coefficients, c is a constant, and $a \neq 0$; also called a quadratic equation. [10.1]

set A collection of objects. [1.1, 1.5]

set-builder notation A method of designating a set that makes use of a variable and a certain property that only elements of that set possess. [1.5]

similar objects Similar objects have the same shape but not necessarily the same size. [8.5]

simplest form of a rational expression A rational expression is in simplest form when the numerator and denominator have no common factors. [8.1]

slope A measure of the slant, or tilt, of a line. The symbol for slope is m. [4.4]

slope-intercept form of a straight line The equation $y = mx + b$, where m is the slope of the line and $(0, b)$ is the y-intercept. [4.4]

solution of a system of equations in three variables An ordered triple that is a solution of each equation of the system. [5.2]

solution of a system of equations in two variables An ordered pair that is a solution of each equation of the system. [5.1]

solution of an equation A number that, when substituted for the variable, results in a true equation. [2.1]

solution of an equation in three variables An ordered triple (x, y, z) whose coordinates make the equation a true statement. [5.2]

solution of an equation in two variables An ordered pair whose coordinates make the equation a true statement. [4.1]

solution set of a system of inequalities The intersection of the solution sets of the individual inequalities. [5.4]

solution set of an inequality A set of numbers, each element of which, when substituted for the variable, results in a true inequality. [2.5]

solving an equation Finding a solution of the equation. [2.1]

sphere A solid in which all points are the same distance from point O, which is called the center of the sphere. [3.3]

square A rectangle with four equal sides. [3.2]

square of a binomial A polynomial that can be expressed in the form $(a + b)^2$. [6.3]

square root A square root of a positive number x is a number a for which $a^2 = x$. [1.2]

square matrix A matrix that has the same number of rows as columns. [5.5 – online]

square root of a perfect square One of the two equal factors of the perfect square. [7.4]

standard form of a quadratic equation A quadratic equation is in standard form when the polynomial is in descending order and equal to zero. [7.5, 10.1]

substitution method An algebraic method of finding an exact solution of a system of linear equations. [5.1]

straight angle An angle whose measure is 180°. [3.1]

sum In addition, the total of two or more numbers. [1.1]

supplementary angles Two angles whose measures have the sum 180°. [3.1]

synthetic division A shorter method of dividing a polynomial by a binomial of the form $x - a$. This method uses only the coefficients of the variable terms. [6.4]

system of equations Two or more equations considered together. [5.1]

system of inequalities Two or more inequalities considered together. [5.4]

terminating decimal A decimal that is formed when division of the numerator of its fractional counterpart by the denominator results in a remainder of zero. [1.2]

terms of a variable expression The addends of the expression. [1.4]

tolerance of a component The amount by which it is acceptable for the component to vary from a given measurement. [2.6]

transversal A line that intersects two other lines at two different points. [3.1]

triangle A three-sided closed figure. [3.1]

trinomial A polynomial of three terms. [6.2]

undefined slope The slope of a vertical line is undefined. [4.4]

uniform motion The motion of an object whose speed and direction do not change. [2.1]

union of two sets The set that contains all elements that belong to either of the sets. [1.5]

upper limit In a tolerance, the greatest acceptable value. [2.6]

value of a function The value of the dependent variable for a given value of the independent variable. [4.2]

variable A letter of the alphabet used to stand for a number that is unknown or that can change. [1.1]

variable expression An expression that contains one or more variables. [1.4]

variable part In a variable term, the variable or variables and their exponents. [1.4]

variable term A term composed of a numerical coefficient and a variable part. When the numerical coefficient is 1 or -1, the 1 is usually not written. [1.4]

vertex Point at which the rays that form an angle meet. [3.1]

vertex of a parabola The point on the parabola with the smallest y-coordinate or the largest y-coordinate. [11.1]

vertical angles Two angles that are on opposite sides of the intersection of two lines. [3.1]

vertical-line test A graph defines a function if any vertical line intersects the graph at no more than one point. [11.2]

volume A measure of the amount of space inside a closed surface. [3.3]

***x*-coordinate** The abscissa in an *xy*-coordinate system. [4.1]

***x*-intercept** The point at which a graph crosses the *x*-axis. [4.3]

***y*-coordinate** The ordinate in an *xy*-coordinate system. [4.1]

***y*-intercept** The point at which a graph crosses the *y*-axis. [4.3]

zero of a function A value of x for which $f(x) = 0$. [11.1]

zero slope The slope of a horizontal line. [4.4]

Index

Q

T

U

V

W

X

Y

Z

Index of Applications

TI-30X IIS

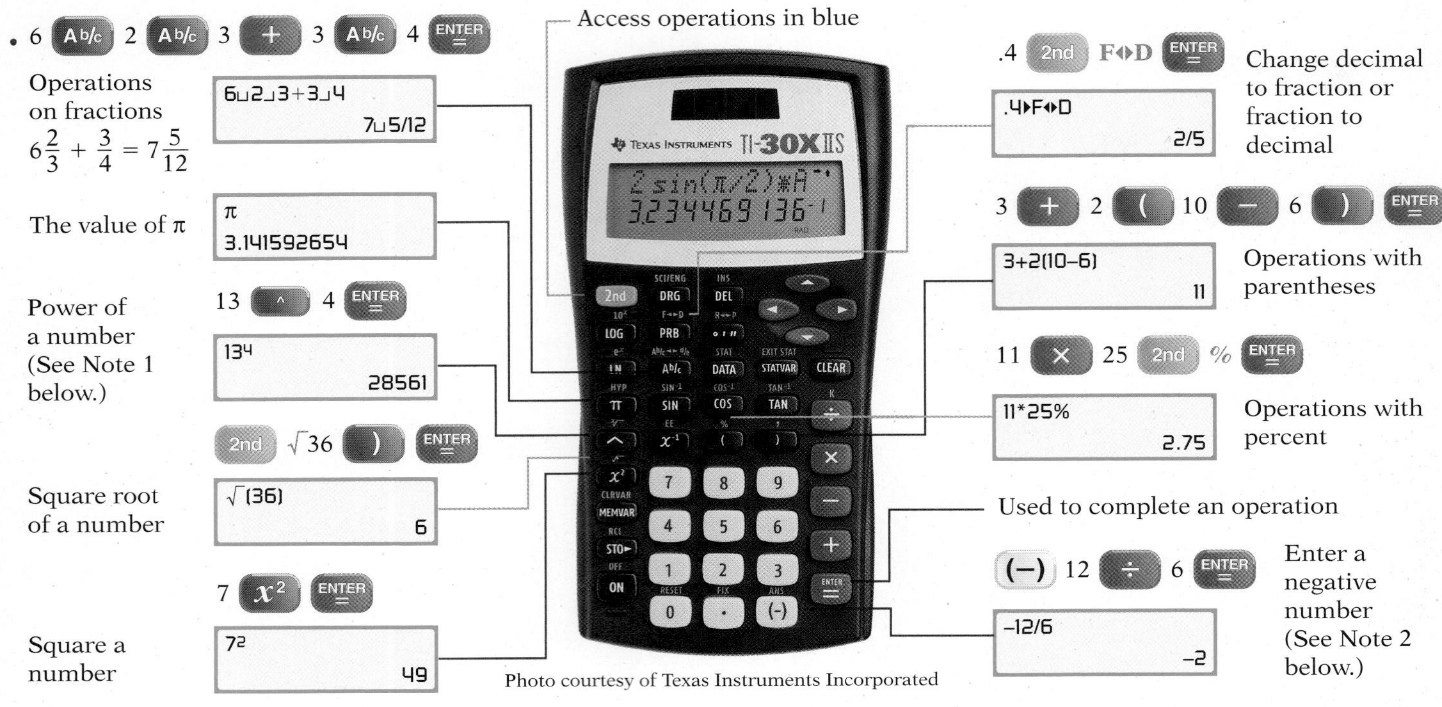

fx-300MS

Access operations in gold

√ 36 =

Square root of a number

√36
6

6 a b/c 2 a b/c 3 + 3 a b/c 4 =

Operations on fractions $6\frac{2}{3} + \frac{3}{4} = 7\frac{5}{12}$

6⌟2⌟3+3⌟4
7⌟5⌟12

7 x^2 =

Square a number

7²
49

(−) 12 ÷ 6 =

Enter a negative number (See Note 2 below.)

−12÷6
−2

Photo courtesy of Casio, Inc.

.4 = SHIFT d/c

Change decimal to fraction

.4
2⌟5

13 ∧ 4 =

Power of a number (See Note 1 below.)

13^4
28561

3 + 2 (10 − 6) =

Operations with parentheses

3+2(10−6)
11

11 × 25 SHIFT % =

Operations with percent

11x25%
2.75

Used to complete an operation

SHIFT π =

The value of π

π
3.141592654

NOTE 1: Some calculators use the y^x key to calculate a power. For those calculators, enter 13 y^x 4 = to evaluate 13^4.

NOTE 2: Some calculators use the +/− key to enter a negative number. For those calculators, enter 12 +/− ÷ 6 = to calculate −12 ÷ 6.